Areas and Volumes

1. Triangle

Area $= (1/2)bh.$

2. Parallelogram

Area $= bh.$

3. Trapezoid

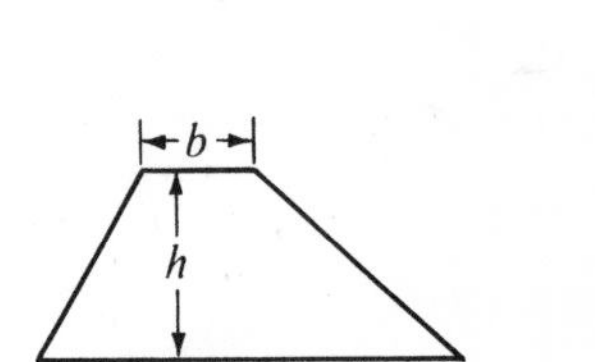

Area $= \dfrac{a+b}{2}h.$

4. Circle

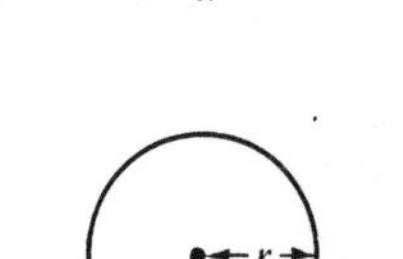

Area $= \pi r^2.$
Circumference $= 2\pi r.$

5. Sector of a Circle

Area $= (1/2)r^2\theta, \quad \theta$ in radians.

6. Right Circular Cylinder

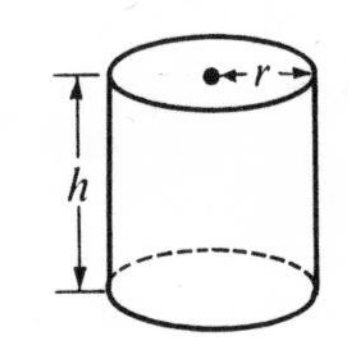

Volume $= \pi r^2 h.$
Lateral surface area $= 2\pi rh.$

7. Right Circular Cone

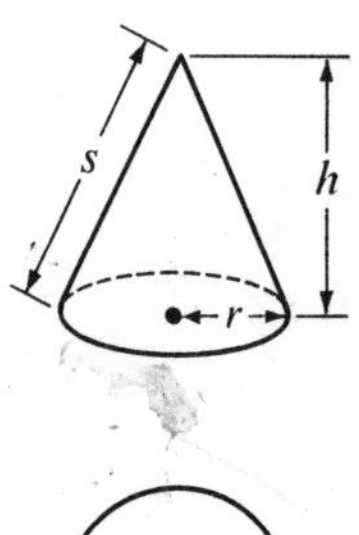

Volume $= (1/3)\pi r^2 h.$
Lateral surface area $= \pi rs.$

8. Sphere

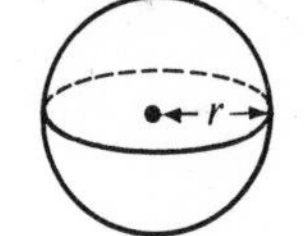

Volume $= (4/3)\pi r^3.$
Surface area $= 4\pi r^2.$

9. Frustrum of a Right Circular Cone

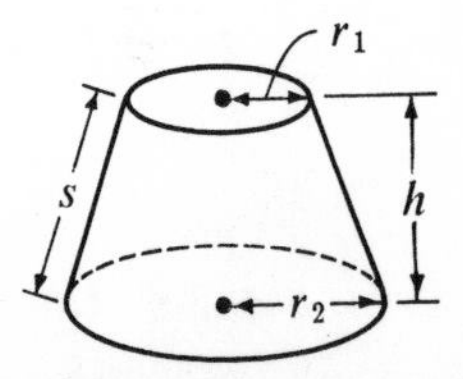

Volume $= (1/3)\pi h(r_1{}^2 + r_1 r_2 + r_2{}^2).$
Lateral surface area $= \pi s(r_1 + r_2).$

Calculus wit
Analytic Geometr

7817

th

·y

nte, Billie Harber, and the production staff of Prindle,
ler the direction of Michael Michaud.
ınd chapter opening details from the brass etching
r. Cover printing by LeHigh Press Lithographers.
oto Times Roman by Syntax International, Ltd. Printed
y & Company.

Calculus
with Analytic Geometry

Second Edition

Prindle, Weber & Schmidt, Inc.
Boston, Massachusetts

Howard E. Campbell   **Paul F. Dierker**

University of Idaho

Dedicated to
Ramona and Arlene

Printed in the United States of America.

Second printing: September 1978

Library of Congress Cataloging in Publication Data

Campbell, Howard E

Calculus with analytic geometry

Includes index.

1. Calculus. 2. Geometry, Analytic. I. Dierker, Paul F., joint author. II. Title.

QA303.C187 515'.15 77-17845
ISBN 0-87150-244-5

Preface

During the past twenty years a revolution has taken place in the standard calculus course. To judge the scope of this revolution, one need only examine the calculus books of two decades ago and compare them with the current books. The older texts tended to dwell primarily on manipulative skills, while many texts today tend toward abstract encyclopedic developments.

We are convinced that the proper presentation for most students lies somewhere between those two extremes. On the one hand, the cookbook approach tends to limit severely the mathematical development of the average student. On the other hand, the average student is not prepared for the shock of an abstract development. Consequently we have sought to develop the essentials of calculus in the most straightforward fashion possible.

It has become common to use the definite integral for defining various quantities. For example, the area under a curve $y=f(x)$ between $x=a$ and $x=b$ is often defined as $\int_a^b f(x)\,dx$. Though the method is logically consistent, it does rob calculus of much of its interest and excitement by making it appear synthetic. Moreover, such an approach makes novel uses of the integral nearly impossible; and, since each application has its own individual definition, no unity is established among the various applications. Accordingly, for example, we have taken care to derive the formula for area from an intuitive understanding of area. In fact, we make three assumptions about area that can actually be taken as the *definition* of area. All applications of the integral may be derived from similar assumptions. Chapter 1 contains preliminary material needed for the study of calculus. Part, or even all of it, can be omitted for students with good backgrounds.

Limits are introduced in Chapter 2, while discussion of limits at infinity is delayed until Chapter 10.

We have included an optional introduction to the differentiation of the trigonometric functions in Section 3.10 and an optional introduction to the integration of the trigonometric functions in Section 4.9. Those who wish to delay the introduction of these functions until the second term need only omit these sections. Chapter 3 also contains a special problem set requiring students to match the graphs of functions with graphs of their derivatives.

Sections in Chapters 5 and 16 have been divided into *Intuitive Developments* and *More Rigorous Developments*. An instructor may wish to cover only some of the *More Rigorous Developments*.

Chapter 9 on numerical methods has been included because we believe that computers have made such a study essential. In that chapter we have included a number of computer-oriented problems.

Chapter 10 (An Extension of the Limit Concept, Improper Integrals, and L'Hospital's Rule), Chapter 11 (Infinite Sequences and Series), and Chapter 12 (Polar Coordinates) are independent of each other.

We have included a brief optional introduction to differential equations in Chapter 13.

Green's Theorem, Stoke's Theorem, and the Divergence Theorem are interesting theorems which are essential to many applications of calculus. These are included in Chapter 17.

Other features new to the second edition include a section on parametric equations (1.9) and one on differentials (3.9), an improved introduction to the integral (4.1–4.2), sections on moments and centroids (5.9) and on hyperbolic functions (6.8), and an improved chapter on numerical methods. Chapter 10 has been divided into two chapters and expanded with a new section on extensions of L'Hospital's Rule (10.3).

Computer graphics have been used when they have aided in exposition. We wish to thank Kenneth I. Joy of Northern Michigan University for preparing the computer artwork and for suggesting the problem set on matching graphs in Chapter 3.

We have included over 4500 problems of varying difficulty. Each chapter contains a set of Technique Review Exercises, in which comprehension of each of the major techniques of the chapter is tested, followed by an extensive set of Additional Exercises, referenced to the relevant sections of the chapter. A set of Challenging Problems designed to interest the better student is contained at the end of each chapter; these problems involve further extensions of theory, as well as applications of mathematics to other disciplines.

There are also two special problem sets to evaluate the student's knowledge of integration techniques (8.7) and the tests for convergence or divergence of series (11.6).

We would like to express our appreciation to the following people who have made many helpful suggestions for improving this edition: John T. White, Texas Technological University; Bruce B. Claflin, Northeastern University; G.G. Bilodeau, Boston College; K.C. Schraut, University of Dayton; Mark P. Hale, Jr., University of Florida; Thomas Schwartzbauer, The Ohio State University; Richard L. Faber, Boston College; William R. Fuller, Purdue University; Clinton B. Gass, DePauw University; Chris Nevison, Colgate University; Joseph Krebs, Boston College; David E. Flesner, Gettysburg College; Michael L. Engquist, Eastern Washington State College; David Ryeburn, Simon Fraser University; Richard C. Orr, State University of New York at Oswego; Philip A. Leonard, Arizona State University; Vyron M. Klassen, California State University at Fullerton; Lester E. Laird, Emporia Kansas State College; and J. B. Fugate, University of Kentucky.

We express our appreciation to Connie Yarborough, Lynda Prather, Pat Wight, Celeste Cummins, and Anna Conditt for typing various versions of this work. We also offer thanks to the students of Mathematics 180, Mathematics 190, and Mathematics 200 at the University of Idaho, who have helped us to class-test the text, and to the members of the Mathematics Department at the University of Idaho, particularly Charles Christenson and Ralph Neuhaus, who also class-tested the manuscript and made helpful suggestions. Finally, we are indebted to Robert Katz for his assistance in readying the completed manuscript for publication.

Howard E. Campbell
Paul F. Dierker

Contents

1

2

3

The Derivative 96

4

Integration 168

5

Applications of the Definite Integral 220

6

Exponential and Logarithmic Functions 278

7

Trigonometric and Inverse Trigonometric Functions 316

8

Other Integration Techniques 348

9

Numerical Methods 384

10

11

17

Vector Calculus 734

1

Preliminaries

1.1 Introduction

The invention of calculus is usually attributed to two seventeenth-century mathematicians, Newton and Leibniz. These two men, one English and one German, independently and almost concurrently developed some of the major theorems and applications of calculus. However, they didn't do it alone; previously, Barrow, Fermat, and others had already developed significant amounts of what we now call calculus. Newton and Leibniz integrated this knowledge into a more coherent package. The fact that two men were able to develop the same theory independently, and essentially at the same time, serves to indicate that calculus was an "idea whose time had come."

Seldom, if ever, has the development of one area of mathematics had such an impact on science. The advent of calculus enabled seventeenth-century scientists to solve problems that had long puzzled the scientific community. Through the succeeding years calculus has contributed significantly to all areas of science. Today calculus is used not only in the physical sciences but also in the biological and social sciences. In this text we shall try to indicate a wide variety of applications.

In any mathematics course it is especially important to have a reasonable understanding of the meanings of the terms used; hence, for completeness, we shall start with some fundamentals.

1.2 Numbers and Coordinate Systems

We shall deal almost exclusively with real numbers, and unless otherwise stated, all numbers mentioned or referred to will be considered to be real numbers.

Definition 1.2.1

> A *real number* is a number that can be expressed as an infinite decimal.

For example: $7 = 7.000\ldots = 7.\overline{0}$, $0 = 0.000\ldots = 0.\overline{0}$, $1/3 = 0.3333\ldots = 0.\overline{3}$, and $68.\overline{271}$ are real numbers. Here the bar over a sequence of digits means that the same sequence repeats forever. Those four examples are all *rational numbers.* That is, each can also be expressed as a quotient of two integers. However, not all real numbers are rational numbers. Those that are not rational are called *irrational.* These are irrational, for example: π, $\sqrt{3}$, and $3.010010001\ldots$, where the pattern continues with four zeros then a 1, followed by five zeros and a 1, etc. In fact, a real number is a rational number if and only if it can be expressed as a *repeating decimal* in which the same sequence of digits repeats forever from some point on.

The real numbers can be associated with the points on a line by taking one point, called the *origin,* and labeling it 0. Then, for each positive number p, we label the point that is p units to the right of the origin as p, and the point that is p units to the left of the origin as $-p$. The unit of measure used is arbitrary and is chosen so that diagrams can be drawn conveniently. An illustration of part of the resulting *number line*, or *real line*, is given in Figure 1.2.1.

Figure 1.2.1

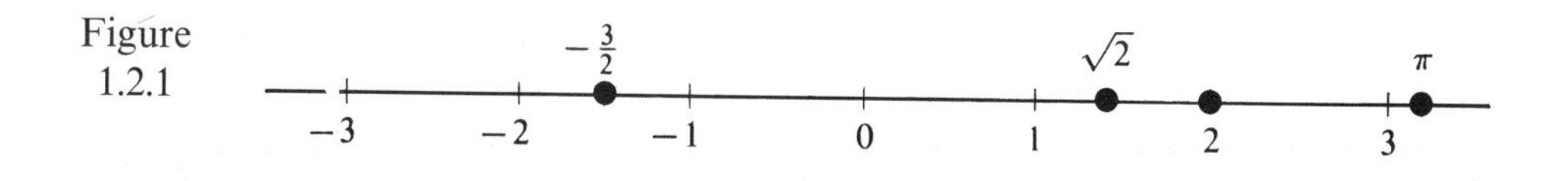

There the indicated points correspond to $-3/2$, $\sqrt{2}$, 2, and π. Now consider the inequality $x > 2$. A number is greater than 2 if and only if it lies to the right of 2 on the number line. Consequently the set of points corresponding to the inequality $x > 2$ is the set of all points to the right of 2. That set is indicated in Figure 1.2.2. Note that a small open circle is used at 2 to indicate that 2 is *not* in the set.

Figure 1.2.2

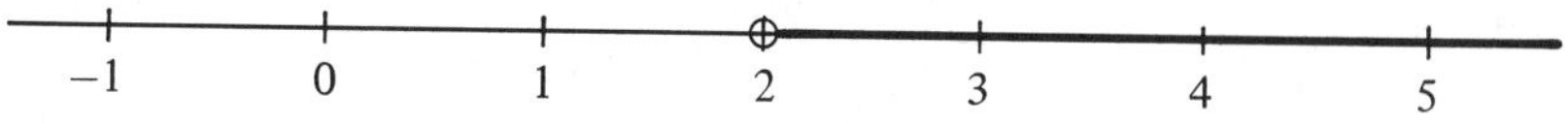

−1 0 1 2 3 4 5

Similarly the set of points corresponding to $-1 \le x < 3$, that is, $x \ge -1$ and $x < 3$, is indicated in Figure 1.2.3. Note the open circle that indicates 3 is not in the set. The heavy dot indicates -1 is in the set.

Figure 1.2.3

We shall now use a pair of number lines to coordinatize a plane; that is, we shall associate each point in the plane with a unique ordered pair of numbers. We proceed as follows. First take two perpendicular real lines that have the same origin. This gives us a *rectangular coordinate system* for locating points in the plane, as shown in Figure 1.2.4. The vertical line is called the *y-axis* and the horizontal line is called the x-axis. Although in Figure 1.2.4 the same unit of measure is used on both axes, we shall find that it is sometimes convenient to use different units of measure.

Each point in the plane is associated with an ordered pair of numbers. The first number in the ordered pair indicates the location of the point horizontally and is called the *x-coordinate*; the second number indicates the location vertically and is called the *y-coordinate*. For example, the point P_1 is 3 units to the right of, and 2 units above the intersection of the axes and hence is labeled (3, 2). The intersection of the axes is called the *origin*, O, and has coordinates (0, 0). A point labeled (a, b) is often called *the point* (a, b).

Other points are also shown in Figure 1.2.4. As indicated, the point P_2 is 2 units to the left of the y-axis and 1 unit above the x-axis and is labeled $(-2, 1)$; the point P_3, whose coordinates are $(-3, -2)$, is 3 units to the left of the y-axis and 2 units below

Figure 1.2.4

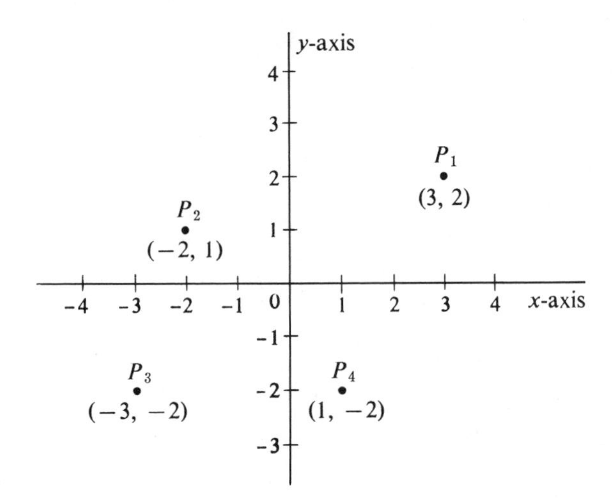

the x-axis. The point $(1, -2)$ is one unit to the right of the y-axis and 2 units below the x-axis.

Note the coordinate axes divide the plane into four parts or *quadrants*. The quadrants are numbered as shown in Figure 1.2.5.

Figure 1.2.5

y-axis

2nd quadrant

1st quadrant

x-axis

3rd quadrant

4th quadrant

Exercises 1.2

In Exercises 1–8, plot the indicated point.

1. $(1, 2)$.
2. $(-1, -2)$.
3. $(2, -3)$.
4. $(-3, 1)$.
5. $(0, 2)$.
6. $(3, 0)$.
7. $(-1, 1)$.
8. $(-1, -1)$.

In Exercises 9–18, indicate the points on the x-axis for which the given statement is true.

9. $2 \leq x < 4$.
10. $-1 < x \leq 3$.
11. $x < 2$ or $x \geq 4$.
12. $x \leq -1$ or $x > 3$.
13. $x < 2$ and $x \geq 4$.
14. $x \leq -1$ and $x > 3$.
15. $-2 \leq x < 3$ and $-1 < x < 4$.
16. $2 < x \leq 5$ and $4 < x < 8$.
17. $-2 \leq x < 3$ or $-1 < x < 4$.
18. $2 < x \leq 5$ or $4 < x < 8$.

In Exercises 19–32, indicate on a coordinate system the points for which the given statement is true.

19. $x > 0$ and $y > 0$.
20. $x < 0$ and $y > 0$.
21. $x < 1$.
22. $y > -1$.
23. $2 < x < 3$ and $1 < y < 2$.
24. $x = 2$ and $-2 < y < -1$.
25. $y = -1$ and $2 < x < 3$.
26. $-1 < x < 0$ and $-2 < y < -1$.

27. $x = 2$.
28. $y = 3$.
29. $x = 2$ and $y = 3$.
30. $x = 0$.
31. $x = 0$ and $-1 < y < 1$.
32. $y = 0$ and $1 < x < 2$.

1.3 Functions

The concept of function is one of the most fundamental ideas of mathematics, and in particular, of calculus.

Definition 1.3.1

> A *function* is a rule that assigns a unique real number to each number in a specified set of real numbers.

In view of this definition we can think of a function as a computer, which, when presented with the right kind of number as *input*, produces exactly one number as *output*. If f denotes the function and x denotes the input, we shall denote the output by $f(x)$. A function f may be pictured as follows:

$$\underset{\text{input}}{x} \rightarrow \underset{\text{computer}}{\boxed{f}} \rightarrow \underset{\text{output}}{f(x)}$$

To specify a particular function we must specify what the computer does to suitable inputs. For example, the computer f might add 3 to each input. In this case we would have

$$2 \rightarrow \boxed{f} \rightarrow 5, \text{ or } f(2) = 2 + 3 = 5;$$

$$-7 \rightarrow \boxed{f} \rightarrow -4, \text{ or } f(-7) = -7 + 3 = -4;$$

$$\frac{3}{4} \rightarrow \boxed{f} \rightarrow \frac{15}{4}, \text{ or } f\left(\frac{3}{4}\right) = \frac{3}{4} + 3 = \frac{15}{4};$$

$$a + 1 \rightarrow \boxed{f} \rightarrow a + 4, \text{ or } f(a + 1) = (a + 1) + 3 = a + 4;$$

and in general,

$$x \rightarrow \boxed{f} \rightarrow x + 3, \text{ or } f(x) = x + 3.$$

Another function g might square each input, then add three times the input, and then subtract 2. In this case,

$$5 \rightarrow \boxed{g} \rightarrow 38, \text{ or } g(5) = 5^2 + 3(5) - 2 = 38;$$

$$a + h \rightarrow g \rightarrow (a + h)^2 + 3(a + h) - 2, \text{ or } g(a + h) = (a + h)^2 + 3(a + h) - 2;$$

and in general,

$$x \rightarrow \boxed{g} \rightarrow x^2 + 3x - 2, \text{ or } g(x) = x^2 + 3x - 2.$$

Usually it is convenient to use only the second notation indicated in each of the illustrations above, and in particular to write only $g(x) = x^2 + 3x - 2$ for the function

g. Thus for inputs of -2, $x+3$, and x^3, we write

$$g(-2) = (-2)^2 + 3(-2) - 2 = 4 - 6 - 2 = -4;$$

$$\begin{aligned} g(x+3) &= (x+3)^2 + 3(x+3) - 2 \\ &= x^2 + 6x + 9 + 3x + 9 - 2 \\ &= x^2 + 9x + 16; \end{aligned}$$

$$g(x^3) = (x^3)^2 + 3(x^3) - 2 = x^6 + 3x^3 - 2.$$

Getting back to the more cumbersome picture with arrows for a moment, it should be noted that $g(x+3)$ could be diagrammed as

$$\begin{aligned} x \to \boxed{f} \to x + 3(=f(x)) \to \boxed{g} \to g(x+3) &= (x+3)^2 + 3(x+3) - 2 \\ &= x^2 + 9x + 16 = g(f(x)), \end{aligned}$$

where we use x as input for f and then take the output $f(x)$ of f and use it as input for g, to get $g(f(x))$ as the final output. This "function of a function," or *composite function*, could also be diagrammed as a single function $g \circ f$ as follows:

$$x \to \boxed{g \circ f} \to (g \circ f)(x).$$

In the more convenient notation we write

$$(g \circ f)(x) = g(f(x)).$$

Similarly, if we let $h(x) = x^3$, then $g(x^3)$ may be written as the composite function $(g \circ h)(x)$. That is,

$$(g \circ h)(x) = g(h(x)) = (x^3)^2 + 3x^3 - 2 = x^6 + 3x^3 - 2.$$

At the beginning of this section we said that when the "right kind of number" is presented to a function as input, it produces exactly one number as output. The functions we have considered so far, however, produced an output for *every* input number. This need not always be the case. For example, let $f(x) = 3/(x-1)$, and consider the input 1. Then $f(1) = 3/(1-1) = 3/0$. But since division by zero has no meaning, there is no output for the input 1. Thus $f(1)$ has no meaning in this case, even though $f(x)$ has a meaning for all other values of x.

Similarly, if $g(x) = \sqrt{x}$, there is no real-number output when the input is a negative number, because there are no real numbers that are square roots of negative numbers. In particular $g(-4) = \sqrt{-4}$ has no meaning in the real number system; it does have a meaning in the complex number system, however, since there, $\sqrt{-4} = 2i$.

The set of all acceptable inputs and the set of all outputs of a function are often referred to and hence have special names.

Definition 1.3.2

(i) The set of all acceptable inputs of a function f is called the *domain* of f.

(ii) The set of all outputs of a function f is called the *range* of f.

For example, the domain of $f(x) = 3/(x-1)$ is the set of all real numbers except 1, and its range is the set of all real numbers except zero. If we require the output to be a real number, both the domain and the range of $g(x) = \sqrt{x}$ are the set of all nonnegative (real) numbers.

It is also the case that a function need not be defined by a single equation. For example a function f may be defined as

$$f(x) = \begin{cases} 2 & \text{if} \quad x < 3 \\ 5 & \text{if} \quad x = 3 \\ 1 & \text{if} \quad x > 3. \end{cases}$$

Notice here, for instance, that $f(0) = 2$, $f(3/2) = 2$, $f(3) = 5$, $f(17/5) = 1$, and $f(457) = 1$.

Exercises 1.3

For Exercises 1–24, let $f(x) = x + 5$, $g(x) = x^4$, and $k(x) = x^2 - 2x + 1$, and find each of the following.

1. $f(3)$.
2. $f(-8)$.
3. $k(0)$.
4. $g(2)$.
5. $k(a + 3)$.
6. $k(b - 2)$.
7. $(f \circ g)(x)$.
8. $(g \circ f)(x)$.
9. $(f \circ g)(2)$.
10. $(g \circ f)(2)$.
11. $(k \circ f)(x)$.
12. $(f \circ k)(x)$.
13. $k(y^3)$.
14. $k(u^4)$.
15. $g(x) + f(x)$.
16. $f(x) + k(x)$.
17. $g(x)k(x)$.
18. $f(x)g(x)$.
19. $k(-3)f(-1)$.
20. $g(2)k(0)$.
21. $f(x + h) - f(x)$.
22. $g(x + h) - g(x)$.
23. $\dfrac{k(x + h) - k(x)}{h}$.
24. $\dfrac{g(x + h) - g(x)}{h}$.

For Exercises 25–32, find (a) the domain, and (b) the range, of each of the functions.

25. $f(x) = \dfrac{1}{x + 2}$.
26. $f(x) = \dfrac{1}{(x - 1)(x + 3)}$.
27. $f(x) = \sqrt{x + 2}$.
28. $f(x) = \sqrt{x - 3}$.
29. $f(x) = \sqrt[3]{\dfrac{1}{x + 2}}$.
30. $f(x) = \sqrt{\dfrac{1}{x - 5}}$.
31. $f(x) = \begin{cases} x & \text{if} \quad x > 0 \\ -3 & \text{if} \quad x = 0 \\ x^2 & \text{if} \quad x < 0. \end{cases}$
32. $f(x) = \begin{cases} 2x & \text{if} \quad x < 3 \\ 7 & \text{if} \quad x = 3 \\ 9 & \text{if} \quad x > 3. \end{cases}$
33. Find a numerical expression for the function f so that $f(x)$ is the largest integer less than or equal to x when $0 \le x \le 4$.
34. The fare of a certain taxi is 50 cents for the first quarter of a mile or any part thereof, and 30 cents for each additional quarter of a mile or part thereof. Find the function f that expresses the fare $f(x)$ for a trip of x miles where $0 \le x \le 2$.
35. A rancher has 600 yards of fencing to be used for three sides of a rectangular corral along the side of a straight canyon wall that will be the fourth side of the corral. Find a function A that expresses the area $A(x)$ of the corral, where x is the length of each of the sides perpendicular to the canyon wall.
36. An open-topped box is to be constructed by cutting equal squares from the corners of a square sheet of aluminum and then folding up the sides. If the sheet has side length of 6 feet, find a

function V that expresses the volume $V(x)$ of the box, where x is the side-length of the squares cut out.

37. Sand is poured on the ground forming a pile in the shape of a right circular cone whose radius is twice its height. Find a function V expressing the volume $V(h)$ of the pile, where h is its height.
38. The material in the aluminum top of a beer can costs \$.0002 per square inch, while the steel for the lateral surface and bottom costs \$.0001 per square inch. If the can is to hold twelve ounces ($1.8 \text{ in}^3 = 1$ oz) find a function C expressing the cost $C(r)$ of the can, where r is its radius.
39. If $f(x) = ax^2 + bx + c$, where a, b, and c are constants, and $f(x-y) = f(x) - f(y)$ for all values of x and y, then what values can a, b, and c have?
40. If $f(x) = ax + b$ and $(f \circ f)(x) = f(x)$ for all x, then what values can a and b have?
41. Let $f(x) = \dfrac{x^n(a-bx)^n}{n!}$. Show that $f\left(\dfrac{a}{b} - x\right) = f(x)$.

1.4 Further Examples of Functions

The functions that we shall most frequently refer to in this text are those based on the following functions:

a Power functions,
b Polynomial functions,
c Rational functions,
d The absolute value function,
e Trigonometric functions,
f Exponential functions,
g Logarithmic functions.

A *power function* has the form $f(x) = x^k$, where k is a real number. A *polynomial function* is a function of the form $f(x) = a_0 + a_1x + a_2x^2 + \cdots + a_nx^n$, where the a_i are particular numbers and n is a nonnegative integer. For example, $1 + 2x - x^2$, $7x^5 - x + 2$, and $3 = 3 + 0 \cdot x + 0 \cdot x^2 + \cdots + 0 \cdot x^n$ are polynomials. A *rational function* is one that can be expressed as the quotient of two polynomials, where the denominator is not zero. In particular, the following are rational functions:

$$\frac{2x^2 - 7x}{x^2 - 1}; \quad \frac{5}{x+3}; \quad \frac{3-x}{8x^5}; \quad x^3 + 5\left(= \frac{x^3+5}{1}\right).$$

The *absolute value* function, $|x|$, is probably less familiar than polynomial and rational functions. Since this function has considerable significance, we shall consider it in more detail.

Definition 1.4.1

The *absolute value*, $|x|$, of x is defined as follows:

$$|x| = \begin{cases} x & \text{if } x \geq 0 \\ -x & \text{if } x < 0. \end{cases}$$

For example, since $-3 < 0$, we have $|-3| = -(-3) = 3$. Also, $|0| = 0$, $|\sqrt{2}| = \sqrt{2}$, and $|-\pi| = -(-\pi) = \pi$. Thus the absolute value of a positive number is the number itself, and the absolute value of a negative number is the negative of the negative number and hence is also a positive number. Note that zero is the only number whose absolute value is not positive. It follows from the definition of absolute value that $|ab| = |a|\,|b|$ and $|a/b| = |a|/|b|$.

It also follows from the definition of $|a|$ that, on the real line, $|a|$ is the distance between the origin and the point a. For example, $|3|$, which equals 3, is the distance between the origin and the point 3, and $|-5|$, which equals 5, is the distance between the origin and the point -5, as indicated in Figure 1.4.1.

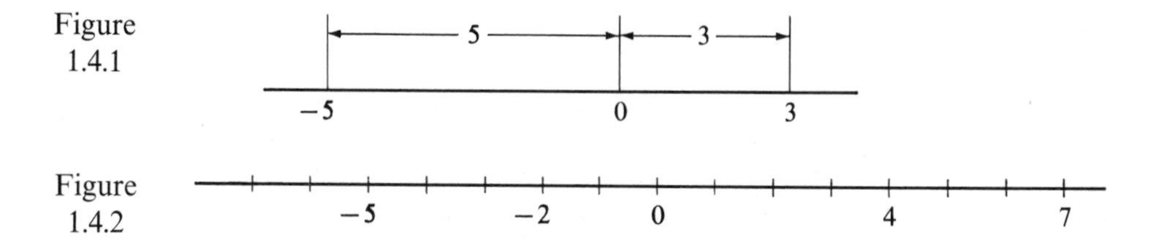

Figure 1.4.1

Figure 1.4.2

As we can see from looking at Figure 1.4.2, the distance between the point 4 and the point 7 is $7 - 4$, or 3. Similarly the distance between -2 and 4 is $4 - (-2) = 4 + 2 = 6$, and the distance between -2 and -5 is $-2 - (-5) = -2 + 5 = 3$. In general, the distance between distinct points a and b is $b - a$ if b is to the right of a, and is $a - b$ if a is to the right of b. In both cases, the distance is the positive one of the two numbers $a - b$ and $b - a$. Since each of the numbers $a - b$ and $b - a$ is the negative of the other, $|a - b|$ (or, what is the same, $|b - a|$) is the positive one. Of course, if neither point is to the right of the other, they are the same point, $a = b$, and the distance between them is $0 = a - b = b - a = |a - b| = |b - a|$. In all cases then, $|a - b| = |b - a|$, and both are the distance between a and b on the real line.

Thus the statement $|x - 2| < 5$ is equivalent to saying that the distance between x and 2 on the real line is less than 5. Similarly, since $|x + 3| = |x - (-3)|$, the statement $|x + 3| > 1$ is equivalent to saying that the distance between x and -3 is greater than 1.

An important inequality involving absolute values is the *triangle inequality*:

$$|a + b| \leq |a| + |b|.$$

In fact, we see that when a and b are both positive or both negative, or when one or both of them is zero, $|a + b| = |a| + |b|$; and when they have opposite signs, $|a + b| < |a| + |b|$. For example,

1. When $a = 2$ and $b = -3$, $|a + b| = |-1| = 1$ and $|a| + |b| = 2 + 3 = 5$.
2. When $a = 5$ and $b = 2$, $|a + b| = |7| = 7$ and $|a| + |b| = 5 + 2 = 7$.
3. When $a = 0$ and $b = -4$, $|a + b| = |-4| = 4$ and $|a| + |b| = 0 + 4 = 4$.

Although we are assuming that the reader has learned a reasonable amount of trigonometry we shall discuss the trigonometric functions as other examples of functions. Before defining these functions, let us briefly consider angles. For our purposes, an *angle* is an amount of rotation from one half-line, called the *initial side*, to another half-line, called the *terminal side*. The two half-lines must start at the same point called the *vertex*. An angle is considered positive if the rotation is in the counterclockwise direction and negative if the rotation is in the clockwise direction. Some pictures of angles are given in Figure 1.4.3. Notice that here α, γ and σ are positive and β and δ are negative.

Figure 1.4.3

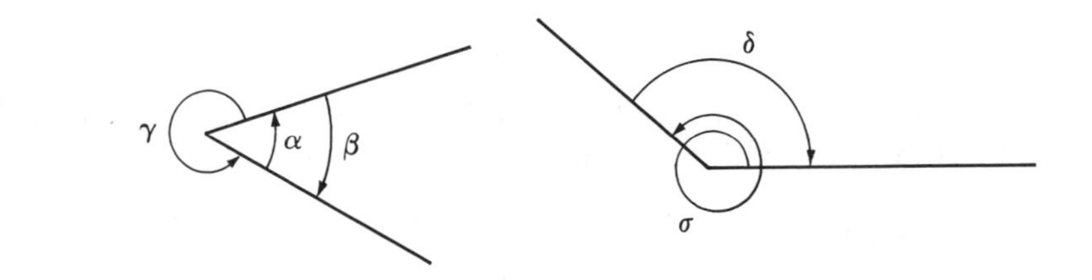

The two most frequently used units of angle measurement are *degrees* and *radians*. Since there are 360 degrees in one full revolution, a degree is 1/360 of a revolution. While degree measurements are more familiar, we shall find that the use of radians will simplify many of the formulas of calculus.

A *radian* is simply the measure of the angle subtended by an arc of length r on a circle of radius r. This relationship is illustrated in Figure 1.4.4. Since the circumference of a circle of radius r is $2\pi r$, 2π arcs of length r can be placed around the circle. Consequently there are 2π radians in one full revolution. Thus,

$$2\pi \text{ radians} = 360 \text{ degrees},$$

and so,

$$1 \text{ radian} = \frac{360}{2\pi} \text{ degrees}$$

and,

$$1 \text{ degree} = \frac{2\pi}{360}, \text{ or } \frac{\pi}{180} \text{ radians}.$$

Note, for example, that π radians $= 180$ degrees, and $\pi/2$ radians $= 90$ degrees.

Figure 1.4.4

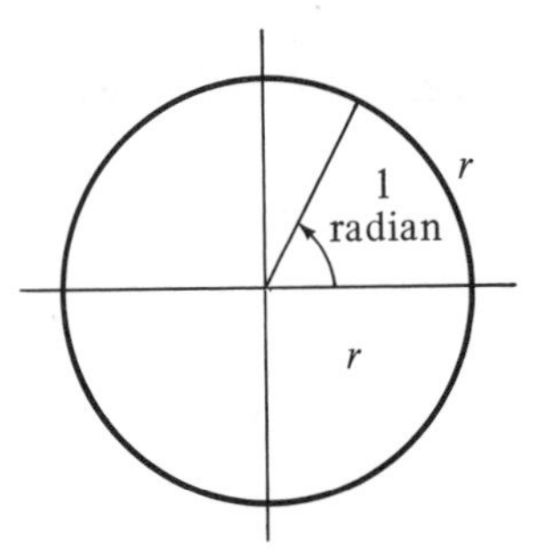

To define the trigonometric functions of an angle θ, we choose a rectangular coordinate system with the same unit of measure on both axes, and then we place the angle in *standard position*, that is, with its vertex at the origin and its initial side along the positive side of the x-axis. A point P different from the origin is then chosen on the terminal side of θ, as shown in Figure 1.4.5.

Figure 1.4.5

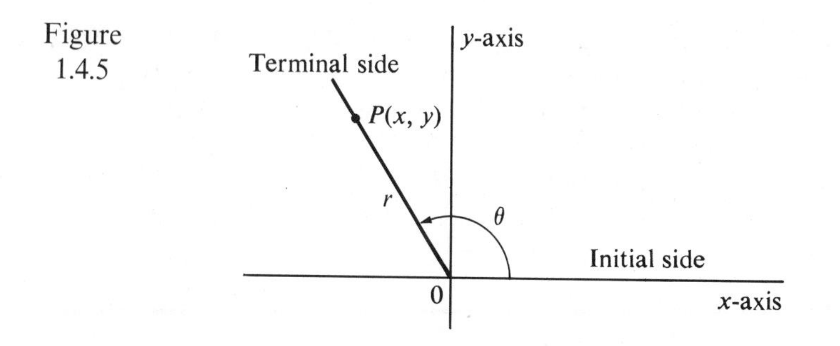

Let the coordinates of P be (x, y) and let r be the distance between P and the origin O. Then we adopt these definitions:

$$\sin \theta = \frac{y}{r} \qquad \tan \theta = \frac{y}{x} \qquad \sec \theta = \frac{r}{x}$$

$$\cos \theta = \frac{x}{r} \qquad \cot \theta = \frac{x}{y} \qquad \csc \theta = \frac{r}{y}.$$

It can be shown using similar triangles that these functions are independent of the location of the point P, as long as P is not at the origin.

For example, the trigonometric functions of 120 degrees, that is, $2\pi/3$ radians, may be determined as follows: (In Figure 1.4.6, a 30°, 60°, 90° triangle, that is, a $\pi/6$, $\pi/3$, $\pi/2$ triangle, with hypotenuse of 2, is shown as an aid.) On choosing the point P with coordinates $(-1, \sqrt{3})$ and $r = 2$ we find that

$$\sin \frac{2\pi}{3} = \frac{y}{r} = \frac{\sqrt{3}}{2}, \qquad \cot \frac{2\pi}{3} = \frac{x}{y} = \frac{-1}{\sqrt{3}} = -\frac{1}{\sqrt{3}},$$

$$\cos \frac{2\pi}{3} = \frac{x}{r} = \frac{-1}{2} = -\frac{1}{2}, \qquad \sec \frac{2\pi}{3} = \frac{r}{x} = \frac{2}{-1} = -2,$$

$$\tan \frac{2\pi}{3} = \frac{y}{x} = \frac{\sqrt{3}}{-1} = -\sqrt{3}, \qquad \csc \frac{2\pi}{3} = \frac{r}{y} = \frac{2}{\sqrt{3}}.$$

An *exponential function* is a function of the form a^x, where a is a fixed positive number, and x is the exponent. For example, if $f(x) = 2^x$, then $f(3) = 2^3 = 8$, and $f(-1/2) = 2^{-1/2} = 1/\sqrt{2}$.

Figure 1.4.6

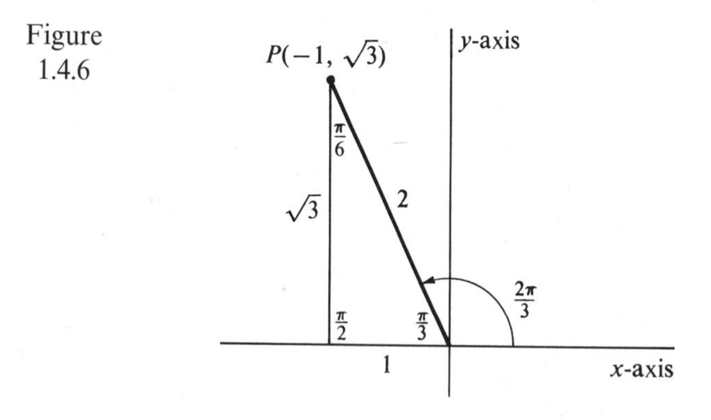

The *logarithm, with base a, of* x, written $\log_a x$, is the power to which a must be raised to obtain x. Here a is a fixed positive number not equal to 1. For example, if $g(x) = \log_3 x$, then

$$g(9) = 2, \text{ since } 3^2 = 9;$$

$$g\left(\frac{1}{81}\right) = -4, \text{ since } 3^{-4} = \frac{1}{81}; \text{ and}$$

$$g(1) = 0, \text{ since } 3^0 = 1.$$

Exercises 1.4

For Exercises 1–12, find the distance between each pair of points on the real line.

1. 3 and 7.
2. 9 and 2.
3. 4 and -6.
4. -5 and 3.
5. -3 and 8.
6. -1 and -7.
7. x and 3.
8. x and 5.
9. x and -6.
10. x and -8.
11. a and b.
12. a and $-b$.

For Exercises 13–20, express each statement in words, in terms of distance on the real line.

13. $|x - 3| < 2$.
14. $|x - 6| < 5$.
15. $|x + 4| \leq 3$.
16. $|x + 1| \leq 2$.
17. $|2 - x| \geq 5$.
18. $|7 - x| \geq 3$.
19. $|x + 6| > 8$.
20. $|x + 5| > 4$.

In Exercises 21–24, express each statement as an inequality involving an absolute value.

21. The distance between x and 4 is less than or equal to 3.
22. The distance between x and 7 is greater than or equal to 10.
23. The distance between x and -6 is greater than 8.
24. The distance between x and -5 is less than 1.

In Exercises 25–40, evaluate each function.

25. $\sin \frac{\pi}{3}$.
26. $\cos \frac{\pi}{3}$.
27. $\tan \frac{\pi}{6}$.
28. $\sec \frac{\pi}{6}$.
29. $\cos \frac{5\pi}{6}$.
30. $\sin \frac{5\pi}{6}$.
31. $\csc \frac{11\pi}{6}$.
32. $\cot \frac{11\pi}{6}$.
33. $\log_2 16$.
34. $\log_3 27$.
35. $\log_3(1/9)$.
36. $\log_2(1/32)$.
37. $\log_5 0.04$.
38. $\log_5 0.008$.
39. $\log_2 0.25$.
40. $\log_2 0.125$.

For Exercises 41–47, find a function f with the indicated property.

41. $f(-x) = f(x)$. (This sort of function is called an *even function.*)
42. $f(-x) = -f(x)$. (This sort of function is called an *odd function.*)
43. $f(x + y) = f(x) + f(y)$. (This sort of function is called an *additive function.*)
44. $f(xy) = f(x)f(y)$. (This sort of function is called a *multiplicative function.*)
45. $f(ax + by) = a \cdot f(x) + b \cdot f(y)$. (This sort of function is called a *linear function.*)
46. $f(x + y) = f(x)f(y)$.
47. $f(xy) = f(x) + f(y)$.
48. Prove that the composition of two even functions is an even function.
49. Prove that the composition of two additive functions is an additive function.

50. Prove that the composition of two multiplicative functions is a multiplicative function.

51. Prove that the composition of two linear functions is a linear function.

In Exercises 52–59, find all values of x for which the given statement is true.

52. $|x| = x$. 53. $|x| = -x$. 54. $|-x| = -x$. 55. $|-x| = x$.

56. $\sqrt{x^2} = x$. 57. $\sqrt{x^2} = -x$. 58. $\sqrt{x^2} = |x|$. 59. $\sqrt{x^2} = -|x|$.

1.5 Graphs of Functions and Equations

If f is a function, we obtain an equation involving x and y by setting y equal to $f(x)$. Thus we obtain the equation $y = f(x)$ from the function f. Of course not all equations arise in this way. Other examples of equations involving x and y but not of the form $y = f(x)$ are $x^2 + y^2 = 25$, and $xy = 1$.

Definition 1.5.1

> A point (a, b) is said to be a *solution* of an equation involving x and y, if, when x is replaced by a, and y is replaced by b, the equation becomes a true statement.

For example, $(-2, 2)$ is a solution of the equation $y = |x|$, because $2 = |-2|$ is a true statement. Similarly, $(2, 2)$, $(-\sqrt{2}, \sqrt{2})$, and $(0, 0)$, along with infinitely-many other points, are also solutions of $y = |x|$. Some solutions of $x^2 + y^2 = 25$ are $(3, 4)$, $(-3, 4)$, $(4, 3)$, and $(-4, -3)$; and again there are infinitely-many other solutions.

Definition 1.5.2

> The set of all points that are solutions of an equation involving x and y is called the *graph of the equation.*

Definition 1.5.3

> The *graph of a function* f is the graph of the equation $y = f(x)$.

The graph of a function or an equation serves much the same purpose as a roadmap giving us an overview of the relationship between x and y. Until we develop some additional techniques we shall sketch a graph by determining some of the points on it and connecting them by a curve or line.

Let us look at some examples.

Example 1 | Sketch the graph of the function $f(x) = x$.

We set $y = f(x)$, that is, $y = x$, and find some solutions of this equation. When $x = 1$, we have $y = 1$; so $(1, 1)$ is a solution. Similarly, $(2, 2)$, $(3, 3)$, $(-1, -1)$, $(0, 0)$, $(-2, -2)$, and $(\sqrt{2}, \sqrt{2})$ are solutions. Those points can be conveniently summarized and plotted as indicated in Figure 1.5.1. ||

We shall see later that the graph of this function is a straight line that goes through the origin at an angle of 45°. The graph is pictured in Figure 1.5.2.

Figure 1.5.1

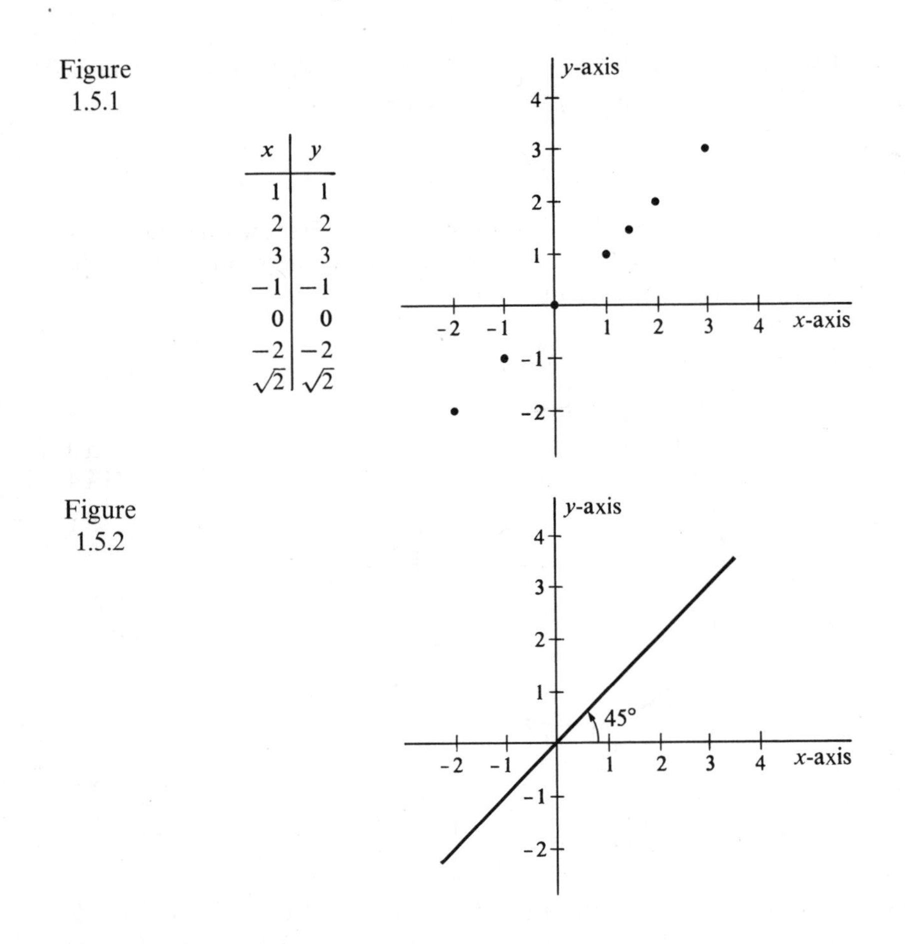

x	y
1	1
2	2
3	3
-1	-1
0	0
-2	-2
$\sqrt{2}$	$\sqrt{2}$

Figure 1.5.2

Example 2 | Sketch the graph of $y = |x|$.

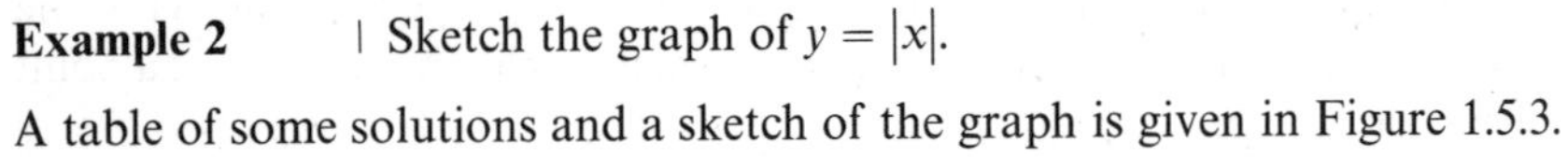

A table of some solutions and a sketch of the graph is given in Figure 1.5.3. ||

Figure 1.5.3

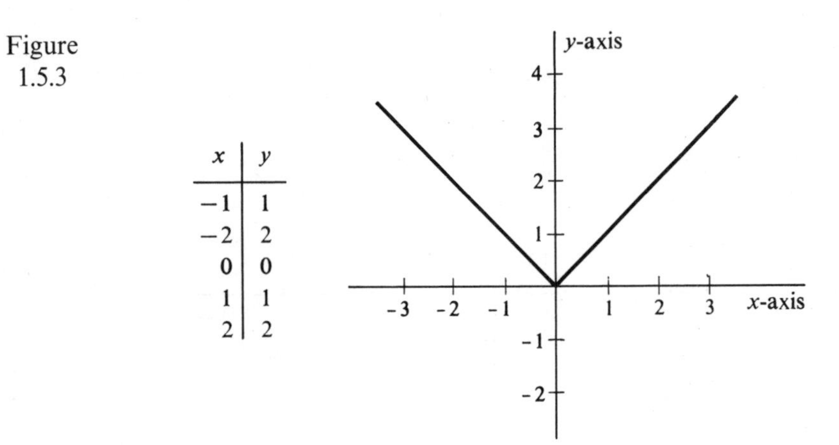

x	y
-1	1
-2	2
0	0
1	1
2	2

Notice that since the graphs in Examples 1 and 2 extend infinitely far out, we cannot completely draw either graph. We can, however, get a good idea of the graphs from what we have drawn. Usually when we sketch a graph we shall have to be content to sketch only a relatively small part of the entire graph.

Example 3 | Sketch the graph of $x^2 + y^2 = 25$.

With the aid of the indicated table of values, we obtain the graph shown in Figure 1.5.4. As we shall see later, the graph in Figure 1.5.4 is a circle of radius 5 with center at the origin. (In this case we have sketched the whole graph.) ||

Figure 1.5.4

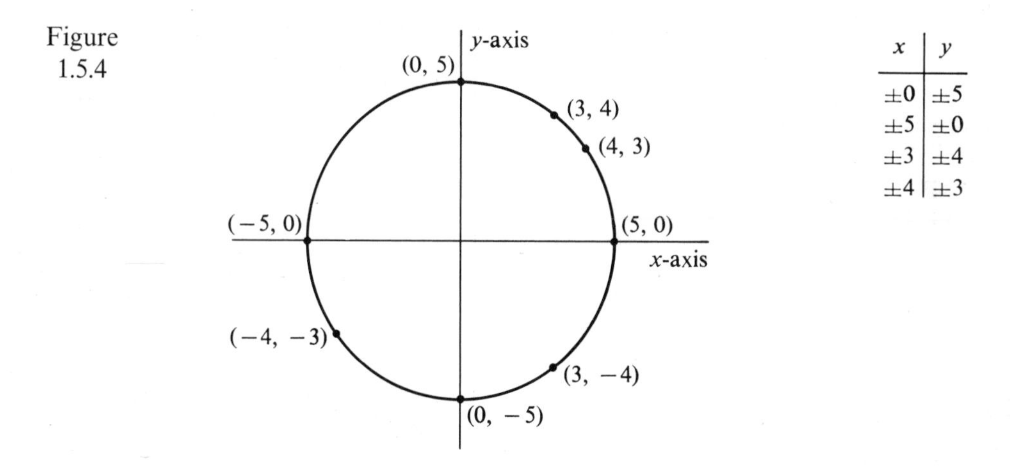

x	y
± 0	± 5
± 5	± 0
± 3	± 4
± 4	± 3

Example 4 | Sketch the graph of $y = \begin{cases} x & \text{if} \quad x \neq 5 \\ 1 & \text{if} \quad x = 5. \end{cases}$

We see that except for $x = 5$, the graph is the same as the one in Example 1. For $x = 5$, however, $y = 1$; so the point (5, 1) is on the graph and the point (5, 5) is not. Thus the graph can be sketched as in Figure 1.5.5, where the small circle ◦ indicates that a point is missing from the graph. ||

Figure 1.5.5

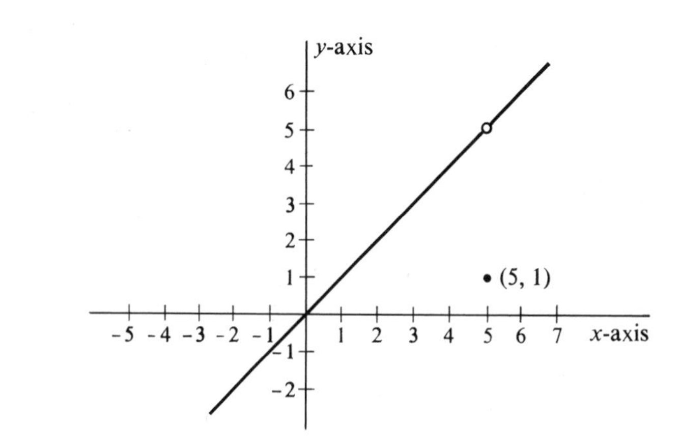

Note that to draw this graph we must lift the pencil off the paper. But the graphs of Examples 1, 2, and 3 have no breaks and so can be drawn without taking the pencil off the paper.

Example 5 Sketch the graph of $y = \begin{cases} x & \text{if} \quad x \le 1 \\ 1 & \text{if} \quad 1 < x \le 2 \\ 3 & \text{if} \quad x > 2. \end{cases}$

Here there is a break at $x = 2$, but no other breaks. There is a corner at $x = 1$, but no break there. ||

Figure 1.5.6

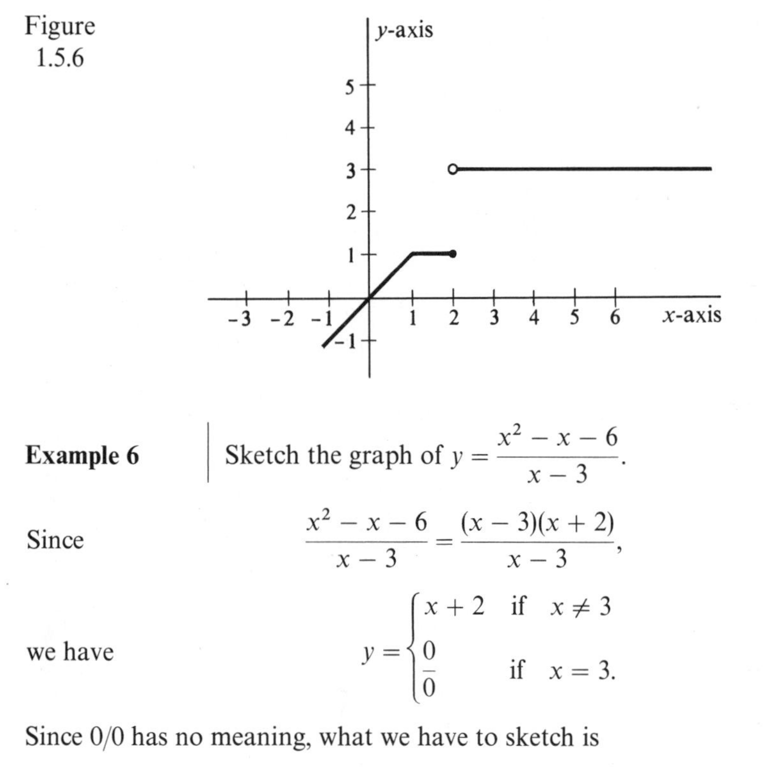

Example 6 Sketch the graph of $y = \dfrac{x^2 - x - 6}{x - 3}$.

Since
$$\frac{x^2 - x - 6}{x - 3} = \frac{(x - 3)(x + 2)}{x - 3},$$

we have
$$y = \begin{cases} x + 2 & \text{if} \quad x \ne 3 \\ \dfrac{0}{0} & \text{if} \quad x = 3. \end{cases}$$

Since 0/0 has no meaning, what we have to sketch is
$$y = \begin{cases} x + 2 & \text{if} \quad x \ne 3 \\ \text{no value} & \text{if} \quad x = 3. \end{cases}$$

With aid of the indicated table of values we obtain the graph shown in Figure 1.5.7.

Figure 1.5.7

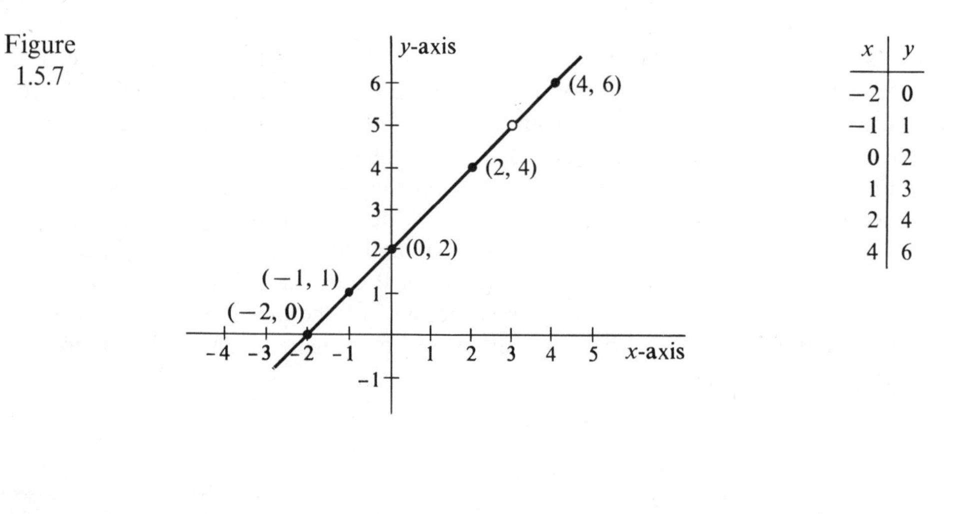

x	y
−2	0
−1	1
0	2
1	3
2	4
4	6

There is a break at $x = 3$, but no other breaks. For $x = 3$ there is no point at all on the graph. ||

Exercises 1.5

Sketch the graph of each function or equation. Also (except for Exercises 1, 4, and 16) either indicate that the graph has no breaks or list the values of x for which breaks occur.

1. $x = 2$.
2. $y = 3$.
3. $y = -4$.
4. $x = -5$.
5. $y = 3x$.
6. $y = 2x$.
7. $y = -2x$.
8. $y = -x$.
9. $f(x) = 2|x|$.
10. $f(x) = 1 + |x|$.
11. $f(x) = x + |x|$.
12. $f(x) = x/|x|$.
13. $x = 1/y$.
14. $y = 1/x$.
15. $x = y + 1$.
16. $y^2 = x$.
17. $y = \sqrt{x^2}$.
18. $y = \sqrt{(x-1)^2}$.
19. $f(x) = \begin{cases} 1 & \text{if } x < 0 \\ 2 & \text{if } x \geq 0. \end{cases}$
20. $f(x) = \begin{cases} 5 & \text{if } x \leq 1 \\ 4 & \text{if } x > 1. \end{cases}$
21. $f(x) = \begin{cases} x & \text{if } x \leq 2 \\ 2 & \text{if } x > 2. \end{cases}$
22. $f(x) = \begin{cases} 1 & \text{if } x < 1 \\ x & \text{if } x \geq 1. \end{cases}$
23. $f(x) = \begin{cases} x & \text{if } x \neq 2 \\ 5 & \text{if } x = 2. \end{cases}$
24. $f(x) = \begin{cases} x^2 & \text{if } x < 2 \\ 2x & \text{if } 2 < x < 5 \\ 2 & \text{if } x \geq 5. \end{cases}$
25. $f(x) = \dfrac{x^2 + 2x}{x}$.
26. $f(x) = \dfrac{x^2 - 1}{x - 1}$.
27. $f(x) = \begin{cases} \dfrac{x^2 + 2x}{x} & \text{if } x \neq 0 \\ 2 & \text{if } x = 0. \end{cases}$
28. $f(x) = \begin{cases} \dfrac{x^2 - 1}{x - 1} & \text{if } x \neq 1 \\ 2 & \text{if } x = 1. \end{cases}$

1.6 Lines

In mathematics the term *line* means *straight line.* If a line is not horizontal, it intersects the x-axis at one and only one point P. The *inclination* of such a line is defined as the positive angle, with P as vertex, from the part of the x-axis to the right of P, to the line. If a line is horizontal we define its inclination to be $0°$ or, equivalently, 0 radians. In Figures 1.6.1 and 1.6.2, the inclinations of the lines shown are α and β, respectively. To avoid distorting the angles, the same unit of measure must be used on both axes.

Note that it follows from the definition of inclination that if θ is the inclination of a line, then $0 \leq \theta < 180°$ if θ is in degrees, or $0 \leq \theta < \pi$ if θ is in radians. If θ is the inclination of a line, then $\tan \theta$ is a measure of the amount the y-coordinate changes per unit change in the x-coordinate. This measure is sufficiently important to be given a special name.

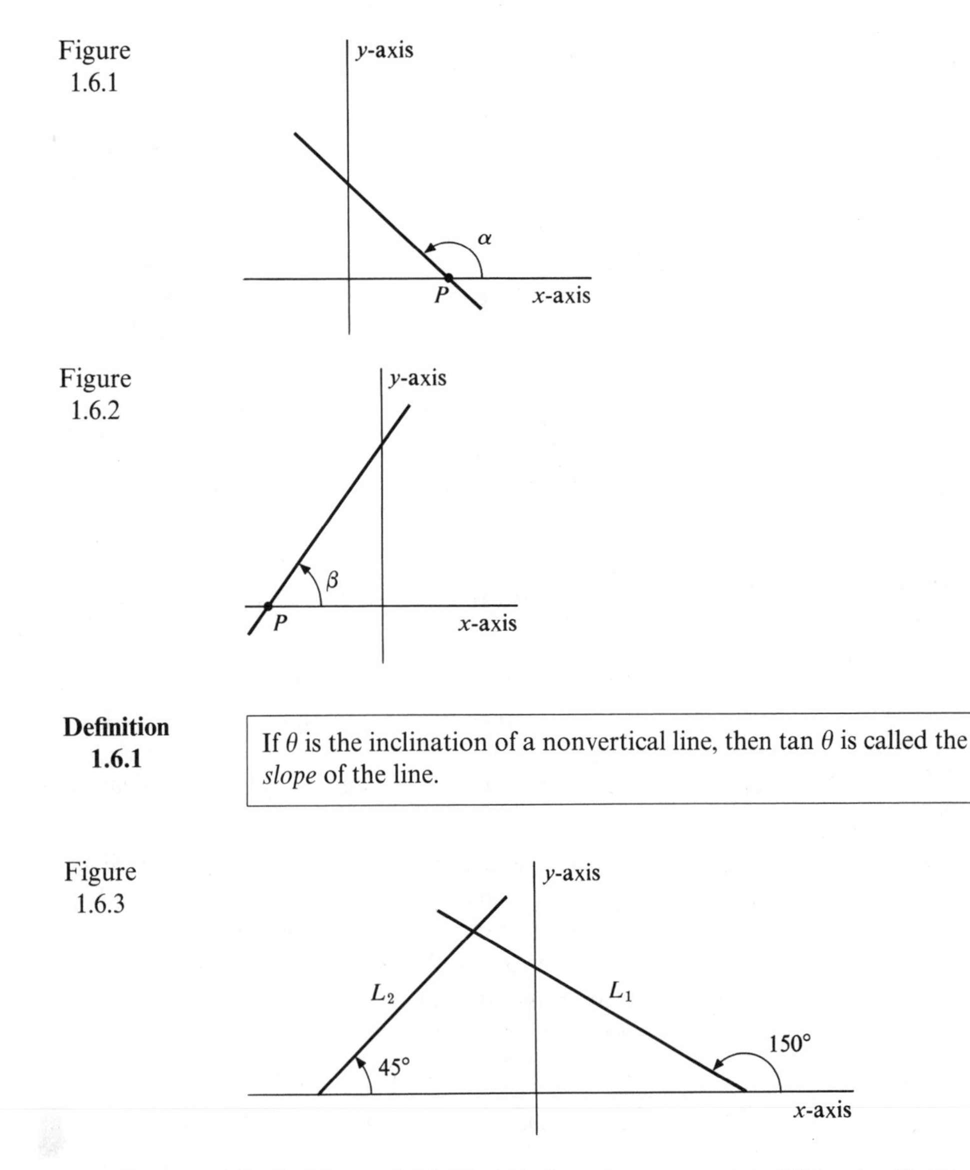

Figure 1.6.1

Figure 1.6.2

Definition 1.6.1

If θ is the inclination of a nonvertical line, then $\tan \theta$ is called the *slope* of the line.

Figure 1.6.3

For example, in Figure 1.6.3, line L_1 has slope $m_1 = \tan 150^\circ = \tan(5\pi/6) = -1/\sqrt{3}$, and line L_2 has slope $m_2 = \tan 45^\circ = \tan(\pi/4) = 1$.

Since vertical lines all have inclination of 90°, or $\pi/2$ radians, and since $\tan 90^\circ$ is undefined, we could not use $\tan \theta$ to define the slope of a vertical line. There is, in fact, no useful way to define the slope of a vertical line; accordingly, vertical lines do not have slopes. On the other hand, horizontal lines have inclination 0°, and since $\tan 0^\circ = 0$, each horizontal line has a slope of 0. Notice the distinction between saying that a line has no slope and saying that a line has slope 0.

The function $\tan \theta$ is positive and increasing as θ varies from 0 to $\pi/2$, and is negative and increasing as θ varies from $\pi/2$ to π. Consequently the slope of a line is positive if the line slopes upward to the right and negative if the line slopes downward to the right. Figure 1.6.4 pictures this relationship.

Figure 1.6.4

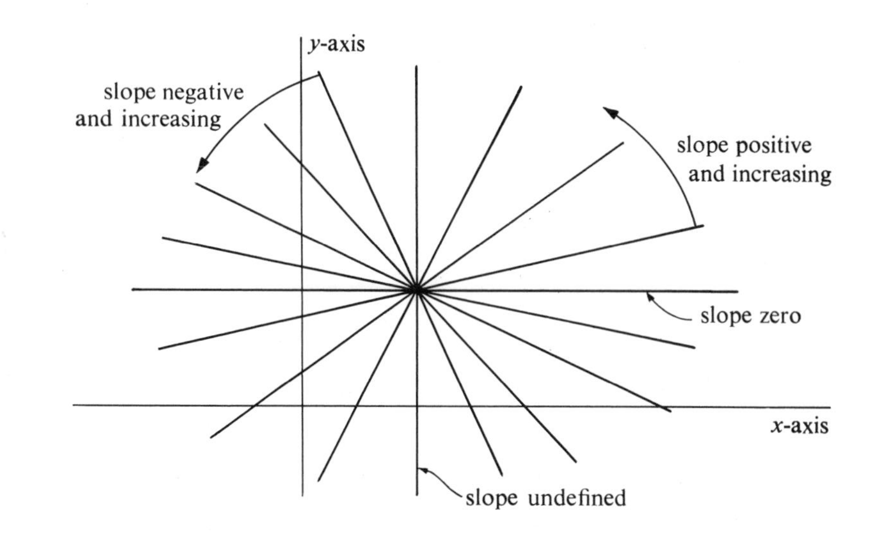

Theorem 1.6.1

> Let L_1 and L_2 be nonvertical lines with slopes m_1 and m_2, respectively. Then
>
> (i) L_1 and L_2 are parallel if and only if $m_1 = m_2$;
>
> (ii) L_1 and L_2 are perpendicular if and only if $m_1 m_2 = -1$.

Proof. Let α_1 be the inclination of L_1 and let α_2 be the inclination of L_2.

(i) L_1 is parallel to L_2 if and only if $\alpha_1 = \alpha_2$ (see Figure 1.6.5). Since both of these angles are greater than or equal to zero and less than 180°, $\alpha_1 = \alpha_2$ if and only if $\tan \alpha_1 = \tan \alpha_2$, which is true if and only if $m_1 = m_2$. Therefore, L_1 is parallel to L_2 if and only if $m_1 = m_2$.

(ii) If L_1 and L_2 are perpendicular, let L_2 be the line with the larger inclination as in Figure 1.6.6. Then, since $\alpha_1 + \beta = 90°$ and $\alpha_2 + \beta = 180°$, we have $\alpha_2 = \alpha_1 + 90°$. Therefore,

$$m_1 m_2 = \tan \alpha_1 \tan \alpha_2 = \tan \alpha_1 \tan(\alpha_1 + 90°) = \tan \alpha_1(-\cot \alpha_1) = -1.$$

Conversely, if $m_1 m_2 = -1$, them $m_2 \neq 0$ (and $m_1 \neq 0$), and so $m_1 = -1/m_2$. But $m_1 = \tan \alpha_1$ and $m_2 = \tan \alpha_2$; so $\tan \alpha_1 = -1/\tan \alpha_2 = -\cot \alpha_2$.

But since $0 \leq \alpha_1 < 180°$ and $0 \leq \alpha_2 < 180°$, that can only happen if $\alpha_1 = \alpha_2 + 90°$ or $\alpha_2 = \alpha_1 + 90°$, and in either case, L_1 and L_2 are perpendicular.

For example, if one line has slope $-2/3$ and a second line of slope m is perpendicular to it, then $(-2/3)m = -1$, and so $m = 3/2$.

Theorem 1.6.1 deals only with lines that have slopes. If one of two perpendicular lines does not have a slope, then it must be vertical, and a line perpendicular to it must be horizontal and have slope 0. Conversely, if one line has no slope and a second line

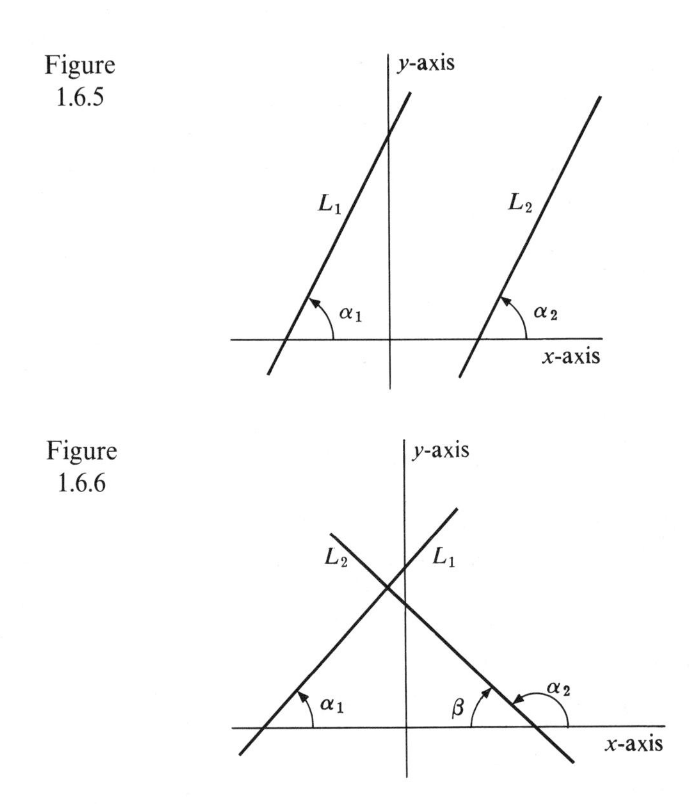

Figure 1.6.5

Figure 1.6.6

has slope 0, the first line is vertical and the second one is horizontal, and hence they are perpendicular.

It is useful to be able to express the slope of a line in terms of the coordinates of any two distinct points on the line. The following theorem indicates how this can be done.

Theorem 1.6.2

> If (x_1, y_1) and (x_2, y_2) are distinct points on a nonvertical line L, and m is slope of L, then
>
> $$m = \frac{y_2 - y_1}{x_2 - x_1}.$$

Proof. It is usually best to refer to a figure when going through a proof. It should be realized, however, that a particular figure represents only one possible situation. A figure is usually drawn to give a reasonably general picture of the situation, but the proof must hold for *all* cases permitted by the hypothesis of the theorem. These ideas should be kept in mind when studying this proof and all others involving diagrams. We now proceed with the proof.

For convenience we consider a new coordinate system with its x'-axis parallel to the x-axis, its y'-axis parallel to the y-axis, and both of the new coordinate axes passing through the point (x_1, y_1). Then, since the x'-axis is parallel to the x-axis, the inclination θ of the line L

relative to the x-axis will be in standard position relative to the $x'y'$-coordinate system.

The point (x_2, y_2) is then a point on the terminal side of θ, and if its coordinates relative to the new coordinate system are (x', y'), we have

$$x' = x_2 - x_1 \quad \text{and} \quad y' = y_2 - y_1.$$

Therefore
$$m = \tan\theta = \frac{y'}{x'} = \frac{y_2 - y_1}{x_2 - x_1}$$

and the theorem is proved.

Figure 1.6.7

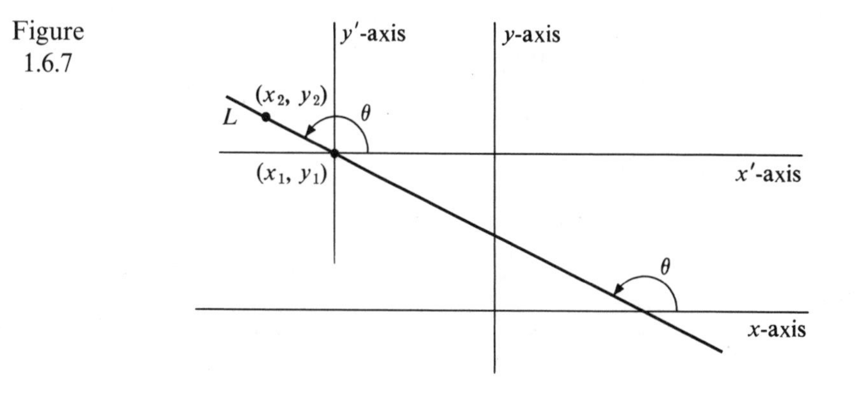

For example, if $(-2, 3)$ and $(7, 5)$ are points on a line, then the slope m of the line is given by

$$m = \frac{5-3}{7-(-2)} = \frac{2}{9}.$$

Here we have considered $(-2, 3)$ to be the point (x_1, y_1) and $(7, 5)$ to be the point (x_2, y_2). Actually we could have made our choices in the opposite way because

$$\frac{5-3}{7-(-2)} = \frac{(-1)[5-3]}{(-1)[7-(-2)]} = \frac{3-5}{-2-7}.$$

In general,
$$\frac{y_2 - y_1}{x_2 - x_1} = \frac{(-1)(y_1 - y_2)}{(-1)(x_1 - x_2)} = \frac{y_1 - y_2}{x_1 - x_2};$$

so it is immaterial which point is chosen as (x_1, y_1).

Now that we have defined the slope of a line and have a way of finding it by using the coordinates of two distinct points on the line, we shall explain how to find an *equation* of a line. First we define what is meant by an *equation of a set*, in a way compatible with our definition, in Section 1.5, of the *graph of an equation.*

Definition 1.6.2

An *equation of a set* is an equation that is satisfied by all points of the set and not satisfied by any other points.

For example, if the set is a vertical line 2 units to the left of the y-axis, then $x = -2$ is an equation of the set because every point on this line has x-coordinate of -2 and points not on this line do not. The equations $x + 2 = 0$ and $3 - x = 5$ are also equations of this set; so we see that, in general, there is no unique equation of a set. That is why we say "*an* equation" of a set rather than "*the* equation" of a set.

Theorem 1.6.3

An equation of the line that has slope m and passes through the point (x_1, y_1) is

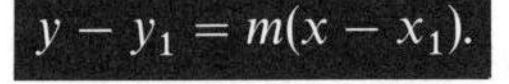

$$y - y_1 = m(x - x_1).$$

Proof. As in Figure 1.6.8, let (x, y) be any point on the line different from (x_1, y_1). Then by Theorem 1.6.2,

$$\frac{y - y_1}{x - x_1} = m, \tag{1}$$

which is an equation satisfied by all points on the line except (x_1, y_1). However, if we multiply both sides of (1) by $x - x_1$, we get

$$y - y_1 = m(x - x_1), \tag{2}$$

which is satisfied by all points that satisfy (1), and which is also satisfied by (x_1, y_1). Thus every point on the line is a solution of (2).

Figure 1.6.8

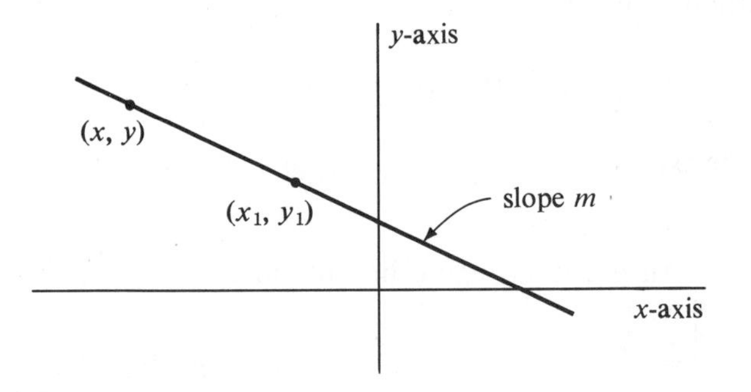

If a point (a, b) is not on the line, then it cannot be a solution of (2); for if it were, we would have

$$b - y_1 = m(a - x_1).$$

Now, either $a - x_1 = 0$ or $a - x_1 \neq 0$. If $a - x_1 = 0$, then $b - y_1 = 0$, and so $b = y_1$, $a = x_1$, and (a, b) is the point (x_1, y_1). But this cannot be the case because (a, b) is not on the line. Therefore $a - x_1 \neq 0$, and so we may divide by it to obtain

$$m = \frac{b - y_1}{a - x_1}.$$

Hence the line passing through the points (a, b) and (x_1, y_1) has slope m. Since there is only one line passing through (x_1, y_1) with slope m, this would force (a, b) to be on the given line, which is a contradiction. Thus we have shown that a point not on the line is not a solution of $y - y_1 = m(x - x_1)$.

We have shown that the equation $y - y_1 = m(x - x_1)$ is satisfied by all points on the given line and is not satisfied by any other points, and so it is an equation of the line.

Example 1 An equation of the line of slope $-1/2$ passing through the point $(1, -3)$ is

$$y - (-3) = -\frac{1}{2}(x - 1) \quad \text{or} \quad y + 3 = -\frac{1}{2}(x - 1).$$

Other equations for the same line are $2y + 6 = -x + 1$ and $x + 2y + 5 = 0$. ||

Example 2 Find an equation of the line passing through the points $(-3, -2)$ and $(1, -1)$.

First we find the slope:

$$m = \frac{-1 - (-2)}{1 - (-3)} = \frac{1}{4}.$$

Then an equation of the line is

$$y - (-2) = \frac{1}{4}(x - (-3)) \quad \text{or} \quad y + 2 = \frac{1}{4}(x + 3).$$

This can also be expressed as $4y + 8 = x + 3$, or as $x - 4y - 5 = 0$. ||

Example 3 An equation of the line of slope m passing through the point $(0, b)$ is

$$y - b = m(x - 0),$$

that is, $y - b = mx$. This can be written as

$$y = mx + b.$$

The number b in this case is called the *y-intercept.* It is the y-coordinate of the point where the line crosses the y-axis, since $y = b$ when $x = 0$. In particular, an equation of the line with slope 2 and y-intercept 5 is $y = 2x + 5$. ||

From Example 3 we see that a line has slope m and y-intercept b if and only if it has an equation of the form

An important consequence of this is that the slope of a line can be read off as the coefficient of the x term after the equation has been solved for y. This fact is used in our next example.

Example 4 | Find an equation of the line passing through the point $(3, -4)$ and perpendicular to the line with equation $3x - 5y = 2$.

To find the slope m_1 of $3x - 5y = 2$, we first write the equation in the form $y = m_1x + b$, getting $5y = 3x - 2$, or $y = (3/5)x - (2/5)$. Hence $m_1 = 3/5$. Thus, if m_2 is the slope of the line in question, $(3/5)m_2 = -1$, and we have $m_2 = -5/3$. Consequently, an equation of the line passing through $(3, -4)$ and perpendicular to the line with equation $3x - 5y = 2$ is

$$y - (-4) = -\frac{5}{3}(x - 3),$$

which can be expressed as $3y + 12 = -5x + 15$, or as

$$5x + 3y - 3 = 0.$$ ||

If a line is vertical, it has no slope and consequently Theorem 1.6.3 is of no use in determining an equation for the line. However, every point on a vertical line has the same x-coordinate and a point not on the line has a different x-coordinate. Thus, a vertical line has an equation of the form $x = k$, where k is the common x-coordinate. For example, a vertical line 3 units to the right of the y-axis has an equation $x = 3$. As we saw earlier in this section, a vertical line 2 units to the left of the y-axis has an equation $x = -2$.

Similarly, a horizontal line has an equation of the form $y = k$, where k is the common y-coordinate of every point on the line. For example, the horizontal line 102 units above the x-axis has an equation $y = 102$.

Example 5 | Find an equation of the line passing through $(-3, 4)$ and perpendicular to the line $y = -5$.

Since the line $y = -5$ is horizontal, any line perpendicular to $y = -5$ is vertical. Thus, the line we seek must have an equation of the form $x = k$, where k is the common x-coordinate of all points on the line. Since the line passes through $(-3, 4)$, $k = -3$ and an equation of the line in question is $x = -3$. ||

In order to sketch a line from its equation, it is usually simplest to find two points on the line, plot them, and then draw the line.

Example 6 | Sketch the graph of the equation $4x - 5y = 8$.

If $x = 0$, then $y = -8/5$; so $(0, -8/5)$ is a point on the line. If $y = 0$, then $x = 2$; so $(2, 0)$ is a second point. Then we can sketch the graph as shown in Figure 1.6.9. ||

Figure 1.6.9

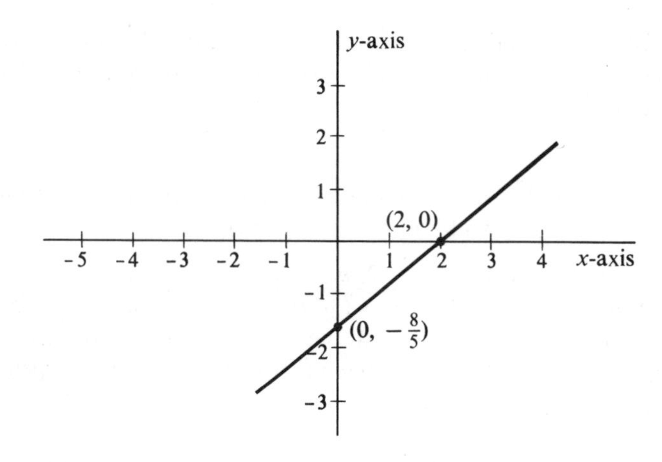

Exercises 1.6

In Exercises 1–4, find the slope of the line passing through the indicated pair of points.

1. (2, 5) and (6, −1).
2. (2, 0) and (1, 4).
3. (1, 3) and (1, −2).
4. (2, −3) and (5, −3).

In Exercises 5–22, find equations of the lines with the given properties.

5. Through (1, −6) with slope 2/5.
6. Through (8, −3) with slope 0.
7. Through (4, 7) and (−1, 3).
8. Through (−4, 1) and (−4, 3).
9. Through (1, 6) and (5, 6).
10. Through (1, 2) and (−3, 7).
11. Parallel to $x = 3$ through (−1, −2).
12. Parallel to $y = 1$ through (−2, 5).
13. Parallel to $3x + 5y = 2$ through (2, −1).
14. Parallel to $8x - 4y = 1$ through (1, 2).
15. Perpendicular to $x - 2y = 5$ through (1, −2).
16. Perpendicular to $x + y = 7$ through (4, 6).
17. Perpendicular to $2y - 3x = 1$ through (−3, 2).
18. Perpendicular to $x + 3y = 2$ through (−1, 2).
19. Perpendicular to $x = -3$ through (1, −5).
20. Perpendicular to $y = -2$ through (2, 5).
21. Perpendicular to $y = 5$ through (6, 7).
22. Perpendicular to $x = 11$ through (−2, −3).
23. Prove that the lines $16x + 20y = 9$ and $15y = 4 - 12x$ are parallel.
24. Prove that the lines $4x - 6y = 7$ and $6x = 7 + 9y$ are parallel.
25. Prove that the lines $6x - 9y = 5$ and $15x + 10y = 7$ are perpendicular.
26. Prove that the lines $x = 3y - 1$ and $2y + 6x = 1$ are perpendicular.
27. Prove that the lines $y + 3 = 0$ and $x - 5 = 0$ are perpendicular.
28. Prove that the lines $x + 1 = 0$ and $y = 7$ are perpendicular.
29. For what value of k does the line $5x + ky = 7$ go through the point (2, 3)?
30. For what value of h does the line $3hx - 2y = 11$ go through the point (−1, 2)?

Sketch the graphs of the equations in Exercises 31–42.

31. $x = 6$.
32. $y = -3$.
33. $y = 0$.
34. $x = 0$.
35. $y = 2x - 8$.
36. $x = 3y + 1$.
37. $x - 3y = 7$.
38. $3x = 4y$.
39. $6x + 3y = 12$.
40. $5y - 2x + 3 = 0$.
41. $y = |x - 2|$.
42. $y = |x + 1|$.

43. There is a linear relationship between temperature measured in degrees Celsius and degrees Fahrenheit. Water boils at 212°F and 100°C. Water freezes at 32°F and 0°C. Find the relationship between Fahrenheit and Celsius temperatures.

44. The relationship between pounds and kilograms is linear. If a 6-kilogram object weighs 13.2 lbs, find the relationship between pounds and kilograms.

1.7 Distance, Circles, Ellipses, Hyperbolas, and Parabolas

Our previous work in Section 1.4 enabled us to determine the distance between any two points on a real line. In this section we shall use that information and The Pythagorean Theorem to determine the distance between two points (a, b) and (c, d) in the plane.

Note that the line segment connecting the points (c, d) and (c, b) is parallel to the y-axis and the line segment connecting the points (a, b) and (c, b) is parallel to the x-axis. Consequently these line segments are perpendicular, and the points (a, b), (c, d), and (c, b) form a right triangle as pictured in Figure 1.7.1. Since (a, b) and (c, b) lie on the same horizontal line, the distance between them is $|a - c|$. Similarly, the distance between (c, d) and (c, b) is $|b - d|$. Then the distance between the points (a, b) and (c, d), which we shall denote by $D((a, b), (c, d))$, is the length of the hypotenuse of a right triangle with sides of length $|a - c|$ and $|b - d|$. Thus, from the Pythagorean Theorem we have this *distance formula*:

$$D((a, b), (c, d)) = \sqrt{|a - c|^2 + |b - d|^2}.$$

Of course, $|a - c|^2 = (a - c)^2$ and $|b - d|^2 = (b - d)^2$, and so we may write

$$D((a, b), (c, d)) = \sqrt{(a - c)^2 + (b - d)^2}.$$

Since $(a - c)^2 = (c - a)^2$ and $(b - d)^2 = (d - b)^2$, we get

$$D((a, b), (c, d)) = \sqrt{(c - a)^2 + (d - b)^2} = D((c, d), (a, b)).$$

To illustrate the use of the distance formula we shall consider a few examples.

Figure 1.7.1

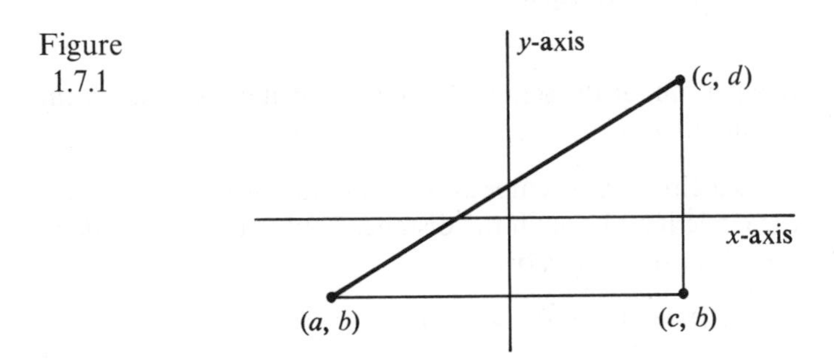

Example 1 | Find the distance between the points $(1, -3)$ and $(-4, 5)$.

Using the distance formula, we have

$$\begin{aligned} D((1, -3), (-4, 5)) &= \sqrt{(1-(-4))^2 + (-3-5)^2} \\ &= \sqrt{25 + 64} \\ &= \sqrt{89}. \end{aligned}$$

||

Example 2 | Show that the point $\left(\frac{a+c}{2}, \frac{b+d}{2}\right)$ is equidistant from the points (a, b) and (c, d).

It is sufficient to show that

$$D\left((a, b), \left(\frac{a+c}{2}, \frac{b+d}{2}\right)\right) = D\left((c, d), \left(\frac{a+c}{2}, \frac{b+d}{2}\right)\right).$$

First note that

$$\begin{aligned} D\left((a, b), \left(\frac{a+c}{2}, \frac{b+d}{2}\right)\right) &= \sqrt{\left(a - \frac{a+c}{2}\right)^2 + \left(b - \frac{b+d}{2}\right)^2} \\ &= \sqrt{\left(\frac{a-c}{2}\right)^2 + \left(\frac{b-d}{2}\right)^2} \\ &= \sqrt{\frac{(a-c)^2}{4} + \frac{(b-d)^2}{4}}. \end{aligned}$$

Now note that

$$\begin{aligned} D\left((c, d), \left(\frac{a+c}{2}, \frac{b+d}{2}\right)\right) &= \sqrt{\left(c - \frac{a+c}{2}\right)^2 + \left(d - \frac{b+d}{2}\right)^2} \\ &= \sqrt{\left(\frac{c-a}{2}\right)^2 + \left(\frac{d-b}{2}\right)^2} \\ &= \sqrt{\frac{(c-a)^2}{4} + \frac{(d-b)^2}{4}} \\ &= \sqrt{\frac{(a-c)^2}{4} + \frac{(b-d)^2}{4}} \end{aligned}$$

Therefore the required two distances are equal. ||

Example 3 | Find an equation for the set of all points (x, y) at a distance r from a given point (h, k).

Note that the described set is a circle with center at (h, k) and radius r. And observe that a point (x, y) is on the circle if and only if the distance between (x, y) and (h, k) is r; that is, if and only if, $D((x, y), (h, k)) = r$, or

$$\sqrt{(x-h)^2 + (y-k)^2} = r, \tag{1}$$

which is an equation of the circle. We may simplify the equation by noting that if a point satisfies equation (1), then it must also satisfy

$$(x - h)^2 + (y - k)^2 = r^2. \tag{2}$$

Conversely, if a point (x, y) satisfies (2) it must also satisfy

$$\pm\sqrt{(x - h)^2 + (y - k)^2} = r.$$

However, since $r > 0$, we may eliminate the minus sign. Hence a point (x, y) that satisfies (2) must also satisfy (1). Consequently, the more convenient form

$$(x - h)^2 + (y - k)^2 = r^2. \tag{3}$$

is also an equation of the circle with center at (h, k) and radius r. ||

When the center of the circle is at the origin, it follows that $h = 0 = k$, and so we have the simpler form

$$x^2 + y^2 = r^2. \tag{4}$$

Figure 1.7.2

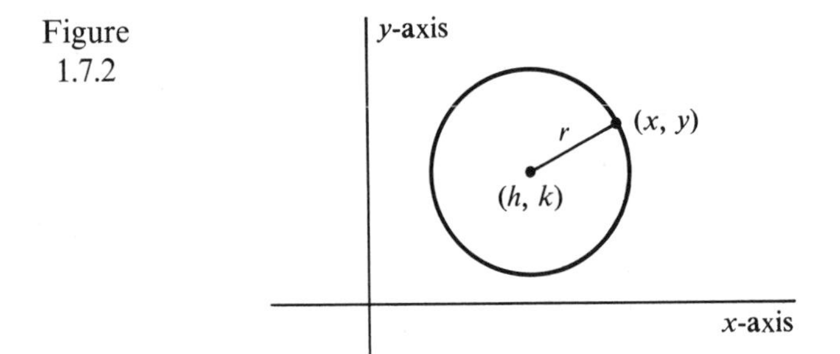

An equation resembling (4) is

$$\frac{x^2}{a^2} + \frac{y^2}{b^2} = 1, \tag{5}$$

where a and b are fixed positive real numbers that are not necessarily different. It is an equation of a circle if and only if $a = b$. The graph of (5) is called an *ellipse.* It becomes more and more circular in shape as a is taken more nearly equal to b. Thus an ellipse can be thought of as a distorted circle. In fact, when looking at a circle such as the top of a drinking glass from an oblique angle, the circle appears to be an ellipse.

The graph of (5) goes through the four points $(\pm a, 0)$ and $(0, \pm b)$ and is pictured in Figure 1.7.3.

The largest diameter of an ellipse is called the *major axis* and the smallest diameter is called the *minor axis.* For an ellipse with an equation of the form (5), the length of the major axis is the larger of the numbers $2a$ and $2b$, and the length of the minor axis is the smaller of those two numbers. The graph can be quickly sketched after plotting the points at the ends of the major and minor axes.

Figure 1.7.3

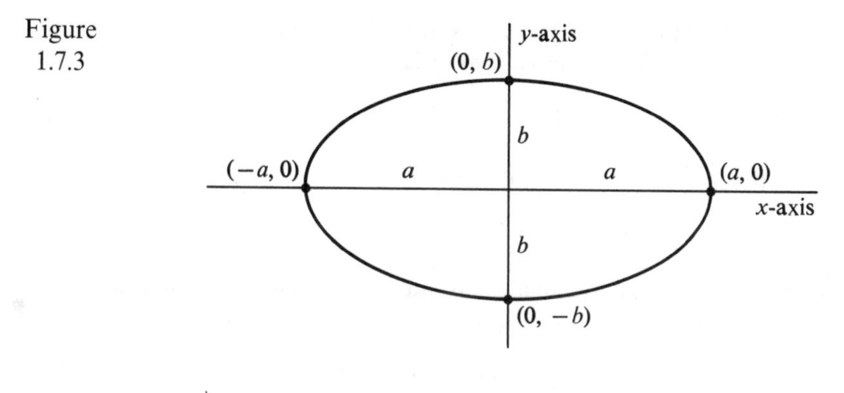

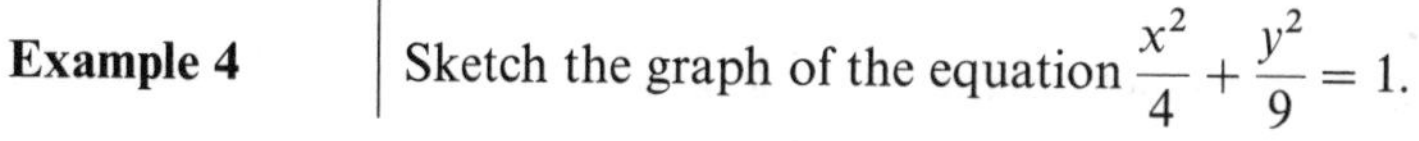

Example 4 Sketch the graph of the equation $\frac{x^2}{4} + \frac{y^2}{9} = 1$.

From our discussion above, the graph of this equation is known to be an ellipse centered at (0, 0), with $a = 2$ and $b = 3$. In addition the graph passes through the points (0, 3), (0, −3), (2, 0), and (−2, 0), which are the ends of the major and minor axes. Using this information we can obtain the sketch indicated in Figure 1.7.4. ||

Figure 1.7.4

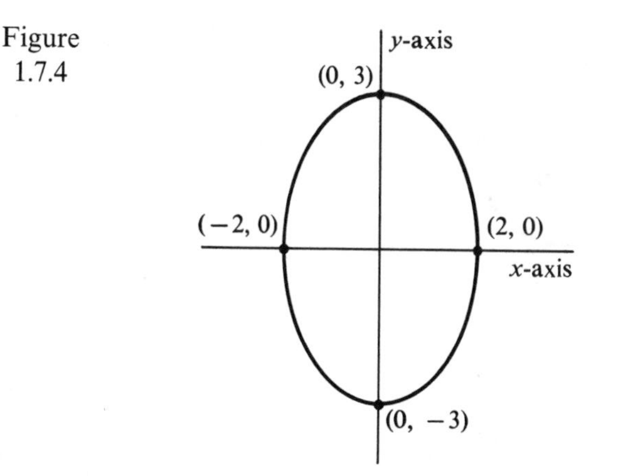

An equation like (5) with a minus sign instead of the plus sign has a graph called a *hyperbola*. Thus the graph of

$$\frac{x^2}{a^2} - \frac{y^2}{b^2} = 1, \tag{6}$$

where a and b are fixed positive real numbers that are not necessarily different, is called a hyperbola.

If we solve (6) for y^2, we obtain

$$y^2 = \frac{b^2}{a^2}(x^2 - a^2),$$

which can be written as

$$y^2 = \frac{b^2}{a^2}x^2\left(1 - \frac{a^2}{x^2}\right).$$

Solving for y, we have

$$y = \pm\frac{b}{a}x\left(1 - \frac{a^2}{x^2}\right)^{1/2}.$$

When $|x|$ is large, a^2/x^2 is small, and the graph is very near the graph of $y = \pm(b/a)x$, that is, very near the lines $y = (b/a)x$ and $y = (-b/a)x$. In fact the larger $|x|$ gets, the nearer the graph is to those lines, but it never touches or crosses them. The graph can be sketched quickly by first plotting the points $(\pm a, 0)$ that satisfy (6), and then by making a smooth curve that gets closer and closer to the lines $y = \pm(b/a)x$ as $|x|$ gets larger and larger, as indicated in Figure 1.7.5.

Figure 1.7.5

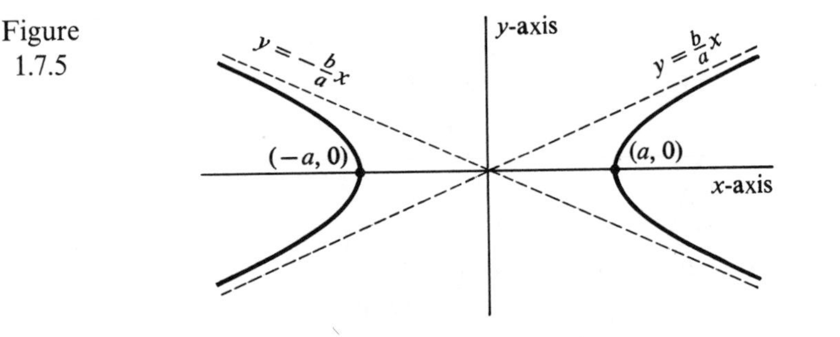

The lines $y = (b/a)x$ and $y = -(b/a)x$ are not part of the graph of the hyperbola (6). However, these lines, called *asymptotes*, are helpful in sketching the graph because it gets arbitrarily close to them. The segment of length $2a$ between the points $(-a, 0)$ and $(a, 0)$ of the graph of (6) is called the *transverse axis* of the hyperbola (6).

Similarly, an equation of the form

$$\frac{y^2}{b^2} - \frac{x^2}{a^2} = 1 \tag{7}$$

also represents a hyperbola with the roles of x and y interchanged. Its graph is indicated in Figure 1.7.6. Note that its transverse axis goes between the points $(0, b)$ and $(0, -b)$ of the graph. However the asymptotes of (7) are the same as the asymptotes of (6). Hyperbolas (6) and (7) are called *conjugate* hyperbolas. Each is called the *conjugate* of the other.

Figure 1.7.6

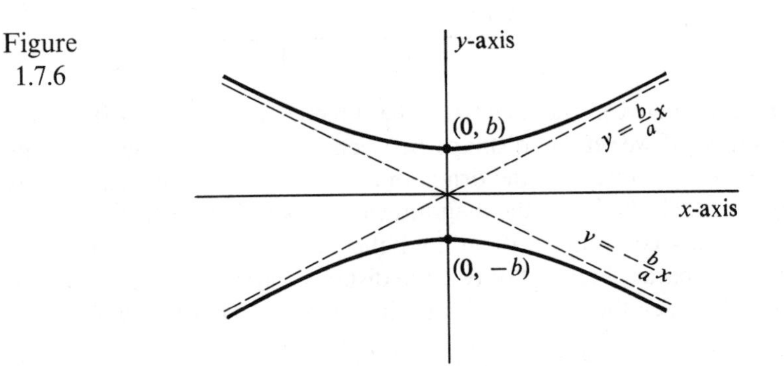

Example 5 | Sketch the graph of the equation $\dfrac{x^2}{9} - \dfrac{y^2}{16} = 1$.

In this case $a = 3$ and $b = 4$. By plotting the asymptotes $y = (4/3)x$ and $y = -(4/3)x$ and noting that the ends of the transverse axes are the points $(3, 0)$ and $(-3, 0)$, we obtain the sketch indicated in Figure 1.7.7. The origin is said to be the *center* of a hyperbola of the form

$$\frac{x^2}{a^2} - \frac{y^2}{b^2} = 1$$

or

$$\frac{y^2}{b^2} - \frac{x^2}{a^2} = 1. \qquad \|$$

Figure 1.7.7

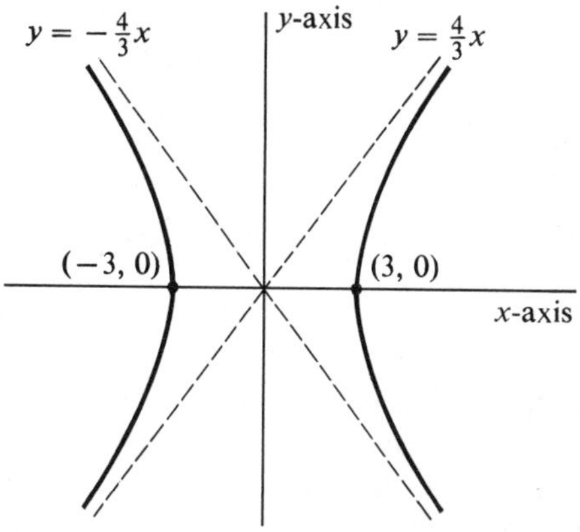

$y = -\frac{4}{3}x$ y-axis $y = \frac{4}{3}x$ $(-3, 0)$ $(3, 0)$ x-axis

The set of points equidistant from a fixed point and a fixed line is called a *parabola.* The fixed point is called the *focus* and the fixed line is called the *directrix.* A picture of such a set is shown in Figure 1.7.8.

Figure 1.7.8

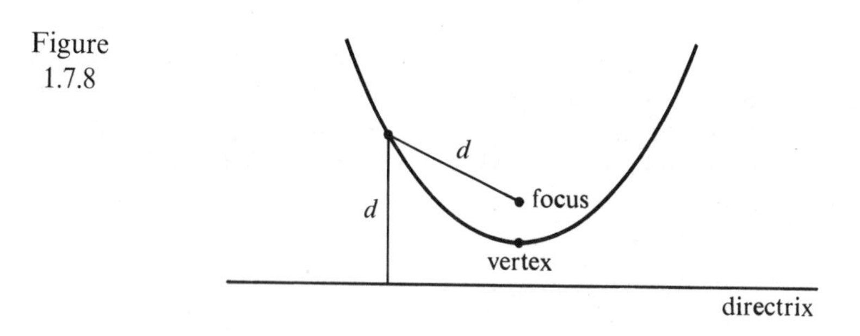

The point midway between the focus and the directrix is a point on the parabola called the *vertex.* If we let the distance between the focus and the vertex be the positive number p, and if we take the origin of a coordinate system at the vertex and put the y-axis through the focus, the diagram will be as in Figure 1.7.9.

The focus then has coordinates $(0, p)$, and the directrix is the line $y = -p$. A point (x, y) is on the parabola if and only if the distance $D((x, y), (0, p))$ is equal to the distance $|y + p|$ between the point (x, y) and the directrix. That is, if and only if

$$|y + p| = \sqrt{x^2 + (y - p)^2}, \tag{8}$$

Figure 1.7.9

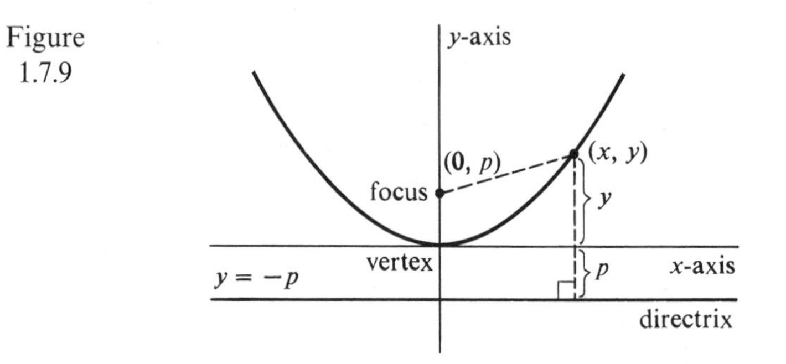

which is an equation of the parabola. We can write a simpler equation for the parabola, however. A point satisfies (8) if and only if

$$(y+p)^2 = x^2 + (y-p)^2, \quad \text{or}$$
$$y^2 + 2py + p^2 = x^2 + y^2 - 2py + p^2.$$

By eliminating y^2 and p^2 and solving for y, we see that

$$y = \frac{x^2}{4p} \tag{9}$$

is also an equation of the parabola. Note that Figure 1.7.9 shows that this parabola "opens upward."

When the focus is at $(0, -p)$ and the directrix is the line $y = p$, with $p > 0$, the vertex is still at the origin, but an equation of the parabola is

$$y = -\frac{x^2}{4p},$$

and the parabola opens downward instead of upward.

Example 6 The graphs of the equations $y = x^2/8$ and $y = 6x^2$ are both parabolas opening upward like the one in Figure 1.7.9. ||

Example 7 The graph of the equation $x = y^2/8$ is also a parabola with the roles of the x-axis and y-axis interchanged. Its graph is illustrated in Figure 1.7.10. ||

Figure 1.7.10

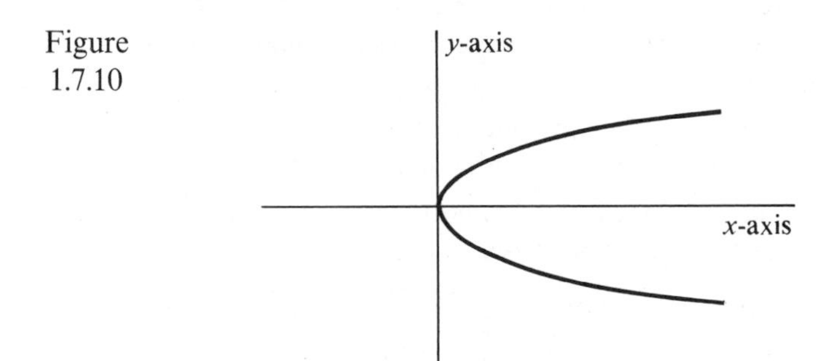

Example 8 | The graph of the equation $y = -2x^2$ is a parabola opening downward and is shown in Figure 1.7.11. ||

Figure 1.7.11

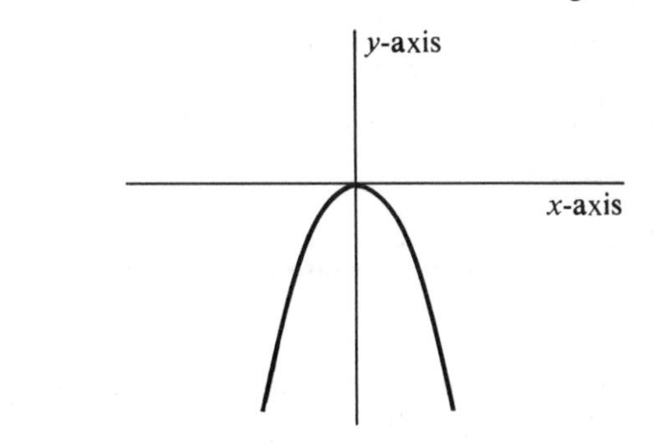

Circles, ellipses, hyperbolas, and parabolas are often called *conic sections*, or *conics*, because each is formed by the intersection of a double cone with an appropriate plane that misses the vertex of the cone. This is illustrated in Figure 1.7.12.

Figure 1.7.12

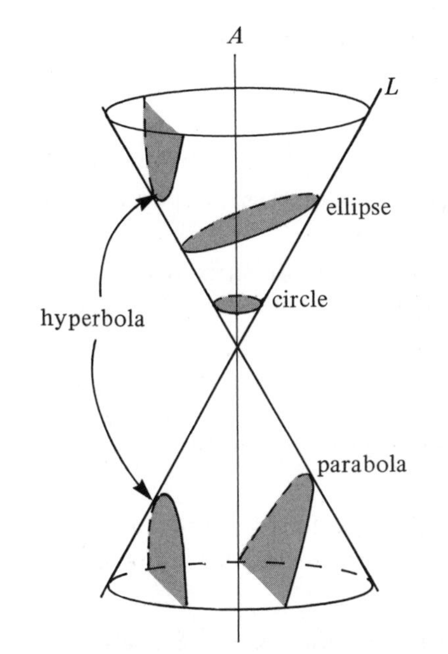

It can be shown that:

a If the cutting plane cuts entirely across one half of the cone, the intersection curve is either an ellipse or a circle. A circle results when the cutting plane is perpendicular to the axis A of the cone.

b If the cutting plane is parallel to exactly one line L on the surface of the cone, the intersection curve is a parabola.

c If the cutting plane intersects both halves of the cone, the intersection curve is a hyperbola.

Every conic has an equation of the form $Ax^2 + Bxy + Cy^2 + Dx + Ey + F = 0$. In this section and the following one, the major axes of the ellipses, the transverse axes of the hyperbolas, and the directrixes of the parabolas are taken to be parallel to one of the coordinate axes. If those conditions are satisfied the conic will have an equation of the form $Ax^2 + Cy^2 + Dx + Ey + F = 0$.

Exercises 1.7

In Exercises 1–6, find the distances between the given points.

1. $(4, 7), (1, 3)$.
2. $(2, 2), (4, -6)$.
3. $(0, 5), (0, -3)$.
4. $(2, 0), (-6, 0)$.
5. $(0, 0), (-5, 12)$.
6. $(3, -4), (0, 0)$.
7. Find all values of x such that the distance between $(x, 7)$ and $(2, 3)$ is 5.
8. Find all values of x such that the distance between $(x, 6)$ and $(1, 3)$ is 5.
9. Find all values of x such that the distance between (x, x) and $(2x, 0)$ is 1.
10. Find all values of y such that the distance between $(y, 1)$ and $(1, y)$ is 2.

In Exercises 11–18, find equations of the circles with the given properties.

11. Radius 6, center $(0, 5)$.
12. Radius 4, center $(-2, 3)$.
13. Center $(2, -4)$, through $(1, 3)$.
14. Center $(-1, 3)$, through $(2, 5)$.
15. Radius 5, through $(3, -4)$ and $(0, 5)$.
16. Through $(0, 0)$, $(0, 3)$ and $(2, 0)$.
17. Tangent to both the x and y axes and passing through the point $(-2, 1)$.
18. Center at $(5, 3)$ and tangent to the x-axis.

For Exercises 19–36, sketch the graph of the given equation. For each ellipse, label the center and four points on the graph. For each hyperbola, use dotted lines to sketch the asymptotes. For each parabola, label the vertex.

19. $\dfrac{x^2}{9} + \dfrac{y^2}{16} = 1$.
20. $\dfrac{x^2}{25} + \dfrac{y^2}{9} = 1$.
21. $9x^2 + 4y^2 = 36$.
22. $4x^2 + 25y^2 = 100$.
23. $\dfrac{x^2}{9} - \dfrac{y^2}{16} = 1$.
24. $\dfrac{x^2}{25} - \dfrac{y^2}{9} = 1$.
25. $\dfrac{y^2}{4} - \dfrac{x^2}{9} = 1$.
26. $\dfrac{y^2}{9} - \dfrac{x^2}{16} = 1$.
27. $12y = x^2$.
28. $y = \dfrac{x^2}{4}$.
29. $x = 4y^2$.
30. $x = 2y^2$.
31. $y = \dfrac{-x^2}{8}$.
32. $y = -6x^2$.
33. $x = -y^2$.
34. $x = \dfrac{-y^2}{8}$.
35. $y = x|x|$.
36. $x = y|y|$.

In Exercises 37–40, use the distance formula to help find equations for the set of all points that satisfy the given condition.

37. Equidistant from the line $y = 1$ and the point $(-1, -1)$.
38. Equidistant from the line $x = -2$ and the point $(2, 3)$.
39. Equidistant from the points $(1, -3)$ and $(-2, 5)$.
40. Equidistant from the points $(-2, 3)$ and $(4, 1)$.

41. Show that the point $((a + c)/2, (b + d)/2)$ is the midpoint of the line segment joining the points (a, b) and (c, d).
42. Use the result of Exercise 41 to find the midpoint of the line segment joining the points $(1, -3)$ and $(-2, 5)$.
43. Use the result of Exercise 41 to find the midpoint of the line segment joining the points $(-2, 3)$ and $(4, 1)$.
44. Find an equation of the line passing through the midpoint of the line segment joining $(1, -3)$ and $(-2, 5)$ and perpendicular to that line segment. Compare your answer to that obtained in Exercise 39.
45. Find an equation of the line passing through the midpoint of the line segment joining $(-2, 3)$ and $(4, 1)$ and perpendicular to that line segment. Compare your answer to that obtained in Exercise 40.

1.8 Translation of Axes

Sometimes it is convenient to consider two coordinate systems in the same plane. In this particular instance we shall consider an xy-coordinate system and an XY-coordinate system, where the X-axis is parallel to the x-axis and the Y-axis is parallel to the y-axis.

As in Figure 1.8.1, let the origin of the XY-system have coordinates (h, k) in the xy-system. Then any point P with coordinates (x, y) in the xy-system has coordinates (X, Y) in the XY-system, where

$$x = X + h, \qquad (1)$$
$$y = Y + k;$$

or

$$X = x - h, \qquad (2)$$
$$Y = y - k.$$

Equations (1) are used to change an equation of a graph from the xy-system to the XY-system, and equations (2) are used to change an equation from the XY-system to the xy-system. Either such change is called a *translation of axes.*

Figure 1.8.1

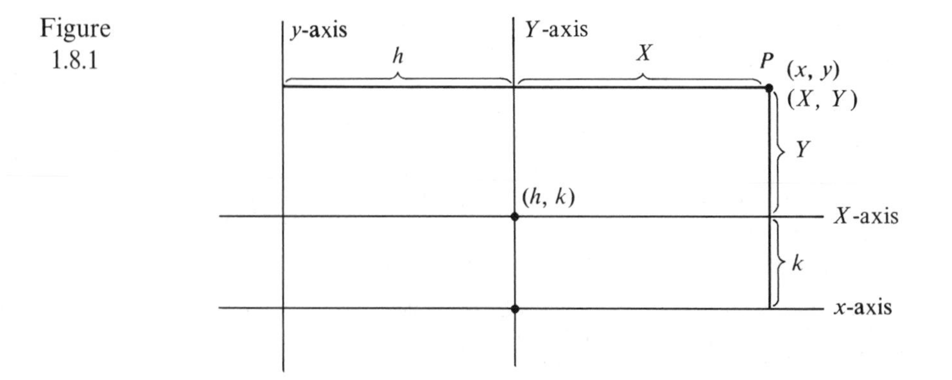

Example 1 | Change the equation $x^2 - 4x + y^2 + 6y = -4$ to one in XY-coordinates, using a translation with $h = 2$ and $k = -3$.

Taking $x = X + 2$ and $y = Y - 3$, we get

$$(X + 2)^2 - 4(X + 2) + (Y - 3)^2 + 6(Y - 3) = -4,$$

or

$$X^2 + 4X + 4 - 4X - 8 + Y^2 - 6Y + 9 + 6Y - 18 = -4.$$

On simplifying we obtain

$$X^2 + Y^2 = 9. \quad ||$$

An important application of translation of axes occurs in the consideration of ellipses and hyperbolas whose centers are not at the origin, and of parabolas whose vertices are not at the origin.

First consider an ellipse whose center is at the point (h, k) and whose major axis is parallel either to the x-axis or to the y-axis. Choose the origin of the XY-system at (h, k), as in Figure 1.8.2.

Figure 1.8.2

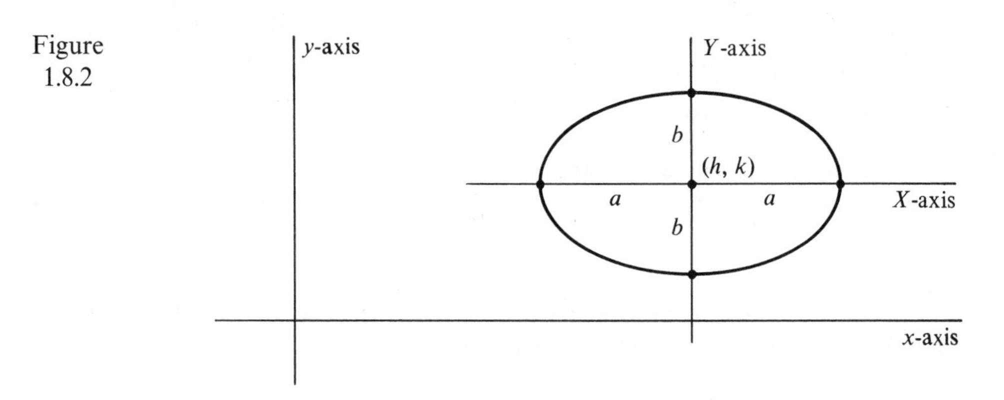

Then, as in Section 1.7, the ellipse has an equation of the form

$$\frac{X^2}{a^2} + \frac{Y^2}{b^2} = 1$$

in the XY-system. Then, by (2), its equation can be expressed as

$$\frac{(x - h)^2}{a^2} + \frac{(y - k)^2}{b^2} = 1$$

in the xy-system.

Example 2 | Sketch the graph of $\dfrac{(x + 2)^2}{16} + \dfrac{(y - 3)^2}{9} = 1.$

From above we know that this is an ellipse with center at $(-2, 3)$, where $a = 4$ and $b = 3$. After first plotting the indicated points 4 units on either side of the center and the points 3 units above and below the center, we sketch the graph pictured in Figure 1.8.3. Note that it was not necessary to show the XY-system, which was used only as an aid in sketching the graph. ||

Figure 1.8.3

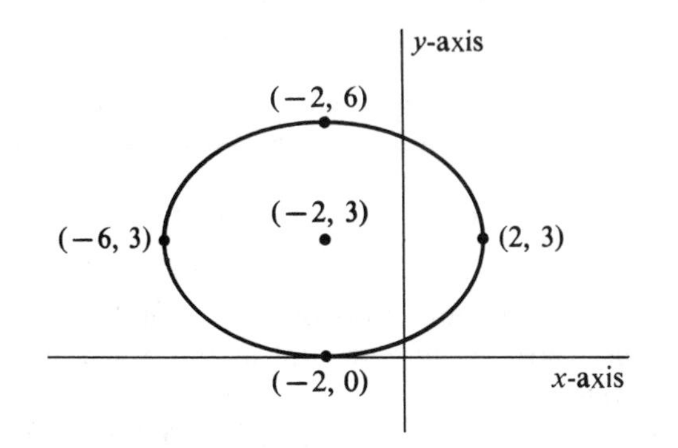

Similarly, for a hyperbola whose center is at (h, k), we take the origin of the XY-system at (h, k). Then the hyperbola has an equation of the form

$$\frac{X^2}{a^2} - \frac{Y^2}{b^2} = 1$$

if its transverse axis is parallel to the x-axis, or

$$\frac{Y^2}{b^2} - \frac{X^2}{a^2} = 1$$

if its transverse axis is parallel to the y-axis. In either case the asymptotes are the lines $Y = \pm(b/a)X$. Thus, in the xy-system the hyperbola has an equation of the form

$$\frac{(x-h)^2}{a^2} - \frac{(y-k)^2}{b^2} = 1 \quad \text{(transverse axis parallel to } x\text{-axis), or}$$

$$\frac{(y-k)^2}{b^2} - \frac{(x-h)^2}{a^2} = 1 \quad \text{(transverse axis parallel to } y\text{-axis),}$$

and the asymptotes are $y - k = \pm(b/a)(x - h)$, that is, they are the lines passing through (h, k) with slopes $\pm b/a$.

Example 3 | Sketch the graph of $\dfrac{(x-1)^2}{25} - \dfrac{(y+2)^2}{9} = 1.$

We know this is a hyperbola centered at $(1, -2)$, with $a = 5$, $b = 3$, and transverse axis parallel to the x-axis. By plotting the ends of the transverse axis 5 units to the left and right of the center $(1, -2)$, and then plotting the asymptotes as the lines passing through $(1, -2)$ with slopes $\pm 3/5$, we can quickly sketch the graph shown in Figure 1.8.4. ||

Figure 1.8.4

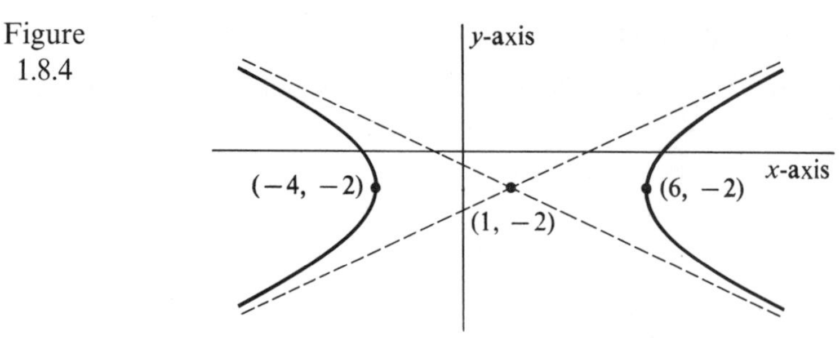

In the case of a parabola with vertex at (h, k) and directrix parallel to the x-axis, again we take the XY-origin at (h, k). We obtain an equation of the form

$$Y = \pm\frac{X^2}{4p}, \quad \text{or}$$

$$y - k = \pm\frac{(x-h)^2}{4p},$$

where $p > 0$ and p is the distance between the vertex and the focus. The plus sign occurs when the focus is above the directrix and the minus sign occurs when the focus is below the directrix.

Similarly, a parabola with vertex at (h, k) and directrix parallel to the y-axis has an equation of the form

$$x - h = \pm\frac{(y-k)^2}{4p},$$

where $p > 0$ and p is the distance between the vertex and the focus. The plus sign occurs when the focus is to the right of the directrix and the minus sign occurs when the focus is to the left of the directrix.

Example 4 | Sketch the graph of $y - 1 = (x + 2)^2$.

This is a parabola with vertex at $(-2, 1)$. The focus is above the directrix and so the parabola opens upward. A sketch is shown in Figure 1.8.5. ||

Figure 1.8.5

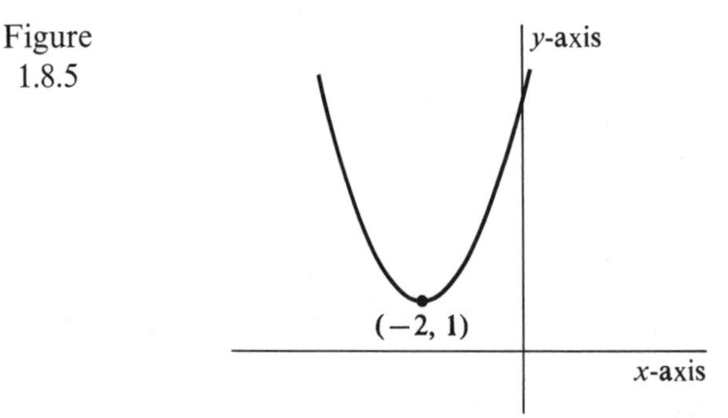

Example 5 | Sketch the graph of $x + 3 = -(1/4)(y - 2)^2$.

This is a parabola with vertex at $(-3, 2)$. Since the minus sign is present, the focus is to the left of the directrix, and so the parabola opens to the left. A sketch is indicated in Figure 1.8.6. ||

Figure 1.8.6

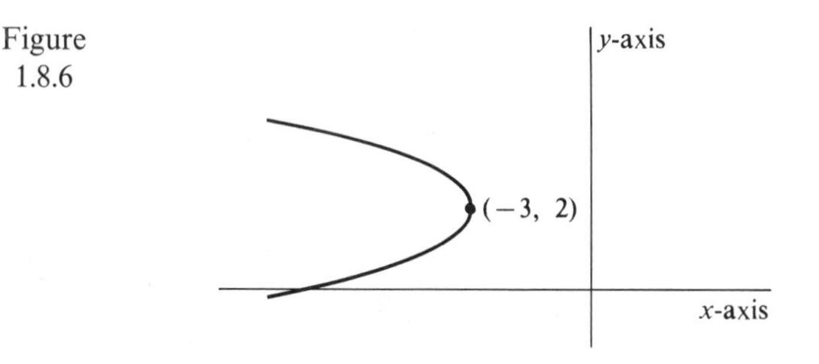

To recognize many second degree equations in x and y as a parabola, ellipse, or hyperbola, it is often necessary to *complete the square* of the second degree term(s). Since

$$(u \pm a)^2 = u^2 \pm 2au + a^2,$$

the quantity that one adds to an expression of the form $u^2 \pm 2au$ to complete the square (make it a perfect square) is the square of half of the $2a$, namely a^2.

Example 6 | Describe the graph of the equation

$$9x^2 + 54x + 4y^2 - 40y = -145.$$

To simplify the process of completing the square, we first factor the coefficient of x^2 out of the part involving x, and similarly with y (leaving space to complete the square); we get

$$9(x^2 + 6x \qquad) + 4(y^2 - 10y \qquad) = -145.$$

To complete the square in x we add the square of half of 6, namely 9, inside the first parenthesis; and to complete the square in y we add 25 inside the second parenthesis; thus we have added $9 \cdot 9 = 81$ and $4 \cdot 25 = 100$ to the left side of the equation. Now we must also add $81 + 100 = 181$ to the right side of the equation. We obtain

$$9(x^2 + 6x + 9) + 4(y^2 - 10y + 25) = -145 + 181, \quad \text{or}$$

$$9(x + 3)^2 + 4(y - 5)^2 = 36.$$

Dividing both sides by 36, we have

$$\frac{(x + 3)^2}{2^2} + \frac{(y - 5)^2}{3^2} = 1.$$

So the graph is an ellipse with $a = 2$, $b = 3$, and center at $(-3, 5)$. ||

Example 7 | Describe the graph of the equation $y = -12x^2 - 72x - 103$.

Here there is only one square to complete. We add 103 to both sides and factor out the coefficient of x^2, to get

$$y + 103 = -12(x^2 + 6x \qquad).$$

To complete the square we add 9 inside the parenthesis; that is, we add $-12 \cdot 9 = -108$ to both sides of the equation, obtaining

$$y - 5 = -12(x^2 + 6x + 9), \quad \text{or}$$

$$y - 5 = -12(x + 3)^2.$$

Thus the graph is a parabola that has vertex at $(-3, 5)$ and opens downward. ||

Exercises 1.8

In Exercises 1–22, sketch the graph of the given equation. For each ellipse label the center and four points on the graph. For each hyperbola label the center and use dotted lines to sketch the asymptotes. For each parabola label the vertex.

1. $\dfrac{(x-2)^2}{16} + \dfrac{(y-1)^2}{9} = 1.$
2. $\dfrac{(x+1)^2}{25} + \dfrac{(y-2)^2}{36} = 1.$
3. $4(x-3)^2 + 9(y+2)^2 = 36.$
4. $25(x+4)^2 + 4(y+1)^2 = 100.$
5. $\dfrac{(x+2)^2}{9} - \dfrac{(y-1)^2}{16} = 1.$
6. $\dfrac{(x-3)^2}{25} - \dfrac{(y+2)^2}{16} = 1.$
7. $\dfrac{y^2}{16} - \dfrac{(x-1)^2}{25} = 1.$
8. $\dfrac{(y+1)^2}{4} - \dfrac{x^2}{9} = 1.$
9. $y - 4 = (x+2)^2.$
10. $y + 3 = 4(x-5)^2.$
11. $y = 2 - (x-3)^2.$
12. $y = -(x+4)^2 - 1.$
13. $x + 1 = (y-2)^2.$
14. $x - 3 = \frac{1}{4}(y+5)^2.$
15. $x = -(y+1)^2.$
16. $x + 2 = -(y+1)^2.$
17. $25x^2 - 150x + 9y^2 + 18y = -9.$
18. $9x^2 + 36x - y^2 + 24y = 1.$
19. $9y^2 - 72y - x^2 - 10x + 110 = 0.$
20. $y^2 + 12y - 8x + 60 = 0.$
21. $x^2 - 8x - 16y - 32 = 0.$
22. $16x^2 - 32x + 9y^2 - 108y + 16 = 0.$
23. Consider the set of all points (x, y) such that the sum of the distances between (x, y) and the points $(-3, 0)$ and $(3, 0)$ is equal to 10. Use the distance formula to show that an equation of this set is $(x^2/25) + (y^2/16) = 1$.
24. Consider the set of all points (x, y) such that the difference of the distances between (x, y) and the points $(-5, 0)$ and $(5, 0)$ is equal to ± 8. Use the distance formula to show that an equation of this set is $(x^2/16) - (y^2/9) = 1$.
25. Let a and c be fixed positive numbers with $a > c$. Consider the set of all points (x, y) such that the sum of the distances between (x, y) and the points $(-c, 0)$ and $(c, 0)$ is equal to $2a$. Use the distance formula to show that an equation of this set is the ellipse $(x^2/a^2) + (y^2/b^2) = 1$, where $b^2 = a^2 - c^2$. (An ellipse is often defined as a graph with the geometric properties indicated.)
26. Let a and c be fixed positive numbers with $a < c$. Consider the set of all points (x, y) such that the difference of the distances between (x, y) and the points $(-c, 0)$ and $(c, 0)$ is equal to $\pm 2a$. Use the distance formula to show that an equation of this set is the hyperbola $(x^2/a^2) - (y^2/b^2) = 1$, where $b^2 = c^2 - a^2$. (A hyperbola is often defined as a graph with the geometric properties indicated.)
27. Let points P and Q have coordinates (x_1, y_1) and (x_2, y_2) in the xy-coordinate system and (X_1, Y_1) and (X_2, Y_2) in the XY-coordinate system. If the XY-coordinate system is obtained

from the xy-coordinate system by a translation, show that

$$\sqrt{(x_1 - x_2)^2 + (y_1 - y_2)^2} = \sqrt{(X_1 - X_2)^2 + (Y_1 - Y_2)^2}.$$

Thus the distance between P and Q is not changed by a translation of axes.

28. Show that the slope of the line $y = mx + b$ is not changed by a translation of axes.

1.9 Parametric Equations

We have seen how a function $y = h(x)$ specifies a value of y for each value of x in its domain. In some circumstances it is more natural to introduce a third variable or *parameter* that in turn determines the values of x and y.

For example, consider a particle moving in a plane with constant velocity of 5 ft/sec in the x-direction and constant velocity of 3 ft/sec in the y-direction. If the particle starts at the origin, its position after t sec will be given by $x = 5t$ ft and $y = 3t$ ft. In this case each value of the parameter t will determine a value of x and y.

More formally, if we denote the parameter by t, we have two functions, $x = f(t)$, $a \leq t \leq b$, and $y = g(t)$, $a \leq t \leq b$, that give the values of x and y that are related to each value of t. The equations $x = f(t)$ and $y = g(t)$ are called *parametric equations*. Note that given a value of t, say $t = t_0$, a point $(x_0, y_0) = (f(t_0), g(t_0))$ is determined. That fact leads to the following definition.

Definition 1.9.1

The graph of the parametric equations $x = f(t)$, $y = g(t)$, $a \leq t \leq b$, is the set of all points (x_0, y_0) such that $x_0 = f(t_0)$ and $y_0 = g(t_0)$ for some t_0, where $a \leq t_0 \leq b$.

The most direct way of graphing the curve representing a set of parametric equations is to calculate the points (x, y) corresponding to various values of t. Those points are then connected in the order of increasing values of t.

Example 1 Sketch the graph of the parametric equations $x = \cos t$, $y = \sin t$, $0 \leq t \leq 2\pi$.

The following table shows selected values of t and the corresponding points.

t	$x = \cos t$	$y = \sin t$	(x, y)
0	1	0	$(1, 0)$
$\pi/4$	$\sqrt{2}/2$	$\sqrt{2}/2$	$(\sqrt{2}/2, \sqrt{2}/2)$
$\pi/3$	$1/2$	$\sqrt{3}/2$	$(1/2, \sqrt{3}/2)$
$\pi/2$	0	1	$(0, 1)$
$3\pi/4$	$-\sqrt{2}/2$	$\sqrt{2}/2$	$(-\sqrt{2}/2, \sqrt{2}/2)$
π	-1	0	$(-1, 0)$
$5\pi/4$	$-\sqrt{2}/2$	$-\sqrt{2}/2$	$(-\sqrt{2}/2, -\sqrt{2}/2)$
$3\pi/2$	0	-1	$(0, -1)$
$7\pi/4$	$\sqrt{2}/2$	$-\sqrt{2}/2$	$(\sqrt{2}/2, -\sqrt{2}/2)$

In Figure 1.9.1 we plot those points and connect the points in the order of increasing t.

Figure 1.9.1

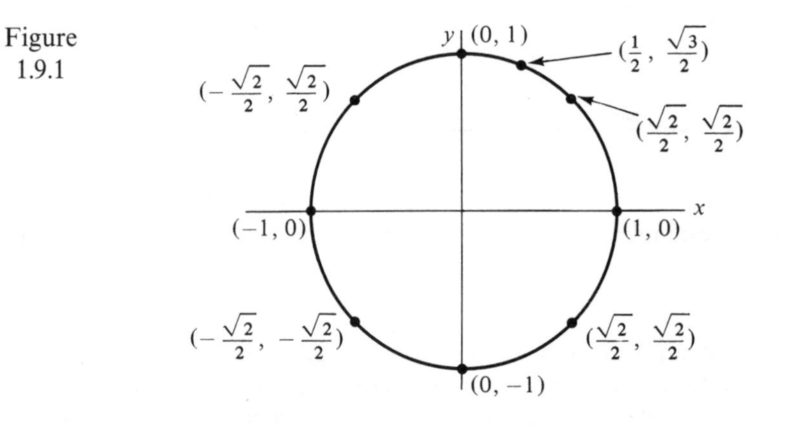

Note that the graph appears to be a circle. But how can we be sure? In this case we need only square the two equations $x = \cos t$ and $y = \sin t$. Adding the results we get

$$x^2 + y^2 = \sin^2 t + \cos^2 t = 1.$$

Conversely if $x^2 + y^2 = 1$, it is possible to select a value of t so that $x = \cos t$ and $y = \sin t$. Consequently a point (x, y) is on the graph of the parametric equations if and only if $x^2 + y^2 = 1$. The graph thus must be a circle of radius one and centered at the origin. ||

At times, elimination of the parameter will result in an equation whose graph is easily identified.

Example 2 | Eliminate the parameter t in the parametric equations $x = -(1/4)(t^2 + 12)$, $y = t + 2$. Graph the resulting equation.

If we solve the second equation for t, we have $t = y - 2$. Substituting this into the first equation gives

$$x = -\frac{1}{4}((y-2)^2 + 12).$$

Thus, we get

$$x + 3 = -\frac{1}{4}(y-2)^2,$$

which is an equation whose graph can be recognized as a parabola. The graph in Figure 1.8.6 is here repeated in Figure 1.9.2. ||

Figure 1.9.2

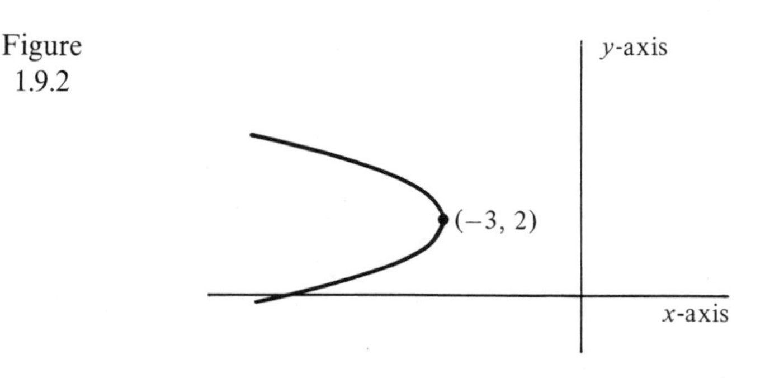

The following example again shows how parametric equations may arise in applications.

Example 3 Let a projectile be fired from the origin at time $t = 0$, with initial velocity v_0 ft/sec. If the projectile is fired at a 45° angle to the horizontal, and air resistance is neglected, develop parametric equations for the position of the projectile at any time t. Eliminate the parameter and identify the graph of the resulting equation.

Since the projectile is fired at a 45° angle, the x and y components of the velocity are both $v_0/\sqrt{2}$. If we neglect air resistance we have

$$x = \frac{v_0}{\sqrt{2}}\,t,$$

and

$$y = \frac{v_0}{\sqrt{2}}\,t - \frac{1}{2}gt^2,$$

where t is the time in seconds and g is the acceleration due to gravity. If we then solve the first equation for t and substitute into the second we obtain

$$y = x - \frac{g}{v_0^{\,2}}x^2,$$

whose graph is a parabola. Thus a projectile fired in a vacuum at 45° follows a parabolic path. ||

Exercises 1.9

In Exercises 1–4, sketch the graph of the parametric equations without eliminating the parameter.

1. $x = t + 3,\quad y = 1 - t.$
2. $x = 2t + 3,\quad y = t - 2,\quad 0 \le t \le 1.$
3. $x = 2u,\quad y = u^2,\quad 0 \le u \le 1.$
4. $x = 3t^2 - 1,\quad y = t + 2,\quad 0 \le t \le 2.$

In Exercises 5–15, eliminate the parameter and identify the graph of the resulting equation if possible.

5. $x = 3v + 7,\quad y = v - 1.$
6. $x = -t + 5,\quad y = -t - 7.$
7. $x = 5t^2 - 7t,\quad y = t + 4.$
8. $x = 7u^2 - 3,\quad y = 4 - 2u.$
9. $x = \sin t,\quad y = \cos t,\quad 0 \le t \le \pi.$
10. $x = 3\sin t,\quad y = 2\cos t,\quad 0 \le t \le \pi.$

11. $x = 4 \sin t, \quad y = 3 \cos t, \quad 0 \leq t \leq \pi.$

12. $x = -1 + \cos t, \quad y = \sin t, \quad 0 \leq t \leq 2\pi.$

13. $x = 2 + \cos t, \quad y = 1 + \sin t, \quad 0 \leq t \leq 2\pi.$

14. $x = v^2, \quad y = v^3.$

15. $x = t^3, \quad y = t^{1/2}.$

16. Show that the equations $x = t/m$, $y = t + b$ describe a straight line with slope m.

17. Show that the equations $x = a \cos \theta$, $y = b \sin \theta$ describe an ellipse.

18. Let a particle move with constant velocity v_0 along the line segment from (1, 1) to (5, 5). Find the parametric equations for the position of the particle at any time t, assuming that the particle is at (1, 1) at time $t = 0$.

19. Show that the equations

$$x = (x_1 - x_0)t + x_0, \quad y = (y_1 - y_0)t + y_0, \quad 0 \leq t \leq 1$$

describe a line segment joining the points (x_0, y_0) and (x_1, y_1).

20. Sketch the graph described by $x = t \cos t, \quad y = t \sin t, \quad 0 \leq t \leq 2\pi.$

21. Find parametric equations for the curve traced by a point P on the circumference of a wheel rolling along the x-axis as indicated in Figure 1.9.3. Use the parameter θ and assume that P is at the origin when $\theta = 0$ radians. Sketch the graph.

Figure 1.9.3

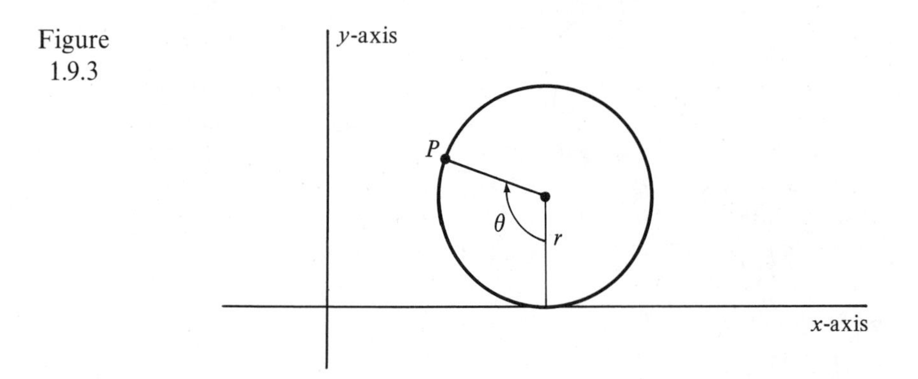

22. Consider a wheel rolling along the x-axis. Find parametric equations for the curve traced by a point P on the radius TQ. This situation is illustrated in Figure 1.9.4.

Figure 1.9.4

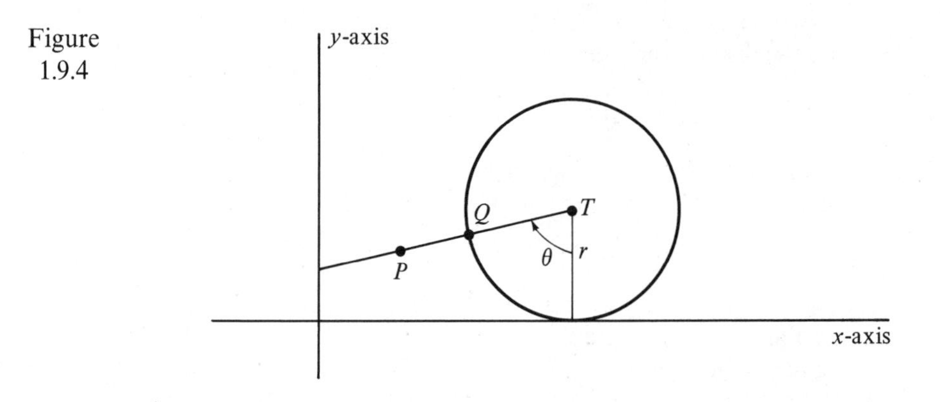

Use the parameter θ and assume Q is at the origin when $\theta = 0$ radians. Let b denote the distance between P and T and consider both cases $b > r$ and $b < r$. Sketch the graphs.

Brief Review of Chapter 1

1. Real Numbers and Rectangular Coordinates

Definition. A *real number* is a number that can be expressed as an infinite decimal.

Real Line. A point is labeled 0, and then for each positive real number p, the point p units to the right of 0 is labeled p, and the point p units to the left of 0 is labeled $-p$. Such a labeled line is called a *real line.*

Rectangular Coordinate System. Two perpendicular real lines intersecting at the origin of each are taken, one vertical, called the y-axis, and one horizontal, called the x-axis. Each point in the plane is labeled with an ordered pair of numbers, the first number is the horizontal location and is called the x-coordinate, and the second is the vertical location and is called the y-coordinate.

2. Functions

A function is considered to be a computer f which, when presented with the right kind of number x as input, produces exactly one number $f(x)$ as output. When f and g are functions an important related function is $f(g(x))$, which is often denoted by $(f \circ g)(x)$. Some examples of functions are power functions, polynomial functions, rational functions, the absolute value function, trigonometric functions, exponential functions, and logarithmic functions.

3. The Absolute Value Function

Definition. The *absolute value* of x, $|x|$, is defined as follows:

$$|x| = \begin{cases} x & \text{if } x \geq 0 \\ -x & \text{if } x < 0. \end{cases}$$

Some Important Properties. $|ab| = |a| \cdot |b|$; $|a/b| = |a|/|b|$; $|a + b| \leq |a| + |b|$; and $|a - b|$ is the distance between a and b on the real line.

4. Graphs

Definition. The set of all points that are solutions of an equation involving x and y is called the *graph of the equation.* The *graph of a function f* is the graph of the equation $y = f(x)$.

Sketching. Although more powerful methods are developed in later chapters, at this point we sketch a graph of an equation by making a table of solutions, plotting the points, and then trying to get from them a general idea of what the graph looks like.

Section 1.4

For Exercises 11–12, find the distance between each pair of points on the real line.

11. -7 and x.

12. -4 and $-x$.

For Exercises 13–14, express each quantity in words, in terms of distance on the real line.

13. $|3 + x| > 1$.

14. $|x + y| < 2$.

For Exercises 15–16, express each statement as an inequality involving absolute value.

15. The distance between -10 and x is less than 12.

16. The distance between 6 and $-x$ is greater than 4.

17. Evaluate $\sec(7\pi/6)$.

18. Evaluate $\log_2 64$.

Section 1.5

Sketch the graph of each function or equation. Also, either indicate that the graph has no breaks or list the values of x for which breaks occur.

19. $y = 1/|x|$.

20. $y = |x| - x$.

21. $f(x) = \begin{cases} 7 & \text{if } x \le 5 \\ 2 & \text{if } x > 5. \end{cases}$

22. $f(x) = \begin{cases} -2x & \text{if } x < 0 \\ 3x & \text{if } x \ge 0. \end{cases}$

23. $f(x) = \begin{cases} 3 & \text{if } x = -2 \\ -1 & \text{if } x \ne -2. \end{cases}$

24. $f(x) = \begin{cases} x & \text{if } x < 1 \\ 2x & \text{if } x \ge 1. \end{cases}$

25. $f(x) = \begin{cases} \dfrac{x^2 + 3x}{x} & \text{if } x \ne 0 \\ 4 & \text{if } x = 0. \end{cases}$

26. $f(x) = \begin{cases} \dfrac{x^2 - 4}{x - 2} & \text{if } x \ne 2 \\ 4 & \text{if } x = 2. \end{cases}$

Section 1.6

Find equations of the lines with the given properties.

27. Through $(5, -2)$ and $(1, 3)$.

28. Through $(6, -2)$ and $(6, 3)$.

29. Parallel to $7x + 3y = 1$ through $(-1, 2)$.

30. Perpendicular to $2x - y = 5$ through $(7, -4)$.

31. Parallel to $y = 4$ through $(-2, 3)$.

32. Perpendicular to $y = 1$ through $(11, 8)$.

33. Prove that the lines $4x = 6y + 1$ and $15y = 10x + 7$ are parallel.

34. Prove that the lines $6y + 10x - 1 = 0$ and $9x = 15y - 2$ are perpendicular.

Section 1.7

35. Find the distance between $(-4, 2)$ and $(-6, -3)$.

36. Find the distance between $(-2, 5)$ and $(-2, -7)$.

37. Find an equation of the circle centered at $(-2, 7)$ and passing through $(1, -3)$.

38. Find an equation of the circle of radius 13 passing through the points $(4, -10)$ and $(11, 7)$.

39. Sketch the graph of $y = 3x^2$.

40. Sketch the graph of $y^2 - x^2/4 = 1$.

5. Lines

Inclination. If a line is not horizontal, it intersects the x-axis at one and only one point P. The *inclination* of such a line is defined as the positive angle, with P as vertex, from the part of the x-axis to the right of P, to the line. If a line is horizontal we define its inclination to be 0° or, equivalently, 0 radians.

Slope. If θ is the inclination of a nonvertical line, then $\tan\theta$ is called the *slope* of the line. Vertical lines do not have slopes. If L_1 and L_2 are lines with slopes m_1 and m_2, then L_1 and L_2 are parallel if and only if $m_1 = m_2$, and they are perpendicular if and only if $m_1 m_2 = -1$. The slope m of a nonvertical line through (x_1, y_1) and (x_2, y_2) is

$$m = \frac{y_2 - y_1}{x_2 - x_1}.$$

Equations. An equation of the line that has slope m and passes through the point (x_1, y_1) is $y - y_1 = m(x - x_1)$. An equation of the line with slope m and y-intercept b is $y = mx + b$. A vertical line has an equation of the form $x = k$, where k is the common x-coordinate.

6. Distances

The distance between points (a, b) and (c, d) is

$$D((a, b), (c, d)) = \sqrt{(a - c)^2 + (b - d)^2}.$$

7. Circles

An equation of the circle with radius r and center at (h, k) is

$$(x - h)^2 + (y - k)^2 = r^2.$$

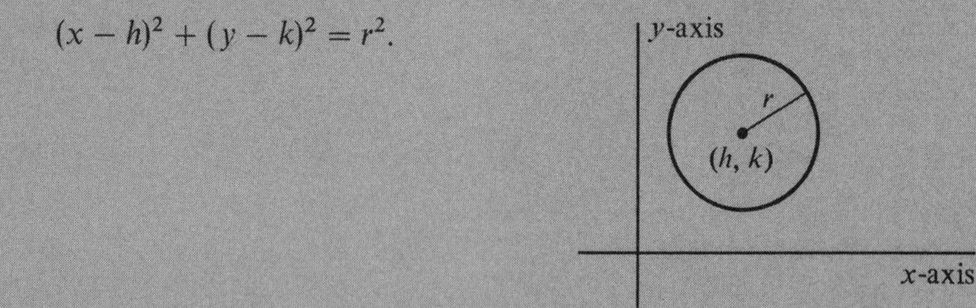

8. Ellipses

An equation of the ellipse with center at (h, k) and axes parallel to the coordinate axes is

$$\frac{(x - h)^2}{a^2} + \frac{(y - k)^2}{b^2} = 1.$$

y-axis
a b a
b (h, k)
x-axis

The larger of the numbers $2a$ and $2b$ is the length of the largest diameter of the ellipse and is called the *major axis*; the smaller of those two numbers is the length of the smallest diameter and is called the *minor axis*.

9. Hyperbolas

Transverse Axis Parallel to x-axis, Center at (h, k)

$$\frac{(x-h)^2}{a^2} - \frac{(y-k)^2}{b^2} = 1.$$

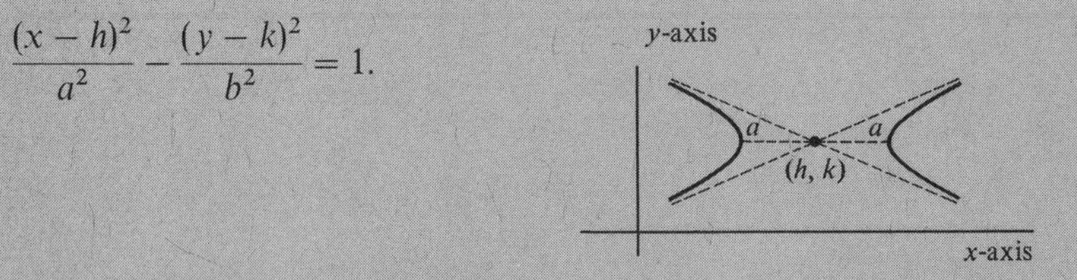

Length of transverse axis: $2a$; asymptotes: $y - k = \pm(b/a)(x - h)$.

Transverse Axis Parallel to y-axis, Center at (h, k)

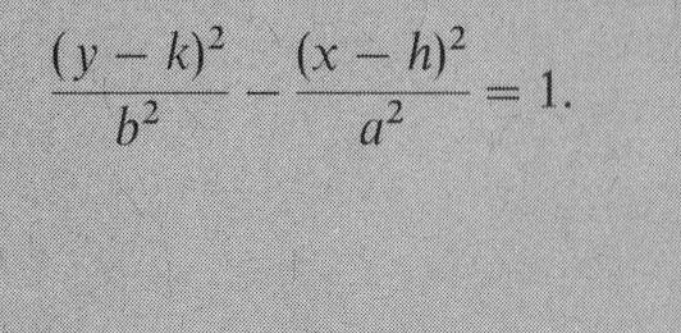

$$\frac{(y-k)^2}{b^2} - \frac{(x-h)^2}{a^2} = 1.$$

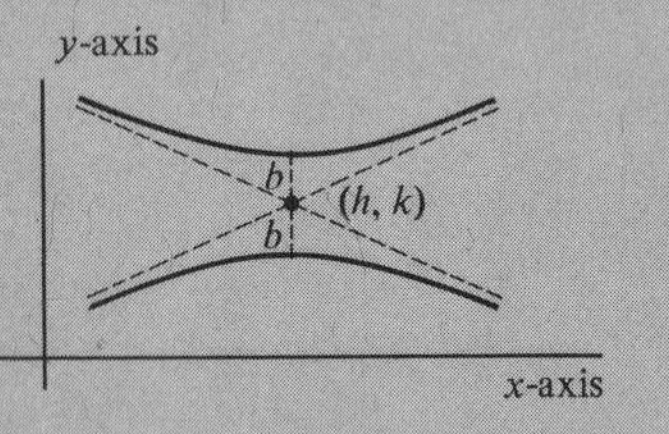

Length of transverse axis: $2b$; asymptotes: $y - k = \pm(b/a)(x - h)$.

10. Parabolas, Vertex at (h, k), Distance between Focus and Directrix = $2p$.

Directrix Parallel to x-axis

$$y - k = \pm\frac{(x-h)^2}{4p}.$$

(Opens upward with + sign, and downward with − sign.)

Directrix Parallel to y-axis

$$x - h = \pm\frac{(y-k)^2}{4p}.$$

(Opens to the right with + sign, and to the left with − sign.)

11. Parametric Equations

The graph of the parametric equations $x = f(t)$, $y = g(t)$, $a \le t \le b$, is the set of all points (x_0, y_0) such that $x_0 = f(t_0)$ and $y_0 = g(t_0)$ for some t_0, where $a \le t_0 \le b$. Such graphs may be sketched directly by locating several points on the graph and connecting them in the order of increasing t. Sometimes it is also possible and helpful to eliminate the parameter t and so obtain an equation directly relating x and y.

Technique Review Exercises, Chapter 1

1. If $f(x) = x^2 + 3x$ and $g(x) = 1 - x^2$, find
   (a) $f(x^3)$, (b) $(f \circ g)(2)$, (c) $g(x + 1)$.
2. Find the range and domain of the function $f(x) = \sqrt{5 - x}$.
3. Explain in words the meaning, on the real line, of each of the following:
   (a) $|x - 5|$, (b) $|6 + x|$.
4. Sketch the graph of
$$f(x) = \begin{cases} x & \text{if } x < 2 \\ 5 & \text{if } x = 2 \\ 1 & \text{if } x > 2. \end{cases}$$
5. Find an equation of the line perpendicular to $5x + 3y = 7$ and passing through $(-2, 3)$.
6. Find an equation of the line passing through $(3, -2)$ and $(-4, -2)$.
7. Find the distance between the points $(-3, 5)$ and $(6, 1)$.
8. Use the distance formula to find an equation of the set of all points equidistant from tl points $(-1, 2)$ and $(3, 5)$. Simplify your result.
9. Find an equation of the circle centered at $(5, -1)$ and passing through the point $(7, 2)$.
10. Sketch the graph of $\frac{(x-4)^2}{25} + \frac{(y+2)^2}{9} = 1$, and label the coordinates of several impor points that aid in the sketching.
11. Sketch the graph of $\frac{(y-1)^2}{16} - \frac{(x+3)^2}{25} = 1$, and include sketches of the asymptot dotted lines.
12. Sketch the graph of $2 - x = (y + 3)^2$.
13. Sketch the graph of $x^2 + y^2 - 4x - 6y + 9 = 0$.
14. Sketch the graph of $y = 3w - 5$, $x = w^2$ without eliminating the parameter.
15. Eliminate the parameter from $y = 5 \sin t$, $x = -2 \cos t$. Identify the graph of the r equation.

Additional Exercises, Chapter 1

Section 1.3

For Exercises 1–6, let $f(x) = x^2 + 2x$ and $g(x) = x^2 - 3x + 5$, and find each of the fo

1. $f(5)$.
2. $g(3)$
3. $(f \circ g)(x)$.
4. $(g \circ f)(x)$.
5. $g(x + h) - g(x)$.
6. $f(x + h) - f(x)$.

For Exercises 7–10, find (a) the domain, and (b) the range of each function.

7. $f(x) = 1/((x - 2)(x + 1))$.
8. $f(x) = \sqrt{x - 5}$.
9. $f(x) = \sqrt{x^2 + 1}$.
10. $f(x) = 1/x(x - 7)$.

41. Find an equation of the set of points equidistant from the line $x = 2$ and the point (6, 1).

42. Find an equation of the set of points equidistant from the points (3, 5) and (−2, 1).

Section 1.8

Sketch the graph of each equation.

43. $\dfrac{(x+2)^2}{25} + \dfrac{(y-1)^2}{4} = 1.$

44. $\dfrac{(x+2)^2}{25} - \dfrac{(y-1)^2}{4} = 1.$

45. $25(y+2)^2 - 9(x-3)^2 = 225.$

46. $25(y+2)^2 + 9(x-3)^2 = 225.$

47. $(y+3)^2 = x - 1.$

48. $y = 3 + (x-5)^2.$

49. $(x+7)^2 + (y-2)^2 = 0.$

50. $(x-3)^2 + (y-6)^2 = 0.$

Section 1.9

In Exercises 51–54, sketch the graph of the given parametric equations without eliminating the parameter.

51. $x = 2t - 3, \quad y = 4t + 10, \quad 0 \le t \le 2.$

52. $x = -6t + 4, \quad y = 3t + 2.$

53. $x = t^2 + 2, \quad y = t - 1.$

54. $x = t^2, \quad y = t^3.$

In Exercises 55–59, eliminate the parameter and try to identify the graph.

55. $x = 7t^2 + 3, \quad y = t - 1.$

56. $x = 7t^2 + 3, \quad y = t^2 - 1.$

57. $x = -\cos 3t, \quad y = 2 \sin 3t.$

58. $x = \sec t, \quad y = \tan t, \quad 0 \le t < \pi/2.$

59. $x = 1 + \cos t, \quad y = 2 \sin t, \quad 0 \le t \le 2\pi.$

Challenging Problems, Chapter 1

1. Let f be a function that satisfies $f(ab) = a \cdot f(b)$ for all a and b. Prove that there is some constant k such that $f(x) = kx$ for all x.

2. Prove analytically that the diagonals of a rhombus are perpendicular.

3. Prove analytically that the line segments joining the midpoints of the opposite sides of any quadralateral bisect each other.

4. Prove that the distance between the line $y = mx + b$ and the origin is $|b|/\sqrt{1+m^2}$

5. Use the result of Problem 4 to help prove that the distance between the line $y = mx + b$ and the point (x_1, y_1) is $|y_1 - mx_1 - b|/\sqrt{1+m^2}$.

6. Let $k \neq 1$ be a constant and let P and Q be fixed points. Show that the set of all points in the plane whose distance from P is k times the distance from Q is a circle. This property, which is sometimes used to define a circle, is credited to Apollonius.

7. We defined a parabola in terms of a focus and a directrix. Ellipses and hyperbolas can be defined similarly. Let e be a positive constant. Consider the set of points (x, y) such that the distance between (x, y) and the focus is equal to e, multiplied by the distance between (x, y) and the directrix. Show that when $e < 1$ the set of points is an ellipse, and when $e > 1$ the set is a hyperbola. (Of course when $e = 1$, it is a parabola.) Use the following as a guide:

   > Let d be the distance between the focus and the directrix, and let $a = ed/|1 - e^2|$. It will be convenient to choose the coordinate system so that the x-axis passes through the focus and the directrix has the equation $x = a/e$. When $e < 1$, take the coordinate system so that the focus is to the left of the directrix; when $e > 1$, take the coordinate system so that the focus is to the right of the directrix. Show that in both cases these choices cause the focus to have coordinates $(ae, 0)$. Next determine an equation for the set of points (x, y) for each case and complete the problem.

8. Show that in Problem 7, the same final equations are obtained when $(-ae, 0)$ is the focus and $x = -a/e$ is the directrix. Therefore both an ellipse and a hyperbola have two foci and two corresponding directrices.

2

Limits, Continuity, and Derivatives

2.1 The Tangent Problem

We begin this chapter by considering a geometric problem with widespread non-geometric consequences.

Tangent Problem

Find the tangent to the graph of $y = f(x)$ at the point $(a, f(a))$.

Since the concept of "tangent line" has not been defined, we shall first attempt to formulate a suitable definition that fits the intuitive requirements of a "tangent line." In particular the tangent line should pass through the point $(a, f(a))$. However, since many lines pass through this point, more information is necessary in order to determine the tangent line. Figure 2.1.1 indicates that a tangent line (or at least what appears to be a tangent line) may cross the curve (in the case of L) or may intersect the curve in more than one point (in the case of L'). Thus the requirements normally associated with the tangent to a circle cannot be used to help specify tangent lines in general. Rather, we shall use a method of successive approximations to obtain the *slope* of the tangent line to the curve at $(a, f(a))$. The tangent line is then the line through $(a, f(a))$ with the slope so obtained.

Figure 2.1.1

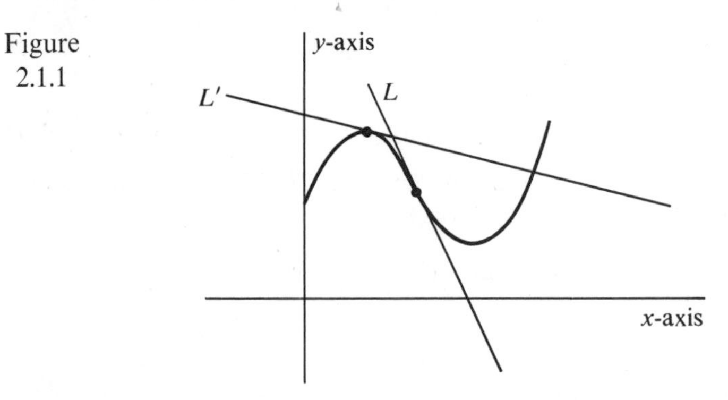

Let h be a positive or negative real number. Both $(a, f(a))$ and $(a + h, f(a + h))$ are points on the graph of $y = f(x)$. It is an easy matter to specify $m(h)$, the slope of the line L_h passing through the points $(a, f(a))$ and $(a + h, f(a + h))$. In fact,

$$m(h) = \frac{f(a + h) - f(a)}{h}.$$

Figure 2.1.2 illustrates that L_h is an approximation (perhaps not a good one) to the tangent line L, and that $m(h)$ is an approximation to m, the slope of the tangent line.

Figure 2.1.2

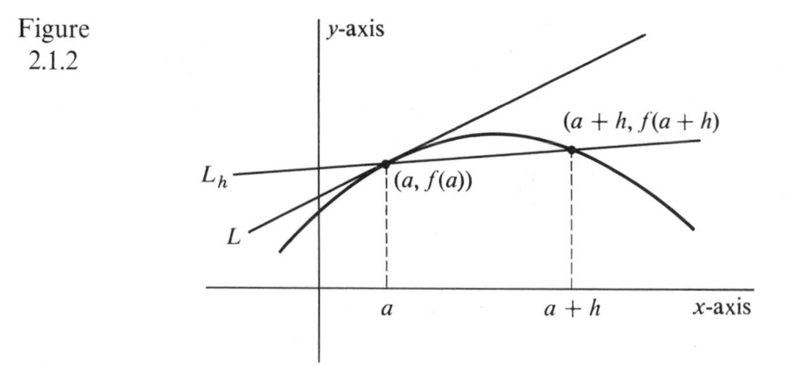

The important point is not that the approximation is close, but rather that the approximation approaches the slope of the tangent line as h approaches zero. This fact is illustrated in Figure 2.1.3 for $h = 1, -1/2, 1/4$, and $-1/8$.

These intuitive considerations lead us to *define* the tangent line L as the line approached by the lines L_h as h tends to zero. Thus the slope m of the tangent line L is the number that

$$m(h) = \frac{f(a+h) - f(a)}{h}.$$

approaches as h gets close to 0. To indicate this fact we write $m(h) \to m$ as $h \to 0$, or

$$\lim_{h \to 0} m(h) = m.$$

(Don't worry too much about the $\lim_{h \to 0}$ notation. We shall consider this more thoroughly in the next section.)

Example 1 | Find the tangent to the graph of $f(x) = x^2 + 3$ at the point (2, 7).

Of course the tangent line passes through the point (2, 7). Moreover, in this case $a = 2$, and so the slope m is the number approached by

$$m(h) = \frac{f(2+h) - f(2)}{h} = \frac{(2+h)^2 + 3 - 7}{h}$$

as h approaches zero.

Simplification yields

$$m(h) = \frac{4 + 4h + h^2 - 4}{h} = \frac{4h + h^2}{h} = 4 + h, \quad h \neq 0.$$

Hence $m(h)$ approaches 4 as h approaches zero, and so $m = 4$. We may now write an equation of the tangent as $y - 7 = 4(x - 2)$, or $y = 4x - 1$. ||

Example 2 | Compute the slope of the line tangent to the graph of $f(x) = x \sin x$ at the point (0, 0).

Figure 2.1.3

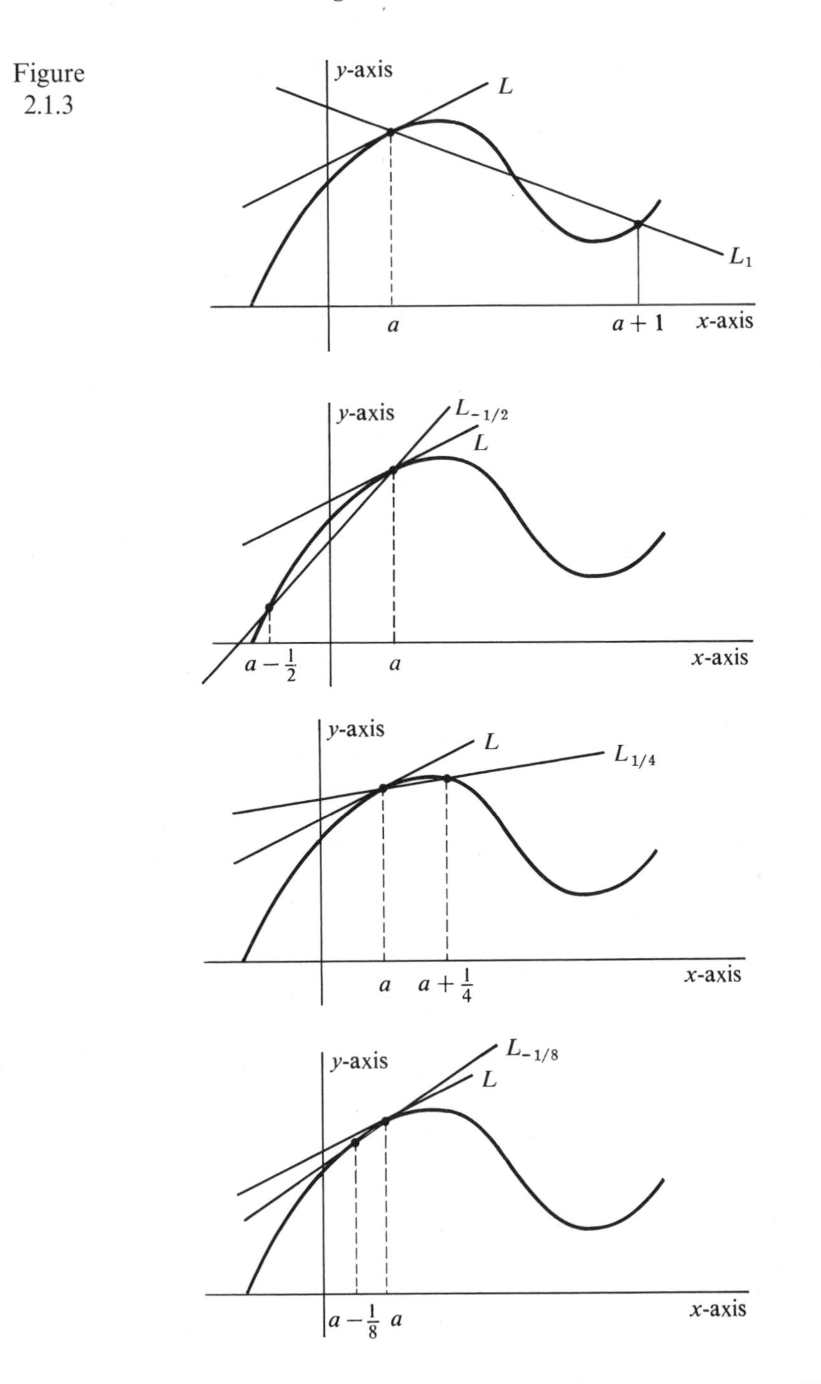

The slope m is the number approached by $m(h)$ as h approaches zero. Here

$$m(h) = \frac{f(0 + h) - f(0)}{h} = \frac{f(h) - f(0)}{h}$$

$$= \frac{h \sin h - 0 \sin 0}{h}.$$

But $0 \sin 0 = 0(0) = 0$; so we get

$$m(h) = \frac{h \sin h}{h} = \sin h, \quad h \neq 0.$$

Since $\sin h$ approaches 0 as h gets close to 0, we get

$$m = \lim_{h \to 0} (\sin h) = 0. \qquad \|$$

We shall return to the tangent problem and some of its consequences later in this chapter.

Exercises 2.1

For Exercises 1–14, use the method of this section to find an equation of the line tangent to the graph of $f(x)$ at the indicated point.

1. $f(x) = x^2$; $(3, 9)$.
2. $f(x) = x^2 - 1$; $(2, 3)$.
3. $f(x) = x^2$; $(0, 0)$.
4. $f(x) = x^2 - 1$; $(-1, 0)$.
5. $f(x) = 4 - x$; $(3, 1)$.
6. $f(x) = x - 2$; $(1, -1)$.
7. $f(x) = 5x + 2$; $(1, 7)$.
8. $f(x) = 6 - x$; $(0, 6)$.
9. $f(x) = x^2 - x$; $(1, 0)$.
10. $f(x) = 3 - x^2$; $(1, 2)$.
11. $f(x) = x^2 - 2x + 3$; $(2, 3)$.
12. $f(x) = x^2 + 4x - 1$; $(0, -1)$.
13. $f(x) = x^2 - 2x + 3$; $(-1, 6)$.
14. $f(x) = x^2 + 4x - 1$; $(-2, -5)$.

For Exercises 15–34, compute the slope of the line tangent to the graph of $f(x)$ at the indicated point.

15. $f(x) = x^2 + 7x$; $(0, 0)$.
16. $f(x) = x^2 - 3$; $(2, 1)$.
17. $f(x) = x^3$; $(2, 8)$.
18. $f(x) = x^3 - 2$; $(3, 25)$.
19. $f(x) = x^3$; $(-1, -1)$.
20. $f(x) = x^3 - 2$; $(-2, -10)$.
21. $f(x) = 1/x$; $(3, 1/3)$.
22. $f(x) = 1/(x + 1)$; $(-2, -1)$.
23. $f(x) = 1/(x - 1)$; $(2, 1)$.
24. $f(x) = 2/(x + 7)$; $(1, 1/4)$.
25. $f(x) = \sqrt{x}$; $(1, 1)$.
    (*Hint*: Multiply the resulting quotient by $(\sqrt{1+h} + \sqrt{1})/(\sqrt{1+h} + \sqrt{1})$.)
26. $f(x) = \sqrt{x - 1}$; $(5, 2)$.
27. $f(x) = xa^x$, $a > 0$; $(0, 0)$.
28. $f(x) = x^2 \sin x$; $(0, 0)$.
29. $f(x) = x^2 \cos x$; $(0, 0)$.
30. $f(x) = x \sin(\pi/2 - x)$; $(0, 0)$.
31. $f(x) = x^7$; $(0, 0)$.
32. $f(x) = x^9$; $(0, 0)$.
33. $f(x) = (x - 2)^2 \sin x$; $(2, 0)$.
34. $f(x) = x \tan x$; $(0, 0)$.
35. Find the slope of the line tangent to $f(x) = x^2 + 2x$ at the point $(a, f(a))$.
36. Find the slope of the line tangent to $f(x) = x^2 - 3x + 1$ at the point $(a, f(a))$.

A line passing through the point $(a, f(a))$ and perpendicular to the tangent line at $(a, f(a))$ is called the *normal* to the graph of $y = f(x)$ at $(a, f(a))$.

37. Find the normal to the graph of $f(x) = 3x^2 + 1$ at $(1, 4)$.
38. Find the normal to the graph of $f(x) = 1 - x^2$ at $(1, 0)$.

2.2 Limits

The process of finding the slope m of the line tangent to the graph of $y = f(x)$ at $(a, f(a))$ involves finding the number m that $m(h)$ approaches as h approaches zero. This is a specific example of the limit process. Since this process recurs throughout calculus, we shall devote this section to a rather general study of limits.

Suppose that $f(x)$ becomes arbitrarily close to the number L as x approaches a. We then say that the limit of $f(x)$ as x approaches a is L, and we write

$$\lim_{x \to a} f(x) = L.$$

The definition of the slope m of a tangent line can now be written in the shorthand form

$$m = \lim_{h \to 0} m(h), \quad \text{or}$$

$$m = \lim_{h \to 0} \frac{f(a+h) - f(a)}{h}.$$

Before looking at some examples we shall restate the meaning of limit in a somewhat more explicit form.

We say that

$$\lim_{x \to a} f(x) = L$$

if $f(x)$ can be made arbitrarily close (that is, closer than any preassigned positive distance) to L by requiring x to be sufficiently close to a but not equal to a. It is important to realize that $f(x)$ must be arbitrarily close to the number L for *all* x that are sufficiently close to a but different from a. With this in mind, let us consider the following examples.

Example 1 | Find $\lim_{x \to 4} (x + 2)$.

In this case $f(x) = x + 2$ can be made arbitrarily close to 6 by requiring x to be sufficiently close to 4 but not equal to 4. For example, $x + 2$ can be made to be within 1/1000 of 6 (that is, $|(x + 2) - 6| < 1/1000$) by requiring that x be within 1/1000 of 4 but not equal to 4 ($0 < |x - 4| < 1/1000$). We get

$$\lim_{x \to 4} (x + 2) = 6. \qquad \|$$

Example 2 | Find $\lim_{x \to 7} \frac{x^2 - 49}{x - 7}$.

Since $\frac{x^2 - 49}{x - 7} = \frac{(x - 7)(x + 7)}{x - 7} = x + 7$ for $x \neq 7$, and we are only concerned about the given function when $x \neq 7$, we have

$$\lim_{x \to 7} \frac{x^2 - 49}{x - 7} = \lim_{x \to 7} \frac{(x - 7)(x + 7)}{x - 7} = \lim_{x \to 7} (x + 7).$$

But $x + 7$ can be made arbitrarily close to 14 by requiring x to be sufficiently close to 7 but different from 7. Thus

$$\lim_{x \to 7} \frac{x^2 - 49}{x - 7} = 14. \qquad \|$$

Example 3 Let $f(x) = \begin{cases} x & \text{if} \quad x \neq 5 \\ 1 & \text{if} \quad x = 5. \end{cases}$ Find $\lim_{x \to 5} f(x)$.

The graph of this function appeared in Example 4 of Section 1.5. That figure is repeated here as Figure 2.2.1. In this case $f(x)$ is arbitrarily close to 5 when x is sufficiently close to, but not equal to, 5. Note that if we were to permit x to be equal to 5, $f(x)$ would take on the value 1 and so would not be arbitrarily close to 5. However, by the definition of limit we are only concerned with values of x that are *not equal to* 5. Thus

$$\lim_{x \to 5} f(x) = 5.$$

Figure 2.2.1

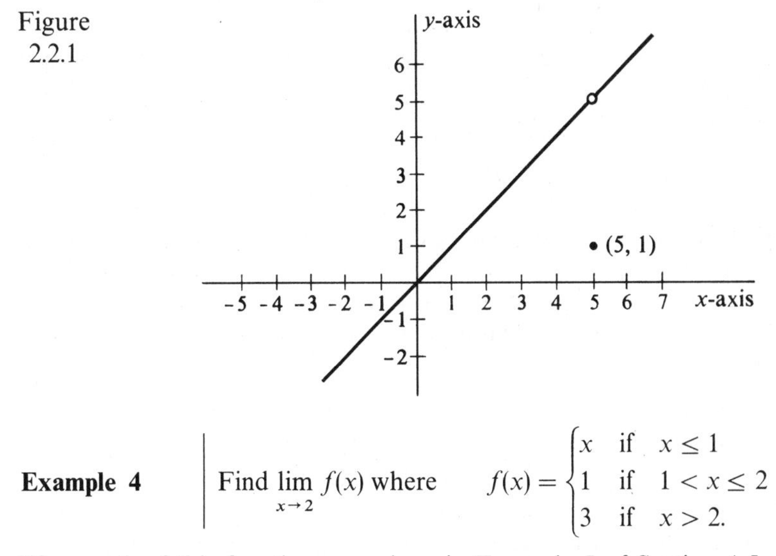

Example 4 Find $\lim_{x \to 2} f(x)$ where $f(x) = \begin{cases} x & \text{if} \quad x \leq 1 \\ 1 & \text{if} \quad 1 < x \leq 2 \\ 3 & \text{if} \quad x > 2. \end{cases}$

The graph of this function was given in Example 5 of Section 1.5 and is repeated here as Figure 2.2.2. ||

Figure 2.2.2

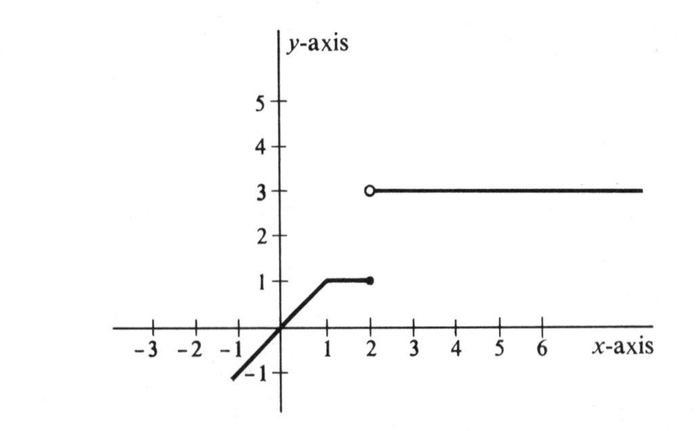

Recall that $\lim_{x \to 2} f(x) = L$ if $f(x)$ can be made arbitrarily close to L by requiring x to be sufficiently close to, but not equal to, 2. If $f(x)$ is to be made arbitrarily close to a number L, it surely must be possible to make $f(x)$ be within 1/10 of L by requiring x to be sufficiently close to 2, but not equal to 2. However, for values of x larger than 2, $f(x) = 3$; and for values of x slightly less than 2, $f(x) = 1$. Consequently L must be within 1/10 of both 1 and 3. Since no number L satisfies this requirement, $\lim_{x \to 2} f(x)$ has no meaning. That is, the limit does not exist. ||

Example 4 illustrates the fact that when $f(x)$ gets arbitrarily close to one number for some values of x sufficiently close to a, but $f(x)$ gets arbitrarily close to a different number for other values of x sufficiently close to a, then $\lim_{x \to a} f(x)$ does not exist. That is, when $\lim_{x \to a} f(x)$ exists it has exactly one value L, not two or more. The proof of that fact is left to the Challenging Problems, Chapter 2. The next example is another such illustration.

Example 5 | Let $f(x) = \begin{cases} 3x & \text{if } x \text{ is rational} \\ 4x & \text{if } x \text{ is irrational.} \end{cases}$ Find $\lim_{x \to 5} f(x)$.

In this case $f(x)$ gets arbitrarily close to 15 for the rational values of x that are sufficiently close to 5, but $f(x)$ gets arbitrarily close to 20, a different number, for the irrational values of x that are sufficiently close to 5. Thus the limit does not exist.

Example 6 | Find $\lim_{x \to -1} x^2$.

Here $f(x) = x^2$ is close to 1 when x is close to -1. In fact x^2 may be made arbitrarily close to 1 by requiring x to be sufficiently close to -1 but not equal to -1. Thus, $\lim_{x \to -1} x^2 = 1$. ||

Example 7 | Let $f(x) = \begin{cases} 2x & \text{if } x \le 1 \\ 2 & \text{if } x > 1. \end{cases}$ Find $\lim_{x \to 1} f(x)$.

Figure 2.2.3 shows the graph of this function. Here, when x is required to be close enough to 1, but not equal to 1, $f(x)$ can be made arbitrarily close to 2. Thus $\lim_{x \to 1} f(x) = 2$. If, for example, we want to make $f(x)$ within a positive distance ε of 2, we need only require x to be within $\varepsilon/2$ of 1 (but not equal to 1). To prove this

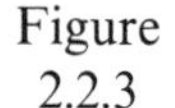
Figure 2.2.3

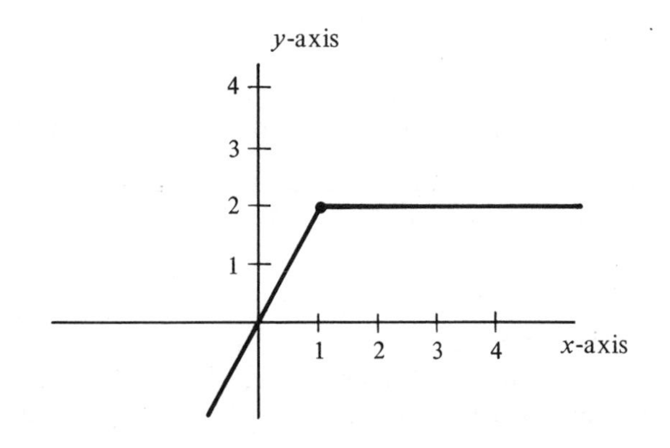

last statement, let $0 < |x - 1| < \varepsilon/2$. Then, when $x < 1$, we have $f(x) = 2x$; so

$$|f(x) - 2| = |2x - 2| = 2|x - 1| < 2\left(\frac{\varepsilon}{2}\right) = \varepsilon;$$

and when $x > 1$, we have $f(x) = 2$; so

$$|f(x) - 2| = |2 - 2| = 0 < \varepsilon.$$

In either case,

$$|f(x) - 2| < \varepsilon.$$

when

$$0 < |x - 1| < \frac{\varepsilon}{2}.$$

Consequently $f(x)$ may be made arbitrarily close to 2, that is, within any preassigned positive distance ε, by requiring x to be close enough to, but not equal to, 1. Hence $\lim_{x \to 1} f(x) = 2$. ||

Example 8 | Find $\lim_{x \to 3} f(x)$ where $f(x) = (x^2 - x - 6)/(x - 3)$.

The graph of this function appeared in Example 6 of Section 1.5. That figure is repeated here as Figure 2.2.4. Using the fact that

$$\frac{x^2 - x - 6}{x - 3} = \frac{(x - 3)(x + 2)}{x - 3},$$

we see that this function can also be written as

$$f(x) = \begin{cases} x + 2 & \text{if } x \neq 3 \\ \text{no value} & \text{if } x = 3. \end{cases}$$

Figure 2.2.4

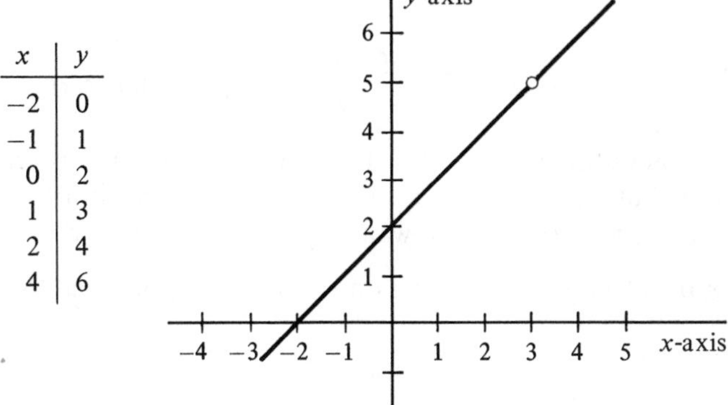

x	y
−2	0
−1	1
0	2
1	3
2	4
4	6

In calculating $\lim_{x \to 3} f(x)$, however, we are *not* concerned with the value of the function *at* $x = 3$. The important point is that $x + 2$ can be made arbitrarily close to 5 by requiring x to be close to 3 but *not* equal to 3. Since this is what is required by the definition of limit, $\lim_{x \to 3} f(x) = 5$. ||

Before we consider some of the major theorems concerning limits, we shall make the definition of limit somewhat more precise. The only possible misconceptions in our present definition arise from the use of the words "arbitrarily close" and "sufficiently close." The phrase "$f(x)$ can be made arbitrarily close to L by choosing x sufficiently close to a" means that $f(x)$ can be made to be within any preassigned positive distance (traditionally called ε) of L by requiring x to be within a specified positive distance (called δ) of a, but not equal to a. That is, $|f(x) - L| < \varepsilon$ when $|x - a| < \delta$ and $x \neq a$. With these changes in mind we can give a precise definition of limit.

Definition 2.2.1

> $\lim_{x \to a} f(x) = L$ means that for each $\varepsilon > 0$ (the preassigned distance) there is a $\delta > 0$ (the measurement of "sufficiently close") such that $|f(x) - L| < \varepsilon$ when $0 < |x - a| < \delta$.

Note that in the statement of this definition we have expressed $|x - a| < \delta$ and $x \neq a$ more compactly as $0 < |x - a| < \delta$. Figure 2.2.5 gives the geometric interpretation of this definition. Note that $f(x)$ lies between $L + \varepsilon$ and $L - \varepsilon$ (that is, $|f(x) - L| < \varepsilon$) for all values of $x \neq a$ between $a - \delta$ and $a + \delta$ (that is, for $0 < |x - a| < \delta$). The value $f(a)$ of the function at a is immaterial in the computation of L.

Since the inequality $|f(x) - L| < \varepsilon$ means that $f(x)$ is between $L - \varepsilon$ and $L + \varepsilon$, this indicates that graphically $f(x)$ is between the lines $y = L - \varepsilon$ and $y = L + \varepsilon$. Similarly the graphical meaning of $0 < |x - a| < \delta$ is that x is between the lines $x = a - \delta$ and $x = a + \delta$ but not on the line $x = a$. Thus Definition 2.2.1 has this graphical interpretation: For each positive number ε (no matter how small) a positive number δ can be chosen so that the graph of $f(x)$ is between the lines $y = L - \varepsilon$ and $y = L + \varepsilon$, when x is between the lines $x = a - \delta$ and $x = a + \delta$ but not on the line $x = a$.

Figure 2.2.5 shows the graphical interpretation for a particular $\varepsilon > 0$. In general, the smaller the ε the smaller the positive number δ must be. That is, the smaller the height of the indicated rectangle the smaller its width must be. It is important to realize that Figure 2.2.5 is only for one particular positive ε and that there must be a similar picture for *every* positive number ε.

Figure 2.2.5

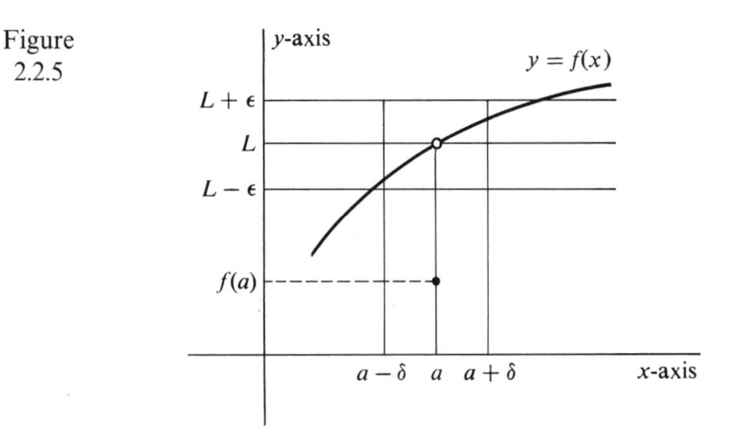

As might be expected, it is sometimes difficult to prove the existence or nonexistence of limits. In later chapters more powerful techniques will be available. However, a calculator is often useful in helping to see if certain limits appear to exist, and in making approximations to the values of those that do exist. The idea is to take values of x closer and closer to a (but not equal to a) and to determine whether the values of $f(x)$ appear to get arbitrarily close to exactly one particular number L. While such procedures do not prove the existence of limits, they do help to obtain a "feel" for the limit concept.

In order to prove that $\lim_{x \to a} f(x)$ exists and has the value L, one must show that $f(x)$ can be made *arbitrarily* close to L for *all* x that are sufficiently close to a, not just rather close for the few values of x we have chosen for the calculations.

The next four examples illustrate the indicated procedure. For each of them we make a table of values with appropriate columns to see if it appears that the given limit exists or not, and if so, we approximate the value of the limit. The calculations were made with a low-priced hand calculator.

Example 9 Approximate $\lim_{x \to 0} \dfrac{x}{\sqrt{4+x}-2}$.

After choosing the values of x shown, we obtain the following table. The functional values for negative x are indented so that it is easier to see how the function is changing.

x	$\dfrac{x}{\sqrt{4+x}-2}$
.1	4.024846
−.1	3.974842
.01	4.002498
−.01	3.997498

x	$\dfrac{x}{\sqrt{4+x}-2}$
.001	4.000250
−.001	3.999750
.0001	4.000025
−.0001	3.999975

The limit appears to exist and to have the value 4. (Actually it can be proved that the limit does exist and does have the value 4.) ||

Example 10 Approximate $\lim_{x \to 1} \dfrac{x^2 + |x-1| - 1}{|x-1|}$.

We obtain the following table, where $f(x) = \dfrac{x^2 + |x-1| - 1}{|x-1|}$.

x	$f(x)$
1.1	3.1
.9	−.9
1.01	3.01
.99	−.09

x	$f(x)$
1.001	3.001
.999	−.009
1.0001	3.0001
.9999	−.0009

It appears that the limit does not exist because $f(x)$ seems to be getting arbitrarily close to 3 for some x and arbitrarily close to 1, a different number, for other values of x. (Actually the limit does not exist.) ||

Example 11 Approximate $\lim_{x \to 0} \dfrac{\sin x}{x}$, where x is in radians.

We obtain the following table:

x	$\sin x$	$\dfrac{\sin x}{x}$
± 1.0	$\pm .841471$	.841471
$\pm .5$	$\pm .479425$	.958850
$\pm .2$	$\pm .198669$	.993345
$\pm .1$	$\pm .099833$	.998330
$\pm .01$	$\pm .009999$	.999900
$\pm .001$	$\pm .001000$	1.000000

It appears from the table that the limit exists and has the value 1. ||

Actually it is proved in Chapter 7 that the limit in Example 11 does exist and does have the value 1.

Example 12 | Approximate $\lim_{x \to 0} (1 + x)^{1/x}$.

We obtain the table below in which the values of the function for negative x are indented so that it is easier to see how the function is changing as the values of x get closer and closer to 0.

x	$1 + x$	$\dfrac{1}{x}$	$(1 + x)^{1/x}$
.5	1.5	2	2.250000
$-.5$	.5	-2	4.000000
.2	1.2	5	2.488311
$-.2$	.8	-5	3.051748
.1	1.1	10	2.593737
$-.1$	.9	-10	2.867959
.01	1.01	100	2.704722
$-.01$	.99	-100	2.731906
.001	1.001	1000	2.715564
$-.001$	.999	-1000	2.719642
.0005	1.0005	2000	2.718280
$-.0005$	.9995	-2000	2.718278

It appears from the table that the limit exists and has a value near 2.71828. ||

The limit of Example 12 arises naturally in the material of Chapter 6, and it can be proved that the limit does indeed exist. An outline of the proof appears in the Challenging Problems, Chapter 11. The limit is so important in calculus that it has been given a special symbol, e. The number e turns out to be an irrational number. (See Challenging Problems, Chapter 11 for an outline of the proof). Since e is irrational, we cannot give an exact decimal expression for it, but a 20 decimal place approximation to e is

$$2.71828182845904523536.$$

It should be noted that none of the functions in Examples 9–12 whose limits we were concerned with as $x \to a$ has any meaning when $x = a$. Of course that does not matter because the definition of a limit as $x \to a$ does not require x to be a.

Exercises 2.2

In Exercises 1–26, find the limit indicated or state that it does not exist.

1. $\lim_{x \to 0} \dfrac{x^2 + 5x}{x}$.

2. $\lim_{x \to 0} \dfrac{x^2 - 3x}{x}$.

3. $\lim_{x \to 2} \dfrac{x^2 - 4}{x - 2}$.

4. $\lim_{x \to 5} \dfrac{x^2 - 25}{x - 5}$.

5. $\lim_{x \to 0} (x/|x|)$.

6. $\lim_{x \to 4} \dfrac{2 - \sqrt{x}}{4 - x}$.

7. $\lim_{x \to 9} \dfrac{3 - \sqrt{x}}{9 - x}$.

8. $\lim_{x \to 3} \dfrac{|x - 3|}{x - 3}$.

9. $\lim_{x \to 0} \dfrac{3x}{x^2 - 8x}$.

10. $\lim_{x \to 0} \dfrac{2x}{x^2 + 6x}$.

11. $\lim_{x \to 3} \dfrac{x - 3}{x^2 - x - 6}$.

12. $\lim_{x \to -4} \dfrac{x + 4}{x^2 + 2x - 8}$.

13. $\lim_{x \to -2} \dfrac{x^2 - x - 6}{x + 2}$.

14. $\lim_{x \to 1} \dfrac{x^2 + 3x - 4}{x - 1}$.

15. $\lim_{x \to 1} f(x)$, where $f(x) = \begin{cases} x^2 & \text{if } x < 1 \\ x & \text{if } x > 1. \end{cases}$

16. $\lim_{x \to 3} f(x)$, where $f(x) = \begin{cases} 3x & \text{if } x < 3 \\ x^2 & \text{if } x > 3. \end{cases}$

17. $\lim_{x \to 0} f(x)$, where $f(x) = \begin{cases} (x^2 - 2x)/x & \text{if } x \neq 0 \\ 1 & \text{if } x = 0. \end{cases}$

18. $\lim_{x \to 5} f(x)$, where $f(x) = \begin{cases} \dfrac{x^2 - 25}{x - 5} & \text{if } x \neq 5 \\ 3 & \text{if } x = 5. \end{cases}$

19. $\lim_{x \to 2} f(x)$, where $f(x) = \begin{cases} x^2 & \text{if } x < 2 \\ 3x & \text{if } x \geq 2. \end{cases}$

20. $\lim_{x \to 0} f(x)$, where $f(x) = \begin{cases} 3 & \text{if } x \geq 0 \\ x - 1 & \text{if } x < 0. \end{cases}$

21. $\lim_{x \to -2} f(x)$, where $f(x) = \begin{cases} 1 & \text{if } x < -2 \\ 5 & \text{if } x = -2 \\ 9 & \text{if } x > -2. \end{cases}$

22. $\lim_{x \to 4} f(x)$, where $f(x) = \begin{cases} x & \text{if } x < 4 \\ 7 & \text{if } x = 4 \\ 2x & \text{if } x > 4. \end{cases}$

23. $\lim_{x \to 0} x \sin\left(\dfrac{1}{x}\right)$.

24. $\lim_{x \to 0} \cos\left(\dfrac{1}{x}\right)$.

25. $\lim_{x\to 0} \sin\left(\frac{1}{x}\right)$.

26. $\lim_{x\to 0} x\cos\left(\frac{1}{x}\right)$.

In Exercises 27–34, make a table of values with appropriate columns to see if it appears that the given limit exists, and if so, approximate the value of the limit.

27. $\lim_{x\to 2} \frac{x-2}{\sqrt{7+x}-3}$.

28. $\lim_{x\to 1} \frac{x-1}{\sqrt{x+24}-5}$.

29. $\lim_{x\to 3} \frac{x^2-|x-3|-9}{|x-3|}$.

30. $\lim_{x\to 2} \frac{x^2+|x-2|-4}{|x-2|}$.

31. $\lim_{x\to 0} \frac{1-\cos x}{x}$.

32. $\lim_{x\to 0} \frac{\sin 2x}{4x}$.

33. $\lim_{x\to 0} \left(\frac{1}{1-x}\right)^{1/x}$.

34. $\lim_{x\to 0} \left(\frac{1+x}{1-x}\right)^{1/x}$.

35. How close to 3 must one restrict x (a) so that $2x$ is within $1/8$ of 6? (b) so that $2x$ is within ε of 6?

36. How close to 5 must one restrict x (a) so that $3x$ is within $1/10$ of 15? (b) so that $3x$ is within ε of 15?

37. How close to -2 must one restrict x (a) so that $4x$ is within $1/10$ of -8? (b) so that $4x$ is within ε of -8?

38. How close to -6 must one restrict x (a) so that $5x$ is within $1/40$ of -30? (b) so that $5x$ is within ε of -30?

39. Find the slope of the tangent to the graph of $\sqrt{x}$ at the point (4, 2).

40. Find the slope of the tangent to the graph of $1/x^2$ at the point $(-1, 1)$.

41. Prove that $\lim_{x\to a} x = a$.

42. Prove that if c is a constant, $\lim_{x\to a} c = c$.

2.3 The Limit Theorem

Calculation of limits directly from the definition can be difficult, but the Limit Theorem will make the computation of many limits a straightforward task.

Theorem 2.3.1

The Limit Theorem

If $\lim_{x\to a} f(x)$ and $\lim_{x\to a} g(x)$ exist, then

(i) $\lim_{x\to a} [f(x)+g(x)] = \lim_{x\to a} f(x) + \lim_{x\to a} g(x)$,

(ii) $\lim_{x\to a} c\cdot f(x) = c\cdot \lim_{x\to a} f(x)$, where c is any constant,

(iii) $\lim_{x\to a} f(x)\cdot g(x) = \lim_{x\to a} f(x)\cdot \lim_{x\to a} g(x)$,

(iv) $\lim_{x\to a} \frac{f(x)}{g(x)} = \frac{\lim_{x\to a} f(x)}{\lim_{x\to a} g(x)}$, provided $\lim_{x\to a} g(x) \neq 0$.

All of the results of the limit theorem are just what one would expect. For example if $f(x)$ is getting arbitrarily close to L_1 and $g(x)$ is getting arbitrarily close to L_2, then surely one would expect that $f(x)g(x)$ must get arbitrarily close to L_1L_2.

To indicate how the various parts of this theorem are proved, we shall prove part (i). The interested reader will be able to find proofs of the remaining parts in most advanced calculus texts.

Proof of Part (i). To simplify the notation we let $\lim_{x \to a} f(x) = L_1$ and $\lim_{x \to a} g(x) = L_2$. Hence $f(x)$ can be made arbitrarily close to L_1 (say within a positive distance ε_1) by requiring x to be sufficiently close to a but not equal to a (say within a distance δ_1). That is, if $0 < |x - a| < \delta_1$, then $|f(x) - L_1| < \varepsilon_1$.

In the same fashion, $g(x)$ can be made arbitrarily close to L_2 (within $\varepsilon_2 > 0$) by requiring $x \neq a$ to be sufficiently close to a (within δ_2 of a). That is, if $0 < |x - a| < \delta_2$, then $|g(x) - L_2| < \varepsilon_2$.

Our task is to show that $f(x) + g(x)$ can be made arbitrarily close to $L_1 + L_2$ (say within a distance ε) by requiring x to be close to a but not equal to a (within δ of a). Hence we want to find a restriction δ such that if $0 < |x - a| < \delta$, then $|f(x) + g(x) - (L_1 + L_2)| < \varepsilon$. To accomplish this, note that

$$|f(x) + g(x) - (L_1 + L_2)| = |f(x) - L_1 + g(x) - L_2|.$$

Now a use of the triangle inequality gives:

$$|f(x) + g(x) - (L_1 + L_2)| \leq |f(x) - L_1| + |g(x) - L_2|. \tag{1}$$

From the discussion above, we know that $|f(x) - L_1| < \varepsilon_1$ when $0 < |x - a| < \delta_1$, and that $|g(x) - L_2| < \varepsilon_2$ when $0 < |x - a| < \delta_2$. Hence, if we require *both* $0 < |x - a| < \delta_1$ and $0 < |x - a| < \delta_2$, we have *both* $|f(x) - L_1| < \varepsilon_1$ and $|g(x) - L_2| < \varepsilon_2$. That is, the restriction $0 < |x - a| < \delta$, where δ is the smaller of δ_1 and δ_2, yields $|f(x) - L_1| < \varepsilon_1$ and $|g(x) - L_2| < \varepsilon_2$. Substitution into (1) gives:

$$|f(x) + g(x) - (L_1 + L_2)| < \varepsilon_1 + \varepsilon_2. \tag{2}$$

Of course we wanted

$$|f(x) + g(x) - (L_1 + L_2)| < \varepsilon.$$

To achieve this we can select any ε_1 and ε_2 that add up to ε. For example, we could let $\varepsilon_1 = \varepsilon_2 = \varepsilon/2$. (This is certainly possible since ε_1 and ε_2 are arbitrary positive numbers.) Then, if $0 < |x - a| < \delta$, where δ is the smaller of δ_1 and δ_2, we can conclude that

$$|f(x) + g(x) - (L_1 + L_2)| < \frac{\varepsilon}{2} + \frac{\varepsilon}{2} = \varepsilon.$$

Consider the following examples of applications of the Limit Theorem.

Example 1 Find $\lim_{x \to a} (x^2 + 3)$.

$$\lim_{x \to a} (x^2 + 3) = \lim_{x \to a} x^2 + \lim_{x \to a} 3 \qquad \text{(by part (i))}$$

$$= \lim_{x \to a} x \cdot \lim_{x \to a} x + \lim_{x \to a} 3 \qquad \text{(by part (iii))}.$$

Then, since $\lim_{x \to a} x = a$ and $\lim_{x \to a} 3 = 3$, we get $\lim_{x \to a} (x^2 + 3) = a \cdot a + 3 = a^2 + 3$. ||

Example 2 Find $\lim_{h \to 0} \dfrac{4h^2 + 3h}{h}$.

Since $(4h^2 + 3h)/h = 4h + 3$, when $h \neq 0$, we have

$$\lim_{h \to 0} \frac{4h^2 + 3h}{h} = \lim_{h \to 0} (4h + 3)$$

$$= \lim_{h \to 0} 4h + \lim_{h \to 0} 3 \qquad \text{(by part (i))}$$

$$= 4 \cdot \lim_{h \to 0} h + \lim_{h \to 0} 3 \qquad \text{(by part (ii))}.$$

Then, since $\lim_{h \to 0} h = 0$ and $\lim_{h \to 0} 3 = 3$, we have

$$\lim_{h \to 0} \frac{4h^2 + 3h}{h} = 4(0) + 3 = 3.$$

Notice that in this example we could not have used part (iv) directly, since the limit of the denominator is zero. (That is, $\lim_{h \to 0} h = 0$.) ||

Example 3 Find $\lim_{h \to 1} \dfrac{4h^2}{h + 5}$.

Since $$\lim_{h \to 1} (h + 5) = \lim_{h \to 1} h + \lim_{h \to 1} 5 = 1 + 5 = 6 \neq 0,$$

we may use part (iv). Then

$$\lim_{h \to 1} \frac{4h^2}{h + 5} = \frac{\lim_{h \to 1} 4h^2}{\lim_{h \to 1} (h + 5)} = \frac{4 \cdot \lim_{h \to 1} h^2}{6}$$

$$= \frac{4\left(\lim_{h \to 1} h\right)\left(\lim_{h \to 1} h\right)}{6} = \frac{4}{6} = \frac{2}{3}.$$ ||

Example 4 Find an equation of the tangent line to the graph of $f(x) = x^2 + 1$ at the point (3, 10).

Recall that the tangent line to the graph of $y = f(x)$ at the point $(a, f(a))$ has slope

$$m = \lim_{h \to 0} m(h) = \lim_{h \to 0} \frac{f(a + h) - f(a)}{h}.$$

In this case the slope of the tangent is

$$m = \lim_{h\to 0} \frac{f(3+h) - f(3)}{h}$$

$$= \lim_{h\to 0} \frac{(3+h)^2 + 1 - 10}{h}$$

$$= \lim_{h\to 0} \frac{9 + 6h + h^2 + 1 - 10}{h} = \lim_{h\to 0} \frac{6h + h^2}{h}.$$

Note that part (iv) does not apply here, since the limit of the denominator is 0. However, we can cancel an h, since $h \neq 0$, to get

$$m = \lim_{h\to 0} (6 + h)$$

$$= \lim_{h\to 0} 6 + \lim_{h\to 0} h$$

$$= 6 + 0$$

$$= 6.$$

Thus the slope of the tangent line is 6, and, since the line also passes through the point (3, 10), its equation is $y - 10 = 6(x - 3)$, or $y = 6x - 8$. ||

Exercises 2.3

In Exercises 1–8, use the limit theorem to calculate the indicated limits.

1. $\lim_{x\to 2} (x^2 + 3x)$.
2. $\lim_{x\to 3} (x^2 - 5x)$.
3. $\lim_{h\to 1} \dfrac{h^3}{h-2}$.
4. $\lim_{t\to 5} \dfrac{t+6}{t^2+1}$.
5. $\lim_{h\to 0} \dfrac{h^2 - 5h}{h}$.
6. $\lim_{x\to 0} \dfrac{x^2 + 6x}{x^2 - x}$.
7. $\lim_{x\to 0} (3x - 2)(x^2 + 7)$.
8. $\lim_{x\to 1} (2x^2 - 1)(x^4 - 5)^2$.

In Exercises 9–16, find an equation of the tangent to the graph of the given function or equation at the indicated point.

9. $y = x^2 - 5x$, at $(3, -6)$.
10. $y = x^2 + 3x$, at $(-1, -2)$.
11. $f(x) = 2x - 3x^2$, at $(1, -1)$.
12. $f(x) = 5 - x^2$, at $(2, 1)$.
13. $f(x) = x^3 - 2x^2 + 6$, at $(2, 6)$.
14. $f(x) = 3x^2 - 5x^3 + 2x$, at $(2, -24)$.
15. $f(x) = 3x^2 + (1/\sqrt{x})$, at $(1, 4)$.
16. $f(x) = x^2 + \sqrt{5 - x}$, at $(1, 3)$.
17. Prove that $\lim_{x\to a} [f(x)]^2 = \left[\lim_{x\to a} f(x)\right]^2$ if $\lim_{x\to a} f(x)$ exists.
18. Prove that $\lim_{x\to a} [f(x)g(x)h(x)] = \lim_{x\to a} f(x) \lim_{x\to a} g(x) \lim_{x\to a} h(x)$ if all limits exist.
19. Prove that $\lim_{x\to a} [f(x) + g(x) + h(x)] = \lim_{x\to a} f(x) + \lim_{x\to a} g(x) + \lim_{x\to a} h(x)$ if all limits exist.
20. Let a, a_0, a_1, and a_2 be constants. Prove that

$$\lim_{x\to a} (a_2x^2 + a_1x + a_0) = a_2a^2 + a_1a + a_0.$$

2.4 Continuity

In Section 1.5, a number of graphs were sketched. Some of them are "continuous" in the sense that there are no breaks in the graphs, while others have breaks or "discontinuities." In this section we shall formalize the concept of continuity and consider a few of the major properties of continuous functions.

Intuitively a function is continuous at a point $x = a$ if the graph of the function has no break or jump at the point $x = a$. Thus, for example, the function whose graph is shown in Figure 2.4.1 is discontinuous at a and continuous at a'.

Figure 2.4.1

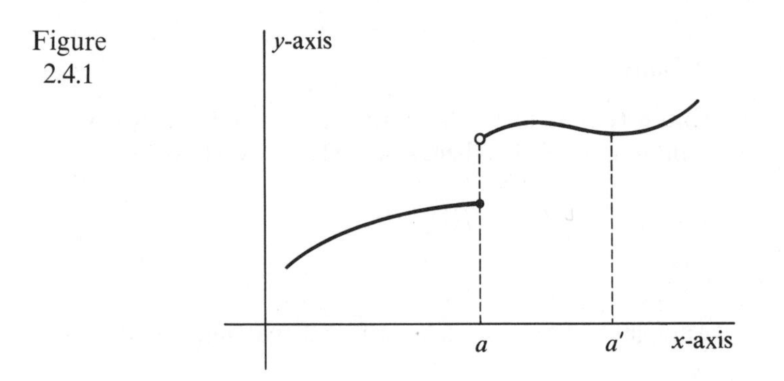

To formalize the definition of continuity we note that in order for the graph of a function *not* to have a break over the point a, it is necessary that

1. f is defined at a, that is, $f(a)$ exists; and
2. values of x close to a must yield values of $f(x)$ close to $f(a)$.

Those two conditions are expressed more precisely in the following:

Definition 2.4.1

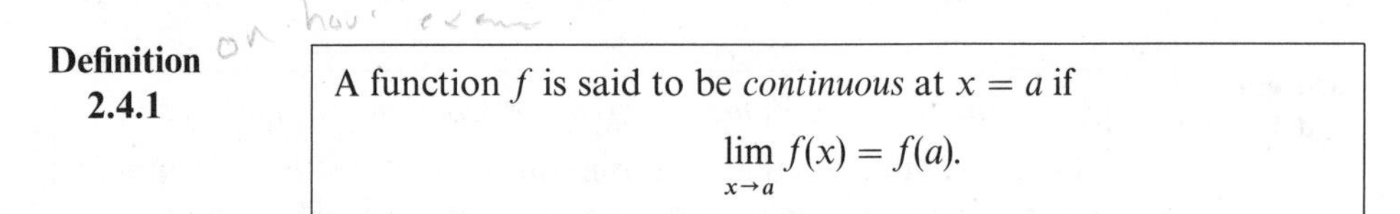

A function f is said to be *continuous* at $x = a$ if

$$\lim_{x \to a} f(x) = f(a).$$

Example 1 | Examine the continuity of $f(x) = x + 2$ at the point $x = 4$.

In Example 1 of Section 2.2, we found that

$$\lim_{x \to 4} (x + 2) = 6.$$

Since $f(4) = 6$, we have

$$\lim_{x \to 4} f(x) = f(4),$$

and so f is continuous at $x = 4$. ||

Example 2 Determine whether $f(x) = \dfrac{x^2 - x - 6}{x - 3}$ is continuous at $x = 3$.

Since $f(3) = 0/0$ is meaningless, there is no value $f(3)$. Hence it cannot be true that $\lim_{x \to 3} f(x) = f(3)$, and so f is discontinuous at $x = 3$. ||

Example 3 Examine the continuity of the function

$$f(x) = \begin{cases} x & \text{if} \quad x \leq 1 \\ 1 & \text{if} \quad 1 < x \leq 2 \\ 3 & \text{if} \quad x > 2 \end{cases}$$

at $x = 1$ and $x = 2$.

Since $f(x)$ can be made arbitrarily close to 1 by requiring x to be sufficiently close to but not equal to 1, we obtain $\lim_{x \to 1} f(x) = 1$. Then, since $f(1) = 1$, we have

$$\lim_{x \to 1} f(x) = f(1),$$

and so f is continuous at $x = 1$.

On the other hand in Example 4 of Section 2.2, we found that $\lim_{x \to 2} f(x)$ does *not* exist. Therefore

$$\lim_{x \to 2} f(x) \neq f(2) = 1.$$

Consequently f is *not* continuous at $x = 2$. ||

The reader may well wonder why we have bothered to define continuous functions. The fact is that continuous functions have many important properties that will be useful in our future study of calculus. Here we mention one of the most fundamental and useful of these properties.

Theorem 2.4.1

Intermediate Value Theorem for Continuous Functions

Let f be a function that is continuous for $a \leq x \leq b$, and suppose that r is some number between $f(a)$ and $f(b)$. Then there is at least one number c between a and b such that $f(c) = r$.

In less rigorous language, the theorem states that a function which is continuous at every point of some interval takes on all values between any two of its values. The proof of the theorem can be found in many advanced calculus texts.

Figure 2.4.2 illustrates The Intermediate Value Theorem. Note that in this case there are several possible choices for c, namely c_1, c_2, or c_3.

If the function f is not continuous, The Intermediate Value Theorem need not hold. For example the function f shown in Figure 2.4.3 has a discontinuity, and for the value of r indicated there is no number c between a and b such that $f(c) = r$.

The polynomial functions afford a wide class of functions that are continuous at every point. Recall the following definition.

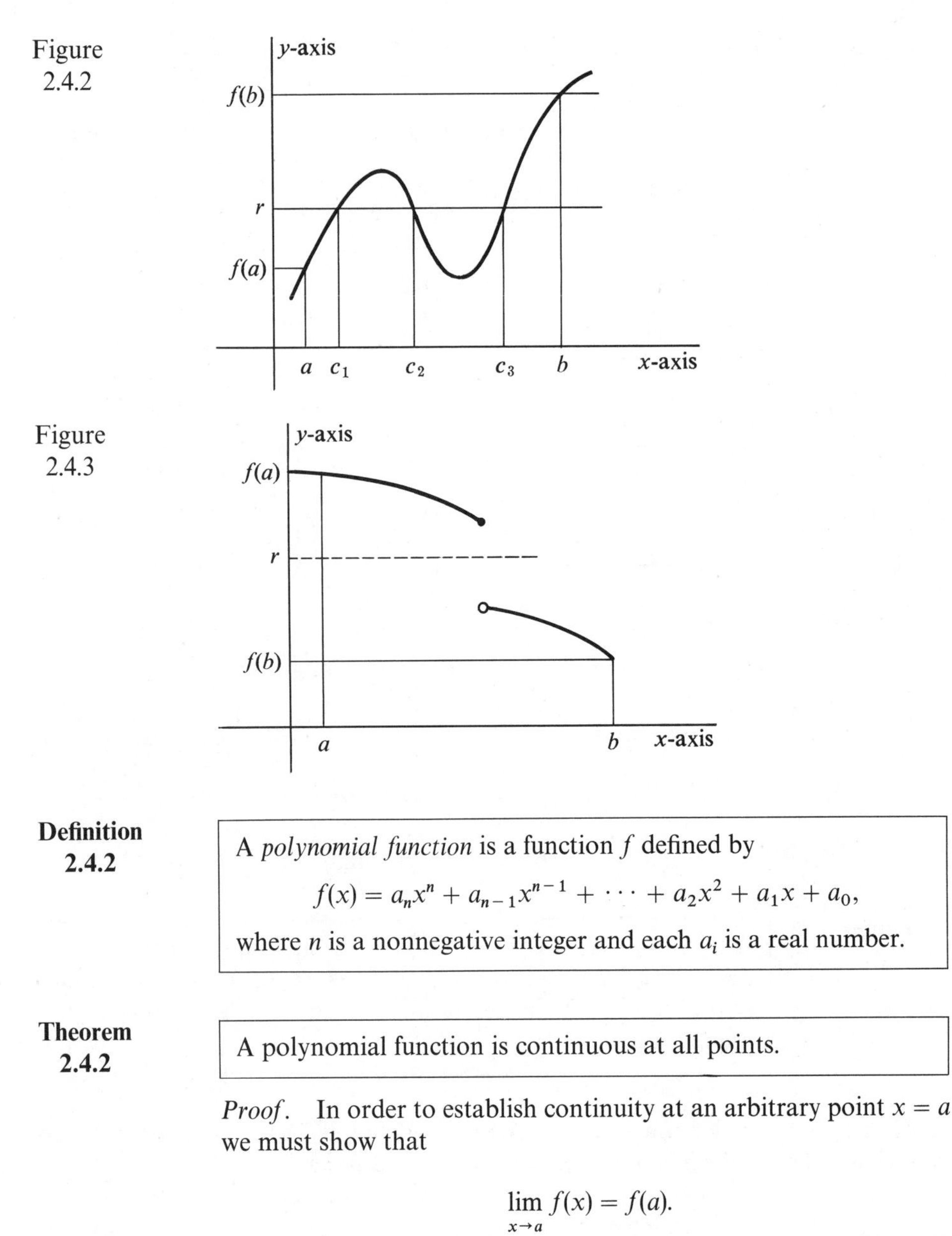

Figure 2.4.2

Figure 2.4.3

Definition 2.4.2

A *polynomial function* is a function f defined by

$$f(x) = a_n x^n + a_{n-1}x^{n-1} + \cdots + a_2x^2 + a_1x + a_0,$$

where n is a nonnegative integer and each a_i is a real number.

Theorem 2.4.2

A polynomial function is continuous at all points.

Proof. In order to establish continuity at an arbitrary point $x = a$ we must show that

$$\lim_{x \to a} f(x) = f(a).$$

The function $f(x)$ under consideration here is

$$f(x) = a_n x^n + a_{n-1}x^{n-1} + \cdots + a_2x^2 + a_1x + a_0.$$

Thus

$$\lim_{x \to a} f(x) = \lim_{x \to a} (a_n x^n + a_{n-1}x^{n-1} + \cdots + a_2x^2 + a_1x + a_0).$$

Now using the "sum part" of The Limit Theorem (Theorem 2.3.1, part (i)), we have

$$\lim_{x \to a} f(x) = \lim_{x \to a} a_n x^n + \lim_{x \to a} a_{n-1}x^{n-1} + \cdots + \lim_{x \to a} a_1x + \lim_{x \to a} a_0.$$

The Limit Theorem allows us to factor the constants a_i out of the limits. We obtain

$$\lim_{x\to a} f(x) = a_n \cdot \lim_{x\to a} x^n + a_{n-1} \cdot \lim_{x\to a} x^{n-1} + \cdots + a_1 \cdot \lim_{x\to a} x + \lim_{x\to a} a_0.$$

Since the limit of a product is the product of the limits (Theorem 2.3.1, part (iii)),

$$\lim_{x\to a} x^2 = \lim_{x\to a} (x \cdot x) = \left(\lim_{x\to a} x\right) \cdot \left(\lim_{x\to a} x\right) = \left(\lim_{x\to a} x\right)^2;$$

and, in general,

$$\lim_{x\to a} x^i = \left(\lim_{x\to a} x\right)^i.$$

Hence,

$$\lim_{x\to a} f(x) = a_n \cdot \left(\lim_{x\to a} x\right)^n + a_{n-1} \cdot \left(\lim_{x\to a} x\right)^{n-1} + \cdots$$
$$+ a_1 \cdot \lim_{x\to a} x + \lim_{x\to a} a_0.$$

Then, substituting $\lim_{x\to a} x = a$ and $\lim_{x\to a} a_0 = a_0$, we get

$$\lim_{x\to a} f(x) = a_n a^n + a_{n-1} a^{n-1} + \cdots + a_1 a + a_0, \quad \text{and so}$$

$$\lim_{x\to a} f(x) = f(a),$$

thus completing the proof of the theorem.

Definition 2.4.3

A function that is continuous at all points of a specified set is said to be *continuous on that set.* Moreover, a function that is continuous on the set of all real numbers is simply called *continuous.*

For example, polynomials are continuous functions. That fact, along with the following results, will expand our stock of continuous functions.

Theorem 2.4.3

Let f and g be functions that are continuous at $x = a$. Then the following functions are continuous at $x = a$:

(i) $f(x) + g(x)$,

(ii) $c \cdot f(x)$, for any constant c,

(iii) $f(x) \cdot g(x)$,

(iv) $\dfrac{f(x)}{g(x)}$ (provided that $g(a) \neq 0$).

Proof. The proof of this theorem is a direct application of The Limit Theorem (Theorem 2.3.1). We are assuming that both f and g are continuous at $x = a$. That is,

$$\lim_{x \to a} f(x) = f(a) \text{ and } \lim_{x \to a} g(x) = g(a).$$

To show that $f(x) + g(x)$ is continuous, for example, we use The Limit Theorem to obtain

$$\begin{aligned}\lim_{x \to a} [f(x) + g(x)] &= \lim_{x \to a} f(x) + \lim_{x \to a} g(x) \\ &= f(a) + g(a).\end{aligned}$$

Hence the function $f(x) + g(x)$ is continuous at $x = a$.

The remaining portions of the proof are simple applications of the corresponding parts of Theorem 2.3.1, and are left as exercises.

Example 4 | Investigate the continuity of $h(x) = \dfrac{x^3 + 7x + 1}{x^2 - x - 2}$.

The functions $f(x) = x^3 + 7x + 1$ and $g(x) = x^2 - x - 2$ are both polynomials and so are continuous for all values of x. By part (iv) of Theorem 2.4.3, the quotient is continuous at all points where the denominator is not zero. At the points where the denominator is zero, $h(x)$ is not continuous because it is not even defined there. Thus $h(x)$ is continuous everywhere except at those points where $x^2 - x - 2 = 0$, or $(x - 2)(x + 1) = 0$. Hence $h(x)$ is continuous except at $x = 2$ and $x = -1$. ||

Since all polynomials are continuous functions, it follows from part (iv) of Theorem 2.4.3 (as in Example 4) that the quotient of any two polynomials is continuous at all points except those where the denominator is zero. That is, rational functions are continuous for all values of x except those where the denominator is zero.

Exercises 2.4

In Exercises 1–14, find all values of x for which the given function is continuous.

1. $f(x) = 7 - \pi x^2$.

2. $f(x) = 3x^5 - 7x$.

3. $f(x) = \dfrac{3x}{x^2 - x - 6}$.

4. $f(x) = \dfrac{x + 2}{x^2 + x - 12}$.

5. $f(x) = \dfrac{x}{x^2 - 3x}$.

6. $f(x) = \dfrac{x + 1}{x^2 - 5x - 6}$.

7. $f(x) = \dfrac{x^2 - x - 6}{x^2 + 7x + 10}$.

8. $f(x) = \dfrac{x^2 + 2x - 8}{x^2 - x - 2}$.

9. $f(x) = \begin{cases} \dfrac{x^2 - x - 6}{x^2 + 7x + 10} & \text{if } x \neq -2 \\ -\frac{5}{3} & \text{if } x = -2. \end{cases}$

10. $f(x) = \begin{cases} \dfrac{x + 1}{x^2 - 5x - 6} & \text{if } x \neq -1 \\ -\frac{1}{6} & \text{if } x = -1. \end{cases}$

11. $f(x) = \begin{cases} \dfrac{x}{x^2 - 3x} & \text{if } x \neq 0 \\ -\frac{1}{2} & \text{if } x = 0. \end{cases}$

12. $f(x) = \begin{cases} \dfrac{x^2 + 2x - 8}{x^2 - x - 2} & \text{if } x \neq 2 \\ 2 & \text{if } x = 2. \end{cases}$

13. $f(x) = \begin{cases} 2x & \text{if } x \text{ is not an integer} \\ 0 & \text{if } x \text{ is an integer.} \end{cases}$

14. $f(x) = \begin{cases} x^2 & \text{if } x \text{ is rational} \\ x & \text{if } x \text{ is irrational.} \end{cases}$

In Exercises 15–20, each function is not defined for the given value of x. Define the function for the given value of x so that it is continuous there or state that this is impossible.

15. $f(x) = \dfrac{x}{x^2 - 3x}$, $x = 0$.

16. $f(x) = \dfrac{x + 1}{x^2 - 5x - 6}$, $x = -1$.

17. $f(x) = \dfrac{x^2 + x - 6}{x^2 - 3x + 2}$, $x = 2$.

18. $f(x) = \dfrac{x^2 + 2x - 3}{x^2 - 2x + 1}$, $x = 1$.

19. $f(x) = \dfrac{x^2 - x - 2}{x^2 + 2x + 1}$, $x = -1$.

20. $f(x) = \dfrac{x^2 - 2x - 15}{x^2 + 2x - 3}$, $x = -3$.

21. An airplane starts from its loading gate in Seattle and reaches a maximum speed of 600 miles per hour before coming to a stop at its loading gate in New York. Use The Intermediate Value Theorem (for continuous functions) to show that the plane must have been traveling at 150 miles per hour for at least two points of its journey.

22. Use The Intermediate Value Theorem to show that the equation $x^3 - 3x + 1 = 0$ has at least three real roots.

23. Use The Intermediate Value Theorem to show that the equation $x^3 - 8x - 5 = 0$ has at least three real roots.

24. Prove part (ii) of Theorem 2.4.3.

25. Prove part (iii) of Theorem 2.4.3.

26. Prove part (iv) of Theorem 2.4.3.

2.5 Graphs of Factored Polynomials and Solution of Inequalities

In the preceding section we saw that a polynomial is continuous at all points. Thus, if $f(x)$ is a polynomial, then the graph of the equation $y = f(x)$ has no breaks. When the polynomial $f(x)$ can be expressed as a product of linear factors, a rough sketch of the graph can be made quickly.

Let us consider a particular case of a factored polynomial:

$$f(x) = (x - 2)(x - 1)(x - 2)(x - 2)(x + 1)(x - 1).$$

First we collect like factors to get

$$f(x) = (x + 1)(x - 1)^2(x - 2)^3.$$

We are therefore concerned with sketching the graph of the equation

$$y = (x + 1)(x - 1)^2(x - 2)^3.$$

Since a product is 0 if and only if one of the factors is 0, we see that $y = 0$ when $x = -1, 1$, or 2, and these are the only values of x for which $y = 0$. So the graph of this equation either crosses or touches the x-axis at $x = -1, 1$, or 2, and these are the only values of x for which this happens.

Now let us consider the intervals determined by the points -1, 1, and 2 on the x-axis. The following four intervals are shown in Figure 2.5.1:

$$x < -1, \quad -1 < x < 1, \quad 1 < x < 2, \quad 2 < x. \tag{1}$$

The polynomial is not zero in any of those intervals, and is continuous, and hence is either everywhere positive or everywhere negative in each of them. It could not be both positive and negative in one of the intervals because then The Intermediate Value Theorem would imply that it is 0 at some point in that interval.

Figure 2.5.1

−1 1 2

We shall next decide whether the graph crosses the x-axis or just touches it at each of the points -1, 1, and 2, where y is 0. If y changes sign at one of these points, the graph crosses the x-axis there; but if y does not change sign, the graph does not cross.

At $x = 2$, $(x - 2)^3$ is 0. But $(x + 1)(x - 1)^2$ keeps the same sign near $x = 2$ because it is not 0 there. However, for values larger than 2, $(x - 2)^3$ is positive, and for values of x less than 2, $(x - 2)^3$ is negative; and so $(x - 2)^3$ changes sign at $x = 2$. We conclude that $y = (x + 1)(x - 1)^2(x - 2)^3$ changes sign at $x = 2$, and so the graph *crosses* the x-axis at $x = 2$.

At $x = 1$, $(x - 1)^2$ is 0 and $(x + 1)(x - 2)^3$ is not 0 at $x = 1$ or any value of x near 1. Hence $(x + 1)(x - 2)^3$ does not change sign at $x = 1$. However, even though $x - 1$ is positive on one side of $x = 1$ and negative on the other side, $(x - 1)^2$ is positive on both sides of 1 and so it too keeps the same sign on both sides. Thus y does not change sign at $x = 1$ and the graph touches but does not cross the x-axis there.

At $x = -1$, $x + 1 = (x + 1)^1$ is 0 and changes sign because it is raised to an odd power. Also $(x - 1)^2(x - 2)^3$ is not 0 at or near $x = -1$ and so does not change sign there. Therefore, $y = (x + 1)(x - 1)^2(x - 2)^3$ changes sign at $x = -1$ and the graph crosses the x-axis at $x = -1$.

In summary, we have found that the graph crosses the x-axis at $x = -1$ and $x = 2$, and touches, but does not cross, at $x = 1$.

We can now make a rough sketch of the graph if we determine whether y is positive or negative for some value of x in any one of the intervals indicated in (1), where y is either everywhere positive or everywhere negative. A large positive value of x is usually easiest. In our case a large positive value of x makes each of the factors of y positive, and so makes y positive. Thus y is positive for all $x > 2$.

Since the graph crosses the x-axis at $x = 2$, y is everywhere negative in the interval $1 < x < 2$. Since it does not cross at $x = 1$, y must be negative for $-1 < x < 1$;

and since the graph crosses the x-axis at $x = -1$, it must be positive for all $x < -1$. A rough sketch of the graph is indicated in Figure 2.5.2.

It is important to realize that we have only obtained a rough sketch of the graph. We don't know much about its real appearance in the intervals stated in (1) above. All we know is the sign of y and the fact that the graph has no breaks. In particular we don't know how low the graph goes in each of the intervals $-1 < x < 1$ and $1 < x < 2$. Later, however, when we study *maximum and minimum points*, we shall be able to determine the lowest points reached in each of those intervals.

Figure 2.5.2

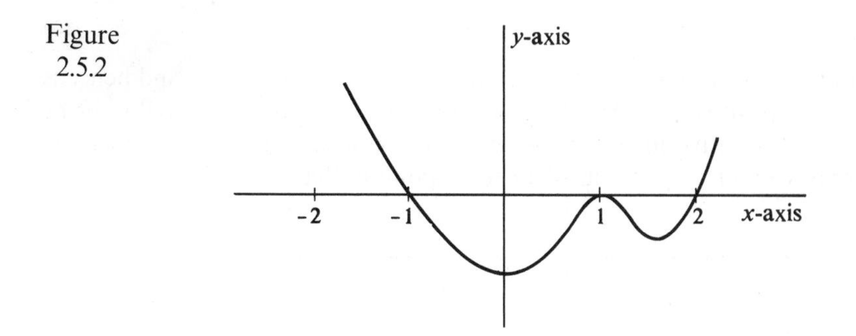

In general, we see that a factor $(x - a)^k$ is 0 at $x = a$ and changes sign at $x = a$ if the exponent k is odd, but does not change sign at $x = a$ if k is even. Of course $(x - a)^k$ does not change sign at $x = b$ if $b \neq a$, because it is not 0 at or near b. Thus, if like factors have been combined, and y is a product of factors of the form $(x - a)^k$, the graph crosses the x-axis at each number a for which k is odd, but does not cross at $x = a$ when k is even.

For example, the graph of

$$y = (x + 2)^2(x - 2)(x - 3)^3(x - 4)^4$$

crosses the x-axis at $x = 2$ and $x = 3$ but does not cross at $x = -2$ and $x = 4$. Moreover, these are the only values of x where the graph either crosses or touches the x-axis.

A rough sketch of the graph can be made quickly. We indicate the steps of the process one at a time.

| *Step 1.* Plot the values of x for which y is zero, marking those where the graph does not cross with a dot $\cdot$ and those where it does cross with $\times$.

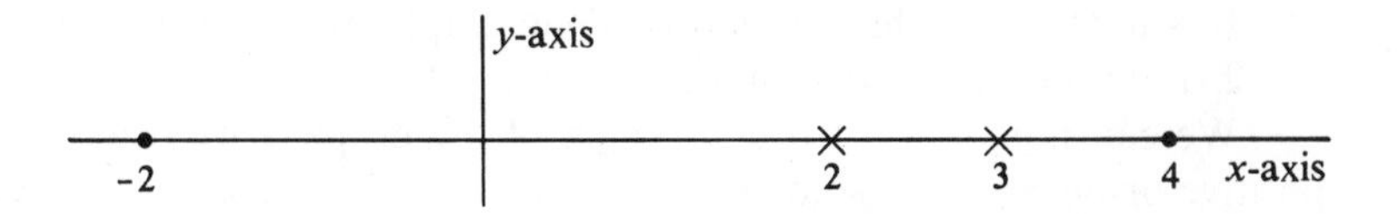

| *Step 2.* Determine whether y is positive or negative for a large value of x and sketch the graph for the right-most interval. In our case every factor of y is positive; so y is positive for all $x > 4$, and we have this situation:

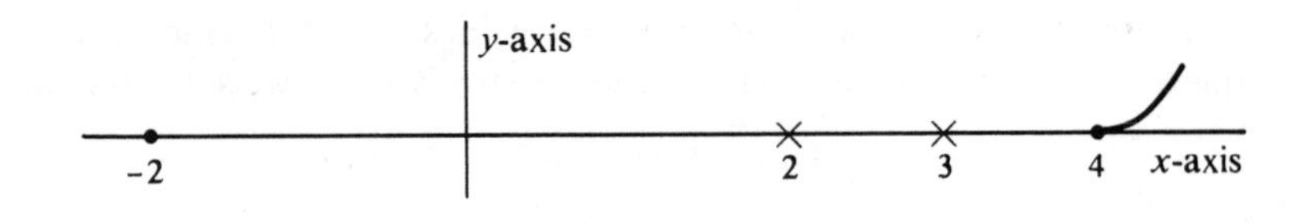

| *Step 3.* Complete the sketch of the graph, going from right to left, using the fact that the graph does not cross but only touches the x-axis at each dot $\cdot$ and crosses at each cross $\times$.

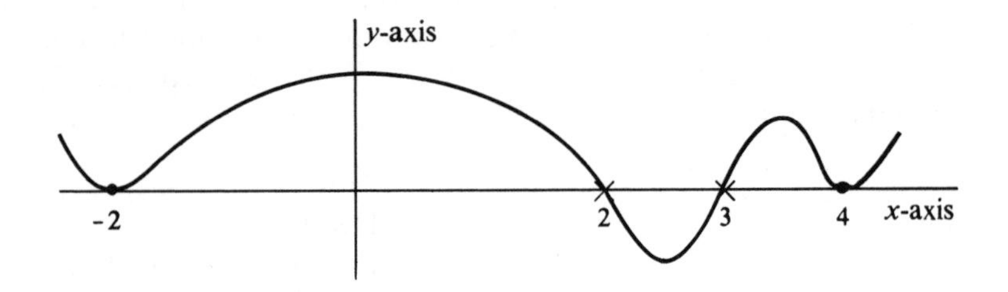

As another example, consider

$$y = (3 - x)^3(x + 5)(1 - x)x^2(x - 3)^2x(x - 1)(1 - x).$$

We first collect like terms, observing that $(x - 3)^2 = (3 - x)^2$ and $1 - x = -(x - 1)$:

$$y = (3 - x)^5(x + 5)x^3(x - 1)^3.$$

Then, using steps 1, 2, and 3, we obtain the sketch shown in Figure 2.5.3. Note in step 2 that $(3 - x)^5$ is negative for a large value of x, while each of the other factors is positive. Consequently the function is negative for large values of x.

Figure 2.5.3

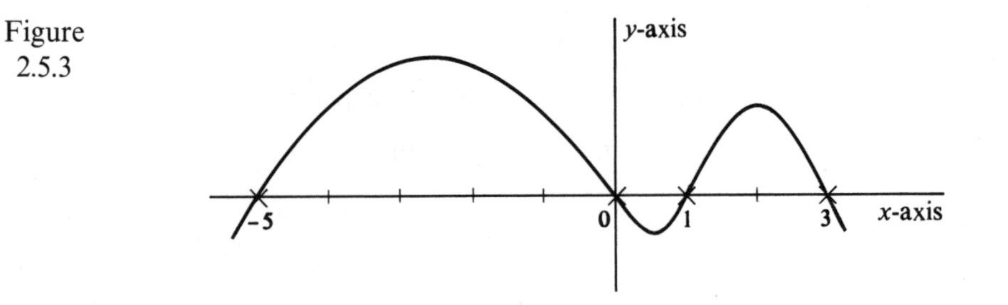

Although the technique used in this section only gives us a rough sketch of the graph of a polynomial that is factored into linear factors, it does determine precisely the values of x for which the polynomial is positive, negative, or zero. Therefore the technique can be used to solve inequalities quickly and easily when one side of the inequality is expressed as a product of linear factors and the other side is 0. Such inequalities can arise in several ways, for example when one wants to determine the values of x for which the square root of a polynomial is real.

Example 1 | Determine the values of x for which the following square root is real:

$$\sqrt{(2x - 5)(x + 1)^3(x + 3)(x - 1)^2(6 - x)}.$$

Of course, to do this we determine the values of x for which the part under the square root sign is greater than or equal to 0. That is, we solve the inequality

$$(2x - 5)(x + 1)^3(x + 3)(x - 1)^2(6 - x) \geq 0.$$

To do this, we set the left side equal to y:

$$y = (2x - 5)(x + 1)^3(x + 3)(x - 1)^2(6 - x).$$

We then sketch the graph of that equation by the methods of this section and look at the graph to find the values of x for which $y \geq 0$. These are the values of x for which the square root is real. We obtain the graph shown in Figure 2.5.4. From the graph we see that $y \geq 0$ for $-3 \leq x \leq -1$ or $x = 1$ or $5/2 \leq x \leq 6$, and so we have the solution to the given problem. ||

Figure 2.5.4

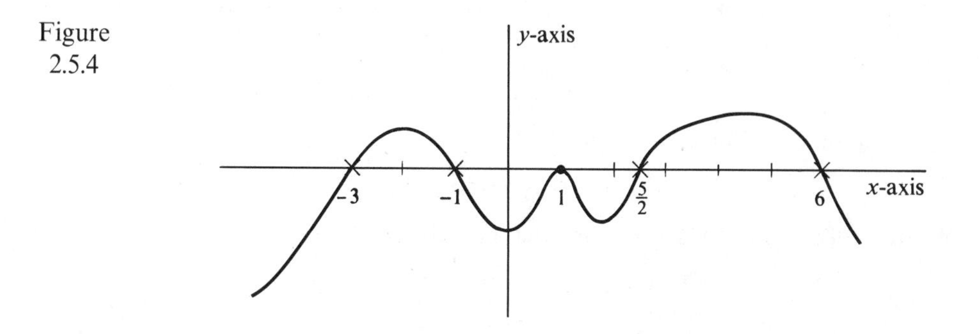

Example 2 | Solve $(4 - x)(x + 3)(x - 2)(6 - x) > 0$.

A sketch of the graph of

$$y = (4 - x)(x + 3)(x - 2)(6 - x)$$

is given in Figure 2.5.5. And so the solution of the inequality is

$$x < -3 \quad \text{or} \quad 2 < x < 4 \quad \text{or} \quad x > 6.$$

||

Figure 2.5.5

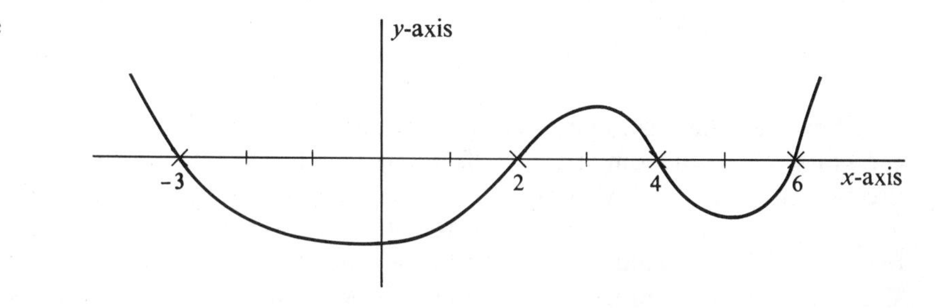

When the inequality involves a quotient N/D, we shall use the following two facts to obtain a solution.

1 N/D is positive if and only if ND is positive.

2 N/D is negative if and only if ND is negative.

These hold because N/D and ND are both positive if and only if N and D are of the same sign, and they are both negative if and only if N and D have opposite signs. When necessary, we also use the fact that $N/D = 0$ if and only if $N = 0$ and $D \neq 0$.

Example 3 | Solve $\dfrac{x(x-2)^2}{(x+3)(x-5)} \geq 0.$

Here we have equality if and only if $x(x-2)^2 = 0$, that is, if and only if $x = 0$ or $x = 2$. To solve the "greater than" part we "move" the denominator to the numerator and proceed as before to sketch the graph of

$$y = x(x-2)^2(x+3)(x-5).$$

A sketch is shown in Figure 2.5.6. Since $x(x-2)^2/(x+3)(x-5) > 0$ if and only if $x(x-2)^2(x+3)(x-5) > 0$, we see from the graph that $x(x-2)^2/(x+3)(x-5) > 0$ if and only if $-3 < x < 0$ or $x > 5$. Combining this result with the fact that the fraction is equal to 0 only when $x = 0$ or $x = 2$, we obtain the solution to the given inequality:

$$-3 < x \leq 0 \quad \text{or} \quad x = 2 \quad \text{or} \quad x > 5. \qquad \|$$

Figure 2.5.6

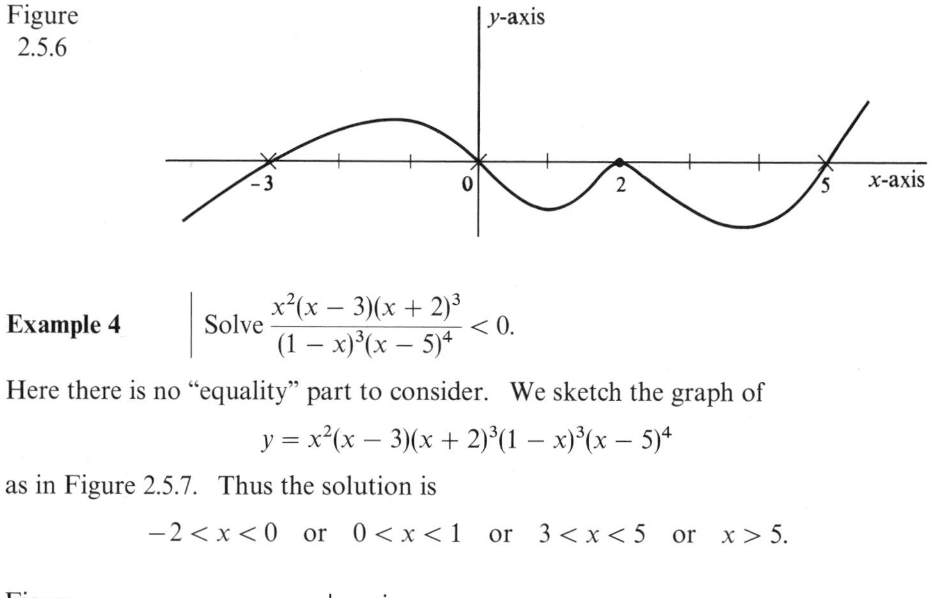

Example 4 | Solve $\dfrac{x^2(x-3)(x+2)^3}{(1-x)^3(x-5)^4} < 0.$

Here there is no "equality" part to consider. We sketch the graph of

$$y = x^2(x-3)(x+2)^3(1-x)^3(x-5)^4$$

as in Figure 2.5.7. Thus the solution is

$$-2 < x < 0 \quad \text{or} \quad 0 < x < 1 \quad \text{or} \quad 3 < x < 5 \quad \text{or} \quad x > 5.$$

Figure 2.5.7

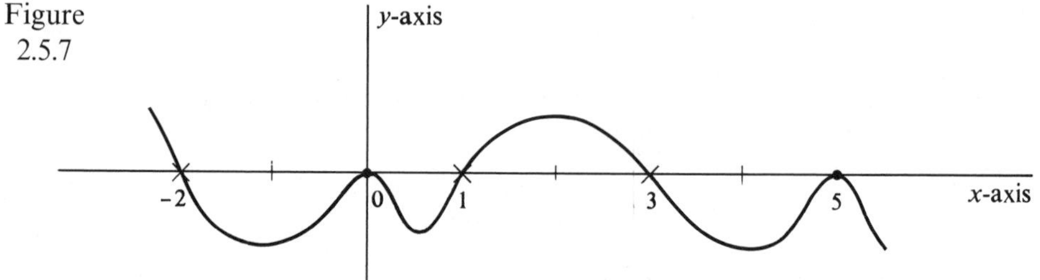

Notice that in this example we do not write $-2 < x < 1$, because that interval includes $x = 0$, which is not a solution. When x is equal to 0, y is also equal to 0, not less than 0. ||

Exercises 2.5

In Exercises 1–10, find the values of x for which the square roots are real.

1. $\sqrt{x-3}$.
2. $\sqrt{x+2}$.
3. $\sqrt{x(x^2+3x-10)}$.
4. $\sqrt{x^2-x-30}$.
5. $\sqrt{6-5x+x^2}$.
6. $\sqrt{(x+3)(x-4)(6-x)}$.
7. $\sqrt{\dfrac{(x+1)(x-3)}{x^2(x-5)^3}}$.
8. $\sqrt{\dfrac{x^2(x-2)(3+x)}{(x-4)(x+2)^2}}$.
9. $\sqrt{\dfrac{(5-x)(2+x)}{(6+x)(x-1)}}$.
10. $\sqrt{\dfrac{x(1+x)}{(x+3)(4+x)}}$.

In Exercises 11–22, solve each inequality.

11. $2x + 11 < 0$.
12. $5x - 7 < 0$.
13. $x^2 + 3x + 2 < 0$.
14. $x^2 - 6x + 5 < 0$.
15. $x^2 + 6 > 5x$.
16. $x^2 + x > 12$.
17. $x(x-1)(x+2) \leq 0$.
18. $(x+3)(x-4)(x-6) \geq 0$.
19. $\dfrac{x(3-x)^3}{(x+1)^2(x-5)} \geq 0$.
20. $\dfrac{(1-x)(2-x)^3(3-x)}{x+3} \leq 0$.
21. $\dfrac{(3+x)(7-x)^3(x+1)}{(1-x)(6+x)(1-x)^3} > 0$.
22. $\dfrac{(x+2)^3(x+1)^3(x+2)^5}{(3x-10)(x-5)^4} > 0$.

Find all values of x for which each of the following expressions is (a) positive, (b) negative.

23. $(3-x)^3(x+7)^2(x+1)$.
24. $\dfrac{x^3(x-2)(x+3)^2}{(x-5)(x+1)}$.
25. $\dfrac{(6+x)^3(x-7)^5}{x(2+x)(1-x)}$.
26. $(8-x)(2-x)(5-x)$.
27. $4x^3 - x^5$.
28. $x^3 - 2x^2 - 8x$.
29. $x^5 + x^4 - 6x^3$.
30. $x^4 - x^3$.

2.6 The Derivative

In Section 2.1, we defined the slope m of the line that is tangent to the graph of a function f at the point $(a, f(a))$ to be

$$\lim_{h\to 0} m(h) = \lim_{h\to 0} \frac{f(a+h)-f(a)}{h}.$$

For each number a for which this limit exists, the limit is a unique number. Hence this association of the number

$$\lim_{h\to 0}\frac{f(a+h)-f(a)}{h}$$

with the number a is a function. If we denote that function by f', we have

$$f'(x)=\lim_{h\to 0}\frac{f(x+h)-f(x)}{h}.$$

Other notations for $f'(x)$ are

$$\frac{df(x)}{dx},\qquad \frac{d}{dx}f(x),\quad \text{and}\quad D_x f(x).$$

If $y=f(x)$, then $f'(x)$ is also denoted by dy/dx or y'. Since the function f' obtained from the function f is one of the most important concepts of calculus, it has a special name:

Definition 2.6.1

If f is a function of x, then the *derivative* of f with respect to x is the function f', where

$$f'(x)=\lim_{h\to 0}\frac{f(x+h)-f(x)}{h}.$$

Geometrically, we see that $f'(x)$ is the slope of the line tangent to the graph of the function f at the point $(x, f(x))$ as illustrated in Figure 2.6.1. Let's look at some examples of derivatives.

Figure 2.6.1

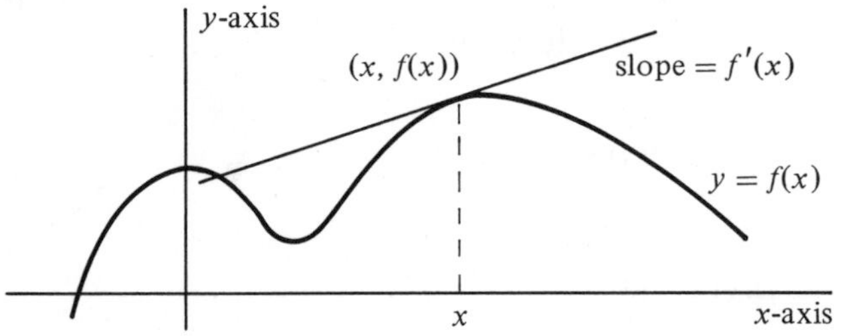

Example 1 | Let $f(x)=x^2+3$.

Then

$$\begin{aligned} f'(x) &= \lim_{h\to 0}\frac{f(x+h)-f(x)}{h}\\ &= \lim_{h\to 0}\frac{(x+h)^2+3-x^2-3}{h}\\ &= \lim_{h\to 0}\frac{x^2+2xh+h^2-x^2}{h}\\ &= \lim_{h\to 0}\frac{2xh+h^2}{h}=\lim_{h\to 0}(2x+h)=2x. \end{aligned}$$

Thus the slope of the tangent to the graph of $y = x^2 + 3$ at the point $(x, x^2 + 3)$ is $2x$. In particular, by letting $x = 2$, and then $x = -3$, we find that the slopes of the tangents to the graph at the points $(2, 7)$ and $(-3, 12)$ are 4 and -6, respectively. ||

Example 2 | Let $f(x) = 6$.

Then
$$f'(x) = \lim_{h \to 0} \frac{f(x+h) - f(x)}{h} = \lim_{h \to 0} \frac{6 - 6}{h}$$
$$= \lim_{h \to 0} \frac{0}{h} = \lim_{h \to 0} 0 = 0.$$

So the slope of the tangent to the graph of $y = 6$ is 0 at all points on the graph. We can also see this directly because the graph of $y = 6$ is a horizontal line 6 units above the x-axis. ||

Example 3 | Let $f(x) = 2 - 3x - 5x^2$.

Then
$$f'(x) = \lim_{h \to 0} \frac{2 - 3(x+h) - 5(x+h)^2 - 2 + 3x + 5x^2}{h}$$
$$= \lim_{h \to 0} \frac{-3x - 3h - 5x^2 - 10xh - 5h^2 + 3x + 5x^2}{h}$$
$$= \lim_{h \to 0} \frac{-3h - 10xh - 5h^2}{h}$$
$$= \lim_{h \to 0} (-3 - 10x - 5h)$$
$$= -3 - 10x.$$
||

Example 4 | Let $f(x) = 1/(x + 1)$.

Then
$$f'(x) = \lim_{h \to 0} \frac{\dfrac{1}{x+h+1} - \dfrac{1}{x+1}}{h}$$
$$= \lim_{h \to 0} \frac{1}{h} \cdot \frac{x + 1 - x - h - 1}{(x+1)(x+h+1)}$$
$$= \lim_{h \to 0} \frac{1}{h} \cdot \frac{-h}{(x+1)(x+h+1)}$$
$$= \lim_{h \to 0} \frac{-1}{(x+1)(x+h+1)}$$
$$= \frac{-1}{(x+1)^2}.$$
||

Example 5 | Let $f(t) = \sqrt{t}$.

Then
$$f'(t) = \lim_{h\to 0} \frac{\sqrt{t+h} - \sqrt{t}}{h}$$
$$= \lim_{h\to 0} \left(\frac{\sqrt{t+h} - \sqrt{t}}{h} \cdot \frac{\sqrt{t+h} + \sqrt{t}}{\sqrt{t+h} + \sqrt{t}}\right)$$
$$= \lim_{h\to 0} \frac{t+h-t}{h(\sqrt{t+h} + \sqrt{t})}$$
$$= \lim_{h\to 0} \frac{h}{h(\sqrt{t+h} + \sqrt{t})} = \lim_{h\to 0} \frac{1}{\sqrt{t+h} + \sqrt{t}}.$$

Now using the fact that the square-root function is continuous, we have

$$\lim_{h\to 0} \sqrt{t+h} = \sqrt{t}, \quad \text{and so}$$
$$f'(t) = \frac{1}{\sqrt{t} + \sqrt{t}} = \frac{1}{2\sqrt{t}}. \qquad \|$$

To find $f'(x)$ directly from the definition, as we have been doing, one must almost always manipulate

$$\frac{f(x+h) - f(x)}{h}$$

until the h cancels out of it, thereby making it easier to find

$$\lim_{h\to 0} \frac{f(x+h) - f(x)}{h}.$$

The reader should be assured that calculation of derivatives is no mere intellectual exercise. Derivatives have many widespread practical uses. For example, it is interesting to know that velocity is a derivative. To help see this, let us consider a car trip of 200 miles that takes 4 hours. On such a trip the distance traveled is a function of the time elapsed. If s is the distance, in miles, from the starting point, and t is the time elapsed, in hours, from the starting time, we then have $s = f(t)$, for some function f. Suppose that the graph of the equation $s = f(t)$ is as indicated in Figure 2.6.2.

We shall discuss the velocity of the car at the point where $t = a$ hours. This will correspond to the point $(a, f(a))$ on the graph. First let us consider the *average velocity* over the interval from a to $a + h$ hours. The distance traveled in this time interval is $f(a + h) - f(a)$ miles, and the time of travel in the interval is h hours. Thus the average velocity over the interval is

$$\frac{f(a+h) - f(a)}{h} \text{ miles per hour,}$$

which is the slope of the line L_h passing through $(a, f(a))$ and $(a + h, f(a + h))$.

Notice that as we take h smaller and smaller,

$$\frac{f(a+h) - f(a)}{h}$$

Figure 2.6.2

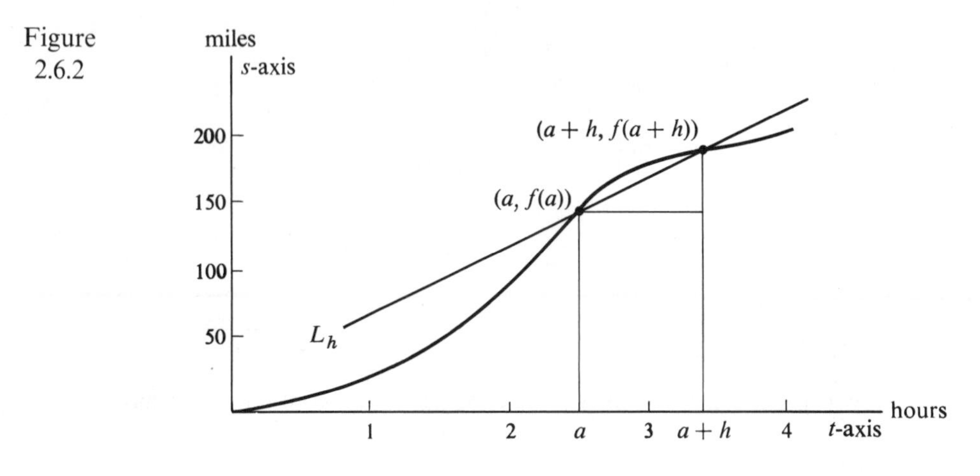

becomes the average velocity over a shorter time interval, and hence it comes closer and closer to the velocity at time $t = a$. That is, the velocity at time a is

$$\lim_{h \to 0} \frac{f(a + h) - f(a)}{h},$$

which, geometrically, is the slope of the tangent to the graph of $s = f(t)$ at the point $(a, f(a))$.

In general, the velocity at any time t is

$$f'(t) = \lim_{h \to 0} \frac{f(t + h) - f(t)}{h}.$$

Thus the velocity at time t is the derivative of $f(t)$ with respect to t. Velocity, the rate of change of distance with respect to time, is only one example of the general concept of "rate of change of one quantity with respect to another." In general, if $y = f(x)$, then the derivative $f'(x)$ $(= dy/dx)$ is the rate of change of y with respect to x.

Example 6 An object is moving along the x-axis according to the equation $x = t^2 - 6t$, where t is the elapsed time in seconds and distance is measured in feet. Find the position of the object and its velocity when t is 2 seconds, 3 seconds, and 7 seconds.

The velocity of the object is

$$\begin{aligned} f'(t) = \frac{dx}{dt} &= \lim_{h \to 0} \frac{f(t + h) - f(t)}{h} \\ &= \lim_{h \to 0} \frac{(t + h)^2 - 6(t + h) - t^2 + 6t}{h} \\ &= \lim_{h \to 0} \frac{t^2 + 2th + h^2 - 6t - 6h - t^2 + 6t}{h} \\ &= \lim_{h \to 0} \frac{2th + h^2 - 6h}{h} = \lim_{h \to 0} (2t + h - 6) = 2t - 6. \end{aligned}$$

Since $f'(t) = 2t - 6$, the velocities when t is 2 seconds, 3 seconds, and 7 seconds are respectively

$$f'(2) = 2(2) - 6 = 4 - 6 = -2 \text{ ft/sec},$$

$$f'(3) = 2(3) - 6 = 6 - 6 = 0 \text{ ft/sec},$$

$$f'(7) = 2(7) - 6 = 14 - 6 = 8 \text{ ft/sec}.$$

The positions of the object at the given times are: $f(2) = 2^2 - 6(2) = -8$ feet, $f(3) = 3^2 - 6(3) = -9$ feet, and $f(7) = 7^2 - 6(7) = 7$ feet. Thus when $t = 2$ seconds, the object is 8 feet to the left of the origin and is traveling at a rate of 2 feet/second to the left. When $t = 3$ seconds, the object is 9 feet to the left of the origin and has a velocity of 0 ft/sec. When $t = 7$ seconds, the object is 7 feet to the right of the origin and is moving at a rate of 8 ft/sec to the right. ||

As another application of the derivative, let us consider the following problem. A rancher has 250 yards of fencing that he wants to use for three sides of the largest rectangular area possible along the side of a straight canyon wall. What dimensions should he use? The problem is illustrated in Figure 2.6.3.

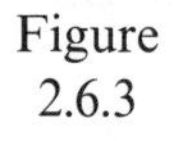
Figure 2.6.3

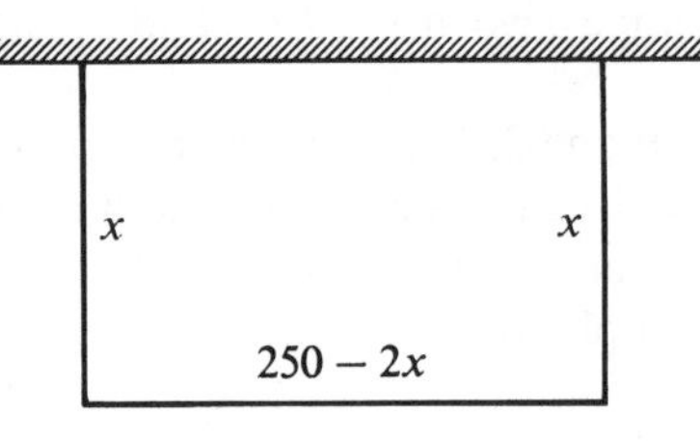

If we let x be the length of each side that is perpendicular to the canyon wall, then the side parallel to the wall has length $250 - 2x$. The area enclosed is then

$$y = A(x) = (250 - 2x)x = 250x - 2x^2.$$

Since this is a factored polynomial we can quickly sketch the graph of the equation by the method of Section 2.5 and obtain the sketch illustrated in Figure 2.6.4.

Figure 2.6.4

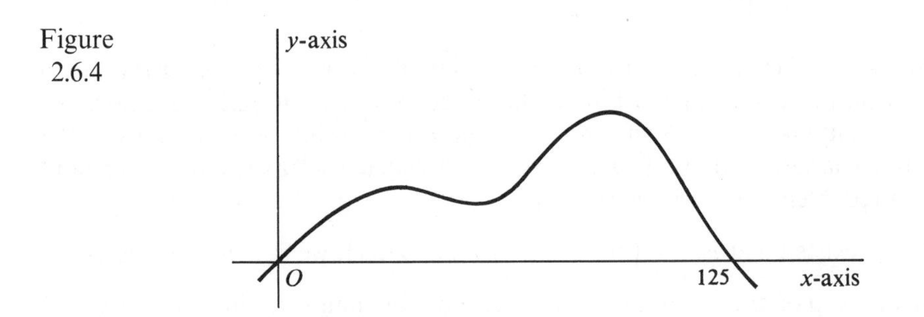

It should be noted that, based on the information we have at this point, we really don't know what the graph actually looks like. We only know that y is 0 when $x = 0$ and when $x = 125$, that y is negative to the left of 0 and to the right of 125, and that y is positive between 0 and 125. However, the derivative $A'(x)$ can be used to give us additional information. By definition,

$$A'(x) = \lim_{h \to 0} \frac{A(x + h) - A(x)}{h}.$$

Hence
$$A'(x) = \lim_{h \to 0} \frac{250(x + h) - 2(x + h)^2 - 250x + 2x^2}{h}$$

$$= \lim_{h \to 0} \frac{250x + 250h - 2x^2 - 4xh - 2h^2 - 250x + 2x^2}{h}$$

$$= \lim_{h \to 0} \frac{250h - 4xh - 2h^2}{h}$$

$$= \lim_{h \to 0} (250 - 4x - 2h)$$

$$= 250 - 4x$$

$$= 4(62.5 - x).$$

We see that the derivative $A'(x)$ is 0 when x is 62.5, is positive for all x less than 62.5, and is negative for all x greater than 62.5. Thus each tangent to the curve has positive slope when $x < 62.5$, has slope 0 when $x = 62.5$, and has negative slope when $x > 62.5$.

Therefore, for values of x less than 62.5, each tangent slopes upward; for values of x greater than 62.5, each tangent slopes downward; and at 62.5 the tangent is horizontal. We conclude that the curve is increasing to the left of 62.5, and is decreasing to the right of 62.5. The graph, then, must look something like Figure 2.6.5.

Figure 2.6.5

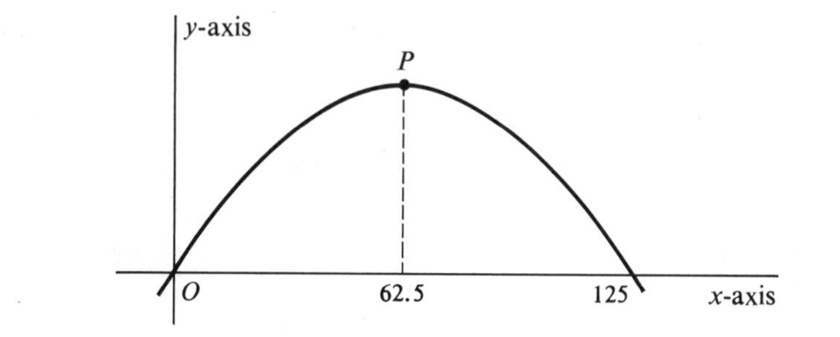

We see that there is a largest value of y, and that it occurs at the point P where x is 62.5. That is, there is a largest area, and it occurs when x is 62.5. Therefore the rancher should take $x = 62.5$ yds and $250 - 2x = 125$ yds as the dimensions of the field with the largest area. In the next chapter we shall consider such maximum (and minimum) problems in much more detail.

If $f'(a)$ exists (that is, $\lim_{h \to 0} (f(a + h) - f(a))/h$ exists), we say that $f(x)$ is *differentiable* at $x = a$ or at the point a. The process of obtaining $f'(x)$ from $f(x)$ is called *differentiation.* It is important to realize that if $f(x)$ is differentiable at $x = a$, then $f(x)$ is continuous at $x = a$. This is proved in the next theorem.

Theorem 2.6.1

> If a function $f(x)$ has a derivative at $x = a$, then $f(x)$ is continuous at $x = a$.

Proof. If $f(x)$ has a derivative at $x = a$ then

$$\lim_{h\to 0} \frac{f(a+h) - f(a)}{h} = f'(a)$$

exists. Thus $f(a)$ exists. Also

$$\lim_{h\to 0} f(a+h) = \lim_{h\to 0} [f(a+h) - f(a) + f(a)]$$

$$= \lim_{h\to 0} \left[\frac{f(a+h) - f(a)}{h} h + f(a)\right].$$

However, we have seen in Theorem 2.3.1 that the limit of the sum of two functions, each of which has a limit, is the sum of the limits of the functions. Hence,

$$\lim_{h\to 0} f(a+h) = \lim_{h\to 0} \left[\frac{f(a+h) - f(a)}{h} h\right] + \lim_{h\to 0} f(a).$$

Using Theorem 2.3.1 again, this time for products, we obtain

$$\lim_{h\to 0} f(a+h) = \lim_{h\to 0} \frac{f(a+h) - f(a)}{h} \cdot \lim_{h\to 0} h + \lim_{h\to 0} f(a)$$

$$= f'(a) \cdot 0 + f(a)$$

$$= f(a).$$

Thus $\lim_{h\to 0} f(a+h) = f(a)$. Let $x = a + h$; then $h \to 0$ is equivalent to $x \to a$, and we have

$$\lim_{x\to a} f(x) = f(a),$$

and so $f(x)$ is continuous at $x = a$.

If $f(x)$ is continuous at $x = a$, however, it is not necessarily differentiable at $x = a$. For example, the function $f(x) = |x|$ is continuous at $x = 0$ but has no derivative at $x = 0$. To see this note that

$$\frac{f(0+h) - f(0)}{h} = \frac{|h|}{h} = \begin{cases} 1 & \text{if } h > 0 \\ -1 & \text{if } h < 0. \end{cases}$$

Thus $\lim_{h\to 0} \frac{f(0+h) - f(0)}{h}$ does not exist. This can also be seen from the graph of $y = |x|$ in Figure 2.6.6. Although there is no break in the graph at $x = 0$, there is no tangent at $x = 0$ and hence no derivative there.

Figure 2.6.6

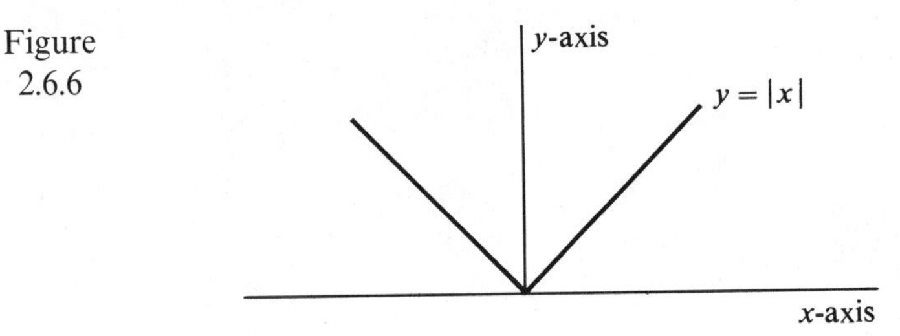

Exercises 2.6

For Exercises 1–16, use the definition of derivative to find the derivative of each function of x.

1. $f(x) = x + 7$.
2. $f(x) = x^2$.
3. $f(x) = x^2 + 5$.
4. $f(x) = x$.
5. $f(x) = -x^2$.
6. $f(x) = 1 - x$.
7. $f(x) = 2x^2 - 4x$.
8. $f(x) = 3 - x^2$.
9. $f(x) = x^3$.
10. $f(x) = x^2 + 3x + 5$.
11. $f(x) = 1/x$.
12. $f(x) = x + (2/x)$.
13. $f(x) = \sqrt{x + 1}$.
14. $f(x) = 1/\sqrt{x}$.
15. $f(x) = 5$.
16. $f(x) = 0$.

If the cost of producing x units of a commodity is $f(x)$, then $f'(x)$ is called the marginal cost, which is the rate of change of the cost with respect to the number of units of the commodity produced. In Exercises 17–20, find the marginal cost for each function $f(x)$.

17. $f(x) = x - x^2$.
18. $f(x) = 1 - (1/x)$.
19. $f(x) = (x^2 - 1)/x$.
20. $f(x) = \sqrt{9 - x}$.

In Exercises 21–22, an object is moving along a straight line according to the given equation where t is the time in seconds and s is the distance from a point P in inches. (a) Find s and the velocity v of the object for each given value of t. (b) Where is the object when $v = 0$?

21. $s = 4t - 9 - t^2$, $t = 0, 1, 6$.
22. $s = t^2 - 8t + 3$, $t = 0, 2, 5$.

In Exercises 23–24, sketch the graph of $f(x)$ and use $f'(x)$ to determine the largest value of the function.

23. $f(x) = 7x - x^2 - 12$.
24. $f(x) = 6 + x - x^2$.
25. What number squared minus itself will produce the least result?
26. What positive number cubed minus itself will produce the least result?
27. Try to calculate $f'(0)$ when $f(x) = |x|$. Be sure to realize that h can be positive or negative.
28. Let $f(x) = x|x|$. Use the definition of derivative to find
    (a) $f'(0)$, and
    (b) $f'(-1)$.

Brief Review of Chapter 2

1. Limits

Definition. $\lim_{x \to a} f(x) = L$ means that for each $\varepsilon > 0$ there is a $\delta > 0$ such that $|f(x) - L| < \varepsilon$ when $0 < |x - a| < \delta$.

Interpretation of Definition. $\lim_{x \to a} f(x) = L$ means that $f(x)$ can be made arbitrarily close to L by requiring x to be sufficiently close to a but not equal to a.

Important Facts. Limit Theorem. (See Section 2.3.)

2. Continuity

Definition. $f(x)$ is continuous at $x = a$ means that $\lim_{x \to a} f(x) = f(a)$; a function is called continuous on a set if it is continuous at all points of the set; a function continuous on the set of all real numbers is called a continuous function.

Interpretation of Definition. $f(x)$ is continuous at $x = a$ when the graph of $f(x)$ does not have a break at the point $x = a$; $f(x)$ is continuous on a set when its graph has no breaks over the set; $f(x)$ is called continuous if it has no breaks anywhere.

Important Facts. Intermediate Value Theorem; all polynomials are continuous; if f and g are continuous at $x = a$ then $f(x) + g(x)$, $c \cdot f(x)$, and $f(x)/g(x)$ (provided $g(a) \neq 0$) are continuous at $x = a$.

3. Graphs of Factored Polynomials

Graphs of these polynomials can be drawn quickly by noting where the polynomials are zero and whether the graph crosses or does not cross the x-axis at such points, and then by finding out if the polynomial is positive or negative for some particular (usually large) value of x.

Uses. Sketching graphs, solving inequalities, determining when square roots are real.

4. Derivatives

Definition.

$$f'(x) = \lim_{h \to 0} \frac{f(x+h) - f(x)}{h}.$$

Interpretation of Definition. $f'(x)$ is a function that expresses the rate of change of $f(x)$ with respect to x; geometrically $f'(x)$ is the slope of the tangent to the graph of $f(x)$ at the point $(x, f(x))$.

Important Facts. Velocity is a derivative; when $f(x)$ has a derivative at $x = a$, $f(x)$ must be continuous at $x = a$; derivatives do not always exist and may even fail to exist at a point where $f(x)$ is continuous.

Technique Review Exercises, Chapter 2

1. Find (a) $\lim_{x \to -2} f(x)$ and (b) $\lim_{x \to 3} f(x)$, if

$$f(x) = \frac{x^2 + x - 2}{x^2 + 5x + 6}.$$

2. Find (a) $\lim_{x \to 5} f(x)$ and (b) $\lim_{x \to -3} f(x)$, if

$$f(x) = \begin{cases} 2|x| & \text{if } x < 5 \\ 7 & \text{if } x = 5 \\ x + 5 & \text{if } x > 5. \end{cases}$$

3. Determine all values of x for which the function $f(x)$ is continuous if

$$f(x) = \begin{cases} \dfrac{x^2 - 2x - 3}{x^2 - x - 6} & \text{if } x \neq 3 \\ \frac{4}{5} & \text{if } x = 3. \end{cases}$$

4. State The Intermediate Value Theorem.
5. Sketch the graph of $y = x^2(x - 4)^3(1 - x)(3 + x)$.
6. For what values of x is the following real-valued:

$$\sqrt{\frac{(x - 1)^3(x + 5)^2}{x(3 - x)(x + 2)}}?$$

7. Use the definition of $f'(x)$ to find $f'(x)$ if $f(x) = 2 - 3x - x^2$.
8. Find an equation of the tangent to the graph of $y = 2 - 3/x$ at the point $(-1, 5)$.
9. An object is moving vertically according to the equation $s = 100t - t^2$, $0 \leq t \leq 100$, where t is the time in seconds and s is the height of the object above the ground. (a) Find the velocity of the object when $t = 5$ sec. (b) What is the time when the object starts to move downward? (c) How high does the object go?

Additional Exercises, Chapter 2

Section 2.1

Use the method of this section to find an equation of the tangent to the graph of $f(x)$ at the point indicated.

1. $f(x) = 3 - x^2$, $(2, -1)$.
2. $f(x) = 7 - 2x$, $(3, 1)$.
3. $f(x) = x^2 + x$, $(3, 12)$.
4. $f(x) = x - x^2$, $(-2, -6)$.
5. $f(x) = 3$, $(5, 3)$.
6. $f(x) = -5$, $(2, -5)$.
7. $f(x) = x^2 - x + 2$, $(1, 2)$.
8. $f(x) = x^2 + 3x - 2$, $(2, 8)$.

Section 2.2

Find the indicated limit or state that it does not exist.

9. $\displaystyle\lim_{x \to 0} \frac{x^2 - 6x}{3x}$.
10. $\displaystyle\lim_{x \to 3} \frac{x^2 - 9}{x - 3}$.
11. $\displaystyle\lim_{x \to 16} \frac{\sqrt{x} - 4}{x - 16}$.
12. $\displaystyle\lim_{x \to 2} \frac{x^2 + 3x - 10}{2 - x}$.
13. $\displaystyle\lim_{x \to -4} \frac{x + 4}{x^2 + 3x - 4}$.
14. $\displaystyle\lim_{x \to 1} \frac{x^2 + 3x - 4}{x - 1}$.
15. $\displaystyle\lim_{x \to 6} f(x)$, where $f(x) = \begin{cases} 7x & \text{if } x \leq 6 \\ 8x & \text{if } x > 6. \end{cases}$
16. $\displaystyle\lim_{x \to 0} f(x)$, where $f(x) = \begin{cases} \dfrac{9x - x^2}{x} & \text{if } x \neq 0 \\ 9 & \text{if } x = 0. \end{cases}$

17. $\lim_{x \to 1} f(x)$, where $f(x) = \begin{cases} \dfrac{1 - x^2}{x - 1} & \text{if } x \neq 1 \\ 2 & \text{if } x = 1. \end{cases}$

18. $\lim_{x \to -3} f(x)$, where $f(x) = \begin{cases} 2 & \text{if } x < -3 \\ 7 & \text{if } x = -3 \\ 5 & \text{if } x > -3. \end{cases}$

Section 2.3

In Exercises 19–22, use The Limit Theorem to calculate the indicated limits.

19. $\lim_{x \to 4} [(x^2 + 3)(x - 1)]$.

20. $\lim_{t \to -1} \dfrac{t^2 - 2}{4t - 7}$.

21. $\lim_{h \to 0} \dfrac{h^3 + 2h^2 - 7h}{h}$.

22. $\lim_{x \to 0} [(3x^2 - 5)(x^3 + 2)]$.

23. Find an equation of the tangent to the graph of $y = 2x - x^2$ at $(2, 0)$.

24. Find an equation of the tangent to the graph of $f(x) = 7 - 2x - 3x^2$ at $(-2, -1)$.

Section 2.4

In Exercises 25–28, find all values of x for which the given function is continuous.

25. $f(x) = 2x^3 - 5x$.

26. $f(x) = \dfrac{x}{x^2 - 7x}$.

27. $f(x) = \dfrac{x + 4}{x^2 + 2x - 8}$.

28. $f(x) = \dfrac{x - 1}{x^2 - 4x + 4}$.

In Exercises 29–32, define the function for the given value of x so that it is continuous there or state that this is impossible.

29. $f(x) = \dfrac{x + 4}{x^2 + 2x - 8}$, $x = -4$.

30. $f(x) = \dfrac{x}{x^2 - 7x}$, $x = 0$.

31. $f(x) = \dfrac{x - 2}{x^2 - 4x + 4}$, $x = 2$.

32. $f(x) = \dfrac{x - 2}{x^2 - 2x - 3}$, $x = 3$.

Section 2.5

In Exercises 33–36, find the values of x for which the square roots are real.

33. $\sqrt{8 - 2x - x^2}$.

34. $\sqrt{x(3 + 2x - x^2)}$.

35. $\sqrt{\dfrac{(x + 5)(1 - x)}{(1 + x)(x - 3)}}$.

36. $\sqrt{\dfrac{x^3(x - 2)}{(x + 4)(2 + x)}}$.

In Exercises 37–40, solve each inequality.

37. $x^2 + 3x > 10$.

38. $x^2 - 21 \leq 3x$.

39. $\dfrac{(5 - x)(2 + x)}{(x - 1)^2(x + 1)} \leq 0$.

40. $\dfrac{(x + 4)^3(x - 5)}{x(x - 1)(x + 2)^5} \geq 0$.

Find all values of x for which each of the following expressions is (a) positive, and (b) negative.

41. $x^3 + 2x^2 - 8x$.

42. $x^2 - x^4$.

Section 2.6

Use the definition of derivative to find the derivative of each function of x.

43. $f(x) = 8x + 6$.

44. $f(x) = 3x^2 - 11$.

45. $f(x) = 7x - 2x^2 - 5$.

46. $f(x) = 2x - 4 - 6x^2$.

47. $f(x) = 17$.

48. $f(x) = 3/x$.

49. $f(x) = 1/(3x)$.

50. $f(x) = (2x)^{1/2}$.

Challenging Problems, Chapter 2

1. A television tower of height h meters is erected at a point P. Suppose that the height, in meters, of the ground above sea level is given by $y = f(x)$ at a distance x meters in the easterly direction from P. Show that a point $(a, f(a))$, $a > 0$, is on a line of sight from the top of the tower if and only if

$$xf(a) - af(x) + h(a - x) > 0 \text{ for all } x \text{ such that } 0 < x < a.$$

2. Let
$$f(x) = \begin{cases} 2 - x & \text{for } x \text{ rational} \\ x & \text{for } x \text{ irrational.} \end{cases}$$

Where, if anywhere, is f continuous?

3. Let f be a function with the following properties:

   1 $f(a + b) = f(a)f(b)$, for all a and b

   2 $f(0) = 1$,

   3 $f'(0) = 1$.

Show that $f'(x) = f(x)$ for all x.

4. A function f is called a *contractive function* if $|f(x_1) - f(x_2)| < |x_1 - x_2|$ for all x_1 and x_2, when $x_1 \neq x_2$. Prove that if f is contractive and differentiable at all points, then $|f'(x)| \leq 1$ for all x.

5. Determine the derivative of $f(x) = \begin{cases} x^2 \sin\left(\dfrac{1}{x}\right) & \text{if } x \neq 0 \\ 0 & \text{if } x = 0 \end{cases}$, at $x = 0$.

6. Every nonzero rational number can be expressed in the form a/b, where a and b are integers with no common integer factor greater than 1. Such a fraction is said to be in *lowest terms*. Let

$$f(x) = \begin{cases} 0 & \text{if } x \text{ is irrational} \\ 1 & \text{if } x = 0 \\ \dfrac{1}{b} & \text{if } x \text{ is a nonzero rational number } \dfrac{a}{b} \text{ in lowest terms.} \end{cases}$$

Prove that f is continuous for every irrational value of x but not continuous for any rational value of x. Use the following outline as a guide to establish that f is continuous at every irrational value of x.

1 Let s be any fixed irrational number and realize we have to show that for each $\varepsilon > 0$ there must be a $\delta > 0$ such that $|f(x)| < \varepsilon$ if $0 < |x - s| < \delta$, and note that only rational values of x need special attention.

2 If given a fixed $\varepsilon > 0$, take a fixed positive integer n such that $\frac{1}{n} < \varepsilon$.

3 Let k be an integer such that $1 \leq k \leq n$ and note that there is a rational number $\frac{h}{k}$ closest to s, that is, for which $d(k) = \left|\frac{h}{k} - s\right| < \frac{1}{k}$.

4 Let δ be the smallest of the numbers $d(1), d(2), \ldots, d(n)$, and show that if $\frac{a}{b}$ is a nonzero rational number in lowest terms for which $0 < \left|\frac{a}{b} - s\right| < \delta$, then $f\left(\frac{a}{b}\right) < \varepsilon$. Use this information to complete the proof.

7. Why is the function f of Problem 6 called the "ruler function"?

8. Let $f(x)$ be continuous over its domain $0 \leq x \leq 3$, and let $0 \leq f(x) \leq 3$. Use The Intermediate Value Theorem to show that $f(x)$ has a *fixed point*; that is, show that there is a number a such that $0 \leq a \leq 3$ and $f(a) = a$.

9. Prove that when $\lim_{x \to a} f(x)$ exists, it has exactly one value. That is, prove that if $\lim_{x \to a} f(x) = L$ and $\lim_{x \to a} f(x) = M$, then $L = M$.

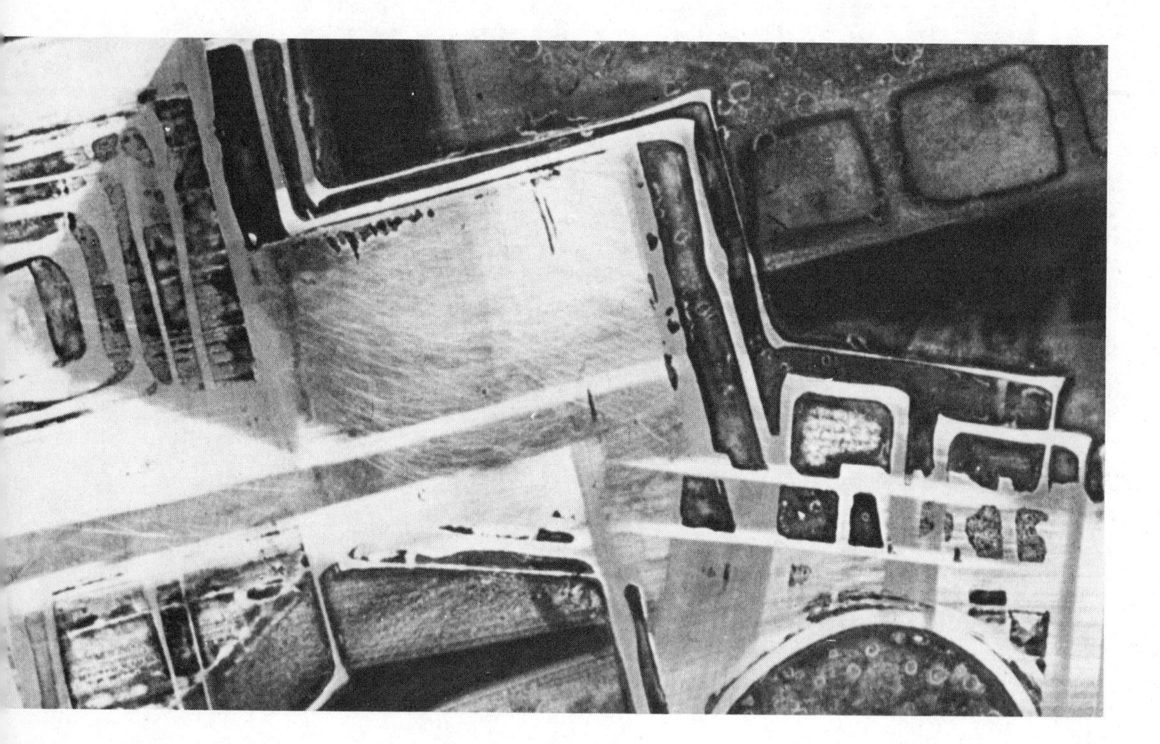

3

The Derivative

Since the derivative has such wide application it is natural that we should now turn our attention to the development of some general rules that will aid us in the calculation of derivatives. This chapter will be devoted to those rules and a few additional applications of the derivative.

3.1 The Derivative of a Polynomial

As a first step in the simplification of calculation of derivatives we shall study the relationship of differentiation to the usual algebraic operations of addition, subtraction, and multiplication by a constant. The information obtained, along with a formula for the derivative of positive-integral powers of x, will enable us to differentiate polynomials.

For example, suppose we know $\dfrac{df(x)}{dx}$ and $\dfrac{dg(x)}{dx}$. Is there then some simple way of expressing $\dfrac{d}{dx}[f(x) + g(x)]$? We know, of course, that by definition

$$\frac{d}{dx}[f(x) + g(x)] = \lim_{h \to 0} \frac{f(x + h) + g(x + h) - [f(x) + g(x)]}{h}.$$

Then

$$\frac{d}{dx}[f(x) + g(x)] = \lim_{h \to 0} \left[\frac{f(x + h) - f(x)}{h} + \frac{g(x + h) - g(x)}{h}\right].$$

From part (i) of Theorem 2.3.1, we know that the limit of a sum is the sum of the separate limits, if the separate limits exist. Consequently,

$$\frac{d}{dx}[f(x) + g(x)] = \lim_{h \to 0} \frac{f(x + h) - f(x)}{h} + \lim_{h \to 0} \frac{g(x + h) - g(x)}{h}.$$

But
$$\lim_{h \to 0} \frac{f(x + h) - f(x)}{h} = \frac{df(x)}{dx},$$

and
$$\lim_{h \to 0} \frac{g(x + h) - g(x)}{h} = \frac{dg(x)}{dx}.$$

Therefore
$$\frac{d}{dx}[f(x) + g(x)] = \frac{df(x)}{dx} + \frac{dg(x)}{dx}.$$

That is, the derivative of the sum of two functions is simply the sum of the separate derivatives.

The same proof, with appropriate changes in the signs, shows that the derivative of the difference of two functions is the difference of the separate derivatives, that is

$$\frac{d}{dx}[f(x) - g(x)] = \frac{df(x)}{dx} - \frac{dg(x)}{dx}.$$

Thus we have the following theorem.

Theorem 3.1.1

If f and g are differentiable functions, then

$$\frac{d}{dx}[f(x) \pm g(x)] = \frac{df(x)}{dx} \pm \frac{dg(x)}{dx}.$$

We now formalize a fact the reader already knows.

Theorem 3.1.2

If c is a constant, then $dc/dx = 0$. (Thus, the derivative of a constant function is 0.)

Proof. The proof is an easy consequence of the definition of the derivative and is left as an exercise.

Theorem 3.1.3

If f is a differentiable function and c is a constant, then

$$\frac{d[c \cdot f(x)]}{dx} = c \cdot \frac{df(x)}{dx}.$$

Proof. The proof follows readily from the definition of derivative and is also left as an exercise.

Before obtaining the derivative of $f(x) = x^n$, for any positive integer n, we first consider the three simplest cases.

If $n = 1$, then $f(x) = x$ and

$$\begin{aligned} \frac{df(x)}{dx} &= \lim_{h \to 0} \frac{f(x+h) - f(x)}{h} \\ &= \lim_{h \to 0} \frac{x + h - x}{h} = \lim_{h \to 0} 1 \\ &= 1. \end{aligned}$$

Thus

$$\frac{dx}{dx} = 1.$$

If $n = 2$, then $f(x) = x^2$ and

$$\begin{aligned}\frac{df(x)}{dx} &= \lim_{h\to 0} \frac{f(x+h) - f(x)}{h} \\ &= \lim_{h\to 0} \frac{(x+h)^2 - x^2}{h} \\ &= \lim_{h\to 0} \frac{x^2 + 2xh + h^2 - x^2}{h} \\ &= \lim_{h\to 0} (2x + h) \\ &= 2x.\end{aligned}$$

Therefore
$$\frac{dx^2}{dx} = 2x.$$

If $n = 3$, then $f(x) = x^3$ and

$$\begin{aligned}\frac{dx^3}{dx} &= \lim_{h\to 0} \frac{(x+h)^3 - x^3}{h} \\ &= \lim_{h\to 0} \frac{x^3 + 3x^2h + 3xh^2 + h^3 - x^3}{h} \\ &= \lim_{h\to 0} (3x^2 + 3xh + h^2) \\ &= 3x^2.\end{aligned}$$

Therefore
$$\frac{dx^3}{dx} = 3x^2.$$

With the knowledge of those special results we shall now attempt the general case. Suppose $f(x) = x^n$, where n is a positive integer. Then

$$\frac{dx^n}{dx} = \lim_{h\to 0} \frac{(x+h)^n - x^n}{h}.$$

By The Binomial Theorem,

$$(x+h)^n = x^n + nhx^{n-1} + \frac{n(n-1)}{2}h^2x^{n-2} + \cdots + h^n,$$

where all terms after the second involve h to the second or higher power. Then

$$\begin{aligned}\frac{dx^n}{dx} &= \lim_{h\to 0} \frac{x^n + nhx^{n-1} + \frac{n(n-1)}{2}h^2x^{n-2} + \cdots + h^n - x^n}{h} \\ &= \lim_{h\to 0} \left(nx^{n-1} + \frac{n(n-1)}{2}hx^{n-2} + \cdots + h^{n-1} \right) \\ &= \lim_{h\to 0} nx^{n-1} + \lim_{h\to 0} \frac{n(n-1)}{2}hx^{n-2} + \cdots + \lim_{h\to 0} h^{n-1}.\end{aligned}$$

Now, since x and n are independent of h,

$$\frac{dx^n}{dx} = nx^{n-1} + \frac{n(n-1)}{2}x^{n-2} \lim_{h\to 0} h + \cdots + \lim_{h\to 0} h^{n-1}$$
$$= nx^{n-1}.$$

Consequently we have the following important result.

Theorem 3.1.4

> If n is a positive integer, then
> $$\frac{dx^n}{dx} = nx^{n-1}.$$

The differentiation rules we have accumulated up to now are summarized in the following list:

1 $\frac{d}{dx}[f(x) \pm g(x)] = \frac{df(x)}{dx} \pm \frac{dg(x)}{dx}$;

2 $\frac{dc}{dx} = 0$, if c is a constant;

3 $\frac{d[c \cdot f(x)]}{dx} = c \cdot \frac{df(x)}{dx}$, if c is a constant;

4 $\frac{dx^n}{dx} = nx^{n-1}$, if n is a positive integer.

Formula (4) says that the derivative of x raised to a power is the exponent times x to a power one less. For example,

$$\frac{dx^5}{dx} = 5x^4 \quad \text{and} \quad \frac{dx^7}{dx} = 7x^6.$$

By (3), the derivative of a constant times a function is the constant times the derivative of the function. For instance,

$$\frac{d}{dx}(4x^7) = 4(7x^6) = 28x^6,$$

$$\frac{d}{dx}(6x^3) = 6(3x^2) = 18x^2.$$

It is usually better to think, but not to write, the first step of the last two examples, and to write immediately

$$\frac{d}{dx}(4x^7) = 28x^6, \quad \text{and} \quad \frac{d}{dx}(6x^3) = 18x^2.$$

Similarly, we immediately write

$$\frac{d}{dx}(11x^4) = 44x^3, \quad \text{and} \quad \frac{d}{dx}(8x^5) = 40x^4.$$

By (1), the derivative of the sum or difference of functions is the sum or difference of their derivatives. Thus, to differentiate a sum or difference, we just go along and differentiate each term of the sum or difference individually. Thus to find $(d/dx)(8x^7 + 2x^4)$ we would write directly $56x^6 + 8x^3$. Similarly, we can write

$$\frac{d}{dx}(4 - 6x^3) = -18x^2,$$

since the derivative of the constant 4 is 0. Using the same technique we find immediately that

$$\frac{d}{dx}(8x^9 - 2x^6 + 14x^2 + 7) = 72x^8 - 12x^5 + 28x.$$

Any polynomial may be differentiated in a similar way.

Exercises 3.1

In Exercises 1–16, find the derivative of each function.

1. $f(x) = 3x^5 + 6x - 3.$
2. $f(x) = 7x^2 + 3x + 1.$
3. $f(x) = 2 - 7x + 6x^8.$
4. $f(x) = 3 + 5x^2 - 3x^6.$
5. $f(x) = x^9 + 3x^4 - 2x + 5.$
6. $f(x) = x^7 - 9x^6 + 10x.$
7. $f(x) = 8x^3 + 3x^2 - 7x + 1.$
8. $f(x) = 6x^4 + 10x^3 + 4x^2 - 7.$
9. $f(x) = (x - 5)(x + 3).$
10. $f(x) = (x + 1)(x - 4).$
11. $f(x) = (x^2 + 8)(x^3 - 2).$
12. $f(x) = (x^4 + 1)(x^5 - 3).$
13. $f(x) = (x^4 + 3x)^2.$
14. $f(x) = (2x^3 - 3)x^2.$
15. $f(x) = x^3(x - 3).$
16. $f(x) = (x^3 - 2x)^2.$
17. Prove Theorem 3.1.2.
18. Prove Theorem 3.1.3.
19. Use the definition of derivative to prove that

$$\frac{d}{dx}[f(x) - g(x)] = \frac{df(x)}{dx} - \frac{dg(x)}{dx}.$$

20. Use the fact that $f(x) - g(x) = f(x) + (-1)g(x)$, and Theorem 3.1.1 with the plus sign, along with Theorem 3.1.3, to prove that

$$\frac{d}{dx}[f(x) - g(x)] = \frac{df(x)}{dx} - \frac{dg(x)}{dx}.$$

3.2 The Product and Quotient Rules

To simplify the differentiation of products and quotients we shall now obtain the product and quotient rules. First consider the problem of finding $(d/dx)[f(x)g(x)]$, where the functions $f(x)$ and $g(x)$ can be differentiated.

From the definition of the derivative,

$$\frac{d}{dx}[f(x)g(x)] = \lim_{h \to 0} \frac{f(x + h)g(x + h) - f(x)g(x)}{h}.$$

To express this limit in a more convenient form we add and subtract $f(x+h)g(x)$ in the numerator. Then

$$\frac{d}{dx}[f(x)g(x)] = \lim_{h\to 0}\frac{f(x+h)g(x+h) - f(x+h)g(x) + f(x+h)g(x) - f(x)g(x)}{h}$$

$$= \lim_{h\to 0}\left[f(x+h)\frac{g(x+h)-g(x)}{h} + g(x)\frac{f(x+h)-f(x)}{h}\right].$$

Again using the fact that the limit of a sum is the sum of the separate limits,

$$\frac{d}{dx}[f(x)g(x)] = \lim_{h\to 0} f(x+h)\frac{g(x+h)-g(x)}{h} + \lim_{h\to 0} g(x)\frac{f(x+h)-f(x)}{h}.$$

And since the limit of a product is the product of the separate limits by Theorem 2.3.1, we have

$$\frac{d}{dx}[f(x)g(x)] = \lim_{h\to 0} f(x+h) \lim_{h\to 0}\frac{g(x+h)-g(x)}{h} + \lim_{h\to 0} g(x) \lim_{h\to 0}\frac{f(x+h)-f(x)}{h}. \tag{1}$$

Since f and g are differentiable functions.

$$\lim_{h\to 0}\frac{f(x+h)-f(x)}{h} = \frac{df(x)}{dx}, \quad \text{and}$$

$$\lim_{h\to 0}\frac{g(x+h)-g(x)}{h} = \frac{dg(x)}{dx}.$$

Moreover, by Theorem 2.6.1, the fact that f is differentiable implies that f is continuous. Consequently, letting $t = x + h$, we have

$$\lim_{h\to 0} f(x+h) = \lim_{t\to x} f(t) = f(x).$$

Since $g(x)$ is independent of h,

$$\lim_{h\to 0} g(x) = g(x).$$

Substitution of those results in (1) yields the desired result:

$$\frac{d}{dx}[f(x)g(x)] = f(x)\frac{dg(x)}{dx} + g(x)\frac{df(x)}{dx}.$$

Note that the derivative of a product is *not* the product of the derivatives. Instead, one multiplies the first function by the derivative of the second and adds the product of the second function by the derivative of the first.

The rule for differentiating quotients,

$$\frac{d}{dx}\left[\frac{f(x)}{g(x)}\right] = \frac{g(x)\dfrac{df(x)}{dx} - f(x)\dfrac{dg(x)}{dx}}{[g(x)]^2},$$

can be proved in a similar way and is left as an exercise. For future reference the results are gathered into a single theorem.

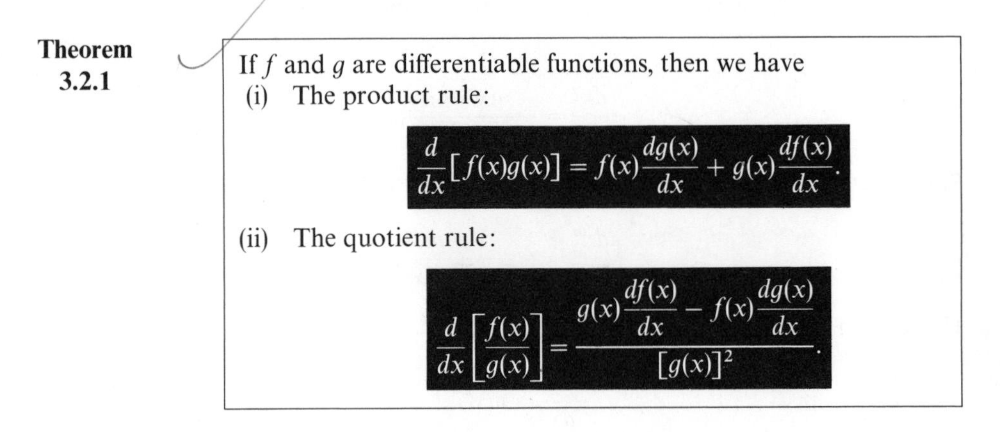

Theorem 3.2.1

If f and g are differentiable functions, then we have

(i) The product rule:

$$\frac{d}{dx}[f(x)g(x)] = f(x)\frac{dg(x)}{dx} + g(x)\frac{df(x)}{dx}.$$

(ii) The quotient rule:

$$\frac{d}{dx}\left[\frac{f(x)}{g(x)}\right] = \frac{g(x)\dfrac{df(x)}{dx} - f(x)\dfrac{dg(x)}{dx}}{[g(x)]^2}.$$

Now let's do some examples by using the product and quotient rules.

Example 1 Calculate $\dfrac{d}{dx}[(5x^3 + 7x^2 + 3)(2x^2 + x + 6)]$.

It is possible (but considerably more tedious) to multiply out the two polynomials involved here and then to calculate the derivative of the resulting polynomial. Instead we shall apply the product rule. We obtain

$$\begin{aligned}\frac{d}{dx}&[(5x^3 + 7x^2 + 3)(2x^2 + x + 6)]\\ &= (5x^3 + 7x^2 + 3)\frac{d}{dx}(2x^2 + x + 6) + (2x^2 + x + 6)\frac{d}{dx}(5x^3 + 7x^2 + 3),\\ &= (5x^3 + 7x^2 + 3)(4x + 1) + (2x^2 + x + 6)(15x^2 + 14x).\end{aligned}$$ ||

Example 2 Calculate $f'(t)$ if $f(t) = \dfrac{t^3 + 5}{t^2 - 7}$.

Application of the quotient rule gives

$$\begin{aligned}f'(t) &= \frac{(t^2 - 7)\dfrac{d}{dt}(t^3 + 5) - (t^3 + 5)\dfrac{d}{dt}(t^2 - 7)}{(t^2 - 7)^2}\\ &= \frac{(t^2 - 7)(3t^2) - (t^3 + 5)(2t)}{(t^2 - 7)^2}\\ &= \frac{3t^4 - 21t^2 - 2t^4 - 10t}{(t^2 - 7)^2}\\ &= \frac{t^4 - 21t^2 - 10t}{(t^2 - 7)^2}.\end{aligned}$$ ||

We have seen in Section 3.1 that when n is a positive integer, $dx^n/dx = nx^{n-1}$. If, on the other hand, n is a negative integer, we let $n = -m$, where $m > 0$. Then, since $x^n = x^{-m} = 1/x^m$, we can use the quotient rule to obtain

$$\frac{dx^n}{dx} = \frac{x^m \dfrac{d1}{dx} - \dfrac{dx^m}{dx}}{[x^m]^2}.$$

Since 1 is a constant, Theorem 3.1.2 yields $d1/dx = 0$. Hence,

$$\frac{dx^n}{dx} = -\frac{1}{[x^m]^2}\frac{dx^m}{dx}.$$

However $m > 0$; so we already know that

$$\frac{dx^m}{dx} = mx^{m-1}.$$

Hence $$\frac{dx^n}{dx} = -\frac{1}{x^{2m}} mx^{m-1} = -mx^{m-1-2m} = -mx^{-m-1}.$$

On substitution of $n = -m$, we have the desired result:

$$\frac{dx^n}{dx} = nx^{n-1}.$$

Since the same formula holds when $n = 0$, that is,

$$\frac{dx^0}{dx} = \frac{d1}{dx} = 0,$$

we have now extended Theorem 3.1.4 to the following more general power-function rule:

Theorem 3.2.2

If n is any integer, then

$$\frac{dx^n}{dx} = nx^{n-1}.$$

Example 3 Using this extension of the power-function rule, we have

$$\frac{dx^{-5}}{dx} = -5x^{-6}, \quad \text{and}$$

$$\frac{d}{dx}[7x^4 - 2x^{-3} + 6x^{-7}] = 28x^3 + 6x^{-4} - 42x^{-8}.$$ ||

Example 4 Calculate $g'(x)$ if $g(x) = \dfrac{x^{-5} - 2x^2}{3x^{-1} + x^3}$.

Application of the quotient rule gives

$$g'(x) = \frac{(3x^{-1} + x^3)(x^{-5} - 2x^2)' - (x^{-5} - 2x^2)(3x^{-1} + x^3)'}{(3x^{-1} + x^3)^2}.$$

Then

$$g'(x) = \frac{(3x^{-1} + x^3)(-5x^{-6} - 4x) - (x^{-5} - 2x^2)(-3x^{-2} + 3x^2)}{(3x^{-1} + x^3)^2}.$$

||

In some problems it is necessary to apply the product rule several times.

Example 5 | Find $\frac{d}{dx}\{(2x-5)(x^5 - x^{-1})(x^2+3)\}$.

Since $(2x-5)(x^5 - x^{-1})(x^2+3) = (2x-5)[(x^5 - x^{-1})(x^2+3)]$, we can apply the product rule to get

$$\frac{d}{dx}\{(2x-5)[(x^5 - x^{-1})(x^2+3)]\} = (2x-5)\frac{d}{dx}[(x^5 - x^{-1})(x^2+3)] + [(x^5 - x^{-1})(x^2+3)]\frac{d}{dx}(2x-5).$$

Using the product rule again, we get

$$\frac{d}{dx}\{(2x-5)[(x^5 - x^{-1})(x^2+3)]\}$$

$$= (2x-5)\left[(x^5 - x^{-1})\frac{d}{dx}(x^2+3) + (x^2+3)\frac{d}{dx}(x^5 - x^{-1})\right] + [(x^5 - x^{-1})(x^2+3)]\frac{d}{dx}(2x-5).$$

Thus,

$$\frac{d}{dx}\{(2x-5)(x^5 - x^{-1})(x^2+3)\} = (2x-5)[(x^5 - x^{-1})(2x) + (x^2+3)(5x^4 + x^{-2})] + (x^5 - x^{-1})(x^2+3)(2).$$

||

Often, solutions to various mathematical problems involve differential equations, that is, equations containing derivatives. A function is a *solution* to a differential equation if the equation becomes a true statement when the function is substituted into the equation.

Example 6 | Show that $y = 2x^4 + 3x - 7$ is a solution to the differential equation $\frac{dy}{dx} - 8x^3 = 3$.

If $y = 2x^4 + 3x - 7$, then $\frac{dy}{dx} = 8x^3 + 3$. Substitution into the differential equation then gives

$$(8x^3 + 3) - 8x^3 = 3, \quad \text{or} \quad 3 = 3.$$

This verifies that $y = 2x^4 + 3x - 7$ is a solution to the differential equation $\frac{dy}{dx} - 8x^3 = 3$. ||

2-24 even

Exercises 3.2

In Exercises 1–24, find the derivative of the given function.

1. $f(x) = 6x^5 + 7x^{-5}$.
2. $f(x) = 14x^{-3} - 2x^{-4}$.
3. $f(t) = 2t^4 - 8t^{-1} + 6t^5$.
4. $f(x) = -4x^3 + 7x^2 - 2x^{-3}$.
5. $f(x) = (x^5 + 3x)(x^2 - 1)$.
6. $f(t) = (t^6 - 2)(t^2 + 8t)$.
7. $f(x) = (x^3 + 3)(x^2 - 5)$.
8. $f(x) = (2x - 11)(x^2 + 5)$.
9. $f(x) = (7x - 3x^{-1})(2x^3 + 4)$.
10. $f(x) = (3x + x^5)(x^2 + 3x - 2)$.
11. $f(x) = (x^3 + x)^2$.
12. $f(x) = (x + 5)(x - 2)$.
13. $f(x) = (x - 4)(x + 3)$.
14. $f(x) = (3x^2 - 2)^2$.
15. $f(u) = \dfrac{u + 5}{u - 3}$.
16. $f(x) = \dfrac{3 - x}{x - 2}$.
17. $f(x) = \dfrac{1}{2x^4 + 1}$.
18. $f(u) = \dfrac{u - 4}{u^2 + 1}$.
19. $f(x) = \dfrac{x^2 - 3x + 5}{x^2 + 3}$.
20. $f(x) = \dfrac{1}{x^3 - 1}$.
21. $f(x) = \dfrac{x^2 + 3x + 2}{x - 1}$.
22. $f(x) = \dfrac{x^3 + 2x - 1}{x}$.
23. $f(x) = (x + 2)(x - 3)(x + 5)$.
24. $f(x) = (x^2 - 3)(x + 4)(x - 5)$.

In Exercises 25–30, verify that the given function is a solution to the given differential equation by substituting the function into the equation and showing that the equation is satisfied. Where the symbol k appears, consider it to be a constant.

25. $\dfrac{dy}{dx} - 2x = 5, \quad y = x^2 + 5x + 7$.
26. $\dfrac{dy}{dx} - 3x^2 = 2x, \quad y = x^3 + x^2 + 5$.
27. $xy' - 2y = 0, \quad y = 3x^2$.
28. $xy' - 3y = -4x, \quad y = kx^3 + 2x$.
29. $y + xy' = 6x, \quad y = kx^{-1} + 3x$.
30. $xy' - y = x^2 + 1, \quad y = x^2 + kx - 1$.
31. Suppose that $y = f(x)$ and $y = g(x)$ both satisfy the differential equation $y' + 7y = 0$. Show that $y = f(x) + g(x)$ also satisfies that equation.
32. Suppose that $y = f(x)$ satisfies the differential equation $y' + 5y = 7$ and $y = g(x)$ satisfies the differential equation $y' + 5y = 0$. Show that $y = f(x) + g(x)$ satisfies the differential equation $y' + 5y = 7$.
33. Let $f(x) = u(x)v(x)$.

    Show
    1 $f''(x) = u(x)v''(x) + 2u'(x)v'(x) + u''(x)v(x)$.
    2 $f'''(x) = u(x)v'''(x) + 3u'(x)v''(x) + 3u''(x)v'(x) + u'''(x)v(x)$.

Compare these results with $[u(x) + v(x)]^2$ and $[u(x) + v(x)]^3$ respectively. Guess the expres-

sion for $f^{(4)}(x)$ ... $'''(x) = \frac{d}{dx}(f''(x))$, and $f^{(4)}(x) = \frac{d}{dx}(f'''(x))$.

ıtena ... s C are related to the overhaul interval x by ... is a constant. Verify that $C(x) = (a/x) + kx^2$

...ring a quantity q of a commodity is related

$$\frac{C(q)}{q} = a,$$

$+ (k/q)$ is a solution.

$$\frac{1}{[g(x)]^2}\frac{dg(x)}{dx}.$$

$$(x)h(x)\frac{dg(x)}{dx} + g(x)h(x)\frac{df(x)}{dx}.$$

Theorem 3.2.1) using the following outline

the fraction.

appropriately.

esult.

Leslie

Thank you for the use of your books. They proved to be most beneficial in my passing this course.

But I thank you even more so for your support and encouragement.

God Bless you

Anthony

B... the derivative we shall discuss one la... his rule gives us an easy method of ca... n (a function of a function).

... d below in Example 4 to calculate th... by viewing this function as $y = f(u) = u^{15}$, where $u = g(x) = x^7 - 2x^{-7}$. That is, we shall view $y = (x^7 - 2x^{-7})^{15}$ as a function of a function.

Theorem 3.3.1

> *The Chain Rule*
>
> If f has a derivative at $g(x_0)$ and g has a derivative at x_0, then $[f \circ g](x)$ has a derivative at x_0 and
>
> $$[f \circ g]'(x_0) = f'(g(x_0)) \cdot g'(x_0).$$

Proof. By definition of the derivative,

$$[f \circ g]'(x_0) = \lim_{h \to 0} \frac{[f \circ g](x_0 + h) - [f \circ g](x_0)}{h},$$

$$= \lim_{h \to 0} \frac{f(g(x_0 + h)) - f(g(x_0))}{h}.$$

Assume that $g(x_0 + h) - g(x_0) \neq 0$. Then we can write

$$[f \circ g]'(x_0) = \lim_{h \to 0} \left[\frac{f(g(x_0 + h)) - f(g(x_0))}{g(x_0 + h) - g(x_0)} \cdot \frac{g(x_0 + h) - g(x_0)}{h} \right],$$

or, assuming that the separate limits exist,

$$[f \circ g]'(x_0) = \lim_{h \to 0} \frac{f(g(x_0 + h)) - f(g(x_0))}{g(x_0 + h) - g(x_0)} \times \lim_{h \to 0} \frac{g(x_0 + h) - g(x_0)}{h}. \tag{1}$$

The second of those two limits is easy to evaluate:

$$\lim_{h \to 0} \frac{g(x_0 + h) - g(x_0)}{h} = g'(x_0).$$

So we turn our attention to the first limit.

Since g is differentiable at x_0 it must be continuous there. Hence,

$$\lim_{h \to 0} g(x_0 + h) = g(x_0), \quad \text{and so}$$

$$\lim_{h \to 0} (g(x_0 + h) - g(x_0)) = 0.$$

Consequently, on letting $g(x_0 + h) - g(x_0) = s$ we have

$$\lim_{h \to 0} \frac{f(g(x_0 + h)) - f(g(x_0))}{g(x_0 + h) - g(x_0)} = \lim_{s \to 0} \frac{f(g(x_0) + s) - f(g(x_0))}{s}.$$

The last limit is simply the definition of $f'(g(x_0))$. On substitution of our results into (1), we have the desired result:

$$[f \circ g]'(x_0) = f'(g(x_0)) \cdot g'(x_0).$$

(The requirement that $g(x_0 + h) - g(x_0) \neq 0$ may be removed by using slightly more advanced techniques, which the interested reader will find in most advanced calculus texts.)

We can conclude from this theorem that if f and g are differentiable functions, then $f \circ g$ is differentiable and

$$[f \circ g]'(x) = f'(g(x)) \cdot g'(x).$$

To express this rule differently we let $u = g(x)$ and $y = f(g(x)) = f(u)$, then

$$g'(x) = \frac{du}{dx}, \quad f'(g(x)) = f'(u) = \frac{dy}{du}, \quad \text{and} \quad [f \circ g]'(x) = \frac{dy}{dx}.$$

Using this notation the chain rule can be written in the following easily remembered forms:

$$\frac{dy}{dx} = \frac{dy}{du}\frac{du}{dx} \tag{2}$$

and

$$\frac{df(u)}{dx} = \frac{df(u)}{du} \cdot \frac{du}{dx}. \tag{3}$$

Formulas (2) and (3) show that the derivative behaves like a fraction. This is the motivation of the *Leibniz notation, dy/dx*, for derivatives and indicates the great convenience of this notation as a memory aid.

Example 1 Find $\frac{dy}{dx}$ when $y = u^5 + 7u$ and $u = 14x^{-1} - 2x$.

Using the chain rule, we obtain

$$\begin{aligned} \frac{dy}{dx} &= \frac{dy}{du}\frac{du}{dx} \\ &= \frac{d}{du}(u^5 + 7u)\frac{d}{dx}(14x^{-1} - 2x) \\ &= (5u^4 + 7)(-14x^{-2} - 2). \end{aligned}$$

Substitution of $u = 14x^{-1} - 2x$ puts the result above in terms of x alone:

$$\frac{dy}{dx} = [5(14x^{-1} - 2x)^4 + 7](-14x^{-2} - 2).$$ ||

Many times the chain rule can be applied without a full knowledge of the functions involved. This technique is illustrated in the next two examples.

Example 2 Let $f(u) = u^3 + 3u$, with $u = g(x)$. If $g'(2) = 7$ and $g(2) = 3$, compute $[f \circ g]'(2)$.

Since $$[f \circ g]'(x) = f'(g(x)) \cdot g'(x)$$

we have $$\begin{aligned} [f \circ g]'(2) &= f'(g(2)) \cdot g'(2) \\ &= f'(3) \cdot g'(2). \end{aligned}$$

Then, since $$f'(u) = 3u^2 + 3,$$

we have $$f'(3) = 30.$$

Consequently $$[f \circ g]'(2) = (30)(7) = 210.$$ ||

As the following example illustrates, practical applications of that technique are often possible.

Example 3 Suppose that the temperature, in degrees, of a particle at a distance x feet from a fixed point P is given by $T(x) = x^3 - x^2 + 3x + 1$. At a particular instant the particle is 1 foot from P and moving away from P at the rate of 10 ft/min. Find the rate of change of the temperature of the particle with respect to time at that instant.

Let $x = g(t)$ be the distance of the particle from P at time t. Then the temperature of the particle at time t is given by $[T \circ g](t)$, and we are to compute

$$[T \circ g]'(t) = T'(g(t)) \cdot g'(t),$$

when $g(t) = 1$ and $g'(t) = 10$.

Since $$T'(x) = 3x^2 - 2x + 3,$$

we have $$T'(1) = 4.$$

Thus $$[T \circ g]'(t) = 4 \cdot 10 = 40 \text{ degrees/minute.}$$ ||

To find $\frac{d}{dx}[g(x)]^n$ we let $u = g(x)$ and use the chain rule. Thus

$$\frac{d}{dx}[g(x)]^n = \frac{du^n}{dx} = \frac{du^n}{du}\frac{du}{dx} = nu^{n-1} \cdot \frac{du}{dx}.$$

On restoring $g(x)$ for u, we get

$$\frac{d}{dx}[g(x)]^n = n[g(x)]^{n-1} \cdot \frac{dg(x)}{dx}.$$

In general, then, to find the derivative of a function raised to a power we form the product of the exponent, the function to one less power, and the derivative of the function.

Example 4 Find $\frac{d}{dx}(x^7 - 2x^{-7})^{15}$.

Use of the previous rule gives

$$\begin{aligned}\frac{d}{dx}(x^7 - 2x^{-7})^{15} &= 15(x^7 - 2x^{-7})^{14} \cdot \frac{d}{dx}(x^7 - 2x^{-7}) \\ &= 15(x^7 - 2x^{-7})^{14}(7x^6 + 14x^{-8}).\end{aligned}$$

Often one only thinks the first step and writes immediately

$$\frac{d}{dx}(x^7 - 2x^{-7})^{15} = 15(x^7 - 2x^{-7})^{14}(7x^6 + 14x^{-8}).$$ ||

Example 5 Calculate $\frac{d}{dx}(3x^2 + 7x - 5x^{-1})^{100}$.

$$\frac{d}{dx}(3x^2 + 7x - 5x^{-1})^{100} = 100(3x^2 + 7x - 5x^{-1})^{99}(6x + 7 + 5x^{-2}).$$ ||

Example 6 | Calculate $\frac{d}{dx}[(x^{-4} - x^5)^{15} + 4x^{-2}]^{10}$.

$$\frac{d}{dx}[(x^{-4} - x^5)^{15} + 4x^{-2}]^{10}$$

$$= 10[(x^{-4} - x^5)^{15} + 4x^{-2}]^9[15(x^{-4} - x^5)^{14}(-4x^{-5} - 5x^4) - 8x^{-3}]. \quad \|$$

The chain rule is also useful in determining $\frac{dy}{dx}$ where both x and y are given as functions of a parameter t. Suppose that the parametric equations $x = f(t)$ and $y = g(t)$ do determine y as a function of x. Then by the chain rule,

$$\frac{dy}{dt} = \frac{dy}{dx}\frac{dx}{dt}$$

or, if $\frac{dx}{dt} \neq 0$,

$$\frac{dy}{dx} = \frac{\frac{dy}{dt}}{\frac{dx}{dt}}.$$

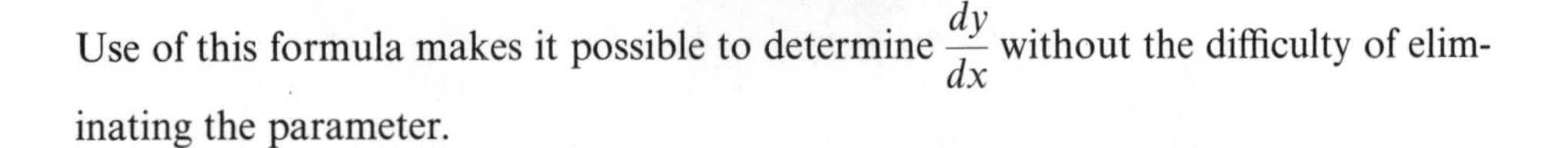

Use of this formula makes it possible to determine $\frac{dy}{dx}$ without the difficulty of eliminating the parameter.

Example 7 | Find the slope of the graph determined by $x = t^4 + 5t^3 - 7$, and $y = t^2 - 3t$ when $t = 1$.

Since $\frac{dx}{dt} = 4t^3 + 15t^2$ and $\frac{dy}{dx} = 2t - 3$, we have

$$\frac{dy}{dx} = \frac{2t - 3}{4t^3 + 15t^2}.$$

Thus when $t = 1$, we have $\frac{dy}{dx} = -\frac{1}{19}$. $\|$

Exercises 3.3

In Exercises 1–32, find dy/dx.

1. $y = 3u^2 - 5u + 2$, where $u = x^2 - 1$.
2. $y = 10u^2 + 3u^{-1}$, where $u = x^3 + 2$.
3. $y = 4u^5 - 7u + u^{-2}$, where $u = x + x^2$.
4. $y = u^{10} - 5(u^3 + 3u)^{14}$, where $u = x^4$.
5. $y = (u + 1)^{-5}$, where $u = x^{-3}$.
6. $y = z(z^3 + 3)^4$, where $z = (x + 3)^2$.

7. $y = \dfrac{u^3 + 3}{u^2 - 1}$, where $u = (x^2 + 4)^3$.

8. $y = \dfrac{u^3 + 5u}{u^5 - 1}$, where $u = x^7 + x^{-1}$.

9. $y = (2x^3 + 3x + 1)^8$.

10. $y = (3x^2 + 7x^{-10})^{-5}$.

11. $y = (4x^{-1} - 2x^{-2})^{-3}$.

12. $y = (10x^3 - 4x)^{-17}$.

13. $y = (x^3 - 2x + 1)^4 + (x^2 + 5)^6$.

14. $y = (x^5 - 6)^3 - (x^3 + 2x)^4$.

15. $y = \left(\dfrac{x + 2}{x - 3}\right)^4$.

16. $y = \left(\dfrac{10x - 2}{3x + 5}\right)^6$.

17. $y = \left(\dfrac{x^{-1} + 3x^{-2}}{4x + 5}\right)^{-6}$

18. $y = \left(\dfrac{x^{-3} - 2x^{-1}}{x^4 - 3}\right)^{-2}$.

19. $y = (x^2 + 3x)^5(x - 1)^{-2}$.

20. $y = (x^3 - 2x)^4(x^7 - 2)^{-3}$.

21. $y = [(3x^2 + 5x)(x - 5)]^3$.

22. $y = [(x^3 + 7)(x^2 + 3x - 1)]^4$.

23. $y = (x^2 + 2)^4(x^3 - 1)^2$.

24. $y = (x^2 - 1)^3(x^3 + 5)$.

25. $y = [(x^2 + 3)^{10} - 5x^{-2}]^6$.

26. $y = [(x^{-3} - 2x)^7 - 7x^2 + 2]^4$.

27. $x = t^3 - 7t, \quad y = 4t - 3$.

28. $x = t^6 - 4t^{-1}, \quad y = t^3 - 2t$.

29. $x = (t - 2)^4, \quad y = (t^3 - 2t)^{-2}$.

30. $x = (t^2 - 7t + 1)^2, \quad y = (t - 3)^{-1}$.

31. $x = (u - 7u^{-2})^4, \quad y = (u^3 + 4u)^5$.

32. $x = (u - 1)^2(u + 1)^{-3}, \quad y = (2u - 3)^4$.

33. Find $[f \circ g]'(2)$ if $f(u) = u^2 - 1$, $g(2) = 3$, and $g'(2) = -1$.

34. Find $[f \circ g]'(-3)$ if $f(u) = -2u^3 + 6u^{-3}$, $g(-3) = 2$, and $g'(-3) = 1/2$.

35. Let $f(u) = -3u^4 + u^{-1}$ and find the derivative of $f \circ g$ at a point a where $g(a) = 2$ and $g'(a) = -3$.

36. Let $f(u) = (6u^{-2} + u)^3$ and find the derivative of $f \circ g$ at a point a where $g(a) = 1/2$ and $g'(a) = -2$.

37. Find an equation of the tangent to $y = (x^2 + 5x + 2)^4$ at the point $(0, 16)$.

38. Find an equation of the tangent to $y = (3x^{-2} - 2x^3)^5$ at the point $(1, 1)$.

39. Find an equation of the tangent to $x = (4v - 2)^2$, $y = (3v^2 - 2v)^{-1}$ when $v = 1$.

40. Find an equation of the tangent to $x = (t^{-3} + 4t)^2$, $y = (t^4 - t^{-4})^2$ when $t = 1$.

41. Find the values of x where the graph of $y = (x^2 - 4)^5$ has a horizontal tangent.

42. Find the values of x where the graph of $y = (x^3 - x)^2$ has a horizontal tangent.

43. The temperature at any point x on the x-axis is $T(x) = x^2 - 3x + 1$, where x is measured in feet. If a particle maintains the temperature of its surroundings, compute the rate of change of the temperature with respect to time of the particle as it moves through the origin with a velocity of -5 ft/sec.

3.4 Derivatives and Curve Sketching

Up to this time our rough sketches of graphs have often ignored some features that might be important. In the past we concentrated mostly on points where the graph of a function crosses the x-axis and where it is above or below this axis. In this section we shall use calculus to determine some additional properties of graphs. Before we continue we shall need the following definitions.

Definition 3.4.1

> A *neighborhood* of a point a is the set o
> $|x - a| < \delta$, where $\delta > 0$. Thus a neighb
> an interval that is centered at a and has no

Definition 3.4.2

> A point a is an *interior point* of a set S if some
> a lies fully in S.

For instance, 2 is an interior point of the interval $0 \leq x \leq 5$, si … igh-borhood of 2, for example $|x - 2| < 1$, lies fully in the interval $0 \leq x \leq 5$. Of course, any smaller interval centered at 2 also lies in the set $0 \leq x \leq 5$. On the other hand, 0 and 5 are not interior points. Indeed, 0 and 5 are called *endpoints* of the interval $0 \leq x \leq 5$.

The sign of $f'(a)$ gives a great deal of information about the behavior of the function f near the point a. The following theorems illustrate the major facts.

Theorem 3.4.1

> If a is an interior point of the domain of f and $f'(a) > 0$, then f is increasing at a. That is, there is a $\delta > 0$ such that if $a - \delta < x_1 < a$, then $f(x_1) < f(a)$, and if $a < x_2 < a + \delta$, then $f(a) < f(x_2)$.

Figure 3.4.1

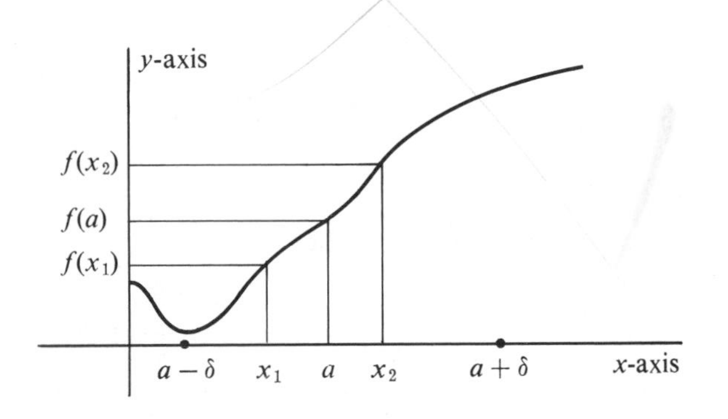

Proof. It will probably be helpful to refer to Figure 3.4.1 throughout the proof. Since $f'(a) > 0$.

$$\lim_{h \to 0} \frac{f(a+h) - f(a)}{h} = f'(a) > 0.$$

Thus,

$$\frac{f(a+h) - f(a)}{h}$$

can be made arbitrarily close to the positive number $f'(a)$ by requiring h to be sufficiently close to zero. (It is understood, of course, that $h \neq 0$.) Since

$$\frac{f(a+h) - f(a)}{h}$$

can be made arbitrarily close to a positive number, the quotient

$$\frac{f(a+h)-f(a)}{h}$$

itself can be made positive by simply choosing h sufficiently small, say $|h| < \delta$. First suppose $a < x_2 < a + \delta$, and let $h = x_2 - a > 0$. Then since $h = x_2 - a < \delta$, we have

$$\frac{f(a+h)-f(a)}{h} > 0,$$

or, using the fact that $x_2 = a + h$,

$$\frac{f(x_2)-f(a)}{h} > 0.$$

Finally, since $h > 0$, we have $f(x_2) - f(a) > 0$ or $f(x_2) > f(a)$, as was to be shown.

If $a - \delta < x_1 < a$, let $h = x_1 - a < 0$. Note that $|h| = a - x_1 < \delta$ and hence

$$\frac{f(a+h)-f(a)}{h} > 0.$$

Substitution of $x_1 = a + h$ yields

$$\frac{f(x_1)-f(a)}{h} > 0.$$

Then since $h < 0$, we must have $f(x_1) - f(a) < 0$ or $f(x_1) < f(a)$, thereby completing the proof.

The following theorem gives the corresponding result for the case where $f'(a) < 0$.

Theorem 3.4.2

> If a is an interior point of the domain of f and $f'(a) < 0$, then f is decreasing at a. That is, there is a $\delta > 0$ such that if $a - \delta < x_1 < a$, then $f(x_1) > f(a)$, and if $a < x_2 < a + \delta$, then $f(a) > f(x_2)$.

Since the proof of the theorem parallels completely that of the previous theorem we leave the proof as an exercise.

The two preceding results are "local" in nature, that is, they tell us something about the nature of a function in the vicinity or neighborhood of a fixed point. These theorems can be generalized to the *global* results of Theorems 3.4.3 and 3.4.4. In general a global result is one that yields information about the behavior of a function over an entire interval. We first define the concepts of *increasing* and *decreasing* on an interval.

Definition 3.4.3

> A function f is said to be *increasing* (*decreasing*) on an interval $a < x < b$ if for all $x_1 < x_2$ in the interval, $f(x_1) < f(x_2)$ $(f(x_1) > f(x_2))$.

Theorem 3.4.3

> If $f'(x) > 0$ for all x such that $a < x < b$, then f is increasing on the interval $a < x < b$.

Of course the corresponding result for decreasing functions is also true.

Theorem 3.4.4

> If $f'(x) < 0$ for all x such that $a < x < b$, then f is decreasing on the interval $a < x < b$.

Those two theorems remain true for the infinite intervals $x > a$ and $x < b$. The proofs of Theorems 3.4.3 and 3.4.4 will be given after a discussion of The Mean Value Theorem in the next section.

Example 1 | Where is the function $f(x) = x^3 + 3x^2 + 2$ increasing? Where is it decreasing?

We have

$$f'(x) = 3x^2 + 6x$$
$$= 3x(x + 2).$$

To apply the previous theorems we shall find where $f'(x)$ is positive and where it is negative. That is, we shall solve the inequalities $3x(x + 2) > 0$ and $3x(x + 2) < 0$. Using the techniques of Chapter 1, we see that $f'(x)$ changes sign at $x = 0$ and $x = -2$. Then, since $f'(x)$ is positive for large values of x, we can sketch a graph of $y = f'(x)$ as in Figure 3.4.2.

Consequently, $f'(x) < 0$ for $-2 < x < 0$

and $f'(x) > 0$ for $x < -2$ or $0 < x$.

Application of Theorems 3.4.3 and 3.4.4 then indicates that f is increasing for $x < -2$ or $0 < x$, and decreasing for $-2 < x < 0$. ||

Figure 3.4.2

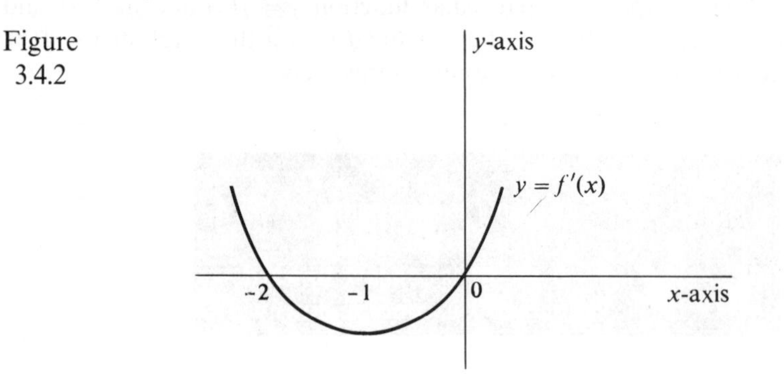

The derivative of $f'(x)$ is called the *second derivative* of f. Various notations are used for the second derivative; for example if $y = f(x)$, the second derivative can be expressed as

$$\frac{d}{dx}\left(\frac{dy}{dx}\right) = \frac{d^2y}{dx^2} = f''(x) = D_x{}^2 f(x) = y''.$$

The sign of the second derivative also reveals useful information about the graph of $y = f(x)$. Assume, for example, that $f''(x) > 0$ for $a < x < b$. Then, since $f''(x)$ is the derivative of f', we may apply Theorem 3.4.3 to find that f' is increasing for $a < x < b$. Since $f'(x)$ is the slope of the tangent to $y = f(x)$ at $(x, f(x))$, and since the slope of a line increases if and only if the line rotates in a counterclockwise direction, we know that the tangent to $y = f(x)$ must rotate in a counterclockwise direction as x increases. Figure 3.4.3(a) illustrates a sequence of such tangents. In Figure 3.4.3(b) the points of tangency are superimposed to exhibit the counterclockwise rotation. Note that the graph of $y = f(x)$ must be concave upward to cause a counterclockwise rotation of the tangent line. Thus, when $f''(x) > 0$ in an interval, the graph of $y = f(x)$ is concave upward in the interval.

Figure 3.4.3

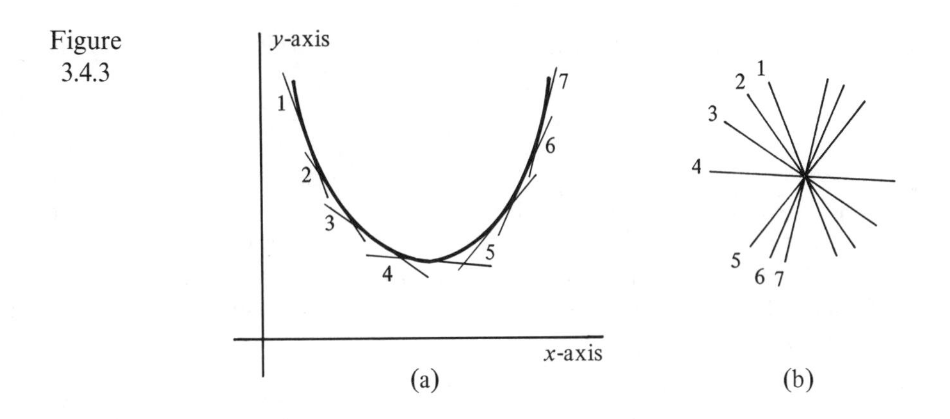

A similar argument would show that the graph of $y = f(x)$ is concave downward on any interval where $f''(x) < 0$.

Figure 3.4.4 shows graphs of a particular function $y = f(x)$ and its first and second derivatives. Note the influence of $f''(x)$ and $f'(x)$ on the graph of $y = f(x)$.

The following table summarizes the results of this section.

Assume $a < x < b$.

(1) If $f'(x) > 0$, then f is increasing.
(2) If $f'(x) < 0$, then f is decreasing.
(3) If $f''(x) > 0$, then f is concave upward.
(4) If $f''(x) < 0$, then f is concave downward.

Figure 3.4.4

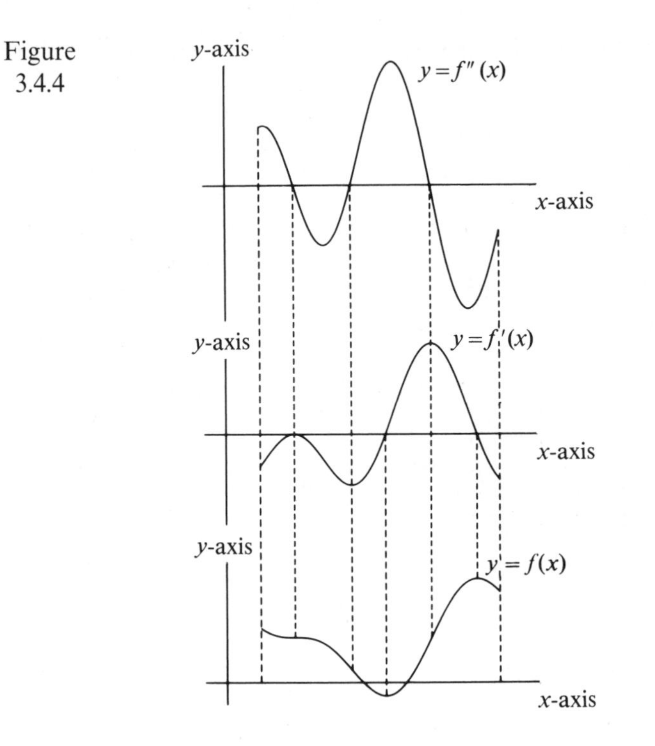

Example 2

Find where the graph of $y = 2x^3 - 9x^2 + 12x$ is increasing, decreasing, concave upward, and concave downward. Use this information to sketch a graph of the equation.

In this case

$$\frac{dy}{dx} = 6x^2 - 18x + 12$$
$$= 6(x - 2)(x - 1).$$

A sketch of the graph of dy/dx indicates that

$$\frac{dy}{dx} > 0 \quad \text{if } x < 1 \text{ or } x > 2, \quad \text{and}$$

$$\frac{dy}{dx} < 0 \quad \text{if } 1 < x < 2.$$

Consequently, $y = 2x^3 - 9x^2 + 12x$ increases for $x < 1$ or $x > 2$, and decreases for $1 < x < 2$. Furthermore

$$\frac{d^2y}{dx^2} = 12x - 18.$$

Hence

$$\frac{d^2y}{dx^2} > 0 \text{ if } x > \frac{3}{2}, \quad \text{and} \quad \frac{d^2y}{dx^2} < 0 \text{ if } x < \frac{3}{2}.$$

We then know that the graph is concave upward for $x > 3/2$ and concave downward for $x < 3/2$.

Since $y = 2x^3 - 9x^2 + 12x$ can be factored as

$$y = x(2x^2 - 9x + 12),$$

and since $2x^2 - 9x + 12$ has no real roots, the graph under consideration crosses the x-axis only at $x = 0$. Using that information and the points (1, 5), (2, 4) and (3/2, 9/2) on the graph where the first or second derivative is zero, enables us to sketch the graph as in Figure 3.4.5. ||

Figure 3.4.5

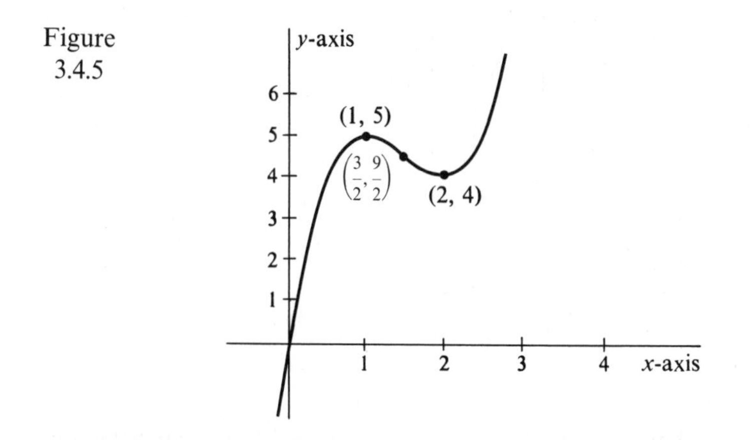

The points where a graph changes concavity are called *inflection points.* The points where a graph has a horizontal tangent or fails to have a tangent are called *critical points.* It should be clear from the preceding example that such points are especially important in sketching graphs. (Note that one inflection point and two critical points are shown in Figure 3.4.5.)

Example 3 | Find where the graph of $y = \dfrac{x+1}{x-5}$ is increasing, decreasing, concave upward, and concave downward. Use this information to sketch the graph of the equation.

In this case

$$\frac{dy}{dx} = \frac{(x-5)-(x+1)}{(x-5)^2} = \frac{-6}{(x-5)^2}.$$

Since $(x - 5)^2 > 0$ for all $x \neq 5$, we have

$$\frac{dy}{dx} < 0 \quad \text{if} \quad x \neq 5.$$

Note that at $x = 5$, both the function and its derivative are undefined. Consequently $y = (x + 1)/(x - 5)$ decreases for all x other than 5.

Since
$$\frac{d^2y}{dx^2} = 12(x - 5)^{-3},$$

we have
$$\frac{d^2y}{dx^2} > 0 \text{ for } x > 5, \quad \text{and} \quad \frac{d^2y}{dx^2} < 0 \text{ for } x < 5.$$

Thus the graph is concave upward when $x > 5$ and concave downward when $x < 5$. This information together with several points on the graph enable us to sketch the graph as in Figure 3.4.6. ||

Figure 3.4.6

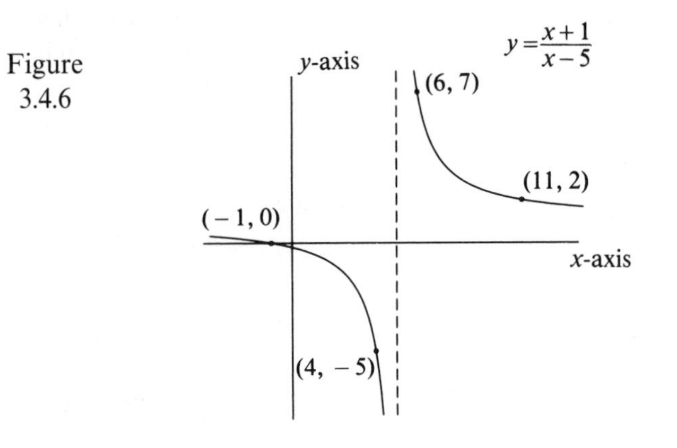

Exercises 3.4

For Exercises 1–6, find y''.

1. $y = x^3 - 5x^2 + 8x - 1.$
2. $y = x^4 + 7x^3 - 6x + 5.$
3. $y = (x^2 + 1)^3.$
4. $y = (x^2 - 3)^2.$
5. $y = \dfrac{x - 2}{x + 1}.$
6. $y = \dfrac{x + 4}{x - 3}.$

For Exercises 7–22, determine where the graph is (a) increasing, (b) decreasing, (c) concave upward, and (d) concave downward. Use this information along with the points of inflection and critical points to sketch the graph of the equation.

7. $y = x^2 - 4x + 2.$
8. $y = x^2 + x - 12.$
9. $y = 3 - 6x - x^2.$
10. $y = -x^2 - 6x + 7.$
11. $y = 12x - x^3 + 1.$
12. $y = x^3 - 3x.$
13. $y = x^4 - 1.$
14. $y = x^3 + 5.$
15. $y = 3x^4 - 4x^3 + 2.$
16. $y = (x + 3)(x^2 + 6x + 6).$
17. $y = (x^2 + 1)^3.$
18. $y = (x^2 - 3)^2.$
19. $y = (x - 3)^4.$
20. $y = (2x + 3)^5.$
21. $y = (x - 2)/(x + 1).$
22. $y = x + (1/x).$
23. Sketch the graph of a function $y = f(x)$ where $f'(x) > 0$ for $x < 5$, $f'(x) < 0$ for $x > 5$, and $f(5) = 1$.
24. Sketch the graph of a function $y = f(x)$ where $f''(x) > 0$ for $x < 5$, $f''(x) < 0$ for $x > 5$, and $f(5) = 1$.

25. Make a sketch similar to Figure 3.4.4, showing $f''(x)$, $f'(x)$, and $f(x)$ for $y = f(x) = x^3 - 3x$.

26. Make a sketch similar to Figure 3.4.4, showing $f''(x)$, $f'(x)$, and $f(x)$ for $y = f(x) = 3x^4 - 4x^3$.

27. Psychologists are interested in expressing the relationship between attainment and practice. Thurston has suggested the "learning curve"

$$G(x) = \frac{a(x + b)}{x + c},$$

where $G(x)$ represents the attainment for x hours of practice, and a, b, and c are positive empirical constants with $b < c$. Plot the graph of $G(x)$. What does a represent?

28. Suppose that the reliability of a test is given by a number r, where $0 < r < 1$, and 0 indicates no reliability, and 1 indicates total reliability. Suppose also that if the test is replaced by a test of equal difficulty but x times as long, the reliability of the new test is given by

$$f(x) = \frac{rx}{1 + (x - 1)r},$$

where r is a constant. Prove that $f'(x) > 0$. What does this fact indicate?

29. One measure of consumer reaction to price changes is *elasticity of demand.* If $x = g(p)$ is the number of units sold at a price p, then

$$E = \frac{p}{x}\frac{dx}{dp}$$

gives the elasticity of demand. Show that an increase in price causes a decrease in *total revenue*, px, if $E < -1$.

30. Prove Theorem 3.4.2.

31. The graphs of six functions are given in Figure 3.4.7(a)–(f). The graphs of the derivatives of those functions are given in a scrambled order in Figure 3.4.8(i)–(vi). Match the graph of each function with the graph of its derivative.

Figure 3.4.7

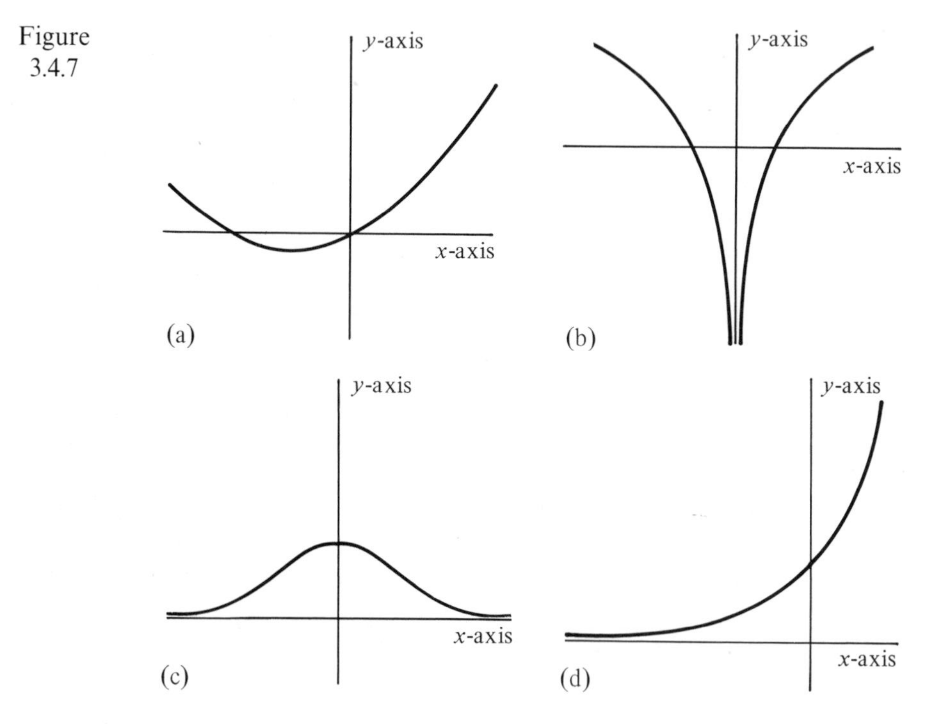

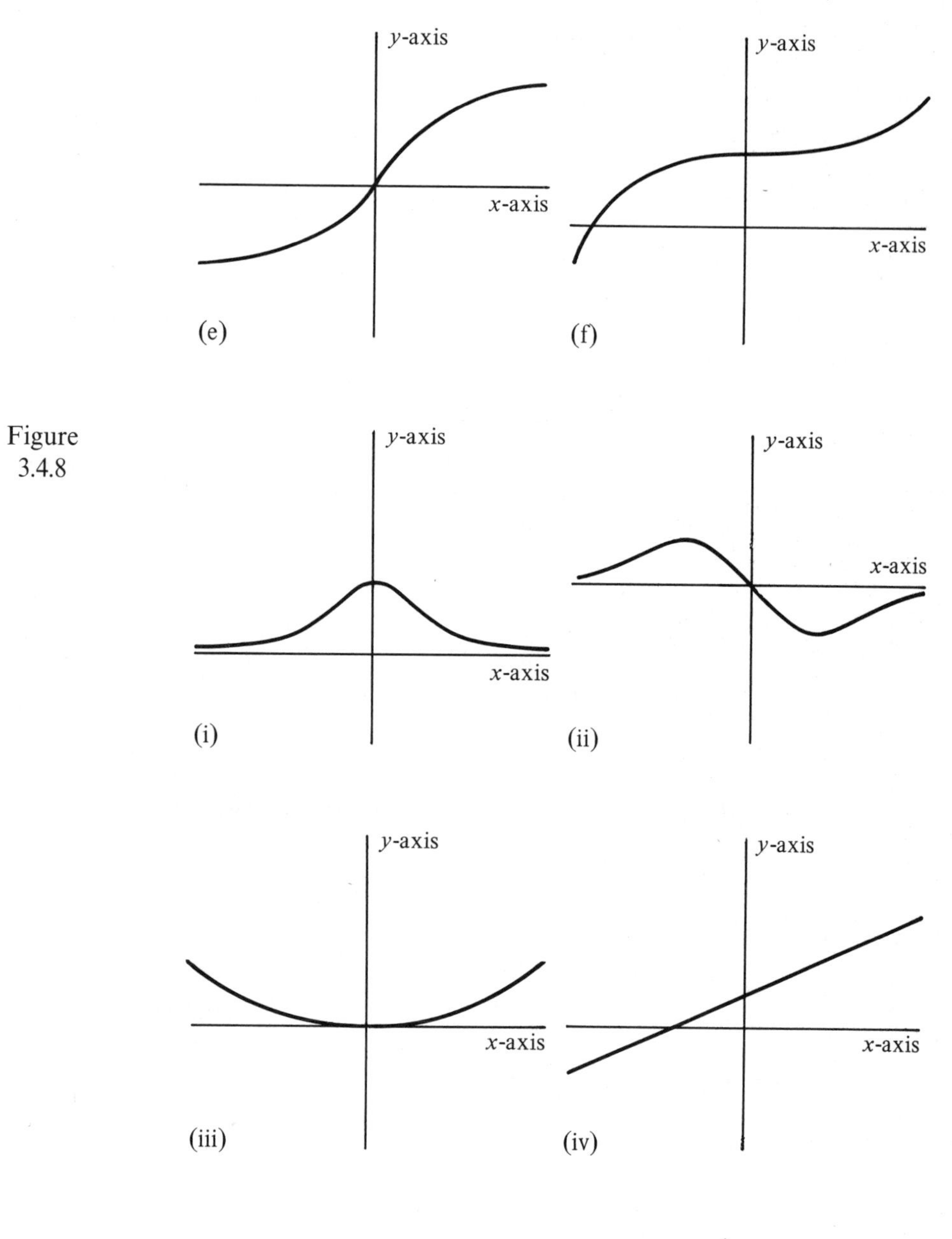

y-axis
x-axis
(e)
y-axis
x-axis
(f)
y-axis
x-axis
(i)
y-axis
x-axis
(ii)
y-axis
x-axis
(iii)
y-axis
x-axis
(iv)

Figure 3.4.8

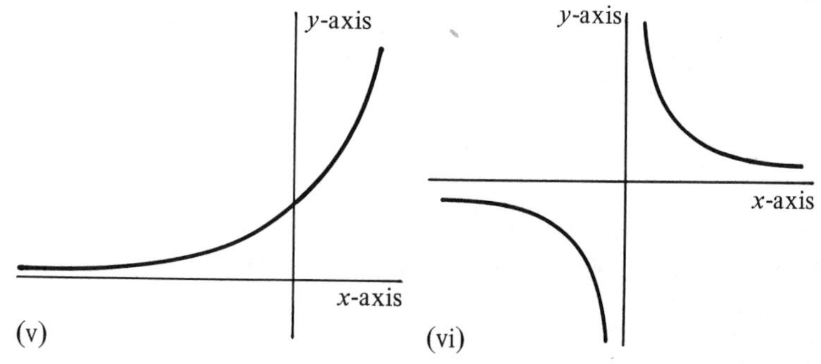

y-axis
x-axis
(v)
y-axis
x-axis
(vi)

3.5 Maximum-Minimum Problems

In Section 2.6 the reader was introduced to the problem of finding maximum values of functions. Since min-max problems arise in many areas of science, we shall discuss them more fully in this section.

Before beginning our discussion we need a few preliminary definitions.

Definition 3.5.1

> A function f is said to have a *local maximum* (or *local minimum*) at a if $f(a)$ is greater (less) than or equal to $f(x)$ in some neighborhood of a and in the domain of f. That is, f has a local maximum (minimum) at a if there is a $\delta > 0$ such that for $|x - a| < \delta$, we have $f(x) \leq f(a)$ ($f(x) \geq f(a)$), where $f(x)$ is defined.
>
> If f has a local maximum (minimum) at a, then $f(a)$ is said to be a *local maximum (minimum) value* of f, and the point $(a, f(a))$ is called a *local maximum (minimum) point.*

Definition 3.5.2

> If c is a point such that $f(c) \geq f(x)$ ($f(c) \leq f(x)$) for all x in the domain of f, then f is said to have an *absolute maximum* (or *absolute minimum*) at c.
>
> If f has an absolute maximum (minimum) at c, then $f(c)$ is said to be an *absolute maximum (minimum) value* of f, and the point $(c, f(c))$ is called an *absolute maximum (minimum) point.*

Figure 3.5.1 illustrates a few of the possibilities for a function whose domain is $a \leq x \leq b$.

Figure 3.5.1

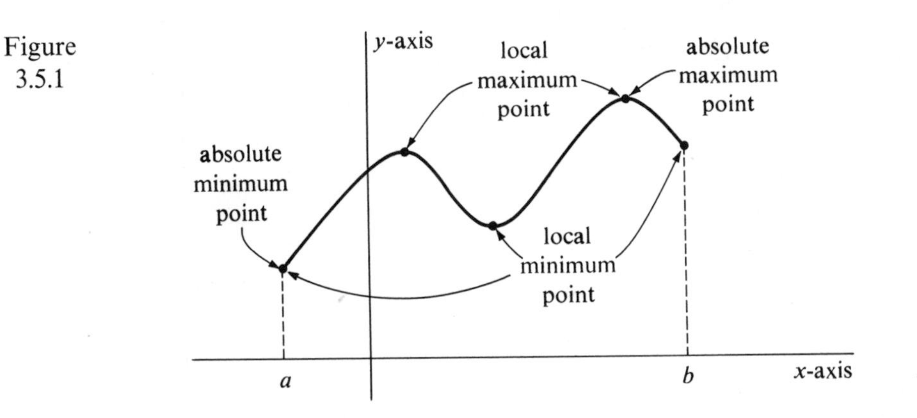

If f has a local maximum or minimum at an interior point a, then f can neither increase nor decrease at a. Consequently, if a is an interior point of the domain of f and $f'(a)$ exists, then it can be neither positive nor negative and so must be zero. This important fact is formalized as follows.

Theorem 3.5.1

> Let a be an interior point of the domain of f. If f has a local maximum or minimum at a and $f'(a)$ exists, then $f'(a) = 0$.

The reader should note that the theorem does *not* require that $f'(a) = 0$ at a local maximum or minimum; nor is it necessary that $(a, f(a))$ be a local maximum or minimum point when $f'(a) = 0$. For example, $f(x) = |x|$ has a local minimum at 0 although $f'(0)$ is not zero; in fact, $f'(0)$ does not exist. On the other hand, if $f(x) = x^3$, then $f'(0) = 0$ but f does not have a local maximum or minimum at 0. Nor does the theorem deal with the endpoints of the domain of f. However, the theorem does indicate that a local maximum or minimum can *only* occur either where $f'(x) = 0$, or where $f'(x)$ does not exist, or at an endpoint of the domain. Since the last statement is so useful we state it as a theorem.

Theorem 3.5.2

> A local maximum or minimum of a function f can occur only at a point a where $f'(a) = 0$, or where $f'(a)$ does not exist, or at an endpoint of the domain of f.

Example 1 Determine all local maximum and minimum points of the function

$$f(x) = 2x^3 - 9x^2 + 12x,$$

where $0 \le x \le 3$.

Here

$$\begin{aligned} f'(x) &= 6x^2 - 18x + 12 \\ &= 6(x - 2)(x - 1). \end{aligned}$$

Consequently $f'(x) = 0$ when $x = 1$ and when $x = 2$. It should also be noted that we must also check the endpoints $x = 0$ and $x = 3$ of the interval of definition. Since all local maxima and minima must occur at points where $f'(x) = 0$, or where $f'(x)$ does not exist, or at an endpoint of the domain, any local maximum or minimum point of this function must occur at (0, 0), (1, 5), (2, 4), or (3, 9). Since $f'(x) > 0$ for $x < 1$ or $x > 2$, and $f'(x) < 0$ for $1 < x < 2$, a sketch of the graph shows that (0, 0) and (2, 4) are local minimum points, while (1, 5) and (3, 9) are local maximum points. ||

Example 2 Determine all local maximum and minimum points of the function

$$f(x) = x^4 + x^3, \quad \text{for} \quad -1 \le x \le 1.$$

Since

$$\begin{aligned} f'(x) &= 4x^3 + 3x^2 \\ &= x^2(4x + 3), \end{aligned}$$

$f'(x) = 0$ when $x = 0$ or $x = -3/4$. Consequently local maximum and minimum points can occur only at (0, 0), $(-3/4, -27/256)$, $(-1, 0)$, or (1, 2). (The last two points are determined by the endpoints of the domain of f.) A sketch of the graph of

$f(x) = x^4 + x^3$, for $-1 \leq x \leq 1$, is given in Figure 3.5.2. The sketch indicates the following facts:

1. $(-1, 0)$ is a local maximum point;
2. $(-3/4, -27/256)$ is a local minimum point; and
3. $(1, 2)$ is a local maximum point.
4. Even though $y = f(x)$ has a horizontal tangent at $x = 0$, it does *not* have a local maximum or minimum there. ||

Figure 3.5.2

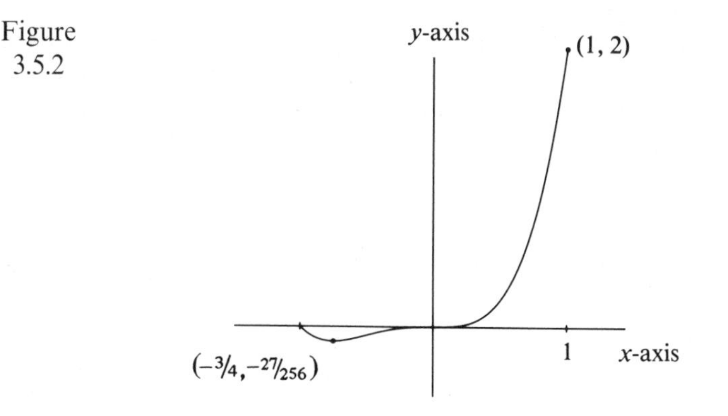

This example again illustrates the fact that a local maximum or minimum need not occur at a simply because $f'(a) = 0$. Rather, our criterion is that the only possible places where a local maximum or minimum can occur are those where $f'(x) = 0$, or where $f'(x)$ does not exist, or at an endpoint of the domain of f.

In the previous two examples we made a quick sketch of the graph of the function in order to determine which points were local maxima, which were local minima, and which were neither. While this technique is the most universally applicable method of sorting local maxima from local minima, there are other techniques available.

Notice that if a graph is concave downward at a point $(a, f(a))$ where $f'(a) = 0$, then $(a, f(a))$ must be a local maximum point. On the other hand, if $f'(a) = 0$ and the graph is concave upward at $(a, f(a))$, then $(a, f(a))$ must be a local minimum point. These facts are illustrated in Figure 3.5.3.

Figure 3.5.3

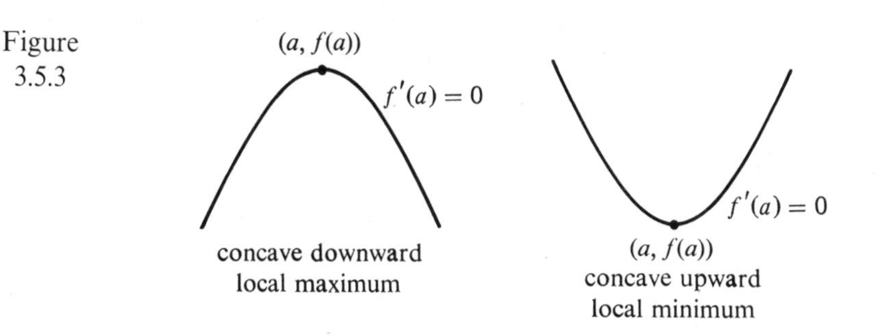

Since $f''(a) < 0$ implies that $y = f(x)$ is concave downward at $x = a$, and since $f''(a) > 0$ implies that $y = f(x)$ is concave upward at $x = a$, we have the following *second derivative test*.

Theorem 3.5.3

> *The Second Derivative Test*
>
> Let a be an interior point of the domain of f at which $f'(a) = 0$. Then,
> (i) if $f''(a) < 0$, then $(a, f(a))$ is a local maximum point;
> (ii) if $f''(a) > 0$, then $(a, f(a))$ is a local minimum point.

Note that in case $f''(a) = 0$, the second derivative test yields *no* conclusion on maximum or minimum, and so another approach, perhaps a sketch of the graph, must be used.

Example 3 | Find all values of x at which $f(x) = 2x^{-2} + x^2$ has a local minimum.

In this case $f'(x) = -4x^{-3} + 2x$, which exists for all x except $x = 0$. Consequently a local minimum can only occur at $x = 0$ or where $-4x^{-3} + 2x = 0$. Solving for x produces

$$4x^{-3} = 2x$$

$$4 = 2x^4$$

$$x = \pm 2^{1/4}.$$

Then, since

$$f''(x) = 12x^{-4} + 2,$$

we have

$$f''(\pm 2^{1/4}) > 0.$$

Consequently, by the second derivative test, f has a local minimum at both $x = 2^{1/4}$ and $x = -2^{1/4}$. Since the function itself is undefined at $x = 0$, it follows that f has neither a local maximum nor a local minimum at $x = 0$. ||

As noted above, the second derivative test fails to give the desired information on maximum or minimum for $x = a$ when $f''(a) = 0$, or when $f''(a)$ does not exist. When the second derivative test fails in that way it is often possible to use our knowledge of the first derivative around $x = a$ to determine whether the function has a local maximum or a local minimum at an interior point a of the domain of f. For example, suppose $f'(x)$ changes sign from negative to positive as x increases through a. Then, as illustrated in Figure 3.5.4, $y = f(x)$ decreases on a small interval

Figure 3.5.4

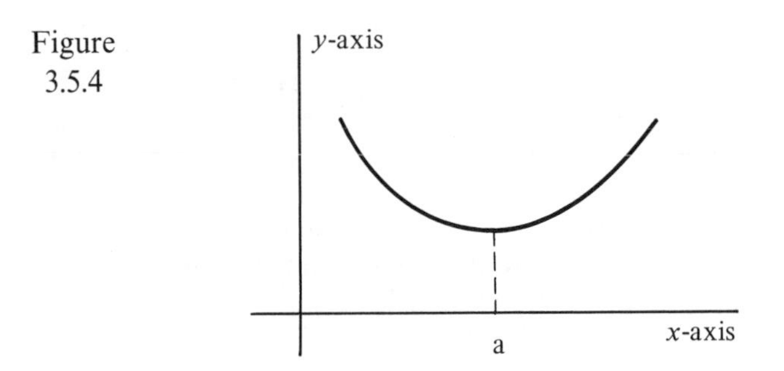

to the left of a and increases on a small interval to the right of a. Consequently the function has a local minimum at a. We have established the first part of the following *First Derivative Test.* The remaining portions are left as an exercise.

Theorem 3.5.4

The First Derivative Test

Let a be an interior point of the domain of f.

(i) If $f'(x)$ changes sign from negative to positive as x increases through a, then $(a, f(a))$ is a local minimum point.

(ii) If $f'(x)$ changes sign from positive to negative as x increases through a, then $(a, f(a))$ is a local maximum point.

(iii) If $f'(x)$ does not change sign at $x = a$, then $(a, f(a))$ is neither a local maximum nor a local minimum point.

The next example illustrates the use of the First Derivative Test.

Example 4 Find all values of x at which the function $f(x) = x^3(x-1)^4$ has a local maximum or minimum.

Use of the product rule gives

$$\begin{aligned} f'(x) &= 3x^2(x-1)^4 + 4x^3(x-1)^3 \\ &= x^2(x-1)^3(7x-3). \end{aligned}$$

Since this derivative exists for all x, and there are no endpoints to consider, any local maximum or minimum may only occur where $f'(x) = 0$, that is, at $x = 0$, $x = 1$, or $x = 3/7$.

Again we use the product rule to calculate $f''(x)$.

$$\begin{aligned} f''(x) &= 2x(x-1)^3(7x-3) + 3x^2(x-1)^2(7x-3) + 7x^2(x-1)^3 \\ &= x(x-1)^2[2(x-1)(7x-3) + 3x(7x-3) + 7x(x-1)]. \end{aligned}$$

Hence,

$$f''(0) = 0, f''(1) = 0, \quad \text{and} \quad f''\left(\frac{3}{7}\right) < 0.$$

Thus, while the second derivative test indicates that f has a local maximum at $x = 3/7$ since $f''(3/7) < 0$, the test gives no maximum-minimum information about the points $x = 0$ and $x = 1$.

Note, however, that the first derivative changes sign from negative to positive at $x = 1$ and does not change sign at $x = 0$. Thus, by the First Derivative Test, $x = 1$ determines a local minimum and the function has neither a local maximum nor a local minimum at $x = 0$. Use of these facts, along with the positiveness of $f(x)$ for large x, enables us to sketch the graph $y = f(x)$ roughly as indicated in Figure 3.5.5. Consequently, f has neither a local minimum nor a local maximum at $x = 0$, and f has a local minimum at $x = 1$. Of course, we saw earlier that f has a local maximum at $x = 3/7$. ||

Figure 3.5.5

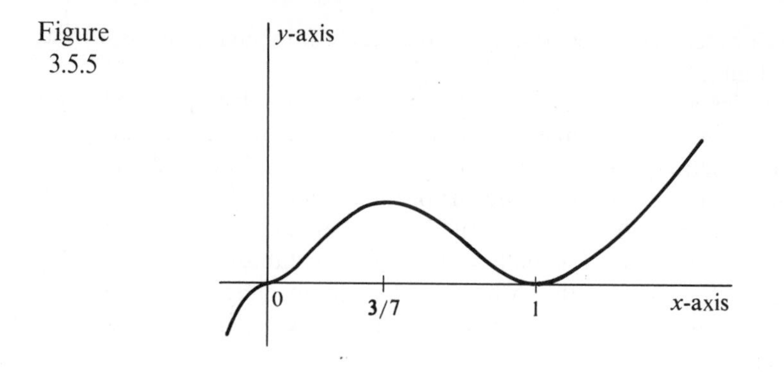

Exercises 3.5

For Exercises 1–20, sketch the graph and determine all local maximum and minimum points of the given functions.

1. $f(x) = 3x^2 - 12x + 1.$
2. $f(x) = -2x^2 + 8.$
3. $f(x) = x^2 - 4x + 2, \quad 3 \leq x \leq 5.$
4. $f(x) = x^2 + 6x - 13, \quad 1 \leq x \leq 2.$
5. $f(x) = x^2 - 4x + 2, \quad 0 \leq x \leq 3.$
6. $f(x) = x^2 + 6x - 13, \quad 0 \leq x \leq 4.$
7. $f(x) = x^3 - 3x^2 + 1.$
8. $f(x) = 2x^3 - 9x^2 + 12x + 7.$
9. $f(x) = -x^4 + 2x^2 + 1, \quad -3 \leq x \leq 3.$
10. $f(x) = x^3 - 3x + 7, \quad -2 \leq x \leq 2.$
11. $f(x) = x^{-2} - x^{-1}, \quad 1 \leq x \leq 3.$
12. $f(x) = x^4 + 8, \quad -1 \leq x \leq 1.$
13. $f(x) = x^{-1} + 4.$
14. $f(x) = x^{-3} + x^{-2}, \quad 1 \leq x \leq 2.$
15. $f(x) = |x|.$
16. $f(x) = 1 - |x|.$
17. $f(x) = \sqrt{x^{10}}, \quad -1 \leq x \leq 2.$
18. $f(x) = \sqrt{x^6}, \quad -2 \leq x \leq 1.$
19. $f(x) = 1/(1 + x^2).$
20. $f(x) = x/(1 + x^2).$
21. Let f and g be positive functions such that f' and g' exist for all x such that $a \leq x \leq b$, and let c be an interior point of the interval. If both f and g have a local minimum at c, prove that $f(x) \cdot g(x)$ has a local minimum at c.
22. Let f and g be positive functions such that f' and g' exist for all x, $a \leq x \leq b$, and let c be an interior point of the interval at which f has a local maximum and g has a local minimum. Prove that $f(x)/g(x)$ has a local maximum at c.

3.6 Absolute Maxima and Minima; The Mean Value Theorem

We are generally more interested in finding the absolute maximum and absolute minimum points of a given function than in finding the local maximum and minimum points. Of course the absolute maxima and minima, if they exist, must occur among the local maxima and minima. Hence, to find the absolute maximum points and absolute minimum points of a function on a closed interval we generally consider the endpoints and all points where the derivative is zero or does not exist; and from these we select the points where the function is greatest or least. When doing this, it is not necessary to determine whether each individual point is a local maximum or local minimum.

Next we consider a theorem that is a classical example of a "pure existence theorem." A pure existence theorem establishes the existence of a certain entity but does not provide a method for finding it. For example, The Intermediate Value Theorem for Continuous Functions assures us of the existence of a number c such that $f(c) = r$, but gives no method of finding it. While these existence theorems may appear to be of little value, they are actually widely used to establish more "practical" results. Theorem 3.6.1 assures us that a function has an absolute maximum value *and* an absolute minimum value under certain conditions. A proof can be found in many advanced calculus texts.

Theorem 3.6.1

> If the function f is continuous for all x such that $a \leq x \leq b$, then f has an absolute maximum value and an absolute minimum value on the interval $a \leq x \leq b$.

It is quite important that the function f of the theorem above be continuous at each point of a *closed interval* (that is, an interval that includes its endpoints). For example, the function $f(x) = 1/x$ is continuous at each point of the interval $0 < x \leq 1$, but is *not* continuous at $x = 0$. Note, however, that the function has no maximum value on the interval $0 < x \leq 1$. Similarly as illustrated in Figure 3.6.1 the function $f(x) = x$, $1 < x < 3$, has no maximum or minimum value; but the function $f(x) = x$, $1 \leq x \leq 3$, does have both.

Figure 3.6.1

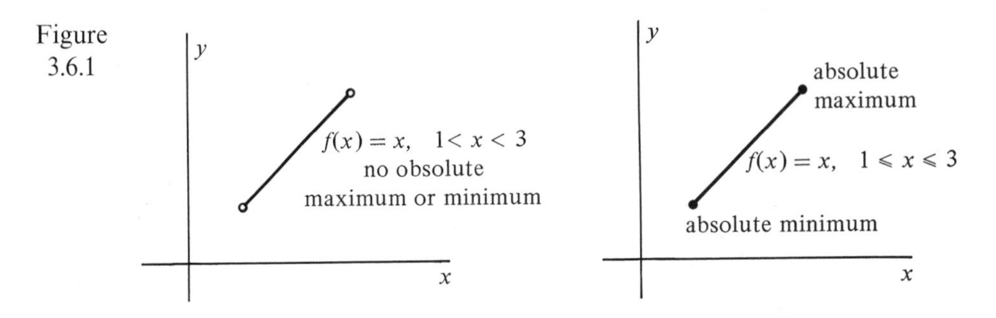

We are now in a position to find the absolute maximum and minimum values whose existence is assured by the preceding theorem.

Example 1 | Find the absolute maximum and minimum points of $f(x) = x^3 - 5x^2 + 8x + 7$ for $-1 \leq x \leq 3$.

Each *local maximum* or *minimum* must occur at a point $(x, f(x))$ where $f'(x) = 0$, or where $f'(x)$ does not exist or at one of the endpoints $x = -1$ and $x = 3$. In this case,

$$\begin{aligned} f'(x) &= 3x^2 - 10x + 8 \\ &= (3x - 4)(x - 2). \end{aligned}$$

Thus, $f'(x) = 0$ if and only if $x = 2$ or $x = 4/3$.

Since, by Theorem 3.6.1, the function f *must* have an absolute maximum and an absolute minimum on the interval $-1 \leq x \leq 3$, one of the four points, (2, 11), (4/3, 301/27), (−1, −7), or (3, 13) must be the absolute maximum point and one of them must be the absolute minimum point. Clearly (3, 13) is the absolute maximum point and (−1, −7) is the absolute minimum point. ||

Since an absolute maximum or minimum does not necessarily exist when the hypothesis of Theorem 3.6.1 is not satisfied, great care must be exercised in attempting to find absolute maxima and minima in such cases.

Example 2 | Find the absolute maximum and absolute minimum value of $f(x) = x^3 - 6x^2 + 9x$.

First one should note that Theorem 3.6.1 *does not* assure us of the existence of an absolute maximum or minimum value, because the function is not restricted to a closed interval. Of course, since there are no endpoints, if either or both the absolute maximum and absolute minimum exist, they can only occur where $f'(x) = 0$ or where $f'(x)$ does not exist. In this case,

$$\begin{aligned} f'(x) &= 3x^2 - 12x + 9 \\ &= 3(x - 1)(x - 3), \end{aligned}$$

and so (1, 4) and (3, 0) are the critical points.

Since $f''(x) = 6x - 12$, we have $f''(1) < 0$ and $f''(3) > 0$. Consequently, f has a local maximum at $x = 1$ and f has a local minimum at $x = 3$. However, a sketch of the graph in Figure 3.6.2 indicates that there is no absolute maximum or absolute minimum value. It should be noted that if $f(x) = x^3 - 6x^2 + 9x$ were restricted to a closed interval, both absolute maximum and absolute minimum values would exist. In particular, for the interval $0 \leq x \leq 5$, the points to consider are (1, 4), (3, 0), (0, 0), and (5, 20). Thus (3, 0) and (0, 0) are both absolute minimum points, and (5, 20) is the absolute maximum point. This could be determined without first finding that (1, 4) is a local maximum point and that (3, 0) is a local minimum point. ||

Figure 3.6.2

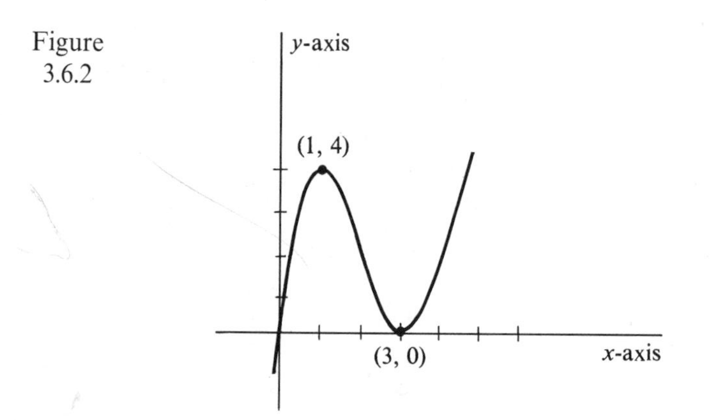

We now have sufficient background information to prove The Mean Value Theorem, which will be used to prove Theorems 3.4.3 and 3.4.4. The Mean Value Theorem will have a wide range of applications in our future development of calculus.

Theorem 3.6.2

The Mean Value Theorem

If the function f has a derivative for each x, $a \leq x \leq b$, then there is some c, $a < c < b$, such that

$$f'(c) = \frac{f(b) - f(a)}{b - a}.$$

Since a geometric interpretation of this theorem leads rather quickly to a proof, we first consider the geometry of the situation. The number

$$\frac{f(b) - f(a)}{b - a}$$

is the slope of the line L passing through the points $(a, f(a))$ and $(b, f(b))$. Since $f'(x)$ is the slope of the tangent to the graph of $y = f(x)$ at $(x, f(x))$, and since lines are parallel if and only if they have equal slopes, The Mean Value Theorem simply states that there is some c, $a \leq c \leq b$ such that the tangent at $(c, f(c))$ is parallel to the line passing through $(a, f(a))$ and $(b, f(b))$. Figure 3.6.3 illustrates this theorem for several situations. Note that in each of these situations it appears that one

Figure 3.6.3

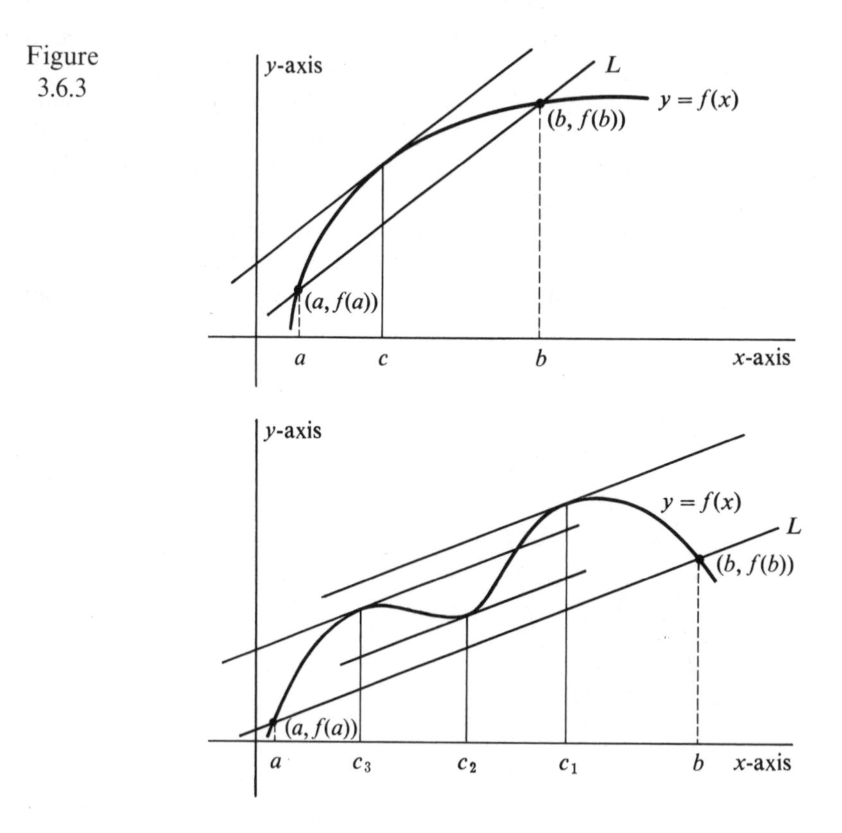

possible choice for c is the point where the vertical distance between the graph of $y = f(x)$ and the line L is a maximum.

Since an equation of the line passing through $(a, f(a))$ and $(b, f(b))$ is

$$y = \frac{f(b) - f(a)}{b - a}(x - b) + f(b),$$

the vertical distance being considered is given by

$$s(x) = f(x) - \frac{f(b) - f(a)}{b - a}(x - b) - f(b).$$

We are thus led to give the following proof of The Mean Value Theorem.

Proof (Mean Value Theorem). Since f is differentiable for each x, $a \leq x \leq b$, f must be continuous for each x in the interval. Consequently the function

$$s(x) = f(x) - \frac{f(b) - f(a)}{b - a}(x - b) - f(b).$$

is continuous for all x, $a \leq x \leq b$, and so must attain an absolute maximum value and an absolute minimum value in this interval, by Theorem 3.6.1. If the absolute maximum value and the absolute minimum value are both zero, then $s(x) = 0$ for $a \leq x \leq b$, and so $s'(c) = 0$ for any c such that $a < c < b$. If either the absolute maximum or the absolute minimum is nonzero, then this absolute maximum or absolute minimum is not attained at the endpoints $x = a$ or $x = b$, since $s(a) = s(b) = 0$. Let c be a value of x, $a < c < b$, at which $s(x)$ attains either an absolute maximum or an absolute minimum. Then we know from Theorem 3.5.1 that $s'(c) = 0$. Then, since

$$s'(x) = f'(x) - \frac{f(b) - f(a)}{b - a}.$$

we have

$$s'(c) = f'(c) - \frac{f(b) - f(a)}{b - a} = 0,$$

and so

$$f'(c) = \frac{f(b) - f(a)}{b - a},$$

as we wanted to prove.

The Mean Value Theorem can be used to give a proof of Theorem 3.4.3.

Proof (Theorem 3.4.3). Suppose that $a \leq x_1 < x_2 \leq b$. Then by The Mean Value Theorem there is some c, $x_1 \leq c \leq x_2$, such that

$$f'(c) = \frac{f(x_2) - f(x_1)}{x_2 - x_1}, \quad \text{or}$$

$$f(x_2) - f(x_1) = f'(c) \cdot (x_2 - x_1).$$

By assumption, $f'(c) > 0$. Then, since $x_2 > x_1$, we have $x_2 - x_1 > 0$, and hence

$$f(x_2) - f(x_1) > 0.$$

Therefore $$f(x_2) > f(x_1),$$

and the function f is increasing, as we wished to prove.

The proof of Theorem 3.4.4 completely parallels the proof above and is left as an exercise.

Exercises 3.6

For Exercises 1–18, determine the absolute maximum and minimum points for each given function $f(x)$, or state that they do not exist.

1. $f(x) = x^2 - 6x - 1, \quad 2 \le x \le 5.$
2. $f(x) = x^2 - 8x + 15, \quad 3 \le x \le 5.$
3. $f(x) = x^2 - 6x - 1, \quad 0 \le x \le 2.$
4. $f(x) = x^2 - 8x + 15, \quad 0 \le x \le 3.$
5. $f(x) = x^2 + 2, \quad -3 \le x \le 1.$
6. $f(x) = -x^2 + 1, \quad 0 \le x \le 1.$
7. $f(x) = x^3 + 6x^2 + 12x, \quad -4 \le x \le 4.$
8. $f(x) = 3x^3 - x + 7, \quad -1 \le x \le 1.$
9. $f(x) = x^4 + 8x^2 + 5, \quad -1 \le x \le 1.$
10. $f(x) = x^4 + 7, \quad -1 \le x \le 1.$
11. $f(x) = x^{-2}, \quad -1 \le x \le 3.$
12. $f(x) = (1 - x)/x, \quad 0 < x \le 1.$
13. $f(x) = x^{-2} - x^{-1}, \quad -1 \le x \le 3.$
14. $f(x) = x^{-3} + x^{-2}, \quad 1 \le x \le 2.$
15. $f(x) = 7, \quad 0 \le x \le 2.$
16. $f(x) = (1/x) + 5.$
17. $f(x) = x^3 + 3x - 2.$
18. $f(x) = x^4 + 3x^2 - 5.$

For Exercises 19–22, find all values of c whose existence is assured by The Mean Value Theorem for the indicated functions and intervals.

19. $f(x) = 3x^2 - 5x + 7, \quad 1 \le x \le 9.$
20. $f(x) = 8 - 6x^2, \quad 3 \le x \le 7.$
21. $f(x) = x^3 - 2, \quad -4 \le x \le 5.$
22. $f(x) = (1/x) + 3, \quad 1/3 \le x \le 6.$
23. When a driver got on a state toll road the attendant stamped the time 8:37 A.M. on the travel ticket he gave to the driver. When the driver arrived at his destination-exit 240 miles down the road, he paid the toll; the exit attendant stamped the exit time 11:17 A.M. on the ticket, collected from the driver, and immediately called his supervisor. How could the police use The Mean Value Theorem in court the next week to prove that the driver was going exactly 90 miles per hour at some point of his trip?
24. Prove that if $f(a) = f(b) = 0$ and f has a derivative for each x, $a \le x \le b$, then $y = f(x)$ has a horizontal tangent at some point of the interval.
25. Use The Mean Value Theorem to prove Theorem 3.4.4.
26. Use The Mean Value Theorem to prove that if $f'(x) = 0$ for all x, $a \le x \le b$, then $f(x)$ is a constant on the interval.
27. Use The Mean Value Theorem to prove that if $F'(x) = G'(x)$ for all x, $a \le x \le b$, then for $a \le x \le b$, it follows that $F(x) - G(x) = C$, where C is a constant.
28. Show that if the hypothesis of The Mean Value Theorem does not hold, the conclusion need not hold. Consider, for example, the function $f(x) = |x|$.

29. Prove that the value of c whose existence is assured by The Mean Value Theorem for a parabola $y = hx^2 + kx + r$ on an interval $a \le x \le b$ is the mid-point $(a + b)/2$ of the interval.

30. Prove the following special case of The Mean Value Theorem, (Theorem 3.6.2).
Rolle's Theorem: If the function f has a derivative for each x, $a \le x \le b$, and $f(a) = f(b) = 0$, then there is some c, $a < c < b$, such that $f'(c) = 0$.

3.7 Applied Maximum-Minimum Problems

The techniques of the previous section are widely applied in almost every area of science. In this section we shall discuss some possible applications.

Example 1 If the cost in dollars per pound of producing x pounds of a certain chemical per hour is given by

$$C(x) = 150 + x(x - 20), \qquad x \ge 0,$$

find the amount of the chemical to be produced per hour in order to minimize the cost. (*Note*: Here the constant 150 represents a fixed cost of operation.)

If an absolute minimum value of the function

$$C(x) = 150 + x^2 - 20x, \qquad x \ge 0$$

exists, it must occur either where $C'(x) = 0$, or where $C'(x)$ does not exist, or at the endpoint $x = 0$. Now

$$C'(x) = 2x - 20.$$

Consequently, $C'(x) = 0$ if and only if $x = 10$. Therefore $(0, 150)$ and $(10, 50)$ are the only candidates for absolute minimum points. Clearly $(0, 150)$ cannot be the absolute minimum point because $150 > 50$. Since

$$C'(x) = 2x - 20 < 0 \quad \text{for all } x < 10, \quad \text{and}$$

$$C'(x) = 2x - 20 > 0 \quad \text{for all } x > 10,$$

the function $C(x)$ is decreasing for all $x < 10$ and increasing for all $x > 10$. Thus, the local minimum point $(10, 50)$ is an absolute minimum, and the absolute minimum cost of 50 dollars per pound occurs when 10 pounds are produced per hour. ||

The preceding example illustrates a very useful variant of the First Derivative Test. Note that in order to show that $(10, 50)$ is an absolute minimum point we needed only to show that $C'(x) < 0$ for all $x < 10$ and $C'(x) > 0$ for all $x > 10$. In general, we may apply the following First Derivative Test for Absolute Extrema.

1 If $f'(x) > 0$ for all $x < a$ and $f'(x) < 0$ for all $x > a$, then $(a, f(a))$ is an absolute maximum point.

2 If $f'(x) < 0$ for all $x < a$ and $f'(x) > 0$ for all $x > a$, then $(a, f(a))$ is an absolute minimum point.

Many times some initial work must be done in order to set up the function to be maximized or minimized.

Example 2 | A rectangular area of 160,000 square feet must be enclosed. How should this be done in order to use the minimum amount L of fencing?

Figure 3.7.1

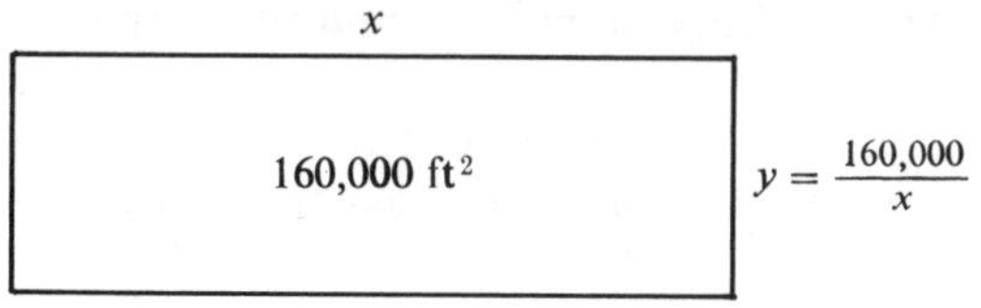

The area to be enclosed can be pictured as in Figure 3.7.1. The total length of fencing used is

$$L = 2x + \frac{320,000}{x}.$$

Since the minimum value of L must occur where $dL/dx = 0$ or where dL/dx does not exist, we examine

$$\frac{dL}{dx} = 2 - 320,000x^{-2}.$$

Then $dL/dx = 0$ if and only if

$$2 - 320,000x^{-2} = 0, \quad \text{or}$$

$$x = \pm 400.$$

Since we are only concerned with $x > 0$, it follows that $x = 400$ is the only possibility for a minimum. In fact, since

$$\frac{dL}{dx} < 0 \quad \text{for all } x \text{ such that } 0 < x < 400, \quad \text{and}$$

$$\frac{dL}{dx} > 0 \quad \text{for all } x > 400,$$

we see that L decreases for all $x < 400$ and increases for all $x > 400$. Thus the local minimum at $x = 400$ is actually the absolute minimum value of L. The corresponding y value is $y = 400$. Consequently, the area to be enclosed should be a square with sides of length 400 feet. ||

Example 3 | What point on the graph of $y = x^2$ is nearest the point (3, 0)?

The distance between the points (x, y) and (3, 0) is

$$L = \sqrt{(x - 3)^2 + y^2}.$$

Since $y = x^2$, we have

$$L = \sqrt{(x-3)^2 + x^4}.$$

Rather than minimizing L we shall minimize L^2. (The reader should verify that this amounts to the same thing.) We have

$$L^2 = (x-3)^2 + x^4, \quad \text{and so}$$

$$\frac{dL^2}{dx} = 2(x-3) + 4x^3.$$

Since dL^2/dx exists for all x, the minimum of L^2, if it exists, must occur where $dL^2/dx = 0$, that is, where

$$2(x-3) + 4x^3 = 0, \quad \text{or}$$

$$4x^3 + 2x - 6 = 0.$$

But $x = 1$ is a zero of this polynomial. Hence we can factor out $(x - 1)$ to get

$$(x-1)(4x^2 + 4x + 6) = 0, \quad \text{or}$$

$$2(x-1)(2x^2 + 2x + 3) = 0.$$

Since the quadratic $2x^2 + 2x + 3$ has no real zeros, $dL^2/dx = 0$ only when $x = 1$. Then, using the fact that

$$\frac{dL^2}{dx} > 0 \quad \text{for all } x > 1, \quad \text{and}$$

$$\frac{dL^2}{dx} < 0 \quad \text{for all } x < 1,$$

we see that L must have an absolute minimum at $x = 1$. Hence $(1, 1)$ is the point on the graph of $y = x^2$ that is closest to the point $(3, 0)$. ||

Example 4 Suppose that the cost in dollars of producing x units of a certain commodity per hour is

$$C(x) = 2x^2 + 8, \qquad 0 \le x \le 10,$$

and the number of units sold per hour at a price p per unit is $x = 18 - p$. Find the number of units that should be produced per hour to maximize the profit. What is the maximum profit?

The total revenue $R(x)$ from the sale of x units at a price p per unit is $R(x) = px$. Since $p = 18 - x$, we have

$$R(x) = (18 - x)x, \qquad 0 \le x \le 10.$$

The profit, $P(x)$, is the difference between the total revenue and the cost. Therefore,

$$\begin{aligned} P(x) &= R(x) - C(x) \\ &= (18 - x)x - (2x^2 + 8) \\ &= -3x^2 + 18x - 8, \quad 0 \le x \le 10. \end{aligned}$$

Since $P(x)$ must have an absolute maximum value in the interval $0 \leq x \leq 10$, this maximum must occur either at one of the endpoints $x = 0$, $x = 10$, or where $P'(x) = 0$. Now

$$P'(x) = -6x + 18$$

is zero only when $x = 3$. Consequently P must have an absolute maximum at $x = 0$, $x = 3$, or $x = 10$. Since $P(0) = -8$, $P(3) = 19$, and $P(10) = -128$, we see that 3 units should be produced per hour to yield a maximum profit of 19 dollars. ||

Exercises 3.7

1. If the sum of two numbers is 30, maximize their product.
2. Find the dimensions of the rectangular area containing 10,000 square meters that can be enclosed with the least amount of fencing.
3. Twelve hundred meters of fencing are used to enclose a rectangular area and to divide it into two equal areas, as shown in Figure 3.7.2. Find the dimensions so that the total area enclosed is the greatest.

Figure 3.7.2

4. An open-topped box is to be constructed by cutting equal squares from each corner of a square sheet of aluminum and folding up the sides. If the sheet has side length of 3 feet, find the volume of the largest such box.
5. A grapefruit grower finds that a grapefruit tree produces 200 grapefruit per year if no more than 16 trees are planted per acre. For each additional tree planted per acre he finds that the yield per tree decreases by 10 grapefruit per year. How many trees per acre should he plant to get the maximum number of grapefruit?
6. Postal regulations require that the sum of the length and girth (perimeter of the smallest cross section) of a package to be sent by parcel post must not exceed 78 inches. What are the dimensions of the largest mailable rectangular package with square cross section?
7. Following the restrictions set forth in the preceding exercise, find the dimensions of the largest mailable cylindrical package having a circular cross section.
8. A wire 36 cm in length is cut into two parts, one of which has length x, $1 \leq x \leq 35$. One piece is bent into a circle, the other into a square. How should the wire be cut (a) to maximize total area enclosed? (b) to minimize the total area?
9. If the wire of Exercise 8 is cut and formed into two squares, where should the cut occur (a) to maximize the enclosed area? (b) to minimize the enclosed area?
10. A v-shaped trough 8 feet long is made from a rectangular sheet of aluminum 20 inches wide by bending it down the middle. Find the width across the top of the trough in order that it have maximum capacity.
11. A herpetologist is designing a rectangular experimental iguana pen to fit the $8' \times 20'$ corner of a Komodo dragon pen as shown in Figure 3.7.3. If only 42 feet of fencing are available for the iguana pen, what dimensions should be used to obtain the largest area for the new pen?

Figure 3.7.3

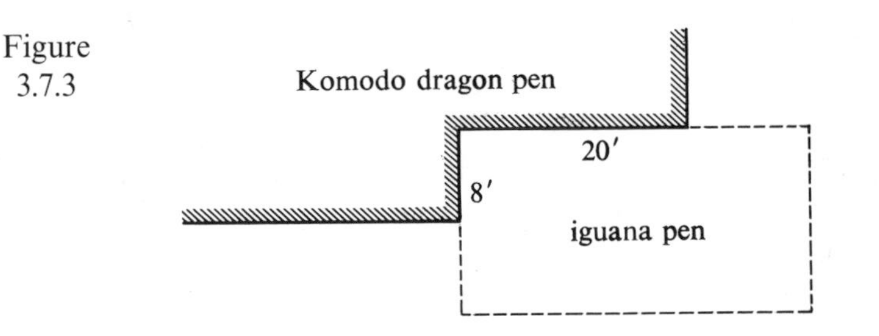

12. Prove that the shortest distance from a point to a straight line is along a perpendicular. (*Hint*: Choose a coordinate system that simplifies the equations of the straight line and the point.)

13. Often equipment maintenance and operating costs C are related to the overhaul-interval x in months by the equation $C(x) = (a/x) + kx$, where a and k are positive constants. Find the optimal overhaul-interval.

14. If the total cost of producing x units of a commodity is given by

$$C(x) = x^3 - 100x + 1500, \qquad x \geq 1,$$

find the value of x that minimizes this cost, and determine the minimal cost.

15. Each month a certain company produces x items at a cost of $(x/2) + 400$ dollars and the price per item at which they can market them is $3 - (x/20{,}000)$ dollars. How many items should the company produce per month to get the greatest profit?

16. If a tax of 10 cents per item is later imposed on the company mentioned in Exercise 15, how much of the tax should the company pass on to the consumer?

17. Assuming that the strength of a rectangular beam varies as the product of the width and the square of the depth, find the dimensions of the strongest rectangular beam that can be cut from a cylindrical log of radius r.

18. A rectangular sheet of paper contains 88 square inches. The margins at the top and bottom are 2 inches and those at the sides are 1 inch. What are the dimensions of the maximum printed area?

19. The space within a 1-km race track is to consist of a rectangle with a semi-circular area at each end. To what dimensions should the track be built in order to maximize the area of the rectangle?

20. A closed cylindrical canister of volume 8π ft^3 is to be made. It must stand between two partitions 8 feet apart in a room with an 8 foot ceiling. If the cost of the top and bottom is \$1/ft^2 and the cost of the lateral surface is \$2/ft^2, find the dimensions of (a) the least expensive canister, and (b) the most expensive canister.

21. A company has 10 duplicating machines, each capable of making 5000 copies per hour. If it costs the company \$1 to make each master copy, and the cost of running n duplicators together is $5 + .03n$ dollars per hour, how many master copies should be made to minimize the cost of making 81,000 copies of a single-page form?

22. A tavern offers dinner and entertainment to an organization for \$6 per person if at least 100, but no more than 300, people attend, with the provision that the price per person will be reduced 20 cents for each 10 people in excess of 100 that come. What number x, $100 \leq x \leq 300$, of people in attendance will maximize the total revenue of the tavern?

23. A rectangle with sides parallel to the coordinates axes is to be inscribed in the region bounded by $y = x^2$, and $y = 4$. Find the dimensions of the rectangle with maximum perimeter.

24. If the velocity v of air through a bronchial tube of radius r under pressure is

$$v = kr^2(a - r),$$

where a is the radius when no pressure applied and k is a constant, find the bronchial-tube radius that will provide the maximum air velocity.

25. The material in the aluminum top of a beer can costs \$.0002 per square inch, while the steel for the lateral surface and bottom costs \$.0001 per square inch. If the can is to hold twelve ounces ($1.8 \text{ in}^3 = 1$ oz) find the dimensions of the least expensive can.

26. An open-topped box is to be constructed by cutting equal squares from each corner of a a 3 ft by 8 ft rectangular sheet of aluminum and folding up the sides. Find the volume of the largest such box.

27. Four hundred meters of fencing is to be added to an existing fence as shown in Figure 3.7.4. How should the fence be located to maximize the enclosed area?

Figure 3.7.4

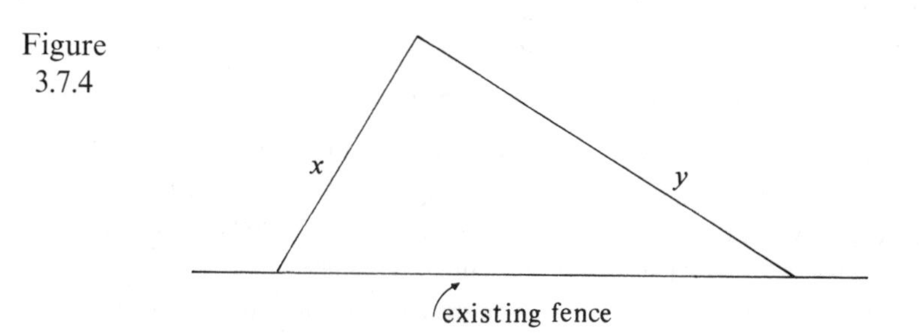

28. A farmer has 5000 meters of fencing with which he is going to fence an area in the shape of a sector of a circle. What is the maximum area he can fence? (The sector is not to be an entire circle.)

29. The method of *least squares* is often used to "fit" a particular type of equation $y = f(x)$ to points $(a_1, b_1), (a_2, b_2), \ldots, (a_m, b_m)$ obtained from experimental data. In this method one selects a function $f(x)$ that minimizes the sum

$$\sum_{i=1}^{m} [b_i - f(a_i)]^2 = [b_1 - f(a_1)]^2 + [b_2 - f(a_2)]^2 + \cdots + [b_m - f(a_m)]^2.$$

Use the method of least squares to find the value of k for which the equation $y = kx + 1$ best "fits" the points (2, 3), (4, 5), (6, 8).

3.8 Implicit Differentiation

Many applications give rise to equations involving both x and y, that is equations expressible in the form $F(x, y) = 0$. For example, an equation of the unit circle is $x^2 + y^2 - 1 = 0$, and the pressure and volume of an ideal gas at constant temperature are related by $xy - k = 0$, where x is pressure, y is the volume, and k is a constant. In considering such equations several questions arise quite naturally.

1 Does the equation $F(x, y) = 0$ always determine y as a function of x, that is, is it always the case that $y = f(x)$?

2 Can one always solve for y and so find the function $y = f(x)$?

We already have an example to indicate that the answer to question (1) is *no*. For example, the graph of $x^2 + y^2 - 1 = 0$ is a circle of radius 1 with center at the origin. Thus every x value, $-1 < x < 1$, determines not one but two y values:

$$y = \pm\sqrt{1 - x^2}.$$

Hence y is not a function of x. Note, however, that a suitable restriction on y, say $y > 0$, together with the equation $x^2 + y^2 - 1 = 0$, does determine y as a function of x, namely,

$$y = \sqrt{1 - x^2}.$$

Question (1) is thus more fruitfully phrased as "Can x and y be suitably restricted to lie in certain intervals so that $F(x, y) = 0$ determines y as a function of x?" Figure 3.8.1 indicates a possible graph of $F(x, y) = 0$. The graph of a function $y = f(x)$ determined by restricting y between c and d and x between a and b is illustrated as the solid curve. In the remainder of this text we shall assume a fact that is proved in most advanced calculus texts: in most (but not all) instances x and y can be restricted to lie in certain intervals so that $F(x, y) = 0$ does determine y as a function of x.

Figure 3.8.1

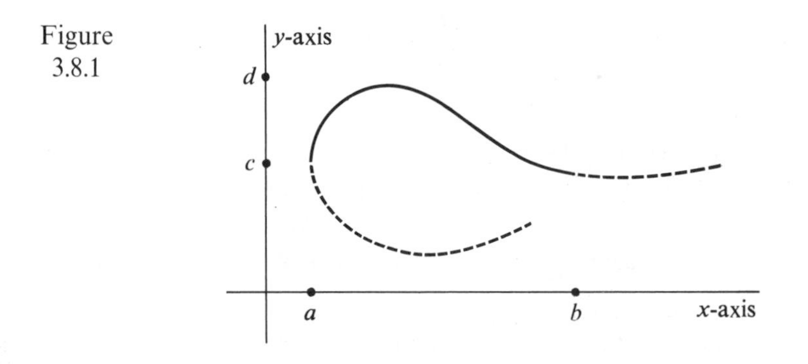

The second question is much easier; a flat *no* is sufficient. Again, however, we must rely on more advanced texts to assure us that equations involving powers of y greater than or equal to 5 are not always solvable for y in terms of x. For example, even though with suitable restrictions on x and y the equation $2y^5x^3 + 3yx^5 + 7 = 0$ does determine y as a function of x, $y = f(x)$, it is *not* possible to "solve for y" in order to find $y = f(x)$.

The negative answer to Question (2) leads to another interesting question. When $F(x, y) = 0$ does determine y as a function of x, is it possible to find $f'(x)$, even when we can't solve for y? Remarkably enough the answer is *yes*. Implicit differentiation enables us to find $f'(x)$ rather easily. For example, though we can't solve

$$2y^5x^3 + 7y^2 + 3yx^5 + 7 = 0$$

for $y = f(x)$, we can assume that y is a differentiable function of x. Then substitution of the unknown function $y = f(x)$ gives

$$2(f(x))^5x^3 + 7(f(x))^2 + 3f(x)x^5 + 7 = 0.$$

Differentiation of both sides produces

$$\frac{d}{dx}[2(f(x))^5x^3 + 7(f(x))^2 + 3f(x)x^5 + 7] = \frac{d0}{dx}, \quad \text{or}$$

$$2\frac{d(f(x))^5x^3}{dx} + 7\frac{d(f(x))^2}{dx} + 3\frac{df(x)x^5}{dx} = 0.$$

Then use of the product rule yields

$$2\left[x^3\frac{d(f(x))^5}{dx} + (f(x))^5\frac{dx^3}{dx}\right] + 7\frac{d(f(x))^2}{dx} + 3\left[x^5\frac{df(x)}{dx} + f(x)\frac{dx^5}{dx}\right] = 0.$$

By the chain rule we know that

$$\frac{d(f(x))^n}{dx} = n(f(x))^{n-1}\frac{df(x)}{dx}.$$

Thus we obtain

$$2\left[5x^3(f(x))^4\frac{df(x)}{dx} + (f(x))^5 3x^2\right] + 14f(x)\frac{df(x)}{dx} + 3\left[x^5\frac{df(x)}{dx} + 5f(x)x^4\right] = 0,$$

or

$$[10x^3(f(x))^4 + 14f(x) + 3x^5]\frac{df(x)}{dx} + 6x^2(f(x))^5 + 15x^4f(x) = 0.$$

We may now solve for $df(x)/dx$:

$$\frac{df(x)}{dx} = -\frac{6x^2(f(x))^5 + 15x^4f(x)}{10x^3(f(x))^4 + 14f(x) + 3x^5} = -\frac{6x^2y^5 + 15x^4y}{10x^3y^4 + 14y + 3x^5}.$$

Note that in order to evaluate $\frac{df(x)}{dx}$ at a particular value of x we must also know the value of y associated with that value of x.

Usually implicit differentiation is done without substitution of $f(x)$ for y. It is merely *assumed* throughout the problem that y *is* a differentiable function of x. Generally, also, it is simpler to use the symbol y' in place of dy/dx. In particular, when a term like $6x^4y^3$ appears in an equation to be differentiated implicitly, we differentiate it to get

$$24x^3y^3 + 18x^4y^2y'.$$

Note that both the product rule and the chain rule were used to get this derivative. (The chain rule was used on y^3 to get $3y^2y'$.) One of the commonest mistakes made in this sort of situation is to omit the y' that the chain rule requires.

Example 1 | Find y' if $x^4 + y^4 = 2$. Also, compute y' at the points (1, 1) and (1, −1).

Differentiating, treating y as a function of x, gives

$$4x^3 + 4y^3y' = 0.$$

Thus

$$y' = -\frac{x^3}{y^3}.$$

At $(1,\ 1)$, $y' = -1$; and at $(1,\ -1)$, $y' = 1$. Figure 3.8.2 shows the graph of $x^4 + y^4 = 2$. ||

Figure 3.8.2

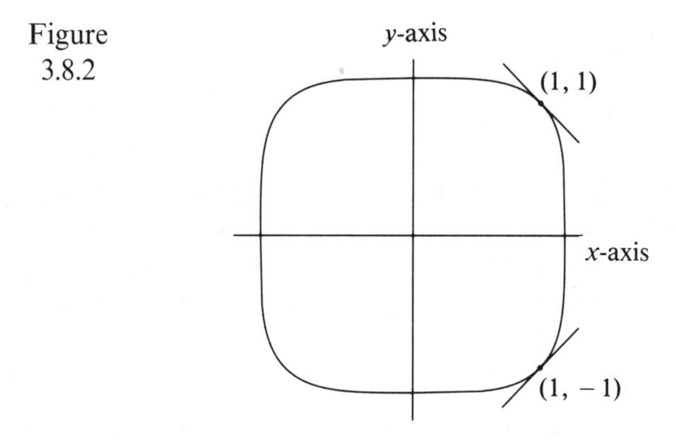

Example 2 | Find y' if $y^2 + x^3y^5 + x = 1$.

Differentiation, treating y as a function of x, yields

$$\frac{d}{dx}(y^2 + x^3y^5 + x) = \frac{d1}{dx}.$$

Hence, being careful to use the chain rule on powers of y, we get

$$2yy' + 3x^2y^5 + 5x^3y^4y' + 1 = 0.$$

Solving for y' gives

$$y' = -\frac{3x^2y^5 + 1}{2y + 5x^3y^4}.$$

||

Example 3 | Find the slopes of the lines tangent to the graph of

$$x^{-2}y^6 + 2y^{-2} - 6 = 0$$

at the point $(1/2,\ 1)$ and at the point $(1/2,\ -1)$.

The slopes of the tangent lines are the values of $dy/dx = y'$ at $(1/2,\ 1)$ and at $(1/2,\ -1)$. Now

$$\frac{d}{dx}(x^{-2}y^6 + 2y^{-2} - 6) = 0, \quad \text{and so}$$

$$6x^{-2}y^5y' - 2x^{-3}y^6 - 4y^{-3}y' = 0.$$

Hence, on solving for y', we have

$$y' = \frac{x^{-3}y^6}{3x^{-2}y^5 - 2y^{-3}}.$$

Thus, when $x = 1/2$ and $y = 1$, we get $y' = 4/5$; and when $x = 1/2$ and $y = -1$, we get $y' = -4/5$. Hence the slope of the tangent at the point $(1/2, 1)$ is $4/5$; and the slope of the tangent at the point $(1/2, -1)$ is $-4/5$. Note that both of these distinct tangent lines occur for the same value of x. ||

In our previous work we found that

$$\frac{dx^n}{dx} = nx^{n-1}, \qquad \text{for all integers } n.$$

We can now use implicit differentiation to prove that the same formula holds even when n is rational. Of course when n is a rational number, x^n is not necessarily a real number for all values of x. To avoid excessive detail in the statement of the next theorem we tacitly assume that it applies only for the appropriate cases where x^n is real.

Theorem 3.8.1

If n is a rational number, then $dx^n/dx = nx^{n-1}$.

Proof. Consider any rational number n, where $n = p/q$ and p and q are integers with $q \neq 0$. Let $y = x^{p/q}$. Then

$$y^q = x^p.$$

Since p and q are both integers, we can differentiate both sides with respect to x to get

$$qy^{q-1}\frac{dy}{dx} = px^{p-1}, \quad \text{and so}$$

$$\frac{dy}{dx} = \frac{px^{p-1}}{qy^{q-1}}.$$

Substitution of $y = x^{p/q}$ then produces

$$\frac{dx^{p/q}}{dx} = \frac{p}{q}\frac{x^{p-1}}{(x^{p/q})^{q-1}}$$

$$= \frac{p}{q}x^{p-1-(p-p/q)}$$

$$= \frac{p}{q}x^{-1+(p/q)}.$$

Then since $n = p/q$, we finally have the desired result:

$$\frac{dx^n}{dx} = nx^{n-1},$$

for every rational number n.

Example 4 Find dy/dx if $y = x^{-3/5} + x^{4/5}$.

$$\frac{dy}{dx} = -\frac{3}{5}x^{(-3/5)-1} + \frac{4}{5}x^{(4/5)-1}$$

$$= -\frac{3}{5}x^{-8/5} + \frac{4}{5}x^{-1/5}. \quad \|$$

Example 5 Find where the graph of $f(x) = x^{2/3}$ is increasing, decreasing, concave upward, and concave downward. Use that information to sketch the graph of the equation and to determine all local extrema.

Since $f'(x) = \frac{2}{3}x^{-1/3} = \frac{2}{3\sqrt[3]{x}}$, we see that $f'(x) > 0$ for $x > 0$, $f'(x) < 0$ for $x < 0$, and the derivative does not exist when $x = 0$. Consequently f increases for $x > 0$, and decreases for $x < 0$.

Since
$$f''(x) = -\frac{2}{9}x^{-4/3} = -\frac{2}{9}(\sqrt[3]{x})^{-4},$$

$f''(x) < 0$ for all $x \neq 0$. Thus, the graph of x is concave downward for all values of x except $x = 0$. Use of this information and the two indicated points allows us to sketch the graph as in Figure 3.8.3. Since $f'(x)$ does not exist at $x = 0$, $x = 0$ is the only critical point. By the First Derivative Test (0, 0) is a local minimum point of f. In fact, since $f'(x) > 0$ for $x > 0$ and $f'(x) < 0$ for $x < 0$, (0, 0) is an absolute minimum. $\|$

Figure 3.8.3

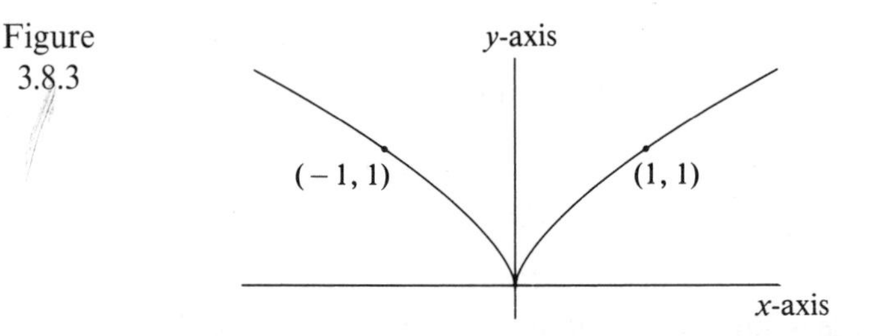

Use of implicit differentiation often reduces the manipulation necessary for the solution of applied maximum-minimum problems. The next example illustrates that fact.

Example 6 A 10 ft^3 oxygen tank is to be constructed in the form of a right circular cylinder. Find the dimensions that minimize the total surface area.

Let h and r denote the height and radius of the right circular cylinder.

Then
$$\pi r^2 h = 10$$

and the total surface area of the tank is

$$S = 2\pi r^2 + 2\pi rh.$$

Differentiation of the first of those equations with respect to r, treating h as a function of r, gives

$$2\pi rh + \pi r^2 \frac{dh}{dr} = 0.$$

Thus,
$$\frac{dh}{dr} = -\frac{2h}{r}.$$

Differentiation of the second equation yields

$$\frac{dS}{dr} = 4\pi r + 2\pi h + 2\pi r \frac{dh}{dr},$$

But
$$\frac{dh}{dr} = -\frac{2h}{r}.$$

Hence
$$\frac{dS}{dr} = 4\pi r + 2\pi h - 4\pi h.$$

We are looking for those values of r and h for which $\frac{dS}{dr} = 0$. That is,

$$4\pi r - 2\pi h = 0.$$

Thus, $h = 2r$. Substitution into $\pi r^2 h = 10$ gives

$$2\pi r^3 = 10.$$

Consequently $r = \sqrt[3]{5/\pi}$ and $h = 2\sqrt[3]{5/\pi}$.

It only remains to show that a cylinder having these dimensions has minimal surface area.

Since
$$\frac{dS}{dr} = 4\pi r - 2\pi h = 4\pi\left(r - \frac{h}{2}\right),$$

we see that

$$\frac{dS}{dr} < 0 \text{ if } r < \frac{h}{2} \quad \text{and} \quad \frac{dS}{dr} > 0 \text{ if } r > \frac{h}{2}.$$

Consequently the absolute minimum of S occurs when $r = h/2$ or $h = 2r$. ||

Exercises 3.8

In Exercises 1–16, find y' by using implicit differentiation.

1. $x^3 + y^5 = 6$.
2. $3x^2y^6 + y^5 + 2 = 0$.
3. $x^2y^2 + 3xy^3 = 7$.
4. $y^7 - 2x^7y^5 = 10$.
5. $\sqrt{xy} = 2 + x^2y^3$.
6. $3x^2y^{2/3} + x^{1/2}y^7 = 14$.
7. $x^2y^{-3} + 3 = y$.
8. $xy^{7/8} + x^{7/8}y = 3$.
9. $x^{2/5} + y^{3/5} = x$.
10. $(x + y)^{2/3} + xy = 0$.
11. $(x^4 - y^5)^7 = (x^3 + y)^5$.
12. $y = (xy - 3x^2y^5)^4$.
13. $(x^2 + y^2)^{-3} = (x^2y^4 - 1)^2$.
14. $(x^{-3} - y^{-2})^2 + xy^4 = 8$.
15. $(x^3y^{-2} + xy^2)^3(xy + 2)^2 = 2$.
16. $(2x^2y^4 - x^3y)(x^{-2}y + 1)^3 = 5$.

17. (a) Show that if $3x^2 - y^2 = 7$ then $y' = 3x/y$.
    (b) Differentiate $y' = 3x/y$ to get $y'' = (3y - 3xy')/y^2$.
    (c) Replace y' by $3x/y$ to show that $y'' = -3(3x^2 - y^2)/y^3$.
    (d) Use $3x^2 - y^2 = 7$ to get $y'' = -21/y^3$.

In Exercises 18–23, use the technique of Exercise 17 to find y''.

18. $x^2 - 3y^2 = 5$.
19. $x^2 + 5y^2 = 7$.
20. $xy = 6$.
21. $xy^2 = 8$.
22. $x^2y = 7$.
23. $x^2y^2 = 17$.
24. Find an equation of the tangent to the graph of $x^2y^4 + x^3y^5 = 2$ at $(1, 1)$.
25. Find an equation of the tangent to the graph of $y^2 = x^2y^3 - 7x + 19$ at $(1, -2)$.
26. Find an equation of the tangent to the graph of $y^6 + 2xy^2 + 3 = 0$ at $(-2, 1)$.
27. Find an equation of the line perpendicular to the tangent to the graph of $x^3 + y^3 = 8$ at $(2, 0)$.
28. Find an equation of the line perpendicular to the tangent to the graph of $(x + y)^{4/3} + xy = 1$ at $(0, 1)$.
29. Find the absolute maximum and minimum of $f(x) = x^{4/5} - x^{-4/5}$ for $-1 \le x \le 1$.
30. Find the absolute maximum and minimum of $f(x) = (x - 2)^{2/3}$ for $0 \le x \le 10$.

In Exercises 31–34, use implicit differentiation to solve the given problem.

31. Show that a rectangle with fixed perimeter has the shortest diagonal when the rectangle is a square.
32. An open trough is to be constructed in the form of a half-cylinder. If the trough is to contain 300 liters of water, what dimensions minimize the total surface area?
33. A conical tent must contain 40π ft^3. Compute the height and radius of the tent with minimal total surface area. (Include the floor material.)
34. Find the height and radius of the cone with largest surface area that can be inscribed in a unit sphere.
35. As shown in Figure 3.8.4 a fence of length L is to be used to establish a storage area in corner of an existing fenced field. Find the values of x and y that maximize the enclosed area.

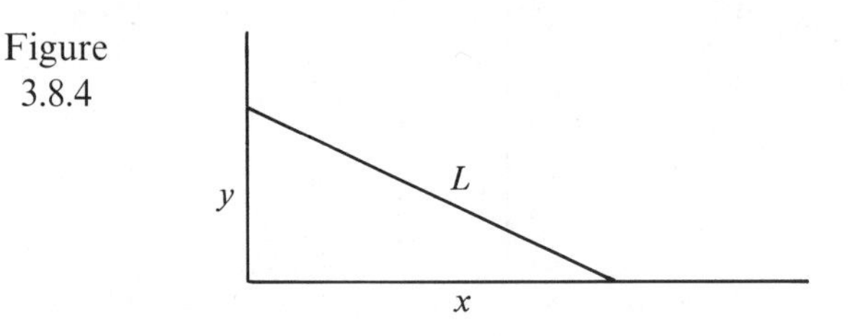

Figure 3.8.4

36. Four hundred meters of fencing is to be added to an existing fence as shown in Figure 3.8.5. Find the value of x that will maximize the enclosed area?

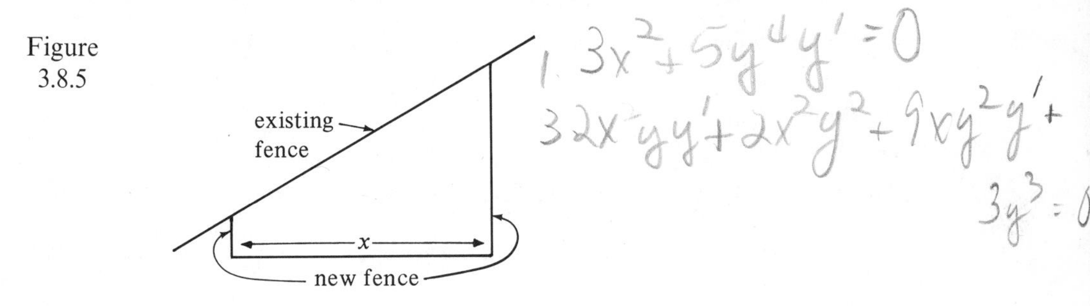

Figure 3.8.5

37. A natural gas line is to be extended to a house 600 feet down a straight road from the present end of the pipe. If the house is set back 300 feet from the road and the cost of laying the line is \$.80 per foot along the road and \$1 per foot through the field next to the road, where should the gas line be set to minimize the cost?

38. Use the chain rule to prove Theorem 3.8.1 as follows:
   (a) Let $y = x^p = (x^{p/q})^q = u^q$ where $u = x^{p/q}$.
   (b) Differentiate to get $dy/dx = px^{p-1} = qu^{q-1}(du/dx)$.
   (c) Solve for du/dx to get $(d/dx)(x^{p/q}) = (p/q)x^{p/q-1}$.

3.9 Differentials

When $y = g(x)$, we have used the symbol $\dfrac{dy}{dx}$ to denote the derivative $g'(x)$. Thus $\dfrac{dy}{dx}$ was defined as a single entity and was not defined as the quotient of the symbols dy and dx. In fact, the symbols dy and dx were not defined as separate entities themselves. We shall now define dy and dx in such a way that our notation is consistent; that is, so that dy divided by dx is equal to the derivative $g'(x)$.

We consider the graph of $y = g(x)$ near a point $(x, g(x))$ on the graph, as shown in Figure 3.9.1. We let dx denote a change in the horizontal coordinate. That change can be positive, negative, or zero. The figure shows a case where dx is positive. Note that $g'(x)$ is the slope of the tangent line to the graph of $y = g(x)$ at the point $(x, g(x))$. Thus $g'(x)dx$ is the vertical change in that tangent line when the horizontal change is dx.

Figure 3.9.1

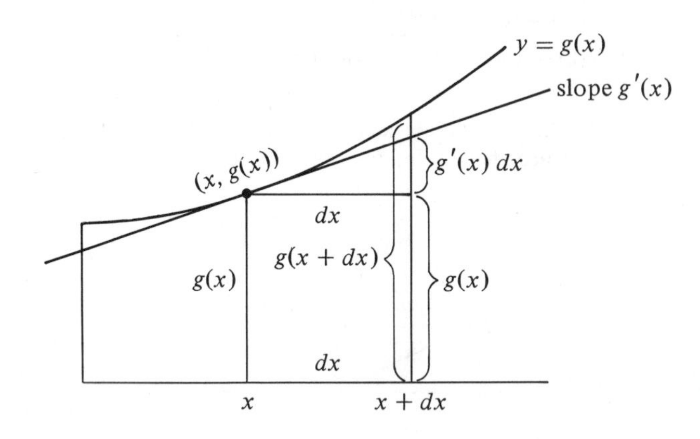

Thus if we let dy denote $g'(x)dx$, then $\dfrac{dy}{dx} = g'(x)$ if $dx \neq 0$.

The following definition summarizes the essential parts of our discussion.

Definition 3.9.1

> Let $y = g(x)$ be a differentiable function of x, and let dx, the *differential* of x, be any real number (positive, negative, or zero). Then the *differential* of $g(x)$, denoted by dy, is given by $dy = g'(x)dx$.

Note that the differential dy depends on both x and dx and hence both x and dx must be given numerical values to determine a numerical value of dy.

Example 1 | Let $g(x) = (x^2 + 3x)^2$.
i) Express dy in terms of x and dx.
ii) Compute the value of dy when $x = 2$ and $dx = .05$.

Since $$g'(x) = 2(x^2 + 3x)(2x + 3),$$

we have $$dy = 2(x^2 + 3x)(2x + 3)dx.$$

If $x = 2$ and $dx = .05$, then

$$dy = 2(10)(7)(.05) = 7. \quad ||$$

In addition to illustrating the geometric significance of the differential, Figure 3.9.1 gives us a powerful approximation technique. Note that dy is approximately the change in $y = g(x)$ between x and $x + dx$.

Thus $$g(x + dx) - g(x) \approx dy. \tag{1}$$

The next example shows a possible use of that approximation.

Example 2 | A square metal plate is expanding due to being heated by the sun. Approximate the change in area as the side length changes from 30″ to 30.1″.

Let $y = g(x)$ be the area of a square of side length x. Then since $g(x) = x^2$, we have $dy = 2x\,dx$ and so,

$$g(x + dx) - g(x) \approx 2x\,dx.$$

If we let $x = 30$ and $dx = 0.1$, we get

$$g(30.1) - g(30) \approx 2(30)(.1) = 6 \text{ square units.} \quad ||$$

Approximation (1) can be written in the form

$$g(x + dx) \approx g(x) + dy,$$

or

$$g(x + dx) \approx g(x) + g'(x)dx. \tag{2}$$

The next two examples illustrate possible uses of that approximation.

Example 3 | Use (2) to approximate $g(2.05)$ where $g(x) = (x^2 + 3x)^2$.

Since $g'(x) = 2(x^2 + 3x)(2x + 3)$, equation (2) gives

$$g(x + dx) = g(x) + 2(x^2 + 3x)(2x + 3)dx.$$

Thus, if we take $x = 2$ and $dx = .05$, we get

$$g(2.05) \approx 10^2 + 2(10)(7)(.05) = 107.$$

Example 4 | Approximate $\sqrt{101}$ by use of (2).

In this case we take $g(x) = \sqrt{x}$. Then since $g'(x) = (1/2)x^{-1/2}$, we have

$$g(x + dx) \approx \sqrt{x} + \frac{1}{2\sqrt{x}}\, dx.$$

Then, if $x = 100$ and $dx = 1$, we obtain

$$\sqrt{101} \approx 10 + \frac{1}{20} = 10.05. \qquad \|$$

In Chapter 9 we shall consider the accuracy of the approximations introduced in this section.

Exercises 3.9

In Exercises 1–11, express dy in terms of x and dx.

1. $y = 7x^2 - 3x$.
2. $y = 3x - 7x^3$.
3. $y = (x^2 - 3)^2$.
4. $y = (x^3 - 5x^2)^3$.
5. $y = \dfrac{x^3 - 3x}{x^2 + 2}$.
6. $y = (x^2 + 7)^3(x - 1)^2$.
7. $y = (x + 3)^{-2}(2x - 4)^3$.
8. $y = \dfrac{x^{-3} + 7x}{x^4 - x^{-2}}$.
9. $y = x^{7/2} - 3x^{1/3}$.
10. $y = x^{4/3} + x^{1/2}$.
11. $y = (x^2 - 7x^3)^{4/5}$.

In Exercises 12–17, use differentials to approximate the indicated functional value.

12. $g(x) = \sqrt{x^2 - 9}$, $g(5.1)$.
13. $g(x) = 3x - 5x^{-3}$, $g(1.06)$.
14. $g(x) = x^{5/2}$, $g(101)$.
15. $g(x) = \sqrt{3 + x^2}$, $g(1.01)$.
16. $g(x) = \sqrt{9 - x^2}$, $g(.3)$.
17. $g(x) = (x^3 - 2)^2$, $g(.1)$.
18. Use differentials to approximate the change in $f(x) = x^{1/2}$ as x increases from 9 to 9.001.
19. Use differentials to approximate the change in $f(x) = (x + 3)^{1/3}$ as x decreases from $x = 24$ to $x = 23.95$.
20. Use differentials to approximate the change in the volume of a cube as its side length decreases from 5.02 cm to 5 cm.
21. Use differentials to approximate the change in the surface area of a sphere as its radius increases from 2 cm to 2.003 cm.
22. Use differentials to approximate the change in the volume of a sphere as its radius increases from 5 cm to 5.03 cm.

3.10 Differentiation; The Derivatives of the Trigonometric Functions†

Up to this point, we have only been able to apply the fundamental techniques and rules of differentiation to functions based on power functions. Thus we have

† The remainder of the text (except for Section 4.9) is independent of this section. This section is included for those classes that prefer an early introduction to differentiation of the trigonometric functions.

not been able to get as much practice in finding deri
derivative formulas for a wider variety of functions.
nity for more practice and to get an early introduct
the trigonometric functions discussed in Chapter
some of the basic differentiation formulas for those

Also, in previous sections of this chapter, pract
in another way because the reader usually knew wh
each exercise and the only challenge was to apply
exercises will contain the additional challenge of de
formula to apply.

The most fundamental differentiation technique
of derivative. That technique however can lead to
Consequently the basic rules that we have developed should be used where possible. Here is a list of those techniques or rules.

Differentiation Techniques or Rules

1. The rule for differentiating a constant.
2. The rule for differentiating a constant times a function.
3. The rule for differentiating a sum or difference.
4. The product rule.
5. The quotient rule.
6. The power function rule.
7. The chain rule.
8. Implicit differentiation.

Before proceeding we need some underlying information. First, to make the differentiation formulas simpler, *all angles will be measured in radians.* Also, we shall use the trigonometric identity.

$$\sin(\alpha + \beta) - \sin(\alpha - \beta) = 2 \cos\alpha \sin\beta, \tag{1}$$

and we shall use the fact that if h is in radians, then

$$\lim_{h \to 0} \frac{\sin(h/2)}{h/2} = 1. \tag{2}$$

Item (1) can be obtained by subtracting these basic identities:

$$\sin(\alpha + \beta) = \sin\alpha \cos\beta + \cos\alpha \sin\beta,$$

$$\sin(\alpha - \beta) = \sin\alpha \cos\beta - \cos\alpha \sin\beta.$$

Item (2) comes from the result indicated in Example 11 of Section 2.2 (and proved in Section 7.1), that if x is in radians then

$$\lim_{x \to 0} \frac{\sin x}{x} = 1.$$

Using that result, with x replaced by $h/2$, we get

$$\lim_{h \to 0} \frac{\sin(h/2)}{h/2} = \lim_{h/2 \to 0} \frac{\sin(h/2)}{h/2} = 1.$$

ctly from the definition of derivative, we obtain

$$\frac{d}{dx}\sin x = \lim_{h\to 0}\frac{\sin(x+h)-\sin x}{h}.$$

We let $\alpha = x + h/2$ and $\beta = h/2$. Then $\alpha + \beta = x + h$ and $\alpha - \beta = x$, and identity (1) becomes

$$\sin(x+h) - \sin x = 2\cos\left(x + \frac{h}{2}\right)\sin\left(\frac{h}{2}\right).$$

Therefore

$$\begin{aligned}\frac{d}{dx}\sin x &= \lim_{h\to 0}\frac{2\cos(x+h/2)\sin(h/2)}{h}\\ &= \lim_{h\to 0}\frac{\cos(x+h/2)\sin(h/2)}{h/2}\\ &= \lim_{h\to 0}\cos\left(x+\frac{h}{2}\right)\lim_{h\to 0}\frac{\sin(h/2)}{h/2}.\end{aligned}$$

But since the cosine function is continuous,

$$\lim_{h\to 0}\cos\left(x + \frac{h}{2}\right) = \cos x.$$

Thus, also using item (2), we get

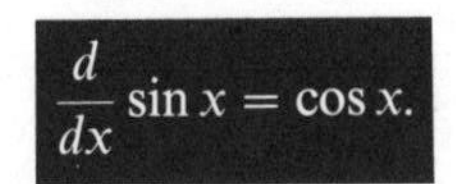

$$\frac{d}{dx}\sin x = \cos x.$$

More generally, if u is a differentiable function of x, the chain rule produces

$$\frac{d}{dx}\sin u = \cos u\,\frac{du}{dx}.$$

Example 1 $\frac{d}{dx}\sin(x^2-5) = \cos(x^2-5)\frac{d}{dx}(x^2-5) = 2x\cos(x^2-5).$

With practice, one usually thinks but does not write the first step and simply writes $\frac{d}{dx}\sin(x^2-5) = 2x\cos(x^2-5)$. ||

Example 2 Find the derivative of $\sin^3 x$.

Use of the power function rule gives

$$\frac{d}{dx}(\sin^3 x) = \frac{d}{dx}(\sin x)^3 = 3\sin^2 x\,\frac{d}{dx}(\sin x) = 3\sin^2 x\cos x.$$ ||

Example 3 | Find $f'(x)$ if $f(x) = x^4 \sin x$.

Use of the product rule gives

$$f'(x) = x^4 \cos x + 4x^3 \sin x. \qquad \|$$

Since $\cos x = \sin(\pi/2 - x)$ and $\cos(\pi/2 - x) = \sin x$, we have

$$\frac{d}{dx}\cos x = \frac{d}{dx}\sin\left(\frac{\pi}{2} - x\right) = (-1)\cos\left(\frac{\pi}{2} - x\right) = -\sin x.$$

That is,

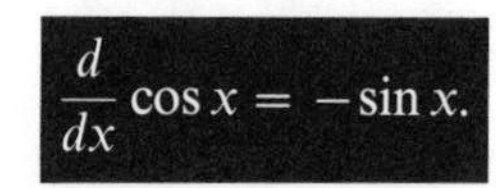

$$\frac{d}{dx}\cos x = -\sin x.$$

More generally, if u is a differentiable function of x, the chain rule gives us

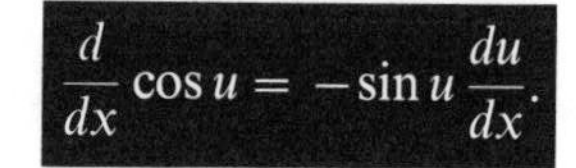

$$\frac{d}{dx}\cos u = -\sin u \frac{du}{dx}.$$

Example 4 | Find $f'(x)$ if $f(x) = \cos(x^2 + 3x)$.

We obtain

$$f'(x) = -\sin(x^2 + 3x)\frac{d}{dx}(x^2 + 3x) = -(2x + 3)\sin(x^2 + 3x).$$

As before, one usually thinks but does not write the first step, and just writes $f'(x) = -(2x + 3)\sin(x^2 + 3x)$. ||

Example 5 | Calculate $\frac{d}{dx}\cos^5(x^3 + 9)$.

Use of the power function rule and the chain rule produces

$$\begin{aligned}\frac{d}{dx}\cos^5(x^3 + 9) &= 5\cos^4(x^3 + 9)\frac{d}{dx}\cos(x^3 + 9)\\ &= -15x^2\cos^4(x^3 + 9)\sin(x^3 + 9).\end{aligned} \qquad \|$$

Example 6 | Calculate y' if $y = \sin^2(8x)\cos^3(7x)$.

Use of the product rule, the power function rule, and the chain rule gives

$$\begin{aligned}y' &= \sin^2(8x) \cdot 3\cos^2(7x) \cdot (-7\sin 7x) + 2\sin(8x) \cdot 8\cos 8x\cos^3(7x)\\ &= -21\sin^2(8x)\cos^2(7x)\sin 7x + 16\sin(8x)\cos(8x)\cos^3(7x).\end{aligned} \qquad \|$$

Example 7 | Find $\frac{d}{dx}\csc x$ and $\frac{d}{dx}\csc u$, where u is a differentiable function of x.

Since $\csc x = \dfrac{1}{\sin x}$, we can use the quotient rule to get

$$\frac{d}{dx}\csc x = \frac{(\sin x)(0) - (1)(\cos x)}{\sin^2 x} = -\frac{1}{\sin x}\frac{\cos x}{\sin x} = -\csc x \cot x.$$

Now, using the chain rule, we get

$$\frac{d}{dx}\csc u = -\csc u \cot u \frac{du}{dx}.$$

||

Example 8 | Find $\dfrac{d}{dx}\csc(7x + x^3)$.

By the result of Example 7, we get

$$\frac{d}{dx}\csc(7x + x^3) = -(7 + 3x^2)\csc(7x + x^3)\cot(7x + x^3).$$

||

Example 9 | Find y' if $x^2 \sin y + 3 \cos xy = \cos 5x + 4$.

We use implicit differentiation to get

$$x^2(\cos y)y' + 2x \sin y - 3 \sin(xy)(xy' + y) = -5 \sin 5x.$$

Now, getting all terms containing y' on the left and the other terms on the right, we have

$$[x^2 \cos y - 3x \sin(xy)]y' = 3y \sin(xy) - 5 \sin 5x - 2x \sin y$$

So

$$y' = \frac{3y \sin(xy) - 5 \sin 5x - 2x \sin y}{x^2 \cos y - 3x \sin(xy)}.$$

||

The derivatives of the other trigonometric functions of a differentiable function u of x are as follows:

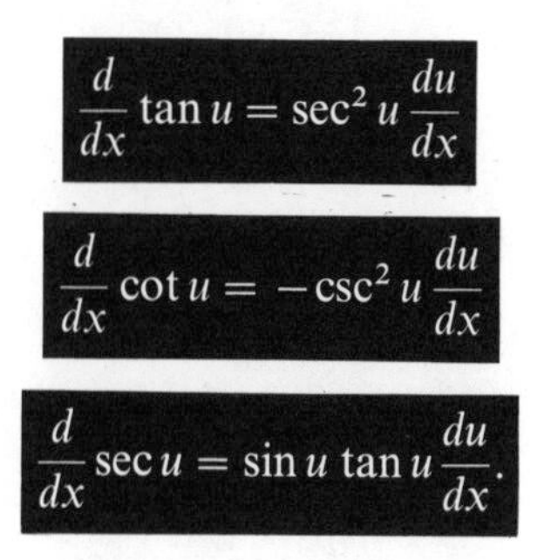

$$\frac{d}{dx}\tan u = \sec^2 u \frac{du}{dx}$$

$$\frac{d}{dx}\cot u = -\csc^2 u \frac{du}{dx}$$

$$\frac{d}{dx}\sec u = \sin u \tan u \frac{du}{dx}.$$

The proofs of those results are left to the exercises.

Exercises 3.10

Skip

Differentiate the functions in Exercises 1–26.

1. $f(x) = \sin(x^5 - 3x)$.
2. $f(x) = \cos(8x + 2)$.
3. $f(x) = x^3 \cos 7x$.
4. $f(x) = x^4 \sin 3x$.
5. $f(x) = \sin 2x \cos 3x$.
6. $f(x) = \sin 8x \cos 2x$.
7. $f(x) = \sin^3 5x \cos^4 6x$.
8. $f(x) = \sin^7 2x \cos^5 3x$.
9. $f(x) = \dfrac{\sin 9x}{\cos 6x}$.
10. $f(x) = \dfrac{\cos 8x}{\sin 4x}$.
11. $f(x) = \tan x = \dfrac{\sin x}{\cos x}$.
12. $f(x) = \cot x = \dfrac{\cos x}{\sin x}$.
13. $f(x) = \tan u$, where $u = g(x)$.
14. $f(x) = \cot u$, where $u = g(x)$.
15. $f(x) = \sec u = \dfrac{1}{\cos u}$, where $u = g(x)$.
16. $f(x) = \sin u \cos u$, where $u = g(x)$.
17. $f(x) = \tan(5 - x^3)$.
18. $f(x) = \sec(x^4 - 2)$.
19. $f(x) = \csc(x - x^3)$.
20. $f(x) = \cot(3x - 2)$.
21. $f(x) = \sec(4x)$.
22. $f(x) = \tan(9x)$.
23. $f(x) = \sec^2(5x)\tan^3(2x)$.
24. $f(x) = \csc^3(7x)\cot^4(6x)$.
25. $f(x) = \dfrac{\cot^3(7x)}{\sec^2(5x)}$.
26. $f(x) = \dfrac{\csc^9(2x)}{\tan^7(5x)}$.

For Exercises 27–32 find y'.

27. $\sin(xy) + \cos x = 7x - \tan y$.
28. $\tan(xy) - 8y = \sec(xy^2)$.
29. $y \cos 3x + 9x = x \sin 2y$.
30. $x \sec 5x = 4y - y \tan 8x$.
31. $\tan^3(x^2y^2) = 5 \sec(xy^3) + 8$.
32. $\sin^2(x^3y^5) + 8x = x \cos^3(3y)$.
33. Find an equation of the tangent to the graph of $y = x \sin x$ at $x = 7\pi/6$.
34. Determine where the graph of $y = \sin x + \cos x$, $-\dfrac{3\pi}{4} \leq x \leq \dfrac{5\pi}{4}$, is
    (a) increasing
    (b) decreasing
    (c) concave upward
    (d) concave downward.
    Also determine
    (e) the local maximum points
    (f) the local minimum points
    (g) the points of inflection
    and
    (h) sketch the graph.
35. What is the slope of the graph of $\sqrt{2}\cos(x + y) = 1$ at the point $(\pi/6, \pi/12)$?
36. Find the points of inflection of the graph of $y = \tan x$.
37. Differentiate both sides of the identity $\sin 2x = 2 \sin x \cos x$ to get an identity for $\cos 2x$.

38. The maximum height h attained by a projectile fired with initial velocity v_0 at an angle of θ to the horizontal is

$$h = \frac{v_0{}^2 \sin^2 \theta}{2g}$$

where g is the acceleration due to gravity. For what angle is h a maximum?

39. An object of weight W is being pulled along a horizontal plane at constant velocity. If the force has magnitude F and is directed at an angle θ to the horizontal then

$$F = \frac{\mu W}{\mu \sin \theta + \cos \theta}$$

where μ is the coefficient of friction. Show that F is a minimum when $\tan \theta = \mu$.

40. A fence 1 m high is 2 m from a wall. Find the length of the shortest beam that can pass over the fence to brace the wall.

41. Find the absolute maximum value of $\sin^2 x \cos x$, $0 \leq x \leq \pi/2$.

42. A line L through the point (0, 1) is rotating about the point (0, 1) at the rate of one revolution per minute in a clockwise direction. At what rate is the point of intersection of L and the x-axis moving along the x-axis when $x = 5$?

43. A v-shaped trough 4 m long is made from a rectangular sheet of aluminum 70 cm wide by bending it down the middle. For maximum capacity, what angle should be used between the sides of the trough?

44. A particle is moving in a plane in such a way that its coordinates are $x = \cos t$, $y = \sin 2t$ at an elapsed time of t seconds. If x and y are measured in meters, find the velocity at which the particle is moving toward the origin when $t = \pi/4$ seconds.

45. An advertising display item is to be made from a circular sheet of paper 40 cm in diameter by cutting out a sector of the circle and attaching the cut edges of the sector together to form a cone. Find the angle of the sector to be cut out that will produce the cone of maximum volume.

3.11 Velocity, Acceleration, and Related Rates

We recall from Section 2.6 that the derivative dy/dx is the rate of change of y with respect to x, and, in general, that the derivative of one quantity with respect to another quantity is the rate of change of the first quantity with respect to the second; that is, if $y = f(x)$, then the rate of change of y with respect to x is $f'(x)$. In this section we shall review and extend such applications of the derivative.

Recall that if a particle moves along a straight line and $s = f(t)$ gives the coordinate of the particle at time t, then the velocity $v(t)$ the particle at time t is

$$v(t) = f'(t).$$

Since *acceleration* is the rate of change of velocity with respect to time, the acceleration $a(t)$ of the particle at time t is

$$a(t) = v'(t) = f''(t).$$

Example 1 Find the velocity and acceleration of a particle that moves along a straight line in such a way that its coordinate $s(t)$ is

$$s(t) = t^2 + (2t - 3)^{1/2}.$$

Since the velocity is $v(t) = s'(t)$, we have

$$v(t) = \frac{d}{dt}[t^2 + (2t - 3)^{1/2}]$$

$$= 2t + (2t - 3)^{-1/2}.$$

The acceleration is given by

$$a(t) = v'(t) = 2 - (2t - 3)^{-3/2}.$$ ||

Solutions to related rate problems are based on the fact that an equation which establishes a relation between two variables may be used together with the chain rule to establish a relation between those variables and their rates of change with respect to time (that is, their derivatives with respect to time.)

It is important to use the chain rule properly when differentiating with respect to the time t. For example, if w is a differentiable function of time t, then

$$\frac{dw^4}{dt} = 4w^3 \frac{dw}{dt}.$$

Always remember to attach the $\frac{dw}{dt}$!

Example 2 The radius of a spherical balloon is increasing at the rate of 2 feet per minute. How fast is the volume of the balloon changing when the radius is 3 feet?

The volume V of the balloon and its radius r are related by

$$V = \frac{4}{3}\pi r^3.$$

If we differentiate both sides with respect to t, we get

$$\frac{dV}{dt} = \frac{4}{3}\pi 3r^2 \frac{dr}{dt}$$

$$= 4\pi r^2 \frac{dr}{dt}.$$

Note that this equation relates V, r, and their rates of change. In this particular case we are given that the radius is increasing at the rate of two feet per minute, that is,

$$\frac{dr}{dt} = 2 \text{ ft/min}.$$

Consequently,
$$\frac{dV}{dt} = 4\pi r^2 2$$
$$= 8\pi r^2.$$

We are asked to find dV/dt when $r = 3$ ft. We have

$$\left[\frac{dV}{dt}\right]_{r=3} = 8\pi 3^2 = 72\pi \text{ ft}^3/\text{min}. \qquad ||$$

Example 3 Water runs from an inverted conical tank at the rate of 2 ft^3/min. If the tank has base radius 10 feet and height 20 feet, find how fast the depth of the water is changing when the depth is 10 feet.

The general situation (for arbitrary time t) is pictured in Figure 3.11.1. Note that h is used to represent the depth of the water, and r the radius of the surface of the water. Use of similar triangles gives

$$\frac{r}{10} = \frac{h}{20}, \quad \text{or}$$
$$r = \frac{1}{2}h.$$

Since the volume V of water is

$$V = \frac{1}{3}\pi r^2 h,$$

and since $r = (1/2)h$, we get
$$V = \frac{\pi h^3}{12}.$$

Consequently,
$$\frac{dV}{dt} = \frac{1}{4}\pi h^2 \frac{dh}{dt}.$$

But we are given that
$$\frac{dV}{dt} = -2 \text{ ft}^3/\text{min}.$$

Figure 3.11.1

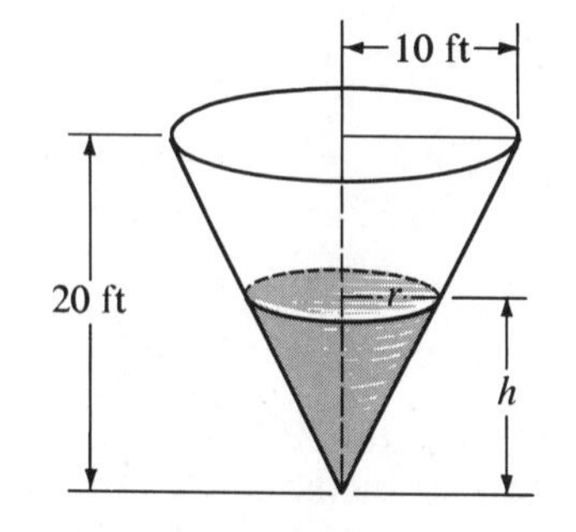

(Note that we have used the minus sign to indicate that the volume V is decreasing as time t increases, because the water is running out of the tank.)

Hence,
$$-2 = \frac{1}{4}\pi h^2 \frac{dh}{dt}, \quad \text{or}$$
$$\frac{dh}{dt} = \frac{-8}{\pi h^2}$$

We are expected to find dh/dt when $h = 10$ ft. In this case,
$$\frac{dh}{dt} = \frac{-8}{100\pi}$$
$$\approx -.025 \text{ ft/min.}$$
||

Example 4 An airplane is flying at the constant elevation of 4000 feet at 400 mi/hr. How fast is it approaching an observer directly beneath its flight path when its horizontal distance from the observer is 3000 feet?

If we let L be the distance between the plane and the observer, and x be the horizontal distance between the plane and the observer, we may picture the situation as shown in Figure 3.11.2.

Figure 3.11.2

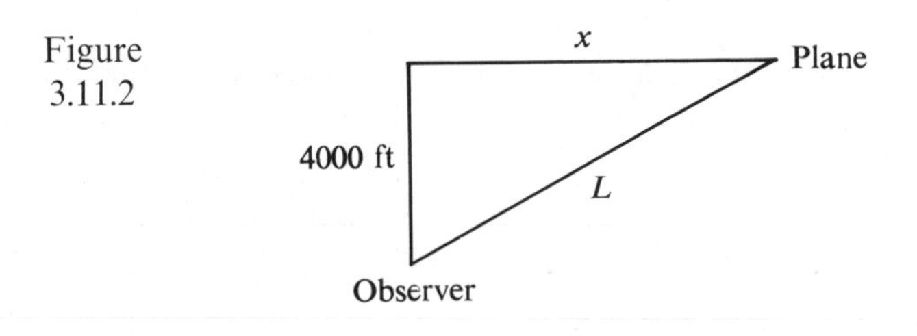

Consequently,
$$L^2 = 4000^2 + x^2,$$
and so
$$2L\frac{dL}{dt} = 2x\frac{dx}{dt}.$$

Here $dx/dt = -400$ mi/hr, since x is decreasing as time increases; so,
$$\frac{dL}{dt} = -400\frac{x}{L} \text{ mi/hr.}$$

We are asked to find dL/dt when $x = 3000$ ft. Since $L^2 = 4000^2 + x^2$, we get $L = 5000$ ft when $x = 3000$ ft. Hence,
$$\frac{dL}{dt} = -400\frac{3000}{5000} \text{ mi/hr}$$
$$= -240 \text{ mi/hr.}$$

That is, the airplane is approaching the observer at 240 mi/hr. ||

The examples above indicate the rather wide diversity of related rate problems. Usually in these problems the rate of change of one variable is given as a constant and the rate of change of a second variable is to be found for a certain value of one of the variables. In addition, sufficient information is given to set up an equation relating the variables. To solve the problem one usually proceeds as follows:

1. Draw a diagram of the general situation.
2. Set up the equation relating the variables, using convenient letters for the variables.
3. Differentiate the equation with respect to the time t, thus obtaining an equation involving the variables and the derivatives (rates of change) of the variables with respect to t.
4. Substitute the given values into the equation, including the given constant rate of change.
5. Solve for the required rate of change.

In drawing a diagram of the situation described by the problem and in establishing the relationship between the variables, *always* distinguish between variables and constants. Also remember to complete the chain rule properly by multiplying by the appropriate derivative with respect to t.

Exercises 3.11

In Exercises 1–6, the coordinate s is given as a function of the time t. Find the velocity and acceleration as functions of t.

1. $s(t) = t^3 - t + 1$.
2. $s(t) = \sqrt{1 + t^2}$.
3. $s(t) = 1/t + t^{1/2}$.
4. $s(t) = (t + 3)^{10}$.
5. $s(t) = t^2(t + 3)$.
6. $s(t) = \dfrac{t^2 - 7t + 5}{t}$.
7. An object is jettisoned upward from the top of a 100 foot gantry so that at time t seconds it is $s = 100 + 160t - 16t^2$ feet above the ground. Find the velocity as a function of time and calculate the following items.
   (a) The velocity at which the object was jettisoned.
   (b) The time required to reach the highest point.
   (c) The distance that the highest point is above the ground.
8. A ball is thrown upward from a 12 foot platform in such a way that at time t seconds it is $s = 12 + 96t - 16t^2$ feet above the ground. Find the velocity as a function of time and calculate the following items.
   (a) The velocity at which the ball was thrown upward.
   (b) The time required to reach the highest point.
   (c) The distance that the highest point is above the ground.
9. The area of a square is decreasing at the rate of 5 cm^2/min. Find the rate of change of the length of a side when the side length is 15 cm.
10. A rectangle remains twice as long as it is wide. If the area of the rectangle is increasing at the rate of 2 cm^2/hr, at what rate is the width increasing when the width is 10 cm?
11. The radius of a thin circular sheet is expanding at the rate of 2 in/min. How fast is the area changing when the radius is 1 foot?

12. Air is escaping from a spherical balloon at the rate of 3 ft^3/min. At what rate is the radius changing when the radius is 3 feet?

13. The sides of an equilateral triangle are increasing at the rate of .2 cm per hour. At what rate is the area changing when the side length is 4 cm?

14. A metal bar with square cross sections is being heated causing its length to increase at .06 cm per minute and its cross-sectional area to increase at 4 square cm per minute. Find a formula for the rate of increase of the volume.

15. Wine is running from an inverted conical tank at the rate of 12π ft^3/min. If the tank has radius 4 feet at the top and a depth of 12 feet, determine how fast the depth of the wine is changing when the depth is 9 feet.

16. Sand is poured on the ground forming a conical pile whose radius is twice its height. If the sand is poured onto the pile at the rate of 64 ft^3/min, how fast is the height changing when it is 8 feet?

17. The base of a 13 m ladder is moved along a horizontal floor at the rate of 2 m/sec away from the wall against which the ladder rests. At what rate is the top of the ladder moving down the wall when the top is 5 m above the floor?

18. Two cars, one traveling east, the other north, approach an intersection. If both cars are traveling 70 km/hr how fast are they approaching each other when they are 1/2 km from the intersection?

19. If a tree trunk adds 1/4 of an inch to its diameter and 1 foot to its height each year, how rapidly is its volume changing when its diameter is 3 feet and its height is 50 feet? (Assume that the tree trunk is a circular cylinder.)

20. The resistance R, measured in ohms, of a copper wire of fixed length varies inversely as its cross-sectional area. Find the rate of change of R with respect to r, the radius of the wire.

21. A weather balloon is released from a point 100 ft horizontally from an observer. If the balloon rises straight upward at the constant rate of 15 ft/sec, how fast is the distance between the observer and the balloon increasing at the end of one second?

22. A spherical ball of ice is melting at the rate of 10 cm^3/min. Find the rate of change of the radius of the ball when the radius is 5 cm.

23. Sometimes foresters assume that certain tree trunks are conical in shape. Assume that a tree trunk is a cone with fixed base radius of 2 ft. The tree increases 5 ft in height each year. Compute the rate of change of the volume when the tree is 150 ft tall.

24. What is the side length of a square whose area is increasing at twice the rate of increase of the side length?

25. What is the radius of a circular plate whose area is increasing at the same rate as its radius?

26. A helium filled balloon is released directly under a light hung 20 m above the ground. The balloon rises at the rate of 2 m/sec and is carried horizontally by a 3 m/sec wind. How fast is the shadow of the balloon moving along the ground after 9 seconds?

27. A spherical drop of water is evaporating at a rate proportional to its surface area. Show that the radius of the drop is decreasing at a constant rate.

28. Newton's law of motion for an object of mass m moving in a straight line says that

$$F = ma$$

where F is the force acting on the object and a is the acceleration of the particle. In relativistic mechanics this law is replaced by

$$F = m_0 \frac{d}{dt} \frac{v}{\sqrt{1 - (v/c)^2}}$$

where m_0 is the mass of the object measured at rest and c is the velocity of light. Show that

$$F = \frac{m_0 a}{(1 - (v/c)^2)^{3/2}}$$

Brief Review of Chapter 3

1. Basic Differentiation Rules

If c is a constant,

$$\frac{dc}{dx} = 0 \quad \text{and} \quad \frac{d[c \cdot f(x)]}{dx} = c\frac{df(x)}{dx}.$$

$$\frac{d}{dx}[f(x) \pm g(x)] = \frac{df(x)}{dx} \pm \frac{dg(x)}{dx}.$$

$$\frac{d}{dx}[f(x)g(x)] = f(x)\frac{dg(x)}{dx} + g(x)\frac{df(x)}{dx}.$$

$$\frac{d}{dx}\left[\frac{f(x)}{g(x)}\right] = \frac{g(x)\frac{df(x)}{dx} - f(x)\frac{dg(x)}{dx}}{[g(x)]^2}.$$

$$\frac{dx^n}{dx} = nx^{n-1}, \quad \text{for any integer } n.$$

2. The Chain Rule

$$\frac{df(u)}{dx} = \frac{df(u)}{du}\frac{du}{dx} \quad \text{or} \quad \frac{dy}{dx} = \frac{dy}{du}\frac{du}{dx},$$

and in particular,

$$\frac{d}{dx}[g(x)]^n = n[g(x)]^{n-1}\frac{dg(x)}{dx}.$$

3. Derivatives and Graphs

Signs of Derivatives. If, for $a < x < b$,
$f'(x) > 0$, then f is increasing;
$f'(x) < 0$, then f is decreasing;
$f''(x) > 0$, then f is concave upward;
$f''(x) < 0$, then f is concave downward.

Inflection Points. These are points where a graph changes concavity, that is, where $f''(x)$ changes sign; at such a point, $f''(x) = 0$ if it exists; but even if $f''(x) = 0$ at a point, it might not be an inflection point because $f''(x)$ might not change sign there.

Critical Points. These are points where the tangent is horizontal or vertical, or fails to exist; a point is a critical point if and only if $f'(x) = 0$ or $f'(x)$ does not exist at the point.

4. Local Maximum and Minimum Points

A function f is said to have a *local maximum* (or *local minimum*) at a if $f(a)$ is greater (less) than or equal to $f(x)$ in some neighborhood of a and in the domain of f; a local maximum or minimum of f can occur only at a point a where $f'(a) = 0$, or where $f'(a)$ does not exist, or at an endpoint of the domain of f. However a point a where $f'(a) = 0$ or $f'(a)$ does not exist does not have to determine a local maximum or minimum point. The second derivative test is often useful in finding local maxima or minima; however this test yields no conclusion about a possible maximum or minimum at a point a where $f''(a) = 0$. The First Derivative Test is often useful in circumstances where The Second Derivative Test yields no conclusion.

5. Absolute Maximum and Minimum Points

Definition. A function f is said to have an *absolute maximum* (or *absolute minimum*) at c if $f(c) \geq f(x)$ $(f(c) \leq f(x))$ for all x in the domain of f.

Existence. When a function f is continuous for all x of a closed interval $a \leq x \leq b$, it must have both an absolute maximum value and an absolute minimum value on the interval. This might not be true when the interval of continuity fails to include both endpoints a and b, or when f has a discontinuity at an interior point of the interval.

Location. To find absolute maximum and absolute minimum values of a function f on a closed interval $a \leq x \leq b$, first find all values of x for which $f'(x)$ is either 0 or does not exist. Then evaluate f at each of those points and at the endpoints a and b. The greatest of such values of $f(x)$ is the absolute maximum value and the smallest is the absolute minimum value.

The First Derivative Test for Absolute Extrema is often useful in showing that one has obtained an absolute extreme.

6. Implicit Differentiation

When an equation of the form $F(x, y) = 0$ determines y as a function of x, it is possible to use implicit differentiation to find y' without solving for y in terms of x. To apply the process of implicit differentiation one simply differentiates $F(x, y) = 0$ with respect to x, while keeping in mind that y is a function of x and so must be differentiated appropriately.

The formula

$$\frac{dx^n}{dx} = nx^{n-1}$$

is extended to all rational numbers n by use of implicit differentiation.

7. Differentials

If $y = f(x)$ is a differentiable function, then

$$dy = f'(x)dx.$$

Since

$$f(x + dx) \approx f(x) + dy,$$

the differential dy approximates the change in y between x and $x + dx$.

8. Velocity, Acceleration, and Related Rates

If the coordinate of a particle moving along a straight line is given by $s = f(t)$, where t denotes time, then the velocity and acceleration of the particle are given by $v(t) = f'(t)$ and $a(t) = f''(t)$ respectively.

Related rate problems are based on the fact that an equation relating two variables may be differentiated to yield an equation relating the variables and their rates of change.

Technique Review Exercises, Chapter 3

1. Find $f'(x)$ if $f(x) = 5x^4 - 3x^2 + 4x^{-7} - x^{-2} + 2$.
2. Find $f'(x)$ if $f(x) = x/(x^2 - 3x + 2)$.
3. Find $f'(x)$ if $f(x) = x^5(x^2 + 3x)^4$.
4. Find $f'(x)$ if $f(x) = [3 + (x^2 - 5x)^3]^2$.
5. For $f(x) = 3x - x^3 + 3$, determine the values of x for which f is increasing, decreasing, concave upward, and concave downward. Also find the coordinates of each critical point and each point of inflection, and sketch the graph.
6. Determine all local maximum and minimum points of $f(x) = x^3 - 3x^2 + 2$, $-2 \le x \le 3$.
7. Determine the absolute maximum and absolute minimum values of $f(x) = 2x^3 + 3x^2 - 12x + 5$, $-3 \le x \le 10$.
8. A rectangular area of 1800 square yards is to be enclosed and divided into 3 equal areas as in the figure. What are the dimensions requiring the least amount of fencing?

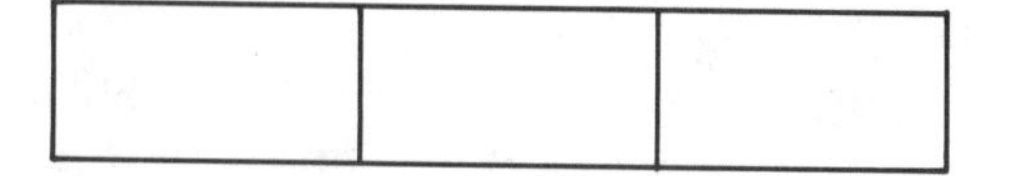

9. The area of a square is decreasing at the rate of 5 square inches per minute. Find the rate of change of the length of a side when the side length is 12 inches.

10. Find y' if $x^4y^2 = x^3 + y^4$.

11. A conical pile of sand is 3 m high; however, the diameter of 5 m is measured to within 2 cm of the correct value. Approximate the error in computing the volume of the sand pile.

12. Find $f'(x)$ if:
    (i) $f(x) = \cos x \sin 5x$
    (ii) $f(x) = \tan(x^3 - 3x)$
    (iii) $f(x) = \sin^2(3x)\cos^3(7x)$

Additional Exercises, Chapter 3

Section 3.1

Find the derivative of each function.

1. $f(x) = 8x^4 - 7x^3 + 2.$
2. $f(x) = 6x - 9x^4 + 2.$
3. $f(x) = (x - 7)(x + 2).$
4. $f(x) = (x + 5)(x + 6).$
5. $f(x) = (x^3 - 2x)^2.$
6. $f(x) = x^3(x^2 - 3)^2.$

Section 3.2

Find $\frac{dy}{dx}$.

7. $y = 11x^{-4} + 6x^4 - 7x^{-2}.$
8. $y = 8x^{-3} - 5x^{-1} + 3.$
9. $y = (x^2 + 5x)(x^3 - 3x).$
10. $y = (6x^{-1} - 5x)(3x^{-2} + 2x).$
11. $y = \frac{x - 7}{x + 1}.$
12. $y = \frac{x}{x - 1}.$
13. $y = \frac{x^3 - 2}{x^2 + 1}.$
14. $y = \frac{x^2 + 7x}{x - 3}.$
15. $y = (x^2 + 1)(x^3 + 1)(x^4 + 1).$
16. $y = (x^3 - 2)(x^5 + 1)(x^2 - 1).$

Section 3.3

Find $\frac{dy}{dx}$.

17. $y = 6u^3 + 3u^2, \quad u = x^{-1} + x^{-2}.$
18. $y = (u + 3)^5, \quad u = x^{-1} - x^7.$
19. $y = (x^2 - 3x)^4(2x + 7)^5.$
20. $y = (x^{-1} + 3x)^2(x^2 - 3x^3)^{-2}.$
21. $x = t^7 - 3t^{-1}, \quad y = t^3 - 2t.$
22. $x = (u^4 + 3)^2, \quad y = u^3 - 1.$
23. $x = u^7 + 3u^{-3}, \quad y = u^3 + 5u^2.$
24. $x = t^{-3}, \quad y = t^2 - t.$
25. $y = \frac{(x + 3)^2}{(2x - 1)^5}.$
26. $y = \frac{(x^2 + 1)^3}{(x^3 - 3x)^5}.$

Section 3.4

Determine where the graph is (a) increasing, (b) decreasing, (c) concave upward, (d) concave downward, and (e) sketch the graph.

27. $y = x^2 - 6x + 1.$
28. $y = 7 - 12x + x^2.$
29. $y = x^3 - 6x^2 + 2.$
30. $y = 8 - 16x^3 + 3x^4.$
31. $y = (x^2 - 1)^3.$
32. $y = 12x - x^3 - 2.$

Section 3.5

Determine all local maximum and minimum points of each function.

33. $f(x) = 7 - 2x - x^2$.
34. $f(x) = 5x^2 + 20x - 3$.
35. $f(x) = x^2 + 4x - 7, \quad 0 \leq x \leq 5$.
36. $f(x) = 3 - 8x - 2x^2, \quad 0 \leq x \leq 3$.
37. $f(x) = x^2 + 4x - 7, \quad -5 \leq x \leq -1$.
38. $f(x) = 3 - 8x - 2x^2, \quad 3 \leq x \leq 10$.
39. $f(x) = x^3 - 2$.
40. $f(x) = x^4 - 1$.

Section 3.6

Determine the absolute maximum and minimum points for each function, or state that they do not exist.

41. $f(x) = x^2 + 4x + 3, \quad 0 \leq x \leq 3$.
42. $f(x) = x^2 + 6x - 1, \quad -4 \leq x \leq 1$.
43. $f(x) = x^2 + 4x + 3, \quad -4 \leq x \leq 1$.
44. $f(x) = x^2 + 6x - 1, \quad 0 \leq x \leq 5$.
45. $f(x) = x^{-1}, \quad -2 \leq x \leq 1$.
46. $f(x) = (x + 3)/x, \quad -1 \leq x \leq 2, x \neq 0$.
47. $f(x) = 6x - x^3 + 2$.
48. $f(x) = x^2 - 3x^4 + 7, \quad -1 \leq x \leq 8$.

Section 3.7

49. A rancher wants to build a rectangular corral containing 10,000 square feet. What should be the dimensions of the corral using the smallest amount of fencing?
50. A cylindrical metal coffee can is to contain a fixed volume V. Show that the can using the least amount of metal has its height equal to its diameter.
51. A plastic container is to have a fixed volume V and have the shape of a cylinder surmounted by a cone. The height of the cylinder must equal the radius r of the cone. Show that the container with the least surface area occurs when $h = r\sqrt{2}$, where h is the height of the cylinder.
52. The center of a circular pond of radius 100 feet is 200 feet from the straight side of an existing fence running along a freeway. A rancher wants to enclose a rectangular field containing the pond by using 2000 feet of fencing for 3 sides of the field and the existing fence along the freeway as the fourth side. To give proper access to the pond he does not want the fence to be closer to the pond than 100 feet. What dimensions should he use for
    (a) The largest such field?
    (b) The smallest such field?

Section 3.8

Find y' using implicit differentiation.

53. $x^3y^4 - 2xy^2 = 3$.
54. $6x^{2/3}y^{3/2} + 4x^2y^{1/2} = 3$.
55. $10x^{3/5} - 15y^{2/5} = y$.
56. $y = (x^2y^3 - 2xy)^2$.
57. Use implicit differentiation to show that the rectangle of maximum area that can be inscribed in a circle is a square.
58. Use implicit differentiation to show that the rectangle of maximum perimeter that can be inscribed in a circle is a square.

Section 3.9

In Exercises 59–62, express dy in terms of x and dx.

59. $y = \dfrac{x - 3}{x + 4}$.
60. $y = (x^{-2} + 3x)^{1/5}$.
61. $y = (x + 2)^3(x^7 - 5x^3)^4$.
62. $y = (x^{-3} - 7x^{-2})^{-1/3}$.

63. Use differentials to approximate $\sqrt{4.03}$.

64. The diameter of a ball bearing is measured as 3 mm, accurate to within .2 mm. Approximately what error may be made in computing the volume?

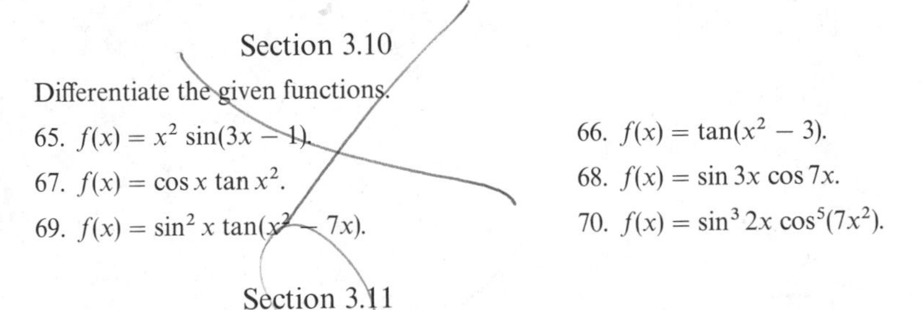

Section 3.10

Differentiate the given functions.

65. $f(x) = x^2 \sin(3x - 1)$.

66. $f(x) = \tan(x^2 - 3)$.

67. $f(x) = \cos x \tan x^2$.

68. $f(x) = \sin 3x \cos 7x$.

69. $f(x) = \sin^2 x \tan(x^2 - 7x)$.

70. $f(x) = \sin^3 2x \cos^5(7x^2)$.

Section 3.11

71. The radius of the base of a cone is increasing at the constant rate of 3 cm per minute and the altitude of the cone is decreasing at the constant rate of 2 cm per minute. What is the rate of change of the volume of the cone when the base radius is 12 cm and the altitude is 8 cm?

72. The base of a triangle is decreasing at the constant rate of 1 inch per minute, but the area of the triangle is increasing at the constant rate of 12 square inches per minute. What is the rate of change of the altitude of the triangle when the altitude is 10 inches and the base is 4 inches?

73. The circumference of a circle is decreasing at a constant rate of 2 feet per second. At what rate is the area changing when the radius is 40 feet?

74. The surface area of a sphere is increasing at a constant rate of 72 square cm per minute. At what rate is the volume changing when the radius is 9 cm?

Challenging Problems, Chapter 3

1. Let f be a function whose domain is the set of all real numbers. Show that if $f(a + b) = f(a) + f(b)$ for all a, b and $f'(0) = c$, then $f'(x) = c$ for all x.

2. Let f be a function with the properties
   (i) $f(x + y) = f(x)f(y)$, for all x, y,
   (ii) $f(x) = 1 + xg(x)$ for all x where $\lim_{x \to 0} g(x) = 1$.

   Show that $f'(x) = f(x)$ for all x.

3. Let x and y be given parametrically as $x = f(t)$ and $y = g(t)$. Show that

$$\frac{d^2y}{dx^2} = \frac{\dfrac{dx}{dt}\dfrac{d^2y}{dt^2} - \dfrac{d^2x}{dt^2}\dfrac{dy}{dt}}{\left(\dfrac{dx}{dt}\right)^3}.$$

4. Use the following outline to establish Schwarz's inequality,

$$(a_1b_1 + a_2b_2 + \cdots + a_nb_n)^2 \leq (a_1{}^2 + a_2{}^2 + \cdots + a_n{}^2)(b_1{}^2 + b_2{}^2 + \cdots + b_n{}^2).$$

(i) Let $f(x) = ax^2 + 2bx + c$, with $a > 0$. Prove that $f(x) \geqslant 0$ for all x, if and only if $b^2 - ac < 0$.
(ii) Apply the result of (i) to the function

$$f(x) = (a_1x + b_1)^2 + (a_2x + b_2)^2 + \cdots + (a_nx + b_n)^2.$$

5. Wildlife management experts are often interested in a reproduction function for a certain animal population. A reproduction function gives the size of a population one year hence in terms of the present population size. Thus if $y = f(x)$ is a reproduction function, x is the present population and y is the population one year hence. Let $y = f(x)$ be a reproduction function and show:
(i) $f(x) - x$ is the sustainable yearly harvest from a population of size $f(x)$.
(ii) The maximal sustainable yearly harvest occurs at x_0 where

$$f'(x_0) = 1 \quad \text{and} \quad f''(x_0) < 0.$$

6. Find two functions $f(x)$ and $g(x)$ that are nowhere differentiable and yet $f(x) + g(x)$ is differentiable everywhere.

7. Let f be a function whose domain includes the interval $a \leq x \leq b$, and let p be a fixed number such that $a < p < b$. Suppose it is known that $f'(x)$ exists for all x, and that $\lim_{x \to p} f'(x) = k$. Use the Mean Value Theorem to prove that $f'(p) = k$.

8. We have seen that if a function has a derivative, then the function must be continuous. It can be proved that the derivative of a function need not be continuous (see Challenging Problems, Chapter 7). Even though they may not be continuous, however, derivatives cannot have "jump" discontinuities and, in fact, do have the intermediate value property. The theorem stating that fact is called Darboux's Theorem. Prove Darboux's Theorem. That is, prove: Let f be a function such that $f'(x)$ exists for $a \leq x \leq b$, and suppose that r is some number between $f'(a)$ and $f'(b)$. Then there is at least one number c, $a < c < b$, such that $f'(c) = r$. (*Hint*: consider the absolute minimum of $g(x) = f(x) - rx$.)

†9. An essential property of parabolic reflectors is that rays parallel to the axis of a parabola are reflected to the focus of the parabola and vice versa. Prove that reflection property, using the following outline as a guide:
(i) Let the focus of the parabola be at $(0, p)$ and let $y = -p$ be the directrix. Then the vertex is at the origin, the y-axis is the axis of the parabola, and an equation of the parabola is $y = x^2/(4p)$.

† From "A Simple Proof of the Reflection Property for Parabolas," R. H. Cowin, *Two-Year College Mathematics Journal*, 1976, pp. 59–60.

(ii) Let a ray parallel to the y-axis meet the parabola at the point (x, y) and note that when the line representing the ray is extended it meets the directrix at the point $(x, -p)$. Show that tangent line at (x, y) is perpendicular to the line through the point $(x, -p)$ and the focus, and use that fact to complete the proof.

4

Integration

4.1 The Area Problem

The problem of determining areas extends back to ancient Greece. At that time even the computation of the area of a circle was a difficult task. It was not until the time of Leibniz and Newton that area problems could be solved easily by using calculus.

Let us consider the region bounded by the graph of $y = f(x)$ and the x-axis between the vertical lines $x = a$ and $x = b$, where $f(x) \geq 0$ for $a \leq x \leq b$. The "area problem" is the problem of determining the area bounded by the graph of $y = f(x)$ and the x-axis between a and b.

Figure 4.1.1

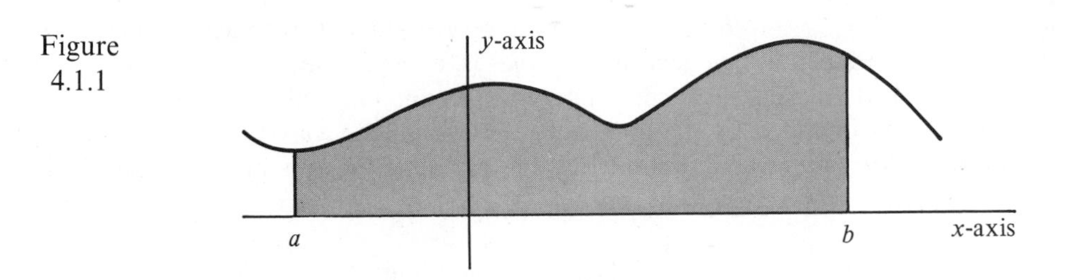

We shall find that the area problem has a solution that is both simple and beautiful! The solution helps to illustrate the great power of mathematics, especially that of calculus. Although the problem can be solved for more general functions, it will be sufficient for our purposes to restrict our attention to functions continuous over an interval with endpoints at a and b.

We shall first approximate the areas involved by slicing the regions into pieces. As shown in Figure 4.1.2, the slices are obtained by dividing the interval from a to b into n equal subintervals, each of width h, by taking equally spaced points x_0 $(=a)$, $x_1, x_2, \ldots, x_n$ $(=b)$. In each of those subintervals we select an arbitrary point. That is, we take any point x_1^* in the first subinterval, any point x_2^* in the second subinterval, . . . , and finally any point x_n^* in the nth subinterval.

Figure 4.1.2

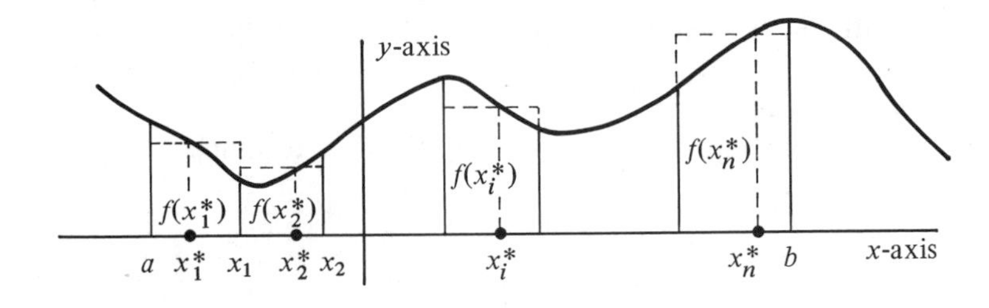

Now, as shown in Figure 4.1.2, $f(x_1^*)h$ is an approximation to the area of the first slice, $f(x_2^*)h$ is an approximation to the area of the second slice, and so forth. Thus, an approximation to the area between a and b is

$$f(x_1^*)h + f(x_2^*)h + f(x_3^*)h + \cdots + f(x_n^*)h.$$

This sum is usually written in more compact form as

$$\sum_{i=1}^{n} f(x_i^*)h.$$

The $\sum_{i=1}^{n} f(x_i^*)h$ is read "the sum from 1 to n of $f(x_i^*)h$" and means that first 1 is substituted for i and the result is added to the result when 2 is substituted for i, etc., until the result obtained by substituting n for i is added. For example,

$$\sum_{i=1}^{4} f(x_i^*)h = f(x_1^*)h + f(x_2^*)h + f(x_3^*)h + f(x_4^*)h.$$

Although $\sum_{i=1}^{n} f(x_i^*)h$ might not be a very good approximation when h is large, it becomes better as h gets smaller, that is, as n gets larger.

To illustrate these ideas further, let's consider the particular problem of calculating the area bounded by the graph of $y = x/2$ and the x-axis between 1 and 5. As illustrated in Figure 4.1.3, we only need to calculate the area of a trapezoid with sides of length 1/2 and 5/2 and of width 4. We obtain

$$\frac{1}{2}\left(\frac{1}{2} + \frac{5}{2}\right)(4) = 6 \text{ square units.}$$

Figure 4.1.3

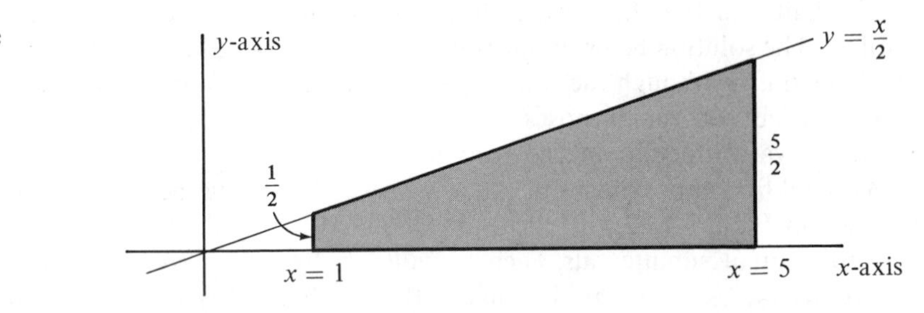

However, to prepare for more complex problems, we shall apply the method discussed above. To use the method we must first subdivide the interval $1 \leq x \leq 5$ into n subintervals, each of width h. Thus, $h = (5 - 1)/n = 4/n$, and so $n = 4/h$. The points $x_1^*, x_2^*, \ldots, x_n^*$ may be selected anywhere in their respective subintervals. We shall, for the sake of convenience, take them at the right-hand endpoints of the subintervals, as illustrated in Figure 4.1.4. Thus $x_1^* = 1 + h$, $x_2^* = 1 + 2h$, $x_3^* = 1 + 3h, \ldots, x_n^* = 1 + nh = 5$.

The contribution of the first approximating rectangle is

$$f(x_1^*)h = \frac{x_1^*}{2}h = \frac{1 + h}{2}h;$$

Figure 4.14

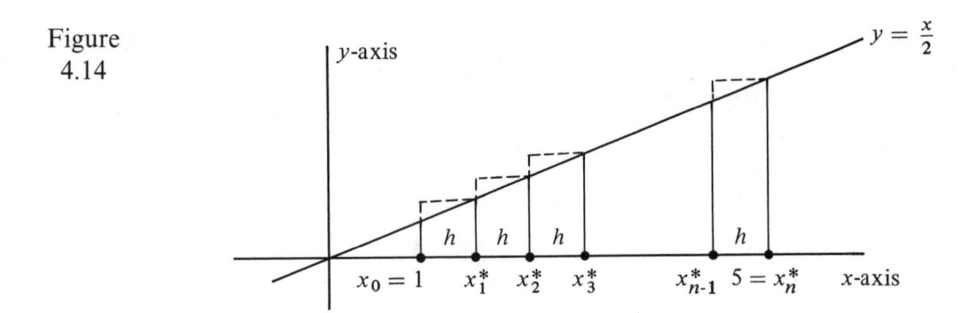

the contribution of the second approximating rectangle is

$$f(x_2^*)h = \frac{x_2^*}{2}h = \frac{1 + 2h}{2}h;$$

and in general, the ith approximating rectangle contributes

$$f(x_i^*)h = \frac{x_i^*}{2}h = \frac{1 + ih}{2}h.$$

Thus, as in our general discussion, the area is approximated by

$$\begin{aligned}\sum_{i=1}^{n} f(x_i^*)h &= f(1 + h)h + f(1 + 2h)h + \cdots + f(1 + nh)h \\ &= h[f(1 + h) + f(1 + 2h) + \cdots + f(1 + nh)] \\ &= h\left[\frac{1 + h}{2} + \frac{1 + 2h}{2} + \cdots + \frac{1 + nh}{2}\right] \\ &= \frac{h}{2}[(1 + h) + (1 + 2h) + \cdots + (1 + nh)] \\ &= \frac{h}{2}[n + (1 + 2 + \cdots + n)h].\end{aligned}$$

But $1 + 2 + \cdots + n = n(n + 1)/2$; so

$$\sum_{i=1}^{n} f(x_i^*)h = \frac{h}{2}\left[n + \frac{n(n + 1)}{2}h\right].$$

Using the fact that $n = 4/h$, this can be expressed as

$$\begin{aligned}\sum_{i=1}^{n} f(x_i^*)h &= \frac{h}{2}\left[\frac{4}{h} + \frac{2}{h}\left(\frac{4}{h} + 1\right)h\right] \\ &= 2 + 4 + h \\ &= 6 + h.\end{aligned}$$

Consequently, as h gets small, this approximation approaches 6 square units, which is the same result we obtained above by elementary means.

Actually, the approximation always approaches the same value (6 square units) no matter how the x_i^*'s are chosen in their respective subintervals. For example, if the x_i^* are taken as the left-hand endpoints of their respective subintervals, we have

$$x_1^* = 1, \quad x_2^* = 1 + h, \quad x_3^* = 1 + 2h, \ldots, x_n^* = 1 + (n-1)h.$$

Consequently an approximation to the area is given by

$$\sum_{i=1}^{n} f(x_i^*)h = \frac{1}{2}h + \frac{1+h}{2}h + \frac{1+2h}{2}h + \cdots + \frac{1+(n-1)h}{2}h$$

$$= \frac{h}{2}[1 + (1+h) + (1+2h) + \cdots + (1+(n-1)h)]$$

$$= \frac{h}{2}[n + (1 + 2 + 3 + \cdots + (n-1))h].$$

As above,
$$1 + 2 + 3 + \cdots + (n-1) = \frac{n(n-1)}{2},$$

and so
$$\sum_{i=1}^{n} f(x_i^*)h = \frac{h}{2}\left[n + \frac{n(n-1)}{2}h\right].$$

Substitution of $n = 4/h$ yields

$$\sum_{i=1}^{n} f(x_i^*)h = \frac{h}{2}\left[\frac{4}{h} + \frac{2}{h}\left(\frac{4}{h} - 1\right)h\right]$$
$$= 2 + 4 - h$$
$$= 6 - h.$$

Thus, again, as h approaches zero, the approximation approaches the value of 6 square units.

Of course, by choosing the x_i^*s to be the right and left-hand endpoints we have chosen those x_i^*'s that yield the approximating rectangles of maximum and minimum heights respectively. Consequently, $6 + h$ is the maximum possible value for

$$\sum_{i=1}^{n} f(x_i^*)h,$$

and $6 - h$ is its minimum possible value. Hence, for *any* choice of the x_i^*'s,

$$6 - h \leq \sum_{i=1}^{n} f(x_i^*)h \leq 6 + h.$$

and, as h approaches zero,

$$\sum_{i=1}^{n} f(x_i^*)h$$

must approach 6. That is,

$$\lim_{h \to 0} \sum_{i=1}^{n} f(x_i^*)h = 6.$$

The quantity

$$\lim_{h \to 0} \sum_{i=1}^{n} f(x_i^*)h$$

is so important, and has so many applications, that we make the following formal definition. The restriction $f(x) \geq 0$ was made to motivate and illustrate that new concept as an area. For many applications the restriction is too confining; so we drop it.

Definition 4.1.1

> Let f be a function defined on the interval $a \leq x \leq b$. Divide the interval $a \leq x \leq b$ into n equal subintervals, each of width h. Take an arbitrary point x_1^* in the first subinterval, an arbitrary point x_2^* in the second subinterval, and so forth. Then, if
>
> $$\lim_{h \to 0} \sum_{i=1}^{n} f(x_i^*)h$$
>
> exists and is independent of the choice of the x_i^*'s, this limit is called the *definite integral* of f from a to b. We write
>
> $$\int_a^b f(x)dx$$
>
> to denote that limit.

In the notation $\int_a^b f(x)dx$, the symbol $\int$ is actually an elongated S representing the limit of the sum. The a and the b are called the *lower* and *upper limits* of integration, respectively, and are the lower and upper ends of the interval $a \leq x \leq b$, over which the sum is taken. The $f(x)$ is called the *integrand.* The differential dx is used as a convenience whose purpose will become apparent later.

Example 1 | Compute the definite integral $\int_{-2}^{1} 4\,dx$.

In this case the interval $-2 \leq x \leq 1$ is subdivided into n subintervals, each of width $h = [1 - (-2)]/n = 3/n$. Thus $n = 3/h$. Since $f(x) = 4$ for all x, $f(x_i^*) = 4$ for any choice of x_i^*. Consequently,

$$\begin{aligned} \sum_{i=1}^{n} f(x_i^*)h &= \sum_{i=1}^{n} 4h \\ &= 4h + 4h + \cdots + 4h \ (n \text{ terms}) \\ &= 4nh = 4\frac{3}{h}h = 12. \end{aligned}$$

Hence,

$$\int_{-2}^{1} 4\,dx = \lim_{h \to 0} \sum_{i=1}^{n} f(x_i^*)h = \lim_{h \to 0} 12 = 12.$$

Note that $\int_{-2}^{1} 4\,dx$ is the area that is bounded above by the graph of $f(x) = 4$ and below by the x-axis, and lies between the vertical lines $x = -2$ and $x = 1$. ||

Example 2 Compute the definite integral $\int_0^7 6x^2\,dx$. Use the fact the

$$1^2 + 2^2 + 3^2 + \cdots + n^2 = \frac{n(n+1)(2n+1)}{6}.$$

Here the interval has width $7 - 0 = 7$, and so each of the n subintervals has width $h = 7/n$. Thus $n = 7/h$. Although the points $x_1^*, x_2^*, \ldots, x_n^*$ may be selected anywhere in their respective subintervals, we shall take them at the right-hand endpoints. Thus $x_1^* = 0 + h$, $x_2^* = 0 + 2h$, $x_3^* = 0 + 3h$, $\ldots, x_i^* = 0 + ih$, $\ldots, x_n^* = 0 + nh = 7$. Since $f(x) = 6x^2$ and $x_i^* = ih$, we get

$$\sum_{i=1}^{n} f(x_i^*)h = \sum_{i=1}^{n} 6(ih)^2 h$$

$$= \sum_{i=1}^{n} 6i^2h^3$$

$$= 6h^3(1^2 + 2^2 + 3^2 + \cdots + n^2).$$

But

$$1^2 + 2^2 + 3^2 + \cdots + n^2 = n(n+1)(2n+1)/6.$$

So

$$\sum_{i=1}^{n} f(x_i^*)h = h^3 n(n+1)(2n+1).$$

Replacing n by $7/h$, we get

$$\sum_{i=1}^{n} f(x_i^*)h = h^3 \frac{7}{h}\left(\frac{7}{h} + 1\right)\left(\frac{14}{h} + 1\right)$$

$$= 7(7 + h)(14 + h).$$

Hence

$$\int_0^7 6x^2\,dx = \lim_{h \to 0} \sum_{i=1}^{n} f(x_i^*)h = \lim_{h \to 0} 7(7 + h)(14 + h) = 686.$$ ||

It should be realized that definite integrals of some functions do not exist, a fact illustrated in the next example.

Example 3 Let f be the function

$$f(x) = \begin{cases} 0 & \text{if } x \text{ is rational} \\ 1 & \text{if } x \text{ is irrational,} \end{cases}$$

and let us attempt to compute $\int_2^3 f(x)dx$.

To compute $\int_2^3 f(x)dx$, we must subdivide the interval $2 \leq x \leq 3$ into n equal subintervals, each of width $h = 1/n$. No matter how small these subintervals are taken, each must contain both rational and irrational numbers. Let us first select a rational x_i^* from each subinterval; then $f(x_i^*) = 0$ for all i and

$$\sum_{i=1}^{n} f(x_i^*)h = 0.$$

On the other hand, if we select an irrational x_i^* from each subinterval, then $f(x_i^*) = 1$ for all i and

$$\begin{aligned}\sum_{i=1}^{n} f(x_i^*)h &= h + h + h + \cdots + h \text{ (n terms)} \\ &= nh \\ &= n\frac{1}{n} \\ &= 1.\end{aligned}$$

Consequently, the quantity

$$\lim_{h \to 0} \sum_{i=1}^{n} f(x_i^*)h$$

depends on the particular choice of the x_i^*. Hence, according to Definition 4.1.1, the definite integral $\int_2^3 f(x)dx$ does not exist. ||

Though the preceding integral does not exist, it is nevertheless true that definite integrals exist under fairly weak conditions. A sufficient condition for their existence is indicated in the next theorem. (A proof can be found in many advanced calculus texts.)

Theorem 4.1.1

If a function f is continuous on the interval $a \leq x \leq b$, then the integral $\int_a^b f(x)dx$ exists; that is, the limit

$$\lim_{h \to 0} \sum_{i=1}^{n} f(x_i^*)h$$

exists and is independent of the choice of the x_i^*'s.

Actually an integral exists under even weaker conditions than indicated in the preceding theorem. In particular, $\int_a^b f(x)dx$ will exist if f has at most a finite number of discontinuities for $a \leq x \leq b$, and there is a number M such that $|f(x)| \leq M$ for $a \leq x \leq b$.

Not only can those conditions be weakened, but the definition of the definite integral can be stated more generally. It is not necessary that the subintervals have equal widths. In the more general definition, $\sum_{i=1}^{n} f(x_i^*)h$ is replaced by the *Riemann Sum* $\sum_{i=1}^{n} f(x_i^*)(x_i - x_{i-1})$ and the limit is taken so that the width of every subinterval approaches zero. Our more restricted definition is much simpler to use (see Problem 4 of Challenging Problems, Chapter 4), and nothing is lost since it is sufficient for all of our purposes.

In order to provide more familiarity with the notation $\sum_{i=1}^{n} f(x_i^*)h$ used in the definition of the definite integral, we present an additional example.

Example 4 | Evaluate $\sum_{i=1}^{4} f(x_i^*)h$ and $\sum_{i=1}^{n} f(x_i^*)h$, where $f(x) = 10x$ and $x_i^* = ih$.

We get

$$\begin{aligned}\sum_{i=1}^{4} f(x_i^*)h &= f(x_1^*)h + f(x_2^*)h + f(x_3^*)h + f(x_4^*)h \\ &= 10x_1^*h + 10x_2^*h + 10x_3^*h + 10x_4^*h \\ &= 10(h)h + 10(2h)h + 10(3h)h + 10(4h)h \\ &= 100h^2.\end{aligned}$$

and

$$\begin{aligned}\sum_{i=1}^{n} f(x_i^*)h &= f(x_1^*)h + f(x_2^*)h + \cdots + f(x_n^*)h \\ &= 10(h)h + 10(2h)h + \cdots + 10(nh)h \\ &= 10h^2(1 + 2 + 3 + \cdots + n) \\ &= 10h^2\frac{n(n+1)}{2} = 5h^2n(n+1).\end{aligned}$$ ||

Exercises 4.1

For Exercises 1–10, evaluate the indicated sums.

1. $\sum_{i=1}^{5} f(x_i^*)h$, where $f(x) = 4x$ and $x_i^* = ih$.
2. $\sum_{i=1}^{6} f(x_i^*)h$, where $f(x) = 6x$ and $x_i^* = (i-1)h$.
3. $\sum_{i=1}^{7} f(x_i^*)h$, where $f(x) = 4x$ and $x_i^* = (i-1)h$.
4. $\sum_{i=1}^{8} f(x_i^*)h$, where $f(x) = 6x$ and $x_i^* = ih$.
5. $\sum_{i=1}^{n} f(x_i^*)h$, where $f(x) = 4x$ and $x_i^* = ih$.

6. $\sum_{i=1}^{n} f(x_i^*)h$, where $f(x) = 6x$ and $x_i^* = (i-1)h$.

7. $\sum_{i=1}^{n} f(x_i^*)h$, where $f(x) = 4x$ and $x_i^* = (i-1)h$.

8. $\sum_{i=1}^{n} f(x_i^*)h$, where $f(x) = 6x$ and $x_i^* = ih$.

9. $\sum_{i=1}^{4} f(x_i^*)h$, where $f(x) = x^2$ and $x_i^* = ih$.

10. $\sum_{i=1}^{n} f(x_i^*)h$, where $f(x) = x^2$ and $x_i^* = (i-1)h$.

For Exercises 11–20, use the definition of the definite integral to compute each integral or state that it does not exist.

11. $\int_2^7 9\,dx$.
12. $\int_{-3}^1 6\,dx$.
13. $\int_{-2}^5 0\,dx$.
14. $\int_{-1}^7 0\,dx$.
15. $\int_3^7 (x-1)dx$.
16. $\int_{-1}^3 (x+2)dx$.
17. $\int_0^3 x^2\,dx$. $\left(\text{Use the fact that } 1^2 + 2^2 + 3^2 + \cdots + n^2 = \frac{n(n+1)(2n+1)}{6}.\right)$
18. $\int_0^2 3x^2\,dx$. (Use the equality of Exercise 17.)
19. $\int_3^8 f(x)dx$, where $f(x) = \begin{cases} 3 & \text{if } x \text{ is rational} \\ 5 & \text{if } x \text{ is irrational.} \end{cases}$
20. $\int_2^4 f(x)dx$, where $f(x) = \begin{cases} 6 & \text{if } x \text{ is rational} \\ 2 & \text{if } x \text{ is irrational.} \end{cases}$

4.2 More on Definite Integrals and Areas

In this section we shall more closely examine the relationship between areas and definite integrals. We shall also consider that relationship for definite integrals of functions that may be negative.

By using Theorem 4.1.1, it is not difficult to show that the definite integral of a continuous nonnegative function f from a to b is equal to the area bounded by the graph of $y = f(x)$ and the x-axis between a and b. We first subdivide the interval $a \le x \le b$ into n subintervals, each of width h. Now, since f is continuous it attains an absolute maximum value M_i and an absolute minimum value m_i on the ith subinterval as indicated in Figure 4.2.1. Let A_i be the area bounded by $y = f(x)$ and the x-axis between x_{i-1} and x_i. Then A_i must be between $m_i h$ and $M_i h$; that is,

$$m_i h \le A_i \le M_i h.$$

Then, since h is positive,

$$m_i \le \frac{A_i}{h} \le M_i.$$

Figure 4.2.1

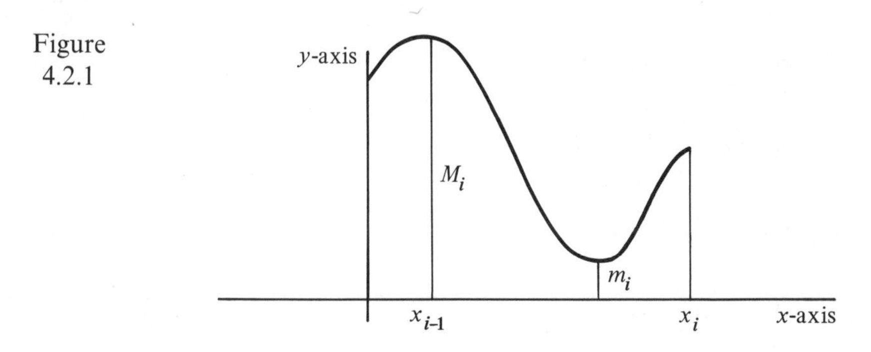

Since the continuous function f has the values m_i and M_i on the ith subinterval, and since A_i/h is between m_i and M_i, The Intermediate Value Theorem for Continuous Functions assures us that there is a point x_i^* such that

$$x_{i-1} \le x_i^* \le x_i \quad \text{and} \quad f(x_i^*) = \frac{A_i}{h}.$$

Therefore, for this choice of x_i^*,

$$A_i = f(x_i^*)h.$$

But the total value A of the area bounded by the graph of $y = f(x)$ and the x-axis between a and b is the sum of the A_i's. That is,

$$A = \sum_{i=1}^{n} A_i = \sum_{i=1}^{n} f(x_i^*)h.$$

Moreover, since A is independent of h and Theorem 4.1.1 guarantees the existence of the limit, we have

$$A = \lim_{h \to 0} A = \lim_{h \to 0} \sum_{i=1}^{n} A_i = \lim_{h \to 0} \sum_{i=1}^{n} f(x_i^*)h.$$

By definition the last expression is $\int_a^b f(x)dx$, and so

$$A = \int_a^b f(x)dx.$$

Thus we have proved the following result.

Theorem 4.2.1

> If a function f is continuous and $f(x) \ge 0$ on the interval $a \le x \le b$, then the area bounded by the graph of $y = f(x)$ and the x-axis between a and b is equal to $\int_a^b f(x)dx$.

The properties of area that we have assumed in the proof of Theorem 4.2.1 are as follows:

1 the usual area formula for rectangles
2 if one plane region is part of another, its area is less than or equal to the area of the other
3 if two plane regions intersect at most on their boundaries, then their total area is the sum of their separate areas.

Frequently the proper intuitive view can be used both as a memory aid and as a guide to additional applications. While the intuitive view is not a proof, it is a simulation of a proof and will serve as a unifying principle in our various applications of the definite integral.

In particular we shall give an intuitive development of the theorem that the area A of the region bounded by the graph of a nonnegative function $f(x)$ and the x-axis between a and b is equal to $\int_a^b f(x)dx$.

In the more rigorous development we subdivided the interval $a \leq x \leq b$ into subintervals each of width h. Those subintervals enabled us to cut the region under consideration into n slices of equal width h. A picture of the ith slice is given in Figure 4.2.2. We then showed that the point x_i^* could be chosen so that the area of the ith slice is exactly $f(x_i^*)h$. We then summed up the areas of the n slices, took the limit as $h \to 0$, and finally used the definition of the definite integral to get

$$A = \sum_{i=1}^{n} f(x_i^*)h = \lim_{h \to 0} \sum_{i=1}^{n} f(x_i^*)h = \int_a^b f(x)dx. \tag{1}$$

Figure 4.2.2

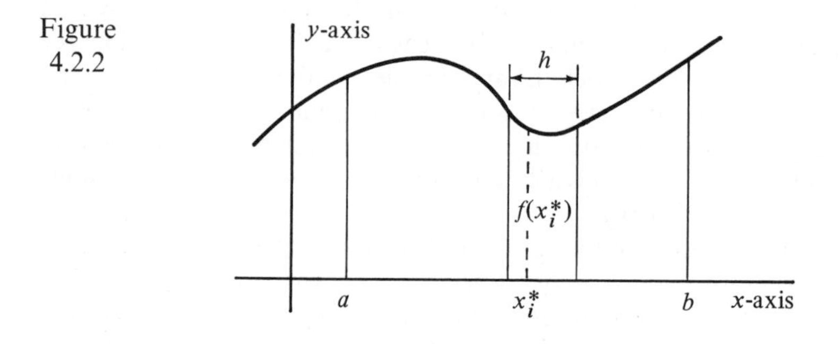

In the intuitive development we simulate the more rigorous development and give a sort of skeleton of the major steps. The ith *slice* of the region is replaced by a subregion of height $f(x)$ and width dx (illustrated in Figure 4.2.3) whose area is called *a typical element of area.* Note that the width h is replaced by dx and the precise $f(x_i^*)$ is replaced simply by $f(x)$. The equalities of item (1) and the reasoning for them is replaced by considering $\int_a^b$ to be an operation summing up the areas of

Figure 4.2.3

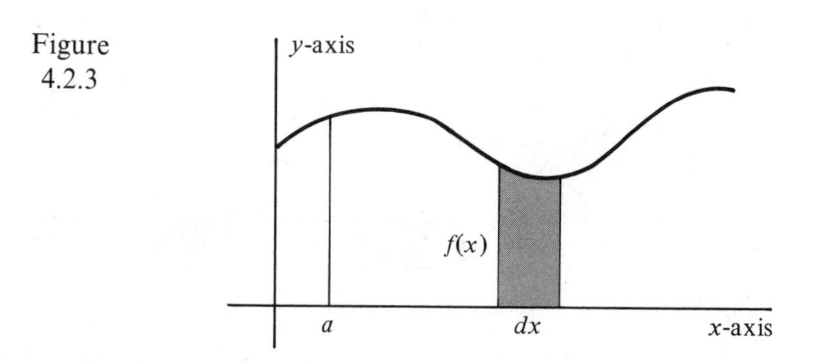

all of the slices over the interval from a to b to get $A = \int_a^b f(x)dx$. Thus the intuitive development might be stated as follows.

> *Intuitive Development.* Sketch a picture where a typical element of area is the area of a region of height $f(x)$ and width dx, as shown in Figure 4.2.3. Then think of $\int_a^b$ as the operation of summing up the elements of area throughout the interval from a to b to get
>
> $$A = \int_a^b f(x)dx.$$

As mentioned above, for most of our proofs involving applications of definite integrals we shall give both an intuitive development similar to the one just given and a more rigorous development.

At our present stage of development even the simplest of definite integrals requires a considerable amount of effort to evaluate. In Section 4.4, however, we shall obtain a beautiful and exceptionally useful theorem called *The Fundamental Theorem of Calculus.* This theorem will enable us to evaluate definite integrals much more easily.

In the next section we shall approach the area problem in a simpler way that will lead us to The Fundamental Theorem. The reason we did not here follow the simpler method of Section 4.3 to attack the area problem without using the concept of definite integral is that definite integrals are valuable in many applications other than finding areas. In particular, definite integrals are valuable in dealing with volumes, lengths of curves, work, pressure, and other applications, including some in statistics, economics, and biology.

Although our motivation for the area problem has been for the case where $f(x) \geq 0$ for $a \leq x \leq b$, we might well ask what happens when $f(x)$ is positive for some of those values of x and negative for others. An illustrative case is shown in Figure 4.2.4. There is a region bounded by the lines $x = a$ and $x = b$ between the graph and the x-axis. In the figure the area of the region is $A_1 + A_2 + A_3 + A_4$. Part of the region is above the x-axis and part of the region is below the x-axis. In Figure 4.2.4, the area of the part above the x-axis is $A_1 + A_3$ and the area of the part below the x-axis is $A_2 + A_4$. It will be convenient to call the area of the part above the x-axis minus the area of the part below the x-axis the *net* area bounded by the graph of $y = f(x)$ and the x-axis between a and b. In the figure the net area bounded by the graph of $y = f(x)$ and the x-axis between a and b is $(A_1 + A_3) - (A_2 + A_4)$. It turns out that $\int_a^b f(x)dx$ is equal to the net area bounded by the graph of

Figure 4.2.4

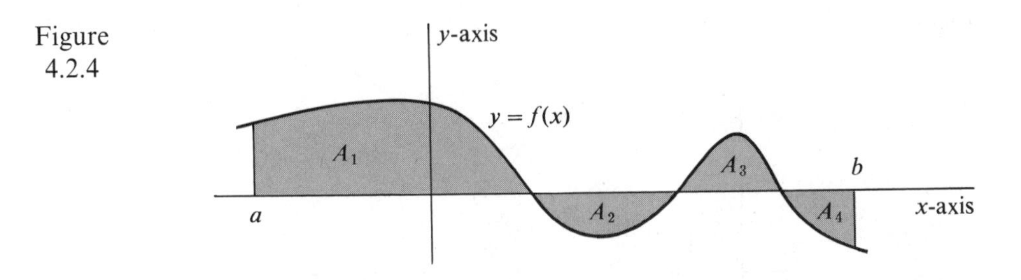

$f(x)$ and the x-axis between a and b. Thus, for the function illustrated in Figure 4.2.4,

$$\int_a^b f(x)dx = (A_1 + A_3) - (A_2 + A_4).$$

The proof of this more general result is almost identical to the proof of Theorem 4.2.1.

Example 1 Use the definite integral to calculate the net area bounded by the graph of $y = 2x$ and the x-axis between $x = -2$ and $x = 3$.

Since the net area A we are to determine is equal to $\int_{-2}^{3} 2x\,dx$, we must evaluate that definite integral. We divide the interval from -2 to 3 into n subintervals, each of width $h = \dfrac{3-(-2)}{n} = \dfrac{5}{n}$. Thus $n = 5/h$. For convenience we take the points $x_1^*, x_2^*, \ldots, x_n^*$ at the right-hand endpoints. Thus $x_1^* = -2 + h$, $x_2^* = -2 + 2h, \ldots,$ $x_n^* = -2 + nh = 3$. Since $f(x) = 2x$ and $x_i^* = -2 + ih$, we get

$$\begin{aligned}\sum_{i=1}^{n} f(x_i^*)h &= \sum_{i=1}^{n} 2(-2 + ih)h \\ &= 2(-2 + h)h + 2(-2 + 2h)h + \cdots + 2(-2 + nh)h \\ &= 2h(-2)n + 2h^2(1 + 2 + \cdots + n) \\ &= -4hn + 2h^2\frac{n(n+1)}{2} \\ &= -4hn + h^2n(n+1).\end{aligned}$$

Since $n = 5/h$, we get

$$\begin{aligned}\sum_{i=1}^{n} f(x_i^*)h &= -4h\left(\frac{5}{h}\right) + h^2\left(\frac{5}{h}\right)\left(\frac{5}{h} + 1\right) \\ &= -20 + 5(5 + h).\end{aligned}$$

Hence

$$\begin{aligned}A = \int_{-2}^{3} 2x\,dx &= \lim_{h\to 0} \sum_{i=1}^{n} f(x_i^*)h \\ &= \lim_{h\to 0} [-20 + 5(5 + h)] \\ &= 5 \text{ square units.}\end{aligned}$$

||

Exercises 4.2

Use the definition of the definite integral to calculate the net area bounded by the graph of $y = f(x)$ and the x-axis between a and b.

1. $f(x) = 2$, $a = -3$ and $b = 5$.
2. $f(x) = 5$, $a = 2$ and $b = 7$.
3. $f(x) = -5$, $a = -1$ and $b = 4$.
4. $f(x) = -2$, $a = -6$ and $b = 1$.
5. $f(x) = x$, $a = 2$ and $b = 6$.
6. $f(x) = 2x$, $a = 1$ and $b = 5$.
7. $f(x) = 4x$, $a = -3$ and $b = 8$.
8. $f(x) = 2x + 1$, $a = -2$ and $b = 5$.

9. $f(x) = 2x - 1, \quad a = -1$ and $b = 3$.

10. $f(x) = 4x, \quad a = -6$ and $b = 1$.

11. $f(x) = 3x^2, \quad a = 0$ and $b = 4$.

12. $f(x) = x^2, \quad a = 0$ and $b = 3$.

4.3 Another Approach to the Area Problem; Antiderivatives

We shall now consider a function $A(x)$ that gives the net area bounded by the graph of $y = f(x)$ and the x-axis between a and x. As before we shall assume that f is continuous on every interval under consideration. Figure 4.3.1 illustrates the situation.

As a particular example, let us take the case shown in Figure 4.3.2, where $f(x) = 3$ and $a = 2$. Here, $A(x)$ is the area of a rectangle of height 3 and width $x - 2$, so $A(x) = 3(x - 2) = 3x - 6$.

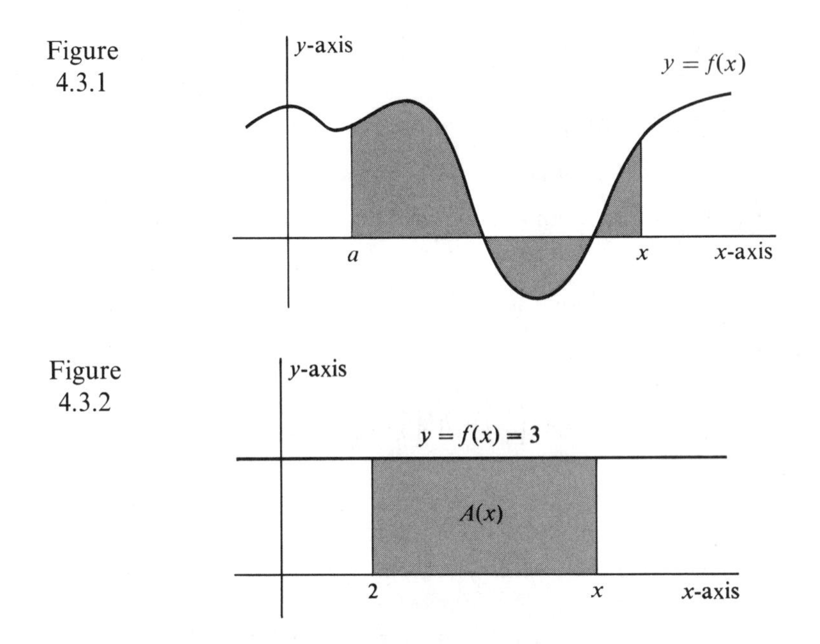

Figure 4.3.1

Figure 4.3.2

Usually, however, $A(x)$ is much more difficult to obtain; therefore let us consider $A(x)$ further. It happens that we can easily find $A'(x)$, the derivative of $A(x)$. In fact we shall prove that $A'(x) = f(x)$.

Before attacking the general proof, let us consider the special case where $f(x) = x/2$ and $a = 4$ as shown in Figure 4.3.3. Here $A(x)$ is the difference of the areas of two triangles and we have

$$A(x) = \frac{1}{2}\left(\frac{x}{2}\right)(x) - \frac{1}{2}(2)(4)$$

$$= \frac{x^2}{4} - 4.$$

Therefore, in this case, $A'(x) = x/2 = f(x)$. We now proceed with the general case.

Figure 4.3.3

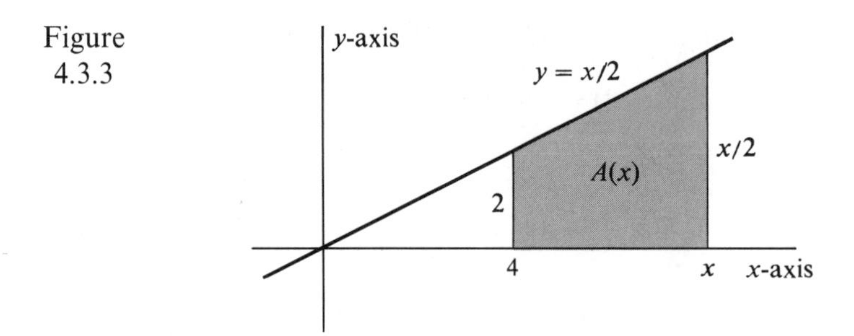

Referring to Figure 4.3.4, note that the net area bounded by the graph of $y = f(x)$ and the x-axis between a and $x + h$ is $A(x + h)$, and the net area between a and x is $A(x)$. Thus, when h is positive, the net area between x and $x + h$ is $A(x + h) - A(x)$. As in Figure 4.3.4, let m be the smallest value and M the largest value of the continuous function $f(x)$ in the closed interval from x to $x + h$. Then we have

$$mh \leq A(x + h) - A(x) \leq Mh,$$

and since $h > 0$,

$$m \leq \frac{A(x + h) - A(x)}{h} \leq M.$$

(Even if h is negative, the preceding inequality is still valid, as the reader can easily verify.)

Figure 4.3.4

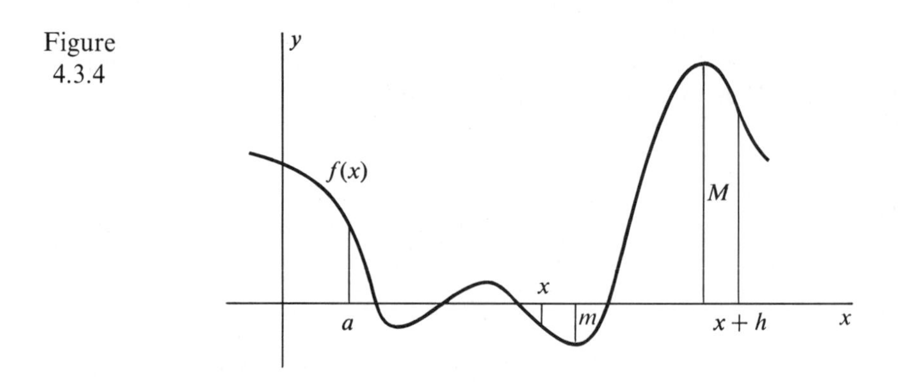

Since f is continuous in the closed interval under consideration, $f(x)$ must take on the values m and M and all values between them (by The Intermediate Value Theorem). Thus there is some x^* such that $x \leq x^* \leq x + h$ and

$$f(x^*) = \frac{A(x + h) - A(x)}{h}.$$

Consequently

$$\lim_{h \to 0} f(x^*) = \lim_{h \to 0} \frac{A(x + h) - A(x)}{h}. \tag{1}$$

Since $x \le x^* \le x + h$, we see that $x^* \to x$ as $h \to 0$. Hence

$$\lim_{h \to 0} f(x^*) = \lim_{x^* \to x} f(x^*).$$

Then, since f is continuous at x,

$$\lim_{x^* \to x} f(x^*) = f(x),$$

and so

$$\lim_{h \to 0} f(x^*) = f(x).$$

Substitution into (1) then gives

$$f(x) = \lim_{h \to 0} \frac{A(x+h) - A(x)}{h}.$$

But by definition of the derivative of $A(x)$, we also have

$$\lim_{h \to 0} \frac{A(x+h) - A(x)}{h} = A'(x),$$

and hence

$$A'(x) = f(x).$$

In summary, we have shown that $A(x)$, the net area bounded by the graph of $y = f(x)$ and the x-axis between a and x, has the property that $A'(x) = f(x)$.

Since we now know that $A'(x) = f(x)$, we are led to the problem of finding a function (in this case $A(x)$) if its derivative is known. That problem is much more difficult than the reverse problem of finding the derivative if the function is known. For one thing, there is never just one function with a given derivative. For example, we know that $(x^3)' = 3x^2$. But also $(x^3 + 2)' = 3x^2$, and in fact, if C is any constant, $(x^3 + C)' = 3x^2$. Thus there are infinitely-many different functions that have $3x^2$ for a derivative.

Definition 4.3.1

If F is a function such that $F'(x) = f(x)$, then $F(x)$ is called an *antiderivative* or *indefinite integral* of $f(x)$.

For example, as we just saw, x^3 and $x^3 + 2$ are both antiderivatives of $3x^2$. The next theorem will be of value in our consideration of antiderivatives.

Theorem 4.3.1

If two functions F and G have the same derivative for $a \le x \le b$, they differ by a constant in the interval. That is, if $F'(x) = G'(x)$ for $a \le x \le b$, then $F(x) - G(x) = C$ where C is a constant, for $a \le x \le b$.

Proof. Let $P(x) = F(x) - G(x)$. Then $P'(x) = F'(x) - G'(x) = 0$ for $a \le x \le b$. Let x_1 and x_2 be any two numbers for which $a \le x_1 < x_2 \le b$. Then by The Mean Value Theorem, there is a number c

such that $x_1 \le c \le x_2$ and

$$P'(c) = \frac{P(x_2) - P(x_1)}{x_2 - x_1}.$$

But $P'(c) = 0$; so $P(x_2) - P(x_1) = 0$. Then $P(x_2) = P(x_1)$, and, since x_1 and x_2 are any two distinct numbers between a and b, it follows that $P(x)$ must be a constant C. We then have

$$F(x) - G(x) = P(x) = C$$

for $a \le x \le b$, which completes the proof of the theorem.

We could write $F(x) - G(x) = C$ in the form $F(x) = G(x) + C$. Therefore, *if $G(x)$ is any antiderivative of a function $f(x)$, then every antiderivative of $f(x)$ has the form $F(x) = G(x) + C$, where C is a constant.*

Example 1 Since $(x^3)' = 3x^2$, every antiderivative of $3x^2$ has the form $x^3 + C$, where C is a constant. ||

Example 2 Since the derivative of $x^2/2$ is x, every antiderivative of x has the form $(x^2/2) + C$ where C is a constant. ||

Example 3 The derivative of x is 1; so every antiderivative of 1 has the form $x + C$, where C is a constant. ||

Example 4 Let a be a constant and let n be any positive integer. Since the derivative of $ax^{n+1}/(n+1)$ is ax^n, every antiderivative of ax^n has the form $ax^{n+1}/(n+1) + C$, where C is a constant. ||

Example 5 Using the result of Example 4, every antiderivative of $3x^7$ has the form $(3x^8/8) + C$ where C is a constant. ||

Now let us apply what we have learned to an area problem.

Example 6 Find the area of the region bounded by the graph of $y = 3x^2$ and the x-axis, between $x = 1$ and $x = 4$.

Since the function $3x^2$ is positive for all x between 1 and 4, the area below the x-axis is 0, and so the net area bounded by $y = 3x^2$ and the x-axis between 1 and 4 is the area of the region bounded by this graph and the x-axis between 1 and 4. Let $A(x)$ be the area of the region under this graph from 1 to x. Then

$$A'(x) = 3x^2, \quad \text{and so}$$

$$A(x) = x^3 + C, \text{ where } C \text{ is a constant.}$$

We can find C, because we know that the area from 1 to 1 is 0. That is, $A(1) = 0$. Thus

$$0 = A(1) = (1)^3 + C = 1 + C,$$

from which we readily get $C = -1$. Therefore

$$A(x) = x^3 - 1.$$

But the area from 1 to 4 is the value of $A(x)$ when $x = 4$:

$$A(4) = 4^3 - 1 = 64 - 1 = 63 \text{ square units.}$$ ||

We shall now do another example in less detail.

Example 7 | Find the net area bounded by the graph of $y = x - 1$ and the x-axis between -2 and 10.

Let $A(x)$ be the net area between -2 and x. Then

$$A'(x) = x - 1, \quad \text{and so}$$

$$A(x) = \frac{x^2}{2} - x + C, \quad \text{where } C \text{ is a constant.}$$

But $A(-2) = 0$; therefore

$$0 = A(-2) = \frac{(-2)^2}{2} - (-2) + C,$$

and hence $C = -4$. Then

$$A(x) = \frac{x^2}{2} - x - 4,$$

and so the result we want is

$$A(10) = \frac{10^2}{2} - 10 - 4 = 36 \text{ square units.}$$ ||

It should be noted that the indicated technique enables us to find the *exact* areas, not just approximations.

It is worthwhile to realize that

$$\int_a^b f(x)dx = \int_a^b f(t)dt$$

since the net area bounded by the graph of $y = f(x)$ and the x-axis between a and b is equal to the net area bounded by the graph of $y = f(t)$ and the t-axis between a and b (using the t-axis in place of the x-axis). This enables us to write $A(x) = \int_a^x f(t)dt$ without confusing symbolism. It is not appropriate to write $A(x) = \int_a^x f(x)dx$, however, because the two x's in the expression $\int_a^x f(x)dx$ would have to have different meanings.

Since $A(x) = \int_a^x f(t)dt$ and $A'(x) = f(x)$, we have the following theorem.

Theorem 4.3.2

Let $f(x)$ be continuous. Then

$$\frac{d}{dx}\left(\int_a^x f(t)dt\right) = f(x).$$

Since that theorem and Theorem 4.4.1 of the next section link the two fundamental processes of calculus, differentiation and integration, each is often called a *Fundamental Theorem of Calculus.* However, since Theorem 4.4.1 is so valuable for the evaluation of definite integrals, we shall call Theorem 4.4.1 *The Fundamental Theorem of Calculus.*

Antiderivatives have uses other than for calculating areas. In particular, they can be applied when studying the effect of gravity on moving objects. Considering the positive direction to be upward, the force due to the gravity of the earth imparts an acceleration of $-g$ on an object close to the earth's surface, where g is approximately 32 ft/sec/sec or 980 cm/sec/sec. We shall neglect air resistance and the fact that the acceleration due to gravity decreases very gradually with the altitude. (See Problem 9 of Challenging Problems, Chapter 4.)

Example 8

A stone is thrown upward with an initial velocity of 64 ft/sec from a point on a ladder 80 ft above the ground. Determine the equation of motion of the stone, the time required for the stone to strike the ground, and the velocity as the stone strikes the ground.

Let s be the height of the stone above the ground and let v be its velocity at elapsed time t. Then

$$\frac{dv}{dt} = -32.$$

Therefore

$$v = -32t + C_1,$$

where C_1 is a constant. Since the initial velocity is 64 ft/sec, $v = 64$ when $t = 0$ and we get

$$64 = -32(0) + C_1.$$

So $C_1 = 64$. Thus

$$v = \frac{ds}{dt} = -32t + 64.$$

Hence

$$s = -16t^2 + 64t + C_2,$$

where C_2 is a constant. But the initial height of the stone is 80 ft, so we know that $s = 80$ when $t = 0$. Therefore

$$80 = -16(0) + 64(0) + C_2,$$

or $C_2 = 80$. Consequently the equation of motion is

$$s = -16t^2 + 64t + 80.$$

To find the time required for the stone to strike the ground we find the value of t for which $s = 0$, that is, we solve the equation

$$-16t^2 + 64t + 80 = 0.$$

We get $t = -1$ or $t = 5$. The solution $t = -1$ is rejected since $t \geq 0$. Thus the time required for the stone to reach the ground is 5 seconds. At that time its velocity is

$$v = -32(5) + 64 = -96 \text{ ft/sec.}$$

||

Exercises 4.3

For Exercises 1–16, find a form for every antiderivative of the given function.

1. $f(x) = 3x$.
2. $f(x) = 5x$.
3. $f(x) = 4x^2$.
4. $f(x) = 7$.
5. $f(x) = 12x^3$.
6. $f(x) = 6x^2$.
7. $f(x) = 2$.
8. $f(x) = 4x^3$.
9. $f(x) = 0$.
10. $f(x) = x^4$.
11. $f(x) = x^5$.
12. $f(x) = x^6$.
13. $f(x) = x + 3$.
14. $f(x) = 5 - x$.
15. $f(x) = x - x^2$.
16. $f(x) = x^2 - x^3$.

For Exercises 17–26, find the area bounded by the graph of $y = f(x)$, and the x-axis between the indicated values of a and b.

17. $f(x) = 2x$, $a = 1$ and $b = 4$.
18. $f(x) = 4x$, $a = 2$ and $b = 5$.
19. $f(x) = 3$, $a = 1$ and $b = 5$.
20. $f(x) = x^2$, $a = 0$ and $b = 3$.
21. $f(x) = x^2 + 2$, $a = 2$ and $b = 5$.
22. $f(x) = 0$, $a = 2$ and $b = 5$.
23. $f(x) = x^3$, $a = 1$ and $b = 2$.
24. $f(x) = x^4$, $a = 1$ and $b = 3$.
25. $f(x) = 10x^4 - 6x^2$, $a = 2$ and $b = 4$.
26. $f(x) = 8x^3 - 6x^5$, $a = 0$ and $b = 1$.

For Exercises 27–32, find the net area bounded by the graph of $y = f(x)$ and the x-axis between the indicated values of a and b.

27. $f(x) = 2x$, $a = -3$ and $b = 2$.
28. $f(x) = 4x$, $a = -5$ and $b = 3$.
29. $f(x) = x - 1$, $a = -7$ and $b = 10$.
30. $f(x) = x - 2$, $a = -4$ and $b = 10$.
31. $f(x) = 4x^3$, $a = -2$ and $b = 1$.
32. $f(x) = 6x^5$, $a = -2$ and $b = 1$.
33. A stone is thrown upward with an initial velocity of 8 ft/sec from a point 24 ft above the ground. Determine the equation of motion of the stone, the time required for the stone to strike the ground, and the velocity as the stone strikes the ground.
34. Do Exercise 33 when the initial velocity is 16 ft/sec and the point is 96 ft above the ground.
35. Do Exercise 33 when the stone is thrown downward at 8 ft/sec.
36. Do Exercise 33 when the stone is thrown downward at 16 ft/sec.
37. A stone is dropped from a height of 96 ft. How long will it take for the stone to hit the ground, and how fast will it be going as it strikes?
38. A stone is dropped from a height of 80 ft. How long will it take for the stone to hit the ground and how fast will it be going when it strikes?

39. A stone was thrown upward at 20 ft/sec and hit the grou
height was the stone thrown?

40. A stone was thrown upward at 30 ft/sec and hit the gro
height was the stone thrown?

4.4 The Fundamental Theorem of

Let $f(x)$ be continuous for all x in the interval $a \leq x \leq$
the definite integral from a to b of $f(x)$ as

$$\int_a^b f(x)dx = \lim_{h \to 0} \sum_{i=1}^{n} f(x_i^*)h,$$

where the interval from a to b is divided into subintervals of equal width h and each x_i^* is an arbitrary point in the ith subinterval.

In addition, we learned that $\int_a^b f(x)dx$ equals the net area bounded by the graph of $y = f(x)$ and the x-axis between a and b. We also saw that definite integrals are very difficult to evaluate.

Using the method of finding areas developed in Section 4.3 we shall now obtain a remarkable theorem that will significantly simplify the calculation of definite integrals. Although our proof requires $f(x)$ to be continuous for $a \leq x \leq b$, that condition can be eliminated by a more theoretical proof.

Theorem 4.4.1

> *The Fundamental Theorem of Calculus*
>
> Let $F(x)$ be *any* function such that $F'(x) = f(x)$ for $a \leq x \leq b$. Then
>
> $$\int_a^b f(x)dx = F(b) - F(a).$$

Proof. Let $A(x)$ be the net area bounded by the graph of $y = f(x)$ and the x-axis between a and x. Then, since $\int_a^b f(x)dx$ is the net area between a and b, it is the value of $A(x)$ when $x = b$, that is,

$$\int_a^b f(x)dx = A(b).$$

Also, since $A'(x) = f(x)$ and $F'(x) = f(x)$ for $a \leq x \leq b$, the two functions differ by at most a constant on the interval. Hence,

$$A(x) = F(x) + C, \quad \text{where } C \text{ is a constant.}$$

Now, since the net area bounded by the graph of $y = f(x)$ and the x-axis between a and a is zero, we have $A(a) = 0$, and so

$$0 = A(a) = F(a) + C.$$

Thus

$$C = -F(a).$$

Therefore,

$$A(x) = F(x) + C = F(x) - F(a).$$

Consequently, $A(b) = F(b) - F(a)$ and we have

$$\int_a^b f(x)dx = A(b) = F(b) - F(a).$$

This completes the proof of the theorem.

For convenience, we often write $F(b) - F(a)$ as $F(x)\big|_a^b$. Using that notation we can say that if F is any function such that $F'(x) = f(x)$, then

$$\int_a^b f(x)dx = F(x)\Big|_a^b$$

Example 1

Since $\frac{d}{dx}(x^2) = 2x$,

$$\int_{-1}^4 2x\,dx = x^2\Big|_{-1}^4 = 4^2 - (-1)^2 = 15. \qquad \|$$

Example 2

Since $\frac{d}{dx}(x^3/3) = x^2$,

$$\int_2^3 x^2\,dx = \frac{x^3}{3}\Big|_2^3 = \frac{3^3}{3} - \frac{2^3}{3} = \frac{27}{3} - \frac{8}{3} = \frac{19}{3}. \qquad \|$$

Example 3

Since $\frac{d}{dx}(x^2 + 3x) = 2x + 3$,

$$\int_{-7}^2 (2x + 3)dx = (x^2 + 3x)\Big|_{-7}^2$$

$$= 2^2 + 3(2) - [(-7)^2 + 3(-7)] = -18. \qquad \|$$

Example 4

Since $\frac{d}{dx}(x^2 + 3) = 2x$,

$$\int_{-1}^4 2x\,dx = (x^2 + 3)\Big|_{-1}^4 = 4^2 + 3 - [(-1)^2 + 3] = 15. \qquad \|$$

Now compare the answers to Examples 1 and 4 and notice how the 3 is eliminated in Example 4. The integrals are the same but are evaluated by using two different choices for $F(x)$, each having $2x$ for a derivative. Both examples illustrate the fact indicated in The Fundamental Theorem that

$$\int_a^b f(x)dx = F(b) - F(a),$$

where $F(x)$ is *any* function such that $F'(x) = f(x)$. We commonly use the simplest possible function for $F(x)$, however. In particular we would evaluate the definite integral of Examples 1 and 4 as in Example 1.

Exercises 4.4

Evaluate the following definite integrals.

1. $\int_2^5 x\,dx.$
2. $\int_0^2 x^2\,dx.$
3. $\int_{-5}^{-3} 6x^2\,dx.$
4. $\int_3^7 6x\,dx.$
5. $\int_{-7}^{-3} 5dx.$
6. $\int_{-1}^4 2dx.$
7. $\int_2^6 x^3\,dx.$
8. $\int_1^3 4x^3\,dx.$
9. $\int_{-3}^2 (x+2)dx.$
10. $\int_{-1}^4 (x^2+4x)dx.$
11. $\int_1^4 (3x^2+4x^3)dx.$
12. $\int_2^3 (10x^4+1)dx.$
13. $\int_3^7 dx\left(=\int_3^7 1\,dx\right).$
14. $\int_0^3 (3+6x^2)dx.$
15. $\int_{-2}^0 (1-x^3)dx.$
16. $\int_{-3}^1 (9-6x^5)dx.$
17. $\int_{-2}^{-1} (6x^2-7x^6)dx.$
18. $\int_{-1}^1 (5x^3-x)dx.$
19. $\int_{-2}^2 (x^5+7x^3)dx.$
20. $\int_{-1}^1 (10x^4-2x^5)dx.$
21. $\int_{-3}^3 x^7\,dx.$
22. $\int_{-4}^4 x^9\,dx.$
23. $\int_{-1}^1 36x^8\,dx.$
24. $\int_{-1}^1 33x^{10}\,dx.$

For Exercises 25–31, let a and b be constants and verify the indicated result by evaluating the definite integral and then differentiating.

25. $\frac{d}{dx}\int_a^x (t^2-3t+2)dt = x^2-3x+2.$
26. $\frac{d}{dx}\int_a^x (7t^3+6t^2-5)dt = 7x^3+6x^2-5.$
27. $\frac{d}{dx}\int_a^{bx} (8-9t^3+t^2)dt = b(8-9(bx)^3+(bx)^2).$
28. $\frac{d}{dx}\int_a^{bx} (6t^2+8t-3)dt = b(6(bx)^2+8(bx)-3).$
29. $\frac{d}{dx}\int_{ax}^{bx} (t^3-8t+6)dt = b[(bx)^3-8(bx)+6]-a[(ax)^3-8(ax)+6].$
30. $\frac{d}{dx}\int_{ax}^{bx} (3t^2+2t-7)dt = b[3(bx)^2+2(bx)-7]-a[3(ax)^2+2(ax)-7].$
31. Show that $\frac{d}{dx}\int_{ax}^{bx} f(t)dt = bf(bx)-af(ax).$
32. An object is moving in a straight line. The velocity $y=v(t)$ is recorded and graphed. What is the physical interpretation of the area of the region under the graph of $y=v(t)$ and above the t-axis for t between a and b?

4.5 Additional Properties of the Definite Integral

When $a \le b$ and a function $f(x)$ is continuous for $a \le x \le b$, we have defined $\int_a^b f(x)dx$, and we have seen that it is equal to the net area bounded by the graph of $y=f(x)$ and the x-axis between a and b.

Actually our original definition of the definite integral can be applied equally well to those cases where $a \ge b$. We simply omit any restriction and give this more general definition.

"Walsh"

Definition 4.5.1

$$\int_a^b f(x)dx = \lim_{h\to 0} \sum_{i=1}^{n} f(x_i^*)h,$$

where $h = (b - a)/n$ and the x_i^*s are arbitrarily selected from the appropriate subintervals.

Let's examine the definite integral further. Suppose that $a = b$. Then

$$h = \frac{b - a}{n} = 0, \qquad \text{and so}$$

$$\int_a^a f(x)dx = \lim_{h\to 0} \sum_{i=1}^{n} f(x_i^*)h = \lim_{h\to 0} 0 = 0. \tag{1}$$

We shall now prove that $\int_a^b f(x)dx = -\int_b^a f(x)dx$. By Definition 4.5.1,

$$\int_a^b f(x)dx = \lim_{h\to 0} \sum_{i=1}^{n} f(x_i^*)h,$$

where $h = (b - a)/n$. Now, again using Definition 4.5.1 to express $\int_b^a f(x)dx$, we obtain

$$\int_b^a f(x)dx = \lim_{r\to 0} \sum_{i=1}^{n} f(x_i^*)r,$$

where
$$r = \frac{a - b}{n} = -\frac{b - a}{n} = -h.$$

Thus,
$$\begin{aligned}\int_a^b f(x)dx &= \lim_{h\to 0} \sum_{i=1}^{n} f(x_i^*)h \\ &= -\lim_{h\to 0} \sum_{i=1}^{n} f(x_i^*)(-h) \\ &= -\lim_{r\to 0} \sum_{i=1}^{n} f(x_i^*)r \\ &= -\int_b^a f(x)dx.\end{aligned}$$

Consequently,

$$\int_a^b f(x)dx = -\int_b^a f(x)dx. \tag{2}$$

We also see that for any constant k,

$$\int_a^b kf(x)dx = \lim_{h\to 0} \sum_{i=1}^{n} kf(x_i^*)h.$$

Factoring out the k and then using The Limit Theorem, we get

$$\lim_{h\to 0} k \sum_{i=1}^{n} f(x_i^*)h = k \cdot \lim_{h\to 0} \sum_{i=1}^{n} f(x_i^*)h = k \int_a^b f(x)dx.$$

Hence

$$\int_a^b kf(x)dx = k \int_a^b f(x)dx. \tag{3}$$

The results (1), (2), and (3) make up the first 3 parts of the following theorem. The fourth part is left as an exercise.

Theorem 4.5.1

If a and b are any real numbers, k is any constant and the indicated integrals exist, then

(i) $\int_a^a f(x)dx = 0,$

(ii) $\int_a^b f(x)dx = -\int_b^a f(x)dx,$

(iii) $\int_a^b kf(x)dx = k \int_a^b f(x)dx,$

(iv) $\int_a^b [f(x) \pm g(x)]dx = \int_a^b f(x)dx \pm \int_a^b g(x)dx.$

It is essential to remember when using part (iii) of this theorem that k must be a constant and so cannot involve x.

We now note that The Fundamental Theorem of Calculus still holds for the more general definition of the definite integral given by Definition 4.5.1. That is, the following proposition is true.

Theorem 4.5.2

If F is any function such that $F'(x) = f(x)$, and $a > b$, then

$$\int_a^b f(x)dx = F(b) - F(a).$$

Proof. Since $a > b$, we already know that

$$\int_b^a f(x)dx = F(a) - F(b), \quad \text{and so}$$

$$-\int_b^a f(x)dx = -[F(a) - F(b)] = F(b) - F(a).$$

But by Theorem 4.5.1,

$$\int_a^b f(x)dx = -\int_b^a f(x)dx.$$

And then by combining the last two steps we get the desired result:

$$\int_a^b f(x)dx = F(b) - F(a).$$

If $a < c < b$, then as shown in Figure 4.5.1, the area from a to c plus the area from c to b is the area from a to b. That is,

$$\int_a^c f(x)dx + \int_c^b f(x)dx = \int_a^b f(x)dx. \tag{4}$$

Figure 4.5.1

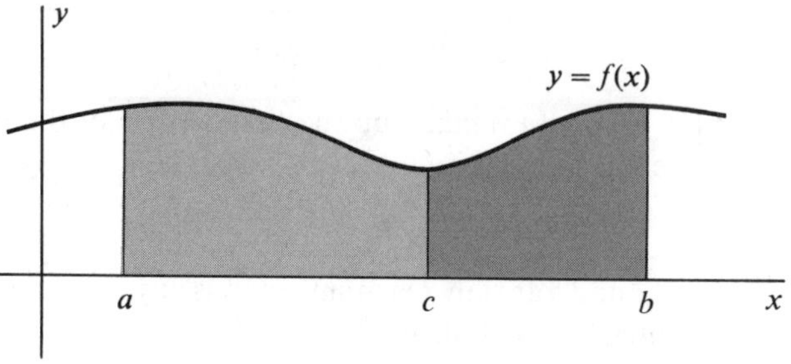

With Definition 4.5.1, equation (4) holds no matter how the real numbers a, b, and c are related, even when we do not have $a < c < b$. This result is stated in the next theorem.

Theorem 4.5.3

If a, b, and c are any real numbers and f is continuous in the closed intervals between a and c, between c and b, and between a and b, then

$$\int_a^c f(x)dx + \int_c^b f(x)dx = \int_a^b f(x)dx.$$

Proof. Suppose that $F'(x) = f(x)$. Then by The Fundamental Theorem of Calculus,

$$\int_a^c f(x)dx + \int_c^b f(x)dx = [F(c) - F(a)] + [F(b) - F(c)] = F(b) - F(a).$$

Also by The Fundamental Theorem, we have

$$\int_a^b f(x)dx = F(b) - F(a).$$

Therefore $$\int_a^c f(x)dx + \int_c^b f(x)dx = \int_a^b f(x)dx.$$

Actually Theorem 4.5.3 can be proved without assuming the existence of a function F such that $F'(x) = f(x)$. Such a proof requires slightly more advanced techniques that can be found in most advanced calculus texts.

Theorem 4.5.3 considerably simplifies the computation of certain integrals, as is illustrated in the next example.

Example 1 | Compute $\int_{-1}^{2} |x^2 - 3x + 2| dx$.

Since $x^2 - 3x + 2 = (x - 2)(x - 1)$, we have

$$x^2 - 3x + 2 \geq 0 \quad \text{if} \quad x \leq 1 \text{ or } x \geq 2, \quad \text{and}$$

$$x^2 - 3x + 2 < 0 \quad \text{if} \quad 1 < x < 2, \quad \text{and so}$$

$$|x^2 - 3x + 2| = \begin{cases} x^2 - 3x + 2 & \text{if} \quad x \leq 1 \text{ or } x \geq 2 \\ -(x^2 - 3x + 2) & \text{if} \quad 1 < x < 2. \end{cases}$$

Then since

$$\int_{-1}^{2} |x^2 - 3x + 2|dx = \int_{-1}^{1} |x^2 - 3x + 2|dx + \int_{1}^{2} |x^2 - 3x + 2|dx,$$

we have

$$\int_{-1}^{2} |x^2 - 3x + 2|dx = \int_{-1}^{1} (x^2 - 3x + 2)dx + \int_{1}^{2} -(x^2 - 3x + 2)dx$$

$$= \left(\frac{1}{3}x^3 - \frac{3}{2}x^2 + 2x\right)\Big|_{-1}^{1} - \left(\frac{1}{3}x^3 - \frac{3}{2}x^2 + 2x\right)\Big|_{1}^{2}$$

$$= 4\frac{5}{6}$$

||

Exercises 4.5

In Exercises 1–26, evaluate the given definite integrals.

1. $\int_2^5 -x\,dx.$
2. $\int_4^4 x^3\,dx.$
3. $\int_3^3 x^9\,dx.$
4. $\int_{-5}^{-5} \sin x\,dx.$
5. $\int_7^7 g(x)dx.$
6. $\int_8^5 (1 - x)dx.$
7. $\int_3^1 x^3\,dx.$
8. $\int_2^1 (x^3 - x)dx.$
9. $\int_2^{-4} (x - 5x^4)dx.$
10. $\int_2^0 (1 - 12x^5)dx.$
11. $\int_{-2}^{5} 0\,dx.$
12. $\int_{-2}^{-1} 8x^7\,dx.$
13. $\int_{-3}^{7} |x|dx.$
14. $\int_0^5 |x - 2|dx.$
15. $\int_{-2}^{3} x|x|dx.$
16. $\int_{-1}^{4} |x^2 - 2x|dx.$
17. $\int_{-3}^{2} |x + x^2|dx.$
18. $\int_0^4 |x^2 - 5x + 6|dx.$
19. $\int_{-4}^{0} |x + 3|dx.$
20. $\int_{-4}^{1} |x^2 + 3x|dx.$
21. $\int_{-1}^{4} |3x - x^2|dx.$
22. $\int_{-10}^{1} |5 + x|dx.$
23. $\int_{-2}^{2} x|x + 1|dx.$
24. $\int_{-2}^{3} x|x - 2|dx.$
25. $\int_{-2}^{3} 6|x + 2 - x^2|dx.$
26. $\int_{-4}^{3} 6|6 - x - x^2|dx.$
27. Assume that for all x such that $a \leq x \leq b$, $f(x)$ and $g(x)$ are continuous and $f(x) \leq g(x)$. It can then be proved that

$$\int_a^b f(x)dx \leq \int_a^b g(x)dx.$$

Using this information, prove that

$$\left|\int_a^b f(x)dx\right| \leq \int_a^b |f(x)|dx.$$

28. Prove part (iv) of Theorem 4.5.1.

29. Let f be continuous for $-a \leq x \leq a$ and let f be an *odd* function (that is, $f(-x) = -f(x)$ for all x). Prove that $\int_{-a}^{a} f(x)dx = 0$.

30. Let f be continuous for $-a \leq x \leq a$ and let f be an *even* function (that is, $f(-x) = f(x)$ for all x). Prove that $\int_{-a}^{a} f(x)dx = 2\int_{0}^{a} f(x)dx$.

31. Use the result of Exercise 29 to evaluate each integral
(a) $\int_{-7}^{7} x^3\sqrt{5x^6 + 3}\, dx$. (b) $\int_{-2}^{2} x^4 \sin 2x\, dx$.

4.6 Computing Areas

We have found that when $a \leq b$, $\int_a^b f(x)dx$ is equal to the net area bounded by the graph of $y = f(x)$ and the x-axis between a and b. Also we have noted that when $f(x) \geq 0$ for $a \leq x \leq b$, the region is entirely above the x-axis, and so $\int_a^b f(x)dx$ is simply the area of the region bounded by the graph of $y = f(x)$ and the x-axis between the vertical lines $x = a$ and $x = b$. Similarly, when $f(x) \leq 0$, the region is entirely below the x-axis, and so the area of the region bounded by $y = f(x)$ and the x-axis between the lines $x = a$ and $x = b$ is $-\int_a^b f(x)dx$.

In the more general case where $f(x)$ is both positive and negative for $a \leq x \leq b$, part of the region bounded by the graph of $y = f(x)$ and the x-axis between the vertical lines $x = a$ and $x = b$ is above the x-axis and part of the region is below the x-axis. We shall call the area of the part of the region above the x-axis *plus* the area of the part of the region below the x-axis the *total* area bounded by the graph of $y = f(x)$ and the x-axis between a and b.

Example 1 Find the total area of the region bounded by the graph of $y = 2x - x^2$ and the x-axis between $x = 1$ and $x = 3$.

Since $2x - x^2 = x(2 - x)$, the graph crosses the x-axis at $x = 0$ and $x = 2$, as in Figure 4.6.1. Hence

$$\int_1^2 (2x - x^2)dx = \text{the area of the part of the region above the } x\text{-axis},$$

and

$$-\int_2^3 (2x - x^2)dx = \text{the area of the part of the region below the } x\text{-axis}.$$

Therefore the total area is

$$\int_1^2 (2x - x^2)dx - \int_2^3 (2x - x^2)dx.$$

Since $\left(x^2 - \dfrac{x^3}{3}\right)' = 2x - x^2$, we can make the following evaluation:

$$\left(x^2 - \frac{x^3}{3}\right)\Bigg|_1^2 - \left(x^2 - \frac{x^3}{3}\right)\Bigg|_2^3 = \left[\left(4 - \frac{8}{3}\right) - \left(1 - \frac{1}{3}\right)\right] - \left[\left(9 - \frac{27}{3}\right) - \left(4 - \frac{8}{3}\right)\right]$$
$$= 2 \text{ square units},$$

which is the desired area. ||

Figure 4.6.1

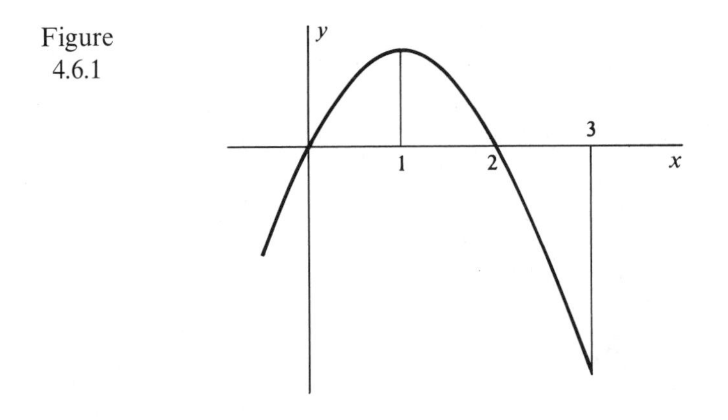

Now we shall show how to find the area of a region bounded by two curves and two vertical lines. Consider the region bounded by the graphs of $y = f(x)$, $y = g(x)$, $x = a$, and $x = b$. The simplest situation is illustrated in Figure 4.6.2. Here we see that the area in question is the area under the upper curve minus the area under the lower curve, that is,

$$\int_a^b f(x)dx - \int_a^b g(x)dx.$$

This holds as long as $y = f(x)$ is the upper curve and $y = g(x)$ is the lower curve, even if one or both of the curves were below or partly below the x-axis. A consideration of cases will verify this fact.

Figure 4.6.2

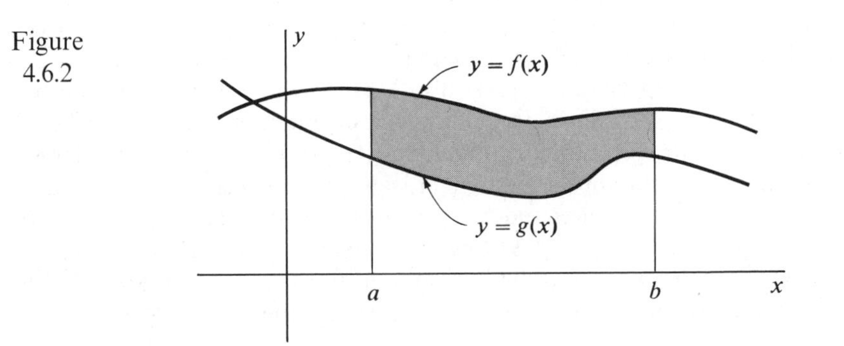

Example 2 | Find the area of the region bounded by $y = x^2 - x - 2$ and $y = x - 2$.

We can rapidly sketch the graph of $y = x^2 - x - 2 = (x - 2)(x + 1)$, as in Figure 4.6.3. In this case the vertical lines between which the region lies are not given and we must determine them by solving the equations $y = x^2 - x - 2$ and $y = x - 2$ simultaneously. Solving the equation $x^2 - x - 2 = x - 2$, we obtain $(0, -2)$ and $(2, 0)$ as the points of intersection. So the region that we are concerned with lies

Figure 4.6.3

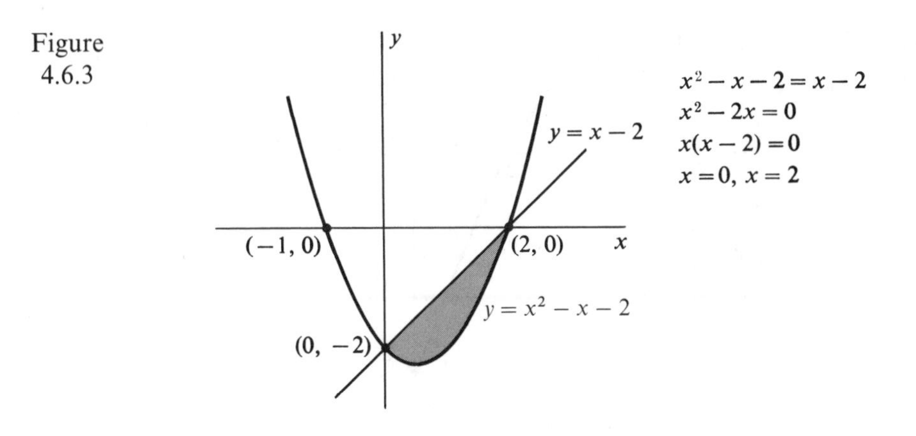

between $x = 0$ and $x = 2$. Since $y = x - 2$ is the upper curve, the area A is

$$A = \int_0^2 (x-2)dx - \int_0^2 (x^2 - x - 2)dx$$

$$= \left(\frac{x^2}{2} - 2x\right)\Bigg|_0^2 - \left(\frac{x^3}{3} - \frac{x^2}{2} - 2x\right)\Bigg|_0^2$$

$$= (-2) - \left(-\frac{10}{3}\right)$$

$$= \frac{4}{3} \text{ square units.}$$

||

Another way to find the area A of the region bounded by the continuous curves $y = f(x)$ and $y = g(x)$ and the vertical lines $x = a$ and $x = b$ is to slice it into n pieces of equal width h and to use a single definite integral. Let $y = f(x)$ be the upper curve and $y = g(x)$ be the lower curve as pictured in Figure 4.6.4. That is, assume that $f(x) \geq g(x)$ for all $a \leq x \leq b$. Since $f(x)$ and $g(x)$ are continuous for $a \leq x \leq b$, we see that $w(x) = f(x) - g(x)$ is continuous and nonnegative for $a \leq x \leq b$. An enlarged picture of the ith slice might appear as in Figure 4.6.5, where m is the smallest value of $w(x)$ for the interval of the slice, and M is the largest value. Then the area A_i of this slice has the property that $mh \leq A_i \leq Mh$, and so $m \leq A_i/h \leq M$.

Since $w(x)$ is continuous in the subinterval and has the values m and M, The Intermediate Value Theorem implies that $w(x)$ assumes the values A_i/h for some

Figure 4.6.4

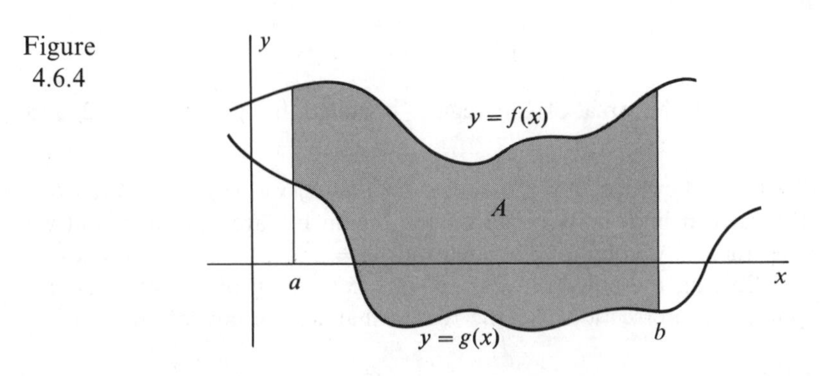

Figure 4.6.5

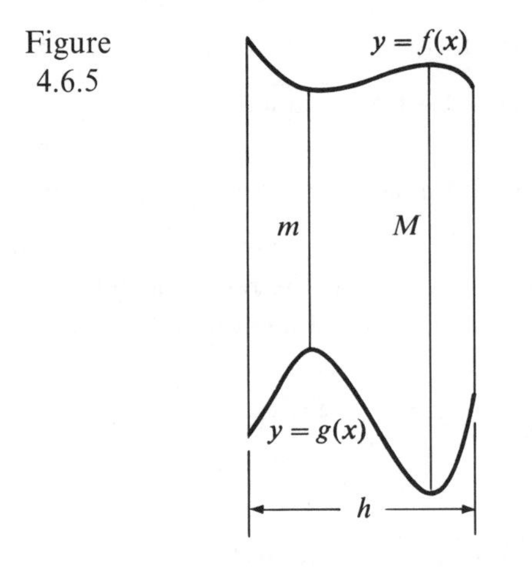

point x_i^* in the subinterval, that is,

$$\frac{A_i}{h} = w(x_i^*),$$

and so

$$A_i = w(x_i^*)h.$$

Then, since the total area is the sum of the areas of the separate slices, we have

$$A = \sum_{i=1}^{n} w(x_i^*)h.$$

Since A is independent of h, then $\lim_{h \to 0} A = A$, and thus A is also equal to the limit of this sum as h approaches zero. Therefore, as before

$$A = \int_a^b w(x)dx = \int_a^b [f(x) - g(x)]dx.$$

Intuitive Development. Sketch a picture where a typical element of area is the area of a region of height $w(x) = f(x) - g(x)$ and width dx, as indicated in Figure 4.6.6. Then as before think of $\int_a^b$ as summing up the elements of area from a to b to get

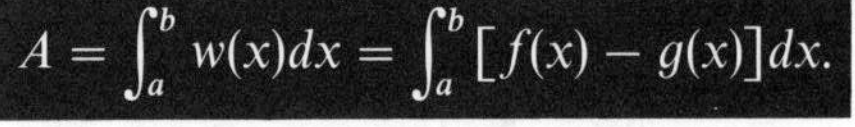

$$A = \int_a^b w(x)dx = \int_a^b [f(x) - g(x)]dx.$$

Figure 4.6.6

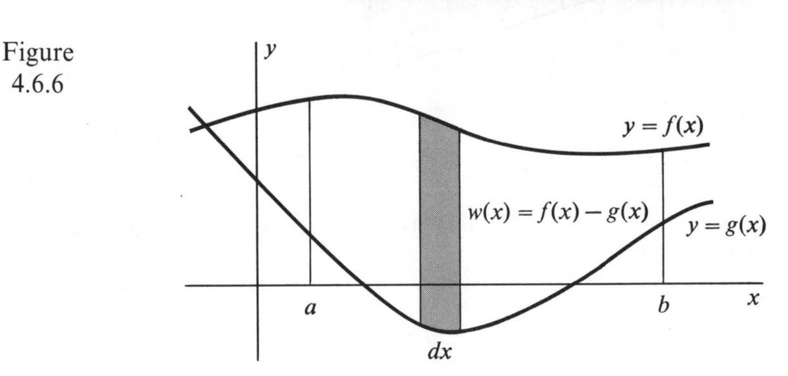

Remember that $w(x)$ has the form

$$w(x) = (\text{upper function}) - (\text{lower function}).$$

Example 3 Find the area of the region bounded by $y = x^2 - x - 2$ and $y = x - 2$.

Note that this is a repeat of Example 2 where we found that the region is between $x = 0$ and $x = 2$ as shown in Figure 4.6.3. Since the upper function is $x - 2$ and the lower function is $x^2 - x - 2$ we obtain

$$w(x) = (x - 2) - (x^2 - x - 2) = 2x - x^2.$$

Therefore the area is

$$\int_0^2 (2x - x^2)dx = \left(x^2 - \frac{x^3}{3}\right)\bigg|_0^2 = 4 - \frac{8}{3} = \frac{4}{3} \text{ square units,}$$

which is the same answer obtained before by a different technique. ||

Example 4 Find the area of the region bounded by the graphs of $y = \frac{x}{2}$ and $y = -\frac{x}{3}$ between $x = 1$ and $x = 5$.

The region is pictured in Figure 4.6.7. Since the upper function minus the lower function is

$$w(x) = \frac{x}{2} - \left(-\frac{x}{3}\right) = \frac{5}{6}x,$$

we get

$$A = \int_1^5 \frac{5}{6} x\, dx = \frac{5x^2}{12}\bigg|_1^5 = \frac{5}{12}(25 - 1) = 10 \text{ square units.}$$ ||

Figure 4.6.7

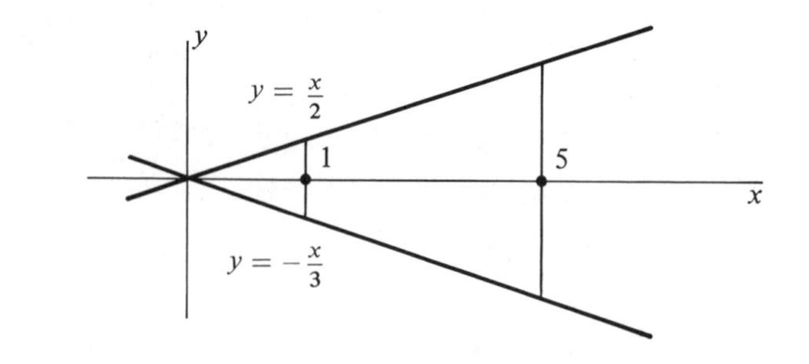

Exercises 4.6

For Exercises 1–8, find the area of the region bounded by the graph of the given equation and the x-axis, between the given values of a and b.

1. $y = x, \quad a = 2, \quad b = 5.$
2. $y = x, \quad a = -7, \quad b = -1.$
3. $y = -x, \quad a = 2, \quad b = 6.$
4. $y = -x, \quad a = -5, \quad b = -1.$

5. $y = -x^2, \quad a = 1, \quad b = 4.$

6. $y = -x^2, \quad a = -2, \quad b = 1.$

7. $y = 3 - x, \quad a = -1, \quad b = 3.$

8. $y = 3 + x, \quad a = 3, \quad b = 7.$

For Exercises 9–18, find the total area bounded by the graph of $y = f(x)$ and the x-axis between $x = a$ and $x = b$.

9. $y = x + 1, \quad a = -5, \quad b = 1.$

10. $y = x - 2, \quad a = -6, \quad b = 3.$

11. $y = 3x^2 - 6x, \quad a = 1, \quad b = 4.$

12. $y = 6x - 6x^2, \quad a = 0, \quad b = 3.$

13. $y = 4x^3, \quad a = -3, \quad b = 2.$

14. $y = -8x^3, \quad a = -4, \quad b = 3.$

15. $y = 6x - 3x^2, \quad a = -2, \quad b = 5.$

16. $y = 6x^2 - 6x, \quad a = -3, \quad b = 4.$

17. $y = 6x^2 + 6x, \quad a = -3, \quad b = 2.$

18. $y = 3x^2 + 6x, \quad a = -4, \quad b = 3.$

For Exercises 19–24, find the area of the region bounded by the graphs of the given equations and the x-axis between the given values of a and b.

19. $y = x, \quad y = 2x, \quad a = 2, \quad b = 8.$

20. $y = 5x, \quad y = 3x, \quad a = 1, \quad b = 4.$

21. $y = x, \quad 4y = x^2, \quad a = 1, \quad b = 3.$

22. $8y = x^2, \quad y = 2x, \quad a = 3, \quad b = 10.$

23. $y = 5 - x^2, \quad y = x^2 - 9, \quad a = -1, \quad b = 2.$

24. $y = x^2 - 6, \quad y = 10 - x^2, \quad a = -2, \quad b = 1.$

For Exercises 25–34, sketch the graphs of the given equations, and then find the area of the region that they enclose.

25. $y = x, \quad y = 3, \quad x = 0.$

26. $y = x, \quad y = 0, \quad x = 4.$

27. $y = 2x, \quad y = 3x, \quad x = 4.$

28. $y = x, \quad y = -x, \quad x = 3.$

29. $y = x^2, \quad y = x.$

30. $y = x^2, \quad y = 2 - x.$

31. $y = 0, \quad y = x^2 - 5x.$

32. $y = 0, \quad y = x^2 - 2x.$

33. $y = x^2 - 2x, \quad y = 2x - 3.$

34. $y = x^2 + 3x, \quad y = -2x - 4.$

4.7 The Indefinite Integral

As we have seen, The Fundamental Theorem of Calculus is a very powerful tool for the evaluation of definite integrals. However to use The Fundamental Theorem of Calculus to evaluate $\int_a^b f(x)dx$ we must be able to determine an antiderivative or indefinite integral of $f(x)$. This naturally leads us to the problem of finding indefinite integrals. Although the problem of finding indefinite integrals is much more difficult than the problem of finding derivatives, we shall be able to find indefinite integrals for many functions.

In Section 4.3 we learned that if $F(x)$ is *any* indefinite integral of $f(x)$, then *every* indefinite integral of $f(x)$ has the form $F(x) + C$, where C is a constant. This idea is so closely related to the evaluation of definite integrals by means of The Fundamental Theorem of Calculus that we express it in symbols as

$$\int f(x)dx = F(x) + C.$$

The expression $\int f(x)dx$ is called *the indefinite integral* of $f(x)$.

Thus in order to determine $\int f(x)dx$ we find a function $F(x)$ for which $F'(x) = f(x)$ and add a constant. This could also be expressed by saying that we add a constant to any function $F(x)$ whose differential is $f(x)dx$.

Example 1 Since $\frac{d}{dx}(x^2) = 2x$, then $\int 2x\, dx = x^2 + C$. ||

Example 2 Since $\frac{d}{dx}\left(\frac{x^3}{3}\right) = x^2$, then $\int x^2\, dx = \frac{x^3}{3} + C$. ||

Example 3 Since $\frac{d}{dx}\left(\frac{x^2}{2} + 5x\right) = x + 5$, then $\int (x + 5)dx = \frac{x^2}{2} + 5x + C$. ||

Note that since

$$\frac{d}{dx}\int f(x)dx = \frac{d}{dx}(F(x) + C) = F'(x) = f(x),$$

we have

$$\frac{d}{dx}\int f(x)dx = f(x).$$

Therefore from The Fundamental Theorem of Calculus we get

$$\int_a^b f(x)dx = \left(\int f(x)dx\right)\Big|_a^b.$$

Let us now develop some formulas for finding indefinite integrals. Actually every differentiation formula can be restated as an integration formula. For example, since $dx/dx = 1$, we have $\int 1\, dx = \int dx = x + C$. Also since $(d/dx)(x^2/2) = x$, then $\int x\, dx = x^2/2 + C$. More generally, when n is rational and $n \neq -1$, we see that $(d/dx)(x^{n+1}/(n + 1)) = (n + 1)x^n/(n + 1) = x^n$, and so we have

$$\int x^n\, dx = \frac{x^{n+1}}{n + 1} + C, \qquad \text{for } n \neq -1.$$

That is, the indefinite integral of a power of x is x to a power one higher, divided by the new exponent, plus a constant. It is worthwhile to remember this important formula in both forms, verbal and symbolic.

Also, if $F'(x) = f(x)$ and $G'(x) = g(x)$, then

$$[F(x) + G(x)]' = F'(x) + G'(x) = f(x) + g(x).$$

Therefore

$$\int [f(x) + g(x)]dx = \int f(x)dx + \int g(x)dx.$$

In similar fashion we can prove that

$$\int [f(x) - g(x)]dx = \int f(x)dx - \int g(x)dx.$$

When k is a constant, $(d/dx)[kF(x)] = k(d/dx)F(x)$, and so we also obtain

$$\int kf(x)dx = k\int f(x)dx.$$

It is important to realize that this does not hold, however, if k is not a constant. For example, since

$$\int x^2\,dx = \frac{x^3}{3} + C \quad \text{and} \quad x^2 \int dx = x^2(x + C)$$

are not equal, we have

$$\int x^2\,dx \neq x^2 \int dx.$$

To summarize, we have proved the following theorem.

Thoerem 4.7.1

(i) $\int dx = x + C$.

(ii) $\int x^n\,dx = \dfrac{x^{n+1}}{n+1} + C$ if n is any rational number other than -1.

(iii) $\int[f(x) + g(x)]dx = \int f(x)dx + \int g(x)dx$.

(iv) $\int[f(x) - g(x)]dx = \int f(x)dx - \int g(x)dx$.

(v) $\int kf(x)dx = k\int f(x)dx$, if k is a constant.

Example 4 By (i) and (v), $\int 3\,dx = 3\int dx = 3(x + C_1) = 3x + 3C_1 = 3x + C$, where $C = 3C_1$. ||

Example 5 By (ii), $\int x^{1/4}\,dx = (4/5)x^{5/4} + C$. ||

Example 6 By (ii) and (iii),

$$\begin{aligned}\int(x^3 + x^4)dx &= \int x^3\,dx + \int x^4\,dx \\ &= \frac{x^4}{4} + C_1 + \frac{x^5}{5} + C_2 \\ &= \frac{x^4}{4} + \frac{x^5}{5} + C \quad \text{where } C = C_1 + C_2.\end{aligned}$$ ||

Example 7 By (ii) and (v),

$$\int 7x^{-9}\,dx = 7\int x^{-9}\,dx = \frac{7x^{-8}}{-8} + C.$$ ||

By combining the results of all parts of the preceding theorem we can immediately write the indefinite integral of any polynomial. For example, we can immediately write

$$\int(3x^4 + 2x^2 - 5x + 6)dx = \frac{3x^5}{5} + \frac{2x^3}{3} - \frac{5x^2}{2} + 6x + C.$$

We shall now consider an integration formula, called the *substitution formula*, that arises from a restatement of the chain rule. Recall that if $y = F(u)$ and $u = g(x)$, then the chain rule can be stated as follows:

$$\frac{dF(u)}{dx} = \frac{dy}{dx} = \frac{dy}{du}\frac{du}{dx} = \frac{dF(u)}{du}\frac{du}{dx}.$$

If we let $dF(u)/du = f(u)$, then we have

$$\frac{d}{dx}F(u) = f(u)\frac{du}{dx}.$$

That is, $F(u)$ is a function whose derivative with respect to x is $f(u)du/dx$, or $F(u)$ is an indefinite integral of $f(u)du/dx$. Thus,

$$\int f(u)\frac{du}{dx}\,dx = F(u) + C.$$

It is important to realize that in order to get $F(u) + C$ for the indefinite integral, the integrand must be $f(u)du/dx$. If the integrand were only $f(u)$ the formula would not hold. Remember that $F(u)$ is a function whose *derivative with respect to u is* $f(u)$.

Example 8 | Find $\int[(x^2 + 1)^2 + 3(x^2 + 1)]2x\,dx$.

Here we have the proper form if we let $u = x^2 + 1$ and $f(u) = u^2 + 3u$, because the $2x$ is exactly the needed du/dx. Our original integral is then equal to

$$\int(u^2 + 3u)\frac{du}{dx}\,dx,$$

which equals $F(u) + C$, where $F(u)$ is a function with the property that $dF(u)/du = u^2 + 3u$. We see that $F(u)$ can be taken to be $u^3/3 + 3u^2/2$. Thus

$$\int(u^2 + 3u)\frac{du}{dx}\,dx = \frac{u^3}{3} + \frac{3u^2}{2} + C.$$

Since $u = x^2 + 1$, we have

$$\int[(x^2 + 1)^2 + 3(x^2 + 1)]2x\,dx = \frac{(x^2 + 1)^3}{3} + \frac{3(x^2 + 1)^2}{2} + C.$$

We write the answer in terms of x. The u was only a symbol used to help us get the result. ||

In that example, if we did not have the $2x$ we could not have used the integration formula above. That is, if we had this integral

$$\int[(x^2 + 1)^2 + 3(x^2 + 1)]dx,$$

we would have to multiply everything out and integrate it as a polynomial:

$$\int[(x^2 + 1)^2 + 3(x^2 + 1)]dx = \int[x^4 + 2x^2 + 1 + 3x^2 + 3]dx$$

$$= \int(x^4 + 5x^2 + 4)dx = \frac{x^5}{5} + \frac{5x^3}{3} + 4x + C.$$

Integration formulas for composite functions can be expressed in simpler form if we use differentials. Recall that if $g(x)$ is a differentiable function of x, then

$$dg(x) = g'(x)dx.$$

If $u = g(x)$ and we rewrite $dg(x) = g'(x)dx$ in terms of u, then, using the fact that $g'(x) = du/dx$, we obtain

$$du = \frac{du}{dx}\,dx.$$

That equation is one of the main reasons for using the differential dx in the notation for the definite integral. Along with the convenient Leibniz notation du/dx for the derivative of u with respect to x, differentials simplify the notation for many integration formulas. In particular, the substitution formula

$$\int f(u)\frac{du}{dx}\,dx = F(u) + C$$

can be expressed more simply as

$$\int f(u)du = F(u) + C,$$

where u is a function of x, and $F(u)$ is any function of u such that $dF(u)/du = f(u)$, and $du = \dfrac{du}{dx}\,dx$ is the differential of u.

An important special instance is the case where $f(u) = u^n$ and $n \neq -1$. In this case $F(u) = u^{n+1}/(n + 1)$, and we have the *power function rule*:

$$\int u^n\,du = \frac{u^{n+1}}{n+1} + C, \quad \text{if } n \neq -1. \tag{1}$$

Again it is well to remember this formula in words as well as in symbols: *the indefinite integral of a power* (different from -1) *of a function times the differential of the function is equal to the function to a power one greater, divided by the new power, plus a constant.*

Using the words helps us to write the answer immediately without writing what u and du are, although we must be sure that the proper differential is there. Let us consider two more examples.

Example 9 | Find $\int(x^2 + 9)^7(2x)dx$.

Here we only have to think to ourselves that the function is $x^2 + 9$, the exponent is 7 and the differential of the function is $2x\,dx$. We can then write the result immediately as

$$\int(x^2 + 9)^7(2x)dx = \frac{(x^2 + 9)^8}{8} + C. \qquad \|$$

If the $2x$, or at least an x, were not present, the integration formula (1) would not be applicable. That is, if we had $\int(x^2 + 9)^7\,dx$, we would have to multiply out the $(x^2 + 9)^7$ and then integrate it as a polynomial.

If an x were present instead of the $2x$, however, we could "fix up" the x by writing $x = (1/2)(2x)$, and then we would move the constant $1/2$ outside the integration sign.

That is we could write:

$$\int (x^2 + 9)^7 x\, dx = \frac{1}{2}\int (x^2 + 9)^7 (2x)dx = \frac{1}{2}\frac{(x^2 + 9)^8}{8} + C$$

$$= \frac{1}{16}(x^2 + 9)^8 + C.$$

Remember that only *constant factors* can be moved outside the integration sign; variables, such as x, cannot. Thus, although $1 = (2x)/(2x)$, we cannot "fix up" $\int (x^2 + 9)^7\, dx$ because

$$\int (x^2 + 9)^7\, dx \neq \frac{1}{2x}\int (x^2 + 9)^7 (2x)dx.$$

Example 10 | Find $\int x\sqrt{3x^2 - 7}\, dx$.

In this case we use the fact that $\sqrt{3x^2 - 7} = (3x^2 - 7)^{1/2}$ and rearrange things to get

$$\int (3x^2 - 7)^{1/2} x\, dx = \frac{1}{6}\int (3x^2 - 7)^{1/2}(6x)dx,$$

where we have also "fixed up" the x to get the proper differential. We can then write the answer immediately as

$$\int x\sqrt{3x^2 - 7}\, dx = \frac{1}{6}\frac{(3x^2 - 7)^{3/2}}{3/2} + C = \frac{1}{9}(3x^2 - 7)^{3/2} + C. \qquad \|$$

Using parts (iii), (iv) and (v) of Theorem 4.7.1, we can integrate many expressions immediately, one term at a time, without extra writing. We give two such examples.

Example 11 | We can write immediately that

$$\int [(x^3 + 3x)^4 + 7(x^3 + 3x)](3x^2 + 3)dx$$

$$= \frac{(x^3 + 3x)^5}{5} + \frac{7(x^3 + 3x)^2}{2} + C. \qquad \|$$

Example 12 | Find $\int [(x^2 - 6x)^3 - 8(x^2 - 6x)^2 + 5(x^2 - 6x) - 8](x - 3)dx$.

Here we first "fix up" the $x - 3$ to be $(1/2)(2x - 6)$ and move the $1/2$ outside the integral sign to get the proper $du = (2x - 6)dx$. We obtain

$$\frac{1}{2}\int [(x^2 - 6x)^3 - 8(x^2 - 6x)^2 + 5(x^2 - 6x) - 8](2x - 6)dx$$

$$= \frac{1}{2}\left[\frac{(x^2 - 6x)^4}{4} - \frac{8(x^2 - 6x)^3}{3} + \frac{5(x^2 - 6x)^2}{2} - 8(x^2 - 6x) + C_1\right]$$

$$= \frac{(x^2 - 6x)^4}{8} - \frac{4(x^2 - 6x)^3}{3} + \frac{5(x^2 - 6x)^2}{4} - 4(x^2 - 6x) + C,$$

where $C = C_1/2$. $\|$

We can also apply the method of substitution directly to definite integrals as follows. Let $u = g(x)$. Since

$$\int f(u)\frac{du}{dx}\,dx = F(u) + C,$$

for any function $F(u)$ such that $dF(u)/du = f(u)$, we have

$$\int_a^b f(u)\frac{du}{dx}\,dx = F(g(b)) - F(g(a)).$$

Thus, if we let $g(a) = c$ and $g(b) = d$, we have

$$\int_a^b f(u)\frac{du}{dx}\,dx = F(d) - F(c).$$

From the more compact form $\int f(u)du = F(u) + C$, we find that

$$\int_c^d f(u)du = F(d) - F(c), \quad \text{and so}$$

$$\int_a^b f(u)\frac{du}{dx}\,dx = \int_c^d f(u)du,$$

where $g(a) = c$ and $g(b) = d$.

Example 13 | Evaluate the definite integral $\int_0^3 x(x^2 + 1)^{1/2}\,dx$.

$$\int_0^3 x(x^2 + 1)^{1/2}\,dx = \frac{1}{2}\int_0^3 (x^2 + 1)^{1/2}2x\,dx.$$

Thus, if we let $u = g(x) = x^2 + 1$, then $du/dx = 2x$. Moreover since $g(0) = 1$ and $g(3) = 10$, we have

$$\int_0^3 x(x^2 + 1)^{1/2}\,dx = \frac{1}{2}\int_1^{10} u^{1/2}\,du$$

$$= \frac{1}{2}\cdot\frac{2}{3}u^{3/2}\Big|_1^{10}$$

$$= \frac{1}{3}(10^{3/2} - 1). \qquad \|$$

Exercises 4.7

Find each of the following integrals.

1. $\int 5\,dx$.
2. $\int x^4\,dx$.
3. $\int t^3\,dt$.
4. $\int \sqrt{x}\,dx$.
5. $\int_1^2 x^{-3}\,dx$.
6. $\int_1^3 \frac{3}{x^2}dx$.
7. $\int \frac{dx}{\sqrt{x}}$.
8. $\int(4x^7 - 11x^2 + 6)dx$.
9. $\int(5 - x^2 + 8x^5)dx$.
10. $\int(x^2 - 9)^2(2x)dx$.

11. $\int t(3 - t^2)^2\, dt$.

12. $\int (x^2 - 9)^2\, dx$.

13. $\int (3 - x^2)^2\, dx$.

14. $\int_{-1}^{0} (x^3 + x)^4(3x^2 + 1)dx$.

15. $\int (1 - v)^3\, dv$.

16. $\int t^2(t^3 + 5)^4\, dt$.

17. $\int x(x^2 + 4)^{1/2}\, dx$.

18. $\int [(x^3 + 2)^4 + 5(x^3 + 2)^2 + 11](3x^2)dx$.

19. $\int (x^4 + 8x^2)^5(x^3 + 4x)dx$.

20. $\int x^2(3 - x^3)^5\, dx$.

21. $\int_2^4 \frac{x}{\sqrt{x^2 - 2}}\, dx$.

22. $\int_1^6 \frac{dw}{\sqrt{w + 3}}$.

23. $\int_0^1 \frac{x^2\, dx}{(x^3 + 1)^5}$.

24. $\int \frac{(x + 2)dx}{(x^2 + 4x - 1)^3}$.

25. $\int (3u - 5)^8\, du$.

26. $\int (4 - 3u)^{10}\, du$.

27. $\int x\sqrt{9 - x^2}\, dx$.

28. $\int_0^1 x^2 \sqrt[5]{x^3 + 1}\, dx$.

29. $\int_0^1 \frac{x}{(5 + 3x^2)^{2/3}}\, dx$.

30. $\int \frac{x^3}{(x^4 + 7)^{4/5}}\, dx$.

31. What is the derivative of $\int h(x)dx$ with respect to x?

4.8 The Mean Value Theorem for Integrals; Average Value of a Function

Another result that will be useful later is one that we might guess from the graph of a positive continuous function $f(x)$, $a \leq x \leq b$.

It appears from Figure 4.8.1 that if c is chosen properly between a and b, then the area of the rectangle whose width is $b - a$ and whose height is $f(c)$ is equal to the area under the graph of $y = f(x)$ between a and b. That is, in fact, true even when $f(x)$ is not positive, as the following Mean Value Theorem for Integrals indicates.

Figure 4.8.1

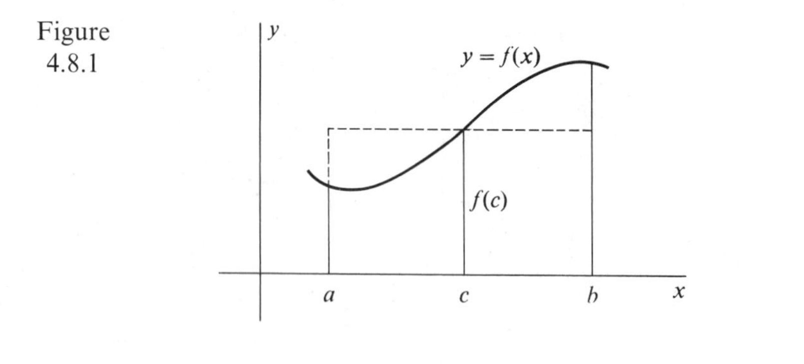

Theorem 4.8.1

Mean Value Theorem for Integrals

Let $f(x)$ be continuous for $a \leq x \leq b$ where $a < b$. Then there is a number c such that

$$\int_a^b f(x)dx = f(c)(b - a), \quad \text{where } a < c < b.$$

Proof. Let $A(x) = \int_a^x f(t)dx$. Then as we saw in Section 4.3, $A'(x) = f(x)$. We shall apply the Mean Value Theorem for derivatives (Theorem 3.6.2) to $A(x)$. By that theorem, there is some c, $a < c < b$ such that

$$A'(c) = \frac{A(b) - A(a)}{b - a},$$

so

$$A(b) - A(a) = A'(c)(b - a).$$

Now, using $A(b) = \int_a^b f(t)dt = \int_a^b f(x)dx$, $A(a) = \int_a^a f(t)dt = 0$, and $A'(c) = f(c)$ in the last equation, we get the desired conclusion that

$$\int_a^b f(x)dx = f(c)(b - a).$$

The number $f(c)$, whose existence is assured by The Mean Value Theorem for Integrals, is called the *average value* (or *mean value*) of $f(x)$ over the interval $a \leq x \leq b$. It is natural to call $f(c)$ the average value of $f(x)$ over the interval because it is the value of $f(x)$ which when multiplied by $b - a$, gives the area bounded by the graph of $y = f(x)$ from a to b. That is, it is the average height of the graph in the interval $a \leq x \leq b$. We make the following definition.

Definition 4.8.1

The *average value* of a continuous function $f(x)$ over an interval $a \leq x \leq b$ is

$$\frac{1}{b - a}\int_a^b f(x)dx \quad \text{if } a \neq b$$

or,

$$f(a) \quad \text{if } a = b.$$

Example 1

A metal plate is heated to $50t$ degrees Fahrenheit at time t. What is the average temperature A of the plate from $t = 0$ to $t = 4$?

$$A = \frac{1}{4}\int_0^4 50t\, dt = \frac{1}{4} \cdot 25t^2 \Big|_0^4 = 100°\text{F}.$$

||

Example 2

Let a particle move along a line so that its position coordinate is $s(t)$ at time t, where $a \leq t \leq b$. Then the velocity is $v(t) = s'(t)$. The average value of $v(t)$ over the interval $a \leq t \leq b$ is

$$\frac{1}{b - a}\int_a^b s'(t)dt = \frac{s(b) - s(a)}{b - a}.$$

On the other hand, in the time interval $a \leq t \leq b$, the particle moves a distance of $s(b) - s(a)$. The average velocity is therefore

$$\frac{s(b) - s(a)}{b - a}$$

Thus the average value of the velocity $v(t)$ over the interval according to Definition 4.8.1 is the same as the usual concept of average velocity! ||

Exercises 4.8

In Exercises 1–10, find the average value of the given function over the indicated interval.

1. $f(x) = 3x, \quad -2 \leq x \leq 6.$
2. $f(x) = -5x, \quad -4 \leq x \leq 2.$
3. $f(x) = x^2, \quad -5 \leq x \leq 2.$
4. $f(x) = -4x, \quad 2 \leq x \leq 8.$
5. $f(x) = 3x^2 - 4x + 5, \quad -2 \leq x \leq 3.$
6. $f(x) = 2x - 6x^2 - 3, \quad -3 \leq x \leq 4.$
7. $f(x) = 6x^2 - 4x^3 + 2, \quad -3 \leq x \leq 2.$
8. $f(x) = 8x^3 - 4x + 1, \quad -1 \leq x \leq 3.$
9. $f(x) = 10x^4 - 2x + 7, \quad -1 \leq x \leq 2.$
10. $f(x) = 4x - 5x^4 - 10, \quad -2 \leq x \leq 2.$

In Exercises 11–18, find a number c such that $\int_a^b f(x)dx = f(c)(b - a)$.

11. $a = 1, \quad b = 7, \quad f(x) = 4x.$
12. $a = 2, \quad b = 5, \quad f(x) = 9x^2.$
13. $a = 1, \quad b = 5, \quad f(x) = 6x^2.$
14. $a = -2, \quad b = 3, \quad f(x) = 2x.$
15. $a = -3, \quad b = 4, \quad f(x) = 1 - 4x^3.$
16. $a = -4, \quad b = 2, \quad f(x) = 1 - 3x^2.$
17. $a = -2, \quad b = 2, \quad f(x) = 6x^2 - 5.$
18. $a = -3, \quad b = 4, \quad f(x) = 8x^3 + 2.$
19. A rod 2 feet long is heated to $60x^2$ degrees Celsius where x is the distance from one end of the rod. Calculate the average temperature of the rod.
20. Find an expression for the average of the linear function $f(x) = cx + d$ over the interval $a \leq x \leq b$.
21. The typing speed w, in words per minute, of a certain secretary is $w = 20t - 5t^2 + 50$, where t is the elapsed time in hours. Find
    (a) the secretary's minimum speed in words per minute,
    (b) the secretary's maximum speed in words per minute,
    (c) the secretary's average speed in words per minute,
    (d) the total number of words the secretary typed,

    over the 3 hour interval $0 \leq t \leq 3$.
22. An object falls from rest for 10 seconds. Use the relations $v = gt$ and $v = \sqrt{2gs}$ to find the average velocity of the object over the interval from $t = 1$ to $t = 4$ (t in seconds):
    (a) if the average is with respect to the time,
    (b) if the average is with respect to the distance.
23. Do Exercise 22 for the interval $t = 2$ to $t = 8$ (t in seconds).

4.9 Integrals of Certain Trigonometric Functions†

Every differentiation formula results in a corresponding integration formula. In this section we study those integration formulas that come from the derivatives of the trigonometric functions obtained in Section 3.10.

For example, since

$$\sin u \frac{du}{dx} = -\frac{d}{dx} \cos u$$

we have

$$\int \sin u \frac{du}{dx} dx = -\cos u + C.$$

† The remainder of the text is independent of this section. This section is included for those classes that desire an early introduction to integration of the trigonometric functions.

This last result is more simply expressed as

$$\int \sin u \, du = -\cos u + C.$$

Example 1 | Compute $\int(\sin x^2)(2x)dx$.

We need only recognize that if $u = x^2$ then $du = 2x\,dx$. We then write immediately

$$\int(\sin x^2)(2x)dx = -\cos x^2 + C. \qquad \|$$

The following table lists the previously obtained differentiation formulas for the trigonometric functions and the corresponding integration formulas.

Differentiation Formula	*Integration Formula*
$\frac{d}{dx}\sin u = \cos u \frac{du}{dx}$	$\int \cos u \, du = \sin u + C$
$\frac{d}{dx}\cos u = -\sin u \frac{du}{dx}$	$\int \sin u \, du = -\cos u + C$
$\frac{d}{dx}\tan u = \sec^2 u \frac{du}{dx}$	$\int \sec^2 u \, du = \tan u + C$
$\frac{d}{dx}\sec u = \sec u \tan u \frac{du}{dx}$	$\int \sec u \tan u \, du = \sec u + C$
$\frac{d}{dx}\csc u = -\csc u \cot u \frac{du}{dx}$	$\int \csc u \cot u \, du = -\csc u + C$

Example 2 | Evaluate $\int x^2 \sec(x^3 - 5)\tan(x^3 - 5)dx$.

In this case we let $u = x^3 - 5$. Then $du = 3x^2\,dx$. Consequently we have

$$\begin{aligned}\int x^2 \sec(x^3 - 5)\tan(x^3 - 5)dx &= \frac{1}{3}\int(\sec(x^3 - 5)\tan(x^3 - 5))3x^2\,dx \\ &= \frac{1}{3}\sec(x^3 - 5) + C.\end{aligned} \qquad \|$$

Example 3 | Evaluate $\int_0^{\pi/4} \cos(\pi - x)dx$.

Let $u = g(x) = \pi - x$. Then $du = -dx$. Moreover, since $g(0) = \pi$ and $g(\pi/4) = 3\pi/4$ we have

$$\begin{aligned}\int_0^{\pi/4} \cos(\pi - x)dx &= -\int_\pi^{3\pi/4} \cos u \, du \\ &= -\sin u \Big|_\pi^{3\pi/4} = -\frac{1}{\sqrt{2}}.\end{aligned} \qquad \|$$

Other integrals involving the trigonometric functions may be evaluated by use of

$$\int u^n \, du = \frac{1}{n+1} u^{n+1} + C, \quad n \neq -1.$$

The next examples illustrate several such cases.

Example 4 | Evaluate $\int \sin^5 x \cos x \, dx$.

If we take $u = \sin x$ then $du = \cos x \, dx$. Consequently

$$\int \sin^5 x \cos x \, dx = \int u^5 \, du = \frac{1}{6} u^6 + C,$$

and so

$$\int \sin^5 x \cos x \, dx = \frac{1}{6} \sin^6 x + C. \qquad \|$$

Example 5 | Evaluate $\int \tan^2 x \sec^2 x \, dx$.

Note that if we let $u = \tan x$ then $du = \sec^2 x \, dx$.

Thus

$$\int \tan^2 x \sec^2 x \, dx = \int u^2 \, du = \frac{1}{3} u^3 + C = \frac{1}{3} \tan^3 x + C. \qquad \|$$

Example 6 | Evaluate $\int x \cos^3(x^2) \sin(x^2) dx$.

If we let $u = \cos(x^2)$ then $du = -2x \sin(x^2) dx$. Thus we "fix up" the integral to get

$$\begin{aligned} \int x \cos^3(x^2) \sin(x^2) dx &= -\frac{1}{2} \int \cos^3(x^2)(-2x) \sin(x^2) dx \\ &= -\frac{1}{2} \int u^3 \, du = -\frac{1}{2}\left(\frac{1}{4} u^4\right) + C \\ &= -\frac{1}{8} \cos^4(x^2) + C. \qquad \| \end{aligned}$$

Example 7 | Evaluate $\int \sec^3 x \tan x \, dx$.

Since

$$\int \sec^3 x \tan x \, dx = \int \sec^2 x \sec x \tan x \, dx$$

we get

$$\int \sec^3 x \tan x \, dx = \frac{1}{3} \sec^3 x + C. \qquad \|$$

Exercises 4.9

For Exercises 1–20, evaluate the integral. .

1. $\int \sin 2x \, dx$.
2. $\int \cos 3x \, dx$.
3. $\int \sec^2 t \, dt$.
4. $\int \csc^2 7x \, dx$.
5. $\int \sec 2x \tan 2x \, dx$.
6. $\int \csc(-3x) \cot(-3x) dx$.
7. $\int x \sec^2(x^2) dx$.
8. $\int (x^2 + 1) \cos(x^3 + 3x) dx$.

9. $\int v^2 \sec(v^3) \tan(v^3) dv$.
10. $\int x^3 \sin(x^4 - 5) dx$.
11. $\int \sin^3 x \cos x \, dx$.
12. $\int \tan^7 t \sec^2 t \, dt$.
13. $\int \cos^5 x \sin x \, dx$.
14. $\int \sec^5 x \sec x \tan x \, dx$.
15. $\int \csc^7 x \csc x \cot x \, dx$.
16. $\int \sec^4 x \tan x \, dx$.
17. $\int \sec^3 t \tan t \, dt$.
18. $\int (\sin^3 x + \sin x) \cos x \, dx$.
19. $\int (7 \cos x - 3 \cos^2 x) \sin x \, dx$.
20. $\int (x^5 - 1) \sin(x^6 - 6x) dx$.
21. Find the area bounded by the graph of $y = \cos x$ and the x-axis between $x = 0$ and $x = \pi/2$.
22. Find the area bounded by the graph of $y = \sec^2 x$ and the x-axis between $x = 0$ and $x = \pi/4$.
23. Find the average value of the function $f(x) = \sin x \cos x$ between $x = 0$ and $x = 2\pi$.
24. Find the area under one arch of the graph of $y = \sin x$.
25. Find the net area bounded by the graph of $y = x \cos(x^2)$ and the x-axis between $x = 0$ and $x = 2\pi$.
26. Find the average value of $f(x) = \sec^2 x \tan^5 x$ between $x = 0$ and $x = \pi/4$.

Brief Review of Chapter 4

1. The Definite Integral

Definition (incomplete).

$$\int_a^b f(x)dx = \lim_{h \to 0} \sum_{i=1}^{n} f(x_i^*)h$$

(a complete definition must include the meaning of x_i^* and h).

Interpretation of Definition. Geometrically, for $f(x) \geq 0$ and $a \leq x \leq b$, $\int_a^b f(x)dx$ is the area of the region bounded by the graph of $y = f(x)$, the x-axis, and the vertical lines $x = a$ and $x = b$.

Important Facts. The Fundamental Theorem of Calculus simplifies the evaluation of $\int_a^b f(x)dx$ when $F(x)$ can be found so that $F'(x) = f(x)$, for then $\int_a^b f(x)dx = F(b) - F(a)$. Definite integrals do not always exist, but they do exist for continuous functions. The Mean Value Theorem for Integrals and the average value of a function are also important.

2. Indefinite Integrals

Definition. If F is a function such that $F'(x) = f(x)$, then $F(x)$ is called an antiderivative or indefinite integral of $f(x)$.

Important Facts. If $F(x)$ is *any* indefinite integral of $f(x)$, then *every* indefinite integral of $f(x)$ has the form $F(x) + C$, where C is a constant. This fact is expressed in symbols as

$$\int f(x)dx = F(x) + C.$$

The expression $\int f(x)dx$ is called the indefinite integral of $f(x)$. Indefinite integration is the "opposite" of differentiation and every differentiation formula can be restated as an integration formula. Indefinite integrals are used mainly to evaluate definite integrals with the help of the Fundamental Theorem of Calculus. A very important integration formula is

$$\int x^n\,dx = \frac{x^{n+1}}{n+1} + C, \qquad n \neq 1.$$

3. Use of Differentials in Integration Formulas

Important Facts. If $u = g(x)$ then $du = (du/dx)dx$. A very important integration formula called the *substitution formula* is

$$\int f(u)\frac{du}{dx}dx = F(u) + C,$$

where u is a function of x and $F(u)$ is any function of u such that $dF(u)/du = f(u)$. Using differentials the substitution formula can be expressed more simply as

$$\int f(u)du = F(u) + C.$$

An important special instance of this is the power function rule:

$$\int u^n\,du = \frac{u^{n+1}}{n+1} + C, \qquad n \neq -1.$$

Technique Review Exercises, Chapter 4

1. Use the *definition* of the definite integral to find the area of the region bounded by $y = 8$, the x-axis, and the lines $x = 3$ and $x = 10$.
2. Use the "$A(x)$ approach" to find the net area bounded by $y = -x^3 + 5$ and the x-axis between -2 and 4.
3. State The Fundamental Theorem of Calculus.
4. Evaluate $\int_{-3}^{2} (x - 6x^2)dx$.
5. Use a definite integral to calculate the following: (a) the net area of the region bounded by the graph of $y = x^3 - 3x$ and the x-axis between 1 and 4; (b) the total area of the region bounded by $y = x^3 - 3x$ and the x-axis, between $x = 1$ and $x = 4$.
6. Find the area of the region bounded by the graphs of the equations $y = -x$ and $y = x^2 - 2$.
7. Evaluate $\int x^2(x^3 - 7)^{10}\,dx$.
8. Evaluate $\int \frac{x^3\,dx}{(2x^4 + 1)^3}$.
9. Evaluate $\int (x^2 + 3x)^2\,dx$.
10. Evaluate $\int_0^1 \frac{x^3\,dx}{(x^4 + 10)^3}$.

11. Evaluate $\int_{-3}^{0} |-x^2 - 2x| dx$.

12. Find the average value of $f(x) = 7 - 8x^3 - 9x^2$ for $-3 \le x \le 1$.

Additional Exercises, Chapter 4

Section 4.1

Use the definition of the definite integral to compute each integral.

1. $\int_{-4}^{2} 17\, dx$.
2. $\int_{-1}^{3} 4\, dx$.

Section 4.2

Use the definition of the definite integral to calculate the net area bounded by the graph of $y = f(x)$ and the x-axis between a and b.

3. $f(x) = x, \quad a = -3, \quad b = 1$.
4. $f(x) = 2x, \quad a = -1, \quad b = 2$.

Section 4.3

In Exercises 5–8, find a form for every antiderivative of the given function.

5. $18x$.
6. 10.
7. $10x^4$.
8. $24x^7$.

In Exercises 9–14, find the net area bounded by the graph of $f(x)$ and the x-axis between the indicated values of a and b.

9. $f(x) = 14x, \quad a = -2$ and $b = 5$.
10. $f(x) = x^3, \quad a = -1$ and $b = 0$.
11. $f(x) = 4, \quad a = 0$ and $b = 9$.
12. $f(x) = x^2, \quad a = -3$ and $b = 1$.
13. $f(x) = 8x^3, \quad a = 2$ and $b = 4$.
14. $f(x) = 10x^4, \quad a = -2$ and $b = 2$.

Section 4.4

Evaluate each definite integral.

15. $\int_{-3}^{1} x\, dx$.
16. $\int_{-1}^{2} 12x^2\, dx$.
17. $\int_{-2}^{3} 19\, dx$.
18. $\int_{2}^{3} 10x\, dx$.
19. $\int_{-1}^{1} 12x^3\, dx$.
20. $\int_{-1}^{4} (1 - x) dx$.
21. $\int_{-2}^{0} -2x^3\, dx$.
22. $\int_{1}^{3} 15x^4\, dx$.

Section 4.5

Evaluate each definite integral.

23. $\int_{15}^{15} x^8\, dx$.
24. $\int_{-2}^{-2} (x^5 - 5x) dx$.
25. $\int_{3}^{-1} x\, dx$.
26. $\int_{4}^{1} 6x^2\, dx$.
27. $\int_{-7}^{2} |x - 3| dx$.
28. $\int_{4}^{1} x|x| dx$.

Section 4.6

29. Find the area bounded by the graph of $y = 5x^4$ and the x-axis between the vertical lines $x = -1$ and $x = 3$.

30. Find the area bounded by the graph of $y = 8x^3$ and the x-axis between the vertical lines $x = -4$ and $x = 0$.

For Exercises 31–34, find the total area bounded by the graph of the given equation and the x-axis between $x = a$ and $x = b$.

31. $y = x - 3, \quad a = -1, b = 4.$
32. $y = 12x - 6x^2, \quad a = -1, b = 4.$
33. $y = 6x^2 + 12x, \quad a = -7, b = 1.$
34. $y = 3x^2 + 3x, \quad a = -2, b = 5.$

For Exercises 35–36 find the area bounded by the graphs of the given equations.

35. $y = x^3, \quad y = |x|.$
36. $y = x^2 - x, \quad y = x + 4.$

Section 4.7

Find each integral.

37. $\int_{-2}^{-1} \frac{24}{x^2}\,dx.$
38. $\int \frac{dx}{\sqrt[3]{x}}.$
39. $\int(x^3 - 4x^7 + 6x^2)dx.$
40. $\int(7 - t^3)^2\,dt.$
41. $\int x\sqrt{x^2 - 1}\,dx.$
42. $\int x^3(x^4 - 2)^5\,dx.$
43. $\int(2x^5 - 1)(x^6 - 3x)^8\,ax.$
44. $\int \frac{u^2\,du}{\sqrt{7 - u^3}}.$
45. $\int_0^1 t\sqrt{3t^2 + 1}\,dt.$
46. $\int \frac{2w + 1}{(w^2 + w - 2)^{3/5}}\,dw.$

Section 4.8

47. Find the average value of $8x(x^2 - 3)^3, \quad -1 \le x \le 2.$
48. Find the average value of $(x^2 + 1)^2, \quad -2 \le x \le 1.$
49. Find a number c such that $1 \le c \le 3$ and $\int_1^3 8x^3\,dx = 16c^3.$
50. Find a number c such that $-2 \le c \le 3$ and $\int_{-2}^3 (5 - 6x^2)dx = (5 - 6c^2)(5).$

Section 4.9

Find each integral.

51. $\int \cos 7x\,dx.$
52. $\int \sin 5x\,dx.$
53. $\int \csc^2 8x\,dx.$
54. $\int \sec^2 3t\,dt.$
55. $\int t^2 \sin(t^3 + 7)dx.$
56. $\int \sin^4 x \cos x\,dx.$
57. $\int \sec^2 u \tan^5 u\,du.$
58. $\int \csc^2 t \cot t\,dt.$

Challenging Problems, Chapter 4

1. Show that $\int_a^b f(x)dx = \int_a^b f(a + b - x)dx.$
2. Use the given outline to prove the following form of Schwarz' Inequality

$$\int_a^b f(x)g(x)dx \le \left(\int_a^b (f(x))^2\,dx\right)^{1/2}\left(\int_a^b (g(x))^2\,dx\right)^{1/2}.$$

(i) Let $h(t) = at^2 + 2bt + c$, with $a > 0$. Prove that $h(t) \ge 0$ for all t if and only if $b^2 - ac \le 0$.

(ii) Apply the result of (i) above to the function

$$h(t) = \int_a^b [tf(x) + g(x)]^2\, dx.$$

3. Let $f(x)$ be a positive continuous function with the property that the area bounded by the graph of $y = f(x)$ between $x = 1$ and $x = t$ is $(t - 1)$ times the area bounded between $x = 1$ and $x = 2$. Prove that $f(x)$ is a constant function.

4. Our definition of the definite integral $\int_a^b f(x)dx$ included dividing the interval from a to b into subintervals of equal width h by taking equally spaced points $x_0\ (=a), x_1, x_2, \ldots, x_n\ (=b)$. The restriction that the points be equally spaced was made for simplification and is not a necessity. In the more general subdivision, $\sum_{i=1}^n f(x_i^*)h$ is replaced by the *Riemann Sum* $\sum_{i=1}^n f(x_i^*)(x_i - x_{i-1})$. The limit is taken in a way that requires the width of every subinterval to approach 0. Calculate $\int_0^1 x\, dx$ using the points $x_0\ (=0), x_1, \ldots, x_n\ (=1)$, where $x_i = (i^2/n^2)$. Then all the subintervals have different widths.

Use the following outline as a guide:

(i) Let $d = \dfrac{1}{n^2}$ and show that $x_i - x_{i-1} = (2i - 1)d$, while $x_i = i^2 d$.

(ii) Take x_i^* to be the right-hand endpoint x_i of the ith subinterval and calculate the Riemann Sum $\sum_{i=1}^n f(x_i^*)(x_i - x_{i-1})$ using $1^2 + 2^2 + 3^2 + \cdots + n^2 = n(n + 1)(2n + 1)/6$, $1^3 + 2^3 + 3^3 + \cdots + n^3 = n^2(n + 1)^2/4$, and $n = 1/d^{1/2}$.

(iii) Note that as $d \to 0$ the width $x_i - x_{i-1}$ of each subinterval approaches 0 and thus calculate

$$\int_0^1 x\, dx = \lim_{d \to 0} \sum_{i=1}^{n} f(x_i^*)(x_i - x_{i-1}).$$

5. Assume that for $a \leq x \leq b$, $f(x)$ is continuous and $0 \leq f(x) \leq p \int_a^x f(t)dt$, where p is a fixed positive number. Prove that $f(x) = 0$ for all x, $a \leq x \leq b$. Use the following outline as a guide:

(i) Let $A(x) = \int_a^x f(t)dt$, $\quad a \leq x \leq b$. Show that $A(x)$ is nondecreasing.

(ii) Let n be a fixed integer, $n \geq 2$, chosen large enough so that $h = \dfrac{b - a}{n} < \dfrac{1}{2p}$.

Consider the n values

$$0 = A(a), \quad A(a + h), \quad A(a + 2h), \ldots, A(a + nh) = A(b).$$

Let $A(a + kh)$ be the last one of those which is zero and take care of the case where $k = n$.

(iii) If $1 \leq k < n$ show that $A(a + kh + h) = hA'(c) = hf(c)$ for some c such that $a + kh < c < a + kh + h$.

(iv) Use (iii) to obtain the contradiction $A(a + kh + h) < (1/2)A(c)$, $c < a + kh + h$, and complete the proof.

6. Prove that if $\int_a^b f(t)dt = \int_{ax}^{bx} f(t)dt$ for all positive a, b and x, then $f(x) = k/x$ for some constant k and all $x > 0$.

7. Use the definition of the definite integral to evaluate

$$\int_1^3 (x^2 + 2)dx$$

8. A parabolic fence with equation $y = x^2$ (x and y measured in meters) has been built. An additional 400 m of fence is to be added as shown in Figure 1. What dimensions for the new fence will maximize the enclosed area A?

Figure 1

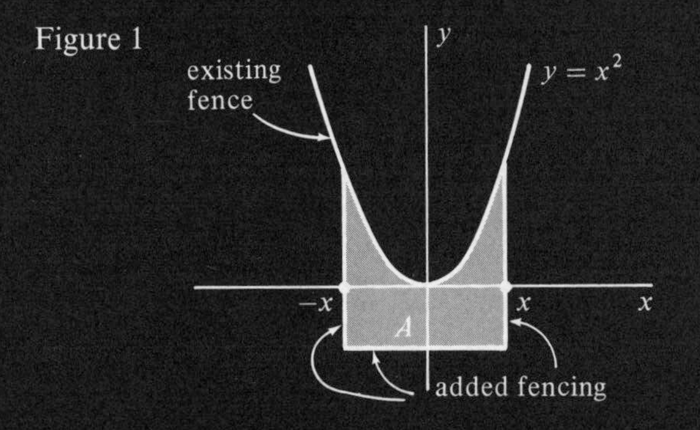

9. Use the indicated outline to establish each of the following results:
   (a) If only the force of the earth's gravity is considered, then an object shot upward with an initial velocity of approximately 7 miles/sec or 11.2 km/sec will never return to the earth! (That velocity is called the *velocity of escape*. Note that air resistance or other gravitational forces such as those of the sun or moon are not considered.)
   (b) If only the force of the earth's gravity and the centrifugal force due to the rotation of the earth are considered, then the extent of the earth's atmosphere at the equator is approximately 36,000 km or 22,000 miles above the earth.

Outline

(i) The acceleration due to gravity at a distance s from the center of the earth is inversely proportional to s^2. Let the positive direction be upward and obtain $a = \dfrac{-gr^2}{s^2}$, where r is the radius of the earth and $-g$ is the acceleration due to gravity at the surface of the earth.

(ii) Multiply the equation from (c) by $2\,\dfrac{ds}{dt}$ and take the antiderivative of each side and show that if the initial velocity of the object is v_0 when $s = d$, $d \geq r$, then

$$v^2 = v_0{}^2 + 2gr^2\left(\frac{1}{s} - \frac{1}{d}\right), \qquad s \geq r.$$

(iii) Show that if $v_0 = \sqrt{2gr}$ when $d = r$, then v is always positive and hence the object does not slow to a stop and then fall back to the earth. That is, it never returns to the earth!

(iv) Calculate $\sqrt{2gr}$ where $g = 32$ ft/sec/sec and $r = 4000$ miles and when $g = 980$ cm/sec/sec and $r = 6400$ km, completing part (a).

(v) Using the equation from (i) and Newton's law, which states that the gravitational force F on an object of mass m, with acceleration due to gravity of a is ma, obtain F for a portion of atmosphere of mass m.

(vi) The centrifugal force due to the earth's rotation on a portion of atmosphere of mass m, at a distance s from the center of the earth and above the equator, is $\frac{mgs}{289r}$, if m is in grams and g in cm/sec/sec. Consider the point of equilibrium of the gravitational force and the centrifugal force. Show that for this point $s^3 = 289r^3$, and obtain $s \approx 6.61r$. Then complete part (b).

10. The acceleration due to gravity of an object inside the earth is directly proportional to the distance of the object from the center of the earth. If a tube ran through the center of the earth and an object fell into the tube, show that its velocity at the center of the earth would be about 5 miles/sec or 8 km/sec. (*Hint*: Use a procedure similar to parts (i) and (ii) of the outline in the previous problem.)

5

Applications of The Definite Integral

In the preceding chapter we saw that the definite integral of a nonnegative function may be interpreted as an area. This is only one of the many ways that definite integrals can be applied. Indeed, definite integrals are used in many disciplines. In this chapter we shall discuss some typical applications.

Each application will be developed from both an intuitive and a more rigorous point of view. Intuitively one may think of the definite integral $\int_a^b f(x)dx$ as the sum of elements $f(x)dx$ between a and b. That interpretation will serve as the basis for our intuitive development of the definite integral.

An intuitive view is particularly valuable in getting a "feel" for the particular application, as a memory aid, and as a source of suggestions for additional applications of the integral. However, it should be recognized that a rigorous development is necessary to insure the validity of any formula.

5.1 Volume

Since the concept of volume is closely related to that of area, it is not surprising that definite integrals can be used to calculate certain volumes. In this section we shall restrict our attention to solids having a known cross-sectional area. That is, let us suppose we have a continuous cross-sectional-area function $A(x)$ of a solid R with respect to a line L. That is, suppose we have a continuous function $A(x)$ that yields the area of the cross section cut from the solid R by a plane $P(x)$ perpendicular to the line L at the point x on L, as in Figure 5.1.1.

Figure 5.1.1

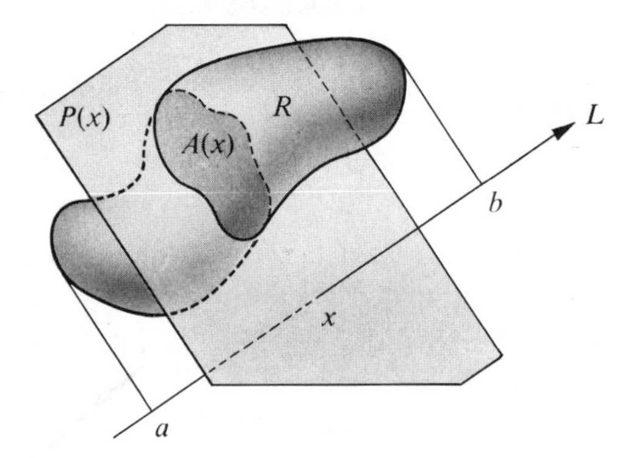

Intuitive Development. The intuitive developments included throughout this chapter follow the pattern established in Section 4.2.

Subdivide the interval $a \leq x \leq b$ (on L) into small subintervals each having width dx. This process yields a subdivision of the solid into a number of slices as shown in Figure 5.1.2. A typical thin

Figure 5.1.2

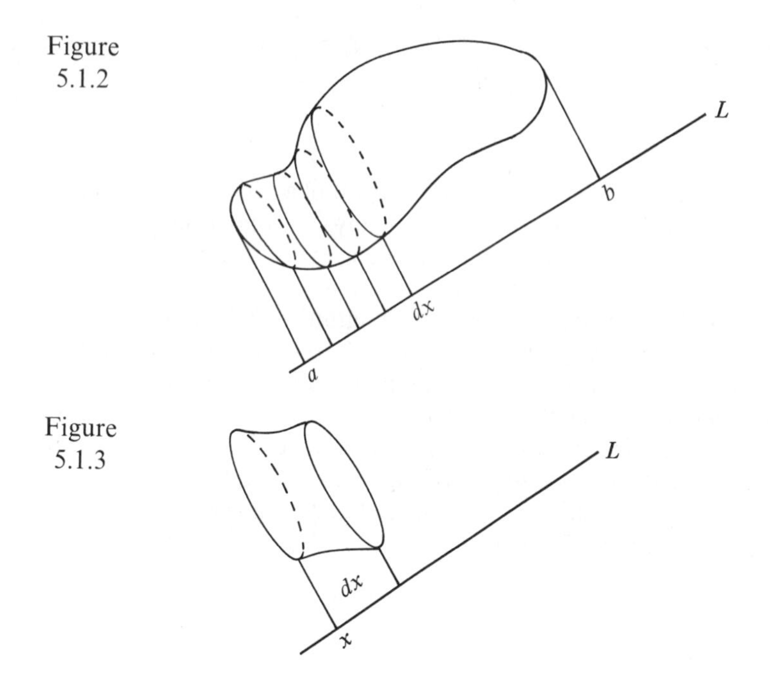

Figure 5.1.3

slice or *element of volume* is shown in a magnified view in Figure 5.1.3. For small values of dx such an element is approximately a cylinder with base area $A(x)$ and height dx. Consequently each element of volume is approximately $A(x)dx$. As in Chapter 4, use of the integral sign to sum these thin slices from $x = a$ to $x = b$ gives the following formula for the volume of the solid:

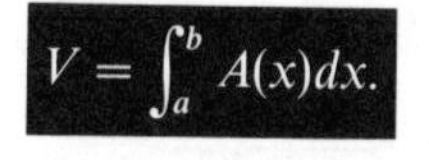

$$V = \int_a^b A(x)dx.$$

More Rigorous Development. A mathematical development must proceed from certain basic assumptions or axioms. Our previous experience with volumes leads us to make the following two general assumptions about volumes.

Assumption 1. If a particular solid has variable cross-sectional area as indicated in Figure 5.1.4, the volume V of the solid is between the minimum cross-sectional area m times the height h and the maximum cross-sectional area M times the height h, that is

$$hm \leq V \leq hM.$$

Figure 5.1.4

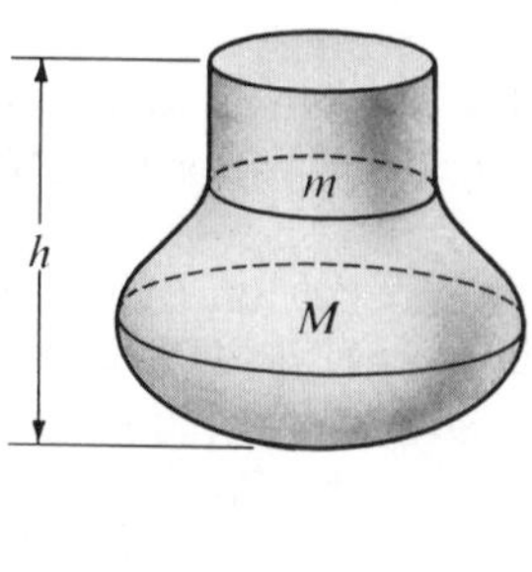

Assumption 2. Volume is additive; that is, if two solids R_1 and R_2 overlap only along their boundaries, or not at all, the volume of the combined solid consisting of R_1 and R_2 is the sum of their separate volumes. Figure 5.1.5 illustrates this situation.

Figure 5.1.5

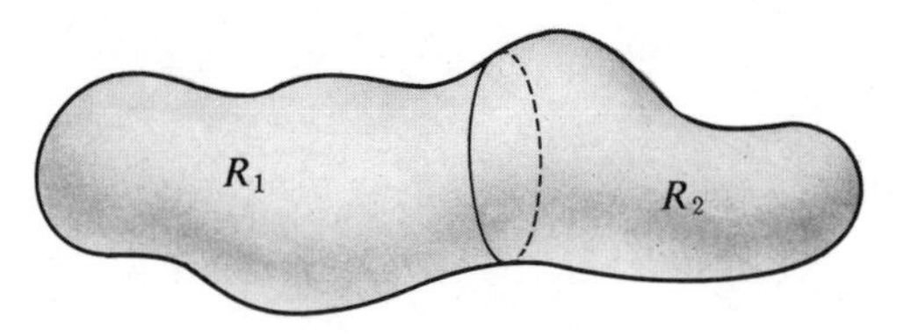

The remarkable fact is that Assumptions 1 and 2 lead to a simple method for calculating certain volumes.

First we express the volume V as a sum of smaller volumes by subdividing the interval $a \leq x \leq b$ (on L) into n smaller subintervals each of width $h = (b - a)/n$. The endpoints of these subintervals are $a = x_0 < x_1 < \cdots < x_{n-1} < x_n = b$, where $x_i = a + ih$. Now consider the volume V_i of the solid bounded by the planes $P(x_{i-1})$ and $P(x_i)$, as illustrated in Figure 5.1.6.

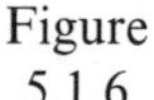
Figure 5.1.6

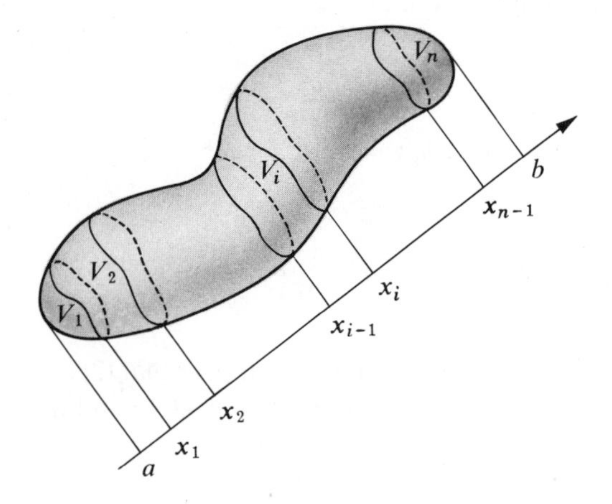

Since volume is additive, the total volume V is

$$V = V_1 + V_2 + V_3 + \cdots + V_n, \quad \text{or}$$

$$V = \sum_{i=1}^{n} V_i. \tag{1}$$

If we could now find each V_i, we would have an expression for V. With this in mind, let's focus our attention on a single V_i. Recall that we have assumed that the cross-sectional-area function $A(x)$ is continuous. Consequently $A(x)$ has a maximum value M and a minimum value m on the subinterval $x_{i-1} \leq x \leq x_i$, as pictured in Figure 5.1.7. Using Assumption 1 concerning volume we can write

$$mh \leq V_i \leq Mh,$$

Figure 5.1.7

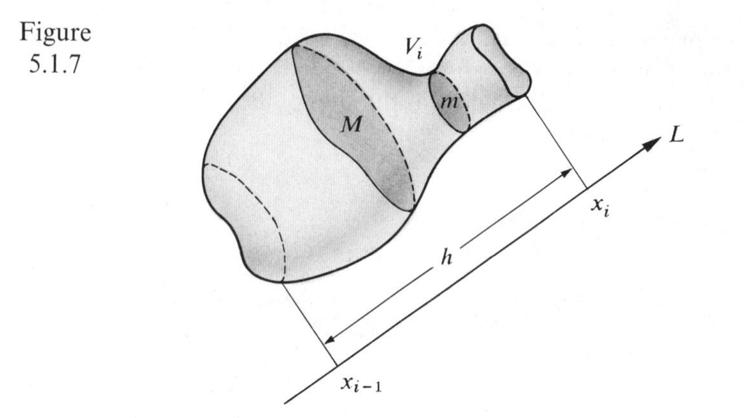

or, since h is positive,

$$m \leq \frac{V_i}{h} \leq M.$$

Since the continuous function $A(x)$ takes on both values m and M in the subinterval, we know by The Intermediate Value Theorem that $A(x)$ must take on the value V_i/h at some point x_i^* in the subinterval; that is,

$$\frac{V_i}{h} = A(x_i^*), \quad \text{or}$$

$$V_i = A(x_i^*)h, \text{ where } x_{i-1} \leq x_i^* \leq x_i.$$

Then, by (1), we get

$$V = \sum_{i=1}^{n} A(x_i^*)h, \qquad x_{i-1} \leq x_i^* \leq x_i. \tag{2}$$

Since we are unable to find each x_i^*, it might appear that (2) is of little use. However, we can apply the same technique used in Section 4.6 to find the area of a region between two curves.

The volume V is independent of h, and so

$$\lim_{h\to 0} V = V. \tag{3}$$

Then, on taking the limit as h approaches zero in (2), we have

$$\lim_{h\to 0} V = V = \lim_{h\to 0} \sum_{i=1}^{n} A(x_i^*)h.$$

But by definition of the definite integral,

$$\lim_{h\to 0} \sum_{i=1}^{n} A(x_i^*)h = \int_a^b A(x)dx, \quad \text{and so}$$

$$V = \int_a^b A(x)dx. \tag{4}$$

Example 1 A solid has cross-sectional area $A(x) = \pi(x^2 - x)^2$ for $0 \le x \le 1$. Find the volume of the solid.

By (4) the volume is given by

$$V = \int_0^1 \pi(x^2 - x)^2\, dx = \pi \int_0^1 (x^4 - 2x^3 + x^2)dx.$$

Then we have

$$V = \pi\left(\frac{1}{5}x^5 - \frac{1}{2}x^4 + \frac{1}{3}x^3\right)\Bigg|_0^1 = \frac{\pi}{30} \text{ cubic units.}$$ ||

In most cases the function $A(x)$ must first be determined from other given information.

Example 2 The base of a solid is the region in the x, y plane bounded by the graph of $y = 1 - x^2$ and the x-axis. The cross sections of the solid perpendicular to the x-axis are semicircles. Find the volume of the solid.

Usually it is helpful to sketch a typical cross section of the volume in question, as in Figure 5.1.8(a). The angle between the axes is distorted to indicate that we are looking at the figure from the right of center. Figure 5.1.8(b) is a computer generated image showing another view of the solid. Portions of a number of cross sections are also indicated. We first find $A(x)$. In this case, $A(x)$ is the area of a

Figure 5.1.8

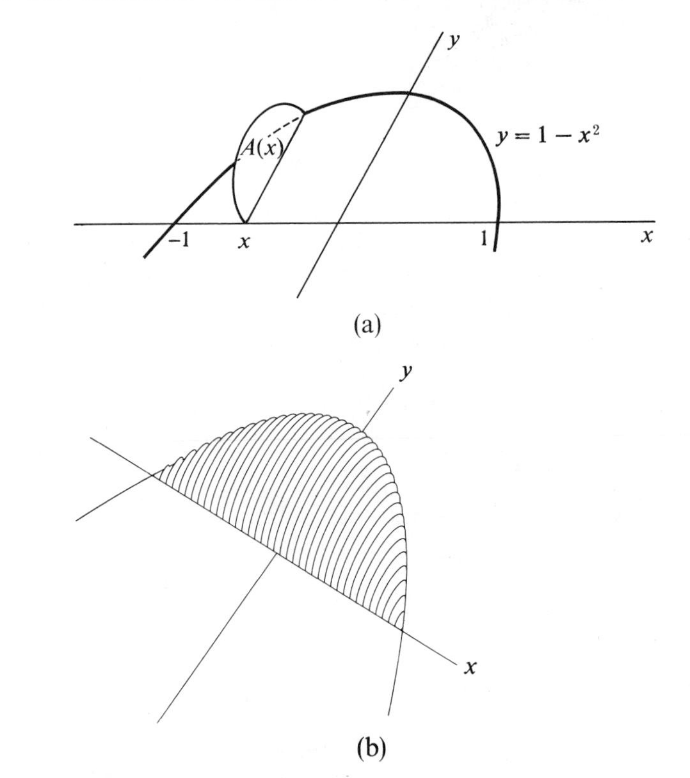

Figure 5.1.9

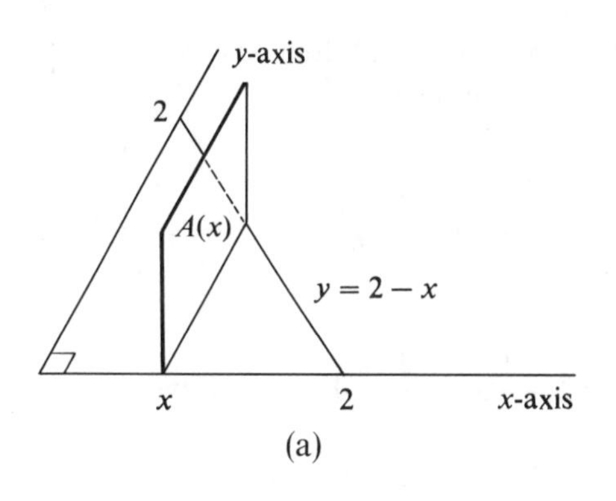

(a)

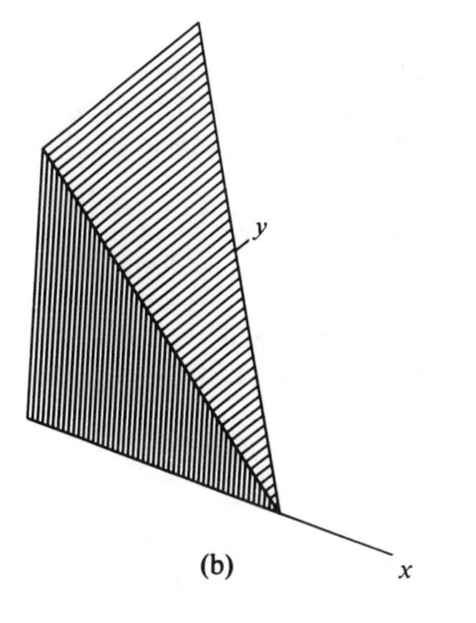

(b)

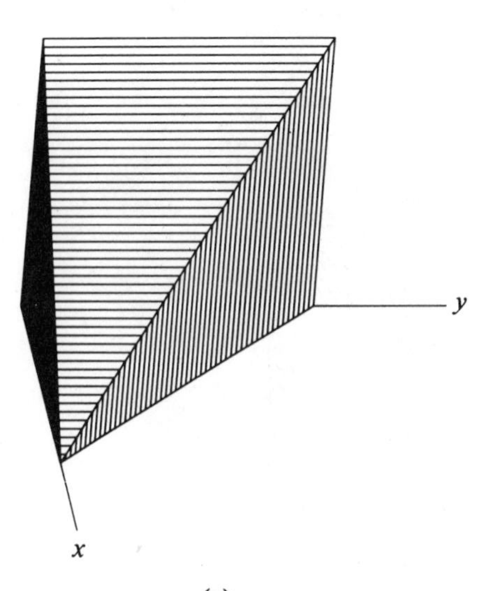

(c)

semicircle with diameter $y = 1 - x^2$. Hence

$$A(x) = \frac{1}{2}\pi\left(\frac{1-x^2}{2}\right)^2 = \frac{1}{8}\pi(1-x^2)^2.$$

$$V = \int_{-1}^{1} \frac{1}{8}\pi(1-x^2)^2\,dx = \frac{1}{8}\pi\int_{-1}^{1}(x^4 - 2x^2 + 1)dx$$

$$= \frac{1}{8}\pi\left(\frac{1}{5}x^5 - \frac{2}{3}x^3 + x\right)\Bigg|_{-1}^{1} = \frac{1}{8}\pi\left(\frac{8}{15} + \frac{8}{15}\right) = \frac{2\pi}{15} \text{ cubic units.} \qquad \|$$

Example 3 | The base of a solid is an isosceles right triangle with two sides of length 2. The cross sections of the solid perpendicular to one of the sides are squares. Find the volume of the solid.

In Figure 5.1.9(a), we give a sketch of the base of the solid and indicate the area $A(x)$ of a typical cross section. Figures 5.1.9(b) and 5.1.9(c) are computer-generated images showing somewhat different views of the solid. Portions of a number of cross sections are also indicated.

If we select the two axes as indicated, the cross-sectional region at x is a square with sides of length $y = 2 - x$. Thus $A(x) = (2-x)^2$, and so

$$V = \int_0^2 (2-x)^2\,dx = \int_0^2 (4 - 4x + x^2)dx$$

$$= \left(4x - 2x^2 + \frac{1}{3}x^3\right)\Bigg|_0^2 = \frac{8}{3} \text{ cubic units.} \qquad \|$$

Solids obtained by rotating regions about an axis are called *solids of revolution.* A specialization of the preceding technique is often used to calculate volumes of such solids. This specialization is called the *disc method.*

Example 4 | The region bounded by the graph of $y = x - x^2$ and the x-axis is rotated about the x-axis. Find the resulting volume.

Since the solid is formed by rotation about the x-axis, cross sections perpendicular to the x-axis are discs (hence the name disc method), as indicated in Figure 5.1.10(a). Figure 5.1.10(b) is a computer-generated image showing another view of the solid. The region to be rotated is shaded in that figure. Portions of a number of cross sections are also indicated. Since the disc at x has radius $y = x - x^2$, the area of the cross section is

$$A(x) = \pi(x - x^2)^2.$$

Then the volume is

$$V = \int_0^1 \pi(x-x^2)^2\,dx = \pi\int_0^1 (x^2 - 2x^3 + x^4)dx$$

$$= \pi\left(\frac{1}{3}x^3 - \frac{1}{2}x^4 + \frac{1}{5}x^5\right)\Bigg|_0^1 = \frac{\pi}{30} \text{ cubic units.} \qquad \|$$

Example 5 | The region bounded by the graphs of $y = 5x^2 - 10x$ and $y = 2x - x^2$ is rotated about the line $y = 1$. Calculate the volume.

Figure 5.1.10

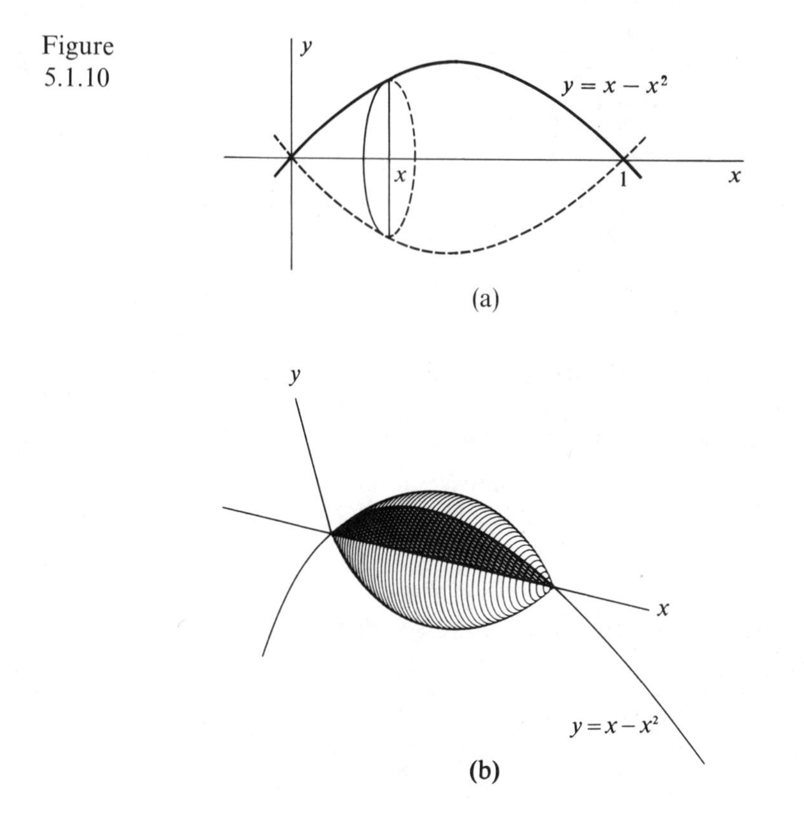

The region to be rotated is indicated in Figure 5.1.11. Note that a typical cross section is a washer with inner radius $r_1 = 1 - 2x + x^2$ and outer radius $r_2 = 1 - 5x^2 + 10x$. Hence

$$A(x) = \pi(1 - 5x^2 + 10x)^2 - \pi(1 - 2x + x^2)^2, \quad \text{and so}$$

$$\begin{aligned} V &= \pi \int_0^2 [(1 - 5x^2 + 10x)^2 - (1 - 2x + x^2)^2]dx \\ &= \pi \int_0^2 (24x^4 - 96x^3 + 84x^2 + 24x)dx \\ &= \pi\left(\frac{24}{5}x^5 - 24x^4 + 28x^3 + 12x^2\right)\Bigg|_0^2 \\ &= \pi\left(\frac{768}{5} - 384 + 224 + 48\right) \\ &= 41\frac{3}{5}\pi \text{ cubic units.} \end{aligned}$$

||

In the previous examples the cross-sectional areas were always taken perpendicular to the x-axis. This need not always be the case, as is illustrated in the next example where the roles of x and y are interchanged; that is, the cross-sectional areas are taken perpendicular to the y-axis.

Figure 5.1.11

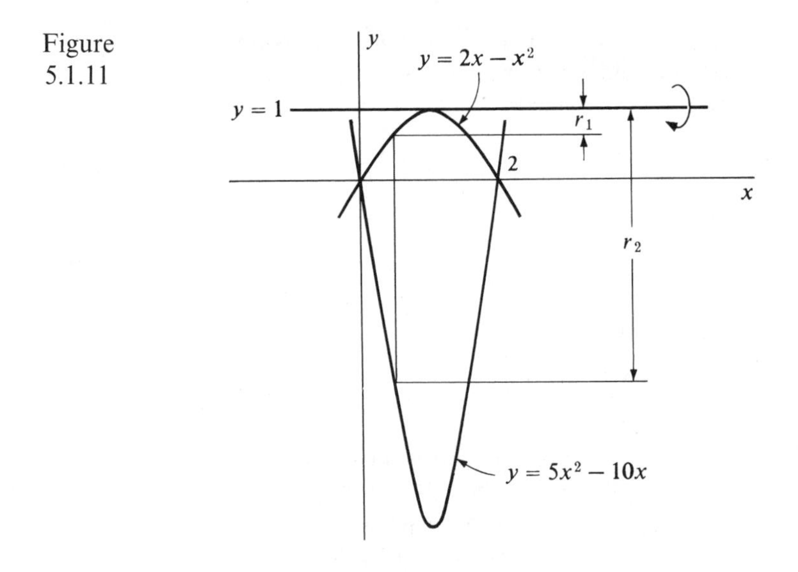

Example 6 The region bounded by the coordinate axes and the graph of $y = x^3 + 1$ is rotated about the y-axis. Calculate the volume of the resulting solid.

Figure 5.1.12(a) shows the region to be rotated and a typical cross section. Figure 5.1.12(b) is a computer-generated image showing a slightly different view of the solid.

Figure 5.1.12

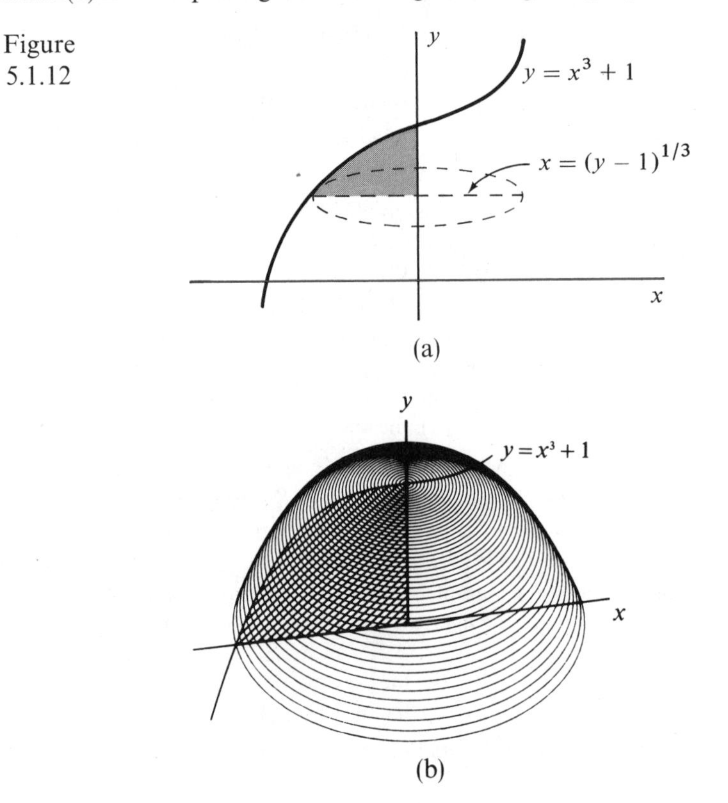

The region to be rotated is cross-hatched in that figure. Portions of a number of cross sections are also indicated. A typical cross section is a circle with radius $(y-1)^{1/3}$. Thus $A(y) = \pi(y-1)^{2/3}$ and

$$V = \int_0^1 \pi(y-1)^{2/3}\,dy = \pi\,\frac{3}{5}(y-1)^{5/3}\Big|_0^1 = \frac{3\pi}{5}. \qquad \|$$

Exercises 5.1

1. The variable cross-sectional area of a solid is given by

$$A(x) = (2x+3)^{1/2}, \qquad 0 \le x \le 1.$$

   Find the volume.

2. The base of a solid is the region bounded by the graphs of $y = x$, $y = 0$, and $x = 3$. The cross sections perpendicular to the x-axis are squares. Find the volume.
3. The base of a solid is the region bounded by the graph of $y = 1 - x^2$ and the x-axis. The cross sections perpendicular to the x-axis are squares. Find the volume.
4. Find the volume of the solid whose base is the region bounded by the graphs of $y = x^3$, $x = 1$, and the x-axis, and whose cross sections perpendicular to the x-axis are semicircles.
5. The base of a solid is the upper semicircle bounded by the x-axis and the graph of $x^2 + y^2 = 1$. The cross sections perpendicular to the x-axis are isosceles right triangles with one side (not the hypotenuse) in the base. Find the volume.
6. Use the methods of this section to prove that the volume of a cube with sides of length s is s^3.
7. Use the methods of this section to find the volume of a right circular cone with base-radius r and height h.
8. The region bounded by $y = 2x$, $y = 6$, and $x = 0$ is rotated about the x-axis. Find the volume.
9. The region bounded by the graphs of $y = x^2$, $x = 5$, and the x-axis is rotated about the x-axis. Find the volume.
10. Let R denote the region bounded by the x-axis, the line $x = 2$, and the graph of $y = x^3$. Set up, but do not evaluate, integrals for the volume swept out as R is rotated about:
    (i) the x-axis, (ii) the y-axis,
    (iii) the line $x = 4$, (iv) the line $y = -2$.
11. Let R denote the region bounded by the y-axis, the line $y = 8$, and the graph of $y = x^3$. Set up, but do not evaluate, integrals for the volume swept out as R is rotated about:
    (i) the x-axis, (ii) the y-axis,
    (iii) the line $x = 4$, (iv) the line $y = -2$.
12. Let R denote the region bounded by the graphs of $y = 4x$ and $y = x^3$ for $x \ge 0$. Set up, but do not evaluate, the integrals for the volume swept out as R is rotated about:
    (i) the x-axis, (ii) the y-axis,
    (iii) the line $x = 3$, (iv) the line $y = -2$.
13. Let R denote the region in the first quadrant bounded by the graphs of $y = 4 - x^2$ and $y = 2 - x$. Set up, but do not evaluate, the integrals for the volume swept out as R is rotated about:
    (i) the x-axis, (ii) the y-axis,
    (iii) the line $y = 7$, (iv) the line $x = -1$.
14. Use the methods of this section to find the volume of a pyramid with square base of side length s and height h.

15. Show that the volume of a sphere of radius r is $(4/3)\pi r^3$.
16. Prove the following theorem. If two solids have equal height and the cross-sectional areas at equal distances from their bases are equal, then the volumes are equal.

5.2 Work Done by a Variable Force

Physics affords many examples of applications of the definite integral, one of which will be discussed in this section and others in Sections 5.9 and 5.10.

It is well known that if a constant force F moves a body in a straight line for a distance d, the work done is simply Fd, which is the product of the force and the distance through which it acts. Suppose, however, that the force in question is not constant and that its magnitude at a point x on the x-axis is given by the continuous function $F(x)$. In this case a rather obvious question presents itself. What is the work $W(a, b)$ done by the variable force $F(x)$ as it moves a body from $x = a$ to $x = b$?

Intuitive Development. Subdivide the interval $a \leq x \leq b$ into small subintervals each with width dx. When dx is small the force $F(x)$ is nearly constant over the distance dx. Consequently the work done in the subinterval is approximately $F(x)dx$. Use of the integral to sum up these "elements of work" between $x = a$ and $x = b$ gives the total work. Thus,

$$W(a, b) = \int_a^b f(x)dx.$$

More Rigorous Development. The following two assumptions about work form the basis of our mathematical development.

Assumption 1. The work done is between the maximum force M times the distance the body is moved and the minimum force m times that distance. That is,

$$m|s - r| \leq W(r, s) \leq M|s - r|.$$

Assumption 2. Work is additive; that is, the work done in moving an object along a line from r to s is the sum of the work done moving it from r to t and then from t to s, where $r < t < s$. Hence,

$$W(r, s) = W(r, t) + W(t, s).$$

Now we can find the total work done by a force $F(x)$ in moving an object from a to b, where $a \leq b$. We first subdivide the interval $a \leq x \leq b$ into n subintervals, each of width $h = (b - a)/n$, by introducing the intermediate points $x_i = a + hi$, $i = 0, 1, \ldots, n$. By Assumption 2, the total work $W(a, b)$ is the sum of the work done in moving the object over all the separate intervals $x_{i-1} \leq x \leq x_i$. Thus

$$W(a, b) = \sum_{i=1}^{n} W(x_{i-1}, x_i). \tag{1}$$

Now since $F(x)$ is continuous for $x_{i-1} \leq x \leq x_i$, it follows that $F(x)$ has both an absolute maximum value M and an absolute minimum value m there. Then by Assumption 1,

$$m(x_i - x_{i-1}) \leq W(x_{i-1}, x_i) \leq M(x_i - x_{i-1}),$$

or, since $h = x_i - x_{i-1}$,

$$mh \leq W(x_{i-1}, x_i) \leq Mh.$$

Thus, since h is positive,

$$m \leq \frac{W(x_{i-1}, x_i)}{h} \leq M.$$

Again using the hypothesis that $F(x)$ is continuous we know that $F(x)$ takes on both values m and M for $x_{i-1} \leq x \leq x_i$. Consequently, by The Intermediate Value Theorem we know that there is some point x_i^*, such that $x_{i-1} \leq x_i^* \leq x_i$ and

$$\frac{W(x_{i-1}, x_i)}{h} = F(x_i^*), \quad \text{or}$$

$$W(x_{i-1}, x_i) = F(x_i^*)h.$$

Substitution into (1) then yields

$$W(a, b) = \sum_{i=1}^{n} F(x_i^*)h,$$

and, since $W(a, b)$ is independent of h, we get

$$W(a, b) = \lim_{h \to 0} \sum_{i=1}^{n} F(x_i^*)h.$$

(This last equation is true whether $a \leq b$ or $a > b$. Its derivation in the case where $a > b$ would only require the reversal of certain inequalities above.)

Then, using the definition of the definite integral,

$$\lim_{h \to 0} \sum_{i=1}^{n} F(x_i^*)h = \int_a^b F(x)dx,$$

we obtain

$$W(a, b) = \int_a^b F(x)dx.$$

Example 1 | Calculate the work done in raising a 2000 pound weight from a depth of 1000 feet, using a cable that weighs 2 lbs/ft.

Let x be the depth of the 2000 pound weight. Then, when x feet of cable are still out, a total weight of $2000 + 2x$ pounds is being lifted by a force directed upward toward decreasing x. That is,

$$F(x) = -(2000 + 2x).$$

The minus sign is used to indicate that the force is directed toward decreasing x. Then

$$W = \int_{1000}^{0} -(2000 + 2x)dx, \quad \text{and so}$$

$$W = (-2000x - x^2)\Big|_{1000}^{0} = 3{,}000{,}000 \text{ foot-pounds.} \qquad ||$$

Example 2 An electron is attracted toward the nucleus of an atom with a force inversely proportional to the square of the distance between them. Calculate the work done by an external force that moves an electron that is c units from the nucleus to $2c$ units from the nucleus.

Select the x-axis in such a way that the nucleus of the atom is located at $x = 0$. Then the attractive force exerted on the electron is directed toward decreasing x with magnitude k/x^2.

The force that moves the electron from c to $2c$ must be equal in magnitude but opposite in direction to the attractive force discussed above. Thus, the force is directed toward increasing x and is the positive quantity $F(x) = k/x^2$. Therefore

$$W = \int_{c}^{2c} \frac{k}{x^2}\,dx = -\frac{k}{x}\Big|_{c}^{2c} = -\frac{k}{2c} + \frac{k}{c} = \frac{k}{2c}. \qquad ||$$

Suppose that a tank as pictured in Figure 5.2.1 is filled with liquid having density k. We want to find the work necessary to pump all the fluid out of the tank. As illustrated, we assume that the area of a cross section cut by a plane perpendicular to the y-axis is $A(y)$. Since the work done does not depend on the way the fluid is moved, we shall think of the fluid as being moved upward by a piston of varying shape. The area of the piston is $A(y)$. Since the depth of the fluid above the piston is $a - y$, the downward pressure on the piston is $k(a - y)$. Consequently the upward force that the piston must exert when it is y units above the origin is

$$F(y) = k(a - y)A(y).$$

Figure 5.2.1

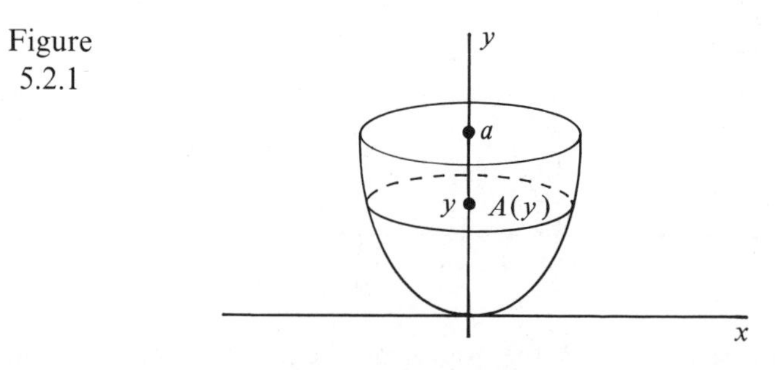

Hence the work done in moving the fluid to the top of the tank is

$$W = \int_0^a k(a - y)A(y)dy.$$

There is a very useful intuitive interpretation of that formula. Suppose that the interval $0 \le y \le a$ is subdivided into subintervals each having width dy. This process cuts the liquid in the tank into thin slabs. A typical slab is illustrated in Figure 5.2.2. Since the volume of the slab is approximately $A(y)dy$, its weight is approximately $kA(y)dy$. Since the slab must be lifted $a - y$ units to pump it out of the tank, $k(a - y)A(y)dy$ is an approximation to the work necessary to pump this typical slab from the tank. Using the integral sign to sum these "elements of work" between 0 and a gives

$$W = \int_0^a k(a - y)A(y)dy.$$

Figure 5.2.2

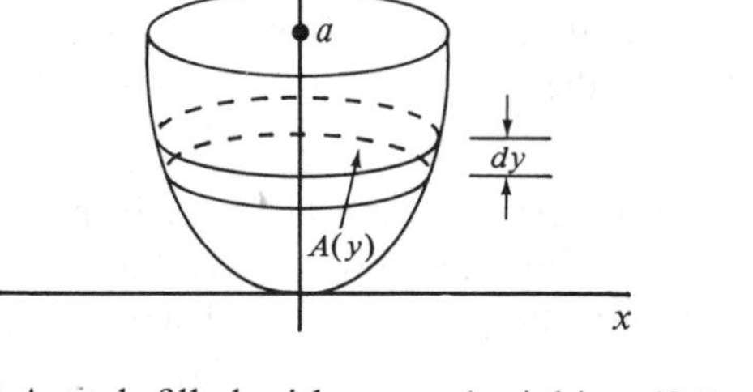

Example 3 A tank filled with water (weighing 62.5 pounds per cubic foot) is in the shape of a right circular cone with a 20 foot vertical axis and a radius of 4 feet at the top. Find the amount of work needed to pump all the water out of the top of the tank.

Take the y-axis as the axis of the cone with the vertex at the origin, as in Figure 5.2.3. Then the radius x of the tank at y feet above the origin is related to y by

$$y = 5x, \quad \text{or} \quad x = \frac{y}{5},$$

Figure 5.2.3

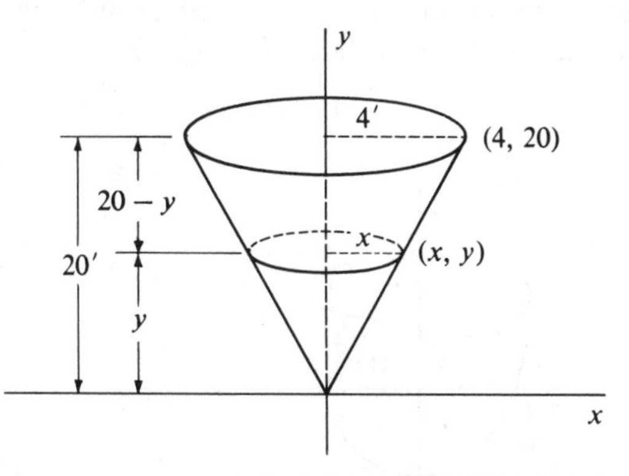

since (x, y) is on the line passing through the origin and the point (4, 20). Thus the cross-sectional area is given by $A(y) = \pi y^2/25$, and the weight of the slab of water illustrated in Figure 5.2.4 is $(62.5)\,\dfrac{\pi y^2}{25}\,dy$. Since the slab must be lifted $20 - y$ feet, the work done in pumping it out of the tank is

$$(62.5)\frac{\pi y^2}{25}(20 - y)dy.$$

Figure 5.2.4

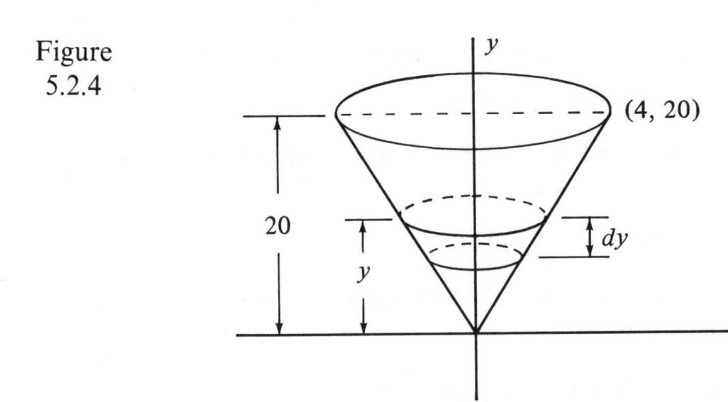

Then, using the integral to sum these increments of work from $y = 0$ to $y = 20$ gives

$$W = \int_0^{20} 2.5\pi y^2(20 - y)dy = 2.5\pi \int_0^{20} (20y^2 - y^3)dy$$

$$= 2.5\pi\left(\frac{20y^3}{3} - \frac{y^4}{4}\right)\Bigg|_0^{20} = 2.5\pi\left(\frac{20^4}{3} - \frac{20^4}{4}\right)$$

$$= 2.5\pi\left(\frac{20^4}{12}\right) \approx 104{,}720 \text{ foot-pounds.}$$ ||

Example 4 | Find the work required to pump the water out of the tank of Example 3 if the discharge of the pump is 10 feet above the top of the tank.

The work required is equal to the work needed to pump the water to the top of the tank plus the work needed to move all of the water up another 10 feet. This additional work is equal to 10 times the weight of the water. Thus the total amount of work is

$$104{,}720 + 10\frac{\pi}{3}(4)^2(20)(62.5) \approx 314{,}160 \text{ foot-pounds.}$$ ||

Example 5 | The parabolic tank illustrated in Figure 5.2.5 is filled with water to a depth of 5 feet. Compute the work done in pumping the water out of a pipe 3 feet above the surface.

Figure 5.2.5

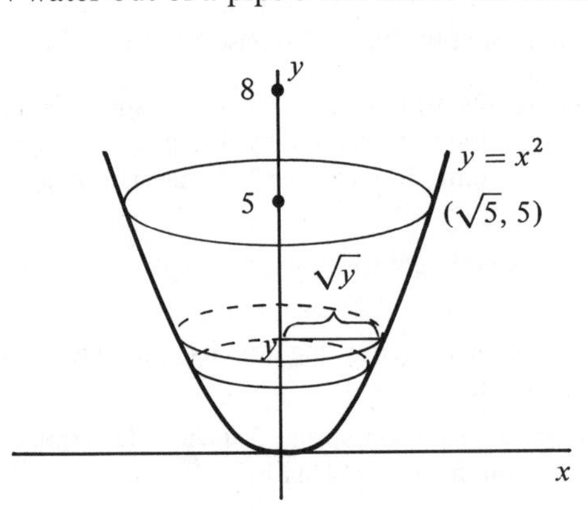

A typical slab located y feet above the x-axis has weight

$$62.5\pi(\sqrt{y})^2\, dy \text{ lbs.}$$

The slab must be lifted $8 - y$ feet. Thus, the work done in pumping it out of a pipe 3 ft above the surface is

$$62.5\pi y(8 - y)dy \text{ ft lbs.}$$

The total work W done is obtained by using the integral to sum from $y = 0$ to $y = 5$. Then

$$W = \int_0^5 62.5\pi y(8 - y)dy = 62.5\pi \int_0^5 (8y - y^2)dy$$

$$= 62.5\pi\left(4y^2 - \frac{1}{3}y^3\right)\Bigg|_0^5 \approx 11454 \text{ ft lbs.}$$

||

Exercises 5.2

1. A chain weighing 5 lbs/ft is unwinding from a drum into a canyon. If 100 feet of chain are already unwound, find the work done by gravity to unwind an additional 30 feet.
2. Find the work done when a chain (weighing 2 pounds per foot), attached to a winch at the top of a canyon 500 feet deep, is used to raise a weight of 800 pounds from the bottom of the canyon to a platform 100 feet above the bottom of the canyon.
3. A cable, 500 feet long and weighing 1 pound per foot, lies coiled at the base of a building 400 feet high. One end of the cable is attached to a lighter cord, which weighs 0.1 pound per foot. Determine the work necessary to raise the upper end of the cable to the top of the building.
4. Water is leaking from a bucket at a constant rate of .1 ft^3/min. If the bucket originally contained 3 cubic feet of water at 62.5 lbs/ft^3, find the work done in raising the bucket 100 feet at the constant rate of 10 ft/min.
5. A piston slowly compresses a monatomic gas. If the gas is maintained at constant temperature, the pressure p in pounds per square foot is

$$p = 20{,}000v^{-3/2},$$

where v is the volume of the gas in cubic feet. Find the work done in compressing the gas from 9 cubic feet to 4 cubic feet.

6. Find the work done in compressing the gas in Exercise 5 from 16 cubic feet to 9 cubic feet.

Hooke's Law states that the elastic restoring force exerted by a spring is directly proportional to the extension of the spring. Thus the force exerted by the spring is $F(x) = kx$, where x is the displacement and k is the spring constant. In Exercises 7–10, use Hooke's Law to find the work done.

7. Determine the work done in extending a spring from its natural length of 5 cm to a length of 8 cm, if $k = 2$ kg per cm.
8. Determine the work done in compressing a spring from its natural length of 4 feet to a length of 2 feet, if $k = 25$ pounds per foot.
9. A weight of 10 pounds stretches a certain spring 2 inches. Determine the work done in stretching the spring 4 inches from its natural length.

10. A force of 12 kg extends a certain spring 5 cm. Determine the work done in stretching the spring from length 6 cm to length 10 cm, if the natural length of the spring is 4 cm.

11. The force exerted by the earth on a body with mass m is $F = k(mM/s^2)$, where k is a proportionality constant determined by the units used, M is the mass of the earth in slugs, and s is the distance of the body in feet from the center of the earth. Find the work required to raise a mass of one slug from the surface of the earth to a distance 400 miles above the surface. Take the radius of the earth to be 4000 miles. (A slug is the unit of mass, in the foot-pound-second system, that can be given an acceleration of 1 foot per second by a force of 1 pound.)

12. A tank full of liquid weighing 70 lb per cubic foot is in the shape of a right circular cone with a 10 ft vertical axis and a radius of 5 ft at the top. Find the amount of work required to pump the liquid out of the top of the tank.

13. A horizontal, cylindrical tank is 4 feet in diameter and 10 feet long. It is half full of oil weighing 50 pounds per cubic foot. Find the work needed to pump the oil out of a pump 6 feet above the top of the tank.

14. Find the work needed if the tank in Exercise 12 is emptied out of a pump 8 m above the top of the tank.

15. An open dipping tank full of solution weighing 60 pounds per cubic foot is in the shape of a rectangular box $3' \times 5' \times 4'$ deep. Find the work needed to pump the solution out of a pump 2 feet above the top of the tank.

16. A vat, 15 feet long, 5 feet wide at the top, and 6 feet deep, has the shape shown in the diagram. The vat contains a liquid weighing 60 pounds per cubic foot to a depth of 4 feet. Determine the work done in pumping out all the liquid at the top of the tank.

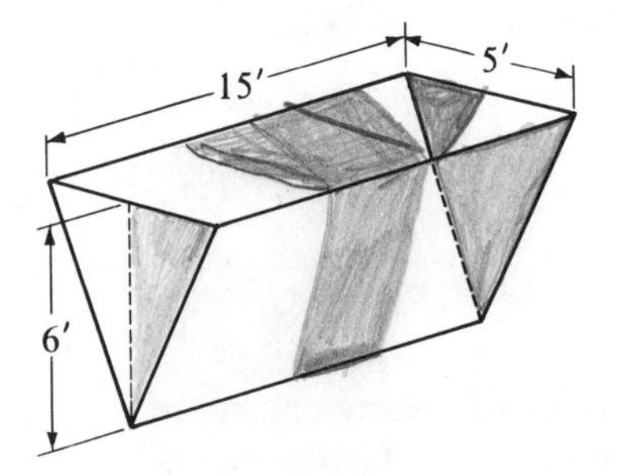

17. A hemispherical reservoir is 20 ft in diameter and contains water with a depth of 5 ft. Find the work done in pumping the water out of a pump 2 feet above the top of the reservoir.

5.3 Arc Length

From our previous work we are already familiar with the concept of the length of a straight line segment. However, we have not yet considered the length of an arc of a curve. In this section we are concerned with the arc length of the graph of $y = f(x)$ for $a \leq x \leq b$, where $f'(x)$ is continuous on the interval.

Intuitive Development. Subdivide the interval $a \leq x \leq b$ into small subintervals each of width dx. This process yields a subdivision of the graph $y = f(x)$ for $a \leq x \leq b$. The subdivided arc and a magnified view of a typical element of arc is shown in Figure 5.3.1.

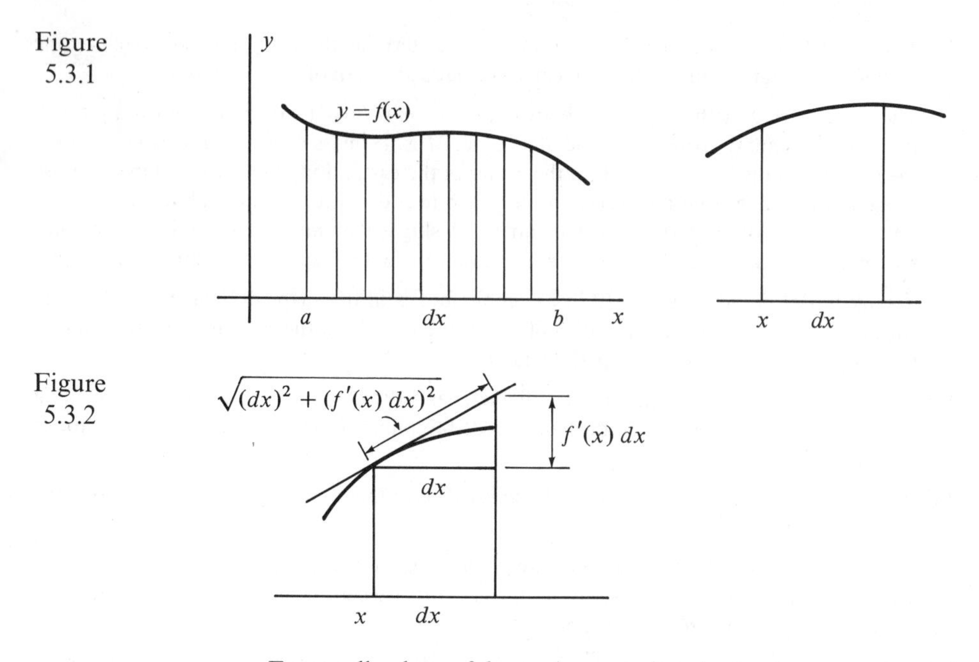

Figure 5.3.1

Figure 5.3.2

For small values of dx an element of arc is closely approximated by the tangent to the graph, as illustrated in Figure 5.3.2.

By the Pythagorean Theorem the length of the tangent line above the segment of length dx is

$$\sqrt{(dx)^2 + (f'(x)dx)^2} = \sqrt{1 + (f'(x))^2}\,dx.$$

Use of the integral to sum these small elements of arc length from a to b gives

$$L = \int_a^b \sqrt{1 + (f'(x))^2}\,dx.$$

More Rigorous Development. If we consider straight line segments, it is clear that the greater the absolute value of the slope the further one must travel along the line segment in order to cover a given horizontal distance h. In fact, if the line has a slope m, as in Figure 5.3.3, the length L of the line segment above an interval of width h is

$$\sqrt{h^2 + (mh)^2}.$$

Figure 5.3.3

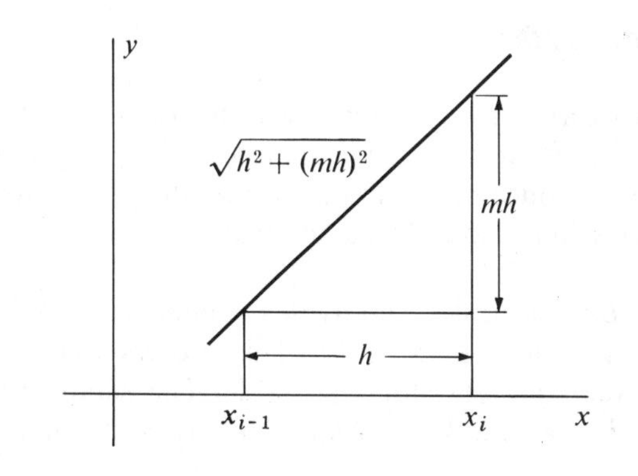

Thus the length of a line segment above an interval of width h is determined by the slope m of the segment.

We shall carry that fact about straight-line lengths over to an assumption concerning arc length in general. Figure 5.3.4 pictures a portion of the graph of $y = f(x)$, along with the tangent lines T_1 and T_2 that have the largest and smallest slope in absolute value. That is, if M is the slope of T_1 and m is the slope of T_2, then

$$|m| \leq |f'(x)| \leq |M|, \quad \text{for} \quad x_{i-1} \leq x \leq x_i.$$

Figure 5.3.4

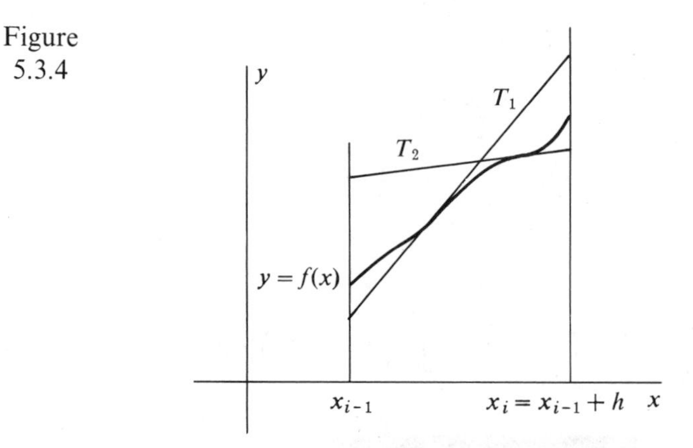

From our consideration of straight line segments it would appear that the greatest arc length would occur if $|f'(x)| = M$ for all x in the interval $x_{i-1} \leq x \leq x_i$. On the other hand, the minimal arc length would occur if $y = f(x)$ were more nearly horizontal, that is, if $|f'(x)| = m$ for all x in the interval $x_{i-1} \leq x \leq x_i$. Thus it is reasonable to assume that the length L of the graph $y = f(x)$ between x_{i-1} and x_i is between the lengths of T_1 and T_2 over the same interval. Hence we assume

$$\sqrt{h^2 + (mh)^2} \leq L \leq \sqrt{h^2 + (Mh)^2}. \tag{1}$$

Secondly, we assume that arc length is additive. That is, we assume that the length of the graph of $y = f(x)$ for $r \leq x \leq s$ is the sum of the length of the graphs of $y = f(x)$ for $r \leq x \leq t$ and $y = f(x)$ for $t \leq x \leq s$.

Those two assumptions lead to a simple formula for the arc length of the graph of $y = f(x)$ for $a \leq x \leq b$. We first subdivide the interval $a \leq x \leq b$ into n subintervals, each of width $h = (b - a)/n$, by introducing the intermediate points $x_i = a + hi$, $i = 0, 1, 2, \ldots, n$. Let L be the arc length of the graph of $y = f(x)$ for $a \leq x \leq b$, and let L_i be the arc length for $x_{i-1} \leq x \leq x_i$. Then the use of the assumption that arc length is additive gives

$$L = \sum_{i=1}^{n} L_i. \tag{2}$$

Since $f'(x)$ is continuous, $|f'(x)|$ has an absolute maximum value M, and an absolute minimum value m, on each interval $x_{i-1} \leq x \leq x_i$. By (1) we know that

$$\sqrt{h^2 + (mh)^2} \leq L_i \leq \sqrt{h^2 + (Mh)^2}, \quad \text{or}$$

$$(\sqrt{1 + m^2})h \leq L_i \leq (\sqrt{1 + M^2})h.$$

Since the continuous function $(\sqrt{1 + [f'(x)]^2})h$ assumes both the values $(\sqrt{1 + m^2})h$ and $(\sqrt{1 + M^2})h$ on the interval $x_{i-1} \leq x \leq x_i$, it must also assume the value L_i there. That is, there is a number x_i^* such that $x_{i-1} \leq x_i^* \leq x_i$ and

$$L_i = (\sqrt{1 + [f'(x_i^*)]^2})h. \tag{3}$$

Substituting into (2), we obtain

$$L = \sum_{i=1}^{n} (\sqrt{1 + [f'(x_i^*)]^2})h.$$

Then, since the arc length is independent of h,

$$L = \lim_{h \to 0} L = \lim_{h \to 0} \sum_{i=1}^{n} (\sqrt{1 + [f'(x_i^*)]^2})h.$$

But by definition,

$$\lim_{h \to 0} \sum_{i=1}^{n} (\sqrt{1 + [f'(x_i^*)]^2})h = \int_a^b \sqrt{1 + [f'(x)]^2}\, dx,$$

and so

$$L = \int_a^b \sqrt{1 + [f'(x)]^2}\, dx.$$

Example 1 Calculate the length of the graph of $f(x) = (2/3)x^{3/2}$ between $x = 0$ and $x = 3$.

Since $f(x) = (2/3)x^{3/2}$, we have $f'(x) = x^{1/2}$. Substituting into the equation for arc length, we get

$$A = \int_0^3 \sqrt{1 + x}\, dx = \frac{2}{3}(1 + x)^{3/2}\Big|_0^3$$

$$= \frac{2}{3}(4^{3/2} - 1) = \frac{14}{3} \text{ units.}$$

A direct change of variable will put the arc length formula,

$$L = \int_a^b \sqrt{1 + \left(\frac{dy}{dx}\right)^2}\, dx, \tag{4}$$

into a form that is easily applied to curves given by the parametric equations $x = f(t)$, $y = g(t)$. Note that

$$\frac{dy}{dx} = \frac{dy/dt}{dx/dt}$$

and

$$dx = \frac{dx}{dt}\, dt.$$

Thus, if $a = f(c)$ and $b = f(d)$ we can substitute into (4) to obtain

$$L = \int_c^d \sqrt{1 + \left(\frac{dy/dt}{dx/dt}\right)^2} \frac{dx}{dt}\, dt,$$

or

$$L = \int_c^d \sqrt{\left(\frac{dx}{dt}\right)^2 + \left(\frac{dy}{dt}\right)^2}\, dt. \qquad \|$$

Example 2 | Find the distance traveled between $t = 0$ and $t = \pi/2$ by a particle whose position at time t is given by $x = \sin^3 t$, $y = \cos^3 t$.

Since

$$\frac{dx}{dt} = 3 \sin^2 t \cos t,$$

and

$$\frac{dy}{dt} = -3 \cos^2 t \sin t,$$

substitution into (5) gives

$$\begin{aligned} L &= \int_0^{\pi/2} \sqrt{9 \sin^4 t \cos^2 t + 9 \cos^4 t \sin^2 t}\, dt \\ &= \int_0^{\pi/2} \sqrt{9 \sin^2 t \cos^2 t(\sin^2 t + \cos^2 t)} dt \\ &= \int_0^{\pi/2} 3 \sin t \cos t\, dt \\ &= \frac{3}{2} \sin^2 t \Big|_0^{\pi/2} = \frac{3}{2} \text{ units.} \end{aligned} \qquad \|$$

Exercises 5.3

In Exercises 1–19, find the length of the graph of the given equation on the indicated interval.

1. $y = (2/3)x^{3/2}, \quad 0 \le x \le 15.$
2. $y = (2/3)x^{3/2}, \quad 15 \le x \le 24.$
3. $y = 6x^{3/2}, \quad 1 \le x \le 4.$
4. $y = (2/3)(x^2 + 1)^{3/2}, \quad -1 \le x \le 2.$
5. $y = (2/3)(x - 5)^{3/2}, \quad 6 \le x \le 8.$
6. $y = (1/3)(x^2 + 2)^{3/2}, \quad 0 \le x \le 3.$
7. $y = (1/3)(x^2 + 2)^{3/2}, \quad 3 \le x \le 6.$
8. $y = \dfrac{x^3}{6} + \dfrac{x^{-1}}{2}, \quad 1 \le x \le 2.$
9. $y = \dfrac{x^3}{12} + x^{-1}, \quad 1 \le x \le 12.$
10. $y = \dfrac{x^3}{4} + \dfrac{x^{-1}}{3}, \quad 2 \le x \le 4.$
11. $y = \dfrac{x^4}{32} + x^{-2}, \quad 2 \le x \le 4.$
12. $y = \dfrac{x^5}{10} + \dfrac{x^{-3}}{6}, \quad 1 \le x \le 2.$
13. $x = y^{3/2}, \quad 0 \le y \le 4.$
14. $y^2 = 4x^3, \quad -2 \le y \le 16.$
15. $y = (1/3)\sqrt{x}(3x - 1), \quad 1 \le x \le 4.$
16. $x = 6(y + 1)^{3/2}, \quad 1 \le y \le 4.$
17. $x = e^t \cos t$, $y = e^t \sin t$, from $t = 0$ to $t = \pi$.
18. $x = 4t^3$, $y = 2t^2$, from $t = 0$ to $t = 1$.
19. $x = \theta^2/2$, $y = (1/3)(2\theta + 1)^{3/2}$, from $\theta = 0$ to $\theta = 9$.
20. Verify that the circumference of a circle of radius r is $2\pi r$.

21. The position of a particle at time t is given by $x = \frac{t^2}{2} + 1$ and $y = \frac{1}{3}(2t + 3)^{3/2}$. Compute the distance the particle travels between $t = 0$ and $t = 2$.

In Exercises 22–24 set up, but do not evaluate, an integral giving the arc length of the indicated graph. The techniques for evaluating these integrals will be developed later.

22. The parabola $y = x^2$ from $x = -1$ to $x = 2$.

23. The first quadrant portion of the ellipse $4x^2 + 9y^2 = 36$.

24. Use the arc length formula developed in this section to verify that the straight-line distance between points (a, b) and (c, d) is given by
$$\sqrt{(a - c)^2 + (b - d)^2}.$$

5.4 General Properties of Applications of Definite Integrals†

We have already discussed several applications of the definite integral. Some of those applications have certain fundamental ideas in common; indeed, many applications of the definite integral share the same ideas. In each case there is a function $f(x)$ that is continuous for $a \leq x \leq b$, and there is a quantity $S(r, s)$ that depends on $f(x)$ and has a value for each subinterval $r \leq x \leq s$ of the interval $a \leq x \leq b$. If $S(r, s)$ also has the other two properties mentioned in Theorem 5.4.1 below, it can be shown that
$$S(a, b) = \int_a^b f(x)dx.$$

$S(a, b)$ might be the area under the graph of $y = f(x)$, the volume of a solid with cross-sectional area $f(x)$, the work done by a force $f(x)$, and so forth.

The relevant theorem will be proved by the same technique used to obtain some of the previous formulas for applications of the definite integral.

Theorem 5.4.1

Let $f(x)$ be continuous for $a \leq x \leq b$, and let $S(r, s)$ be a quantity that depends on $f(x)$ and has a value for each subinterval $r \leq x \leq s$ of the interval $a \leq x \leq b$. Assume further that S has the following two properties:

(i) S is additive, that is,
$$S(r, t) + S(t, s) = S(r, s) \qquad \text{for any } t \text{ such that } r \leq t \leq s.$$

(ii) Where m is the absolute minimum value of $f(x)$ and M is the absolute maximum value of $f(x)$ for $s \leq x \leq t$:
$$m(t - s) \leq S(s, t) \leq M(t - s) \qquad \text{for all } s, t \text{ between } a \text{ and } b.$$

Then
$$S(a, b) = \int_a^b f(x)dx.$$

† The material in this section is optional for classes wishing to use the *Intuitive Development* in the remaining sections of Chapter 5.

Proof. Divide the interval $a \le x \le b$ into n subintervals, each of width h, by selecting these equally spaced points:

$$x_0 = a, \quad x_1 = a + h, \quad x_2 = a + 2h, \quad \ldots, x_n = a + nh = b.$$

Then by (i), we get

$$S(a, b) = S(x_0, x_1) + S(x_1, x_2) + \cdots + S(x_{n-1}, x_n) \tag{1}$$

$$= \sum_{i=1}^{n} S(x_{i-1}, x_i).$$

And from (ii), we have

$$m(x_i - x_{i-1}) \le S(x_{i-1}, x_i) \le M(x_i - x_{i-1}),$$

or, replacing the positive number $x_i - x_{i-1}$, by h, we get

$$mh \le S(x_{i-1}, x_i) \le Mh.$$

Dividing by the positive number h, we obtain

$$m \le \frac{S(x_{i-1}, x_i)}{h} \le M.$$

Since the continuous function $f(x)$ takes both of the values m and M in the interval $x_{i-1} \le x \le x_i$, The Intermediate Value Theorem for continuous functions assures that there is a number x_i^* such that $x_{i-1} \le x_i^* \le x_i$ and

$$\frac{S(x_{i-1}, x_i)}{h} = f(x_i^*), \quad \text{or}$$

$$S(x_{i-1}, x_i) = f(x_i^*)h.$$

Then by (1),

$$S(a, b) = \sum_{i=1}^{n} S(x_{i-1}, x_i) = \sum_{i=1}^{n} f(x_i^*)h. \tag{2}$$

However, since $S(a, b)$ is independent of h,

$$\lim_{h \to 0} S(a, b) = S(a, b).$$

Then, taking the limit as h approaches zero in (2), we get

$$\lim_{h \to 0} S(a, b) = S(a, b) = \lim_{h \to 0} \sum_{i=1}^{n} f(x_i^*)h.$$

But by definition of the definite integral,

$$\lim_{h \to 0} \sum_{i=1}^{n} f(x_i^*)h = \int_a^b f(x)dx,$$

and hence $$S(a, b) = \int_a^b f(x)dx.$$

In order to apply this theorem, we only need to know that $f(x)$ is continuous for $a \le x \le b$ and that S satisfies properties (i) and (ii). We can then conclude that

$$S(a, b) = \int_a^b f(x)dx.$$

In later sections, additional applications of the definite integral are obtained directly from Theorem 5.4.1 instead of going through a derivation similar to the proof of Theorem 5.4.1 in each individual case.

The material in this section is an illustration of the great power of generality and abstraction in mathematics, where the essential aspects of many problems are distilled into a single unifying idea. Results are then obtained for the general, abstract situation. These general results are applied to the individual problems, making it unnecessary to derive the results for each separate case. This not only saves a great deal of effort, but gives an insight into the really important aspects of the problems.

Theorem 5.4.1 can be used with proper selection of $f(x)$ and $S(r, s)$ to obtain directly the individual integrals for volume by cross-sectional area, work, length, and consumer surplus.

5.5 Perfect Price Discrimination and Consumer Surplus

In many situations a producer will lower the unit price of a product to increase sales. Generally decreased price corresponds to increased sales. Let x denote the quantity of a commodity sold, and p the price of the commodity per unit. The discussion above indicates that often a decrease in p causes an increase in x. As we shall see later that fact implies that there is a function, $p = f(x)$, which gives the price as a function of the number of units sold.

For example, the following might be a function that gives the price of electricity in cents per kilowatt hour, where x is the number of kilowatt hours used per month:

$$f(x) = \begin{cases} 4 & \text{if} \quad 0 \le x \le 50 \\ 3 & \text{if} \quad 50 < x \le 100 \\ 2 & \text{if} \quad 100 < x \le 150 \\ 1.1 & \text{if} \quad 150 < x \le 350 \\ 1 & \text{if} \quad 350 < x \le 600 \\ 0.9 & \text{if} \quad 600 < x. \end{cases} \tag{1}$$

In an ideal market situation the price p per unit is considered to be a continuous function $f(x)$ of the number x of units previously sold. For example, the cost of electricity, in cents per kilowatt hour, might be given by

$$f(x) = \frac{75}{x+1} + 2, \qquad \text{for } x > 0.$$

If, instead of selling a certain number of units at a fixed price, a producer charges $f(x)$ for unit x, we say that perfect price discrimination has been attained.

A natural problem is the determination of the revenue $R(a, b)$ that results as x changes from a to b when perfect price discrimination has been attained.

Intuitive Development. Subdivide the interval $a \leq x \leq b$ into small subintervals each having width dx. Over small subintervals, $p = f(x)$ is nearly constant. The resulting revenue from the sale of dx items at a price of $f(x)$ per unit is $f(x)dx$. Use of the integral to sum this revenue from $x = a$ to $x = b$ gives

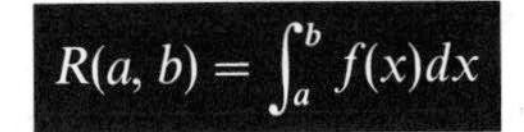

$$R(a, b) = \int_a^b f(x)dx$$

More Rigorous Development. We assume that $f(x)$ is continuous and that for any two demands r and s,

(a) $$R(r, t) + R(t, s) = R(r, s), \quad \text{for } r \leq t \leq s, \quad \text{and}$$

(b) $$m(s - r) \leq R(r, s) \leq M(s - r),$$

where m and M are the largest and smallest values of $f(x)$ for $r \leq x \leq s$. Then, by using Theorem 5.4.1 with $R(r, s) = S(r, s)$, we have

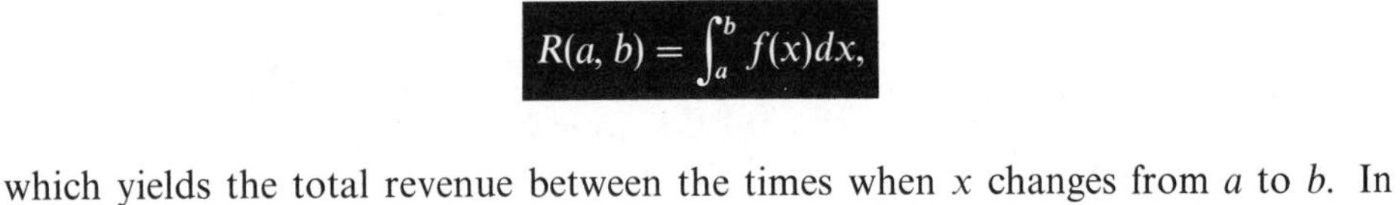

$$R(a, b) = \int_a^b f(x)dx,$$

which yields the total revenue between the times when x changes from a to b. In particular,

$$R(0, b) = \int_0^b f(x)dx. \tag{2}$$

Thus we have an expression for the total revenue from the sale of the first b units.

Of course, in most instances perfect price discrimination is not practiced. Rather, a seller will simply sell b units at a fixed price p. Then the *consumer surplus C* is the difference between the revenue possible under perfect price discrimination and the actual revenue pb. That is,

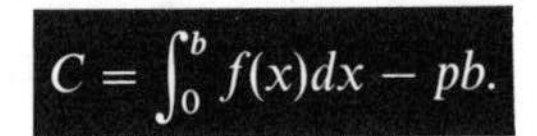

$$C = \int_0^b f(x)dx - pb.$$

C is a measure of the advantage (when C is positive) or disadvantage (when C is negative) to the consumer when perfect price discrimination is not practiced.

Example 1 | If the demand function is $p = 11 - x^2/100$, find the consumer surplus when the actual selling price is $p = 2$.

We get the number x of units sold when $p = 2$ by substituting $p = 2$ in the demand function. Thus $2 = 11 - x^2/100$, and therefore $x^2 = 900$, or $x = 30$. Hence

$$\begin{aligned} C &= \int_0^{30}\left(11 - \frac{x^2}{100}\right)dx - 2(30) \\ &= 11x - \frac{x^3}{300}\Big|_0^{30} - 60 \\ &= 180. \end{aligned}$$

||

Exercises 5.5

For each demand function $p = f(x)$, find the consumer surplus for the given actual price p per unit sold or number x of units sold.

1. $p = 90 - x^3, \quad p = 26.$
2. $p = 103 - \dfrac{x}{200}, \quad p = 98.$
3. $p = 20 - \dfrac{x}{50}, \quad x = 100.$
4. $p = 40 - \dfrac{x^2}{50}, \quad x = 40.$
5. $p = 544 - x^2, \quad p = 400.$
6. $p = 20 - x^4, \quad p = 4.$
7. $p = 200 - (x + 2)^2, \quad x = 8.$
8. $p = 100 - \dfrac{25}{(x+1)^2}, \quad x = 4.$
9. $p = 200 - \dfrac{100}{(x+1)^2} - x^2, \quad x = 4.$
10. $p = 5000 - x^3, \quad p = 4973,$
11. $p = 50 - \sqrt{x}, \quad p = 47.$
12. $p = 500 - \sqrt{x+1}, \quad x = 15.$
13. $p = 100 - x\sqrt{x^2 + 9}, \quad x = 4.$
14. $p = 30 - x\sqrt{x^2 + 16}, \quad x = 3.$

5.6 The Theorem of Bliss

Before we discuss any further applications it will be helpful to have some additional theory. Recall that the definite integral is defined as follows:

$$\int_a^b f(x)dx = \lim_{h \to 0} \sum_{i=1}^{n} f(x_i^*)h,$$

where the interval $a \leq x \leq b$ has been divided into n subintervals of equal width $h = (b - a)/n$ by introduction of the points $x_i = a + hi$ for $i = 1, 2, \ldots, n$, and where x_i^* is an arbitrarily chosen point between x_{i-1} and x_i. According to that definition we see that

$$\int_a^b f(x)g(x)dx = \lim_{h \to 0} \sum_{i=1}^{n} f(x_i^*)g(x_i^*)h.$$

The Theorem of Bliss tells us that we need not evaluate the func
same point x_i^* in each subinterval, but we may evaluate f at one
x_i^*, and g at another, x_i^{**}. Of course both points x_i^* and x_i^{**} m
the ith subinterval.

Theorem 5.6.1

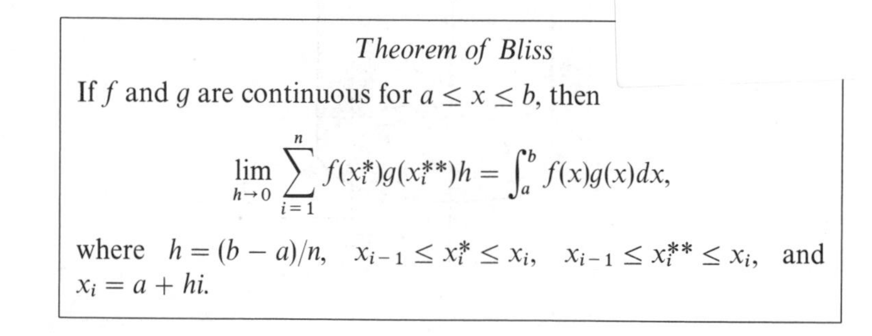

Theorem of Bliss

If f and g are continuous for $a \leq x \leq b$, then

$$\lim_{h \to 0} \sum_{i=1}^{n} f(x_i^*)g(x_i^{**})h = \int_a^b f(x)g(x)dx,$$

where $h = (b - a)/n$, $x_{i-1} \leq x_i^* \leq x_i$, $x_{i-1} \leq x_i^{**} \leq x_i$, and $x_i = a + hi$.

A proof of Bliss' Theorem can be found in some advanced texts.

5.7 The Method of Cylindrical Shells

For finding volumes the method of cylindrical shells can often be used advantageously in cases where the disc method would involve an integral that is difficult to evaluate. In this section we shall suppose that the width of a region above the point x is represented by a continuous functions $w(x)$. Let that region be rotated about the y-axis as indicated in Figure 5.7.1.

Figure 5.7.1

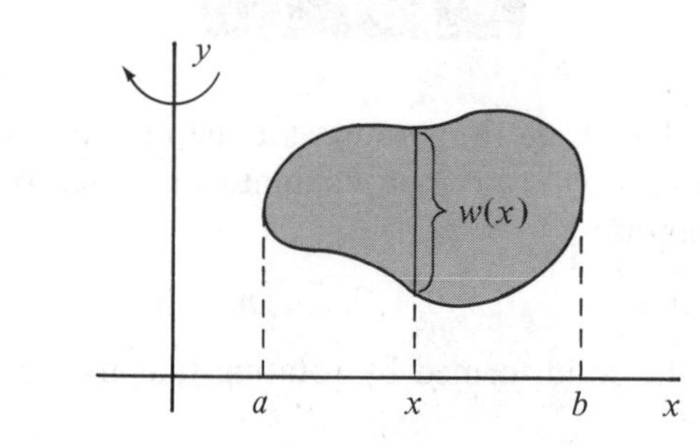

Intuitive Development. Subdivide the interval $a \leq x \leq b$ into small subintervals each having width dx. Such subdivision results in a subdivision of the given region into a number of thin elements of area. Rotation of a typical element of area about the y-axis yields the element of volume pictured in Figure 5.7.2. Note that this element of volume is a cylindrical shell, which accounts for the name of the method. In order to estimate the volume of the cylindrical shell, cut it vertically and lay it out flat as illustrated in Figure 5.7.3. Thus the volume of the element is approximately $2\pi x w(x)dx$. Use

igure
5.7.2

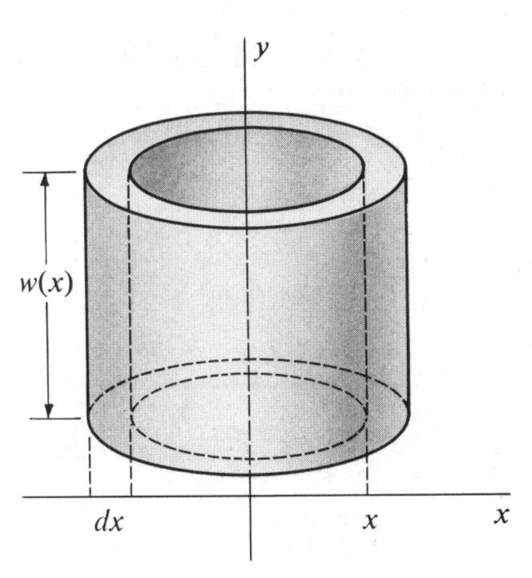

Figure
5.7.3

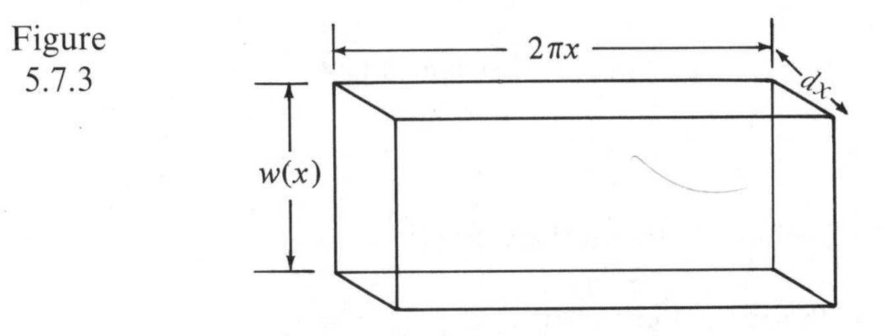

of the integral to sum the elements of volume from $x = a$ to $x = b$ gives

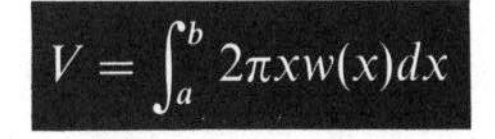

$$V = \int_a^b 2\pi x w(x) dx$$

More Rigorous Development. To derive the formula for the method of cylindrical shells we first divide the interval $a \leq x \leq b$ into n subintervals, each of width $h = (b - a)/n$, by introducing the points

$$x_i = a + hi, \qquad i = 0, 1, 2, \ldots, n.$$

Let V_i be the volume of the solid formed by rotating that part of the region

Figure
5.7.4

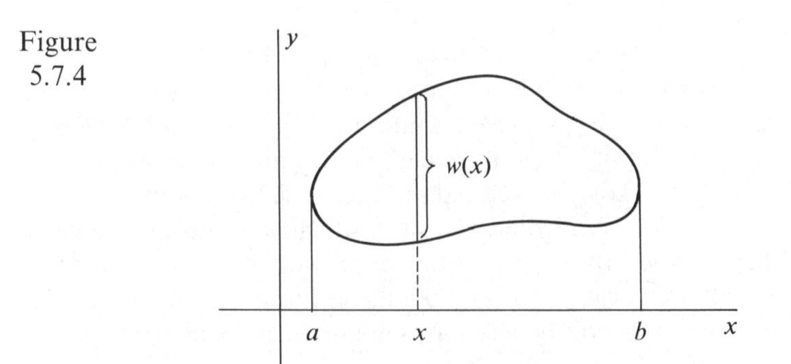

between x_{i-1} and x_i about the y-axis. Then the total volume is given by

$$V = \sum_{i=1}^{n} V_i.$$

In Figure 5.7.5 a typical "element of volume" is pictured. The maximal cross section of that cylindrical shell made by planes perpendicular to the y-axis is the washer-shaped area indicated in Figure 5.7.6. The area A of this cross section is given by

$$A = \pi x_i^2 - \pi x_{i-1}^2.$$

Figure 5.7.5

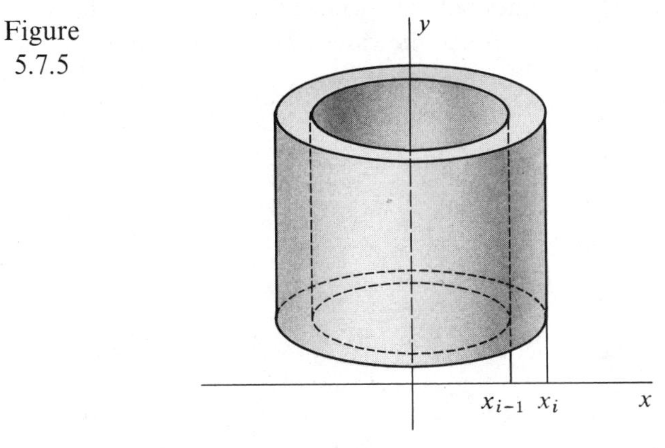

Figure 5.7.6

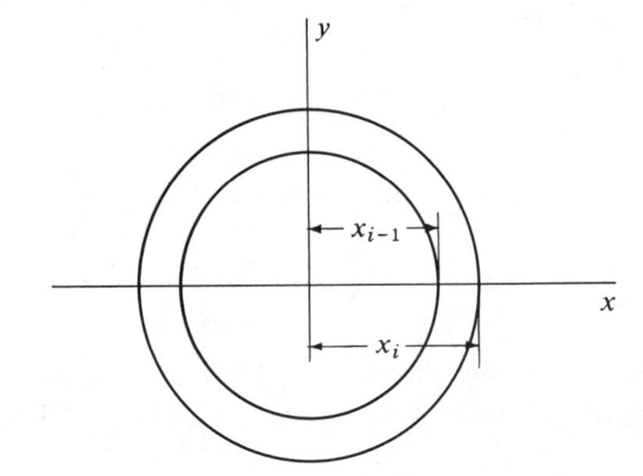

For our further work we will modify the expression for A as follows:

$$A = \pi(x_i - x_{i-1})(x_i + x_{i-1}).$$

Then, since $x_i - x_{i-1} = h$,

$$A = \pi h(x_i + x_{i-1}) = 2\pi h \frac{x_i + x_{i-1}}{2}.$$

Now let x_i^* be the average of x_i and x_{i-1}; that is, let $x_i^* = (x_i + x_{i-1})/2$. Then $x_{i-1} \leq x_i^* \leq x_i$ and

$$A = 2\pi h x_i^*.$$

Since we now have an expression for the maximal cross-sectional area we can concentrate on finding an expression for the height of the cylindrical shell. Note that the height above the point $(x, 0)$ is $w(x)$. Because $w(x)$ is continuous, it has a maximum value K and a minimum value k on $x_{i-1} \leq x \leq x_i$. The volume V_i of the cylindrical shell is between the cross-sectional area A times the minimal height k and the cross-sectional area A times the maximal height K. That is,

$$Ak \leq V_i \leq AK, \quad \text{or}$$

$$2\pi h x_i^* k \leq V_i \leq 2\pi h x_i^* K.$$

Since the function $w(x)$ is continuous, so is the function $g(x) = 2\pi h x_i^* \cdot w(x)$. Moreover, since the function $w(x)$ assumes both the values k and K on the interval $x_{i-1} \leq x \leq x_i$, we know that $g(x)$ must assume the values $2\pi h x_i^* k$ and $2\pi h x_i^* K$ and all values between them on this interval.

Then, using the fact that

$$2\pi h x_i^* k \leq V_i \leq 2h x_i^* K,$$

we find that there is some point x_i^{**} such that $x_{i-1} \leq x_i^{**} \leq x_i$ and

$$V_i = g(x_i^{**}), \quad \text{or}$$

$$V_i = 2\pi x_i^* w(x_i^{**})h.$$

Hence
$$V = \sum_{i=1}^{n} V_i = \sum_{i=1}^{n} 2\pi x_i^* w(x_i^{**})h.$$

Taking the limit of both sides, we get

$$\lim_{h \to 0} V = \lim_{h \to 0} \sum_{i=1}^{n} 2\pi x_i^* w(x_i^{**})h.$$

Since the left side equals V (because V is independent of h), we get

$$V = \lim_{h \to 0} \sum_{i=1}^{n} 2\pi x_i^* w(x_i^{**})h.$$

Since the points x_i^* and x_i^{**} are both from the ith subinterval, we can apply the Theorem of Bliss (Theorem 5.6.1) to the two functions x and $w(x)$. Then

$$\lim_{h \to 0} \sum_{i=1}^{n} 2\pi x_i^* w(x_i^{**})h = \int_a^b 2\pi x w(x)\,dx$$

and so
$$V = 2\pi \int_a^b x w(x)\,dx.$$

Example 1 The region bounded by the graphs of $y = -x^2 + 4x$ and $y = x^2 - 4x + 6$ is rotated about the y-axis. Calculate the resulting volume.

The region bounded by these graphs is pictured in Figure 5.7.7. Note that since $-x^2 + 4x \geq x^2 - 4x + 6$ for $1 \leq x \leq 3$,

$$w(x) = (-x^2 + 4x) - (x^2 - 4x + 6) = -2x^2 + 8x - 6.$$

Figure 5.7.7

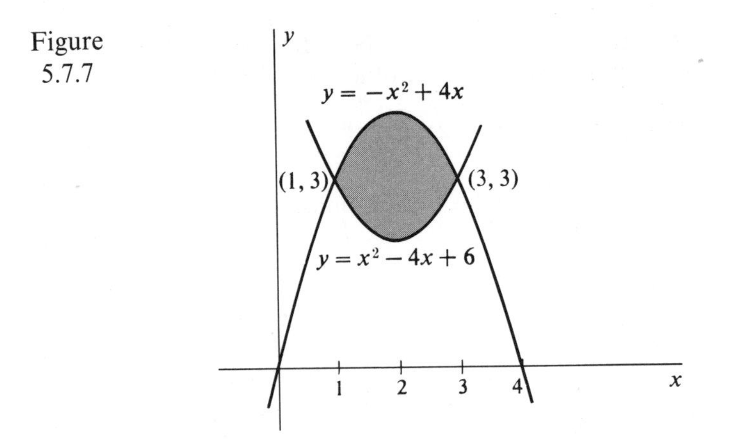

Then substituting into

$$V = 2\pi \int_a^b xw(x)dx, \quad \text{we get}$$

$$V = 2\pi \int_1^3 x(-2x^2 + 8x - 6)dx = 2\pi \int_1^3 (-2x^3 + 8x^2 - 6x)dx$$

$$= 2\pi\left(-\frac{1}{2}x^4 + \frac{8}{3}x^3 - 3x^2\right)\Bigg|_1^3 = \frac{32\pi}{3} \text{ cubic units.}$$

||

Of course not all solids of revolution are obtained by rotation about the y-axis. Suppose, for example, that a region A is rotated about a line L. That situation is illustrated in Figure 5.7.8, where an equation of L is $t = r$. In this case the volume is given by

$$V = \int_a^b 2\pi(t - r)w(t)dt.$$

Figure 5.7.8

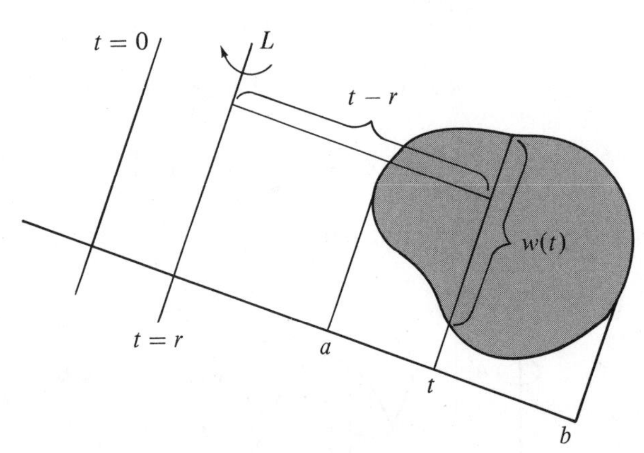

To develop an intuitive view of this formula we think of a small element formed by the rotation of the element of area $w(t)dt$ through an orbital distance $2\pi(t - r)$. Then when the integral is used to sum these elements from $t = a$ and $t = b$ we obtain the formula above.

Example 2 | The region bounded by the graph of $x = y - y^2$ and the y-axis is rotated about the x-axis. Find the resulting volume.

The region in question is pictured in Figure 5.7.9. Note that at a distance y from the x-axis the region has a cross-sectional width of $x = y - y^2$. Thus

$$V = \int_0^1 2\pi y(y - y^2)dy = 2\pi \int_0^1 (y^2 - y^3)dy$$

$$= 2\pi\left(\frac{1}{3}y^3 - \frac{1}{4}y^4\right)\Bigg|_0^1 = \frac{\pi}{6} \text{ cubic units.} \quad \|$$

Figure 5.7.9

Example 3 | The region bounded by the graphs of $y = -x^2 + 4x$ and $y = x^2 - 4x + 6$ is rotated about the line $x = -1$. Find the resulting volume.

The region to be rotated is the same as that considered in Example 1; only the axis of rotation has been changed, as indicated in Figure 5.7.10. The cross section at x is at a distance $x + 1$ from the axis of rotation $x = -1$. At x the cross-sectional width is

$$(-x^2 + 4x) - (x^2 - 4x + 6) = -2x^2 + 8x - 6, \quad \text{and so}$$

$$V = 2\pi \int_1^3 (x + 1)(-2x^2 + 8x - 6)dx$$

$$= 2\pi \int_1^3 (-2x^3 + 6x^2 + 2x - 6)dx$$

$$= 2\pi\left(-\frac{1}{2}x^4 + 2x^3 + x^2 - 6x\right)\Bigg|_1^3$$

$$= 16\pi \text{ cubic units.} \quad \|$$

Figure 5.7.10

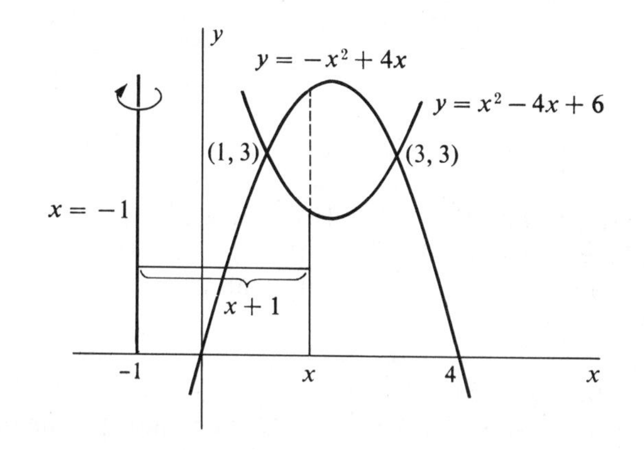

Exercises 5.7

In Exercises 1–22, use the method of cylindrical shells to find the volume obtained when the given region is rotated about the indicated axis.

1. The region bounded by $y = 2x$, $x = 3$, and the x-axis is rotated about the y-axis.
2. The region bounded by $y = 2x$, $y = 8$, and the y-axis is rotated about the y-axis.
3. The region of Exercise 1 is rotated about the x-axis.
4. The region of Exercise 2 is rotated about the x-axis.
5. The region of Exercise 1 is rotated about the line $x = -2$.
6. The region of Exercise 2 is rotated about the line $y = -4$.
7. The region of Exercise 1 is rotated about the line $y = -3$.
8. The region of Exercise 2 is rotated about the line $x = -3$.
9. The region of Exercise 1 is rotated about the line $x = 5$.
10. The region of Exercise 2 is rotated about the line $y = 9$.
11. The region of Exercise 1 is rotated about the line $y = 8$.
12. The region of Exercise 2 is rotated about the line $x = 7$.
13. The region bounded by $y = (x - 1)^2$ and the line $y = 1$ is rotated about the y-axis.
14. The region bounded by $y = x^3$, $y = 1$, and the y-axis is rotated about the x-axis.
15. The region bounded by $y = x^3$, $y = 1$, and the y-axis is rotated about the x-axis.
16. The region bounded by $y = x - x^2$ and the x-axis is rotated about the line $x = -1$.
17. The region bounded by $y = (x - 1)^2$ and the line $y = -x + 1$ is rotated about the line $x = 4$.
18. The region bounded by the lines $x = 0$, $x = 1$, the x-axis, and the graph of $y = \sqrt{3 - x^2}$ is rotated about the y-axis.
19. The region is bounded by $y = x^2$ and $y = x + 2$. The axis of rotation is the line $x = -3$.
20. The region is bounded by $x = y^2$ and $x = 6 - y$. The axis of rotation is the line $x = -1$.
21. The region of Exercise 19 is rotated about the line $y = -3$.
22. The region of Exercise 20 is rotated about the line $y = 3$.
23. Set up but do not evaluate a definite integral for the volume of Exercise 16 using the disc method.
24. Find the volume generated when an isosceles right triangle with two sides of length 1 is rotated about a line that is parallel to and 3 units from one side, as indicated in Figure 5.7.11.

Figure 5.7.11

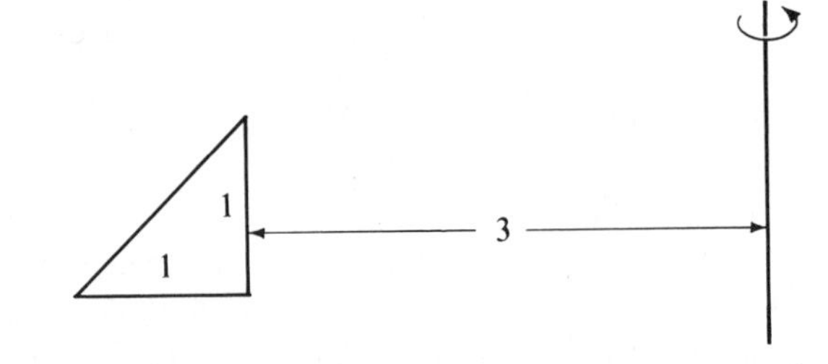

25. Use the method of cylindrical shells to prove that the volume of a right circular cone is given by $(1/3)\pi r^2 h$, where r is the base-radius and h is the height of the cone.

5.8 Area of a Surface of Revolution

If the graph of a continuous function $f(x)$, $a \leq x \leq b$, is revolved about the x-axis, a *surface of revolution* is obtained. As expected, the definite integral provides a tool for calculating the area of such a surface.

Intuitive Development. Subdivide the interval $a \leq x \leq b$ into small subintervals each of width dx. Rotation of the resulting "element of arc" about the x-axis results in the "element of surface" pictured in Figure 5.8.1. When that element of surface is cut and laid out flat it appears as the rectangle illustrated in Figure 5.8.2. Thus the area of the element of surface is $2\pi|f(x)|\sqrt{1 + (f'(x))^2}\,dx$. Use of the integral to sum up the elements of surface area for $a \leq x \leq b$ gives

$$A = 2\pi \int_a^b |f(x)|\sqrt{1 + (f'(x))^2}\,dx.$$

Figure 5.8.1

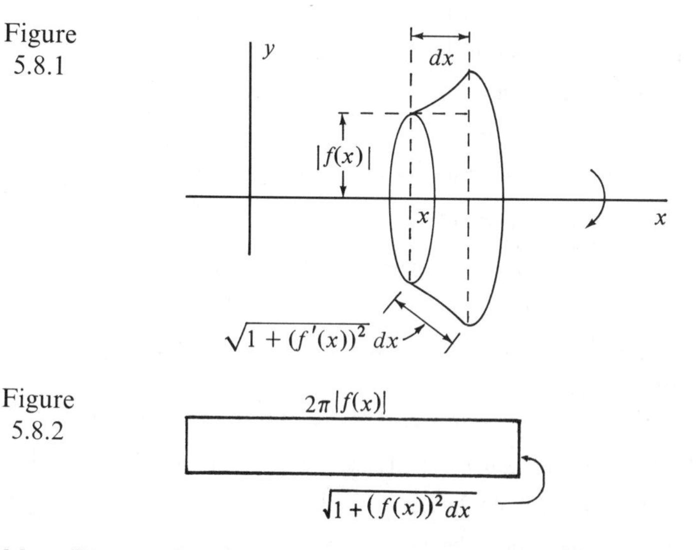

Figure 5.8.2

$2\pi|f(x)|$

$\sqrt{1+(f(x))^2}dx$

More Rigorous Development. Our two basic assumptions concerning the area of a surface of revolution are similar to the assumptions made in our previous applications of the definite integral.

Assumption 1. The area of a surface of revolution is "additive." That is, the area of the surface formed by the revolution of the graph of $y = f(x)$, $r \leq x \leq s$, is the sum of the areas of the surfaces formed by the revolution of the graphs of $y = f(x)$, $r \leq x \leq t$, and $y = f(x)$, $t \leq x \leq s$. This assumption is illustrated in Figure 5.8.3.

Assumption 2. If L is the length of the arc to be rotated, then the area A of the resulting surface is somewhere between L times the minimum circumference and L times the maximum circumference. Thus, if m and M are the minimum and maximum radii, respectively, then

$$2\pi mL \leq A \leq 2\pi ML.$$

Figure 5.8.3

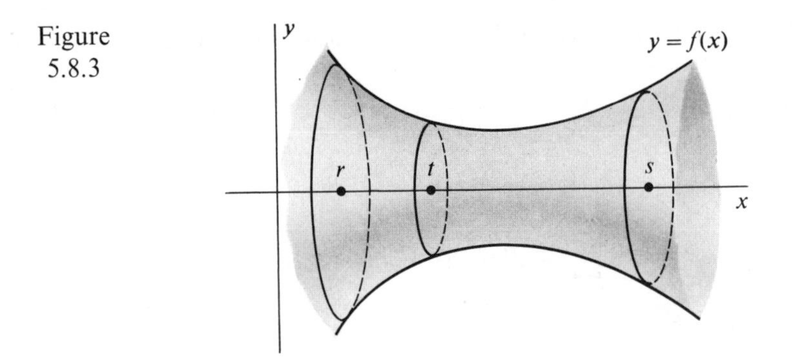

Just as in our previous applications of the definite integral we subdivide the interval $a \leq x \leq b$ into n subintervals, each of width $h = (b - a)/n$, by introducing the points $x_i = a + hi$, $i = 0, 1, 2, \ldots, n$. Then, if A_i is the area of the surface formed by revolving the graph of $y = f(x)$, $x_{i-1} \leq x \leq x_i$, about the x-axis, we may use Assumption 1 to get

$$A = \sum_{i=1}^{n} A_i,$$

where A is the total area of the surface of revolution.

Now concentrating on the area A_i of a typical slice, pictured in Figure 5.8.4, we may use Assumption 2 to obtain

$$2\pi m L_i \leq A_i \leq 2\pi M L_i,$$

where m is the absolute minimum of $|f(x)|$ on the interval $x_{i-1} \leq x \leq x_i$, and M is its absolute maximum there, and L_i is the length of arc to be revolved.

Figure 5.8.4

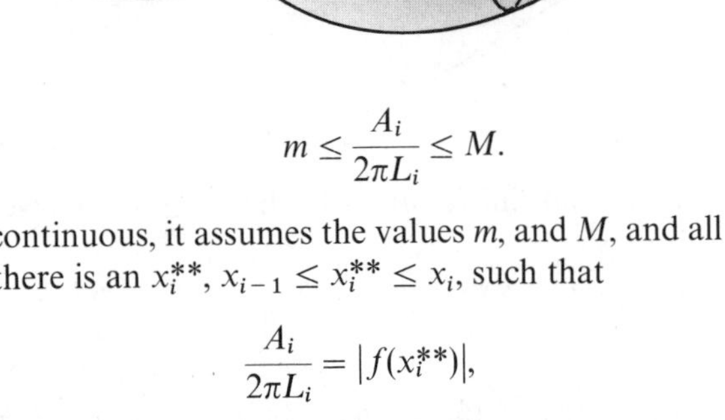

Then

$$m \leq \frac{A_i}{2\pi L_i} \leq M.$$

Since $|f(x)|$ is continuous, it assumes the values m, and M, and all the values between them. Hence there is an x_i^{**}, $x_{i-1} \leq x_i^{**} \leq x_i$, such that

$$\frac{A_i}{2\pi L_i} = |f(x_i^{**})|,$$

and so

$$A_i = 2\pi |f(x_i^{**})| L_i.$$

Equation (3) of Section 5.3 assures us that there is some point x_i^*, $x_{i-1} \leq x_i^* \leq x_i$, such that

$$L_i = (\sqrt{1 + [f'(x_i^*)]^2})h.$$

Hence
$$A_i = 2\pi |f(x_i^{**})| \sqrt{1 + [f'(x_i^*)]^2}\, h.$$

So
$$A = \sum_{i=1}^{n} A_i = \sum_{i=1}^{n} 2\pi |f(x_i^{**})| \sqrt{1 + [f'(x_i^*)]^2}\, h.$$

Since A is independent of h,

$$A = \lim_{h \to 0} \sum_{i=1}^{n} 2\pi |f(x_i^{**})| \sqrt{1 + [f'(x_i^*)]^2}\, h.$$

If we now apply the Theorem of Bliss (Theorem 5.6.1) to the functions $|f(x)|$ and $g(x) = \sqrt{1 + [f'(x)]^2}$, we obtain the final result:

$$A = 2\pi \int_a^b |f(x)| \sqrt{1 + [f'(x)]^2}\, dx.$$

Example 1 Use calculus to show that the lateral surface area of a right circular cone of base-radius r and height h is $\pi r \sqrt{r^2 + h^2}$.

The surface area in question is formed by rotating the line segment from (0, 0) to (h, r) about the x-axis, as indicated in Figure 5.8.5. Hence we take $f(x) = (r/h)x$, $0 \leq x \leq h$, and obtain

$$A = 2\pi \int_0^h \left|\frac{r}{h} x\right| \sqrt{1 + \left(\frac{r}{h}\right)^2}\, dx.$$

Figure 5.8.5

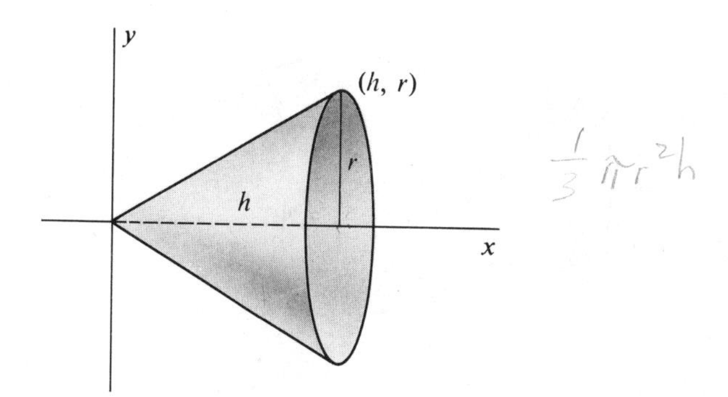

Since h and r are constants, and since $(r/h)x \geq 0$ when $x \geq 0$, we have

$$A = \frac{2\pi r}{h} \sqrt{1 + \frac{r^2}{h^2}} \int_0^h x\, dx = \frac{2\pi r}{h} \sqrt{1 + \frac{r^2}{h^2}} \cdot \frac{1}{2} h^2 = \pi r \sqrt{h^2 + r^2}.$$ ||

Example 2 Find the area of the surface obtained when the graph of $f(x) = x + 3$, $-5 \leq x \leq 2$, is revolved about the x-axis.

Since $f'(x) = 1$, we have

$$A = 2\pi \int_{-5}^{2} |x + 3|\sqrt{2}\, dx$$
$$= 2\sqrt{2}\pi \int_{-5}^{-3} |x + 3| dx + 2\sqrt{2}\pi \int_{-3}^{2} |x + 3| dx.$$

For $-5 \leq x \leq -3$, we have $|x + 3| = -(x + 3)$; and for $-3 \leq x \leq 2$, we have $|x + 3| = x + 3$. Thus,

$$A = -2\sqrt{2}\pi \int_{-5}^{-3} (x + 3) dx + 2\sqrt{2}\pi \int_{-3}^{2} (x + 3) dx$$
$$= -2\sqrt{2}\pi \left(\frac{1}{2}x^2 + 3x\right)\Bigg|_{-5}^{-3} + 2\sqrt{2}\pi \left(\frac{1}{2}x^2 + 3x\right)\Bigg|_{-3}^{2}$$
$$= 4\sqrt{2}\pi + 25\sqrt{2}\pi = 29\sqrt{2}\pi \text{ square units.}$$

||

Of course lines other than the x-axis may be used as axes of rotation. Suppose, for example, that the line $y = k$ is used as the axis of rotation. Then, as illustrated in Figure 5.8.6, an element of arc is rotated through the orbital distance $2\pi|k - f(x)|$. Consequently the area of the surface generated as the arc $y = f(x)$, $a \leq x \leq b$, is rotated about the line $y = k$ is

$$A = 2\pi \int_{a}^{b} |k - f(x)| \sqrt{1 + [f'(x)]^2}\, dx.$$

Figure 5.8.6

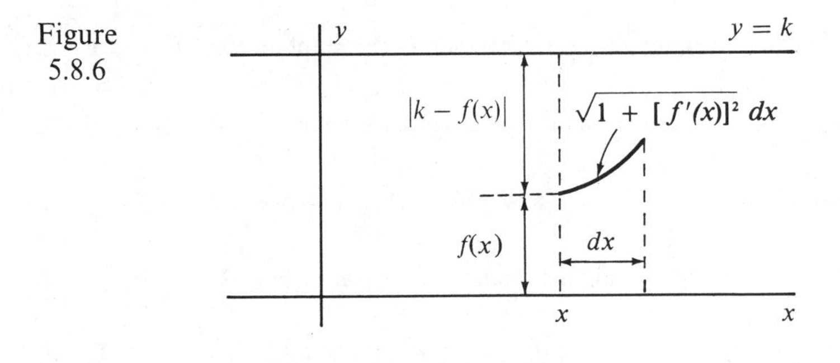

In a similar fashion, if the arc $x = g(y)$, $c \leq y \leq d$, is rotated about the line $x = k$, the area of the resulting surface is

$$A = 2\pi \int_{c}^{d} |k - g(y)| \sqrt{1 + [g'(y)]^2}\, dy.$$

Example 3 | Find the area of the surface formed when the graph of $x = y^3$, $0 \leq y \leq 1$, is rotated about the y-axis.

$$A = 2\pi \int_{0}^{1} |y^3| \sqrt{1 + [3y^2]^2}\, dy = 2\pi \int_{0}^{1} y^3 \sqrt{1 + 9y^4}\, dy$$
$$= \frac{2\pi}{36} \int_{0}^{1} (1 + 9y^4)^{1/2}(36y^3\, dy) = \frac{\pi}{18} \cdot \frac{2}{3}(1 + 9y^4)^{3/2}\Bigg|_{0}^{1}$$
$$= \frac{\pi}{27}(10\sqrt{10} - 1) \text{ square units.}$$

||

Exercises 5.8

1. Find the area of the surface obtained when the graph of $y = 2x - 1$, $1 \le x \le 4$, is revolved about the x-axis.
2. Find the area of the surface obtained when the graph of $y = 4 - x$, $2 \le x \le 4$, is revolved about the x-axis.
3. Find the area generated when the graph of Exercise 1 is revolved about the y-axis.
4. Find the area generated when the graph of Exercise 2 is revolved about the y-axis.
5. Find the area of the surface resulting from the rotation of the graph of $x = y^3$, $1 \le y \le 2$, about the y-axis.
6. Find the area of the surface resulting from the rotation of the graph of $y = (1/3)x^3$, $0 \le x \le 3$, about the x-axis.
7. Find the area of the surface resulting from the rotation of the graph of $y = (x^3/12) + x^{-1}$, $1 \le x \le 2$, about the x-axis.
8. Find the area of the surface resulting from the rotation of the graph of $x = (1/6)y^3 + 1/(2y)$, $1 \le y \le 2$, about the y-axis.
9. Find the area of the surface resulting from the rotation of the graph of $y = \sqrt{x}$, $1 \le x \le 2$, about the x-axis.
10. Find the area of the surface resulting from the rotation of the graph of $y = 3x^{1/3}$, $0 \le y \le 3$, about the y-axis.
11. Find the area of the surface resulting from the rotation of the graph $y = x^2$, $0 \le y \le 2$, about the y-axis.
12. Find the area of the surface resulting from the rotation of the graph of $y^2 = 4x$, $0 \le x \le 4$, about the x-axis.
13. Find the area of the surface resulting from the rotation of the graph of $y^2 + x = 4$, $0 \le x \le 4$, about the x-axis.
14. Find the area of the surface resulting from the rotation of the graph of $y = x + 3$, $1 \le x \le 2$, about the line $y = 5$.
15. Find the area of the surface resulting from the rotation of the graph of $y = 2x - 1$, $0 \le x \le 3$, about the line $y = 4$.
16. Find the area of the surface resulting from the rotation of the graph of $y = (x^3/12) + x^{-1}$, $2 \le x \le 3$, about the line $y = 1$.
17. Find the area of the surface resulting from the rotation of the graph of $y = 2x - 1$, $0 \le x \le 3$, about the line $x = 4$.
18. Prove that the surface area of a sphere of radius r is $4\pi r^2$.
19. Prove that the surface area of a frustrum of a right circular cone of height h and radii r_1 and r_2 is given by

$$\pi(r_1 + r_2)\sqrt{h^2 + (r_1 - r_2)^2}.$$

20. A paraboloid is generated by revolving the curve $x = \sqrt{y}$ about the y-axis. A *zone* of the paraboloid is that portion of the surface between two parallel planes which are perpendicular to the y-axis. Determine a formula for the area of a zone of altitude h which lies between the planes $y = a$ and $y = a + h$, $a \ge 0$.
21. That portion of the surface of a sphere between two parallel planes a distance h apart is called a zone of altitude h. Determine a formula for the area of a zone of altitude $h < 2r$ on

a surface of a sphere of radius r. Does the area of a zone of altitude h depend on the location of the zone?

5.9 Moments and Centroids

Suppose several objects are located at various points along the x-axis. Specifically suppose that n objects are located along the x-axis as illustrated in Figure 5.9.1. That is, suppose that a weight w_i is located at x_i. The moment of that system of weights about x_0 is

$$M_{x_0} = w_1(x_1 - x_0) + w_2(x_2 - x_0) + \cdots + w_n(x_n - x_0).$$

Figure 5.9.1

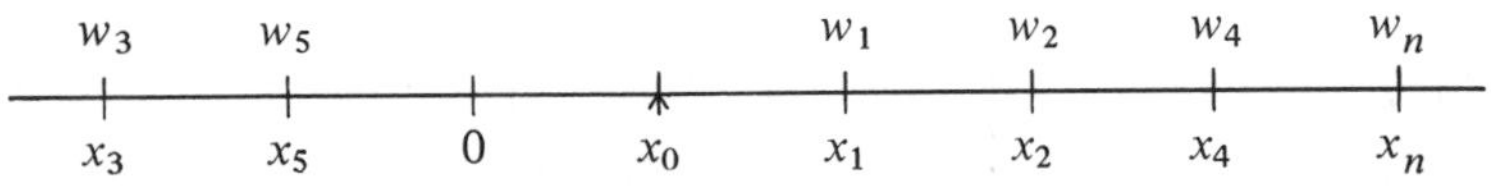

w_3 w_5 w_1 w_2 w_4 w_n

x_3 x_5 0 x_0 x_1 x_2 x_4 x_n

If one views the x-axis as a seesaw supported at the point x_0, then M_{x_0} measures the tendency of the seesaw to rotate about x_0. The point $\bar{x}$ for which $M_{\bar{x}} = 0$ is called the centroid of the system. Note that when a system is supported at its centroid it is balanced and will not rotate.

Example 1 Find the centroid of the system shown in Figure 5.9.2. The indicated weights are given in grams.

Figure 5.9.2

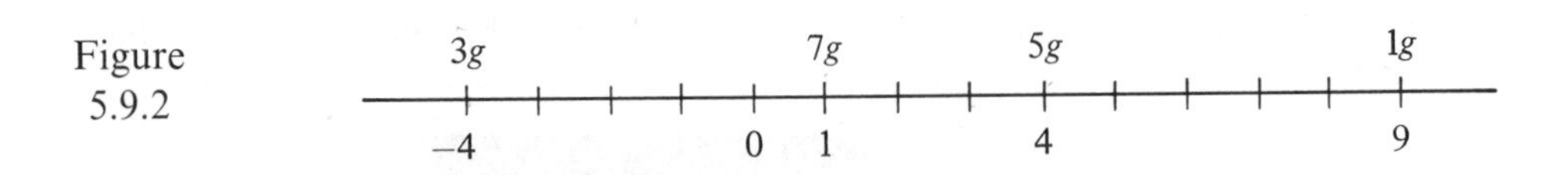

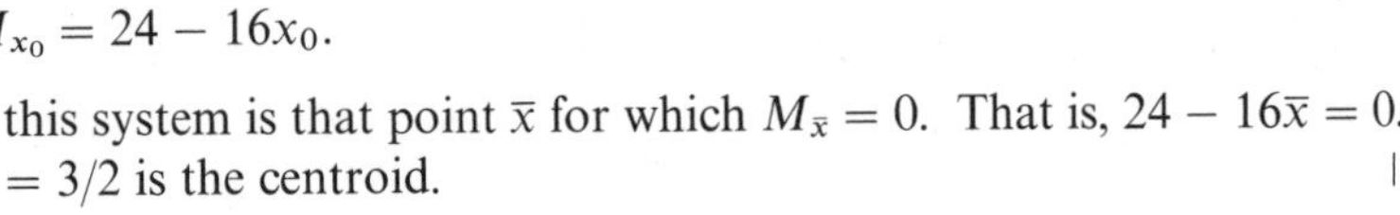

The moment of this system about the point x_0 is

$$M_{x_0} = 3(-4 - x_0) + 7(1 - x_0) + 5(4 - x_0) + 1(9 - x_0)$$

$$M_{x_0} = 24 - 16x_0.$$

The centroid of this system is that point $\bar{x}$ for which $M_{\bar{x}} = 0$. That is, $24 - 16\bar{x} = 0$. Consequently $\bar{x} = 3/2$ is the centroid. ||

Suppose that a set of n weights are distributed in the plane. Let (x_i, y_i) be the position of a weight w_i. In much the same way as before, one may consider moments. Now however the moments measure the tendency of the system to rotate about specified lines. For example, M_{x_0}, is

$$M_{x_0} = w_1(x_1 - x_0) + w_2(x_2 - x_0) + \cdots + w_n(x_n - x_0),$$

and measures the tendency of the given system to rotate about the line $x = x_0$.

The moment about the line $y = y_0$ is

$$M_{y_0} = w_1(y_1 - y_0) + w_2(y_2 - y_0) + \cdots + w_n(x_n - x_0).$$

It measures the tendency of the system to rotate about the line $y = y_0$.

The point $(\bar{x}, \bar{y})$ for which $M_{\bar{x}} = 0$ and $M_{\bar{y}} = 0$ is called the *centroid* of the system. If supported at the centroid the system will not rotate and thus will balance. The centroid need not be within the system, but if the entire weight of the system were located at the centroid, the tendency to rotate about any line would remain the same.

We wish to extend the concepts of moments and centroids to areas. For this purpose we wish to consider the region as being covered by a thin lamina of uniform density one. This will allow us to extend the physical concepts of moment and centroid to areas.

Consider a region whose width above the points x is given by $w(x)$. Such a region is illustrated in Figure 5.9.3.

Figure 5.9.3

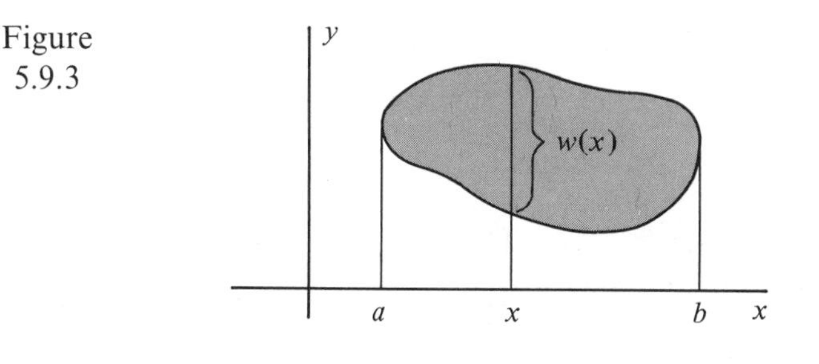

Intuitive Development. Subdivide the interval $a \le x \le b$ into a number of small subintervals each having width dx. A typical thin slice of area is pictured in Figure 5.9.4. The weight of this thin slice is approximately $w(x)dx$. Consequently the moment of the mass about $x = x_0$ is $(x - x_0)w(x)dx$. Use of the integral to sum these elements of moment for all x between $x = a$ and $x = b$ gives

$$M_{x_0} = \int_a^b (x - x_0)w(x)dx.$$

Figure 5.9.4

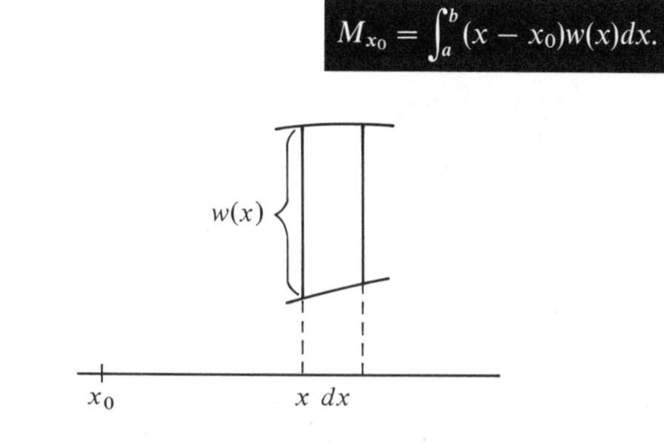

More Rigorous Development. Let $M_{x_0}(r, s)$ denote the moment of that portion of the area between $x = r$ and $x = s$. We make the following two assumptions about moments:

Assumption 1. Moments are additive; that is,

$$M_{x_0}(r, t) + M_{x_0}(t, s) = M_{x_0}(r, s), \qquad \text{for } r \le t \le s.$$

Assumption 2. Let A denote the portion of the area between $x = r$ and $x = s$. Then if m is the minimum value of $x - x_0$ for $r \le x \le s$ and M is the maximum value of $x - x_0$ for $r \le x \le s$, we have

$$mA \le M_{x_0}(r, s) \le MA.$$

These assumptions lead directly to the following formula for the moment about the line $x = x_0$.

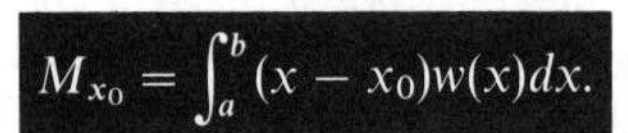

$$M_{x_0} = \int_a^b (x - x_0)w(x)dx.$$

In order to find the position of a line perpendicular to the x-axis upon which the area will balance, we set $M_{\bar{x}} = 0$ and solve for $\bar{x}$. Since $\int_a^b (x - \bar{x})w(x)dx = 0$ we have

$$\int_a^b xw(x)dx - \bar{x}\int_a^b w(x)dx = 0.$$

Thus,

$$\bar{x} = \frac{\int_a^b xw(x)dx}{\int_a^b w(x)dx}.$$

Since the area A of the region is given by

$$A = \int_a^b w(x)dx,$$

the equation for $\bar{x}$ may be put into the simpler form

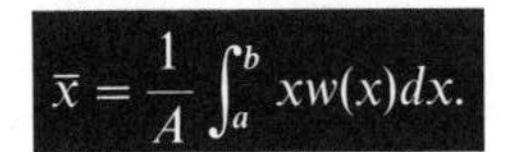

$$\bar{x} = \frac{1}{A}\int_a^b xw(x)dx.$$

Suppose that the width of a region opposite a point y is given by $l(y)$, as pictured in Figure 5.9.5.

Figure 5.9.5

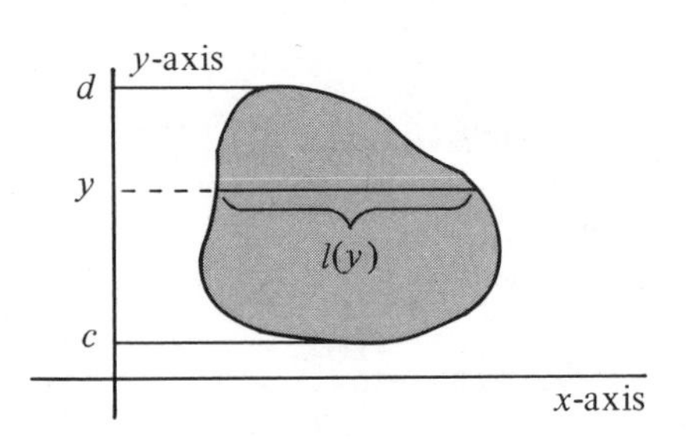

In a manner very similar to the above, it can be shown that the moment of the region with respect to the line $y = y_0$ is given by

$$M_{y_0} = \int_c^d (y - y_0)l(y)dy.$$

If we now set $M_{\bar{y}} = 0$ and solve for $\bar{y}$, we have

$$\bar{y} = \frac{\int_c^d yl(y)dy}{\int_c^d l(y)dy}.$$

As before the area of the region is given by

$$A = \int_c^d l(y)dy.$$

Consequently

$$\bar{y} = \frac{1}{A}\int_c^d yl(y)dy.$$

Example 2 | Compute the centroid of the triangle bounded by the coordinate axes and the line $3x + 7y = 21$.

The region is illustrated in Figure 5.9.6.

Figure 5.9.6

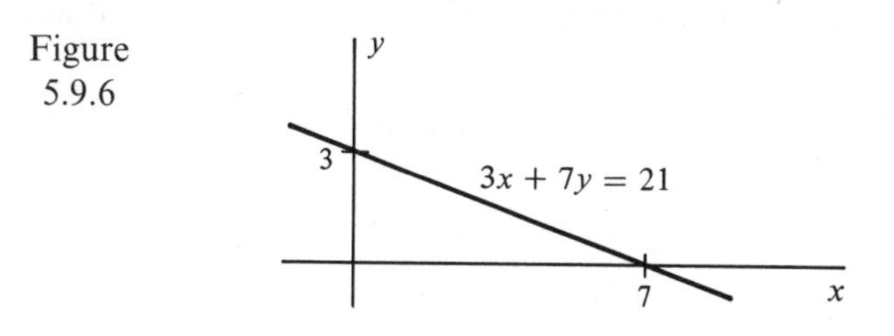

The area of this triangle is $A = 1/2 \cdot 7 \cdot 3 = 21/2$. Since the width of the triangle above x is given by $w(x) = 3 - 3x/7$,

$$\begin{aligned}\bar{x} &= \frac{1}{A}\int_0^7 x\left(3 - \frac{3}{7}x\right)dx \\ &= \frac{1}{A}\int_0^7 \left(3x - \frac{3}{7}x^2\right)dx = \frac{1}{A}\left(\frac{3}{2}x^2 - \frac{1}{7}x^3\right)\Bigg|_0^7 \\ &= \frac{1}{A}\left(\frac{147}{2} - 49\right).\end{aligned}$$

Then since $A = 21/2$,

$$\bar{x} = \frac{2}{21}\left(\frac{147}{2} - 49\right) = \frac{7}{3}.$$

The width opposite a point y is $l(y) = 7 - (7/3)y$. Thus,

$$\begin{aligned}\bar{y} &= \frac{1}{A}\int_0^3 y\left(7 - \frac{7}{3}y\right)dy = \frac{1}{A}\left(\frac{7}{2}y^2 - \frac{7}{9}y^3\right)\Bigg|_0^3 \\ &= \frac{1}{A}\left(\frac{63}{2} - 21\right).\end{aligned}$$

Then, since $A = 21/2$,

$$\bar{y} = \frac{2}{21}\left(\frac{63}{2} - 21\right) = 1.$$

Consequently the centroid is $\left(\frac{7}{3}, 1\right)$. ||

Example 3 | Compute the centroid of the region bounded by the curves $y = x^3$ and $x = y^2$.

The region in question is illustrated in Figure 5.9.7.

Figure 5.9.7

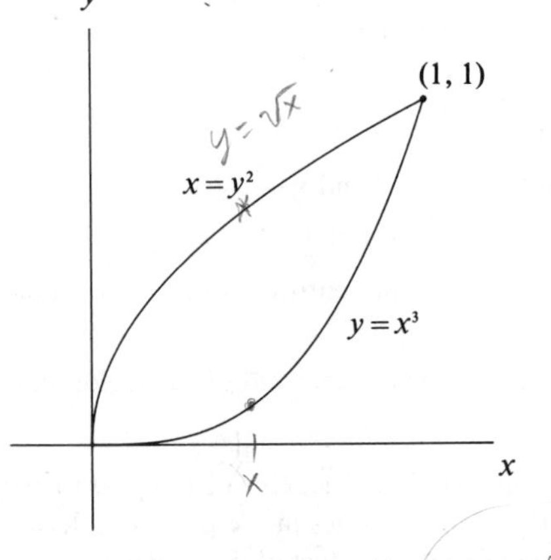

Note that the width above x is given by $w(x) = \sqrt{x} - x^3$. Thus the area is

$$A = \int_0^1 (\sqrt{x} - x^3)dx = \frac{2}{3}x^{3/2} - \frac{1}{4}x^4\Big|_0^1 = \frac{5}{12}.$$

Then
$$\bar{x} = \frac{1}{A}\int_0^1 x(\sqrt{x} - x^3)dx = \frac{12}{5}\int_0^1 (x^{3/2} - x^4)dx$$

$$= \frac{12}{5}\left(\frac{2}{5}x^{5/2} - \frac{1}{5}x^5\right)\Big|_0^1 = \frac{12}{25}.$$

Since the width opposite y is given by $l(y) = \sqrt[3]{y} - y^2$, we have

$$\bar{y} = \frac{1}{A}\int_0^1 y(\sqrt[3]{y} - y^2)dy = \frac{12}{5}\int_0^1 (y^{4/3} - y^3)dy$$

$$= \frac{12}{5}\left(\frac{3}{7}y^{7/3} - \frac{1}{4}y^4\right)\Big|_0^1 = \frac{3}{7}.$$

Consequently the centroid of the given region is $\left(\frac{12}{25}, \frac{3}{7}\right)$. ||

Exercises 5.9

In Exercises 1–15, compute the centroid of the region bounded by the given curves.

1. The coordinate axes and the line $3x + 4y = 12$.
2. The coordinate axes and the line $2x + 8y = 9$.
3. The curves $y = x^2$ and $x = y^2$.
4. The x-axis and the curve $y = x^2 - x$.
5. The y-axis and the curves $y = x^2$ and $y = 5$, $x \geq 0$.
6. The curve $y = x^2$ and the line $y = x$.

7. The curve $x = y^2$ and the line $y = x$.
8. The curves $y = 4 - x^2$ and $y = x^2 - 4$.
9. The curves $y = x^3$ and $y = x^2$.
10. The curves $y = 4x$ and $y = x^2$.
11. The line $ax + by = ab$ and the coordinate axes.
12. The curve $y = x^2$ and the line $y = 1$.
13. The curve $y = x^4$ and the line $y = 1$.
14. The coordinate axes and the lines $y = x + 2$ and $x + y = 6$.
15. The coordinate axes and the lines $y = 3x + 1$ and $x + y = 5$.
16. Set up, but do not evaluate, integrals for the centroid of the region bounded by the positive coordinate axes and the curve $y = 1 - x^2$.
17. Set up, but do not evaluate, integrals for the centroid of the region bounded by the x-axis and the curve $y = 1 - x^2$.

In Exercises 18–23, use the following fact (The Theorem of Pappus) to compute the volume: If a region in a plane is rotated about a line that lies in the plane but does not intersect the region, then the volume generated is equal to the product of the area of the region and the distance traveled by its centroid. (See Problem 2, in the Challenging Problems, Chapter 5)

18. The region bounded by the coordinate axes and the line $3x + 4y = 12$ is rotated about the line $y = -1$.
19. The region bounded by the coordinate axes and the line $2x + 8y = 9$ is rotated about the line $x = 5$.
20. The region bounded by $y = x^2$ and $y = x^3$ is rotated about the line $y = -4$.
21. The region of Exercise 9 is rotated about the line $x = 3$.
22. The region bounded by the x-axis and the curve $y = x^2 - 2x$ is rotated about the line $x = -1$.
23. Because of symmetry the centroid of a circular disc is at the center of the disc. Derive a formula for the volume of the torus illustrated in Figure 5.9.8.

Figure 5.9.8

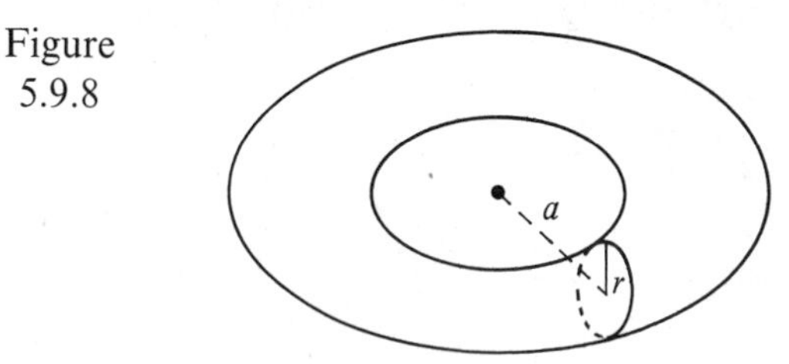

5.10 Force Exerted by a Fluid on a Vertical Surface

The pressure (force per unit area) exerted by a liquid at a depth of y units is given by $p(y) = ky$, where k is the weight of the fluid per unit volume. Since we are considering a fluid, the pressure at depth y is equal in all directions.

Suppose that a vertical plate is submerged in a fluid in such a way that the width of the plate at depth y is $w(y)$, as pictured in Figure 5.10.1. (In these problems it is most convenient to consider the positive y-axis as pointing downward.) We wish to find the total force exerted on one side of the plate by the fluid.

Figure 5.10.1

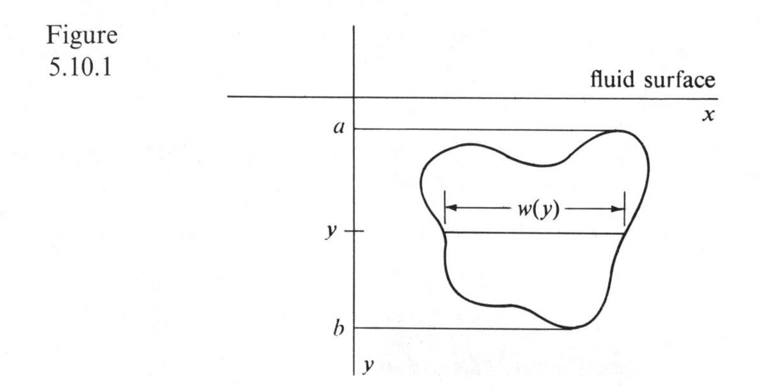

Intuitive Development. Subdivide the interval $a \leq y \leq b$ into small subintervals each of width dy. This subdivision results in a subdivision of the plate into a number of thin slices. A typical slice is illustrated in Figure 5.10.2. The area of that slice is approximately $w(y)dy$.

Figure 5.10.2

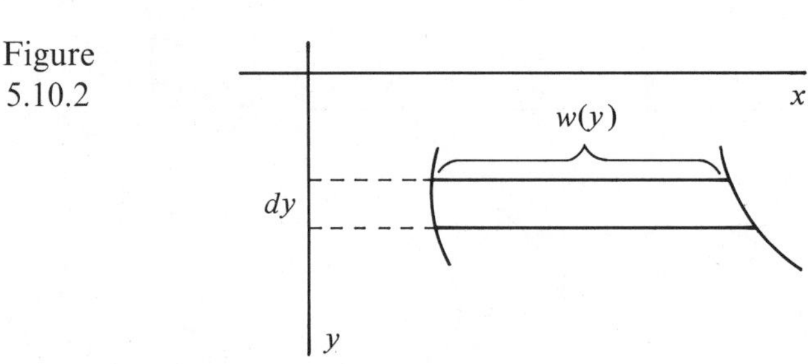

Since the pressure exerted by the liquid at depth y is ky, the total force exerted on the slice is approximately $kyw(y)dy$. Use of the integral to sum the forces from $y = a$ to $y = b$ gives

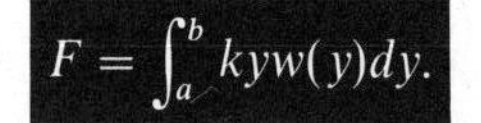

$$F = \int_a^b kyw(y)dy.$$

More Rigorous Development. To find the total force exerted on one side of the plate we make the following assumptions concerning $F(r, s)$, the total force exerted on one side of the plate between any two depths of r and s units.

Assumption 1. $F(r, t) + F(t, s) = F(r, s)$, for $r \leq t \leq s$.

Assumption 2. Let $A(r, s)$ be the area of the plate between the depths r and s. Then the force $F(r, s)$ on this section of the plate is between the area times the density k times the minimal depth r, and the area times the density k times the maximal depth s. That is,

$$A(r, s)kr \leq F(r, s) \leq A(r, s)ks.$$

To put Assumption 2 into a more workable form, we note that

$$A(r, s) = \int_r^s w(y)dy,$$

and so we are assured by The Mean Value Theorem for Integrals that there is some y^*, $r \le y^* \le s$, such that $A(r, s) = w(y^*)(s - r)$. Consequently we have

$$w(y^*)(s - r)kr \le F(r, s) \le w(y^*)(s - r)ks. \tag{1}$$

Use of (1), Assumption 1, and the Theorem of Bliss gives the final result:

$$F(a, b) = \int_a^b kyw(y)dy.$$

Example 1 Find the force due to water pressure on one side of the triangular plate illustrated in Figure 5.10.3. Assume that the units are given in feet and that water weighs 62.5 pounds per cubic foot.

In this case, $w(y) = (2 - y) - \dfrac{y - 2}{2} = 3 - \dfrac{3}{2}y$, $0 < y < 2$.

Figure 5.10.3

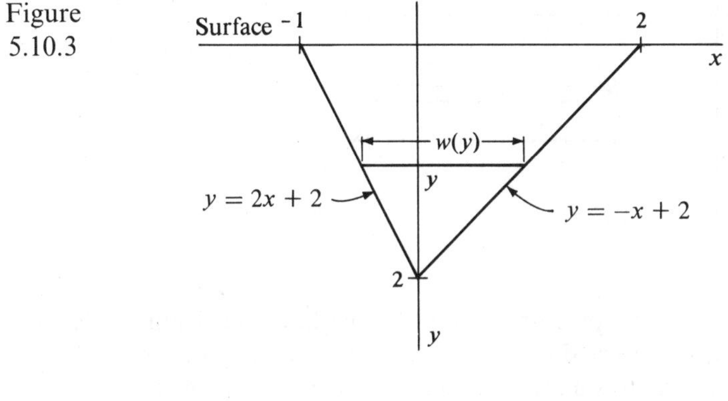

Then $$F = \int_0^2 (62.5)y\left(3 - \frac{3}{2}y\right)dy = 62.5\int_0^2 \left(3y - \frac{3}{2}y^2\right)dy$$

$$= 62.5\left(\frac{3}{2}y^2 - \frac{1}{2}y^3\right)\Bigg|_0^2 = 62.5(2) = 125 \text{ lbs.}$$ ||

Exercises 5.10

In Exercises 1–10, find the force due to water pressure on one side of a vertical plate of the given shape situated under water as indicated. Assume the water weighs 62.5 pounds per cubic foot. Assume that the units on the coordinate axes are in feet.

1. An isosceles triangle whose 8′ base is parallel to the surface and 5′ down, if the altitude of the triangle is 10′ and the vertex is down.
2. An isosceles right triangle with sides 3′ long, with one of its 3′ sides, parallel to the surface and 2′ below the surface.

3. A semicircle whose 10′ diameter is at the surface.
4. The region bounded by $x = (y - 5)^2$ and $x = 9$, where the x-axis is at the surface and the y-axis is downward.
5. The semi-ellipse shown in Figure 5.10.4.

Figure 5.10.4

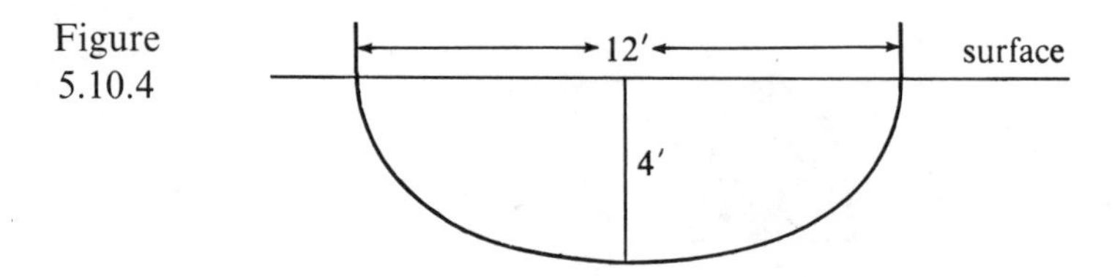

6. A 9′ by 8′ rectangle with its upper 9′ edge parallel to and 4′ below the surface.
7. The region bounded by $x = (y - 4)^2$ and $x = y + 2$, where the x-axis is at the surface and the y-axis is downward.
8. A square of side length 4′ with its upper edge parallel to and 10′ below the surface.
9. A vertical end of a water trough having the dimensions indicated.

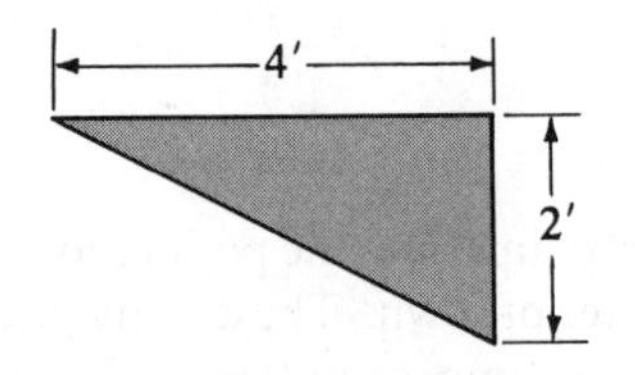

10. A vertical end of a water trough having the dimensions indicated.

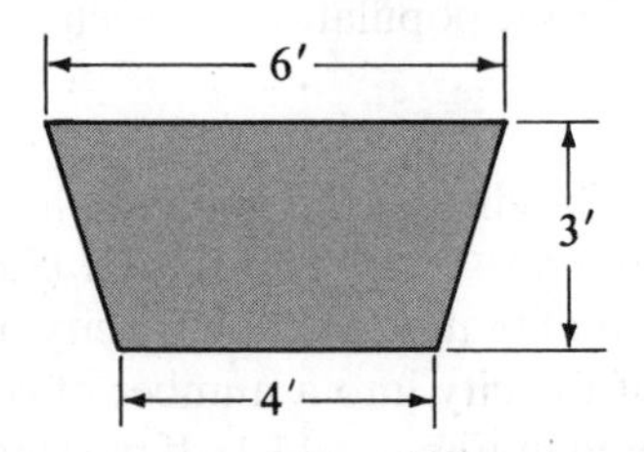

11. A plate in the shape of the region bounded by the parabola $y = 2x^2$ and the line $y = 3$ is submerged vertically in a fluid with its upper edge 3 feet below the surface of the water, as indicated. Find the force exerted on one side of the plate if the density of the fluid is 50 pounds per cubic foot.

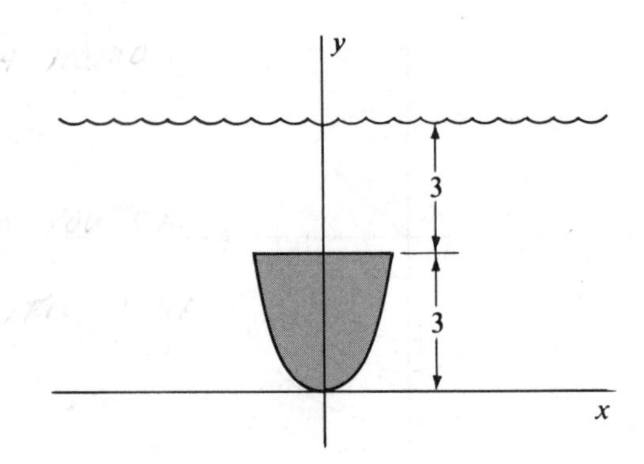

12. Determine the force on one side of the plate described in Exercise 11 if it is submerged vertically with its vertex 3 feet below the surface of the fluid, as indicated.

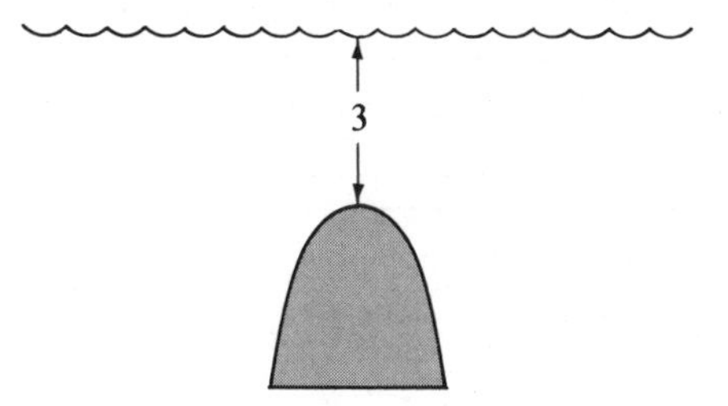

13. Use (1) and Assumption 1 of this section to prove that

$$F(a, b) = \int_a^b kyw(y)dy.$$

14. Show that the force on one side of a vertically submerged plate of area A is $k\bar{y}A$, where $\bar{y}$ is the depth of the centroid of the plate and k is the fluid density.

5.11 Population

Sociologists have found that, as a first approximation, the population density of a city depends only on the distance from the center of town. The density generally is higher near the center of town and decreases as one moves outward. Suppose that a continuous function $f(x)$ gives the population density at a distance x from the center of town. We want to find a formula for $P(a, b)$, the population between any two distances a and b from the center of town.

Intuitive Development. Subdivide the interval $a \leq x \leq b$ into a number of small subintervals each of width dx. Such subdivision results in the subdivision of that portion of the city between a and b units from the center of the city into a number of concentric rings. A typical ring is illustrated in Figure 5.11.1. If the ring is cut and laid out flat to form a rectangle as shown in Figure 5.11.2, it becomes evident that its area is approximately $2\pi x\, dx$.

Figure 5.11.1

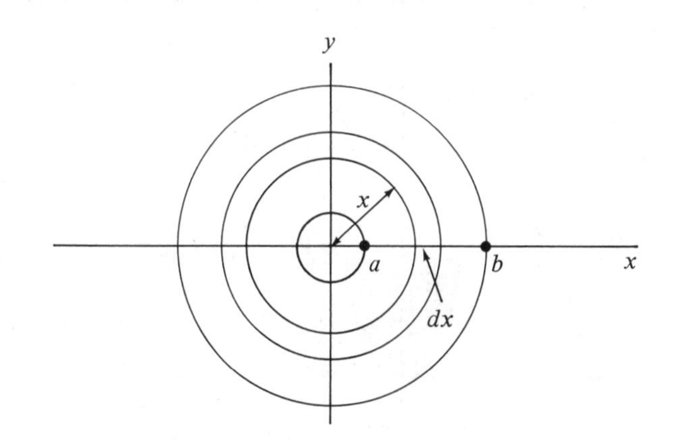

Figure 5.11.2

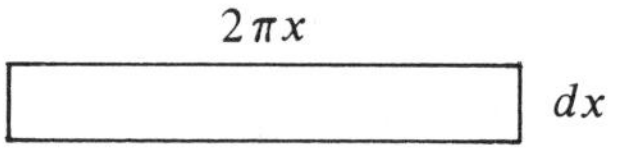

$2\pi x$ ... dx

For small values of dx, $f(x)$, the population density, will be nearly constant over the ring of Figure 5.11.1. Consequently the population in that ring is approximately $2\pi x f(x)dx$. Use of the integral to sum from $x = a$ to $x = b$ gives

$$P(a, b) = 2\pi \int_a^b xf(x)dx.$$

More Rigorous Development. We make the following two assumptions:

Assumption 1. $P(r, t) + P(t, s) = P(r, s)$, for $r \le t \le s$.

Assumption 2. $\pi(s^2 - r^2)m \le P(r, s) \le \pi(s^2 - r^2)M$, where m and M are the absolute minimum and maximum of $f(x)$ for $r \le x \le s$.

Those assumptions, along with the Theorem of Bliss, lead directly to the following formula for the population living between a and b miles from the center of the city:

$$P(a, b) = 2\pi \int_a^b xf(x)\,dx.$$

Example 1 Find the population within 4 miles of the center of a city whose population density is given by $f(x) = 10{,}000/(x^2 + 1)^4$, where x is the number of miles from the center of the city.

We have

$$\begin{aligned} P(0, 4) &= 2\pi \int_0^4 \frac{10{,}000x}{(x^2 + 1)^4}\,dx \\ &= 10{,}000\pi \int_0^4 (x^2 + 1)^{-4}\,2x\,dx \\ &= 10{,}000\pi \left[-\frac{1}{3}(x^2 + 1)^{-3} \right]_0^4 \\ &= 10{,}000\pi \left(\frac{1}{3} - \frac{1}{3(17)^3} \right) \\ &\approx 10{,}470. \end{aligned}$$

||

Exercises 5.11

In Exercises 1–12, find the population between the given numbers of miles from the center of a city for the indicated population density $f(x)$ at a distance x miles from the center.

1. $f(x) = 100 - x, \quad 0 \le x \le 1.$
2. $f(x) = 1000 - x^2, \quad 0 \le x \le 3.$
3. $f(x) = 100 - x, \quad 1 \le x \le 3.$
4. $f(x) = 1000 - x^2, \quad 3 \le x \le 6.$

5. $f(x) = x^2 - 100x + 1000, \quad 0 \leq x \leq 3.$
6. $f(x) = 2000 - x^3, \quad 0 \leq x \leq 2.$
7. $f(x) = x^2 - 100x + 1000, \quad 3 \leq x \leq 6.$
8. $f(x) = 2000 - x^3, \quad 4 \leq x \leq 6.$
9. $f(x) = 10{,}000/(x^2 + 2)^3, \quad 0 \leq x \leq 3.$
10. $f(x) = 20{,}000/(x^2 + 1)^2, \quad 0 \leq x \leq 1.$
11. $f(x) = 10{,}000/(x^2 + 2)^3, \quad 3 \leq x \leq 8.$
12. $f(x) = 20{,}000/(x^2 + 1)^2, \quad 1 \leq x \leq 3.$
13. Explain why Assumptions 1 and 2 given in the text are reasonable.
14. Supply an intuitive explanation of this formula:

$$P(a, b) = 2\pi \int_a^b xf(x)dx.$$

Brief Review of Chapter 5

1. Volumes

Volumes with Known Cross-Sectional Area. If R is a solid and $A(x)$ is a continuous function giving the area of the cross section cut from R by a plane perpendicular to the x-axis at point x, then the volume V of R between the planes perpendicular to the x-axis at a and b, where $a \leq b$, is

$$V = \int_a^b A(x)dx.$$

When the cross section is made by a plane perpendicular to the y-axis and the appropriate notational changes are made, the volume of R is

$$V = \int_c^d A(y)dy.$$

Disc Method. This is a particular case of a solid of known cross-sectional area when the solid is obtained by rotation. The cross sections are then discs, or discs with center holes ("washer-shaped" regions).

Cylindrical Shell Method. Let A be an area of a region that is rotated about a line L, and let $w(t)$ be the width of a cross section parallel to L and at a distance t from L. Then the volume obtained when the part of the region between $t = c$ and $t = d$ is rotated about L is

$$V = \int_c^d 2\pi t w(t)dt.$$

2. Work

The work $W(a, b)$ done by a force $f(x)$ when it moves a body along the x-axis from $x = a$ to $x = b$ is

$$W(a, b) = \int_a^b f(x)dx.$$

3. Arc Length

The arc length L of the graph of $y = f(x)$, for $a \leq x \leq b$, is

$$L = \int_a^b \sqrt{1 + [f'(x)]^2}\, dx.$$

The arc length of the graph of $x = f(t)$, $y = g(t)$, $c \leq t \leq d$ is

$$L = \int_c^d \sqrt{\left(\frac{dx}{dt}\right)^2 + \left(\frac{dy}{dt}\right)^2}\, dt.$$

4. Consumer Surplus

For a demand function $f(x)$, the consumer surplus C, when b units are sold at a fixed price p, is

$$C = \int_0^b f(x)dx - pb.$$

5. Surface Area

When the graph of a function $f(x)$, $a \leq x \leq b$, is revolved about the x-axis, the area A of the resulting surface of revolution is

$$A = 2\pi \int_a^b |f(x)| \sqrt{1 + [f'(x)]^2}\, dx.$$

6. Fluid Force

Let a plate be vertically submerged in a fluid whose weight per unit volume is k, and let $w(y)$ be the width of the plate at depth y. Then the total force exerted by the fluid on one side of the plate between two depths $y = a$ and $y = b$ is

$$F(a, b) = \int_a^b kyw(y)dy.$$

7. Moments and Centroids

Let a region have width $w(x)$, $a \leq x \leq b$, above the point x, and width $l(y)$, $c \leq y \leq d$, opposite the point y. Then if A is the area of the region,

$$\overline{x} = \frac{1}{A} \int_a^b xw(x)dx \quad \text{and} \quad \overline{y} = \frac{1}{A} \int_c^d yl(y)dy$$

are the x and y coordinates of the centroid of A.

8. Population Density

Let $f(x)$ be the population density of a city at a distance x from the center of town. Then the total population between two distances a and b from the center of town is

$$P(a, b) = 2\pi \int_a^b xf(x)dx.$$

Technique Review Exercises, Chapter 5

1. The base of a solid is the region bounded by $y = x^2 - 2x$ and the x-axis. The cross sections perpendicular to the x-axis are squares. Find the volume of the solid.
2. Find the volume obtained when the region bounded by $y = x^2$, $x = 2$, and the x-axis is rotated about the line $x = -3$: (a) by the disc method, (b) by the shell method.
3. Find the work done in raising a 1500 pound weight from a depth of 200 feet, using a chain weighing 4 pounds per foot.
4. Find the length of the graph of $y = 2(x + 1)^{3/2} - 5$ between $x = 6$ and $x = 10$.
5. Find the length of the graph of $x = t^2$, $y = (2/3)t^3$ between $t = 1$ and $t = 2$.
6. If the demand function is $100 - x^{1/2}$, find the consumer surplus if units are sold at a fixed price of 70.
7. Find the area obtained when the graph of $3y = (x + 2)^3$, $-2 \leq x \leq -1$, is revolved about the x-axis.
8. Find the centroid of the region bounded by $y = x^2$, $x = 1$, and $y = 0$.
9. A semicircular plate of diameter 4′ is vertically submerged in oil weighing 50 pounds per cubic foot so that its diameter is at the surface. Find the total force exerted by the oil on one side of the plate.
10. The population density of a certain town is $2000 - 10x^3$ at a distance x miles from the center of town, for $0 \leq x \leq 5$. Find the total population between 1 and 3 miles from the center.

Additional Exercises, Chapter 5

Section 5.1

1. The base of an object is the region bounded by the graphs of $y = x^2$ and $y = 2x + 3$. The cross sections perpendicular to the x-axis are squares. Find the volume of the object.
2. The region bounded by $y = x^2$, and $y = 3$, is rotated about the x-axis. Find the resulting volume.
3. Find the volume of the solid obtained when the region of Exercise 2 is rotated about the line $y = -2$.
4. Find the volume of the solid obtained when the region of Exercise 2 is rotated about the line $x = -3$.
5. The region bounded by $y = x^2$ and $y = 6 - x$ is revolved about the line $y = -3$. Find the resulting volume.
6. The region of Exercise 5 is revolved about the line $x = 5$. Find the resulting volume.
7. The region of Exercise 5 is revolved about the line $y = 10$. Find the resulting volume.
8. The region bounded by $y = x^2$ and $y = 0$ and $x = 5$ is rotated about the y-axis. Find the resulting volume.
9. The region of Exercise 8 is revolved about the line $y = -3$. Find the resulting volume.
10. The region of Exercise 8 is revolved about the line $x = 7$. Find the resulting volume.

Section 5.2

11. Find the work done when a cable (weighing 3 pounds per foot) attached to a winch at the top of a 200-foot building is used to raise a weight of 500 pounds from the ground to a window 60 feet above the ground.

12. The force exerted by a spring when its natural length is increased by amount x is $F = kx$, where k is a constant. Find the work done in extending a spring of natural length 6 cm from a length of 8 cm to a length of 11 cm, if a force of 24 kg extends it to 9 cm.

13. Find the work done is compressing the spring of Exercise 12 from its natural length to a length of 4 inches.

14. A tank full of liquid weighing 60 pounds per cubic foot is in the shape of a right circular cone with a 12′ vertical axis and a radius of 4′ at the top. Find the work needed to pump the liquid out of the top of the tank.

15. Find the work needed to pump the liquid of Exercise 14 out of a pump 5′ above the top of the tank.

16. An open settling tank is 3/4 full of liquid weighing 70 pounds per cubic foot and is in the shape of a rectangular box 10′ × 8′ × 12′ deep. Find the work needed to pump the liquid out of a pump 4′ above the top of the tank.

Section 5.3

Find the length of the graph of the given equation between the indicated values.

17. $y = (2/3)x^{3/2}, \quad 3 \le x \le 35.$

18. $y = (2/3)(x + 2)^{3/2}, \quad 6 \le x \le 13.$

19. $y = \dfrac{x^3}{2} + \dfrac{x^{-1}}{6}, \quad 1 \le x \le 2.$

20. $y = \dfrac{x^3}{3} + \dfrac{x^{-1}}{4}, \quad 3 \le x \le 6.$

21. $y = (1/3)(x^2 - 2)^{3/2}, \quad 0 \le x \le 6.$

22. $y = (1/3)(x^2 - 2)^{3/2}, \quad 2 \le x \le 10.$

23. $x = 1 + t^2, \quad y = 1 - t^2, \quad 1 \le t \le 2.$

24. $x = t^3, \quad y = t^2, \quad 0 \le t \le 1.$

25. $x = a(1 - \cos\theta), \quad y = a\sin\theta, \quad 0 \le \theta \le 2\pi.$

26. $x = (1/4)t^4, \quad y = (1/5)t^5, \quad 0 \le t \le 2.$

Section 5.5

For each demand function $p = f(x)$, find the consumer surplus for the given actual price p per unit or number x of units sold.

27. $p = 155 - x^3, \quad p = 30.$

28. $p = 50 - x/100, \quad x = 400.$

29. $p = 300 - 10\sqrt{x + 3}, \quad x = 6.$

30. $p = 100 - x^4, \quad p = 19.$

Section 5.7

Use the method of cylindrical shells to find the volume obtained when the given region is rotated about the indicated axis.

31. The region bounded by $y = 3x$, $y = 6$, and the y-axis is rotated about the y-axis.

32. The region of Exercise 31 is rotated about the x-axis.

33. The region of Exercise 31 is rotated about the line $x = 5$.

34. The region of Exercise 31 is rotated about the line $y = -2$.

35. The region of Exercise 31 is rotated about (a) the line $x = -2$, (b) the line $y = 9$.

36. The region bounded by $y = x^2$ and $y = 2 - x$ is rotated about the x-axis.

37. The region of Exercise 36 is rotated about the line $y = 5$.

38. The region of Exercise 36 is rotated about the line $y = -2$.

39. The region of Exercise 36 is rotated about the line $x = -3$.

40. The region of Exercise 36 is rotated about the line $x = 3$.

Section 5.8

41. Find the area of the surface resulting when the graph of $y = \sqrt{7x - 10}$, $2 \le x \le 5$, is rotated about the x-axis.

42. Do Exercise **41** with the rotation about the y-axis.

43. Find the area of the surface resulting when the graph of $y = (2/3)x^3$, $0 \le x \le 2^{1/4}$, is revolved about the x-axis.

44. Find the area of the surface resulting from the revolution of the graph of $y = x^{1/2}$, $2 \le x \le 12$, about the x-axis.

45. Find the area of the surface resulting from the revolution of the graph of $y = x^{1/3}$, $0 \le y \le 1$, about the y-axis.

46. Find the area of the surface resulting from the revolution of the graph of $x = y^3/12 + y^{-1}$, $1 \le y \le 2$, about the y-axis.

Section 5.9

In Exercises 47–50 find the centroid of the region bounded by the given curves.

47. The y-axis and the lines $y = 2x + 1$ and $y = 7$.

48. The positive coordinate axes and the lines $y = x + 1$ and $x = 5$.

49. The curves $y = x^4$ and $x = y^2$.

50. The curves $y = x^4$ and $x = y^4$.

51. Use the Theorem of Pappus to compute the volume that results when the region of Exercise 47 is rotated about the line $x = -1$.

52. Use the Theorem of Pappus to compute the volume that results when the region of Exercise 49 is rotated about the line $y = -2$.

Section 5.10

Find the force due to water pressure on one side of a vertical plate of the given shape situated under water as indicated. Assume the water weighs 62.5 pounds per cubic foot.

53. A $10'$ by $12'$ rectangle with its upper $10'$ edge parallel to and $6'$ below the surface.

54. A right triangle with its upper $8'$ edge parallel to and $4'$ below the surface, if it has a vertical edge extending to a depth $6'$ below its upper edge.

55. A semicircle whose $12'$ diameter is at the surface.

56. The region bounded by $x = (y - 8)^2$ and $x = 25$, where the x-axis is at the surface and the y-axis is downward.

Section 5.11

Find the population between the given numbers of miles from the center of a city for the indicated population density $f(x)$ at a distance x miles from the center.

57. $f(x) = 1000 - x^3$: (a) $0 \le x \le 2$, (b) $1 \le x \le 3$.

58. $f(x) = 10{,}000 - x^4$: (a) $0 \le x \le 1$, (b) $1 \le x \le 2$.

59. $f(x) = 50{,}000/(x^2 + 3)^2$: (a) $0 \le x \le 1$, (b) $1 \le x \le 2$.

60. $f(x) = 4000/(x^2 + 1)^2$: (a) $0 \le x \le 1$, (b) $0 \le x \le 3$.

Challenging Problems, Chapter 5.

1. Water is evaporating from an open container. The rate of evaporation is directly proportional to the surface area. Show that the depth of the water in the container decreases at a constant rate no matter what the shape of the container.

2. Prove the following *Theorem of Pappus*.
Theorem: If a region in the plane is rotated about a line that lies in the plane but does not intersect the region, then the volume generated is equal to the product of the area of the region and the distance traveled by its centroid.

3. A cow is tied to a circular silo by a rope of length L. Assume that $L < \pi r$, where r is the radius of the silo. Use the following outline to calculate a formula for the area over which the cow can graze.
   (i) The grazing area is shown in Figure 1. Note that the grazing area consists of a semicircle of radius L and two congruent regions labeled R and S. Verify that the indicated angle has radian measure L/r.

Figure 1

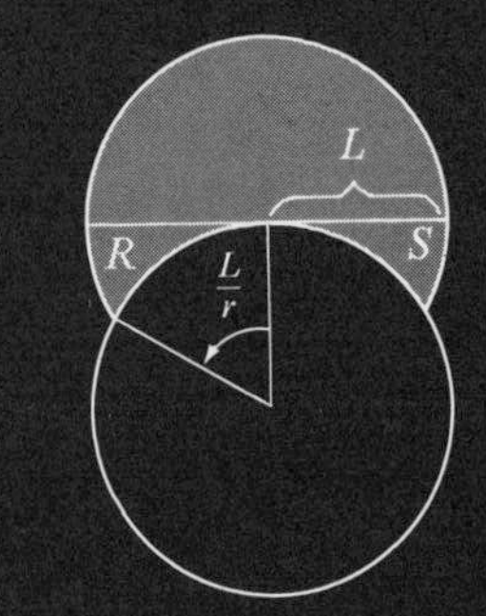

   (ii) Show that the area of the region illustrated in Figure 2 is $A = (1/2)k^2\alpha$.

Figure 2

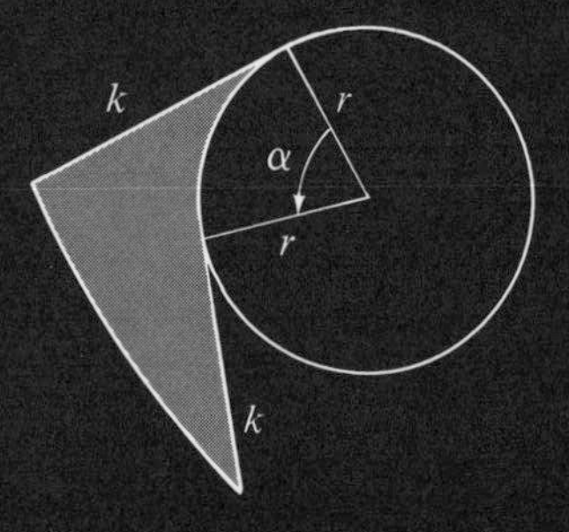

   Hint: Note that if $\alpha = 2\pi$, the area is $\pi(k^2 + r^2) - \pi r^2$.
   (iii) Subdivide the angle L/r into n subangles each of size $h = L/nr$, by introducing the subangles $\theta_i = Li/nr$, $i = 0, 1, 2, \ldots, n$. That process causes a subdivision of the region R into n subregions with areas $A_1, A_2, \ldots, A_n$. A typical subregion is shown in Figure 3. Verify the lengths shown in Figure 3.

Figure 3

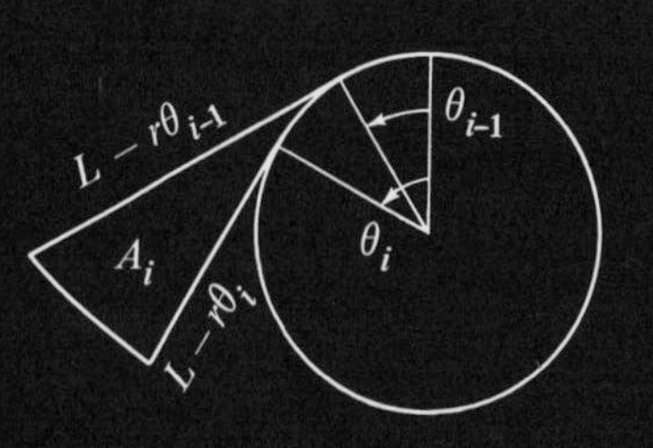

(iv) Use step (ii) to show that

$$\frac{1}{2}(L - r\theta_i)^2 h \leq A_i \leq \frac{1}{2}(L - r\theta_{i-1})^2 h.$$

(v) Show that there is a θ_i^*, $\theta_{i-1} \leq \theta_i^* \leq \theta_i$ such that

$$A_i = \frac{1}{2}(L - r\theta_i^*)^2 h.$$

(vi) Prove that the area A of R is given by

$$A = \int_0^{L/r} \frac{1}{2}(L - r\theta)^2 \, d\theta.$$

(vii) Show that the total grazing area is

$$L^2\left(\frac{L}{3r} + \frac{\pi}{2}\right).$$

6

Exponential and Logarithmic Functions

6.1 The Exponential Function

We recall the familiar definition of a^x for rational numbers x as follows:

Definition 6.1.1

> Let a be a positive real number and let m and n be positive integers. Then
>
> $$\text{(i)}\quad a^{m/n} = (\sqrt[n]{a})^m,$$
> $$\text{(ii)}\quad a^0 = 1,$$
> $$\text{(iii)}\quad a^{-m/n} = \frac{1}{a^{m/n}}.$$

This definition indicates that $f(x) = a^x$ is a function for rational x. That is, for each rational number x there is a unique output $f(x) = a^x$. We have required a to be positive, since if a is not positive there are values of x for which a^x either has no meaning or is not a real number. For example 0^0 has no meaning and $(-1)^{1/2}$ is not real.

Definition 6.1.1 can be extended to produce a continuous function whose domain consists of all real numbers. For example, let us consider a^π. The number π is an irrational number and has this nonrepeating decimal representation: $\pi = 3.14159265\ldots$. Since all these numbers are rational,

$$3.1 = \frac{31}{10}, \quad 3.14 = \frac{314}{10^2}, \quad 3.141 = \frac{3141}{10^3}, \quad 3.1415 = \frac{31415}{10^4}, \ldots,$$

the following numbers have been defined in Definition 6.1.1:

$$a^{3.1}, \quad a^{3.14}, \quad a^{3.141}, \quad a^{3.1415}, \ldots.$$

The more decimal places we take, the nearer the exponent x gets to π. Moreover, it can be proved that as x approaches π, a^x approaches a unique number, which is defined to be a^π. More generally, and in more precise language, a^x is defined for all real numbers x as follows:

Definition 6.1.2

> Let a be a positive real number, r be any fixed real number, and t be rational. Then we define a^r by
>
> $$a^r = \lim_{t \to r} a^t.$$

It can be proved that every limit occurring in Definition 6.1.2 exists and that the following familar properties for rational powers also hold for all real powers:

$$a^r \cdot a^s = a^{r+s},$$

$$(a^r)^s = a^{rs},$$

$$(ab)^r = a^r b^r,$$

$$\frac{a^r}{a^s} = a^{r-s}.$$

Notice that according to the definition of a^r, we almost have the statement that a^x is continuous at each real number r. In order to show continuity for each r, we must establish that $\lim_{t \to r} a^t = a^r$ is true not only when t is rational, but also when t is irrational, that is, when t is any real number. Actually this follows from the case where t is rational because the rational numbers occur so densely on the real line. (Like the proofs of certain other statements not given in this section, the proof is more suitable for a course in advanced calculus and hence will not be given here.) Since the result is important, we do state it formally as follows:

Theorem 6.1.1

> Let a and r be real numbers, where a is positive. Then a^x is continuous at $x = r$. That is, a^x is a continuous function.

Let us investigate a^x further, first for the case where $a > 1$. We know that $a^0 = 1$. If $a^k = 1$ and $k \neq 0$, then

$$a = (a^k)^{1/k} = 1^{1/k} = 1.$$

This is impossible since $a > 1$. Thus $a^k \neq 1$ if $k \neq 0$.

Since $a^1 = a > 1$, $a^x > 1$ for all $x > 0$. For if $a^p < 1$ for some $p > 0$, then by The Intermediate Value Theorem for continuous functions, there would be a number q between 1 and p for which $a^q = 1$.

Now if $x < 0$, let $x = -k$ where $k > 0$. Then

$$a^x = a^{-k} = \frac{1}{a^k}.$$

But since $a^k > 1$, $0 < 1/a^k < 1$. That is, $0 < a^x < 1$.

Similarly, it can be proved that if $0 < a < 1$, then $a^x > 1$ for all $x < 0$, and $0 < a^x < 1$ for all $x > 0$. We summarize the results just obtained in the following theorem.

Theorem 6.1.2

> (i) If $a > 1$, then $0 < a^x < 1$ for all $x < 0$, and $a^x > 1$ for all $x > 0$.
> (ii) If $0 < a < 1$, then $a^x > 1$ for all $x < 0$, and $0 < a^x < 1$ for all $x > 0$.
>
> Note in particular that if $a \neq 0$, then $a^x > 0$ for all x.

From this theorem we see that when $r > s$, that is, when $r - s > 0$,

$$a^{r-s} > 1 \text{ if } a > 1, \quad \text{and} \quad a^{r-s} < 1 \text{ if } 0 < a < 1.$$

Multiplying both sides of those inequalities by the positive number a^s, we obtain

$$a^r > a^s \text{ if } a > 1, \quad \text{and} \quad a^r < a^s \text{ if } 0 < a < 1.$$

So we have proved another theorem.

Theorem 6.1.3

> If $r > s$, then
> (i) $a^r > a^s$ when $a > 1$, that is, $f(x) = a^x$ is an increasing function, and
> (ii) $a^r < a^s$ when $0 < a < 1$, that is $f(x) = a^x$ is a decreasing function.

The information obtained so far about the function a^x is helpful in sketching the graph of $y = a^x$. If we evaluated the function for various values of x and various fixed values of a, we would see that the graph has essentially three different shapes, depending on whether $0 < a < 1$, $a = 1$, or $a > 1$. Typical cases are sketched in Figure 6.1.1. Of course when $a = 1$, then $a^x = 1^x = 1$ for all x; so in this case the graph is a horizontal line that is 1 unit above the x-axis, as indicated.

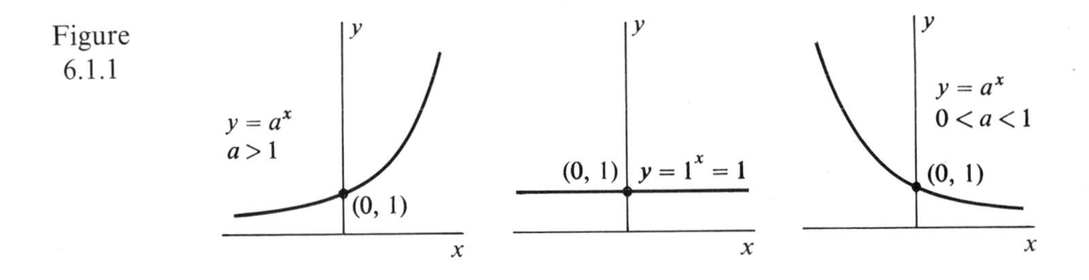

Figure 6.1.1

Exercises 6.1

For Exercises 1–8, find the value of each expression.

1. $8^{1/3}$.
2. $(8)^{-2/3}$.
3. $(32)^{-3/5}$.
4. $81^{3/4}$.
5. $(1/9)^{-3/2}$.
6. $(27)^{-2/3}$.
7. $(27/8)^{-2/3}$.
8. $(36/25)^{-3/2}$.

For Exercises 9–20, sketch the graph of each equation by plotting three well-chosen points.

9. $y = 2^x$.
10. $y = (1/2)^x$.
11. $y = 3^{-x}$.
12. $y = 5^x$.
13. $y = (1/5)^x$.
14. $y = 2^{-x}$.
15. $y = 2^x - 1$.
16. $y = 1 + 2^x$.
17. $y = 3^{x-2}$.
18. $y = 2^{x-1}$.
19. $y = 2^{x+3} + 1$.
20. $y = 2^{x+2} - 3$.

6.2 Monotone Functions, Inverse Functions

We have found that the function $f(x) = a^x$ is increasing for all x if $a > 1$ and is decreasing for all x if $0 < a < 1$. If a function is either increasing on an interval or decreasing on the interval, it is said to be *monotone* on the interval.

Thus, if $a > 0$ and $a \neq 1$, then $f(x) = a^x$ is monotone on every interval. When a function $f(x)$ is continuous and monotone on an interval $a \leq x \leq b$, the equation $y = f(x)$ relates x and y so that for each suitable y there is exactly one x. More precisely, we have the following theorem.

Theorem 6.2.1

> Let $f(x)$ be continuous and monotone for $a \leq x \leq b$, and let d be any number in the closed interval between $f(a)$ and $f(b)$. Then there is a *unique* number c such that $a \leq c \leq b$ and $f(c) = d$.

Figure 6.2.1

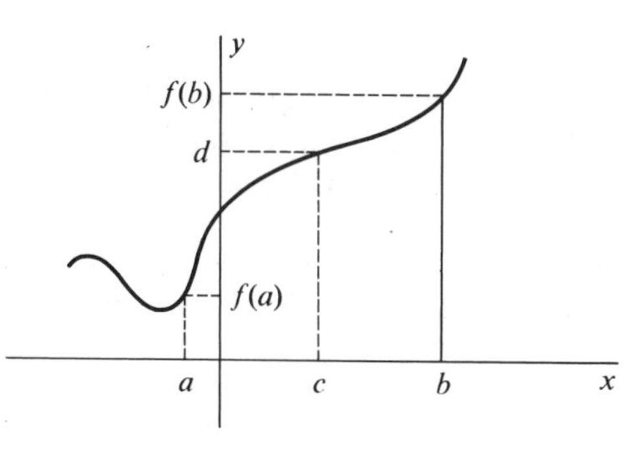

Proof. We shall assume that $f(x)$ is increasing. The proof for the case where it is decreasing is similar. Let d be a number such that $f(a) \leq d \leq f(b)$. Then by The Intermediate Value Theorem for continuous functions, there is at least one number c such that $a \leq c \leq b$ and $f(c) = d$. If, in addition, there were another number c_1 such that $a \leq c_1 \leq b$ and $f(c_1) = d$, then $f(c_1) = f(c)$. However, since f is increasing we can have neither $c_1 > c$ nor $c > c_1$, since these facts would imply $f(c_1) > f(c)$ or $f(c) > f(c_1)$. Consequently we must have $c_1 = c$, and so c is the *unique* number such that $a \leq c \leq b$ and $f(c) = d$, thus completing the proof.

If $f(x)$ is continuous and increasing for $a \leq x \leq b$, then for each input y such that $f(a) \leq y \leq f(b)$, there is a unique output x such that $a \leq x \leq b$ and $f(x) = y$. Thus we have a function $g(y)$, with $g(y) = x$, for all y such that $f(a) \leq y \leq f(b)$. Since a point (x, y) is on the graph of $x = g(y)$ if and only if it is on the graph of $y = f(x)$, the graph of the equation $x = g(y)$ for $f(a) \leq y \leq f(b)$ is the same as the graph of the equation $y = f(x)$ for $a \leq x \leq b$. We also have

$$f(g(y)) = f(x) = y \quad \text{for all} \quad f(a) \leq y \leq f(b), \quad \text{and}$$

$$g(f(x)) = x \quad \text{for all} \quad a \leq x \leq b.$$

Thus, in a sense, "g undoes the work done by f" and "f undoes the work done by g." Corresponding statements hold when $f(x)$ is decreasing for $a \leq x \leq b$. In either case, the function g is called the *inverse* of function f, according to the next definition.

Definition 6.2.1

> Let $f(x)$ be a function with domain $a \leq x \leq b$. A function $g(y)$ such that
>
> $$g(f(x)) = x \quad \text{for} \quad a \leq x \leq b, \quad \text{and}$$
>
> $$f(g(y)) = y \quad \text{for} \quad y \text{ in the range of } f$$
>
> is called the *inverse* of $f(x)$ on the interval $a \leq x \leq b$.

Returning to our computer analogy for functions, we illustrate the relationship between inverse functions f and g in Figure 6.2.2.

Figure 6.2.2

$$x \to \boxed{f} \to f(x) \to \boxed{g} \to g(f(x)) = x$$

$$y \to \boxed{g} \to g(y) \to \boxed{f} \to f(g(y)) = y$$

Not all functions have inverse functions; in fact if $f(x)$ is continuous and not monotone for $a \leq x \leq b$, it cannot have an inverse on that interval. In this case, for some d such that $f(a) \leq d \leq f(b)$, there will be at least two different numbers c_1 and c_2 such that $f(c_1) = f(c_2) = d$, as illustrated in Figure 6.2.3.

Figure 6.2.3

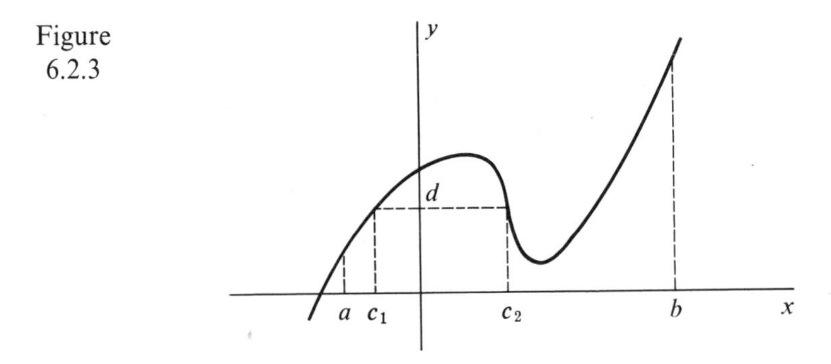

Since the inverse of a continuous function is also continuous, as is proved in many advanced calculus texts, we have the following theorem.

Theorem 6.2.2

> If $f(x)$ is continuous and monotone on the interval between a and b, it has an inverse function $g(y)$ that is continuous and monotone on the interval between $f(a)$ and $f(b)$. Moreover, $g(y)$ is increasing or decreasing accordingly as $f(x)$ is increasing or decreasing.

Example 1 | Let $y = f(x) = 2x$ for $1 \leq x \leq 5$.

This function is continuous and increasing over its domain. Solving for x in terms of y, we have $x = g(y) = y/2$ for $2 \leq y \leq 10$. As the theorem indicates, $g(y)$ is continuous and increasing for $2 \le y \le 10$. Note also that in accord with Definition 6.2.1,

$$g(f(x)) = g(2x) = \frac{2x}{2} = x \quad \text{for} \quad 1 \le x \le 5, \quad \text{and}$$

$$f(g(y)) = f\left(\frac{y}{2}\right) = 2\left(\frac{y}{2}\right) = y \quad \text{for} \quad 2 \le y \le 10.$$

||

Example 2 | Let $y = f(x) = x^2$ for $x \ge 0$.

Again the function is continuous and increasing on the given interval. Solving for x in terms of y, we have $x = g(y) = y^{1/2}$ for $y \ge 0$, where $g(y)$ is continuous and increasing. (Note that if there were no restriction on x, then $f(x) = x^2$ would not be monotone, but it could be separated into two monotone functions, because the function is decreasing when $x \le 0$ and increasing when $x \ge 0$.) In the next example we shall consider the case where $x \le 0$. ||

Example 3 | Let $y = f(x) = x^2$ for $x \le 0$.

This function is continuous and decreasing. Notice that for $x \le 0$, we have $y \ge 0$; so when we solve for x we have $x = g(y) = -(y)^{1/2}$ for $y \ge 0$. Here $g(y)$ is continuous and decreasing. ||

Now let's consider a function $f(x)$ with an inverse $g(y)$. If f has a derivative at a point x_0, must g have a derivative at y_0, where $y_0 = f(x_0)$? To find out we must decide on the existence of

$$\lim_{h\to 0} \frac{g(y_0 + h) - g(y_0)}{h} = g'(y_0).$$

Since g is the inverse of f, $g(y_0) = x_0$ and so

$$\frac{g(y_0 + h) - g(y_0)}{h} = \frac{g(y_0 + h) - x_0}{h}. \tag{1}$$

Let $g(y_0 + h) - x_0 = k$. Then $x_0 + k = g(y_0 + h)$, and hence $y_0 + h = f(x_0 + k)$; so we have $h = f(x_0 + k) - y_0 = f(x_0 + k) - f(x_0)$. Therefore

$$\frac{g(y_0 + h) - x_0}{h} = \frac{k}{f(x_0 + k) - f(x_0)} = \frac{1}{\dfrac{f(x_0 + k) - f(x_0)}{k}}. \tag{2}$$

Also, since f has a derivative at x_0, f must be continuous at x_0 and so by Theorem 6.2.2, g must be continuous at y_0. Therefore $\lim_{h\to 0} g(y_0 + h) = g(y_0)$, and thus when $h \to 0$, $k \to 0$. Consequently by (1) and (2), if $f'(x_0) = 0$,

$$\lim_{h\to 0} \frac{g(y_0 + h) - g(y_0)}{h} = \lim_{k\to 0} \frac{1}{\dfrac{f(x_0 + k) - f(x_0)}{k}} = \frac{1}{f'(x_0)}.$$

Thus if $f'(x_0) \neq 0$, g does have a derivative at y_0; and moreover $g'(y_0) = \dfrac{1}{f'(x_0)}$.

We have proved the following theorem.

Theorem 6.2.3

> Let $f(x)$ be a differentiable function with an inverse $g(y)$ for $a \leq x \leq b$ such that $f'(x) \neq 0$, $a \leq x \leq b$. Then $g(y)$ is differentiable and
>
> $$g'(y) = \frac{1}{f'(x)}, \qquad a \leq x \leq b.$$

Using the Leibniz notation, that equality can be expressed as

$$\frac{dx}{dy} = \frac{1}{\dfrac{dy}{dx}}.$$

Again we see the great value of the Leibniz notation because of its consistency with fractional notation!

We shall use Theorem 6.2.3 in Chapter 7 to obtain the derivatives of the inverse trigonometric functions. The next example, while producing no new results, illustrates the technique of using that theorem to get the derivative of an inverse function from the derivative of a function.

Example 4 | Let $y = f(x) = x^n$, where $x > 0$ and n is a positive integer.

Then $x = y^{1/n}$ and

$$\frac{dx}{dy} = \frac{d}{dy}(y^{1/n}) = \frac{1}{\dfrac{dy}{dx}} = \frac{1}{nx^{n-1}} = \frac{1}{n} \cdot \frac{1}{x^{n-1}}.$$

But
$$\frac{1}{x^{n-1}} = x^{1-n} = (y^{1/n})^{(1-n)} = y^{(1/n)(1-n)} = y^{(1/n)-1}.$$

So
$$\frac{dx}{dy} = \frac{d}{dy}(y^{1/n}) = \frac{1}{n}y^{(1/n)-1},$$

as we have already established. ||

The next example illustrates the great value of Theorem 6.2.3 when the inverse function is difficult or impossible to express in a convenient form.

Example 5 | Find the derivative of the inverse of $y = (7x^3 - 3)^{-3}$ at $x = 1$.

Since
$$\frac{dy}{dx} = -3(7x^3 - 3)^{-2}(21x^2),$$

the derivative of the inverse function is

$$\frac{dx}{dy} = \frac{1}{\frac{dy}{dx}} = -\frac{(7x^3 - 3)^2}{63x^2}, \qquad \text{if } x \neq 0.$$

Thus, if $x = 1$ we have $\frac{dx}{dy} = -\frac{16}{63}$. ||

Exercises 6.2

In Exercises 1–10, find the inverse of each function and find the domain of the inverse function.

1. $f(x) = 2x + 5, \quad -2 \leq x \leq 7.$
2. $f(x) = 3x - 2, \quad 0 \leq x \leq 6.$
3. $f(x) = x^2 - 4, \quad x \leq 0.$
4. $f(x) = x^2 + 1, \quad x \geq 0.$
5. $f(x) = x^2 - 4, \quad x \geq 0.$
6. $f(x) = x^2 + 1, \quad x \leq 0.$
7. $f(x) = \sqrt{x - 5}, \quad 5 < x < 30.$
8. $f(x) = \sqrt{x + 7}, \quad -3 < x < 10.$
9. $f(x) = \begin{cases} 2x & \text{if } 0 \leq x < 1 \\ 5 - x & \text{if } 1 \leq x \leq 3. \end{cases}$
10. $f(x) = \begin{cases} 2 - x & \text{if } 0 \leq x < 1 \\ x - 1 & \text{if } 1 \leq x \leq 2. \end{cases}$

Restrict the domain of each of the following functions in two ways so that for each restriction the function has an inverse function.

11. $f(x) = |x|.$
12. $f(x) = x^4.$
13. $f(x) = (x - 3)^2.$
14. $f(x) = |x + 2|.$
15. $f(x) = |5 - x|.$
16. $f(x) = (8 - x)^6.$
17. $f(x) = 6 - x^4.$
18. $f(x) = 3 - x^2.$
19. $f(x) = x^2 - 3.$
20. $f(x) = x^4 + 5.$

In Exercises 21–28 find the derivative of the inverse of the given function at the given point.

21. $y = (x - 1)^5, \quad x = 2.$
22. $y = (2x + 3)^{-2}, \quad x = 1.$
23. $y = x^4 + 7x - 8, \quad x = 1$
24. $y = x^5 - 7x^2 + 4, \quad x = 2$
25. $y = x^{-3} + x^2 + 3, \quad x = 2$
26. $y = x^4 + 2x^{-1} + 9, \quad x = 1$
27. $y = x^7 + 2x, \quad x = 0$
28. $y = x^5 - 3x, \quad x = 0$

6.3 The Logarithmic Function

We have seen that when $a > 0$, the exponential function a^x is continuous for all x; when $a > 1$, a^x is increasing; and when $0 < a < 1$, a^x is decreasing. Therefore, when $a > 0$ and $a \neq 1$, the function a^x has an inverse function.

Definition 6.3.1

> When $a > 0$ and $a \neq 1$, the inverse of the exponential function $y = a^x$ is called the *logarithmic function with base a.* The logarithm of y with base a is denoted by $\log_a y$.

So we see that if $y = a^x$, then $x = \log_a y$. From the definition of inverse functions we know that

$$a^{\log_a y} = y.$$

Thus $\log_a y$ is the power to which a must be raised to obtain y. For example $\log_{10} 100 = 2$, $\log_{10} 0.1 = -1$, $\log_2 8 = 3$, $\log_{1/3} 9 = -2$, and $\log_{1/3}(1/9) = 2$. Usually when we consider functions and their graphs it is most convenient to consider them in the form $y = f(x)$. Thus we shall consider the function $y = f(x) = \log_a x$ for positive values of a other than 1. The domain of $y = \log_a x$ is the set of positive real numbers and its range is the set of all real numbers. If $a = 10$, then $\log_a x$ is the familiar *common logarithm* and is denoted by $\log x$ with the 10 omitted.

By Theorem 6.1.3 and Theorem 6.2.2, we have the following theorem.

Theorem 6.3.1

> The function $\log_a x$ is continuous for all positive x, increasing when $a > 1$, and is decreasing when $0 < a < 1$.

Since the exponential and logarithmic functions are inverses of each other, the graphs of $x = \log_a y$ and $y = a^x$ are the same. Consequently the graphs of $x = \log_a y$ for the cases where $a > 1$ and $0 < a < 1$ are as shown in Figure 6.3.1.

Figure 6.3.1

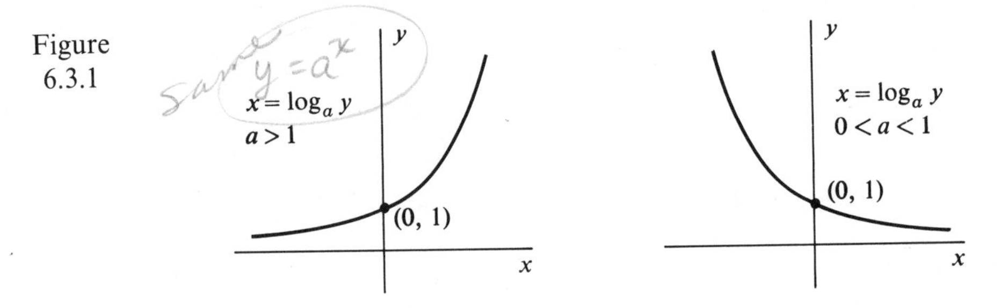

Of course we are usually interested in the graph of $y = \log_a x$ in the cases where $a > 1$ and $0 < a < 1$. Thus we must interchange the roles of x and y. This is most easily accomplished by rotating the graphs of Figure 6.3.1 one-half revolution about the line $y = x$. The graphs we obtain after such a rotation are illustrated in Figure 6.3.2.

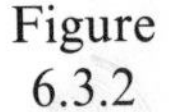

Figure 6.3.2

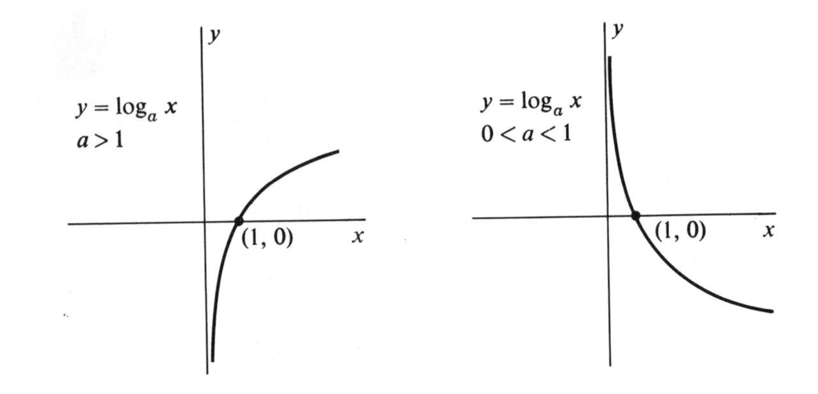

It can be proved from the properties of powers that for all positive real numbers r and s:

$$\log_a(rs) = \log_a r + \log_a s,$$

$$\log_a \frac{r}{s} = \log_a r - \log_a s,$$

$$\log_a(r^s) = s \log_a r.$$

Exercises 6.3

Evaluate each expression in Exercises 1–12.

1. $\log .01$.
2. $\log_2 64$.
3. $\log_3(1/27)$.
4. $\log_{1/3} 9$.
5. $\log_{1/3}(1/27)$.
6. $\log .001$.
7. $\log_3 81$.
8. $\log_{1/2} 16$.
9. $\log_{1/2}(1/32)$.
10. $\log_2 32$.
11. $\log 10^{17}$
12. $10^{\log 37}$.

Sketch the graphs of the equations in Exercises 13–16.

13. $y = \log_2 x$.
14. $y = \log_{1/2} x$.
15. $y = \log_3 x$.
16. $y = \log_{1/3} x$.
17. Show that $\log_a(x + h) - \log_a x = \log_a(1 + h/x)$.
18. Show that $\frac{1}{h} \log_a\left(1 + \frac{h}{x}\right) = \frac{1}{x} \log_a\left(1 + \frac{h}{x}\right)^{x/h}$.
19. Show that for any positive real numbers a and b different from 1, $\log_a b = 1/\log_b a$.
20. Explain why it follows directly from the definition of $\log_a r$ that (a) $a^{\log_a r} = r$, (b) $\log_a(a^r) = r$.

6.4 Differentiation of Logarithmic Functions

Next we shall obtain the formula for the derivative of $\log_a x$, and then the formula for the derivative of $\log_a u$ where u is a differentiable function of x. Of course the formulas will only hold for values of x for which the logarithms are defined. That is, for $\log_a x$ the differentiation formula will only hold when x is positive. Since every differentiation formula gives rise to a corresponding integration formula, we shall also obtain the related integration formulas.

Let $f(x) = \log_a x$ where $x > 0$. Then

$$f'(x) = \lim_{k \to 0} \frac{f(x + h) - f(x)}{h} = \lim_{h \to 0} \frac{1}{h} [\log_a(x + h) - \log_a x].$$

Since

$$\log_a(x + h) - \log_a x = \log_a\left(\frac{x + h}{x}\right) = \log_a\left(1 + \frac{h}{x}\right),$$

we get

$$f'(x) = \lim_{h \to 0} \left[\frac{1}{h} \log_a\left(1 + \frac{h}{x}\right)\right] = \lim_{h \to 0} \left[\frac{1}{x} \cdot \frac{x}{h} \log_a\left(1 + \frac{h}{x}\right)\right].$$

Using the facts that $s \log_a r = \log_a(r^s)$ and x is independent of h, we obtain

$$f'(x) = \lim_{h \to 0} \frac{1}{x} \log_a\left(1 + \frac{h}{x}\right)^{x/h} = \frac{1}{x} \lim_{h \to 0} \log_a\left(1 + \frac{h}{x}\right)^{x/h}.$$

Now let $t = h/x$. Then $h \to 0$ is equivalent to $t \to 0$; so

$$f'(x) = \frac{1}{x} \lim_{t \to 0} \log_a (1 + t)^{1/t}.$$

However, since the logarithmic function is continuous, we have

$$\lim_{t \to 0} [\log_a (1 + t)^{1/t}] = \log_a \left[\lim_{t \to 0} (1 + t)^{1/t} \right]. \tag{1}$$

It can be proved that $\lim_{t \to 0} (1 + t)^{1/t}$ exists and has a value between 2 and 3. (See Challenging Problems, Chapter 11 for an outline of the proof.) Note, that in Example 12 of Section 2.2, a table of values was given to help approximate this limit (with t replaced by x). There we obtained the approximation 2.71828.

The limit $\lim_{t \to 0} (1 + t)^{1/t}$, which arose naturally from the consideration of the derivative of the logarithmic function, is very important in calculus because its use permits considerable simplification. As indicated in Section 2.2, this limit is given the special symbol e. The number e is irrational (an outline of the proof is given in the Challenging Problems, Chapter 11), and e is not even a solution to a polynomial equation with integral coefficients. A twenty-place decimal approximation to e is $e \approx 2.71828182845904523536$.

Getting back to $f'(x) = (d/dx) \log_a x$, we now can write

$$\lim_{t \to 0} \log_a (1 + t)^{1/t} = \log_a e, \quad \text{and so by (1)}$$

$$\frac{d}{dx} \log_a x = \frac{1}{x} \log_a e.$$

This formula is simplest for the case where $a = e$, since $\log_e e = 1$. In that case we have $(d/dx) \log_e x = 1/x$. Because of this simplification the base e is almost always used for logarithmic functions in calculus. The logarithmic function with base e is given the special symbol $\ln x$ in place of the more cumbersome $\log_e x$. Thus

$$\frac{d}{dx} \ln x = \frac{1}{x}. \tag{2}$$

It follows from (2) or from Theorem 6.3.1 that $\ln x$ is an increasing function for all positive x.

In case we have $y = \log_a u$ or $y = \ln u$, where u is a differentiable, positive-valued function of x, we use the chain rule, $dy/dx = (dy/du)(du/dx)$, to obtain

$$\frac{d}{dx} \log_a u = \frac{d}{du} \log_a u \frac{du}{dx}.$$

But $(d/du) \log_a u = (1/u) \log_a e$, and hence we have

$$\frac{d}{dx} \log_a u = \frac{1}{u} \frac{du}{dx} \log_a e. \tag{3}$$

When $a = e$, we get

$$\frac{d}{dx} \ln u = \frac{1}{u} \frac{du}{dx} = \frac{du/dx}{u}. \tag{4}$$

In applications it is often useful to think of formula (4) in words: the derivative with respect to x of the natural logarithm of a function of x is the derivative of the function with respect to x divided by the function. For example,

to find
$$\frac{d\ln(x^2+1)}{dx},$$

we think
$$\frac{d(x^2+1)/dx}{x^2+1}$$

and write directly
$$\frac{d\ln(x^2+1)}{dx}=\frac{2x}{x^2+1}.$$

Example 1 $\frac{d}{dx}\ln(x^2-2x+5)=\frac{2x-2}{x^2-2x+5}.$ ||

Example 2 $\frac{d}{dx}\ln(x^4+3x^2)=\frac{4x^3+6x}{x^4+3x^2}=\frac{4x^2+6}{x^3+3x}.$ ||

Example 3 $\frac{d}{dx}\ln(x^2+5)^4=\frac{4(x^2+5)^3\cdot 2x}{(x^2+5)^4}=\frac{8x}{x^2+5}.$ ||

Example 4 $\frac{d}{dx}\log_5(x^2+1)=\frac{2x}{x^2+1}\log_5 e.$ ||

Since $\ln x$ is only defined when $x>0$, the function $\ln|x|$ has a larger domain because it is also defined when $x<0$. Let us find the derivative of $\ln|x|$. We have

$$\frac{d}{dx}\ln|x|=\frac{d|x|/dx}{|x|}.$$

Since $|x|=\begin{cases}x & \text{if } x\geq 0\\ -x & \text{if } x<0,\end{cases}$ we have $\frac{d}{dx}|x|=\begin{cases}1 & \text{if } x>0\\ -1 & \text{if } x<0.\end{cases}$

Hence
$$\frac{d}{dx}\ln|x|=\begin{cases}1/x & \text{if } x>0\\ -1/-x=1/x & \text{if } x<0.\end{cases}$$

That is, for all $x\neq 0$,

$$\frac{d\ln|x|}{dx}=\frac{1}{x}. \qquad (5)$$

If u is a function of x, then for $u\neq 0$, we also have

$$\frac{d\ln|u|}{dx}=\frac{d\ln|u|}{du}\frac{du}{dx}.$$

But by (5)

$$\frac{d\ln|u|}{du}=\frac{1}{u}.$$

So we have extended (4) as follows:

$$\frac{d}{dx}\ln|u| = \frac{1}{u}\frac{du}{dx} = \frac{du/dx}{u}, \; u \neq 0.$$

Similarly, by essentially the same argument we can extend (3) thus:

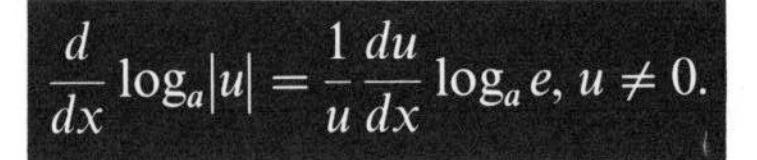

$$\frac{d}{dx}\log_a|u| = \frac{1}{u}\frac{du}{dx}\log_a e, \; u \neq 0.$$

Example 5 $$\frac{d}{dx}\ln|1 - x^2| = \frac{-2x}{1 - x^2}.$$ ||

Example 6 $$\frac{d}{dx}\ln|x^3 + 4x - 11| = \frac{3x^2 + 4}{x^3 + 4x - 11}.$$ ||

Example 7 $$\frac{d}{dx}\log_a|3x - x^4| = \frac{3 - 4x^3}{3x - x^4}\log_a e.$$ ||

We have seen that corresponding to every differentiation formula there is an integration formula. When u is a function of x, we know that $\ln|u|$ is a function whose derivative is $(1/u)(du/dx)$. Hence

$$\int \frac{1}{u}\frac{du}{dx}\,dx = \ln|u| + C$$

where C is any constant. Since $(du/dx)dx = du$, we can rewrite the formula above more simply as

$$\int \frac{du}{u} = \ln|u| + C.$$

That is, when the integrand is in the form of a fraction whose numerator is the differential of the denominator, the integral is the natural log of the absolute value of the denominator. For example,

$$\int \frac{2x}{x^2 - 7}\,dx = \int \frac{2x\,dx}{x^2 - 7} = \int \frac{d(x^2 - 7)}{x^2 - 7} = \ln|x^2 - 7| + C,$$

because the numerator is the differential of the denominator. Similarly,

$$\int \frac{3x^2 + 6x}{x^3 + 3x^2 + 2}\,dx = \ln|x^3 + 3x^2 + 2| + C.$$

Often we "fix up" the numerator by multiplying by a suitable constant k to make it the differential of the denominator. When this is done, of course, we also multiply by $1/k$. We then use the fact that a constant may be moved from one side of the integral sign to the other in order to move the $1/k$ outside the integral sign. This process is used in the next two examples.

Example 8 Evaluate $\int \frac{x^2 + 2x}{x^3 + 3x^2 + 2} dx$.

$$\int \frac{x^2 + 2x}{x^3 + 3x^2 + 2} dx = \frac{1}{3} \int \frac{3(x^2 + 2x)}{x^3 + 3x^2 + 2} dx$$

$$= \frac{1}{3} \int \frac{d(x^3 + 3x^2 + 2)}{x^3 + 3x^2 + 2} = \frac{1}{3} \ln|x^3 + 3x^2 + 2| + C. \qquad \|$$

Example 9 Evaluate $\int \frac{7x\, dx}{x^2 - 7}$.

$$\int \frac{7x\, dx}{x^2 - 7} = \frac{7}{2} \int \frac{2x\, dx}{x^2 - 7} = \frac{7}{2} \int \frac{d(x^2 - 7)}{x^2 - 7} = \frac{7}{2} \ln|x^2 - 7| + C. \qquad \|$$

Remember, however, that only constants, not x's, can be moved outside the integral sign.

The formula $(d/dx) \ln x = 1/x$ can be used to evaluate certain integrals involving the natural logarithm.

Example 10 Evaluate $\int_1^e \frac{1}{x} (\ln x)^5 dx$.

Since

$$\frac{dx}{x} = d(\ln x),$$

we have

$$\int_1^e \frac{1}{x} (\ln x)^5 dx = \int_1^e (\ln x)^5 d(\ln x)$$

$$= \frac{1}{6} (\ln x)^6 \Big|_1^e = \frac{1}{6}. \qquad \|$$

It is interesting to note that we have now filled the gap in the power function rule:

$$\int u^n\, du = \frac{u^{n+1}}{n+1} + C, \qquad \text{if } n \neq -1.$$

For when $n = -1$ we have

$$\int u^{-1}\, du = \int \frac{du}{u} = \ln|u| + C.$$

Exercises 6.4

Differentiate each function in Exercises 1–12.

1. $f(x) = \ln(x^6 + 3x^2 + 5)$.
2. $f(x) = \ln(7 + 5x^2)$.
3. $f(x) = x^2 \log(x - 1)$.
4. $f(x) = x^4 \log(3x^2 + 11)$.
5. $f(x) = x \log_5(4x + 5)$.
6. $f(x) = x \log_2(8x + 3)$.

7. $f(x) = \ln|x^2 - 3x + 2|$.
8. $f(x) = \ln|1 - x^7|$.
9. $f(x) = \ln|x^3 - 3x + 1|^3$.
10. $f(x) = \ln|x^2 - 2|^7$.
11. $f(x) = \ln(\ln|x^3 + 2|)$.
12. $f(x) = \ln(\ln|1 - x|)$.

Evaluate each integral in Exercises 13–26.

13. $\int_1^3 \frac{1}{x}\,dx$.
14. $\int \frac{2x}{x^2 + 3}\,dx$.
15. $\int \frac{x}{x^2 + 3}\,dx$.
16. $\int \frac{2x - 6}{x^2 - 6x + 7}\,dx$.
17. $\int \frac{x - 3}{x^2 - 6x + 7}\,dx$.
18. $\int \frac{x^2 + 4x}{x^3 + 6x^2 - 7}\,dx$.
19. $\int \frac{x^3}{1 - x^4}\,dx$.
20. $\int \frac{x^7}{x^8 + 5}\,dx$.
21. $\int \frac{dx}{x \ln x}$.
22. $\int_2^e \frac{\ln x}{x}\,dx$.
23. $\int_1^{e^2} \frac{(\ln x)^3}{x}\,dx$.
24. $\int \frac{dx}{x(\ln x)^4}$.
25. $\int \frac{\ln x}{x}\,dx$.
26. $\int \frac{(\ln x - 1)^2}{x}\,dx$.

27. Find the local maximum and minimum points and the points of inflection of the graph of $y = x \ln x$.
28. Find the local maximum and minimum points and the points of inflection of the graph of $y = \frac{\ln x}{x}$.
29. Find the area of the region bounded by the graph of $y = x^3/(x^4 - 8)$, the x-axis, and the lines $x = 2$ and $x = 3$.
30. Find the volume obtained when the region bounded by the x-axis, the lines $x = 1$ and $x = 2$, and the graph of $y = [x/(5x^2 + 1)]^{1/2}$ is rotated about the x-axis.
31. For a certain kind of telegraph cable the speed s of the signal is given by $s = kx^2 \ln(1/x)$, where k is a positive constant and x is the ratio of the radius of the core to the thickness of the covering. Determine the value of x for which the maximum speed occurs.
32. Show that the area under $y = 1/x$ and above the x-axis between a and b is the same as the area under $y = 1/x$ and above the x-axis between ka and kb, for all positive numbers k, a, and b.

6.5 The Derivative of a^x; Logarithmic Differentiation

We shall first find the derivative of a^x, where a is a positive constant not equal to 1. Let $y = a^x$. Then y is necessarily positive and so

$$\ln y = \ln(a^x).$$

But $\ln(a^x) = x \ln a$, and therefore we have

$$\ln y = x \ln a.$$

Hence

$$\frac{d \ln y}{dx} = \frac{d(x \ln a)}{dx}.$$

But

$$\frac{d \ln y}{dx} = \frac{1}{y}\frac{dy}{dx} \quad \text{and} \quad \frac{d(x \ln a)}{dx} = \ln a.$$

Thus
$$\frac{1}{y}\frac{dy}{dx} = \ln a,$$

and so
$$\frac{dy}{dx} = y \ln a.$$

Replacing y by a^x we get

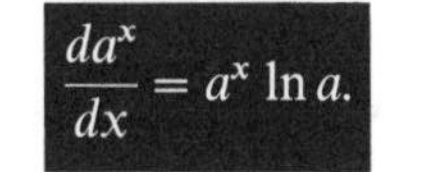

$$\frac{da^x}{dx} = a^x \ln a. \tag{1}$$

In particular, since $\ln e = 1$, we also have

$$\frac{de^x}{dx} = e^x. \tag{2}$$

Thus e^x has the interesting property of being equal to its derivative!

When u is a differentiable function of x, formulas (1) and (2) can be extended by using the chain rule. The extended results are stated in the next theorem.

Theorem 6.5.1

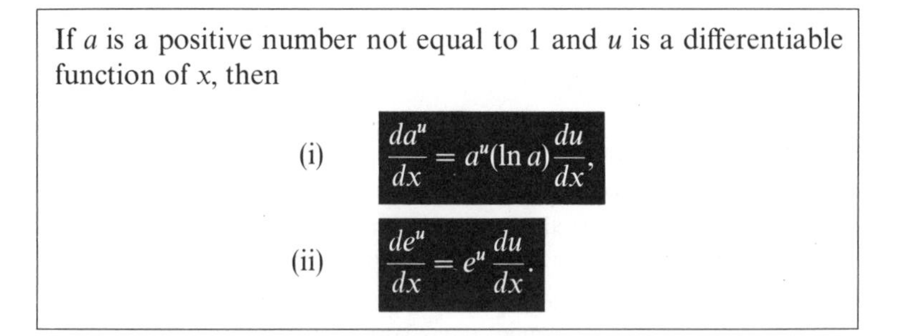

If a is a positive number not equal to 1 and u is a differentiable function of x, then

(i) $$\frac{da^u}{dx} = a^u(\ln a)\frac{du}{dx},$$

(ii) $$\frac{de^u}{dx} = e^u\frac{du}{dx}.$$

For example,

$$\frac{d}{dx}3^{x^2-3x+2} = (2x-3)3^{x^2-3x+2}\ln 3,$$

$$\frac{d}{dx}e^{5x-x^3+7} = (5-3x^2)e^{5x-x^3+7}.$$

The technique we used to find the derivative of a^x is also used for other functions. It is used for a function $f(x)$ when $\ln f(x)$ is easier to differentiate than $f(x)$. The technique is called *logarithmic differentiation.* We now illustrate logarithmic differentiation with some examples.

Example 1 | Find dy/dx if $y = (x^2+1)^{x+5}$.

From $y = (x^2+1)^{x+5}$ we obtain

$$\ln y = \ln(x^2+1)^{x+5} = (x+5)\ln(x^2+1).$$

Differentiating with respect to x gives

$$\frac{1}{y}\frac{dy}{dx} = (x+5)\frac{2x}{x^2+1} + \ln(x^2+1).$$

Then, multiplying by y and replacing it by $(x^2+1)^{x+5}$, we get

$$\frac{dy}{dx} = (x^2+1)^{x+5}\left[\frac{2x^2+10x}{x^2+1} + \ln(x^2+1)\right]. \qquad \|$$

When f(x) is not positive, ln $f(x)$ is not meaningful. However when $f(x) \neq 0$, $|f(x)|$ is positive and so $\ln|f(x)|$ is meaningful. Thus when $y = f(x)$ and $f(x)$ is negative for some x, we use an additional step in logarithmic differentiation. We use the fact that if $y = f(x)$ then $|y| = |f(x)|$. The process still does not produce a derivative for those values of x such that $f(x) = 0$, because $\ln|f(x)|$ has no meaning then.

Example 2 | Find dy/dx if $y = (x^2-3)^2(1-3x)^5(x^3-2x)^3$.

Here y is negative for some x and so we first write

$$|y| = |x^2-3|^2|1-3x|^5|x^3-2x|^3.$$

Hence, using the properties of logarithms,

$$\ln|y| = 2\ln|x^2-3| + 5\ln|1-3x| + 3\ln|x^3-2x|.$$

Therefore $\quad \dfrac{1}{y}\dfrac{dy}{dx} = \dfrac{2(2x)}{x^2-3} + \dfrac{5(-3)}{1-3x} + \dfrac{3(3x^2-2)}{x^3-2x},\quad$ and so

$$\frac{dy}{dx} = (x^2-3)^2(1-3x)^5(x^3-2x)^3\left[\frac{4x}{x^2-3} - \frac{15}{1-3x} + \frac{9x^2-6}{x^3-2x}\right]. \qquad \|$$

Example 3 | Find $\dfrac{dy}{dx}$ if $y = \dfrac{(3-x)^{1/3}(x^2-2)^{1/2}}{(x^3+1)^2}$.

We have

$$\ln|y| = \ln[|3-x|^{1/3}|x^2-2|^{1/2}] - \ln|x^3+1|^2$$

$$= \frac{1}{3}\ln|3-x| + \frac{1}{2}\ln|x^2-2| - 2\ln|x^3+1|.$$

Thus $\quad \dfrac{1}{y}\dfrac{dy}{dx} = \dfrac{1}{3}\cdot\dfrac{-1}{3-x} + \dfrac{1}{2}\cdot\dfrac{2x}{x^2-2} - 2\cdot\dfrac{3x^2}{x^3+1}.$

Multiplying both sides by the expression for y we obtain

$$\frac{dy}{dx} = \frac{(3-x)^{1/3}(x^2-2)^{1/2}}{(x^3+1)^2}\left[\frac{-1}{9-3x} + \frac{x}{x^2-2} - \frac{6x^2}{x^3+1}\right]. \qquad \|$$

As another application of logarithmic differentiation, we extend Theorem 3.1.4 to the case where n is any real number. (At the stage of development when Theorem 3.1.4 was proved we were only able to consider the case where n is a rational number.)

Theorem 6.5.2

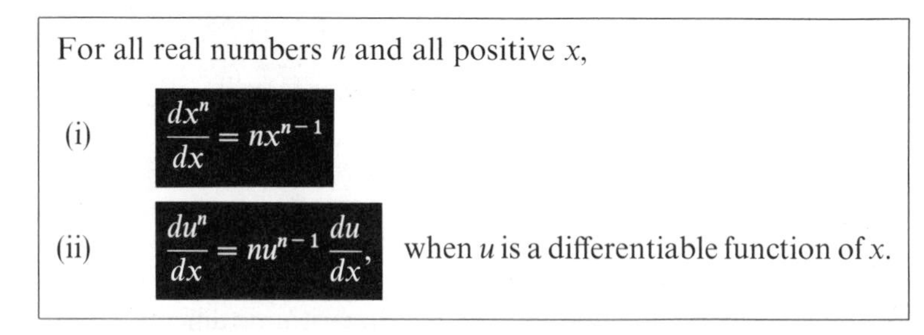

For all real numbers n and all positive x,

(i) $\dfrac{dx^n}{dx} = nx^{n-1}$

(ii) $\dfrac{du^n}{dx} = nu^{n-1}\dfrac{du}{dx}$, when u is a differentiable function of x.

Proof. (i) Let $y = x^n$ where n is any real number. Then $\ln y = n \ln x$. Therefore

$$\frac{1}{y}\frac{dy}{dx} = n\frac{1}{x}, \qquad \text{and so}$$

$$\frac{dy}{dx} = \frac{dx^n}{dx} = y\frac{n}{x} = x^n\frac{n}{x} = nx^{n-1}.$$

(ii) If $y = u^n$, where u is a differentiable function of x, then by part (i), we have $dy/du = nu^{n-1}$. But $dy/dx = (dy/du)(du/dx)$. Hence

$$\frac{dy}{dx} = \frac{du^n}{dx} = nu^{n-1}\frac{du}{dx}.$$

Let C and k be constants and consider the function $f(x) = Ce^{kx}$. Since $\dfrac{df}{dx} = kCe^{kx} = kf(x)$, that function has the property that its derivative is a constant times the function. Stated another way, we have a function of x whose rate of change with respect to x is proportional to the function.

Instances of this sort of behavior occur in the world around us. In particular, under certain conditions, life forms increase in number at a rate proportional to the number present. In some cases the weight of a living organism increases at a rate proportional to the weight present. Radioactive materials decay at a rate proportional to the amount of radioactive material present.

There are many other examples of this phenomenon and we might well wonder whether there is any other function $g(x)$ such that $\dfrac{dg}{dx} = kg(x)$. If $g(x)$ is such a function, let $h(x) = g(x)e^{-kx}$. Then

$$\begin{aligned}\frac{dh}{dx} &= \frac{dg}{dx}e^{-kx} - kg(x)e^{-kx}\\ &= e^{-kx}\left(\frac{dg}{dx} - kg(x)\right)\\ &= 0,\end{aligned}$$

since $\dfrac{dg}{dx} = kg(x)$. Thus $\dfrac{dh}{dx} = 0$, and so $h(x)$ must be a constant, say C_1. Therefore

$$g(x)e^{-kx} = C_1,$$

and so $g(x) = C_1 e^{kx}$. That is, $g(x)$ must also be a constant times e^{kx}. Thus, a constant multiple of e^{kx} is the only function $f(x)$ such that $\frac{df(x)}{dx} = kf(x)$. It should be noted that $f(x)$ is increasing if and only if kC is a positive constant, while $f(x)$ is decreasing if and only if kC is a negative constant.

In summary, if k is a constant and $\frac{df(x)}{dx} = kx$, then $f(x) = Ce^{kx}$, where C is a constant.

Example 4 | A baby whale is growing at a rate proportional to its weight. If the whale weighed 100 lbs at birth and weighs 150 lbs in 1 month, how much will it weigh when it is 1 year old?

Let W be the weight of the whale in pounds and let t be the time in months. Then for some constant k, $\frac{dW}{dt} = kW$; so

$$W = Ce^{kt},$$

where C is a constant. We know that when $t = 0$, $W = 100$; so

$$100 = Ce^0 = C.$$

Thus,

$$W = 100e^{kt}.$$

We also know that $W = 150$ when $t = 1$; so we have

$$150 = 100e^k.$$

Therefore,

$$e^k = \frac{3}{2},$$

and since $e^{kt} = (e^k)^t$, we get

$$W = 100\left(\frac{3}{2}\right)^t.$$

When the whale is 1 year old, $t = 12$ and hence

$$W = 100\left(\frac{3}{2}\right)^{12} \approx 12{,}975 \text{ lbs.}$$ ||

Exercises 6.5

Differentiate each function in Exercises 1–16.

1. $f(x) = (x^2 + 9)^{\sqrt{2}}$.
2. $f(x) = (1 - x^3)^{-\pi}$.
3. $f(x) = x^2(3x^4 - 1)^{1/\sqrt{3}}$.
4. $f(x) = x^3(x + 2)^{-1/\pi}$.
5. $f(x) = 7^x$.
6. $f(t) = e^{7t}$.
7. $f(t) = e^{5t^2 - 3t + 2}$.
8. $f(x) = e^{5 - x^4}$.
9. $f(x) = e^{x^3 + 8}$.
10. $f(x) = e^{(x^2 + 1)^2}$.
11. $f(x) = e^{(x^3 + 2)^2}$.
12. $f(x) = e^{(x - 1)/(x + 1)}$.

13. $f(x) = x^x$.
14. $f(x) = x^{(x^2)}$.
15. $f(x) = (x + 1)^{2x-1}$.
16. $f(x) = x^{(x^x)}$.

Use logarithmic differentiation to differentiate each function in Exercises 17–22.

17. $f(x) = (x + 5)^4(x - 1)^3(x + 3)^2$.
18. $f(x) = \dfrac{(2x - 1)^3(1 - x)^4}{(x + 5)^5}$.
19. $f(x) = \dfrac{(x^2 + 5)^3(1 - x)^5}{(x^3 + 1)^2(x + 4)}$.
20. $f(x) = \dfrac{(x - 3)(x^2 + 7)^5}{(x^4 - 1)(x + 6)}$.
21. $f(x) = \dfrac{(2x + 7)^{3/2}(x^2 + 1)^{3/4}}{(x + 2)^3}$.
22. $f(x) = (4 - x)^{1/3}(x^2 + 1)^{3/5}(2x + 5)^4$.

23. A certain culture of bacteria is increasing at a rate proportional to the number present. If there were 10,000 bacteria at 1 P.M. and 12,000 at 3 P.M., how many would there be at 8 P.M.?

24. The population of a certain city is assumed to be increasing at a rate proportional to the number present. If the population was 30,000 in 1950 and 35,000 in 1970, what will be the city's population in the year 2000?

25. Radium decays at a rate proportional to the mass of the radium that is present. It takes 1700 years for a mass of radium to be reduced to half of its original mass; that is, the *half-life* of radium is 1700 years. A certain supply of radium contained 10 grams in 1940. How many grams were left in 1970?

26. A piece of equipment costing \$100,000 when new is assumed to depreciate with time at a rate proportional to its value. If its value is \$20,000 at the end of 8 years, what would the value of the item be when it is 12 years old?

27. Assume that a certain metal casting is cooling at a rate proportional to the difference in temperature of the casting and the air surrounding it, which is assumed to remain at 70°. If the casting cools from an initial 800° to 700° in one minute, how long will it take to cool to 100°?

28. When a bank pays *continuous interest*, a sum of money in the bank increases at a rate proportional to the amount present, with the constant of proportionality equal to the interest rate expressed as a decimal. If the interest rate is 5% per year (that is, $k = .05$), how much will an initial deposit of \$1000 appreciate to in (a) 1 year, (b) 5 years?

29. Living things absorb radioactive carbon 14. When they die the absorption stops and the carbon 14 decays at a fixed rate. If an animal skeleton has only 1/5 as much carbon 14 as it did when the animal died, how long ago did the animal die? (Assume the half-life of carbon 14 is 5600 years.)

30. Determine all local and absolute extrema and the inflection points of the graph of $y = e^{(-x^2)}$ and sketch the graph.

31. Find the absolute minimum value of $f(x) = x^2 \ln x$.

32. A function, called a *Gompertz function*, that is useful in certain growth or decay problems is $g(t) = ca^{(b^t)}$, where a, b, and c are positive constants, a and b are less than 1, and t is the time. Find $g'(t)$ and $g''(t)$.

33. Suppose that the number N of bacteria in a culture is $N = 50{,}000(.5)^{t/3}$, where t is the time in hours. Find N when $t = 0$, 9, and 12 hours. Find the rate of change of N with respect to the time when $t = 0$ and $t = 3$ hours.

34. Let $f(x) = x^n$, where n is any real number greater than 1. Use the definition of $f'(0)$ to prove that $f'(0) = 0$.

6.6 Integration of Exponential Functions

Let u be a differentiable function of x. Corresponding to the differentiation formula

$$\frac{de^u}{dx} = e^u \frac{du}{dx},$$

we have the integration formula

$$\int e^u \frac{du}{dx}\, dx = e^u + C.$$

Replacing $(du/dx)dx$ by du, we have

$$\int e^u\, du = e^u + C.$$

Similarly, since

$$\frac{da^u}{dx} = a^u \frac{du}{dx} \ln a,$$

we have

$$\frac{d}{dx}\left[\frac{a^u}{\ln a}\right] = a^u \frac{du}{dx}.$$

The corresponding integration formula is

$$\int a^u \frac{du}{dx}\, dx = \frac{a^u}{\ln a} + C, \quad \text{or}$$

$$\int a^u\, du = \frac{a^u}{\ln a} + C.$$

Example 1 $\int e^x\, dx = e^x + C.$ ||

Example 2 $\int e^{2x}\, dx = \frac{1}{2}\int e^{2x}(2\, dx) = \frac{1}{2}e^{2x} + C.$ ||

Example 3 $\int e^{x^2-5} 2x\, dx = e^{x^2-5} + C.$ ||

Example 4 $\int 7xe^{x^2-5}\, dx = \frac{7}{2}\int e^{x^2-5} 2x\, dx = \frac{7}{2}e^{x^2-5} + C.$ ||

Example 5 $\int x^2 e^{x^3}\, dx = \frac{1}{3}\int e^{x^3} 3x^2\, dx = \frac{1}{3}e^{x^3} + C.$ ||

Example 6 $\int (x-1)e^{x^2-2x+3}\, dx = \frac{1}{2}e^{x^2-2x+3} + C.$ ||

Example 7 $\int e^{-3x}\,dx = -\frac{1}{3}\int e^{-3x}(-3)\,dx = -\frac{e^{-3x}}{3} + C.$ ||

Example 8 $\int \frac{e^{1/x}}{x^2}\,dx = -\int e^{1/x}\left(-\frac{1}{x^2}\right)dx = -e^{1/x} + C.$ ||

Example 9 $\int 5^{x^2+7x}(2x+7)\,dx = \frac{1}{\ln 5}5^{x^2+7x} + C.$ ||

Exercises 6.6

Evaluate each integral in Exercises 1–20.

1. $\int e^{3x}\,dx.$
2. $\int e^{x^2+1}2x\,dx.$
3. $\int e^{-t}\,dt.$
4. $\int_0^2 xe^{5-x^2}\,dx.$
5. $\int_0^2 e^{6x-3x^2}(1-x)dx.$
6. $\int e^{-11x}\,dx.$
7. $\int(x^2+2x)e^{x^3+3x^2}\,dx.$
8. $\int(5x^2+10x)e^{x^3+3x^2}\,dx.$
9. $\int \frac{e^{1/x^2}}{x^3}\,dx.$
10. $\int 3^x\,dx.$
11. $\int_0^1 5^{3x}\,dx.$
12. $\int(e^x - e^{-x})^2\,dx.$
13. $\int_0^2 \frac{dx}{e^{x/2}}.$
14. $\int \frac{e^x\,dx}{e^x+3}.$
15. $\int \frac{(e^{3x}+1)^2}{e^{6x}}\,dx.$
16. $\int \frac{e^{x^{1/2}}}{x^{1/2}}\,dx.$
17. $\int \frac{x\,dx}{e^{(x^2)}}.$
18. $\int \frac{e^{1/x^2}}{x^3}\,dx.$
19. $\int \frac{e^{-x}}{5-e^{-x}}\,dx.$
20. $\int \frac{(e^x-1)^2}{e^{2x}}\,dx.$
21. Find the area of the region bounded by the graph of $y = (e^x - e^{-x})/2$, the x-axis, and the lines $x = 0$ and $x = \ln 3$.
22. Find the volume obtained when the region bounded by $y = e^x$, the x-axis, and the lines $x = 0$ and $x = \ln 5$ is rotated about the x-axis.
23. Find the area of the largest rectangle with one side on the x-axis that can be inscribed under the graph of $y = e^{(-x^2)}$.
24. Find the area of the surface generated when the graph of $y = (e^x + e^{-x})/2$, $0 \le x \le 2$, is revolved about the x-axis.
25. Do Exercise 24 for the case where the revolution is about the y-axis.

6.7 Integration by Parts

The integration formula corresponding to the differentiation formula for a product is called the *integration by parts* formula. Let u and v be differentiable functions of x.

Then

$$\frac{d}{dx}(uv) = u\frac{dv}{dx} + v\frac{du}{dx}, \quad \text{and so}$$

$$\int \left[u\frac{dv}{dx} + v\frac{du}{dx} \right] dx = uv + C.$$

Since the integral of a sum of two functions is the sum of the integrals of the functions, we can express the preceding formula thus:

$$\int u\frac{dv}{dx}\,dx + \int v\frac{du}{dx}\,dx = uv + C.$$

Since

$$\frac{dv}{dx}dx = dv \quad \text{and} \quad \frac{du}{dx}dx = du,$$

we have

$$\int u\,dv + \int v\,du = uv + C.$$

We now subtract $\int v\,du$ from both sides to obtain

$$\int u\,dv = uv - \int v\,du + C. \tag{1}$$

That formula is important because it enables us to find $\int u\,dv$ by finding $\int v\,du$, which is very helpful when $\int v\,du$ is easier to find than $\int u\,dv$.

Example 1 | Find $\int xe^x\,dx$.

We have no formula to evaluate this directly, but we can do it by integration by parts. Let $u = x$ and $dv = e^x\,dx$; then $du = dx$ and we take $v = e^x$ in order to have $dv = e^x\,dx$.

Thus, $uv = xe^x$ and $v\,du = e^x\,dx$. Using (1), we get

$$\int xe^x\,dx = xe^x - \int e^x\,dx + C.$$

Now $\int e^x\,dx = e^x + C_1$, where C_1 is a constant; so we have

$$\int xe^x\,dx = xe^x - e^x - C_1 + C.$$

But $-C_1 + C = C_2$, where C_2 is a constant; hence

$$\int xe^x\,dx = xe^x - e^x + C_2. \qquad ||$$

Notice how the constant C of formula (1) combines with the constant C_1 obtained when we evaluate $\int v\,du$ to produce another constant C_2. In such cases, the same result would be obtained if the C were omitted, because a constant occurs when $\int v\,du$ is evaluated. Consequently, for simplicity, formula (1) is usually written without the C as follows:

$$\int u\,dv = uv - \int v\,du. \tag{2}$$

However, we shall see cases in the next chapter where the C is needed. In such cases we shall still use formula (2) and add the C when it is convenient.

Since $dv = v'\,dx$ and $du = u'\,dx$, the formula can also be expressed in the form

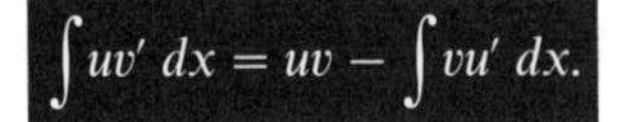

$$\int uv'\,dx = uv - \int vu'\,dx.$$

Example 2 | Find $\int (x + 3)e^{x-2}\,dx$.

Let $u = x + 3$ and $dv = e^{x-2}\,dx$. Then $du = dx$ and we take $v = e^{x-2}$. We then have

$$\begin{aligned}\int (x + 3)e^{x-2}\,dx &= (x + 3)e^{x-2} - \int e^{x-2}\,dx \\ &= (x + 3)e^{x-2} - e^{x-2} + C.\end{aligned}$$

||

It is important to realize that when using integration by parts to integrate a product, one has a choice for u and dv. To illustrate the choice, let us consider the integral of Example 1 again: $\int xe^x\,dx$. We were able to evaluate this integral fairly easily by taking $u = x$ and $dv = e^x\,dx$. But we could have also chosen $u = e^x$ and $dv = x\,dx$. With that choice we have $du = e^x\,dx$ and $v = x^2/2$. Thus, using (2) we have

$$\int xe^x\,dx = \frac{x^2}{2}e^x - \int \frac{x^2}{2}e^x\,dx.$$

Here, instead of $\int (x^2/2)e^x\,dx$ being easier to evaluate than the original integral, it is at least as difficult.

When using integration by parts, we should try to have $\int v\,du$ turn out to be simpler to evaluate than $\int u\,dv$. If one choice does not produce such a result, then another choice should be made for u and dv. Of course it could happen that no choice helps in the integration. In that situation the technique of integration by parts fails, and some other technique should be attempted.

Example 3 | Find $\int \ln|x|dx$.

Let $u = \ln|x|$ and $dv = dx$. Then $du = (1/x)dx$, $v = x$, and consequently $v\,du = x(1/x)dx = dx$,

So we have
$$\int \ln|x|dx = x\ln|x| - \int dx.$$

Thus
$$\int \ln|x|dx = x\ln|x| - x + C.$$

||

Example 4 | Find $\int x\ln|x^2|dx$.

Let $u = \ln|x^2|$ and $dv = x\,dx$. Then $du = (2x/x^2)dx$, $v = x^2/2$, and $v\,du = x\,dx$.

Therefore
$$\int x \ln|x^2| dx = \frac{x^2}{2} \ln|x^2| - \int x\, dx$$
$$= \frac{x^2}{2} \ln|x^2| - \frac{x^2}{2} + C. \qquad \|$$

Example 5 | Find $\int x \ln|x| dx$.

Let $u = \ln|x|$ and $dv = x\, dx$. Then $du = (1/x)dx$, $v = x^2/2$, and $v\, du = (x/2)dx$. Hence

$$\int x \ln|x| dx = \frac{x^2}{2} \ln|x| - \int \frac{x}{2} dx$$
$$= \frac{x^2}{2} \ln|x| - \frac{x^2}{4} + C. \qquad \|$$

When going from dv to v it is not necessary to add a constant of integration, although there are occasions where adding a particular constant is helpful. Let us see how a constant added at the point of going from dv to v is eliminated, by re-evaluating the integral of Example 5.

Example 6 | Find $\int x \ln|x| dx$.

Let $u = \ln|x|$ and $dv = x\, dx$. Then $du = (1/x)dx$ and $v = x^2/2 + C_1$; hence

$$\int x \ln|x| dx = \frac{x^2}{2} \ln|x| + C_1 \ln|x| - \int \left(\frac{x}{2} + \frac{C_1}{x}\right) dx$$
$$= \frac{x^2}{2} \ln|x| + C_1 \ln|x| - \frac{x^2}{4} - C_1 \ln|x| + C.$$

The $C_1 \ln|x|$ is eliminated and we have the same result as in Example 5,

$$\int x \ln|x| dx = \frac{x^2}{2} \ln|x| - \frac{x^2}{4} + C. \qquad \|$$

Sometimes it is necessary to use integration by parts more than once to evaluate an integral, as illustrated in the next example.

Example 7 | Find $\int x^2 e^x\, dx$.

Let $u = x^2$ and $dv = e^x\, dx$. Then $du = 2x\, dx$ and $v = e^x$; so

$$\int x^2 e^x\, dx = x^2 e^x - 2 \int x e^x\, dx.$$

Now, even though $\int x e^x\, dx$ is simpler than the original integral, we still need integration by parts to evaluate it. As in Example 1, we let $u = x$ and $dv = e^x\, dx$. Then

$du = dx$ and $v = e^x$; so we have

$$\int x^2e^x\,dx = x^2e^x - 2xe^x + 2\int e^x\,dx$$
$$= x^2e^x - 2xe^x + 2e^x + C. \qquad \|$$

Example 7 is a specific case of a more general technique where several successive integrations by parts can be used to evaluate an integral. The technique applies to integrals of the form $\int p(x)q(x)dx$, where p is a polynomial of degree n in x, and q is a function whose indefinite integral is easily obtained. Specifically, it can be shown by repeated application of the integration by parts formula that

$$\int pq\,dx = pq_1 - p'q_2 + p''q_3 - \cdots + (-1)^n p^{(n)}q_{n+1} + C, \tag{3}$$

where $q_1' = q$, $q_2' = q_1, \ldots, q_{n-1}' = q_n$, and $p^{(n)}$ is the nth derivative of p.

That is, $\int pq\,dx$ is equal to a sum of terms starting with pq_1, where each successive term is obtained by differentiating p one more time, integrating q one more time, and alternating the signs between terms.

Example 8 | Find $\int(x^3 - 2x^2 - 3)e^{3x}\,dx$.

Applying (3) with $p(x) = x^3 + 2x^2 - 3$ and $q(x) = e^{3x}$, we obtain

$$(x^3 + 2x^2 - 3)\frac{e^{3x}}{3} - (3x^2 + 4x)\frac{e^{3x}}{9} + (6x + 4)\frac{e^{3x}}{27} - 6\frac{e^{3x}}{81} + C.$$

Note that here, $q_1 = \dfrac{e^{3x}}{3}$, $q_2 = \dfrac{1}{3}\dfrac{e^{3x}}{3} = \dfrac{e^{3x}}{9}$, and so forth. $\|$

Exercises 6.7

Use integration by parts to evaluate each of the integrals in Exercises 1–14.

1. $\int x^7 \ln|x|dx$.
2. $\int x^2 \ln|x|dx$.
3. $\int x^3e^x\,dx$.
4. $\int xe^{5x}\,dx$.
5. $\int xe^{-2x}\,dx$.
6. $\int x^2 \ln|x^5|dx$.
7. $\int x \ln^2|x|dx\ (=\int x(\ln|x|)^2\,dx)$.
8. $\int x \ln^3|x|dx\ (=\int x(\ln|x|)^3\,dx)$.
9. $\int x^{-2} \ln|x|dx$.
10. $\int(x + 7)e^{2x-1}\,dx$.
11. $\int x^2e^{3x}\,dx$.
12. $\int \dfrac{x^2}{e^x}\,dx$.
13. $\int \ln x\,dx$.
14. $\int x^3 \ln(x^4)dx$.

Use formula (3) to evaluate each of the integrals in Exercises 15–20.

15. $\int x^5e^x\,dx$.
16. $\int x^4e^x\,dx$.
17. $\int(x^4 - x^3)e^{2x}\,dx$.
18. $\int(x^5 - 2x^4 + 3x^2 - 7)e^{-x}\,dx$.
19. $\int(x^7 + 3x^4)e^{-2x}\,dx$.
20. $\int(x^3 - 8x^2 + 5)e^{3x}\,dx$.

6.8 The Hyperbolic Functions

Hyperbolic functions are certain combinations of exponential functions and occur often in various applications. For example, the shape of a hanging cable is represented by a hyperbolic function. These functions are similar in many respects to the trigonometric functions. In fact, they play the same role relative to triangles in Hyperbolic Geometry that the trigonometric functions play in Euclidean Geometry. Because of the analogous role, the symbols for the hyperbolic function are also similar: sinh, cosh, tanh, coth, sech, and csch. In this section we shall discuss the basic properties of these functions.

The hyperbolic functions are defined as follows.

Definition 6.8.1

(a) hyperbolic sine: $\sinh x = \dfrac{e^x - e^{-x}}{2}$.

(b) hyperbolic cosine: $\cosh x = \dfrac{e^x + e^{-x}}{2}$.

(c) hyperbolic tangent: $\tanh x = \dfrac{e^x - e^{-x}}{e^x + e^{-x}} = \dfrac{\sinh x}{\cosh x}$.

(d) hyperbolic cotangent: $\coth x = \dfrac{\cosh x}{\sinh x}, \quad x \neq 0$.

(e) hyperbolic secant: $\operatorname{sech} x = \dfrac{1}{\cosh x}$.

(f) hyperbolic cosecant: $\operatorname{csch} x = \dfrac{1}{\sinh x}, \quad x \neq 0$.

The hyperbolic function for the shape of a hanging cable, called a *catenary*, is $a \cosh(x/a)$, where a is a constant.

Various identities involving the hyperbolic functions follow directly from the definitions. In the following theorem we list the most basic ones.

Theorem 6.8.1

(a) $\cosh^2 x - \sinh^2 x = 1$
(b) $\operatorname{sech}^2 x = 1 - \tanh^2 x$
(c) $\operatorname{csch}^2 x = \coth^2 x - 1$
(d) $\cosh(x + y) = \cosh x \cosh y + \sinh x \sinh y$
(e) $\sinh(x + y) = \sinh x \cosh y + \cosh x \sinh y$
(f) $\sinh(-x) = -\sinh x$
(g) $\cosh(-x) = \cosh x$.

Proof.

(a) From the definitions it follows that

$$\cosh^2 x = \frac{1}{4}(e^x + e^{-x})^2 = \frac{1}{4}(e^{2x} + 2 + e^{-2x})$$

and, $\sinh^2 x = \frac{1}{4}(e^x - e^{-x})^2 = \frac{1}{4}(e^{2x} - 2 + e^{-2x}).$

Consequently $\cosh^2 x - \sinh^2 x = 1.$

The proofs of the remaining identities will be left as exercises.

The derivatives of the hyperbolic functions also follow directly from their definitions. For example,

$$\frac{d}{dx}\sinh x = \frac{d}{dx}\frac{e^x - e^{-x}}{2} = \frac{e^x + e^{-x}}{2} = \cosh x,$$

and

$$\frac{d}{dx}\cosh x = \frac{d}{dx}\frac{e^x + e^{-x}}{2} = \frac{e^x - e^{-x}}{2} = \sinh x.$$

Thus if u is a differentiable function of x, we have:

$$\frac{d}{dx}\sinh u = \cosh u \frac{du}{dx},$$

$$\frac{d}{dx}\cosh u = \sinh u \frac{du}{dx}.$$

For convenience we list the differentiation formulas for all hyperbolic functions in the next theorem. The proofs of the last four are left as exercises.

Theorem 6.8.2

If u is a differentiable function of x, then:

(i) $\frac{d}{dx}\sinh u = \cosh u \frac{du}{dx}$

(ii) $\frac{d}{dx}\cosh u = \sinh u \frac{du}{dx}$

(iii) $\frac{d}{dx}\tanh u = \operatorname{sech}^2 u \frac{du}{dx}$

(iv) $\frac{d}{dx}\coth u = -\operatorname{csch}^2 u \frac{du}{dx}, \quad u \neq 0$

(v) $\frac{d}{dx}\operatorname{sech} u = -\operatorname{sech} u \tanh u \frac{du}{dx}$

(vi) $\frac{d}{dx}\operatorname{csch} u = -\operatorname{csch} u \coth u \frac{du}{dx}, \quad u \neq 0.$

Example 1 | Compute the derivative of $\cosh(x^2 + 3x)$.

$$\frac{d}{dx}\cosh(x^2 + 3x) = \sinh(x^2 + 3x)\frac{d(x^2 + 3x)}{dx} = (2x + 3)\sinh(x^2 + 3x).$$ ||

Example 2 Compute the derivative of $\tanh^3(\sin x)$.

$$\frac{d}{dx}\tanh^3(\sinh x) = 3\tanh^2(\sinh x)\frac{d}{dx}\sinh x$$

$$= 3(\cosh x)(\tanh^2(\sinh x)). \qquad \|$$

Our familiarity with the exponential function enables us to sketch the graphs of the hyperbolic cosine and the hyperbolic sine. For example, note that since $\cosh x = (1/2)(e^x + e^{-x})$, $\cosh x$ is simply the average of e^x and e^{-x}. Consequently the graph of $y = \cosh x$ falls exactly midway between the graphs of $y = e^x$ and $y = e^{-x}$. That fact is illustrated in Figure 6.8.1.

Figure 6.8.1

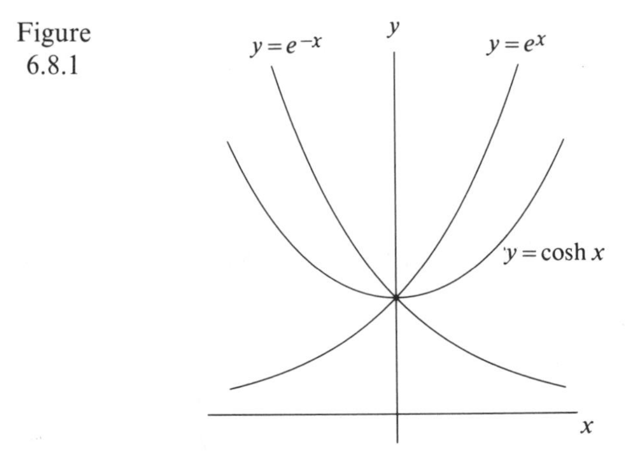

The graph of $y = \sinh x$ is illustrated in Figure 6.8.2. This graph may be obtained by observing that $\sinh x$ is the average of e^x and $-e^{-x}$.

Figure 6.8.2

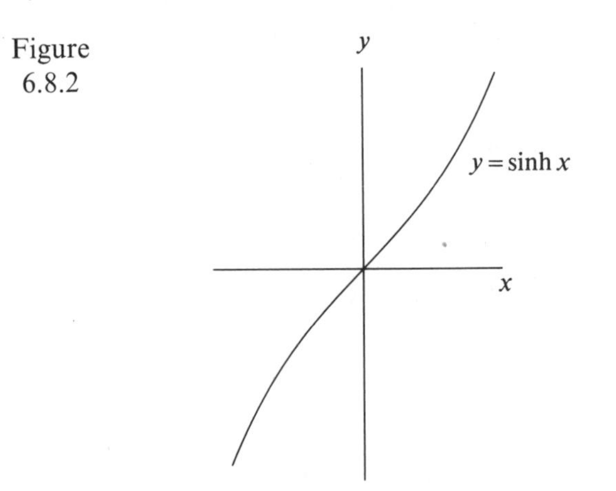

Each of the differentiation formulas of Theorem 6.8.2 has a corresponding integration formula. For example, since $\frac{d}{dx}\sinh u = \cosh u\frac{du}{dx}$, we have

$$\int \cosh u\, du = \sinh u + C.$$

The remaining integration formulas are listed for convenience:

$$\int \sinh u\, du = \cosh u + C$$

$$\int \operatorname{sech}^2 u\, du = \tanh u + C$$

$$\int \operatorname{csch}^2 u\, du = -\coth u + C$$

$$\int \operatorname{sech} u \tanh u\, du = -\operatorname{sech} u + C$$

$$\int \operatorname{csch} u \coth u\, du = -\operatorname{csch} u + C.$$

Example 3 $\int \sinh(2x + 7)dx = \frac{1}{2}\int \sinh(2x + 7)2\, dx = \frac{1}{2}\cosh(2x + 7) + C.$ ||

Example 4 $\int 2x \cosh(x^2 - 7)dx = \sinh(x^2 - 7) + C.$ ||

Example 5 $\int \tanh x \operatorname{sech}^2 x\, dx = \int \tanh x\, d(\tanh x) = \frac{1}{2}\tanh^2 x + C.$ ||

Example 6 $\int \sinh^3 x \cosh x\, dx = \int \sinh^3 x\, d(\sinh x) = \frac{1}{4}\sinh^4 x + C.$ ||

Example 7 $\int \coth x\, dx = \int \frac{\cosh x}{\sinh x} dx = \int \frac{d(\sinh x)}{\sinh x} = \ln|\sinh x| + C.$ ||

Integration by parts can be used to evaluate certain integrals involving hyperbolic functions.

Example 8 Evaluate $\int x \sinh x\, dx$. Let $u = x$ and $dv = \sinh x\, dx$. Then $du = dx$ and $v = \cosh x$; so

$$\begin{aligned}\int x \sinh x\, dx &= x \cosh x - \int \cosh x\, dx \\ &= x \cosh x - \sinh x + C.\end{aligned}$$ ||

Exercises 6.8

In Exercises 1–8 prove the given identity.

1. $\operatorname{sech}^2 x = 1 - \tanh^2 x.$
2. $\operatorname{csch}^2 x = \coth^2 x - 1.$
3. $\cosh(x + y) = \cosh x \cosh y + \sinh x \sinh y.$
4. $\sinh(x + y) = \sinh x \cosh y + \cosh x \sinh y.$
5. $\sinh(-x) = -\sinh x.$
6. $\cosh(-x) = \cosh x.$
7. $\sinh 2x = 2 \sinh x \cosh x.$
8. $\cosh 2x = \cosh^2 x + \sinh^2 x.$
9. Prove parts (iii), (iv), (v), and (vi) of Theorem 6.8.2.

In Exercises 10–21 compute the derivative of the given function.

10. $f(x) = \sinh(3 - x^3)$.
11. $f(x) = \sinh(x^2 + 5x)$.
12. $f(x) = \cosh^2(x^2 - 1)$.
13. $f(x) = \cosh^3(3x + 2)$.
14. $f(x) = \tanh^2(x + 5)$.
15. $f(x) = \coth(4x^2 + 7x)$.
16. $f(x) = \operatorname{csch}^2(\ln x)$.
17. $f(x) = \operatorname{sech}^5(\sqrt{2x - 1})$.
18. $f(x) = e^x \sinh(2x)$.
19. $f(x) = e^{2x} \cosh(3x)$.
20. $f(x) = \sinh^2 2x \cosh^3 x$.
21. $f(x) = \cosh x \coth x$.

In Exercises 22–39 compute the given integral.

22. $\int_1^2 \cosh(x - 1)dx$.
23. $\int \sinh(2x + 4)dx$.
24. $\int e^x \sinh(e^x)dx$.
25. $\int \frac{1}{x} \cosh(\ln x)dx$.
26. $\int x \operatorname{sech}^2(3x^2 + 1)dx$.
27. $\int \tanh x \operatorname{sech}^2 x\, dx$.
28. $\int x \cosh x\, dx$.
29. $\int_0^1 x \sinh x^2\, dx$.
30. $\int \coth x \operatorname{csch}^2 x\, dx$.
31. $\int \sinh^3 x \cosh x\, dx$.
32. $\int x^2 \cosh x^3\, dx$.
33. $\int x^2 \cosh x\, dx$.
34. $\int \frac{1}{x} \operatorname{sech}^2(\ln x)dx$.
35. $\int e^x \cosh(e^x)dx$.
36. $\int \frac{\sinh x}{\cosh^5 x} dx$.
37. $\int_1^2 \frac{\cosh x}{\sinh^4 x} dx$.
38. $\int_0^1 \tanh x\, dx$.
39. $\int \coth x\, dx$.

40. Show that $y = a\cosh(x/a)$ satisfies the differential equation $y'' = \frac{1}{a}\sqrt{1 + (y')^2}$. This differential equation expresses the equilibrium condition for the forces acting on a homogenous hanging cable. For this reason a hanging cable will have the shape of the graph of $y = a\cosh(x/a)$. The graph of $y = a\cosh(x/a)$ is called a catenary.

41. Evaluate the integral $\int \sinh ax \cosh bx\, dx$, $a \neq b, a \neq 0, b \neq 0$. (*Hint*: Apply integration by parts twice and use the resulting equation.)

42. Show that the graph of the parametric equations $x = \cosh t$, $y = \sinh t$ is the right half of the hyperbola $x^2 - y^2 = 1$.

Brief Review of Chapter 6

1. The Exponential Function

Definition. Let a be a positive real number, let r be any fixed real number, and restrict t to the set of rational numbers. Then

$$a^r = \lim_{t \to r} a^t.$$

Important Facts. The function a^x is continuous for all x. If $a > 1$, then a^x is increasing for all x, and $0 < a^x < 1$ for all $x < 0$, and $a^x > 1$ for all $x > 0$. If $0 < a < 1$, then a^x is decreasing for all x, and $a^x > 1$ for all $x < 0$, and $0 < a^x < 1$ for all $x > 0$. The function Ce^{kx} is useful in certain growth and decay problems.

Differentiation Formulas.

$$\frac{da^u}{dx} = a^u(\ln a)\frac{du}{dx}; \qquad \frac{de^u}{dx} = e^u \frac{du}{dx}.$$

Integration Formulas.

$$\int e^u \, du = e^u + C; \qquad \int a^u \, du = \frac{a^u}{\ln a} + C.$$

2. Inverse Functions

Definition. Let $f(x)$ be a function with domain $a \leq x \leq b$. A function $g(y)$ such that $g(f(x)) = x$, for $a \leq x \leq b$, and $f(g(y)) = y$, for y in the range of f, is called the *inverse* of $f(x)$ on $a \leq x \leq b$.

Important Facts. With suitable restrictions on the domains, the inverse of $y = f(x)$, if it exists, is the solution of $y = f(x)$ for x in terms of y. If $f(x)$ is continuous and monotone on the interval $a \leq x \leq b$, it has an inverse function $g(y)$ that is continuous and monotone on the interval between $f(a)$ and $f(b)$. The inverse function $g(y)$ is increasing or decreasing according as $f(x)$ is increasing or decreasing.

3. The Logarithmic Function

Definition. When $a > 0$ and $a \neq 1$, the inverse of the exponential function $y = a^x$ is called the logarithmic function with base a. The logarithm, with base a, of x is denoted by $\log_a x$.

Important Facts. The function $\log_a x$ is continuous for all positive x, and is increasing for $a > 1$ and decreasing for $0 < a < 1$. For all positive real numbers r and s,

$$\log_a(rs) = \log_a r + \log_a s; \quad \log_a \frac{r}{s} = \log_a r - \log_a s; \quad \log_a(r^s) = s \log_a r.$$

Differentiation Formulas.

$$\frac{d}{dx}\ln|u| = \frac{1}{u}\frac{du}{dx}; \qquad \frac{d}{dx}\log_a|u| = \frac{1}{u}\frac{du}{dx}\log_a e.$$

Integration Formula.

$$\int \frac{du}{u} = \ln|u| + C.$$

4. Logarithmic Differentiation

Meaning. The method of finding dy/dx when $y = f(x)$ by first writing $\ln|y| = \ln|f(x)|$, then differentiating to get $((1/y)(dy/dx)) = (d/dx)\ln|f(x)|$, and finally multiplying both sides by y to get dy/dx.

Uses. Especially useful for differentiating functions of the form $p(x)^{q(x)}$ or for differentiating complicated products or quotients.

5. Integration by Parts

Meaning. The use of the integration formula

$$\int u\,dv = uv - \int v\,du.$$

Uses. To find $\int u\,dv$ by finding $\int v\,du$ when $\int v\,du$ is more easily evaluated. The choices of u and v are important. Sometimes integration by parts must be used more than once to evaluate an integral.

6. The Hyperbolic Functions

Definitions.

$$\sinh x = \frac{e^x - e^{-x}}{2} \qquad \coth x = \frac{\cosh x}{\sinh x}, \quad x \neq 0$$

$$\cosh x = \frac{e^x + e^{-x}}{2} \qquad \operatorname{sech} x = \frac{1}{\cosh x}$$

$$\tanh x = \frac{\sinh x}{\cosh x} \qquad \operatorname{csch} x = \frac{1}{\sinh x}, \quad x \neq 0$$

Basic Identities.

$$\cosh^2 x - \sinh^2 x = 1 \qquad \operatorname{sech}^2 x = 1 - \tanh^2 x$$

$$\operatorname{csch}^2 x = \coth^2 x - 1$$

$$\cosh(x + y) = \cosh x \cosh y + \sinh x \sinh y$$

$$\sinh(x + y) = \sinh x \cosh y + \cosh x \sinh y$$

$$\sinh(-x) = -\sinh x \qquad \cosh(-x) = \cosh x$$

Differentiation Formulas.

$$\frac{d}{dx}\sinh u = \cosh u\frac{du}{dx} \qquad \frac{d}{dx}\coth u = -\operatorname{csch}^2 u\frac{du}{dx}, \quad u \neq 0$$

$$\frac{d}{dx}\cosh u = \sinh u\frac{du}{dx} \qquad \frac{d}{dx}\operatorname{sech} u = -\operatorname{sech} u \tanh u\frac{du}{dx}$$

$$\frac{d}{dx}\tanh u = \operatorname{sech}^2 u\frac{du}{dx} \qquad \frac{d}{dx}\operatorname{csch} u = -\operatorname{csch} u \coth u\frac{du}{dx}, \quad u \neq 0$$

Integration Formulas.

$\int \sinh u \, du = \cosh u + C$ $\int \cosh u \, du = \sinh u + C$

$\int \operatorname{sech}^2 u \, du = \tanh u + C$ $\int \operatorname{csch}^2 u \, du = -\coth u + C$

$\int \operatorname{sech} u \tanh u \, du = -\operatorname{sech} u + C$ $\int \operatorname{csch} u \coth u \, du = -\operatorname{csch} u + C$

Technique Review Exercises, Chapter 6

1. Find the inverse function (and its domain) of $x^2 + 3$, $-7 \le x \le -1$.
2. Find y' if (a) $y = e^{(3x^2 - 2x + 1)}$, (b) $y = 6^{1-3x}$.
3. Find y' if $y = x^{3x-1}$.
4. Find y' if $y = \ln|x^3 - 2|^5$.
5. Evaluate $\int \frac{e^{2/x}}{x^2} dx$.
6. Evaluate $\int x^3 \ln(x^2) dx$.
7. Evaluate $\int x^2 e^{3x} dx$.
8. Evaluate $\int \frac{x \, dx}{3x^2 + 5}$.
9. Find y' if $y = \sinh^3(x^2 + 5)$.
10. Evaluate $\int \sinh^3(7x) \cosh(7x) dx$.

Additional Exercises, Chapter 6

Section 6.1

Find the value of each expression.

1. $64^{2/3}$.
2. $(27)^{2/3}$.
3. $(8/64)^{-2/3}$.
4. $(16)^{-3/4}$.

Sketch the graph of each equation by plotting three well-chosen points.

5. $y = 3^x$.
6. $y = 2^{-x}$.
7. $y = (1/3)^x$.
8. $y = 3^x + 1$.

Section 6.2

Find the inverse of each function, and find the domain of the inverse function.

9. $6x - 1$, $-1 \le x \le 3$.
10. $x^2 + 5$, $x \le 0$.
11. $x^2 + 5$, $x \ge 0$.
12. $\sqrt{6 - x}$, $-3 < x < 5$.

Section 6.3

Evaluate each expression in Exercises 13–15.

13. $\log .0001$.
14. $\log_3 27$.
15. $\log_2(1/16)$.
16. Sketch the graph of $y = \log_5 x$.
17. Sketch the graph of $y = \log_{1/5} x$.

Section 6.4

Differentiate each function in Exercises 18–23.

18. $f(x) = \ln(8x^3 - 3x + 1)$.
19. $f(x) = \ln(16x^7 + 2x^5 - 8x)$.
20. $f(x) = \log(x^2 - x + 1)$.
21. $f(x) = \log_3|7 - x|$.
22. $f(x) = \ln|x^2 - x - 18|$.
23. $f(x) = \ln|2 - x - x^3|^4$.

Evaluate each integral in Exercises 24–27.

24. $\int \frac{4x^3 + 3}{x^4 + 3x - 2}\, dx$.
25. $\int_2^5 \frac{dx}{1 - x}$.
26. $\int \frac{(\ln x)^2}{x}\, dx$.
27. $\int \frac{dx}{x(\ln x)^2}$.

Section 6.5

Differentiate each function.

28. $f(x) = (x^5 - x)^\pi$.
29. $f(x) = 16^x$.
30. $f(x) = e^{7x - 8x^2}$.
31. $f(x) = x^{9x}$.
32. $f(x) = (x - 7)^2(x + 5)^4(x - 3)^3$.
33. $f(x) = \frac{(x + 17)^4}{(x - 6)^5(x - 8)^9}$.
34. $(3 + x)^{x-2}$.

Section 6.6

Evaluate each integral.

35. $\int e^{15x}\, dx$.
36. $\int xe^{1-x^2}\, dx$.
37. $\int x^3 e^{x^4+5}\, dx$.
38. $\int 9^x\, dx$.
39. $\int \frac{e^{1/x^3}}{x^4}\, dx$.
40. $\int \frac{e^{5x}}{e^{5x} - 2}\, dx$.
41. $\int_0^1 \frac{t^2\, dt}{e^{(t^3)}}$.
42. $\int \frac{(e^{-t} + 3)^2}{e^{2t}}\, dt$.

Section 6.7

Evaluate each integral.

43. $\int xe^{4x}\, dx$.
44. $\int xe^{-3x}\, dx$.
45. $\int x^{10} \ln|x| dx$.
46. $\int x^2 e^{2x}\, dx$.
47. $\int x^4 \ln|x^3| dx$.
48. $\int x \ln^2|3x| dx$.
49. $\int x^3 e^{3x}\, dx$.
50. $\int (x - 2)e^{1-2x}\, dx$.

Section 6.8

Differentiate each function in Exercises 51–56.

51. $f(x) = \cosh(4x - 3)$.
52. $f(x) = \cosh(x^2 + 7x)$.
53. $f(x) = \sinh^5(x^2 - x)$.
54. $f(x) = \sinh^7(x^3 + 5)$.
55. $f(x) = \tanh^4(2 - x^2)$.
56. $f(x) = \operatorname{sech}^3(x^2 + 3)$.

Evaluate each integral in Exercises 57–60.

57. $\int x \sinh(5 - x^2) dx$.
58. $\int x^2 \cosh(x^3 + 2) dx$.
59. $\int \operatorname{sech}(3 - 2x) \tanh(3 - 2x) dx$.
60. $\int \sinh(5x) \cosh^3(5x) dx$.

Challenging Problems, Chapter 6

1. Show that $\lim\limits_{t\to 0} \dfrac{x^t - 1}{t} = \ln x, \quad x > 0.$

2. Prove that the function $f(x) = \left(1 + \dfrac{1}{x}\right)^x$ is increasing for all $x > 0$. (*Hint*: To show that $\ln\left(1 + \dfrac{1}{x}\right) > \dfrac{1/x}{1 + 1/x}$ for all $x > 0$, show that $\dfrac{1}{z} + \ln z > 1$ for all $z > 1$.)

3. Let $f(x)$ be continuous for $a \le x \le b$, and suppose g is the inverse of the function f. Prove that

$$\int_a^b f(x)dx = bf(b) - af(a) - \int_{f(a)}^{f(b)} g(y)dy,$$

and use that result to evaluate $\int_1^e \ln x\, dx$.

4. Explain the result of Problem 3 graphically.

5. Although we defined $\ln x$ to be the inverse of the exponential function with base e, it could have been defined by $\ln x = \int_1^x \dfrac{dt}{t}$, $x > 0$. Using that definition, prove each of the following (assume $a > 0$, $b > 0$):
   (a) $\ln 1 = 0$.
   (b) $\dfrac{d}{dx} \ln|x| = \dfrac{1}{x}, \quad x \neq 0.$
   (c) $\int_a^b \dfrac{dt}{t} = \ln b - \ln a.$
   (d) $\ln(ab) = \ln a + \ln b$. (*Hint*: Use (c) and consider $\ln(ab) - \ln a$.)
   (e) $\ln(a^p) = p \ln a$. (*Hint*: Let $t = z^p$.)

6. Show that if $y = \dfrac{e^x - e^{-x}}{2}$, then $x = \ln(y + \sqrt{y^2 + 1})$.

7

Trigonometric and Inverse Trigonometric Functions

7.1 The Derivative of the Sine Function

In order to find the derivative of $\sin x$ we shall need the fact that if h is in radians, then

$$\lim_{h \to 0} \frac{\sin h}{h} = 1. \tag{1}$$

To obtain that result, the familiar formula $A = \pi r^2$ for the area A of the region interior to a circle of radius r will be used. That area formula is most easily obtained by using the methods of calculus. However, it is easy to fall into a reasoning trap when evaluating an integral for A by using integration formulas that depend on the formula for the derivative of $\sin x$. The reasoning would be faulty because the formula for A would then be used to get the formula for A and the reasoning would be truly "circular." The derivation of the formula $A = \pi r^2$ without faulty reasoning is somewhat long and involved. It is outlined in Challenging Problems 3 and 4, Chapter 7, for the interested reader.

Another result that we shall need for differentiating $\sin x$ is the trigonometric identity

$$\sin(\alpha + \beta) - \sin(\alpha - \beta) = 2 \cos \alpha \sin \beta. \tag{2}$$

It can be obtained by subtracting these familiar identities:

$$\sin(\alpha + \beta) = \sin \alpha \cos \beta + \cos \alpha \sin \beta,$$

$$\sin(\alpha - \beta) = \sin \alpha \cos \beta - \cos \alpha \sin \beta.$$

Item (1) is discussed in the next theorem.

Theorem 7.1.1

If h is in radians, then $\lim_{h \to 0} \frac{\sin h}{h} = 1$.

Proof. First we shall show why we can restrict our attention to the case where h is positive. For if h is negative we let $k = -h$. Then k is positive and

$$\frac{\sin h}{h} = \frac{\sin(-k)}{-k} = \frac{-\sin k}{-k} = \frac{\sin k}{k}.$$

Thus we may restrict our consideration to positive values of h. Let h be a small positive number and consider Figure 7.1.1, where $\widehat{PQ}$ is an arc of a circle of radius 1, PR is perpendicular to OQ, and $\widehat{SR}$ is an arc of a circle of radius $OR = \cos h$. Then $PR = \sin h$ and we

Figure 7.1.1

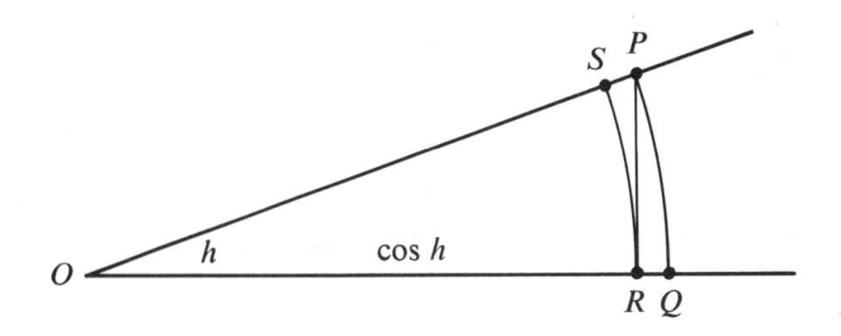

have (3)

$$\text{area of sector } ORS \leq \text{area of triangle } ORP \leq \text{area of sector } OPQ.$$

Since the area of a circle of radius r is πr^2, and a sector with central angle h is the fractional part $h/2\pi$ of a complete circle, the sector has area

$$\left(\frac{h}{2\pi}\right)\pi r^2 = \frac{1}{2}hr^2.$$

So by (3) $$\frac{1}{2}h\cos^2 h \leq \frac{1}{2}\sin h\cos h \leq \frac{1}{2}h.$$

We multiply those inequalities by the positive number $2/(h\cos h)$ to obtain

$$\cos h \leq \frac{\sin h}{h} \leq \frac{1}{\cos h}.$$

Since the cosine function is continuous,

$$\lim_{h\to 0}\cos h = \cos 0 = 1 \quad \text{and} \quad \lim_{h\to 0}\frac{1}{\cos h} = \frac{1}{1} = 1,$$

and so $$\lim_{h\to 0}\frac{\sin h}{h} = 1.$$

If h is in degrees rather than in radians, the area of a sector of a circle of radius r with central angle h is the fractional part $h/360$ of the area of the whole circle. Hence the sector has area $(h/360)\pi r^2$. It is left as an exercise to show, by the same technique used to prove Theorem 7.1.1, that when h is in degrees,

$$\lim_{h\to 0}\frac{\sin h}{h} = \frac{\pi}{180} \tag{4}$$

It is because $\lim_{h\to 0}(\sin h)/h = 1$ when h is in radians that radians are used in calculus. From now on, unless otherwise indicated, all angles will be understood to be measured in radians.

We are now able to obtain the formula for the derivative of $\sin x$, when x is in radians. By definition of the derivative we have

$$\frac{d}{dx}\sin x = \lim_{h\to 0}\frac{\sin(x+h) - \sin x}{h}.$$

Now, letting $\alpha + \beta = x + h$ and $\alpha - \beta = x$, we have $\alpha = x + h/2$ and $\beta = h/2$. Then by identity (2),

$$\sin(x + h) - \sin x = 2 \cos\left(x + \frac{h}{2}\right) \sin\left(\frac{h}{2}\right).$$

Hence,

$$\frac{d}{dx} \sin x = \lim_{h \to 0} \frac{2 \cos(x + h/2) \sin(h/2)}{h}$$

$$= \lim_{h \to 0} \frac{\cos(x + h/2) \sin(h/2)}{h/2}$$

$$= \lim_{h \to 0} \cos\left(x + \frac{h}{2}\right) \lim_{h \to 0} \frac{\sin(h/2)}{h/2}.$$

But since the cosine function is continuous,

$$\lim_{h \to 0} \cos\left(x + \frac{h}{2}\right) = \cos x.$$

And by Theorem 7.1.1,

$$\lim_{h \to 0} \frac{\sin(h/2)}{h/2} = \lim_{h/2 \to 0} \frac{\sin(h/2)}{h/2} = 1.$$

Hence

$$\frac{d}{dx} \sin x = \cos x.$$

If u is a differentiable function of x the chain rule gives

$$\frac{d}{dx} \sin u = \cos u \frac{du}{dx}.$$

We have now proved the following theorem.

Theorem 7.1.2

When x and u are in radians, and u is a differentiable function of x,

(i) $$\frac{d}{dx} \sin x = \cos x,$$

(ii) $$\frac{d}{dx} \sin u = \cos u \frac{du}{dx}.$$

Example 1

$$\frac{d}{dx} \sin(x^3 + 3x + 5) = \cos(x^3 + 3x + 5) \frac{d(x^3 + 3x + 5)}{dx}$$

$$= (3x^2 + 3) \cos(x^3 + 3x + 5).$$ ||

Example 2

$$\frac{d}{dx} \sin(e^{(x^2)}) = \cos(e^{(x^2)}) \frac{de^{(x^2)}}{dx} = 2xe^{(x^2)} \cos(e^{(x^2)}).$$ ||

If x were in degrees, it would follow from (4) and the derivation of the formula for the derivative of $\sin x$ that $(d/dx) \sin x = (\pi/180) \cos x$.

By use of Theorem 7.1.1, various other limits can be obtained, as indicated in the following examples.

Example 3

$$\lim_{x\to 0} \frac{\sin 3x}{x} = 3 \cdot \lim_{x\to 0} \frac{\sin 3x}{3x} = 3 \cdot \lim_{3x\to 0} \frac{\sin 3x}{3x} = 3 \cdot 1 = 3. \quad \|$$

Example 4

$$\lim_{x\to 0} \frac{1-\cos x}{x} = \lim_{x\to 0} \frac{1-\cos x}{x} \frac{1+\cos x}{1+\cos x} = \lim_{x\to 0} \frac{1-\cos^2 x}{x(1+\cos x)}$$

$$= \lim_{x\to 0} \frac{\sin^2 x}{x(1+\cos x)} = \lim_{x\to 0} \frac{\sin x}{x} \cdot \lim_{x\to 0} \frac{\sin x}{1+\cos x}$$

$$= 1 \cdot \frac{0}{2} = 0. \quad \|$$

Example 5

$$\lim_{x\to 0} \frac{x}{\sin 5x} = \frac{1}{5} \cdot \lim_{x\to 0} \frac{5x}{\sin 5x} = \frac{1}{5} \cdot \lim_{x\to 0} \frac{1}{\dfrac{\sin 5x}{5x}}$$

$$= \frac{1}{5} \frac{1}{\lim\limits_{5x\to 0} \dfrac{\sin 5x}{5x}} = \frac{1}{5} \cdot 1 = \frac{1}{5}. \quad \|$$

Example 6

$$\lim_{x\to 0} \frac{\sin 5x}{\sin 2x} = \lim_{x\to 0} \left[\frac{\sin 5x}{5x} \cdot \frac{2x}{\sin 2x} \cdot \frac{5}{2}\right]$$

$$= \frac{5}{2} \cdot \lim_{5x\to 0} \frac{\sin 5x}{5x} \cdot \lim_{2x\to 0} \frac{2x}{\sin 2x} = \frac{5}{2} \cdot 1 \cdot 1 = \frac{5}{2}. \quad \|$$

Exercises 7.1

Find the limits in Exercises 1–14.

1. $\lim\limits_{x\to 0} \dfrac{x}{\sin x}$.
2. $\lim\limits_{x\to 0} \dfrac{\sin 7x}{x}$.
3. $\lim\limits_{x\to 0} \dfrac{\sin 9x}{2x}$.
4. $\lim\limits_{x\to 0} (x \csc 3x)$.
5. $\lim\limits_{x\to 0} \dfrac{1}{2x \csc 8x}$.
6. $\lim\limits_{x\to 0} \dfrac{1-\cos x}{x^2}$.
7. $\lim\limits_{x\to 0} \dfrac{3x - 2\sin 5x}{x}$.
8. $\lim\limits_{x\to 0} \dfrac{\sin^2 x}{1-\cos x}$.
9. $\lim\limits_{x\to 0} \dfrac{\tan 3x}{5x}$.
10. $\lim\limits_{x\to 0} \dfrac{\sin 2x}{\sqrt{x}}$.
11. $\lim\limits_{x\to 0} \dfrac{1-\cos x}{\sin x}$.
12. $\lim\limits_{x\to 0} \dfrac{x}{x+\sin x}$.
13. $\lim\limits_{x\to 0} \dfrac{x \cos x}{1-\cos x}$.
14. $\lim\limits_{x\to \pi} \dfrac{\sin x}{\pi - x}$.

Differentiate the functions in Exercises 15–26.

15. $f(x) = \sin 7x$.
16. $f(x) = x^3 \sin 4x$.
17. $f(x) = \sin^4 3x$.
18. $f(x) = x \sin^2 5x$.
19. $f(x) = \sin(x^2 + x - 1)$.
20. $f(x) = \sin(1/x)$.
21. $f(x) = \csc x \left(= \dfrac{1}{\sin x} \right)$.
22. $f(x) = \dfrac{x^3}{\sin x}$.
23. $f(x) = e^{\sin x}$.
24. $f(x) = \ln(\sin x)$.
25. $f(x) = x^{\sin x}$.
26. $f(x) = (\sin x)^x$.

27. If a rocket is launched at an angle θ to the horizontal and with initial velocity v_0, its range R is

$$R = \frac{{v_0}^2}{g} \sin 2\theta, \qquad 0 \le \theta \le \frac{\pi}{2},$$

where g is the acceleration due to gravity. Find the angle that gives the greatest range.

28. A light is to be placed over the center of a square table six feet on each side. The intensity of light reaching any point on the table is directly proportional to the sine of the angle of incidence and inversely proportional to the square of the distance between the point and the light. How high above the table should the light be hung in order to give maximum illumination to the corners of the table? (The angle of incidence is the angle between the light ray and the horizontal table.)

29. A vertical circular disc of radius 2′ is partly submerged in a chemical solution and rotated slowly, as shown in Figure 7.1.2. How high above the solution should the center be to maximize the wetted surface of the disc above the solution?

Figure 7.1.2

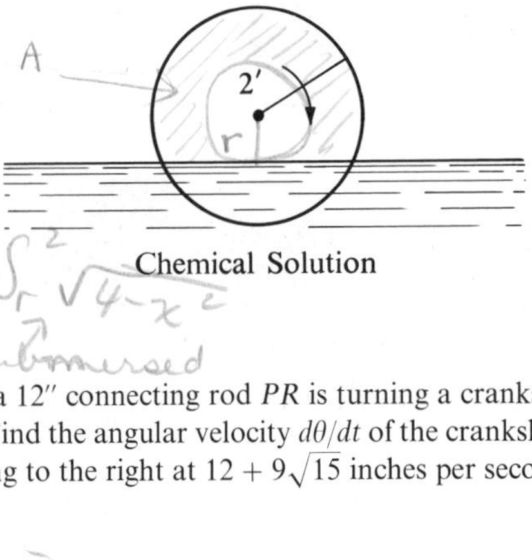

2′

Chemical Solution

30. A piston with a 12″ connecting rod PR is turning a crankshaft OR of radius 5″, as shown in Figure 7.1.3. Find the angular velocity $d\theta/dt$ of the crankshaft at the instant when $\sin \theta = 4/5$ and P is moving to the right at $12 + 9\sqrt{15}$ inches per second.

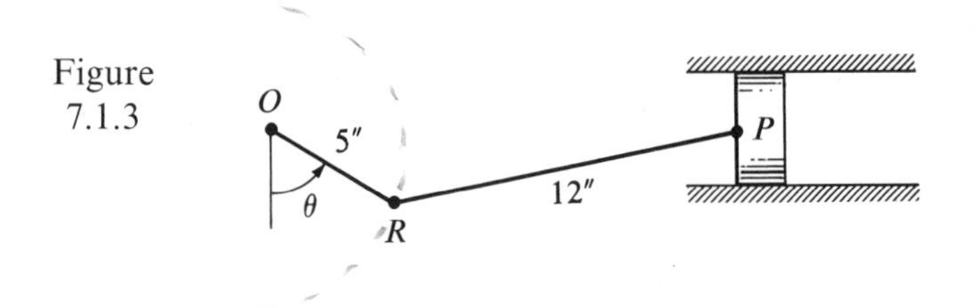

Figure 7.1.3

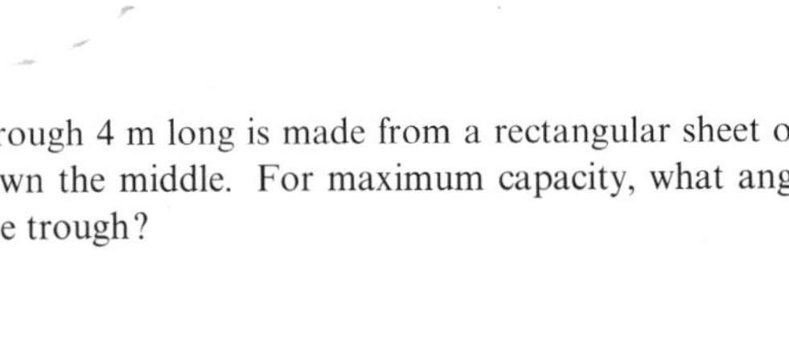

31. A V-shaped trough 4 m long is made from a rectangular sheet of aluminum 70 cm wide by bending it down the middle. For maximum capacity, what angle should be used between the sides of the trough?

7.2 The Derivatives of the Other Trigonometric Functions

First we derive the formula for the derivative of $\cos x$ with the aid of the identities

$$\cos x = \sin\left(\frac{\pi}{2} - x\right) \quad \text{and} \quad \cos\left(\frac{\pi}{2} - x\right) = \sin x.$$

We have

$$\frac{d}{dx}\cos x = \frac{d}{dx}\sin\left(\frac{\pi}{2} - x\right) = (-1)\cos\left(\frac{\pi}{2} - x\right) = -\sin x.$$

If u is a differentiable function of x, we can use the chain rule to get

$$\frac{d}{dx}\cos u = -\sin u\frac{du}{dx}.$$

Now we derive the formula for the derivative of $\tan x$ by using the identity $\tan x = \sin x/\cos x$ and the quotient rule:

$$\frac{d}{dx}\tan x = \frac{d}{dx}\frac{\sin x}{\cos x} = \frac{\cos x\cos x - \sin x(-\sin x)}{\cos^2 x}$$

$$= \frac{\cos^2 x + \sin^2 x}{\cos^2 x} = \frac{1}{\cos^2 x} = \sec^2 x.$$

If u is a differentiable function of x, then with the aid of the chain rule we get

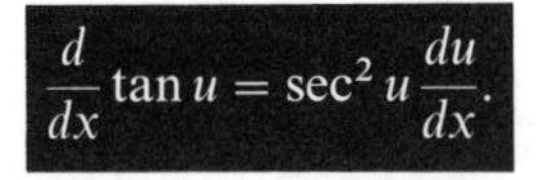

$$\frac{d}{dx}\tan u = \sec^2 u\frac{du}{dx}.$$

By expressing the other functions in terms of sine and cosine, we can also obtain the following formulas.

$$\frac{d}{dx}\cot u = -\csc^2 u\frac{du}{dx}. \qquad \frac{d}{dx}\sec u = \sec u\tan u\frac{du}{dx}.$$

$$\frac{d}{dx}\csc u = -\csc u\cot u\frac{du}{dx}.$$

Example 1

$$\frac{d}{dx}\tan(x^2 - 1) = \sec^2(x^2 - 1)\frac{d}{dx}(x^2 - 1) = 2x\sec^2(x^2 - 1). \qquad \|$$

Example 2

$$\frac{d}{dx}\csc^3(x^2 + 3x) = 3\csc^2(x^2 + 3x)\frac{d}{dx}\csc(x^2 + 3x)$$

$$= 3\csc^2(x^2 + 3x)[-\csc(x^2 + 3x)\cot(x^2 + 3x)](2x + 3)$$

$$= -(6x + 9)\csc^3(x^2 + 3x)\cot(x^2 + 3x). \qquad \|$$

Example 3

$$\frac{d}{dx}\cot(1 - x^3) = -\csc^2(1 - x^3)\frac{d}{dx}(1 - x^3) = 3x^2\csc^2(1 - x^3). \qquad \|$$

Exercises 7.2

Differentiate the functions in Exercises 1–16.

1. $f(x) = \cos(3 - x^2)$.
2. $f(x) = \cos(x^2 + 8)$.
3. $f(x) = \cos^3(5 - 2x)$.
4. $f(x) = \cos^2(3x - 1)$.
5. $f(x) = \sec^2(3x^2 - 1)$.
6. $f(x) = \sec 3x \cot 5x$.
7. $f(x) = \tan\sqrt{2x - 1}$.
8. $f(x) = e^x \tan 3x$.
9. $f(x) = \csc^2(e^x + x)$.
10. $f(x) = \tan(1 - x)$.
11. $f(x) = e^{2x} \sec 3x$.
12. $f(x) = \sec^3(x + e^x)$.
13. $f(x) = \cot^3(x^2 - 1)$.
14. $f(x) = \csc^2 5x \cot^3 4x$.
15. $f(x) = \sec^2 3x \tan^5 3x$.
16. $f(x) = e^{5x} \csc 7x$.

Find the differentials of the functions in Exercises 17–24.

17. $f(x) = e^{\csc 3x}$.
18. $f(x) = \ln|\tan 2x|$.
19. $f(x) = \ln|\sec 5x|$.
20. $f(x) = e^{\cot 7x}$.
21. $f(x) = \cos 5x$.
22. $f(x) = \tan^2 6x$.
23. $f(x) = \sec 2x \tan 2x$.
24. $f(x) = \cos^2(x^3 + 2x)$.
25. Prove that $(d/dx) \cot u = -\csc^2 u(du/dx)$.
26. Prove that $(d/dx) \sec u = \sec u \tan u(du/dx)$.
27. Prove that $(d/dx) \csc u = -\csc u \cot u(du/dx)$.
28. Show that $(d/dx)(\cos^2 x - \sin^2 x - \cos 2x) = 0$. What conclusion can be drawn from this fact?
29. A light lies 1/4 mile directly off shore from a point P. If the light rotates at the rate of 60 radians per hour, calculate how fast the beam of light passes a point 1/4 mile down the straight shore from P.
30. A fence 4′ high is 5′ from a wall. Find the length of the shortest beam that can pass over the fence to brace the wall.
31. A 4′ wide hallway runs perpendicularly into a corridor 10′ wide. Find the length of the longest straight thin board that can be moved horizontally around the corner.
32. Assume that when an airplane wing with a flat bottom is inclined at an angle α from the horizontal, its lifting power is proportional to $\sin^2\alpha \cos\alpha$. For what value of α is the lifting power greatest?
33. Consider an inclined plane making an angle θ with the horizontal. The maximum weight W that can be pulled up this plane with a force F is

$$W = \frac{F}{\mu}(\cos\theta + \mu \sin\theta), \qquad 0 \le \theta \le \frac{\pi}{2},$$

where μ is the coefficient of friction. For what value of θ is W a maximum?

34. Three points A, B, and C are joined by line segments as shown in Figure 7.2.1. Find the length CD that minimizes the total length of the line segments.

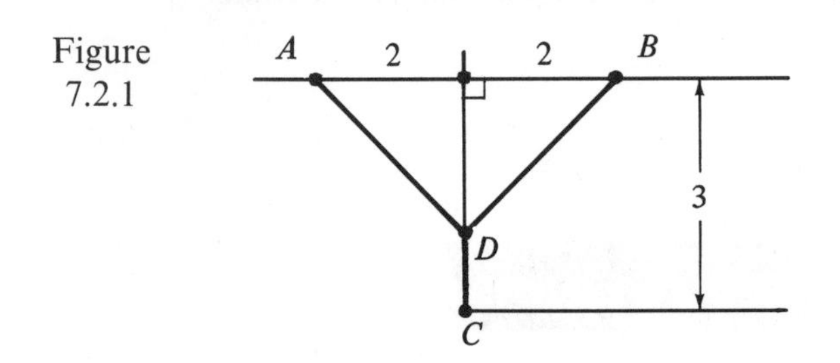

Figure 7.2.1

35. Use trigonometric functions to compute the radius and height of the right circular cylinder of largest lateral surface area that can be inscribed in a sphere of radius 1.

36. A phone cable is to be laid from point A on one side of a river to point B on the opposite side. The cost of laying the cable along the river is \$15/ft, while the cost of submerging the cable in the river is \$20/ft. If A and B are located as in Figure 7.2.2, find the length AC that minimizes the total cost.

Figure 7.2.2

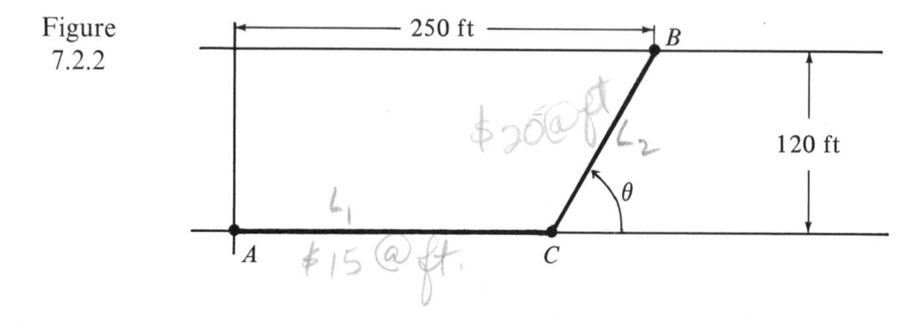

37. Differentiate both sides of the identity $\sin 2x = 2 \sin x \cos x$ to get an identity for $\cos 2x$.

38. Determine where the graph of $y = \sin x + \cos x$, $-\frac{3\pi}{4} \leq x \leq \frac{5\pi}{4}$, is (a) increasing, (b) decreasing, (c) concave upward, (d) concave downward. Also determine (e) the local maximum points, (f) the local minimum points, (g) the points of inflection, and (h) sketch the graph.

39. Find the points of inflection of the graph of $y = \tan x$.

40. What is the slope of the graph of $\sqrt{2} \cos(x + y) = 1$ at the point $(\pi/6, \pi/12)$?

41. An object of weight W is being pulled along a horizontal plane at constant velocity. If the force has magnitude F and is directed at an angle θ to the horizontal then

$$F = \frac{\mu W}{\mu \sin \theta + \cos \theta}$$

where μ is the coefficient of friction. Show that F is a minimum when $\tan \theta = \mu$.

42. A particle is moving in a plane in such a way that its coordinates are $x = \cos t$, $y = \sin 2t$ at an elapsed time of t seconds. If x and y are measured in meters, find the velocity at which the particle is moving toward the origin when $t = \pi/4$ seconds.

7.3 The Integrals of the Trigonometric Functions

Since $d(\sin u) = \cos u \, du$, we have

$$\int \cos u \, du = \int d(\sin u) = \sin u + C.$$

Thus

$$\int \cos u \, du = \sin u + C.$$

Similarly, since $d(\cos u) = -\sin u\,du$, we have $\sin u\,du = -d(\cos u)$, and so

$$\int \sin u\,du = \int -d(\cos u) = -\int d(\cos u) = -\cos u + C.$$

Thus

$$\int \sin u\,du = -\cos u + C.$$

Furthermore,

$$\int \tan u\,du = \int \frac{\sin u\,du}{\cos u} = -\int \frac{-\sin u\,du}{\cos u}$$

$$= -\int \frac{d(\cos u)}{\cos u} = -\ln|\cos u| + C.$$

And since $-\ln|\cos u| = \ln|\sec u|$, we have

$$\int \tan u\,du = -\ln|\cos u| + C = \ln|\sec u| + C.$$

Similarly,

$$\int \cot u\,du = \ln|\sin u| + C = -\ln|\csc u| + C.$$

Example 1 $\int \cos x\,dx = \sin x + C.$ ||

Example 2 $\int \sin 3x(3\,dx) = \int \sin 3x\,d(3x) = -\cos 3x + C.$ ||

Example 3

$$\int x\cos(x^2)dx = \frac{1}{2}\int \cos(x^2)(2x\,dx)$$

$$= \frac{1}{2}\int \cos(x^2)\,d(x^2) = \frac{1}{2}\sin(x^2) + C.$$ ||

Example 4 $\int \tan x\,dx = \ln|\sec x| + C.$ ||

Example 5

$$\int x^2\tan(x^3 + 2)dx = \frac{1}{3}\int \tan(x^3 + 2)(3x^2\,dx)$$

$$= \frac{1}{3}\int \tan(x^3 + 2)d(x^3 + 2)$$

$$= \frac{1}{3}\ln|\sec(x^3 + 2)| + C.$$ ||

Example 6

$$\int \sin x \cos^2 x \, dx = -\int (\cos^2 x)(-\sin x \, dx)$$

$$= -\int \cos^2 x \, d(\cos x) = -\frac{\cos^3 x}{3} + C. \quad \|$$

Example 7

$$\int \sin 3x \cos^5 3x \, dx = -\frac{1}{3}\int (\cos^5 3x)(-3 \sin 3x \, dx)$$

$$= -\frac{1}{3}\int (\cos^5 3x) \, d(\cos 3x)$$

$$= -\frac{1}{3}\frac{\cos^6 3x}{6} + C = -\frac{1}{18}\cos^6 3x + C. \quad \|$$

Example 8

$$\int e^x(\cot e^x)dx = \int (\cot e^x)(e^x \, dx) = \int (\cot e^x) \, d(e^x)$$

$$= -\ln|\csc e^x| + C = \ln|\sin e^x| + C. \quad \|$$

Integration by parts can often be used on integrals involving trigonometric functions. The next example illustrates that fact.

Example 9 | Evaluate $\int x \cos x \, dx$.

Let $u = x$ and $dv = \cos x \, dx$. Then $du = dx$ and $v = \sin x$; so

$$\int x \cos x \, dx = x \sin x - \int \sin x \, dx$$

$$= x \sin x + \cos x + C. \quad \|$$

To obtain $\int \sec u \, du$, we proceed as follows:

$$\int \sec u \, du = \int \frac{(\sec u)(\sec u + \tan u)}{\sec u + \tan u} du$$

$$= \int \frac{(\sec^2 u + \sec u \tan u)du}{\sec u + \tan u}.$$

Since $d(\sec u) = \sec u \tan u \, du$ and $d(\tan u) = \sec^2 u \, du$, the numerator is the differential of the denominator; hence we obtain the natural log of the absolute value of the denominator plus, a constant; that is,

$$\boxed{\int \sec u \, du = \ln|\sec u + \tan u| + C.}$$

Similarly,

$$\int \csc u \, du = -\ln|\csc u + \cot u| + C.$$

Example 10

$$\int (e^x + 2x) \sec(x^2 + e^x) dx = \int \sec(x^2 + e^x)(e^x + 2x) dx$$
$$= \int \sec(x^2 + e^x) \, d(x^2 + e^x)$$
$$= \ln|\sec(x^2 + e^x) + \tan(x^2 + e^x)| + C.$$

||

We now obtain an integration formula corresponding to each of the following differentiation formulas:

$$\frac{d}{dx} \tan u = \sec^2 u \frac{du}{dx},$$
$$\frac{d}{dx} \cot u = -\csc^2 u \frac{du}{dx},$$
$$\frac{d}{dx} \sec u = \sec u \tan u \frac{du}{dx},$$
$$\frac{d}{dx} \csc u = -\csc u \cot u \frac{du}{dx}.$$

from pg 322

We obtain these formulas:

$$\int \sec^2 u \, du = \tan u + C,$$
$$\int \csc^2 u \, du = -\cot u + C.$$

$$\int \sec u \tan u \, du = \sec u + C,$$
$$\int \csc u \cot u \, du = -\csc u + C.$$

Example 11

$$\int x \sec^2(x^2 + 5) dx = \frac{1}{2} \int \sec^2(x^2 + 5)(2x \, dx)$$
$$= \frac{1}{2} \tan(x^2 + 5) + C.$$

||

Example 12

$$\int \csc^2 3x \, dx = \frac{1}{3} \int \csc^2 3x(3 \, dx) = -\frac{1}{3} \cot 3x + C.$$

||

Example 13 $\int \sec 2x \tan 2x\, dx = \frac{1}{2}\int \sec 2x \tan 2x(2\, dx) = \frac{1}{2}\sec 2x + C.$ ||

Example 14 $\int e^x(\csc e^x)(\cot e^x)dx = \int(\csc e^x)(\cot e^x)(e^x\, dx) = -\csc e^x + C.$ ||

Exercise 7.3

In Exercises 1–32 evaluate each integral.

1. $\int \sin 5x\, dx.$
2. $\int \cos 7x\, dx.$
3. $\int x \tan(x^2)dx.$
4. $\int x^2 \sec(1 - x^3)dx.$
5. $\int \frac{1}{x}\sin(\ln x)dx.$
6. $\int \sin^3 x \cos x\, dx.$
7. $\int \sin x \cos^2 x\, dx.$
8. $\int e^x \cos(e^x)dx.$
9. $\int \sec^2 x\, dx.$
10. $\int \sec^2 6x\, dx.$
11. $\int x \sin x\, dx.$
12. $\int x \sec(x^2)\tan(x^2)dx.$
13. $\int \csc^2(5x - 11)dx.$
14. $\int x \sec^2 x\, dx.$
15. $\int \sec 5x\, dx.$
16. $\int x \csc(x^2)dx.$
17. $\int_0^{\pi/2} \frac{\cos 5x}{1 + \sin 5x}dx.$
18. $\int_0^{\pi/3}(\sin x + \cos x)^2\, dx.$
19. $\int \sec^2 2x \tan^5 2x\, dx.$
20. $\int \frac{\sec^2 5x}{2 + \tan 5x}dx.$
21. $\int_0^{\pi/12}\sec 3x \tan 3x\, dx.$
22. $\int \sec^3 x \tan x\, dx.$
23. $\int \sec^5 x \tan x\, dx.$
24. $\int \sin^3 x \cos^3 x\, dx.$
25. $\int \sin x\, e^{\cos x}\, dx.$
26. $\int \sec^2 3x\, e^{\tan 3x}\, dx.$
27. $\int x \sin 2x\, dx.$
28. $\int x \cos 3x\, dx.$
29. $\int x^2 \sin x\, dx.$
30. $\int x^2 \cos x\, dx.$
31. $\int \sin(\ln x)\cos^2(\ln x)\frac{1}{x}\, dx.$
32. $\int \frac{1}{x}\tan(\ln x)\sec^2(\ln x)dx.$
33. Find the area of the region under one arch of the curve $y = 2 \sin 3x$.
34. Find the area of the region under the curve $y = \sec x \sin x$ between $x = 0$ and $x = \pi/4$.
35. The region under one arch of the curve $y = \sin x$ is revolved about the y-axis. Compute the volume of the resulting solid.
36. Compute the arc length of $y = \ln(\cos x)$ between $x = 0$ and $x = \pi/4$.
37. A force of $\sin x \cos^3 x$ lbs is exerted on an object a distance x feet from the origin. Compute the work done in moving the object from 0 to $\pi/2$.
38. The region under $y = \sin(x^2/\pi)$ between $x = 0$ and $x = \pi$ is revolved about the y-axis. Compute the volume of the resulting solid.

7.4 Inverse Trigonometric Functions

Consider the function $f(x) = \sin x$, whose graph is indicated in Figure 7.4.1. Let c be between -1 and 1. Since the horizontal line $y = c$ intersects the graph in many different places, the output c corresponds to many different inputs. Consequently the function $f(x) = \sin x$ has no inverse function; and since similar remarks hold for all the trigonometric functions, none of the trigonometric functions has an inverse.

Figure 7.4.1

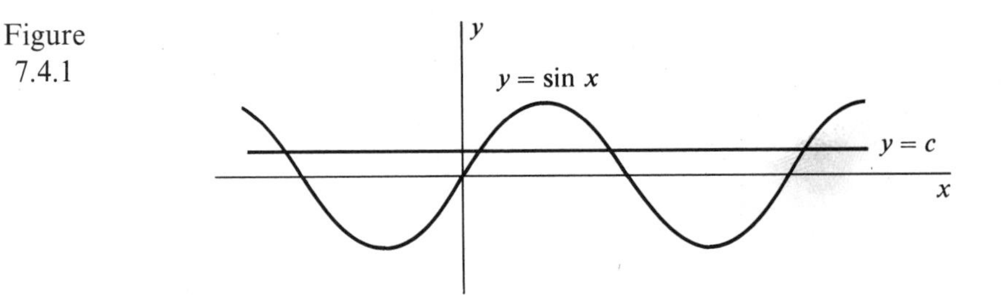

However the function $f(x) = \sin x$, $-\pi/2 \leq x \leq \pi/2$, whose graph is indicated in Figure 7.4.2, is both continuous and increasing. Therefore, by Theorem 6.2.2, that function has a continuous, increasing inverse function, which is called the *arcsine* function. Since the arcsine function is the inverse of $\sin x$, $-\pi/2 \leq x \leq \pi/2$,we have

$$y = \arcsin x \quad \text{if and only if} \quad \sin y = x \textit{ and } -\frac{\pi}{2} \leq y \leq \frac{\pi}{2}.$$

Figure 7.4.2

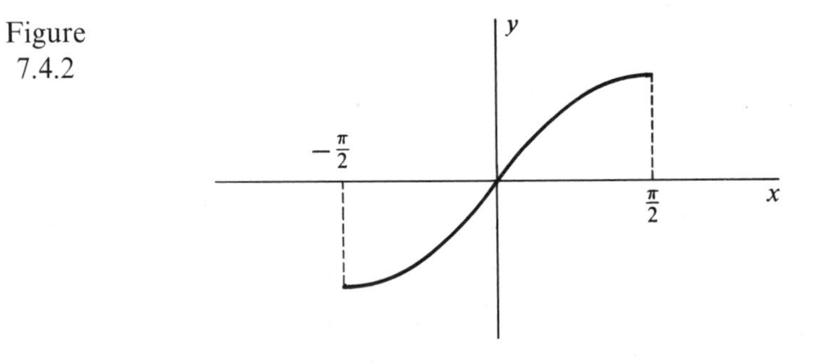

The graph of $y = \arcsin x$ is pictured in Figure 7.4.3. Note that the graph of $y = \arcsin x$ is obtained by rotating the graph of $y = \sin x$ one-half revolution about the line $y = x$.

Figure 7.4.3

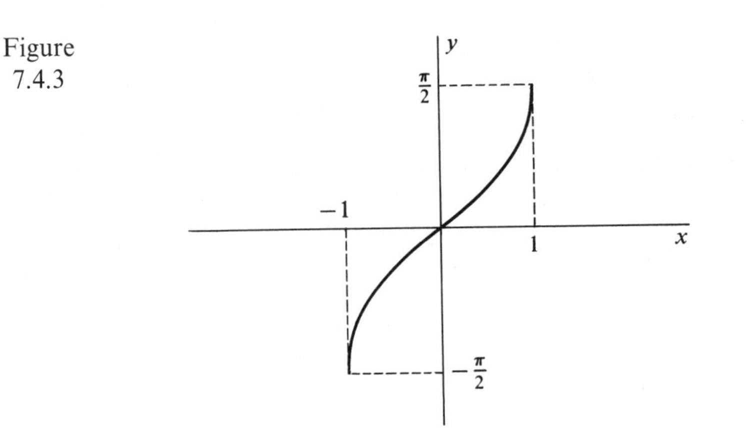

Example 1 | Evaluate arcsin 1/2.

$y = \arcsin 1/2$ if and only if $1/2 = \sin y$ and $-\pi/2 \leq y \leq \pi/2$. Since $\pi/6$ meets both requirements ($1/2 = \sin \pi/6$ and $-\pi/2 \leq \pi/6 \leq \pi/2$), we have $\arcsin 1/2 = \pi/6$. ||

Also, according to the definition of inverse function (Definition 6.2.1), the arcsin function has these two basic properties:

$$\arcsin(\sin x) = x, \qquad \text{for } -\frac{\pi}{2} \le x \le \frac{\pi}{2}. \tag{1}$$

$$\sin(\arcsin y) = y, \qquad \text{for } -1 \le y \le 1. \tag{2}$$

Example 2 | Evaluate sin(arcsin 2/3).

By (2) we have $\sin(\arcsin 2/3) = 2/3$. ||

Example 3 | Evaluate $\arcsin(\sin 5\pi/4)$.

We cannot use (1) directly to obtain this because $5\pi/4$ is not in the interval $-\pi/2 \le x \le \pi/2$. However, $\sin 5\pi/4 = \sin(-\pi/4)$, and $-\pi/4$ is in the proper interval. Therefore

$$\arcsin\left(\sin\frac{5\pi}{4}\right) = \arcsin\left[\sin\left(-\frac{\pi}{4}\right)\right] = -\frac{\pi}{4}.$$ ||

This example illustrates the fact the $\arcsin(\sin x)$ might not equal x. In fact, according to (1) above,

$$\arcsin(\sin x) = x \quad \text{if and only if} \quad -\frac{\pi}{2} \le x \le \frac{\pi}{2}.$$

The other inverse trigonometric functions are motivated and defined similarly to arcsin. Since none of the trigonometric functions is monotone and since it is necessary for a continuous function to be monotone to have an inverse, it is necessary to restrict the domain of each trigonometric function to an interval where it is monotone. Of course the most convenient interval should be chosen. For example, the arccosine function is the inverse of the cosine function restricted to the interval $0 \le x \le \pi$. Thus

$$y = \arccos x \quad \text{if and only if} \quad \cos y = x \text{ and } 0 \le y \le \pi.$$

Next we list the definitions of the remaining inverse trigonometric functions and we indicate their graphs in Figure 7.4.4. As we have previously noted, the graph of an inverse function may be obtained by rotating the graph of the function one-half revolution about the line $y = x$.

$$y = \arctan x \quad \text{if and only if} \quad \tan y = x \text{ and } -\frac{\pi}{2} < y < \frac{\pi}{2}.$$

$$y = \operatorname{arccot} x \quad \text{if and only if} \quad \cot y = x \text{ and } 0 < y < \pi.$$

$$y = \operatorname{arcsec} x \quad \text{if and only if} \quad \sec y = x \text{ and } 0 \le y \le \pi, \quad y \ne \frac{\pi}{2}.$$

$$y = \operatorname{arccsc} x \quad \text{if and only if} \quad \csc y = x \text{ and } -\frac{\pi}{2} \le y \le \frac{\pi}{2}, \quad y \ne 0.$$

Figure 7.4.4

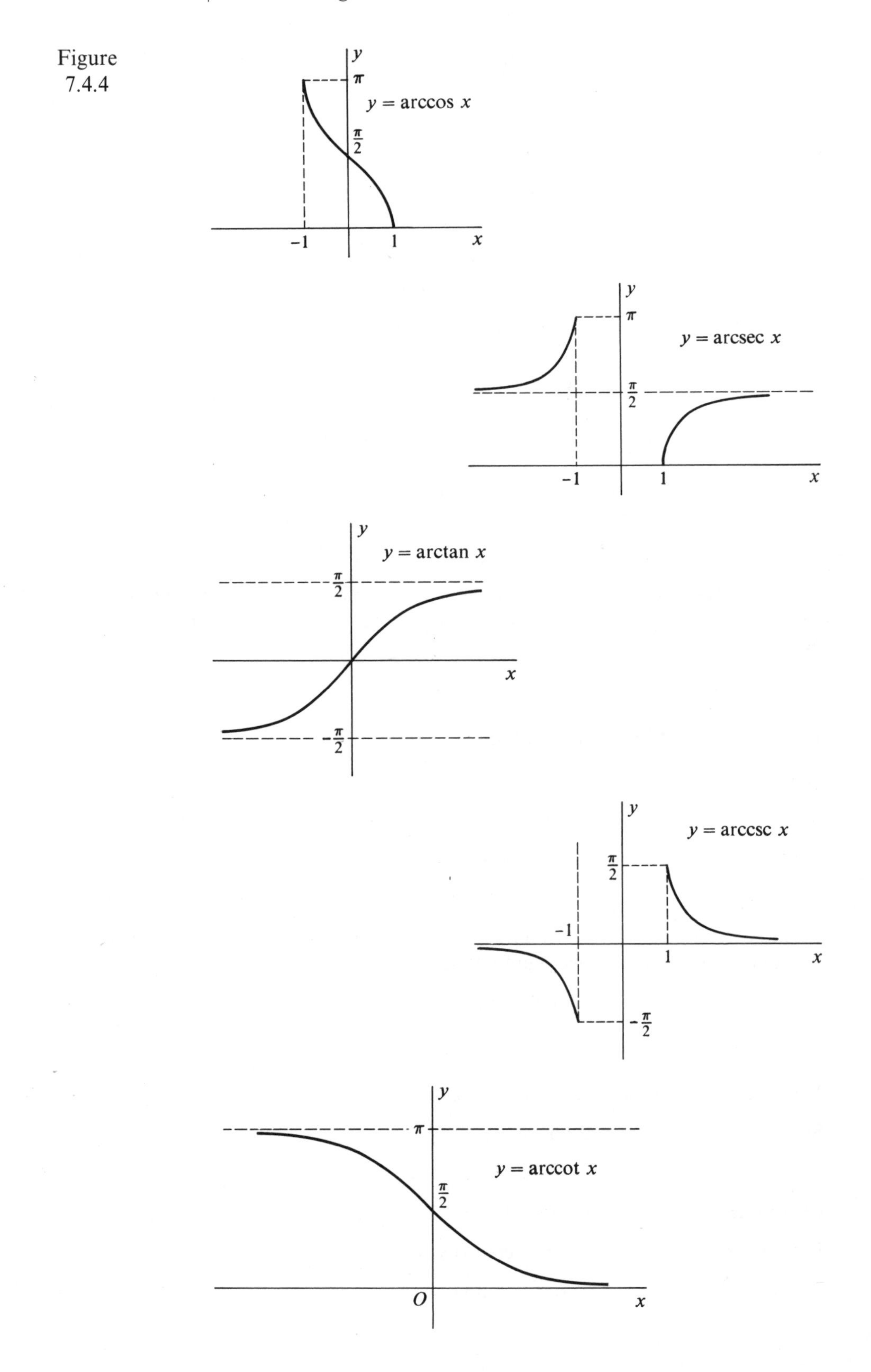
y
π
y = arccos x
π/2
−1
1
x
y
π
y = arcsec x
π/2
−1
1
x
y
y = arctan x
π/2
x
−π/2
y
y = arccsc x
π/2
−1
1
x
−π/2
y
π
y = arccot x
π/2
O
x

The following properties are similar to those indicated in (1) and (2) above.

$$\arccos(\cos x) = x, \quad \text{for } 0 \le x \le \pi. \tag{3}$$

$$\cos(\arccos y) = y, \quad \text{for } -1 \le y \le 1. \tag{4}$$

$$\arctan(\tan x) = x, \quad \text{for } -\frac{\pi}{2} < x < \frac{\pi}{2}. \tag{5}$$

$$\tan(\arctan y) = y, \quad \text{for all } y. \tag{6}$$

$$\operatorname{arccot}(\cot x) = x, \quad \text{for } 0 < x < \pi. \tag{7}$$

$$\cot(\operatorname{arccot} y) = y, \quad \text{for all } y. \tag{8}$$

$$\operatorname{arcsec}(\sec x) = x, \quad \text{for } 0 \le x \le \pi, \quad x \ne \frac{\pi}{2}. \tag{9}$$

$$\sec(\operatorname{arcsec} y) = y, \quad \text{for } y \le -1 \text{ or } y \ge 1. \tag{10}$$

$$\operatorname{arccsc}(\csc x) = x, \quad \text{for } -\frac{\pi}{2} \le x \le \frac{\pi}{2}, \quad x \ne 0. \tag{11}$$

$$\csc(\operatorname{arccsc} y) = y, \quad \text{for } y \le -1 \text{ or } y \ge 1. \tag{12}$$

To familiarize ourselves with the inverse trigonometric functions and to derive some results necessary for our further work, we shall consider three additional examples.

Example 4 | Calculate $\cos(\arcsin x)$.

Since

$$\cos \alpha = \pm\sqrt{1 - \sin^2 \alpha},$$

we have

$$\cos(\arcsin x) = \pm\sqrt{1 - [\sin(\arcsin x)]^2}.$$

But from (2),

$$\sin(\arcsin x) = x,$$

and so

$$\cos(\arcsin x) = \pm\sqrt{1 - x^2}.$$

Furthermore, since $-\pi/2 \le \arcsin x \le \pi/2$, we have $\cos(\arcsin x) \ge 0$; and therefore we can eliminate the negative sign to obtain

$$\cos(\arcsin x) = \sqrt{1 - x^2}.$$

||

Example 5 | Calculate $\sec(\arctan x)$.

Since $\sec^2 \alpha = 1 + \tan^2 \alpha$, we have

$$\sec \alpha = \pm\sqrt{1 + \tan^2 \alpha}.$$

Consequently,

$$\sec(\arctan x) = \pm\sqrt{1 + \tan^2(\arctan x)}.$$

But from (6),

$$\tan(\arctan x) = x,$$

and so

$$\sec(\arctan x) = \pm\sqrt{1 + x^2}.$$

Since $-\pi/2 < \arctan x < \pi/2$, we have $\sec(\arctan x) \geq 0$; and therefore we can eliminate the negative sign to get

$$\sec(\arctan x) = \sqrt{1 + x^2}.$$

||

Example 6 | Calculate $\tan(\operatorname{arcsec} x)$.

Since $\tan^2 \alpha = \sec^2 \alpha - 1$, we have

$$\tan \alpha = \pm\sqrt{\sec^2 \alpha - 1}.$$

Hence,

$$\tan(\operatorname{arcsec} x) = \pm\sqrt{\sec^2(\operatorname{arcsec} x) - 1}.$$

Then, using the fact that $\sec(\operatorname{arcsec} x) = x$, we have

$$\tan(\operatorname{arcsec} x) = \pm\sqrt{x^2 - 1}.$$

In this case, however, we cannot eliminate the negative sign; but we can find out when to use it.

When $x \geq 1$, we have $0 \leq \operatorname{arcsec} x < \pi/2$ and so $\tan(\operatorname{arcsec} x) \geq 0$. If, on the other hand, $x \leq -1$, we have $\pi/2 < \operatorname{arcsec} x \leq \pi$ and so $\tan(\operatorname{arcsec} x) \leq 0$. Hence we have

$$\tan(\operatorname{arcsec} x) = \begin{cases} \sqrt{x^2 - 1} & \text{if } x \geq 1 \\ -\sqrt{x^2 - 1} & \text{if } x \leq -1. \end{cases}$$

||

Exercises 7.4

Simplify the expressions in Exercises 1–30.

1. $\arcsin \sqrt{3}/2$.
2. $\arctan 1$.
3. $\operatorname{arccot} \sqrt{3}$.
4. $\arccos \sqrt{3}/2$.
5. $\arctan(-\sqrt{3})$.
6. $\operatorname{arccot}(-1)$.
7. $\arccos 1/2$.
8. $\arcsin(-1/2)$.
9. $\operatorname{arcsec}(-\sqrt{2})$.
10. $\operatorname{arcsec}(-2)$.
11. $\operatorname{arccsc} 2$.
12. $\operatorname{arccsc}(-\sqrt{2})$.
13. $\sec(\operatorname{arcsec} 5)$.
14. $\arctan(\tan \pi/7)$.
15. $\arctan(\tan 10\pi)$.
16. $\arcsin(\sin 5\pi/6)$.
17. $\arccos(\cos 7\pi/6)$.
18. $\operatorname{arccot}[\cot(-\pi/3)]$.
19. $\operatorname{arcsec}(\sec 4\pi/3)$.
20. $\arctan(\tan 2\pi/3)$.
21. $\cos(\arcsin 1/2)$.
22. $\cos(\arcsin 1/4)$.
23. $\tan(\operatorname{arcsec} 5)$.
24. $\tan(\operatorname{arcsec}(-3))$.
25. $\tan(\operatorname{arcsec}(-5))$.
26. $\tan(\operatorname{arcsec}(3))$.
27. $\sec(\arctan 10)$.
28. $\sec(\arctan 2)$.
29. $\arcsin\left(\sin \frac{7\pi}{8}\right)$.
30. $\arccos\left(\cos \frac{3\pi}{2}\right)$.

Find the inverse of each function and the domain of the inverse function in Exercises 31–38. The domains of the functions will have to be appropriately restricted.

31. $y = 3 \sin 5x$.
32. $y = 2 \cos(x + 3)$.
33. $y = 3 \arcsin 2x$.
34. $y = 5 \arccos 3x$.
35. $y = 5 \sec(2 - 3x)$.
36. $y = 3 \tan(5 - 2x)$.
37. $y = \sin x \cos x$.
38. $y = \tan x - \cot x$.

7.5 Derivatives of the Inverse Trigonometric Functions

Since we know the derivatives of the trigonometric functions, Theorem 6.2.3 will make it an easy matter to obtain the derivatives of the inverse trigonometric functions. Suppose that $y = f(x)$ and $x = g(y)$ are inverse functions. Then if $dx/dy \neq 0$, Theorem 6.2.3 assures us that

$$\frac{dy}{dx} = \frac{1}{dx/dy}.$$

In particular, to find $d(\arcsin x)/dx$, let $y = \arcsin x$; then $x = \sin y$, $-\pi/2 \leq y \leq \pi/2$, and $dx/dy = \cos y$. Thus

$$\frac{d(\arcsin x)}{dx} = \frac{dy}{dx} = \frac{1}{dx/dy} = \frac{1}{\cos y} = \frac{1}{\cos(\arcsin x)}.$$

Using the result of Example 4, Section 7.4, we have $\cos(\arcsin x) = \sqrt{1 - x^2}$; so

$$\frac{d(\arcsin x)}{dx} = \frac{1}{\sqrt{1 - x^2}}, \qquad |x| < 1.$$

Similarly, to find $d(\arctan x)/dx$, we will let $y = \arctan x$. Then $x = \tan y$, $-\pi/2 < y < \pi/2$, and $dx/dy = \sec^2 y$. Therefore

$$\frac{d}{dx}\arctan x = \frac{dy}{dx} = \frac{1}{dx/dy} = \frac{1}{\sec^2 y} = \frac{1}{1 + \tan^2 y} = \frac{1}{1 + x^2}.$$

Hence

$$\frac{d}{dx}\arctan x = \frac{1}{x^2 + 1}.$$

Finally, we obtain $d(\operatorname{arcsec} x)/dx$ by letting $y = \operatorname{arcsec} x$. Then $x = \sec y$, $0 \leq y \leq \pi$, and $dx/dy = \sec y \tan y$. Hence

$$\frac{d}{dx}\operatorname{arcsec} x = \frac{1}{\sec y \tan y} = \frac{1}{\sec(\operatorname{arcsec} x)\tan(\operatorname{arcsec} x)}$$

From (10) and Example 6 of Section 7.4, we have $\sec(\operatorname{arcsec} x) = x$ and

$$\tan(\operatorname{arcsec} x) = \begin{cases} \sqrt{x^2 - 1} & \text{if } x \geq 1 \\ -\sqrt{x^2 - 1} & \text{if } x \leq -1. \end{cases}$$

Hence

$$\frac{d}{dx}\operatorname{arcsec} x = \begin{cases} \dfrac{1}{x\sqrt{x^2 - 1}} & \text{if } x > 1 \\ \dfrac{1}{-x\sqrt{x^2 - 1}} & \text{if } x < -1. \end{cases}$$

This can be written in the simpler form

$$\frac{d}{dx}\operatorname{arcsec} x = \frac{1}{|x|\sqrt{x^2 - 1}}, \qquad \text{for all } |x| > 1.$$

In a similar way one can derive the following three results:

$$\frac{d}{dx}\arccos x = -\frac{1}{\sqrt{1-x^2}}, \qquad \text{for all } |x| < 1;$$

$$\frac{d}{dx}\operatorname{arccot} x = -\frac{1}{x^2+1};$$

$$\frac{d}{dx}\operatorname{arccsc} x = -\frac{1}{|x|\sqrt{x^2-1}}, \qquad \text{for all } |x| > 1.$$

Using these results in conjunction with the chain rule, we get the following theorem.

Theorem 7.5.1

If u is a differentiable function of x, then

(i) $$\frac{d}{dx}\arcsin u = \frac{1}{\sqrt{1-u^2}}\frac{du}{dx},$$

(ii) $$\frac{d}{dx}\arccos u = -\frac{1}{\sqrt{1-u^2}}\frac{du}{dx},$$

(iii) $$\frac{d}{dx}\arctan u = \frac{1}{u^2+1}\frac{du}{dx},$$

(iv) $$\frac{d}{dx}\operatorname{arccot} u = -\frac{1}{u^2+1}\frac{du}{dx},$$

(v) $$\frac{d}{dx}\operatorname{arcsec} u = \frac{1}{|u|\sqrt{u^2-1}}\frac{du}{dx},$$

(vi) $$\frac{d}{dx}\operatorname{arccsc} u = -\frac{1}{|u|\sqrt{u^2-1}}\frac{du}{dx}.$$

Note that the derivative of each function differs from the derivative of its corresponding "co" function only by a negative sign.

Example 1 | Calculate $d[\arctan(2x^2+1)]/dx$.

$$\frac{d}{dx}\arctan(2x^2+1) = \frac{1}{(2x^2+1)^2+1}(4x) = \frac{2x}{2x^4+2x^2+1}. \qquad \|$$

Example 2 | Calculate $d[\operatorname{arcsec}(e^x)]/dx$.

$$\frac{d}{dx}\operatorname{arcsec}(e^x) = \frac{1}{|e^x|\sqrt{(e^x)^2-1}}e^x.$$

Then, since $e^x > 0$, we have $|e^x| = e^x$, and so

$$\frac{d}{dx}\operatorname{arcsec}(e^x) = \frac{1}{\sqrt{e^{2x}-1}}. \qquad \|$$

Example 3 Calculate $\frac{d}{dx}[\arcsin(\ln x)]^2$.

$$\frac{d}{dx}[\arcsin(\ln x)]^2 = 2\arcsin(\ln x)\frac{1}{\sqrt{1-(\ln x)^2}}\frac{1}{x}$$

$$= \frac{2\arcsin(\ln x)}{x\sqrt{1-(\ln x)^2}}.$$

||

Exercises 7.5

Find the derivative of each function in Exercises 1–14.

1. $f(x) = \arcsin 5x$.
2. $f(x) = \arctan(x - 1)$.
3. $f(x) = \operatorname{arcsec}(x^{-2})$.
4. $f(x) = (4 - x^2)^{-1/2} - \arcsin(x/a)$.
5. $f(x) = \arctan(e^x - 1)$.
6. $f(x) = \arccos(\tan x + 1)$.
7. $f(x) = \ln(\operatorname{arccsc} x)$.
8. $f(x) = e^{\arcsin x}$.
9. $f(x) = e^{\arctan 3x}$.
10. $f(x) = \arccos(\ln 5x)$.
11. $f(x) = \arcsin(x^3)$.
12. $f(x) = \arctan(x^2 + 3)$.
13. $f(x) = \operatorname{arcsec}(e^x)$.
14. $f(x) = \arcsin(x^4)$.
15. Compute the derivative of $y = \arcsin x + \arccos x$. What conclusion can you draw from this result?
16. Compute the derivative of $y = \arctan x - \operatorname{arccot}(1/x)$. What conclusion can you draw from this result?
17. Compute the derivative of $y = \operatorname{arccot}(\sec x + \tan x)$. What conclusion can you draw from this result?

Use integration by parts to find each indefinite integral in Exercises 18–21.

18. $\int \arcsin x\, dx$.
19. $\int \arctan x\, dx$.
20. $\int \operatorname{arccot} x\, dx$.
21. $\int \arccos x\, dx$.
22. At what point along the x-axis does the line segment joining (0, 1) and (1, 2) subtend the maximum angle?
23. A road sign 5 feet in height is suspended so that its lower edge is 20 feet above a roadway. If a driver's eyes are 5 feet above the roadway, at what distance down the road can the driver see the sign best? Assume that the driver will see the sign best when the angle subtended by the sign at his eye is largest.
24. An airplane 1 mile high is flying horizontally at 400 mi/hr in a line that will pass directly over an observer. At what rate is the angle of elevation of the plane changing when the plane is over a point on the ground 1/2 mile from the observer?

7.6 Integration Formulas Involving the Inverse Trigonometric Functions

Since every differentiation formula gives rise to a corresponding integration formula we can obtain several new integration formulas from Theorem 7.5.1. In particular we have the following formulas:

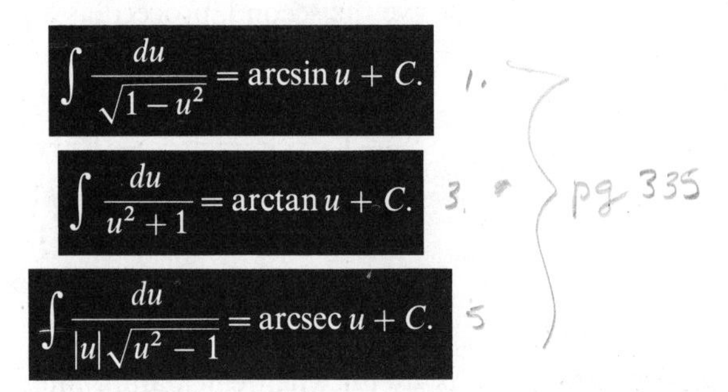

$$\int \frac{du}{\sqrt{1-u^2}} = \arcsin u + C.$$

$$\int \frac{du}{u^2+1} = \arctan u + C.$$

$$\int \frac{du}{|u|\sqrt{u^2-1}} = \operatorname{arcsec} u + C.$$

Those new formulas can be used alone or in conjunction with the techniques of substitution and integration by parts.

Example 1

$$\int_0^3 \frac{dx}{x^2+9} = \frac{1}{9}\int_0^3 \frac{dx}{(x/3)^2+1} = \frac{1}{3}\int_0^3 \frac{d(x/3)}{(x/3)^2+1}$$

$$= \frac{1}{3}\arctan\frac{x}{3}\Big|_0^3 = \frac{1}{3}\left(\frac{\pi}{4} - 0\right) = \frac{\pi}{12}. \qquad \|$$

Example 2

$$\int \frac{x\,dx}{\sqrt{1-x^4}} = \frac{1}{2}\int \frac{d(x^2)}{\sqrt{1-(x^2)^2}} = \frac{1}{2}\arcsin(x^2) + C. \qquad \|$$

Example 3

$$\int_0^{1/3} \frac{dx}{\sqrt{4-9x^2}} = \frac{1}{2}\int_0^{1/3} \frac{dx}{\sqrt{1-(3x/2)^2}} = \frac{1}{3}\int_0^{1/3} \frac{d(3x/2)}{\sqrt{1-(3x/2)^2}}$$

$$= \frac{1}{3}\arcsin\frac{3x}{2}\Big|_0^{1/3} = \frac{1}{3}\left(\frac{\pi}{6} - 0\right) = \frac{\pi}{18}. \qquad \|$$

Example 4

$$\int \frac{e^x\,dx}{1+e^{2x}} = \int \frac{d(e^x)}{1+(e^x)^2} = \arctan(e^x) + C. \qquad \|$$

Example 5

$$\int \frac{\sec^2 y\,dy}{\sqrt{1-\tan^2 y}} = \int \frac{d(\tan y)}{\sqrt{1-\tan^2 y}} = \arcsin(\tan y) + C. \qquad \|$$

Two generalizations of the formulas above are often useful:

$$\int \frac{du}{u^2+a^2} = \frac{1}{a}\arctan\frac{u}{a} + C,$$

$$\int \frac{du}{\sqrt{a^2-u^2}} = \arcsin\frac{u}{a} + C.$$

The derivation of the first formula is very similar to Example 1 above and will be left as an exercise. To derive the second, proceed as follows.

$$\int \frac{du}{\sqrt{a^2 - u^2}} = \int \frac{du}{a\sqrt{1 - \left(\dfrac{u}{a}\right)^2}} = \int \frac{d\left(\dfrac{u}{a}\right)}{\sqrt{1 - \left(\dfrac{u}{a}\right)^2}}$$

$$= \arcsin\left(\frac{u}{a}\right) + C.$$

Integration by parts is useful when calculating integrals involving the inverse trigonometric functions.

Example 6 | Find $\int \arccos x \, dx$.

Let $u = \arccos x$ and $dv = dx$. Then $du = -dx/\sqrt{1 - x^2}$ and $v = x$; so

$$\int \arccos x \, dx = x \arccos x + \int \frac{x}{\sqrt{1 - x^2}} \, dx$$

$$= x \arccos x - \frac{1}{2} \int (1 - x^2)^{-1/2} \, d(1 - x^2)$$

$$= x \arccos x - (1 - x^2)^{1/2} + C.$$ ||

Exercises 7.6

In Exercises 1–22 evaluate the given integral.

1. $\int_{-1}^{1} \dfrac{dx}{3 + x^2}$.
2. $\int \dfrac{dx}{\sqrt{9 - x^2}}$.
3. $\int \dfrac{dx}{9x^2 + 4}$.
4. $\int \dfrac{e^x \, dx}{\sqrt{1 - e^{2x}}}$.
5. $\int \dfrac{e^x \, dx}{e^x \sqrt{e^{2x} - 4}}$.
6. $\int \dfrac{\sin 3x \, dx}{9 + \cos^2 3x}$.
7. $\int \dfrac{dx}{|x|\sqrt{9x^2 - 16}}$.
8. $\int \dfrac{x^2 \, dx}{\sqrt{1 - x^6}}$.
9. $\int \dfrac{dx}{\sqrt{25 - x^2}}$.
10. $\int \dfrac{dx}{3x^2 + 5}$.
11. $\int \dfrac{dx}{x\sqrt{x^4 - 9}}$.
12. $\int \dfrac{\sec^2 x \, dx}{\sqrt{1 - 4\tan^2 x}}$.
13. $\int \dfrac{dx}{x\sqrt{4 - 9(\ln x)^2}}$.
14. $\int \dfrac{dx}{x^{1/2} + x^{3/2}}$.
15. $\int \dfrac{x \, dx}{\sqrt{16 - x^2}}$.
16. $\int \dfrac{x^3 \, dx}{x^4 + 4}$.
17. $\int \dfrac{x \, dx}{x^4 + 4}$.
18. $\int \dfrac{x^2 \, dx}{\sqrt{4 - x^6}}$.
19. $\int \dfrac{dx}{\sqrt{x}(1 + x)}$.
20. $\int \dfrac{dx}{\sqrt{x - x^2}}$.
21. $\int \dfrac{dx}{x + x(\ln x)^2}$.
22. $\int \dfrac{dx}{x\sqrt{1 - (\ln x)^2}}$.
23. Find the area of the region bounded by the graph of $y = 1/(x^2 + 1)$ and the x-axis, between the vertical lines $x = -1$ and $x = 1$.

24. Find the length of the graph of $y = -(1 - x^2)^{1/2}$, $0 \le x \le \sqrt{2}/2$.
25. The region bounded by the graph of $y = [1/(1 + x^2)]^{1/2}$, the x-axis, and the lines $x = 0$ and $x = 1$ is rotated about the x-axis. Find the resulting volume.
26. Prove that

$$\int \frac{du}{u^2 + a^2} = \frac{1}{a} \arctan \frac{u}{a} + C.$$

27. Write the integration formulas resulting from these equations:

$$\frac{d \arccos u}{dx} = \frac{-1}{\sqrt{1 - u^2}} \frac{du}{dx},$$

$$\frac{d \operatorname{arccot} u}{dx} = \frac{-1}{u^2 + 1} \frac{du}{dx},$$

$$\frac{d \operatorname{arccsc} u}{dx} = \frac{-1}{|u|\sqrt{u^2 - 1}} \frac{du}{dx}.$$

Why weren't those integration formulas included in this section?

Brief Review of Chapter 7

1. The Derivatives of the Trigonometric Functions

It has been proved that if x is measured in radians, then

$$\lim_{x \to 0} \frac{\sin x}{x} = 1.$$

That fact is needed to obtain the derivative of the sine function and is helpful in evaluating other limits involving trigonometric functions.

$$\frac{d}{dx} \sin u = \cos u \frac{du}{dx}; \qquad \frac{d}{dx} \cos u = -\sin u \frac{du}{dx}.$$

$$\frac{d}{dx} \tan u = \sec^2 u \frac{du}{dx}; \qquad \frac{d}{dx} \cot u = -\csc^2 u \frac{du}{dx}.$$

$$\frac{d}{dx} \sec u = \sec u \tan u \frac{du}{dx}; \qquad \frac{d}{dx} \csc u = -\csc u \cot u \frac{du}{dx}.$$

2. The Integrals of the Trigonometric Functions

$$\int \sin u \, du = -\cos u + C; \qquad \int \cos u \, du = \sin u + C.$$

$$\int \tan u \, du = \ln|\sec u| + C; \qquad \int \cot u \, du = \ln|\sin u| + C.$$

$$\int \sec u \, du = \ln|\sec u + \tan u| + C; \qquad \int \csc u \, du = -\ln|\csc u + \cot u| + C.$$

3. Integrals Arising from Trigonometric Derivatives

$$\int \sec^2 u\, du = \tan u + C; \qquad \int \csc^2 u\, du = -\cot u + C.$$

$$\int \sec u \tan u\, du = \sec u + C; \qquad \int \csc u \cot u\, du = -\csc u + C.$$

4. Definitions of Inverse Trigonometric Functions

$y = \arcsin x$ if and only if $\sin y = x$ and $-\frac{\pi}{2} \leq y \leq \frac{\pi}{2}$.

$y = \arccos x$ if and only if $\cos y = x$ and $0 \leq y \leq \pi$.

$y = \operatorname{arccot} x$ if and only if $\cot y = x$ and $0 < y < \pi$.

$y = \arctan x$ if and only if $\tan y = x$ and $-\frac{\pi}{2} < y < \frac{\pi}{2}$.

$y = \operatorname{arcsec} x$ if and only if $\sec y = x$ and $0 \leq y \leq \pi$, $y \neq \frac{\pi}{2}$.

$y = \operatorname{arccsc} x$ if and only if $\csc y = x$ and $-\frac{\pi}{2} \leq y \leq \frac{\pi}{2}$, $y \neq 0$.

5. Derivatives of the Inverse Trigonometric Functions

$$\frac{d(\arcsin u)}{dx} = \frac{1}{\sqrt{1-u^2}}\frac{du}{dx}. \qquad \frac{d(\arccos u)}{dx} = \frac{-1}{\sqrt{1-u^2}}\frac{du}{dx}.$$

$$\frac{d(\arctan u)}{dx} = \frac{1}{u^2+1}\frac{du}{dx}. \qquad \frac{d(\operatorname{arccot} u)}{dx} = \frac{-1}{u^2+1}\frac{du}{dx}.$$

$$\frac{d(\operatorname{arcsec} u)}{dx} = \frac{1}{|u|\sqrt{u^2-1}}\frac{du}{dx}. \qquad \frac{d(\operatorname{arccsc} u)}{dx} = \frac{-1}{|u|\sqrt{u^2-1}}\frac{du}{dx}.$$

6. Integrals Arising from Derivatives of Inverse Trigonometric Functions

$$\int \frac{du}{\sqrt{1-u^2}} = \arcsin u + C,$$

$$\int \frac{du}{u^2+1} = \arctan u + C,$$

$$\int \frac{du}{|u|\sqrt{u^2-1}} = \operatorname{arcsec} u + C.$$

Technique Review Exercises, Chapter 7

1. Find y' if $y = \sec(3x - 1)$.
2. Find y' if $y = \cot^2(1 - x^2)$.

3. Find y' if $y = \cos^3(x^3 - 2x)^2$.

4. Evaluate $\int x \sin(x^2 + 3)dx$.

5. Evaluate $\int \tan(2x - 3)dx$.

6. Evaluate $\int \csc(1 - 2x)dx$.

7. Evaluate $\int x \sec^2(x^2 - 2)dx$.

8. Evaluate $\int \sec 5x \tan 5x\, dx$.

9. Evaluate (a) $\arcsin(\sin 2\pi/3)$, (b) $\arctan(-1)$.

10. Find y' if $y = \arcsin(x^2 - 5)$.

11. Find y' if $y = \arctan^2(3x)$.

12. Find y' if $y = \operatorname{arcsec}(x^2 + 1)$.

13. Evaluate $\int \frac{dx}{(16 - x^2)^{1/2}}$.

14. Evaluate $\int \frac{x^2\, dx}{1 + x^6}$.

15. Evaluate $\int \frac{dx}{x(x^4 - 1)^{1/2}}$.

Additional Exercises, Chapter 7

Section 7.1

Find the limits in Exercises 1–6.

1. $\lim_{x\to 0} \frac{x}{\sin 2x}$.

2. $\lim_{x\to 0} \frac{\sin 8x}{9x}$.

3. $\lim_{t\to 0} \frac{4 - \cos 2t}{3t}$.

4. $\lim_{t\to 0} \frac{4t}{\tan 7t}$.

5. $\lim_{x\to 0} \frac{2x}{5x - 3\sin x}$.

6. $\lim_{x\to 0} \frac{\sin 4x}{5x - 2\sin 3x}$.

Differentiate each function in Exercises 7–12.

7. $f(x) = \sin 12x$.

8. $f(x) = x^5 \sin(7x^2 - 3)$.

9. $f(x) = \sin^7 4x$.

10. $f(x) = e^{5 \sin 3x}$.

11. $f(x) = \ln^3(8x)$.

12. $f(x) = x^{\sin 2x}$.

Section 7.2

Differentiate each function.

13. $f(x) = \cos(x^3 - 2)$.

14. $f(x) = \cos^4(x^2 + 4)$.

15. $f(x) = \tan^2(x^2 - 1)$.

16. $f(x) = \sec^3 x \tan^2 x$.

17. $f(x) = e^{2x} \csc 3x$.

18. $f(x) = \sec^2(e^{3x} + 5x)$.

19. $f(x) = \csc^5(3 - x^2)$.

20. $f(x) = e^{\tan 7x}$.

Section 7.3

Evaluate each integral.

21. $\int \sin 15x\, dx$.

22. $\int x \cos(x^2 + 2)dx$.

23. $\int x^2 \tan(2 - x^3)dx$.

24. $\int \sec^2 8x\, dx$.

25. $\int x^2 \sec(x^3) \tan(x^3)dx$.

26. $\int \sec 12x\, dx$.

27. $\int x^2 \csc(x^3)dx$.

28. $\int (x+3)\cot(x^2+6x)dx$.

29. $\int x \sin(3x)dx$.

30. $\int e^{\tan x} \sec^2 x \, dx$.

Section 7.4

Simplify each expression in Exercises 31–34.

31. $\arcsin(-1/2)$.

32. $\arccos(-1/2)$.

33. $\arctan 0$.

34. $\operatorname{arcsec}(2^{1/2})$.

Find the inverse of each function and the domain of the inverse function for suitable domains of the given functions.

35. $y = 2 \cos 4x$.

36. $y = 7 \arcsin 3x$.

Section 7.5

Find the derivative of each function.

37. $f(x) = \arcsin 10x$.

38. $f(x) = \arctan(x^2 - 1)$.

39. $f(x) = \operatorname{arcsec}(e^{2x})$.

40. $f(x) = e^{\arccos 5x}$.

41. $f(x) = \ln(\arcsin x^2)$.

42. $f(x) = \arctan(\ln 7x)$.

Section 7.6

Evaluate each integral.

43. $\int \frac{dx}{\sqrt{49 - x^2}}$.

44. $\int \frac{dx}{4x^2 + 25}$.

45. $\int \frac{dx}{|x|\sqrt{36x^2 - 9}}$.

46. $\int \frac{x\,dx}{\sqrt{9 - x^4}}$.

47. $\int \frac{dx}{\sqrt{e^{2x} - 1}}$.

48. $\int \frac{dx}{8 + 5x^2}$.

49. $\int \frac{e^{2x}\,dx}{e^{4x} + 49}$.

50. $\int \frac{e^{3x}\,dx}{\sqrt{100 - e^{6x}}}$.

Challenging Problems, Chapter 7

1. Evaluate

$$\lim_{h \to 0} \frac{1}{h}\left(\int_1^{2+h} \frac{\sin t}{t}\,dt - \int_1^2 \frac{\sin t}{t}\,dt\right).$$

2. Use the technique of the proof of Theorem 7.1.1 to show that if h is in degrees, $\lim_{h \to 0} (\sin h)/h = \pi/180$.

3. Obtain the familiar formula $L = 2\pi r$ for the circumference (or length) L of a circle of radius r by using the following outline.

   (i) Define π to be the ratio of the circumference of a circle of radius 1 to the length of its diameter.

(ii) Use the arc length formula to show that

$$L = 8r \int_0^{r/\sqrt{2}} (r^2 - x^2)^{-1/2}\, dx.$$

(iii) Use your definition of π and the formula of step (ii) to deduce that

$$8 \int_0^{1/\sqrt{2}} (1 - x^2)^{-1/2}\, dx = 2\pi.$$

(iv) Evaluate the integral for L by using the transformation $x = rz$ and the result of step (iii) to get $L = 2\pi r$.

(v) Why wasn't the integral of (ii) expressed as

$$L = 2r \int_{-r}^{r} (r^2 - x^2)^{-1/2}\, dx?$$

4. Obtain the familiar formula $A = \pi r^2$ for the area A of the region interior to a circle of radius r by using the following outline.

(i) Note that

$$\frac{A}{8} = \int_0^{r/\sqrt{2}} (r^2 - x^2)^{1/2}\, dx - \frac{r^2}{4}.$$

(ii) Apply integration by parts with

$$u = (r^2 - x^2)^{1/2} \quad \text{to} \quad \int (r^2 - x^2)^{1/2}\, dx.$$

(iii) Obtain

$$\int (r^2 - x^2)^{1/2}\, dx = \frac{x}{2}(r^2 - x^2)^{1/2} + \frac{r}{2} \int (r^2 - x^2)^{-1/2}\, dx.$$

(iv) Use the result of Problem 3 along with the results of (i) and (iii) to complete the proof.

(v) Why wasn't the integral of step (i) expressed as

$$A = 2 \int_{-r}^{r} (r^2 - x^2)^{1/2}\, dx?$$

5. Show that the function

$$f(x) = \begin{cases} x^2 \sin(1/x) & \text{if } x \neq 0 \\ 0 & \text{if } x = 0 \end{cases}$$

has a derivative for all x. (Be careful to obtain $f'(0)$ correctly.) Show that its derivative is not continuous at $x = 0$. $f'(x)$ is an example of a discontinuous function having the intermediate value property on every interval. (See Problem 8 in the Challenging Problems, Chapter 3.)

6. Let $f(x) = \begin{cases} 2x \sin(1/x) - \cos(1/x) & \text{if } x \neq 0 \\ 0 & \text{if } x = 0. \end{cases}$

(i) Evaluate $\int f(x)dx$.

(ii) Use the result of (i) to evaluate $\int_0^{2/\pi} f(x)dx$. Note that this is an example of the application of the Fundamental Theorem of Calculus to the integral of a discontinuous function.

7. A rectangular piece of sheet steel 3 m long and 80 cm wide is to be made into a trough for a concrete mixing truck. The sheet is to be bent so that its cross section is in the shape of a segment of a circle. Find the radius that will maximize the cross sectional area of the trough.

8. A pipeline is to be laid between two opposite points, A and B, on a circular lake. The cost of laying the pipeline along the shore is c dollars per meter, and the cost of laying the pipeline under water is kc dollars per meter. Company policy has been to lay the pipeline either entirely under water or entirely along the shore, depending on which is least costly. It is suggested that it may be cheaper to lay the pipeline under water to some point, C, on the shore and then follow the shoreline as shown in the figure. Evaluate that suggestion.

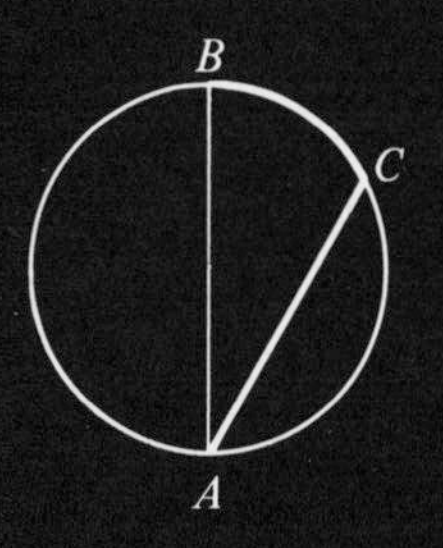

9. The curvature K of the graph of $y = f(x)$ is defined as the absolute value of the rate of change of the angle ϕ with respect to arc length.

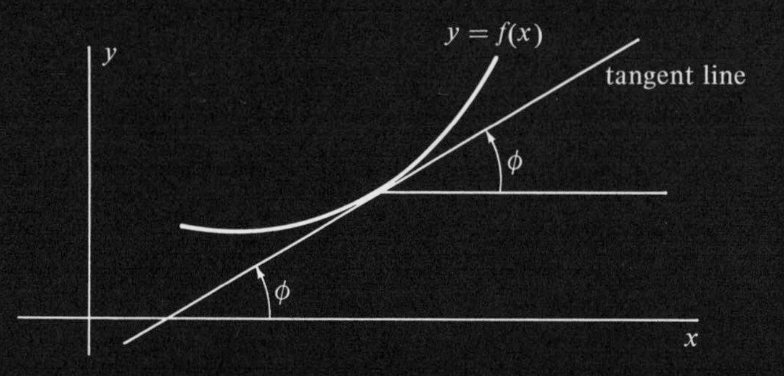

(i) Prove that

$$K = \frac{|f''(x)|}{[1 + (f'(x))^2]^{3/2}}$$

(ii) Compute the curvature of a circle of radius a.

10. Consider the following graph. Suppose that a particle slides along this frictionless graph starting from rest at the origin. If the particle moves only under the influence of gravity and if v denotes the speed of the particle at (x, y) then conservation of energy dictates that

$$-mgy = \frac{1}{2}mv^2$$

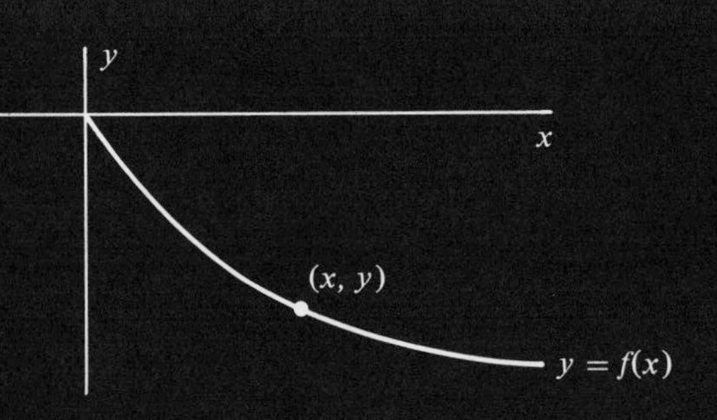

where m is the mass of the particle and g the acceleration due to gravity. Use that equation to show that the time required for the particle to slide to the point (x_1, y_1) is

$$t_1 = \int_0^{x_1} \sqrt{\frac{1 + (f'(x))^2}{-2gf(x)}}\, dx.$$

$\left(\textit{Hint}: \text{First obtain } \frac{ds}{dt} = \frac{ds}{dx}\frac{dx}{dt} \text{ then compute } \frac{dt}{dx}.\right)$

11. Use the result of the previous problem to prove the following remarkable property of the graph of $x = a(\phi - \sin \phi)$, $y = a(\cos \phi - 1)$. Particles starting at (x_0, y_0) and (x_1, y_1) will require the *same* time to slide to $(a\pi, -2a)$ for *any* points (x_0, y_0) and (x_1, y_1) different from $(a\pi, -2a)$.

    Because of that property this graph is called a tautochrone ("same time"). In addition this curve has the brachistochrone ("shortest time") property. That is, it is the curve along which a particle sliding only under the influence of gravity will move from (0.0) to $(a\pi, -2a)$ in the shortest time.

12. The speed of light depends on the medium. In fact, the more optically dense the medium, the lower the speed of light is. When light passes from one medium to another its direction abruptly changes. The phenomenon is called *refraction*. An important property of light called Fermat's principle applies to both reflection and refraction. The principal states that when light travels from one point to another, it moves along the path that requires the least time. Use Fermat's principle to prove each of the following:

    (a) If light is reflected as shown in its path from P_1 to P_2, then $\alpha_1 = \alpha_2$.

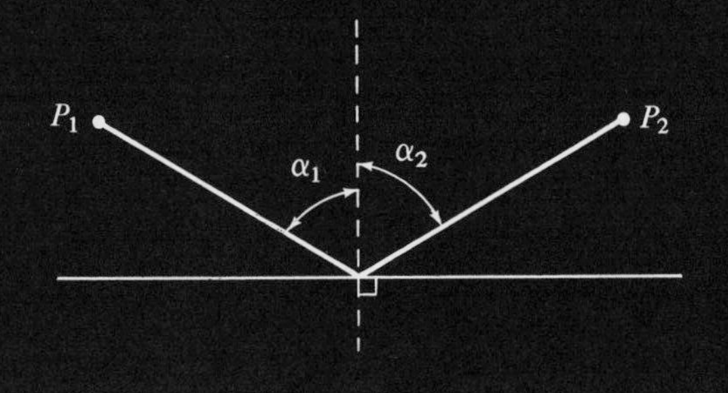

(b) (*Snell's Law of Refraction*) If light is refracted as shown in its path from P_1 to P_2, then $\dfrac{\sin \alpha_1}{\sin \alpha_2} = \dfrac{s_1}{s_2}$, where s_1 is the speed of light in the first medium and s_2 is the speed of light in the second medium.

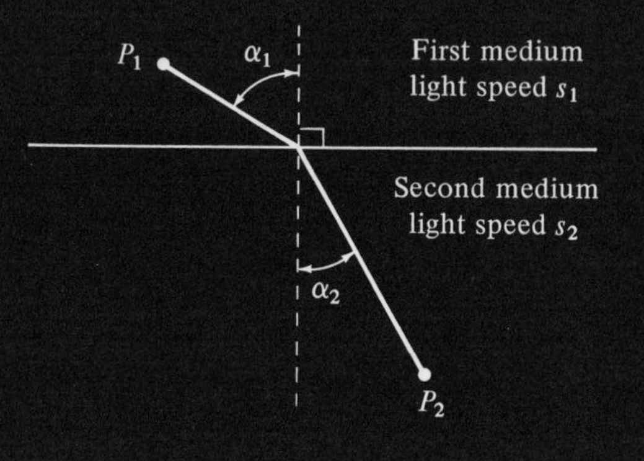

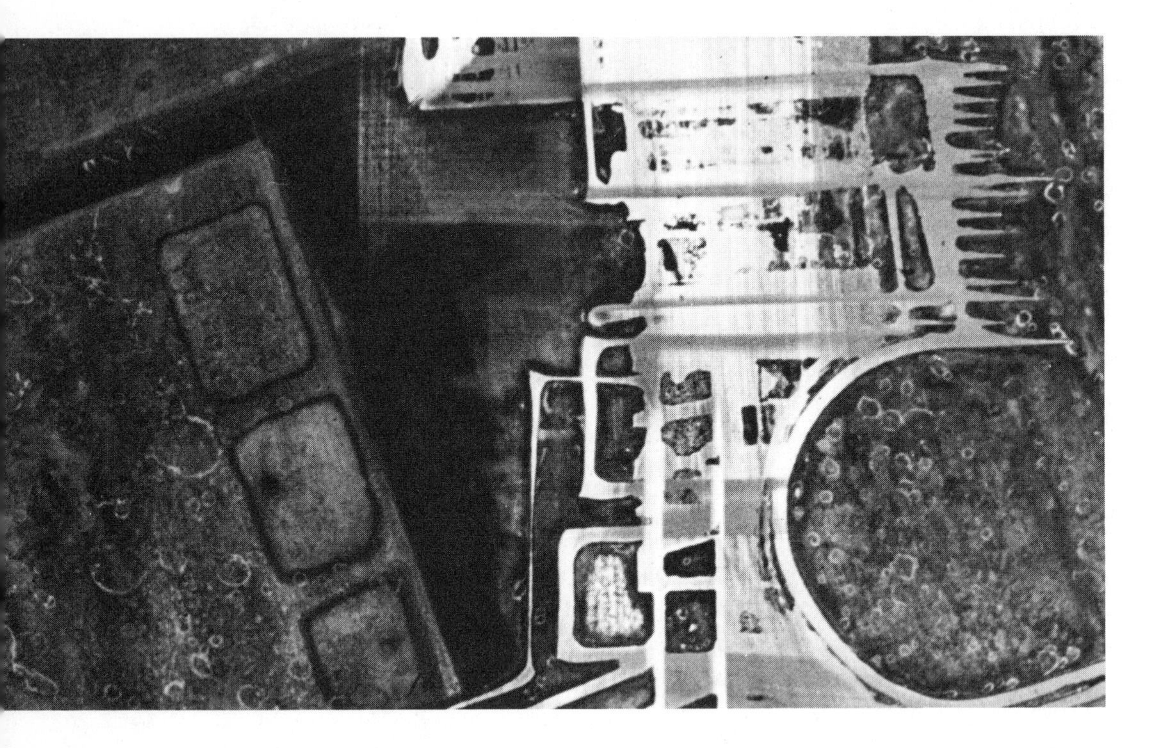

8

Other Integration Techniques

8.1 Further Application of Previous Techniques

In addition to formulas for integrating specific types of functions such as polynomial, logarithmic, trigonometric, and inverse trigonometric functions, we have developed these two general techniques.

The substitution formula (Section 4.7):

$$\int f(u)du = F(u) + C, \tag{1}$$

where u is a differentiable function and $F(u)$ is any function of u such that $dF(u)/du = f(u)$.

Integration by parts (Section 6.7):

$$\int u\, dv = uv - \int v\, du \tag{2}$$

where u and v are differentiable functions.

As we found in Section 4.7, it is often best to apply the substitution formula by thinking, but not writing, what the symbol u represents.

Example 1 $\displaystyle \int x^2(x^3 + 3)^4\, dx = \frac{1}{3}\int (x^3 + 3)^4\, 3x^2\, dx = \frac{(x^3 + 3)^5}{15} + C.$ ||

Example 2 $\displaystyle \int xe^{x^2+3}\, dx = \frac{1}{2}\int e^{x^2+3}(2x)dx = \frac{e^{x^2+3}}{2} + C.$ ||

However, sometimes it *is* best to write the function that u represents, to express the whole integral in terms of u, to integrate it, and then to express the result in terms of x. A particular case of this occurs when the integral contains a fractional power of a linear function of x.

Example 3 $\int (x + 3)\sqrt{x - 7}\, dx.$

Let $u = (x - 7)^{1/2}$. Then $u^2 = x - 7$, $x = u^2 + 7$, $dx = 2u\, du$, and $x + 3 = u^2 + 10$. Expressing the whole integral in terms of u, we have

$$\int (u^2 + 10)u \cdot 2u\, du = 2\int (u^4 + 10u^2)du = \frac{2u^5}{5} + \frac{20u^3}{3} + C.$$

Replacing u by $(x - 7)^{1/2}$, we obtain the final result:

$$\int (x + 3)\sqrt{x - 7}\, dx = \frac{2(x - 7)^{5/2}}{5} + \frac{20(x - 7)^{3/2}}{3} + C.$$ ||

The technique of substitution can also be used on definite integrals, as the next example illustrates.

Example 4 | $\int_{-1}^{6} \frac{x-1}{(x+2)^{2/3}} dx.$

Let $u = (x+2)^{1/3}$. Then $u^3 = x + 2$, $x = u^3 - 2$, $dx = 3u^2\, du$, and $x - 1 = u^3 - 3$. Moreover, since $u = 1$ when $x = -1$ and $u = 2$ when $x = 6$, we have,

$$\int_{-1}^{6} \frac{x-1}{(x+2)^{2/3}} dx = \int_{1}^{2} \frac{u^3 - 3}{u^2} 3u^2\, du$$

$$= 3\int_{1}^{2} (u^3 - 3) du = \frac{3u^4}{4} - 9u \Big|_{1}^{2} = \frac{9}{4}. \qquad \|$$

In Examples 3 and 4, the substitutions $u = (x-7)^{1/2}$ and $u = (x+2)^{1/3}$ lead to integrals containing no fractional exponents. In those and other integrals the integration can also be done with different substitutions that produce integrals containing fractional exponents. In particular the integrals of Examples 3 and 4 can be found using the substitutions $u = x - 7$ and $u = x + 2$, as we show next.

For Example 3, let $u = x - 7$. Then $x = u + 7$, $dx = du$, and so

$$\int (x+3)\sqrt{x-7}\, dx = \int (u+10)u^{1/2}\, du = \int (u^{3/2} + 10u^{1/2}) du$$

$$= \frac{2u^{5/2}}{5} + \frac{20u^{3/2}}{3} + C$$

$$= \frac{2(x-7)^{5/2}}{5} + \frac{20(x-7)^{3/2}}{3} + C.$$

For Example 4, let $u = x + 2$. Then $x = u - 2$ and $dx = du$. Moreover, since $u = 1$ when $x = -1$ and $u = 8$ when $x = 6$, we have

$$\int_{-1}^{6} \frac{x-1}{(x+2)^{2/3}} dx = \int_{1}^{8} \frac{u-3}{u^{2/3}} du = \int_{1}^{8} (u^{1/3} - 3u^{-2/3}) du$$

$$= \frac{3u^{4/3}}{4} - 9u^{1/3} \Big|_{1}^{8} = \frac{9}{4}.$$

However, in many cases it is necessary to avoid fractional exponents in order to complete the integration. In such cases an additional substitution should be made to eliminate the fractional exponents, or the problem should be reworked using the technique of Examples 3 and 4.

Although we have used integration by parts in several ways, the technique was introduced before we obtained integration formulas for the trigonometric functions. There are some clever applications of integration by parts to certain integrals involving trigonometric functions, as we illustrate next.

Example 5 | $\int e^x \sin x\, dx.$

Let $u = e^x$ and $dv = \sin x\, dx$. Then $du = e^x\, dx$ and $v = -\cos x$. Hence

$$\int e^x \sin x\, dx = -e^x \cos x + \int e^x \cos x\, dx.$$

Now apply integration by parts again by letting $u = e^x$ and $dv = \cos x\,dx$. Then $du = e^x\,dx$, and $v = \sin x$. Thus

$$\int e^x \sin x\,dx = -e^x \cos x + e^x \sin x - \int e^x \sin x\,dx.$$

Our effort to obtain a simpler integral than the given one has resulted in the same integral we started with. However, if we add $\int e^x \sin x\,dx$ to both sides of the last equation, the problem can be solved! We obtain

$$2\int e^x \sin x\,dx = -e^x \cos x + e^x \sin x.$$

In Section 6.7, it was mentioned that the constant of integration is usually not added until we evaluate the last integral obtained. In the present example we did not evaluate the last integral obtained on the right side, but we moved it to the left side. Hence we should add a constant now to get

$$2\int e^x \sin x\,dx = -e^x \cos x + e^x \sin x + C_1.$$

Then, dividing both sides by 2, we get

$$\int e^x \sin x\,dx = \frac{1}{2}e^x(\sin x - \cos x) + C,$$

where $C = C_1/2$. ||

The technique of Example 5 is applicable to integrals of the form

$$\int e^{ax} \sin bx\,dx \quad \text{or} \quad \int e^{ax} \cos bx\,dx,$$

where a and b are constants. Similarly, the technique of using integration by parts twice and then transposing a term from the right side of the resulting equation to the left side, can be used for integrals of the form

$$\int \sin ax \sin bx\,dx, \quad \int \sin ax \cos bx\,dx, \quad \int \cos ax \cos bx\,dx,$$

when a and b are distinct constants.

Example 6 | $\int \sin 2x \cos 5x\,dx$.

Let $u = \sin 2x$ and $dv = \cos 5x\,dx$. Then $du = 2\cos 2x\,dx$, $v = (\sin 5x)/5$, and so

$$\int \sin 2x \cos 5x\,dx = \frac{\sin 2x \sin 5x}{5} - \frac{2}{5}\int \cos 2x \sin 5x\,dx.$$

Now let $u = \cos 2x$ and $dv = \sin 5x\,dx$. Then $du = -2\sin 2x\,dx$ and $v = -(\cos 5x)/5$. We obtain

$$\int \sin 2x \cos 5x\,dx = \frac{\sin 2x \sin 5x}{5} + \frac{2\cos 2x \cos 5x}{25} + \frac{4}{25}\int \sin 2x \cos 5x\,dx.$$

As before we are back to the same integral we started with. If we subtract $(4/25)\int \sin 2x \cos 5x\, dx$ from both sides and add the integration constant C_1, we have

$$\frac{21}{25}\int \sin 2x \cos 5x\, dx = \frac{1}{5}\sin 2x \sin 5x + \frac{2}{25}\cos 2x \cos 5x + C_1.$$

Thus

$$\int \sin 2x \cos 5x\, dx = \frac{5}{21}\sin 2x \sin 5x + \frac{2}{21}\cos 2x \cos 5x + C,$$

where $C = (25/21)C_1$. ||

It is essential here that, in the second integration by parts, u be the portion obtained from the u in the first integration by parts. Notice what happens to the integral of Example 5 when this is not done. Proceeding from the equation

$$\int e^x \sin x\, dx = -e^x \cos x + \int e^x \cos x\, dx$$

in Example 5, we let $u = \cos x$ and $dv = e^x\, dx$. Then we have $du = -\sin x\, dx$ and $v = e^x$, and so

$$\int e^x \sin x\, dx = -e^x \cos x + e^x \cos x + \int e^x \sin x\, dx.$$

Then, since $-e^x \cos x + e^x \cos x = 0$, we arrive at

$$\int e^x \sin x\, dx = \int e^x \sin x\, dx,$$

a fact we knew before starting the problem! What happened was that the second integration by parts reversed the work of the first integration by parts, so that we arrived back at the starting point.

Exercises 8.1

In Exercises 1–30, evaluate the integrals.

1. $\int (1 + x)\sqrt{x + 3}\, dx.$
2. $\int (x - 3)\sqrt{2x + 5}\, dx.$
3. $\int (x^2 - 3)(x + 5)^{2/3}\, dx.$
4. $\int (x + 6)(x - 1)^{3/2}\, dx.$
5. $\int \frac{2x + 3}{\sqrt{4x - 1}}\, dx.$
6. $\int \frac{1 - x}{\sqrt{x + 9}}\, dx.$
7. $\int_1^2 \frac{x + 2}{(x + 1)^{3/2}}\, dx.$
8. $\int_0^2 \frac{x\, dx}{(x + 2)^{1/4}}.$
9. $\int e^{2x} \sin x\, dx.$
10. $\int e^x \cos x\, dx.$
11. $\int_0^{\pi/2} e^{2x} \sin x\, dx.$
12. $\int_0^{\pi/2} e^x \cos x\, dx.$
13. $\int e^{2x} \sin 2x\, dx.$
14. $\int e^{2x} \cos x\, dx.$
15. $\int e^{3x} \cos x\, dx.$
16. $\int e^{2x} \cos 3x\, dx.$
17. $\int_{-\pi}^{\pi} \sin x \cos 2x\, dx.$
18. $\int_{-\pi}^{\pi} \sin 2x \cos 3x\, dx.$
19. $\int \cos 2x \cos 5x\, dx.$
20. $\int \sin 3x \cos 2x\, dx.$
21. $\int \sin(\ln x)dx.$
22. $\int \cos(\ln x)dx.$
23. $\int x^2 \sin x\, dx.$
24. $\int x^2 \cos x\, dx.$
25. $\int_0^{\pi/2} x^2 \sin x\, dx.$
26. $\int_0^{\pi/2} x^2 \cos x\, dx.$
27. $\int e^{ax} \sin bx\, dx.$
28. $\int \sin ax \cos bx\, dx.$
29. $\int \sec^3 x\, dx.$
30. $\int \csc^3 x\, dx.$

In Exercises 31–34 use the indicated substitution to evaluate each integral.

31. $\int \frac{dx}{x^2\sqrt{4-x^2}}$, $u = \frac{2}{x}$.

32. $\int \frac{dx}{x^2\sqrt{x^2+9}}$, $u = \frac{3}{x}$.

33. $\int \frac{x^2\,dx}{(x+2)^3}$, $u = x + 2$.

34. $\int \frac{\sqrt{x^2-25}}{x^4}\,dx$, $u = \frac{5}{x}$.

35. Find the net area of the region bounded by the graph of $y = \sin 2x \sin 3x$ and the x-axis from $x = 0$ to $x = \pi/2$.

36. Find the net area of the region bounded by the graph of $y = \cos 4x \cos 2x$ and the x-axis from $x = \pi/2$ to $x = \pi$.

37. The region under the graph of $y = \sin x$ between $x = 0$ and $x = \pi$ is rotated about the y-axis. Find the resulting volume.

38. Evaluate $\int \cos x \sin x\,dx$ by using
 (a) $\int \cos x \sin x\,dx = -\int \cos x\,d(\cos x)$,
 (b) $\int \cos x \sin x\,dx = \int \sin x\,d(\sin x)$.
 Do you note any discrepancy?

39. Use integration by parts three times to show that

$$\int_0^1 x^5(1-x)^3\,dx = \frac{3!5!}{9!}.$$

8.2 Integrals of Powers of Certain Trigonometric Functions

An integral of the form

$$\int \sin^r x \cos^s x\,dx,$$

where either r or s is a positive odd integer, can be evaluated by using the following technique. If r is positive and odd, the integral is first written as

$$-\int \sin^{r-1} x \cos^s x(-\sin x\,dx).$$

Then $r - 1$ is either zero or a positive even integer; so $\sin^{r-1} x$ either equals 1 or can be changed to cosines using the fact that $\sin^2 x = 1 - \cos^2 x$. Since $d(\cos x) = -\sin x\,dx$, the resulting integral can be evaluated easily.

Example 1

$$\begin{aligned}\int \sin^3 x \cos^\pi x\,dx &= -\int \sin^2 x \cos^\pi x(-\sin x\,dx)\\ &= -\int (1-\cos^2 x)(\cos^\pi x)(-\sin x\,dx)\\ &= -\int (\cos^\pi x - \cos^{\pi+2} x)d(\cos x)\\ &= -\frac{\cos^{\pi+1} x}{\pi+1} + \frac{\cos^{\pi+3} x}{\pi+3} + C.\end{aligned}$$

||

When s is a positive odd integer, the same procedure is used with the roles of sine and cosine reversed.

Example 2

$$\begin{aligned}\int_0^{\pi/2} \sin^2 x \cos^5 x\,dx &= \int_0^{\pi/2} \sin^2 x \cos^4 x(\cos x\,dx)\\ &= \int_0^{\pi/2} \sin^2 x(1-\sin^2 x)^2(\cos x\,dx)\\ &= \int_0^{\pi/2} \sin^2 x(1-2\sin^2 x+\sin^4 x)(\cos x\,dx)\\ &= \int_0^{\pi/2} (\sin^2 x - 2\sin^4 x + \sin^6 x)d(\sin x)\\ &= \left(\frac{\sin^3 x}{3} - \frac{2\sin^5 x}{5} + \frac{\sin^7 x}{7}\right)\Big|_0^{\pi/2}\\ &= \left(\frac{1}{3} - \frac{2}{5} + \frac{1}{7}\right) = \frac{8}{105}.\end{aligned}$$

||

Since $d(\sec x) = \sec x\ \tan x\,dx$, $d(\tan x) = \sec^2 x\,dx$, and $\sec^2 x = 1 + \tan^2 x$, a similar technique may be used to evaluate integrals of the form

$$\int \sec^r x \tan^s x\,dx,$$

where either r is even and positive or s is odd and positive. If r is even and positive, we first write

$$\int \sec^r x \tan^s x\,dx = \int \sec^{r-2} x \tan^s x(\sec^2 x)dx,$$

and then change $\sec^{r-2} x$ to a polynomial in $\tan^2 x$ by using $\sec^2 x = 1 + \tan^2 x$. If s is odd and positive, we first write

$$\int \sec^r x \tan^s x\,dx = \int \sec^{r-1} x \tan^{s-1} x(\sec x \tan x\,dx),$$

and then change $\tan^{s-1} x$ to a polynomial in $\sec^2 x$ by using $\tan^2 x = \sec^2 x - 1$.

Example 3

$$\begin{aligned}\int \sec^4 x \tan^{3/2} x\,dx &= \int \sec^2 x \tan^{3/2} x(\sec^2 x\,dx)\\ &= \int (1+\tan^2 x)\tan^{3/2} x(\sec^2 x\,dx)\\ &= \int (\tan^{3/2} x + \tan^{7/2} x)d(\tan x)\\ &= \frac{2\tan^{5/2} x}{5} + \frac{2\tan^{9/2} x}{9} + C.\end{aligned}$$

||

Example 4

$$\int \sec^3 x \tan^5 x\, dx = \int \sec^2 x \tan^4 x(\sec x \tan x\, dx)$$
$$= \int \sec^2 x(\sec^2 x - 1)^2(\sec x \tan x\, dx)$$
$$= \int (\sec^6 x - 2\sec^4 x + \sec^2 x) d(\sec x)$$
$$= \frac{\sec^7 x}{7} - \frac{2\sec^5 x}{5} + \frac{\sec^3 x}{3} + C. \qquad \|$$

Since $d(\csc x) = -\csc x \cot x\, dx$, $d(\cot x) = -\csc^2 x\, dx$, and $\csc^2 x = 1 + \cot^2 x$, the technique is almost the same for integrals of the form

$$\int \csc^r x \cot^s x\, dx,$$

where either r is even and positive or s is odd positive.

Example 5

$$\int_{\pi/4}^{\pi/2} \csc^4 x \cot^{3/2} x\, dx = -\int_{\pi/4}^{\pi/2} \csc^2 x \cot^{3/2} x(-\csc^2 x\, dx)$$
$$= -\int_{\pi/4}^{\pi/2} (1 + \cot^2 x)\cot^{3/2} x(-\csc^2 x\, dx)$$
$$= -\int_{\pi/4}^{\pi/2} (\cot^{3/2} x + \cot^{7/2} x) d(\cot x)$$
$$= -\frac{2\cot^{5/2} x}{5} - \frac{2\cot^{9/2} x}{9}\Bigg|_{\pi/4}^{\pi/2}$$
$$= \frac{2}{5} + \frac{2}{9} = \frac{28}{45}. \qquad \|$$

When both r and s are positive even integers, we can evaluate

$$\int \sin^r x \cos^s x\, dx$$

by using these trigonometric identities:

$$\sin\theta\cos\theta = \frac{\sin 2\theta}{2}, \tag{1}$$

$$\sin^2\theta = \frac{1 - \cos 2\theta}{2}. \tag{2}$$

$$\cos^2\theta = \frac{1 + \cos 2\theta}{2}. \tag{3}$$

The identity (1) is used first if it applies. Let us consider some examples.

Example 6

$$\int \sin^4 x \cos^4 x\, dx = \int \left(\frac{\sin 2x}{2}\right)^4 dx = \frac{1}{16}\int \sin^4 2x\, dx$$
$$= \frac{1}{16}\int (\sin^2 2x)^2\, dx = \frac{1}{16}\int \left(\frac{1-\cos 4x}{2}\right)^2 dx$$
$$= \frac{1}{64}\int (1 - 2\cos 4x + \cos^2 4x)dx$$
$$= \frac{x}{64} - \frac{1}{32}\int \cos 4x\, dx + \frac{1}{64}\int \cos^2 4x\, dx$$
$$= \frac{x}{64} - \frac{1}{32}\cdot\frac{1}{4}\sin 4x + \frac{1}{64}\int \frac{1+\cos 8x}{2} dx$$
$$= \frac{x}{64} - \frac{\sin 4x}{128} + \frac{1}{128}\left(x + \frac{1}{8}\sin 8x\right) + C$$
$$= \frac{1}{128}\left(3x - \sin 4x + \frac{1}{8}\sin 8x\right) + C. \quad \|$$

If identity (1) is not applicable, then identies (2) or (3) are applied immediately.

Example 7

$$\int \cos^2 3x\, dx = \int \frac{1+\cos 6x}{2} dx = \frac{x}{2} + \frac{\sin 6x}{12} + C. \quad \|$$

Exercises 8.2

1. $\int \sin^4 x \cos^3 x\, dx$.
2. $\int \sin^3 x \cos^4 x\, dx$.
3. $\int \cos^2 x \sin^3 x\, dx$.
4. $\int \sin^2 x \cos^3 x\, dx$.
5. $\int_0^{\pi/4} \sec^2 x \tan^2 x\, dx$.
6. $\int_0^{\pi/4} \sec^2 x \tan^6 x\, dx$.
7. $\int \sec 2x \tan^3 2x\, dx$.
8. $\int \sec^3 x \tan^3 x\, dx$.
9. $\int \tan^5 x\, dx$.
10. $\int \tan^3 2x\, dx$.
11. $\int \sec^6 x\, dx$.
12. $\int \sec^4 x\, dx$.
13. $\int \sec^4 x \sqrt{\tan x}\, dx$.
14. $\int \sec^{-3} 4x \tan^5 4x\, dx$.
15. $\int_{-\pi}^{\pi} \sqrt[3]{\cos x} \sin^3 x\, dx$.
16. $\int_0^{\pi} \sqrt{\sin x} \cos^5 x\, dx$.
17. $\int \cos^7 5x \sin^{-2} 5x\, dx$.
18. $\int \sin^5 x \cos^{-4} x\, dx$.
19. $\int \csc^4 5x \cot^2 5x\, dx$.
20. $\int \csc^3 x \cot^5 x\, dx$.
21. $\int \csc^2 x \cot^{-5} x\, dx$.
22. $\int \csc^{2/3} x \cot^3 x\, dx$.
23. $\int_0^{\pi/2} \cos^2(-3x)dx$.
24. $\int_0^{\pi/2} \sin^2 5x\, dx$.
25. $\int \sin^2 x \cos^4 x\, dx$.
26. $\int \sin^2 x \cos^2 x\, dx$.
27. $\int_0^{\pi} \sin^6 x \cos^2 x\, dx$.
28. $\int_0^{\pi} \sin^6 x \cos^6 x\, dx$.
29. The region under the graph of $y = \sin 2x$ between $x = 0$ and $x = \pi/2$ is rotated about the x-axis. Find the resulting volume.
30. Due to influences of food supply and predator population the size of certain populations are periodic functions of time t. If the size of a population is

$$N = k \sin^2 2\pi t \cos^2 2\pi t + a,$$

where k and a are constants, determine the average population from $t = 0$ to $t = 1$.

8.3 Trigonometric Substitutions

With the identities

$$1 - \sin^2\theta = \cos^2\theta, \tag{1}$$

$$1 + \tan^2\theta = \sec^2\theta, \tag{2}$$

integrals containing $\sqrt{a^2 - u^2}$, $\sqrt{a^2 + u^2}$, and $\sqrt{u^2 - a^2}$, where u is a differentiable function and a is a *positive* constant, can often be evaluated. The technique is to let u equal a times an appropriate trigonometric function. The substitutions are made according to the following table:

occurrence	*substitution*	
$\sqrt{a^2 - u^2}$	$u = a\sin\theta,$	$-\pi/2 \leq \theta \leq \pi/2.$
$\sqrt{a^2 + u^2}$	$u = a\tan\theta,$	$-\pi/2 < \theta < \pi/2.$
$\sqrt{u^2 - a^2}$	$u = a\sec\theta,$	$\begin{cases} 0 \leq \theta < \pi/2 & \text{if } u \geq a \\ \pi/2 < \theta \leq \pi & \text{if } u \leq -a. \end{cases}$

Notice what happens to the first two square roots with these substitutions.

For $\sqrt{a^2 - u^2}$, let $u = a\sin\theta$, $-\pi/2 \leq \theta \leq \pi/2$. We obtain

$$\sqrt{a^2 - u^2} = \sqrt{a^2 - a^2\sin^2\theta} = a\sqrt{\cos^2\theta}.$$

Since $-\pi/2 \leq \theta \leq \pi/2$, we have $\cos\theta \geq 0$ and hence $\sqrt{\cos^2\theta} = \cos\theta$. Thus

$$\sqrt{a^2 - u^2} = a\cos\theta.$$

For $\sqrt{a^2 + u^2}$, let $u = a\tan\theta$, $-\pi/2 < \theta < \pi/2$. We get

$$\sqrt{a^2 + u^2} = \sqrt{a^2 + a^2\tan^2\theta} = a\sqrt{\sec^2\theta}.$$

Since $-\pi/2 < \theta < \pi/2$, we have $\sec\theta \geq 0$; and so $\sqrt{\sec^2\theta} = \sec\theta$. Thus

$$\sqrt{a^2 + u^2} = a\sec\theta.$$

Example 1 | Evaluate $\int x^3\sqrt{4 - x^2}\,dx$.

From the table above it is apparent that we should let $x = 2\sin\theta$, $-\pi/2 \leq \theta \leq \pi/2$. Then $dx = 2\cos\theta\,d\theta$ and $\sqrt{4 - x^2} = 2\cos\theta$; so we have

$$\begin{aligned}
\int x^3\sqrt{4 - x^2}\,dx &= \int (2^3\sin^3\theta)(2\cos\theta)(2\cos\theta\,d\theta) \\
&= 32\int \sin^3\theta\cos^2\theta\,d\theta \\
&= -32\int (1 - \cos^2\theta)(\cos^2\theta)(-\sin\theta\,d\theta) \\
&= 32\int (-\cos^2\theta + \cos^4\theta)d(\cos\theta) \\
&= 32\left(-\frac{\cos^3\theta}{3} + \frac{\cos^5\theta}{5}\right) + C.
\end{aligned}$$

However, since the original integral is in terms of x, we should write the result in terms of x. The θ is only a tool for obtaining the result. Since $2\cos\theta = \sqrt{4-x^2}$, we have $\cos\theta = \sqrt{4-x^2}/2$. Therefore,

$$\int x^3\sqrt{4-x^2}\,dx = 32\left[\frac{(4-x^2)^{5/2}}{5\cdot 2^5} - \frac{(4-x^2)^{3/2}}{3\cdot 2^3}\right] + C$$

$$= \frac{(4-x^2)^{5/2}}{5} - \frac{4(4-x^2)^{3/2}}{3} + C. \qquad \|$$

Trigonometric substitutions can be applied equally well to definite integrals, as indicated in the next two examples.

Example 2 | Evaluate $\int_0^3 x^3\sqrt{9+x^2}\,dx$.

Let $x = 3\tan\theta$, $-\pi/2 < \theta < \pi/2$. Then

$$dx = 3\sec^2\theta\,d\theta \quad \text{and} \quad \sqrt{9+x^2} = 3\sec\theta.$$

Moreover, since $\theta = \arctan x/3$, $\theta = 0$ when $x = 0$ and $\theta = \pi/4$ when $x = 3$. Then

$$\int_0^3 x^3\sqrt{9+x^2}\,dx = \int_0^{\pi/4}(3^3\tan^3\theta)(3\sec\theta)(3\sec^2\theta)d\theta$$

$$= 3^5\int_0^{\pi/4}\tan^3\theta\sec^3\theta\,d\theta$$

$$= 3^5\int_0^{\pi/4}\tan^2\theta\sec^2\theta(\sec\theta\tan\theta\,d\theta)$$

$$= 3^5\int_0^{\pi/4}(\sec^4\theta - \sec^2\theta)d(\sec\theta)$$

$$= 3^5\left[\frac{\sec^5\theta}{5} - \frac{\sec^3\theta}{3}\right]\Bigg|_0^{\pi/4}$$

$$= 3^5\left(\frac{1}{5}\cdot 2^{5/2} - \frac{1}{3}\cdot 2^{3/2} - \frac{1}{5} + \frac{1}{3}\right)$$

$$= 3^4\cdot\frac{2}{5}(2^{1/2}+1). \qquad \|$$

A similar technique is used when an integral contains $\sqrt{u^2-a^2}$. Note that $\sqrt{u^2-a^2}$ is a real number only when $u \le -a$ or when $u \ge a$. Using identity (2) in the form

$$\sec^2\theta - 1 = \tan^2\theta,$$

we let $u = a\sec\theta$, where $0 \le \theta < \pi/2$ for $u \ge a$, and $\pi/2 < \theta \le \pi$ for $u \le -a$. Then

$$\sqrt{u^2-a^2} = \sqrt{a^2\sec^2\theta - a^2} = a\sqrt{\tan^2\theta}.$$

But since for $0 \le \theta < \pi/2$, $\tan\theta \ge 0$, and since for $\pi/2 < \theta \le \pi$, $\tan\theta \le 0$, we have

$$\sqrt{u^2 - a^2} = \begin{cases} a\tan\theta & \text{when } u \ge a \\ -a\tan\theta & \text{when } u \le -a. \end{cases}$$

Example 3 Evaluate $\int_{8/\sqrt{3}}^{8} \frac{dx}{\sqrt{x^2 - 16}}$.

Since $x > 4$ we let $x = 4\sec\theta$, $0 \le \theta < \pi/2$. Then $dx = 4\sec\theta\tan\theta\, d\theta$ and $\sqrt{x^2 - 16} = \sqrt{16\sec^2\theta - 16} = 4\tan\theta$. Since $\theta = \operatorname{arcsec} x/4$, $\theta = \pi/6$ when $x = 8/\sqrt{3}$, and $\theta = \pi/3$ when $x = 8$. Thus

$$\begin{aligned}\int_{8/\sqrt{3}}^{8} \frac{dx}{\sqrt{x^2 - 16}} &= \int_{\pi/6}^{\pi/3} \frac{4\sec\theta\tan\theta\, d\theta}{4\tan\theta} = \int_{\pi/6}^{\pi/3} \sec\theta\, d\theta \\ &= \ln|\sec\theta + \tan\theta|\Big|_{\pi/6}^{\pi/3} \\ &= \ln|2 + \sqrt{3}| - \ln\left|\frac{2}{\sqrt{3}} + \frac{1}{\sqrt{3}}\right| \\ &= \ln|2 + \sqrt{3}| - \ln|\sqrt{3}| \\ &= \ln(2 + \sqrt{3}) - \ln(\sqrt{3}).\end{aligned}$$

||

It is sometimes necessary to complete the square first in order to put the integral into the proper form.

Example 4 Evaluate $\int \frac{x\, dx}{\sqrt{5 - x^2 + 4x}}$.

Here we first complete the square to get

$$5 - x^2 + 4x = 5 - (x^2 - 4x) = 5 - (x^2 - 4x + 4) + 4 = 9 - (x - 2)^2.$$

We then let $x - 2 = 3\sin\theta$, $-\pi/2 \le \theta \le \pi/2$, obtaining $dx = 3\cos\theta\, d\theta$ and $\sqrt{5 - x^2 + 4x} = \sqrt{9 - (x - 2)^2} = 3\cos\theta$. Then

$$\begin{aligned}\int \frac{x\, dx}{\sqrt{5 - x^2 + 4x}} &= \int \frac{(2 + 3\sin\theta)3\cos\theta\, d\theta}{3\cos\theta} = \int (2 + 3\sin\theta)d\theta \\ &= 2\theta - 3\cos\theta + C.\end{aligned}$$

Then since $\sin\theta = (x - 2)/3$ and $-\pi/2 \le \theta \le \pi/2$, we know that $\theta = \arcsin(x - 2)/3$. Substituting that result and $3\cos\theta = \sqrt{5 - x^2 + 4x}$ gives

$$\int \frac{x\, dx}{\sqrt{5 - x^2 + 4x}} = 2\arcsin\frac{x - 2}{3} - \sqrt{5 - x^2 + 4x} + C.$$

||

We shall now use the substitution $u = a\sec\theta$ to obtain an integration formula for $\int \frac{du}{\sqrt{u^2 - a^2}}$ where a is a positive constant. We shall only derive the formula for the case where $u > a$, although the formula also holds for $u < -a$. The derivation is similar but a little more complicated for $u < -a$, and is left as an exercise.

Let $u = a \sec\theta$, then $du = a \sec\theta \tan\theta \, d\theta$; and when $u > a$, $\sqrt{u^2 - a^2} = a \tan\theta$. So we get

$$\begin{aligned}
\int \frac{du}{\sqrt{u^2 - a^2}} &= \int \frac{a \sec\theta \tan\theta \, d\theta}{a \tan\theta} \\
&= \int \sec\theta \, d\theta \\
&= \ln|\sec\theta + \tan\theta| + C_1 \\
&= \ln\left|\frac{u}{a} + \frac{\sqrt{u^2 - a^2}}{a}\right| + C_1 \\
&= \ln\left|u + \sqrt{u^2 - a^2}\right| - \ln a + C_1 \\
&= \ln\left|u + \sqrt{u^2 - a^2}\right| + C,
\end{aligned}$$

where C is the constant $C_1 - \ln a$. Thus, we have the formula

$$\int \frac{du}{\sqrt{u^2 - a^2}} = \ln\left|u + \sqrt{u^2 - a^2}\right| + C.$$

Exercises 8.3

In Exercises 1–20, evaluate the integrals.

1. $\int \frac{dx}{\sqrt{4 + x^2}}$.
2. $\int \frac{dx}{\sqrt{7 + x^2}}$.
3. $\int \sqrt{9 - x^2} \, dx$.
4. $\int \sqrt{4 - x^2} \, dx$.
5. $\int_0^1 \frac{x \, dx}{(9 - x^2)^{3/2}}$.
6. $\int_0^1 x\sqrt{4 + x^2} \, dx$.
7. $\int \frac{x^3}{\sqrt{9 - 2x^2}} \, dx$.
8. $\int \frac{x^3}{\sqrt{7 - x^2}} \, dx$.
9. $\int_0^2 \frac{dx}{(9 - x^2)^{3/2}}$.
10. $\int_0^{1/2} \frac{dx}{(1 - x^2)^{3/2}}$.
11. $\int \frac{dx}{(x^2 - 16)^{3/2}}, \quad x < -4$.
12. $\int \frac{dx}{(x^2 - 4)^{3/2}}, \quad x > 2$.
13. $\int \sqrt{7 - 16x^2} \, dx$.
14. $\int \sqrt{8 - 3x^2} \, dx$.
15. $\int_0^1 \frac{dx}{\sqrt{x^2 + 8x + 32}}$.
16. $\int_{-1}^1 \frac{x \, dx}{\sqrt{8 - 2x - x^2}}$.
17. $\int (x + 7)^5 \sqrt{-x^2 - 14x - 45} \, dx$.
18. $\int (x - 3)^3 \sqrt{6x - x^2} \, dx$.
19. $\int (x - 3)^3 \sqrt{x^2 - 6x + 13} \, dx$.
20. $\int (x + 2)^3 \sqrt{x^2 + 4x + 13} \, dx$.

In Exercises 21–24, evaluate the integrals directly without using a trigonometric substitution.

21. $\int \frac{dx}{\sqrt{x^2 - 16}}$. 22. $\int \frac{dx}{\sqrt{x^2 - 9}}$. 23. $\int \frac{x\,dx}{\sqrt{x^4 - 25}}$. 24. $\int \frac{x^2\,dx}{\sqrt{x^6 - 4}}$.

25. Compute the area of the region under the graph of $y = \sqrt{4 - x^2}$ for $-2 \le x \le 2$.

26. The region of Exercise 25 is rotated about the line $x = 2$. Find the resulting volume.

27. Show that if $x = 2 \arctan u$, then $u = \tan \frac{x}{2}$, $dx = \frac{2\,du}{1 + u^2}$, $\sin x = 2 \sin \frac{x}{2} \cos \frac{x}{2} = \frac{2u}{1 + u^2}$, and $\cos x = \frac{1 - u^2}{1 + u^2}$.

In Exercises 28–31, use the substitution $x = 2 \arctan u$ and the results of Exercise 27 to evaluate the integrals.

28. $\int \frac{dx}{1 + \sin x}$. 29. $\int \frac{dx}{\sin x + \tan x}$.

30. $\int \frac{dx}{2 + \cos x}$. 31. $\int \frac{dx}{1 + \cos x}$.

32. Complete the derivation of the integration formula given at the end of this section for the case where $u < -a$.

8.4 Integrals of Functions with Quadratics in the Denominator

When the numerator of a function to be integrated is a constant or is linear, and the denominator is either a quadratic expression or the square root of a quadratic expression, the integral can then be put in a form that enables us to use one of the following formulas. It may be necessary to "complete the square" in the quadratic.

$$\int \frac{du}{u^2 + a^2} = \frac{1}{a} \arctan \frac{u}{a} + C$$

$$\int \frac{du}{u^2 - a^2} = \frac{1}{2a} \ln \left| \frac{u - a}{u + a} \right| + C$$

$$\int \frac{du}{\sqrt{a^2 - u^2}} = \arcsin \frac{u}{a} + C$$

$$\int \frac{du}{\sqrt{u^2 - a^2}} = \ln\left|u + \sqrt{u^2 - a^2}\right| + C.$$

All but the second of those formulas have appeared in previous sections. To obtain the second formula we use the fact that

$$\frac{1}{2a}\left[\frac{1}{u - a} - \frac{1}{u + a}\right] = \frac{1}{u^2 - a^2}$$

to get

$$\int \frac{du}{u^2 - a^2} = \frac{1}{2a}\left[\int \frac{du}{u-a} - \int \frac{du}{u+a}\right]$$

$$= \frac{1}{2a}[\ln|u-a| - \ln|u+a|] + C$$

$$= \frac{1}{2a}\ln\left|\frac{u-a}{u+a}\right| + C.$$

The procedure is simplest when the numerator is a constant.

Example 1

$$\int \frac{dx}{x^2 - 8x + 30} = \int \frac{dx}{x^2 - 8x + 16 + 14}$$

$$= \int \frac{dx}{(x-4)^2 + 14}$$

$$= \frac{1}{\sqrt{14}} \arctan \frac{x-4}{\sqrt{14}} + C. \qquad \|$$

Example 2

$$\int \frac{dx}{2x^2 - 5x + 2} = \frac{1}{2}\int \frac{dx}{x^2 - (5/2)x + 1}$$

$$= \frac{1}{2}\int \frac{dx}{x^2 - (5/2)x + 25/16 + 1 - 25/16}$$

$$= \frac{1}{2}\int \frac{dx}{(x - 5/4)^2 - 9/16}$$

$$= \frac{1}{4(3/4)}\ln\left|\frac{x - 5/4 - 3/4}{x - 5/4 + 3/4}\right| + C$$

$$= \frac{1}{3}\ln\left|\frac{x-2}{x-1/2}\right| + C. \qquad \|$$

Example 3

$$\int \frac{dx}{\sqrt{16 + 6x - x^2}} = \int \frac{dx}{\sqrt{16 - (x^2 - 6x)}}$$

$$= \int \frac{dx}{\sqrt{16 + 9 - (x^2 - 6x + 9)}}$$

$$= \int \frac{dx}{\sqrt{25 - (x-3)^2}}$$

$$= \arcsin \frac{x-3}{5} + C. \qquad \|$$

When the numerator is linear, the integral is first expressed as the sum of two integrals, one with a numerator that is the derivative of the quadratic denominator and one with a constant numerator.

Example 4

$$\begin{aligned}\int \frac{3x-2}{x^2-4}\,dx &= 3\int \frac{x-2/3}{x^2-4}\,dx\\ &= \frac{3}{2}\int \frac{2x-4/3}{x^2-4}\,dx\\ &= \frac{3}{2}\int \frac{2x}{x^2-4}\,dx - \frac{3}{2}\cdot\frac{4}{3}\int \frac{dx}{x^2-4}\\ &= \frac{3}{2}\ln|x^2-4| - \frac{1}{2}\ln\left|\frac{x-2}{x+2}\right| + C.\end{aligned}$$

||

Example 5

$$\begin{aligned}\int \frac{x+5}{\sqrt{x^2-9}}\,dx &= \frac{1}{2}\int \frac{2x+10}{\sqrt{x^2-9}}\,dx\\ &= \frac{1}{2}\int \frac{2x}{(x^2-9)^{1/2}}\,dx + \frac{10}{2}\int \frac{dx}{\sqrt{x^2-9}}\\ &= \frac{1}{2}\int (x^2-9)^{-1/2}(2x\,dx) + 5\int \frac{dx}{\sqrt{x^2-9}}\\ &= \frac{1}{2}\cdot\frac{2}{1}(x^2-9)^{1/2} + 5\ln|x+\sqrt{x^2-9}| + C\\ &= (x^2-9)^{1/2} + 5\ln|x+\sqrt{x^2-9}| + C.\end{aligned}$$

||

Example 6

$$\begin{aligned}\int \frac{5x+2}{x^2-4x+10}\,dx &= \frac{5}{2}\int \frac{2x+4/5}{x^2-4x+10}\,dx\\ &= \frac{5}{2}\int \frac{2x-4+4+4/5}{x^2-4x+10}\,dx\\ &= \frac{5}{2}\int \frac{2x-4}{x^2-4x+10}\,dx + 12\int \frac{dx}{x^2-4x+10}\\ &= \frac{5}{2}\int \frac{2x-4}{x^2-4x+10}\,dx + 12\int \frac{dx}{x^2-4x+4+6}\\ &= \frac{5}{2}\ln|x^2-4x+10| + 12\int \frac{dx}{(x-2)^2+6}\\ &= \frac{5}{2}\ln|x^2-4x+10| + \frac{12}{\sqrt{6}}\arctan\frac{x-2}{\sqrt{6}} + C\\ &= \frac{5}{2}\ln|x^2-4x+10| + 2\sqrt{6}\arctan\frac{x-2}{\sqrt{6}} + C.\end{aligned}$$

||

Example 7

$$\int \frac{3x-4}{\sqrt{x^2-6x+1}}\,dx = 3\int \frac{x-4/3}{\sqrt{x^2-6x+1}}\,dx$$

$$= \frac{3}{2}\int \frac{2x-8/3}{\sqrt{x^2-6x+1}}\,dx$$

$$= \frac{3}{2}\int \frac{2x-6+6-8/3}{\sqrt{x^2-6x+1}}$$

$$= \frac{3}{2}\int \frac{(2x-6)dx}{\sqrt{x^2-6x+1}} + \frac{3}{2}\cdot\frac{10}{3}\int \frac{dx}{\sqrt{x^2-6x+1}}$$

$$= \frac{3}{2}\int (x^2-6x+1)^{-1/2}(2x-6)dx$$

$$+ 5\int \frac{dx}{\sqrt{x^2-6x+9+1-9}}$$

$$= \frac{3}{2}\cdot\frac{2}{1}(x^2-6x+1)^{1/2} + 5\int \frac{dx}{\sqrt{(x-3)^2-8}}$$

$$= 3(x^2-6x+1)^{1/2}$$
$$+ 5\ln\left|x-3+\sqrt{(x-3)^2-8}\right| + C. \qquad \|$$

Exercises 8.4

In Exercises 1–36, evaluate the given integrals.

1. $\int \frac{dx}{x^2+4}$.
2. $\int \frac{dx}{x^2-25}$.
3. $\int \frac{dx}{x^2-9}$.
4. $\int \frac{dx}{x^2+36}$.
5. $\int \frac{dx}{12-x^2}$.
6. $\int \frac{dx}{x^2+10}$.
7. $\int \frac{dx}{x^2+19}$.
8. $\int \frac{dx}{7-x^2}$.
9. $\int \frac{3x-7}{x^2+16}\,dx$.
10. $\int \frac{5x+6}{x^2-4}\,dx$.
11. $\int \frac{2x+1}{\sqrt{4-(x+3)^2}}\,dx$.
12. $\int \frac{3x-1}{\sqrt{(x-2)^2-1}}\,dx$.
13. $\int \frac{dx}{x^2+4x+8}$.
14. $\int \frac{dx}{x^2-2x+7}$. ← complete the square
15. $\int \frac{dx}{7-x^2-6x}$.
16. $\int \frac{dx}{2-x^2+6x}$.
17. $\int \frac{5\,dx}{x^2-3x+9}$.
18. $\int \frac{4\,dx}{6-x+x^2}$.
19. $\int \frac{3\,dx}{\sqrt{x^2+4x}}$.
20. $\int \frac{2\,dx}{\sqrt{7-6x-x^2}}$.
21. $\int \frac{5\,dx}{\sqrt{16+6x-x^2}}$.
22. $\int \frac{7\,dx}{\sqrt{x^2+8x+7}}$.
23. $\int_0^1 \frac{x\,dx}{x^2-4x+13}$.
24. $\int_0^1 \frac{x\,dx}{x^2-2x+7}$.
25. $\int \frac{x-3}{20-x^2+4x}\,dx$.
26. $\int \frac{1+x}{3-x^2+5x}\,dx$.
27. $\int_0^2 \frac{3-2x}{7-x+x^2}\,dx$.

28. $\int_0^1 \frac{3x-2}{4-x+x^2}\,dx.$ 29. $\int \frac{x+3}{7-x^2-6x}\,dx.$ 30. $\int \frac{1-x}{2x-x^2+7}\,dx.$

31. $\int \frac{6x-5}{3+4x^2-x}\,dx.$ 32. $\int \frac{7x+1}{2+x^2+3x}\,dx.$ 33. $\int \frac{x+3}{\sqrt{5-x^2-4x}}\,dx.$

34. $\int \frac{x-2}{\sqrt{x^2+6x}}\,dx.$ 35. $\int \frac{5x-2}{\sqrt{x^2-4x-5}}\,dx.$ 36. $\int \frac{7x+3}{\sqrt{3+2x-x^2}}\,dx.$

37. The region below the graph of $y = 1/(x^2+4x+8)$ between $x = 0$ and $x = 1$ is rotated about the y-axis. Find the resulting volume.

38. Planes perpendicular to the y-axis cut a given solid in such a way that the cross-sectional area is given by

$$A(y) = \frac{2y+1}{y^2+3y+4}, \qquad \text{for } 1 \le y \le 2.$$

Compute the volume.

8.5 Integration Tables

Although we have discussed a number of techniques for finding indefinite integrals and have developed many integration formulas, there are still many functions whose indefinite integrals are difficult or impossible to obtain. An exhaustive study of those that can be obtained would be too tiring and time-consuming for our purposes.

Actually, it is not possible to write the indefinite integral of certain functions in terms of rational functions, power functions, exponential functions, logarithmic functions, and trigonometric functions. Here are some examples:

$$\int \frac{\sin x}{x}\,dx, \qquad \int \frac{e^x}{x+2}\,dx, \qquad \int x^x\,dx.$$

However, we shall see in the next chapter that there are various methods of effectively approximating definite integrals, whether or not the corresponding indefinite integrals can be found.

Also, there are some good lists of integration formulas that go beyond those of this book. For example, integration tables are included in Burington, *Mathematical Tables and Formulas* (McGraw-Hill), or in Selby, *Standard Mathematical Tables* (Chemical Rubber Company). Another example is Peirce, *A Short Table of Integrals* (Ginn and Company).

In a table of integrals one finds essentially all of the integral formulas we have developed (for simplification the constants of integration are omitted) and many more such as

$$\int \frac{dx}{a+b\sin x} = \begin{cases} \dfrac{2}{\sqrt{a^2-b^2}} \arctan\left(\dfrac{a\tan(x/2)+b}{\sqrt{a^2-b^2}}\right) & \text{if } a^2 > b^2 \\[2ex] \dfrac{1}{\sqrt{b^2-a^2}} \ln\left|\dfrac{a\tan(x/2)+b-\sqrt{b^2-a^2}}{a\tan(x/2)+b+\sqrt{b^2-a^2}}\right| & \text{if } b^2 > a^2. \end{cases}$$

That formula could be used, for example, to find $\int \frac{dx}{7+2\sin x}$.

Also included in a table of integrals are *reduction formulas.* A reduction formula can be used one or more times to reduce successively an appropriate integral until an integral is obtained that is easy to evaluate. One such reduction formula is

$$\int \frac{dx}{(x^2 + a^2)^{k+1}} = \frac{x}{2ka^2(x^2 + a^2)^k} + \frac{2k - 1}{2ka^2} \int \frac{dx}{(x^2 + a^2)^k}. \tag{1}$$

We shall illustrate its use by evaluating $\int \frac{dx}{(x^2 + 4)^3}$. Applying the reduction formula once with $a = 2$ and $k = 2$, we get

$$\int \frac{dx}{(x^2 + 4)^3} = \frac{x}{16(x^2 + 4)^2} + \frac{3}{16} \int \frac{dx}{(x^2 + 4)^2}.$$

A second application with $a = 2$ and $k = 1$ yields

$$\int \frac{dx}{(x^2 + 4)^2} = \frac{x}{8(x^2 + 4)} + \frac{1}{8} \int \frac{dx}{x^2 + 4}, \quad \text{and so}$$

$$\int \frac{dx}{(x^2 + 4)^3} = \frac{x}{16(x^2 + 4)^2} + \frac{3x}{128(x^2 + 4)} + \frac{3}{128} \int \frac{dx}{x^2 + 4}.$$

Then, on evaluating the final integral we obtain

$$\int \frac{dx}{(x^2 + 4)^3} = \frac{x}{16(x^2 + 4)^2} + \frac{3x}{128(x^2 + 4)} + \frac{3}{256} \arctan \frac{x}{2} + C.$$

In this case we had to apply the reduction formula twice; but to evaluate

$$\int \frac{dx}{(x^2 + 4)^7},$$

we would have to apply the reduction formula six times.

Although reduction formulas can usually be obtained by applying integration by parts, the application may not be simple. For example, item (1) above can be obtained as follows.

Let $u = (x^2 + a^2)^{-k}$ and $dv = dx$; then $du = -2kx(x^2 + a^2)^{-k-1}\, dx$ and $v = x$. Hence,

$$\int \frac{dx}{(x^2 + a^2)^k} = \frac{x}{(x^2 + a^2)^k} + 2k \int \frac{x^2\, dx}{(x^2 + a^2)^{k+1}}.$$

Then, since

$$\frac{x^2}{(x^2 + a^2)^{k+1}} = \frac{x^2 + a^2 - a^2}{(x^2 + a^2)^{k+1}} = \frac{1}{(x^2 + a^2)^k} - \frac{a^2}{(x^2 + a^2)^{k+1}},$$

we have

$$\int \frac{dx}{(x^2 + a^2)^k} = \frac{x}{(x^2 + a^2)^k} + 2k \int \frac{dx}{(x^2 + a^2)^k} - 2ka^2 \int \frac{dx}{(x^2 + a^2)^{k+1}}.$$

Subtracting $\int \frac{dx}{(x^2 + a^2)^k}$ from both sides and adding $2ka^2 \int \frac{dx}{(x^2 + a^2)^{k+1}}$, we get

$$2ka^2 \int \frac{dx}{(x^2 + a^2)^{k+1}} = \frac{x}{(x^2 + a^2)^k} + (2k - 1) \int \frac{dx}{(x^2 + a^2)^k}.$$

Then, dividing both sides by $2ka^2$, we have

$$\int \frac{dx}{(x^2 + a^2)^{k+1}} = \frac{x}{2ka^2(x^2 + a^2)^k} + \frac{2k - 1}{2ka^2} \int \frac{dx}{(x^2 + a^2)^k}.$$

That formula will be used in the next section.

Exercises 8.5

Evaluate the integrals in Exercises 1–8.

1. $\int \frac{dx}{-2 + 7 \sin x}$.
2. $\int_0^{\pi/2} \frac{dx}{3 + 4 \sin x}$.
3. $\int_0^{\pi/2} \frac{dx}{7 - 4 \sin x}$.
4. $\int \frac{dx}{4 + 3 \sin x}$.
5. $\int \frac{dx}{(x^2 + 3)^2}$.
6. $\int \frac{dx}{(x^2 + 4)^2}$.
7. $\int \frac{dx}{(x^2 + 7)^3}$.
8. $\int \frac{dx}{(x^2 + 9)^3}$.
9. Use integration by parts to verify the reduction formula

$$\int \sin^n x \, dx = -\frac{1}{n} \sin^{n-1} x \cos x + \frac{n-1}{n} \int \sin^{n-2} x \, dx.$$

10. Use integration by parts to verify the reduction formula

$$\int \cos^n x \, dx = \frac{1}{n} \cos^{n-1} x \sin x + \frac{n-1}{n} \int \cos^{n-2} x \, dx.$$

11. Show that $\int x^n \cos x \, dx = x^n \sin x - n \int x^{n-1} \sin x \, dx$.
12. Show that $\int x^n \sin x \, dx = -x^n \cos x + n \int x^{n-1} \cos x \, dx$.

In Exercises 13–16, use the formulas developed in Exercises 9–12 to evaluate the given integral.

13. $\int \cos^3 x \, dx$.
14. $\int x^2 \sin x \, dx$.
15. $\int \cos^4 x \, dx$.
16. $\int x^3 \cos x \, dx$.
17. Use the result of Exercise 9 to show that

$$\int_0^{\pi/2} \sin^n x \, dx = \frac{n-1}{n} \int_0^{\pi/2} \sin^{n-2} x \, dx.$$

18. Use the result of Exercise 10 to show that

$$\int_0^{\pi/2} \cos^n x \, dx = \frac{n-1}{n} \int_0^{\pi/2} \cos^{n-2} x \, dx.$$

19. Use the result of Exercise 17 to obtain Wallis' formula that for any positive integer n,

$$\int_0^{\pi/2} \sin^{2n} x\, dx = \frac{1 \cdot 3 \cdot \cdots \cdot (2n-1)}{2 \cdot 4 \cdot \cdots \cdot (2n)} \cdot \frac{\pi}{2}.$$

20. Use the result of Exercise 18 to obtain Wallis' formula that for any positive integer n,

$$\int_0^{\pi/2} \cos^{2n} x\, dx = \frac{1 \cdot 3 \cdot \cdots \cdot (2n-1)}{2 \cdot 4 \cdot \cdots \cdot (2n)} \cdot \frac{\pi}{2}.$$

21. Use the result of Exercise 17 to obtain Wallis' formula that for any positive integer n,

$$\int_0^{\pi/2} \sin^{2n-1} x\, dx = \frac{2 \cdot 4 \cdot \cdots \cdot (2n-2)}{3 \cdot 5 \cdot \cdots \cdot (2n-1)}.$$

22. Use the result of Exercise 18 to obtain Wallis' formula that for any positive integer n,

$$\int_0^{\pi/2} \cos^{2n-1} x\, dx = \frac{2 \cdot 4 \cdot \cdots \cdot (2n-2)}{3 \cdot 5 \cdot \cdots \cdot (2n-1)}.$$

8.6 Partial Fractions

The method of *partial fractions* can be used to integrate rational functions by decomposing them into forms that can be handled by the formulas we have developed so far. We recall that a rational function is one that can be expressed as the quotient of two polynomials, the denominator of which is not 0. When the degree of the numerator is greater than or equal to the degree of the denominator, a rational function is called an *improper fraction*; and when the degree of the numerator is less than the degree of the denominator, it is called a *proper fraction.*

When a rational function is an improper fraction, we can divide the numerator by the denominator to obtain a quotient in the form of a polynomial plus a remainder in the form of a proper fraction. That is, an improper fraction can be expressed as the sum of a polynomial and a proper fraction.

Example 1

$$\frac{3x^2+7}{x^2+x-2} = 3 + \frac{-3x+13}{x^2+x-2}.$$ ||

Example 2

$$\frac{x^4+3x-1}{x^2+x-2} = x^2 - x + 3 + \frac{-2x+5}{x^2+x-2}.$$ ||

To integrate an improper fraction, we first express it as the sum of a polynomial and a proper fraction. Since we know how to integrate polynomials, we shall now concentrate on integration of proper fractions. The method of partial fractions applies to proper fractions and enables us to express them as sums of other fractions that can be integrated.

It can be proved that every polynomial with real coefficients is equal to a product of factors each of which is either a linear or a quadratic expression that cannot be factored into linear factors with real coefficients.

Example 3 | $x^3 - 8 = (x - 2)(x^2 + 2x + 4)$.

Here $x^2 + 2x + 4$ cannot be factored into real linear factors since it has no real zeros. ||

Thus a proper fraction can be expressed so that its denominator is a product of linear and quadratic factors, then like factors are combined. Of course, it could happen that the factoring is difficult.

Example 4

$$\frac{x^4 - 2x^2 + 3x - 5}{x^7 + 5x^5 + 2x^4 - 8x^3 + 16x^2 - 48x + 32}$$

$$= \frac{x^4 - 2x^2 + 3x - 5}{(x^2 + 4)(x - 1)(x + 2)(x - 1)(x^2 + 4)}$$

$$= \frac{x^4 - 2x^2 + 3x - 5}{(x^2 + 4)^2(x - 1)^2(x + 2)}.$$ ||

It can be proved that every proper fraction can be expressed as the sum of other fractions called *partial fractions* of the following types.

Corresponding to each power $[L(x)]^r$ of a linear factor $L(x)$ occurring in the denominator of the original fraction, we get (1)

$$\frac{a_1}{L(x)} + \frac{a_2}{[L(x)]^2} + \cdots + \frac{a_r}{[L(x)]^r},$$

where $a_1, a_2, \ldots, a_r$ are constants to be determined. In particular, if $r = 1$, the linear factor $L(x)$ occurs only to the first power and so there is only one partial fraction corresponding to it; namely, one of the form $a/L(x)$, where a is a constant to be determined.

Corresponding to each power $[q(x)]^s$ of a quadratic factor $q(x)$ occurring in the denominator of the original fraction, we get (2)

$$\frac{c_1x + d_1}{q(x)} + \frac{c_2x + d_2}{[q(x)]^2} + \cdots + \frac{c_sx + d_s}{[q(x)]^s},$$

where the c's and d's are constants to be determined. In particular, when $s = 1$, the quadratic factor $q(x)$ occurs only to the first power and so there is only one partial fraction corresponding to it; namely, one of the form $(cx + d)/q(x)$, where c and d are constants to be determined.

Example 5

$$\frac{x^4 - 2x^2 + 3x - 5}{(x^2 + 4)^2(x - 1)^2(x + 2)} = \frac{a}{x - 1} + \frac{b}{(x - 1)^2} + \frac{c}{x + 2} + \frac{dx + e}{x^2 + 4} + \frac{gx + h}{(x^2 + 4)^2},$$

where a, b, c, d, e, g, and h are constants to be evaluated. ||

Example 6

$$\frac{x^3 + 3x + 1}{(x + 3)(x - 2)(x + 1)^3} = \frac{a}{x + 1} + \frac{b}{(x + 1)^2} + \frac{c}{(x + 1)^3} + \frac{d}{x + 3} + \frac{e}{x - 2}$$

where a, b, c, d, and e are constants to be evaluated. ||

Example 7

$$\frac{2x^2 - x + 5}{(x + 2)(x^2 + 1)} = \frac{a}{x + 2} + \frac{bx + c}{x^2 + 1},$$

where a, b, and c are constants to be evaluated. ||

Example 8

$$\frac{3x + 4}{(x - 1)(x + 2)} = \frac{a}{x - 1} + \frac{b}{x + 2},$$

where a and b are constants to be determined. ||

As indicated above, every proper fraction can be expressed as a sum of partial fractions. Consequently the *existence* of the constants such as a, b, c, and d, is not in question. However, we must obtain a method of finding those constants. First we multiply both sides of the equality by the denominator of the proper fraction. That eliminates all the denominators on both sides of the equation and so produces polynomials on both sides. Then the constants may be evaluated by using the fact that these polynomials must have the same value for all values of x. This procedure is illustrated in the next example.

Example 9 As in Example 8, $\dfrac{3x + 4}{(x - 1)(x + 2)} = \dfrac{a}{x - 1} + \dfrac{b}{x + 2}$.

To find the values of a and b, first multiply both sides by $(x - 1)(x + 2)$ to obtain

$$3x + 4 = a(x + 2) + b(x - 1).$$

Since the polynomials on both sides of the equation have the same value for each value of x, we use the most convenient values of x, namely, $x = -2$ and $x = 1$. (These values make $(x + 2)$ and $(x - 1)$ equal to zero, respectively.) For $x = -2$, we have

$$3(-2) + 4 = a(-2 + 2) + b(-2 - 1),$$

which yields $\quad -2 = -3b, \quad \text{or} \quad b = \frac{2}{3}.$

For $x = 1$, we have

$$3(1) + 4 = a(1 + 2) + b(1 - 1),$$

which gives $\quad 7 = 3a, \quad \text{or} \quad a = \frac{7}{3}.$

Thus
$$\frac{3x+4}{(x-1)(x+2)} = \frac{7/3}{x-1} + \frac{2/3}{x+2}.$$

Another way to evaluate the constants is to use the fact that if two polynomials have the same value for each value of x, then their corresponding coefficients are equal. Proceeding from $3x + 4 = a(x + 2) + b(x - 1)$ we multiply and collect like terms to get

$$3x + 4 = (a + b)x + (2a - b).$$

Setting corresponding coefficients equal we get the system of equations

$$a + b = 3$$
$$2a - b = 4.$$

On solving this system we get $a = 7/3$ and $b = 2/3$ as before. ||

After finding the constants for a partial fraction decomposition, one can check the work by recombining the partial fractions to be sure that the original fraction is obtained.

Example 10 | Check the result of Examples 8 and 9 that

$$\frac{7/3}{x-1} + \frac{2/3}{x+2} = \frac{3x+4}{(x-1)(x-2)}.$$

We recombine the partial functions to get

$$\frac{7/3}{x-1} + \frac{2/3}{x+2} = \frac{\frac{7}{3}(x+2)}{(x-1)(x+2)} + \frac{\frac{2}{3}(x-1)}{(x-1)(x+2)}$$
$$= \frac{3x+4}{(x-1)(x+2)}.$$

So our result checks as it should. ||

We now look at some examples of the evaluation of integrals by using partial fractions.

Example 11 | Use the result of Examples 8 and 9 to evaluate $\int \frac{3x+4}{(x-1)(x+2)}\,dx$.

By Examples 8 and 9

$$\frac{3x+4}{(x-1)(x-2)} = \frac{7/3}{x-1} + \frac{2/3}{x+2}.$$

So $$\int \frac{3x+4}{(x-1)(x+2)}\,dx = \frac{7}{3}\int \frac{dx}{x-1} + \frac{2}{3}\int \frac{dx}{x+2}$$
$$= \frac{7}{3}\ln|x-1| + \frac{2}{3}\ln|x+2| + C. \qquad \|$$

Example 12 Evaluate $\int \frac{x^4+5x^3+9x^2+4x+11}{(x+2)(x^2+1)}\,dx$.

We first note that the degree of the numerator exceeds that of the denominator. So we use long division to get

$$\frac{x^4+5x^3+9x^2+4x+11}{(x+2)(x^2+1)} = x+3+\frac{2x^2-x+5}{(x+2)(x^2+1)},$$
$$\int \frac{x^4+5x^3+9x^2+4x+11}{(x+2)(x^2+1)}\,dx = \int (x+3)dx + \int \frac{2x^2-x+5}{(x+2)(x^2+1)}\,dx.$$

Since the polynomial $(x+3)$ is easily integrated, we concentrate on the evaluation of

$$\int \frac{2x^2-x+5}{(x+2)(x^2+1)}\,dx.$$

As in Example 7, we write

$$\frac{2x^2-x+5}{(x+2)(x^2+1)} = \frac{a}{x+2} + \frac{bx+c}{x^2+1},$$

and then determine the values of a, b, and c.

First multiply both sides of the equation by $(x+2)(x^2+1)$ to get

$$2x^2-x+5 = a(x^2+1) + (bx+c)(x+2).$$

Since both sides must be equal for all values of x, they must be equal when $x=-2$. So we have $15=5a$, or $a=3$. It is not evident that other real values of x cause a factor on the right hand side to be zero; however $x=0$ and $x=1$ are good values to use since they simplify the calculations. When $x=0$, we have $5=a+2c$. Then, since $a=3$, we get $c=1$. If we let $x=1$, we obtain $6=2a+3b+3c$. Then, since $a=3$ and $c=1$, we have $b=-1$. Hence

$$\int \frac{2x^2-x+5}{(x+2)(x^2+1)}\,dx = 3\int \frac{dx}{x+2} + \int \frac{-x+1}{x^2+1}\,dx.$$

Now $$3\int \frac{dx}{x+2} = 3\ln|x+2| + C_1, \quad \text{and}$$

$$\int \frac{-x+1}{x^2+1}\,dx = -\frac{1}{2}\int \frac{2x}{x^2+1}\,dx + \int \frac{dx}{x^2+1}$$
$$= -\frac{1}{2}\ln|x^2+1| + \arctan x + C_2.$$

Since $$\int (x+3)dx = \frac{1}{2}x^2 + 3x + C_3,$$

we have $\int \frac{x^4 + 5x^3 + 9x^2 + 4x + 11}{(x+2)(x^2+1)} dx$

$$= \frac{1}{2}x^2 + 3x + 3 \ln|x+2| - \frac{1}{2}\ln|x^2+1| + \arctan x + C,$$

where $C = C_1 + C_2 + C_3$. ||

At times, use of nonreal, complex values for x will considerably simplify the calculations involved in the evaluation of the constants. This is illustrated in the next example.

Example 13 $\int \frac{x^4 + 2x^3 + 5x^2 + 5x + 2}{(x^2+4)^2(x-2)} dx.$

Let $$\frac{x^4 + 2x^3 + 5x^2 + 5x + 2}{(x^2+4)^2(x-2)} = \frac{a}{x-2} + \frac{bx+c}{(x^2+4)^2} + \frac{dx+e}{x^2+4}.$$

Multiply both sides by $(x^2+4)^2(x-2)$ to get

$$x^4 + 2x^3 + 5x^2 + 5x + 2 = a(x^2+4)^2 + (bx+c)(x-2) + (dx+e)(x-2)(x^2+4).$$

Two convenient values of x to substitute in that equation are $x = 2$ and $x = 2i$. We choose $x = 2i$, because when x is replaced by $2i$, $x^2 + 4$ becomes $(2i)^2 + 4 = 4i^2 + 4 = 4(-1) + 4 = 0$.

For $x = 2$, we obtain $\quad 64 = 64a, \quad$ or $\quad a = 1$.

For $x = 2i$, we have $\quad -2 - 6i = (-4b - 2c) + (2c - 4b)i$.

Since two complex numbers are equal if and only if their real parts are equal and their imaginary parts are equal, we get

$$-4b - 2c = -2, \quad \text{and}$$

$$-4b + 2c = -6;$$

so $b = 1$ and $c = -1$.

There are no other convenient values of x that make factors on the right equal to zero, except $x = -2i$. But that will only give us $b = 1$ and $c = -1$ again. However, $x = 0$ and $x = 1$ are good because they make the calculations simple.

For $x = 0$, we have $\quad 2 = 16a - 2c - 8e$,

and for $x = 1$, $\quad 15 = 25a - b - c - 5d - 5e$.

Since $a = 1$, $b = 1$, and $c = -1$, we can reduce those two equations to

$$-16 = -8e, \quad \text{and}$$

$$-10 = -5d - 5e.$$

Thus $e = 2$ and $d = 0$. Consequently,

$$\int \frac{x^4 + 2x^3 + 5x^2 + 5x + 2}{(x^2+4)^2(x-2)} dx = \int \frac{dx}{x-2} + \int \frac{x-1}{(x^2+4)^2} dx + 2\int \frac{dx}{x^2+4} \qquad (3)$$

Since
$$\int \frac{dx}{x-2} = \ln|x-2| + C_1$$

and
$$2\int \frac{dx}{x^2+4} = \arctan\frac{x}{2} + C_2,$$

we are left with the second integral, which can be written as

$$\int \frac{x-1}{(x^2+4)^2}\,dx = \frac{1}{2}\int \frac{2x\,dx}{(x^2+4)^2} - \int \frac{dx}{(x^2+4)^2}.$$

The first of those two integrals can be evaluated immediately, and we may use formula (1) of Section 8.5, with $a = 2$ and $k = 1$, to evaluate the second integral. We have

$$\int \frac{x-1}{(x^2+4)^2}\,dx = \frac{1}{2}\frac{(x^2+4)^{-1}}{-1} - \frac{x}{8(x^2+4)} - \frac{1}{8}\int \frac{dx}{x^2+4}$$
$$= \frac{-1}{2(x^2+4)} - \frac{x}{8(x^2+4)} - \frac{1}{16}\arctan\frac{x}{2} + C_3.$$

On substitution into (3), we get

$$\int \frac{x^4+2x^3+5x^2+5x+2}{(x^2+4)^2(x-2)}\,dx = \ln|x-2| + \arctan\frac{x}{2} - \frac{1}{2(x^2+4)} - \frac{x}{8(x^2+4)}$$
$$- \frac{1}{16}\arctan\frac{x}{2} + C$$
$$= \ln|x-2| + \frac{15}{16}\arctan\frac{x}{2} - \frac{x+4}{8(x^2+4)} + C,$$

where $C = C_1 + C_2 + C_3$. ||

Exercises 8.6

Evaluate the integrals in Exercise 1–26.

1. $\int_{-1}^{1} \frac{5x-11}{(x+5)(x-2)}\,dx.$
2. $\int_{0}^{2} \frac{x+3}{(x-3)(x+1)}\,dx.$
3. $\int \frac{3x^2-8x+12}{x^3-x^2-12x}\,dx.$
4. $\int \frac{x^2+x+2}{x^3-x^2-6x}\,dx.$
5. $\int \frac{x-3}{x+1}\,dx.$
6. $\int \frac{x+5}{x+2}\,dx.$
7. $\int \frac{3x^2-8x-4}{x^3-4x}\,dx.$
8. $\int \frac{x+5}{x^3-x}\,dx.$
9. $\int_{1}^{2} \frac{2x^2-5x-10}{x^2(x+2)}\,dx.$
10. $\int_{2}^{3} \frac{x^2+3}{x^2(x-1)}\,dx.$
11. $\int \frac{2x+1}{(x+1)^2}\,dx.$ do not use partical fractions.
12. $\int \frac{2x-3}{(x-2)^2}\,dx.$

13. $\int \frac{x^3+1}{(x^2+1)^2}\,dx.$

14. $\int \frac{x^3-x^2+2x+1}{(x^2+1)^2}\,dx.$

15. $\int \frac{x^3-x^2-16x-4}{x^2-2x-15}\,dx.$

16. $\int \frac{x^2+3x+1}{x^2-1}\,dx.$

17. $\int \frac{x^4-2x^2-7x+2}{x^2-4}\,dx.$

18. $\int \frac{x^3+3x-4}{x^2-1}\,dx.$

19. $\int \frac{5x^2+3x+16}{(x^2+2x+5)(x-1)}\,dx.$

20. $\int \frac{x^2+2x-5}{(x^2+4)(x-2)}\,dx.$

21. $\int_0^1 \frac{x^2+12x-5}{(x+1)^2(x-7)}\,dx.$

22. $\int_0^1 \frac{2x^2-3x+1}{(x-2)^2(x+1)}\,dx.$

23. $\int \frac{2x^2+x+8}{(x^2+4)^2}\,dx.$

24. $\int \frac{x-5}{(x^2+4)^2}\,dx.$

25. $\int \frac{x^4+5x+3}{(x-1)(x^2+2)^2}\,dx.$

26. $\int \frac{x^4-3x^2-3x+3}{(x^2+1)^2(x-2)}\,dx.$

In Exercises 27–28 use the indicated substitution to help evaluate each integral.

27. $\int \frac{dx}{x(3x+5)^2}, \quad u = 3x+5.$

28. $\int \frac{dx}{x(7x-4)^3}, \quad u = 7x-4.$

29. Use partial fractions to prove this integration formula:

$$\int \frac{dx}{x(a+bx)} = -\frac{1}{a}\ln\left|\frac{a+bx}{x}\right| + C.$$

30. Use partial fractions to prove this integration formula:

$$\int \frac{dx}{x^2(a+bx)} = -\frac{1}{ax} + \frac{b}{a^2}\ln\left|\frac{a+bx}{x}\right| + C.$$

31. Population size x is sometimes related to time t by the equation $dx/dt = kx(a-x)(1-m/x)$, where m and a are constants and $m < a$. By integration we get

$$\int \frac{dx}{kx(a-x)(1-m/x)} = t.$$

Evaluate this integral. What happens to populations with less than m members?

8.7 Recognition of Techniques

Do all odds.

In this chapter we have discussed a number of techniques for finding indefinite integrals. Each of these techniques, however, is applicable only to a specific form of integrand. It is not only important to be able to use each technique, but also to be able to *recognize which technique to use*. In the previous sections of this chapter, the reader usually knew which technique to apply for each exercise, and the challenge was to apply it correctly. In this section the main challenge for each exercise will be to recognize which technique to apply. Also included will be exercises on techniques and standard forms from previous chapters. Therefore we should be particularly careful to recognize a standard form that can be integrated directly, such as the power function or a form leading to a logarithm.

Here is a rough classification of all the techniques we have discussed so far:

Techniques of Integration

1. Standard form (such as power function, log function, etc.).
2. Integration by parts (one or more times).
3. Substitution (usually for a fractional power of a linear function).
4. Products of powers of sines and cosines, secants and tangents, or cosecants and cotangents.
5. Trigonometric substitution.
6. Completing the square in the denominator.
7. Partial fractions.
8. Integration tables.

Practice is the best way to gain facility in recognizing which technique to apply. For that reason we have included many examples and exercises.

Example 1 $\displaystyle \int \frac{x\,dx}{\sqrt{x^2-9}}.$

This is a standard form (a power function) and can be done immediately:

$$\frac{1}{2}\int (x^2-9)^{-1/2}2x\,dx = (x^2-9)^{1/2} + C. \qquad \|$$

Example 2 $\displaystyle \int e^{2x} \cos 5x\,dx.$

The technique is to integrate by parts twice and then to move an integral from the right side of the resulting equation to the left side. ||

Example 3 $\displaystyle \int \frac{dx}{\sqrt{x^2-9}}.$

The technique is to make the trigonometric substitution $x = 3\sec\theta$, where $0 \le \theta < \pi/2$ if $x > 3$, or $\pi/2 < \theta \le \pi$ if $x < -3$. ||

Example 4 $\displaystyle \int \frac{x^3\,dx}{x^2-4}.$

The technique is to use partial fractions, after dividing the denominator into the numerator to get a polynomial plus a proper fraction. ||

Example 5 $\displaystyle \int \sin 7x \cos 2x\,dx.$

The technique is to integrate by parts twice. ||

Example 6 $\displaystyle \int \frac{x\,dx}{9-x^2}.$

This is a standard form (a logarithm). The integral equals

$$-\frac{1}{2}\int\frac{-2x\,dx}{9-x^2} = -\frac{1}{2}\ln|9-x^2| + C. \qquad \|$$

Example 7 $\displaystyle\int\frac{dx}{\sqrt{9-x^2}}.$

This can be converted to a standard form, an arcsine. ||

Example 8 $\displaystyle\int x^3 \ln x\,dx.$

Integration by parts is the technique to use here. ||

Example 9 $\displaystyle\int\sqrt{x-3}\,dx.$

A standard form (power function):

$$\int\sqrt{x-3}\,dx = \int(x-3)^{1/2}\,dx = \frac{2(x-3)^{3/2}}{3} + C. \qquad \|$$

Example 10 $\displaystyle\int\frac{dx}{x\ln x}.$

A standard form (producing a logarithm):

$$\int\frac{dx}{x\ln x} = \int\frac{dx/x}{\ln x} = \ln|\ln x| + C. \qquad \|$$

Example 11 $\displaystyle\int\frac{x^2\,dx}{1+\sqrt{x+2}}.$

The technique is substitution. Let $u = \sqrt{x+2}$. ||

Example 12 $\displaystyle\int xe^{(x^2)}\,dx.$

A standard form:

$$\int xe^{(x^2)}\,dx = \frac{1}{2}\int e^{(x^2)}(2x\,dx) = \frac{1}{2}e^{(x^2)} + C. \qquad \|$$

Example 13 $\displaystyle\int x^2e^x\,dx.$

Use integration by parts. ||

Example 14 $\displaystyle\int\frac{dx}{x^2+x+7}.$

The technique is to complete the square in the denominator and then use a standard form. ||

Example 15 $\displaystyle\int \frac{\sin x - \cos x}{\sin x + \cos x}\, dx.$

This produces a logarithm because the numerator is a constant (-1) times the derivative of the denominator. ||

Exercises 8.7

Each of the first fifty-two exercises has two parts; part (a) is to determine the technique to use, and part (b) is to use the technique and to find the final result.

1. $\displaystyle\int \frac{x+5}{x^2+10x}\, dx.$
2. $\displaystyle\int \frac{dx}{x^2-5}.$
3. $\displaystyle\int \frac{x^2\, dx}{x^2-4}.$
4. $\displaystyle\int_4^5 \frac{x\, dx}{\sqrt{2x-3}}.$
5. $\displaystyle\int \frac{dx}{x^2-6x}.$
6. $\int e^x \sin 2x\, dx.$
7. $\displaystyle\int_0^1 \frac{x\, dx}{x^4+1}.$
8. $\int \sin^2 x \cos^3 x\, dx.$
9. $\int x^2 \sin x\, dx.$
10. $\displaystyle\int_{-\pi}^{\pi} \cos x \sin 2x\, dx.$
11. $\displaystyle\int \frac{x^3+1}{x+2}\, dx.$
12. $\displaystyle\int \frac{x\, dx}{\sqrt{5+3x^2}}.$
13. $\displaystyle\int_0^{\pi/2} \frac{\cos x}{\sin x + 3}\, dx.$
14. $\displaystyle\int \frac{dx}{x^2-x-6}.$
15. $\displaystyle\int \frac{x^2+5x+18}{(x+3)(x^2+2x+3)}\, dx.$
16. $\displaystyle\int \frac{dx}{x^2+6x+13}.$
17. $\displaystyle\int_{-\pi}^{\pi} \sin x \sin 7x\, dx.$
18. $\int x^2 e^{3x}\, dx.$
19. $\displaystyle\int \frac{dx}{(x^2+9)^2}.$
20. $\int \sin^3 2x\, dx.$
21. $\int \cos x\, e^{\sin x}\, dx.$
22. $\int x^2 \ln x\, dx.$
23. $\displaystyle\int \frac{e^{4x}\, dx}{4+e^{4x}}.$
24. $\displaystyle\int \frac{dx}{\sqrt{5+4x-x^2}}.$
25. $\int x^6 \ln 2x\, dx.$
26. $\int x(x^2-16)^4\, dx.$
27. $\displaystyle\int_3^4 x\sqrt{x-3}\, dx.$
28. $\int \sec^4 2x\, dx.$
29. $\displaystyle\int \frac{x^4-38x^2+73x-64}{(x-5)(x+7)}\, dx.$
30. $\int x^4(x^4-16)dx.$
31. $\displaystyle\int \frac{\sec^2 x}{\tan x}\, dx.$
32. $\int \sec^3 2x \tan^3 2x\, dx.$
33. $\int \sin^2 5x\, dx.$
34. $\int x \sec^2 x\, dx.$
35. $\displaystyle\int \frac{dx}{(2-x^2)^{3/2}}.$
36. $\int \csc^3 2x \cot 2x\, dx.$
37. $\displaystyle\int_0^{\pi/3} \cos^2 3x\, dx.$
38. $\displaystyle\int_0^1 e^{3x} \cos x\, dx.$
39. $\int x \sec x \tan x\, dx.$
40. $\displaystyle\int \frac{7x^2-18x+5}{(x-3)(2x^2-5x+4)}\, dx.$

41. $\int \frac{x\,dx}{1+x^4}$.

42. $\int \frac{x}{(x-3)^{2/3}}\,dx$.

43. $\int \cos^3 7x\,dx$.

44. $\int \frac{\cos x\,dx}{1+\sin^2 x}$.

45. $\int x(3x^2+1)^{1/2}\,dx$.

46. $\int_{-1}^{1} \frac{x^3+3x^2-26x-41}{x^2+3x-28}\,dx$.

47. $\int \sin\sqrt{x}\,dx$.

48. $\int e^{\sqrt{x+2}}\,dx$.

49. $\int e^{\sqrt[3]{x}}\,dx$.

50. $\int \ln\sqrt[5]{x}\,dx$.

51. $\int \ln\sqrt{x+3}\,dx$.

52. $\int \cos\sqrt[3]{x}\,dx$.

53. The Theory of Relativity indicates that the mass of an object is given by

$$m = \frac{m_0}{\sqrt{1-v^2/c^2}}.$$

where m_0 is the mass of the object at rest, v its velocity, and c the velocity of light. If the velocity of an object is given as $v = kct$, where t is time measured in seconds, find the average mass of the object relative to the time from $t = 0$ to $t = 1/(2k)$.

54. A simplified mathematical model of the spread of a disease is based on the assumption that the rate of spread is jointly proportional to the fraction of those infected and the fraction of those healthy. Thus if $p(t)$ is the fraction infected at time t, then $1 - p(t)$ is the fraction of the population that is healthy. The fundamental assumption is then

$$\frac{dp(t)}{dt} = kp(t)[1-p(t)],$$

where k is a constant of proportionality. Consequently

$$\frac{dp}{p(1-p)} = k\,dt.$$

Integrate both sides of the equation and solve for p as a function of t.

55. Prove the reduction formula

$$\int x^m e^{ax}\,dx = \frac{x^m e^{ax}}{a} - \frac{m}{a}\int x^{m-1}e^{ax}\,dx.$$

56. Prove the reduction formula

$$\int x^a(\ln x)^n\,dx = \frac{x^{a+1}(\ln x)^n}{a+1} - \frac{n}{a+1}\int x^a(\ln x)^{n-1}\,dx, \qquad \text{if } a \neq -1.$$

Brief Review of Chapter 8

1. Integrals Containing $(ax + b)^{p/q}$

The substitution $u = (ax + b)^{1/q}$ should be used.

2. Integration by Parts

Integration by parts applied twice can be used to evaluate the integrals $\int e^{ax}\sin bx\,dx$, $\int e^{ax}\cos bx\,dx$, and $\int \cos mx \sin nx\,dx$.

3. Integrals of Powers of Trigonometric Functions

The integral $\int \sin^r x \cos^s x \, dx$ is evaluated as follows:

(a) If r is odd, the integral is written as

$$-\int \sin^{r-1} x \cos^s x \, d(\cos x).$$

Then one uses $\sin^2 x = 1 - \cos^2 x$ to express the integral in terms of $\cos x$. If s is odd, the approach is similar.

(b) If both r and s are even, the following identities are used in the order given:

$$\sin\theta\cos\theta = \frac{\sin 2\theta}{2},$$

$$\sin^2\theta = \frac{1-\cos 2\theta}{2},$$

$$\cos^2\theta = \frac{1+\cos 2\theta}{2}.$$

(c) Similar techniques are useful for

$$\int \csc^r x \cot^s x \, dx \quad \text{and} \quad \int \sec^r x \tan^s x \, dx.$$

4. Trigonometric Substitutions

Trigonometric substitutions made according to the following table are often useful.

occurrence	*substitution*	
$\sqrt{a^2-u^2}$	$u = a\sin\theta,$	$-\pi/2 \le \theta \le \pi/2$
$\sqrt{a^2+u^2}$	$u = a\tan\theta,$	$-\pi/2 < \theta < \pi/2$
$\sqrt{u^2-a^2}$	$u = a\sec\theta,$	$\begin{cases} 0 \le \theta < \pi/2 & \text{if } u \ge a \\ \pi/2 < \theta \le \pi & \text{if } u \le -a \end{cases}$

5. Integrals with Quadratics in the Denominator

The integration formulas

$$\int \frac{du}{u^2-a^2} = \frac{1}{2a}\ln\left|\frac{u-a}{u+a}\right| + C, \qquad \int \frac{du}{\sqrt{u^2-a^2}} = \ln\left|u+\sqrt{u^2-a^2}\right| + C,$$

$$\int \frac{du}{u^2+a^2} = \frac{1}{a}\arctan\frac{u}{a} + C, \qquad \int \frac{du}{\sqrt{a^2-u^2}} = \arcsin\frac{u}{a} + C$$

enable us to evaluate integrals of the form

$$\int \frac{rx+s}{ax^2+bx+c}\,dx \quad \text{or} \quad \int \frac{rx+s}{\sqrt{ax^2+bx+c}}\,dx.$$

Completing the square is often necessary.

6. Integration Tables

These tables contain many useful integration formulas not included in this text. The reduction formula

$$\int \frac{dx}{(x^2 + a^2)^{k+1}} = \frac{x}{2ka^2(x^2 + a^2)^k} + \frac{2k - 1}{2ka^2} \int \frac{dx}{(x^2 + a^2)^k}$$

is particularly useful when applying the method of partial fractions.

7. Partial Fractions

The method of partial fractions can be used to integrate rational functions where the degree of the numerator is less than the degree of the denominator. If the degree of the numerator is greater than or equal to that of the denominator, long division is used first.

Technique Review Exercises, Chapter 8

1. $\int \frac{3x^3 - 2x^2 - 10x + 9}{x^2 - x - 6}\, dx.$
2. $\int \frac{x\, dx}{(2x + 1)^{3/2}}.$
3. $\int \sin^2 x \cos^5 x\, dx.$
4. $\int \sin^2 \pi t \cos^2 \pi t\, dt.$
5. $\int_0^1 \frac{dx}{4 + x^2}.$
6. $\int_0^{3/2} \frac{x^2\, dx}{\sqrt{9 - x^2}}.$
7. $\int \sec^4 2x \tan^3 2x\, dx.$
8. $\int \frac{dx}{\sqrt{5 + 4x - x^2}}.$
9. $\int \sin 3x \cos x\, dx.$
10. $\int \frac{x\, dx}{x^2 - 4x - 5}.$
11. $\int \frac{e^{\sqrt{x+1}}}{\sqrt{x + 1}}\, dx.$

Additional Exercises, Chapter 8

Section 8.1

Evaluate the integrals.

1. $\int_4^5 (x - 1)\sqrt{x - 3}\, dx.$
2. $\int (1 - 3x)(x + 1)^{1/3}\, dx.$
3. $\int e^x \cos 3x\, dx.$
4. $\int_0^1 e^{2x} \sin x\, dx.$
5. $\int_0^{\pi/2} \sin x \cos 3x\, dx.$
6. $\int \sin 2x \sin 5x\, dx.$
7. $\int x \sin(2x) dx.$
8. $\int_0^1 x^2 \cos(2x + 1) dx.$

Section 8.2

Evaluate the integrals.

9. $\int x \sec^2(x^2)\tan^5(x^2)dx.$

10. $\int \tan^3 3x\, dx.$

11. $\int_0^{\pi/2} \sin^6 x \cos^3 x\, dx.$

12. $\int \sec^4 3x\, dx.$

13. $\int \csc^{1/4} x \cot^5 x\, dx.$

14. $\int \cos^4 x\, dx.$

15. $\int \sec^4 x \tan^{5/2} x\, dx.$

16. $\int \sin^3 x \cos^4 x\, dx.$

Section 8.3

Evaluate the integrals.

17. $\int \frac{dx}{\sqrt{9+x^2}}.$

18. $\int_0^1 \sqrt{16-x^2}\, dx.$

19. $\int x\sqrt{7-x^2}\, dx.$

20. $\int \frac{dx}{(16-x^2)^{3/2}}.$

21. $\int_0^1 \frac{dx}{\sqrt{x^2+4x+16}}.$

22. $\int \frac{x\, dx}{\sqrt{9-4x-x^2}}.$

23. $\int (x-4)^3\sqrt{10-x^2+8x}\, dx.$

24. $\int (x+4)^5\sqrt{x^2+8x+7}\, dx.$

Section 8.4

Evaluate the integrals.

25. $\int \frac{dx}{x^2-6x+5}.$

26. $\int \frac{dx}{3-x^2-5x}.$

27. $\int \frac{x\, dx}{x^2-6x+5}.$

28. $\int \frac{x\, dx}{3-x^2-5x}.$

29. $\int \frac{3x+4}{4-2x+x^2}\, dx.$

30. $\int_{-1}^{0} \frac{x-1}{x^2+4x+5}\, dx.$

31. $\int \frac{2x-1}{3+4x^2-8x}\, dx.$

32. $\int \frac{x+3}{7-x^2-6x}\, dx.$

33. $\int \frac{dx}{\sqrt{40+6x-x^2}}.$

34. $\int \frac{x+4}{\sqrt{-x^2-14x-24}}\, dx.$

Section 8.5

Evaluate the integral.

35. $\int \frac{dx}{2+\sin x}.$

36. $\int \frac{dx}{1-\sin x}.$

37. $\int_1^2 \frac{dx}{(x^2+4)^2}.$

38. $\int \frac{dx}{(x^2+8)^2}.$

39. $\int \frac{dx}{(x^2+4)^3}.$

40. $\int \frac{dx}{(x^2+8)^3}.$

Section 8.6

Evaluate the integrals.

41. $\int \frac{dx}{(x-2)(x+1)}.$

42. $\int_{-2}^{-1} \frac{x+1}{x(x-1)}\, dx.$

43. $\int \frac{x^3}{(x-3)(x+1)} dx.$

44. $\int \frac{x\,dx}{x^2-3x-4}.$

45. $\int_1^2 \frac{6x^2+x+24}{x(x^2+6)} dx.$

46. $\int \frac{8x^3-18x^2+14x-16}{(x-1)^2(x^2+3)} dx.$

47. $\int \frac{x^2-x+1}{x(x-2)^2} dx.$

48. $\int \frac{-x^3+x^2+7x-5}{(x+3)^2(x^2+1)} dx.$

Section 8.7

Evaluate the integrals.

49. $\int_0^{\pi/4} \frac{\sin^3 x}{\cos x} dx.$

50. $\int \frac{dx}{x^2+4x+5}.$

51. $\int \frac{x^2}{(x^2+5)^2} dx.$

52. $\int \frac{dx}{\sqrt{x^2-4}}.$

53. $\int \frac{x}{x^2+4x+5} dx.$

54. $\int \frac{x}{\sqrt{8-x^2+2x}} dx.$

55. $\int \frac{x^3+1}{x^3-9x} dx.$

56. $\int_0^1 \frac{2x-1}{(x^2-4)^2} dx.$

57. Verify the reduction formula

$$\int \sin^n x\,dx = -\frac{\sin^{n-1} x \cos x}{n} + \frac{n-1}{n}\int \sin^{n-2} x\,dx.$$

Challenging Problems, Chapter 8

1. What's wrong with the following "proof" that $0 = 1$? Apply integration by parts to $\int x^{-1}\,dx$ by letting $u = x^{-1}$ and $dv = dx$. Then $du = -x^{-2}\,dx$ and $v = x$. Thus

$$\int x^{-1}\,dx = x \cdot x^{-1} - \int x(-x^{-2})dx.$$

Hence
$$\int x^{-1}\,dx = 1 + \int x^{-1}\,dx,$$

and on substracting $\int x^{-1}\,dx$ from both sides, we have $0 = 1$.

2. Show that $\int_0^1 x^r(1-x)^s\,dx = \frac{r!s!}{(r+s+1)!}.$

9

Numerical Methods

9.1 Introduction

In this chapter we consider two interrelated topics, numerical methods and approximations. While the term "numerical methods" applies equally well to precise and approximate computations, in actual practice approximations must almost always be used, since precise methods of computation are available only for *rational* numbers. Thus, while the precise area of a circle of radius 2 units is 4π square units, we must use some rational approximation to π in order to obtain a numerical answer.

Of course the precision demanded of the final result dictates the precision that should be used in finding an approximation. Although society developed up to a certain point by using the biblical approximation $\pi \approx 3$ ($\approx$ is read as "is approximately equal to"), a far better approximation is necessary today. This brings up the central point concerning approximations. *In order to make use of an approximation, it is necessary to know how much the approximation might differ from the exact value.* That is, we must have some bound on the error of the approximation.

Generally, numerical methods are applied to problems that lack a straightforward solution. For example, though the quadratic formula gives a direct and simple method for solving any quadratic equation, we can prove that no such algorithm exists for the general fifth-degree equation. Similarly, although The Fundamental Theorem of Calculus affords a method of evaluating many definite integrals, it is of little use in evaluating integrals such as

$$\int_1^2 \frac{\sin x}{x}\,dx, \qquad \int_0^2 \frac{4^x}{x+1}\,dx, \qquad \text{and} \int_1^3 x^x\,dx,$$

because the corresponding indefinite integrals cannot be expressed in any convenient form.

We begin by discussing two methods for the numerical solution of equations: Newton's Method and the bisection method.

9.2 Newton's Method

Tangents to a curve are often useful approximations to the curve in cases where the curve itself is difficult to work with. Newton's Method uses tangents to the graph of a function to generate successive approximations to a zero of the given function. Suppose that c_1 is an initial approximation to the number α where $f(\alpha) = 0$. In Figure 9.2.1, we indicate the desired point α, the initial approximation c_1, and the tangent to the graph of $y = f(x)$ at $(c_1, f(c_1))$. The fundamental idea of Newton's Method is that the point c_2 where the indicated tangent crosses the x-axis is usually a better approximation to α than c_1 is.

Figure 9.2.1

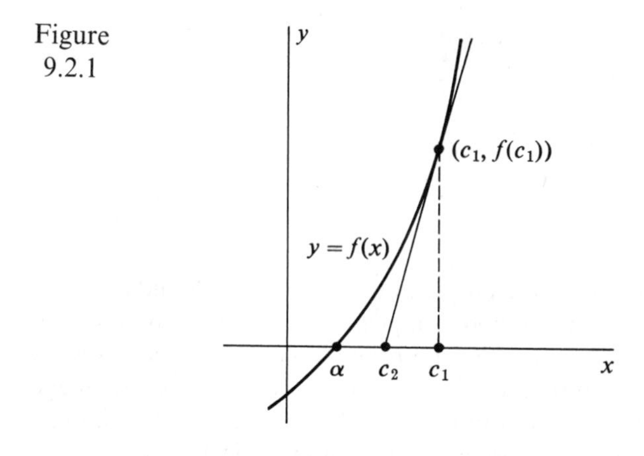

If we now repeat the procedure above, using c_2 in place of c_1, we obtain an even better approximation, as indicated in Figure 9.2.2. That is, the point c_3 where the tangent to the graph at $(c_2, f(c_2))$ crosses the x-axis is a better approximation to the point α. By continuing this process we can often obtain an approximation to α that is as accurate as we wish. In that case, we say that Newton's Method *converges* to the zero α.

Figure 9.2.2

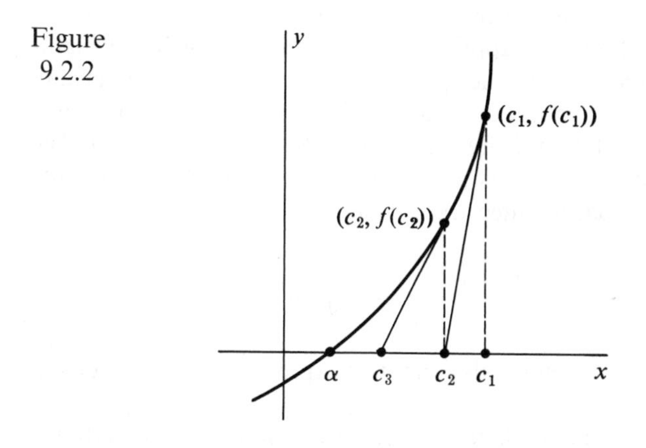

To obtain an algebraic expression corresponding to the preceding geometric discussion, we can proceed as follows. An equation of the line tangent to the graph of $y = f(x)$ at $(c_1, f(c_1))$ is

$$y - f(c_1) = f'(c_1)(x - c_1).$$

On that tangent, $x = c_2$ when $y = 0$; hence

$$-f(c_1) = f'(c_1)(c_2 - c_1), \quad \text{and so}$$

$$c_2 = c_1 - \frac{f(c_1)}{f'(c_1)}.$$

In general, for any positive integer i the line tangent to $y = f(x)$ at $(c_i, f(c_i))$ has an equation

$$y - f(c_i) = f'(c_i)(x - c_i).$$

On that tangent, $x = c_{i+1}$ when $y = 0$. Therefore we get

$$c_{i+1} = c_i - \frac{f(c_i)}{f'(c_i)}. \tag{1}$$

This is the recursion relation governing Newton's Method.

Example 1 | Beginning with $c_1 = 1$, find the third approximation to the cube root of 3, that is, to the zero of $f(x) = x^3 - 3$.

Since $f'(x) = 3x^2$, the recursion relation (1) becomes

$$c_{i+1} = c_i - \frac{c_i^{\,3} - 3}{3c_i^{\,2}} = \frac{2c_i^{\,3} + 3}{3c_i^{\,2}}.$$

Thus,
$$c_2 = \frac{2c_1^{\,3} + 3}{3c_1^{\,2}}.$$

and so
$$c_2 = \frac{5}{3} = 1.\overline{6}.$$

(Here we are using the common convention that a bar over one or more digits indicates that they repeat forever.) The third approximation is therefore

$$c_3 = \frac{2(5/3)^3 + 3}{3(5/3)^2} \approx 1.47. \qquad \|$$

At this point, we don't know how close the approximation c_3 is to the actual zero $\alpha = \sqrt[3]{3}$. Before we attempt to estimate the error, we should note that, in some cases, Newton's Method may actually fail to converge to a zero! Figure 9.2.3 illustrates such a case.

Figure 9.2.3

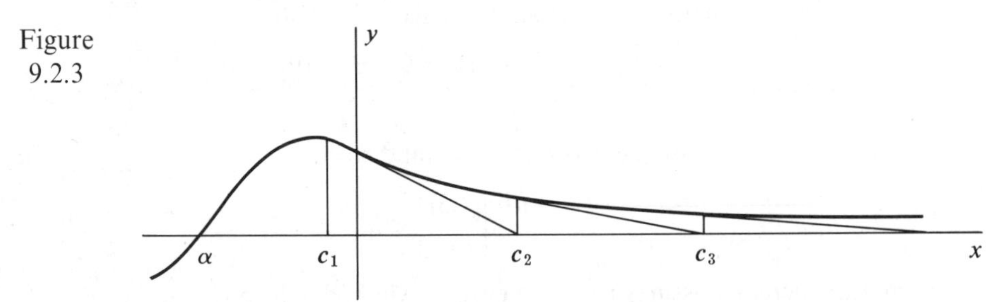

In view of the example illustrated in Figure 9.2.3 and Example 1, two questions present themselves.

1. Are there any criteria that will ensure the convergence of Newton's Method to a zero of the function?
2. Is there a reasonable bound, B_n, on the error $E_n = |c_n - \alpha|$?

The first of those questions is answered by the following theorem. (For a proof the reader is referred to a suitable numerical-analysis text such as Henrici, *Elements of Numerical Analysis* (Wiley, 1964).)

Theorem 9.2.1

If $f(x)$ is continuous for $a \le x \le b$ and

(i) $f(x)$ has a zero between a and b,
(ii) $f(x)$ is either increasing on the interval $a \le x \le b$ or decreasing on the interval $a \le x \le b$,
(iii) $f(x)$ has the same concavity at each point of the interval $a \le x \le b$,
(iv) $\left|\frac{f(c)}{f'(c)}\right| \le b - a$ for the endpoint c of the interval $a \le x \le b$ at which $|f'(x)|$ is smaller,

then Newton's Method converges to the zero of $f(x)$ between a and b, for any choice of c_1, $a \le c_1 \le b$.

The last hypothesis, (iv), assures us that any tangent line to $y = f(x)$, $a \le x \le b$, intersects the x-axis at a point between a and b.

Example 2 Prove that Newton's Method converges to a zero of the function $f(x) = x^2 - x - 10$ for any choice of c_1, $3 \le c_1 \le 4$.

We need only check that the hypotheses of Theorem 9.2.1 are satisfied for the continuous function $f(x) = x^2 - x - 10$.

(i) Since $f(3) = -4 < 0$ and $f(4) = 2 > 0$ The Intermediate Value Theorem assures us that $f(x)$ does have a zero between 3 and 4.
(ii) Since $f'(x) = 2x - 1 > 0$ for $3 \le x \le 4$, $f(x)$ is increasing for all x, $3 \le x \le 4$.
(iii) Since $f''(x) = 2 > 0$, $f(x)$ is concave upward at all points of the interval.
(iv) Since $|f'(x)| = |2x - 1|$, $|f'(3)| = 5$, and $|f'(4)| = 7$, we know that $|f'(x)|$ is smaller when $x = 3$. Also,

$$\left|\frac{f(3)}{f'(3)}\right| = \left|\frac{9 - 3 - 10}{5}\right| = \frac{4}{5}.$$

Consequently, since $a = 3$ and $b = 4$,

$$\left|\frac{f(3)}{f'(3)}\right| < b - a = 1.$$

Theorem 9.2.1 therefore assures us that Newton's Method will converge. ||

Note, however, that this analysis does not yield a bound on the error. The following theorem will yield a bound on the error if some initial error is known.

Theorem 9.2.2

If $f(x)$ has a zero, α, in the interval $a \le x \le b$, and if for $a \le x \le b$ we have $|f''(x)| \le M$ and $|f'(x)| \ge m > 0$, then the error bounds, $B_i \ge |c_i - \alpha|$, are related as follows:

$$B_{n+1} \le B_n^2 \frac{M}{2m}. \tag{2}$$

The proof of this theorem will be left to the Challenging Problems, Chapter 9.

It should be noted that the theorem does not give any specific bound, B_{n+1}, on the error $E_{n+1} = |c_{i+1} - \alpha|$, but it does establish a relation between any bound B_n and the successive bound B_{n+1}.

Example 3 | Use Newton's Method to find the third approximation to the zero of $f(x) = x^2 - x - 10$ that lies between $x = 3$ and $x = 4$. Analyze the error at each stage of the computation.

In this case, $f'(x) = 2x - 1$; so, in general,

$$c_{i+1} = c_i - \frac{c_i^2 - c_i - 10}{2c_i - 1} = \frac{c_i^2 + 10}{2c_i - 1}.$$

Thus, if we let $c_1 = 3$, we have

$$c_2 = \frac{9 + 10}{5} = \frac{19}{5} = 3.8, \quad \text{and}$$

$$c_3 = \frac{(3.8)^2 + 10}{2(3.8) - 1} \approx 3.703.$$

To analyze the error we use Theorem 9.2.2. Since $f''(x) = 2$, we have $|f''(x)| \leq 2$, and so we can take $M = 2$. Using $f'(x) = 2x - 1$, we see that

$$f'(x) = 2x - 1 \geq 5, \qquad \text{for } 3 \leq x \leq 4.$$

Consequently, we take $m = 5$. Since the zero in question lies between $x = 3$ and $x = 4$, the initial error in selecting $c_1 = 3$ is less than or equal to 1. That is, $E_1 \leqslant 1$, so we may take $B_1 = 1$. Then using this information in (2), we obtain

$$B_2 \leq (1)^2 \frac{2}{10} = \frac{1}{5}.$$

Then, again using $B_2 \leq 1/5$ in (2), we get

$$B_3 \leq \left(\frac{1}{5}\right)^2 \frac{2}{10} = \frac{1}{125}.$$

Hence the approximation $c_3 \approx 3.703$ is within 1/125 of the desired zero. ||

The reader should be careful to note that Theorem 9.2.1 gives conditions sufficient to insure that Newton's Method converges, while Theorem 9.2.2 gives a method of establishing error bounds. It is possible that Newton's Method may converge (from Theorem 9.2.1) yet the error bounds established by Theorem 9.2.2 may not tend to zero because of an unsuitable choice of B_1. This fact will be illustrated in the exercises. (See Exercises 5, 11, and 17 in particular.)

Newton's Method is often used in computers for calculating various simple functions such as square roots, cube roots, and reciprocals. These will be examined in greater detail in the exercises.

As an indication of a possible computer application of Newton's Method we shall consider the problem of finding a number α such that

$$k\alpha - e^{-\alpha} = 0,$$

where k is a fixed number and $k \geq 1$. That is, we shall use Newton's Method to find the zero of the function

$$f(x) = kx - e^{-x}.$$

Note first that $f(0) = -1 < 0$ and $f(1) = k - 1/e > 0$, since $k \geq 1$. Thus we must have $0 < \alpha < 1$, and an initial approximation of $c_1 = 1/2$ will result in an initial error, E_1, bounded above by 1/2, that is, $E_1 \leq 1/2$, or $B_1 = 1/2$.

Since
$$f'(x) = k + e^{-x},$$

the recursion relation given by Newton's Method is

$$c_{i+1} = c_i - \frac{kc_i - e^{-c_i}}{k + e^{-c_i}}.$$

To obtain bounds on the successive errors, note that

$$|f'(x)| = |k + e^{-x}| = k + e^{-x} \geq k;$$

so we may take $m = k$. In addition,

$$|f''(x)| = |-e^{-x}| = e^{-x} < 1, \qquad \text{for } x > 0,$$

and hence we can take $M = 1$. Then, since

$$B_{n+1} \leq B_n{}^2 \frac{M}{2m},$$

we have
$$B_{n+1} \leq B_n{}^2 \frac{1}{2k}.$$

Note that since $B_1 = 1/2$ and $k \geq 1$, the errors will continually decrease.

The flow chart pictured in Figure 9.2.4 is the first step toward a computer program. Note that the constant k, along with a maximal allowable error F, must be given as input.

Figure 9.2.4

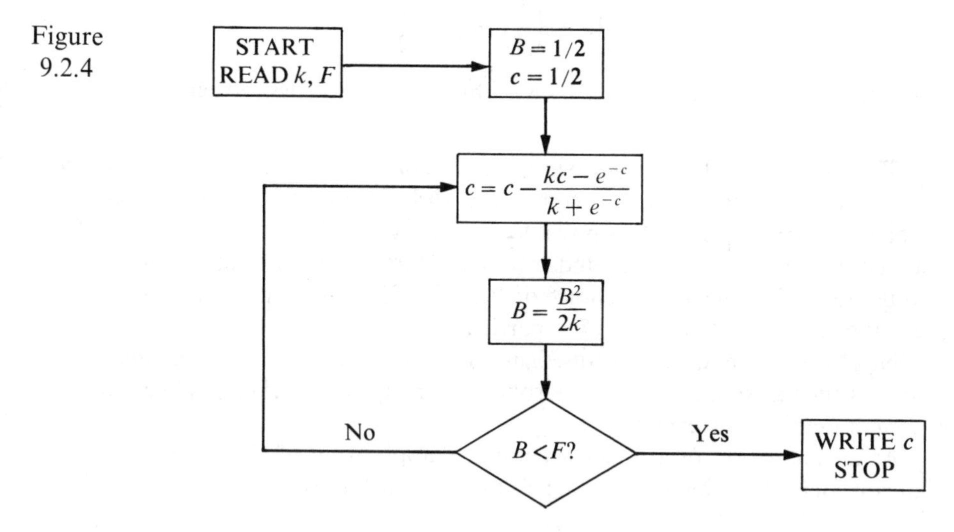

Also note that the flow chart uses the "assignment" equal sign commonly used in computers. Rather than indicating equality, this sign tells the computer that the right-hand side of the equation is to be assigned to the place indicated on the left-hand side.

Exercises 9.2

In Exercises 1–10, use the given initial approximation to calculate c_3, the third approximation to a zero.

1. $x^2 - 5 = 0, \quad c_1 = 2.$
2. $x^2 - 14 = 0, \quad c_1 = 4.$
3. $x^2 - x - 3 = 0, \quad c_1 = -1.$
4. $x^2 + x - 1 = 0, \quad c_1 = 1/2.$
5. $x^4 - 5 = 0, \quad c_1 = 1.$
6. $x^4 - 2 = 0, \quad c_1 = 1.$
7. $3x^3 - x^2 - 4 = 0, \quad c_1 = 1.$
8. $x^4 - x - 1 = 0, \quad c_1 = 0.$
9. $100 - 1/x = 0, \quad c_1 = 1.$
10. $10 - 1/x = 0, \quad c_1 = 1.$
11. Prove that Newton's Method converges when applied in Exercise 5 above. (Note that if $f(x) = x^4 - 5$, then $f(1) < 0$ and $f(2) > 0$.)
12. Prove that Newton's Method converges when applied in Exercise 6 above.
13. As in Example 3, analyze the errors in Exercise 1 for the interval $2 \le x \le 3$. (Note that $E_1 \le 1$.)
14. As in Example 3, analyze the errors in Exercise 2 for the interval $3 \le x \le 4$. (Note that $E_1 \le 1$.)
15. Analyze the errors in Exercise 3 for the interval $-2 \le x \le -1$. (Note that $E_1 \le 1$.)
16. Analyze the errors in Exercise 4 for the interval $0 \le x \le 1$. (Note that $E_1 \le 1$.)
17. Analyze the errors in Exercise 5 for the interval $1 \le x \le 2$. How do you explain this result in view of Exercise 11?
18. Newton's Method is often used by computers to approximate reciprocals. That is, since $1/a$ is a zero of $f(x) = 1/x - a$, Newton's Method may be applied to $f(x) = 1/x - a$ in order to approximate $1/a$. Develop the recursion formulas *not* involving division that can be used for this approximation.
19. Develop a recursion relation that can be used for approximating $a^{1/2}$.
20. Use Newton's Method to develop a recursion formula to approximate α where $\cos\alpha = \alpha$. Prove that Newton's Method will converge for this case.
21. Determine the x value of the point of intersection of the graphs of $y = x^2 + 1$ and $y = x^3$ with an error of less than $1/5$.
22. Determine the x value of the point of intersection of the graphs of $y = x^3$ and $y = x/2 - 1$ with an error of less than $1/100$.
23. Write a computer program to compute $a^{1/2}$ with an error of less than 10^{-6}. Use this program to compute $5^{1/2}$ and $(.15)^{1/2}$ with an error of less than 10^{-6}.
24. Write a computer program to compute $a^{1/3}$ with an error of less than 10^{-6}. Use this program to compute $5^{1/3}$ and $(.15)^{1/3}$ with an error of less than 10^{-6}.
25. Write a computer program to find all values of x for which $2\sin x = x$ with an error of less than 10^{-5}.
26. The following three graphs illustrate instances for which Newton's Method fails to converge. Indicate which of the hypotheses of Theorem 9.2.1 are violated in each case.

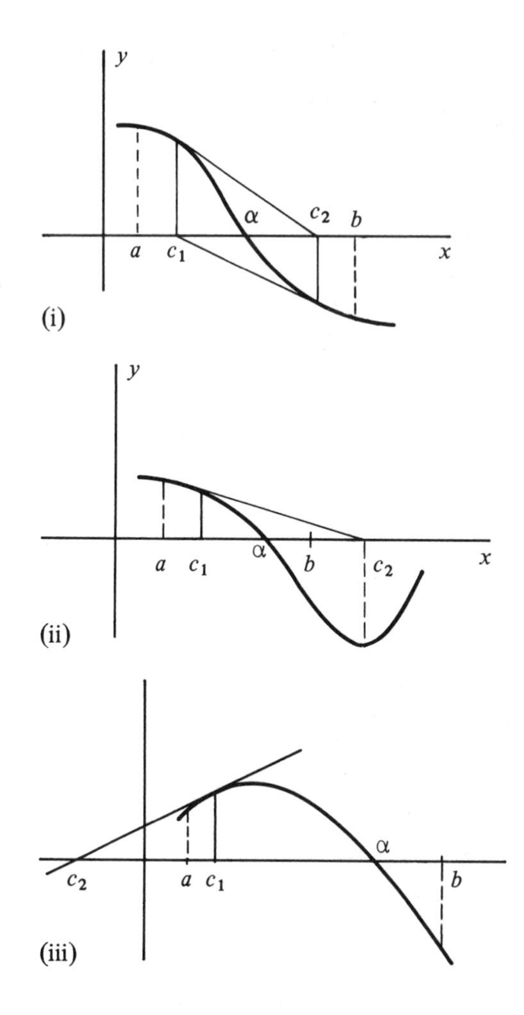

9.3 Bisection Method

In our discussion of Newton's Method we noted that there are some instances where it fails. On the other hand, the bisection method that we shall discuss shortly is safe to use on all continuous functions. However, the large amount of necessary computation makes this method useful primarily on high speed computers.

The central idea is to isolate the desired zero in an interval, and then by repeated bisection of the interval to confine the zero to smaller and smaller intervals. Suppose, for example, that f is a continuous function with $f(a) < 0$ and $f(b) > 0$. Since f is continuous, The Intermediate Value Theorem for continuous functions assures us that there is an α such that $a < \alpha < b$ and $f(\alpha) = 0$. We now evaluate f at the point $x_1 = (b + a)/2$, midway between a and b. Of course if $f(x_1) = 0$, we're finished. If $f(x_1) \neq 0$, then $f(x_1) > 0$ or $f(x_1) < 0$. If $f(x_1) > 0$, then α is between a and x_1; while if $f(x_1) < 0$, then α is between x_1 and b. These two situations are pictured in Figure 9.3.1(a) and (b), respectively. In either case the zero, α, is then captured within an interval half the size of the previous interval. Continued repetition of this method will determined α to any desired accuracy.

Figure 9.3.1

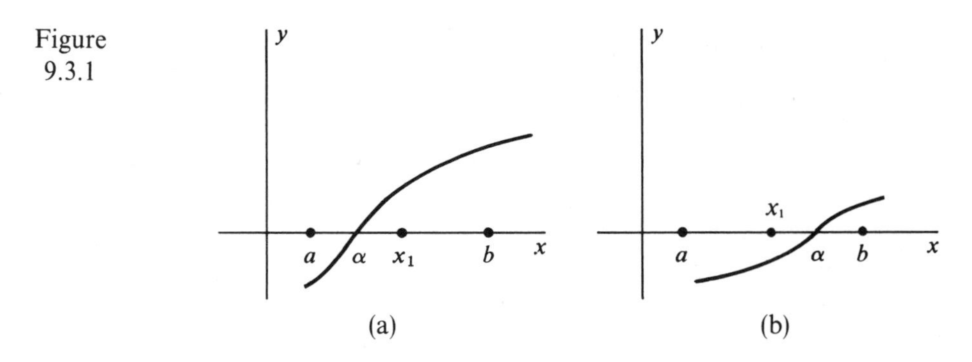

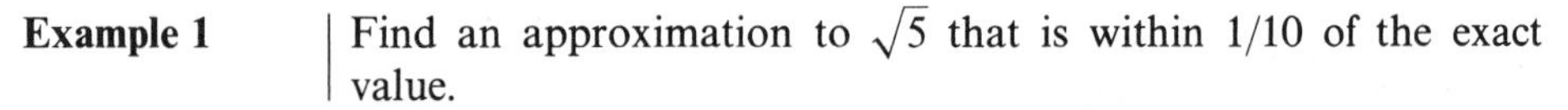

Example 1 | Find an approximation to $\sqrt{5}$ that is within 1/10 of the exact value.

Since $\sqrt{5}$ is the positive zero of $f(x) = x^2 - 5$ and since $f(2) = -1 < 0$ and $f(3) = 4 > 0$, we know that $2 < \sqrt{5} < 3$.

For $x_1 = (2 + 3)/2 = 2.5$, we have $f(2.5) = 1.25 > 0$. Since $f(2.5)$ and $f(2) = -1$ differ in sign, $2 < \sqrt{5} < 2.5$. Evaluation of f midway between 2 and 2.5 gives $f(2.25) = .0625 > 0$. Then, since $f(2.25)$ and $f(2)$ differ in sign, $2 < \sqrt{5} < 2.25$. Evaluation of f midway between 2 and 2.25 gives $f(2.125) = -.484375$. Since $f(2.25)$ and $f(2.125)$ differ in sign, $2.125 < \sqrt{5} < 2.25$. Consequently the point 2.1875, midway between 2.125 and 2.25, is within 1/10 of the exact value of $\sqrt{5}$. ||

Figure 9.3.2 pictures a flow chart that applies the bisection method to the calculation of $b^{1/2}$ where $b \geq 1$. In this flow chart we read-in both b and F, the maximal allowable error. Note that since $b \geq 1$, we have $1 \leq b^{1/2} \leq b$. That inequality establishes the initial interval confining $b^{1/2}$.

Figure 9.3.2

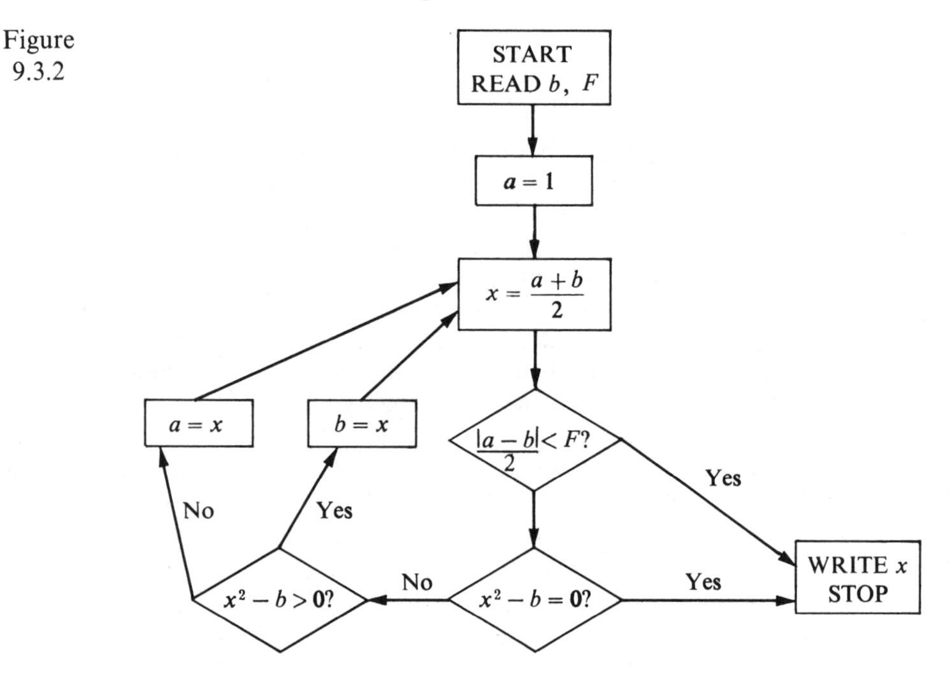

Exercises 9.3

1. Use the bisection method to find an approximation to $23^{1/2}$ that is within 1/10 of the exact value.
2. Use the bisection method to find an approximation to $11^{1/2}$ that is within 1/10 of the exact value.
3. Use the bisection method to find an approximation to $31^{1/3}$ that is within 1/10 of the exact value.
4. Use the bisection method to find an approximation to $12^{1/3}$ that is within 1/10 of the exact value.
5. Using the bisection method, write a computer program to compute to any desired accuracy an approximation to $a^{1/2}$, where $a \geq 1$. Design the program to write the number of iterations used. Use the program to approximate $5^{1/2}$ and $(123.1)^{1/2}$ with an error less than 10^{-5}.
6. Using the bisection method, write a computer program to find, with an error less than 10^{-5}, all values of x such that $2 \sin x = x$.
7. Using the bisection method, write a computer program to find all values of x for which $\tan x = \cos x$ with an error less than 10^{-5}.
8. Write a computer program by using the bisection method to compute $a^{1/2}$ with an error of less than 10^{-6}. Use this program to compute $5^{1/2}$ and $(.15)^{1/2}$ with an error of less than 10^{-6}. Compare this program with that written for Exercise 23, Section 9.2.
9. Write a computer program by using the bisection method to compute $a^{1/3}$ with an error of less than 10^{-6}. Use this program to compute $5^{1/3}$ and $(.15)^{1/3}$ with an error of less than 10^{-6}. Compare this program with that written for Exercise 24, Section 9.2.
10. If $1 < \alpha < 1.6$, how many bisections are needed to find an approximation that is within .0001 of the exact value?

9.4 Trapezoid Rule

As we pointed out in the introduction to this chapter, it is often difficult or impossible to evaluate certain definite integrals with The Fundamental Theorem. In those cases we usually turn to some numerical technique such as the *trapezoid rule.*

Consider the definite integral $\int_a^b f(x)dx$, where $f(x) \geq 0$ for $a \leq x \leq b$. Since this integral represents the area of the region under the graph of $y = f(x)$ between $x = a$ and $x = b$, an approximation to the area will be an approximation to the integral. To approximate the area, first subdivide the interval $a \leq x \leq b$ into n subintervals, each of width $h = (b - a)/n$, by introducing the points $x_0 = a$, $x_1 = a + h, \ldots, x_i = a + ih, \ldots, x_n = a + nh = b$. Each of the subintervals is used as the base of a trapezoid whose area approximates the area of the region under the curve and above the subinterval, as indicated in Figure 9.4.1.

Since the area of the ith trapezoid is $h[f(x_{i-1}) + f(x_i)]/2$, the total area under the graph is approximated by

$$\sum_{i=1}^{n} \frac{h}{2}[f(x_{i-1}) + f(x_i)], \quad \text{and so}$$

$$\int_a^b f(x)dx \approx \frac{h}{2} \sum_{i=1}^{n} [f(x_{i-1}) + f(x_i)].$$

Figure
9.4.1

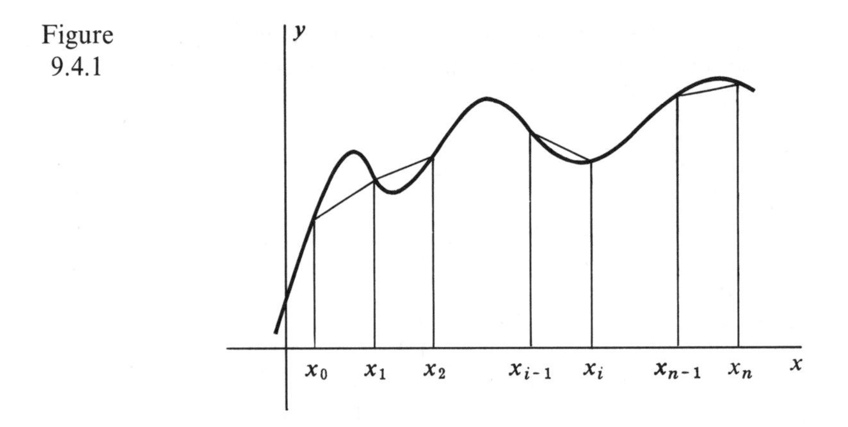

This formula may be simplified as follows:

$$\int_a^b f(x)dx \approx \frac{h}{2}[f(x_0)+f(x_1)+f(x_1)+f(x_2)+f(x_2)+f(x_3)+\cdots+f(x_{n-1})+f(x_n)]$$

$$\approx \frac{h}{2}[f(x_0)+2f(x_1)+2f(x_2)+\cdots+2f(x_{n-1})+f(x_n)].$$

Thus, for $h=(b-a)/n$,

$$\int_a^b f(x)dx \approx h\left[\frac{1}{2}f(x_0)+f(x_1)+f(x_2)+\cdots+f(x_{n-1})+\frac{1}{2}f(x_n)\right].$$

To remember this formula, it is best to think of it in this form:

$$\int_a^b f(x)dx \approx h\left[\frac{1}{2}(\textit{zero}\text{th term})+\text{intermediate terms}+\frac{1}{2}(\text{last term})\right].$$

Example 1 | Use the trapezoid rule, with $n=2$, to find an approximation to

$$\int_{\pi/2}^{\pi} \frac{\sin x}{x}\,dx.$$

Since $h=(b-a)/n$, we have $h=(\pi-\pi/2)/2=\pi/4$. A convenient way to organize the computations involved in the trapezoid rule is to construct the following table.

i	x_i	$f(x_i)$	*multiplier*	*term*
0	$\pi/2$	$2/\pi$	1/2	$1/\pi$
1	$3\pi/4$	$2\sqrt{2}/3\pi$	1	$2\sqrt{2}/3\pi$
2	π	0	1/2	0

Then,

$$\int_{\pi/2}^{\pi} \frac{\sin x}{x}\,dx \approx \frac{\pi}{4}\left(\frac{1}{\pi}+\frac{2\sqrt{2}}{3\pi}+0\right)$$

$$\approx \frac{1}{4}\left(1+\frac{2\sqrt{2}}{3}\right).$$

||

Example 2 Use the trapezoid rule, with $n = 4$, to approximate $\int_0^1 x^2\,dx$. How does the approximation compare with the exact value?

Here $h = 1/4$, and we have the following table:

i	x_i	$f(x_i)$	*multiplier*	*term*
0	0	0	1/2	0
1	1/4	1/16	1	1/16
2	1/2	1/4	1	1/4
3	3/4	9/16	1	9/16
4	1	1	1/2	1/2

Hence,

$$\int_0^1 x^2\,dx \approx \frac{1}{4}\left(0 + \frac{1}{16} + \frac{1}{4} + \frac{9}{16} + \frac{1}{2}\right) = \frac{11}{32}.$$

Consequently,

$$\int_0^1 x^2\,dx \approx .344.$$

The precise value is

$$\int_0^1 x^2\,dx = \frac{1}{3}x^3\Big|_0^1 = \frac{1}{3} = .33\overline{3},$$

and so the error is less than .02. ||

As for all approximations, it is natural to look for a bound on the error

$$E_n = \left|\int_a^b f(x)dx - h\left[\frac{1}{2}f(x_0) + f(x_1) + \cdots + f(x_{n-1}) + \frac{1}{2}f(x_n)\right]\right|.$$

The following theorem, whose proof can be found in many texts on numerical analysis, gives such a bound.

Theorem 9.4.1

If $|f''(x)| \le M$ for $a \le x \le b$, then

$$E_n \le \frac{b-a}{12}Mh^2. \tag{1}$$

Example 3 Use the trapezoid rule with $n = 5$ to approximate $\ln 2 = \int_1^2 dx/x$, and find a bound for the error in the approximation.

Here $h = 1/5$, and we have this table:

i	x_i	$f(x_i)$	*multiplier*	*term*
0	1	1	1/2	1/2
1	6/5	5/6	1	5/6
2	7/5	5/7	1	5/7
3	8/5	5/8	1	5/8
4	9/5	5/9	1	5/9
5	2	1/2	1/2	1/4

Hence, $$\ln 2 \approx \frac{1}{5}\left(\frac{1}{2} + \frac{5}{6} + \frac{5}{7} + \frac{5}{8} + \frac{5}{9} + \frac{1}{4}\right) \approx .6956.$$

To find a bound on the error E_5, we need to find the maximum M of $f''(x)$ for $1 \leq x \leq 2$. In this case, $f(x) = x^{-1}$ and so $f''(x) = 2x^{-3}$. Hence $|f''(x)|$ assumes its maximum value when $x = 1$, and we may take $M = 2$. Substitution of $M = 2$, $h = 1/5$, $a = 1$, and $b = 2$ into (1) gives $E_5 \leq 1/150$. ||

In Example 3 we were given a value of n and asked to compute a bound on the error for that particular value of n. In many instances the process must be reversed. That is, we are given the maximal allowable error, F, and must determine the smallest value of n for which $E_n \leq F$. In that case we use (1) along with $h = (b - a)/n$ to determine the smallest acceptable value of n.

Example 4 What is the smallest value of n that can be used so that the error in approximating $\ln 2 = \int_1^2 \frac{dx}{x}$ by the trapezoid rule is less than or equal to 10^{-6}?

Again we have $M = 2$, $a = 1$, and $b = 2$. However, in this case, $h = 1/n$ is unknown. Substitution into (1) gives $E_n \leq 1/6n^2$, and hence we need to choose n so large that $1/6n^2 < 10^{-6}$, that is,

$$n^2 > \frac{1}{6} \cdot 10^6 = 166{,}666\,\frac{2}{3}.$$

Since $(409)^2 = 167{,}281$, we have $n = 409$. ||

In practical situations, we often observe the value of an unknown function, $f(x)$, at certain specified and equally spaced points. For example, one might record the temperature T at each hour during a day. While it is natural to suppose that this temperature T is a function of time, $T = f(t)$, the precise function is unknown. Nevertheless, it might be useful to have an approximation to $\int_0^{24} f(t)dt$ or, in general, an approximation to $\int_a^b f(x)dx$. The trapezoid rule gives one method for obtaining such an approximation.

Example 5 A bar of metal is being cooled and the temperature of the bar is observed every 40 minutes. Over a 120-minute period the following temperatures are observed: 450°F, 221°F, 180°F, 164°F. Use the trapezoid rule to find an approximation to the average temperature,

$$\frac{1}{120}\int_0^{120} f(t)dt.$$

In this case, $n = 3$ and $h = 40$; so using $f(0) = 450$, $f(40) = 221$, $f(80) = 180$, and $f(120) = 164$, we obtain

$$\frac{1}{120}\int_0^{120} f(t)dt \approx \frac{40}{120}\left[\frac{1}{2}(450) + 221 + 180 + \frac{1}{2}(164)\right] = 236°\text{F}.$$

It should be noted here that the average of the *four* temperatures is 253.75°F, not 236°F. To what do you attribute the difference? ||

Exercises 9.4

Use the trapezoid rule, with $n = 4$, to approximate the integrals in Exercises 1–14.

1. $\int_1^5 \frac{dx}{x} = \ln 5.$
2. $\int_1^3 \frac{dx}{x} = \ln 3.$
3. $\int_0^2 \frac{dx}{1+x^2} = \arctan 2.$
4. $\int_0^1 \frac{dx}{1+x^2} = \frac{\pi}{4}.$
5. $\int_0^1 81^{-x}\,dx.$
6. $\int_0^1 16^{-x}\,dx.$
7. $\int_0^1 \frac{16^x}{x+2}\,dx.$
8. $\int_0^2 \frac{4^x}{x+1}\,dx.$
9. $\int_\pi^{2\pi} \frac{\sin x}{x}\,dx.$
10. $\int_\pi^{2\pi} \frac{\cos x}{x}\,dx.$
11. $\int_0^2 \frac{dx}{x^3-9}.$
12. $\int_0^1 \frac{dx}{1+x^3}.$
13. $\int_1^3 (1+x^2)^{1/2}\,dx.$
14. $\int_0^{1.6} (1+x^2)^{1/2}\,dx.$
15. Find a bound on the error in Exercise 1.
16. Find a bound on the error in Exercise 2.
17. Find a bound on the error in Exercise 5. (Use the fact that $\ln 81 < 5$.)
18. Find a bound on the error in Exercise 6. (Use the fact that $\ln 16 < 3$.)
19. Find a bound on the error made by using the trapezoid rule with $n = 10$ to approximate $\int_1^4 \ln(x^2)dx$.
20. Find a bound on the error made by using the trapezoid rule with $n = 10$ to approximate $\int_0^1 e^{(-x^2)}\,dx$.
21. Find the smallest value of n so that the error in Exercise 1 is less than (a) 1/100, (b) 1/1000.
22. Find the smallest value of n so that the error in Exercise 2 is less than (a) 1/100, (b) 1/1000.
23. Find the smallest value of n so that the error in Exercise 5 is less than 1/100.
24. Find the smallest value of n so that the error in Exercise 6 is less than 1/100.
25. Use the trapezoid rule with $n = 4$ to approximate the arc length of the graph of $y = \sin x$ from $x = 0$ to $x = \pi$.
26. The region under the graph of $y = \sin x$, $0 \le x \le \pi/2$, is rotated about the y-axis. Use the trapezoid rule with $n = 2$ to approximate the volume. Find a bound on the error.
27. The tank pictured in Figure 9.4.2 has the indicated cross-sectional areas. Use the trapezoid rule to approximate the volume.

Figure 9.4.2

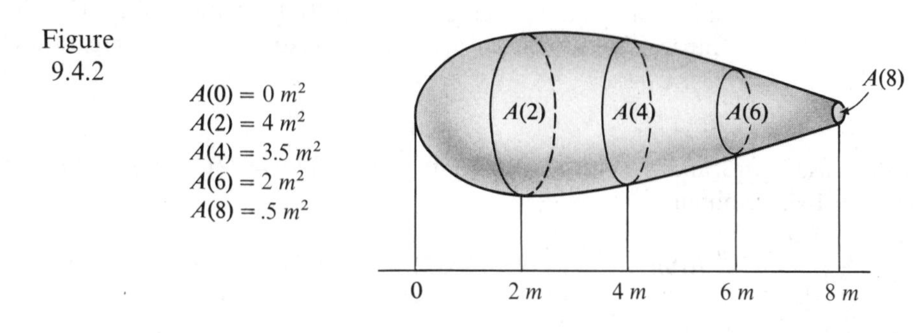

28. As an object is moved twenty meters, the following readings of the force (in kg) applied to the object were made at 5-m intervals: 20, 27, 26, 23, and 20. Use the trapezoid rule to approximate the work done in moving the object.

29. An object is in motion along a line. The velocity, as measured at several instants of time, is given in the following table:

t (sec)	0	1	2	3	4	5	6
v (m/sec)	3.2	2.7	2.9	4.0	4.7	5.6	5.7

Use the trapezoid rule to approximate the distance traveled from $t = 0$ to $t = 6$ sec.

30. The width of a swimming pool at five-foot intervals is indicated in Figure 9.4.3. Use the trapezoid rule to approximate the surface area of the water.

Figure 9.4.3

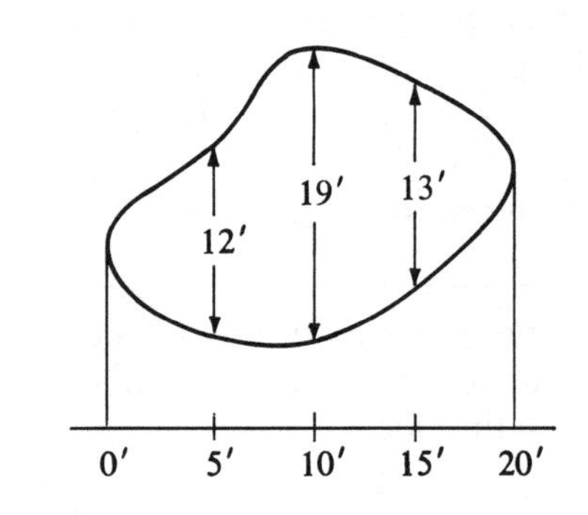

31. Show by use of a graph that the trapezoid rule will overestimate $\int_a^b f(x)dx$ if $f''(x) > 0$ for $a \leq x \leq b$.

9.5 Simpson's Rule

To develop the trapezoid rule we approximated a given curve by a sequence of line segments and used the area of the region beneath the segments as an approximation to the definite integral of the given function.

Simpson's Rule is much the same; however, in this case one approximates the given curve by a sequence of parabolic arcs. To accomplish this, we first subdivide the interval $a \leq x \leq b$ into an *even* number of subintervals, each of width h, by introducing the points $a = x_0 < x_1 < \cdots < x_n = b$. A parabolic arc, that is, a portion of the graph of $y = ax^2 + bx + c$ for some a, b, and c, is then passed through each successive triple of points. The situation for four subintervals and a general pair of subintervals is shown in Figure 9.5.1. In that figure the parabolic arcs are indicated by the dotted curves.

As with the trapezoidal rule the area of the region under these approximating arcs is used as an approximation to the definite integral $\int_a^b f(x)dx$. While it might appear that such an approximation would involve a great deal of calculation, the next theorem will considerably simplify the situation by giving us an expression for the area under a parabolic arc.

Figure 9.5.1

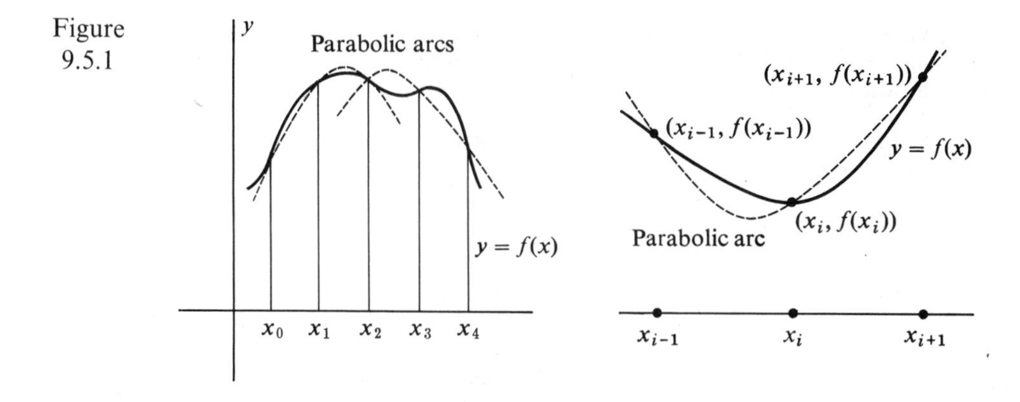

Theorem 9.5.1

If the parabolic segment $y = ax^2 + bx + c$, $x_{i-1} \le x \le x_{i+1}$, passes through the points $(x_{i-1}, f(x_{i-1}))$, $(x_i, f(x_i))$, and $(x_{i+1}, f(x_{i+1}))$, where $x_i - x_{i-1} = x_{i+1} - x_i = h$, then

$$\int_{x_{i-1}}^{x_{i+1}} (ax^2 + bx + c)dx = \frac{h}{3}[f(x_{i-1}) + 4f(x_i) + f(x_{i+1})].$$

Proof. Though the proof of this theorem is not difficult, it does involve a considerable amount of algebraic manipulation and so will only be given in outline form.

Since the graph of $y = ax^2 + bx + c$ is to pass through the points $(x_{i-1}, f(x_{i-1}))$, $(x_i, f(x_i))$, and $(x_{i+1}, f(x_{i+1}))$, we must have

$$f(x_{i-1}) = a(x_{i-1})^2 + bx_{i-1} + c,$$
$$f(x_i) = a(x_i)^2 + bx_i + c,$$
$$f(x_{i+1}) = a(x_{i+1})^2 + bx_{i+1} + c.$$

If we solve those three equations for the unknowns a, b, and c, and then evaluate the integral

$$\int_{x_{i-1}}^{x_{i+1}} (ax^2 + bx + c)dx,$$

we obtain the desired result.

With that information at hand let's return to the general situation depicted in Figure 9.5.2, where again the dotted curves are the parabolic arcs. In view of Theorem 9.5.1, the area beneath the first parabolic arc, between x_0 and x_2, is

$$\frac{h}{3}[f(x_0) + 4f(x_1) + f(x_2)],$$

and the area beneath the second parabolic arc, between x_2 and x_4 is

$$\frac{h}{3}[f(x_2) + 4f(x_3) + f(x_4)].$$

Figure 9.5.2

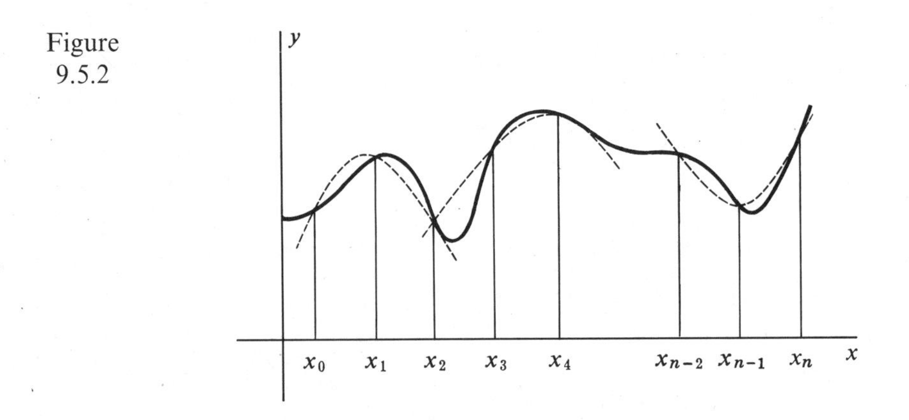

In general, the area beneath the parabolic arc between x_i and x_{i+2}, for each even value of i, is

$$\frac{h}{3}[f(x_i) + 4f(x_{i+1}) + f(x_{i+2})].$$

Consequently, the total area under the parabolic arcs is

$$\frac{h}{3}[f(x_0) + 4f(x_1) + f(x_2)] + \frac{h}{3}[f(x_2) + 4f(x_3) + f(x_4)] + \cdots + \frac{h}{3}[f(x_{n-2}) + 4f(x_{n-1}) + f(x_n)].$$

Therefore, on simplifying this expression, we have

$$\int_a^b f(x)dx \approx \frac{h}{3}[f(x_0) + 4f(x_1) + 2f(x_2) + 4f(x_3) + 2f(x_4) + \cdots + 2f(x_{n-2}) + 4f(x_{n-1}) + f(x_n)],$$

where n is even, and the x_i's are the endpoints of n subintervals, each of width h.

As with many formulas, it is best to remember this in words:

$$\int_a^b f(x)dx \approx \frac{h}{3}[\textit{zero}\text{th term} + \text{last term} + \text{twice even-numbered terms} + \text{four times odd-numbered terms}].$$

Example 1 Use Simpson's Rule with $n = 2$ to find an approximation to $\ln 2 = \int_1^2 dx/x$.

Since $h = (b - a)/n$, we have $h = 1/2$. A convenient way to organize the computations involved in Simpson's Rule is to construct the following table.

i	x_i	$f(x_i)$	*multiplier*	*term*
0	1	1	1	1
1	3/2	2/3	4	8/3
2	2	1/2	1	1/2

Substitution into
$$\int_a^b f(x)dx \approx \frac{h}{3} \; [f(x_0) + 4f(x_1) + f(x_2)]$$

gives
$$\ln 2 = \int_1^2 \frac{dx}{x} \approx \frac{1}{6}\left[1 + \frac{8}{3} + \frac{1}{2}\right] = .69\overline{4}. \qquad \|$$

Example 2 | Use Simpson's Rule with $n = 4$ to approximate $\int_0^1 x^2\, dx$.

In this case $h = 1/4$, and so we have the following table.

i	x_i	$f(x_i)$	*multiplier*	*term*
0	0	0	1	0
1	1/4	1/16	4	1/4
2	1/2	1/4	2	1/2
3	3/4	9/16	4	9/4
4	1	1	1	1

Consequently,
$$\int_0^1 x^2\, dx \approx \frac{1}{12}\left[0 + \frac{1}{4} + \frac{1}{2} + \frac{9}{4} + 1\right] = \frac{1}{3}.$$

Thus Simpson's Rule yielded the exact result 1/3, while the trapezoid rule gave the approximation .344. (See Example 2 of Section 9.4.) ||

Again we must, as with all approximations, develop some bound on the error

$$E_n = \left| \int_a^b f(x)dx - \frac{h}{3}[f(x_0) + 4f(x_1) + 2f(x_2) + 4f(x_3) + \cdots \right.$$
$$\left. + 2f(x_{n-2}) + 4f(x_{n-1}) + f(x_n)] \right|.$$

Though we shall not present a proof, the following theorem provides such a bound.

Theorem 9.5.2

> If $|f^{(4)}(x)| \leq M$ for all x in the interval $a \leq x \leq b$, and if $h = (b - a)/n$, then
> $$E_n \leq \frac{h^4(b - a)}{180} M.$$

Notice that by choosing n larger, h is made smaller and h^4 is made much smaller. Hence the bound on the error decreases rapidly as n increases.

As an immediate consequence of Theorem 9.5.2, we note that Simpson's Rule will give the precise result when applied to any polynomial involving only powers of x less than 4. In this case, $f(x) = a_3x^3 + a_2x^2 + a_1x + a_0$ and direct computation gives $f^{(4)}(x) = 0$ for all x. Consequently we can take $M = 0$ in Theorem 9.5.2 to get $E_n = 0$, for every even integer n.

Example 3 | Find a bound on the error in the approximation $\ln 2 \approx .69\overline{4}$, given in Example 1.

Since $f(x) = 1/x = x^{-1}$, we have $f'(x) = -x^{-2}$, $f''(x) = 2x^{-3}$, $f'''(x) = -6x^{-4}$, and finally $f^{(4)}(x) = 24x^{-5}$. Since

$$|f^{(4)}(x)| = \left|\frac{24}{x^5}\right| \le 24, \qquad \text{for } 1 \le x \le 2,$$

we may take $M = 24$ when applying Theorem 9.5.2. Then, using $h = 1/2$, $n = 2$, and $b - a = 1$, we get

$$E_2 \le \frac{(1/2)^4}{180} 24, \quad \text{or} \quad E_2 \le \frac{1}{120}.$$

In Example 3 of Section 9.4, we found that the trapezoidal rule with $n = 5$ gave an error bounded by $1/150$. ||

As in the case of the trapezoid rule, we must often determine the smallest value of n that can be used if the error is to be smaller than some preassigned number.

Example 4 | What is the smallest value of n that can be used so that the error in approximating $\ln 2$ by Simpson's Rule is less than 10^{-6}?

As in Example 3 above, we may take $M = 24$. Then, since $b - a = 1$,

$$E_n \le \frac{h^4}{180} 24 = \frac{2}{15} h^4.$$

Since $h = (b - a)/n = 1/n$, we have

$$E_n \le \frac{2}{15n^4}.$$

Thus we need only choose n so large that

$$\frac{2}{15n^4} < 10^{-6}, \quad \text{or} \quad n^4 > \frac{2}{15} 10^6 = 133{,}333\frac{1}{3}.$$

Since $(20)^4 = 160{,}000$, it is sufficient to use Simpson's Rule with $n = 20$. In Example 4 of Section 9.4 we found that to achieve the same accuracy with the trapezoid rule required $n = 409$. ||

It is instructive to compare the error bounds that result when Simpson's Rule and the Trapezoid Rule are applied to a specific integral.

Example 5 | For $n = 2, 4, 6, 8$, and 10 compute the error bounds that result from the application of Simpson's Rule and the Trapezoid Rule to $\int_1^2 \frac{dx}{x}$.

By Example 4 of Section 9.4 we know that $1/(6n^2)$ is a bound on the error resulting from an application of the Trapezoid Rule to $\int_1^2 \frac{dx}{x}$. From Example 4 above we note that $2/(15n^4)$ is a bound on the error resulting from an application of Simpson's Rule to $\int_1^2 \frac{dx}{x}$. Thus we have the following table of error bounds:

n	*Trapezoid Rule* $1/(6n^2)$	*Simpson's Rule* $2/(15n^4)$
2	1/24	1/120
4	1/96	1/1920
6	1/216	1/9720
8	1/348	1/30720
10	1/600	1/75000

||

It should be clear that even more accurate methods of approximate integration are possible. One could, for example, pass higher order polynomials through more points on the given curve. The interested student will find entire chapters of numerical analysis texts devoted to this topic. See, for example, F. B. Hildebrand, *Introduction to Numerical Analysis* (McGraw-Hill, 1956).

Exercises 9.5

Use Simpson's Rule, with $n = 4$, to approximate the integrals in Exercises 1–14.

1. $\int_1^5 \frac{dx}{x} = \ln 5.$
2. $\int_1^3 \frac{dx}{x} = \ln 3.$
3. $\int_0^2 \frac{dx}{1 + x^2} = \arctan 2.$
4. $\int_0^1 \frac{dx}{1 + x^2} = \frac{\pi}{4}.$
5. $\int_0^1 81^{-x}\, dx.$
6. $\int_0^1 16^{-x}\, dx.$
7. $\int_0^1 \frac{16^x}{x + 2}\, dx.$
8. $\int_0^2 \frac{4^x}{x + 1}\, dx.$
9. $\int_\pi^{2\pi} \frac{\sin x}{x}\, dx.$
10. $\int_\pi^{2\pi} \frac{\cos x}{x}\, dx.$
11. $\int_0^2 \frac{dx}{x^3 - 9}.$
12. $\int_0^1 \frac{dx}{1 + x^3}.$
13. $\int_1^3 (1 + x^2)^{1/2}\, dx.$
14. $\int_0^{1.6} (1 + x^2)^{1/2}\, dx.$
15. Find a bound on the error in Exercise 1. Compare your result to that in Exercise 15 of Section 9.4.
16. Find a bound on the error in Exercise 2. Compare your result to that in Exercise 16 of Section 9.4.
17. Find a bound on the error in Exercise 5. (Use the fact that $\ln 81 < 5$.) Compare your result to that found in Exercise 17 of Section 9.4.
18. Find a bound on the error in Exercise 6. (Use the fact that $\ln 16 < 3$.) Compare your result to that found in Exercise 18 of Section 9.4.
19. Find a bound on the error made by using Simpson's Rule with $n = 10$ to approximate $\int_1^4 \ln(x^2)dx$. Compare your result to that found in Exercise 19 of Section 9.4.

20. Find the smallest value of n so that the error in Exercise 2 is less than (a) 1/100, (b) 1/1000. Compare your results to those obtained in Exercise 22 of Section 9.4.

21. Find the smallest value of n so that the error in Exercise 1 is less than (a) 1/100, (b) 1/1000. Compare your results to those obtained in Exercise 21 of Section 9.4.

22. Find the smallest value of n so that the error in Exercise 6 is less than 1/100. Compare your result to that obtained in Exercise 24 of Section 9.4.

23. Find the smallest value of n so that the error in Exercise 5 is less than 1/100. Compare your result to that obtained in Exercise 23 of Section 9.4.

24. Use Simpson's Rule with $n = 4$ to approximate the arc length of the graph of $y = \sin x$ from $x = 0$ to $x = \pi$.

25. The region under the graph of $y = \sin x$, $0 \le x \le \pi/2$, is rotated about the y-axis. Use Simpson's Rule with $n = 2$ to approximate the volume. Find a bound on the error.

26. The tank pictured in Figure 9.5.3 has the indicated cross-sectional areas. Use Simpson's Rule to approximate the volume.

Figure 9.5.3

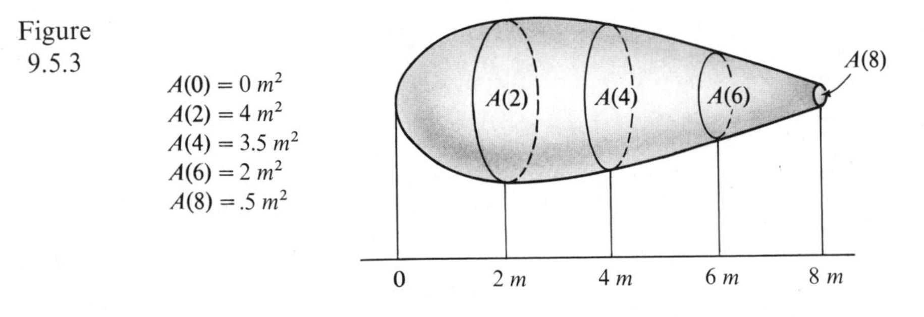

27. As an object was moved thirty meters, the following readings of the force (in kg) applied to the object were made at 5-m intervals: 20, 27, 26, 23, 20, 18, and 12. Use Simpson's Rule to approximate the work done in moving the object.

28. An object is in motion along a line. The velocity, as measured at several instants of time, is given in the following table:

t (sec)	0	1	2	3	4	5	6
v (m/sec)	3.2	2.7	2.9	4.0	4.7	5.6	5.7

Use Simpson's Rule to approximate the distance traveled from $t = 0$ to $t = 6$ sec.

29. The width of a swimming pool at five-foot intervals is as indicated in Figure 9.5.4. Use Simpson's Rule to approximate the surface area of the water.

Figure 9.5.4

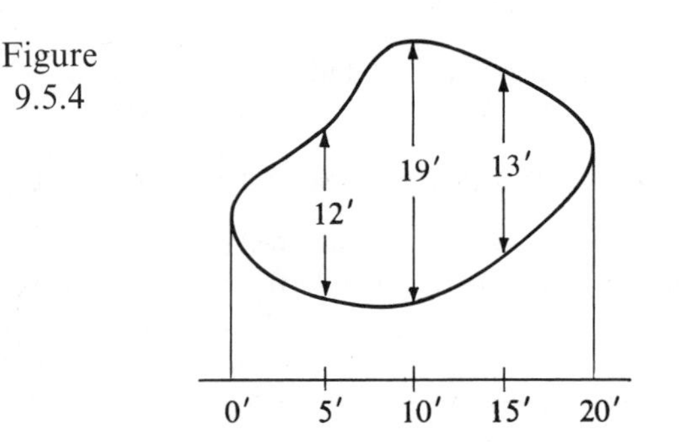

30. The arc length of the ellipse $(x^2/a^2) + (y^2/b^2) = 1$ is given by

$$L = a \int_0^{2\pi} \sqrt{1 - \varepsilon^2 \sin^2 t}\, dt,$$

where $\varepsilon = (a^2 - b^2)/a^2$. Approximate the integral by Simpson's Rule with $n = 8$. (The Fundamental Theorem is not useful here.)

31. Prove that Simpson's Rule will give the precise result when applied to $\int_a^b f(x)dx$ where $f(x)$ is a cubic polynomial.

9.6 Linear Approximations

In Section 3.9 we studied the use of differentials in the approximation of functions. In this section we will investigate this process more closely.

As shown in Figure 9.6.1

$$f(x) \approx f(a) + f'(a)dx,$$

or, since $dx = x - a$,

$$f(x) \approx f(a) + f'(a)(x - a).$$

Figure 9.6.1

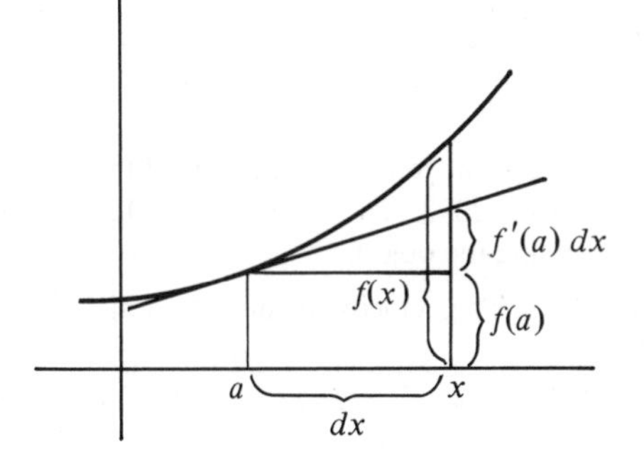

Viewed in another way, that formula simply tells us that the tangent line to $y = f(x)$ at $x = a$, $y = f(a) + f'(a)(x - a)$, approximates the graph of $y = f(x)$ near the point of tangency. Thus, to approximate $f(x)$ for some particular value of x, we select a in such a way that

(i) a is near x, and
(ii) $f(a)$ and $f'(a)$ can be easily calculated.

Example 1 | Use the linear approximation formula to approximate $\sqrt{101}$.

In this case let $f(x) = x^{1/2}$. Then $f'(x) = (1/2)x^{-1/2}$. The selection of $a = 100$ fits both criteria above. Then substitution of $x = 101$ and $a = 100$ into the approximation formula

$$x^{1/2} \approx a^{1/2} + \frac{1}{2}a^{-1/2}(x - a)$$

gives $$\sqrt{101} \approx \sqrt{100} + \frac{101 - 100}{2\sqrt{100}} = 10 + \frac{1}{20} = 10.05.$$ ||

Example 2 | Find the linear approximation to $\sin x$ for values of x close to 0.

Since $f(x) = \sin x$, we have $f'(x) = \cos x$ and so

$$\sin x \approx \sin a + (x - a)\cos a.$$

Since we are interested in an approximation for values of x close to 0, we let $a = 0$. Then $\sin x \approx x$. ||

Now we establish a bound on the error

$$E(x) = f(x) - [f(a) + f'(a)(x - a)]$$

obtained when using the linear approximation formula to approximate $f(x)$ in a neighborhood N of a.

Note that $E'(t) = f'(t) - f'(a)$, and so

$$\frac{E'(t)}{t - a} = \frac{f'(t) - f'(a)}{t - a}. \tag{1}$$

If we now assume that $f''(t)$ exists for all t in N, we may apply the Mean Value Theorem to $y = f'(t)$ to obtain

$$\frac{f'(t) - f'(a)}{t - a} = f''(p), \qquad \text{for some } p \text{ in } N.$$

Consequently if $|f''(x)| \leq M$ for all x in N, we have

$$\left|\frac{f'(t) - f'(a)}{t - a}\right| \leq M, \qquad \text{for all } t \text{ in } N, \text{ or}$$

$$-M \leq \frac{f'(t) - f'(a)}{t - a} \leq M, \qquad \text{for all } t \text{ in } N.$$

Thus in view of (1) we have

$$-M \leq \frac{E'(t)}{t - a} \leq M, \qquad \text{for all } t \text{ in } N.$$

Then if $t - a > 0$ we obtain

$$-M(t - a) \leq E'(t) \leq M(t - a), \qquad \text{for all } t \text{ in } N. \tag{2}$$

(If $t - a < 0$, the inequalities above are reversed, but the proof is essentially the same.)

Since $E(a) = 0$, we have

$$E(x) = E(x) - E(a) = \int_a^x E'(t)dt.$$

Thus from (2) we get

$$-\int_a^x M(t - a)dt \leq E(x) \leq \int_a^x M(t - a)dt.$$

Then, since M is a constant, we may integrate to obtain

$$-\frac{1}{2}M(x-a)^2 \le E(x) \le \frac{1}{2}M(x-a)^2, \quad \text{or}$$

$$|E(x)| \le \frac{1}{2}M(x-a)^2.$$

In summary we have this result:

Theorem 9.6.1

> If $f''(x)$ is continuous in some neighborhood N of a, then for each x in N,
>
> $$f(x) = f(a) + f'(a)(x-a) + E_1(x).$$
>
> Moreover, if $|f''(x)| \le M$ for x in N, then
>
> $$|E_1(x)| \le \frac{1}{2}M(x-a)^2.$$

Figure 9.6.2 illustrates the geometric significance of Theorem 9.6.1.

Figure 9.6.2

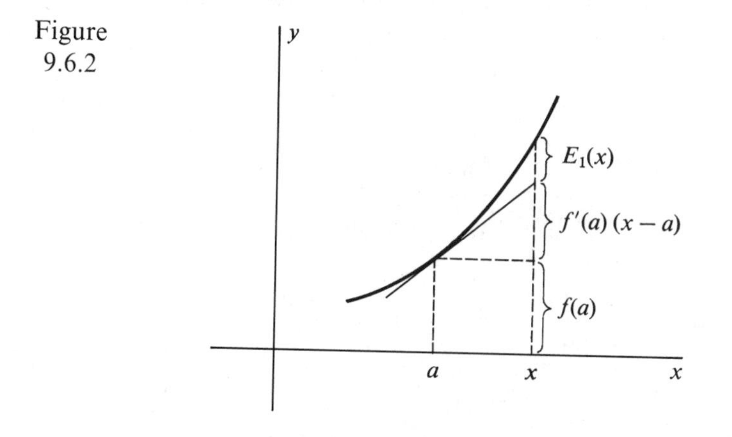

Example 3 Find a bound for the error in the approximation $\sin(.2) \approx .2$ obtained from Example 2.

Since

$$|f''(x)| = |-\sin x| \le 1,$$

we may take $M = 1$. Hence, since $a = 0$, we have

$$|E(x)| \le \frac{1}{2}x^2, \quad \text{and so}$$

$$|E(.2)| \le \frac{1}{2}(.2)^2 = .02.$$

Note here that a large value of x could result in a large error and consequently a poor approximation. ||

Example 4 | Find the linear approximation to $f(x) = x^{1/2}$ for $a = 1$, and establish a bound on the error for $x \geq 1/2$.

In this case $f(x) = x^{1/2}$, $f'(x) = (1/2)x^{-1/2}$, and $f''(x) = -(1/4)x^{-3/2}$. Thus, since $f(a) = 1$ and $f'(a) = 1/2$, we have

$$x^{1/2} \approx 1 + \frac{1}{2}(x - 1) = \frac{1}{2}(x + 1).$$

Since

$$|f''(x)| = \left| -\frac{1}{4}x^{-3/2} \right| = \frac{1}{4}x^{-3/2},$$

we know $|f''(x)|$ is a decreasing function. Since we are interested only in those values of x for which $x \geq 1/2$, we may take

$$M = \left| f''\left(\frac{1}{2}\right) \right| = \frac{1}{4}\left(\frac{1}{2}\right)^{-3/2} = \frac{2^{3/2}}{4} = 2^{-1/2}.$$

Then, since

$$|E(x)| \leq \frac{1}{2}M(x - a)^2,$$

we have

$$|E(x)| \leq 2^{-3/2}(x - 1)^2,$$

and so

$$|E(x)| \leq 2^{-3/2}(x - 1)^2, \qquad \text{for all } x \geq \frac{1}{2}.$$

Note again that this linear approximation results in a small error if x is close to 1, but could result in a large error for those values of x far from 1. ||

Exercises 9.6

In Exercises 1–20, find the linear approximation to the given function near the given point a. If no such approximation exists, state so.

1. $f(x) = x^2 - 6x - 2, \quad a = 0.$
2. $f(x) = x^2 + 3x + 2, \quad a = 0.$
3. $f(x) = x^2 - 6x - 2, \quad a = -1.$
4. $f(x) = x^2 + 3x + 2, \quad a = 1.$
5. $f(x) = -2x + 7, \quad a = 0.$
6. $f(x) = -3x + 17, \quad a = 0.$
7. $f(x) = -2x + 7, \quad a = 1.$
8. $f(x) = -3x + 17, \quad a = -1.$
9. $f(x) = \ln x, \quad a = 2.$
10. $f(x) = \ln x, \quad a = 1.$
11. $f(x) = e^x, \quad a = 0.$
12. $f(x) = e^x, \quad a = 1.$
13. $f(x) = \tan x, \quad a = \pi/4.$
14. $f(x) = \sec x, \quad a = \pi/4.$
15. $f(x) = |x|, \quad a = 0.$
16. $f(x) = \sqrt{\ln x}, \quad a = 1.$
17. $f(x) = x \ln x, \quad a = 1.$
18. $f(x) = \sin x \cos x, \quad a = 0.$
19. $f(x) = e^{\sin x}, \quad a = 0.$
20. $f(x) = \tan^2 x, \quad a = 0.$
21. Use the result of Exercise 11 to find an approximation to e.
22. Graph the function $f(x) = \sin x$ and its linear approximation at $a = 0$.
23. Graph the function $f(x) = x^2 + 3x + 2$ and its linear approximation at $a = 0$.
24. Use Theorem 9.6.1 to establish a bound on the error in the approximation of Exercise 2. What is the actual error?

25. Use Theorem 9.6.1 to establish a bound on the error in the approximation of Exercise 3. What is the actual error?
26. Find a bound on the error in the approximation of Exercise 10 if $x \geq 1/2$.
27. Find a bound on the error in the approximation of Exercise 11 if $|x| \leq 1$.
28. Find an approximation to $\sin(.03)$ and find a bound on the error.
29. Find an approximation to $\tan(\pi/4 - .01)$.
30. Explain why the approximation

$$\int_0^{.01} x \sin x \, dx \approx \int_0^{.01} x^2 \, dx$$

may be used. Find a bound on the error made in using such an approximation.

31. Explain why the approximation

$$\int_{.9}^{1} x \ln x \, dx \approx \int_{.9}^{1} (x^2 - x) dx$$

might be used.

32. Find a bound on the error in the approximation

$$\int_a^b x \sin x \, dx \approx \frac{1}{3}(b^3 - a^3).$$

9.7 Second and Higher Degree Approximations; Taylor's Theorem

The tangent line to a graph is the line that best conforms to the graph near the point of tangency. In fact, if $y = p_1(x)$ is an equation of the line tangent to $y = f(x)$ at $(a, f(a))$, then the functions p_1 and f agree in two ways:

$$p_1(a) = f(a), \quad \text{and}$$
$$p_1'(a) = f'(a).$$

It seems reasonable that a second-degree equation $y = p_2(x)$ which agrees with f in these three ways,

$$\text{(i)} \quad p_2(a) = f(a),$$
$$\text{(ii)} \quad p_2'(a) = f'(a),$$
$$\text{(iii)} \quad p_2''(a) = f''(a),$$

might well be a better approximation to f. It is most convenient to express $p_2(x)$ as a second-degree polynomial in $x - a$. We leave as an exercise the proof that the desired second-degree polynomial is

$$p_2(x) = f(a) + f'(a)(x - a) + \frac{1}{2}f''(a)(x - a)^2.$$

Hence we have the approximation

$$f(x) \approx f(a) + f'(a)(x - a) + \frac{1}{2}f''(a)(x - a)^2. \tag{1}$$

Example 1 | Use (1) to find a second-degree approximation to $\sqrt{101}$.

In this case $f(x) = x^{1/2}$, $f'(x) = (1/2)x^{-1/2}$, and $f''(x) = -(1/4)x^{-3/2}$.

Hence $$x^{1/2} \approx a^{1/2} + \frac{1}{2}a^{-1/2}(x - a) - \frac{1}{8}a^{-3/2}(x - a)^2.$$

Then, letting $x = 101$ and $a = 100$, we get

$$\sqrt{101} \approx 10 + \frac{1}{20} - \frac{1}{8000} = 10.0498750.$$

We shall show later that the error in this approximation is less than $4(10)^{-5}$. ||

Example 2 | Develop a second-degree approximation to $\cos x$ for small x.

Here $f(x) = \cos x$, $f'(x) = -\sin x$, and $f''(x) = -\cos x$. Thus,

$$\cos x \approx \cos a - (x - a)\sin a - \frac{1}{2}(x - a)^2 \cos a.$$

If we now let $a = 0$, we obtain

$$\cos x \approx 1 - \frac{1}{2}x^2.$$ ||

As with the linear approximations of the previous section it is natural to attempt to find a bound on the error in a neighborhood N of a. In this case we denote the error by $E_2(x)$. Thus

$$E_2(x) = f(x) - \left[f(a) + f'(a)(x - a) + \frac{1}{2}f''(a)(x - a)^2\right], \quad \text{and}$$

$$E_2'(x) = f'(x) - f'(a) - f''(a)(x - a), \quad \text{and} \tag{2}$$
$$E_2''(x) = f''(x) - f''(a).$$

Hence

$$\frac{E_2''(x)}{x - a} = \frac{f''(x) - f''(a)}{x - a}. \tag{3}$$

Assuming that f''' exists in N, we can apply The Mean Value Theorem to $y = f''(x)$ to obtain

$$\frac{f''(x) - f''(a)}{x - a} = f'''(c)$$

for each x in N and some c between x and a. Hence, in view of (3),

$$E_2''(x) = f'''(c)(x - a), \qquad \text{for some } c \text{ between } x \text{ and } a.$$

An argument similar to the one used in developing an expression for the error in a linear approximation can be used to show that

$$E_2'(x) = \frac{1}{2}f'''(c)(x - a)^2,$$

for some c between a and x.

Using the same procedure again, we get

$$E_2(x) = \frac{1}{6} f'''(c)(x - a)^3$$

for some c between x and a.

As before, while the expression for $E_2(x)$ is not useful for a precise calculation of the error, it is sufficient to establish a bound on the error. Thus, if M is the maximum of $|f'''(t)|$ for t in N, then

$$|E_2(x)| \leq \frac{1}{6} M |x - a|^3.$$

In summary we have the following result:

Theorem 9.7.1

If $f'''(x)$ is continuous in some neighborhood N of a, then for each x in N,

$$f(x) = f(a) + f'(a)(x - a) + \frac{1}{2} f''(a)(x - a)^2 + E_2(x).$$

Moreover, if M is the maximum of $|f'''(x)|$ for x in N, then

$$|E_2(x)| \leq \frac{1}{6} M |x - a|^3.$$

Example 3 Use of the approximation developed in Example 2 gives $\cos(.1) \approx .995$. Establish a bound on the error in this approximation.

Since $f(x) = \cos x$, we have $f'''(x) = \sin x$, and so $|f'''(x)| \leq 1$. Thus we can take $M = 1$, and since $a = 0$, we have

$$|E_2(x)| \leq \frac{|x|^3}{6}.$$

In this case $x = .1$, and so

$$|E_2(.1)| \leq \frac{|.1|^3}{6} = .0001\bar{6}.$$

That is, the approximation $\cos(.1) \approx .995$ is correct to three decimal places. ||

Before continuing along this line, let's review what has been done in the last two sections. In Section 9.6, we approximated a given function $f(x)$ by a linear polynomial in $(x - a)$ that passes through the point $(a, f(a))$ and has the same derivative as $f(x)$ at this point. Theorem 9.7.1 allows us to approximate the function f with a second-degree polynomial in $(x - a)$. This second-degree polynomial is not only forced to pass through the point $(a, f(a))$ and to have the same derivative as $f(x)$ at a, but also to have the same second derivative at a. Let's carry this idea further and try to find an nth-degree polynomial in $(x - a)$, that is, a polynomial of the form

$$p_n(x) = a_n(x - a)^n + \cdots + a_1(x - a) + a_0,$$

such that

$$p_n(a) = f(a)$$
$$p_n'(a) = f'(a)$$
$$p_n''(a) = f''(a)$$
$$\cdots\cdots\cdots\cdots$$
$$p_n^{(n)}(a) = f^{(n)}(a).$$

Under those conditions it can be readily shown that

$$p_n(x) = f(a) + f'(a)(x - a) + \frac{1}{2}f''(a)(x - a)^2$$
$$+ \frac{1}{3!}f'''(a)(x - a)^3 + \cdots + \frac{1}{n!}f^{(n)}(a)(x - a)^n.$$

The proof is left as an exercise.

The next theorem gives the precise information about the approximation of $f(x)$ by $p_n(x)$. Its proof, with the assumption that $f^{(n+1)}(x)$ is continuous, is similar to the one given in the preceding section. A proof with the less restrictive assumption that $f^{(n+1)}(x)$ exists can be found in many advanced calculus texts.

Theorem 9.7.2

Taylor's Theorem

If $f^{(n+1)}(x)$ exists in some neighborhood N of a, then for each x in N,

$$f(x) = f(a) + f'(a)(x - a) + \frac{1}{2}f''(a)(x - a)^2 + \frac{1}{3!}f'''(a)(x - a)^3$$
$$+ \cdots + \frac{1}{n!}f^{(n)}(a)(x - a)^n + E_n(x).$$

Moreover, if M_{n+1} is the maximum of $|f^{(n+1)}(t)|$ for t in N, then

$$|E_n(x)| \leq \frac{1}{(n + 1)!}M_{n+1}|x - a|^{n+1}.$$

When we are interested in approximating a function near a, it is best to express the approximating polynomial in powers of $x - a$ because $x - a$ is small when x is near a.

Example 4 Apply Taylor's Theorem, with $n = 3$, to the polynomial $f(x) = 5x^3 - 7x^2 + 2x - 3$ for $a = 1$. Determine a bound for the error.

$f(x) = 5x^3 - 7x^2 + 2x - 3$; so $f(1) = -3$.

$f'(x) = 15x^2 - 14x + 2$; so $f'(1) = 3$.

$f''(x) = 30x - 14$; so $f''(1) = 16$.

$f'''(x) = 30$; so $f'''(1) = 30$.

Furthermore, $f^{(4)}(x) = 0$. Then we substitute into

$$f(x) \approx f(a) + f'(a)(x - a) + \frac{1}{2}f''(a)(x - a)^2 + \frac{1}{3!}f'''(a)(x - a)^3$$

to get

$$f(x) \approx -3 + 3(x - 1) + \frac{1}{2}(16)(x - 1)^2 + \frac{1}{6}(30)(x - 1)^3. \tag{4}$$

From Theorem 9.7.2 we know that

$$|E_3| \leq \frac{1}{4!}M_4|x - a|^4,$$

where M_4 is the maximum of $|f^{(4)}(x)|$. Since $f^{(4)}(x) = 0$, we can take $M_4 = 0$ and so $E_3 = 0$. Hence the approximation given in (4) is the exact value, and simplification of the right-hand side of (4) will actually yield $5x^3 - 7x^2 + 2x - 3$. ||

There is a very interesting analogy between the approximations given by Taylor's Theorem and the terminating decimal approximations to real numbers. Consider the real number π. Below we indicate successive terminating decimal approximations to π, and we show a bound on the possible error.

terminating decimal approximation	*bound on error*
3	.5
3.1	.05
3.14	.005
3.142	.0005
3.1416	.00005

Taylor's Theorem allows us to treat functions in much the same way. Rather than approximating a real number with rational numbers of longer and longer length, we approximate a given function with successively higher degree polynomials in $x - a$ as indicated in the table below.

polynomial approximation	*bound on error*
$f(a) + f'(a)(x - a)$	$\frac{1}{2}M_2\lvert x - a\rvert^2$
$f(a) + f'(a)(x - a) + \frac{1}{2}f''(a)(x - a)^2$	$\frac{1}{3!}M_3\lvert x - a\rvert^3$
$f(a) + f'(a)(x - a) + \frac{1}{2}f''(a)(x - a)^2 + \frac{1}{3!}f'''(a)(x - a)^3$	$\frac{1}{4!}M_4\lvert x - a\rvert^4$

Example 5 | Construct successive polynomial approximations to $f(x) = e^x$, for $a = 0$. In each case establish a bound on the error.

Since $f(x) = e^x$, we have $f^{(k)}(x) = e^x$ and $f^{(k)}(0) = 1$ for all integers k. If M_k is the maximum of $f^{(k)}(x)$ when x is restricted to some neighborhood N of 0, we can construct the following table:

polynomial approximation	*bound on error*
$1 + x$	$\frac{1}{2} M_2 \lvert x \rvert^2$
$1 + x + \frac{1}{2}x^2$	$\frac{1}{3!} M_3 \lvert x \rvert^3$
$1 + x + \frac{1}{2}x^2 + \frac{1}{3!}x^3$	$\frac{1}{4!} M_4 \lvert x \rvert^4$
$1 + x + \frac{1}{2}x^2 + \frac{1}{3!}x^3 + \frac{1}{4!}x^4$	$\frac{1}{5!} M_5 \lvert x \rvert^5$
.....	.....
$1 + x + \frac{1}{2}x^2 + \frac{1}{3!}x^3 + \frac{1}{4!}x^4 + \cdots + \frac{1}{n!}x^n$	$\frac{1}{(n+1)!} M_{n+1} \lvert x \rvert^{n+1}$

To be more specific, suppose that we wanted an approximation to $f(1) = e$. In that case we could let the neighborhood N of $a = 0$ be the interval $-2 < x < 2$. Since $f^{(k)}(x) = e^x$ is less than $e^2 \leq 9$ on this interval, we may take $M_k = 9$ for all k. Then, on substitution of $x = 1$ in our previous approximations, we obtain:

approximation to e	*bound on error*
$1 + 1 = 2$	$\frac{9}{2}$
$1 + 1 + \frac{1}{2} = 2.5$	$\frac{1}{3!}(9) = \frac{3}{2}$
$1 + 1 + \frac{1}{2} + \frac{1}{3!} = 2.\overline{6}$	$\frac{1}{4!}(9) = \frac{3}{8}$
$1 + 1 + \frac{1}{2} + \frac{1}{3!} + \frac{1}{4!} = 2.708\overline{3}$	$\frac{1}{5!}(9) = \frac{3}{40}$

Continuing in this way, we could calculate e to any desired accuracy

In Figure 9.7.1 the graphs of $f(x) = e^x$ and the first three approximating polynomials are shown. Note that the polynomials conform more and more closely to the graph of $f(x) = e^x$ as the degree of the polynomial increases. ||

In our previous examples we did not have to worry about the neighborhood N of a; we were able to take N to be the entire real line in each case. That is not

Figure 9.7.1

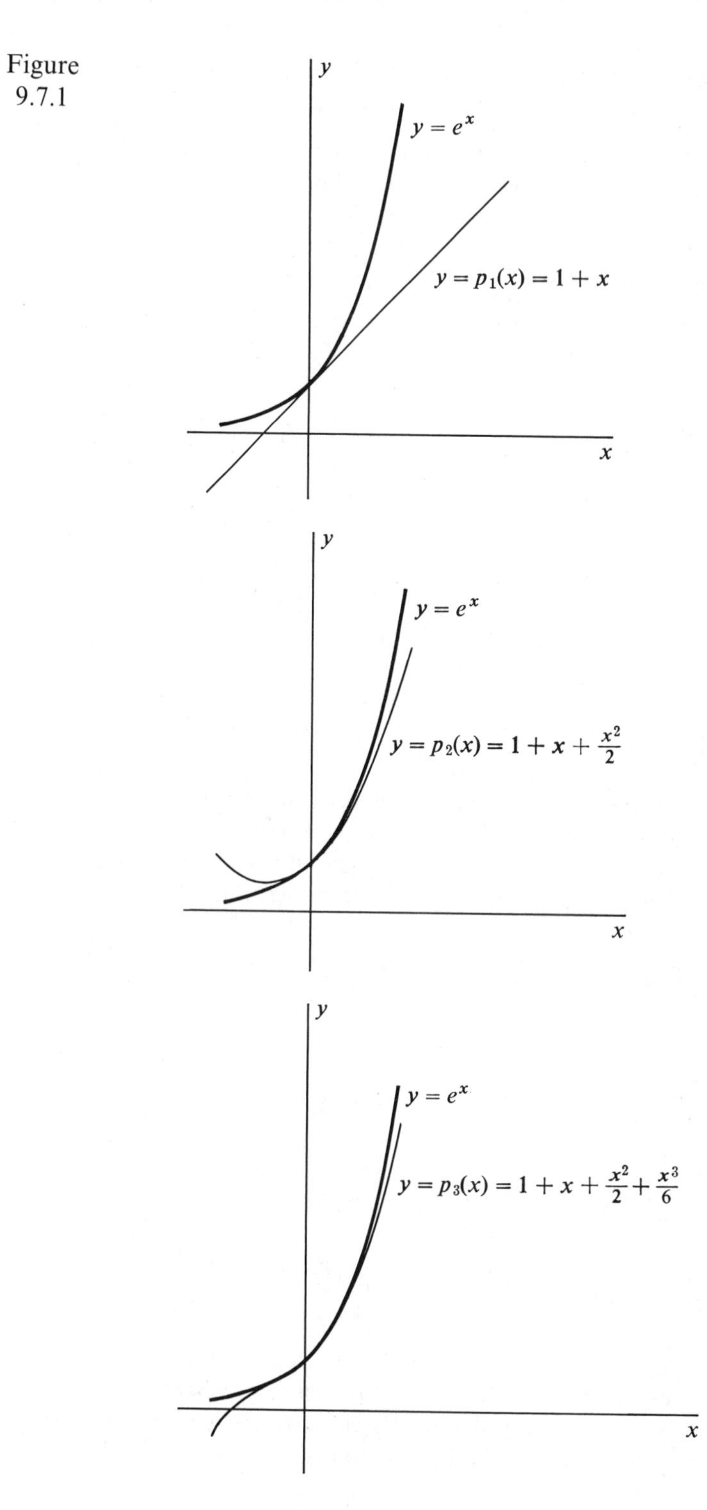

always possible. Suppose we were to apply Taylor's Theorem to the function $f(x) = (1 - x^2)^{1/2}$ with $a = 0$. Then the first derivative

$$f'(x) = -x(1 - x^2)^{-1/2}$$

does not exist when $x = \pm 1$, and hence no higher order derivatives exist for $x = \pm 1$. But they do exist when x is in the interval $-1 < x < 1$, however. Since the hypothesis of Taylor's Theorem demands the existence of $f^{(n+1)}(x)$, we must restrict our considerations to the interval $-1 < x < 1$.

Example 6 | Apply Taylor's Theorem with $a = 1$ to $f(x) = x^{-1}$, and make an analysis of the error terms.

We have

$$\begin{aligned} f(x) &= x^{-1}, & f(1) &= 1; \\ f'(x) &= -x^{-2}, & f'(1) &= -1; \\ f''(x) &= 2x^{-3}, & f''(1) &= 2; \\ f'''(x) &= -(3!)x^{-4}, & f'''(1) &= -3! \end{aligned}$$

And, in general, since

$$f^{(k)}(x) = (-1)^k(k!)x^{-(k+1)},$$

we have

$$f^{(k)}(1) = (-1)^k k!.$$

Note that since $f(x)$ and its derivatives do not exist at $x = 0$, we must restrict ourselves to some neighborhood N of $a = 1$ that excludes 0. For example, we might take N to be the set of all *positive* real numbers, that is, the set of all x for which $x > 0$. We may substitute the information above into

$$f(x) = f(a) + f'(a)(x - a) + \frac{1}{2}f''(a)(x - a)^2 + \cdots + \frac{1}{n!}f^{(n)}(a)(x - a)^n + E_n(x)$$

to obtain

$$x^{-1} = 1 - (x - 1) + \frac{1}{2}(2)(x - 1)^2 - \frac{1}{3!}(3!)(x - 1)^3 + \cdots$$
$$+ \frac{1}{n!}(-1)^n(n!)(x - 1)^n + E_n(x).$$

So

$$x^{-1} = 1 - (x - 1) + (x - 1)^2 - (x - 1)^3 + (x - 1)^4 - \cdots$$
$$+ (-1)^n(x - 1)^n + E_n(x).$$

From Taylor's Theorem we know that

$$|E_n(x)| \leq \frac{1}{(n + 1)!}M_{n+1}|x - a|^{n+1},$$

where M_{n+1} is the maximum of $|f^{(n+1)}(x)|$ for x in the neighborhood N. In this particular case,

$$f^{(n+1)}(x) = (-1)^{n+1}(n+1)! \cdot x^{-(n+2)}.$$

Hence if we take the neighborhood N as the set of all x with $x > 0$, then $f^{(n+1)}(x)$ has *no* maximum. That is, $x^{-(n+2)}$ can be made arbitrarily large just by taking x close enough to 0. Consequently, to establish a bound on the error, we must reduce the size of N to keep x greater than or equal to a fixed positive number. For example, if N is now taken to be the set of all x for which $x > 1/2$, we may take

$$M_{n+1} = (n+1)! \cdot 2^{n+2}.$$

So in the neighborhood N, we have

$$|E_n(x)| \le \frac{1}{(n+1)!}(n+1)! \cdot 2^{n+2}|x-1|^{n+1}, \quad \text{or}$$

$$|E_n(x)| \le 2|2x-2|^{n+1}, \qquad \text{if } x > \frac{1}{2}.$$

This analysis immediately presents an additional problem. The bound $2|2x-2|^{n+1}$ does not, in general, get smaller as n becomes larger. However, if $2|x-1| < 1$, then the error does decrease with each increase in n. Thus we also restrict our attention to values of x for which

$$2|x-1| < 1, \quad \text{or} \quad |x-1| < \frac{1}{2}, \quad \text{or} \quad \frac{1}{2} < x < \frac{3}{2}.$$

Let's summarize the results of this rather complicated example.

1. For all x such that $x > 0$, Taylor's Theorem yields
$$x^{-1} = 1 - (x-1) + (x-1)^2 - (x-1)^3 + \cdots + (-1)^n(x-1)^n + E_n(x).$$
2. The error term $E_n(x)$ can only be bounded if we also require x to be greater than or equal to a fixed positive number. In the example, the distance was 1/2; however, any positive number could serve.
3. The bounds on the error term $E_n(x)$ are made to decrease with increasing n by the additional restriction that $1/2 < x < 3/2$.

Exercises 9.7

Apply Taylor's Theorem, with $a = 0$ and $n = 4$, to each function in Exercises 1–14. In each case (except Exercise 14) also establish a bound on the error $E_4(x)$.

1. $f(x) = x^4 - x + 7$.
2. $f(x) = x^4 + 2x^2 + 3$.
3. $f(x) = 2x^5 - 3x^4 + x^3 - 3$.
4. $f(x) = x^5 + 2x^2 + 3$.
5. $f(x) = \sin x$.
6. $f(x) = \cos x$.
7. $f(x) = e^x$, $|x| \le 1$.
8. $f(x) = \ln(x+1)$, $|x| \le 1/2$.

9. $f(x) = \sqrt[3]{x+2}, \quad |x| \leq 1.$

10. $f(x) = \sqrt{x+1}, \quad |x| \leq 1/2.$

11. $f(x) = \dfrac{1}{x-3}, \quad |x| \leq 1.$

12. $f(x) = \dfrac{1}{x+2}, \quad |x| \leq 1.$

13. $f(x) = e^{(x^2)}. \quad |x| < 1.$

14. $f(x) = \tan^2 x, \quad |x| < 1.$

Apply Taylor's Theorem, with $a = 1$ and $n = 4$, to the functions in Exercises 15–22. In each case also establish a bound on the error $E_4(x)$.

15. $f(x) = 2x^4 - 7x^3 + 3.$

16. $f(x) = \ln x, \quad 1/2 \leq x \leq 3/2.$

17. $f(x) = \cos(\pi x/2).$

18. $f(x) = e^x, \quad 0 \leq x \leq 2.$

19. $f(x) = x^{1/2}, \quad 1/2 \leq x \leq 3/2.$

20. $f(x) = \sin \pi x.$

21. $f(x) = x \ln x, \quad 1/2 \leq x \leq 3/2.$

22. $f(x) = \sin \pi x \cos \pi x.$

23. Let f be a function and let $p_2(x) = a_0 + a_1(x-a) + a_2(x-a)^2$, where a, a_0, a_1, a_2 are numbers. Let $p_2(a) = f(a)$, $p_2'(a) = f'(a)$, and $p_2''(a) = f''(a)$. Show by direct substitution of a for x in the appropriate polynomials that $a_0 = f(a)$, $a_1 = f'(a)$, and $a_2 = (1/2)f''(a)$. Show conversely that for the polynomial

$$p_2(x) = f(a) + f'(a)(x-a) + \frac{1}{2}f''(a)(x-a)^2,$$

it follows that $p_2(a) = f(a)$, $p_2'(a) = f'(a)$, and $p_2''(a) = f''(a)$.

24. Let f be a function and let $p_3(x) = a_0 + a_1(x-a) + a_2(x-a)^2 + a_3(x-a)^3$, where a, a_0, a_1, a_2, a_3 are numbers. Let $p_3(a) = f(a)$, $p_3'(a) = f'(a)$, $p_3''(a) = f''(a)$, and $p_3'''(a) = f'''(a)$. Show by direct substitution of a for x in the appropriate polynomials that

$$a_0 = f(a), \quad a_1 = f'(a), \quad a_2 = \frac{1}{2}f''(a), \quad \text{and} \quad a_3 = \frac{1}{3!}f'''(a).$$

Show, conversely, that for the polynomial

$$p_3(x) = f(a) + f'(a)(x-a) + \frac{1}{2}f''(a)(x-a)^2 + \frac{1}{3!}f'''(a)(x-a)^3,$$

$p_3(a) = f(a)$, $p_3'(a) = f'(a)$, $p_3''(a) = f''(a)$, and $p_3'''(a) = f'''(a)$.

25. Let f be a function and let

$$(*) \qquad p_n(x) = a_0 + a_1(x-a) + a_2(x-a)^2 + \cdots + a_n(x-a)^n,$$

where $a, a_0, a_1, a_2, \ldots, a_n$ are numbers. Let $p_n(a) = f(a)$, $p_n'(a) = f'(a)$, $p_n''(a) = f''(a), \ldots,$ $p_n^{(n)}(a) = f^{(n)}(a)$. Show by direct substitution of a for x in the appropriate polynomials that

$$a_0 = f(a), \quad a_1 = f'(a), \quad a_2 = \frac{1}{2!}f''(a), \ldots, a_n = \frac{1}{n!}f^{(n)}(a).$$

Show, conversely, that the polynomial (*) has the properties $p_n(a) = f(a)$, $p_n'(a) = f'(a)$, $p_n''(a) = f''(a), \ldots, p_n^{(n)}(a) = f^{(n)}(a)$.

26. Explain why the approximation

$$\int_0^{.01} x \sin x \, dx \approx \int_0^{.01} \left(x^2 - \frac{x^4}{6}\right) dx$$

may be used. What is a bound on the error in this approximation?

27. Explain why the approximation

$$\int_0^{.01} e^{x^2}\,dx = \int_0^{.01}\left(1 + x^2 + \frac{x^4}{2}\right)dx$$

may be used. What is a bound on the error in this approximation?

28. Establish a bound on the error in this approximation

$$\int_a^b x\sin x\,dx \approx \frac{1}{3}(b^3 - a^3) - \frac{1}{30}(b^5 - a^5).$$

29. Let m_0 be the rest mass of an object. If the object has velocity v, then the theory of relativity indicates that its mass is a function of its velocity and is given by

$$m = \frac{m_0}{\sqrt{1 - v^2/c^2}},$$

where c is the velocity of light. Obtain this approximation:

$$m \approx m_0 + \frac{m_0}{2}\left(\frac{v}{c}\right)^2.$$

Brief Review of Chapter 9

method	*use*	*formula*	*error*
Newton's Method	Approximation of zeros.	$c_{i+1} = c_i - \dfrac{f(c_i)}{f'(c_i)}$.	$B_{n+1} \le B_n^2 \dfrac{M}{2m}$.
Bisection Method	Approximation of zeros.	Successively bisect intervals containing the zero.	Half of last interval length.
Trapezoid Rule	Approximation of definite integrals.	$h[\frac{1}{2}f(x_0) + f(x_1) + \cdots + f(x_{n-1}) + \frac{1}{2}f(x_n)]$.	$E_n \le \dfrac{b-a}{12} Mh^2$.
Simpson's Rule	Approximation of definite integrals.	$\dfrac{h}{3}[f(x_0) + 4f(x_1) + 2f(x_2) + 4f(x_3) + \cdots + 2f(x_{n-2}) + 4f(x_{n-1}) + f(x_n)]$.	$E_n \le \dfrac{h^4(b-a)}{180} M$.
Taylor's Theorem	Approximation of functions by polynomials.	$f(a) + f'(a)(x-a) + \dfrac{1}{2!}f''(a)(x-a)^2 + \dfrac{1}{3!}f'''(a)(x-a)^3 + \cdots + \dfrac{1}{n!}f^{(n)}(a)(x-a)^n$.	$\lvert E_n \rvert \le \dfrac{M_{n+1}\lvert x-a\rvert^{n+1}}{(n+1)!}$

Since this outline does not include an explanation of the symbols used, the reader must refer to the appropriate theorems for a full statement of the details.

Linear approximations are the special case of Taylor's Theorem when $n = 1$. Consequently, only Taylor's Theorem is included in the outline above.

Technique Review Exercises, Chapter 9

1. Use Newton's Method to obtain a recursion relation that will generate successive approximations to $3^{1/2}$.
2. Let $c_1 = 2$ and use the recursion relation in Exercise 1 to find c_3. Find a bound on the error E_3.
3. Note that $1 < 3^{1/2} < 2$, and use the bisection method to approximate $3^{1/2}$ with an error of less than 1/10.
4. Use the Trapezoid Rule with $n = 4$ to approximate $\int_0^1 16^x\, dx$.
5. Use Simpson's Rule with $n = 4$ to approximate $\int_0^1 16^x\, dx$.
6. (a) Find a bound on the error in Exercise 4. (Use $\ln 16 < 3$.)
   (b) Find a bound on the error in Exercise 5. (Use $\ln 16 < 3$.)
7. Apply Taylor's Theorem, with $n = 4$ and $a = 0$, to approximate the function $f(x) = e^{-x}$. $|x| \leq 1$. Find a bound on the error.

Additional Exercises, Chapter 9

Section 9.2

In Exercises 1–6, use Newton's Method with the given initial approximation to calculate c_3, the third approximation to a zero.

1. $x^2 - 6 = 0, \quad c_1 = 2.$
2. $x^2 + 3x - 1 = 0, \quad c_1 = 0.$
3. $x^4 - 2x - 3 = 0, \quad c_1 = 1.$
4. $x^3 - 2x^2 - 1 = 0, \quad c_1 = 2.$
5. $x^5 - 2 = 0, \quad c_1 = 1.$
6. $x^5 - x^2 + x + 1 = 0, \quad c_1 = 0.$
7. Prove that Newton's Method converges when applied in Exercise 2 above.
8. Analyze the errors in Exercise 2 above. (Note that since $f(0) < 0$ and $f(1) > 0$, $E_1 < 1$.)
9. Prove that Newton's Method converges when applied in Exercise 4 above.
10. Analyze the errors in Exercise 4 above. (Note that since $f(2) < 0$ and $f(3) > 0$, $E_1 < 1$.)

Section 9.3

11. Use the bisection method to find an approximation to $5^{1/2}$ that is within 1/10 of the exact value.
12. Use the bisection method to find an approximation to $10^{1/2}$ that is within 1/10 of the exact value.
13. Use the bisection method to find an approximation to $4^{1/3}$ that is within 1/10 of the exact value.

14. Use the bisection method to find an approximation to $7^{1/3}$ that is within 1/10 of the exact value.

Section 9.4

In Exercises 15–18, use the Trapezoid Rule with $n = 4$ to compute approximations to the given integrals.

15. $\int_0^1 \frac{dx}{1+x} = \ln 2.$

16. $\int_{-1}^2 \sqrt{2+x^3}\,dx.$

17. $\int_0^1 e^{x^2}\,dx.$

18. $\int_0^\pi \ln(1+\sin x)dx.$

19. Find a bound on the error in Exercise 15.

20. Find a bound on the error in Exercise 17.

21. Find the smallest value of n so that the error in Exercise 17 is less than 1/1000.

22. Find the smallest value of n so that the error in Exercise 15 is less than 1/1000.

Section 9.5

In Exercises 23–26, use Simpson's Rule with $n = 4$ to compute approximations to the given integrals.

23. $\int_0^1 \frac{dx}{1+x} = \ln 2.$

24. $\int_{-1}^3 \sqrt{2+x^3}\,dx.$

25. $\int_0^1 e^{x^2}\,dx.$

26. $\int_0^\pi \ln(1+\sin x)dx.$

27. Find a bound on the error in Exercise 23.

28. Find a bound on the error in Exercise 25.

29. Find the smallest value of n so that the error in Exercise 25 is less than 1/1000.

30. Find the smallest value of n so that the error in Exercise 23 is less than 1/1000.

Section 9.6

In Exercises 31–34, find the linear approximation of the given function near the given point a. If no such approximation exists, state so.

31. $f(x) = \cos x, \quad a = \pi/2.$

32. $f(x) = e^{(x^2)}, \quad a = 1.$

33. $f(x) = x^{5/6}, \quad a = 0.$

34. $f(x) = \sin(\ln x), \quad a = 1.$

35. Find a bound on the error in Exercise 31.

36. Find a bound on the error in Exercise 34.

Section 9.7

In Exercises 37–42, apply Taylor's Theorem with the specified values of a and n to obtain an approximation to the given function.

37. $f(x) = \cos x, \quad a = \pi/2, n = 4.$

38. $f(x) = \sin(e^x), \quad a = 0, n = 2.$

39. $f(x) = \cos^2 x, \quad a = 0, n = 4.$

40. $f(x) = \sin(\ln x), \quad a = 1, n = 2.$

41. $f(x) = x^4 + 3x^2 + 1, \quad a = 2, n = 4.$

42. $f(x) = x^4 - x^2, \quad a = 2, n = 4.$

43. Find a bound on the error in Exercise 37.

44. Find a bound on the error in Exercise 38.
45. Find a bound on the error in Exercise 39.
46. Find a bound on the error in Exercise 40.

Challenging Problems, Chapter 9

1. There is another method of approximating zeros that closely resembles Newton's Method. This method requires two starting points c_1 and c_2, and uses the line through $(c_1, f(c_1))$ and $(c_2, f(c_2))$ instead of the tangent line. Figure 1 illustrates how one can obtain c_3 from c_1 and c_2. Derive the expression

$$c_{i+1} = c_i + \frac{f(c_i)(c_i - c_{i-1})}{f(c_{i-1}) - f(c_i)},$$

which gives the $(i + 1)$st approximation in terms of the preceding two approximations, c_i and c_{i-1}.

Figure 1

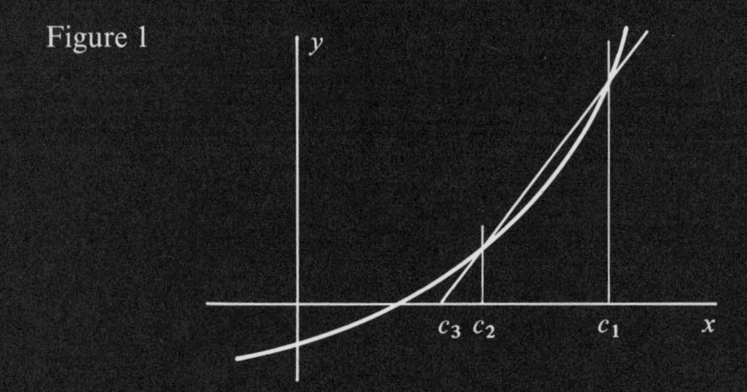

2. Prove that Newton's Method converges if $(b - a)\frac{M}{2m} < 1$, where a, b, M, and m have the meaning assigned in Theorem 9.2.2, and $c_1 = \frac{a + b}{2}$.

3. Use the following outline to establish a proof of Theorem 9.2.2, which gives a bound for the error in Newton's Method.

   (i) Show that

$$f(\alpha) = f(c_i) + f'(c_i)(\alpha - c_i) + E, \quad \text{and}$$

$$|E| \leq \frac{M}{2}|\alpha - c_i|^2, \quad \text{where}$$

$$M \geq |f''(t)|, \quad \text{for all } t \text{ between } \alpha \text{ and } c_i.$$

(ii) Show that

$$c_i - \frac{f(c_i)}{f'(c_i)} - \alpha = \frac{E}{f'(c_i)}, \quad \text{and so}$$

$$c_{i+1} - \alpha = \frac{E}{f'(c_i)}.$$

(iii) Then use the bound on E and the assumption that $|f'(c_i)| \geq m > 0$ to show that

$$E_{i+1} \leq E_i^2 \cdot \frac{M}{2m}.$$

4. One of the simplest methods of approximating integrals is to approximate the area under a curve by a series of rectangles. This "rectangle rule" is illustrated in Figure 2 for $n = 6$. Since the area of the jth rectangle is $f(x_{j-1})h$, where $h = (b - a)/n$, we have the following approximation

$$\int_a^b f(x)dx \approx h[f(x_0) + f(x_1) + \cdots + f(x_{n-1})].$$

Use the following outline to show that the error made in using this approximation is less than or equal to

$$\frac{M(b-a)^2}{n}, \qquad \text{where } M \geq |f'(x)| \text{ for } a \leq x \leq b.$$

(i) Note that

$$\left|\int_{x_i}^{x_{i+1}} f(x)dx - f(x_i)h\right| \leq [f(x_i^*) - f(x_i^{**})]h,$$

where $f(x_i^*)$ and $f(x_i^{**})$ are the maximum and minimum, respectively, of $f(x)$ for $x_i \leq x \leq x_{i+1}$.

(ii) Use The Mean Value Theorem to show that

$$[f(x_i^*) - f(x_i^{**})]h \leq Mh^2.$$

Figure 2

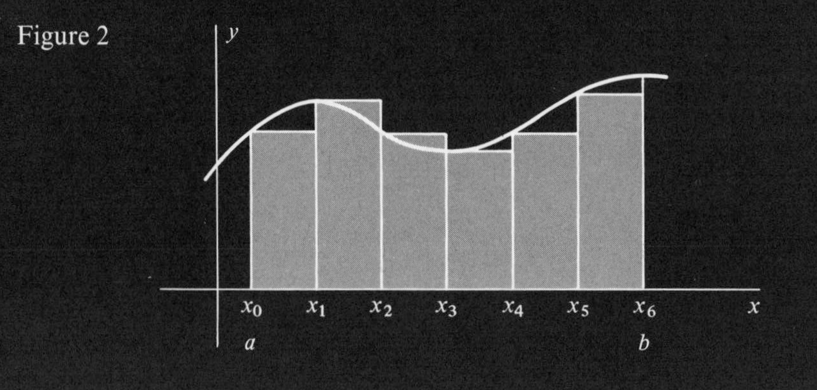

5. Let $f(x)$ be continuous for $a \le x \le b$, and assume that $f(a)f(b) < 0$. Let $x_1 < x_2 < \cdots < x_n$ be the zeros of $f(x)$ in the interval $a \le x \le b$. Suppose that the graph of $y = f(x)$ actually crosses the x-axis at each x_i. Show that the bisection method must either:

   a) hit a zero exactly or,
   b) converge to a zero with an odd subscript.

6. (*Round-Off Error*) As has been shown, $f(n) = \left(1 + \frac{1}{n}\right)^n$ is an increasing function. However a computer generated the following results:

$$f(460) = 2.714723$$

$$f(450) = 2.715071.$$

Note that $2.714723 < 2.715071$. Round-off error contributes to this discrepancy as the following outline shows:

$$f(460) = \left(1 + \frac{1}{460}\right)^{460} \approx (1.002173913)^{460}.$$

However, due to round off, the computer actually calculates

$$(1.002174)^{460},$$

and so causes an error of approximately

$$E = (1.002174)^{460} - (1.002173913)^{460}.$$

Approximate E as follows:

(i) Let $g(x) = (1.002174 - x)^{460}$.
Obtain the linear approximation for $g(x)$ about $x = 0$.
(ii) Use this result to approximate $g(0) - g(x)$.
(iii) Note that $E = g(0) - g(.000000087)$.
(iv) Show that $E \approx .0001$.

10

An Extension of the Limit Concept, Improper Integrals, and l'Hospital's Rule

10.1 An Extension of the Limit Concept

In our previous work we found the limit concept to be particularly useful. In this section we shall consider the following extension of the limit concept:

$$\lim_{x \to \infty} f(x) = L.$$

That is, we shall consider the meaning of the limit of $f(x)$ as x increases without bound. The definition is similar to that of the earlier concept.

We say that $\lim_{x \to \infty} f(x) = L$ if the values of the function $f(x)$ can be made arbitrarily close (that is, closer than any preassigned positive distance) to L by requiring x to be sufficiently large.

Note that we did not say that "x is sufficiently close to but not equal to infinity," because infinity is not a real number. In fact, in our considerations, infinity is not a number of any kind but is used only to indicate that x increases without bound.

As in our previous discussion of limits, we now give a precise definition.

Definition 10.1.1

> $\lim_{x \to \infty} f(x) = L$ if for each positive number ε (the preassigned distance) there is a positive number N (the measurement of *sufficiently large*) such that $|f(x) - L| < \varepsilon$ when $x > N$.

Since the inequality $|f(x) - L| < \varepsilon$ means that $f(x)$ is between $L - \varepsilon$ and $L + \varepsilon$, we note that this means graphically that $f(x)$ is between the lines $y = L - \varepsilon$ and $y = L + \varepsilon$. Since the graphical meaning of $x > N$ is that x is to the right of N, Definition 10.1.1 has this graphical interpretation: For each positive number ε (no matter how small) a positive number N can be chosen so that the graph of $f(x)$ is between the lines $y = L - \varepsilon$ and $y = L + \varepsilon$, when x is to the right of N.

Figure 10.1.1 shows the graphical interpretation for a particular $\varepsilon > 0$. In general, the smaller the ε, the larger the N must be; that is, the closer the lines $y = L - \varepsilon$ and $y = L + \varepsilon$, the farther to the right N must be. It is important to realize

Figure 10.1.1

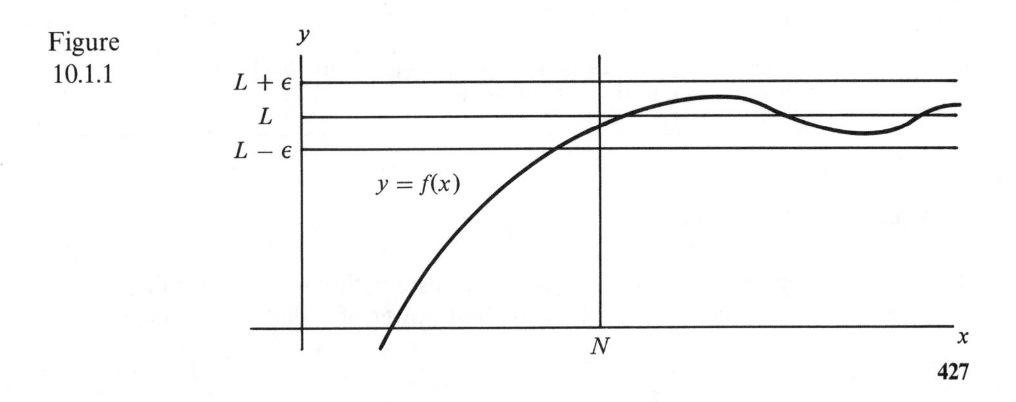

that Figure 10.1.1 is only for one particular positive ε and that there must be a similar picture for *every* positive number ε.

Example 1 | Find $\lim_{x\to\infty} \frac{1}{x+1}$.

The function $f(x) = 1/(x+1)$ can be made arbitrarily close to 0 by requiring x to be large enough. For instance, if $\varepsilon = 10^{-3} = .001$, we see that by taking $N = 10^3 = 1000$ we have

$$|f(x) - 0| = \left|\frac{1}{x+1}\right| = \frac{1}{x+1} < \frac{1}{1000+1} < .001, \qquad \text{if } x > 10^3.$$

If $\varepsilon = 10^{-8}$, we need a larger N, and we see that 10^8 will do; that is

$$|f(x) - 0| = \frac{1}{x+1} < \frac{1}{10^8+1} < 10^{-8}, \qquad \text{if } x > 10^8.$$

Clearly, no matter how small a positive number we take for ε, we can find a large enough N so that $|f(x) - 0| = \frac{1}{x+1} < \varepsilon$ if $x > N$. Thus we have

$$\lim_{x\to\infty} \frac{1}{x+1} = 0. \qquad \|$$

As in Chapter 2, where x approaches a finite number, there is a limit theorem for our new limit concept. Both the statement and proof of the theorem closely parallel that of The Limit Theorem (Theorem 2.3.1).

Theorem 10.1.1

If both $\lim_{x\to\infty} f(x)$ and $\lim_{x\to\infty} g(x)$ exist, then

(i) $\lim_{x\to\infty} [f(x) + g(x)] = \lim_{x\to\infty} f(x) + \lim_{x\to\infty} g(x)$,

(ii) $\lim_{x\to\infty} [cf(x)] = c \lim_{x\to\infty} f(x)$, if c is a constant,

(iii) $\lim_{x\to\infty} [f(x) \cdot g(x)] = \lim_{x\to\infty} f(x) \cdot \lim_{x\to\infty} g(x)$,

(iv) $\lim_{x\to\infty} \frac{f(x)}{g(x)} = \frac{\lim_{x\to\infty} f(x)}{\lim_{x\to\infty} g(x)}$, provided $\lim_{x\to\infty} g(x) \neq 0$.

Example 2 | Find $\lim_{x\to\infty} \frac{2x^2 - 7x + 2}{3x^2 + 8x - 11}$.

To see what happens to this function as x increases without bound, we first divide the numerator and denominator by the highest power of x that occurs, namely x^2.

Then

$$\lim_{x\to\infty} \frac{2x^2 - 7x + 2}{3x^2 + 8x - 11} = \lim_{x\to\infty} \frac{2 - \dfrac{7}{x} + \dfrac{2}{x^2}}{3 + \dfrac{8}{x} - \dfrac{11}{x^2}}.$$

Now $7/x$, $2/x^2$, $8/x$, and $11/x^2$ each approach 0 as x increases without bound; but the constants 2 and 3 remain unchanged. Thus, on applying parts (iv) and (i) of Theorem 10.1.1, we have

$$\begin{aligned}\lim_{x\to\infty} \frac{2x^2 - 7x + 2}{3x^2 + 8x - 11} &= \lim_{x\to\infty} \frac{2 - \dfrac{7}{x} + \dfrac{2}{x^2}}{3 + \dfrac{8}{x} - \dfrac{11}{x^2}} \\ &= \frac{2 - \lim\limits_{x\to\infty} \dfrac{7}{x} + \lim\limits_{x\to\infty} \dfrac{2}{x^2}}{3 + \lim\limits_{x\to\infty} \dfrac{8}{x} - \lim\limits_{x\to\infty} \dfrac{11}{x^2}} \\ &= \frac{2 - 0 + 0}{3 + 0 - 0} \\ &= \frac{2}{3}.\end{aligned}$$

||

As illustrated in the next example, we need not always show so much detail in applying Theorem 10.1.1.

Example 3 Find $\lim\limits_{x\to\infty} \dfrac{4x^5 + 2x^7 - x + 6}{8x^3 + 11x^8 + 5}$.

Again we divide the numerator and denominator by the highest power of x that occurs, x^8, to get

$$\lim_{x\to\infty} \frac{4x^5 + 2x^7 - x + 6}{8x^3 + 11x^8 + 5} = \lim_{x\to\infty} \frac{\dfrac{4}{x^3} + \dfrac{2}{x} - \dfrac{1}{x^7} + \dfrac{6}{x^8}}{\dfrac{8}{x^5} + 11 + \dfrac{5}{x^8}} = \frac{0}{11} = 0.$$

||

Example 4 Suppose that the population of a nation as a function of time is given by

$$f(t) = \frac{aC}{e^{-at} + C},$$

where a and C are positive constants. What is the limit of the population as t increases without bound?

We are expected to compute

$$\lim_{t \to \infty} f(t) = \lim_{t \to \infty} \frac{aC}{e^{-at} + C}.$$

An application of Theorem 10.1.1 yields

$$\lim_{t \to \infty} f(t) = \frac{aC}{\left(\lim\limits_{t \to \infty} e^{-at}\right) + C}.$$

Now, since a is a positive constant, e^{at} increases without bound as t increases without bound. Hence

$$\lim_{t \to \infty} e^{-at} = \lim_{t \to \infty} \frac{1}{e^{at}} = 0, \quad \text{and so}$$

$$\lim_{t \to \infty} f(t) = \frac{aC}{C} = a.$$

Consequently, a is the limit of the population. ||

Example 5 | Find $\lim\limits_{x \to \infty} \sin x$.

For large values of x, the sine function continues to attain periodically all values between and including -1 and 1. Thus there is no unique number approached by $\sin x$ as x increases without bound; so $\lim\limits_{x \to \infty} \sin x$ does not exist. ||

Example 6 | Find $\lim\limits_{x \to \infty} \dfrac{2x^3 - 3x + 8}{4x^2 + 6x - 1}$.

$$\lim_{x \to \infty} \frac{2x^3 - 3x + 8}{4x^2 + 6x - 1} = \lim_{x \to \infty} \frac{2 - \frac{3}{x^2} + \frac{8}{x^3}}{\frac{4}{x} + \frac{6}{x^2} - \frac{1}{x^3}}.$$

Hence The Limit Theorem does not apply because the denominator approaches 0 as x increases without bound. However, as x increases without bound, the numerator approaches 2, while the denominator becomes positive and approaches 0. Hence, as x increases without bound, the fraction increases without bound. Therefore the given limit does not exist. When the limit fails to exist in this particular way we say that the limit is infinite and we write

$$\lim_{x \to \infty} \frac{2x^3 - 3x + 8}{4x^2 + 6x - 1} = \infty.$$ ||

In general we will write $\lim\limits_{x \to \infty} f(x) = \infty$ to indicate that $f(x)$ increases without bound as x increases without bound. Similarly we will write $\lim\limits_{x \to a} f(x) = \infty$ to indicate that $f(x)$ increases without bound as x approaches a. In these cases the limit of $f(x)$ *does not exist*, even though we often say that the limit is infinite.

Similarly, the statements $\lim_{x\to\infty} f(x) = -\infty$ and $\lim_{x\to a} f(x) = -\infty$ express the fact that the limit does not exist because the function decreases without bound.

Example 7 Find $\lim_{x\to\infty} \frac{3-x^3}{2+x}$.

$$\lim_{x\to\infty} \frac{3-x^3}{2+x} = \lim_{x\to\infty} \frac{\frac{3}{x^3}-1}{\frac{2}{x^3}+\frac{1}{x^2}} = -\infty.$$ ||

As in Definition 10.1.1, we define

$$\lim_{x\to-\infty} f(x) = L$$

by replacing $x > N$ by $x < -N$ in that definition. We shall not go into this idea in any more detail, but we wish to point out that the corresponding limit theorem holds.

Exercises 10.1

Evaluate the limits in Exercises 1–28.

1. $\lim_{x\to\infty} \frac{3x^4-x^2+1}{6+40x^3}$.
2. $\lim_{x\to\infty} \frac{8x^3+2x-7}{100x^2+6}$.
3. $\lim_{x\to\infty} \frac{7x^4-100x^5+17}{6x^7-x^2+5}$.
4. $\lim_{x\to\infty} \frac{81x^3+9x^4+11}{6x^2-9+7x^5}$.
5. $\lim_{x\to\infty} \frac{\cos x}{x}$.
6. $\lim_{x\to\infty} \frac{\sin x}{x}$.
7. $\lim_{x\to\infty} \frac{5}{7^x}$.
8. $\lim_{x\to\infty} \frac{3}{e^x}$.
9. $\lim_{x\to\infty} \frac{1}{x^3-x^2}$.
10. $\lim_{x\to\infty} (x^3-x^2)$.
11. $\lim_{x\to\infty} \tan x$.
12. $\lim_{x\to\infty} \cos x$.
13. $\lim_{x\to\infty} \frac{5x+3}{x-70}$.
14. $\lim_{x\to\infty} \frac{x+2}{x-1}$.
15. $\lim_{x\to\infty} \left[\frac{x^2-9}{x+3} - x\right]$
16. $\lim_{x\to\infty} \left[x - \frac{x^2-1}{x-1}\right]$.
17. $\lim_{n\to\infty} \frac{10n}{n-6}$.
18. $\lim_{n\to\infty} \frac{n}{n+3}$.
19. $\lim_{n\to\infty} \frac{2n^3-6n+1}{4-n^3}$.
20. $\lim_{n\to\infty} \frac{3n^4-7n+2}{6-n^4}$.
21. $\lim_{n\to\infty} \left[\frac{1}{n} - \frac{1}{n^2}\right]$.
22. $\lim_{n\to\infty} \left[\frac{1}{n^4} - \frac{1}{n}\right]$.
23. $\lim_{n\to\infty} [n(1-n)]$.
24. $\lim_{n\to\infty} [n^2-n^3]$.
25. $\lim_{x\to\infty} \frac{\sqrt{4x^2-1}}{5x}$.
26. $\lim_{x\to\infty} \frac{\sqrt{3x^2+4x-2}}{7-2x}$.
27. $\lim_{n\to\infty} \frac{\sqrt[3]{n^6-3n}}{n^2-7}$.
28. $\lim_{n\to\infty} \frac{\sqrt[3]{n^3+7n^2+3n}}{5n-1}$.
29. Suppose that the reliability of a test is given by a number r, where $0 \le r \le 1$. Suppose also that if the test is replaced by a test of equal difficulty but n times as long, the reliability of

the new test is given by

$$f_n(r) = \frac{nr}{1 + (n-1)r}.$$

Compute $\lim_{n\to\infty} f_n(r)$ and $\lim_{r\to 1} f_n(r)$. How do you interpret your results?

30. Psychologists are interested in expressing the relation between attainment and practice. Thurston has suggested the "learning curve"

$$G(x) = \frac{a(x+b)}{x+c},$$

where $G(x)$ is the attainment, x the hours of practice, and a, b, and c positive empirical constants with $b < c$. Compute $\lim_{x\to\infty} G(x)$. What does a represent?

31. The value of

$$\lim_{n\to\infty} \frac{x^n}{x^n + 1}$$

depends on the value of x. Let

$$f(x) = \lim_{n\to\infty} \frac{x^n}{x^n + 1}.$$

Sketch the graph of $y = f(x)$ for $x \geq 0$.

10.2 l'Hospital's Rule

Previously when faced with the calculation of limits such as

$$\lim_{x\to 0} \frac{\sin x}{x} \quad \text{and} \quad \lim_{x\to 1} \frac{\ln x}{x-1},$$

we supplied an argument that applied only to the particular limit under consideration. In this section we shall prove *l'Hospital's Rule*, which provides a simple method of computing many limits of the form

$$\lim_{x\to a} \frac{f(x)}{g(x)}, \qquad \text{where both } \lim_{x\to a} f(x) = 0 \text{ and } \lim_{x\to a} g(x) = 0.$$

Before we discuss l'Hospital's Rule, however, we need an extension of The Mean Value Theorem for derivatives.

Theorem 10.2.1

If the functions f and g are such that
(i) $f'(x)$ and $g'(x)$ exist for $a \leq x \leq b$,
(ii) $g'(x) \neq 0$, for $a \leq x \leq b$,
(iii) and $g(a) \neq g(b)$,

then
$$\frac{f(b) - f(a)}{g(b) - g(a)} = \frac{f'(c)}{g'(c)},$$

for some c such that $a < c < b$.

Proof. We intend to show the existence of a number c such that $a < c < b$ and

$$\frac{f(b) - f(a)}{g(b) - g(a)} = \frac{f'(c)}{g'(c)}, \quad \text{or}$$

$$[f(b) - f(a)]g'(c) = [g(b) - g(a)]f'(c), \quad \text{or}$$

$$[f(b) - f(a)]g'(c) - [g(b) - g(a)]f'(c) = 0.$$

To accomplish this, consider the new function

$$h(x) = [f(b) - f(a)]g(x) - [g(b) - g(a)]f(x).$$

Note that

$$h'(x) = [f(b) - f(a)]g'(x) - [g(b) - g(a)]f'(x).$$

Thus, to complete the proof we need only find c, such that $a < c < b$ and $h'(c) = 0$. Note now that

$$h(a) = [f(b) - f(a)]g(a) - [g(b) - g(a)]f(a) = f(b)g(a) - g(b)f(a)$$

and

$$h(b) = [f(b) - f(a)]g(b) - [g(b) - g(a)]f(b) = -f(a)g(b) + g(a)f(b).$$

Hence $$h(a) = h(b).$$

By The Mean Value Theorem we know that there is a number c such that $a < c < b$ and

$$h'(c) = \frac{h(b) - h(a)}{b - a}.$$

Since $h(a) = h(b)$, we have $h'(c) = 0$; and so the proof is completed.

Theorem 10.2.2

l'Hospital's Rule

If (i) $\lim_{x \to a} f(x) = 0$ and $\lim_{x \to a} g(x) = 0$,

(ii) $f'(x)$ and $g'(x)$ exist in a neighborhood of a, except possibly at a itself,

(iii) and $g'(x) \neq 0$ in this neighborhood, except possibly at a itself,

then $$\lim_{x \to a} \frac{f(x)}{g(x)} = \lim_{x \to a} \frac{f'(x)}{g'(x)}.$$

Proof. Since the proof of the theorem as stated would take us too far afield, we shall prove only the special case where $f'(x)$ and $g'(x)$ are continuous at $x = a$. By the preceding theorem there is a number c between a and x such that

$$\frac{f(x) - f(a)}{g(x) - g(a)} = \frac{f'(c)}{g'(c)}. \tag{1}$$

Now, since both $f'(a)$ and $g'(a)$ exist, f and g are continuous at $x = a$. Consequently

$$f(a) = \lim_{x \to a} f(x) = 0 \quad \text{and} \quad g(a) = \lim_{x \to a} g(x) = 0.$$

Substitution into (1) then gives

$$\frac{f(x)}{g(x)} = \frac{f'(c)}{g'(c)}, \qquad \text{for some } c \text{ between } x \text{ and } a. \tag{2}$$

But since c is between x and a,

$$\lim_{x \to a} \frac{f'(c)}{g'(c)} = \lim_{x \to a} \frac{f'(x)}{g'(x)}.$$

Taking the limit on both sides of (2), we have the desired result:

$$\lim_{x \to a} \frac{f(x)}{g(x)} = \lim_{x \to a} \frac{f'(x)}{g'(x)}.$$

Example 1 | Use l'Hospital's Rule to verify that $\lim_{x \to 0} \frac{\sin x}{x} = 1$.

In this case the functions $f(x) = \sin x$ and $g(x) = x$ satisfy all the hypotheses of the theorem. Hence

$$\lim_{x \to 0} \frac{\sin x}{x} = \lim_{x \to 0} \frac{\cos x}{1} = 1.$$

As with all theorems, great care must be exercised to assure that the hypotheses are satisfied. For example, it is true that

$$\lim_{x \to 1} \frac{x^2 + x - 2}{x + 1} = \lim_{x \to 1} \frac{(x - 1)(x + 2)}{x + 1} = 0.$$

||

However, an unguarded application of l'Hospital's Rule would yield

$$\lim_{x \to 1} \frac{x^2 + x - 2}{x + 1} = \lim_{x \to 1} \frac{2x + 1}{1} = 3.$$

This apparent contradiction arises from the fact that not all of the hypotheses of l'Hospital's Rule are satisfied. In fact

$$\lim_{x \to 1} g(x) = \lim_{x \to 1} (x + 1) = 2 \neq 0.$$

At times, several applications of l'Hospital's Rule might be necessary to arrive at an easily evaluated limit.

Example 2 | Evaluate $\lim_{x \to 1} \frac{e^{x-1} - x}{(x - 1)^2}$.

First we verify that $\lim_{x\to a} f(x) = \lim_{x\to a} g(x) = 0$. In this case,

$$\lim_{x\to a} f(x) = \lim_{x\to 1} (e^{x-1} - x) = 1 - 1 = 0, \quad \text{and}$$

$$\lim_{x\to a} g(x) = \lim_{x\to 1} (x-1)^2 = 0.$$

Since both functions possess continuous nonzero derivatives when $x \neq 1$, we may apply l'Hospital's Rule to obtain

$$\lim_{x\to 1} \frac{e^{x-1} - x}{(x-1)^2} = \lim_{x\to 1} \frac{e^{x-1} - 1}{2(x-1)}.$$

Since the value of the last limit is not apparent, we apply l'Hospital's Rule once again. This time we use $f(x) = e^{x-1} - 1$ and $g(x) = 2(x-1)$, which also satisfy the hypotheses. Hence

$$\lim_{x\to 1} \frac{e^{x-1} - x}{(x-1)^2} = \lim_{x\to 1} \frac{e^{x-1} - 1}{2(x-1)} = \lim_{x\to 1} \frac{e^{x-1}}{2} = \frac{1}{2}. \qquad \|$$

So far our examples have been concerned only with the case where a is a real number. However, it can be proved that l'Hospital's Rule is also applicable to the case where a is ∞. We need only make certain that there is a number N such that

(i) $\lim_{x\to\infty} f(x) = \lim_{x\to\infty} g(x) = 0$,

(ii) $f'(x)$ and $g'(x)$ exist when $x > N$,

(iii) $g'(x) \neq 0$ when $x > N$.

Example 3 | Evaluate $\displaystyle\lim_{x\to\infty} \frac{\ln(1 + x^{-1})}{x^{-1}}$.

An application of l'Hospital's Rule with $f(x) = \ln(1 + x^{-1})$ and $g(x) = x^{-1}$ gives

$$\lim_{x\to\infty} \frac{\ln(1 + x^{-1})}{x^{-1}} = \lim_{x\to\infty} \frac{\dfrac{-x^{-2}}{(1 + x^{-1})}}{(-x^{-2})}.$$

On simplifying the last limit we get

$$\lim_{x\to\infty} \frac{\ln(1 + x^{-1})}{x^{-1}} = \lim_{x\to\infty} \frac{1}{1 + x^{-1}} = 1. \qquad \|$$

When the numerator and the denominator of a limit both approach 0, we often say that the limit *has the form* 0/0. Similarly when the numerator and denominator both increase without bound, we say the limit *has the form* ∞/∞. Theorem 10.2.2 deals with limits of the form 0/0. The following version of l'Hospital's Rule deals with limits of the form ∞/∞.

Theorem 10.2.3

l'Hospital's Rule

If (i) $\lim_{x \to a} f(x) = \infty$ and $\lim_{x \to a} g(x) = \infty$.

(ii) $f'(x)$ and $g'(x)$ exist in a neighborhood of a, except possibly at a itself,

(iii) and $g'(x) \neq 0$ in this neighborhood, except possibly at a itself,

then

$$\lim_{x \to a} \frac{f(x)}{g(x)} = \lim_{x \to a} \frac{f'(x)}{g'(x)}.$$

Example 4 Evaluate $\lim_{x \to \infty} \frac{x}{e^x}$.

Since $\lim_{x \to \infty} x = \infty$ and $\lim_{x \to \infty} e^x = \infty$, the given limit has the form ∞/∞. An application of Theorem 10.2.3 gives

$$\lim_{x \to \infty} \frac{x}{e^x} = \lim_{x \to \infty} \frac{1}{e^x} = 0.$$ ||

Exercises 10.2

Evaluate the limits in Exercises 1–32.

1. $\lim_{x \to 0} \frac{1 - \cos x}{x^2}$.
2. $\lim_{x \to 0} \frac{1 - \cos x}{x}$.
3. $\lim_{x \to 0} \frac{\sin 3x}{x}$.
4. $\lim_{x \to 0} \frac{\sin 3x}{\sqrt{x}}$.
5. $\lim_{x \to -2} \frac{x + 2}{x^2 - 4}$.
6. $\lim_{x \to 3} \frac{x - 3}{x^2 - 9}$.
7. $\lim_{x \to \infty} \frac{-5x^3 + 7x}{3 - x^3}$.
8. $\lim_{x \to \infty} \frac{x^2 - 7x}{3x^2 + 2}$.
9. $\lim_{t \to 2} \frac{\sqrt{t^2 + 12} - 2t}{t - 2}$.
10. $\lim_{t \to 0} \frac{\sqrt{2 + t} - \sqrt{2 - t}}{t}$.
11. $\lim_{x \to \infty} \frac{\ln(x + 1)}{x}$.
12. $\lim_{x \to 2} \frac{\ln(x/2)}{x/2 - 1}$.
13. $\lim_{x \to \infty} \frac{e^x}{\ln x}$.
14. $\lim_{x \to \infty} \frac{x + \ln x}{x \ln x}$.
15. $\lim_{x \to \infty} \frac{10x^3}{e^x}$.
16. $\lim_{x \to \infty} \frac{x^2}{e^x}$.
17. $\lim_{x \to 0} \frac{x^3 + 3x^2 - x}{2x^3 - 3x + 1}$.
18. $\lim_{x \to 2} \frac{x^2 + 3x - 10}{x^2 + 2x + 1}$.
19. $\lim_{x \to 0} \frac{\ln|\cos x|}{\ln\left|\frac{x + 1}{x - 1}\right|}$.
20. $\lim_{x \to 0} \frac{\ln|\sin x|}{\ln|\tan x|}$.
21. $\lim_{x \to 0} \frac{\ln|x|}{\cos x}$.
22. $\lim_{x \to 0} \frac{\ln|x|}{x^2 + 2x + 1}$.
23. $\lim_{x \to 0} \frac{\arcsin 3x}{x}$.
24. $\lim_{x \to 0} \frac{\arctan x - x}{x^3}$.
25. $\lim_{x \to \infty} \frac{3x^3 + 7x}{e^x}$.
26. $\lim_{x \to \infty} \frac{x^2 + 3x}{e^x}$.
27. $\lim_{x \to \infty} \frac{\ln(2x + 1)}{\ln x}$.
28. $\lim_{x \to \infty} \frac{e^{3x} - 1}{x - \sin x}$.
29. $\lim_{x \to 0} \frac{x \ln|x|}{x + \ln|x|}$.
30. $\lim_{x \to \infty} \frac{a^x}{x}$, $a > 1$.

31. $\lim_{x \to 0} \frac{\cot 2x}{\cot 3x}$.

32. $\lim_{\theta \to \pi/2} \frac{\tan 3\theta}{\tan 5\theta}$.

33. Evaluate the limit

$$\lim_{x \to \infty} \frac{8x^2 + 3x + 1}{100x^2 + 2x},$$

using both the methods of Section 10.1 and l'Hospital's Rule.

34. What is wrong with the following repeated application of l'Hospital's Rule?

$$\lim_{x \to 0} \frac{x^2 + 3x}{2x^2 - 4x} = \lim_{x \to 0} \frac{2x + 3}{4x - 4} = \lim_{x \to 0} \frac{2}{4} = \frac{1}{2}.$$

35. Show that Theorem 10.2.1 implies The Mean Value Theorem by making the proper choice of a function $g(x)$.

10.3 Extensions of l'Hospital's Rule

In the previous section we applied l'Hospital's Rule to compute limits of the form $0/0$ and ∞/∞. At times, limits that are not of one of those forms can be manipulated into one of them. This is best illustrated by means of examples.

The first example has the form $\infty - \infty$.

Example 1 Evaluate $\lim_{x \to 0} \left[\frac{1}{\sin x} - \frac{1}{x}\right]$.

$$\lim_{x \to 0} \left[\frac{1}{\sin x} - \frac{1}{x}\right] = \lim_{x \to 0} \frac{x - \sin x}{x \sin x}.$$

This second limit is of the form $0/0$. Two applications of l'Hospital's Rule yield

$$\lim_{x \to 0} \left[\frac{1}{\sin x} - \frac{1}{x}\right] = \lim_{x \to 0} \frac{1 - \cos x}{\sin x + x \cos x} = \lim_{x \to 0} \frac{\sin x}{2 \cos x - x \sin x} = 0.$$ ||

The next example is of the $0 \cdot \infty$ form.

Example 2 Evaluate $\lim_{x \to 0} x \ln|x|$.

$$\lim_{x \to 0} x \ln|x| = \lim_{x \to 0} \frac{\ln|x|}{1/x}.$$

The second limit is now of the form ∞/∞. An application of l'Hospital's Rule gives

$$\lim_{x \to 0} x \ln|x| = \lim_{x \to 0} \frac{1/x}{-1/x^2} = \lim_{x \to 0} (-x) = 0.$$ ||

Sometimes use of the logarithm function will enable us to put a limit in the desired form. That technique is particularly valuable when calculating limits of the form 1^∞, 0^0, and ∞^0.

It is tempting but incorrect to reason that since 1 to any finite power is 1, any limit of the form 1^∞ must also be 1. That a limit of the form 1^∞ need not be 1 is illustrated in the following example.

Example 3 | Evaluate $\lim_{x\to 0} (1+x)^{1/x}$.

This limit is of the form 1^∞. Note that since

$$\ln(1+x)^{1/x} = \frac{1}{x}\ln(1+x),$$

we have

$$\lim_{x\to 0} \ln(1+x)^{1/x} = \lim_{x\to 0} \frac{\ln(1+x)}{x}.$$

The last of these limits is of the form ∞/∞. An application of l'Hospital's Rule gives

$$\lim_{x\to 0} \ln(1+x)^{1/x} = \lim_{x\to 0} \frac{1}{1+x} = 1.$$

Then since the logarithm function is continuous, we have

$$\ln\left(\lim_{x\to 0} (1+x)^{1/x}\right) = 1.$$

Using the fact that $\ln r = s$ implies $r = e^s$, we get

$$\lim_{x\to 0} (1+x)^{1/x} = e^1 = e. \qquad \|$$

The next example illustrates a limit having the form 0^0.

Example 4 | Evaluate $\lim_{x\to 0} |x|^x$.

$$\lim_{x\to 0} \ln(|x|^x) = \lim_{x\to 0} (x \cdot \ln|x|) = \lim_{x\to 0} \frac{\ln|x|}{1/x}.$$

We can now apply l'Hospital's Rule to get

$$\lim_{x\to 0} \ln(|x|^x) = \lim_{x\to 0} \frac{1/x}{-1/x^2} = \lim_{x\to 0} (-x) = 0.$$

As in Example 3 we use the continuity of the logarithm function. We obtain

$$\ln\left(\lim_{x\to 0} |x|^x\right) = 0.$$

Then, since $\ln r = s$ implies $r = e^s$, we have

$$\lim_{x\to 0} |x|^x = e^0 = 1. \qquad \|$$

Finally we illustrate a limit of the form ∞^0.

Example 5 Evaluate $\lim_{x\to\pi/2} |\tan x|^{\cos x}$.

$$\lim_{x\to\pi/2} \ln|\tan x|^{\cos x} = \lim_{x\to\pi/2} \cos x \ln|\tan x| = \lim_{x\to\pi/2} \frac{\ln|\tan x|}{\sec x}.$$

Since the last limit is of the form ∞/∞, we apply l'Hospital's Rule and get

$$\lim_{x\to\pi/2} \ln|\tan x|^{\cos x} = \lim_{x\to\pi/2} \frac{\dfrac{\sec^2 x}{\tan x}}{\sec x \tan x} = \lim_{x\to\pi/2} \frac{\sec x}{\tan^2 x}$$
$$= \lim_{x\to\pi/2} \frac{\cos x}{\sin^2 x} = 0.$$

Thus, since the logarithm function is continuous,

$$\ln\left(\lim_{x\to\pi/2} |\tan x|^{\cos x}\right) = 0,$$

and so

$$\lim_{x\to\pi/2} |\tan x|^{\cos x} = e^0 = 1. \quad \|$$

Exercises 10.3

Evaluate the limits in Exercises 1–28.

1. $\lim_{x\to 0} (\csc x - \cot x)$.
2. $\lim_{x\to\pi/2} (\sec x - \tan x)$.
3. $\lim_{x\to 0} (\ln|x| - e^x \ln|x|)$.
4. $\lim_{x\to 1} \left(\frac{1}{\ln x} - \frac{x}{\ln x}\right)$.
5. $\lim_{x\to 0} \left(\frac{1}{x} - \frac{1}{e^x - 1}\right)$.
6. $\lim_{x\to 1} \left(\frac{x}{x-1} - \frac{1}{\ln x}\right)$.
7. $\lim_{x\to 0} \left(\frac{1}{x} - \frac{1}{xe^{ax}}\right)$.
8. $\lim_{x\to\infty} [\ln(3x+1) - \ln(2x-5)]$.
9. $\lim_{x\to 0} x^2 \ln|x|$.
10. $\lim_{x\to 0} x \cot x$.
11. $\lim_{x\to 0} (\sin x) \ln|x|$.
12. $\lim_{x\to\infty} (x-1)e^{(-x^2)}$.
13. $\lim_{x\to\pi/2} \left(\frac{\pi}{2} - x\right) \tan x$.
14. $\lim_{x\to\infty} x^2 e^{-x}$.
15. $\lim_{x\to\infty} x^{2/x}$.
16. $\lim_{x\to\infty} x^{\sin(1/x)}$.
17. $\lim_{x\to 0} |x^{3x}|$.
18. $\lim_{x\to 0} |x|^{\sin x}$.
19. $\lim_{x\to 0} x(\ln|x|)^2$.
20. $\lim_{x\to\infty} \left(1 + \frac{1}{x}\right)^{x^2}$.
21. $\lim_{x\to\infty} \left(\frac{x}{x+1}\right)^x$.
22. $\lim_{x\to\infty} x^{(e^{-x})}$.
23. $\lim_{x\to 0} (1 + 3x^2)^{2/x^2}$.
24. $\lim_{x\to\infty} \left(1 + \frac{1}{x^2}\right)^{x^3}$.
25. $\lim_{x\to\infty} x(3^{1/x} - 2^{1/x})$.
26. $\lim_{x\to 0} x(a^{1/x} - b^{1/x})$.
27. $\lim_{y\to\infty} \left(\frac{1+y}{y-1}\right)^y$.
28. $\lim_{y\to 0} \left(\frac{1-y}{1+y}\right)^{1/y}$.

29. Use l'Hospital's Rule to show that

$$\lim_{x\to\infty}\left[\frac{a^{1/x}+b^{1/x}}{2}\right]^x = \sqrt{ab}.$$

10.4 Improper Integrals

In defining the definite integral $\int_a^b f(x)dx$ in Section 4.1, we assumed that a and b are real numbers and that $f(x)$ is defined for $a \le x \le b$. We also pointed out that the definite integral always exists when $f(x)$ is continuous for $a \le x \le b$.

It is often necessary to consider integrals that do not meet those basic requirements. For example, the integral

$$\int_{-1}^{1}\sqrt{\frac{1+x}{1-x}}\,dx$$

is useful in the study of aerodynamics. However, the function

$$\sqrt{\frac{1+x}{1-x}}$$

is not continuous at $x = 1$.

Since that type of integral appears so often in applications of mathematics, we shall now extend the definition of $\int_a^b f(x)dx$ to include cases where $f(x)$ is not defined or not continuous for a value c, where $a \le c \le b$, but is continuous for all other x in the interval. We shall also extend the definition to include other useful cases where $a = -\infty$ or $b = \infty$. Integrals of these types, or mixtures of these types, are called *improper integrals.*

Let c be a number such that $a \le c \le b$, and let $f(x)$ be a function that is defined and continuous for $a \le x \le b$ but $x \ne c$. When $x = c$, let $f(x)$ be undefined, or defined but not continuous. We consider three types of integrals: Type I, $c = a$; Type II, $c = b$; Type III, $a < c < b$.

Type I. $c = a$

Let p be a small positive number. Then the function $f(x)$ is defined and continuous for all x such that $a + p \le x \le b$. Thus the following integral has a meaning:

$$\int_{a+p}^{b} f(x)dx.$$

We then define

$$\int_a^b f(x)dx = \lim_{p\to 0}\int_{a+p}^{b} f(x)dx.$$

when the limit exists and we leave it undefined if the limit does not exist. Figure 10.4.1 illustrates the geometric significance of this definition. Note that $\int_{a+p}^{b} f(x)dx$ equals the net area bounded by the graph of $y = f(x)$ and the x-axis between $a + p$ and b. Consequently, if $\int_a^b f(x)dx$ exists, it may be thought of as the net area bounded by the graph of $y = f(x)$ and the x-axis between a and b.

Figure 10.4.1

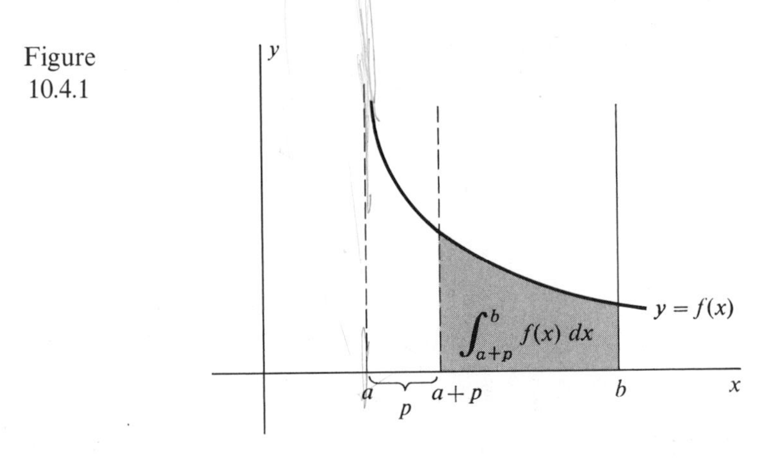

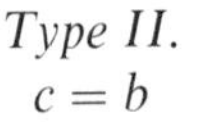

Type II. $c = b$

Let q be a small positive number. Then $f(x)$ is continuous for all x such that $a \leq x \leq b - q$, and so the following integral has a meaning:

$$\int_{a}^{b-q} f(x)dx.$$

We then define

$$\int_{a}^{b} f(x)dx = \lim_{q \to 0} \int_{a}^{b-q} f(x)dx$$

when the limit exists, and we leave it undefined if the limit does not exist. Figure 10.4.2 illustrates the geometry of this definition. Again, the improper integral $\int_a^b f(x)dx$ (when it exists) can be thought of as the net area bounded by the graph of $y = f(x)$ and the x-axis between a and b.

Figure 10.4.2

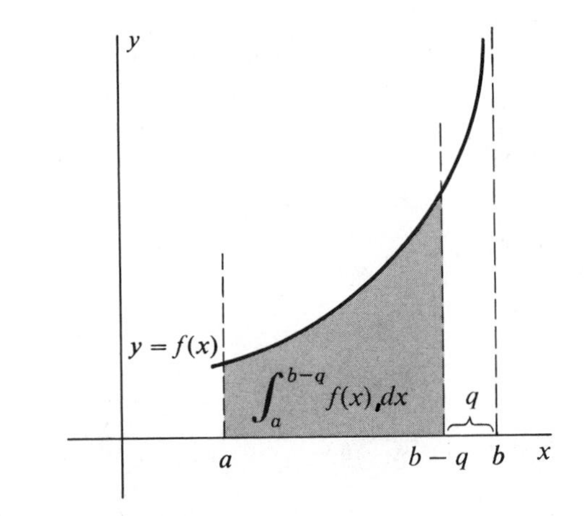

Type III. $a < c < b$

In this case we define

$$\int_{a}^{b} f(x)dx = \int_{a}^{c} f(x)dx + \int_{c}^{b} f(x)dx,$$

provided the Type II integral, $\int_a^c f(x)dx$, and the Type I integral, $\int_c^b f(x)dx$, both exist; otherwise we leave $\int_a^b f(x)dx$ undefined.

When an improper integral exists we often say that it *converges*, and when it does not exist we say that it *diverges*.

Example 1 | Find $\int_0^1 x^{-1/2}\,dx$.

This is a Type I integral.

$$\int_{0+p}^{1} x^{-1/2}\,dx = 2x^{1/2}\Big|_{0+p}^{1} = 2 - 2p^{1/2}.$$

Since $\lim_{p\to 0}(2 - 2p^{1/2}) = 2$, the given integral exists and equals 2. Thus

$$\int_0^1 x^{-1/2}\,dx = 2.$$ ||

Example 2 | Find $\int_0^1 x^{-2}\,dx$.

This is a Type I integral.

$$\int_{0+p}^{1} x^{-2}\,dx = (-x^{-1})\Big|_{0+p}^{1} = -1 + p^{-1}.$$

Since $\lim_{p\to 0}(-1 + p^{-1})$ does not exist, the given integral does not exist; it diverges. ||

Example 3 | Find $\int_0^1 (1 - x)^{-1/3}\,dx$.

This is a Type II integral.

$$\int_0^{1-q} (1 - x)^{-1/3}\,dx = -\frac{3}{2}(1 - x)^{2/3}\Big|_0^{1-q} = -\frac{3}{2}q^{2/3} + \frac{3}{2}.$$

Since $\lim_{q\to 0}[-(3/2)q^{2/3} + (3/2)] = 3/2$, the given integral converges and equals 3/2:

$$\int_0^1 (1 - x)^{-1/3}\,dx = \frac{3}{2}.$$ ||

Example 4 | Find $\int_{-1}^{2} x^{-3}\,dx$.

This is a Type III integral with $c = 0$.

$$\begin{aligned}\int_{-1}^{2} x^{-3}\,dx &= \int_{-1}^{0} x^{-3}\,dx + \int_0^2 x^{-3}\,dx \\ &= \lim_{q\to 0}\int_{-1}^{0-q} x^{-3}\,dx + \lim_{p\to 0}\int_{0+p}^{2} x^{-3}\,dx \\ &= \lim_{q\to 0}\frac{x^{-2}}{-2}\Big|_{-1}^{-q} + \lim_{p\to 0}\frac{x^{-2}}{-2}\Big|_{p}^{2} \\ &= \lim_{q\to 0}\left(\frac{(-q)^{-2}}{-2} + \frac{1}{2}\right) + \lim_{p\to 0}\left(-\frac{1}{8} + \frac{p^{-2}}{2}\right).\end{aligned}$$

Since neither of those limits exists, the given integral does not exist.

It should be noted that an incorrect result will be obtained if this integral is not recognized as an improper integral. Thus, if we simply integrate without taking note of the discontinuity at $x = 0$, we obtain

$$\int_{-1}^{2} x^{-3}\,dx = -\frac{1}{2}x^{-2}\Big|_{-1}^{2} = -\frac{1}{8} + \frac{1}{2} = \frac{3}{8},$$

which cannot be correct since the given integral does not exist. ||

The situations above are extended as outlined below to the cases where there are any finite number of values of x for which the function $f(x)$ is either undefined or not continuous. For example, suppose that

$$a = c_1 < c_2 < c_3 < b,$$

and that $f(x)$ is discontinuous only at $x = c_1, c_2$, and c_3. We then choose any numbers d_1 and d_2 such that

$$a = c_1 < d_1 < c_2 < d_2 < c_3 < b.$$

We then define the original integral as the sum of integrals of Types I and II, as follows:

$$\int_a^b f(x)dx = \int_{c_1}^{d_1} f(x)dx + \int_{d_1}^{c_2} f(x)dx + \int_{c_2}^{d_2} f(x)dx$$
$$+ \int_{d_2}^{c_3} f(x)dx + \int_{c_3}^{b} f(x)dx.$$

Of course the original integral exists only if each of the new integrals used in the definition exists.

There are three other types of improper integrals.

Type IV. $\int_a^\infty f(x)dx$ where $f(x)$ is continuous for all $x \geq a$.

We define $$\int_a^\infty f(x)dx = \lim_{b\to\infty} \int_a^b f(x)dx,$$

if the limit exists; and we leave it undefined if the limit does not exist. Figure 10.4.3 pictures the geometry of this definition. Note that if $f(x)$ is continuous, $\int_a^b f(x)dx$ exists for all numbers b and represents the net area bounded by the graph of $y = f(x)$ and the x-axis between a and b. Consequently, if it exists, $\int_a^\infty f(x)dx$ may

Figure 10.4.3

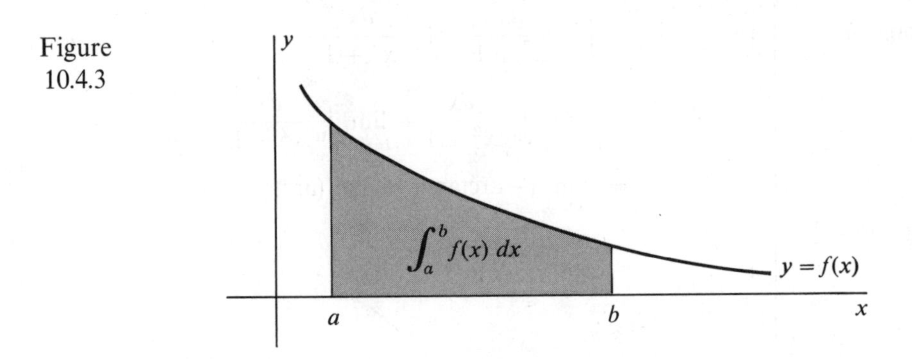

be thought of as the net area bounded by the graph of $y = f(x)$ and the x-axis between a and ∞.

Type V. $\int_{-\infty}^{b} f(x)dx$ where $f(x)$ is continuous for all $x \leq b$.

We define
$$\int_{-\infty}^{b} f(x)dx = \lim_{a \to -\infty} \int_{a}^{b} f(x)dx,$$

if the limit exists; and we leave it undefined otherwise. Figure 10.4.4 illustrates the geometry of this definition. Similar to the previous cases, when it exists, $\int_{-\infty}^{b} f(x)dx$ may be thought of as the net area bounded by the graph of $y = f(x)$ and the x-axis between $-\infty$ and b.

Figure 10.4.4

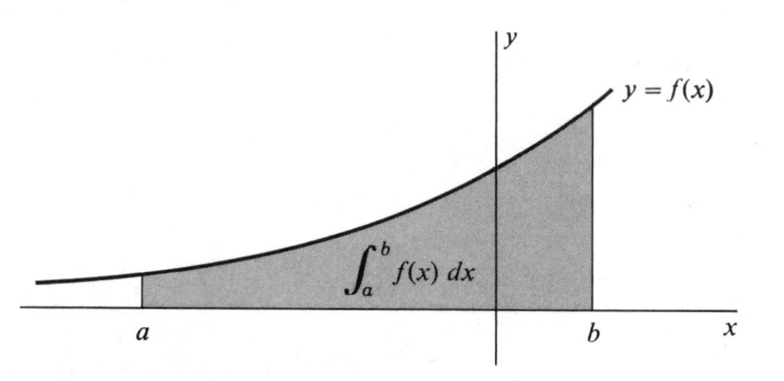

Type VI. $\int_{-\infty}^{\infty} f(x)dx$ where $f(x)$ is continuous for all x.

We choose any convenient real number c and then we define

$$\int_{-\infty}^{\infty} f(x)dx = \int_{-\infty}^{c} f(x)dx + \int_{c}^{\infty} f(x)dx,$$

if both integrals on the right exist; and we leave it undefined otherwise. The value of the integral is independent of the choice of c.

Example 5

$$\int_{1}^{\infty} x^{-2}\,dx = \lim_{b \to \infty} \int_{1}^{b} x^{-2}\,dx = \lim_{b \to \infty} -x^{-1}\Big|_{1}^{b}$$
$$= \lim_{b \to \infty} \left(\frac{-1}{b} + 1\right) = 1.$$ ||

Example 6

$$\int_{-\infty}^{\infty} \frac{dx}{x^2+1} = \int_{-\infty}^{0} \frac{dx}{x^2+1} + \int_{0}^{\infty} \frac{dx}{x^2+1}$$
$$= \lim_{a \to -\infty} \int_{a}^{0} \frac{dx}{x^2+1} + \lim_{b \to \infty} \int_{0}^{b} \frac{dx}{x^2+1}$$
$$= \lim_{a \to -\infty} (-\arctan a) + \lim_{b \to \infty} (\arctan b)$$
$$= \frac{\pi}{2} + \frac{\pi}{2}$$
$$= \pi.$$ ||

It's possible, of course, that a single integral will be a combination of several types. For example,

$$\int_{-1}^{\infty} \frac{dx}{x} = \int_{-1}^{0} \frac{dx}{x} + \int_{0}^{1} \frac{dx}{x} + \int_{1}^{\infty} \frac{dx}{x},$$

and the integral is expressed as a sum of integrals of Types I, II, and IV. The original integral exists if and only if each of the three component integrals exists.

As the next four examples indicate, improper integrals are used in fields as diverse as physics and economics.

Example 7 It is possible to lift a one-pound payload from the surface of the earth to any given distance from the surface? If so, compute the work necessary to lift a one-pound payload to "infinity."

The force exerted by gravity is inversely proportional to the square of the distance of the object from the center of the earth. Thus, the force is given by

$$f(x) = \frac{c}{x^2},$$

where c is the constant of proportionality and x is the distance from the center of the earth.

Since a force of one pound is exerted when the object is on the surface of the earth (4000 miles from the center), we have $1 = c/(4000)^2$, or $c = (4000)^2$; and so the force exerted by gravity at a distance x from the center of the earth is given by

$$f(x) = \frac{(4000)^2}{x^2}.$$

Hence the work done in lifting the payload to a distance r from the surface is given by

$$W = \int_{4000}^{r} \frac{(4000)^2}{x^2}\, dx.$$

Integrating, we find that

$$W = -\frac{(4000)^2}{x}\bigg|_{4000}^{r} = 4000 - \frac{(4000)^2}{r} \text{ mile-lbs.}$$

Consequently it is possible to lift a one-pound payload to a distance r from the surface of the earth by expending work of $4000 - (4000)^2/r$ mile-lbs.

To lift the one-pound payload to "infinity," the necessary work is

$$\begin{aligned} W &= \int_{4000}^{\infty} \frac{(4000)^2}{x^2}\, dx \\ &= \lim_{r\to\infty} \int_{4000}^{r} \frac{(4000)^2}{x^2}\, dx. \\ &= \lim_{r\to\infty} 4000 - \frac{(4000)^2}{r} \\ &= 4000 \text{ mile-lbs.} \end{aligned}$$

Hence the total work is represented by a convergent integral and 4000 mile-pounds of work is necessary to lift a one-pound payload to "infinity." ||

Example 8 Suppose that a constant yield R is received at regular intervals from some capital asset. If the yield on the asset is assumed to be compounded continuously at a constant rate r, then the *present value* of the capital asset C is given by

$$C = \int_0^\infty Re^{-rt}\,dt.$$

Find a simpler expression for C.

Since the yield R is constant,

$$\begin{aligned} C &= R\int_0^\infty e^{-rt}\,dt \\ &= R \lim_{b\to\infty} \int_0^b e^{-rt}\,dt \\ &= R \lim_{b\to\infty} -\frac{1}{r}\int_0^b e^{-rt}(-r\,dt) \\ &= R \lim_{b\to\infty} \left(-\frac{1}{r}e^{-rt}\right)\Bigg|_0^b \\ &= R \lim_{b\to\infty} \left(-\frac{1}{r}e^{-rb} + \frac{1}{r}\right). \end{aligned}$$

Then, since $r > 0$, $\lim_{b\to\infty} e^{-rb} = 0$; and so $C = R/r$. ||

Example 9 As a result of a model of national income distribution suggested by economist and sociologist Vilfredo Pareto, the total personal income greater than level y is given by

$$f(y) = \int_y^\infty abx^{-b}\,dx,$$

where a and b are positive constants with $b > 1$. Find another expression for $f(y)$.

Since a and b are constants, we have

$$\begin{aligned} f(y) &= ab \lim_{c\to\infty} \int_y^c x^{-b}\,dx \\ &= ab \lim_{c\to\infty} \frac{1}{1-b}x^{1-b}\Bigg|_y^c \\ &= ab \lim_{c\to\infty} \frac{1}{1-b}(c^{1-b} - y^{1-b}). \end{aligned}$$

Now since $b > 1$, we have $1 - b < 0$. Therefore

$$\lim_{c\to\infty} c^{1-b} = 0,$$

and so
$$f(y) = \frac{ab}{b-1} y^{1-b}.$$
||

Intuition can often be misleading when dealing with improper integrals. What might appear to be self-evident can, in fact, turn out to be false, as illustrated in the next example.

Example 10 Compute the volume of a container having vertical cross-sectional area $A(x) = |1 - 1/\sqrt{x}|$ ft^2, where x is in feet and $0 \le x \le 1$. Such a container is pictured in Figure 10.4.5. What is the surface area of this container?

Figure 10.4.5

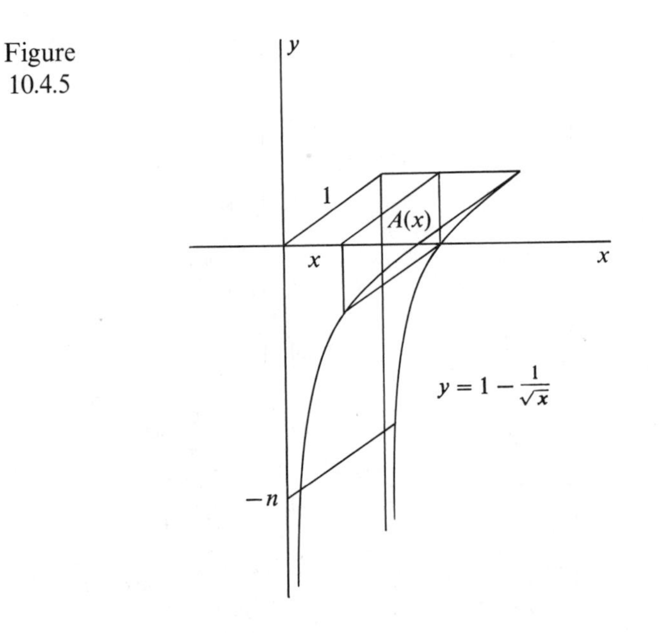

Finding the volume involves an improper integral:

$$V = \int_0^1 \left|1 - \frac{1}{\sqrt{x}}\right| dx = \int_0^1 \left(\frac{1}{\sqrt{x}} - 1\right) dx$$
$$= \lim_{p \to 0} \int_{0+p}^1 \left(\frac{1}{\sqrt{x}} - 1\right) dx = \lim_{p \to 0} (2x^{1/2} - x)\Big|_p^1$$
$$= \lim_{p \to 0} (1 - 2p^{1/2} + p) = 1 \text{ ft}^3.$$

However, the container does not have a finite surface area. To see this, just note that a rectangle of width 1 and height n fits on the surface of the container, as indicated in Figure 10.4.5. Consequently, the surface area is greater than every n; that is, the surface area is unbounded. What an unusual container! It is only large enough to hold 1 cubic foot of paint, but its inside does not have a finite surface area; so we could never paint the inside!! ||

Exercises 10.4

Evaluate the integrals in Exercises 1–37.

1. $\int_0^1 x^{-1/2}\,dx.$
2. $\int_0^1 x^{-1/3}\,dx.$
3. $\int_{-1}^0 \frac{dx}{x^2}.$
4. $\int_0^2 \frac{dx}{x}.$
5. $\int_{-\infty}^0 \frac{dx}{x-5}.$
6. $\int_0^\infty \frac{dx}{(x-3)^3}.$
7. $\int_0^3 (x-3)^{-3/5}\,dx.$
8. $\int_\infty^0 \frac{dx}{4+x^2}.$
9. $\int_{-\infty}^0 \frac{e^x}{1+e^{2x}}\,dx.$
10. $\int_{-\infty}^0 \frac{e^t}{e^t+1}\,dt.$
11. $\int_3^0 \frac{dx}{\sqrt{9-x^2}}.$
12. $\int_0^{-2} (x+1)^{-3/2}\,dx.$
13. $\int_0^4 (x-3)^{-5/3}\,dx.$
14. $\int_{-\infty}^0 \frac{dx}{2x+5}.$
15. $\int_{-\infty}^{-3} \frac{t}{1-t^2}\,dt.$
16. $\int_{-\infty}^0 \frac{x}{(4x^2+3)^2}\,dx.$
17. $\int_0^\infty \frac{dx}{3-x}.$
18. $\int_0^\infty \sin x\,dx.$
19. $\int_\infty^0 e^{-x}\,dx.$
20. $\int_{-\infty}^1 \frac{dx}{(x+3)^2}.$
21. $\int_6^\infty \frac{dt}{t^2-9}.$
22. $\int_2^\infty \frac{t}{\sqrt{3t^2-5}}\,dt.$
23. $\int_2^\infty \frac{dx}{x\sqrt{x^2+4}}.$
24. $\int_0^\infty \frac{dx}{(x^2+1)^{3/2}}.$
25. $\int_0^\infty \frac{dx}{(x-3)(x+2)x}.$
26. $\int_{-4}^\infty \frac{x^2}{4x^2+1}\,dx.$
27. $\int_{-\infty}^\infty \frac{r}{r^2+9}\,dr.$
28. $\int_0^{\pi/2} \frac{\cos\theta}{\sqrt{1-\sin\theta}}\,d\theta.$
29. $\int_0^{\pi/2} \tan\theta\,d\theta.$
30. $\int_{-\infty}^\infty \frac{x^3}{x^4+4}\,dx.$
31. $\int_{-\infty}^\infty \frac{x}{\sqrt{3x^2+2}}\,dx.$
32. $\int_{-2}^4 \frac{dt}{\sqrt{16-t^2}}.$
33. $\int_0^a \frac{1}{\sqrt{a^2-x^2}}\,dx, \quad a>0.$
34. $\int_{-\infty}^\infty \frac{t}{\sqrt{(t^2+4)^3}}\,dt.$
35. $\int_0^1 \frac{dx}{\sqrt{x}(x+1)}.$
36. $\int_0^5 f(x)dx$, where $f(x)=\begin{cases} x^2 & \text{if } 0<x<2 \\ 6x & \text{if } x>2. \end{cases}$
37. $\int_1^3 f(x)dx$, where $f(x)=\begin{cases} 2x & \text{if } x<2 \\ 7 & \text{if } x=2 \\ x^3 & \text{if } x>2. \end{cases}$
38. Find the area of the region under the graph of $y = xe^{-x}$ from 0 to ∞.
39. By interpreting the given integrals as areas, prove that

$$\int_1^\infty \frac{dx}{x^2} = \int_0^1 \frac{dy}{\sqrt{y}} - 1.$$

40. What is wrong with the following "proof" that $-2 > 0$? Since $1/x^2 > 0$ for all x, $\int_{-1}^{1} dx/x^2 > 0$. However,

$$\int_{-1}^{1} \frac{dx}{x^2} = -\frac{1}{x}\Big|_{-1}^{1} = -1 - 1 = -2,$$

and so $-2 > 0$.

41. Compute $\int_0^\infty 1000e^{-.06t}\,dt$ to find the present value of \$1000 per year compounded continuously at the rate of 6%.

42. Assume that the reaction to a given quantity of medication t hours after the drug is given is $r(t) = kte^{-t^2}$, where k is a constant and the reaction between times a and b is given by

$$\int_a^b kte^{-t^2}\,dt.$$

Compute the total reaction:

$$\int_0^\infty kte^{-t^2}\,dt.$$

Laplace Transforms are useful in solving special types of differential equations. The Laplace Transform of a function $f(x)$, $L[f(x)]$, is defined as

$$L[f(x)] = \int_0^\infty e^{-tx}f(x)dx.$$

In Exercises 43–47 compute the Laplace Transform of the given function.

43. $f(x) = 1.$

44. $f(x) = x.$

45. $f(x) = \sin x.$

46. $f(x) = \cos x.$

47. $f(x) = e^{ax}.$

48. Show that $L[x^n] = \dfrac{n}{t} L[x^{n-1}]$.

Brief Review of Chapter 10

1. Extending the Definition of Limits

Definition. $\lim_{x\to\infty} f(x) = L$, if given $\varepsilon > 0$ there is a positive number N such that $|f(x) - L| < \varepsilon$ when $x > N$.

Interpretation of Definition. $f(x)$ can be made arbitrarily close to L by choosing x sufficiently large. The definition and interpretation of $\lim_{x\to-\infty} f(x) = L$ are analogous.

Methods of Calculation. The Limit Theorem and l'Hospital's Rule are aids in calculating many limits.

2. l'Hospital's Rule

This rule is helpful when calculating limits of the form $\lim_{x \to a} f(x)/g(x)$, where either

(i) $\lim_{x \to a} f(x) = \lim_{x \to a} g(x) = 0$, or

(ii) $\lim_{x \to a} f(x) = \lim_{x \to a} g(x) = \pm\infty$.

In those cases, if f and g satisfy additional hypotheses, we have

$$\lim_{x \to a} \frac{f(x)}{g(x)} = \lim_{x \to a} \frac{f'(x)}{g'(x)}.$$

3. Extensions of l'Hospital's Rule

Certain limits of the form $\infty - \infty$, 1^∞, 0^0, and ∞^0 can be changed into the form $0/0$ or ∞/∞ with suitable manipulations. When the form is 1^∞, 0^0, or ∞^0 use of the logarithm function is often beneficial.

4. Improper Integrals

Improper integrals are divided into two major types.

(i) $\int_a^b f(x)dx$, where $f(x)$ is discontinuous somewhere in the interval $a \le x \le b$.

(ii) $\int_a^\infty f(x)dx$, $\int_{-\infty}^b f(x)dx$, and $\int_{-\infty}^\infty f(x)dx$, where the integration is over an infinite "interval."

In each case the integrals are defined as limits of previously defined integrals.

Technique Review Exercises, Chapter 10

1. Compute the limits

(a) $\lim_{x \to 2} \frac{x^2 + 3x - 10}{2x^2 - 10x + 12}$. (b) $\lim_{x \to 2} \frac{x^2 + 3x - 10}{2x^2 - 10x + 11}$. (c) $\lim_{x \to \infty} \frac{x^2 + 3x - 10}{2x^2 - 10x + 11}$.

2. Compute the limits

(a) $\lim_{x \to 0} \frac{1 - \cos 3x}{x^2}$. (b) $\lim_{x \to 0} 2x \ln|x|$.

3. Compute the limits

(a) $\lim_{x \to \infty} (e^x/\ln x)$. (b) $\lim_{x \to \infty} (\ln x)^{1/x}$.

4. Evaluate the integral $\int_1^\infty xe^{-x^2}\, dx$.

5. Evaluate the integral $\int_{-2}^0 dx/(x + 1)^2$.

Additional Exercises, Chapter 10

Section 10.1

In Exercises 1–8, compute the given limit.

1. $\lim_{x \to \infty} \frac{7x^3 + 3x}{4x^2 - 10x^3}$. 2. $\lim_{x \to \infty} \frac{\sin^5 x}{x}$. 3. $\lim_{x \to \infty} \frac{x^4 - x^3}{x^2 + 5}$.

4. $\lim_{x\to\infty} (x^5 - x^2)$.

5. $\lim_{x\to\infty} \dfrac{3x^2 + 7x}{2x^2 - 10}$.

6. $\lim_{x\to\infty} \dfrac{7x - 4x^3}{2x^3 + 5}$.

7. $\lim_{x\to\infty} \dfrac{10x^3 - 7x^2 + 2}{5x^3}$.

8. $\lim_{x\to\infty} \cot x$.

Section 10.2

In Exercises 9–20, compute the given limit.

9. $\lim_{x\to 0} \dfrac{\ln(\cos x)}{x}$.

10. $\lim_{x\to 0} \dfrac{1 - e^x}{x}$.

11. $\lim_{x\to -2} \dfrac{x^2 + 3x + 1}{x^2 - x - 6}$.

12. $\lim_{x\to\infty} \dfrac{\ln x}{x}$.

13. $\lim_{x\to\infty} \dfrac{\ln(\ln|x|)}{x}$.

14. $\lim_{x\to 0} \dfrac{(\sin x)^2}{x}$.

15. $\lim_{x\to 1} \dfrac{\ln x}{x^3 - 1}$.

16. $\lim_{x\to 0} \dfrac{7^x - 1}{3^x - 1}$.

17. $\lim_{x\to \pi/2} \dfrac{(2x/\pi) - \sin x}{x - \pi/2}$.

18. $\lim_{x\to \pi/2} \dfrac{x - (\pi/2)}{\ln(\sin x)}$.

19. $\lim_{x\to 0} \dfrac{1 - \cos x}{\sin x}$.

20. $\lim_{x\to \pi/2} \dfrac{1 - \sin x}{\cos x}$.

Section 10.3

In Exercises 21–34, compute the given limit.

21. $\lim_{x\to\infty} x^{5/x}$.

22. $\lim_{x\to 0} |x|^{5x}$.

23. $\lim_{x\to 0} x^3 \ln|x|$.

24. $\lim_{x\to \pi/2} (\cos x)(\ln|x - \pi/2|)$.

25. $\lim_{x\to\infty} (x^2 + 3x)e^{-x}$.

26. $\lim_{x\to \pi/2} (x - \pi/2) \tan x$.

27. $\lim_{x\to\infty} x^{\sin x}$.

28. $\lim_{x\to\infty} (1 - x^{-1})^{x^3}$.

29. $\lim_{x\to 0} xa^{1/x}$, $a > 1$.

30. $\lim_{x\to\infty} (x^3 - \ln x)$.

31. $\lim_{x\to\infty} (a^x - \ln x)$, $a > 1$.

32. $\lim_{x\to\infty} \dfrac{\ln(2x - 5)}{\ln x}$.

33. $\lim_{x\to 0} \left[\dfrac{1}{x} - \dfrac{1}{\sin x}\right]$.

34. $\lim_{x\to 0} \left[\dfrac{1}{x \tan x} - \dfrac{1}{x^2}\right]$.

Section 10.4

In Exercises 35–40, compute the given integral.

35. $\int_0^1 \dfrac{dx}{x^{3/2}}$.

36. $\int_1^\infty \dfrac{dx}{x^{3/2}}$.

37. $\int_1^\infty \dfrac{dx}{1 + x^2}$.

38. $\int_0^1 \dfrac{dx}{1 - x^2}$.

39. $\int_0^2 \dfrac{dx}{x^2 - 1}$.

40. $\int_0^1 \dfrac{dx}{(1 - x^2)^{1/2}}$.

Challenging Problems, Chapter 10

1. Show that $\lim_{n\to\infty} n(\sqrt[n]{x} - 1) = \ln x$.

2. Let $f''(x)$ exist for all x. Show that

$$\lim_{h\to 0} \frac{f(x+2h) - 2f(x+h) + f(x)}{h^2} = f''(x).$$

3. Let $f(x)$ be a differentiable function, and suppose $\lim_{x\to\infty} f(x) = L$. Show that if $\lim_{x\to\infty} f'(x)$ exists then $\lim_{x\to\infty} f'(x) = 0$.

4. Let $f(x)$ be a continuous function. Express

$$\lim_{n\to\infty} \frac{1}{n}\left[f\left(\frac{1}{n}\right) + f\left(\frac{2}{n}\right) + \cdots + f\left(\frac{n}{n}\right)\right]$$

as a definite integral, and use that result to evaluate

(i) $\lim_{n\to\infty} \frac{1}{n^4}[1 + 2^3 + 3^3 + \cdots + n^3]$,

(ii) $\lim_{n\to\infty} \frac{1}{n^{5/2}}[1 + 2^{3/2} + 3^{3/2} + \cdots + n^{3/2}]$.

5. A certain continuous function $f(x)$ has the following properties:
   (i) There are numbers m and M such that $m \le f(x) \le M$ for all x.
   (ii) $f(a)$ is the average of $f(x)$ over the interval $a - h \le x \le a + h$ for all a and h.

   Show that $f(x)$ is a constant function. (*Hint*: Differentiate $f(x) = \frac{1}{2h}\int_{x-h}^{x+h} f(t)dt$, and let h increase without bound.)

6. Construct a continuous function $f(x)$ that demonstrates that it is not necessary that $\lim_{x\to\infty} f(x) = 0$ for $\int_0^\infty f(x)dx$ to exist. (*Hint*: Consider the segment of a graph shown below, where n is a fixed positive integer.)

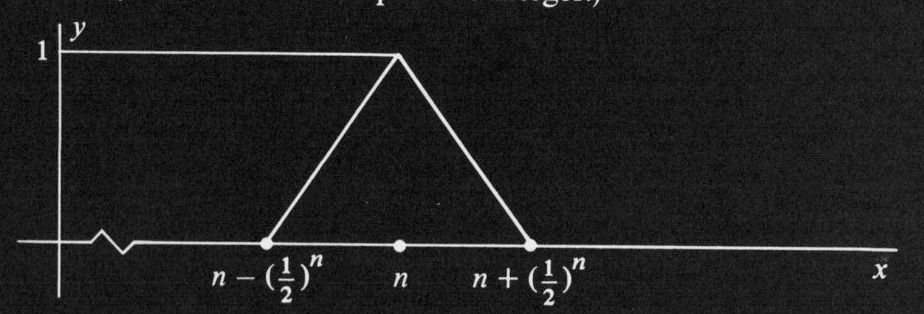

7. Solve the problem proposed by Louis Rotando of Westchester Community College, N.Y. (*American Mathematical Monthly*, Feb. 1976). Let $b > 1$. Find the values of b for which the equation $\log_b x = x$ has: (a) exactly one solution, (b) exactly two solutions, (c) no solution. Use the following hint of a solution by Robert Plumber of Cincinnati Country Day School, Ohio: Note that the equation is equivalent to $b = x^{1/x}$, and consider the graphs of $y = b$ and $y = x^{1/x}$.

8. The gamma function $\Gamma(x)$ is defined as follows:

$$\Gamma(x) = \int_0^\infty t^{x-1}e^{-t}\,dt, \qquad \text{for } x > 0.$$

(a) Show that $\Gamma(x+1) = x\Gamma(x)$.
(b) Show that when n is a positive integer, $\Gamma(n) = (n-1)!$ (Remember that $m!$ is defined to be 1 if $m = 0$.)

9. Prove that if $f(x)$ is a continuous function for $0 \le x \le 1$, then

$$\int_0^{\pi/2} f(\sin x)dx = \int_0^{\pi/2} f(\cos x)dx.$$

Hint: Consider the integral

$$\int_0^1 \frac{f(u)}{\sqrt{1-u^2}}\,du.$$

10. The integration formulas

$$\int \frac{dx}{x^p} = \begin{cases} \dfrac{1}{1-p}x^{1-p} + C & \text{if } p \neq 1 \\ \ln x + C & \text{if } p = 1 \end{cases}$$

appear strange because of the seeming dichotomy. Prove that this dichotomy is only illusory by showing that

$$\lim_{p\to 1} \int_1^b \frac{dx}{x^p} = \int_1^b \frac{dx}{x}.$$

11. A famous theorem in the theory of numbers called The Prime Number Theorem states that if $\pi(x)$ denotes the number of positive prime numbers that are less than or equal to a real number x, then $x/\ln x$ approximates $\pi(x)$ in the sense that

$$\lim_{x\to\infty} \frac{\pi(x)}{\dfrac{x}{\ln x}} = 1.$$

Use the following outline to prove that $n \ln n$ approximates the nth positive prime number p_n in a similar sense. That is, show that

$$\lim_{n\to\infty} \frac{p_n}{n \ln n} = 1.$$

(i) Use The Prime Number Theorem to evaluate $\displaystyle\lim_{x\to\infty}\left(\frac{\ln \pi(x)}{\ln x} + \frac{\ln \ln x}{\ln x} - 1\right)$.

(ii) Use that result to help evaluate $\displaystyle\lim_{x\to\infty} \frac{\ln \pi(x)}{\ln x}$.

(iii) Use (ii) and The Prime Number Theorem to compute $\displaystyle\lim_{x\to\infty} \frac{x}{\pi(x)\ln \pi(x)}$.

(iv) Replace x by p_n and complete the proof.

11

Infinite Sequences and Series

11.1 Infinite Sequences

Intuitively an infinite sequence is simply a nonending list of numbers. For example, when applying Newton's Method the recursion relation

$$c_{i+1} = c_i - \frac{f(c_i)}{f'(c_i)}$$

and a value of c_1 gives an infinite sequence $c_1, c_2, c_3, \ldots, c_i, c_{i+1}, \ldots$ of approximations to a zero of $f(x)$.

More formally, an infinite sequence can be viewed as a rule that assigns a real number a_i to each positive integer i. Thus we have the following definition.

Definition 11.1.1

> An *infinite sequence* is a function whose domain is the set of all integers greater than or equal to a fixed integer h. (Usually $h = 0$ or 1.)

To indicate an infinite sequence we shall usually write it as a function of n with braces around it. Thus we shall write $\{f(n)\}$ when the function is f. The values of the function are called the *terms* of the infinite sequence. In particular, $f(1)$ is called the first term, $f(2)$ is called the second term, $f(3)$ is called the third term, and $f(n)$ is called the nth term. We often write a_1 for $f(1)$, a_2 for $f(2)$, a_3 for $f(3)$, and a_n for $f(n)$. With such notation the sequence can be written as $\{a_n\}$, or as $\{a_1, a_2, a_3, \ldots\}$.

For example the sequence $\{1/n\}$ has first term $a_1 = 1$, second term $a_2 = 1/2$, third term $a_3 = 1/3$, 81st term $a_{81} = 1/81$, and nth term, $a_n = 1/n$.

When the notation $\{a_1, a_2, a_3, \ldots\}$ is used, there should be an evident pattern for determining the nth term. For example, the nth term of the sequence

$$\{1, 3, 5, 7, 9, \ldots\}$$

is $2n - 1$, and the nth term of

$$\left\{1, -\frac{1}{3}, \frac{1}{5}, -\frac{1}{7}, \ldots\right\}$$

is $(-1)^{n+1}/(2n - 1)$.

Infinite sequences arise quite naturally in many applications of mathematics.

Example 1

A sum of money introduced into a local economy, such as the spending of tourists or convention delegates, has more than just an initial effect. Suppose, for example, that each person or business spends 80% of its income locally. (The remaining 20% is partly saved and partly spent outside the local economy). Then, if \$1,000,000 is introduced into the economy, 80% of \$1,000,000, or \$800,000, is again spent and so reintroduced into the economy. Of course the process will continue, 80% of the \$800,000 will again be spent, and so forth.

An infinite sequence is useful in describing the phenomenon. If n is the number of times the money has changed hands, and $f(n)$ is the amount of money spent at that stage, we have

$$f(1) = \$1{,}000{,}000,$$

$$f(2) = \$(.80)(1{,}000{,}000),$$

$$f(3) = \$(.80)(.80)(1{,}000{,}000) = \$(.80)^2(1{,}000{,}000),$$

$$f(4) = \$(.80)(.80)^2(1{,}000{,}000) = \$(.80)^3(1{,}000{,}000),$$

and, in general,

$$f(n) = \$(.80)^{n-1}(1{,}000{,}000).$$

||

Consequently, the situation is described by the infinite sequence

$$\{f(n)\} = \{\$(.80)^{n-1}(1{,}000{,}000)\}.$$

Previously we discussed limits such as $\lim_{x\to\infty} f(x)$. Since an infinite sequence *is* a function, it makes sense to define and discuss $\lim_{n\to\infty} f(n)$, or $\lim_{n\to\infty} a_n$.

Definition 11.1.2

> Let $\{a_n\}$ be an infinite sequence. Then
>
> $$\lim_{n\to\infty} a_n = L$$
>
> if for each positive number ε, there is a positive number N such that
>
> $$|a_n - L| < \varepsilon$$
>
> whenever n is an integer greater than N.

The Limit Theorem also holds for this kind of limit and can be proved in much the same fashion as Theorem 11.1.1.

Definition 11.1.3

> When $\lim_{n\to\infty} a_n$ exists, the infinite sequence $\{a_n\}$ is called *convergent* and the number $\lim_{n\to\infty} a_n$ is called the *limit* of the sequence. When $\lim_{n\to\infty} a_n$ does not exist, the sequence is called *divergent*. When $\lim_{n\to\infty} a_n = a$, we also say that the sequence *converges* to a.

Let us test some sequences for convergence.

Example 2 | $\{1/n\}$.

Since $\lim_{n\to\infty} \frac{1}{n} = 0$, the sequence converges and has limit 0. ||

Example 3 | $\{(-1)^n\}$.

$\lim_{n\to\infty}(-1)^n$ does not exist because $(-1)^n$ alternates between 1 and -1 and hence does not get arbitrarily close to a unique number. Therefore, $\{(-1)^n\}$ diverges. ||

Example 4 | $\{a_n\}$, where $a_n = (n^2 - 2)/(3n - 4n^2)$.

$$\lim_{n\to\infty} \frac{n^2 - 2}{3n - 4n^2} = \lim_{n\to\infty} \frac{1 - 2/n^2}{3/n - 4} = -\frac{1}{4}.$$

Thus this sequence converges, and its limit is $-1/4$. ||

Example 5 | $\left\{\dfrac{5n^2 + 2n^3 - 1}{n^2 + 10n}\right\}$.

$$\lim_{n\to\infty} \frac{5n^2 + 2n^3 - 1}{n^2 + 10n} = \lim_{n\to\infty} \frac{\dfrac{5}{n} + 2 - \dfrac{1}{n^3}}{\dfrac{1}{n} + \dfrac{10}{n^2}} = \infty.$$

So the sequence diverges. ||

Sometimes we shall be concerned with a sequence $\{a_n\}$ for which there is a real number B such that $a_i \leq B$ for all i. More generally, we have the following definition.

Definition 11.1.4

> An *upper bound of a set S* of numbers is any number B that is greater than or equal to every number in S. (An upper bound of S does not have to be in S.) When S is the set of terms of a sequence $\{a_n\}$, we call B an *upper bound of the sequence.*

For example, 5 is an upper bound of the set of all numbers x such that $x^2 < 2$. Also, 32, 3, and $\sqrt{2}$ are upper bounds of that set. The number $\sqrt{2}$ turns out to be the least number that is an upper bound of that set.

In general suppose that there is an upper bound of a non-empty set S of numbers. Then obviously there are infinitely many upper bounds of S. But not so obvious is the fact that of all the upper bounds for S there must be a smallest or least, which we call the least upper bound of S.

Definition 11.1.5

> If there is an upper bound for a set S of numbers, then the least of all the upper bounds of S is called the *least upper bound of S.*

As we pointed out above every non-empty set of real numbers with an upper bound has a least upper bound. That property of the real numbers is called the *completeness property*. It will be left as an exercise to show that this property might not hold in other systems.

We shall use the completeness property to prove the following theorem about sequences.

Theorem 11.1.1

> Let $\{a_n\}$ be a sequence such that $a_i \leq a_{i+1}$ for all i, and assume that $\{a_n\}$ has an upper bound. Then $\{a_n\}$ converges and its limit is its least upper bound.

Proof. Since $\{a_n\}$ has an upper bound, it has a least upper bound B, by the completeness property. Let ε be any preassigned positive number and consider $B - \varepsilon$. Since $B - \varepsilon < B$ and since B is the least upper bound of $\{a_n\}$, $B - \varepsilon$ is not an upper bound of $\{a_n\}$. Therefore there is a number N such that $B - \varepsilon < a_N$. But according to the hypotheses of the theorem,

$$a_N \leq a_{N+1} \leq a_{N+2} \leq \cdots,$$

and hence $B - \varepsilon < a_n$ for all $n > N$. Also $a_n \leq B$ for all n, because B is an upper bound of $\{a_n\}$. Therefore,

$$B - \varepsilon < a_n \leq B, \qquad \text{when } n > N.$$

Thus

$$|a_n - B| < \varepsilon, \qquad \text{when } n > N,$$

and so by the definition of such limits, we have $\lim_{n \to \infty} a_n = B$. That is, $\{a_n\}$ converges and its limit is B.

Example 6

The sequence $\{(2n - 1)/n\} = \{1/1, 3/2, 5/3, 7/4, \ldots\}$ is such that $a_i \leq a_{i+1}$ for all i and it has 5 as an upper bound. Since $(2n - 1)/n = 2 - 1/n$, the least upper bound is 2; and hence the sequence converges and its limit is 2. ||

On the other hand, if $\{a_n\}$ is any sequence that has no upper bound (regardless of whether $a_i \leq a_{i+1}$ for all i), then $\{a_n\}$ diverges. This is true because the sequence contains arbitrarily large terms. Thus we have the following theorem.

Theorem 11.1.2

> If a sequence has no upper bound, it diverges.

It follows from this theorem that if a sequence converges, it must have an upper bound; for if it had no upper bound, it would have to diverge. This fact is important enough to be stated as a separate theorem.

Theorem 11.1.3

> If a sequence converges, it has an upper bound.

Example 7

The sequence

$$\{n[1 + (-1)^n]\} = \{0, 4, 0, 8, 0, 12, \ldots\}$$

has no upper bound and hence diverges. Notice also that here we do not have $\lim_{n \to \infty} a_n = \infty$, because the terms of the sequence alternate between 0 and increasingly larger positive numbers. ||

Example 8 The sequence $\{n\} = \{1, 2, 3, 4, 5, \ldots\}$ has no upper bound and hence diverges. Note, however, that we have $\lim_{n\to\infty} a_n = \infty$. ||

The concepts of lower bound and greatest lower bound are defined in a way that is completely analogous to the definitions of upper bound and least upper bound.

Definition 11.1.6

A *lower bound of a set S* of numbers is any number that is less than or equal to every number in S. (A lower bound of S does not have to be in S.)

Definition 11.1.7

The greatest of all lower bounds of a set S of numbers is called the *greatest lower bound of S*.

The completeness property and theorems analogous to Theorems 11.1.1, 11.1.2, and 11.1.3 also hold if the terms lower bound and greatest lower bound are used. The statement of these facts is left as an exercise.

Exercises 11.1

In Exercises 1–22, find the limit of the given sequence or state that it is divergent.

1. $\left\{\dfrac{2n^2+3}{3n^2+7}\right\}$.
2. $\left\{\dfrac{n^2+1}{n^2-2}\right\}$.
3. $\left\{\dfrac{2n^2+3n}{n^3+5}\right\}$.
4. $\left\{\dfrac{n+1}{3n^2-1}\right\}$.
5. $\left\{\dfrac{n^4-100n}{n^3+50n^2}\right\}$.
6. $\left\{\dfrac{n^3-7n}{n^2+2}\right\}$.
7. $\left\{1+\dfrac{(-1)^n}{n}\right\}$.
8. $\left\{\dfrac{(-1)^n}{n}\right\}$.
9. $\{\sin(n\pi)\}$.
10. $\left\{\sin\dfrac{n\pi}{2}\right\}$.
11. $\left\{\dfrac{1+(-1)^n}{2}\right\}$.
12. $\left\{(-1)^n-\dfrac{1}{n}\right\}$.
13. $\left\{\dfrac{\ln n}{n}\right\}$.
14. $\left\{\dfrac{n-1}{e^{2n+1}}\right\}$.
15. $\left\{\dfrac{\ln\left(\dfrac{1}{n}\right)}{n^2}\right\}$.
16. $\{n^2e^{-n}\}$.
17. $\{n^2e^{(-n^2)}\}$.
18. $\left\{\dfrac{1}{n^2}\ln n\right\}$.
19. $\{n^{(-1)^n}\}$.
20. $\left\{1+(-1)^n\dfrac{1}{n}\right\}$.
21. $\{x_n\}$, where $x_n = \begin{cases} 1 & \text{if } n < 1000 \\ 5 & \text{if } n \geq 1000. \end{cases}$
22. $\{x_n\}$, where $x_{2n-1} = 3$, $x_{2n} = 2$.
23. State the theorems analogous to Theorems 11.1.1, 11.1.2, and 11.1.3, using the terms lower bound and greatest lower bound.

For each of the sequences in Exercises 24–35,

(a) Determine the nth term.

(b) Find two upper bounds or state that none exists.

(c) Find the least upper bound or state that none exists.

(d) Find two lower bounds or state that none exists.

(e) Find the greatest lower bound or state that none exists.

(f) Find the limit of the sequence or state that it is divergent.

24. $\left\{\frac{1}{2}, \frac{1}{3}, \frac{1}{4}, \frac{1}{5}, \ldots\right\}$.

25. $\left\{\frac{1}{2}, \frac{2}{3}, \frac{3}{4}, \frac{4}{5}, \ldots\right\}$.

26. $\{-1, -1, -1, \ldots\}$.

27. $\{5, 5, 5, 5, \ldots\}$.

28. $\{0, 1, 0, 1, 0, 1, \ldots\}$.

29. $\{1, 2, 1, 2, 1, 2, \ldots\}$.

30. $\left\{\frac{1}{2}, -\frac{1}{3}, \frac{1}{4}, -\frac{1}{5}, \frac{1}{6}, -\frac{1}{7}, \ldots\right\}$.

31. $\left\{1, -1, \frac{1}{2}, -\frac{1}{2}, \frac{1}{4}, -\frac{1}{4}, \frac{1}{8}, -\frac{1}{8}, \ldots\right\}$.

32. $\{1, -1, 2, -2, 3, -3, \ldots\}$.

33. $\{1, 2, 3, 4, 5, \ldots\}$.

34. $\left\{2, \frac{5}{2}, 2, \frac{7}{3}, 2, \frac{9}{4}, 2, \frac{11}{5}, \ldots\right\}$.

35. $\left\{1, \frac{1}{2}, 1, \frac{2}{3}, 1, \frac{3}{4}, \ldots\right\}$.

36. A ball is dropped from a height of 6 feet and always rebounds .7 of the height it has fallen. Find a sequence that describes how high it bounces on each successive bounce. Does the resultant sequence converge? If so, what is its limit?

37. Due to birth, death, immigration, and emigration the population of a certain country increases in such a way that the population at the end of any ten-year period is 1.2 times the population at the beginning of the ten-year period. Given an initial population of N, find a sequence that yields the population at the end of each successive ten-year period. Does this sequence converge?

38. Does the completeness property hold in the set of integers? Why?

39. Prove that a sequence cannot converge to two different limits.

40. Let $\{a_n\}$ be a convergent sequence with $a_n \leq M$ for all n. Then show that $\lim_{n\to\infty} a_n \leq M$. Is it possible to have $a_n < M$ for all n and yet have $\lim_{n\to\infty} a_n = M$?

41. Does the existence of $\lim_{n\to\infty} |a_n|$ imply the existence of the $\lim_{n\to\infty} a_n$? Prove your assertion.

42. The sides of an equilateral triangle are bisected and the midpoints are connected to form a new equilateral triangle contained in the original triangle. If the perimeter of the original triangle is P and x_n is the perimeter of the nth triangle constructed in this process, write an expression for x_n in terms of n and P. Does the sequence converge?

11.2 Infinite Series

The process of addition enables us to find the sum of a *finite* number of terms. Under certain conditions it is desirable to extend this process to the sum of *infinitely many* terms.

Example 1 In Example 1 of the preceding section we considered what would happen if \$1,000,000 is introduced into a local economy in which each person or business spends 80% of its income locally. The result was that

$$\{a_n\} = \{(.80)^{n-1}(1{,}000{,}000)\}$$

is an infinite sequence in which a_n is the amount that changes hands during the nth transaction.

A concise description of the total effect of the \$1,000,000 might be achieved by determining the total amount spent due to the initial introduction of the \$1,000,000. To find the total sum spent we would have to add all the terms of the infinite sequence. Thus we would need to find

$$1{,}000{,}000 + (.80)1{,}000{,}000 + (.80)^2 1{,}000{,}000 + \cdots + (.80)^{n-1}1{,}000{,}000 + \cdots.$$

After some preliminary considerations we shall find that sum. ||

Definition 11.2.1 The indicated sum

$$\sum_{i=1}^{\infty} a_i = a_1 + a_2 + a_3 + a_4 + \cdots + a_n + \cdots$$

of the terms of an infinite sequence $\{a_1, a_2, a_3, \ldots\}$ is called an *infinite series.*

Example 2 The indicated sum of the terms of $\{1, 1/2, 1/4, 1/8, \ldots\}$ is the infinite series

$$1 + \frac{1}{2} + \frac{1}{4} + \frac{1}{8} + \cdots.$$ ||

We have not yet defined the sum of infinitely many terms. However, each of the following sums has only a finite number of terms and is therefore meaningful:

$$s_1 = a_1$$

$$s_2 = a_1 + a_2$$

$$s_3 = a_1 + a_2 + a_3$$

$$s_4 = a_1 + a_2 + a_3 + a_4$$

$$s_5 = a_1 + a_2 + a_3 + a_4 + a_5$$

. . . .

$$s_n = a_1 + a_2 + a_3 + \cdots + a_n$$

. . . .

Those sums are called the *partial sums* of the infinite series $\sum_{i=1}^{\infty} a_i$; and in particular, s_1 is the first partial sum, s_2 is the second partial sum, . . . , and s_n is the nth *partial sum.* The infinite sequence

$$\{s_n\} = \{s_1, s_2, s_3, \ldots\}$$

of partial sums of $\sum_{i=1}^{\infty} a_i$ is used to give a meaning to $\sum_{i=1}^{\infty} a_i$ in certain cases, as indicated in the next definition.

Definition 11.2.2

The infinite series

$$\sum_{i=1}^{\infty} a_i = a_1 + a_2 + a_3 + \cdots$$

is called *convergent* or *divergent* accordingly, as its sequence $\{s_n\}$ of partial sums is convergent or divergent. When $\{s_n\}$ converges, its limit S is called the *sum* of the infinite series $\sum_{i=1}^{\infty} a_i$, and we shall often say that the series converges to S.

Example 3 | Consider the series $\sum_{i=0}^{\infty}(1/2^i) = 1 + 1/2 + 1/4 + 1/8 + \cdots$.

Then

$$s_n = 1 + \frac{1}{2} + \left(\frac{1}{2}\right)^2 + \left(\frac{1}{2}\right)^3 + \cdots + \left(\frac{1}{2}\right)^{n-1}.$$

Now

$$\frac{1}{2}s_n = \frac{1}{2} + \left(\frac{1}{2}\right)^2 + \left(\frac{1}{2}\right)^3 + \cdots + \left(\frac{1}{2}\right)^{n-1} + \left(\frac{1}{2}\right)^n,$$

and so

$$s_n - \frac{1}{2}s_n = 1 - \left(\frac{1}{2}\right)^n = 1 - \frac{1}{2^n}.$$

But also

$$s_n - \frac{1}{2}s_n = \frac{1}{2}s_n,$$

and hence

$$s_n = 2\left(1 - \frac{1}{2^n}\right).$$

Therefore

$$\lim_{n \to \infty} s_n = 2(1) = 2.$$

Consequently the series converges and its sum is 2. ||

In general, a series of the form

$$\sum_{i=0}^{\infty} ar^i = a + ar + ar^2 + ar^3 + \cdots,$$

where $a \neq 0$ and r are constants, is called a *geometric series.* Note that the series of Example 3 is a geometric series with $a = 1$ and $r = 1/2$.

The method of Example 3 can be extended to find the sum of certain geometric series. In the general case,

$$s_n = a + ar + ar^2 + \cdots + ar^{n-1}, \quad \text{and}$$

$$rs_n = ar + ar^2 + \cdots + ar^{n-1} + ar^n.$$

Then $$s_n - rs_n = (1 - r)s_n = a - ar^n = a(1 - r^n).$$

So $$s_n = \frac{a(1 - r^n)}{1 - r} = a\left(\frac{1}{1 - r} - \frac{r^n}{1 - r}\right), \qquad \text{if } r \neq 1,$$

and $$s_n = na, \qquad \text{if } r = 1.$$

However, if $|r| < 1$, then $\lim_{n \to \infty} r^n = 0$; and if $|r| > 1$ or $r = -1$, then that limit does not exist. If $r = 1$, then $\lim_{n \to \infty} na = \pm\infty$. Consequently, the geometric series converges and its sum is $a/(1 - r)$ if $|r| < 1$; and it diverges if $|r| \geq 1$. We state this result as the next theorem.

Theorem 11.2.1

> The geometric series $\sum_{i=0}^{\infty} ar^i$ converges and its sum is $a/(1 - r)$ if $|r| < 1$; and it diverges if $|r| \geq 1$.

Example 4 In Example 1 of this section it was noted that the infinite series

$$1{,}000{,}000 + (.80)1{,}000{,}000 + (.80)^2 1{,}000{,}000 + \cdots + (.80)^{n-1} 1{,}000{,}000 + \cdots$$

gives the total spending that results from the introduction of \$1,000,000 into an economy in which each person or business spends 80% of its income. This series is a geometric series with $a = 1{,}000{,}000$ and $r = .80$. Consequently the series converges and its sum is

$$\frac{1{,}000{,}000}{1 - .8} = 5{,}000{,}000.$$

||

Thus the original \$1,000,000 ultimately has the effect of causing an increased total gross income of \$5,000,000 to the people of the locality. This is an example of the *multiplier effect* of spending. In this case the multiplying factor is 5.

Example 5 The series $\sum_{i=0}^{\infty} 2(-4/3)^i$ is a geometric series with $a = 2$ and $r = -4/3$. Since $|r| > 1$, it diverges. ||

Now notice that for the series

$$\sum_{i=1}^{\infty} a_i = a_1 + a_2 + a_3 + \cdots,$$

we have $$s_n - s_{n-1} = a_n.$$

Thus, if the series converges and its sum is A, we have

$$\lim_{n\to\infty} a_n = \lim_{n\to\infty} (s_n - s_{n-1}) = \lim_{n\to\infty} s_n - \lim_{n\to\infty} s_{n-1} = A - A = 0.$$

We have therefore proved that if a series converges, then its nth term must approach 0 as n increases without bound. Consequently, if the nth term does not approach zero, then the series diverges. This test for divergence is important enough to be stated as the next theorem, which we call The Divergence Test. For convenience we shall often write

$$\sum \quad \text{instead of} \quad \sum_{i=1}^{\infty}.$$

Theorem 11.2.2

The Divergence Test

If $\lim_{n\to\infty} a_n$ does not exist or exists and is not 0, then the series Σa_i diverges.

Example 6

The series $\Sigma(i/i + 1) = 1/2 + 2/3 + 3/4 + \cdots$ diverges because

$$\lim_{n\to\infty} a_n = \lim_{n\to\infty} \frac{n}{n+1} = \lim_{n\to\infty} \frac{1}{1 + 1/n} = 1 \neq 0. \qquad \|$$

It is very important to realize that Theorem 11.2.2 *does not say* that if $\lim_{n\to\infty} a_n = 0$, then the series converges. In fact when $\lim_{n\to\infty} a_n = 0$, the series may either converge or diverge. Consider, for example, the following series:

$$\sum_{i=1}^{\infty} \frac{1}{\sqrt{i}} = 1 + \frac{1}{\sqrt{2}} + \frac{1}{\sqrt{3}} + \frac{1}{\sqrt{4}} + \cdots.$$

Here $a_n = 1/\sqrt{n}$ and $\lim_{n\to\infty} a_n = \lim_{n\to\infty} 1/\sqrt{n} = 0$, but the series does not converge! We shall prove that this series diverges by showing that its sequence of partial sums has no upper bound.

In this case

$$s_n = 1 + \frac{1}{\sqrt{2}} + \frac{1}{\sqrt{3}} + \cdots + \frac{1}{\sqrt{n}}.$$

Then, since

$$1 \geq \frac{1}{\sqrt{n}}, \quad \frac{1}{\sqrt{2}} \geq \frac{1}{\sqrt{n}}, \quad \frac{1}{\sqrt{3}} \geq \frac{1}{\sqrt{n}}, \ldots, \frac{1}{\sqrt{n}} \geq \frac{1}{\sqrt{n}},$$

we have

$$s_n \geq \frac{1}{\sqrt{n}} + \frac{1}{\sqrt{n}} + \cdots + \frac{1}{\sqrt{n}} = \frac{n}{\sqrt{n}} = \sqrt{n}.$$

From this we see that $\{s_n\}$ is unbounded. Consequently, by Theorem 11.1.2, the sequence $\{s_n\}$ of partial sums of the series diverges, and therefore the series diverges. In fact, in this case, $\lim_{n\to\infty} s_n = \infty$.

We emphasize again that the relationship $\lim_{n\to\infty} a_n = 0$ provides us with no information on convergence or divergence of the infinite series Σa_i. However if $\lim_{n\to\infty} a_n \neq 0$, we know that Σa_i diverges.

Another important thing to notice about an infinite series is that if a *finite* number of its terms are changed or removed, then its convergence or divergence is not affected. Of course, if it converges and a finite number of terms are changed or removed, its sum is probably affected.

Example 7 We saw in Example 2 that

$$1 + \frac{1}{2} + \frac{1}{4} + \frac{1}{8} + \frac{1}{16} + \cdots$$

converges and its sum is 2. If we remove the first 3 terms $1 + 1/2 + 1/4$, the series still converges but its sum is $2 - (1 + 1/2 + 1/4) = 1/4$. ||

In general, suppose that we remove the first m terms of a series

$$a_1 + a_2 + a_3 + \cdots + a_m + a_{m+1} + \cdots$$

that converges to A. Then, since

$$a_1 + a_2 + a_3 + \cdots + a_m = H$$

is an ordinary finite sum of m numbers, it can be shown that the new infinite series

$$a_{m+1} + a_{m+2} + a_{m+3} + \cdots$$

converges to $A - H$. This, of course, is true even if m is large. The m might be a million, or a billion, or any number, large or small.

Similarly, if every term of an infinite series is multiplied by a nonzero constant k, then its convergence or divergence is not affected. To show this, consider

$$\sum_{i=1}^{\infty} a_i \quad \text{and} \quad \sum_{i=1}^{\infty} ka_i, \quad k \neq 0.$$

Let

$$s_n = \sum_{i=1}^{n} a_i \quad \text{and} \quad S_n = \sum_{i=1}^{n} ka_i.$$

Then $S_n = ks_n$. So $\lim_{n\to\infty} S_n$ exists if and only if $\lim_{n\to\infty} s_n$ exists. That is, Σa_i converges if and only if Σka_i converges. In addition, we note that when $\lim_{n\to\infty} s_n$ exists,

$$\lim_{n\to\infty} S_n = \lim_{n\to\infty} ks_n = k \cdot \lim_{n\to\infty} s_n.$$

Thus, when Σa_i converges to a number S, it follows that Σka_i converges to kS.

Since this section contains many of the fundamental properties of series that are necessary for future developments, we summarize the major points.

1. A series is said to converge if and only if its sequence of partial sums converges.
2. The geometric series $\sum_{i=0}^{\infty} ar^i$ converges to the sum $a/(1-r)$ if $|r| < 1$, and diverges if $|r| \geq 1$.
3. The Divergence Test: If $\lim_{n\to\infty} a_n$ does not exist or $\lim_{n\to\infty} a_n \neq 0$, then Σa_n diverges.
4. Even though $\lim_{n\to\infty} (1/\sqrt{n}) = 0$, $\sum_{i=1}^{\infty} 1/\sqrt{i}$ diverges. Therefore we must be careful to understand what The Divergence Test actually says.
5. Any finite number of terms may be changed or removed from a series without affecting the convergence or divergence of the series. However, the sum of the series may change when this is done.
6. If every term of a series is multiplied by a nonzero constant, its convergence or divergence is not affected. However, the sum of the series may change if this is done.

Exercises 11.2

Decide whether each of the series in Exercises 1–18 converges or diverges. Find the sum of each convergent series.

1. $1 + 1 + 1 + 1 + 1 + \cdots.$

2. $3 - 3 + 3 - 3 + 3 - 3 + \cdots.$

3. $5 + \frac{5}{3} + \frac{5}{9} + \frac{5}{27} + \cdots.$

4. $2 + \frac{2}{5} + \frac{2}{25} + \frac{2}{125} + \cdots.$

5. $1 + \frac{6}{5} + \left(\frac{6}{5}\right)^2 + \left(\frac{6}{5}\right)^3 + \cdots.$

6. $\sum_{i=0}^{\infty} \left(\frac{5}{4}\right)^i.$

7. $\sum_{i=0}^{\infty} (-1)^i \left(\frac{1}{3}\right)^i.$

8. $1 - \frac{1}{2} + \frac{1}{4} - \frac{1}{8} + \cdots.$

9. $\sum_{i=0}^{\infty} \frac{1}{11}\left(-\frac{9}{7}\right)^i.$

10. $\frac{1}{5} + \frac{2}{7} + \frac{3}{9} + \frac{4}{11} + \cdots.$

11. $\frac{1}{3} + \frac{2}{7} + \frac{3}{11} + \frac{4}{15} + \frac{5}{19} + \cdots.$

12. $\frac{1}{\sqrt{300}} + \frac{1}{\sqrt{301}} + \frac{1}{\sqrt{302}} + \cdots.$

13. $\sum_{i=1}^{\infty} (-1)^{i+1}.$

14. $\sum_{i=1}^{\infty} [1 + (-1)^i].$

15. $r^2 + r^4 + r^6 + \cdots,$ r any real number.

16. $\frac{1}{1+b^2} + \frac{1}{(1+b^2)^2} + \frac{1}{(1+b^2)^3} + \cdots,$ b any real number different from 0.

17. $\sum_{i=0}^{\infty} (2x)^i, \quad |x| < \frac{1}{2}.$

18. $\sum_{i=0}^{\infty} (-x)^i, \quad |x| < 1.$

For each of the Exercises 19–22, the nth partial sum (not the nth term) of a series is given. Decide whether each series is convergent or divergent and find the sum of each convergent one.

19. $\dfrac{n+2}{n-1}$.
20. $\dfrac{2+n^2}{3n^2-7}$.
21. $\dfrac{(-1)^n}{2n}$.
22. $3+(-1)^n$.

23. Find an expression for the nth partial sum of the series.

$$\sum_{i=1}^{\infty}\left(\frac{1}{i}-\frac{1}{i+1}\right),$$

and then decide whether the series converges or diverges. (*Hint*: Do not combine the fractions $1/i$ and $1/(i+1)$ over a common denominator.) Why is this series called a telescopic series?

24. Find an expression for the nth partial sum of the series

$$\sum_{i=2}^{\infty}\ln\left(\frac{i}{i-1}\right),$$

and then decide if the series converges or diverges.

25. A ball is dropped from a height of 6 feet and always rebounds .7 of the height it falls. Find a series that describes the total distance travelled by the ball, and also find that total distance.

26. The ball described in Exercise 25 will bounce infinitely often, but will it bounce forever? If not, how long will it bounce? (*Hint*: The time required for the ball to rise or fall a distance h is given by $\sqrt{2h/g}$, where g is the gravitation constant 32.2 ft/sec^2.)

27. For every dollar loaned by a bank, R cents is ultimately redeposited in that bank. Assuming that the bank must keep at least m dollars in deposits and has $M > m$ dollars in deposits, what is the maximum the bank may ultimately loan? (Assume $R < 100$.)

28. In a certain population of N bacteria, 3/4 of each generation produces a single offspring, but the remainder produces none. Assuming no deaths, find the final population.

29. Two people play a game in which they successively flip a coin. The first person to flip a head is the winner. Probability Theory indicates that the first person to flip has a probability of winning equal to

$$\frac{1}{2}+\frac{1}{8}+\frac{1}{32}+\cdots+\frac{1}{2^{2n-1}}+\cdots.$$

Find the sum of this series.

30. Suppose that the series Σa_n converges to the sum A and suppose that

$$a_1+a_2+\cdots+a_m=H.$$

Prove that the series

$$a_{m+1}+a_{m+2}+a_{m+3}+\cdots$$

converges to $A-H$.

31. Prove that if Σa_n and Σb_n both converge, then so does the series $\Sigma(a_n+b_n)$. What is the sum of the new series? Can anything be concluded if both series diverge?

32. Show that a series of the type

$$(a_1+a_2)+(a_3+a_4)+(a_5+a_6)+\cdots$$

may converge, though the series

$$a_1+a_2+a_3+\cdots$$

diverges. This indicates that it is not always possible to "group" terms of a series.

33. Prove that if the series Σa_i converges, then so does the series $\Sigma(a_i + a_{i+1})$. Compare the sums of these two series.

11.3 Series of Positive Terms

When every term of a series is positive we call the series a *positive series*. Let Σp_n be a positive series. We can apply Theorem 11.1.1 to the sequence $\{s_n\}$ of partial sums of this series because $s_{i+1} = s_i + p_{i+1} > s_i$ for all i. Thus, the series converges if $\{s_n\}$ has an upper bound and its sum is the least upper bound of $\{s_n\}$. On the other hand, if $\{s_n\}$ does not have an upper bound, then the series diverges by Theorem 11.1.2. Consequently the series converges if and only if $\{s_n\}$ has an upper bound, and when it converges, it converges to the least upper bound of $\{s_n\}$. Hence we have proved the following theorem.

Theorem 11.3.1

A positive series converges if and only if its sequence of partial sums $\{s_n\}$ has an upper bound. When it does converge, its sum is the least upper bound of $\{s_n\}$.

We shall use that theorem to develop some tests for convergence and divergence of positive series.

Suppose that Σp_n is a positive series that is to be tested for convergence or divergence. Suppose also that Σa_n is a convergent series with $a_i \geq p_i$ for all i. Then if $\{t_n\}$ and $\{s_n\}$ are the sequence of partial sums of Σa_n and Σp_n, respectively, we have $t_n \geq s_n$ for all n. Since the series Σa_n converges, its sequence of partial sums, $\{t_n\}$, is bounded above. Thus the sequence of partial sums $\{s_n\}$ is also bounded above. Consequently by Theorem 11.3.1, the series Σp_n converges. We have thus proved the first part of the following theorem, called The Comparison Test. The remaining portion of the theorem can be established in a similar fashion.

Theorem 11.3.2

The Comparison Test

Let Σp_n be a positive series to be tested for convergence or divergence.

(i) If there exists a convergent positive series Σa_n such that $p_i \leq a_i$ for all i, then Σp_n converges.

(ii) If there exists a divergent positive series Σb_n such that $b_i \leq p_i$ for all i, then Σp_n diverges.

Example 1

Test the series $\displaystyle\sum_{n=1}^{\infty} \frac{n^{3/2} + 1}{n^2}$ for convergence.

Note that for all n we have

$$\frac{n^{3/2}+1}{n^2} = \frac{n^{3/2}}{n^2} + \frac{1}{n^2} = \frac{1}{\sqrt{n}} + \frac{1}{n^2} \geq \frac{1}{\sqrt{n}}.$$

Then, since we know that the series $\Sigma 1/\sqrt{n}$ diverges, we can conclude that the given series diverges. ||

Example 2 Test the following series for convergence:

$$\sum_{n=1}^{\infty} \frac{1}{n2^n} = \frac{1}{2} + \frac{1}{2 \cdot 2^2} + \frac{1}{3 \cdot 2^3} + \frac{1}{4 \cdot 2^4} + \cdots.$$

Since for all n,

$$\frac{1}{n2^n} \leq \frac{1}{2^n},$$

and since we know that the positive series $\Sigma 1/2^n$ converges (because it is a geometric series with $|r| < 1$), we can conclude that the given series converges. ||

It is important to realize that Theorem 11.3.2 provides no information about the convergence or divergence of the positive series Σp_n if $p_i \leq b_i$ for all i and Σb_i diverges, or if $a_i \leq p_i$ for all i and Σa_i converges.

We next obtain a powerful convergence test involving integration.

Theorem 11.3.3

The Integral Test

Let Σp_n be a positive series to be tested for convergence or divergence. Suppose there is a decreasing function $f(x)$ that is positive and continuous for all x such that $x \geq 1$, and suppose that $f(i) = p_i$ for all positive integers i. Then the series Σp_n converges or diverges according as the improper integral $\int_1^{\infty} f(x)dx$ converges or diverges. When the integral converges with value L, the series converges to a sum that is less than or equal to $L + p_1$.

Proof. Figure 11.3.1 illustrates the relationship between the function $f(x)$ and the series Σp_n. In Figure 11.3.2 additional lines have

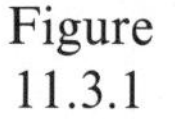
Figure 11.3.1

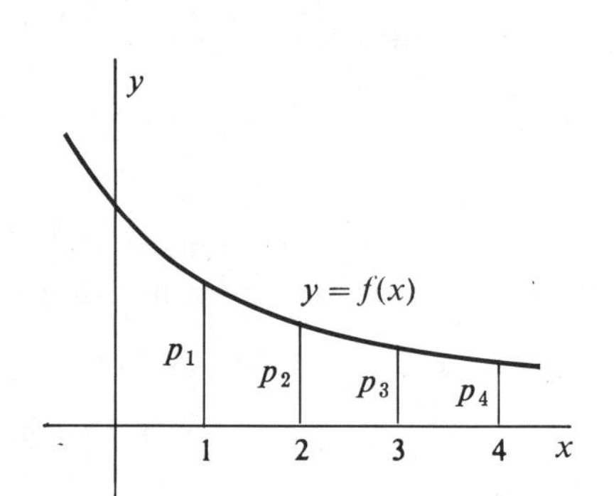

Figure 11.3.2

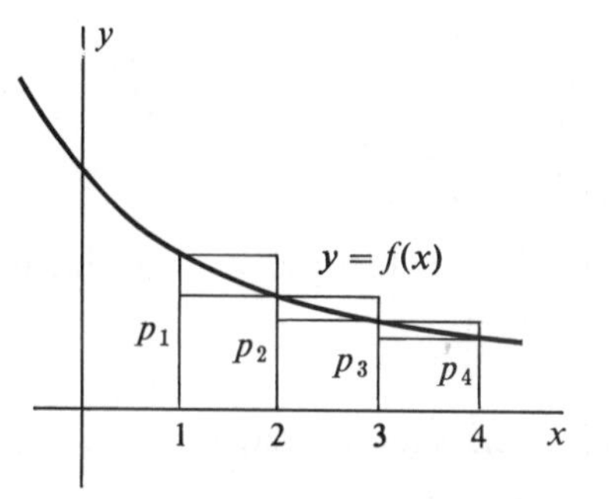

been drawn to form two rectangles over each of the indicated unit intervals. Referring to Figure 11.3.2, note that the largest rectangle above the first interval has area p_1, the largest rectangle above the second interval has area p_2, and so forth; while the smaller rectangles have areas p_2, p_3, and so forth. So we see from the areas involved that the partial sum s_n of Σp_n is related to the integral from 1 to n as follows:

$$p_2 + p_3 + \cdots + p_n \leq \int_1^n f(x)dx \leq p_1 + p_2 + p_3 + \cdots + p_{n-1},$$

or

$$s_n - p_1 \leq \int_1^n f(x)dx \leq s_{n-1}.$$

There are two cases to consider:

| *Case 1.* $\int_1^\infty f(x)dx$ converges.

| *Case 2.* $\int_1^\infty f(x)dx$ diverges.

Since in Case 1, $\int_1^\infty f(x)dx$ converges, there is a number L such that $\lim_{n\to\infty} \int_1^n f(x)dx = L$. Since $f(x) \geq 0$ for all x, $\int_1^n f(x)dx \leq L$. Consequently $L + p_1$ is an upper bound for $\{s_n\}$, and so Σp_n converges to a sum less than or equal to $L + p_1$. In Case 2, the integral diverges, and so must increase without bound because $f(x)$ is positive for all $x \geq 1$. But since

$$\int_1^n f(x)dx \leq s_{n-1},$$

we must also have $\lim_{n\to\infty} s_n = \infty$; and so the series Σp_n diverges.

Example 3 Test the following series for convergence:

$$\sum_{n=1}^{\infty} \frac{1}{n^2} = 1 + \frac{1}{4} + \frac{1}{9} + \frac{1}{16} + \cdots.$$

The function $f(x) = x^{-2}$ is positive, decreasing, and continuous for all $x \geq 1$, and $f(i) = 1/i^2 = p_i$ for all positive integers i. So we can apply the integral test using that function. We obtain

$$\lim_{b\to\infty} \int_1^b x^{-2}\,dx = \lim_{b\to\infty} (-x^{-1})\Big|_1^b = \lim_{b\to\infty} \left(-\frac{1}{b} + 1\right) = 1.$$

Since the integral converges to 1, the series converges; and its sum is less than or equal to $1 + p_1 = 1 + 1 = 2$. ||

Example 4 Test the *harmonic series* for convergence:

$$\sum_{n=1}^{\infty} \frac{1}{n} = 1 + \frac{1}{2} + \frac{1}{3} + \frac{1}{4} + \cdots.$$

The function $f(x) = x^{-1}$ is a positive, decreasing, continuous function with $f(i) = 1/i = p_i$. Then applying the integral test we get

$$\lim_{b\to\infty} \int_1^b x^{-1}\,dx = \lim_{b\to\infty} (\ln|b| - \ln|1|) = \lim_{b\to\infty} \ln b = \infty.$$

Therefore the harmonic series diverges. ||

Examples 3 and 4 are particular cases of series of the form

$$\sum_{n=1}^{\infty} \frac{1}{n^k} = 1 + \frac{1}{2^k} + \frac{1}{3^k} + \frac{1}{4^k} + \cdots,$$

where k is a real number. Such a series is often called a *k-series.* We consider three cases: $k = 0$, $k < 0$, and $k > 0$.

In case $k = 0$, the series is $1 + 1 + 1 + \cdots$. Thus $s_n = n$, and so the series diverges.

In case $k < 0$, let $k = -p$ where $p > 0$. Then the series is

$$\sum_{n=1}^{\infty} \frac{1}{n^k} = \sum_{n=1}^{\infty} n^p = 1 + 2^p + 3^p + \cdots.$$

Since $n^p \geq 1$ for all n, $\lim_{n\to\infty} a_n = \lim_{n\to\infty} n^p \neq 0$. Therefore by The Divergence Test, Theorem 11.2.2, the series diverges for all $k < 0$.

In case $k > 0$, the function $f(x) = 1/x^k$ is positive, decreasing, and continuous for all $x \geq 1$; so we may apply the integral test. For $k \neq 1$, we get

$$\int_1^{\infty} x^{-k}\,dx = \lim_{b\to\infty} \int_1^b x^{-k}\,dx = \lim_{b\to\infty} \left.\frac{x^{1-k}}{1-k}\right|_1^b = \lim_{b\to\infty} \left(\frac{b^{1-k}}{1-k} - \frac{1}{1-k}\right).$$

Now, if $k > 1$, then

$$\lim_{b\to\infty} b^{1-k} = \lim_{b\to\infty} \frac{1}{b^{k-1}} = 0, \quad \text{and so}$$

$$\int_1^{\infty} x^k\,dx = \frac{-1}{1-k} = \frac{1}{k-1}.$$

Therefore the series converges, and

$$\sum_{n=1}^{\infty} \frac{1}{n^k} \leq \frac{1}{k-1} + 1 = \frac{k}{k-1}.$$

If $0 < k < 1$, then $\lim_{b \to \infty} b^{1-k} = \infty$; so the integral and the series diverge. Finally, if $k = 1$, we have the divergent harmonic series considered in Example 4. We summarize these results in the next theorem.

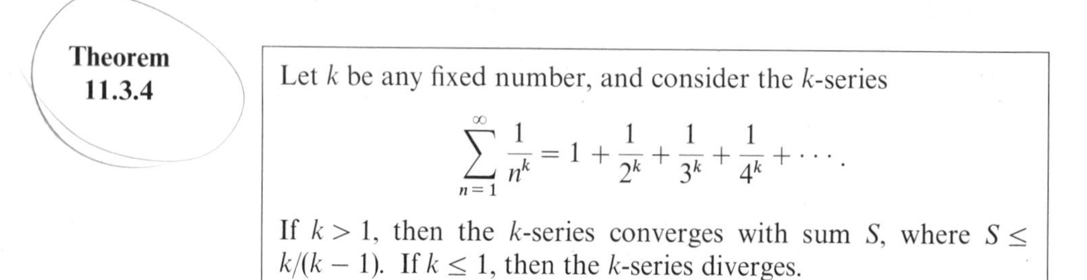

Theorem 11.3.4

Let k be any fixed number, and consider the k-series

$$\sum_{n=1}^{\infty} \frac{1}{n^k} = 1 + \frac{1}{2^k} + \frac{1}{3^k} + \frac{1}{4^k} + \cdots.$$

If $k > 1$, then the k-series converges with sum S, where $S \leq k/(k-1)$. If $k \leq 1$, then the k-series diverges.

Exercises 11.3

For Exercises 1–35 decide whether each series converges or diverges, and give a reason why.

1. $1 + \frac{1}{\sqrt[3]{2}} + \frac{1}{\sqrt[3]{3}} + \frac{1}{\sqrt[3]{4}} + \cdots.$

2. $\frac{1}{\sqrt{2}} + \frac{1}{\sqrt{3}} + \frac{1}{\sqrt{4}} + \frac{1}{\sqrt{5}} + \cdots.$

3. $\frac{1}{2^2} + \frac{1}{3^2} + \frac{1}{4^2} + \frac{1}{5^2} + \cdots.$

4. $1 + \frac{1}{2\sqrt{2}} + \frac{1}{3\sqrt{3}} + \frac{1}{4\sqrt{4}} + \cdots.$

5. $\frac{1}{5} + \frac{1}{10} + \frac{1}{17} + \cdots + \frac{1}{n^2+1} + \cdots.$

6. $\sum_{n=1}^{\infty} \frac{1}{n\sqrt{n+1}}.$

7. $\frac{1}{101} + \frac{1}{201} + \frac{1}{301} + \frac{1}{401} + \cdots.$

8. $\sum_{n=1}^{\infty} \frac{1}{1000n+5}.$

9. $\frac{1}{2} + \frac{1}{5} + \frac{1}{8} + \frac{1}{11} + \cdots.$

10. $1 + \frac{1}{3} + \frac{1}{5} + \frac{1}{7} + \frac{1}{9} + \cdots.$

11. $1 + \frac{1}{3} + \frac{1}{6} + \frac{1}{9} + \frac{1}{12} + \cdots.$

12. $1 + \frac{1}{5} + \frac{1}{10} + \frac{1}{15} + \cdots.$

13. $\frac{2}{3} + \frac{3}{13} + \frac{4}{23} + \frac{5}{33} + \cdots.$

14. $\frac{1}{6} + \frac{2}{11} + \frac{3}{16} + \frac{4}{21} + \cdots.$

15. $\sum_{k=0}^{\infty} \left(\frac{13}{15}\right)^k.$

16. $\sum_{n=1}^{\infty} \frac{1}{\sqrt{n^2+3n}}.$

17. $\sum_{n=0}^{\infty} (1.0001)^n.$

18. $\sum_{n=1}^{\infty} \frac{1}{\sqrt{n^2+4}}.$

19. $\sum_{i=0}^{\infty} \frac{1}{\sqrt{i+3}}.$

20. $\sum_{i=1}^{\infty} \frac{1}{\sqrt{i+2}}.$

21. $\sum_{n=1}^{\infty} \frac{1}{\sqrt[3]{n^4+5}}.$

22. $\sum_{n=1}^{\infty} \frac{1}{\sqrt{n^3+1}}.$

23. $\sum_{n=1}^{\infty} \frac{n^2+3n}{2+5n^2}.$

24. $\sum_{n=1}^{\infty} \frac{n+1}{n+2}.$

25. $\sum_{n=4}^{\infty} \left(\frac{1}{n-3} - \frac{1}{n}\right).$

26. $\sum_{k=1}^{\infty} e^{-k}.$

27. $\sum_{n=1}^{\infty} ne^{-n^2}.$

28. $\sum_{n=2}^{\infty} \frac{1}{n(\ln n)^2}.$

29. $\sum_{n=2}^{\infty} \frac{1}{n(\ln n)}.$

30. $\sum_{k=1}^{\infty} \frac{1}{\sqrt{k}+100}$.

31. $\sum_{k=1}^{\infty} \frac{k}{2k^2-1}$.

32. $\sum_{k=1}^{\infty} \frac{k+5}{k^3+3k-2}$.

33. $\sum_{n=1}^{\infty} \frac{1}{n^3+n^2}$.

34. $\sum_{n=1}^{\infty} ne^{-n}$.

35. $\sum_{n=1}^{\infty} \frac{1}{n+n^2}$.

36. Prove the following test.
Cauchy's Root Test: Let Σp_n be a positive series. Then
(1) If there is an integer N such that $\sqrt[n]{p_n} \le r < 1$ for all $n > N$, the series converges; and
(2) If $\sqrt[n]{p_n} \ge 1$ for infinitely many n, the series diverges.

37. Let Σp_n be a positive series, and let $f(x)$ be a function that is continuous and decreasing for all $x \ge N$. Moreover, assume that $f(n) = p_n$ for all $n \ge N$. Show that the error made in replacing the series by the sum $\sum_{n=1}^{N} p_n$ is less than $\int_N^\infty f(x)dx$.

38. Based on the results of Exercise 37, determine the smallest number of terms of the series $\sum_{n=1}^{\infty} 1/n^2$ that can be added to determine the sum with an error of less than 10^{-3}.

39. Prove that $\sum_{k=2}^{\infty} 1/(k(\ln k)^p)$ converges if $p > 1$ and diverges if $p \le 1$.

40. Prove part (ii) of Theorem 11.3.2.

11.4 The Limit Form of The Comparison Test and The Ratio Test

In this section we shall first obtain another test, which is called *The Limit Form of The Comparison Test.* The Limit Form of The Comparison Test is often easier to apply than The Comparison Test of the preceding section.

Theorem 11.4.1

The Limit Form of The Comparison Test

Let Σp_n be a positive series to be tested for convergence or divergence.

(i) If there exists a convergent positive series Σb_n such that $\lim_{n\to\infty} (p_n/b_n)$ exists, then Σp_n converges.

(ii) If there exists a divergent positive series Σa_n such that $\lim_{n\to\infty} (a_n/p_n)$ exists, then Σp_n diverges.

Proof. (i) The statement that $\lim_{n\to\infty} (p_n/b_n)$ exists is equivalent to the statement that the sequence $\{p_n/b_n\}$ converges. But then by Theorem 11.1.3, the sequence must have an upper bound B (which must be positive since every p_n/b_n is positive), that is, $p_n/b_n \le B$ for all n, or $p_n \le Bb_n$ for all n. But Σb_n converges, and hence ΣBb_n is a convergent positive series. Therefore, by The Comparsion Test, Σp_n converges.

(ii) As in the proof of part (i), since $\lim_{n\to\infty} (a_n/p_n)$ exists, the sequence $\{a_n/p_n\}$ must have an upper bound H (which must be positive since every a_n/p_n is positive), that is, $a_n/p_n \le H$ for all n, or $a_n \le Hp_n$ for all n. But Σa_n is a divergent positive series. Therefore, by The

Comparison Test, ΣHp_n diverges, and so Σp_n diverges. This completes the proof.

When testing a positive series Σp_n, where p_n is a rational function of n and the degree of the denominator of p_n minus the degree of the numerator of p_n is k, the k-series $\Sigma 1/n^k$ should be used.

Example 1 | Test this series for convergence: $1 + 1/3 + 1/5 + 1/7 + 1/9 + \cdots$.

The nth term of this series is $1/(2n - 1)$. Then since the degree of the denominator minus the degree of the numerator is 1, we will use the divergent harmonic series, $\Sigma 1/n$, for comparison. We get

$$\lim_{n\to\infty} \frac{a^n}{p^n} = \lim_{n\to\infty} \frac{1}{n} \cdot \frac{2n-1}{1} = \lim_{n\to\infty} \frac{2n-1}{n} = \lim_{n\to\infty} \left(2 - \frac{1}{n}\right) = 2.$$

Since the limit exists, the given series diverges by Theorem 11.4.1, part (ii). ||

Example 2 | Test for convergence: $\sum_{n=1}^{\infty} \frac{n+1}{n^3} = 2 + \frac{3}{8} + \frac{4}{27} + \cdots$.

Since the power of n in the denominator is 2 more than the power of n in the numerator, the series ought to be compared with the k-series $\Sigma 1/n^2$, which converges by Theorem 11.3.4. We have

$$\lim_{n\to\infty} \frac{p_n}{b_n} = \lim_{n\to\infty} \left(\frac{n+1}{n^3}\right)\left(\frac{n^2}{1}\right) = \lim_{n\to\infty} \left(1 + \frac{1}{n}\right) = 1.$$

Since the limit exists, the given series converges. ||

It is important to remember that when using The Limit Form of The Comparison Test, the terms of a known convergent, positive series always appear in the denominator and the terms of a known divergent, positive series always appear in the numerator.

Next we discuss another kind of test for which known series are not needed.

Theorem 11.4.2

The Ratio Test

Let Σp_n be a positive series to be tested for convergence or divergence, and assume that

$$\lim_{n\to\infty} \frac{p_{n+1}}{p_n} = L.$$

(i) If $L < 1$, the given series converges.
(ii) If $L > 1$ or $L = \infty$, the given series diverges.
(iii) If $L = 1$, the test fails to determine whether the series converges or diverges.

Proof.

(i) $L < 1$. Let r be a number such that $L < r < 1$. Since $\lim_{n\to\infty} (p_{n+1}/p_n) = L$, p_{n+1}/p_n can be made arbitrarily close to L if n is sufficiently large. In particular, $p_{n+1}/p_n < r$ if n is greater than or equal to some fixed number N. Therefore

$$p_{n+1} < rp_n \qquad \text{for all } n \geq N.$$

Thus

$$p_{N+1} < rp_N$$

$$p_{N+2} < rp_{N+1} < r^2 p_N$$

$$p_{N+3} < rp_{N+2} < r^3 p_N.$$

In general, $\quad p_{N+k} < r^k p_N \quad$ for every positive integer k.

Hence every term of the series

$$p_N + p_{N+1} + p_{N+2} + p_{N+3} + \cdots \tag{1}$$

is less than or equal to the corresponding term of the series

$$p_N + rp_N + r^2 p_N + r^3 p_N + \cdots. \tag{2}$$

But series (2) is a geometric series with $|r| = r < 1$, and so it converges. Therefore series (1) converges by The Comparison Test. However series (1) is the original series with a finite number, $N - 1$, of its terms removed. But we have seen previously that removing a finite number of terms does not change the convergence or divergence of the series. Therefore the original positive series converges.

(ii) $L > 1$ or $L = \infty$. Let s be a number such that $1 < s < L$. Since $\lim_{n\to\infty} (p_{n+1}/p_n) = L$, p_{n+1}/p_n can be made arbitrarily close to L if n is sufficiently large. In particular, $p_{n+1}/p_n > s > 1$ if n is greater than or equal to some fixed number M. Therefore $p_{M+1} > p_M$, $p_{M+2} > p_{M+1}$, $p_{M+3} > p_{M+2}, \ldots$, and in general,

$$p_{M+k+1} > p_{M+k} \qquad \text{for all } k.$$

Clearly the terms beyond the Mth term are positive and increasing. Hence $\lim_{n\to\infty} p_n \neq 0$. Therefore, by The Divergence Test, the series diverges.

(iii) $L = 1$. In this case some positive series converge and some diverge, so that the test fails to distinguish convergence or divergence. To show this, consider the convergent positive series $\Sigma 1/n^2$ and the divergent positive series $\Sigma 1/n$.

For $\Sigma 1/n^2$ we have

$$\lim_{n\to\infty} \frac{p_{n+1}}{p_n} = \lim_{n\to\infty} \frac{1}{(n+1)^2} \cdot \frac{n^2}{1} = \lim_{n\to\infty} \frac{n^2}{n^2 + 2n + 1}$$

$$= \lim_{n\to\infty} \frac{1}{1 + \dfrac{2}{n} + \dfrac{1}{n^2}} = 1.$$

For $\Sigma 1/n$ we have

$$\lim_{n\to\infty}\frac{p_{n+1}}{p_n}=\lim_{n\to\infty}\frac{1}{n+1}\cdot\frac{n}{1}=\lim_{n\to\infty}\frac{n}{n+1}=\lim_{n\to\infty}\frac{1}{1+1/n}=1.$$

Clearly when $L=1$, the series may either converge or diverge.

Example 3 | Test $\Sigma 5^{n+1}/n!$ for convergence.

$$\lim_{n\to\infty}\frac{p_{n+1}}{p_n}=\lim_{n\to\infty}\frac{5^{n+2}}{(n+1)!}\cdot\frac{n!}{5^{n+1}}=\lim_{n\to\infty}\frac{5}{n+1}=0.$$

Since $\lim_{n\to\infty} p_{n+1}/p_n=0<1$, the series converges. ||

Example 4 | Test the following series for convergence:

$$1+\frac{1\cdot 3}{2!}+\frac{1\cdot 3\cdot 5}{3!}+\frac{1\cdot 3\cdot 5\cdot 7}{4!}+\cdots+\frac{1\cdot 3\cdot 5\cdots(2n-1)}{n!}+\cdots.$$

We have

$$\lim_{n\to\infty}\frac{p_{n+1}}{p_n}=\lim_{n\to\infty}\left(\frac{1\cdot 3\cdot 5\cdots(2n-1)(2n+1)}{(n+1)!}\cdot\frac{n!}{1\cdot 3\cdot 5\cdots(2n-1)}\right)$$

$$=\lim_{n\to\infty}\frac{2n+1}{n+1}=\lim_{n\to\infty}\frac{2+1/n}{1+1/n}=2>1.$$

So the series diverges. This example also illustrates the fact that it is not always wise to try to obtain a compact expression for the nth term when applying The Ratio Test. ||

Exercises 11.4

Determine whether each of the series given in Exercises 1–32 converges or diverges.

1. $\sum_{n=2}^{\infty}\frac{n!}{(n-2)!}$.
2. $\sum_{n=1}^{\infty}\frac{(n-1)!}{n!}$.
3. $\sum_{i=1}^{\infty}\frac{i!}{100^i}$.
4. $\sum_{i=1}^{\infty}\frac{10^i}{i^2}$.
5. $\sum_{n=1}^{\infty}\frac{6^{n-1}}{n5^{n+3}}$.
6. $\sum_{n=1}^{\infty}\frac{4^{n+1}}{3^n5^{n+2}}$.
7. $\sum_{n=1}^{\infty}\frac{n^3-5}{n^4+7n}$.
8. $\sum_{n=1}^{\infty}\frac{n^2+1}{n^3+3}$.
9. $\sum_{n=1}^{\infty}\frac{n^3+3n^2+1}{2n^5+4n^3+6}$.
10. $\sum_{n=0}^{\infty}\frac{n^2+3n-1}{2n^4-7n+1}$.
11. $\sum_{n=1}^{\infty}\frac{1}{n(n+2)^{1/2}}$.
12. $\sum_{n=1}^{\infty}\frac{1}{n^{1/2}(3n-2)}$.
13. $\sum_{n=1}^{\infty}\frac{n^7 6^{n+2}}{2^{3n}}$.
14. $\sum_{n=1}^{\infty}\frac{n!}{100^n}$.
15. $\sum_{n=1}^{\infty}2^{1/n}$.
16. $\sum_{n=1}^{\infty}\frac{n^2 5^n}{2^{2n}}$.
17. $\frac{2}{3\cdot 4}+\frac{3}{4\cdot 5}+\frac{4}{5\cdot 6}+\frac{5}{6\cdot 7}+\cdots$.

18. $\frac{1}{1\cdot 2}+\frac{1}{2\cdot 3}+\frac{1}{3\cdot 4}+\frac{1}{4\cdot 5}+\cdots.$

19. $\frac{1}{\sqrt{1\cdot 2\cdot 3}}+\frac{1}{\sqrt{2\cdot 3\cdot 4}}+\frac{1}{\sqrt{3\cdot 4\cdot 5}}+\cdots.$

20. $\frac{1}{\sqrt{3}}+\frac{1}{\sqrt{8}}+\frac{1}{\sqrt{15}}+\frac{1}{\sqrt{24}}+\frac{1}{\sqrt{35}}+\cdots.$

21. $\frac{2}{2\cdot 4}+\frac{3}{4\cdot 6}+\frac{4}{6\cdot 8}+\cdots.$

22. $\frac{1}{3}+\frac{1\cdot 3}{3\cdot 6}+\frac{1\cdot 3\cdot 5}{3\cdot 6\cdot 9}+\cdots.$

23. $\frac{2}{2}+\frac{2\cdot 5}{2\cdot 4}+\frac{2\cdot 5\cdot 8}{2\cdot 4\cdot 6}+\frac{2\cdot 5\cdot 8\cdot 11}{2\cdot 4\cdot 6\cdot 8}+\cdots.$

24. $\frac{5}{2}+\frac{5\cdot 9}{2\cdot 7}+\frac{5\cdot 9\cdot 13}{2\cdot 7\cdot 12}+\frac{5\cdot 9\cdot 13\cdot 17}{2\cdot 7\cdot 12\cdot 17}+\cdots.$

25. $\sum_{k=1}^{\infty}\frac{3^{k-1}}{k2^k}.$

26. $\sum_{k=1}^{\infty}\frac{2^k}{3^{k+1}}.$

27. $\sum_{n=0}^{\infty}\frac{n^5 6^n}{(n+1)!}.$

28. $\sum_{n=1}^{\infty}\frac{5^{n+1}n^2}{n!}.$

29. $\sum_{n=1}^{\infty}\frac{n^3 n!}{(3n)!}.$

30. $\sum_{n=1}^{\infty}\frac{n!}{(2n)!}.$

31. $\sum_{n=1}^{\infty}\frac{2^n}{(2n)!n^2}.$

32. $\sum_{n=1}^{\infty}\frac{n^3}{2^n(n+1)!}.$

Determine whether the series given in Exercises 33–38 converge for all the indicated values of x.

33. $\sum_{n=1}^{\infty}\frac{x^n}{n},\quad 0<x<1.$

34. $\sum_{n=1}^{\infty}\frac{2^n}{n}x^n,\quad 0<x<\frac{1}{2}.$

35. $\sum_{n=1}^{\infty}\frac{x^n}{n},\quad 0<x\le 1.$

36. $\sum_{n=1}^{\infty}2^n x^n,\quad 0<x\le\frac{1}{2}.$

37. $\sum_{n=0}^{\infty}\frac{x^n}{n!},\quad$ all $x>0.$

38. $\sum_{n=1}^{\infty}\frac{(x-5)^n}{n},\quad 5<x<6.$

39. Let Σp_n be a positive series for which $p_{n+1}/p_n<r<1$ for all $n\ge N$. Show that the error made in replacing the series by $\sum_{n=1}^{N}p_n$ is less than $rp_N/(1-r)$.

40. Use the result of Exercise 39 to show that

$$\left|\sum_{n=1}^{\infty}\frac{2^n}{n!}-\sum_{n=1}^{5}\frac{2^n}{n!}\right|<.1\overline{3}.$$

11.5 Series of Positive and Negative Terms

If an infinite series

$$\sum_{n=1}^{\infty}a_n=a_1+a_2+a_3+\cdots$$

has both positive and negative terms but no zero terms, then a positive series

$$\sum_{n=1}^{\infty}|a_n|=|a_1|+|a_2|+|a_3|+\cdots$$

can be obtained from it by taking the absolute value of each of its terms.

Definition 11.5.1

A series Σa_n is called *absolutely convergent* if the series $\Sigma|a_n|$ converges.

Theorem 11.5.1

If a series is absolutely convergent, then it is convergent.

Proof. Let Σa_n be an absolutely convergent series. Then the series $\Sigma|a_n|$ converges. Let

$$s_n = a_1 + a_2 + \cdots + a_n \quad \text{and} \quad S_n = |a_1| + |a_2| + \cdots + |a_n|.$$

Then $\{s_n\}$ is the sequence of partial sums of the series Σa_n, and $\{S_n\}$ is the sequence of partial sums of the series $\Sigma|a_n|$. We know that $\{S_n\}$ converges, and we must prove that $\{s_n\}$ converges. Let P_n be the sum of the positive terms occurring among $a_1, a_2, \ldots, a_n$; and let N_n be the sum of the absolute values of the negative terms occurring among $a_1, a_2, \ldots, a_n$. Then P_n, N_n, and S_n are positive for all n, and

$$s_n = P_n - N_n \quad \text{and} \quad S_n = P_n + N_n.$$

Let $\lim_{n\to\infty} S_n = S$. Since $\{S_n\}$ is an increasing sequence and $S_n = P_n + N_n$, it follows that for all n,

$$P_n \leq S_n \leq S \quad \text{and} \quad N_n \leq S_n \leq S.$$

That is, S is an upper bound for both $\{P_n\}$ and $\{N_n\}$. Since $P_i \leq P_{i+1}$ and $N_i \leq N_{i+1}$ for all i, and since the sequences $\{P_n\}$ and $\{N_n\}$ have upper bounds, they must converge by Theorem 11.1.1. Let

$$\lim_{n\to\infty} P_n = P \quad \text{and} \quad \lim_{n\to\infty} N_n = N.$$

Then by The Limit Theorem,

$$\lim_{n\to\infty} s_n = \lim_{n\to\infty} (P_n - N_n) = \lim_{n\to\infty} P_n - \lim_{n\to\infty} N_n = P - N.$$

Hence the series Σa_n converges, and the theorem is proved.

Example 1 We know that this series converges: $\sum \frac{1}{n^2} = 1 + \frac{1}{2^2} + \frac{1}{3^2} + \cdots$.

Therefore, by definition, the following series is absolutely convergent:

$$1 - \frac{1}{2^2} + \frac{1}{3^2} - \frac{1}{4^2} + \frac{1}{5^2} - \cdots.$$

By the last theorem, that series must also be convergent. ||

We note that now The Ratio Test (Theorem 11.4.2) can be extended to series whose terms are arbitrary, that is, not necessarily positive.

Theorem 11.5.2

The Ratio Test

Let Σa_n be a series to be tested for convergence or divergence, and assume that

$$\lim_{n\to\infty}\left|\frac{a_{n+1}}{a_n}\right| = L.$$

(i) If $L < 1$, the given series is absolutely convergent.
(ii) If $L > 1$ or $L = \infty$, the given series diverges.
(iii) If $L = 1$, the test fails to determine whether the series converges or diverges.

Proof. Let $\lim_{n\to\infty} |a_{n+1}|/|a_n| = L$, and consider the following three cases:

(i) $L < 1$. In this case

$$\lim_{n\to\infty}\frac{|a_{n+1}|}{|a_n|} < 1.$$

Thus $\Sigma|a_n|$ converges by The Ratio Test, and so the given series is absolutely convergent.

(ii) $L > 1$. In this case it can be shown (as in Theorem 11.4.2) that $\lim_{n\to\infty} |a_n| \neq 0$. But then $\lim_{n\to\infty} a_n \neq 0$; so the given series diverges by The Divergence Test.

(iii) $L = 1$. In this case the series may converge or diverge. In fact $L = 1$ for the convergent series $\Sigma 1/n^2$ and also for the divergent series $\Sigma 1/n$.

Example 2

Test $\displaystyle\sum_{n=1}^{\infty}\frac{(-1)^{n+1}n}{3^n}$ for convergence.

$$\lim_{n\to\infty}\left|\frac{a_{n+1}}{a_n}\right| = \lim_{n\to\infty}\left|\frac{n+1}{3^{n+1}}\cdot\frac{3^n}{n}\right| = \lim_{n\to\infty}\frac{1}{3}\left(\frac{n+1}{n}\right)$$

$$= \frac{1}{3}\lim_{n\to\infty}\left(1+\frac{1}{n}\right) = \frac{1}{3} < 1.$$

So the series is absolutely convergent and hence convergent. ||

Example 3

The series

$$1 - \frac{1}{2} + \frac{1}{3} - \frac{1}{4} + \frac{1}{5} - \cdots$$

is not absolutely convergent, because the series of absolute values is the harmonic series, which we know is divergent. However, we shall show that the given series is *conditionally convergent*, a concept that is defined next. ||

Definition 11.5.2

> A *conditionally convergent* series is a convergent series that is not absolutely convergent.

Figure 11.5.1 illustrates the relationships between the various definitions relating to convergence.

Figure 11.5.1

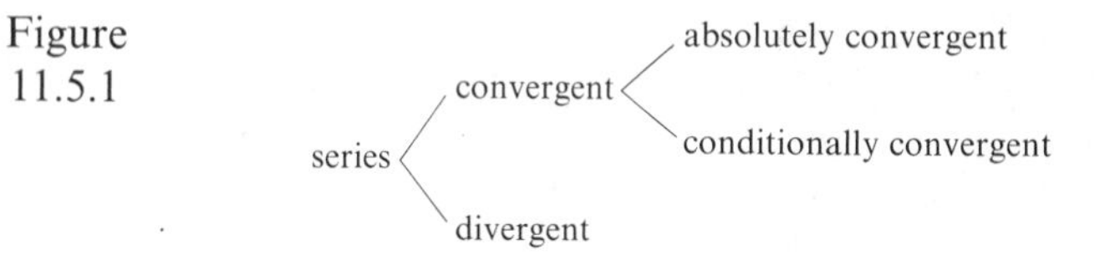

The terms of the series of Examples 2 and 3 alternate in sign; such series are called *alternating series.* There is a fairly simple test for convergence that applies to many alternating series.

Theorem 11.5.3

> *The Alternating Series Test*
>
> Consider the following alternating series in which each a_n is positive:
>
> $$a_1 - a_2 + a_3 - a_4 + \cdots.$$
>
> (i) If $\lim_{n \to \infty} a_n \neq 0$, or the limit does not exist, then the series diverges.
>
> (ii) If $a_{n+1} < a_n$ for all n and $\lim_{n \to \infty} a_n = 0$, then the series converges.

Before proving the theorem let's review what it says about alternating series.

(i) If $\lim_{n \to \infty} a_n$ differs from 0 or fails to exist then the alternating series diverges.

(ii) If $\lim_{n \to \infty} a_n = 0$ and the absolute value of each term is less than the absolute value of the preceding term, then the alternating series converges.

Now we prove the theorem.

Proof. We already know by The Divergence Test that if $\lim_{n \to \infty} a_n \neq 0$, the series diverges. Consequently we have established part (i) of the theorem. In order to establish part (ii) we need only prove that if $\lim_{n \to \infty} a_n = 0$, then the series converges.

Let s_n be the nth partial sum of the series, that is, let

$$s_n = a_1 - a_2 + a_3 - a_4 + \cdots + (-1)^{n-1} a_n.$$

Let us first consider s_n for *even* values of n. Then we can write

$$s_n = (a_1 - a_2) + (a_3 - a_4) + (a_5 - a_6) + \cdots + (a_{n-1} - a_n), \quad \text{or} \qquad (1)$$

$$s_n = a_1 - (a_2 - a_3) - (a_4 - a_5) - \cdots - (a_{n-2} - a_{n-1}) - a_n. \qquad (2)$$

Since $a_{i+1} < a_i$ for all i, each of the expressions enclosed in parentheses is positive. Thus we see from (1) that the sequence $\{s_n\}$ is increasing. From (2) we see that $s_n \le a_1$. Therefore when n is even, $\{s_n\}$ is an increasing sequence with an upper bound and therefore converges. Let

$$\lim_{\substack{n\to\infty \\ n \text{ even}}} s_n = S.$$

Now let n be *odd*. Then $n - 1$ is even and

$$s_n = s_{n-1} + a_n.$$

We have assumed that $\lim\limits_{n\to\infty} a_n = 0$, and since $n - 1$ is even,

$$\lim_{\substack{n\to\infty \\ n \text{ odd}}} s_n = \lim_{\substack{(n-1)\to\infty \\ (n-1) \text{ even}}} (s_{n-1} + a_n) = S + \lim_{\substack{n\to\infty \\ n \text{ odd}}} a_n = S + 0.$$

Therefore when n is odd,

$$\lim_{\substack{n\to\infty \\ n \text{ odd}}} s_n = S.$$

We have now completed the proof that the series converges because, when n is either even or odd, $\lim\limits_{n\to\infty} s_n = S$.

Example 4 Test the following series for convergence:

$$\sum_{n=1}^{\infty} \frac{(-1)^{n+1}}{n} = 1 - \frac{1}{2} + \frac{1}{3} - \frac{1}{4} + \frac{1}{5} - \cdots.$$

Here we have an alternating series with $a_n = 1/n$; so $a_{i+1} < a_i$ for all i, and hence The Alternating Series Test applies. Since $\lim\limits_{n\to\infty} 1/n = 0$, the series converges. As we saw in Example 3, this series is not absolutely convergent; hence it is conditionally convergent. ||

Example 5 Test the following series for convergence:

$$\sum_{n=1}^{\infty} \frac{(-1)^{n+1}(n+1)}{5n} = \frac{2}{3} - \frac{3}{10} + \frac{4}{15} - \frac{5}{20} + \frac{6}{25} - \cdots.$$

Since this is an alternating series in which the terms continually decrease in absolute value, The Alternating Series Test applies. We have

$$\lim_{n\to\infty} \frac{n+1}{5n} = \lim_{n\to\infty} \frac{1 + 1/n}{5} = \frac{1}{5} \neq 0.$$

So the series diverges. ||

The next theorem concerns the error made when a particular approximation is made to the sum of a certain kind of convergent alternating series.

Theorem 11.5.4

> Let each a_i be positive and let
>
> $$a_1 - a_2 - a_3 - a_4 + \cdots$$
>
> be a convergent alternating series with sum S. If $a_{i+1} < a_i$ for all i, then the nth partial sum has the property that
>
> $$|S - s_n| < a_{n+1}, \qquad \text{for all } n.$$

Proof. First let n be even. Then

$$\begin{aligned} S - s_n &= a_{n+1} - a_{n+2} + a_{n+3} - a_{n+4} + \cdots \\ &= (a_{n+1} - a_{n+2}) + (a_{n+3} - a_{n+4}) + \cdots. \end{aligned}$$

Thus, since $S - s_n$ is a sum of positive terms, $S - s_n$ is positive. Also

$$\begin{aligned} S - s_n &= a_{n+1} - (a_{n+2} - a_{n+3}) - (a_{n+4} - a_{n+5}) - \cdots \\ &< a_{n+1}. \end{aligned}$$

Now let n be odd. Then

$$\begin{aligned} s_n - S = a_{n+1} - a_{n+2} + \cdots &= (a_{n+1} - a_{n+2}) + (a_{n+3} - a_{n+4}) + \cdots \\ &= a_{n+1} - (a_{n+2} - a_{n+3}) - \cdots \\ &< a_{n+1}. \end{aligned}$$

Hence, since n is odd, $s_n - S$ is also positive and less than a_{n+1}. Therefore, for all n,

$$|S - s_n| < a_{n+1}.$$

Example 6 If we use the sum, s_{19}, of the first 19 terms of the series

$$1 - \frac{1}{2} + \frac{1}{3} - \frac{1}{4} + \cdots$$

to approximate its sum S, the error is less than $1/20 = .05 = 5 \times 10^{-2}$; that is, as an approximation to S, s_{19} is accurate to one decimal place. ||

Exercises 11.5

Decide whether the series in Exercises 1–16 are convergent or divergent.

1. $-1 + \frac{1}{8} - \frac{1}{15} + \frac{1}{22} - \cdots.$
2. $1 - \frac{1}{4} + \frac{1}{9} - \frac{1}{14} + \frac{1}{19} - \cdots.$
3. $\frac{2}{13} - \frac{4}{23} + \frac{6}{33} - \frac{8}{43} + \cdots.$
4. $\frac{1}{4} - \frac{3}{6} + \frac{5}{8} - \frac{7}{10} + \frac{9}{12} - \cdots.$

5. $\frac{5}{2} - \frac{10}{4} + \frac{15}{8} - \frac{20}{16} + \cdots.$

6. $\frac{1}{2} - \frac{3}{4} + \frac{5}{8} - \frac{7}{16} + \frac{9}{32} - \cdots.$

7. $-\frac{1}{5} + \frac{2}{7} - \frac{3}{9} + \frac{4}{11} - \cdots.$

8. $\frac{1}{3} - \frac{2}{4} + \frac{3}{5} - \frac{4}{6} + \cdots.$

9. $1 - \frac{1}{\sqrt[3]{2}} + \frac{1}{\sqrt[3]{3}} - \frac{1}{\sqrt[3]{4}} + \cdots.$

10. $1 - \frac{1}{\sqrt{2}} + \frac{1}{\sqrt{3}} - \frac{1}{\sqrt{4}} + \frac{1}{\sqrt{5}} - \cdots.$

11. $\sum_{n=2}^{\infty} \frac{(-1)^n}{\ln n}.$

12. $\sum_{n=1}^{\infty} (-1)^n \sin\left(\frac{1}{n}\right).$

13. $\sum_{n=1}^{\infty} (-1)^n \frac{7n+2}{n^2+1}.$

14. $\sum_{n=1}^{\infty} (-1)^n \frac{\sin(n\pi/2)}{n}.$

15. $\sum_{n=1}^{\infty} (-1)^n \frac{3\sqrt{n}}{2n+5}.$

16. $\sum_{n=1}^{\infty} (-1)\frac{n^2}{e^n}.$

In Exercises 17–20, find the sum of the convergent series to the indicated accuracy.

17. $1 - \frac{2}{4} + \frac{3}{4^2} - \frac{4}{4^3} + \frac{5}{4^4} - \cdots$; error less than $5 \cdot 10^{-2}$.

18. $1 - \frac{1}{2} + \frac{1}{3} - \frac{1}{4} + \frac{1}{5} - \cdots$; error less than 10^{-1}.

19. $1 - \frac{1}{10} + \frac{2}{10^2} - \frac{3}{10^3} + \cdots$; error less than $5 \cdot 10^{-4}$.

20. $\sum_{n=0}^{\infty} (-1)^n \frac{n}{3^n}$; error less than $5 \cdot 10^{-3}$.

Decide whether the series in Exercises 21–30 are absolutely convergent, conditionally convergent, or divergent.

21. $-1 + \frac{1}{6} - \frac{2}{8} + \frac{3}{10} - \frac{4}{12} + \cdots.$

22. $1 - \frac{1}{2} + \frac{1}{3} - \frac{1}{4} + \frac{1}{5} - \cdots.$

23. $1 - \frac{1}{\sqrt{2}} + \frac{1}{\sqrt{4}} - \frac{1}{\sqrt{6}} + \frac{1}{\sqrt{8}} - \cdots.$

24. $\sum_{n=1}^{\infty} (-1)^n \frac{n(n+1)}{2(n+2)(n+3)}.$

25. $\sum_{n=1}^{\infty} (-1)^n \frac{n^3+1}{3n^{9/2}+7n^2}.$

26. $\sum_{n=1}^{\infty} (-1)^n \frac{n+1}{n^2}.$

27. $\sum_{n=1}^{\infty} (-1)^n 7^{1/n}.$

28. $\sum_{n=2}^{\infty} (-1)^n \frac{\sqrt{n}}{n-1}.$

29. $\sum_{n=1}^{\infty} \frac{(-1)^n}{\ln(n+1)}.$

30. $\sum_{n=1}^{\infty} \frac{(-1)^n}{n} \sin\left(\frac{1}{n}\right).$

In Exercises 31–34, determine if the given series converges for the indicated values of x.

31. $\sum_{n=1}^{\infty} \frac{x^n}{n}, \quad -1 \leq x < 1.$

32. $\sum_{n=1}^{\infty} \frac{x^n}{n^2}, \quad |x| \leq 1.$

33. $\sum_{n=1}^{\infty} \frac{x^n}{n!}, \quad$ all x.

34. $\sum_{n=0}^{\infty} 2^n x^n, \quad |x| < \frac{1}{2}.$

35. Does the series $1 + 1/2 - 1/4 + 1/8 + 1/16 - 1/32 + 1/64 + 1/128 - 1/256 + \cdots$ converge? (Note that this is not an alternating series.)

36. Prove that $\lim_{n\to\infty} |a_n| = 0$ if and only if $\lim_{n\to\infty} a_n = 0$.

11.6 Testing Series

In the preceding sections we presented a variety of tests for convergence or divergence of series. However, since the exercises at the end of each of those sections concerned the tests just discussed, the reader virtually knew in advance which test or tests to apply. We shall now turn our attention to the problem of deciding which test to apply.

In the following summary we list the tests that were discussed in this chapter.

Tests for Convergence or Divergence

1. A series converges or diverges according to the existence or non-existence of the limit of the sequence of partial sums. At times a series may be tested by examining its sequence of partial sums.
2. Geometric series and k-series may be tested for convergence by reference to Theorem 11.2.1 or Theorem 11.3.4.
3. The Alternating Series Test.
4. The Divergence Test.
5. The Ratio Test.
6. The Integral Test.
7. The Comparison Test (limit form or direct comparison form).

If the series is a geometric series or a k-series, we know immediately whether it converges or diverges: the geometric series

$$\sum_{n=0}^{\infty} ar^n \quad \text{converges if } |r| < 1, \text{ and diverges if } |r| \geq 1;$$

and the k-series

$$\sum_{n=1}^{\infty} \frac{1}{n^k} \quad \text{converges if } k > 1, \text{ and diverges if } k \leq 1.$$

Clearly, we should first decide whether the series being tested is a geometric series or a k-series.

We know from the definition of convergence that if the limit of the sequence of partial sums (not the nth term) of a series exists, the series converges; but if the limit does not exist, the series diverges. When an expression for the nth partial sum is known or easily found, this test can be applied.

For an alternating series, we should be careful to verify that the alternating series test is applicable. That is, we should be sure that each term is, in absolute value, less than the absolute value of the preceding term. If the alternating series test applies, it should be used, since it definitely determines whether a series converges or diverges.

If none of the tests above can be used and the limit of the nth term can be obtained, The Divergence Test should be applied. Remember, however, that this test is only conclusive (and indicates divergence) when the limit of the nth term fails to exist or differs from 0. From only the fact that the limit of the nth term is 0, we *cannot* conclude that the series converges.

The Ratio Test is particularly suitable for series whose terms involve products, factorials, or powers. The absolute value form of the test (Theorem 11.5.2) should be used because it is just as applicable as the other form (Theorem 11.4.2), and in the case where the limit is less than 1, produces the stronger conclusion of absolute convergence rather than just convergence. Of course, The Ratio Test fails if the limits is 1, in which case another test must be used.

For positive series The Integral Test, The Limit Form of The Comparison Test, or The Comparison Test are candidates to consider. For series that involve both positive and negative terms those tests can be attempted on the positive series obtained by replacing each term by its absolute value. If the absolute value series is shown to converge, then the given series converges absolutely; but if the absolute value series diverges, then the original series is either conditionally convergent or divergent.

The Integral Test is good to use because it is always conclusive when the test can be completed. However, it must be remembered that The Integral Test applies only to positive series for which a positive, decreasing, continuous function can be found so that for every n, the function has a value equal to the nth term of the series.

The Limit Form of The Comparison Test is especially useful when the nth term of the series is in the form of a quotient in which the numerator and denominator are both sums (or differences) of powers of n. In this case the series should be compared with k-series $\Sigma 1/n^k$, where k is chosen as the degree of the denominator of the nth term minus the degree of the numerator.

If all the other tests fail, we should try to apply a direct comparison form of The Comparison Test or try to transform the terms of the series so that one of the tests applies.

It should be remembered that the convergence or divergence of a series is not changed if

(a) a *finite* number of its terms are changed or removed, or

(b) each of its terms is multiplied by the same nonzero constant.

Now let's look at some series and decide which tests to use. We shall also consider each alternating series for absolute convergence.

Example 1 $$\sum \frac{10n^3 + 7n^2}{n^{9/2} + 3}.$$

Use The Limit Form of The Comparison Test, and compare with the convergent k-series $\Sigma 1/n^{3/2}$. ||

Example 2 $$1 - \frac{1}{\sqrt{3}} + \frac{1}{\sqrt{5}} - \frac{1}{\sqrt{7}} + \cdots.$$

The Alternating Series Test applies. We see that the pattern indicates that the nth term is $1/\sqrt{2n-1}$. For the absolute-value series use The Limit Form of The Comparison Test with a k-series, where $k = 1/2$. ||

Example 3 | Σne^{-n^2}.

Use The Integral Test. ||

Example 4 | $\dfrac{5}{1^2} - \dfrac{5^2}{2^2} + \dfrac{5^3}{3^2} - \dfrac{5^4}{4^2} + \cdots$.

One might try The Divergence Test, but if the limit is difficult to find, The Ratio Test should be used. ||

Example 5 | $\dfrac{2}{21} - \dfrac{3}{28} + \dfrac{4}{35} - \dfrac{5}{42} + \cdots$.

The Alternating Series Test does not apply since it is not true that each term is less in absolute value than the absolute value of the preceding term. Use The Divergence Test. ||

Example 6 | $1 - \dfrac{1 \cdot 2}{1 \cdot 3} + \dfrac{1 \cdot 2 \cdot 3}{1 \cdot 3 \cdot 5} - \dfrac{1 \cdot 2 \cdot 3 \cdot 4}{1 \cdot 3 \cdot 5 \cdot 7} + \cdots$.

Use The Ratio Test. ||

Example 7 | $\displaystyle\sum_{n=1}^{\infty} \left(\frac{1}{n+1} - \frac{1}{n}\right)$.

The nth partial sum s_n can be obtained by writing the first few terms and the nth term:

$$s_n = \left(\frac{1}{2} - \frac{1}{1}\right) + \left(\frac{1}{3} - \frac{1}{2}\right) + \left(\frac{1}{4} - \frac{1}{3}\right) + \cdots + \left(\frac{1}{n+1} - \frac{1}{n}\right),$$

and then noticing that we have $s_n = [1/(n+1)] - 1$ as a result of a collapsing. The definition of convergence should then be used. ||

Example 8 | $3 + 2 + \dfrac{4}{3} + \dfrac{8}{9} + \dfrac{16}{27} + \cdots$.

This should be recognized as a geometric series with $r = 2/3$ and $a = 3$. ||

Example 9 | $\displaystyle\sum \frac{2 - n + n^2}{70n^2 + 3}$.

Use The Divergence Test. ||

Example 10 $\quad \dfrac{1}{\sqrt{3}} + \dfrac{1}{\sqrt{4}} + \dfrac{1}{\sqrt{5}} + \dfrac{1}{\sqrt{6}} + \cdots.$

This is the result of removing the first two terms of a k-series. ||

Exercises 11.6

Determine whether these series converge or diverge. In addition, decide whether each convergent alternating series is absolutely convergent or conditionally convergent.

1. $\sum \dfrac{10n^3 + 7n^2}{n^{9/2} + 3}$.

2. $1 - \dfrac{1}{\sqrt{3}} + \dfrac{1}{\sqrt{5}} - \dfrac{1}{\sqrt{7}} + \cdots.$

3. $\Sigma n^2 e^{-n^3}$.

4. $\dfrac{5}{1^2} - \dfrac{5^2}{2^2} + \dfrac{5^3}{3^2} - \dfrac{5^4}{4^2} + \cdots.$

5. $\dfrac{2}{7} - \dfrac{3}{14} + \dfrac{4}{21} - \dfrac{5}{28} + \cdots.$

6. $1 - \dfrac{1 \cdot 2}{1 \cdot 3} + \dfrac{1 \cdot 2 \cdot 3}{1 \cdot 3 \cdot 5} - \dfrac{1 \cdot 2 \cdot 3 \cdot 4}{1 \cdot 3 \cdot 5 \cdot 7} + \cdots.$

7. $\sum \left(\dfrac{1}{n+1} - \dfrac{1}{n} \right)$.

8. $1 - \dfrac{3}{4} + \dfrac{5}{8} - \dfrac{9}{16} + \dfrac{17}{32} - \dfrac{33}{64} + \cdots.$

9. $\sum \dfrac{n + 100}{n^3}$.

10. $1 + \dfrac{1}{3} + \dfrac{1}{5} + \dfrac{1}{7} + \dfrac{1}{9} + \cdots.$

11. $\dfrac{1}{1001} + \dfrac{1}{2002} + \dfrac{1}{3003} + \dfrac{1}{4004} + \cdots.$

12. $\sum \dfrac{i+3}{100+i}$.

13. $\sum \dfrac{1 - n^2}{50n^2 - 7n + 2}$.

14. $\dfrac{1}{3} + \dfrac{2}{5} + \dfrac{3}{7} + \dfrac{4}{9} + \cdots.$

15. $\dfrac{1}{\sqrt{1 \cdot 2}} + \dfrac{1}{\sqrt{2 \cdot 3}} + \dfrac{1}{\sqrt{3 \cdot 4}} + \cdots.$

16. $\dfrac{1}{3} + \dfrac{1 \cdot 3}{3 \cdot 6} + \dfrac{1 \cdot 3 \cdot 5}{3 \cdot 6 \cdot 9} + \dfrac{1 \cdot 3 \cdot 5 \cdot 7}{3 \cdot 6 \cdot 9 \cdot 12} + \cdots.$

17. $3 + 2 + \dfrac{4}{3} + \dfrac{8}{9} + \dfrac{16}{27} + \cdots.$

18. $\dfrac{1}{\sqrt{3}} + \dfrac{1}{\sqrt{4}} + \dfrac{1}{\sqrt{5}} + \dfrac{1}{\sqrt{6}} + \cdots.$

19. $\sum \dfrac{n!}{(2n)!}$.

20. $\sum (-1)^n \dfrac{n+1}{n^2}$.

21. $\sum_{k=2}^{\infty} \dfrac{1}{k \ln k}$.

22. $\dfrac{1}{4\sqrt{4}} + \dfrac{1}{5\sqrt{5}} + \dfrac{1}{6\sqrt{6}} + \dfrac{1}{7\sqrt{7}} + \cdots.$

23. $\sum \dfrac{1}{\sqrt{n^3 + 1}}$.

24. $\sum (-1)^n \dfrac{n(n+2)}{3(n+1)(n+3)}$.

25. $\dfrac{1}{3} - \dfrac{2}{4} + \dfrac{3}{5} - \dfrac{4}{6} + \cdots.$

26. $\sum \dfrac{n^3}{3^n (n+2)!}$.

27. $1 - \dfrac{2!}{1 \cdot 3} + \dfrac{3!}{1 \cdot 3 \cdot 5} - \dfrac{4!}{1 \cdot 3 \cdot 5 \cdot 7} + \cdots.$

28. $\sum \dfrac{n^2 + 5}{198 + 3n^3}$.

29. $\sum \dfrac{2^{i+1}}{3^i}$.

30. $\sum \left[\dfrac{1}{n^2} - \dfrac{1}{(n+2)^2} \right]$.

31. $\sum_{n=2}^{\infty} \dfrac{1}{n(\ln n)^5}$.

32. $\sum_{n=0}^{\infty} \dfrac{3n^2}{\sqrt{n^5 + 3n + 1}}$.

33. $\sum \frac{n^2}{3^n}$.

34. $\sum \frac{1}{k^{2/3}}$.

35. $\sum_{n=0}^{\infty} \frac{3^{n+1}}{4^{n+3}}$.

36. $1 - \frac{10}{2} + \frac{20}{4} - \frac{30}{8} + \frac{40}{16} - \cdots$.

37. $\sum \frac{2^n}{(2n)!}$.

38. $\Sigma n^3 e^{-n^4}$.

39. $\sum \frac{n^n}{n!}$.

11.7 Power Series

Let a be a constant and let a_i be a constant for each nonnegative integer i. For each value of x, the following expression determines an infinite series:

$$a_0 + a_1(x - a) + a_2(x - a)^2 + \cdots + a_n(x - a)^n + \cdots. \tag{1}$$

We call the expression in (1) a *power series in* $x - a$. A power series may converge for some values of x and diverge for others.

Example 1 Find the values of x for which this power series converges and those for which it diverges:

$$\sum_{n=0}^{\infty} \frac{(x-1)^n}{2^n(n+1)} = 1 + \frac{1}{2(2)}(x-1) + \frac{1}{2^2(3)}(x-1)^2 + \cdots.$$

Let $u_n = \frac{(x-1)^n}{2^n(n+1)}$ and apply the ratio test. We get

$$\lim_{n\to\infty}\left|\frac{u_{n+1}}{u_n}\right| = \lim_{n\to\infty}\left|\frac{(x-1)^{n+1}}{2^{n+1}(n+2)}\frac{2^n(n+1)}{(x-1)^n}\right| = \lim_{n\to\infty}\frac{|x-1|}{2}\frac{n+1}{n+2}$$

$$= \frac{|x-1|}{2}\lim_{n\to\infty}\frac{n+1}{n+2} = \frac{|x-1|}{2}\lim_{n\to\infty}\frac{1+1/n}{1+2/n} = \frac{|x-1|}{2}.$$

Thus we know that if $|x - 1|/2 < 1$, or $|x - 1| < 2$, the series is absolutely convergent. We also know that if $|x - 1|/2 > 1$, or $|x - 1| > 2$, the series diverges. We see then that the series converges for all values of x for which the distance between x and the point 1 is less than 2, and diverges if that distance is greater than 2. So the series converges for all x in the interval $-1 < x < 3$, and diverges if $x < -1$ or $x > 3$. When $x = -1$ or when $x = 3$, $|x - 1|/2 = 1$; and hence The Ratio Test fails for those values. Accordingly we must test further.

When $x = -1$, we obtain

$$1 - \frac{1}{2} + \frac{1}{3} - \frac{1}{4} + \frac{1}{5} - \cdots,$$

which converges by The Alternating Series Test.

When $x = 3$, we get

$$1 + \frac{1}{2} + \frac{1}{3} + \frac{1}{4} + \frac{1}{5} + \cdots,$$

which is the divergent harmonic series.

Therefore the series converges for $-1 \leq x < 3$ and diverges for all other values of x. That interval is pictured in Figure 11.7.1. ||

Figure 11.7.1

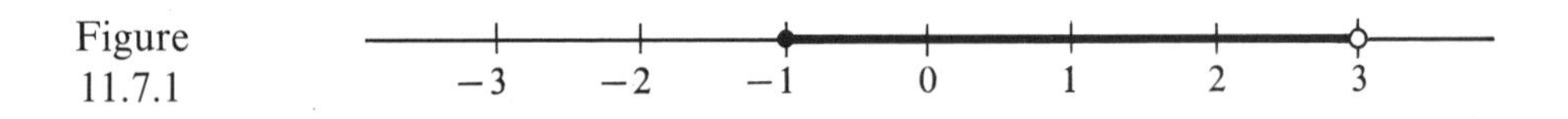

It is easy to see that the power series (1) converges to the sum a_0 when $x = a$, because in this case $s_n = a_0$ for all n, and so $\lim_{n \to \infty} s_n = a_0$. Therefore a power series always converges for at least one value of x. In general the following result can be proved. (A proof can be found in many advanced calculus texts.)

Theorem 11.7.1

A power series

$$a_0 + a_1(x - a) + a_2(x - a)^2 + \cdots + a_n(x - a)^n + \cdots$$

either

(i) converges only for $x = a$ and diverges for all other x, or

(ii) converges absolutely for $|x - a| < r$ and diverges for $|x - a| > r$, for some positive number r, or

(iii) converges absolutely for all x.

In case (ii), the series converges for all x in the interval $|x - a| < r$, called the *interval of convergence*. As shown in Figure 11.7.2, the point a is at the center of this interval because $|x - a| < r$ if and only if the distance between x and a is less than r. The number r is called the *radius of convergence* because a circle of radius r centered at a goes through the endpoints of the interval of convergence. For case (i), the radius of convergence is considered to be 0, and for case (iii), the radius of convergence is said to be ∞.

Figure 11.7.2

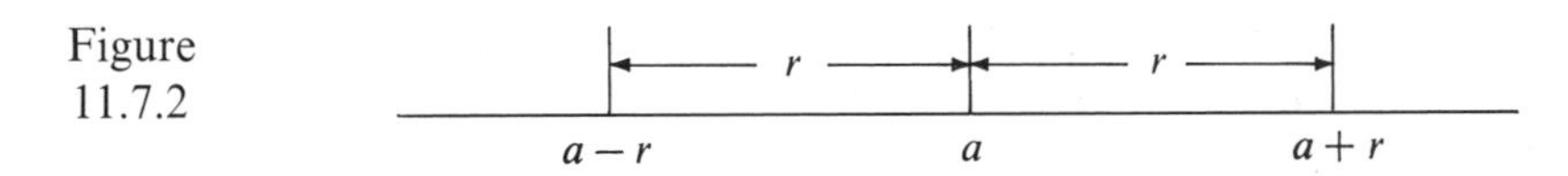

Note that for case (ii), Theorem 11.7.1 does not mention convergence or divergence at the two points where $|x - a| = r$. Those two points are the endpoints, $a - r$ and $a + r$, of the interval, and at those points the series may converge or diverge. Some series converge at both endpoints and some converge at neither endpoint, while others converge at one endpoint but not the other.

For many power series, The Ratio Test can be used to determine the radius of convergence and hence the interval of convergence, as in Example 1. *For the case where the radius of convergence is positive The Ratio Test fails for the endpoints, which must be tested separately*, as they were in Example 1. Note that in Example 1 the radius of convergence is 2.

Example 2 The power series

$$1 + (x - a) + 2!(x - a)^2 + 3!(x - a)^3 + \cdots + n!(x - a)^n + \cdots$$

diverges for all $x \neq a$ and converges for $x = a$.

The Ratio Test shows this, since

$$\begin{aligned}\lim_{n\to\infty}\left|\frac{u_{n+1}}{u_n}\right| &= \lim_{n\to\infty}\left|\frac{(n+1)!(x-a)^{n+1}}{n!(x-a)^n}\right| \\ &= \lim_{n\to\infty}|(n+1)(x-a)| \\ &= \begin{cases}\infty & \text{if } x \neq a \\ 0 & \text{if } x = a.\end{cases}\end{aligned}$$

In this example the radius of convergence is clearly zero. ||

Example 3 Find the values of x for which this series converges:

$$\sum_{n=0}^{\infty}\frac{x^n}{n!} = 1 + x + \frac{x^2}{2!} + \frac{x^3}{3!} + \cdots.$$

(Recall that by definition $0! = 1$.) For all x, we have

$$\lim_{n\to\infty}\left|\frac{u_{n+1}}{u_n}\right| = \lim_{n\to\infty}\left|\frac{x^{n+1}}{(n+1)!}\frac{n!}{x^n}\right| = \lim_{n\to\infty}\left|\frac{x}{n+1}\right| = 0 < 1;$$

hence the series converges for all x, and so the radius of convergence is ∞. ||

The previous two examples clearly illustrate the power of the Ratio Test when applied to power series. The next theorem uses the Ratio Test to obtain an expression for the radius of convergence of a power series.

Theorem 11.7.2

Let r be the radius of convergence of $\sum_{n=0}^{\infty} a_n(x-a)^n$. Then

(i) $r = \dfrac{1}{\lim_{n\to\infty}\left|\dfrac{a_{n+1}}{a_n}\right|}$, if $\lim_{n\to\infty}\left|\dfrac{a_{n+1}}{a_n}\right|$ exists and is nonzero.

(ii) $r = \infty$, if $\lim_{n\to\infty}\left|\dfrac{a_{n+1}}{a_n}\right| = 0$.

(iii) $r = 0$, if $\lim_{n\to\infty}\left|\dfrac{a_{n+1}}{a_n}\right| = \infty$.

Proof.

$$\lim_{n\to\infty}\left|\frac{a_{n+1}(x-a)^{n+1}}{a_n(x-a)^n}\right| = |x-a|\lim_{n\to\infty}\left|\frac{a_{n+1}}{a_n}\right|.$$

Thus by the Ratio Test, the series $\sum_{n=0}^{\infty} a_n(x-a)^n$ converges absolutely if and only if

$$|x-a| \lim_{n\to\infty} \left|\frac{a_{n+1}}{a_n}\right| < 1.$$

Parts (ii) and (iii) follow directly from this inequality. If $\lim_{n\to\infty} \left|\frac{a_{n+1}}{a_n}\right| \neq 0$, division by $\lim_{n\to\infty} \left|\frac{a_{n+1}}{a_n}\right|$ gives part (i).

Example 4 Compute the radius of convergence of $\sum_{n=0}^{\infty} n^2(-3)^n(x-1)^n$.

$$\lim_{n\to\infty} \left|\frac{a_{n+1}}{a_n}\right| = \lim_{n\to\infty} \left|\frac{(n+1)^2(-3)^{n+1}}{n^2(-3)^n}\right|$$

$$= \lim_{n\to\infty} 3\left(\frac{n+1}{n}\right)^2 = 3.$$

Thus by Theorem 11.7.2 the radius of convergence is 1/3. ||

For each x for which a power series converges, a unique sum is determined. That is, for each input x, a unique output, the sum of the series, is obtained. Accordingly, a power series is a function f whose domain is the set of all x for which the series converges. Thus the domain of

$$f(x) = a_0 + a_1(x-a) + a_2(x-a)^2 + \cdots$$

is either the single value a, a finite interval, or the set of all real numbers.

Suppose that the series

$$f(x) = a_0 + a_1(x-a) + a_2(x-a)^2 + \cdots + a_n(x-a)^n + \cdots$$

converges when $|x-a| < r$. It is then natural to ask for differentiation and integration formulas for $f(x)$. The next theorem, whose proof is found in many advanced calculus texts, answers those questions.

Theorem 11.7.3

Suppose that the power series

$$f(x) = a_0 + a_1(x-a) + a_2(x-a)^2 + \cdots + a_n(x-a)^n + \cdots$$

converges when $|x-a| < r$. Then

(i) The function $f(x)$ is continuous on the interval $|x-a| < r$.

(ii) $f'(x) = a_1 + 2a_2(x-a) + 3a_3(x-a)^2 + \cdots + na_n(x-a)^{n-1} + \cdots$, when $|x-a| < r$. Moreover, the radius of convergence of this series is r.

(iii) $$\int f(x)dx = C + a_0(x-a) + \frac{a_1}{2}(x-a)^2 + \frac{a_2}{3}(x-a)^3 + \cdots + \frac{a_n}{n+1}(x-a)^{n+1} + \cdots, \quad \text{when } |x-a| < r.$$

Moreover, the radius of convergence of this series is r.

Note that Theorem 11.7.3 assures us that if we differentiate a convergent power series "term for term," the resulting series converges to the derivative of the sum of the power series. It is important to recognize that Theorem 11.7.3 does not mention the endpoints of the interval of convergence. Even though the given series converges at both endpoints, the term-by-term derivative of the series may diverge at those points.

Example 5 Consider the power series

$$f(x) = \sum_{n=1}^{\infty} \frac{x^n}{n} = x + \frac{x^2}{2} + \frac{x^3}{3} + \cdots.$$

We have

$$\lim_{n\to\infty} \left|\frac{u_{n+1}}{u_n}\right| = \lim_{n\to\infty} \left|\frac{x^{n+1}}{n+1} \cdot \frac{n}{x^n}\right| = \lim_{n\to\infty} \frac{n}{n+1}|x|$$

$$= \lim_{n\to\infty} \frac{1}{1+1/n}|x| = |x|.$$

So the series converges for $|x| < 1$. For the endpoints 1 and -1 of the interval $|x| < 1$, we have these two series:

$$1 + \frac{1}{2} + \frac{1}{3} + \frac{1}{4} + \cdots,$$

$$-1 + \frac{1}{2} - \frac{1}{3} + \frac{1}{4} - \cdots.$$

Thus the power series converges for $-1 \leq x < 1$ and diverges for all other values of x.

The series obtained by differentiating each term of the given power series is

$$f'(x) = \sum_{n=1}^{\infty} x^{n-1} = 1 + x + x^2 + x^3 + \cdots.$$

This is a geometric series, which converges by Theorem 11.2.1 if $|x| < 1$ and diverges if $|x| \geq 1$. We already know by Theorem 11.7.1 that the derivative series has radius of convergence 1, as the given series does. However, note that although the given series converges for $x = -1$, the derivative series does not. This example illustrates the fact that the derivative series may act differently at the end points of the interval of convergence than the given series does. ||

Example 6 We saw in Example 1 that the following power series has radius of convergence 2:

$$f(x) = \sum_{n=0}^{\infty} \frac{(x-1)^n}{2^n(n+1)}$$

$$= 1 + \frac{1}{2 \cdot 2}(x-1) + \frac{1}{2^2 \cdot 3}(x-1)^2 + \frac{1}{2^3 \cdot 4}(x-1)^3 + \cdots.$$

Hence the series is a continuous function when $|x-1|<2$. Furthermore, when $|x-1|<2$, the series is differentiable, and its derivative is

$$f'(x)=\sum_{n=0}^{\infty}\frac{n(x-1)^{n-1}}{2^n(n+1)}=\frac{1}{2\cdot 2}+\frac{2}{2^2\cdot 3}(x-1)+\frac{3}{2^3\cdot 4}(x-1)^2+\cdots. \qquad ||$$

In some instances we are able to find the sum of a power series. The following two examples illustrate this.

Example 7 | Find the function to which the series $\sum_{n=0}^{\infty}(2x)^n$ converges.

The series under consideration is simply a geometric series with $r=2x$, and $a=1$. Thus

$$\sum_{n=0}^{\infty}(2x)^n=\frac{1}{1-2x}, \qquad \text{if } |2x|<1.$$

Consequently the given series converges to $f(x)=\dfrac{1}{1-2x}$ for $|x|<1/2$. ||

Example 8 | Find the function to which the series $\sum_{n=1}^{\infty} n(2x)^{n-1}$ converges.

Note that this series is simply the term for term derivative of the series

$$\sum_{n=0}^{\infty}(2x)^n=\frac{1}{1-2x}, \qquad \text{for } |x|<\frac{1}{2}.$$

Thus, by Theorem 11.7.3 we have

$$\sum_{n=1}^{\infty} n(2x)^{n-1}=\frac{2}{(1-2x)^2}, \qquad \text{for } |x|<\frac{1}{2}. \qquad ||$$

Exercises 11.7

In Exercises 1–26, find all values of x for which the series converges and state the radius of convergence. Also (where possible) find the derivative series, and state its radius of convergence.

1. $1+x+\dfrac{x^2}{2!}+\dfrac{x^3}{3!}+\dfrac{x^4}{4!}+\cdots.$

2. $x-\dfrac{x^3}{3!}+\dfrac{x^5}{5!}-\dfrac{x^7}{7!}+\cdots.$

3. $1+\dfrac{x^3}{2!}+\dfrac{x^6}{3!}+\dfrac{x^9}{4!}+\dfrac{x^{12}}{5!}+\cdots.$

4. $1-\dfrac{x^2}{2!}+\dfrac{x^4}{4!}-\dfrac{x^6}{6!}+\cdots.$

5. $1+\dfrac{3}{7}x+\dfrac{5}{7^2}x^2+\dfrac{7}{7^3}x^3+\dfrac{9}{7^4}x^4+\cdots.$

6. $1+\dfrac{2x}{5}+\dfrac{3x^2}{5^2}+\dfrac{4x^3}{5^3}+\cdots.$

7. $\displaystyle\sum_{n=1}^{\infty}(-1)^n\frac{x^n}{3^n n^{3/2}}.$

8. $\displaystyle\sum_{n=1}^{\infty}(-1)^n\frac{2^n}{5^n n^2}x^n.$

9. $1+(x+5)+\dfrac{1}{2}(x+5)^2+\dfrac{1}{3}(x+5)^3+\cdots.$

10. $1+2(x-3)+3(x-3)^2+4(x-3)^3+\cdots.$

11. $\sum_{n=0}^{\infty} (-1)^n \frac{n}{3^n}(x+1)^n$.

12. $(x+2) - \frac{(x+2)^2}{2} + \frac{(x+2)^3}{3} - \frac{(x+2)^4}{4} + \cdots$.

13. $\sum_{k=0}^{\infty} (-1)^k \frac{k!}{(20)^k} x^k$.

14. $1 + 2x + 2^2x^2 + 2^3x^3 + 2^4x^4 + \cdots$.

15. $\frac{(x+4)}{5^2} + \frac{(x+4)^2}{5^3 2^2} + \frac{(x+4)^3}{5^4 3^2} + \frac{(x+4)^4}{5^5 4^2} + \frac{(x+4)^5}{5^6 5^2} + \cdots$.

16. $1 + \frac{(x-2)}{3 \cdot 2} + \frac{(x-2)^2}{3^2 \cdot 3} + \frac{(x-2)^3}{3^3 \cdot 4} + \cdots$.

17. $\sum_{k=1}^{\infty} k^k(x-5)^k$.

18. $\sum_{k=0}^{\infty} \frac{k}{x^k}$.

19. $\sum_{k=1}^{\infty} \frac{(-1)^k(x+3)^k}{\sqrt{k}}$.

20. $\sum_{n=0}^{\infty} \frac{x^n}{n^2+5}$.

21. $\sum_{n=0}^{\infty} \frac{n^3}{x^{2n}}$.

22. $\sum \frac{2 \cdot 4 \cdot 6 \cdots (2n)}{1 \cdot 3 \cdot 5 \cdots (2n-1)} x^n$.

23. $\sum_{n=1}^{\infty} \frac{(2x-3)^n}{n}$.

24. $\sum_{n=1}^{\infty} \frac{(5-3x)^n}{n^2}$.

25. $\sum_{n=1}^{\infty} \frac{n^3}{n!} x^n$.

26. $\sum_{n=1}^{\infty} \left(\frac{x}{n}\right)^n$.

In Exercises 27–34, find the function to which the given series converges.

27. $\sum_{n=0}^{\infty} x^n$.

28. $\sum_{k=0}^{\infty} (-1)^k x^k$.

29. $\sum_{n=1}^{\infty} nx^{n-1}$.

30. $\sum_{n=1}^{\infty} n(-1)^n x^{n-1}$.

31. $\sum_{n=0}^{\infty} \frac{x^{n+1}}{n+1}$.

32. $\sum_{n=0}^{\infty} \frac{(-1)^n}{n+1} x^{n+1}$.

33. $\sum_{n=0}^{\infty} (n+2)(n+1)x^n$.

34. $\sum_{n=0}^{\infty} (n+2)(n+1)(-1)^n x^n$.

35. If $f(x) = \sum_{n=0}^{\infty} x^n/n!$, find $f'(x)$. Use this information to guess a simple expression for $f(x)$.

11.8 Taylor Series

In the last section we saw that when a power series

$$a_0 + a_1(x-a) + a_2(x-a)^2 + \cdots + a_n(x-a)^n + \cdots$$

converges for more than just $x = a$, it converges on an interval or on the entire real line. We also noted that the series is a differentiable function for all values of x in its interval of convergence, except possibly at the end points.

In this section we shall consider the converse question. That is, given a function $f(x)$, under what conditions can we find a power series such that for some interval $|x - a| < r$,

$$f(x) = a_0 + a_1(x - a) + a_2(x - a)^2 + \cdots + a_n(x - a)^n + \cdots ?$$

When we can find a power series that converges with sum $f(x)$ in some interval, we say that the function $f(x)$ is *expressible* as a power series in that interval.

To begin our discussion we will suppose that $f(x)$ can be expressed as a power series in some interval and try to determine what that power series must be. That is, suppose that when $|x - a| < r$, we have

$$f(x) = a_0 + a_1(x - a) + a_2(x - a)^2 + \cdots + a_n(x - a)^n + \cdots. \tag{1}$$

Then by Theorem 11.7.3 we must have

$$f'(x) = a_1 + 2a_2(x - a) + 3a_3(x - a)^2 + 4a_4(x - a)^3 + \cdots. \tag{2}$$

Similarly, we also have the following results:

$$f''(x) = 2a_2 + 3 \cdot 2a_3(x - a) + 4 \cdot 3a_4(x - a)^2 + \cdots. \tag{3}$$

$$f'''(x) = 3 \cdot 2a_3 + 4 \cdot 3 \cdot 2a_4(x - a) + 5 \cdot 4 \cdot 3a_5(x - a)^2 + \cdots. \tag{4}$$

$$f''''(x) = 4 \cdot 3 \cdot 2a_4 + 5 \cdot 4 \cdot 3 \cdot 2a_5(x - a) + \cdots. \tag{5}$$

. . . .

$$f^{(n)}(x) = n!a_n + (n + 1)!a_{n+1}(x - a) + \cdots. \tag{6}$$

By (1), $f(a) = a_0 = 0!a_0$.

By (2), $f'(a) = a_1 = 1!a_1$.

By (3), $f''(a) = 2a_2 = 2!a_2$.

By (4), $f'''(a) = 3 \cdot 2a_3 = 3!a_3$.

By (5), $f''''(a) = 4 \cdot 3 \cdot 2a_4 = 4!a_4$.

. . . .

By (6), $f^{(n)}(a) = n!a_n$.

Therefore, $a_n = f^n(a)/n!$ for all n, and we have

$$f(x) = f(a) + f'(a)(x - a) + \frac{f''(a)}{2!}(x - a)^2 + \frac{f'''(a)}{3!}(x - a)^3 + \cdots.$$

So, if $f(x)$ can be expressed as a power series in $x - a$, the power series is uniquely determined.

When can $f(x)$ be expressed as a power series? To investigate this question we recall Theorem 9.7.2 (Taylor's Theorem), which states that if $f^{(n+1)}$ exists in some neighborhood N of a, then

$$f(x) = f(a) + f'(a)(x - a) + \frac{1}{2!}f''(a)(x - a)^2$$

$$+ \frac{1}{3!}f'''(a)(x - a)^3 + \cdots + \frac{1}{n!}f^{(n)}(a)(x - a)^n + E_n(x),$$

for x in N; and if $|f^{(n+1)}(x)| \le M_{n+1}$ for x in N, then

$$|E_n(x)| \le \frac{1}{(n+1)!} M_{n+1}|x-a|^{n+1}.$$

Let

$$s_n(x) = f(a) + f'(a)(x-a) + \cdots + \frac{f^n(a)}{n!}(x-a)^n.$$

Then

$$f(x) = s_n(x) + E_n(x), \tag{7}$$

and $s_n(x)$ is the nth partial sum of the power series

$$f(a) + f'(a)(x-a) + \frac{f''(a)}{2!}(x-a)^2 + \frac{f'''(a)}{3!}(x-a)^3 + \cdots + \frac{f^{(n)}(a)}{n!}(x-a)^n + \cdots. \tag{8}$$

We see from (7) that

$$\lim_{n\to\infty} s_n = \lim_{n\to\infty} [f(x) - E_n(x)] = \lim_{n\to\infty} f(x) - \lim_{n\to\infty} E_n(x) = f(x) - \lim_{n\to\infty} E_n(x).$$

Thus $\lim_{n\to\infty} s_n$ exists if and only if $\lim_{n\to\infty} E_n(x)$ exists. That is, the series (8) converges if and only if $\lim_{n\to\infty} E_n$ exists. Moreover, $\lim_{n\to\infty} s_n = f(x)$ if and only if $\lim_{n\to\infty} E_n(x) = 0$. That is, the series (8) converges to $f(x)$ if and only if $\lim_{n\to\infty} E_n(x) = 0$.

If the series

$$f(a) + f'(a)(x-a) + \frac{f''(a)}{2!}(x-a)^2 + \frac{f'''(a)}{3!}(x-a)^3 + \cdots + \frac{f^{(n)}(a)}{n!}(x-a)^n + \cdots$$

converges to $f(x)$ when $|x-a| < r$, we call it *The Taylor Series* for $f(x)$ about the point a.

We have now established the following theorem.

Theorem 11.8.1

(i) If a function $f(x)$ is expressible as a power series about $x = a$, then there is only one such power series:

$$f(x) = f(a) + f'(a)(x-a) + \frac{f''(a)}{2!}(x-a)^2 + \cdots + \frac{f^{(n)}(a)}{n!}(x-a)^n + \cdots.$$

(ii) A function $f(x)$ with derivatives of all orders is expressible as a power series about $x = a$ if and only if

$$\lim_{n\to\infty} E_n(x) = 0.$$

In mathematical language, part (i) of the theorem establishes the uniqueness of a power series representation about $x = a$, while part (ii) gives necessary and sufficient conditions for the existence of such a power series.

Example 1 | Find the Taylor Series for $f(x) = \sin x$ about $x = 0$.

We have

$$\begin{aligned} f(x) &= \sin x, & f(0) &= 0, \\ f'(x) &= \cos x, & f'(0) &= 1, \\ f''(x) &= -\sin x, & f''(0) &= 0, \\ f'''(x) &= -\cos x, & f'''(0) &= -1, \\ &\cdots\cdots & &\cdots\cdots \end{aligned}$$

We see that the pattern is an endless repetition of 0, 1, 0, and -1. Thus the Taylor Series for $\sin x$ is

$$0 + x + 0 - \frac{x^3}{3!} + 0 + \frac{x^5}{5!} + 0 - \frac{x^7}{7!} + \cdots.$$

Since $f^{(n)}(x) = \pm\sin x$ or $\pm\cos x$, for every n, we have $|f^{(n)}(x)| \le 1$ for all x and all n. Hence we can take $M_{n+1} = 1$ in Taylor's Theorem to obtain

$$|E_n(x)| \le \frac{|x|^{n+1}}{(n+1)!}.$$

In Example 3 of Section 11.7 we saw that the series $\Sigma x^n/n!$ converges for all x. Therefore its nth term and its $(n + 1)$st term must approach 0. Hence

$$\lim_{n\to\infty} E_n(x) = \lim_{n\to\infty} \left|\frac{x^{n+1}}{(n+1)!}\right| = 0.$$

Therefore the radius of convergence is ∞, and moreover the Taylor Series converges with sum $\sin x$ for all x. Thus, for all x we have

$$\sin x = x - \frac{x^3}{3!} + \frac{x^5}{5!} - \frac{x^7}{7!} + \cdots. \qquad \|$$

According to Theorem 11.7.3, the series obtained by differentiating the series for $\sin x$ converges to the sum $d(\sin x)/dx = \cos x$ and also has radius of convergence ∞. Therefore, for all x, we have

$$\cos x = 1 - \frac{x^2}{2!} + \frac{x^4}{4!} - \frac{x^6}{6!} + \cdots.$$

Example 2 | Find the Taylor Series for $f(x) = e^x$ about $x = 0$.

We have

$$\begin{aligned} f(x) &= e^x, & f(0) &= 1, \\ f'(x) &= e^x, & f'(0) &= 1, \\ f''(x) &= e^x, & f''(0) &= 1, \\ &\cdots\cdots & &\cdots\cdots \\ f^{(n)}(x) &= e^x, & f^{(n)}(0) &= 1, \\ f^{(n+1)}(x) &= e^x, & f^{(n+1)}(0) &= 1, \\ &\cdots\cdots & &\cdots\cdots \end{aligned}$$

and
$$|E_n(x)| \leq \frac{M_{n+1}}{(n+1)!}|x|^{n+1}.$$

Since by Theorems 6.1.2 and 6.1.3, the function e^x is increasing and positive for all x, $M_{n+1} = e^{|x|}$. Thus

$$\lim_{n\to\infty} |E_n(x)| \leq \lim_{n\to\infty} \frac{e^{|x|}|x^{n+1}|}{(n+1)!} = e^{|x|} \lim_{n\to\infty} \frac{|x^{n+1}|}{(n+1)!}.$$

Now, as in the preceding example,

$$\lim_{n\to\infty} \left|\frac{x^{n+1}}{(n+1)!}\right| = 0, \quad \text{and so}$$

$$\lim_{n\to\infty} |E_n(x)| \leq e^{|x|} \lim_{n\to\infty} \frac{|x^{n+1}|}{(n+1)!} = 0.$$

Consequently $\lim_{n\to\infty} E_n(x) = 0$, and the Taylor Series for e^x must converge with sum e^x for all x. We obtain

$$e^x = 1 + x + \frac{x^2}{2!} + \frac{x^3}{3!} + \frac{x^4}{4!} + \cdots.$$

Note that when each term of the series is differentiated, we obtain the same series. Of course, since $de^x/dx = e^x$, Theorem 11.7.3(ii) guarantees that the differentiated series has the same sum as the original series. ||

It should be clear by now that it is difficult to find the Taylor Series representation of a function by taking successive derivatives and then showing that $\lim_{n\to\infty} E_n(x) = 0$. Consequently we often make use of various theorems to ease this difficulty. For example, an application of Theorem 11.7.3 in Example 1 enabled us to obtain the Taylor Series for $\cos x$ quite simply. The following examples will illustrate possible courses of action more fully.

Example 3 | Find the Taylor Series for $\sin(x^2)$ about $x = 0$.

For all x we know that

$$\sin x = x - \frac{x^3}{3!} + \frac{x^5}{5!} - \frac{x^7}{7!} + \cdots.$$

Hence, for all x,

$$\sin(x^2) = x^2 - \frac{(x^2)^3}{3!} + \frac{(x^2)^5}{5!} - \frac{(x^2)^7}{7!} + \cdots, \quad \text{or}$$

$$\sin(x^2) = x^2 - \frac{x^6}{3!} + \frac{x^{10}}{5!} - \frac{x^{14}}{7!} + \cdots.$$

This *is* a power series that converges to $\sin(x^2)$ for all x. Consequently, by the uniqueness part of Theorem 11.8.1, it *must* be the desired Taylor Series. ||

Example 4 | Find the Taylor Series for the polynomial $f(x) = x^4 - 5x^3 + 7x - 10$ about $x = 0$.

Note that

$$f(x) = -10 + 7x + 0x^2 - 5x^3 + x^4 + 0x^5 + 0x^6 + \cdots$$

is already expressed as a power series (with a lot of zero terms). Since there is only one way to express $f(x)$ as a power series about $x = 0$, that power series must be the Taylor Series for $f(x)$. In the same way any polynomial in x can be viewed as a Taylor Series about $x = 0$. ||

At times, geometric series provide an easy way of calculating Taylor Series.

Example 5 | Find the Taylor Series expansion of $f(x) = 1/(1 - x)$ about $x = 0$.

Recall that the geometric series $\sum_{n=0}^{\infty} ar^n$ converges with sum $a/(1 - r)$ if $|r| < 1$. Thus, if we take $a = 1$ and $r = x$, we see that $\sum_{n=0}^{\infty} x^n$ converges to $1/(1 - x)$ if $|x| < 1$. So, if $|x| < 1$,

$$\frac{1}{1-x} = \sum_{n=0}^{\infty} x^n = 1 + x + x^2 + x^3 + \cdots,$$

which is, by uniqueness, the Taylor Series for $f(x) = 1/(1 - x)$ about $x = 0$. ||

Example 6 | Find the Taylor Series for $f(x) = \ln|1 - x|$ about $x = 0$.

By Theorem 11.7.3, we can integrate this series term by term when $|x| < 1$:

$$\frac{1}{1-x} = 1 + x + x^2 + x^3 + \cdots.$$

Since

$$\int \frac{dx}{1-x} = -\ln|1 - x| + C,$$

we have, for $|x| < 1$,

$$-\ln|1 - x| + C = x + \frac{1}{2}x^2 + \frac{1}{3}x^3 + \cdots. \tag{9}$$

Note that if $x = 0$,

$$-\ln|1 - x| = -\ln 1 = 0.$$

Substituting $x = 0$ in both sides of equation (9) gives $C = 0$. Consequently, if $|x| < 1$, then

$$\ln|1 - x| = -x - \frac{1}{2}x^2 - \frac{1}{3}x^3 - \cdots.$$ ||

Addition and multiplication of power series can also simplify the calculation of Taylor Series. For example, if we add the corresponding terms of these two power series

$$f(x) = a_0 + a_1(x - a) + a_2(x - a)^2 + a_3(x - a)^3 + \cdots,$$

$$g(x) = b_0 + b_1(x - a) + b_2(x - a)^2 + b_3(x - a)^3 + \cdots,$$

we obtain

$$(a_0 + b_0) + (a_1 + b_1)(x - a) + (a_2 + b_2)(x - a)^2 + (a_3 + b_3)(x - a)^3 + \cdots.$$

It can be proved that the series converges to $f(x) + g(x)$ for every value of x for which the two original series converge.

It is also possible to multiply the two given series to obtain

$$a_0b_0 + (a_0b_1 + a_1b_0)(x - a) + (a_0b_2 + a_1b_1 + a_2b_0)(x - a)^2 + (a_0b_3 + a_1b_2 + a_2b_1 + a_3b_0)(x - a)^3 + \cdots,$$

for which the nth term is

$$(a_0b_n + a_1b_{n-1} + a_2b_{n-2} + \cdots + a_{n-1}b_1 + a_nb_0)(x - a)^n.$$

Note that in each summand of each coefficient the sum of the subscripts adds up to the exponent of $(x - a)$.

It can be proved that the product series converges to $f(x)g(x)$ for every value of x that is interior to both intervals of convergence.

Example 7 Find the first four nonzero terms of the Taylor Series for $e^x \sin x$ about $x = 0$.

Since, for all x,

$$e^x = 1 + x + \frac{x^2}{2!} + \frac{x^3}{3!} + \frac{x^4}{4!} + \cdots$$

and

$$\sin x = x - \frac{x^3}{3!} + \frac{x^5}{5!} - \frac{x^7}{7!} + \cdots,$$

we need only multiply the two series to obtain the desired terms. Thus

$$e^x \sin x = \left(1 + x + \frac{x^2}{2!} + \frac{x^3}{3!} + \frac{x^4}{4!} + \cdots\right)\left(x - \frac{x^3}{3!} + \frac{x^5}{5!} - \frac{x^7}{7!} + \cdots\right)$$

$$= 1x + x^2 + \left(\frac{1}{2!} - \frac{1}{3!}\right)x^3 + \left(\frac{1}{3!} - \frac{1}{3!}\right)x^4 + \left(\frac{1}{4!} - \frac{1}{2!3!} + \frac{1}{5!}\right)x^5 + \cdots.$$

Therefore the desired result is

$$e^x \sin x = x + x^2 + \left(\frac{1}{2!} - \frac{1}{3!}\right)x^3 + \left(\frac{1}{4!} - \frac{1}{2!3!} + \frac{1}{5!}\right)x^5 + \cdots.$$ ||

Taylor Series are also very useful for solving differential equations, a fact that we shall illustrate later. For now, however, we shall show that Taylor Series can be used to solve problems that we could not solve before.

Example 8 Consider the continuous function

$$f(x) = \begin{cases} \dfrac{\sin x}{x} & \text{if } x \neq 0 \\ 1 & \text{if } x = 0. \end{cases}$$

Compute the integral $\int_0^1 f(x)dx$.

Since

$$\sin x = x - \frac{x^3}{3!} + \frac{x^5}{5!} - \frac{x^7}{7!} + \cdots,$$

we have

$$\frac{\sin x}{x} = 1 - \frac{x^2}{3!} + \frac{x^4}{5!} - \frac{x^6}{7!} + \cdots, \qquad \text{if } x \neq 0.$$

In fact, since the series

$$1 - \frac{x^2}{3!} + \frac{x^4}{5!} - \frac{x^6}{7!} + \cdots$$

converges to 1 when $x = 0$, we have

$$f(x) = 1 - \frac{x^2}{3!} + \frac{x^4}{5!} - \frac{x^6}{7!} + \cdots, \qquad \text{for all } x.$$

Then by Theorem 11.7.3, we have

$$\begin{aligned} \int_0^1 f(x)dx &= \int_0^1 dx - \int_0^1 \frac{x^2}{3!}dx + \int_0^1 \frac{x^4}{5!}dx - \int_0^1 \frac{x^6}{7!}dx + \cdots \\ &= x\Big|_0^1 - \frac{x^3}{3 \cdot 3!}\Big|_0^1 + \frac{x^5}{5 \cdot 5!}\Big|_0^1 - \frac{x^7}{7 \cdot 7!}\Big|_0^1 + \cdots \\ &= 1 - \frac{1}{3 \cdot 3!} + \frac{1}{5 \cdot 5!} - \frac{1}{7 \cdot 7!} + \cdots. \end{aligned}$$

||

Exercises 11.8

1. Find the Taylor Series expansion for $f(x) = x^3 + 1$: (a) about $x = 0$, (b) about $x = 1$, (c) about $x = -1$. Where does each series converge with sum $f(x)$? How can there be three power series expressions for $f(x)$ in view of the uniqueness of power series?
2. Find the Taylor Series expansion for $f(x) = x^4 - 3x^2 + 5$: (a) about $x = 0$, (b) about $x = 1$, (c) about $x = -1$. Where does each series converge with sum $f(x)$? How can there be three power series expressions for $f(x)$ in view of the uniqueness of power series?
3. Find the first three nonzero terms of the Taylor Series for $f(x) = \tan x$ about $x = \pi/4$.
4. Find the first four nonzero terms of the Taylor Series for $f(x) = e^{\sin x}$ about $x = 0$.
5. Show that if one uses the approximation $1 - x^2/2 + x^4/4!$ for $\cos x$ on the interval $|x| < 1$, the error is less than $1/6!$.

6. Show that if one uses the approximation $1 + x + x^2/2! + x^3/3!$ for e^x on the interval $|x| < 1$, the error is less than $e/4!$.
7. Find the Taylor Series for $\cos 2x$ about $x = 0$.
8. Find the Taylor Series for x^2e^{-x} about $x = 0$.
9. Find the Taylor Series for $x \sin x$ about $x = 0$.
10. Find the Taylor Series for $1/(x - 1)$ about $x = 0$.
11. Find the Taylor Series for $1/(x + 1)$ about $x = 0$.
12. Find the first five nonzero terms of the Taylor Series for $f(x) = \sin x \cos x$ about $x = 0$ by multiplying the series for $\sin x$ and $\cos x$. Where does the resulting series converge to $\sin x \cos x$?
13. Find the first five nonzero terms of the Taylor Series for $f(x) = e^x \sin x$ about $x = 0$ by multiplying the series for e^x and $\sin x$. Where does the resulting series converge to $e^x \sin x$?
14. Find the Taylor Series for $f(x) = \sin x$ about $x = \pi/2$. (*Hint*: Use the series for $\cos x$, the identity $\cos(x - \pi/2) = \sin x$, and uniqueness.)
15. Find the Taylor Series for $1/(1 + x^2)$ about $x = 0$. (*Hint*: Consider a geometric series.)
16. Use the result of Exercise 15 to find the Taylor Series for $\arctan x$ about $x = 0$.
17. Square both sides of $(1 - x)^{-1} = \sum_{n=0}^{\infty} x^n$, $|x| < 1$, to show that $(1 - x)^{-2} = \sum_{n=0}^{\infty}(n + 1)x^n$, $|x| < 1$.
18. Differentiate both sides of $(1 - x)^{-1} = \sum_{n=0}^{\infty} x^n$, $|x| < 1$, to obtain a Taylor Series for $(1 - x)^{-2}$. Compare the result with that of Exercise 17.
19. Show that
    (a) $\ln|x + 1| = x - (1/2)x^2 + (1/3)x^3 - (1/4)x^4 + \cdots$, for $|x| < 1$.
    (b) $\ln x = (x - 1) - (1/2)(x - 1)^2 + (1/3)(x - 1)^3 - \cdots$, for $0 < x < 2$.
20. Use $\sin_2 x = (1 - \cos 2x)/2$ to obtain the Taylor Series for $\sin_2 x$ about $x = 0$.
21. Use a Taylor Series to compute e^{-1} with an error of less than $5 \cdot 10^{-2}$.
22. Use a Taylor Series to compute $\int_0^1 \sin(x^2)dx$ with an error of less than $5 \cdot 10^{-3}$.
23. Use a Taylor Series to compute $\int_0^1 \cos(x^3)dx$ with an error of less than $5 \cdot 10^{-4}$.
24. Use the results of Exercise 19 to compute $\int_0^1(\ln|1 + x|)/x\,dx$ with an error of less than $5 \cdot 10^{-2}$.
25. Use the Taylor Series

$$(1 - x^2)^{-1/2} = 1 + \frac{1}{2}x^2 + \frac{1 \cdot 3}{2 \cdot 4}x^4 + \frac{1 \cdot 3 \cdot 5}{2 \cdot 4 \cdot 6}x^6 + \cdots, \qquad |x| \leq 1$$

to obtain Taylor Series for $\arcsin x$ and $(1 - x^2)^{-3/2}$ about $x = 0$.

26. Prove that if $\sum_{n=1}^{\infty} a_nx^n = 0$ for all x, $a \leq x \leq b$, then $a_n = 0$ for all n.

Brief Review of Chapter 11

1. Infinite Sequences

Definition. An *infinite sequence* is a function f whose domain is the set of all integers greater than or equal to a fixed integer h.

Notation. The terms of an infinite sequence are often denoted by writing a_n instead of $f(n)$. The sequence is often expressed as $\{f(n)\}$, $\{a_n\}$, or $\{a_1, a_2, a_3, \ldots, a_n, \ldots\}$.

Convergence. An infinite sequence is called *convergent* if $\lim_{n\to\infty} a_n$ exists. Otherwise the sequence is called *divergent*.

2. Bounded Sequences

Definition. A real number B such that $x \leq B$ for all x in a set S is called an *upper bound* of S. The least of all upper bounds is called the *least upper bound*.

Completeness Property. If a non-empty set of real numbers has an upper bound, then it has a least upper bound.

The definitions of lower bound and greatest lower bound are similar. The completeness property also holds for lower bounds and greatest lower bounds.

The major uses of the completeness property in this chapter are in the proofs of the following theorems.

Theorem. If $\{a_n\}$ is a sequence such that $a_i \leq a_{i+1}$ for all i, and if $\{a_n\}$ has an upper bound, then $\{a_n\}$ converges and its limit is its least upper bound.

Theorem. If a sequence $\{a_n\}$ has no upper bound, it diverges.

3. Convergence and Divergence of Infinite Series

$\sum_{i=1}^{\infty} a_i$ is an infinite series and its sequence $\{s_n\}$ of *partial sums* is defined by

$$s_n = a_1$$

$$s_2 = a_1 + a_2$$

$$\cdots\cdots\cdots\cdots$$

$$s_n = a_1 + a_2 + \cdots + a_n$$

$$\cdots\cdots\cdots\cdots\cdots\cdots$$

An infinite series *converges* if its sequence of partial sums converges; the limit of the sequence of partial sums (if it exists) is called the *sum* of the series. There are two major questions that concern us:

1 Does a series converge?

2 If a series does converge, what is its sum?

The first of those two questions is often more easily answered than the second. The Divergence Test will sometimes disclose that a series is divergent.

Divergence Test. If $\lim_{n\to\infty} a_n \neq 0$, then $\sum_{i=1}^{\infty} a_i$ is divergent. However, there are divergent series for which $\lim_{n\to\infty} a_n = 0$.

Removal, change, or addition of a finite number of terms will not affect the convergence or divergence of a series, but the sum will usually be changed.

4. Geometric Series

A series of the form $\sum_{i=0}^{\infty} ar^i$ is called a *geometric series.*

Theorem. The geometric series $\sum_{i=0}^{\infty} ar^i$ converges to the sum $a/(1-r)$ if $|r| < 1$, and diverges if $|r| \geq 1$.

5. Tests for Positive Series

(a) Comparison Test
(b) Integral Test
(c) Limit Form of the Comparison Test
(d) Ratio Test

6. Series of Positive and Negative Terms

(*a*) *Absolute Convergence and Conditional Convergence*

Definition. A series Σa_n is called *absolutely convergent* if and only if the series $\Sigma|a_n|$ converges.

Theorem. If a series is absolutely convergent, then it is convergent.

Definition. A series that is convergent but not absolutely convergent is called *conditionally convergent.*

(*b*) *Tests*

1 All tests applicable to positive series may be applied to $\Sigma|a_n|$ to determine absolute convergence.

2 For alternating series we can often apply The Alternating Series Test.

7. Power Series

A series of the form $\sum_{n=0}^{\infty} a_n(x-a)^n$ is called a *power series.* These series may converge for some values of x but diverge for others. In general the power series $\sum_{n=0}^{\infty} a_n(x-a)^n$ converges

(a) only at $x = a$, or
(b) absolutely for all x such that $|x-a| < r$, or
(c) absolutely for all x.

The number r is called the radius of convergence. The Ratio Test can often be used to find the radius of convergence of a power series. Other tests must be applied at the end points of the interval of convergence.

If the power series $f(x) = \sum_{n=0}^{\infty} a_n(x-a)^n$ has radius of convergence r, then

$$f'(x) = \sum_{n=1}^{\infty} na_n(x-a)^{n-1}, \qquad \text{for } |x-a| < r,$$

and the following series is an indefinite integral of $f(x)$:

$$\sum_{n=0}^{\infty} \frac{a_n}{n+1}(x-a)^{n+1}, \qquad \text{for } |x-a| < r,$$

8. Taylor Series

There is only one power series that converges to a given function. That is, if

$$f(x) = \sum_{n=0}^{\infty} a_n(x-a)^n,$$

then a_n is uniquely determined to be $a_n = f^{(n)}(a)/n!$. When the power series $\sum_{n=0}^{\infty}(f^{(n)}(a)/n!)(x-a)$ converges to $f(x)$, the series is called The Taylor Series for the function $f(x)$. That series converges if $\lim_{n\to\infty} E_n(x)$ exists, and it converges to $f(x)$ if and only if $\lim_{n\to\infty} E_n(x) = 0$. The inequality

$$|E_n(x)| \le \frac{1}{(n+1)!} M_{n+1} |x-a|^{n+1}$$

is often useful in trying to establish that a Taylor Series converges.

Suppose that a Taylor Series converges to $f(x)$ for $|x-a| < r$. If the series is differentiated term for term, the resulting series converges to $f'(x)$ when $|x-a| < r$. Similarly, if the series is integrated term for term, the resulting series converges to an indefinite integral of $f(x)$ when $|x-a| < r$.

Also, Taylor Series may be added or multiplied as follows. If

$$f(x) = \sum_{n=0}^{\infty} a_n(x-a)^n \quad \text{and} \quad g(x) = \sum_{n=0}^{\infty} b_n(x-a)^n, \qquad \text{for } |x-a| < r,$$

then

$$f(x) + g(x) = \sum_{n=0}^{\infty} (a_n + b_n)(x-a)^n, \qquad \text{for } |x-a| < r,$$

and

$$f(x)g(x) = \sum_{n=0}^{\infty} c_n(x-a)^n, \qquad \text{for } |x-a| < r,$$

where

$$c_n = a_0b_n + a_1b_{n-1} + a_2b_{n-2} + \cdots + a_{n-1}b_1 + a_nb_0.$$

Technique Review Exercises, Chapter 11

1. Find the limit of the sequence or state that it diverges.
   (a) $\{-1/4, 2/5, -3/6, 4/7, -5/8, \ldots\}$
   (b) $\{3/2, -3/4, 3/8, -3/16, \ldots\}$
2. Use the definition of convergence to determine if the series $\sum_{i=1}^{\infty} \ln((i+1)/i)$ converges. If the series converges, find its sum.

3. Test the series $\sum_{n=1}^{\infty}(n+3)/(1000+n)$ for convergence.
4. Find the sum of the series $2/9 + 2/27 + 2/81 + \cdots$.
5. Test the series $\sum_{n=1}^{\infty}(\sqrt{n}+3)/(1-n+n^2)$ for convergence.
6. Test the series $\sum_{n=2}^{\infty} 1/(n(\ln n)^3)$ for convergence.
7. Find all values of x for which the series $\sum_{n=1}^{\infty}(x+3)^n/(n \cdot 2^n)$ converges.
8. Is the following series absolutely convergent, conditionally convergent, or divergent: $\sum_{n=30}^{\infty}(-1)^n/(\sqrt{n}-5)$?
9. Find the Taylor Series for $f(x) = e^{-x^3}$ about $x = 0$. Show that the series converges to $f(x)$ for all x.
10. Find the Taylor Series for $f(x) = 2\sin x + \cos x$. Where does the series converge to $f(x)$?

Additional Exercises, Chapter 11

Section 11.1

In Exercises 1–6, find the limit of the given sequence.

1. $\{3 - 1/(-2)^n\}$.
2. $\{\cos(2^n\pi)\}$.
3. $\{\cos(n\pi)\}$.
4. $\{ne^{-n}\}$.
5. $\left\{\dfrac{n^2+5n^3}{13n^3-7}\right\}$.
6. $\left\{n^2 \ln\left(1-\dfrac{1}{n}\right)\right\}$.
7. State the completeness property.

Section 11.2

Decide whether each of the series given in Exercises 8–27 converges. In Exercises 8–15, find the sum of all convergent series.

8. $\displaystyle\sum_{n=1}^{\infty}\frac{1}{(n+1000)^{1/2}}$.
9. $\displaystyle\sum_{n=1}^{\infty}\frac{1}{5}\left(\frac{1}{8}\right)^n$.
10. $\dfrac{5}{4}+\dfrac{5}{16}+\dfrac{5}{64}+\cdots$.
11. $\displaystyle\sum_{n=1}^{\infty}\left(\frac{1}{1000}-\frac{1}{n}\right)$.
12. $\displaystyle\sum_{n=1}^{\infty}\frac{2n^2-3n}{14+10n^2}$.
13. $\dfrac{10}{\sqrt{401}}+\dfrac{10}{\sqrt{402}}+\dfrac{10}{\sqrt{403}}+\cdots$.
14. $\displaystyle\sum_{n=1}^{\infty}\ln\left(\frac{n+1}{n}\right)$.
15. $\displaystyle\sum_{n=1}^{\infty}\frac{1}{e^{2n}}$.

Section 11.3

16. $\displaystyle\sum_{n=1}^{\infty}\frac{1}{n^3+n}$.
17. $\displaystyle\sum_{n=1}^{\infty}\frac{2n+1}{n^2(n+1)^2}$.
18. $\displaystyle\sum_{n=1}^{\infty}\frac{1}{(n+3)^{1/3}}$.
19. $\displaystyle\sum_{n=2}^{\infty}\frac{1}{n^2-n}$.
20. $\displaystyle\sum_{n=2}^{\infty}\frac{1}{n(\ln n)^5}$.
21. $\displaystyle\sum_{n=1}^{\infty}\frac{1}{10n+n^{1/2}}$.

Section 11.4

22. $\sum_{n=2}^{\infty} \frac{3^{n-1}}{n!2n}$.

23. $\sum_{n=1}^{\infty} \frac{7n^2+1}{n^2(n+1)!}$.

24. $\sum_{n=1}^{\infty} \frac{1}{n(n+10)^{1/2}}$.

25. $\sum_{n=1}^{\infty} \frac{n^3-3n^2+1}{n^5+10n^3-1}$.

26. $\sum_{n=1}^{\infty} \frac{13n}{n!}$.

27. $\sum_{n=1}^{\infty} \frac{2^n}{(n-1)!3^{n-1}}$.

Section 11.5

In Exercises 28–33, determine whether the given series is absolutely convergent, conditionally convergent, or divergent.

28. $1 - \frac{1}{3} + \frac{1}{5} - \frac{1}{7} + \frac{1}{9} - \cdots$.

29. $\sum_{n=1}^{\infty} (-1)^n \frac{1}{\sqrt{n}}$.

30. $\sum_{n=1}^{\infty} (-1)^n \frac{n^2+3n}{n^4+1}$.

31. $\sum_{n=1}^{\infty} (-1)^n \frac{(n!)^2}{(2n)!}$.

32. $\sum_{n=0}^{\infty} (-1)^n \frac{n+1}{n}$.

33. $\frac{1}{7} - \frac{1}{3}\left(\frac{1}{7}\right)^3 + \frac{1}{5}\left(\frac{1}{7}\right)^5 - \frac{1}{7}\left(\frac{1}{7}\right)^7 + \cdots$.

34. Find an approximation to the sum of the series $-1 + \frac{1}{10} - \frac{1}{100} + \frac{1}{1000} - \cdots$ that is correct to within $5(10)^{-6}$.

Section 11.6

Determine whether the series given in Exercises 35–40 converge or diverge.

35. $\sum_{n=1}^{\infty} \frac{3n}{n!}$.

36. $\sum_{n=1}^{\infty} n^3 e^{(-n^4)}$.

37. $-\frac{5}{2} + \frac{10}{4} - \frac{15}{8} + \frac{20}{16} - \cdots$.

38. $\sum_{n=1}^{\infty} \frac{1}{(n^3+7)^{1/2}}$.

39. $\sum_{n=1}^{\infty} \frac{\ln n}{n}$.

40. $\sum_{n=1}^{\infty} \frac{n^7}{n!}$.

Section 11.7

In Exercises 41–46, find all values of x for which the given series converges.

41. $\sum_{n=1}^{\infty} \frac{n+1}{\sqrt{n}}(x+3)^n$.

42. $\sum_{n=1}^{\infty} \frac{(-1)^n}{3} x^n$.

43. $x - \frac{x^3}{3!} + \frac{x^5}{5!} - \frac{x^7}{7!} + \cdots$.

44. $(x+1)^2 + (x+1)^3 + (x+1)^4 + \cdots$.

45. $\sum_{n=1}^{\infty} \frac{x^n}{n(n+1)}$.

46. $\sum_{n=1}^{\infty} n!x^n$.

Section 11.8

47. Find the Taylor Series for $1/(1-x)$, $|x|<1$. (*Hint*: Consider a geometric series.)
48. Use Exercise 47 to find a Taylor Series for $f(x)=\ln|1-x|$, $x<1$.
49. Find the first four terms of the Taylor Series for $f(x)=\ln(1+e^x)$ about $x=0$.
50. Find the first four nonzero terms of the Taylor Series for $f(x)=(\ln|1-x|)(\cos x)$ about $x=0$ by multiplying the series for $\ln|1-x|$ and $\cos x$.

Challenging Problems, Chapter 11

1. Let $f_k(x)$ denote the kth iteration of the continuous function $f(x)$. That is, let $f_1(x)=f(x)$, $f_2(x)=f(f(x))$, and in general, $f_k(x)=f(f_{k-1}(x))$. Assume that

$$\lim_{k\to\infty} f_k(x)=L.$$

Show that L is a fixed point of f; that is, $f(L)=L$.

2. Let $f_k(x)$ be the kth interation of the sine function. Let $0\le x\le\pi$. Show $\lim_{k\to\infty} f_k(x)$ exists and use the results of Problem 1 to determine the value of the limit. (*Hint*: To prove the existence of the limit prove that $f_1(x)\ge f_2(x)\ge\cdots\ge f_k(x)\ge\cdots\ge 0$.)

3. If you have a hand-held calculator with the cosine function, enter 1 radian and take the cosine function repeatedly (say about 20 times). What happens? Make a conjecture and prove it.

4. Show that a function defined as the sum of an infinite series of continuous functions is not necessarily continuous. (*Hint*: Consider

$$f(x)=\sum_{n=1}^{\infty}(\cos^n x-\cos^{n-1}x),\quad -\frac{\pi}{2}\le x\le\frac{\pi}{2}.\Big)$$

5. The computation of π is an interesting application of Taylor series. The following sequence of steps illustrates some of the problems that arise in this computation and it suggests solutions.
   (i) Use the Taylor series for $\arctan x$ to show that

$$\pi=4\sum_{i=0}^{\infty}(-1)^i\frac{1}{2i+1}.$$

   (ii) Show that to compute π with an error of less than $5(10)^{-10}$ one must use $5(10)^{-8}-1$ terms of this series.

The remaining steps of this problem illustrate a method of avoiding the use of such a large number of terms.

(iii) Use the identity

$$\tan(\alpha + \beta) = \frac{\tan\alpha + \tan\beta}{1 - \tan\alpha \tan\beta}$$

to prove

$$\pi = 4\left(\arctan\frac{1}{2} + \arctan\frac{1}{3}\right).$$

(iv) Use the result of (iii) to show that

$$\pi = 4\sum_{i=0}^{\infty}(-1)^i\frac{1}{(2i+1)}\left(\frac{1}{2^{2i+1}} + \frac{1}{3^{2i+1}}\right)$$

(v) Show that 15 terms of the series in (iv) are sufficient for the computation of π with an error of less than $5(10)^{-10}$.

6. A paradox of Zeno describes a race between Achilles and a tortoise. In the race the tortoise was given a head start. Zeno reasoned as follows that the tortoise must win the race. To overtake the tortoise, Achilles must at some point cover half the distance between him and the tortoise. At this point he still has half the distance to make up. When he then covers half the remaining distance he still has 1/4 of the original distance to make up, and so on. Consequently, Achilles can never catch the tortoise. Assume that the difference in speed between Achilles and the tortoise is a constant, c ft/sec. Then use series to explain the paradox.

7. Prove that the Completeness Property does not hold for the set of rational numbers.

8. Construct a divergent alternating series $a_1 - a_2 + a_3 - a_4 + \cdots$ for which $\lim_{n\to\infty} a_n = 0$.

9. If the probability of success of an experiment is p, then the expected number of trials required for a successful experiment is

$$\sum_{n=1}^{\infty} npq^{n-1}, \qquad \text{where } q = 1 - p.$$

(i) Find the sum of this series.
(ii) The probability of rolling a seven on a pair of dice is 1/6. What is the expected number of rolls required to obtain a 7?

10. Infinite series may be used to show that the graph of a continuous function on a closed interval may fail to be finite in length. Consider, for example, the function

$$f(x) = \begin{cases} x\sin(\pi/x) & \text{if } 0 < x \le 1 \\ 0 & \text{if } x = 0. \end{cases}$$

This function is continuous for $0 \le x \le 1$.

In Figure 1, one arch of the graph is pictured. By using the illustration first show that the arc length of the graph of $y = f(x)$, $1/n \leq x \leq 1(n-1)$, is greater than or equal to $2/(2n-1)$. Use this fact to conclude that the graph does not have finite arc length for $0 \leq x \leq 1$.

Figure 1

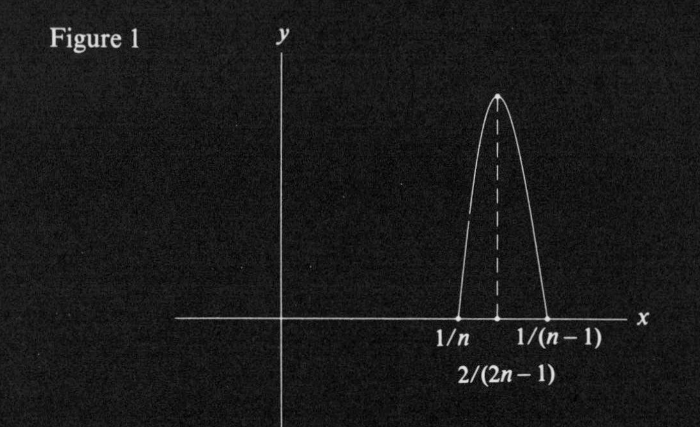

11. Use the result in Problem 11 of Challenging Problems, Chapter 10 to help prove the divergence of the infinite series $\sum_{n=1}^{\infty}(1/P_n)$, where P_n is the nth positive prime number.

12. Show that the Taylor Series with $a = 0$ for

$$f(x) = \begin{cases} 0 & \text{if } x = 0 \\ e^{(-1/x^2)} & \text{if } x \neq 0 \end{cases}$$

converges to $f(x)$ only at $x = 0$. (*Hint*: Use the definition of the derivative to show $f^{(n)}(0) = 0$ for all n.)

13. Use the following outline to prove that $\lim_{h \to 0} (1 + h)^{1/h}$ (the limit defined to be e) exists.

(i) Let $t = 1/h$. Then $f(t) = \left(1 + \frac{1}{t}\right)^t = (1 + h)^{1/h}$.

(ii) First let $t = n$, where n is a positive integer. Use the binomial theorem to show that

$$f(n) = 1 + 1 + \frac{1}{2!} \cdot \frac{n-1}{n} + \frac{1}{3!} \frac{n-1}{n} \cdot \frac{n-2}{n} + \cdots$$

$$+ \frac{1}{(n-1)!} \cdot \frac{n-1}{n} \cdot \frac{n-2}{n} \cdots \frac{2}{n} + \frac{1}{n!} \frac{n-1}{n} \cdot \frac{n-2}{n} \cdots \frac{1}{n}$$

and

$$f(n+1) = 1 + 1 + \frac{1}{2!}\cdot\frac{n}{n+1} + \frac{1}{3!}\frac{n}{n+1}\cdot\frac{n-1}{n+1} + \cdots$$
$$+ \frac{1}{n!}\cdot\frac{n}{n+1}\cdot\frac{n-1}{n+1}\cdots\frac{2}{n+1} + \left(\frac{1}{n+1}\right)^{n+1}.$$

(iii) Show that if $k < n$, then $\frac{k}{n} < \frac{k+1}{n+1}$ and conclude from step (ii) that $f(n) < f(n+1)$.

(iv) Use the first part of step (ii) to obtain

$$f(n) < 1 + 1 + \frac{1}{2!} + \frac{1}{3!} + \cdots + \frac{1}{n!} < 1 + 1 + \frac{1}{2} + \left(\frac{1}{2}\right)^2 + \cdots + \left(\frac{1}{2}\right)^{n-1}.$$

(v) Then show that $f(n) < 3 - \left(\frac{1}{2}\right)^{n-1} < 3$ for all n.

(vi) Conclude that the infinite sequence $\{f(n)\}$ must converge to a number $L \le 3$; that is, $\lim_{n\to\infty}\left(1 + \frac{1}{n}\right)^n = L$, when n is an integer.

(vii) Next let t be any positive real number. Then there is a positive integer $n(t)$ such that $n(t) \le t \le n(t) + 1$, and so

$$1 + \frac{1}{n(t)+1} \le 1 + \frac{1}{t} \le 1 + \frac{1}{n(t)}.$$

(viii) Show that

$$\left(1 + \frac{1}{n(t)+1}\right)^{n(t)} \le \left(1 + \frac{1}{t}\right)^t \le \left(1 + \frac{1}{n(t)}\right)^{n(t)+1}$$

and use that inequality in the form

$$\frac{\left(1 + \frac{1}{n(t)+1}\right)^{n(t)+1}}{1 + \frac{1}{n(t)+1}} \le \left(1 + \frac{1}{t}\right)^t \le \left(1 + \frac{1}{n(t)}\right)^{n(t)}\left(1 + \frac{1}{n(t)}\right).$$

(ix) Use the last inequality to show that $\lim_{t\to\infty}\left(1 + \frac{1}{t}\right)^t = L$.

(x) Let r be a negative real number and let $r = -s$, where $s > 0$; and show that

$$\left(1 + \frac{1}{r}\right)^r = \left(1 + \frac{1}{s-1}\right)^s = \left(1 + \frac{1}{s-1}\right)^{s-1}\left(1 + \frac{1}{s-1}\right),$$

and use that equality to show that $\lim_{r\to-\infty}\left(1 + \frac{1}{r}\right)^r = L$.

(xi) Conclude that $\lim_{h\to 0}(1+h)^{1/h} = L$ and so that limit exists.

14. Use the following outline to prove that e is irrational (not the quotient of two integers).
   (i) Assume that e is rational, say $e = p/q$, where p and q are integers and $q > 1$.
   (ii) Then, since

$$e = 1 + 1 + \frac{1}{2!} + \frac{1}{3!} + \frac{1}{4!} + \cdots,$$

   we have

$$\frac{p}{q} = 1 + 1 + \frac{1}{2!} + \frac{1}{3!} + \cdots + \frac{1}{q!} + \frac{1}{(q+1)!} + \cdots.$$

   (iii) Multiplication by $q!$ produces

$$p(q-1)! = q!\left(1 + 1 + \frac{1}{2!} + \frac{1}{3!} + \cdots + \frac{1}{q!}\right) + q!\left(\frac{1}{(q+1)!} + \frac{1}{(q+2)!} + \cdots\right)$$

   or

$$\begin{aligned} p(q-1)! - q!\left(1 + 1 + \frac{1}{2!} + \frac{1}{3!} + \cdots + \frac{1}{q!}\right) &= q!\left(\frac{1}{(q+1)!} + \frac{1}{(q+2)!} + \cdots\right) \\ &= \frac{1}{q+1} + \frac{1}{(q+1)(q+2)} + \frac{1}{(q+1)(q+2)(q+3)} + \cdots. \end{aligned}$$

   (iv) Now show that the left side of the preceding equation is an integer, while the right side is positive and less than 1. This contradiction will complete the proof. (To show that the right side is less than 1 show that each of its terms is less than or equal to a corresponding term of a geometric series whose sum is less than 1.)

15. Use the following outline of an elegant proof by Ivan Niven (*Bulletin of the American Mathematical Society*, Vol. 53, p. 509) to prove that π is irrational.
   (i) Assume that π is rational, say $\pi = a/b$, where a and b are positive integers.
   (ii) Let n be a positive integer, $f(x) = \dfrac{x^n(a - bx)^n}{n!}$,

   and

$$F(x) = f(x) - f''(x) + f^{(4)}(x) - \cdots + (-1)^n f^{(2n)}(x).$$

   (iii) Show that $f\left(\dfrac{a}{b} - x\right) = f(x)$.

(iv) Observe that $n!f(x)$ is a polynomial with integer coefficients, and that each of its terms has degree n or higher in x.

(v) Use step (iv) to conclude that f and all its derivatives are integers when $x = 0$, and then use step (iii) to conclude that this is also true when $x = \pi = a/b$. Conclude that $F(0)$ and $F(\pi)$ are integers.

(vi) Show that the derivative of $F'(x)\sin x - F(x)\cos x$ is $f(x)\sin x$.

(vii) Use step (vi) to get $\int_0^\pi f(x)\sin x\,dx = F(\pi) + F(0)$.

(viii) Show that for $0 \le x \le \pi$, $f(x)$ has its absolute maximal value when $x = a/2b$.

(ix) Show that for $0 < x < \pi$,

$$0 < f(x)\sin x \le \frac{(\pi a)^n}{2^{2n}n!} < \frac{(\pi a)^n}{n!}, \quad n > 1.$$

(x) Use the fact that $\sum \frac{x^n}{n!}$ converges to show that $\lim_{n\to\infty} \frac{(\pi a)^n}{n!} = 0$, and so conclude that if n is large enough, $0 < \frac{(\pi a)^n}{n!} < \frac{1}{\pi}$.

(xi) Use steps (ix) and (x) to show that $0 < \int_0^\pi f(x)\sin x\,dx < 1$.

(xii) Use steps (v) and (vii) along with step (xi) to arrive at a contradiction and thus complete the proof.

12

Polar Coordinates

12.1 Polar Coordinates

Previously we always specified the position of a point in a plane by giving its x and y coordinates, that is, by specifying the directed distances of the point from a horizontal axis and a vertical axis. There are several other ways of "coordinatizing" the plane. In this chapter we shall make a detailed study of one such new method, the polar coordinate system.

To establish a polar coordinate system in a plane, first select a point O, which is called the *pole* or *origin* of the coordinate system, and then from point O extend a ray R. The ray R is normally taken to be horizontal and extending to the right, as in Figure 12.1.1.

Figure 12.1.1

O R

The point P whose polar coordinates are (r, θ) is obtained by considering the angle θ with vertex at O and initial side as the ray R. When r is positive or zero, the point P is the point on the terminal side of θ, at a distance r from O. When r is negative, the point P is the point in the opposite direction of the terminal side of θ, at a distance $|r|$ from O.

Figures 12.1.2 and 12.1.3 illustrate the points determined by several pairs of polar coordinates.

Figure 12.1.2

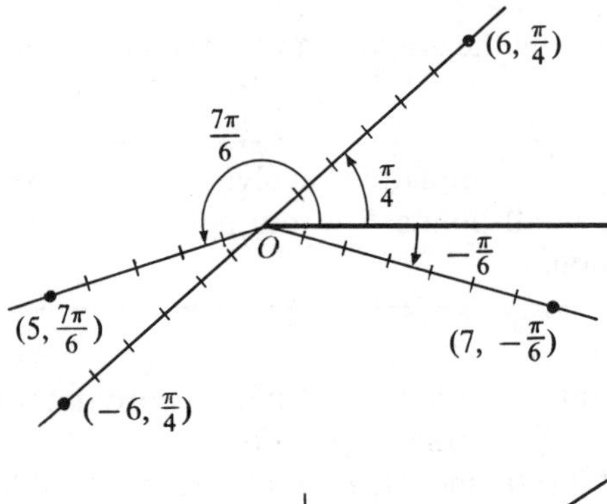

$(6, \frac{\pi}{4})$ $\frac{7\pi}{6}$ $\frac{\pi}{4}$ O $-\frac{\pi}{6}$ R $(5, \frac{7\pi}{6})$ $(7, -\frac{\pi}{6})$ $(-6, \frac{\pi}{4})$

Figure 12.1.3

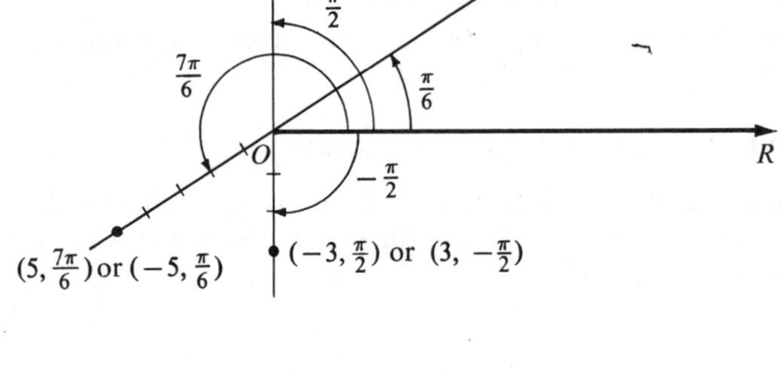

$\frac{\pi}{2}$ $\frac{7\pi}{6}$ $\frac{\pi}{6}$ O R $-\frac{\pi}{2}$ $(5, \frac{7\pi}{6})$ or $(-5, \frac{\pi}{6})$ $(-3, \frac{\pi}{2})$ or $(3, -\frac{\pi}{2})$

From the example of Figure 12.1.3 it should be clear that *every point P has many different pairs of polar coordinates.* For example, both $(-5, \pi/6)$ and $(5, 7\pi/6)$ represent the same point. Indeed, no matter what values are given to r and θ, the coordinates (r, θ), $(r, \theta + 2\pi)$, and $(-r, \pi + \theta)$ always represent the same point. That fact makes polar coordinates somewhat more difficult to work with than rectangular coordinates. However, we shall find that polar coordinates can be used to simplify many problems.

Exercises 12.1

For Exercises 1–20 plot the following points on a polar coordinate system.

1. $(3, \pi/4)$.
2. $(5, \pi/2)$.
3. $(-3, \pi/4)$.
4. $(-5, \pi/2)$.
5. $(-7, -2\pi/3)$.
6. $(7, \pi/3)$.
7. $(7, 7\pi/3)$.
8. $(-7, -8\pi/3)$.
9. $(-2, \pi)$.
10. $(2, 3\pi/2)$.
11. $(2, 0)$.
12. $(2, -\pi/2)$.
13. $(-5, 4\pi/3)$.
14. $(3, 5\pi/6)$.
15. $(3, -7\pi/6)$.
16. $(2, -4\pi/3)$.
17. $(5, -\pi)$.
18. $(-2, -\pi)$.
19. $(-3, -11\pi/6)$.
20. $(-5, 7\pi/6)$.
21. Show that the distance between the points (r_1, θ_1) and (r_2, θ_2) is $[r_1{}^2 + r_2{}^2 - 2r_1r_2 \cos(\theta_2 - \theta_1)]^{1/2}$. (*Hint*: Use the "Law of Cosines.")

12.2 Graphs of Polar Equations

We begin this section by defining the graph of a polar equation.

Definition 12.2.1

> The *graph* of an equation involving polar coordinates r and θ is the set of all points P whose polar coordinates (r, θ) satisfy the equation.

Thus a point P need only have *one* set of polar coordinates (r, θ) that satisfy the given equation in order to be on the graph. For example, the point represented by $(1, 2\pi)$ is on the graph of $\theta = 0$, since $(1, 2\pi)$ is also represented by $(1, 0)$, a pair of polar coordinates that satisfy the given equation.

In the following examples we shall graph a few polar equations.

Example 1 | Graph these equations: (a) $r = 2$; (b) $\theta = -\pi/4$.

(a) The graph of $r = 2$ is the set of all points (r, θ) with $r = 2$, that is, the set of all points at a distance 2 from the pole O. So the graph is a circle of radius 2 centered at the pole.

(b) The graph of $\theta = -\pi/4$ is the set of all points (r, θ) with $\theta = -\pi/4$, that is, the set of all points on the line making an angle $-\pi/4$ radians with the initial ray R. Both graphs are indicated in Figure 12.2.1. ||

Figure 12.2.1

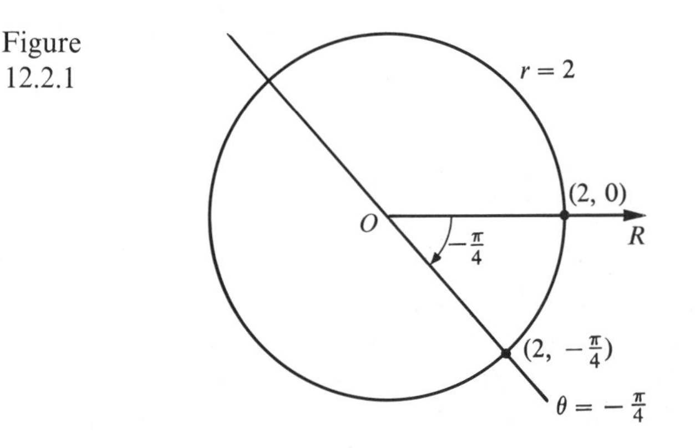

Example 2 | Sketch the graph of the spiral $r = \theta$, $\theta \geq 0$.

The following table gives a few points on the graph of $r = \theta$ for $\theta \geq 0$. If we plot those points and connect them with a smooth curve, we obtain the approximate graph indicated in Figure 12.2.2. The graph continues to spiral outward around the pole, since r increases as θ increases.

Figure 12.2.2

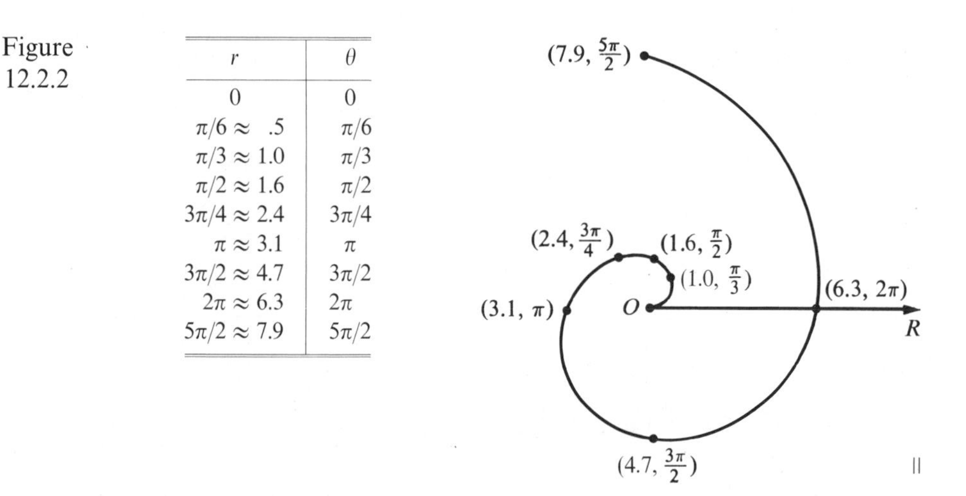

r	θ
0	0
$\pi/6 \approx .5$	$\pi/6$
$\pi/3 \approx 1.0$	$\pi/3$
$\pi/2 \approx 1.6$	$\pi/2$
$3\pi/4 \approx 2.4$	$3\pi/4$
$\pi \approx 3.1$	π
$3\pi/2 \approx 4.7$	$3\pi/2$
$2\pi \approx 6.3$	2π
$5\pi/2 \approx 7.9$	$5\pi/2$

||

Example 3 | Sketch the graph of $r = 1 + \sin\theta$.

Before determining some points on the graph it is often best to determine the general shape of the graph. Since $\sin\theta$ increases from 0 to 1 as θ ranges from 0 to $\pi/2$, it follows that $r = 1 + \sin\theta$ must increase from 1 to 2 as θ goes from 0 to $\pi/2$. Similarly, as θ varies from $\pi/2$ to π, we see that $\sin\theta$ decreases from 1 to 0 and so $r = 1 + \sin\theta$ decreases from the value 2 to the value 1. That information enables us to make a rough sketch of the graph for $0 \leq \theta \leq \pi$, as indicated in Figure 12.2.3.

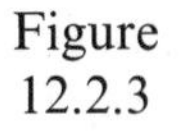

Figure 12.2.3

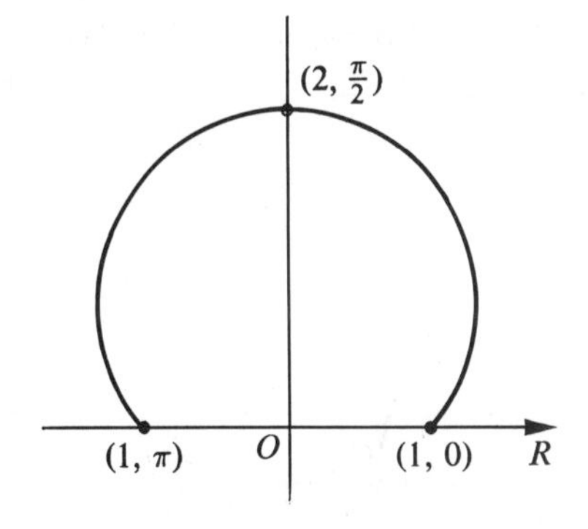

Continuing in the same way, we find that since $\sin\theta$ decreases from 0 to -1 as θ ranges from π to $3\pi/2$, $r = 1 + \sin\theta$ must decrease from 1 to 0 in that range. Finally, $\sin\theta$ increases from -1 to 0 as θ goes from $3\pi/2$ to 2π, so that $r = 1 + \sin\theta$ must increase from 0 to 1 there. The information can be conveniently summarized in the table below.

θ	$\sin\theta$	$r = 1 + \sin\theta$
$0 \to \pi/2$	$0 \to 1$	$1 \to 2$
$\pi/2 \to \pi$	$1 \to 0$	$2 \to 1$
$\pi \to 3\pi/2$	$0 \to -1$	$1 \to 0$
$3\pi/2 \to 2\pi$	$-1 \to 0$	$0 \to 1$

From that table a more complete sketch can be made, as shown in Figure 12.2.4. ||

Figure 12.2.4

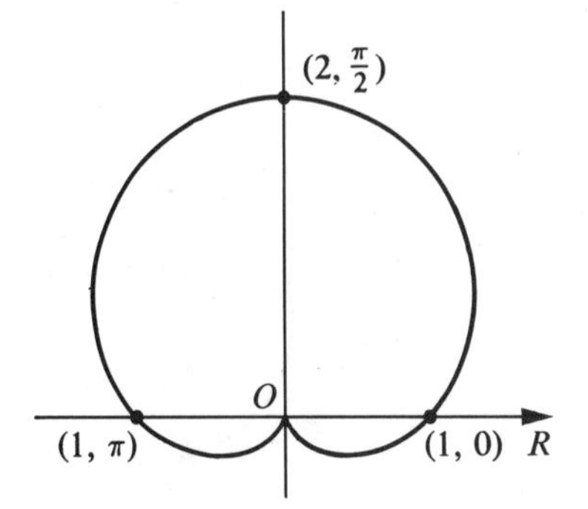

Example 3 illustrates one of the forms of symmetry that is valuable in sketching graphs of polar equations. Since $\sin(\pi - \theta) = \sin\theta$, we know $r = 1 + \sin\theta$ if and only if $r = 1 + \sin(\pi - \theta)$. Thus, a point (r, θ) is on the graph of $r = 1 + \sin\theta$ if and only if the point $(r, \pi - \theta)$ is also on the graph. As shown in Figure 12.2.5, the point $(r, \pi - \theta)$ can be obtained from the point (r, θ) by reflecting through the line $\theta = \pi/2$. Hence we have symmetry about the line $\theta = \pi/2$.

Since each point has many different pairs of polar coordinates, symmetry about the line $\theta = \pi/2$ may occur in many different ways. For example, we see from

Figure 12.2.5

Figure 12.2.6

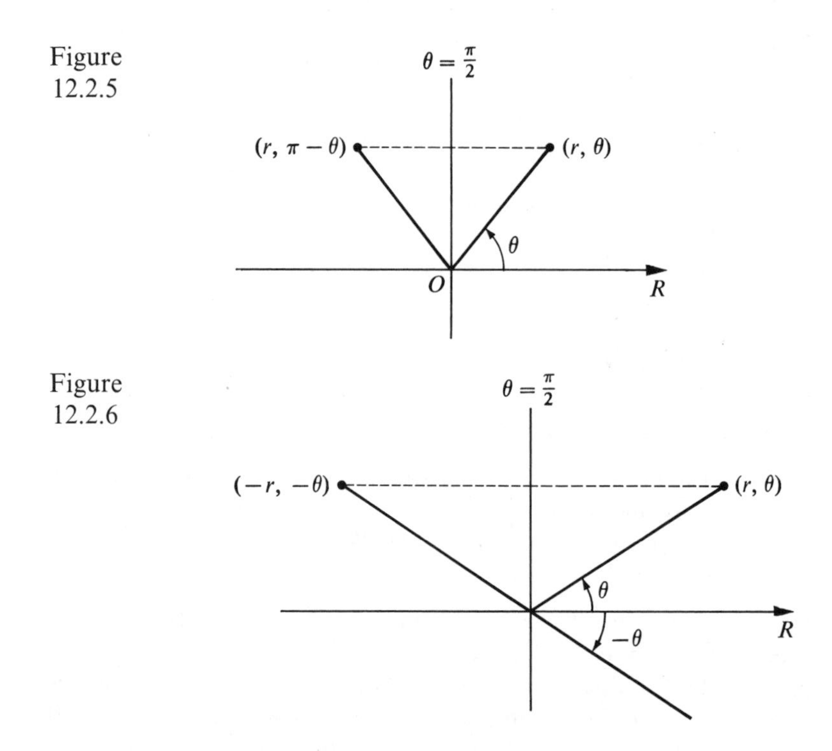

Figure 12.2.6 that a graph also has symmetry about the line $\theta = \pi/2$ provided that the point $(-r, -\theta)$ is on the graph if and only if the point (r, θ) is on the graph.

Other symmetries are also useful. For example suppose a point (r, θ) is on a graph if and only if the point $(-r, \theta)$ is on the graph. In that case the graph has *polar symmetry*, as indicated in Figure 12.2.7.

Figure 12.2.7

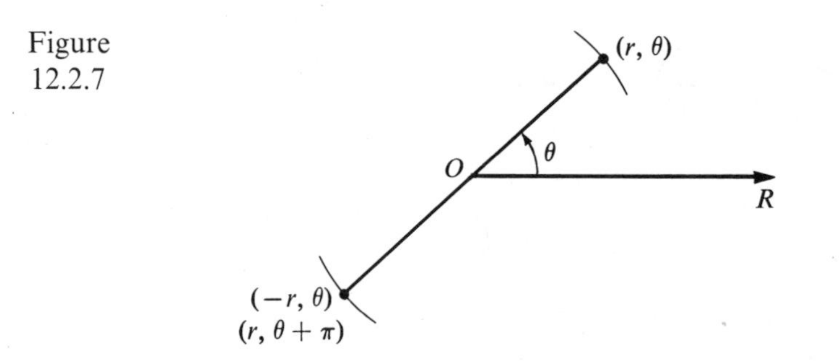

As with symmetry about the line $\theta = \pi/2$, polar symmetry can occur in other ways. For example, a graph will also have polar symmetry provided that (r, θ) is on the graph if and only if $(r, \theta + \pi)$ is on the graph.

Finally, a graph has symmetry around the horizontal line $\theta = 0$ provided that (r, θ) is on the graph if and only if $(r, -\theta)$ is on the graph, as illustrated in Figure 12.2.8.

Some of the other ways that a graph may have symmetry about the line $\theta = 0$ will be examined in the exercises.

Figure 12.2.8

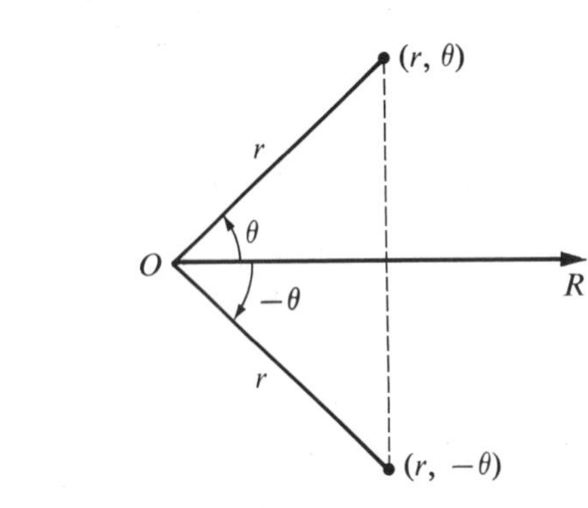

Example 4 | Sketch the graph of $r = \cos 2\theta$.

Since $\cos 2\theta = \cos 2(-\theta)$, a point (r, θ) is on the graph if and only if the point $(r, -\theta)$ is also on the graph. Consequently the graph is symmetric about the line $\theta = 0$. Moreover since $\cos 2(\pi - \theta) = \cos(2\pi - 2\theta) = \cos(-2\theta) = \cos 2\theta$, the graph also has symmetry about the vertical line $\theta = \pi/2$.

To sketch the graph, note that $\cos 2\theta$ decreases from 1 to 0 as 2θ ranges from 0 to $\pi/2$, that is, as θ goes from 0 to $\pi/4$. This enables us to sketch the portion of the graph indicated in Figure 12.2.9. Continuing in the same fashion we note that as θ ranges from $\pi/4$ to $\pi/2$, we see that 2θ goes from $\pi/2$ to π and so $\cos 2\theta$ decreases from 0 to -1. That information enables us to sketch the additional portion of the graph indicated in Figure 12.2.10. By use of symmetry we can complete the sketch as

Figure 12.2.9

Figure 12.2.10

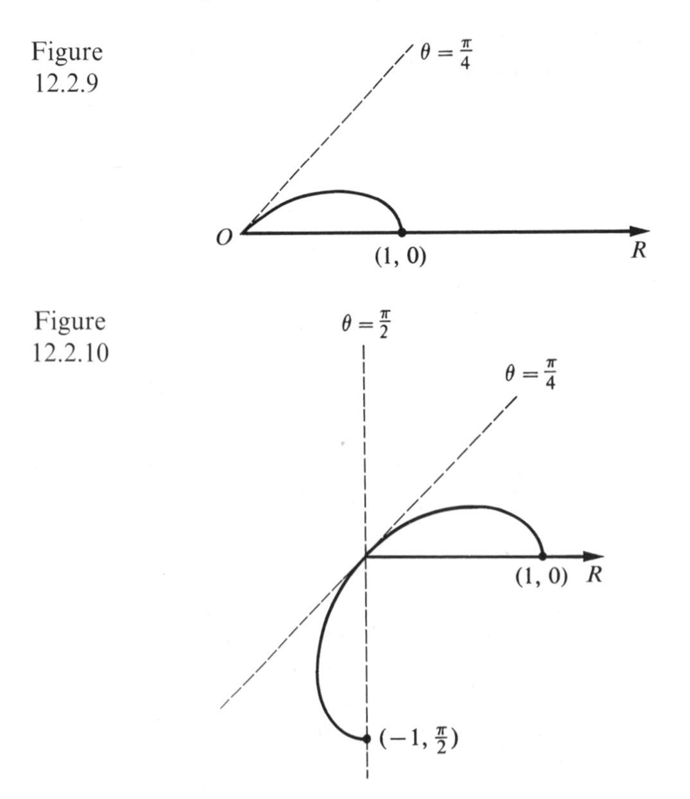

indicated in Figure 12.2.11. Of course if we plotted a few more points on the graph we would determine the graph more precisely. ||

Figure 12.2.11

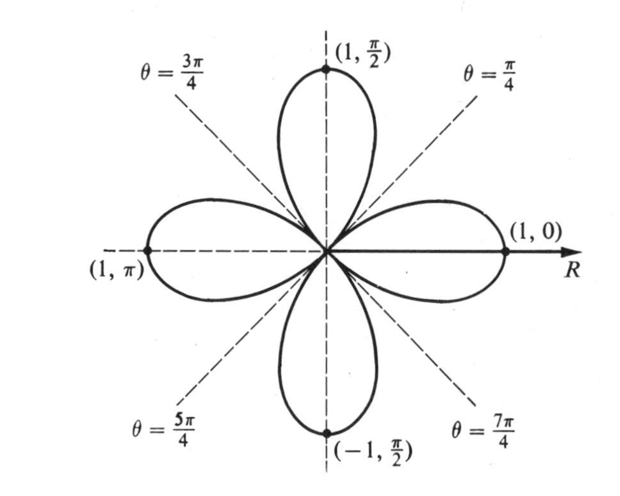

Example 5 | Sketch the graph of $r = 2 + 3 \sin \theta$.

A summary of the way r changes as θ changes is indicated in the following table.

θ	$\sin \theta$	$r = 2 + 3 \sin \theta$
$0 \to \pi/2$	$0 \to 1$	$2 \to 5$
$\pi/2 \to \pi$	$1 \to 0$	$5 \to 2$
$\pi \to 3\pi/2$	$0 \to -1$	$2 \to -1$
$3\pi/2 \to 2\pi$	$-1 \to 0$	$-1 \to 2$

As θ goes from 0 to $\pi/2$ and then from $\pi/2$ to π we obtain the part of the graph indicated in Figure 12.2.12. As we continue to sketch the graph while θ goes from π to $3\pi/2$, we see that r goes from 2 to -1. As r changes in that way, it must go from 2 through 0 to -1. Thus the graph goes through the origin and up to the point $(-1, 3\pi/2)$. After r is 0, it is negative, and so the points on the graph are in the opposite direction of the terminal sides of the associated angles. The graph then

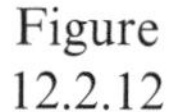
Figure 12.2.12

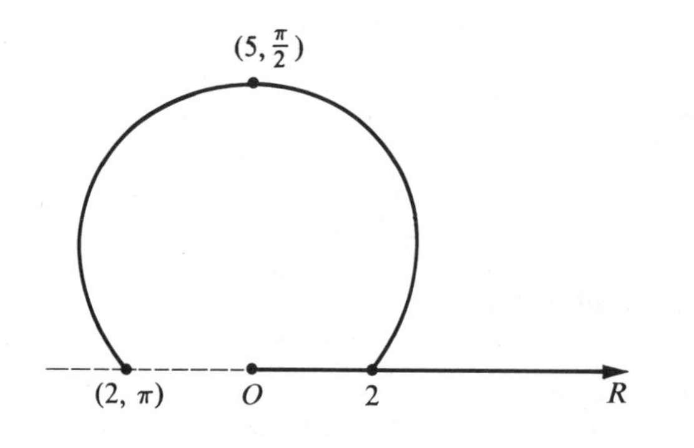

continues, as indicated in Figure 12.2.13, to the point $(-1, 3\pi/2)$. To complete the sketch of the graph we go from $(-1, 3\pi/2)$ to $(2, 2\pi)$, and hence again the graph passes through the pole, as indicated in Figure 12.2.14.

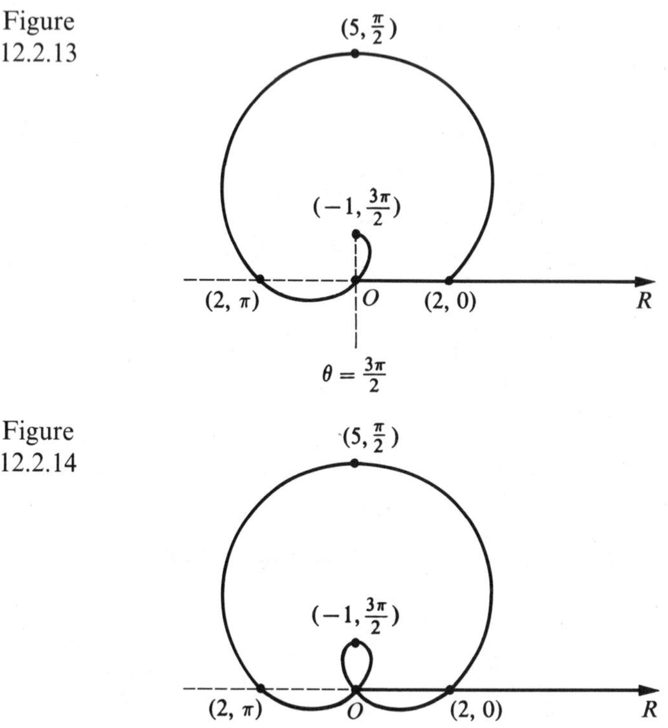

Figure 12.2.13

Figure 12.2.14

||

Exercises 12.2

In Exercises 1–22, sketch the graph of the given equation in polar coordinates. In each case be sure to use any symmetry of the graph as an aid in sketching the graph.

1. $r \sin\theta = 1$.
2. $r \cos\theta = 1$.
3. $r = \sin\theta$.
4. $r = \cos\theta$.
5. $r = 1 - \cos\theta$.
6. $r = 1 + \cos\theta$.
7. $r = 4 - \sin\theta$.
8. $r = 4 - \cos\theta$.
9. $r = \cos 3\theta$.
10. $r = \sin 2\theta$.
11. $r = 2 + 3\cos\theta$.
12. $r = 1 + 2\sin\theta$.
13. $r = \sin 3\theta$.
14. $r = \cos 2\theta$.
15. $r = \cos(\theta/2)$.
16. $r = \sin(\theta/2)$.
17. $r = 1 + \sin(\theta/2)$.
18. $r = 1 + \cos(\theta/2)$.
19. $r = e^{\theta/\pi}, \quad \theta \geq 0$.
20. $r = e^{\theta/\pi}, \quad \theta \leq 0$.
21. $r^2 = a^2 \cos 2\theta, \quad a > 0$.
22. $r = 2\sin^2(\theta/2)$.
23. Show that a graph has symmetry about the line $\theta = 0$ in either of the following situations.
    (a) (r, θ) is on the graph if and only if $(-r, \pi - \theta)$ is on the graph.
    (b) (r, θ) is on the graph if and only if $(r, 2\pi - \theta)$ is on the graph.
    Find two other possible ways that a graph may have symmetry about the line $\theta = 0$.
24. Solve the two equations $r = \theta$ and $\theta = \pi/4$ simultaneously. Give a complete list of all points the graphs have in common. How do you explain the apparent contradiction?

25. Let a and b be positive constants. As illustrated in Figure 12.2.15, let Q be a point on the line $r\cos\theta = a$ and consider the line through the pole O and the point Q. On this line select a point P such that the distance from P to Q is b. Find an equation for the set of all such points P as Q moves along the line $r\cos\theta = a$, and sketch the graph.

Figure 12.2.15

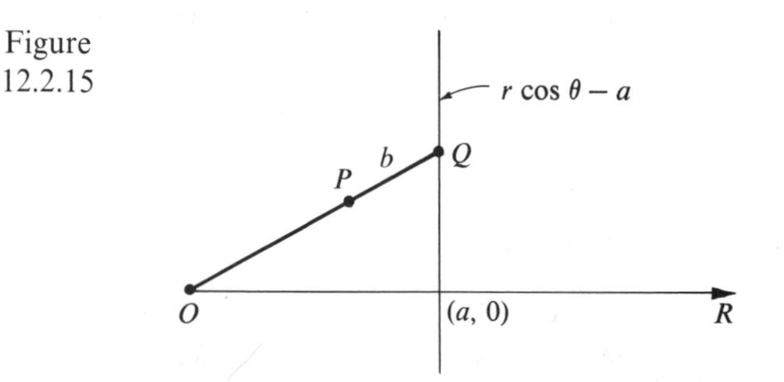

26. As in the previous exercise, let Q be a point on the circle $r = a\cos\theta$; and consider the line through the pole O and the point Q. On this line select a point P such that the distance from P to Q is b. Find an equation for the set of all points P and sketch the graph.

27. A conic may be defined as follows: Let F be a fixed point (called the *focus*), let D be a fixed line (called the *directrix*), and let ε be a positive constant (called the *eccentricity*). Let P be a point and let d_F be the distance between P and F; and let d_D be the distance between P and D. The set of all points P such that $d_F = \varepsilon d_D$ is called a *conic*. In fact, the conic is an ellipse if $\varepsilon < 1$, a parabola if $\varepsilon = 1$, and a hyperbola if $\varepsilon > 1$. Let D be the vertical line p units to the left of the pole (where p is a fixed positive constant), and let F be at the pole. Show that a polar equation of the conic is

$$r = \frac{\varepsilon p}{1 - \varepsilon\cos\theta}.$$

12.3 Relations between Polar and Rectangular Coordinates

It is often useful to transform an equation expressed in one coordinate system into an equation expressed in another. To see how this can be done, we shall superimpose a polar coordinate system on a rectangular coordinate system, with the ray R corresponding to the positive x-axis. Then every point P in the plane will not only have a set of polar coordinates (r, θ) but also a set of rectangular coordinates (x, y). From Figure 12.3.1 we see that

$$x = r\cos\theta \quad \text{and} \quad y = r\sin\theta. \tag{1}$$

Figure 12.3.1

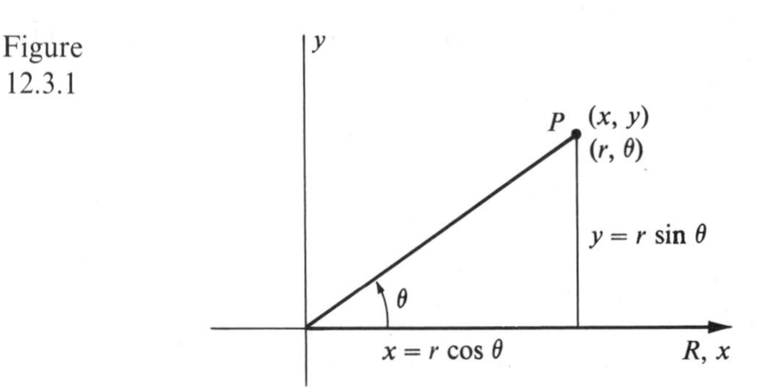

Those relations can be used to transform any equation in rectangular form to one in polar form.

Example 1 | Transform the equation $x^2 + y^2 = 4$ to polar form.

From (1) above we have

$$r^2 \cos^2 \theta + r^2 \sin^2 \theta = 4.$$

Using the fact that $\cos^2 \theta + \sin^2 \theta = 1$, we obtain

$$r^2 = 4, \quad \text{or} \quad |r| = 2. \qquad \|$$

To reverse that process and transform polar equations into rectangular form we again superimpose a polar coordinate system on a rectangular coordinate system, as in Figure 12.3.2.

Figure 12.3.2

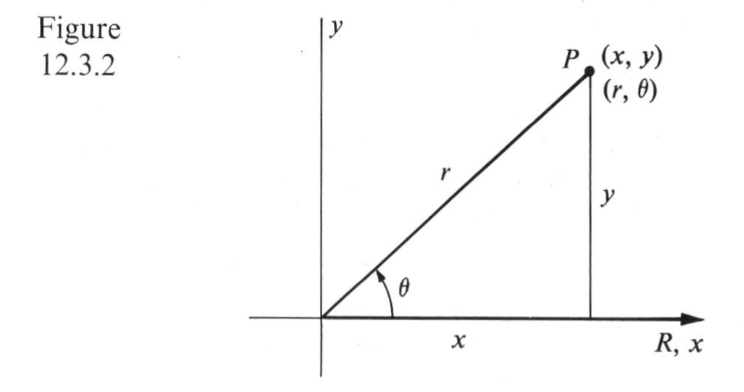

Reference to Figure 12.3.2 indicates that

$$r^2 = x^2 + y^2 \quad \text{and} \quad \tan \theta = \frac{y}{x}. \tag{2}$$

Even when transforming from polar to rectangular coordinates it is wise to keep in mind the relations $x = r \cos \theta$ and $y = r \sin \theta$, which often lead to simplifications not apparent from a direct use of equations (2). Thus, in converting the equation $r \cos \theta = 4$ to rectangular coordinates, we immediately have $x = 4$.

Example 2 | Transform the polar equation $r = 1/(1 - \sin \theta)$ into rectangular form.

The equation

$$r = \frac{1}{1 - \sin \theta}$$

is equivalent to the equation

$$r - r \sin \theta = 1. \tag{3}$$

Since $r = \pm\sqrt{x^2 + y^2}$ and $r \sin\theta = y$, substitution into (3) yields the rectangular form

$$\pm\sqrt{x^2 + y^2} - y = 1, \quad \text{or}$$
$$\pm\sqrt{x^2 + y^2} = 1 + y.$$

If we then square both sides and simplify, we obtain the equation of a parabola:

$$y = \frac{1}{2}x^2 - \frac{1}{2}.$$

||

y = r sin θ
x = r cos θ

Exercises 12.3

In Exercises 1–16, convert the rectangular equation to polar form.

1. $x^2 + y^2 = 8$.
2. $x^2 + y^2 = 4$.
3. $x^2 + y^2 = 4x$.
4. $x^2 + y^2 = 3y$.
5. $y = 2x + x^2 + y^2$.
6. $y = 2x$.
7. $x = -1$.
8. $y = 5$.
9. $y^2 = -x^2 + 2x - 3$.
10. $y = x^2 + 2x$.
11. $(x^2 + y^2)^2 = 2xy$.
12. $(x^2 + y^2)^3 = y^2$.
13. $(x^2 + y^2)^3 = x^2$.
14. $(x^2 + y^2)^2 = x^2 - y^2$.
15. $x^2 + y^2 + 7x = 3y$.
16. $y^2 = 6x + 9$.

In Exercises 17–30, convert the polar equation to rectangular form.

17. $r = -8$.
18. $r = 5$.
19. $r = 1/(\cos\theta + \sin\theta)$.
20. $r = 1/(1 - \cos\theta)$.
21. $r \sin\theta = -5$.
22. $r \cos\theta = 7$.
23. $r = 2\cos\theta, \quad r \neq 0$.
24. $r^2 = 3$.
25. $r^2 = \sin\theta$.
26. $r^2 = \sin 2\theta$.
27. $r^2 = \cos 2\theta$.
28. $r^2 = \cos\theta$.
29. $r = 3/(1 + \cos\theta)$.
30. $r = 3\sin\theta - 7\cos\theta$.

In Exercises 31–34, sketch the graph of the given equation.

31. $(x^2 + y^2)^2 = 2xy$.
32. $(x^2 + y^2)^3 = y^2$.
33. $(x^2 + y^2)^3 = x^2$.
34. $(x^2 + y^2)^2 = x^2 - y^2$.

12.4 Area in Polar Coordinates

Omit

If we are given a polar equation $r = f(\theta)$ that specifies r as a continuous function of θ, it is reasonable to ask for the area A of the region bounded by $r = f(\theta)$ between the two rays $\theta = \alpha$ and $\theta = \beta$. Such a region is indicated in Figure 12.4.1.

To find that area we make the following two fundamental assumptions about area.

Assumption 1. Area is additive. That is, if two regions with areas A_1 and A_2 either do not overlap or only touch along their boundaries, then the area of the combined region is $A_1 + A_2$.

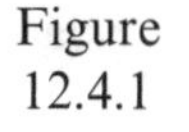
Figure 12.4.1

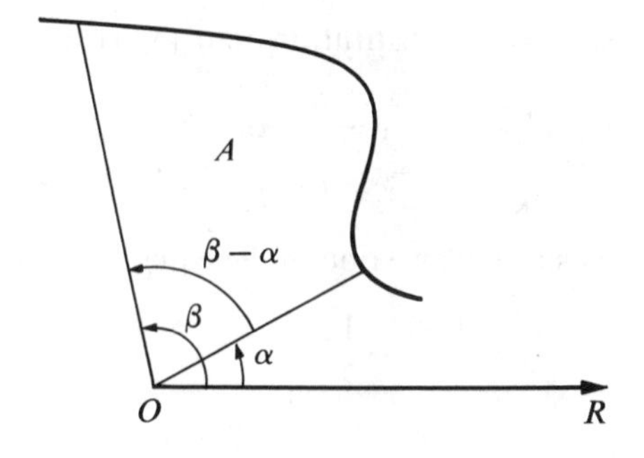

Assumption 2. Let the regions R_1 and R_2 have areas A_1 and A_2, respectively. If R_1 contains R_2, that is R_2 is inside R_1, then $A_2 \leq A_1$.

We now proceed in the same way as with all our previous applications of the definite integral. First subdivide the angle $\beta - \alpha$ into n equal subangles, each with measure $h = (\beta - \alpha)/n$. The subdivision is accomplished by introducing the new angles $\theta_0, \theta_1, \ldots, \theta_{n-1}, \theta_n$, where

$$\theta_0 = \alpha, \quad \theta_1 = \alpha + h, \quad \theta_2 = \alpha + 2h, \ldots, \theta_n = \alpha + nh = \beta.$$

The rays $\theta = \theta_0, \theta = \theta_1, \ldots, \theta = \theta_n$ cut the region into n subregions with areas $A_1, A_2, \ldots, A_n$ as shown in Figure 12.4.2. Since the smaller regions only overlap along their boundaries, we may use Assumption 1 to get

$$A = \sum_{i=1}^{n} A_i. \tag{1}$$

Figure 12.4.2

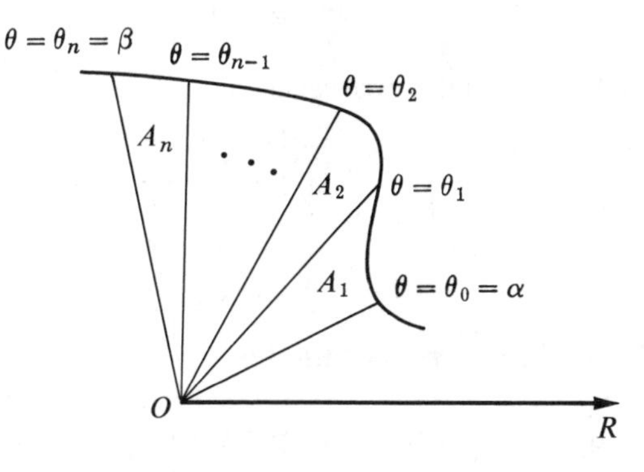

Hence we may concentrate our efforts on finding an expression for the typical subarea A_i pictured in Figure 12.4.3. Since $[f(\theta)]^2$ is a continuous function it takes on a maximum and a minimum value for $\theta_{i-1} \leq \theta \leq \theta_i$. Let M_i be the maximum and let m_i be the minimum of $[f(\theta)]^2$ for $\theta_{i-1} \leq \theta \leq \theta_i$. We then have the situation shown in Figure 12.4.4.

From Assumption 2 we now have

$$\text{area of sector } OPQ \leq A_i \leq \text{area of sector } ORS. \tag{2}$$

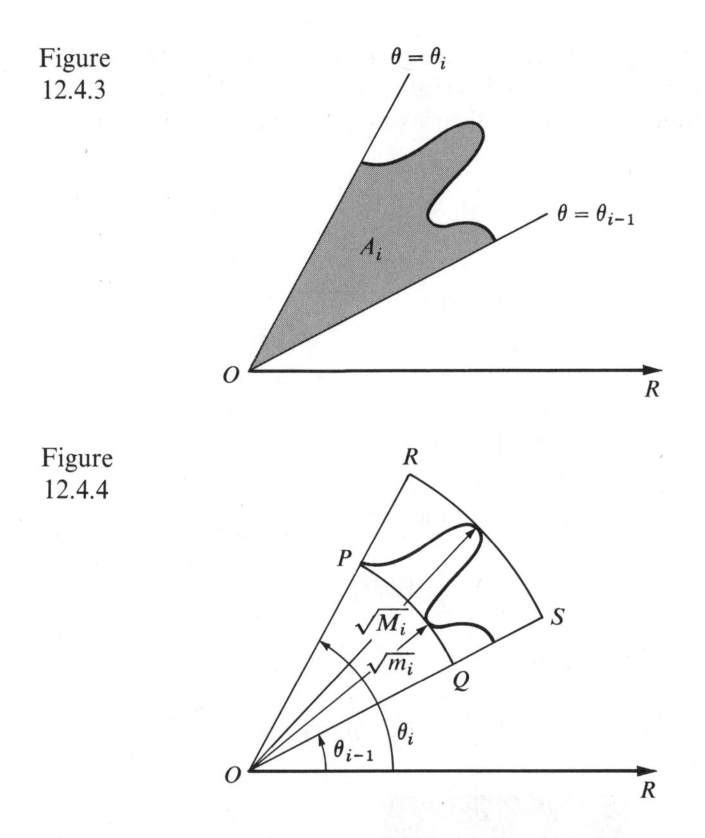

Figure 12.4.3

Figure 12.4.4

Since the area of a sector of a circle of radius r subtended by an angle θ measured in radians is $(1/2)r^2\theta$, we know that

$$\text{area of sector } ORS = \frac{1}{2} M_i(\theta_i - \theta_{i-1}),$$

and

$$\text{area of sector } OPQ = \frac{1}{2} m_i(\theta_i - \theta_{i-1}).$$

Then, using the fact that $\theta_i - \theta_{i-1} = h$, we have

$$\text{area of sector } ORS = \frac{1}{2} M_i h$$

and

$$\text{area of sector } OPQ = \frac{1}{2} m_i h.$$

Consequently, substitution into (2) gives

$$\frac{1}{2} m_i h \leq A_i \leq \frac{1}{2} M_i h,$$

or, since h is positive,

$$m_i \leq \frac{2A_i}{h} \leq M_i.$$

Hence the number $2A_i/h$ lies somewhere between the two values m_i and M_i assumed by the continuous function $[f(\theta)]^2$ on the interval $\theta_{i-1} \leq \theta \leq \theta_i$. The Intermediate Value Theorem for continuous functions then assures us that there is some θ_i^*, $\theta_{i-1} \leq \theta_i^* \leq \theta_i$, such that

$$[f(\theta_i^*)]^2 = \frac{2A_i}{h}, \quad \text{or}$$

$$A_i = \frac{1}{2}[f(\theta_i^*)]^2 h.$$

Now substitution into (1) gives

$$A = \sum_{i=1}^{n} \frac{1}{2}[f(\theta_i^*)]^2 h.$$

Since the area A is independent of the size h of the subdivisions,

$$A = \lim_{h \to 0} A = \lim_{h \to 0} \sum_{i=1}^{n} \frac{1}{2}[f(\theta_i^*)]^2 h.$$

But by definition of the definite integral,

$$\lim_{h \to 0} \sum_{i=1}^{n} \frac{1}{2}[f(\theta_i^*)]^2 h = \int_\alpha^\beta \frac{1}{2}[f(\theta)]^2 \, d\theta. \quad \text{Thus}$$

$$A = \frac{1}{2}\int_\alpha^\beta [f(\theta)]^2 \, d\theta.$$

Example 1 Find the area of the region bounded by the spiral $r = \theta$ and the rays $\theta = 0$, $\theta = \pi/2$.

The graph of $r = \theta$ is sketched in Example 2 of Section 12.2. The area to be found is sketched in Figure 12.4.5. Since $f(\theta) = \theta$, substitution into the area formula gives

$$A = \frac{1}{2}\int_0^{\pi/2} \theta^2 \, d\theta = \frac{1}{6}\theta^3 \bigg|_0^{\pi/2} = \frac{\pi^3}{48} \text{ square units.}$$

||

Figure 12.4.5

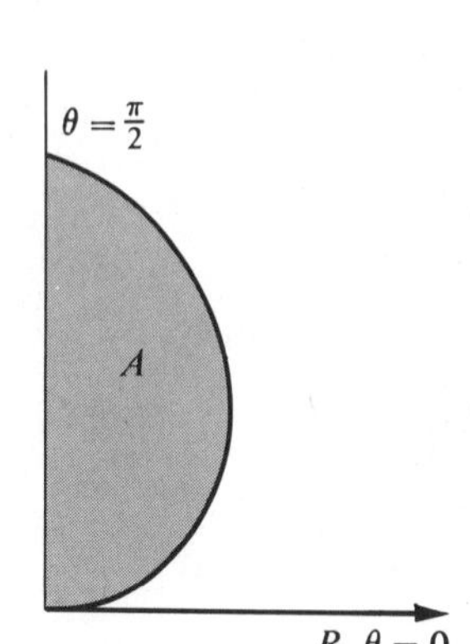

Example 2 Find the area of the region that is inside the circle $r = 1/2$ and also inside the graph of $r = \cos 2\theta$.

The area to be found is sketched in Figure 12.4.6. By symmetry we need only find the area A_1 between $\theta = 0$ and $\theta = \pi/4$ and then multiply by 8 to obtain the total area A. Hence

$$A = 8 \cdot \frac{1}{2} \int_0^{\pi/4} [f(\theta)]^2 \, d\theta.$$

Figure 12.4.6

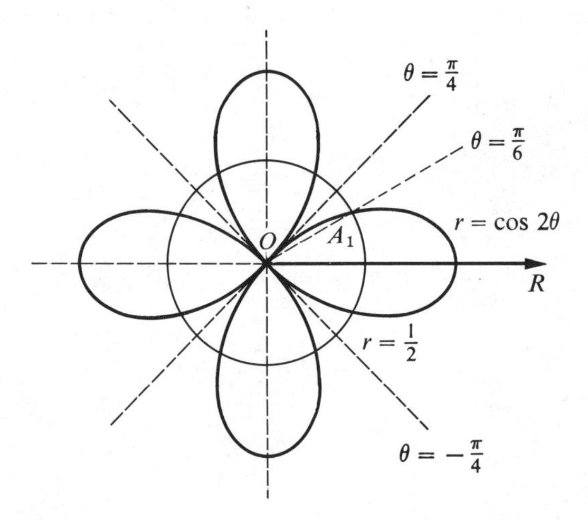

Note, however, that the $f(\theta)$ we are interested in is not defined by a single equation for $0 \le \theta \le \pi/4$. In that interval the graph of $r = 1/2$ intersects the graph of $r = \cos 2\theta$ when $\cos 2\theta = 1/2$ or $\theta = \pi/6$. Thus the $f(\theta)$ we need is

$$f(\theta) = \begin{cases} \dfrac{1}{2} & \text{if } 0 \le \theta \le \dfrac{\pi}{6} \\[2ex] \cos 2\theta & \text{if } \dfrac{\pi}{6} \le \theta \le \dfrac{\pi}{4}. \end{cases}$$

Consequently, since

$$A = 8\left[\frac{1}{2}\int_0^{\pi/6} [f(\theta)]^2 \, d\theta + \frac{1}{2}\int_{\pi/6}^{\pi/4} [f(\theta)]^2 \, d\theta\right],$$

we have

$$A = 4\left[\int_0^{\pi/6} \frac{1}{4} \, d\theta + \int_{\pi/6}^{\pi/4} \cos^2 2\theta \, d\theta\right].$$

Then

$$\int_0^{\pi/6} \frac{1}{4} \, d\theta = \frac{\pi}{24}, \quad \text{and}$$

$$\int_{\pi/6}^{\pi/4} \cos^2 2\theta \, d\theta = \int_{\pi/6}^{\pi/4} \frac{1 + \cos 4\theta}{2} \, d\theta = \left[\frac{1}{2}\theta + \frac{1}{8}\sin 4\theta\right]_{\pi/6}^{\pi/4}$$

$$= \frac{\pi}{8} - \frac{\pi}{12} - \frac{1}{8}\frac{\sqrt{3}}{2} = \frac{\pi}{24} - \frac{\sqrt{3}}{16}.$$

So
$$A = 4\left(\frac{\pi}{24} + \frac{\pi}{24} - \frac{\sqrt{3}}{16}\right) = \frac{\pi}{3} - \frac{\sqrt{3}}{4} \text{ square units.}$$ ||

Example 3 | Find the area of the region that is exterior to the circle $r = 1/2$ and bounded by $r = \cos 2\theta,\ -\pi/4 < \theta < \pi/4$.

From Figure 12.4.6 and the symmetry of the figure we see that this area can be found by subtracting the area bounded by the circle $r = 1/2$ between $\theta = -\pi/6$ and $\theta = \pi/6$ from the area bounded by $r = \cos 2\theta$ between $\theta = -\pi/6$ and $\theta = \pi/6$. That is,

$$A = \frac{1}{2}\int_{-\pi/6}^{\pi/6} \cos^2 2\theta \, d\theta - \frac{1}{2}\int_{-\pi/6}^{\pi/6} \left(\frac{1}{2}\right)^2 d\theta$$

But
$$\frac{1}{2}\int_{-\pi/6}^{\pi/6} \left(\frac{1}{2}\right)^2 d\theta = \frac{1}{8}\int_{-\pi/6}^{\pi/6} d\theta = \frac{1}{8}\left(\frac{\pi}{6} + \frac{\pi}{6}\right) = \frac{\pi}{24}.$$

Also
$$\begin{aligned}\frac{1}{2}\int_{-\pi/6}^{\pi/6} \cos^2 2\theta \, d\theta &= \frac{1}{2}\int_{-\pi/6}^{\pi/6} \frac{1 + \cos 4\theta}{2} \, d\theta \\ &= \frac{1}{4}\left[\theta + \frac{1}{4}\sin 4\theta\right]_{-\pi/6}^{\pi/6} \\ &= \frac{1}{4}\left(\frac{\pi}{3} + \frac{\sqrt{3}}{4}\right) \\ &= \frac{\pi}{12} + \frac{\sqrt{3}}{16}.\end{aligned}$$

Hence
$$A = \frac{\pi}{12} + \frac{\sqrt{3}}{16} - \frac{\pi}{24} = \frac{\pi}{24} + \frac{\sqrt{3}}{16} \text{ square units.}$$ ||

Exercises 12.4

1. Find the area of the region bounded by the graph of $r = \cos\theta$ between $\theta = 0$ and $\theta = \pi/4$.
2. Find the area of the region bounded by the graph of $r = \sin\theta$ between $\theta = 0$ and $\theta = \pi/4$.
3. Find the area of the region bounded by the graph of $r = e^{3\theta}$ between $\theta = 0$ and $\theta = \pi/2$.
4. Find the area of the region bounded by the graph of $r = \theta$ between $\theta = 0$ and $\theta = \pi/4$.
5. Find the area of the entire region bounded by the graph of $r = 2 + \cos\theta$.
6. Find the area of the entire region bounded by the graph of $r = 1 + \sin\theta$.
7. Find the area of the entire region bounded by the graph of $r = 2\sin^2(\theta/2)$.
8. Find the area of the entire region bounded by the graph of $r = 4\sin\theta$.
9. Find the area of the region bounded by one loop of the graph of $r = 2\sin 3\theta$.
10. Find the area of the region bounded by one loop of the graph of $r = 3\cos 2\theta$.
11. Find the total area of the region bounded by the graph of $r = 1 + 2\cos\theta$. Be careful not to count the area within the smaller loop twice.

12. Find the area of the region bounded by the spiral $r = \theta$ for $0 \leq \theta \leq 5\pi/2$. Be careful not to count twice the area bounded by $r = \theta$, $0 \leq \theta \leq \pi/2$.

13. Find the area of the region bounded by the graph of $r = 1 + \sin\theta$ outside $r = 1$.

14. Find the area of the region bounded by the graph of $r = 2\sin\theta$ outside $r = 1$.

15. Find the area of the region inside the graph of $r = 4\cos\theta$ and to the right of the line $r = \sec\theta$.

16. Find the area of the smallest of the regions bounded by the graphs of $r = \sin\theta$ and $r = \cos\theta$.

17. Find the area of the region bounded by the small loop of the graph of $r = 1 + 2\cos\theta$.

18. Find the area of the region bounded by the small loop of the graph of $r = 1 + 2\sin\theta$.

19. Find the area of the region inside the circle $r = \cos\theta$ and outside the graph of $r = 1 + \sin\theta$.

20. Find the area of the smallest of the regions bounded by the graphs of $r = \sin\theta$ and $r\sin\theta = 1/4$.

21. String is being unwound from a spool of radius a in such a way that the string is always perpendicular to the spool as it unwinds. Compute the area of the region swept out by the string as the end of the string moves through the first one-quarter of a revolution. See Figure 12.4.7 for an illustration of the desired area A.

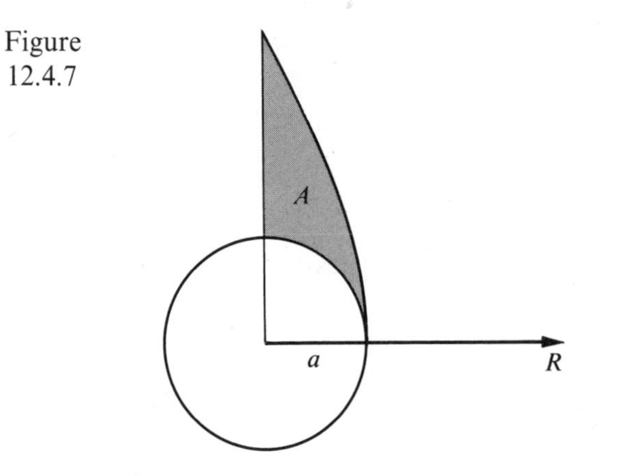

Figure 12.4.7

12.5 Arc Length in Polar Coordinates

In Section 5.3 we derived the equation

$$L = \int_a^b \sqrt{1 + \left(\frac{dy}{dx}\right)^2}\, dx \tag{1}$$

for the arc length of the graph of $y = g(x)$ where $a \leq x \leq b$. We shall now develop a formula that can be used to get the arc length of the polar graph of a differentiable function $r = f(\theta)$ for $\alpha \leq \theta \leq \beta$. The change of variables $x = r\cos\theta$, $y = r\sin\theta$ applied to (1) will enable us to obtain such a formula. In order to give a relatively simple derivation of the arc length formula, we shall assume that $dx/d\theta \neq 0$. Actually it is sufficient to assume that $f'(\theta)$ is continuous.

Since $x = r\cos\theta$, $y = r\sin\theta$, and $r = f(\theta)$, we have $x = f(\theta)\cos\theta$ and $y = f(\theta)\sin\theta$. Consequently, both x and y are differentiable functions of θ. The chain rule then assures us that

$$\frac{dy}{dx}\frac{dx}{d\theta} = \frac{dy}{d\theta}.$$

Thus, if $\frac{dx}{d\theta} \neq 0$,

$$\frac{dy}{dx} = \frac{dy}{d\theta}\Big/\frac{dx}{d\theta}.$$

Hence, if $a = f(\alpha)\cos\alpha$ and $b = f(\beta)\cos\beta$, we have

$$L = \int_a^b \sqrt{1 + \left(\frac{dy}{dx}\right)^2}\,dx = \int_\alpha^\beta \sqrt{1 + \left(\frac{dy}{d\theta}\Big/\frac{dx}{d\theta}\right)^2}\,\frac{dx}{d\theta}\,d\theta.$$

Transfering $dx/d\theta$ under the radical sign simplifies the formula to

$$L = \int_\alpha^\beta \sqrt{\left(\frac{dx}{d\theta}\right)^2 + \left(\frac{dy}{d\theta}\right)^2}\,d\theta. \tag{2}$$

One final simplification will put the formula into a more useful form. We know that

$$x = f(\theta)\cos\theta, \quad \text{and}$$

$$\frac{dx}{d\theta} = f'(\theta)\cos\theta - f(\theta)\sin\theta.$$

In the same way, $y = f(\theta)\sin\theta$ implies that

$$\frac{dy}{d\theta} = f'(\theta)\sin\theta + f(\theta)\cos\theta.$$

Consequently

$$\left(\frac{dx}{d\theta}\right)^2 + \left(\frac{dy}{d\theta}\right)^2 = [f'(\theta)]^2\cos^2\theta - 2f'(\theta)f(\theta)\sin\theta\cos\theta + [f(\theta)]^2\sin^2\theta$$
$$+ [f'(\theta)]^2\sin^2\theta + 2f'(\theta)f(\theta)\sin\theta\cos\theta + [f(\theta)]^2\cos^2\theta.$$

Then, using the fact that $\sin^2\theta + \cos^2\theta = 1$, we have

$$\left(\frac{dx}{d\theta}\right)^2 + \left(\frac{dy}{d\theta}\right)^2 = [f'(\theta)]^2 + [f(\theta)]^2.$$

Substitution of that result into (2) then gives

$$L = \int_\alpha^\beta \sqrt{[f'(\theta)]^2 + [f(\theta)]^2}\,d\theta.$$

Example 1 Find the arc length of the polar graph of the spiral $r = e^\theta$ between $\theta = 0$ and $\theta = \ln 6$.

Since $f(\theta) = e^\theta$, we have $f'(\theta) = e^\theta$ and

$$L = \int_0^{\ln 6} \sqrt{e^{2\theta} + e^{2\theta}}\,d\theta = \sqrt{2}\int_0^{\ln 6} e^\theta\,d\theta$$
$$= \sqrt{2}e^\theta\Big|_0^{\ln 6} = 6\sqrt{2} - \sqrt{2} = 5\sqrt{2} \text{ units.}$$

||

Example 2 | Find the length of the entire graph of $r = 1 + \cos\theta$.

The graph is sketched in Figure 12.5.1. In this case, $f(\theta) = 1 + \cos\theta$ and $f'(\theta) = -\sin\theta$. Moreover, since θ must range from 0 to 2π to cover the entire curve,

$$L = \int_0^{2\pi} \sqrt{\sin^2\theta + 1 + 2\cos\theta + \cos^2\theta}\, d\theta$$

$$= \sqrt{2}\int_0^{2\pi} \sqrt{1 + \cos\theta}\, d\theta.$$

Figure 12.5.1

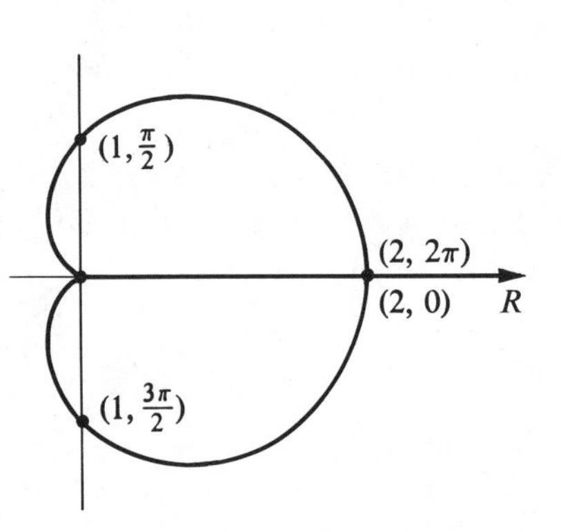

Then, using the identity $1 + \cos\theta = 2\cos^2(\theta/2)$, we have

$$L = 2\int_0^{2\pi} \left|\cos\frac{\theta}{2}\right| d\theta.$$

Since $\cos(\theta/2) \geq 0$ for $0 \leq \theta \leq \pi$, and $\cos(\theta/2) \leq 0$ for $\pi \leq \theta \leq 2\pi$, we have

$$L = 2\int_0^{\pi} \cos\frac{\theta}{2}\, d\theta - 2\int_{\pi}^{2\pi} \cos\frac{\theta}{2}\, d\theta$$

$$= 4\sin\frac{\theta}{2}\Bigg|_0^{\pi} - 4\sin\frac{\theta}{2}\Bigg|_{\pi}^{2\pi} = 8 \text{ units.}$$

That arc length could also be obtained by using the symmetry of the original graph: the entire length is twice the length from $\theta = 0$ to $\theta = \pi$. Consequently

$$L = 2\sqrt{2}\int_0^{\pi} \sqrt{1 + \cos\theta}\, d\theta = 4\int_0^{\pi} \cos\frac{\theta}{2}\, d\theta = 8 \text{ units.}$$ ||

Exercises 12.5

In each of the following exercises find the arc length of the given polar graph. Where no restriction is given the length of the entire curve should be found.

1. $r = 5$.
2. $\theta = \pi, \quad 3 \leq r \leq 9$.
3. $\theta = 0, \quad 1 \leq r \leq 5$.
4. $r = -3$.
5. $r = \theta^2, \quad 0 \leq \theta \leq \pi$.
6. $r = 5\sin\theta$.
7. $r = 9\cos\theta$.
8. $r = -3\theta^2, \quad 0 \leq \theta \leq 2\pi$.
9. $r = e^{5\theta}, \quad 0 \leq \theta \leq \ln 3$.
10. $r = e^{6\theta}, \quad 0 \leq \theta \leq \ln 5$.
11. $r = 12\sin\theta$.
12. $r = 2\tan\theta\sin\theta, \quad 0 \leq \theta \leq \pi/3$.

13. $r = 3 \sec\theta,\quad 0 \leq \theta \leq \pi/4$.

14. $r = 1 - \cos\theta$.

15. $r = 1 + \sin\theta$. (*Hint*: Use the substitution $\theta = \alpha + \pi/2$.)

16. $r = \cos^2(\theta/2),\quad 0 \leq \theta \leq \pi$.

17. $r = \sin^3(\theta/3)$.

Brief Review of Chapter 12

1. The Polar Coordinate System

To set up a polar coordinate system we first select a ray R with initial point O. The polar coordinates (r, θ) of a point P are then determined as indicated. Note that in any case the point (r, θ) is a distance $|r|$ from O.

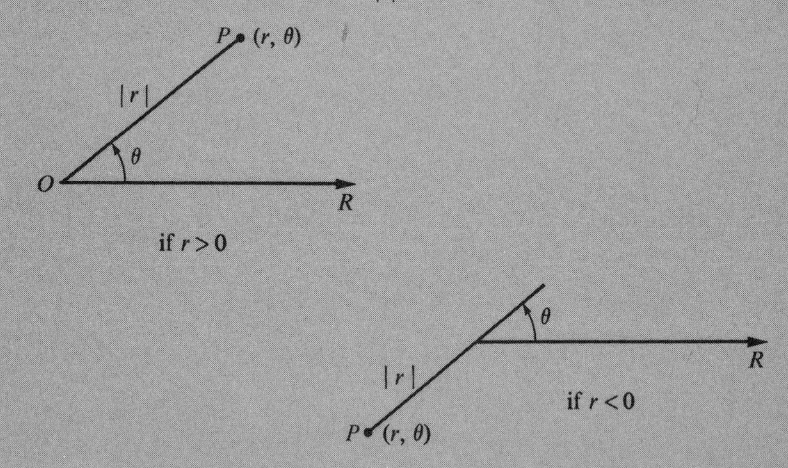

2. Relations between Polar and Rectangular Coordinates

If the ray R is taken to be the positive x-axis each point in the plane will have rectangular coordinates (x, y) and polar coordinates (r, θ), which are related by the equations $x^2 + y^2 = r^2$, $x = r\cos\theta$, and $y = r\sin\theta$.

3. Area in Polar Coordinates

The area of the region bounded by the lines $\theta = \alpha$, $\theta = \beta$, where $\alpha \leq \beta$, and the graph of $r = f(\theta)$ is given by

$$A = \frac{1}{2}\int_\alpha^\beta [f(\theta)]^2\, d\theta.$$

4. Arc Length in Polar Coordinates

The arc length of the graph of $r = f(\theta)$ between $\theta = \alpha$ and $\theta = \beta$ is

$$L = \int_\alpha^\beta \sqrt{[f'(\theta)]^2 + [f(\theta)]^2}\, d\theta.$$

Technique Review Exercises, Chapter 12

1. Show that the graph of $r = 3 + 5 \sin\theta$ is symmetric with respect to the line $\theta = \pi/2$.
2. Convert the polar equation $r = 3 + 5 \sin\theta$ to rectangular form.
3. Sketch the graph of $r = 3 + 5 \sin\theta$.
4. Find the length of $r = e^{2\theta}$ if $0 \le \theta \le 2\pi$.
5. Find the total length of $r = 1 + \cos\theta$.
6. Find the area of the region bounded by $r^2 = 9 \cos 2\theta$.

Additional Exercises, Chapter 12

Section 12.1

In Exercises 1–6, plot the given point and give another set of polar coordinates that refer to the same point.

1. $(-1, \pi/6)$.
2. $(1, -\pi/6)$.
3. $(2, 5\pi/4)$.
4. $(2, 9\pi/4)$.
5. $(-2, -\pi/4)$.
6. $(2, 7\pi/2)$.

Section 12.2

In Exercises 7–16, sketch the graph of the given equation in polar coordinates.

7. $r = \sec\theta$.
8. $r = 1 - \sin\theta$.
9. $r = 2 - 2 \cos\theta$.
10. $r = 3 + \cos\theta$.
11. $r = \sin 4\theta$.
12. $r = \cos 4\theta$.
13. $r = 3 + 4 \cos\theta$.
14. $r = 1 - \sin(\theta/2)$.
15. $r = \theta/10$.
16. $r = \cos^2\theta$.

Section 12.3

In Exercises 17–20, convert the given equation to polar form.

17. $x^2 + y^2 = 2x$.
18. $x^2 + y^2 = -y$.
19. $(x - 2)^2 + y^2 = 1$.
20. $x = 3$.

In Exercises 21–24, convert the given equation to rectangular form.

21. $\theta = \pi/6$.
22. $r = -1$.
23. $r = 1/(\cos\theta - \sin\theta)$.
24. $r = 2 \sin\theta, \quad r \ne 0$.

Section 12.4

In Exercises 25–32, find the area of the region specified.

25. Bounded by the graph of $r = e^{2\theta}$, θ between $\pi/2$ and $3\pi/4$.
26. Bounded by the graph of $r = \theta^2$ between $\theta = \pi/2$ and $\theta = \pi$.
27. Bounded by the graph of $r = 2 - \sin\theta$.
28. Bounded by the graph of $r = 1 - 2 \cos\theta$.
29. Bounded by the graph of $r = 2 - \sin\theta$ outside $r = 1$.
30. Bounded by the graph of $r = 3$ and the graph of $r = 1 - 2 \cos\theta$.

31. Bounded by the smaller loop of $r = 1 - 2\sin\theta$.

32. Inside the circle $r = \sin\theta$ and outside the graph of $r = 1 - \cos\theta$.

Section 12.5

In Exercises 33–40, find the arc length of the indicated polar graph. Where no restriction is given the length of the entire curve should be found.

33. $r = a\theta^2, \quad a > 0, \quad 0 \le \theta \le 2\pi$.

34. $r = \cos^3(\theta/3)$.

35. $r = 1 - \sin\theta$.

36. $r = a\theta^2, \quad 0 \le \theta \le \pi$.

37. $r = \cos^3(\theta/3), \quad 0 \le \theta \le \pi$.

38. $r = \sin^2(\theta/2), \quad 0 \le \theta \le \pi/2$.

39. $r = 2\sec\theta, \quad 0 \le \theta \le \pi/4$.

40. $r = \tan\theta/\cos\theta, \quad 0 \le \theta \le \pi/4$.

Challenging Problems, Chapter 12

1. Determine the coordinates of the points of intersection of the graphs of $r = 2$ and $r = 2\sin\theta - 3$.
2. Prove that an equation of the line through the points (r_1, θ_1) and (r_2, θ_2) is $rr_1\sin(\theta_1 - \theta) + rr_2\sin(\theta - \theta_2) + r_1r_2\sin(\theta_2 - \theta_1) = 0$.

13

Differential Equations

13.1 Differential Equations

Equations involving derivatives of functions are called *differential equations* and arise quite naturally in many applications of mathematics. For example, the mass x of a radioactive substance is known to decrease at a rate proportional to the amount present. Since dx/dt is the rate of change of the mass with respect to the time t, we are led quite naturally to the differential equation

$$\frac{dx}{dt} = -kx, \tag{1}$$

where k is a positive constant of proportionality that depends on the material involved and the units used.

In addition to many physical applications, equations involving quantities and rates of change of these quantities arise in areas as diverse as economics, biology, and sociology. That fact explains the importance of differential equations in applied mathematics.

We begin our discussion of differential equations with a few general comments and definitions.

A number is said to be a *solution* to an equation if upon substitution of the number into the equation it becomes a true statement. For example, 1 is a solution to the equation

$$5x^4 - 3x^3 + x^2 - 10x + 7 = 0.$$

Of course there are other numbers, perhaps not real, that also satisfy that equation, and ordinarily we are interested in finding a complete list of solutions to an equation.

In a completely analogous fashion a *function* is said to be a *solution* to a differential equation for $a \le x \le b$ if the differential equation becomes a true statement for $a \le x \le b$ upon substitution of this function into the differential equation. For example, $y = \sin x$ is a solution to the differential equation

$$y'' + y = 0 \qquad \text{for all } x.$$

At times familiar techniques can be used to solve differential equations. For example, consider equation (1), with $k = 1$:

$$\frac{dx}{dt} = -x. \tag{2}$$

If the solution function $x = f(t)$ is never zero, we may divide by x to obtain

$$\frac{1}{x}\frac{dx}{dt} = -1.$$

On integrating with respect to time t, we obtain

$$\int\left(\frac{1}{x}\frac{dx}{dt}\right)dt = -\int dt, \quad \text{or}$$

$$\int\frac{dx}{x} = -\int dt.$$

Consequently $$\ln|x| + C_1 = -t + C_2,$$

where C_1 and C_2 are arbitrary constants. We can solve for x, the mass at time t, as follows:

$$\ln|x| = -t + (C_2 - C_1),$$
$$|x| = e^{-t+(C_2-C_1)},$$

and since $x \geq 0$,

$$x = e^{-t}e^{(C_2-C_1)}.$$

Since C_1 and C_2 are arbitrary constants, $e^{C_2-C_1}$ is an arbitrary positive constant that we may denote by C. Thus

$$x = Ce^{-t} \tag{3}$$

is a solution to equation (2) for all choices of a positive constant C. However, we assumed that $x \neq 0$ in arriving at those solutions. In fact, the equation $x = 0$, for all t, is itself a solution, which may be incorporated into (3) by simply allowing C to be 0. Since (3) gives a solution for any choice of $C \geq 0$, the equation (2) has infinitely many solutions.

Since differential equations generally have infinitely many solutions, we often seek a particular solution that satisfies additional conditions.

For example, suppose we know that $x = 5$ when $t = 0$; that is, the mass is 5 units initially. Then we must have

$$5 = Ce^{-0}, \quad \text{or} \quad C = 5.$$

The particular solution we seek is then $x = 5e^{-t}$.

In Figure 13.1.1 we illustrate a few of the solutions of (2) for four values of C: $C_4 > C_3 > C_2 > C_1 > 0$.

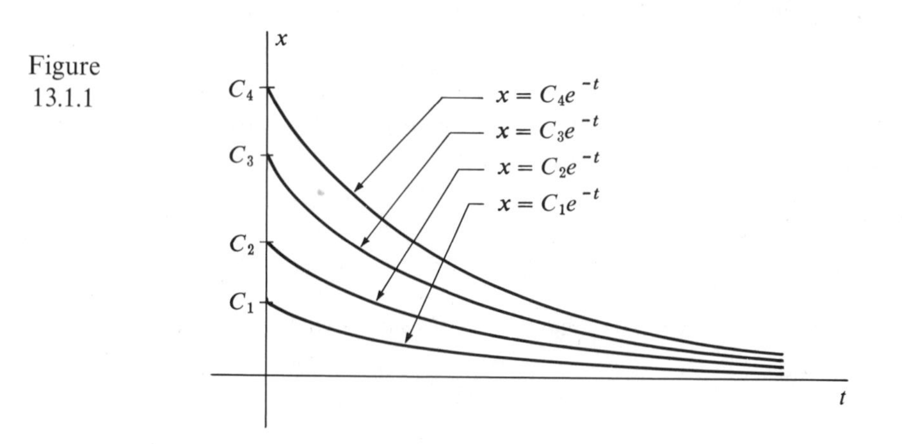

Figure 13.1.1

The following definition will be useful in our study of differential equations.

Definition 13.1.1

> The *order* of a differential equation is the order of the highest derivative appearing in the equation.

For example, both $y''' + y = x^4$ and $x^2y''' + \sin x = y''$ are third order differential equations, while $(y'')^3 - (y')^2 + xy = 0$ is a second order differential equation.

Generally speaking the solutions to a differential equation form a family of curves. For a differential equation of order n, the family of solution curves is an n-parameter family, that is, it contains n arbitrary constants, and n conditions are needed to determine a particular one of the solution curves.

A detailed study of differential equations would involve questions of existence and uniqueness of solutions. Our aims in this chapter will be considerably more modest. We shall be concerned primarily with first order differential equations that illustrate applications of previous material. The more general questions of existence and uniqueness will be reserved for a course in differential equations.

Exercises 13.1

In Exercises 1–10 verify that the given function is a solution to the given differential equation. Assume that k, C, C_1 and C_2 are arbitrary constants.

1. $\dfrac{dy}{dx} + ky = 0, \quad y = Ce^{-kx}.$
2. $\dfrac{dy}{dx} + \dfrac{(x+2)y}{x} = 2, \quad y = 2 - \dfrac{4}{x} + \dfrac{4}{x^2} + \dfrac{C}{x^2}e^{-x}.$
3. $y'' + 2y' - 3y = 5e^{2x}, \quad y = e^{2x} + C_1e^x + C_2e^{-3x}.$
4. $\dfrac{d^2s}{dt^2} + 4s = 3\sin t, \quad s = \sin t + C_1 \sin 2t + C_2 \cos 2t.$
5. $\dfrac{d^2x}{dt^2} + x = t^2 + e^{2t}, \quad x = C_1 \sin t + C_2 \cos t + \dfrac{1}{5}e^{2t} + t^2 - 2.$
6. $y''y''' - 1 = 0, \quad y = \pm(1/15)(2x + C_1)^{5/2} + C_2x + C_3.$
7. $y'' + 6y' + 8y = 0, \quad y = C_1e^{-2x} + C_2e^{-4x}.$
8. $y'' + 6y' + 9y = 0, \quad y = e^{-3x}(C_1x + C_2).$
9. $y'' + 6y' + 10y = 0, \quad y = e^{-3x}(C_1 \cos x + C_2 \sin x).$
10. $y'' - 5y' + 4y = x^2 + 1, \quad y = C_1e^x + C_2e^{4x} + x^2/4 + 5x/8 + 29/32.$
11. In Exercise 1 find the particular solution satisfying the condition $y(0) = 2$.
12. In Exercise 2 find the particular solution satisfying the condition $y(1) = 0$.
13. In Exercise 3 find the particular solution satisfying the conditions $y(0) = 0$ and $y'(0) = 0$.
14. In Exercise 4 find the particular solution satisfying the conditions $s(0) = 1$ and $s(\pi/4) = 0$.
15. In Exercise 7 find the particular solution satisfying the conditions $y(0) = 0$ and $y'(0) = 2$.
16. In Exercise 8 find the particular solution satisfying the conditions $y(0) = y'(0) = 0$.

17. Determine the values of λ for which $e^{\lambda x}$ is a solution to the differential equation $y'' - cy = 0$, where c is a positive constant.

13.2 Separable Differential Equations

A first order differential equation is called *separable* if it can be written in the form

$$g(y)\frac{dy}{dx} = f(x). \tag{1}$$

Integration of both sides of this equation with respect to x gives

$$\int g(y)\frac{dy}{dx}dx = \int f(x)dx, \quad \text{or}$$

$$\int g(y)dy = \int f(x)dx.$$

Such integration yields a solution as illustrated in the next example.

Example 1 Solve the differential equation

$$2y\frac{dy}{dx} - (x + xy^2) = 0.$$

A simple algebraic manipulation results in an equation with the form of equation (1):

$$\frac{2y}{1 + y^2}\frac{dy}{dx} = x.$$

On integration we have

$$\int \frac{2y}{1 + y^2}dy = \int x\,dx, \quad \text{or}$$

$$\ln(1 + y^2) = \frac{1}{2}x^2 + C_1, \tag{2}$$

where C_1 is an arbitrary constant. Of course for some applications it is best to solve (2) for y in terms of x. Use of both sides of (2) as exponents for e gives

$$1 + y^2 = e^{x^2/2}e^{C_1}, \quad \text{or}$$

$$y = \pm(e^{C_1}e^{x^2/2} - 1)^{1/2}.$$

Since C_1 is an arbitrary constant, e^{C_1} is an arbitrary positive constant, and letting $e^{C_1} = C$, we have

$$y = \pm(Ce^{x^2/2} - 1)^{1/2}.$$

||

Example 2 In the simplest cases the growth of a population is directly proportional to the population. A more complex, realistic model is obtained by supposing that the population is limited by food supply and space to some maximal size a and that the growth of the population is proportional to the product $N(a - N)$, where N represents the population at time t. In that case we arrive at the differential equation

$$\frac{dN}{dt} = kN(a - N), \tag{3}$$

where k is the constant of proportionality.

To solve (3) we first separate the variables N and t to get

$$\frac{1}{N(a - N)} \frac{dN}{dt} = k$$

if N is never zero and N is never a. In such case we have

$$\int \frac{dN}{N(a - N)} = \int k\, dt. \tag{4}$$

Since

$$\frac{1}{N(a - N)} = \frac{1}{a}\left(\frac{1}{N} + \frac{1}{a - N}\right).$$

we get

$$\int \frac{dN}{N(a - N)} = \frac{1}{a}\int \frac{dN}{N} + \frac{1}{a}\int \frac{dN}{a - N}.$$

Then from (4) we have

$$\frac{1}{a}\ln N - \frac{1}{a}\ln(a - N) + C_1 = kt + C_2,$$

or, letting $C_3 = C_2 - C_1$,

$$\frac{1}{a}\ln\left(\frac{N}{a - N}\right) = kt + C_3.$$

Hence, if we let $C_4 = aC_3$, we have

$$\ln\left(\frac{N}{a - N}\right) = akt + C_4.$$

Therefore

$$\frac{N}{a - N} = e^{akt}e^{C_4}.$$

Setting $C = e^{C_4}$, which is an arbitrary positive constant, we get

$$\frac{N}{a - N} = Ce^{akt}. \tag{5}$$

Solving (5) for N we obtain the *logistic equation*

$$N = \frac{aCe^{akt}}{1 + Ce^{akt}}.$$

We can now multiply numerator and denominator by e^{-akt} to get

$$N = \frac{aC}{e^{-akt} + C}. \tag{6}$$

With the equation in such form it is clear that as t increases without bound the term e^{-akt} approaches zero, and so the population approaches the limiting value a. If the population is N_0 when $t = 0$, we must have

$$N_0 = \frac{aC}{1 + C}.$$

Solving for C, we get

$$C = \frac{N_0}{a - N_0},$$

and then substituting into (6) we have

$$N = \frac{\dfrac{aN_0}{a - N_0}}{e^{-akt} + \dfrac{N_0}{a - N_0}}.$$

Finally, multiplying numerator and denominator by $a - N_0$, we obtain

$$N = \frac{aN_0}{(a - N_0)e^{-akt} + N_0}.$$

A sketch of the graph of that equation is given in Figure 13.2.1. In solving this differential equation we assumed that N was never zero and N was never a. In fact the constant functions $N(t) = 0$ and $N(t) = a$ are also solutions to equation (3). ||

Figure 13.2.1

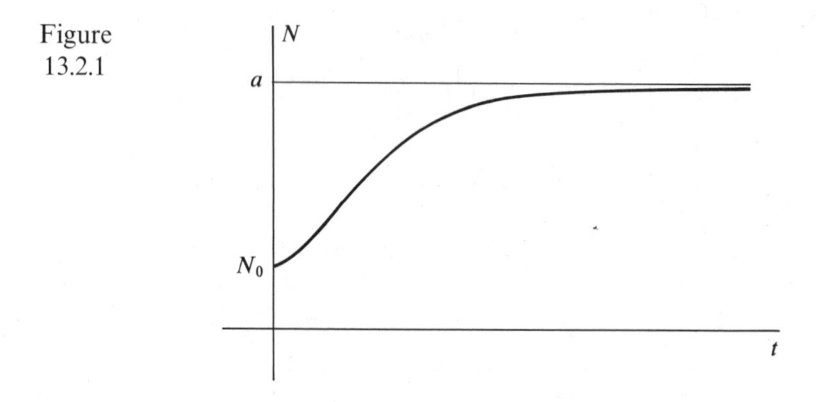

Exercises 13.2

In Exercises 1–14, find the general solution (the one involving an arbitrary constant) to each of the differential equations, by separating the variables. If the variables are not separable, answer "not separable." In those cases where a point is given, also find the particular solution that passes through the given point.

1. $\dfrac{dy}{dx} = 2xy^2$.
2. $dy = 3x^2y\,dx$, $(0, 1)$.
3. $\dfrac{dy}{dx} = e^{-2x} - y$, $(0, 5)$.
4. $\dfrac{ds}{dt} - 3t = \dfrac{3t}{s}$.
5. $x^3\dfrac{dy}{dx} = y - x^2y$.
6. $x^3\dfrac{dy}{dx} = y^2 - x^2y$.
7. $\sin y\,dx = (x^2 + 1)\cos y\,dy$.
8. $\dfrac{y\,dy}{x\,dx} + 1 = 2y^2$.
9. $\dfrac{dy}{dx} = -\dfrac{y+3}{x-2}$, $(3, 4)$.
10. $\dfrac{dx}{dt} = (7 - x)^2$, $(0, 0)$.
11. $y' - 10x^4 = 0$, $(1, 8)$.
12. $y' = (y^2 + y)/(x^2 + x)$, $(-2, -3)$.
13. $y' = e^x(1 + y^2)$.
14. $y' = y + y^3$.
15. The growth of some populations (in particular if the population is well below any limiting value) is directly proportional to the population. Thus, if N represents the population, then $dN/dt = kN$. Solve that equation and compare the solution with the results of Example 2.
16. World population was estimated to be 1550 million in 1900 and 2500 million in 1950. Use that information and the solution to Exercise 15 to estimate the population of the world in the year 2000.
17. A radioactive substance is assumed to decay at a rate proportional to its mass. Write a differential equation describing that situation. If 1 pound of a substance will reduce to 1/2 pound in 90 years, find an expression for the amount of the substance present at any time t.
18. Plants absorb radioactive carbon dioxide containing carbon 14. After the plants die the absorption stops and the carbon 14 decays at a fixed rate. Those facts enable one to date an artifact by determining the amount of carbon 14 present.

    A sample of wood has 1/3 as much carbon 14 as it would have contained initially. How old is the sample? (The half life of carbon 14 is 5600 years; that is, in 5600 years 1/2 of the carbon 14 has decayed.)
19. The rate at which a body cools is proportional to the difference in temperature between the body and its surroundings. If a given body will cool from 600° to 300° in 30 minutes when put in 20° air, find an expression giving the temperature of the body at any time t.
20. Chemicals often dissolve in water at a rate that is proportional to the product of the concentration of the undissolved chemical and the difference between the concentration in a saturated solution and the concentration of the actual solution. If none of the chemical is dissolved initially and if the concentration in a saturated solution is M, find a function giving the concentration of dissolved chemical as a function of time.
21. Sociologists often study "social diffusion" or the spread of a particular idea or fad through a population. The rate of spread is assumed to be proportional to the product of the number of individuals presently holding the idea and the number of people not holding the idea. Find a function giving the number of people holding a particular idea at time t.

22. The rate at which a fluid evaporates is proportional to the surface area of the fluid exposed to the air. If the conical tank pictured in Figure 13.2.2 is initially filled with water, find an expression for the volume as a function of time t. At what time will the tank be empty?

Figure 13.2.2

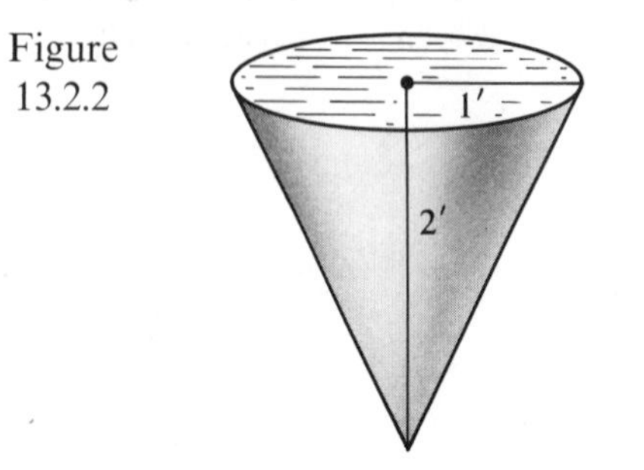

23. Let $f(t)$ be the weight of a limb of an animal body at time t. The ratio $f'(t)/f(t)$ is called the specific rate of growth of the limb. The specific rate of growth is often inversely proportional to t. Use that information to find $f(t)$ if $f(1) = 1$ and $f(2) = 3$ (lbs).
24. For what population size is the graph of the logistic equation concave upward?
25. Suppose that glucose solution is fed into the bloodstream in such a way that k grams of glucose is introduced each minute. Assume that the body removes glucose from the bloodstream at a rate proportional to the amount present. Find an expression for the amount of glucose present at time t.

13.3 First Order Linear Differential Equations

A *linear* differential equation has the form

$$f_n(x)\frac{d^n y}{dx^n} + f_{n-1}(x)\frac{d^{n-1} y}{dx^{n-1}} + \cdots + f_1(x)\frac{dy}{dx} + f_0(x)y = g(x).$$

Since a first order differential equation does not involve derivatives of higher order than the first, a first order linear differential equation must have the form

$$f_1(x)\frac{dy}{dx} + f_0(x)y = g(x). \tag{1}$$

If the left-hand side of (1) is a derivative, finding a solution is simply a matter of integration, as is illustrated in the next example.

Example 1 Find a solution to the differential equation

$$x^3\frac{dy}{dx} + 3x^2y = \sin x.$$

The left-hand side is a derivative. In fact,

$$\frac{d(x^3y)}{dx} = x^3\frac{dy}{dx} + 3x^2y.$$

Thus we may rewrite the original equation as

$$\frac{d(x^3y)}{dx} = \sin x.$$

Integrating both sides, we obtain

$$x^3y = -\cos x + C,$$

and so for $x \neq 0$,

$$y = -\frac{\cos x}{x^3} + \frac{C}{x^3}. \qquad \|$$

Certainly we can't expect that the left-hand side of (1) will always be a derivative. For example, in the equation

$$\frac{dy}{dx} + \frac{3y}{x} = \frac{1}{x^3}\sin x$$

the left-hand side is simply not the derivative of a product. However, if we multiply both sides of the equation by x^3 we have

$$x^3\frac{dy}{dx} + 3x^2y = \sin x.$$

Now, as we observed in Example 1, the left-hand side is the derivative of x^3y, and so the equation can be solved easily.

Consequently, to solve an equation of the form

$$\frac{dy}{dx} + P(x)y = Q(x) \tag{2}$$

we might try to find a function $\rho(x)$ such that

$$\rho(x)\frac{dy}{dx} + \rho(x)P(x)y$$

is the derivative of a product. In fact, if things are to work out as above, we would like to find $\rho(x)$ such that

$$\frac{d}{dx}[\rho(x)y] = \rho(x)\frac{dy}{dx} + \rho(x)P(x)y,$$

or, using the product rule,

$$\rho(x)\frac{dy}{dx} + y\frac{d\rho(x)}{dx} = \rho(x)\frac{dy}{dx} + \rho(x)P(x)y.$$

Consequently we are seeking a function $\rho(x)$ that satisfies

$$\frac{d\rho(x)}{dx} = \rho(x)P(x).$$

This is a separable differential equation and so

$$\int \frac{d\rho(x)}{\rho(x)} = \int P(x)dx,$$

$$\ln \rho(x) = \int P(x)dx + C,$$

$$\rho(x) = e^{\int P(x)dx + C}.$$

Since we were only looking for *one* function $\rho(x)$ that could be used to turn the left-hand side into the derivative of a product, we might just as well take $C = 0$ and use

$$\rho(x) = e^{\int P(x)dx}.$$

The function $\rho(x)$ is called an *integrating factor* for (2).

Example 2 Find a solution $y = f(x)$ to the differential equation

$$\frac{dy}{dx} + \frac{1}{x}y = \frac{\sin x}{x}$$

such that $f(1) = 0$.

Since

$$P(x) = \frac{1}{x},$$

$$\rho(x) = e^{\int P(x)dx} = e^{\int dx/x} = e^{\ln x} = x$$

is an integrating factor. Thus, on multiplying through by x, we have

$$x\frac{dy}{dx} + y = \sin x, \quad \text{or}$$

$$\frac{d(xy)}{dx} = \sin x.$$

On integrating we obtain

$$xy = -\cos x + C, \quad \text{or}$$

$$y = -\frac{\cos x}{x} + \frac{C}{x}.$$

Since y is to be zero when $x = 1$, we must have

$$0 = -\cos 1 + C, \quad \text{or} \quad C = \cos 1.$$

Thus the solution we seek is

$$y = -\frac{1}{x}\cos x + \frac{1}{x}\cos 1.$$

||

A word of warning is in order. The integrating factor

$$\rho(x) = e^{\int P(x)dx}$$

only works for linear equations in which the coefficient of dy/dx is 1. At times, as is illustrated in the next example, the equation must first be put in that form.

Example 3 Find a solution $y = f(x)$ to the differential equation

$$\frac{1}{\tan x}\frac{dy}{dx} + 2y = \tan x$$

such that $f(0) = 0$.

Multiplication by $\tan x$ puts the equation in the proper form for finding an integrating factor:

$$\frac{dy}{dx} + 2y \tan x = \tan^2 x.$$

Here the integrating factor is

$$\rho(x) = e^{\int 2 \tan x \, dx} = e^{2 \ln(\sec x)} = e^{\ln(\sec^2 x)} = \sec^2 x.$$

On multiplication by $\rho(x)$ the equation becomes

$$\sec^2 x \frac{dy}{dx} + 2(\tan x \sec^2 x)y = \tan^2 x \sec^2 x, \quad \text{or}$$

$$\frac{d}{dx}(y \sec^2 x) = \tan^2 x \sec^2 x.$$

Thus

$$y \sec^2 x = \int \tan^2 x \sec^2 x \, dx = \int \tan^2 x \, d(\tan x),$$

or

$$y \sec^2 x = \frac{1}{3} \tan^3 x + C.$$

Consequently

$$y = (\cos^2 x)\left(\frac{1}{3} \tan^3 x + C\right).$$

Since y must be zero when x is zero, $C = 0$. Thus the solution we seek is

$$y = \frac{1}{3} \cos^2 x \tan^3 x. \qquad \|$$

Example 4 Suppose that a body of mass m has a force $f(t) = e^{-t}$ applied to it and that its motion is being resisted by a frictional force proportional to its velocity. If the body is at rest at time $t = 0$, find the velocity of the body as a function of time.

Newton's Law tells us that $F = ma$, where F is the total force applied to a body of mass m, and a is its acceleration. Using $a = dv/dt$, we may write

$$F = m\frac{dv}{dt}. \tag{3}$$

Since the force of friction is opposite to the direction of motion, the total force is given by

$$F = e^{-t} - kv,$$

where e^{-t} is the external force applied to the body and $-kv$ is the frictional force. Substitution into (3) then gives

$$m\frac{dv}{dt} = e^{-t} - kv, \quad \text{or}$$

$$\frac{dv}{dt} + \frac{k}{m}v = \frac{1}{m}e^{-t},$$

which is a first order linear equation.

The integrating factor for the equation is

$$\rho(t) = e^{\int (k/m)dt} = e^{kt/m}.$$

Multiplication by that factor yields

$$e^{kt/m}\frac{dv}{dt} + \frac{kv}{m}e^{kt/m} = \frac{1}{m}e^{t(k/m-1)}, \quad \text{or}$$

$$\frac{d(ve^{kt/m})}{dt} = \frac{1}{m}e^{t(k/m-1)}.$$

Thus

$$ve^{kt/m} = \frac{1}{m}\int e^{t(k/m-1)}\,dt$$

and so

$$ve^{kt/m} = \frac{1}{m(k/m-1)}e^{t(k/m-1)} + C,$$

or

$$v = \frac{1}{k-m}e^{-t} + Ce^{-kt/m}.$$

Since the body is at rest at time $t = 0$, we have

$$0 = \frac{1}{k-m} + C, \quad \text{or} \quad C = -\frac{1}{k-m}$$

Hence the velocity of the body is given by

$$v = \frac{1}{k-m}(e^{-t} - e^{-kt/m}).$$

||

Exercises 13.3

For Exercises 1–14, obtain the general solution to the given differential equation if it is linear; and if it is not linear, simply answer "not linear." If a point is given, also find the particular solution passing through the given point.

1. $\dfrac{dy}{dx} + \dfrac{5y}{x} = x^2 + 2x, \quad (1, 0).$
2. $\dfrac{dy}{dx} - 2y = e^{3x}.$
3. $\dfrac{ds}{dt} - e^{\sin t} = s\cos t.$
4. $y\dfrac{dy}{dx} - 2\dfrac{y}{x} = x^2.$
5. $x^2y' - 2xy = 2.$
6. $y' + 2xy + x = e^{-x^2}.$

7. $\dfrac{dy}{dx} - y^2 = x^2 y.$

8. $\dfrac{dy}{dt} = e^{-2t} - y, \quad (0, 5).$

9. $y' + y = x^2, \quad (0, 5).$

10. $y' - 2y = e^x, \quad (0, 4).$

11. $y' - (\tan x)y = x, \quad (\pi/4, 1).$

12. $y' + xy^2 = 1.$

13. $y' = x \sin y, \quad (1, \pi/4).$

14. $y' + xy = 4x^3.$

Many times a differential equation that is not linear can be put into linear form by a simple transformation. Exercises 15–18 illustrate that fact.

15. Though the equation $dx/dy - 2x/y = 2y^3$ is not linear in y' and y, it is linear in x' and x. Use the technique of this section to solve that equation.

16. An equation of the form

$$\frac{dy}{dx} + P(x)y = Q(x)y^n, \qquad n \neq 0, 1$$

is called a *Bernoulli equation.* Show that the substitution $y^{1-n} = z$ produces an equation that is linear in dz/dx and z.

17. Use the technique of Exercise 16 to solve the differential equation $y' - y = xy^{1/2}$.

18. Use the technique of Exercise 16 to solve the differential equation $xy' + y = x^2y^2$.

19. Find a curve $y = f(x)$ such that the trapezoid formed by the tangent to the curve at any point, the vertical line through the point, and the coordinate axes, has constant area A.

20. At each point (x, y) of a curve the tangent to the curve intersects the y-axis at $(0, 2x^2)$. Find an equation of the curve.

21. If ω is the angular velocity of a planet in its path around the sun and if r is the distance of the planet from the sun, one can obtain the equation

$$2\frac{dr}{d\omega} + r = 0.$$

Solve that equation for ω.

13.4 Two Special Types of Second Order Differential Equations

In general a second order differential equation will involve d^2y/dx^2, dy/dx, y, and x. In this section we shall show how to solve a second order equation that does not involve y or x.

Suppose first that the dependent variable y is missing. Then the equation involves only d^2y/dx^2, dy/dx, and x. If we make the substitution $w = dy/dx$, then $dw/dx = d^2y/dx^2$ and the resulting equation will involve dw/dx, w, and x. That is, it will be a first order equation to which previous techniques may be applied.

Example 1 Find a solution $y = f(x)$ to the differential equation

$$\frac{d^2y}{dx^2} - \frac{1}{x}\left(\frac{dy}{dx} + 5\right) = 0$$

such that $f(1) = 0$ and $f'(1) = 0$.

We let $w = dy/dx$. Then $d^2y/dx^2 = dw/dx$. On substitution the differential equation becomes

$$\frac{dw}{dx} - \frac{w+5}{x} = 0.$$

Separating variables yields

$$\int \frac{dw}{w+5} = \int \frac{dx}{x}, \quad \text{or}$$

$$\ln(w+5) = \ln x + C.$$

Since $w = dy/dx = f'(x)$ is to be zero when $x = 1$, we must have

$$\ln 5 = \ln 1 + C, \quad \text{and so} \quad C = \ln 5.$$

Consequently

$$\ln(w+5) = \ln x + \ln 5,$$

$$\ln(w+5) = \ln 5x,$$

and so

$$w + 5 = 5x, \quad \text{or} \quad w = 5x - 5.$$

Then, since $w = dy/dx$,

$$\frac{dy}{dx} = 5x - 5.$$

Integration gives

$$y = \frac{5}{2}x^2 - 5x + C.$$

Since y is to be zero when $x = 1$, we must have

$$0 = \frac{5}{2} - 5 + C, \quad \text{and so} \quad C = \frac{5}{2}.$$

Hence the desired solution is

$$y = \frac{5}{2}x^2 - 5x + \frac{5}{2}.$$

||

Example 2 A body of mass m is released from rest. Find the distance that the body has fallen in the elapsed time t if the motion of the body through the air is resisted by a force proportional to its velocity.

Two forces act on this falling body. First, the constant force exerted by gravity is proportional to m and is given by mg, where g is the gravitational constant. In addition to that force, air resistance is a force in the opposite (upward) direction given by kv, where v is the velocity of the body and k is a constant of proportionality. The net force in the downward direction is therefore

$$F = mg - kv.$$

By Newton's Law, $F = ma$, where a is the acceleration of the body. Consequently

$$ma = mg - kv. \tag{1}$$

If we let x be the distance the body has fallen in the elapsed time t, then

$$v = \frac{dx}{dt} \quad \text{and} \quad a = \frac{d^2x}{dt^2}.$$

Then (1) becomes

$$m\frac{d^2x}{dt^2} = mg - k\frac{dx}{dt}, \quad \text{or}$$

$$m\frac{d^2x}{dt^2} + k\frac{dx}{dt} = mg. \tag{2}$$

To solve that equation let $w = dx/dt = v$. Then $d^2x/dt^2 = dv/dt$ and (2) becomes

$$m\frac{dv}{dt} + kv = mg.$$

Though that equation is a first order linear equation, we shall find it easier to separate variables. We obtain

$$\frac{m}{mg - kv}\frac{dv}{dt} = 1,$$

and so

$$\int \frac{mdv}{mg - kv} = \int dt.$$

Consequently

$$-\frac{m}{k}\ln(mg - kv) = t + C_1,$$

and so,

$$\begin{aligned} mg - kv &= e^{-tk/m - C_1k/m} \\ &= e^{-tk/m}e^{-C_1k/m} \\ &= C_2e^{-tk/m}, \end{aligned}$$

where $C_2 = e^{-C_1k/m}$.

Then, on solving for $v = dx/dt$, we have

$$\frac{dx}{dt} = \frac{1}{k}(mg - C_2e^{-kt/m}). \tag{3}$$

Since at time $t = 0$ the body is dropped from rest, we must have

$$\frac{dx}{dt} = 0, \qquad \text{when } t = 0.$$

Thus

$$0 = \frac{1}{k}(mg - C_2), \quad \text{or} \quad C_2 = mg.$$

Equation (3) then becomes

$$\frac{dx}{dt} = \frac{mg}{k}(1 - e^{-kt/m}).$$

Integration then yields

$$x = \frac{mg}{k}\left(\int dt - \int e^{-kt/m}\,dt\right), \quad \text{or}$$

$$x = \frac{mg}{k}\left(t + \frac{m}{k}e^{-kt/m} + C\right).$$

Since we are to find the distance the body has fallen in the elapsed time t, we have $x = 0$ when $t = 0$. Thus

$$0 = \frac{mg}{k}\left(\frac{m}{k} + C\right), \quad \text{or} \quad C = -\frac{m}{k}.$$

Finally we get

$$x = \frac{mg}{k}\left(t + \frac{m}{k}e^{-kt/m} - \frac{m}{k}\right).$$

||

One other type of second order differential equation is easily reducible to first order. If the independent variable x is missing, the equation is a function of d^2y/dx^2, dy/dx, and y alone. In that case the substitution $w = dy/dx$ is used again. In addition we note that

$$\frac{d^2y}{dx^2} = \frac{dw}{dx} = \frac{dw}{dy}\frac{dy}{dx} = w\frac{dw}{dy}.$$

Use of the substitutions $dy/dx = w$ and $d^2y/dx^2 = w\,dw/dy$ will reduce the equation to first order.

Example 3 | Solve the differential equation

$$\frac{d^2y}{dx^2} + \frac{dy}{dx} = 0.$$

Substitution of $dy/dx = w$ and $d^2y/dx^2 = w\,dw/dy$ gives

$$w\frac{dw}{dy} + w = 0.$$

One solution is clearly $w = 0$. If w is not zero, we may divide by w to obtain

$$\frac{dw}{dy} + 1 = 0.$$

Separation of variables gives

$$\int dw = -\int dy,$$

$$w = -y + C_1.$$

Since $w = dy/dx$, we then have

$$\frac{dy}{dx} = -y + C_1, \quad \text{or}$$

$$\int \frac{dy}{y - C_1} = -\int dx.$$

Then

$$\ln|y - C_1| = -x + C_2,$$

or

$$|y - C_1| = e^{-x+C_2} = e^{-x}e^{C_2} = C_3e^{-x},$$

where $C_3 = e^{C_2}$. Then

$$y - C_1 = \pm C_3e^{-x}.$$

Letting $C = \pm C_3$, we arrive finally at

$$y = Ce^{-x} + C_1,$$

where C and C_1 are arbitrary constants. ||

Exercises 13.4

In Exercises 1–16, solve the indicated differential equation. Where conditions are given, find the solution that satisfies them.

1. $\dfrac{d^2x}{dt^2} + \dfrac{1}{t}\dfrac{dx}{dt} = 1.$
2. $y'' - 2y' = e^{3x}, \quad y'(0) = 0, \quad \mathbf{y(0) = 0}.$
3. $(y')^2 + 2yy'' = 0, \quad y'(0) = y(0) = 1.$
4. $\dfrac{d^2x}{dt^2} + \left(\dfrac{dx}{dt}\right)^2 + 1 = 0, \quad \dfrac{dx}{dt} = 0$ and $x = 0$ when $t = 0$.
5. $x^2\dfrac{d^2y}{dx^2} + x^2 = 2x\dfrac{dy}{dx}.$
6. $y^2y'' = y', \quad y = 1$ and $y' = -1$ when $x = 0$.
7. $x^2\dfrac{d^2y}{dx^2} - 2x\dfrac{dy}{dx} = 2, \quad y'(1) = y(1) = 0.$
8. $\dfrac{d^2y}{dx^2} + e^{-y}\dfrac{dy}{dx} = 0.$
9. $y'' + y = 0, \quad y(0) = 0, \quad y'(0) = 1.$
10. $(1 + t)\dfrac{d^2s}{dt^2} + \dfrac{ds}{dt} = 0.$
11. $t\dfrac{d^2s}{dt^2} = \dfrac{ds}{dt}, \quad s(1) = 0, \quad s'(1) = 1.$
12. $y'y'' + (y')^2 = 2, \quad y(0) = y'(0) = 1.$
13. $xy'' - (y')^2 = -4.$
14. $y'' - (y')^3 - y' = 0.$
15. $t\dfrac{d^2s}{dt^2} - \dfrac{ds}{dt} = \dfrac{2}{t} - \ln t.$
16. $s\dfrac{d^2s}{dt^2} + \left(\dfrac{ds}{dt}\right)^3 = 0.$
17. Find a solution to $(y')^2 + 2yy'' = 0$ such that $y(0) = y'(0) = 0$. (See Exercise 3.)
18. Find a solution to $y'' + e^{-y}y' = 0$ such that $y(0) = y'(0) = 0$. (See Exercise 8.)
19. A 50 kilogram mass is resting on a frictionless surface. The mass is then attached to a spring and displaced 5 centimeters from the equilibrium position. The body is released from rest

and allowed to oscillate. If the spring has a spring constant of 2 newtons/cm., find the displacement from the equilibrium position x as a function of time. (*Hint*: By Hooke's Law the force exerted by the spring is $2x$, where x is the displacement in centimeters.)

20. Suppose that each of two quantities grows at a rate proportional to its own size. That is, $dx_1/dt = k_1x_1$ and $dx_2/dt = k_2x_2$. Prove that the sum $x = x_1 + x_2$ satisfies the differential equation

$$\frac{d^2x}{dt^2} - (k_1 + k_2)\frac{dx}{dt} + k_1k_2x = 0.$$

13.5 Series Solutions

There are many simple looking differential equations that we cannot solve by using the techniques discussed so far. For example, none of our previous techniques applies to these equations:

$$\frac{d^2y}{dx^2} + x^2\frac{dy}{dx} = xy \qquad \text{and} \qquad x\frac{d^2y}{dx^2} - 2\frac{dy}{dx} + xy = 0.$$

The series technique to be introduced in this section will enable us to generate solutions to those and many other difficult differential equations.

In its simplest form the series method is used to find a function of the form

$$y = \sum_{i=0}^{\infty} a_ix^i = a_0 + a_1x + a_2x^2 + \cdots + a_nx^n + \cdots \tag{1}$$

that satisfies the differential equation for all x in some neighborhood of the origin. By Theorem 11.7.3 we can differentiate a convergent power series term by term within the interval of convergence. On doing that we obtain series for y' and then for y'':

$$\frac{dy}{dx} = \sum_{i=0}^{\infty} ia_ix^{i-1} = a_1 + 2a_2x + 3a_3x^2 + \cdots + na_nx^{n-1} + \cdots, \tag{2}$$

$$\frac{d^2y}{dx^2} = \sum_{i=0}^{\infty} i(i-1)a_ix^{i-2}$$

$$= 2a_2 + 6a_3x + 12a_4x^2 + \cdots + n(n-1)a_nx^{n-2} + \cdots. \tag{3}$$

The expressions (1), (2), and (3) are then substituted into the given equation in an attempt to determine the coefficients a_i, $i = 0, 1, \ldots$.

To illustrate the method let's consider an equation that we had solved previously.

Example 1 | Use the series method to find the general solution of $y'' + y = 0$.

We try to find a power series solution in a neighborhood N of the origin. For all x in some neighborhood N of the origin we have

$$y = a_0 + a_1x + a_2x^2 + \cdots + a_nx^n + \cdots,$$

$$\frac{dy}{dx} = a_1 + 2a_2x + 3a_3x^2 + \cdots + na_nx^{n-1} + \cdots,$$

and $$\frac{d^2y}{dx^2} = 2a_2 + 6a_3x + 12a_4x^2 + \cdots + n(n-1)a_nx^{n-2} + \cdots.$$

Then, on substitution into the original equation, we obtain

$$2a_2 + 6a_3x + 12a_4x^2 + \cdots + n(n-1)a_nx^{n-2} + \cdots$$
$$+ a_0 + a_1x + a_2x^2 + \cdots + a_nx^n + \cdots = 0.$$

Gathering like terms we get

$$(2a_2 + a_0) + (6a_3 + a_1)x + (12a_4 + a_2)x^2 + \cdots$$
$$+ [(n+2)(n+1)a_{n+2} + a_n]x^n + \cdots = 0. \tag{4}$$

Note that in order to gather like terms we were forced to determine the coefficient of x^n in the series expansion for d^2y/dx^2. In the natural course of events the coefficient $n(n-1)a_n$ of x^{n-2} was given. However, simply adding 2 to each appearance of n yields the coefficient $(n+2)(n+1)a_{n+2}$ of x^n.

Since the series given in (4) must sum to 0 for all x in N, we can use the uniqueness of power series expansions to see that each coefficient must be 0. Therefore

$$\begin{aligned} 2a_2 + a_0 &= 0, \\ 6a_3 + a_1 &= 0, \\ 12a_4 + a_2 &= 0, \\ &\cdots\cdots \\ (n+2)(n+1)a_{n+2} + a_n &= 0. \\ &\cdots\cdots \end{aligned} \tag{5}$$

Since the family of solutions to $y'' + y = 0$ should contain two arbitrary constants, we shall let a_0 and a_1 be arbitrary; and we shall solve the equations (5) for the remaining constants in terms of a_0 and a_1. In succession, those equations yield

$$\begin{aligned} a_2 &= -\frac{1}{2}a_0, \\ a_3 &= -\frac{1}{6}a_1, \\ a_4 &= -\frac{1}{12}a_2 = \frac{1}{24}a_0, \\ &\cdots\cdots \end{aligned}$$

Thus, $$y = a_0 + a_1x - \frac{1}{2}a_0x^2 - \frac{1}{6}a_1x^3 + \frac{1}{24}a_0x^4 + \cdots,$$

or $$y = a_0\left(1 - \frac{1}{2}x^2 + \frac{1}{24}x^4 + \cdots\right) + a_1\left(x - \frac{1}{6}x^3 + \cdots\right)$$

is an expression for the general solution. However, it is reasonable to ask for more terms in the solution. Generally more terms are most efficiently found by referring to the general relation

$$(n+2)(n+1)a_{n+2} + a_n = 0, \quad \text{or}$$

$$a_{n+2} = \frac{-a_n}{(n+2)(n+1)}. \tag{6}$$

Using a_0 as our starting point and successively applying (6), we obtain

(setting $n = 0$) $$a_2 = \frac{-a_0}{2 \cdot 1} = -\frac{a_0}{2!},$$

(setting $n = 2$) $$a_4 = \frac{-a_2}{4 \cdot 3} = \frac{a_0}{4 \cdot 3 \cdot 2 \cdot 1} = \frac{a_0}{4!},$$

(setting $n = 4$) $$a_6 = \frac{-a_4}{6 \cdot 5} = -\frac{a_0}{4!}\frac{1}{6 \cdot 5} = -\frac{a_0}{6!}.$$

In general, for even n,

$$a_n = (-1)^{n/2} \cdot \frac{a_0}{n!}. \tag{7}$$

Application of (6), beginning with a_1, gives

(setting $n = 1$) $$a_3 = -\frac{a_1}{3 \cdot 2} = -\frac{a_1}{3!},$$

(setting $n = 3$) $$a_5 = \frac{-a_3}{5 \cdot 4} = \frac{a_1}{3!}\frac{1}{5 \cdot 4} = \frac{a_1}{5!},$$

and in general for odd n,

$$a_n = (-1)^{(n-1)/2} \cdot \frac{a_1}{n!}. \tag{8}$$

Hence, substitution into

$$y = a_0 + a_1 x + a_2 x^2 + a_3 x^3 + \cdots + a_n x^n + \cdots,$$

using $n = 2k$ in (7) for the kth even integer, and $n = 2k - 1$ in (8) for the kth odd integer, gives

$$y = a_0 + a_1 x - \frac{a_0}{2!}x^2 - \frac{a_1}{3!}x^3 + \cdots$$
$$+ (-1)^{k-1} \cdot \frac{a_1}{(2k-1)!}x^{2k-1} + (-1)^k \frac{a_0}{(2k)!}x^{2k} + \cdots,$$

or

$$y = a_0\left(1 - \frac{x^2}{2!} + \frac{x^4}{4!} - \cdots + (-1)^k \frac{x^{2k}}{(2k)!} + \cdots\right)$$
$$+ a_1\left(x - \frac{x^3}{3!} + \frac{x^5}{5!} - \cdots + (-1)^{k-1} \frac{x^{2k-1}}{(2k-1)!} + \cdots\right).$$

From our previous work we recognize that those series represent the sine and cosine functions, so that

$$y = a_0 \sin x + a_1 \cos x.$$

However, usually we can't recognize the final series as being that of some elementary function. ||

Example 2 | Find the particular solution to the differential equation

$$\frac{d^2y}{dx^2} + x^2\frac{dy}{dx} + xy = 0$$

satisfying the conditions $y(0) = y'(0) = 1$.

Again we try to find a solution y that is expressible as a power series in some neighborhood N of the origin. Thus, for all x in N,

$$y = a_0 + a_1x + a_2x^2 + \cdots + a_nx^n + \cdots,$$

$$\frac{dy}{dx} = a_1 + 2a_2x + 3a_3x^2 + \cdots + na_nx^{n-1} + \cdots$$

and $$\frac{d^2y}{dx^2} = 2a_2 + 6a_3x + 12a_4x^2 + \cdots + n(n-1)a_nx^{n-2} + \cdots.$$

Since both y and dy/dx are to be 1 when x is 0, we see immediately that $a_0 = 1$ and $a_1 = 1$. Substituting that information into the original differential equation, we obtain

$$\begin{aligned}(2a_2 + 6a_3x + 12a_4x^2 + \cdots + n(n-1)a_nx^{n-2} + \cdots)&\\ + x^2(1 + 2a_2x + 3a_3x^2 + \cdots + na_nx^{n-1} + \cdots)&\\ + x(1 + x + a_2x^2 + \cdots + a_nx^n + \cdots) = 0,& \quad \text{or}\end{aligned}$$

$$\begin{aligned}(2a_2 + 6a_3x + 12a_4x^2 + \cdots + n(n-1)a_nx^{n-2} + \cdots)&\\ + (x^2 + 2a_2x^3 + 3a_3x^4 + \cdots + na_nx^{n+1} + \cdots)&\\ + (x + x^2 + a_2x^3 + \cdots + a_nx^{n+1} + \cdots) = 0.&\end{aligned}$$

Gathering like terms we get

$$2a_2 + (6a_3 + 1)x + (12a_4 + 2)x^2 + \cdots + ((n+3)(n+2)a_{n+3} + na_n + a_n)x^{n+1} + \cdots = 0.$$

Consequently

$$2a_2 = 0,$$

$$6a_3 + 1 = 0,$$

$$12a_4 + 2 = 0,$$

and, in general,

$$(n+3)(n+2)a_{n+3} + (n+1)a_n = 0.$$

Hence

$$a_2 = 0,$$

$$a_3 = -\frac{1}{6},$$

$$a_4 = -\frac{1}{6},$$

and, in general,

$$a_{n+3} = -\frac{n+1}{(n+3)(n+2)} a_n. \tag{9}$$

Using that general relation we obtain

$$a_5 = -\frac{3}{20} a_2 = 0,$$

$$a_6 = -\frac{4}{30} a_3 = \frac{1}{45}.$$

Hence the desired particular solution is

$$y = 1 + x - \frac{1}{6} x^3 - \frac{1}{6} x^4 + \frac{1}{45} x^6 + \cdots.$$

Though we cannot recognize that series as an elementary function, we could use (9) to generate as many terms of the series for y as we wish, and so obtain as accurate an approximation to y as we wish. ||

In the preceding examples the differential equations have solutions that are expressible as a power series in some neighborhood of the origin. However, the solution of a given differential equation might not possess a power series expansion at the origin, or we may be interested in the solution near some point $a \neq 0$. In either case we should attempt to find a power series of the form

$$y = \sum_{k=0}^{\infty} a_k (x - a)^k.$$

Exercises 13.5

In Exercises 1–10 find a series solution to the given differential equation in some neighborhood of the origin.

1. $y' + 3x^2 y = 1.$
2. $y' - x^3 y = 1.$
3. $y' = 2xy.$
4. $y'' - xy' + y = 0.$
5. $(x^2 - 1)y'' + 6xy' + 3y = 0.$
6. $y'' + 2xy' - x^2 y = 0.$
7. $y'' - xy' - y = 0.$
8. $(x^2 + 1)y'' + xy' + x^2 y = 0.$
9. $y'' - xy = 0.$
10. $y'' - xy = x^4.$
11. Find a series solution of the form $y = \sum_{k=0}^{\infty} a_k (x-1)^k$ to the differential equation $y' = 2xy + x^2$, such that the solution passes through the point $(1, 0)$.

12. Find a series solution of the form $y = \sum_{k=0}^{\infty} a_k(x+3)^k$ to the differential equation $y' + x^2y = 2$.
13. Find a series solution of the form $y = \sum_{k=0}^{\infty} a_k(x-1)^k$ to the differential equation $y'' + (x-1)y' + y = 0$.
14. Find a series solution of the form $y = \sum_{k=0}^{\infty} a_k(x+1)^k$ to the differential equation $(x^2+2x)y'' + (x+1)y' - 4y = 0$.

13.6 Numerical Solutions

Since some differential equations are quite difficult to solve and yet appear often in practical situations, it is not surprising that a number of numerical techniques have been developed that yield approximations to solutions. In this section we shall introduce the most elementary of these techniques, *The Cauchy Polygon Method.*

Suppose we are concerned with a first order differential equation that gives dy/dx as a function of x and y. Then dy/dx, the slope of the solution function, is given in terms of x and y. For example, if

$$\frac{dy}{dx} = x - y, \tag{1}$$

then the slope of the solution function $y = f(x)$ at $x = 3$ and $y = 2$ is

$$\frac{dy}{dx} = 3 - 2 = 1.$$

In Figure 13.6.1 we illustrate the *direction field* specified by the differential equation (1). That is, we exhibit the tangents to the solution curves at various points of the plane.

Figure 13.6.1

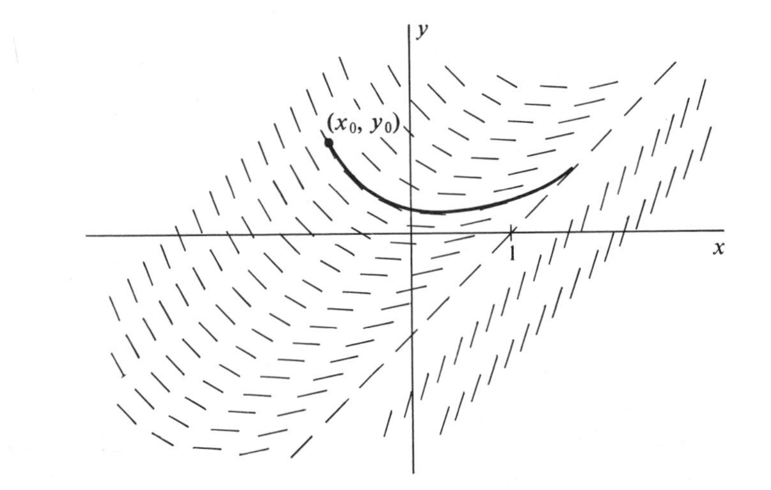

The idea of The Cauchy Polygon Method is to use the line segments of Figure 13.6.1 to build an approximate solution. For example, if we want a solution through the point (x_0, y_0), we could use the line segments illustrated. The general technique should be clear from the following example.

Example 1 Use The Cauchy Polygon Method with five subintervals to obtain an approximate solution, for $0 \le x \le 1$, to the differential equation $dy/dx = 1/y$. The solution is to pass through the point (0, .5).

We first subdivide the interval $0 \le x \le 1$ into 5 equal subintervals by introducing the points $0 \le .2 \le .4 \le .6 \le .8 \le 1$. When $x = 0$, we know that $y = .5$ and that the slope of the solution curve is $dy/dx = 1/.5 = 2$. The line that has slope 2 and passes through the point (0, .5) is used as an approximation to the solution curve on the interval $0 \le x \le .2$, as indicated in Figure 13.6.2. Since that approximating line has the equation $y = 2x + .5$, it passes through the point (.2, .9), which is used as a starting point for our next linear approximation of the solution curve.

Figure 13.6.2

x	*exact* y	*approximate* y
0	.5	.5
.2	.81	.9
.4	1.03	1.12
.6	1.20	1.30
.8	1.36	1.45
1.0	1.5	1.59

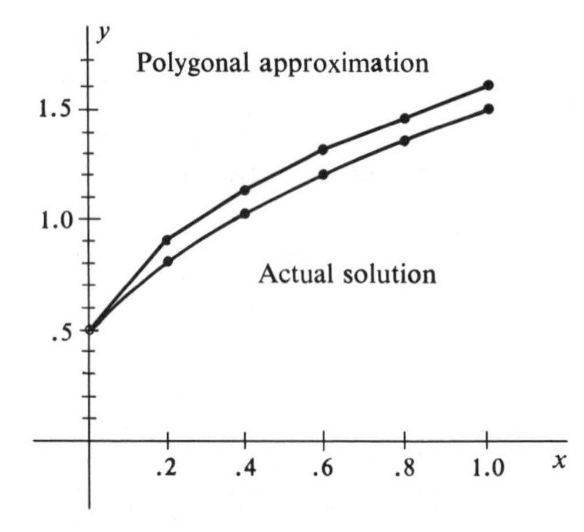

When $y = .9$, the solution curve has slope

$$\frac{dy}{dx} = \frac{1}{.9} \approx 1.11.$$

Then we use the straight line $y - .9 = 1.11(x - .2)$ passing through the point (.2, .9) with slope 1.11 as an approximation to the solution curve on the interval $.2 \le x \le .4$. That straight line passes through the point (.4, 1.12).

When $y = 1.12$, the solution curve has slope

$$\frac{dy}{dx} = \frac{1}{1.12} \approx .89.$$

Using the straight line passing through the point (.4, 1.12) with slope .89 as an approximation to the solution curve, we find that the approximate solution passes through the point

$$(.6, 1.12 + (.2)(.89)) = (.6, 1.30).$$

Continuing in this fashion we find that the polygonal approximation curve must pass through the points (.8, 1.45) and (1, 1.59).

In this instance the original differential equation is easy to solve. The exact particular solution is $(1/2)y^2 = x + 1/8$. Figure 13.6.2 gives a comparison between the exact solution and the approximation obtained by The Cauchy Polygonal Method. ||

This crude method for solving differential equations has at least one serious drawback. As one attempts to extend the approximate solution over larger intervals the errors may be compounded so that the approximation becomes less and less accurate. Of course subdivision of the interval into smaller subintervals will tend to increase the accuracy of the approximation, but any analysis of the error is quite difficult.

More accurate numerical methods for the solution of differential equations have been developed and can be found in most texts on numerical analysis.

Exercises 13.6

Starting at the given point, use The Cauchy Polygonal Method to obtain an approximate solution to the following equations. In each case subdivide the interval into three parts. For Exercise 1, also compare the approximate results with the exact results.

1. $y' = (x + 1)/y$, $(0, 1)$, $0 \leq x \leq 1$.
2. $y' = xy$, $(1, 1)$, $1 \leq x \leq 2$.
3. $y' = x^2/y$, $(0, 1)$, $0 \leq x \leq 1$.
4. $y' = x - y^2$, $(1, 1)$, $1 \leq x \leq 1.3$.
5. $y' = (1 - y^2)^{1/2}$, $(0, 0)$, $0 \leq x \leq 0.3$.
6. $y' = x2^{xy}$, $(1, -1)$, $1 \leq x \leq 4$.

Brief Review of Chapter 13

1. Preliminaries

Definition. The *order* of a differential equation is the order of the highest derivative appearing in the equation.

In general a differential equation will have many solutions; so we often seek a solution that will also satisfy additional conditions.

2. Separable Differential Equations

A differential equation is called separable if it can be written in the form $g(y)dy/dx = f(x)$. In that case, integration will yield a solution.

3. First Order Linear Equations

A first order linear equation has the form $dy/dx + P(x)y = Q(x)$. Multiplication by the integrating factor $\rho(x) = e^{\int P(x)dx}$ turns the left-hand side into the derivative of a product. Integration then yields a solution.

4. Second Order Equations with One Variable Missing

(a) If y, the dependent variable, is missing, the substitutions $dy/dx = w$ and $d^2y/dx^2 = w'$ will reduce the equation to first order.
(b) If x, the independent variable, is missing, the substitutions $dy/dx = w$ and $d^2y/dx^2 = w(dw/dy)$ will reduce the equation to first order.

5. Series Solutions

To determine a series solution to a differential equation near $x = a$, try substituting

$$y = \sum_{n=0}^{\infty} a_n(x-a)^n.$$

The constants a_n, $n = 0, 1, 2, \ldots$, are then determined from the resulting equation.

6. The Cauchy Polygon Method

If an equation gives y' as a function of x and y, this method may be used to generate an approximate solution.

Technique Review Exercises, Chapter 13

1. Find a solution to $x^2y' + 2xy = 1$ that passes through the point $(1, 0)$.
2. Find a solution to the differential equation $dx/dt - 2x^2 = x^2 \sin t$.
3. Find a solution $y = f(x)$ to the different equation $y'' + y'/x = x$ such that $f'(1) = f(1) = 0$.
4. Solve the differential equation $2x\dfrac{d^2x}{dt^2} + \left(\dfrac{dx}{dt}\right)^2 = 0$.
5. Use the series method with $a = 0$ to find a solution to $y'' - 2xy' - 4y = 0$.
6. Use The Cauchy Polygon Method with $h = .1$ to find an approximation to the solution of $dy/dx = x + y$ on the interval $0 \le x \le .3$. The solution is to pass through $(0, 1)$.

Additional Exercises, Chapter 13

Section 13.1

In Exercises 1–6, verify that the given function is a solution to the given differential equation.

1. $y'' + 4y = 0$, $y = C_1 \cos 2x + C_2 \sin 2x$.
2. $y'' + a^2y = 0$, $y = C_1 \cos ax + C_2 \sin ax$.
3. $1/(1 - x^2)^{1/2} + y'/y = 0$, $y = Ce^{-\arcsin x}$.
4. $y' = (1 - y)/(1 - x)$, $y = 1 + C(x - 1)$.

5. $y'' + 2y' - 15y = 0, \quad y = C_1e^{3x} + C_2e^{-5x}$.
6. $y'' - (\alpha + \beta)y' + \alpha\beta y = 0, \quad y = C_1e^{\alpha x} + C_2e^{\beta x}$.
7. In Exercise 4, find the particular solution satisfying the condition $y(0) = 0$.
8. In Exercise 1, find the particular solution satisfying the conditions $y(0) = y'(0) = 0$.
9. In Exercise 4, find the particular solution satisfying the condition $y(2) = 1$.
10. In Exercise 1, find the particular solution satisfying the conditions $y(0) = 1$, $y'(0) = 1/2$.

Section 13.2

In Exercises 11–18, solve the given differential equation by separation of variables. If the equation is not separable, say so. In those cases where a specific point is given, also find the particular solution through that point.

11. $y' = \dfrac{y-1}{x-1}$.
12. $x^2 \dfrac{dy}{dx} = y - x^3y, \quad (1, 1)$.
13. $\dfrac{dy}{dx} - y = x^2y$.
14. $\sin x\, dy - \cos y\, dx = 0$.
15. $\dfrac{dx}{dt} = (1 - x^2)t, \quad (0, 0)$.
16. $y' = \sin y - \cos x$.
17. $y\dfrac{dy}{dx} - x^2 = xy$.
18. $x(\sin t)dt = \dfrac{x^2}{t}\, dx$.

Section 13.3

In Exercises 19–24, obtain the general solution to the given differential equation if it is linear. If the equation is not linear, say so. If a point is given, also find the particular solution passing through that point.

19. $y' + \dfrac{3y}{x} = x^3 + 3x, \quad (1, 0)$.
20. $\dfrac{dy}{dt} = e^{-3t} + 6y$.
21. $\dfrac{ds}{dt} + s \tan t = \sec t$.
22. $\dfrac{dy}{dx} = -y \sec x + \cos x$.
23. $t\dfrac{ds}{dt} + 3\dfrac{s}{t^2} = t^{-2}$.
24. $y' + \dfrac{y}{x} = \sin(x^2), \quad (1, 0)$.
25. Use the substitution $u = y^{1/2}$ to solve the equation $y' - y = xy^{1/2}$. (This equation is of the Bernoulli type.)
26. Solve the equation $ds/dt + s/t = s^2t$ by use of the substitution $s = u^{-1}$. (This is another example of a Bernoulli equation.)

Section 13.4

In Exercises 27–34, solve the indicated differential equation. Where conditions are given, find the particular solution that satisfies those conditions.

27. $\dfrac{d^2y}{dx^2} + 2\dfrac{dy}{dx} = 0, \quad y'(0) = y(0) = 1$.
28. $xy'' + y' = 0$.
29. $yy'' = 1 + (y')^2$.
30. $y'' - (1 + (y')^2)^{3/2} = 0$.

31. $y'' - x^3 = -\frac{1}{x}y'$.

32. $\frac{d^2y}{dx^2} - y\frac{dy}{dx} = y$.

33. $t^2\frac{d^2s}{dt^2} = \left(\frac{ds}{dt}\right)^2$.

34. $\frac{d^2x}{dt^2} - t^2 = t\frac{dx}{dt}$, $x'(0) = 1$, $x(0) = 0$.

Section 13.5

In Exercises 35–42, find a series solution to the given differential equation in some neighborhood of the origin.

35. $y' - xy = x$.

36. $y' - x^2y = 1 + x^3$.

37. $2y' + (x^2 - 1)y = x$.

38. $y' - (1 + x)y = -x^2$.

39. $y'' + xy' = 1$.

40. $y'' - 2xy' + x^2y = 0$.

41. $(1 - x)y'' + xy' - x = 0$.

42. $(1 + x^2)y'' + xy = 1$.

43. Find a series solution of the form $\sum_{k=0}^{\infty} a_k(x - 1)^k$ to the equation $y' - xy = x$.

Section 13.6

Use The Cauchy Polygon Method to find approximate solutions to the equations given in Exercises 44–46. In each case subdivide the interval into three parts.

44. $y' = xy + 1$, $(0, 1)$, $0 \le x \le 1$.

45. $y' = \sin\left(\frac{x}{y}\right)$, $(0, 1)$, $0 \le x \le \frac{\pi}{2}$.

46. $y' = x^2 \sin x$, $(\pi, 1)$, $\pi \le x \le 3\pi$.

Challenging Problems, Chapter 13

1. (i) A tank initially holds 50 gallons of pure water. A brine solution containing one pound of salt per gallon is pumped into the tank at a rate of 5 gal/min, while the new solution is allowed to run out at the same rate. Find a function giving the number of pounds of salt in the tank as a function of time.
   (ii) Suppose that the weight of salt in the incoming solution increases in such a way that at time t the incoming solution contains $1 - e^{-t}$ pounds of salt per gallon. Find a function giving the number of pounds of salt in the tank as a function of time.

14

Vectors and 3-Space

14.1 The 3-Dimensional Coordinate System

In our earlier work we established coordinate systems on the real line and in the plane in order to facilitate the study of geometric properties in an algebraic or "analytic" manner. In much the same way we shall establish a coordinate system in 3-space to aid in the analytic study of 3-dimensional geometric properties. To establish such a coordinate system we take three mutually perpendicular real lines having the same origin, O. That gives us a 3-dimensional coordinate system for locating points in space, as shown in Figure 14.1.1.

Figure 14.1.1

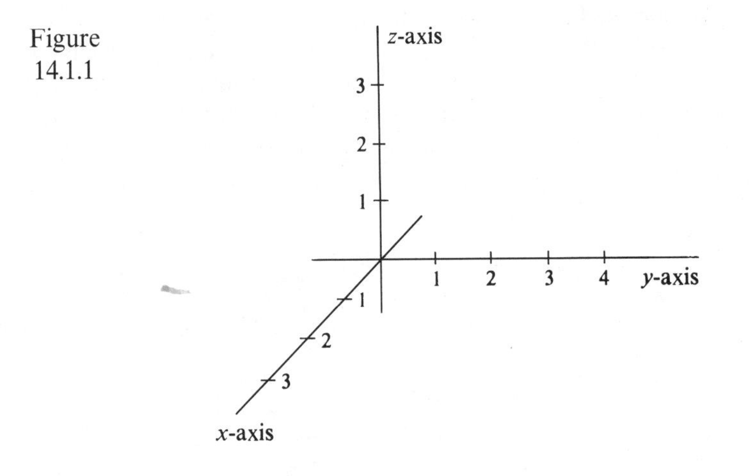

The three perpendicular lines are usually called the *x-axis*, the *y-axis*, and the *z-axis*, as indicated in Figure 14.1.1. The plane determined by the x-axis and the y-axis is called the *xy-plane*; the plane determined by the x-axis and the z-axis is called the *xz-plane*; and the plane determined by the y-axis and the z-axis is called the *yz-plane*. Although we shall assume that the same unit of measure is used on all three axes, it is sometimes convenient to use different units of measure on the axes.

With each point in 3-space there is associated an ordered triple of numbers, called coordinates. The first number in the ordered triple indicates the location of the point in the direction of the x-axis, the second one indicates its location in the direction of the y-axis, and the third number indicates the location of the point in the direction of the z-axis. For example, in Figure 14.1.2, the point P_1 is 1 unit in the x-direction, 3 units in the y-direction, and 2 units in the z-direction, and so has coordinates (1, 3, 2). The point P_2 in the xz-plane 3 units behind the yz-plane and 2 units above the xy-plane has coordinates $(-3, 0, 2)$. The point of intersection of all three axes is called the *origin*, which has coordinates (0, 0, 0).

The three coordinate planes, namely the xy-plane, the yz-plane, and the xz-plane, divide the 3-space into 8 parts, often called *octants*. The octant consisting of all points whose coordinates are all nonnegative is called the *first octant*.

Figure 14.1.2

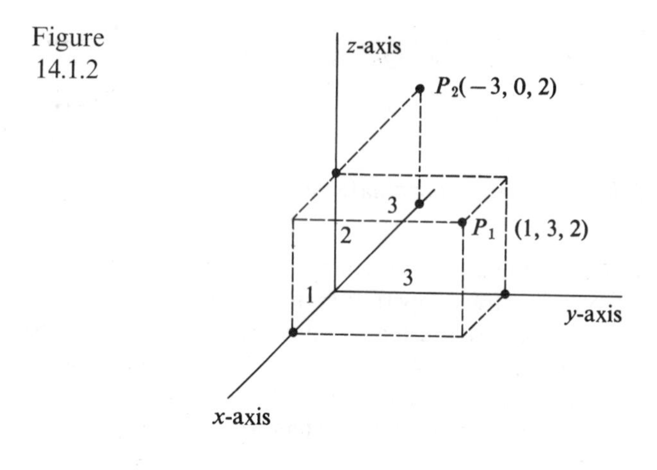

The coordinate system we have chosen is described as a *right-handed* coordinate system. If we were to interchange the x- and y-axes, the resulting coordinate system would be described as a *left-handed* coordinate system. The description "right-handed" comes from the right hand as follows: Place the right hand in the configuration shown in Figure 14.1.3 so that the fingers curve in the direction necessary to rotate the x-axis to the y-axis through an angle of $\pi/2$. Then the thumb points in the direction of the z-axis.

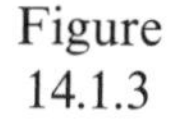
Figure 14.1.3

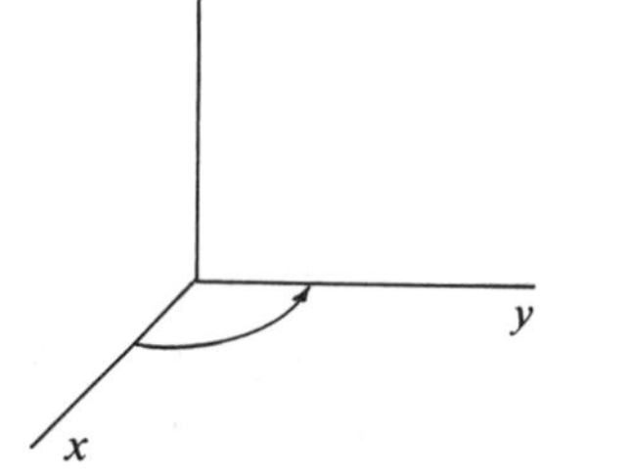

Definition 14.1.1

> A point (a, b, c) is said to be a *solution* of an equation involving x, y, and z, if when x is replaced by a, y is replaced by b, and z is replaced by c, the equation becomes a true statement.

For example, $(1, -2, 2)$ is a solution of the equation $x^2 + y^2 + z^2 = 9$. Similarly, $(0, 0, 3)$, $(1, 2, -2)$, $(2, -1, -2)$, and $(-2, -2, -1)$, along the infinitely many other points, are also solutions of $x^2 + y^2 + z^2 = 9$.

Definition 14.1.2

> The set of all points that are solutions of an equation involving x, y, and z is called the *graph of the equation.*

Example 1 We see that the graph of the equation $z = 2$ is the set of all points with z-coordinate 2, and so it is the plane that is parallel to the xy-plane and 2 units above it. ||

Example 2 The graph of the equation $y = -3$ is the set of all points with y-coordinate -3, and hence it is the plane that is parallel to the xz-plane and 3 units to the left of it. ||

Note that in our present context we are considering the equations $z = 2$ and $y = -3$ to be equations involving x, y, and z. This could be emphasized by writing $z = 2$ as $0 \cdot x + 0 \cdot y + z = 2$, and $y = -3$ as $0 \cdot x + y + 0 \cdot z = -3$.

As in the two examples above, it is generally true that the graph of an equation involving x, y, and z is a surface. In particular, as illustrated in Examples 1 and 2, the graph of any equation of the form $x =$ a constant, $y =$ a constant, or $z =$ a constant is a plane parallel to one of the coordinate planes.

Exercises 14.1

For Exercises 1–6, opposite vertices of a rectangular box whose edges are parallel to the coordinate axes are given. Find the coordinates of the other 6 vertices.

1. $(0, 0, 0)$, $(1, 2, 3)$.
2. $(0, 0, 0)$, $(2, 3, 5)$.
3. $(3, 7, 1)$, $(1, 6, 4)$.
4. $(8, 2, 3)$, $(3, 5, 7)$.
5. $(-1, 3, -2)$, $(-3, -1, 5)$.
6. $(4, -2, 1)$, $(-3, -1, -6)$.

In Exercises 7–27, describe in words the graph in 3 dimensions of each equation.

7. $z = 0$.
8. $x = 0$.
9. $y = 10$.
10. $z = 6$.
11. $x = -2$.
12. $y = -5$.
13. $z = -1$.
14. $x = y$.
15. $y = z$.
16. $z = x$.
17. $x^2 + y^2 = 0$.
18. $y^2 + z^2 = 0$.
19. $x^2 + z^2 = 0$.
20. $x^2 + y^2 + z^2 = 0$.
21. $(x - 1)^2 + (y + 2)^2 + z^2 = 0$.
22. $|x| = 3$.
23. $|y| = 2$.
24. $|z| = 0$.
25. $|x - 2| = 3$.
26. $|y + 1| = 1$.
27. $|z - 3| = 4$.

14.2 Distance

In Section 1.7 we obtained the formula for the distance between two points in the xy-plane. In three dimensions the formula and its derivation are similar to the 2-dimensional case.

Consider two points P_1 and P_2 with coordinates (x_1, y_1, z_1) and (x_2, y_2, z_2), as in Figure 14.2.1.

The segment P_1P_2 can be considered to be the diagonal of a rectangular box as indicated. The point Q, whose x and y-coordinates are the same as those of P_2 and

Figure 14.2.1

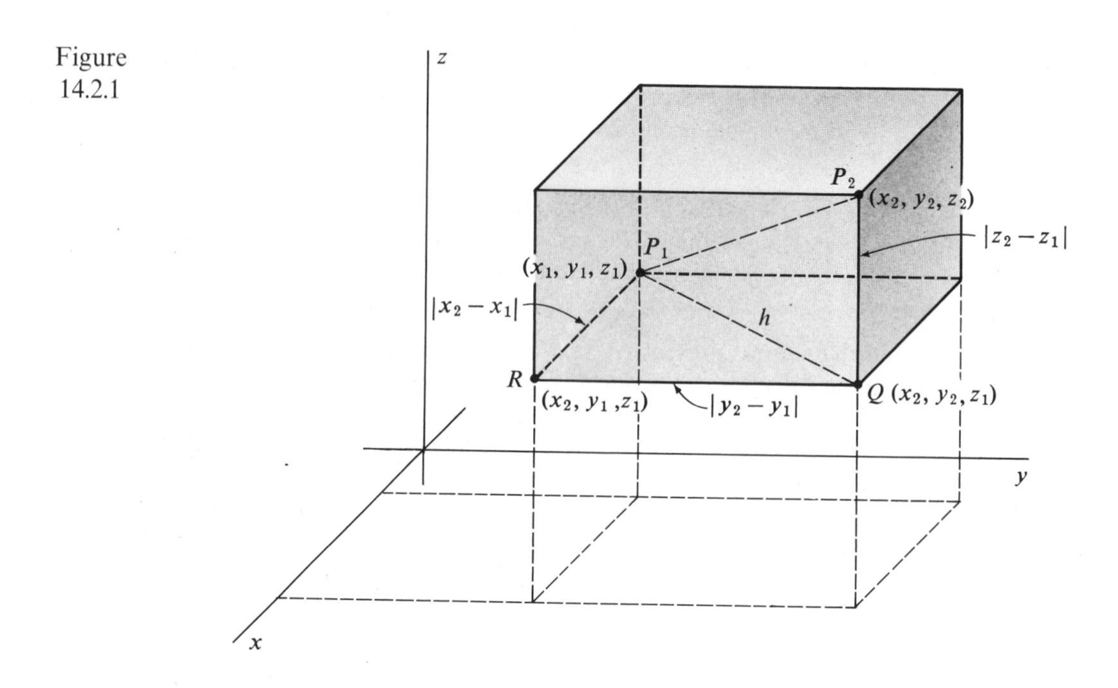

whose z-coordinate is the same as that of P_1, is then one end of a diagonal of length h of a horizontal face of the box, the other end being P_1. The point R has coordinates (x_2, y_1, z_1). Since R and P_1 are on a line parallel to the x-axis, the distance between them is $|x_2 - x_1|$. Similarly, the distance between R and Q is $|y_2 - y_1|$, and the distance between Q and P_2 is $|z_2 - z_1|$.

By applying the Pythagorean Theorem to the right triangle P_1RQ we see that

$$h^2 = |x_2 - x_1|^2 + |y_2 - y_1|^2 = (x_2 - x_1)^2 + (y_2 - y_1)^2.$$

But the triangle P_1QP_2 is also a right triangle, and hence the distance between P_1 and P_2, which we denote by

$$D((x_1, y_1, z_1), (x_2, y_2, z_2)),$$

is

$$\sqrt{h^2 + |z_2 - z_1|^2} = \sqrt{h^2 + (z_2 - z_1)^2}.$$

Finally, substituting for h^2, we have the *distance formula* for any points (x_1, y_1, z_1) and (x_2, y_2, z_2):

$$D((x_1, y_1, z_1), (x_2, y_2, z_2)) = \sqrt{(x_2 - x_1)^2 + (y_2 - y_1)^2 + (z_2 - z_1)^2}.$$

distance

Example 1 The distance between $(-2, 3, 1)$ and $(4, 1, -3)$ is

$$D((-2, 3, 1), (4, 1, -3)) = \sqrt{(4+2)^2 + (1-3)^2 + (-3-1)^2} = \sqrt{56}. \quad \|$$

It is left as an exercise to use the distance formula to prove that the midpoint of the segment joining the points (x_1, y_1, z_1) and (x_2, y_2, z_2) is the point

$$\left(\frac{x_1 + x_2}{2}, \frac{y_1 + y_2}{2}, \frac{z_1 + z_2}{2}\right).$$

mid pt.

Using the distance formula we can find an equation of the sphere with center at the point (x_0, y_0, z_0) and radius r. Since a point is on this sphere if and only if its distance from the point (x_0, y_0, z_0) is r, a point (x, y, z) is on the sphere if and only if

$$\sqrt{(x - x_0)^2 + (y - y_0)^2 + (z - z_0)^2} = r. \tag{1}$$

But a point satisfying that equation must also satisfy the equation

$$(x - x_0)^2 + (y - y_0)^2 + (z - z_0)^2 = r^2. \tag{2}$$

Conversely, any point that satisfies equation (2) must satisfy

$$\pm\sqrt{(x - x_0)^2 + (y - y_0)^2 + (z - z_0)^2} = r.$$

But since $r \geq 0$, the minus sign can be eliminated, and so any point satisfying (2) must also satisfy (1). Thus we see that an equation of the sphere of radius r with center at (x_0, y_0, z_0) is

$$(x - x_0)^2 + (y - y_0)^2 + (z - z_0)^2 = r^2.$$

Example 2 An equation of the sphere of radius 6 with center at $(-2, 5, 1)$ is

$$(x + 2)^2 + (y - 5)^2 + (z - 1)^2 = 36. \quad ||$$

Exercises 14.2

For Exercises 1–4, find the distance between the given points.

1. $(-4, 5, -1)$ and $(-6, 3, 2)$.
2. $(2, 0, -3)$ and $(-1, -3, -2)$.
3. $(0, -2, 3)$ and $(-2, 5, 1)$.
4. $(1, -2, 7)$ and $(3, -5, -1)$.

For Exercises 5–10, find an equation of the sphere indicated.

5. Center at $(1, -2, 3)$ and radius 7.
6. Center at $(4, 1, -3)$ and radius 5.
7. Center at $(-4, 2, 3)$ and passing through the point $(-2, 5, 1)$.
8. Center at $(1, -5, 6)$ and passing through the point $(0, 3, -2)$.
9. Ends of one diameter at $(-2, 3, 5)$ and $(4, 1, 3)$.
10. Ends of one diameter at $(3, -7, -4)$ and $(9, 1, -8)$.

For Exercises 11–16, find an inequality for the set indicated.

11. The set of all points inside the sphere with radius 2 and center at $(4, -1, 3)$.
12. The set of all points outside the sphere with radius 3 and center at $(-2, 5, -1)$.
13. The set of all points outside the sphere of Exercise 7.
14. The set of all points inside the sphere of Exercise 8.
15. The set of points inside the sphere of Exercise 5 and outside the sphere of Exercise 7.
16. The set of points inside the sphere of Exercise 8 and outside the sphere of Exercise 6.
17. Find an equation of the set of all points equidistant from the plane $y = 4$ and the origin.
18. Find an equation of the set of all points equidistant from the plane $z = -1$ and the point $(0, 0, 1)$.

19. Find an equation of the set of all points equidistant from the points $(-1, 2, 3)$ and $(-2, -3, 5)$.
20. Find an equation of the set of all points equidistant from the points $(3, 4, -2)$ and $(-1, 1, 3)$.
21. Prove that the midpoint of the segment joining the points (x_1, y_1, z_1) and (x_2, y_2, z_2) is the point

$$\left(\frac{x_1 + x_2}{2}, \frac{y_1 + y_2}{2}, \frac{z_1 + z_2}{2}\right).$$

14.3 Sketching Graphs

In general the best approach to sketching a graph in three dimensions is to determine information about curves of intersection of the graph with planes parallel to each of the three coordinate planes. In particular we find the curves of intersection of the 3-dimensional graph with planes parallel to the xy-plane, that is, planes of the form $z = t$, where t is a constant. Similarly we use planes of the form $x = r$, where r is a constant (parallel to the yz-plane), or of the form $y = s$, where s is a constant (parallel to the xz-plane).

Example 1 | Sketch the graph of the equation $x^2 + y^2 = z$.

We see that a plane of the form $z = t$ intersects the graph in a circle $x^2 + y^2 = t$ of radius $\sqrt{t}$. Clearly, only planes of the form $z = t$, where t is not negative, intersect the graph; and no part of the graph is below the xy-plane. When $t = 0$, the "circle" of intersection is the point (0, 0) in the xy-plane; and as t increases, the radius of the circle increases. By setting $y = 0$, we see that the graph intersects the xz-plane in the parabola $x^2 = z$; and by setting $x = 0$ we see that the graph intersects the yz-plane in the parabola $y^2 = z$.

By sketching those two parabolas and a circle or two above the xy-plane we obtain the sketch of the graph indicated in Figure 14.3.1(a). Figure 14.3.1(b) is a computer-generated image showing a slightly different view of the graph. ||

As in two dimensions a graph is often unbounded, and so we can only sketch a portion of the graph, as in Figure 14.3.1.

Often a graph can be sketched quickly by finding its curves of intersection with the three coordinate planes and mentally noting the shape of the curves of intersection of the graph with planes parallel to the coordinate planes.

Example 2 | Sketch the graph of the equation $\frac{x^2}{4} + \frac{y^2}{25} + \frac{z^2}{9} = 1$.

The graph intersects the xy-plane ($z = 0$) in the ellipse

$$\frac{x^2}{4} + \frac{y^2}{25} = 1.$$

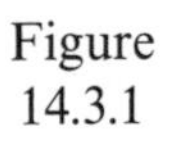

Figure 14.3.1

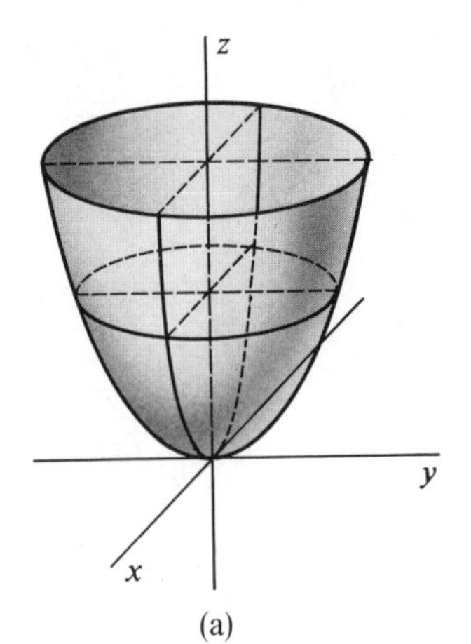

(a)

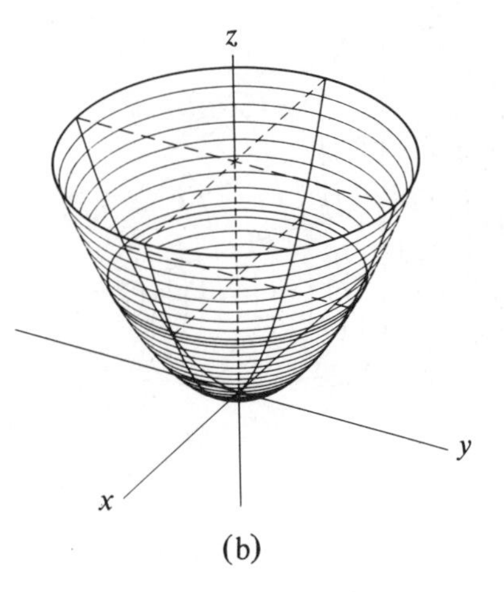

(b)

The graph also cuts the xz-plane and the yz-plane in ellipses with equations

$$\frac{x^2}{4} + \frac{z^2}{9} = 1 \quad \text{and} \quad \frac{y^2}{25} + \frac{z^2}{9} = 1.$$

In fact, we mentally note that every plane parallel to a coordinate plane that intersects the surface in more than one point intersects it in an ellipse. We sketch the three ellipses and obtain a melon-shaped figure, called an ellipsoid, shown in Figure 14.3.2(a). Figure 14.3.2(b) is a computer-generated image showing a slightly different view of the graph. ||

Example 3 | Sketch the graph of the equation $\dfrac{x^2}{4} + \dfrac{y^2}{9} - \dfrac{z^2}{16} = 1$.

In the xy-plane the intersection is the ellipse

$$\frac{x^2}{4} + \frac{y^2}{9} = 1.$$

Figure 14.3.2

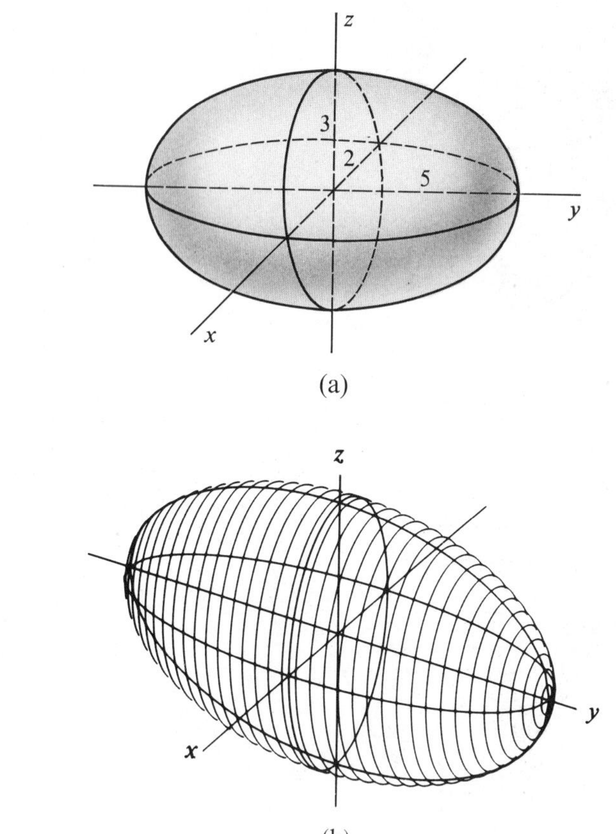

(a)

(b)

In the xz-plane and yz-plane the intersections are the hyperbolas

$$\frac{x^2}{4} - \frac{z^2}{16} = 1 \quad \text{and} \quad \frac{y^2}{9} - \frac{z^2}{16} = 1.$$

We also notice that a plane of the form $z = t$, t constant, intersects the graph in the ellipse

$$\frac{x^2}{4} + \frac{y^2}{9} = 1 + \frac{t^2}{16}.$$

That ellipse is smallest when $t = 0$ and gets larger as $|t|$ increases. Using that information we obtain the graph indicated in Figure 14.3.3. ||

Example 4 | Sketch the graph of $\frac{x^2}{9} + \frac{y^2}{16} = \frac{z^2}{4}$.

In the xz-plane the intersection has equation $x^2/9 = z^2/4$, which is equivalent to $z = \pm 2x/3$, and therefore represents the two lines with equations $z = 2x/3$ and

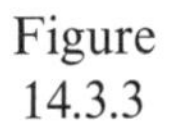
Figure
14.3.3

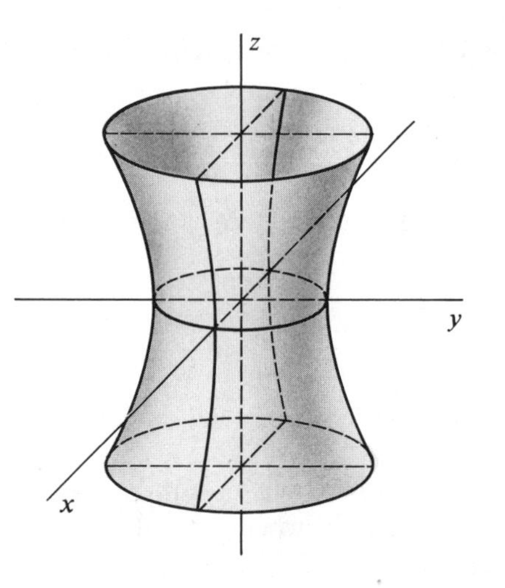

$z = -2x/3$. In the yz-plane we also get two lines through the origin; the equations are $y = \pm 2z$. When $z = 0$, we have

$$\frac{x^2}{9} + \frac{y^2}{16} = 0,$$

which is satisfied only by the point (0, 0). A plane with equation $z = t \neq 0$ intersects the graph of the given equation in the ellipse

$$\frac{x^2}{9} + \frac{y^2}{16} = \frac{t^2}{4}.$$

The resulting graph is called an elliptic cone, part of which is shown in Figure 14.3.4. ||

Figure
14.3.4

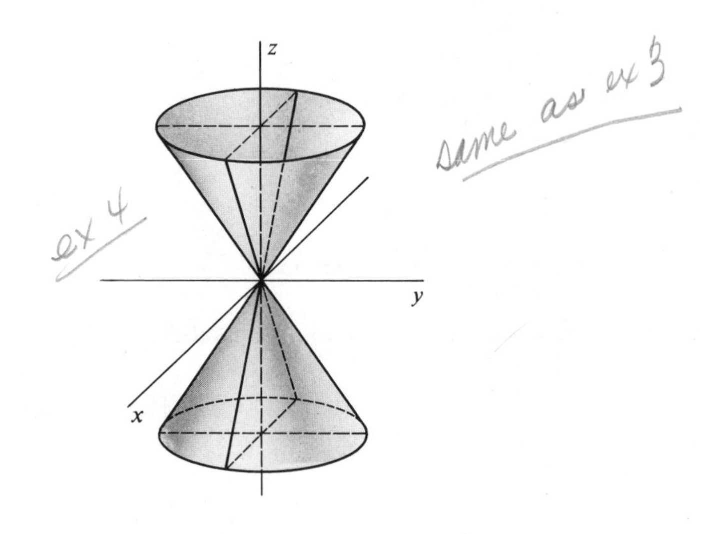

Example 5 | Sketch the graph of $\dfrac{z^2}{9} - \dfrac{x^2}{16} - \dfrac{y^2}{25} = 1$.

In the xz-plane and the yz-plane the intersections are the hyperbolas:

$$\frac{z^2}{9} - \frac{x^2}{16} = 1 \quad \text{and} \quad \frac{z^2}{9} - \frac{y^2}{25} = 1.$$

A plane with equation $z = t$ intersects the graph in

$$\frac{x^2}{16} + \frac{y^2}{25} = \frac{t^2}{9} - 1, \quad \text{or}$$

$$\frac{x^2}{16} + \frac{y^2}{25} = \frac{t^2 - 9}{9}. \qquad (1)$$

Since the left side of equation (1) cannot be negative, there are only curves of intersection when $t^2 - 9$ is not negative; that is, when $t \leq -3$ or $t \geq 3$. Thus no part of the graph of the original equation exists when $-3 < z < 3$.

When $t < -3$ or $t > 3$, the graph of (1) is an ellipse. Consequently we have the graph indicated in Figure 14.3.5. ||

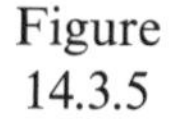
Figure 14.3.5

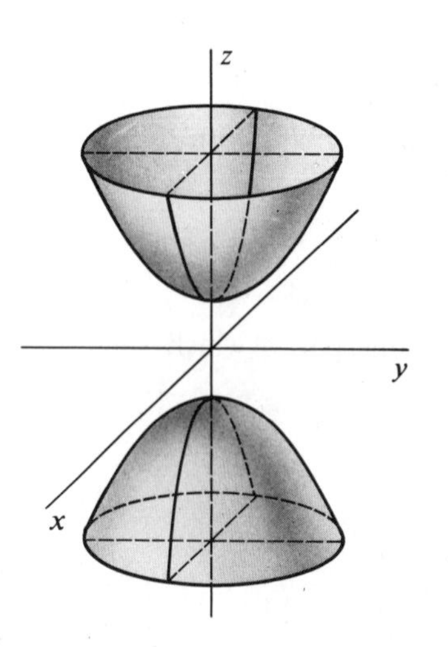

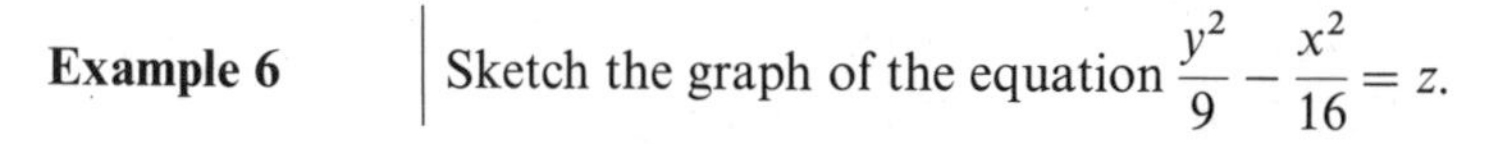
Example 6 | Sketch the graph of the equation $\frac{y^2}{9} - \frac{x^2}{16} = z$.

In the xz-plane and yz-plane the intersections are the parabolas

$$x^2 = -16z \quad \text{and} \quad y^2 = 9z.$$

For a plane $z = t$, the intersection is the hyperbola

$$\frac{y^2}{9t} - \frac{x^2}{16t} = 1.$$

On sketching those curves we get the saddle-shaped graph shown in Figure 14.3.6. ||

Figure 14.3.6

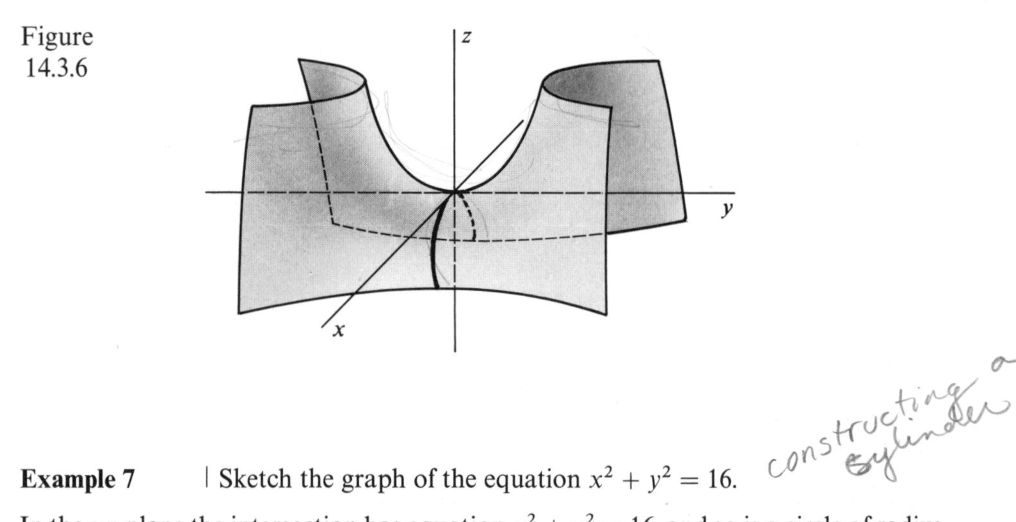

Example 7 | Sketch the graph of the equation $x^2 + y^2 = 16$.

In the xy-plane the intersection has equation $x^2 + y^2 = 16$, and so is a circle of radius 4 centered at the origin. In fact in any plane with equation $z = t$, the intersection is the circle with equation $x^2 + y^2 = 16$, and so is a circle of radius 4 centered on the z-axis. Thus we have the graph illustrated in Figure 14.3.7. Note that the graph can be obtained by moving a line around the circle $x^2 + y^2 = 16$ in the xy-plane so that the line remains parallel to the z-axis. ||

If the equation had been $y = x^2$, the graph could be obtained by moving a line around the parabola $y = x^2$ in the xy-plane so that the line remains parallel to the z-axis. Similar remarks hold for any equation that does not contain z. Surfaces that can be obtained by moving a line so that it remains parallel to a fixed line are called *cylindrical surfaces*. The graph of an equation that does not contain y is also a cylindrical surface. In that case the generating line is moved parallel to the y-axis. Similar remarks hold for the graph of an equation that does not contain x.

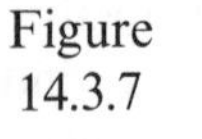
Figure 14.3.7

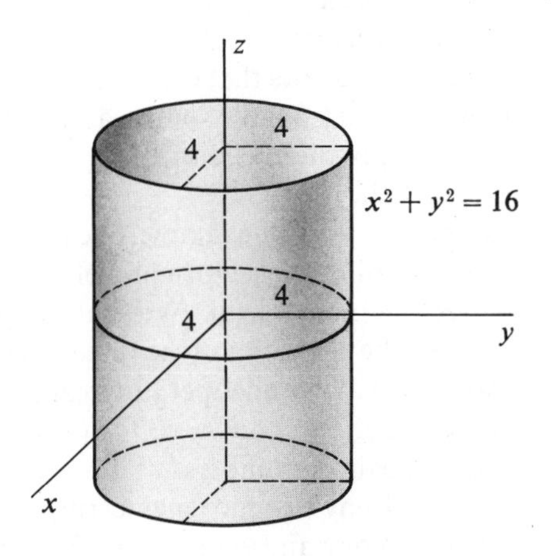

Exercises 14.3

Sketch in 3-space the graphs of the following equations.

1. $\frac{x^2}{9} + \frac{y^2}{4} + \frac{z^2}{16} = 1.$
2. $x^2 + y^2 = 2z.$
3. $x^2 + z^2 = y.$
4. $\frac{x^2}{16} - \frac{y^2}{9} + \frac{z^2}{9} = 1.$
5. $9x^2 + y^2 = z^2.$
6. $x^2 + \frac{y^2}{16} + \frac{z^2}{9} = 1.$
7. $\frac{x^2}{25} + \frac{y^2}{16} - \frac{z^2}{36} = 1.$
8. $x^2 + 4y^2 = z^2.$
9. $z^2 - x^2 - y^2 = 1.$
10. $y^2 - x^2 = z.$
11. $y^2 - 4x^2 = z.$
12. $y^2 - x^2 - z^2 = 1.$
13. $4y^2 + z^2 = x.$
14. $y^2 + z^2 - x^2 = 4.$
15. $x^2 + y^2 + z^2 = 25.$
16. $x^2 + y^2 + z^2 = 1.$
17. $(x + 2)^2 + (y - 3)^2 + z^2 = 9.$
18. $(x - 1)^2 + (y - 5)^2 + (z - 2)^2 = 36.$
19. $(x - 3)^2 + (y + 2)^2 + (z - 4)^2 = 0.$
20. $(x + 2)^2 + (y - 5)^2 + (z - 3)^2 = 0.$
21. $x^2 + y^2 = 4.$
22. $y^2 + z^2 = 9.$
23. $x^2 + z^2 = 16.$
24. $(x + 2)^2 + (y + 3)^2 = 1.$
25. $y = x^2.$
26. $x = y^2.$
27. $4x^2 + y^2 = 16.$
28. $9x^2 + z^2 = 9.$
29. $y = 2x.$
30. $z = -x.$
31. $2z = y + 4.$
32. $x = z - 2.$
33. $y^2 - x^2 = 0.$
34. $z^2 - y^2 = 0.$

14.4 Vectors

Vectors have been used for many years to describe physical quantities that have both magnitude and direction. For example, vectors are often used to describe velocities and forces. In Figure 14.4.1 we picture the vectors that can be used to describe a wind from the west at 20 miles per hour and a wind from the south-east at 40 miles per hour. Of course the length of the vector is simply the speed of the wind and the direction of the wind is the direction indicated by the arrow.

A vector can be described in several ways. For example, one might simply specify the length and direction in order to determine the vector uniquely. However we shall find another method of specifying vectors more useful. We shall simply assume that the vector starts at the origin and specify the end point of the vector. Then as indicated in Figure 14.4.1, a wind from the southeast at 40 miles per hour can be fully described by the ordered pair of numbers $(20\sqrt{2}, 20\sqrt{2})$. Thus a vector (at least in the plane) can simply be thought of as an ordered pair of numbers.

The use of vectors is not limited to physics and engineering. As we shall see, vectors are now used in areas such as biology and economics. Before we investigate

Figure 14.4.1

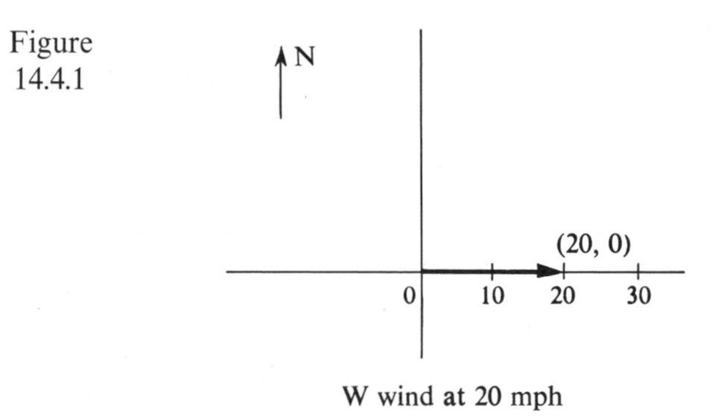

W wind at 20 mph

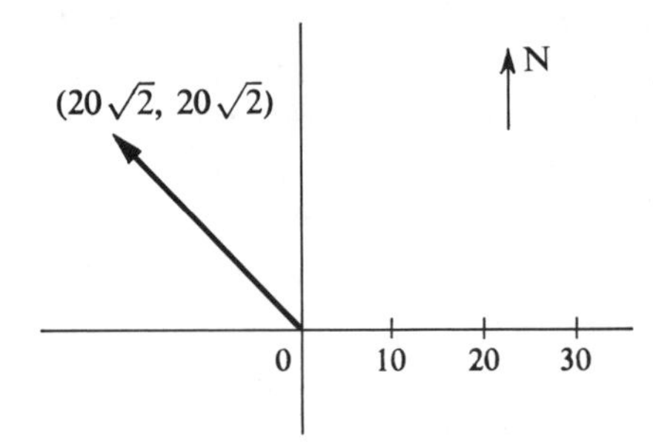

SE wind at 40 mph

such uses, however, we need to give a formal definition of vectors. Recalling that a vector in the plane is fully described by an ordered pair of numbers, we give the following definition.

Definition 14.4.1

> An *n-dimensional vector* is an ordered n-tuple
>
> $$(a_1, a_2, \ldots, a_n)$$
>
> of numbers $a_1, a_2, \ldots, a_n$, where $(a_1, a_2, \ldots, a_n) = (b_1, b_2, \ldots, b_n)$ if and only if $a_1 = b_1, a_2 = b_2, \ldots, a_n = b_n$.

For example, the ordered pairs (3, 5), (0, 0) $(8, \pi)$, and $(7, \pi)$ are 2-dimensional vectors. Also $(3, 5) = (3, 5)$ but $(8, \pi) \neq (7, \pi)$ because $8 \neq 7$. As another example, with n equal to 3, the following ordered 3-tuples or ordered triples, are vectors: $(\pi, 0, 3)$, $(2, 1, -5)$, and $(3, 0, \pi)$. Note that $(\pi, 0, 3) \neq (3, 0, \pi)$ because $\pi \neq 3$ (or $3 \neq \pi$).

While most of the familiar vector quantities are represented by vectors of dimensions 2 and 3, the use of vectors is not limited to these dimensions. For example, a description of the use of 71 pounds of beef, 20 pounds of pork, 27 pounds of fish, and 14 dozen eggs by a restaurant can be indicated by the 4-dimensional vector (71, 20, 27, 14).

Vectors are now even used in biology for ecological systems to help describe the population, growth, and interaction of plants and animals. For example if there are 17 deer, 11 elk, 80 snakes, 0 cougars, and 160 trout in a certain region, we might use the 5-dimensional vector (17, 11, 80, 0, 160) to describe that population.

We shall use the symbol R^n to denote the set of all n-dimensional vectors. Vectors will be denoted by ordered n-tuples or boldface capitals. The individual numbers $a_1, a_2, \ldots, a_n$ of a vector

$$\mathbf{A} = (a_1, a_2, \ldots, a_n)$$

are called the *components* or *coordinates* of **A**, a_1 being the first component, a_2 the second component, and so forth.

Geometrically it is often useful to interpret a vector of R^2 or R^3 as an arrow, in the plane or in space, from the origin to the point whose coordinates are the same as those of the vector. When this is done the vector is often called the *position vector* of the point and the point is called the *head* of the vector. In this way each vector in R^2 or R^3 corresponds to a unique point, and conversely each point corresponds to a unique vector. Because of this correspondence it is often unnecessary to distinguish between points and vectors. Some vectors are illustrated by arrows in the next two examples.

Example 1 The vectors $(-3, 2)$, $(4, 3)$, and $(3, 1)$ of R^2 may be pictured as the arrows in Figure 14.4.2. ||

Figure 14.4.2

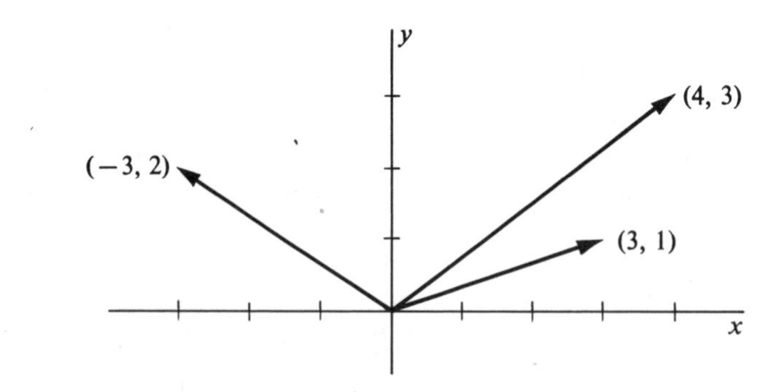

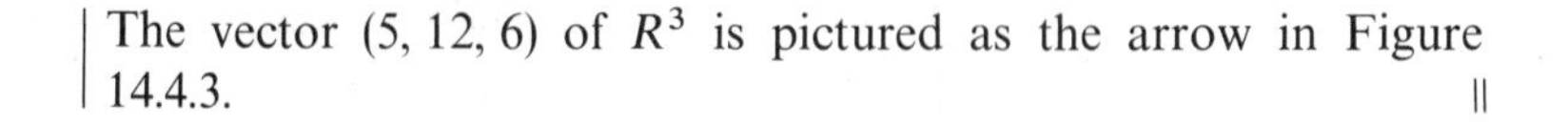

Example 2 The vector $(5, 12, 6)$ of R^3 is pictured as the arrow in Figure 14.4.3. ||

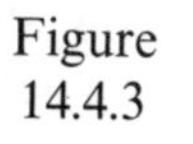

Figure 14.4.3

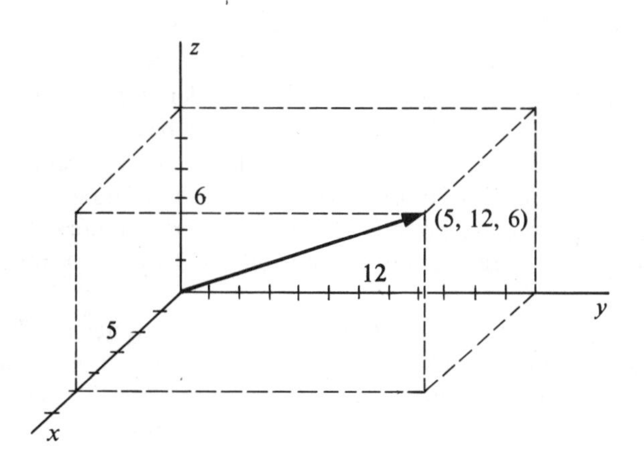

As we stated above, in our geometric interpretation of vectors as arrows, all of the arrows start at the origin and two vectors are equal if and only if they are the same arrow. Often in physics or engineering it is useful to consider arrows that do not start at the origin. If we let **A** be a vector and consider any point P, then there is exactly one arrow that starts at P and has the same length and direction as **A**. We call this arrow *the replica of* **A** *at P*. Some illustrations of replicas of a vector **A** are shown in Figure 14.4.4.

Figure 14.4.4

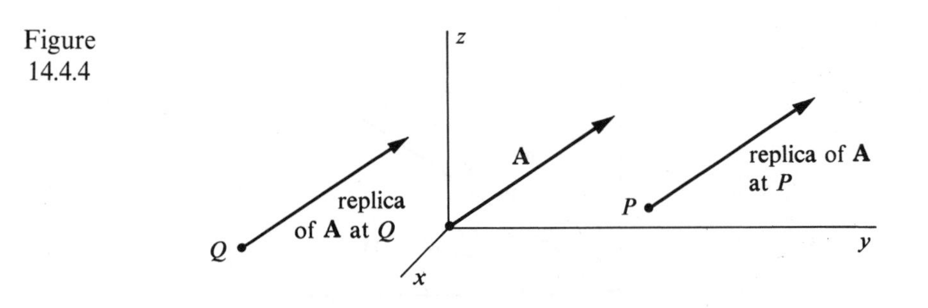

A force on an object or the velocity of an object can be described by a vector at the origin or as a replica of the vector at the object. The length of the arrow used is the magnitude of the force or velocity, and the direction of the arrow used is the direction of the force or velocity. For example a turning force of 3 pounds on the steering wheel of a car might be described by the replica at P shown in Figure 14.4.5.

Of course the force indicated by the replica at P has a much different effect on the direction of the car than the force indicated by the replica at Q, even though these two arrows have the same length and direction.

It is useful to be able to add and subtract vectors and to multiply a vector by a number. We define these operations in the following definition.

Definition 14.4.2

Let $\mathbf{A} = (a_1, a_2, \ldots, a_n)$ and $\mathbf{B} = (b_1, b_2, \ldots, b_n)$ be two vectors of R^n, and let k be any number. We define:

Addition

$(a_1, a_2, \ldots, a_n) + (b_1, b_2, \ldots, b_n) = (a_1 + b_1, a_2 + b_2, \ldots, a_n + b_n).$

Subtraction

$(a_1, a_2, \ldots, a_n) - (b_1, b_2, \ldots, b_n) = (a_1 - b_1, a_2 - b_2, \ldots, a_n - b_n).$

Scalar Multiplication

$k(a_1, a_2, \ldots, a_n) = (ka_1, ka_2, \ldots, ka_n).$

For example:

$$(\pi, 0, 3) + (2, 1, -5) = (\pi + 2, 1, -2),$$
$$(\pi, 0, 3) - (2, 1, -5) = (\pi - 2, -1, 8),$$
$$6(\pi, 0, 3) = (6\pi, 0, 18).$$

Geometrically, the sum $\mathbf{A} + \mathbf{B}$ of vectors **A** and **B** is a diagonal of the parallelogram having **A** and **B** as adjacent sides, as illustrated for R^2 and R^3 in Figures 14.4.6 and 14.4.7.

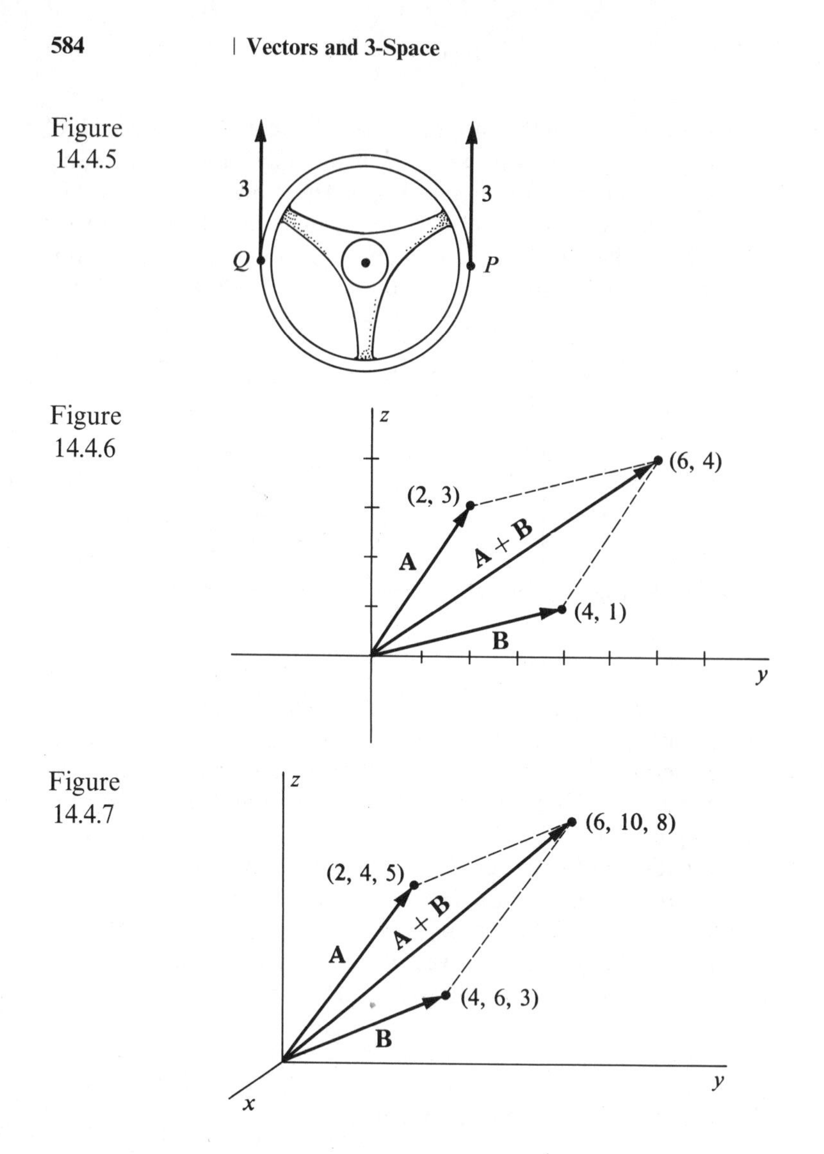

Figure 14.4.5

Figure 14.4.6

Figure 14.4.7

This "parallelogram law" for adding vectors **A** and **B** can also be conceived as attaching a replica of **B** to the head of **A**. The third side of the resulting triangle is then **A** + **B**. Similarly **A** + **B** can be obtained by attaching a replica of **A** to the head of **B**. These replica-attaching interpretations of addition are illustrated in Figure 14.4.8.

To get a geometric picture of subtraction, consider two vectors **A** and **B**, both of which are either in R^2 or in R^3. It is easy to see that $\mathbf{B} + (\mathbf{A} - \mathbf{B}) = \mathbf{A}$. Thus

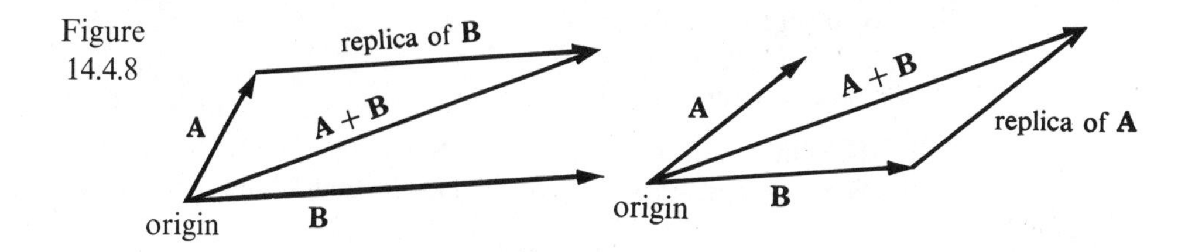

Figure 14.4.8

A − **B** is the vector which when added to **B** produces **A**. Using that fact and the replica-attaching picture of addition, we can get a picture of **A** − **B**. Indeed a replica of **A** − **B** goes from the head of **B** to the head of **A** as shown in Figure 14.4.9.

Figure 14.4.9

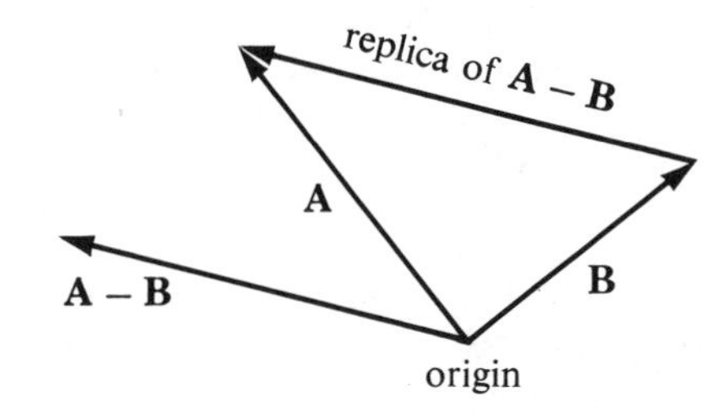

In an actual physical situation, if **W** is a vector describing the velocity of the water in a river and **B** is a vector describing the velocity of a boat in still water, then **W** + **B** is the vector describing the actual velocity of the boat. That is, the boat moves in the direction of **W** + **B** with a speed to the length of **W** + **B**.

Figure 14.4.10

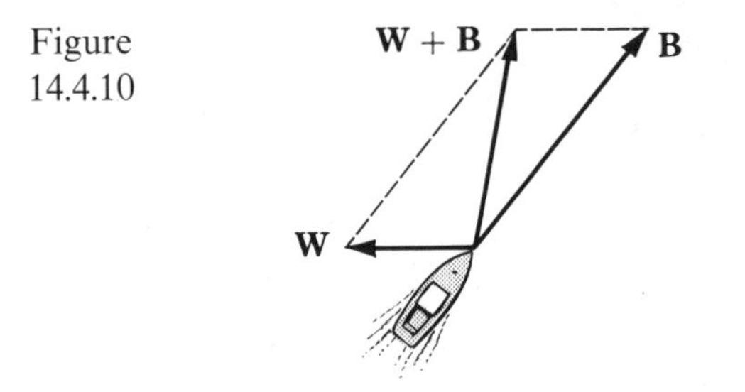

As an example of vector addition in economics, let the components of a vector indicate the number of pounds of beef, the number of pounds of pork, the number of pounds of fish, and the number of dozens of eggs used by a restaurant. Then, if

$$\mathbf{A} = (71, 20, 27, 14) \quad \text{and} \quad \mathbf{B} = (36, 18, 53, 11)$$

describe the use of these items on two successive days, their sum

$$\mathbf{A} + \mathbf{B} = (107, 38, 80, 25)$$

describes the total use of the items for the two days.

If the restaurant had the same usage vector

$$\mathbf{A} = (71, 20, 27, 14)$$

every day for a week, then the vector describing the total use of the items for a week is the scalar product 7**A**. Thus the usage vector for the week could be found as follows:

$$\begin{aligned} 7\mathbf{A} &= (7 \cdot 71, 7 \cdot 20, 7 \cdot 27, 7 \cdot 14) \\ &= (497, 140, 189, 98). \end{aligned}$$

In R^2 the length of a vector (a, b) is $(a^2 + b^2)^{1/2}$. In R^3 the length of a vector (a, b, c) is $(a^2 + b^2 + c^2)^{1/2}$. In those cases, the length of a vector is the distance between the origin and the point whose coordinates are the same as those of the vector. In general, we define the length (also called magnitude or absolute value) of a vector in R^n as follows.

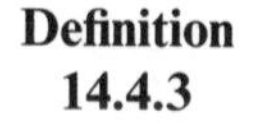

Definition 14.4.3

> The length $|\mathbf{A}|$ of a vector $\mathbf{A} = (a_1, a_2, \ldots, a_n)$ is
> $$|\mathbf{A}| = \sqrt{a_1{}^2 + a_2{}^2 + \cdots + a_n{}^2}.$$

It is easy to prove that for any number k and any vector $\mathbf{A}$, we have $|k\mathbf{A}| = |k| \cdot |\mathbf{A}|$. Consequently, the scalar product $k\mathbf{A}$ is a vector in the same direction as $\mathbf{A}$ if $k > 0$, but in the opposite direction of $\mathbf{A}$ if $k < 0$. Moreover, the length of $k\mathbf{A}$ is $|k|$ times the length of $\mathbf{A}$, as illustrated in Figure 14.4.11.

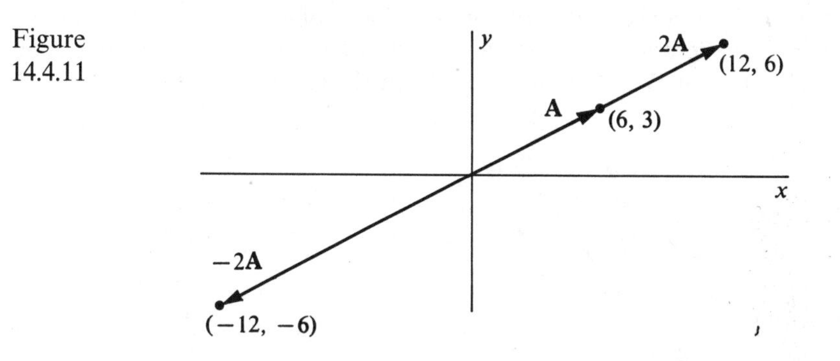

Figure 14.4.11

It follows from our replica-attaching interpretation of subtraction that if $\mathbf{A} = (x_1, y_1, z_1)$ and $\mathbf{B} = (x_2, y_2, z_2)$ are vectors in R^3, then the length of the vector $\mathbf{A} - \mathbf{B}$ is equal to the distance between the points (x_1, y_1, z_1) and (x_2, y_2, z_2), which are the heads of vectors $\mathbf{A}$ and $\mathbf{B}$. That is,

$$|\mathbf{A} - \mathbf{B}| = \sqrt{(x_2 - x_1)^2 + (y_2 - y_1)^2 + (z_2 - z_1)^2}, \quad \text{or}$$
$$|\mathbf{A} - \mathbf{B}|^2 = (x_2 - x_1)^2 + (y_2 - y_1)^2 + (z_2 - z_1)^2.$$

In general, if $\mathbf{A} = (a_1, a_2, \ldots, a_n)$ and $\mathbf{B} = (b_1, b_2, \ldots, b_n)$ are any two vectors of R^n, then

$$\mathbf{A} - \mathbf{B} = (a_1 - b_1, a_2 - b_2, \ldots, a_n - b_n), \quad \text{and}$$
$$|\mathbf{A} - \mathbf{B}| = \sqrt{(a_1 - b_1)^2 + (a_2 - b_2)^2 + \cdots + (a_n - b_n)^2}.$$

A *nonzero vector* is any vector whose components are not all 0. Two nonzero vectors $\mathbf{A}$ and $\mathbf{B}$ of R^2 or R^3 with the same or opposite directions are called *parallel*. Thus $\mathbf{A}$ and $\mathbf{B}$ are parallel if and only if each is a nonzero scalar multiple of the other.

A *unit vector* is a vector of length 1. Note that if $|\mathbf{A}| \neq 0$, the vector $\mathbf{A}/|\mathbf{A}|$ is a unit vector because

$$\left|\frac{\mathbf{A}}{|\mathbf{A}|}\right| = \left|\frac{1}{|\mathbf{A}|}\mathbf{A}\right| = \frac{1}{|\mathbf{A}|}|\mathbf{A}| = 1.$$

Let **A**, **B**, and **C** be any vectors of R^n, and let a and b be any real numbers. Then the following results can easily be proved.

1. $\mathbf{A} + \mathbf{B}$ and $a\mathbf{A}$ are in R^n. (closure properties)
2. $\mathbf{A} + \mathbf{B} = \mathbf{B} + \mathbf{A}$. (commutative property)
3. $(\mathbf{A} + \mathbf{B}) + \mathbf{C} = \mathbf{A} + (\mathbf{B} + \mathbf{C})$. (associative property)
4. There is a zero vector $\mathbf{0}$ such that $\mathbf{0} + \mathbf{A} = \mathbf{A} + \mathbf{0} = \mathbf{A}$. ($\mathbf{0}$ is the additive identity.)
5. There is a vector $-\mathbf{A}$, which depends on $\mathbf{A}$, such that $\mathbf{A} + (-\mathbf{A}) = (-\mathbf{A}) + \mathbf{A} = \mathbf{0}$. (additive inverse property)
6. $(a + b)\mathbf{A} = a\mathbf{A} + b\mathbf{A}$ and $a(\mathbf{A} + \mathbf{B}) = a\mathbf{A} + a\mathbf{B}$. (distributive properties)
7. $(ab)\mathbf{A} = a(b\mathbf{A})$. (associative property)
8. $1\mathbf{A} = \mathbf{A}$. (1 is the scalar identity.)

Every component of the zero vector $\mathbf{0}$ is 0; that is, $\mathbf{0} = (0, 0, \ldots, 0)$. Also $\mathbf{A} = \mathbf{0}$ if and only if $|\mathbf{A}| = 0$. The additive inverse of a vector $\mathbf{A} = (a_1, a_2, \ldots, a_n)$ is the vector $-\mathbf{A} = (-a_1, -a_2, \ldots, -a_n)$. Note that each component of $-\mathbf{A}$ is the negative of the corresponding component of **A**. In the 2- and 3-dimensional cases, $-\mathbf{A}$ is a vector that has the same length as **A** but has a direction that is opposite to that of **A**, as illustrated in Figure 14.4.12.

Figure 14.4.12

In R^n there are n especially useful unit vectors:

$$\mathbf{E}_1 = (1, 0, 0, \ldots, 0),$$
$$\mathbf{E}_2 = (0, 1, 0, \ldots, 0),$$
$$\cdots\cdots\cdots\cdots\cdots\cdots$$
$$\mathbf{E}_n = (0, 0, 0, \ldots, 1).$$

Note that for $\mathbf{E}_i$, the ith component is 1 and all other components are 0. Those n unit vectors have the important property that every vector of R^n can be expressed as a *linear combination* of them, in the following sense. Let $(a_1, a_2, \ldots, a_n)$ be any vector of R^n. Then

$$\begin{aligned}(a_1, a_2, \ldots, a_n) &= a_1(1, 0, 0, \ldots, 0) + a_2(0, 1, 0, \ldots, 0) + \cdots + a_n(0, 0, 0, \ldots, 1)\\ &= a_1\mathbf{E}_1 + a_2\mathbf{E}_2 + \cdots + a_n\mathbf{E}_n.\end{aligned}$$

In R^3 the unit vectors $\mathbf{E}_1$, $\mathbf{E}_2$, and $\mathbf{E}_3$ are often denoted by **i**, **j**, and **k**, respectively, that is

$$\mathbf{i} = (1, 0, 0), \quad \mathbf{j} = (0, 1, 0), \quad \text{and } \mathbf{k} = (0, 0, 1).$$

Thus any vector (a, b, c) of R^3 can be written as

$$(a, b, c) = a\mathbf{i} + b\mathbf{j} + c\mathbf{k}.$$

Exercises 14.4

For Exercises 1–12, perform the indicated operations.

1. $(5, 7, -3) + (2, 4, 1)$.
2. $(1, 7, 6) + (3, -10, 5)$.
3. $4(-1, 2, 5, 3)$.
4. $3(5, -2, 6, 0)$.
5. $(0, 0, 0) + (6, 2, -1)$.
6. $(5, -3, 2) + (0, 0, 0)$.
7. $6[(1, -2) + (4, 5)]$.
8. $4[(3, 6) + (5, -2)]$.
9. $3[(1, -2) - (4, 5)]$.
10. $2[(3, 6) - (5, -2)]$.
11. $(2 + 3)(-4, 2, 1)$
12. $(4 + 2)(5, 3, -2)$.

For Exercises 13–16, find the vector $\mathbf{X}$.

13. $(5, 1, 3) + \mathbf{X} = (2, -3, 4)$.
14. $\mathbf{X} + (7, 3) = (5, -4)$.
15. $3\mathbf{X} + (1, 5) = (4, 2) + \mathbf{X}$.
16. $4\mathbf{X} + (1, 3, 6) = (3, -2, 1) + 2\mathbf{X}$.

For Exercises 17–24, find the scalars x and y.

17. $(4, x, 2y) = (4, 6, 10)$.
18. $(2x, 9, y) = (6, 9, -3)$.
19. $x(1, 7) + y(2, 1) = (5, 4)$.
20. $x(5, 2) + y(1, 3) = (7, 5)$.
21. $(2x + 5, 7) = (8, y - 1)$.
22. $(9, x + 3) = (3y - 4, 5)$.
23. $x(y, 2) = (7, 6)$.
24. $(8, 5) = x(4, y)$.

For Exercises 25–30, find the length of each vector.

25. $(4, 6)$.
26. $(-2, 5)$.
27. $(1, 2, -2)$.
28. $(3, -1, 4)$.
29. $(4, 0, 2, 1, -3)$.
30. $(1, -2, 3, 5)$.
31. Prove that the length of $k\mathbf{A}$ is $|k|$ times the length of $\mathbf{A}$; that is, $|k\mathbf{A}| = |k| \cdot |\mathbf{A}|$.
32. If a and b are any two real numbers and $\mathbf{A}$ and $\mathbf{B}$ are any two vectors of R^n, prove that (a) $\mathbf{A} + \mathbf{B} = \mathbf{B} + \mathbf{A}$, (b) $(a + b)\mathbf{A} = a\mathbf{A} + b\mathbf{A}$.

In Exercises 33–36, find a unit vector whose direction is the same as the direction of the given vector.

33. $(3, -6, 6)$.
34. $(-2, 3, 6)$.
35. $(6, 4, -12)$.
36. $(3, -2, -5)$.

In Exercises 37–40, let $\mathbf{A}$, $\mathbf{B}$, and $\mathbf{C}$ be the position vectors of the points given. Prove that the three points lie on the same line by proving that the vectors $\mathbf{A} - \mathbf{B}$ and $\mathbf{B} - \mathbf{C}$ are parallel.

37. $(1, 2, 3)$, $(3, 1, 6)$, $(-3, 4, -3)$.
38. $(3, -2, -7)$, $(9, -6, 1)$, $(12, -8, 5)$.
39. $(-3, 8, 2)$, $(3, -13, 8)$, $(-11, 36, -6)$.
40. $(5, 0, -2)$, $(8, -6, -17)$, $(4, 2, 3)$.
41. Do Exercise 37 by proving that $\mathbf{A} - \mathbf{B}$ and $\mathbf{A} - \mathbf{C}$ are parallel.
42. Do Exercise 38 by proving that $\mathbf{A} - \mathbf{B}$ and $\mathbf{A} - \mathbf{C}$ are parallel.

43. Use position vectors to prove that the points $(2, -3, 5)$, $(6, -1, 4)$, $(-3, 7, 1)$, and $(1, 9, 0)$ are the vertices of a parallelogram.

44. Use position vectors to prove that the midpoint of the segment joining the points (x_1, y_1, z_1) and (x_2, y_2, z_2) is

$$\left(\frac{x_1 + x_2}{2}, \frac{y_1 + y_2}{2}, \frac{z_1 + z_2}{2}\right).$$

14.5 Dot and Cross Products

In this section we shall discuss two useful ways of multiplying vectors, the dot product and the cross product. While *the cross product can only be defined for vectors in* R^3, *the dot product is defined for vectors in* R^n.

In the preceding section we saw that the vector $\mathbf{A} = (71, 20, 27, 14)$ could be used to describe the use by a restaurant of 71 pounds of beef, 20 pounds of pork, 27 pounds of fish, and 14 dozen eggs on a particular day. Suppose the costs per pound of the beef, pork, and fish are \$.90, \$.80, and \$.70, in that order, and suppose the eggs cost \$.55 per dozen. Then the vector $\mathbf{B} = (.90, .80, .70, .55)$ would conveniently describe those costs. The total cost of the items can be obtained by multiplying corresponding components of the vectors and adding the products. Thus the total cost is

$$(71)(.90) + (20)(.80) + (27)(.70) + (14)(.55) \text{ or } \$106.50.$$

That number is called the dot product of the vectors **A** and **B** according to the next definition.

Definition 14.5.1

> The *dot product* $\mathbf{A} \cdot \mathbf{B}$ of vectors $\mathbf{A} = (a_1, a_2, \ldots, a_n)$ and $\mathbf{B} = (b_1, b_2, \ldots, b_n)$ of R^n is the number
>
> $$\mathbf{A} \cdot \mathbf{B} = a_1 b_1 + a_2 b_2 + \cdots + a_n b_n.$$

In particular for R^3 we have:

Definition 14.5.2

> The *dot product* $\mathbf{A} \cdot \mathbf{B}$ of vectors $\mathbf{A} = (x_1, y_1, z_1)$ and $\mathbf{B} = (x_2, y_2, z_2)$ of R^3 is the number
>
> $$\mathbf{A} \cdot \mathbf{B} = x_1 x_2 + y_1 y_2 + z_1 z_2.$$

Example 1

$$(2, 1, 3) \cdot (5, 4, 2) = 2 \cdot 5 + 1 \cdot 4 + 3 \cdot 2 = 20.$$

$$\begin{aligned}(1, 6, 2, 3) \cdot (3, -2, 1, -5) &= 1 \cdot 3 + 6(-2) + 2 \cdot 1 + 3(-5) \\ &= -22.\end{aligned}$$

||

It follows directly from the preceding definition that the dot product has the properties indicated in the next theorem, whose proof is left to the exercises.

Theorem 14.5.1

Let **A**, **B**, and **C** be vectors of R^n and let a be any real number. Then

(i) $$\mathbf{A} \cdot \mathbf{B} = \mathbf{B} \cdot \mathbf{A},$$

(ii) $$a(\mathbf{A} \cdot \mathbf{B}) = (a\mathbf{A}) \cdot \mathbf{B} = \mathbf{A} \cdot (a\mathbf{B})$$

(iii) $$\mathbf{A} \cdot (\mathbf{B} + \mathbf{C}) = \mathbf{A} \cdot \mathbf{B} + \mathbf{A} \cdot \mathbf{C}$$

(iv) $$\mathbf{A} \cdot (\mathbf{B} - \mathbf{C}) = \mathbf{A} \cdot \mathbf{B} - \mathbf{A} \cdot \mathbf{C}$$

(v) $$\mathbf{A} \cdot \mathbf{A} = |\mathbf{A}|^2.$$

It is important to notice that the dot product of two vectors is a *real number* and not a vector.

In R^3 the law of cosines and the dot product can be used to obtain an expression for the cosine of the angle between two nonzero vectors. As in Figure 14.5.1, let $\mathbf{A} = (x_1, y_1, z_1)$ and $\mathbf{B} = (x_2, y_2, z_2)$ be two nonzero vectors in R^3, and let θ, $0 \leq \theta \leq \pi$, be the angle between them.

Figure 14.5.1

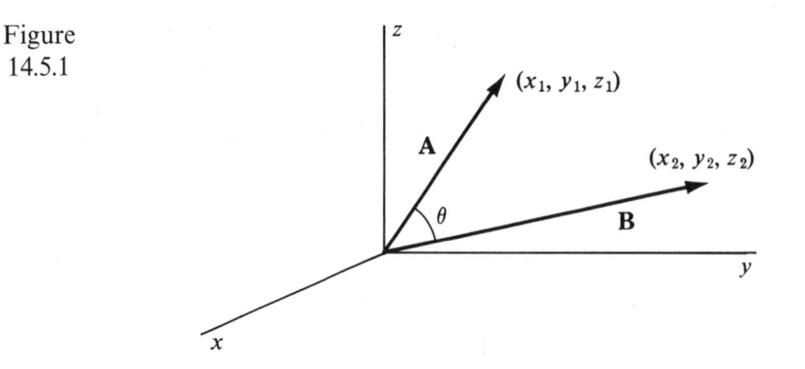

Then using parts (v), (iv), and then (i) of Theorem 14.5.1, we obtain

$$|\mathbf{A} - \mathbf{B}|^2 = (\mathbf{A} - \mathbf{B}) \cdot (\mathbf{A} - \mathbf{B}) = \mathbf{A} \cdot \mathbf{A} + \mathbf{B} \cdot \mathbf{B} - 2\mathbf{A} \cdot \mathbf{B}.$$

But $\mathbf{A} \cdot \mathbf{A} = |\mathbf{A}|^2$ and $\mathbf{B} \cdot \mathbf{B} = |\mathbf{B}|^2$ by part (v) of Theorem 14.5.1. Hence

$$|\mathbf{A} - \mathbf{B}|^2 = |\mathbf{A}|^2 + |\mathbf{B}|^2 - 2\mathbf{A} \cdot \mathbf{B}. \tag{1}$$

Also, using the fact that $|\mathbf{A} - \mathbf{B}|$ is the distance between the points (x_1, y_1, z_1) and (x_2, y_2, z_2), and applying the law of cosines, we get

$$|\mathbf{A} - \mathbf{B}|^2 = |\mathbf{A}|^2 + |\mathbf{B}|^2 - 2|\mathbf{A}|\,|\mathbf{B}| \cos\theta. \tag{2}$$

Then from (1) and (2) we get

$$\mathbf{A} \cdot \mathbf{B} = |\mathbf{A}|\,|\mathbf{B}| \cos\theta.$$

Since $\mathbf{A} \neq \mathbf{0}$ and $\mathbf{B} \neq \mathbf{0}$, it follows that $|\mathbf{A}|\,|\mathbf{B}| \neq 0$; so we can divide by $|\mathbf{A}|\,|\mathbf{B}|$ to obtain

$$\cos\theta = \frac{\mathbf{A}\cdot\mathbf{B}}{|\mathbf{A}|\,|\mathbf{B}|}. \tag{3}$$

We have proved the following useful theorem.

Theorem 14.5.2 The cosine of the angle θ between two nonzero vectors $\mathbf{A} = (x_1, y_1, z_1)$ and $\mathbf{B} = (x_2, y_2, z_2)$ of R^3 can be expressed as

$$\cos\theta = \frac{\mathbf{A}\cdot\mathbf{B}}{|\mathbf{A}|\,|\mathbf{B}|}.$$

Using Theorem 14.5.2 as a guide we define the angle between any two nonzero vectors in R^n, $n > 3$, as follows:

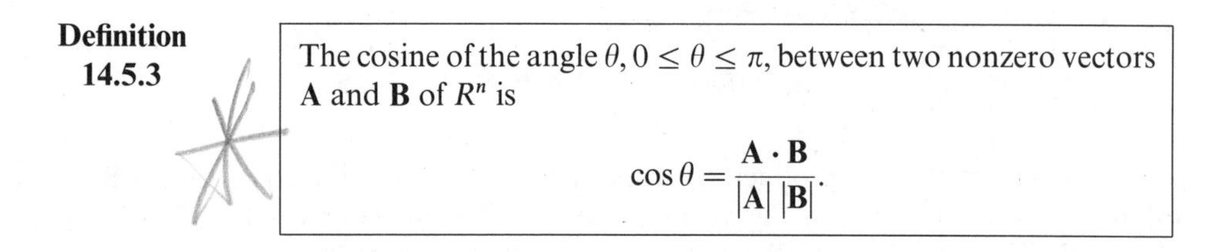

Definition 14.5.3 The cosine of the angle θ, $0 \leq \theta \leq \pi$, between two nonzero vectors $\mathbf{A}$ and $\mathbf{B}$ of R^n is

$$\cos\theta = \frac{\mathbf{A}\cdot\mathbf{B}}{|\mathbf{A}|\,|\mathbf{B}|}.$$

Example 2 | Find the angle θ between the vectors $(1, 2, -3)$ and $(-1, -2, 3)$.

We use Theorem 14.5.2 to obtain

$$\cos\theta = \frac{(1)(-1) + (2)(-2) + (-3)(3)}{\sqrt{1^2 + 2^2 + (-3)^2}\sqrt{(-1)^2 + (-2)^2 + 3^2}} = \frac{-14}{14} = -1.$$

Since, furthermore, $0 \leq \theta \leq \pi$, the angle θ between the two vectors has radian measure π. ||

Theorem 14.5.2 enables us to obtain a simple criterion for determining whether or not two nonzero vectors are perpendicular (orthogonal). Vectors $\mathbf{A}$ and $\mathbf{B}$ are perpendicular if and only if the angle θ between them has radian measure $\pi/2$. But $\cos \pi/2 = 0$. Hence $\mathbf{A} \perp \mathbf{B}$ if and only if

$$\cos\theta = \frac{\mathbf{A}\cdot\mathbf{B}}{|\mathbf{A}|\,|\mathbf{B}|} = 0.$$

But a fraction is zero if and only if its numerator is zero. Thus we have proved the theorem.

Theorem 14.5.3 Two nonzero vectors $\mathbf{A}$ and $\mathbf{B}$ of R^3 are perpendicular if and only if $\mathbf{A}\cdot\mathbf{B} = 0$.

Example 3 The vectors $\mathbf{A} = (2, -2, 5)$ and $\mathbf{B} = (3, 8, 2)$ are perpendicular since

$$\mathbf{A} \cdot \mathbf{B} = (2)(3) + (-2)(8) + (5)(2) = 0;$$

but the vectors $\mathbf{C} = (1, 3, 2)$ and $\mathbf{D} = (-4, 2, 7)$ are not perpendicular since

$$\mathbf{C} \cdot \mathbf{D} = (1)(-4) + (3)(2) + (2)(7) = 16 \neq 0. \qquad \|$$

Let $\mathbf{A}$ and $\mathbf{B}$ be vectors of R^3, as in Figure 14.5.2. Let L be the line containing the vector $\mathbf{B}$. Let P be the point of intersection of L and the line perpendicular to L through the head of $\mathbf{A}$. Then the position vector of P is called the *projection of* $\mathbf{A}$ *on* $\mathbf{B}$ or the *component vector of* $\mathbf{A}$ *in the direction of* $\mathbf{B}$. We shall often use $\mathbf{A_B}$ to denote the component vector of $\mathbf{A}$ in the direction of $\mathbf{B}$.

Figure 14.5.2

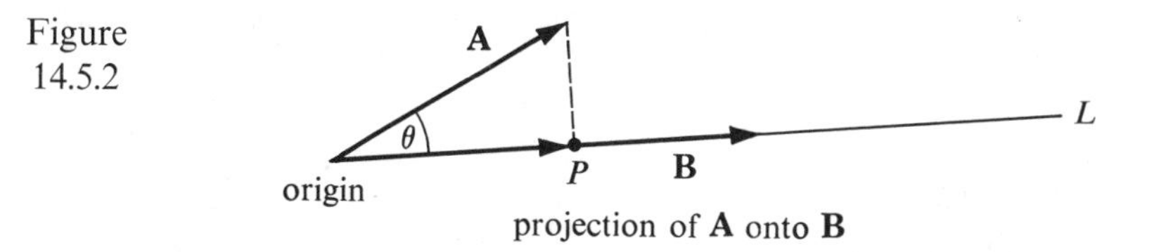

projection of **A** onto **B**

To interpret the dot product $\mathbf{A} \cdot \mathbf{B}$ geometrically we use the equation $\mathbf{A} \cdot \mathbf{B} = |\mathbf{A}|\,|\mathbf{B}| \cos\theta$. As indicated in Figure 14.5.3, $|\mathbf{A}|\,|\cos\theta|$ is the length of the projection of $\mathbf{A}$ on $\mathbf{B}$. Thus $|\mathbf{A} \cdot \mathbf{B}| = |\mathbf{A}|\,|\mathbf{B}|\,|\cos\theta|$ is the length of the projection of $\mathbf{A}$ on $\mathbf{B}$ times the length of $\mathbf{B}$. Similarly, $|\mathbf{A} \cdot \mathbf{B}|$ is also the length of the projection of $\mathbf{B}$ on $\mathbf{A}$ times the length of $\mathbf{A}$.

Figure 14.5.3

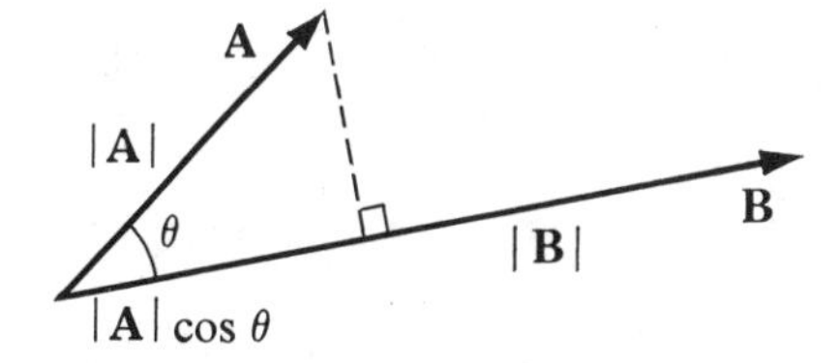

We now obtain a formula for $\mathbf{A_B}$, the component vector of $\mathbf{A}$ in the direction of $\mathbf{B}$, when $\mathbf{B} \neq \mathbf{0}$. Since $|\mathbf{A}|\,|\cos\theta|$ is the length of $\mathbf{A_B}$, $\mathbf{A_B}$ is $|\mathbf{A}| \cos\theta$ times the unit vector in the direction of $\mathbf{B}$. Thus

$$\mathbf{A_B} = |\mathbf{A}|(\cos\theta)\frac{\mathbf{B}}{|\mathbf{B}|}.$$

Then, since $\cos\theta = (\mathbf{A} \cdot \mathbf{B})/|\mathbf{A}|\,|\mathbf{B}|$ and $|\mathbf{B}|^2 = \mathbf{B} \cdot \mathbf{B}$, we obtain

$$\begin{aligned}\mathbf{A_B} &= |\mathbf{A}|(\cos\theta)\frac{\mathbf{B}}{|\mathbf{B}|} \\ &= |\mathbf{A}|\frac{\mathbf{A} \cdot \mathbf{B}}{|\mathbf{A}|\,|\mathbf{B}|}\frac{\mathbf{B}}{|\mathbf{B}|} = \frac{\mathbf{A} \cdot \mathbf{B}}{|\mathbf{B}|^2}\mathbf{B} = \frac{\mathbf{A} \cdot \mathbf{B}}{\mathbf{B} \cdot \mathbf{B}}\mathbf{B}.\end{aligned}$$

We have now proved the following theorem.

Theorem 14.5.4

Let **A** and **B** be any vectors of R^3 with $\mathbf{B} \neq 0$. Then the component vector of **A** in the direction of **B** is given by

$$\mathbf{A_B} = \frac{\mathbf{A} \cdot \mathbf{B}}{\mathbf{B} \cdot \mathbf{B}} \mathbf{B}.$$

Example 4

If $\mathbf{A} = (2, 1, -3)$ and $\mathbf{B} = (-1, 3, 4)$, then $\mathbf{A_B}$, the component vector of **A** in the direction of **B**, is found as follows:

$$\begin{aligned} \mathbf{A_B} = \frac{\mathbf{A} \cdot \mathbf{B}}{\mathbf{B} \cdot \mathbf{B}} \mathbf{B} &= \frac{(2, 1, -3) \cdot (-1, 3, 4)}{(-1, 3, 4) \cdot (-1, 3, 4)} (-1, 3, 4) \\ &= \frac{(2)(-1) + (1)(3) + (-3)(4)}{(-1)(-1) + (3)(3) + (4)(4)} (-1, 3, 4) \\ &= \frac{-11}{26} (-1, 3, 4) \\ &= \left(\frac{11}{26}, \frac{-33}{26}, \frac{-22}{13} \right). \end{aligned}$$

||

Example 5

If $\mathbf{A} = (2, 1, -3)$ and $\mathbf{B} = (1, 0, 0)$, then

$$\begin{aligned} \mathbf{A_B} &= \frac{(2, 1, -3) \cdot (1, 0, 0)}{(1, 0, 0) \cdot (1, 0, 0)} (1, 0, 0) \\ &= \frac{2}{1} (1, 0, 0) \\ &= (2, 0, 0). \end{aligned}$$

||

In contrast to the dot product, the cross product of two vectors of R^3 is a vector of R^3, as the next definition indicates.

Definition 14.5.4

The *cross product* $\mathbf{A} \times \mathbf{B}$ of two vectors $\mathbf{A} = (x_1, y_1, z_1)$ and $\mathbf{B} = (x_2, y_2, z_2)$ of R^3 is the vector

$$\mathbf{A} \times \mathbf{B} = (y_1 z_2 - y_2 z_1, \quad z_1 x_2 - z_2 x_1, \quad x_1 y_2 - x_2 y_1).$$

A useful way to help remember the definition of the cross product is to express it as a determinant in terms of the unit vectors **i**, **j**, and **k**:

$$\begin{aligned} \mathbf{A} \times \mathbf{B} &= (y_1 z_2 - y_2 z_1)\mathbf{i} + (z_1 x_2 - z_2 x_1)\mathbf{j} + (x_1 y_2 - x_2 y_1)\mathbf{k} \\ &= \begin{vmatrix} \mathbf{i} & \mathbf{j} & \mathbf{k} \\ x_1 & y_1 & z_1 \\ x_2 & y_2 & z_2 \end{vmatrix}. \end{aligned}$$

That can be easily verified by expanding the determinant by minors along the top row.

Example 6

$$(3, 4, 1) \times (6, 7, 5) = \begin{vmatrix} \mathbf{i} & \mathbf{j} & \mathbf{k} \\ 3 & 4 & 1 \\ 6 & 7 & 5 \end{vmatrix}$$

$$= (4 \cdot 5 - 7 \cdot 1)\mathbf{i} + (1 \cdot 6 - 5 \cdot 3)\mathbf{j} + (3 \cdot 7 - 4 \cdot 6)\mathbf{k}$$

$$= 13\mathbf{i} - 9\mathbf{j} - 3\mathbf{k}$$

$$= (13, -9, -3).$$ ||

Remember that the cross product of two vectors is a vector, while their dot product is a number!

If **A** and **B** are nonzero vectors of R^3 that are not parallel, then there is exactly one plane containing the two vectors. The major reason for defining $\mathbf{A} \times \mathbf{B}$ is stated in the next theorem.

Theorem 14.5.5 Let **A** and **B** be nonzero vectors of R^3 that are not parallel. Then $\mathbf{A} \times \mathbf{B}$ is perpendicular (normal) to the plane containing **A** and **B**.

Proof. Let $\mathbf{A} = (x_1, y_1, z_1)$ and $\mathbf{B} = (x_2, y_2, z_2)$ be nonzero, nonparallel vectors or R^3. Then

$$\mathbf{A} \cdot (\mathbf{A} \times \mathbf{B}) = (x_1, y_1, z_1) \cdot (y_1 z_2 - y_2 z_1, \quad z_1 x_2 - z_2 x_1, \quad x_1 y_2 - x_2 y_1)$$

$$= x_1 y_1 z_2 - x_1 y_2 z_1 + y_1 z_1 x_2 - y_1 z_2 x_1 + z_1 x_1 y_2 - z_1 x_2 y_1$$

$$= 0.$$

Similarly $\mathbf{B} \cdot (\mathbf{A} \times \mathbf{B}) = 0$; therefore $\mathbf{A} \times \mathbf{B}$ is normal to both **A** and **B** and so is normal to the plane containing **A** and **B**. We have therefore completed the proof.

Both $\mathbf{A} \times \mathbf{B}$ and $-(\mathbf{A} \times \mathbf{B})$ are perpendicular to the plane containing **A** and **B**. Consequently there are two possible directions for a vector that is normal to the plane containing **A** and **B**. It can be proved that in our right-handed coordinate system the direction of $\mathbf{A} \times \mathbf{B}$ is the direction determined by the *right-hand rule*.

Right Hand Rule. Place the right hand in the configuration shown in Figure 14.5.4 so that the fingers curve in the direction necessary

Figure 14.5.4

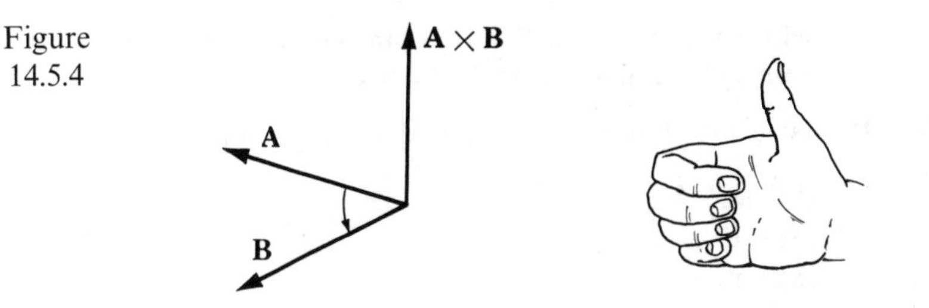

to rotate vector **A** to vector **B** through the angle between **A** and **B**. Then the thumb points in the direction of $\mathbf{A} \times \mathbf{B}$.

Example 7

$$\mathbf{i} \times \mathbf{j} = (1, 0, 0) \times (0, 1, 0) = \begin{vmatrix} \mathbf{i} & \mathbf{j} & \mathbf{k} \\ 1 & 0 & 0 \\ 0 & 1 & 0 \end{vmatrix} = \mathbf{k}.$$

||

Figure 14.5.5

z
k
j
y
i
x

It is left as an exercise to prove that $|\mathbf{A} \times \mathbf{B}|$ is equal to the area of the parallelogram having **A** and **B** as adjacent sides. Figure 14.5.6 illustrates this fact. It follows that we also have

$$|\mathbf{A} \times \mathbf{B}| = |\mathbf{A}|\,|\mathbf{B}| \sin\theta,$$

where θ is the angle between **A** and **B**.

Figure 14.5.6

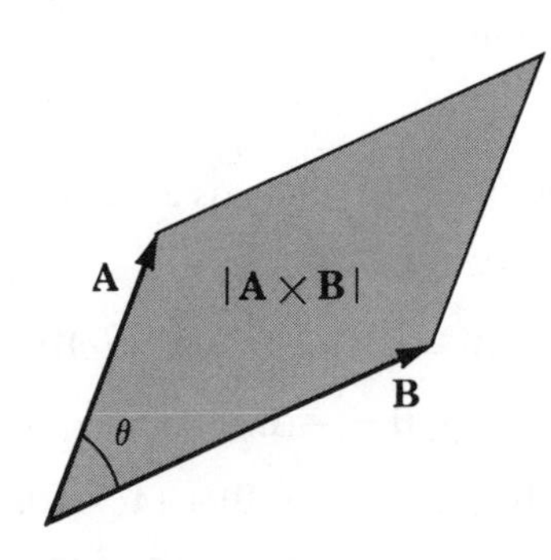

Vectors are also useful in calculating volumes of parallelopipeds. Consider the parallelopiped shown in Figure 14.5.7 having the vectors **A**, **B**, and **C** as adjacent edges. Let θ be the angle between **C** and the vector $\mathbf{A} \times \mathbf{B}$. Then since $\mathbf{A} \times \mathbf{B}$ is perpendicular to the plane containing **A** and **B**, the altitude h of the parallelopiped is $|\mathbf{C}|\,|\cos\theta|$. Since the area of its base is $|\mathbf{A} \times \mathbf{B}|$, the volume V of the parallelopiped is

$$V = |\mathbf{A} \times \mathbf{B}|\,|\mathbf{C}|\,|\cos\theta|.$$

But

$$\cos\theta = \frac{(\mathbf{A} \times \mathbf{B}) \cdot \mathbf{C}}{|\mathbf{A} \times \mathbf{B}|\,|\mathbf{C}|}.$$

Figure 14.5.7

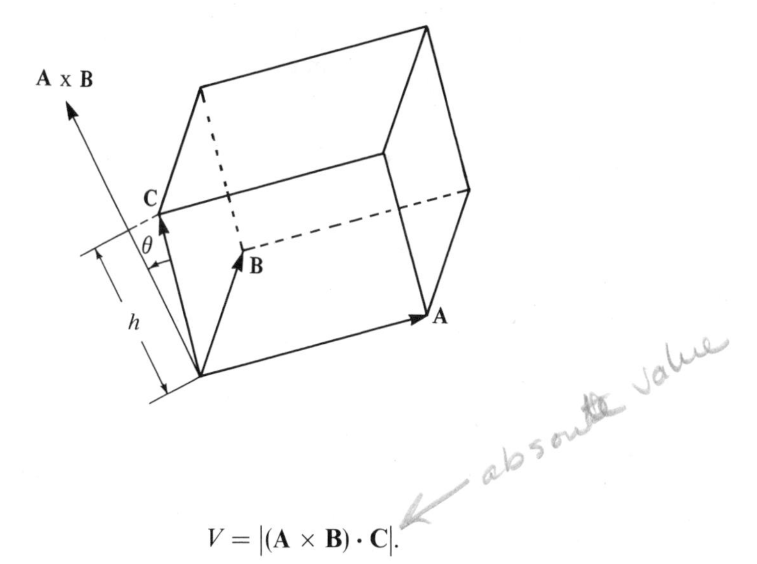

So we get

$$V = |(\mathbf{A} \times \mathbf{B}) \cdot \mathbf{C}|.$$

It is left as an exercise to show that if $\mathbf{A} = (x_1, y_1, z_1)$, $\mathbf{B} = (x_2, y_2, z_2)$, and $\mathbf{C} = (x_3, y_3, z_3)$, then

$$(\mathbf{A} \times \mathbf{B}) \cdot \mathbf{C} = \begin{vmatrix} x_1 & y_1 & z_1 \\ x_2 & y_2 & z_2 \\ x_3 & y_3 & z_3 \end{vmatrix}.$$

Hence V is equal to the absolute value of that determinant.

It follows directly from the definition of the cross product that cross products have the properties listed in the next theorem.

Theorem 14.5.6

If **A**, **B**, and **C** are vectors of R^3, and a is a real number, then

(i) $$\mathbf{A} \times \mathbf{A} = \mathbf{0},$$

(ii) $$a(\mathbf{A} \times \mathbf{B}) = (a\mathbf{A}) \times \mathbf{B} = \mathbf{A} \times (a\mathbf{B}),$$

(iii) $$\mathbf{A} \times \mathbf{B} = -(\mathbf{B} \times \mathbf{A}),$$

(iv) $$\mathbf{A} \times (\mathbf{B} + \mathbf{C}) = (\mathbf{A} \times \mathbf{B}) + (\mathbf{A} \times \mathbf{C}).$$

Exercises 14.5

For each pair of vectors **A** and **B** in Exercises 1–6, find (a) $\mathbf{A} \cdot \mathbf{B}$, (b) $\mathbf{A} \times \mathbf{B}$, (c) $\mathbf{B} \times \mathbf{A}$, (d) $|\mathbf{A} - \mathbf{B}|$, (e) $|\mathbf{B} - \mathbf{A}|$, (f) the angle between **A** and **B**, (g) the component vector of **A** in the direction of **B**.

1. $\mathbf{A} = (1, 0, 0)$, $\mathbf{B} = (0, 0, 1)$.
2. $\mathbf{A} = (1, 2, 3)$, $\mathbf{B} = (-1, -2, -3)$.
3. $\mathbf{A} = (2, -1, 2)$, $\mathbf{B} = (5, 2, -4)$.
4. $\mathbf{A} = (-2, 4, 3)$, $\mathbf{B} = (-2, 4, 3)$.
5. $\mathbf{A} = (1, 1, \sqrt{6})$, $\mathbf{B} = (-1, -1, \sqrt{6})$.
6. $\mathbf{A} = (3, 3, \sqrt{6})$, $\mathbf{B} = (-3, -3, \sqrt{6})$.
7. Use vectors to prove that the line through the points $(-2, 5, 2)$ and $(3, 4, -1)$ is perpendicular to the line through the points $(6, -1, 5)$ and $(2, -30, 8)$.

8. Use vectors to prove that the line through the points $(3, -1, 7)$ and $(4, 2, 5)$ is not perpendicular to the line through the points $(-1, 0, 2)$ and $(2, -3, 4)$.

For each pair of vectors in Exercises 9–12, find a vector that is perpendicular to both vectors.

9. $(1, -2, 3)$, $(2, 1, 4)$.
10. $(2, 0, -3)$, $(3, 1, -1)$.
11. $(-1, 5, 2)$, $(4, 1, -1)$.
12. $(2, -1, 3)$, $(-3, 2, 1)$.

For Exercises 13–16, use the definition of the dot product to prove that if $\mathbf{A} = (a_1, a_2, \ldots, a_n)$, $\mathbf{B} = (b_1, b_2, \ldots, b_n)$, $\mathbf{C} = (c_1, c_2, \ldots, c_n)$ are any vectors of R^n, and if a is any real number, then the indicated identity is satisfied.

13. $\mathbf{A} \cdot \mathbf{B} = \mathbf{B} \cdot \mathbf{A}$.
14. $a(\mathbf{A} \cdot \mathbf{B}) = (a\mathbf{A}) \cdot \mathbf{B} = \mathbf{A} \cdot (a\mathbf{B})$.
15. $\mathbf{A} \cdot (\mathbf{B} + \mathbf{C}) = \mathbf{A} \cdot \mathbf{B} + \mathbf{A} \cdot \mathbf{C}$.
16. $\mathbf{A} \cdot \mathbf{A} = |\mathbf{A}|^2$.

For Exercises 17–20, use the definition of the cross product to prove that if $\mathbf{A} = (x_1, y_1, z_1)$, $\mathbf{B} = (x_2, y_2, z_2)$, $\mathbf{C} = (x_3, y_3, z_3)$ are any vectors of R^3, and if a is any real number, then the indicated identity is satisfied.

17. $\mathbf{A} \times \mathbf{A} = \mathbf{0}$.
18. $a(\mathbf{A} \times \mathbf{B}) = (a\mathbf{A}) \times \mathbf{B} = \mathbf{A} \times (a\mathbf{B})$.
19. $\mathbf{A} \times \mathbf{B} = -(\mathbf{B} \times \mathbf{A})$.
20. $\mathbf{A} \times (\mathbf{B} + \mathbf{C}) = (\mathbf{A} \times \mathbf{B}) + (\mathbf{A} \times \mathbf{C})$.
21. Use a numerical example of particular vectors **A**, **B**, and **C** of R^3 to show that it is not true in general that

$$\mathbf{A} \times (\mathbf{B} \times \mathbf{C}) = (\mathbf{A} \times \mathbf{B}) \times \mathbf{C}.$$

22. Verify that the vectors you chose for Exercise 21 do satisfy the identity

$$\mathbf{A} \times (\mathbf{B} \times \mathbf{C}) = (\mathbf{A} \times \mathbf{B}) \times \mathbf{C} - (\mathbf{A} \times \mathbf{C}) \times \mathbf{B}.$$

23. Prove that if $\mathbf{A} = (x_1, y_1, z_1)$, $\mathbf{B} = (x_2, y_2, z_2)$, and $\mathbf{C} = (x_3, y_3, z_3)$ are any three vectors of R^3, then

$$\mathbf{A} \times (\mathbf{B} \times \mathbf{C}) = (\mathbf{A} \times \mathbf{B}) \times \mathbf{C} - (\mathbf{A} \times \mathbf{C}) \times \mathbf{B}.$$

24. Let $\mathbf{A} = (x_1, y_1, z_1)$ and $\mathbf{B} = (x_2, y_2, z_2)$, and prove that $|\mathbf{A} \times \mathbf{B}|^2 = |\mathbf{A}|^2|\mathbf{B}|^2 - (\mathbf{A} \cdot \mathbf{B})^2$.
25. Use the result of Exercise 24 to prove that $|\mathbf{A} \times \mathbf{B}| = |\mathbf{A}|\,|\mathbf{B}| \sin \theta$, and hence that $|\mathbf{A} \times \mathbf{B}|$ is equal to the area of the parallelogram having **A** and **B** as adjacent sides.
26. Use the result of Exercise 25 to find the area of a parallelogram having three of its vertices at $(3, -2, 4)$, $(5, -1, 7)$, and $(-1, 1, 2)$.
27. Use the result of Exercise 25 to find the area of the triangle whose vertices are at $(-1, 4, 2)$, $(3, 2, 1)$, and $(0, 3, 5)$.
28. Show that two nonzero vectors **A** and **B** are parallel if and only if $\mathbf{A} \times \mathbf{B} = 0$.
29. Show that if $\mathbf{A} = (x_1, y_1, z_1)$, $\mathbf{B} = (x_2, y_2, z_2)$, and $\mathbf{C} = (x_3, y_3, z_3)$, then

$$(\mathbf{A} \times \mathbf{B}) \cdot \mathbf{C} = \begin{vmatrix} x_1 & y_1 & z_1 \\ x_2 & y_2 & z_2 \\ x_3 & y_3 & z_3 \end{vmatrix}.$$

30. Find the volume of the parallelepiped having the vectors $(2, 1, 3)$, $(-1, 4, 2)$, and $(4, -3, 5)$ as adjacent edges.
31. Three adjacent edges of a parallelepiped meet at the point $(2, -1, 4)$. The other ends of the three edges are at $(5, 2, -1)$, $(-2, 5, 1)$, and $(1, 0, -2)$. Find the volume of the parallelepiped.

14.6 Equations of Lines and Planes

Let $\mathbf{D} = (a, b, c)$ be a fixed non-zero vector in R^3 and let (x_0, y_0, z_0) be a fixed point. Then there is one and only one line L through the point (x_0, y_0, z_0) parallel to $\mathbf{D}$. We shall find several types of equations for L. As shown in Figure 14.6.1, let $\mathbf{P}_0$ be the position vector of the point (x_0, y_0, z_0) and let $\mathbf{P}$ be the position vector of any point (x, y, z). Then the arrow from the point (x_0, y_0, z_0) to the point (x, y, z) is a replica of $\mathbf{P} - \mathbf{P}_0$. Thus the point (x, y, z) is on L if and only if the vector $\mathbf{P} - \mathbf{P}_0$ is parallel to $\mathbf{D}$, that is, if and only if

$$\mathbf{P} - \mathbf{P}_0 = t\mathbf{D}, \tag{1}$$

for some real number t. Therefore, as t varies over the set of all real numbers, the points (x, y, z) satisfying equation (1) will be the set of those points and only those points on line L. Hence (1) is an equation of the line L, which we shall call a *vector equation* for L. The variable t is called the *parameter*.

Figure 14.6.1

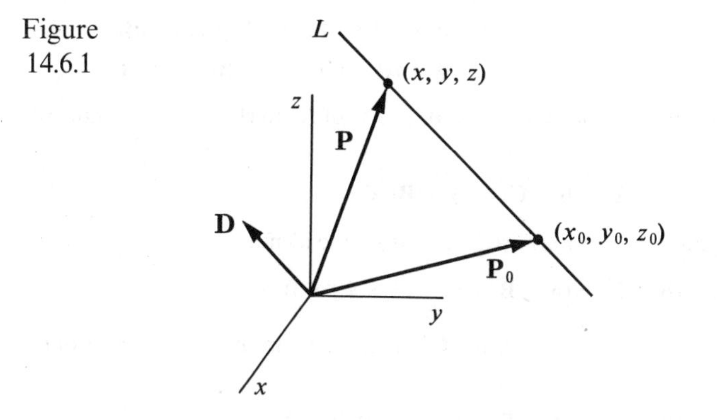

By writing the components of the vector equation (1), we get parametric equations for L. First, the vector equation $\mathbf{P} - \mathbf{P}_0 = t\mathbf{D}$ becomes

$$(x, y, z) - (x_0, y_0, z_0) = t(a, b, c), \quad \text{or}$$

$$\begin{aligned}(x, y, z) &= (x_0, y_0, z_0) + t(a, b, c)\\ &= (x_0 + ta, \quad y_0 + tb, \quad z_0 + tc).\end{aligned}$$

Then, using the definition of equality of vectors, we get *parametric equations* for L:

$$x = x_0 + ta,$$

$$y = y_0 + tb,$$

$$z = z_0 + tc.$$

If none of the numbers a, b, c is zero, the parameter can be eliminated from those equations by solving each of them for t. We obtain

$$\frac{x - x_0}{a} = t, \qquad \frac{y - y_0}{b} = t, \qquad \frac{z - z_0}{c} = t.$$

Therefore $$\frac{x - x_0}{a} = \frac{y - y_0}{b} = \frac{z - z_0}{c}.$$

The equations in the preceding line are called *symmetric equations* for the line L.

The results obtained so far are summarized in the next theorem.

Theorem 14.6.1

Equations for the line that passes through a point with position vector $\mathbf{P}_0 = (x_0, y_0, z_0)$ and is parallel to a fixed vector $\mathbf{D} = (a, b, c)$ can be expressed in any of the following three ways.

(1) Vector Equation:

$\mathbf{P} - \mathbf{P}_0 = t\mathbf{D}$, where $\mathbf{P} = (x, y, z)$.

(2) Parametric Equations:

$x = x_0 + ta,$
$y = y_0 + tb,$
$z = z_0 + tc.$

(3) Symmetric Equations (if $a \neq 0$, $b \neq 0$, and $c \neq 0$):

$$\frac{x - x_0}{a} = \frac{y - y_0}{b} = \frac{z - z_0}{c}.$$

Next we find equations for a line L passing through two given points (x_0, y_0, z_0) and (x_1, y_1, z_1). Let $\mathbf{P}_0$ and $\mathbf{P}_1$ be the position vectors of these points, as in Figure 14.6.2, and let $\mathbf{P}$ be the position vector of an arbitrary point (x, y, z). Then the arrow from the point (x_0, y_0, z_0) to the point (x_1, y_1, z_1) is a replica of the vector $\mathbf{P}_1 - \mathbf{P}_0$. Therefore the line L, which passes through the point (x_0, y_0, z_0), is parallel to the fixed vector $\mathbf{P}_1 - \mathbf{P}_0 = (x_1 - x_0, y_1 - y_0, z_1 - z_0)$. Consequently by Theorem 14.6.1 we get the following theorem.

Theorem 14.6.2

Equations for the line passing through the two points with position vectors $\mathbf{P}_0 = (x_0, y_0, z_0)$ and $\mathbf{P}_1 = (x_1, y_1, z_1)$ can be expressed in any of the following three ways.

(1) Vector Equation:

$\mathbf{P} - \mathbf{P}_0 = t(\mathbf{P}_1 - \mathbf{P}_0).$

(2) Parametric Equations:

$x = x_0 + t(x_1 - x_0),$
$y = y_0 + t(y_1 - y_0),$
$z = z_0 + t(z_1 - z_0).$

(3) Symmetric Equations (if $x \neq x_0$, $y_1 \neq y_0$, and $z_1 \neq z_0$):

$$\frac{x - x_0}{x_1 - x_0} = \frac{y - y_0}{y_1 - y_0} = \frac{z - z_0}{z_1 - z_0}.$$

For *parametric equations* note that if $0 \leq t \leq 1$, then x lies between x_0 and x_1. Since similar remarks hold for y and z, we have the following theorem.

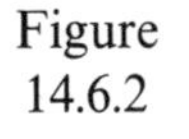

Figure 14.6.2

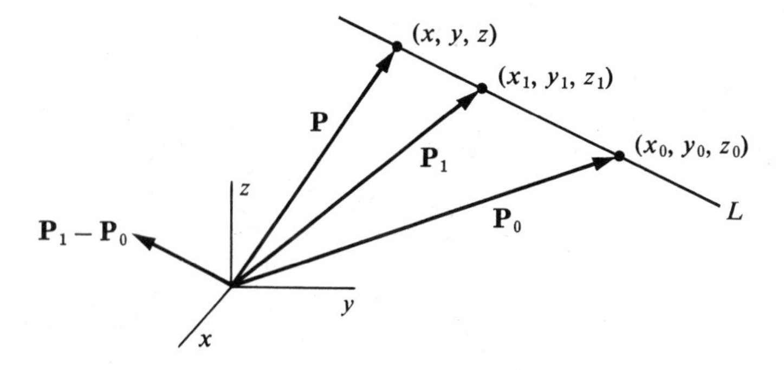

Theorem 14.6.3

> Equations for the *line segment* joining the points (x_0, y_0, z_0) and (x_1, y_1, z_1) can be expressed in this parametric form:
>
> $$x = x_0 + t(x_1 - x_0),$$
> $$y = y_0 + t(y_1 - y_0),$$
> $$z = z_0 + t(z_1 - z_0), \qquad \text{for } 0 \le t \le 1.$$

Example 1 Find parametric and symmetric equations for the line passing through $(-2, 1, 3)$ and $(4, 5, 0)$.

Let $(x_0, y_0, z_0) = (-2, 1, 3)$ and $(x_1, y_1, z_1) = (4, 5, 0)$. Then parametric equations are

$$x = -2 + 6t,$$
$$y = 1 + 4t,$$
$$z = 3 - 3t.$$

Symmetric equations are

$$\frac{x+2}{6} = \frac{y-1}{4} = \frac{z-3}{-3}.$$

||

Example 2 Parametric equations for the line segment joining $(5, 1, -2)$ and $(7, -3, 4)$ are

$$x = 5 + 2t,$$
$$y = 1 - 4t,$$
$$z = -2 + 6t,$$

where $0 \le t \le 1$. ||

If $\mathbf{V} = (a, b, c) \neq (0, 0, 0)$ is a fixed vector in R^3 and (x_0, y_0, z_0) is a fixed point, then there is one and only one plane H that is perpendicular to $\mathbf{V}$ and passes through (x_0, y_0, z_0), as illustrated in Figure 14.6.3. We shall now develop an equation for this plane.

Figure 14.6.3

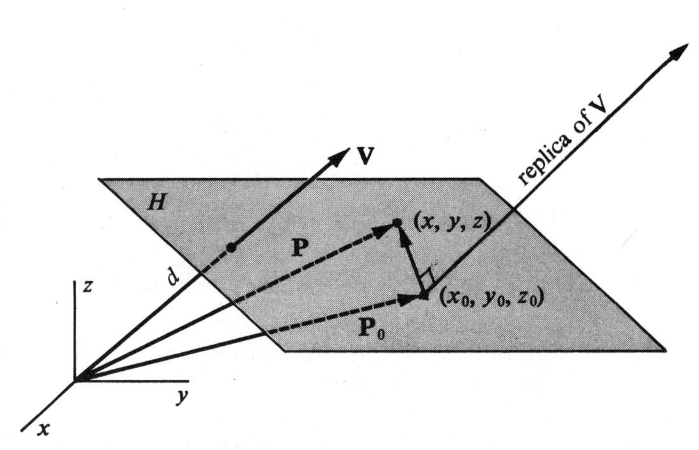

Let $\mathbf{P}_0$ be the position vector of the point (x_0, y_0, z_0), and let $\mathbf{P}$ be the position vector of any point (x, y, z). Then the point (x, y, z) is in the plane H if and only if the arrow from the point (x_0, y_0, z_0) to the point (x, y, z) is perpendicular to the vector $\mathbf{V}$, that is, if and only if the vector $\mathbf{P} - \mathbf{P}_0$ is perpendicular to $\mathbf{V}$. But by Theorem 14.5.3, $(\mathbf{P} - \mathbf{P}_0) \perp \mathbf{V}$ if and only if normal

$$\mathbf{V} \cdot (\mathbf{P} - \mathbf{P}_0) = 0. \tag{2}$$

So (2) is a vector equation for the plane. If we write equation (2) in terms of the components of the vectors, we obtain

$$(a, b, c) \cdot (x - x_0, \quad y - y_0, \quad z - z_0) = 0, \quad \text{or}$$

$$a(x - x_0) + b(y - y_0) + c(z - z_0) = 0. \tag{3}$$

Thus we have proved the following theorem.

Theorem 14.6.4

An equation of the plane that is perpendicular to a nonzero vector $\mathbf{V} = (a, b, c)$ and passes through a point (x_0, y_0, z_0) is

$$a(x - x_0) + b(y - y_0) + c(z - z_0) = 0.$$

An equation of the plane that is perpendicular to the vector $(2, 6, -5)$ and passes through the point $(3, -2, 4)$ is

$$2(x - 3) + 6(y + 2) - 5(z - 4) = 0$$

or, by multiplying out and moving the constant to the right side

$$2x + 6y - 5z = -26. \qquad \|$$

An equation of the form

$$a(x - x_0) + b(y - y_0) + c(z - z_0) = 0,$$

where a, b, and c are not all zero can be changed to

$$ax + by + cz = ax_0 + by_0 + cz_0.$$

Or, if we let $d = ax_0 + by_0 + cz_0$, the equation can be expressed as

$$ax + by + cz = d.$$

If a, b, and c are not all zero, that equation is called a *linear equation* in x, y, and z. Since every plane is perpendicular to many vectors and passes through many points, every plane has a linear equation; and we have proved part of the next theorem.

Theorem 14.6.5

> Every plane has an equation of the form
>
> $$ax + by + cz = d$$
>
> where a, b, c, and d are real numbers and a, b, c are not all zero. Conversely, every such equation is an equation of a plane.

To complete the proof we need to show that every equation of the form indicated in the theorem is an equation of a plane. Let (x_0, y_0, z_0) be any point satisfying that equation. Then

$$ax_0 + by_0 + cz_0 = d.$$

So the given equation can be written in the form

$$ax + by + cz = ax_0 + by_0 + cz_0, \quad \text{or}$$

$$a(x - x_0) + b(y - y_0) + c(z - z_0) = 0$$

which is an equation for the plane that is perpendicular to the vector $\mathbf{V} = (a, b, c)$ and passes through the point (x_0, y_0, z_0). Therefore the proof is complete.

We note that in the proof of Theorem 14.6.5 we also established that if an equation of a plane is $ax + by + cz = d$, then the nonzero vector $\mathbf{V} = (a, b, c)$ is perpendicular to the plane. $\mathbf{V}$ is often called a *normal* to the plane. For example, the vector $(2, -1, 3)$ is perpendicular to (and is a normal to) the plane with equation $2x - y + 3z = 7$.

Since there is one and only one plane H that passes through three fixed points that are not on the same straight line, it is natural to try to determine an equation for H if the three points are known. To see how to proceed, let $\mathbf{A}$, $\mathbf{B}$, and $\mathbf{C}$ be the position vectors of the three points as indicated in Figure 14.6.4. By considering the replicas of $\mathbf{B} - \mathbf{A}$ and $\mathbf{C} - \mathbf{A}$ as shown in the figure, we see that the plane containing the vectors $\mathbf{B} - \mathbf{A}$ and $\mathbf{C} - \mathbf{A}$ is parallel to the plane H. Thus, $\mathbf{V} = (\mathbf{B} - \mathbf{A}) \times (\mathbf{C} - \mathbf{A})$ is a vector perpendicular to H, and we can use it along with any one of the three given points to determine an equation for H.

Figure 14.6.4

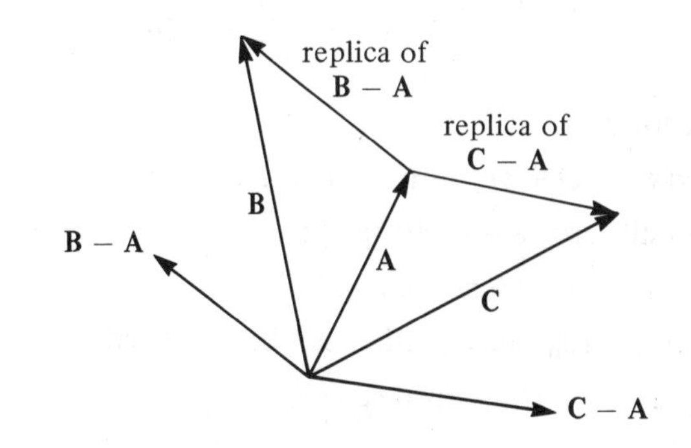

Example 4 | Find an equation of the plane passing through the points $(0, 1, -1)$, $(1, 2, 3)$, and $(2, -1, 0)$.

Let $\mathbf{A} = (0, 1, -1)$, $\mathbf{B} = (1, 2, 3)$, and $\mathbf{C} = (2, -1, 0)$ be the position vectors of the given points. Then $\mathbf{B} - \mathbf{A} = (1, 1, 4)$ and $\mathbf{C} - \mathbf{A} = (2, -2, 1)$, and

$$\mathbf{V} = (\mathbf{B} - \mathbf{A}) \times (\mathbf{C} - \mathbf{A}) = \begin{vmatrix} \mathbf{i} & \mathbf{j} & \mathbf{k} \\ 1 & 1 & 4 \\ 2 & -2 & 1 \end{vmatrix}$$

$$= 9\mathbf{i} + 7\mathbf{j} - 4\mathbf{k} = (9, 7, -4).$$

Since the plane is perpendicular to $\mathbf{V}$ and passes through the point $(0, 1, -1)$ we can express an equation of the plane as

$$9x + 7(y - 1) - 4(z + 1) = 0, \quad \text{or}$$

$$9x + 7y - 4z = 11. \qquad \|$$

Using normals to two planes, we can determine whether the planes are parallel or perpendicular. When they are neither parallel or perpendicular, we can also use normals to calculate the acute angle between the two planes. For if $\mathbf{V}_1$ and $\mathbf{V}_2$ are normals to two planes, then the planes are perpendicular if and only if $\mathbf{V}_1$ and $\mathbf{V}_2$ are perpendicular, and the planes are parallel if and only if $\mathbf{V}_1$ and $\mathbf{V}_2$ are parallel. Also, the acute angle between the two planes is the acute angle between $\mathbf{V}_1$ and $\mathbf{V}_2$.

Example 5 | Show that the planes with equations $2x - 4y + 10z = 7$ and $3x - 6y + 15z = 11$ are parallel.

Normals to the planes are $\mathbf{V}_1 = (2, -4, 10)$ and $\mathbf{V}_2 = (3, -6, 15)$. Since $\mathbf{V}_2 = (3/2)\mathbf{V}_1$, we may conclude that $\mathbf{V}_1$ and $\mathbf{V}_2$ are parallel and so the planes are parallel. ||

Example 6 | Show that the planes with equations $2x + 5y - 4z = 5$ and $9x - 6y - 3z = 2$ are perpendicular.

Normals to the planes are $\mathbf{V}_1 = (2, 5, -4)$ and $\mathbf{V}_2 = (9, -6, -3)$. Since $\mathbf{V}_1 \cdot \mathbf{V}_2 = 18 - 30 + 12 = 0$, we may conclude that $\mathbf{V}_1$ and $\mathbf{V}_2$ are perpendicular and so the planes are perpendicular. ||

Example 7 | Find the acute angle between the planes with equations $x + y - \sqrt{6}z = 3$. and $x + y + \sqrt{6}z = 5$.

Normals to the planes are $\mathbf{V}_1 = (1, 1, -\sqrt{6})$ and $\mathbf{V}_2 = (1, 1, \sqrt{6})$. Let θ be the angle between $\mathbf{V}_1$ and $\mathbf{V}_2$. Then

$$\cos\theta = \frac{\mathbf{V}_1 \cdot \mathbf{V}_2}{|\mathbf{V}_1|\,|\mathbf{V}_2|} = \frac{1 + 1 - 6}{\sqrt{8}\sqrt{8}} = -\frac{1}{2}.$$

Thus, $\theta = 2\pi/3$, and since $2\pi/3$ is not acute, the acute angle between the planes is $\pi - 2\pi/3 = \pi/3$. ||

Exercises 14.6

For Exercises 1–4, find equations for the lines passing through the given points.

1. $(6, 1, -2)$ and $(8, 1, 3)$.
2. $(2, -2, 3)$ and $(2, 1, 6)$.
3. $(2, -2, 3)$ and $(2, 3, 3)$.
4. $(3, -2, 4)$ and $(5, -2, 4)$.

For Exercises 5–8, find equations for the lines passing through the given points: (a) in parametric form, (b) in symmetric form.

5. $(8, -2, 4)$ and $(5, 4, -1)$.
6. $(-2, 5, 6)$ and $(4, 2, -1)$.
7. $(3, 0, -1)$ and $(-2, 5, 7)$.
8. $(0, 0, 1)$ and $(3, -2, 4)$.
9. Find parametric equations for the segment joining each pair of points in Exercises 1, 3, 5, and 7.
10. Find parametric equations for the segment joining each pair of points in Exercises 2, 4, 6, and 8.

For Exercises 11–16, find equations of the lines indicated.

11. Passing through the point $(2, -1, 4)$ and parallel to the line through the points $(5, 2, 1)$ and $(1, 3, -2)$.
12. Perpendicular to the plane $2x - 3y + 5z = 9$ and passing through the point $(-4, 2, 6)$.
13. Perpendicular to the plane $5x + 2y - z = 7$ and passing through the point $(8, -2, 3)$.
14. Passing through the point $(-3, 5, 1)$ and parallel to the line through the points $(0, 1, 2)$ and $(2, -2, 6)$.
15. Passing through the point $(1, -2, 5)$ and parallel to the line through the points $(3, 1, 7)$ and $(2, -1, 3)$.
16. Passing through the point $(3, 2, -1)$ and parallel to the line through the points $(0, -1, 5)$ and $(6, 1, -2)$.

For Exercises 17–18, find an equation of the plane indicated.

17. Passing through the points $(2, 0, -1)$, $(3, 1, 4)$, and $(0, -2, 5)$.
18. Passing through the points $(3, -1, 2)$, $(1, 2, 4)$, and $(4, 2, -1)$.

In Exercises 19–22, find the point of intersection of the line with the given parametric equations and the plane with the given equation.

19. $x = t + 2,\quad y = 2t - 1,\quad z = t;\quad x - y + 3z = 9.$
20. $x = t - 1,\quad y = 3t,\quad z = t + 3;\quad 2x + y - z = 15.$
21. $x = 2t + 1,\quad y = 5t,\quad z = 1 - t;\quad 3x - 2y - z = 8.$
22. $x = 4t + 2,\quad y = t - 2,\quad z = 3t;\quad x + 3y - 4z = 21.$
23. Without using vectors, prove that the line with equations $x = 2t - 3$, $y = t + 2$, $z = t - 5$ is parallel to the plane with equation $x - 6y + 4z = 7$.
24. Use vectors to help prove that the line through the points $(2, -1, 7)$ and $(3, 2, -1)$ is parallel to the plane with equation $2x + 2y + z = 8$.

For Exercises 25–28, find an equation of the plane that is perpendicular to the given vector and passes through the given point.

25. Vector $(5, 6, 1)$, point $(3, -2, 4)$.
26. Vector $(0, 2, 5)$, point $(1, 3, -6)$.
27. Vector $(1, 0, 6)$, point $(-3, 2, 0)$.
28. Vector $(-3, 2, 0)$, point $(-2, 0, 3)$.

29. Find an equation of the plane that passes through the point $(-2, 6, 1)$ and is perpendicular to the line through the points $(2, 4, 5)$ and $(1, -2, 3)$.
30. Find an equation of the plane that passes through the point $(1, 2, -3)$ and is perpendicular to the line through the points $(1, 0, 2)$ and $(8, 1, 7)$.
31. Find an equation of the plane that passes through the point $(1, 5, -3)$ and is parallel to the plane $3x - 5y + 7z = 8$.
32. Find an equation of the plane that passes through the point $(0, -1, 2)$ and is parallel to the plane $7x + 4y - 2z = 6$.
33. Find an equation of the plane passing through the points $(0, 0, 3)$, $(0, -2, 0)$, and $(1, 0, 0)$. (*Hint*: A point is in the plane if and only if it satisfies an equation of the plane.)
34. Find an equation of the plane passing through the points $(2, 0, 0)$, $(0, 3, 0)$, and $(0, 0, 5)$.
35. Find an equation of the vertical plane passing through $(-1, 3, 2)$ and $(2, 9, 5)$. (*Hint*: A vector perpendicular to a vertical plane must be in the xy-plane and hence must have the form $(a, b, 0)$.)
36. Find an equation of the vertical plane passing through $(3, 2, 1)$ and $(-1, 4, 7)$.
37. Find the acute angle between the planes $x + y + \sqrt{6}z = 2$ and $x + y - \sqrt{6}z = 5$. (*Hint*: Find the angle between normals to the planes.)
38. Find the acute angle between the planes $3x + 3y + \sqrt{6}z = 7$ and $3x + 3y - \sqrt{6}z = -2$.
39. Prove that the planes $15x - 3y + 9z = 4$ and $20x - 4y + 12z = 3$ are parallel.
40. Prove that the planes $x + 5y - 2z = 8$ and $-4x + 2y + 3z = 9$ are perpendicular.
41. Prove that the planes $4x - 3y + z = 1$ and $2x + 6y + 10z = 7$ are perpendicular.
42. Prove that the planes $7x - 2y + 5z = 9$ and $-14x + 4y - 5z = 2$ are parallel.

14.7 Functions from R^n to R^m

In this section we shall consider several examples of functions with inputs in R^n and outputs in R^m. Recall that we have thought of a function as a computer which when presented with an element of its domain as input produces a unique output. Previously we were concerned with functions whose range and domain were subsets of the real numbers. In this section we extend that idea by considering examples of functions which when presented with the right kind of vector $(x_1, \ldots, x_n)$ of R^n as input, produce exactly one vector $(y_1, \ldots, y_m)$ of R^m as output. We may picture such a function f as follows:

$$(x_1, \ldots, x_n) \to \boxed{f} \to (y_1, \ldots, y_m).$$

Much of our previous terminology and notation carries over to these new considerations. For example, the output of f associated with the input $(x_1, \ldots, x_n)$ is denoted by $f(x_1, \ldots, x_n)$. Moreover, as before, the set of acceptable inputs of a function is called its *domain* and the set of all outputs is called the *range* of the function.

We shall indicate the fact that a function f has R^n as its domain and has a subset of R^m as its range by saying that f maps R^n into R^m and by writing $f: R^n \to R^m$. Similarly the notation $f: R^n \to R$ is used to indicate that the range of f is a subset of the real numbers, and the expression $f: R \to R^m$ indicates that the domain of f is the set of all real numbers.

To specify a particular function we must specify what the computer does to suitable inputs. For example, a function f such that $f: R^4 \to R^3$ might be given as

$$(x_1, x_2, x_3, x_4) \to \boxed{f} \to (x_1 + x_2, x_4{}^2, x_1 x_3),$$

or

$$f(x_1, x_2, x_3, x_4) = (x_1 + x_2, x_4{}^2, x_1 x_3).$$

In particular, for that function f,

$$f(2, 5, 3, 6) = (2 + 5, 6^2, 2 \cdot 3) = (7, 36, 6).$$

The function g defined as

$$g(x_1, x_2, \ldots, x_n) = \sqrt{x_1{}^2 + x_2{}^2 + \cdots + x_n{}^2}$$

is a function $g: R^n \to R$ for every positive integer n. Another way of expressing the function g can be obtained by letting $\mathbf{X} = (x_1, x_2, \ldots, x_n)$. Then, since

$$\sqrt{x_1{}^2 + x_2{}^2 + \cdots + x_n{}^2} = |\mathbf{X}| = \sqrt{\mathbf{X} \cdot \mathbf{X}},$$

g can be expressed as $g(\mathbf{X}) = |\mathbf{X}|$, or as $g(\mathbf{X}) = \sqrt{\mathbf{X} \cdot \mathbf{X}}$.

If $f: R^n \to R^m$, it may be more convenient at times to use the notation $f(\mathbf{X})$, where $\mathbf{X} = (x_1, x_2, \ldots, x_n)$, rather than $f(x_1, x_2, \ldots, x_n)$.

Vector-valued functions of vectors (or functions from R^n to R^m) arise quite naturally. We shall give several of the more common examples.

Example 1

Time-Independent Fluid Flow

Consider the problem of specifying the velocity of a fluid flowing in some region H. If the flow is time-independent, it is simply necessary to specify the velocity vector of the fluid at each point P of H. This can be pictured by drawing the replica at P of the appropriate velocity vector. Figure 14.7.1 illustrates replicas of velocity vectors at a few points of a region H.

Figure 14.7.1

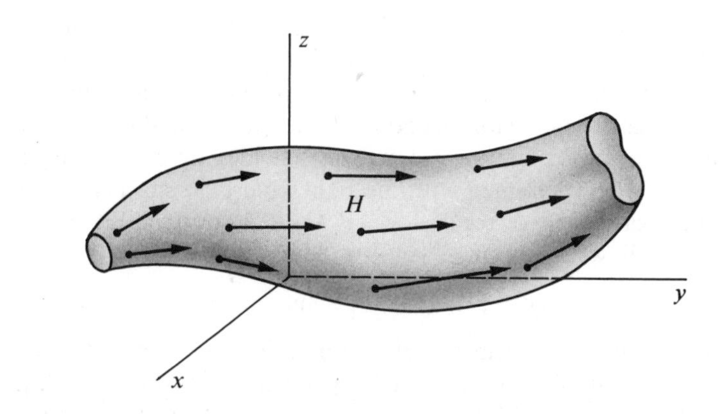

We can also view the specification of velocity vectors as a function F that assigns a velocity vector $F(\mathbf{X})$ to each position vector $\mathbf{X}$ of a point in region H. F is then a function whose domain is a subset of R^3 and whose range is a subset of R^3. ||

In Example 1, both $\mathbf{X}$ and $F(\mathbf{X})$ are 3-dimensional vectors. A case where $\mathbf{X}$ and $F(\mathbf{X})$ do not have the same dimension is illustrated in the next example.

Example 2

Time-Dependent Fluid Flow

This is very similar to the preceding case. However, now the association of velocity vectors with position vectors is not only a function of position, but is also a function of time t. Here we may denote the velocity vector associated with the point (x_1, x_2, x_3) of R^3 at time t by $F(x_1, x_2, x_3, t)$. In this case the velocity vector $F(\mathbf{X})$ is 3-dimensional, while the vector $\mathbf{X} = (x_1, x_2, x_3, t)$ is 4-dimensional. ||

Force fields yield common examples of functions F that map R^3 into R^3. Force fields may be induced in many ways, for instance, by an electrical charge, by a magnetized particle, or by a unit mass, as in the next example.

Example 3

The Gravitational Field Induced by a Particle of Unit Mass

Suppose that a unit mass is located at the origin. We wish to describe the gravitational force exerted on a particle of mass m by the unit mass at the origin. The force thus induced is directed toward the origin and has a magnitude equal to gm/d^2, where g is the gravitational constant and d is the distance between the two masses. If $\mathbf{X}$ is the position vector of the mass m, the force $F(\mathbf{X})$ induced by the unit mass has the direction $-\mathbf{X}$, with $|F(\mathbf{X})| = gm/|\mathbf{X}|^2$. All that information is contained in this simple vector equation:

$$F(\mathbf{X}) = \frac{gm}{|\mathbf{X}|^2}\left(\frac{-\mathbf{X}}{|\mathbf{X}|}\right) = -\frac{gm}{|\mathbf{X}|^3}\mathbf{X}.$$ ||

Example 4

Total Cost

Suppose that n different materials are used in a certain production process. Let c_i be the cost per unit of the ith material. Then the cost of x_i units of the ith material is c_ix_i. Consequently, the total cost of using x_1 units of the first material, x_2 units of the second material, etc., is

$$c_1x_1 + c_2c_2 + \cdots + c_nx_n.$$

Let $\mathbf{X} = (x_1, x_2, \ldots, x_n)$, and let $\mathbf{C} = (c_1, \ldots, c_n)$. The total cost $F(\mathbf{X})$ of using x_1 units of the first material, x_2 units of the second material, etc., can then be given as $F(\mathbf{X}) = \mathbf{C} \cdot \mathbf{X}$.

In this case the domain F is a subset of R^n, and the range of F is a subset of R. ||

Example 5

Position in Space as a Function of Time

Consider a particle that is moving on a circle in the xy-plane with constant angular speed ω. Suppose that the circle has radius r and has its center at the origin, as pictured in Figure 14.7.2.

Figure 14.7.2

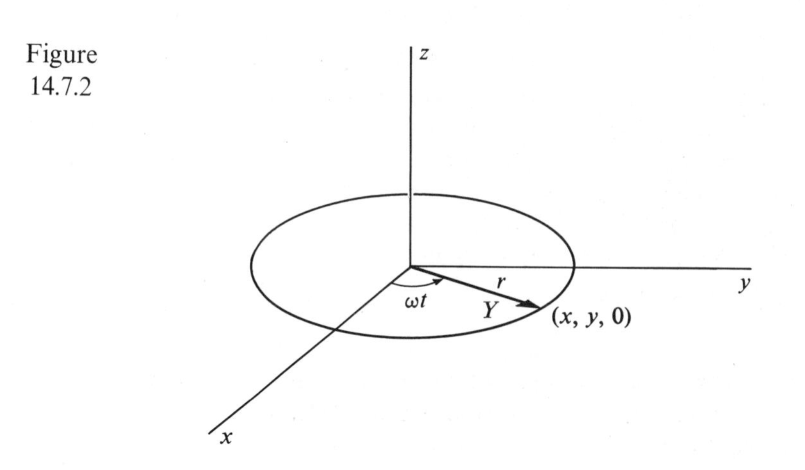

Let t be the elapsed time in minutes from a starting point on the x-axis and let $X(t) = (x, y, z)$ be the position vector of the particle. Then the angle from the x-axis to $X(t)$ is ωt and so

$$x = r \cos \omega t,$$

$$y = r \sin \omega t,$$

$$z = 0.$$

Thus
$$X(t) = (r \cos \omega t, r \sin \omega t, 0).$$

Note that X is a function mapping R (the time axis) into R^3, that is, $X: R \to R^3$.

The x, y, and z components of velocity are the derivatives of the corresponding components of $X(t)$ with respect to the time t. Consequently, the velocity components are

$$\frac{dx}{dt} = -r\omega \sin \omega t,$$

$$\frac{dy}{dt} = r\omega \cos \omega t,$$

$$\frac{dz}{dt} = 0,$$

and the velocity $V(t)$ at elapsed time t is given by

$$V(t) = (-r\omega \sin \omega t, r\omega \cos \omega t, 0).$$

Note that $V: R \to R^3$.

The components of the acceleration vector $A(t)$ are the derivatives of the corresponding components of velocity with respect to t and hence

$$A(t) = (-r\omega^2 \cos \omega t, -r\omega^2 \sin \omega t, 0).$$

It is interesting to notice that in this case

$$A(t) = -\omega^2 X(t).$$

Consequently the direction of the acceleration vector $A(t)$ is always opposite that of $X(t)$; that is, the direction of acceleration is always toward the origin! ||

We are led by Example 5 to the definition of the derivative of a function F mapping a subset of R into R^m.

Definition 14.7.1

> Let $F(t) = (x_1(t), x_2(t), \ldots, x_m(t))$ be a function that maps a subset of R into R^m. Then the derivative of F with respect to t is the function F', defined by
>
> $$F'(t) = (x_1'(t), x_2'(t), \ldots, x_m'(t)).$$

Note that the domain of F' is a subset of R and the range of F' is contained in R^m. Another notation for $F'(t)$ is dF/dt. In particular, if $F(t) = (t^4, t, 3, 4t)$, then

$$\frac{dF}{dt} = (4t^3, 1, 0, 4).$$

Similar to Example 5, let $X(t) = (x(t), y(t), z(t))$ be the position vector of a particle in three dimensional space, where t is the elapsed time from a starting point. Then the velocity vector $V(t)$ of the particle is $X'(t)$ and the acceleration vector $A(t)$ of the particle is $V'(t)$. That is, $V(t) = X'(t)$ and $A(t) = V'(t)$.

Functions of Several Variables

The functions considered in Chapters 1–12 are all functions of one variable. For such functions, both the domain and range are subsets of the real number system R. When the domain is a subset of R^n and the range is a subset of R, it is common to call the function a *real-valued function of n variables.*

In particular, when the domain is a subset of R^2 and the range is a subset of R, the function is called a *real-valued function of two variables.* From now on the functions that we shall consider most are functions of two variables. We shall usually think of a function of two variables as having inputs that are points in the plane (rather than the corresponding position vectors). For example, the function f defined as $f(x, y) = 2x/(x - y)$ can be thought of as a function whose domain is the set of all points (x, y) in the plane with $x \neq y$.

Functions of two variables are common. For instance, the area A of a triangle with base x and altitude y is $A = xy/2$; and the volume V of a right-circular cone of base radius x and altitude y is $V = \pi x^2 y/3$.

Although there is no useful way of graphing functions of more than two variables, it is possible to graph real-valued functions of two variables. If f is a real-valued function of two variables, its graph is the graph of the equation $z = f(x, y)$. For example, the graph of $f(x, y) = (x^2 + y^2)^{1/2}$ is the graph of the equation $z = (x^2 + y^2)^{1/2}$. Consequently, the graph of $f(x, y) = (x^2 + y^2)^{1/2}$ is the upper portion of the cone determined by $z^2 = x^2 + y^2$.

Similarly, the graph of $f(x, y) = \sqrt{4 - x^2 - y^2}$ is the upper portion of the sphere with equation $x^2 + y^2 + z^2 = 4$.

Exercises 14.7

For each $F(t)$ in Exercises 1–6, find $F'(t)$.

1. $F(t) = (t^2, t^3, 5t, 7)$.
2. $F(t) = (4t, t^5, t^4, 6t)$.
3. $F(t) = (e^{2t}, \sin t, \ln 5t)$.
4. $F(t) = (\cos 3t, e^{3t}, \tan 4t)$.
5. $F(t) = (\sin^2 3t, \cos^2 3t, t^{-3})$.
6. $F(t) = (e^t \sin t, e^t \cos t, \sec 6t)$.

Find the range and domain of each function in Exercises 7–14.

7. $f(x, y) = (x^2 - y^2)^{1/2}$.
8. $f(x, y) = (y/x)^{1/2}$.
9. $f(x, y) = (9 - x^2 - y^2)^{1/2}$.
10. $f(x, y) = (xy)^{1/2}$.
11. $f(x, y) = \left(\dfrac{x + y}{x - y}\right)^{1/2}$.
12. $f(x, y) = \dfrac{x}{x^2 - y^2}$.
13. $f(x, y) = (xy)^{1/2} + y^{1/2}$.
14. $f(x, y) = 1/xy^2$.
15. Show that in Example 5, $V(t)$ is perpendicular to $Y(t)$. What conclusion can be drawn from this?
16. The position vector of a moving particle is $\mathbf{X} = (t, -t^2, t^3)$. (a) Find the velocity vector of the particle when $t = 2$. (b) Also, when $t = 2$, find the vector component of the velocity vector of the particle in the direction of the vector $(2, 1, -1)$.
17. The position vector of a moving particle P is $\mathbf{X} = (2 \cos t, 2 \sin t, 5t)$. (a) Find the velocity vector $\mathbf{V}$ and the acceleration vector $\mathbf{A}$; and show that they both have constant magnitude. (b) Find the position vector of P and its velocity vector and acceleration vectors when $t = \pi$. (c) The path of P is called a *helix*. Describe the path. (Note that P must be on the cylinder $x^2 + y^2 = 4$.)
18. A particle is moving with velocity vector $\mathbf{V} = (2t, 1 - 6t^2, 4t^3)$. If the particle is at the point $(-2, 3, 4)$ when $t = 1$, find the position vector of the particle.
19. A particle is moving with velocity vector $\mathbf{V} = (9t^2, 12t^2, 6t + 5)$. If the particle is at the point $(4, -1, 2)$ when $t = 2$, find the position vector of the particle.
20. The position vector of a moving particle is $\mathbf{X} = (a \cos kt, b \sin kt, 0)$, where k is a constant. Find the velocity vector and the acceleration vector; and show that the acceleration vector is always directed toward the origin.
21. The position vector of a moving particle P is $\mathbf{X} = (\cos \omega t, \cos \omega t, \sqrt{2} \sin \omega t)$, where t is in seconds and ω is in radians per second.
    (a) Find the velocity vector $\mathbf{V}$ and the acceleration vector $\mathbf{A}$.
    (b) Find the position vector of P and its velocity vector and acceleration vector when $t = 1/2$ second, if $\omega = 3\pi$ radians per second.
    (c) Describe the path of P.
22. A particle P is moving with position vector $\mathbf{X} = (a \sin 2t, a + a \cos 2t, 2a \sin t)$, $0 \le t \le \pi/2$.
    (a) Find the velocity and acceleration vectors.
    (b) Find the position vector of P when $t = 0$, $t = \pi/4$ and $t = \pi/2$.
    (c) Show that P is on the intersection of the sphere $x^2 + y^2 + x^2 = 4a^2$ and the cylinder $x^2 + (y - a)^2 = a^2$.

14.8 Limits and Continuity

We shall extend the concept of the limit of a function of a single variable to the concept of the limit of a function $f: R^n \to R^m$. That is, we shall define

$$\lim_{\mathbf{X} \to \mathbf{A}} f(\mathbf{X}) = \mathbf{L}$$

where $\mathbf{A} = (a_1, a_2, \ldots, a_n)$ is a fixed vector of R^n, $\mathbf{L}$ is a fixed vector of R^m, and $\mathbf{X} = (x_1, x_2, \ldots, x_n)$.

Recall that $\lim_{x \to a} f(x) = L$ means that $f(x)$ can be made arbitrarily close to L by requiring x to be sufficiently close to but not equal to a. We saw that another way of saying this is "$f(x)$ can be made to be within any preassigned positive distance (called ε of L by requiring x to be within a small enough positive distance (called δ) of a, but not equal to a." In that way we obtained the precise definition of limit, Definition 2.2.1, which we repeat here for convenience.

Definition 2.2.1

> $\lim_{x \to a} f(x) = L$ means that for each $\varepsilon > 0$ (the preassigned distance) there is a $\delta > 0$ (the measurement of sufficient closeness) such that $|f(x) - L| < \varepsilon$ when $0 < |x - a| < \delta$.

Similarly, $\lim_{\mathbf{X} \to \mathbf{A}} f(\mathbf{X}) = \mathbf{L}$ means that $f(\mathbf{X})$ can be made to be within any preassigned positive distance (called ε) of $\mathbf{L}$ by requiring $\mathbf{X}$ to be within a small enough positive distance (called δ) of $\mathbf{A}$, but not equal to $\mathbf{A}$.

Since the distance between $\mathbf{X}$ and $\mathbf{A}$ is $|\mathbf{X} - \mathbf{A}|$ and the distance between $f(\mathbf{X})$ and $\mathbf{L}$ is $|f(\mathbf{X}) - \mathbf{L}|$, we extend Definition 2.2.1 as follows:

Definition 14.8.1

> $\lim_{\mathbf{X} \to \mathbf{A}} f(\mathbf{X}) = \mathbf{L}$ means that for each $\varepsilon > 0$ (the preassigned distance) there is a $\delta > 0$ (the measurement of sufficient closeness) such that $|f(\mathbf{X}) - \mathbf{L}| < \varepsilon$ when $0 < |\mathbf{X} - \mathbf{A}| < \delta$.

It should be noted that since $0 < |\mathbf{X} - \mathbf{A}|$ in this definition, $\mathbf{X}$ is not permitted to be the vector $(a_1, a_2, \ldots, a_n)$. Thus the value of the function f at $\mathbf{A}$, namely $f(\mathbf{A})$, is immaterial in the computation of $\mathbf{L}$. This is similar to the situation for the case of $\lim_{x \to a} f(x) = L$, where the value of the function at a, namely $f(a)$, is immaterial in the computation of L.

Example 1 | Find $\lim_{(x_1, x_2, x_3, x_4) \to (1, 3, 5, 2)} (x_1 + x_2 + x_3 + x_4 + 6)$.

In this case $f(x_1, x_2, x_3, x_4) = x_1 + x_2 + x_3 + x_4 + 6$ can be made arbitrarily close to 17 by requiring (x_1, x_2, x_3, x_4) to be sufficiently close to $(1, 3, 5, 2)$.

Thus

$$\lim_{(x_1, x_2, x_3, x_4) \to (1, 3, 5, 2)} (x_1 + x_2 + x_3 + x_4 + 6) = 17. \qquad \|$$

Example 2

Let $f(x_1, x_2, x_3) = (x_1 + x_2, x_2x_3)$ be a function from R^3 to R^2, and find

$$\lim_{(x_1, x_2, x_3)\to(1, 2, 3)} f(x_1, x_2, x_3).$$

In this case $f(x_1, x_2, x_3) = (x_1 + x_2, x_2x_3)$ can be made arbitrarily close to (3, 6) by requiring (x_1, x_2, x_3) to be sufficiently close to (1, 2, 3).

Thus

$$\lim_{(x_1, x_2, x_3)\to(1, 2, 3)} f(x_1, x_2, x_3) = (3, 6). \qquad \|$$

We have noted earlier that when $n = 1$, $f:R \to R^m$ and the domain of f is a subset of the real numbers. Thus f has the form $f(t) = (x_1(t), x_2(t), \ldots, x_m(t))$. In this case **A** is a real number which we denote by a, and a limit has the form

$$\lim_{t\to a} f(t) = \mathbf{L},$$

where **L** is a fixed vector of R^m.

Example 3

Let $f(t) = (t^2, 3t, 7, 1 - t)$. Then $f:R \to R^4$ and

$$\lim_{t\to 2} f(t) = (4, 6, 7, -1). \qquad \|$$

As in the case of limits of functions of a single variable (Theorem 2.3.1), there is a *Limit Theorem* for functions from R^n to R^m that is an aid in calculating many limits. Its proof is rather long and tedious and will not be given, but the proof is very similar to the single-variable case.

Theorem 14.8.1

The Limit Theorem

If $\lim_{\mathbf{X}\to\mathbf{A}} f(\mathbf{X})$ and $\lim_{\mathbf{X}\to\mathbf{A}} g(\mathbf{X})$ exist in R^m, then

(i) $\lim_{\mathbf{X}\to\mathbf{A}} [f(\mathbf{X}) + g(\mathbf{X})] = \lim_{\mathbf{X}\to\mathbf{A}} f(\mathbf{X}) + \lim_{\mathbf{X}\to\mathbf{A}} g(\mathbf{X})$;

(ii) $\lim_{\mathbf{X}\to\mathbf{A}} cf(\mathbf{X}) = c \lim_{\mathbf{X}\to\mathbf{A}} f(\mathbf{X})$, where c is any constant;

(iii) $\lim_{\mathbf{X}\to\mathbf{A}} f(\mathbf{X})g(\mathbf{X}) = \lim_{\mathbf{X}\to\mathbf{A}} f(\mathbf{X}) \lim_{\mathbf{X}\to\mathbf{A}} g(\mathbf{X})$, if $m = 1$;

(iv) $\lim_{\mathbf{X}\to\mathbf{A}} \dfrac{f(\mathbf{X})}{g(\mathbf{X})} = \dfrac{\lim_{\mathbf{X}\to\mathbf{A}} f(\mathbf{X})}{\lim_{\mathbf{X}\to\mathbf{A}} g(\mathbf{X})}$, provided $m = 1$ and $\lim_{\mathbf{X}\to\mathbf{A}} g(\mathbf{X}) \neq 0$;

(v) $\lim_{\mathbf{X}\to\mathbf{A}} (f_1(\mathbf{X}), f_2(\mathbf{X}), \ldots, f_n(\mathbf{X}))$

$$= \left(\lim_{\mathbf{X}\to\mathbf{A}} f_1(\mathbf{X}), \lim_{\mathbf{X}\to\mathbf{A}} f_2(\mathbf{X}), \ldots, \lim_{\mathbf{X}\to\mathbf{A}} f_n(\mathbf{X})\right).$$

It is left as an exercise to use (v) to show that if $F(t) = (x_1(t), x_2(t), \ldots, x_m(t))$, then $F'(t)$, which is defined by Definition 14.7.1, can be expressed as

$$F'(t) = \lim_{h \to 0} \frac{F(t+h) - F(t)}{h}.$$

We define continuity for functions from R^n to R^m of n variables in much the same way that we defined it for functions of one variable.

Definition 14.8.2

> A function $f: R^n \to R^m$ is said to be *continuous at* **A** if
>
> $$\lim_{\mathbf{X} \to \mathbf{A}} f(\mathbf{X}) = f(\mathbf{A}).$$

Definition 14.8.3

> A function $f: R^n \to R^m$ is said to be *continuous on a set S* if and only if f is continuous at every point of S.

Limits and Continuity for Functions of Two Variables

Since we shall be most interested in functions of two variables, we shall restate the definitions of limit and continuity and the Limit Theorem specifically for such functions.

For $n = 2$ and $m = 1$, that is for a real-valued function $f(x, y)$ of two variables (whose inputs are considered to be points in the plane, rather than the corresponding position vectors), **X** has the form (x, y), **L** is a real number, **A** has the form (a, b) and

$$|\mathbf{X} - \mathbf{A}| = \sqrt{(x-a)^2 + (y-b)^2}.$$

So Definition 14.8.1 (the definition of limit) takes this form:

Definition 14.8.4

> $\lim_{(x, y) \to (a, b)} f(x, y) = L$ means that for each $\varepsilon > 0$ there is a $\delta > 0$ such that $|f(x, y) - L| < \varepsilon$ when $0 < \sqrt{(x-a)^2 + (y-b)^2} < \delta$.

Example 4

> Let $f(x, y) = \begin{cases} x + y & \text{if} \quad (x, y) \neq (4, 6) \\ 4 & \text{if} \quad (x, y) = (4, 6). \end{cases}$
>
> Find $\lim_{(x, y) \to (4, 6)} f(x, y)$.

We see that $f(x, y)$ can be made arbitrarily close to 10 by requiring (x, y) to be close enough to the point $(4, 6)$, *but not at* $(4, 6)$. Thus

$$\lim_{(x, y) \to (4, 6)} f(x, y) = 10.$$

Note that

$$\lim_{(x, y) \to (4, 6)} f(x, y) \neq 4$$

even though $f(4, 6) = 4$. The graph of $z = f(x, y)$ is shown in Figure 14.8.1. ||

Figure 14.8.1

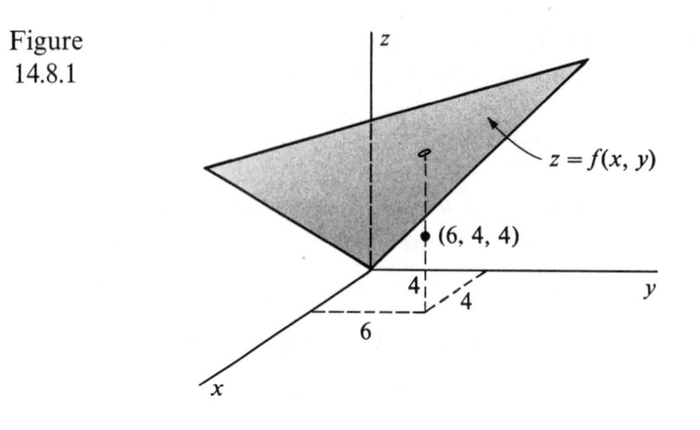

Example 5 | Show that $\lim_{(x, y)\to(0, 0)} \dfrac{xy}{x^2 + y^2}$ does not exist.

In any small disc about (0, 0) the function $f(x, y) = xy/(x^2 + y^2)$ assumes many real values. In particular, suppose the point (x, y) is on the line $y = kx$, for any fixed real number k. Then

$$\frac{xy}{x^2 + y^2} = \frac{kx^2}{x^2 + k^2x^2} = \frac{k}{1 + k^2}.$$

Thus $f(x, y) = xy/(x^2 + y^2)$ assumes the value $k/(1 + k^2)$ whether or not the point (x, y) is near (0, 0), as long as (x, y) is on the line $y = kx$ and not at (0, 0). Of course (x, y) is not permitted to be at (0, 0) by definition of limit. Because the function cannot be made arbitrarily close to a unique number, the limit does not exist. ||

For functions of two variables the Limit Theorem is:

Theorem 14.8.2

If $\lim_{(x, y)\to(a, b)} f(x, y)$ and $\lim_{(x, y)\to(a, b)} g(x, y)$ exist, then

(i) $$\lim_{(x, y)\to(a, b)} [f(x, y) + g(x, y)] = \lim_{(x, y)\to(a, b)} f(x, y) + \lim_{(x, y)\to(a, b)} g(x, y),$$

(ii) $$\lim_{(x, y)\to(a, b)} cf(x, y) = c \lim_{(x, y)\to(a, b)} f(x, y),$$

(iii) $$\lim_{(x, y)\to(a, b)} [f(x, y)g(x, y)] = \lim_{(x, y)\to(a, b)} f(x, y) \lim_{(x, y)\to(a, b)} g(x, y),$$

(iv) $$\lim_{(x, y)\to(a, b)} \frac{f(x, y)}{g(x, y)} = \frac{\lim_{(x, y)\to(a, b)} f(x, y)}{\lim_{(x, y)\to(a, b)} g(x, y)},$$

provided $\lim_{(x, y)\to(a, b)} g(x, y) \neq 0$.

For a real-valued function $f(x, y)$ of two variables, the continuity definitions (Definitions 14.8.2 and 14.8.3) are as follows:

Definition 14.8.5

A function $f(x, y)$ is said to be *continuous at* (a, b) if and only if

$$\lim_{(x, y)\to(a, b)} f(x, y) = f(a, b).$$

Definition 14.8.6

A function $f(x, y)$ is said to be *continuous on a set* S if and only if $f(x, y)$ is continuous at every point of S.

Example 6

The function

$$f(x, y) = \begin{cases} x + y & \text{if} \quad (x, y) \neq (4, 6) \\ 4 & \text{if} \quad (x, y) = (4, 6) \end{cases}$$

of Example 4 has the property that $f(4, 6) = 4$ and

$$\lim_{(x, y)\to(4, 6)} f(x, y) = 10.$$

Thus, since $4 \neq 10$, the function is not continuous at (4, 6). However, that function is continuous on the set of all points different from (4, 6). In particular, it is continuous at (5, 2) because $f(5, 2) = 7$ and

$$\lim_{(x, y)\to(5, 2)} f(x, y) = 7. \qquad \|$$

The geometric interpretation of continuity for a function $f(x, y)$ is also similar to the single-variable case. That is, the function $f(x, y)$ is continuous at (a, b) if and only if the graph of $z = f(x, y)$ does not have a break or discontinuity above the point (a, b). It is continuous on a set S if it has no breaks or discontinuities above S.

The next theorem indicates one of the most important properties of continuous real-valued functions of two variables. It is similar to Theorem 2.4.1, The Intermediate Value Theorem for Continuous Functions (of one variable).

Theorem 14.8.3

Intermediate Value Theorem for Continuous Functions of Two Variables

Let $f(x, y)$ be a function that is continuous in a connected region R. Let (a, b) and (c, d) be any two points of R, and suppose that r is some number between $f(a, b)$ and $f(c, d)$. Then there is at least one point (p, q) in R such that $f(p, q) = r$.

In less rigorous language the theorem states that a function $f(x, y)$ which is continuous at every point of an unbroken or single-pieced region takes on all values between any two of its values. A proof of this theorem and a more rigorous discussion of connectivity would take us too far afield.

Exercises 14.8

1. Let $\mathbf{X} = (x_1, x_2, x_3)$, $\mathbf{A} = (2, -1, 3)$, $f(\mathbf{X}) = (x_1x_2, x_2x_3, x_1x_3, x_2)$, and $g(\mathbf{X}) = (x_1^2, x_2^2, x_1 + x_3, x_3 - x_2)$. Find
(a) $\lim_{\mathbf{X}\to\mathbf{A}} [2f(\mathbf{X}) + 3g(\mathbf{X})]$, (b) $\lim_{\mathbf{X}\to\mathbf{A}} [6f(\mathbf{X}) + g(\mathbf{X})]$.

2. Let $\mathbf{X} = (x_1, x_2, x_3, x_4)$, $\mathbf{A} = (3, 0, -2, -1)$, $f(\mathbf{X}) = (x_1x_2, x_3^2 + x_4)$, and $g(\mathbf{X}) = (x_2x_3^2, x_4)$. Find
(a) $\lim_{\mathbf{X}\to\mathbf{A}} [5f(\mathbf{X}) + 4g(\mathbf{X})]$, (b) $\lim_{\mathbf{X}\to\mathbf{A}} [7f(\mathbf{X}) + g(\mathbf{X})]$.

3. Let $f(x, y) = |x|/(|x| + |y|)$. Find
(a) $\lim_{(x, y)\to(3, 5)} f(x, y)$, (b) $\lim_{(x, y)\to(0, 0)} f(x, y)$.

4. Let $f(x, y) = \begin{cases} x + y + 8 & \text{if } (x, y) = (2, 3) \\ x^2 + y^2 & \text{if } (x, y) \neq (2, 3). \end{cases}$

Find $\lim_{(x, y)\to(2, 3)} f(x, y)$.

5. Let $f(x, y) = \begin{cases} x^2 + y & \text{if } y = 3x \\ 2x + y & \text{if } y \neq 3x. \end{cases}$

Find (a) $\lim_{(x, y)\to(2, 5)} f(x, y)$, (b) $\lim_{(x, y)\to(1, 3)} f(x, y)$.

6. Let $f(x, y) = \begin{cases} xy & \text{if } x \neq y \\ x + y & \text{if } x = y. \end{cases}$

Find (a) $\lim_{(x, y)\to(3, 3)} f(x, y)$, (b) $\lim_{(x, y)\to(2, 2)} f(x, y)$.

7. For what points (x, y) is the function of Exercise 3 not continuous?
8. For what points (x, y) is the function of Exercise 4 not continuous?
9. For what points (x, y) is the function of Exercise 5 continuous?
10. For what points (x, y) is the function of Exercise 6 continuous?
11. Show that $f(\mathbf{X}) = (f_1(\mathbf{X}), f_2(\mathbf{X}), \ldots, f_n(\mathbf{X}))$ is continuous at $\mathbf{A}$ if and only if each of the $f_i(\mathbf{X})$ is continuous at $\mathbf{A}$.
12. Use the method of Example 5 to prove that $\lim_{(x, y)\to(0, 0)} x^2/(x^2 + y^2)$ does not exist.
13. Use the method of Example 5 to prove that $\lim_{(x, y)\to(0, 0)} (x + y)/(x - y)$ does not exist.
14. Show that $F'(t)$ as defined by Definition 14.7.1 can be expressed as

$$F'(t) = \lim_{h\to 0} \frac{F(t + h) - F(t)}{h}.$$

Brief Review of Chapter 14

1. Three Dimensional Coordinates and Graphs

Three Dimensional Coordinate System. We take three mutually perpendicular real lines, called the x-axis, the y-axis, and the z-axis, intersecting at the origin of each. With each point in space there is associated an ordered triple of numbers, the first number indicates the location of the point in the direction of the x-axis, the second indicates its location in the direction of the y-axis, and the third indicates the location of the point in the direction of the z-axis.

Graph of an Equation. The set of all points that are solutions of an equation involving x, y, and z is called the *graph of the equation.*

Usually the graph of an equation involving x, y, and z is a surface.

2. Distance

The distance between points (x_1, y_1, z_1) and (x_2, y_2, z_2) is

$$D((x_1, y_1, z_1), (x_2, y_2, z_2)) = \sqrt{(x_2 - x_1)^2 + (y_2 - y_1)^2 + (z_2 - z_1)^2}.$$

3. Spheres

An equation of the sphere of radius r with center at (x_0, y_0, z_0) is

$$(x - x_0)^2 + (y - y_0)^2 + (z - z_0)^2 = r^2.$$

4. Sketching Graphs

To sketch the graph of an equation in three dimensions, first find the intersection of the graph with each of the three coordinate planes, that is, with the plane $x = 0$, the plane $y = 0$, and the plane $z = 0$. It is then usually best to find the curves of intersection with at least some of the planes parallel to the coordinate planes, that is, planes of the form $x = r$ (parallel to the yz-plane), $y = s$ (parallel to the xz-plane), and $z = t$ (parallel to the xy-plane), where r, s, and t are constants.

5. Vectors

Definition. An *n-dimensional vector* is an ordered n-tuple $(a_1, a_2, \ldots, a_n)$ of real numbers $a_1, a_2, \ldots, a_n$ where equality, addition, and scalar multiplication are defined as follows:

$$(a_1, a_2, \ldots, a_n) = (b_1, b_2, \ldots, b_n) \quad \text{if and only if} \quad a_1 = b_1, a_2 = b_2, \ldots, a_n = b_n.$$

$$(a_1, a_2, \ldots, a_n) + (b_1, b_2, \ldots, b_n) = (a_1 + b_1, a_2 + b_2, \ldots, a_n + b_n),$$

$$k(a_1, a_2, \ldots, a_n) = (ka_1, ka_2, \ldots, ka_n), \quad \text{for any real number } k.$$

Length, Definition. The length $|\mathbf{A}|$ of a vector $\mathbf{A} = (a_1, a_2, \ldots, a_n)$ is

$$|\mathbf{A}| = \sqrt{{a_1}^2 + {a_2}^2 + \cdots + {a_n}^2}.$$

Geometric Interpretation. An n-dimensional vector (an element of R^n) is thought of geometrically as a directed line segment (arrow), in Euclidian n-space, from the origin to the point whose coordinates are the same as those of the vector. For $n = 2$ or $n = 3$, the arrows can be pictured conveniently. When pictured, two vectors are equal if and only if they are the same arrow, and the sum of two vectors $\mathbf{A}$ and $\mathbf{B}$ is a diagonal of the parallelogram having $\mathbf{A}$ and $\mathbf{B}$ as adjacent sides. The "replica-attaching" interpretation of addition is also useful.

The scalar multiplication $k\mathbf{A}$ produces a vector of length $|k|$ times the length of $\mathbf{A}$. Also, $k\mathbf{A}$ is in the same direction as $\mathbf{A}$ if $k > 0$, and in the opposite direction of $\mathbf{A}$ if $k < 0$.

Unit Vectors. A vector of length 1 is called a *unit vector*. For any vector $\mathbf{A}$, $\mathbf{A}/|\mathbf{A}|$ is a unit vector. In R^3, the unit vectors (1, 0, 0), (0, 1, 0), and (0, 0, 1) are usually denoted by **i**, **j**, and **k**, respectively.

6. The Dot Product

Definition. The *dot product* $\mathbf{A} \cdot \mathbf{B}$ of vectors $\mathbf{A} = (a_1, a_2, \ldots, a_n)$ and $\mathbf{B} = (b_1, b_2, \ldots, b_n)$ of R^n is

$$\mathbf{A} \cdot \mathbf{B} = a_1 b_1 + a_2 b_2 + \cdots + a_n b_n.$$

When $\mathbf{A} = (x_1, y_1, z_1)$ and $\mathbf{B} = (x_2, y_2, z_2)$ this definition gives

$$\mathbf{A} \cdot \mathbf{B} = x_1 x_2 + y_1 y_2 + z_1 z_2.$$

Uses. When neither of the vectors $\mathbf{A}$ or $\mathbf{B}$ is $\mathbf{0}$, the cosine of the angle θ between $\mathbf{A}$ and $\mathbf{B}$ is

$$\cos \theta = \frac{\mathbf{A} \cdot \mathbf{B}}{|\mathbf{A}|\,|\mathbf{B}|}.$$

Also, two nonzero vectors $\mathbf{A}$ and $\mathbf{B}$ are perpendicular if and only if $\mathbf{A} \cdot \mathbf{B} = 0$. The component vector of $\mathbf{A}$ in the direction of $\mathbf{B}$ is

$$\frac{\mathbf{A} \cdot \mathbf{B}}{\mathbf{B} \cdot \mathbf{B}} \mathbf{B}$$

7. The Cross Product

Definition. The *cross product* $\mathbf{A} \times \mathbf{B}$ of two vectors $\mathbf{A} = (x_1, y_1, z_1)$ and $\mathbf{B} = (x_2, y_2, z_2)$ of R^3 is

$$\mathbf{A} \times \mathbf{B} = \begin{vmatrix} \mathbf{i} & \mathbf{j} & \mathbf{k} \\ x_1 & y_1 & z_1 \\ x_2 & y_2 & z_2 \end{vmatrix}.$$

Uses. The vector $\mathbf{A} \times \mathbf{B}$ is perpendicular to the vector $\mathbf{A}$ and to the vector $\mathbf{B}$ and so is perpendicular to the plane determined by $\mathbf{A}$ and $\mathbf{B}$. The area of the parallelogram having $\mathbf{A}$ and $\mathbf{B}$ as adjacent sides is $|\mathbf{A} \times \mathbf{B}|$. The volume of the parallelopiped having $\mathbf{A}$, $\mathbf{B}$ and $\mathbf{C}$ as adjacent edges is $|(\mathbf{A} \times \mathbf{B}) \cdot \mathbf{C}|$.

8. Equations of Lines

Passing through a point (x_0, y_0, z_0) with position vector $\mathbf{P}_0$ and parallel to a fixed vector $\mathbf{D} = (a, b, c)$:

Vector Equation. $\mathbf{P} - \mathbf{P}_0 = t\mathbf{D}$, where $\mathbf{P}$ is the position vector of the point (x, y, z).

Parametric Equations.

$$\begin{cases} x = x_0 + ta, \\ y = y_0 + tb, \\ z = z_0 + tc. \end{cases}$$

Symmetric Equations. (If $a \neq 0$, $b \neq 0$, and $c \neq 0$),

$$\frac{x - x_0}{a} = \frac{y - y_0}{b} = \frac{z - z_0}{c}.$$

Passing through two points (x_0, y_0, z_0) and (x_1, y_1, z_1) with position vectors $\mathbf{P}_0$ and $\mathbf{P}_1$, respectively:

Vector Equation. $\mathbf{P} - \mathbf{P}_0 = t(\mathbf{P}_1 - \mathbf{P}_0)$, where $\mathbf{P}$ is the position vector of the point (x, y, z).

Parametric Equations.

$$\begin{cases} x = x_0 + t(x_1 - x_0), \\ y = y_0 + t(y_1 - y_0), \\ z = z_0 + t(z_1 - z_0). \end{cases}$$

Symmetric Equations. (If $x_1 \neq x_0$, $y_1 \neq y_0$, $z_1 \neq z_0$),

$$\frac{x - x_0}{x_1 - x_0} = \frac{y - y_0}{y_1 - y_0} = \frac{z - z_0}{z_1 - z_0}.$$

Parametric equations for the line segment joining the points (x_0, y_0, z_0) and (x_1, y_1, z_1) are

$$\left.\begin{aligned} x &= x_0 + t(x_1 - x_0) \\ y &= y_0 + t(y_1 - y_0) \\ z &= z_0 + t(z_1 - z_0) \end{aligned}\right\} \qquad \text{where } 0 \leq t \leq 1.$$

9. Equations of Planes

An equation of the plane that is perpendicular to a nonzero vector (a, b, c) and passes through a point (x_0, y_0, z_0) is

$$a(x - x_0) + b(y - y_0) + c(z - z_0) = 0.$$

Every plane has an equation of the form

$$ax + by + cz = d,$$

where a, b, c, and d are real constants and a, b, and c are not all 0. And every such equation is an equation of a plane.

10. Functions from R^n to R^m

Meaning. A function $f: R^n \to R^m$ is considered to be a computer which when presented with the right kind of vector $(x_1, x_2, \ldots, x_n)$ of R^n as input, produces exactly one vector $(y_1, y_2, \ldots, y_m)$ of R^m as output.

Derivative of $F: R \to R^m$. When $F(t) = (x_1(t), x_2(t), \ldots, x_m(t))$, the derivative of F with respect to t is the function $F': R \to R^m$ where

$$F'(t) = (x_1'(t), x_2'(t), \ldots, x_m'(t)).$$

11. Limits and Continuity

Definition. Let $f: R^n \to R^m$ and let $\mathbf{A}$ and $\mathbf{L}$ be fixed vectors of R^n and R^m, respectively. Then $\lim_{\mathbf{X} \to \mathbf{A}} f(\mathbf{X}) = \mathbf{L}$ means that for each real number $\varepsilon > 0$ there is a number $\delta > 0$ such that $|f(\mathbf{X}) - \mathbf{L}| < \varepsilon$ when $0 < |\mathbf{X} - \mathbf{A}| < \delta$.

Interpretation of Definition. $\lim_{\mathbf{X} \to \mathbf{A}} f(\mathbf{X}) = \mathbf{L}$ means that $f(\mathbf{X})$ can be made arbitrarily close to $\mathbf{L}$ by requiring $\mathbf{X}$ to be sufficiently close to $\mathbf{A}$ but not equal to $\mathbf{A}$.

Definition. $f(\mathbf{X})$ is continuous at $\mathbf{A}$ means that $\lim_{\mathbf{X} \to \mathbf{A}} f(\mathbf{X}) = f(\mathbf{A})$.

12. Functions of Two Variables

Meaning. A function $f(x, y)$ from R^2 into R is called a *function of two variables.*

Limits. $\lim_{(x, y) \to (a, b)} f(x, y) = L$ means that for each $\varepsilon > 0$ there is a $\delta > 0$ such that $|f(x, y) - L| < \varepsilon$ when $0 < \sqrt{(x-a)^2 + (y-b)^2} < d$.

Continuity, Definitions. A function $f(x, y)$ is said to be *continuous at* (a, b) if and only if $\lim_{(x, y) \to (a, b)} f(x, y) = f(a, b)$. A function $f(x, y)$ is said to be *continuous on a set* S if and only if $f(x, y)$ is continuous at every point of S.

Continuity, Geometric Interpretation. A function $f(x, y)$ is continuous at (a, b) if and only if the surface $z = f(x, y)$ does not have a break above the point (a, b). It is continuous on a set S if it has no breaks above S.

Technique Review Exercises, Chapter 14

1. In words describe the graph of the equation $y = -10$.
2. Find the distance between the points $(-5, 2, 1)$ and $(-2, -3, 3)$.
3. Find an equation of the sphere centered at $(2, -3, 1)$ and passing through the point $(4, 1, -2)$.
4. Sketch the graph of $x^2 + 4z^2 = 4y^2$.
5. If $\mathbf{A} = (1, 2, -3, 4)$ and $\mathbf{B} = (-2, 1, 2, 5)$, find $(\mathbf{A} - 2\mathbf{B}) \cdot \mathbf{A}$.
6. Prove that the vectors $(2, -5, 8)$ and $(-2, 4, 3)$ are perpendicular.
7. Find a unit vector that is perpendicular to both of the vectors $(2, -1, -3)$ and $(3, 2, 4)$.
8. Find the vector component of the vector $(-2, 3, 5)$ in the direction of the vector $(1, -2, 3)$.
9. Find the angle between the vectors $(4, -6, 12^{1/2})$ and $(4, 6, -(12)^{1/2})$.
10. Find parametric equations for the line through the points $(2, 5, -1)$ and $(-3, 5, 2)$.
11. Find symmetric equations for the line that is perpendicular to the plane $4y - 3x + z = 7$ and passes through the point $(5, 8, -6)$.

12. Find an equation of the plane that passes through the point $(5, -4, -6)$ and is perpendicular to the line through the points $(7, -1, 2)$ and $(6, 4, -1)$.

13. If the position vector of a moving object is $\mathbf{P} = (5t^2, t^3, e^{2t})$, find the velocity vector of the object when $t = 1$.

14. Let $f(x, y) = \begin{cases} x^2 - y^2 & \text{if } x \neq y \\ x^2 + y^2 & \text{if } x = y. \end{cases}$

    Find (a) $\lim_{(x, y) \to (0, 0)} f(x, y)$, (b) $\lim_{(x, y) \to (-2, -2)} f(x, y)$.

15. For what points (x, y) is the function of Exercise 14 continuous?

Additional Exercises, Chapter 14

Section 14.1

In words, describe the graph of each equation in 3-space.

1. $x = 7$.
2. $|z| = 9$.
3. $y = 2x$.
4. $(x - 1)^2 + y^2 = 0$.

Section 14.2

5. Find the distance between the points $(-3, 5, -2)$ and $(8, 3, 1)$.
6. Find the distance between the points $(4, 0, -3)$ and $(6, -4, 7)$.

For Exercises 7–9, find an equation for the sphere indicated.

7. Center at $(5, 6, -2)$ and radius 9.
8. Center at $(7, -6, 5)$ and passing through the point $(3, 0, -1)$.
9. Ends of one diameter at $(5, -3, 8)$ and $(3, 1, -2)$.
10. Find an inequality for the set of all points inside the sphere with radius 7 and centered at $(0, -5, 4)$.
11. Find an inequality for the set of all points outside the sphere with center at $(-1, 7, 9)$, if $(2, 6, -1)$ is one point on the sphere.
12. Find an equation of the set of all points equidistant from the points $(4, 7, -3)$ and $(1, -2, -4)$.

Section 14.3

Sketch the graph of the following equations in 3-space.

13. $z^2 = 3x^2 + 3y^2$.
14. $x = y^2 + z^2$.
15. $16x^2 + 9y^2 + 36z^2 = 144$.
16. $y^2 = x^2 + z^2$.
17. $x^2 + (y - 2)^2 + (z - 3)^2 = 9$.
18. $4z = x^2 + 4y^2$.
19. $4x^2 + y^2 - 16z^2 = 16$.
20. $z^2 = x^2$.
21. $x^2 + y^2 = 25$.
22. $9x^2 + 4y^2 = 36$.

Section 14.4

23. Find a vector $\mathbf{X}$ so that $(1, -2, 3) - \mathbf{X} = (-3, 5, 7) + 2\mathbf{X}$.
24. Find $4(1, -3, 2) - 5(6, 1, -8)$.
25. Find the length of the vector $2(-3, 2, 4)$.

26. Find a unit vector whose direction is the same as the direction of the vector $(3, -7, -2)$.
27. Use vectors to prove that the points $(1, -4, 7)$, $(-2, -2, 2)$, and $(13, -12, 27)$ lie on the same line.
28. Use vectors to prove that the points $(7, 3, -1)$, $(-2, 4, 5)$, $(4, 3, 7)$, and $(13, 2, 1)$ are the vertices of a parallelogram.

Section 14.5

29. Find $\mathbf{A} \cdot \mathbf{B}$ if $\mathbf{A} = (2, -1, 3, 8)$ and $\mathbf{B} = (5, 6, -4, 1)$.
30. Find $\mathbf{A} \times \mathbf{B}$ if $\mathbf{A} = (-2, 3, 4)$ and $\mathbf{B} = (5, -2, 4)$.
31. Find the component vector of $(3, -2, -1)$ in the direction of the vector $(2, -1, 4)$.
32. Find the angle between the vectors $(1, -1, 1)$ and $(1 + 2^{1/2}, 1 - 2^{1/2}, 2^{1/2})$.
33. Find a unit vector perpendicular to both of the vectors $(1, 2, -3)$ and $(-2, -1, 1)$.
34. Use vectors to find the area of the triangle whose vertices are at $(3, -2, 0)$, $(-3, 1, 1)$, and $(4, -1, 5)$.

Section 14.6

In Exercises 35–38, find equations for the lines indicated.

35. Passing through the points $(5, -7, 2)$ and $(1, 2, -3)$.
36. Passing through the points $(3, -4, 5)$ and $(3, 2, -1)$.
37. Passing through the point $(8, 5, 1)$ and perpendicular to the plane $4x - 7y + 3z = 2$.
38. Passing through the point $(-2, 5, 7)$ and parallel to the line through the points $(8, 9, 4)$ and $(1, 6, 2)$.

In Exercises 39–42, find an equation of the plane indicated.

39. Perpendicular to the vector $(-2, 3, 5)$ and passing through the point $(9, 1, 3)$.
40. Passing through the point $(7, 0, -1)$ and perpendicular to the line through the points $(-4, 5, 6)$ and $(-2, 1, 7)$.
41. Passing through the point $(5, -3, 2)$ and parallel to the plane $5x + 2y - z = 7$.
42. Passing through the points $(-3, 0, 0)$, $(0, 2, 0)$, and $(0, 0, 6)$.

Section 14.7

43. Find $F'(t)$ if $F(t) = (\sin 3t, e^{7t}, t^4)$.
44. Find the range and domain of the function $f(x, y) = (4 - x^2 - y^2)^{1/2}$.
45. A particle is moving with velocity vector $\mathbf{V} = (e^t, 4t, 3t^2)$. If the particle is at the point $(4, -1, 3)$ when $t = 2$, find the position vector of the particle.

Section 14.8

46. Find $\lim_{(x, y, z)\to(1, 2, 3)} (x + y, 2yz, xyz)$.
47. Let $f(x, y) = \begin{cases} x + y & \text{if } x \neq y \\ x + 2y & \text{if } x = y. \end{cases}$

    Find (a) $\lim_{(x, y)\to(5, 0)} f(x, y)$; (b) $\lim_{(x, y)\to(0, 0)} f(x, y)$.
48. For what points (x, y) is the function of Exercise 47 continuous?

Challenging Problems, Chapter 14

1. A wheel of radius 1 foot is rolling on a horizontal surface with a constant angular velocity of 6 radians per second. If $\mathbf{Y}$ is the position vector of a point P on the rim, show that we can take $\mathbf{Y} = (6t - \sin 6t, 1 - \cos 6t)$, where t is the time in seconds. Find velocity vectors for P when (a) P is at the top of the wheel, (b) P is at the bottom of the wheel. How fast is the axle moving forward?

2. Prove that $\lim_{(x,y)\to(0,0)} y/(x^2 + y)$ does not exist.

3. Kepler's Second Law of planetary motion states that the segment joining the sun and a planet sweeps out equal areas in equal time intervals. Use Kepler's Second Law along with the following outline to show that the acceleration vector of the planet is always directed toward the sun.
   (i) Let $r = f(\theta)$ be the polar equation of the path of the planet with the sun located at the origin. Let $\theta = g(t)$ where t denotes the elapsed time.
   (ii) Differentiate $A = \frac{1}{2}\int_{\theta_1}^{\theta} (f(u))^2\, du$ with respect to time and obtain
   $$r^2 \frac{d\theta}{dt} = c, \qquad c \text{ a constant.}$$
   (iii) Differentiate again to get
   $$2r\frac{dr}{dt}\frac{d\theta}{dt} + r^2\frac{d^2\theta}{dt^2} = 0$$
   (iv) Use $x = r\cos\theta$, $y = r\sin\theta$ and the result of (iii) to obtain
   $$\frac{d^2x}{dt^2} = \left[\frac{1}{r}\frac{d^2r}{dt^2} - \left(\frac{d^2\theta}{dt^2}\right)^2\right]x$$
   and
   $$\frac{d^2y}{dt^2} = \left[\frac{1}{r}\frac{d^2r}{dt^2} - \left(\frac{d^2\theta}{dt^2}\right)^2\right]y.$$
   Note that the desired result follows from these final equations.

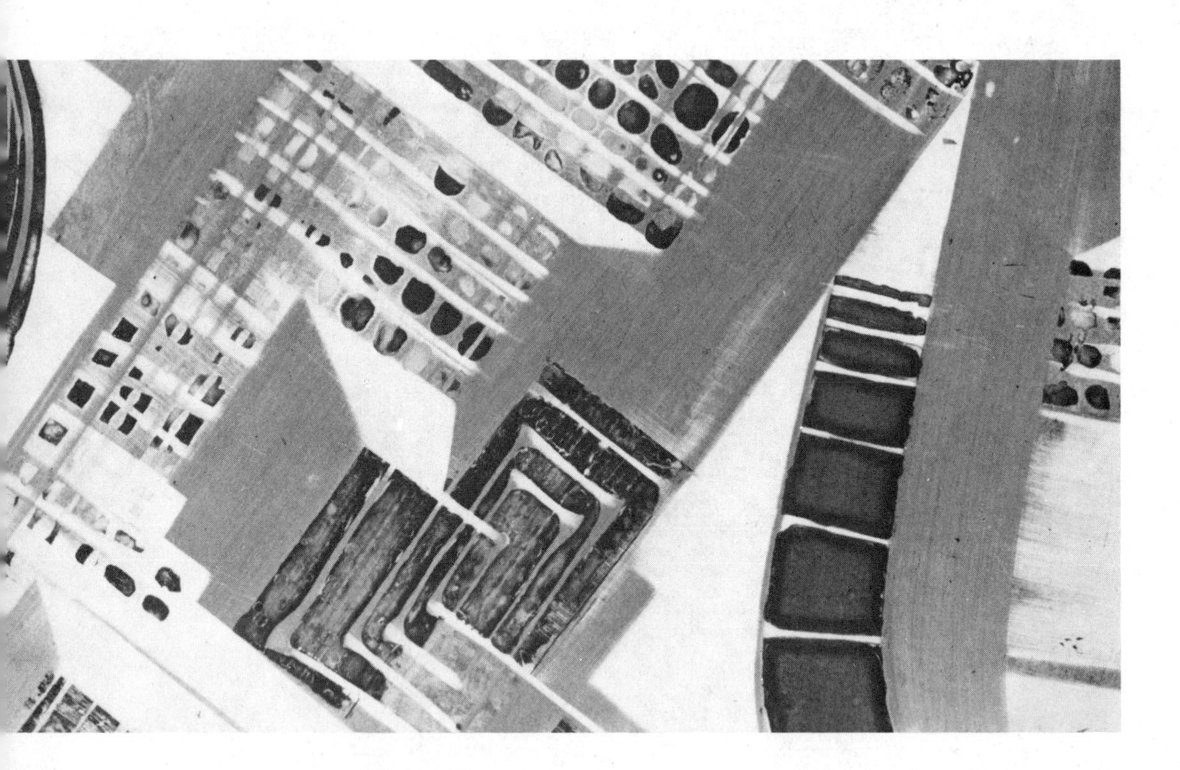

15

Differential Calculus of Functions of Several Variables

15.1 Partial Derivatives

In our previous work the idea of the rate of change of a function led to the concept of a derivative. However, if unmodified, the concept of rate of change has little application to functions of two or more variables. Consider, for example, the function $f(x, y) = x + 2y$, a portion of whose graph is sketched in Figure 15.1.1. With a little reflection it should be clear that the rate of change of $f(x, y)$ as the point (x, y) moves parallel to the y-axis is quite different from the rate of change as (x, y) moves parallel to the x-axis. Thus the rate of change of a function of two variables x and y depends on the "direction of motion" of the point (x, y). Hence, when asking for the rate of change of $f(x, y)$ at a point (x_0, y_0), it is also necessary to specify a direction. When the direction of motion of (x, y) is parallel to the x-axis or to the y-axis, the concept of a partial derivative arises.

Figure 15.1.1

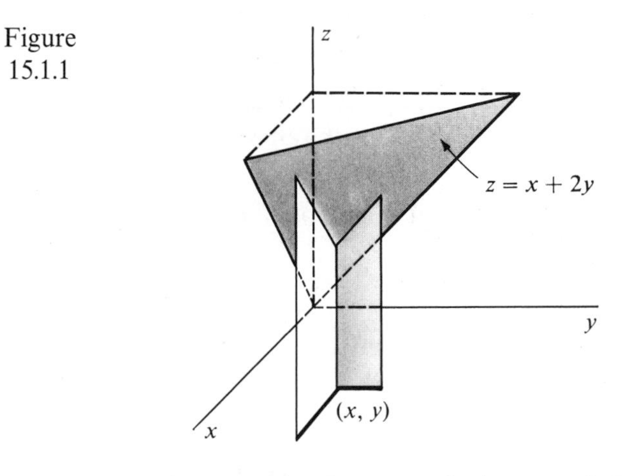

Suppose, for example, that we seek the rate of change of a function f as the point (x_0, y_0) is approached in a direction parallel to the x-axis. As pictured in Figure 15.1.2, we consider the limit of the slope of the line passing through the points $(x_0, y_0, f(x_0, y_0))$ and $(x_0 + h, y_0, f(x_0 + h, y_0))$. That is, we consider

$$\lim_{h \to 0} \frac{f(x_0 + h, y_0) - f(x_0, y_0)}{h}.$$

Since that limit is quite important we make the following definition.

Definition 15.1.1

The *partial derivative of* $f(x, y)$ *with respect to* x *at* (x_0, y_0) is denoted by $f_x(x_0, y_0)$ and is defined as

$$f_x(x_0, y_0) = \lim_{h \to 0} \frac{f(x_0 + h, y_0) - f(x_0, y_0)}{h}.$$

Figure 15.1.2

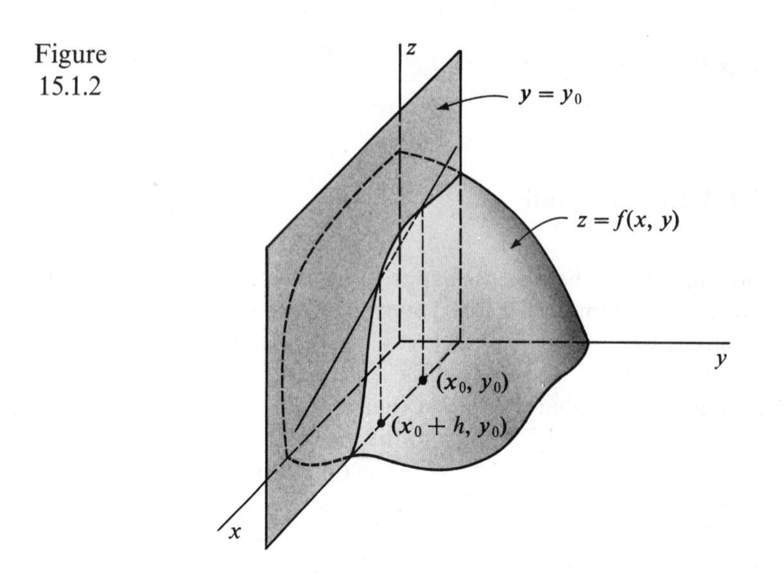

Similarly, if the direction of approach is parallel to the y-axis, we get the concept of a partial derivative with respect to y.

Definition 15.1.2

The *partial derivative of* $f(x, y)$ *with respect to* y *at* (x_0, y_0) is denoted by $f_y(x_0, y_0)$ and is defined as

$$f_y(x_0, y_0) = \lim_{h \to 0} \frac{f(x_0, y_0 + h) - f(x_0, y_0)}{h}.$$

Fortunately, the calculation of partial derivatives is quite simple. Consider the function $f(x, y_0)$. Since y_0 is a constant, $f(x, y_0)$ is a function of x alone. By definition of the derivative we have

$$\left.\frac{df(x, y_0)}{dx}\right|_{x=x_0} = \lim_{h \to 0} \frac{f(x_0 + h, y_0) - f(x_0, y_0)}{h}.$$

Comparison of that equation with Definition 15.1.1 yields

$$f_x(x_0, y_0) = \left.\frac{df(x, y_0)}{dx}\right|_{x=x_0}.$$

Hence, to find $f_x(x, y)$ we apply the usual rules of differentiation but treat y as a constant. Similarly, when calculating $f_y(x, y)$, we treat x as a constant and differentiate $f(x, y)$ with respect to y.

Example 1 | Find $f_x(x, y)$ and $f_y(x, y)$, if $f(x, y) = e^x \sin y + 2xy + y$.

Differentiation of $f(x, y)$, treating y as a constant, yields

$$f_x(x, y) = e^x \sin y + 2y.$$

Similarly, treating x as a constant and differentiating with respect to y, we have

$$f_y(x, y) = e^x \cos y + 2x + 1. \qquad \|$$

Other notations similar to the Leibniz notation for the ordinary derivative are often used. That is, the symbol $\partial f/\partial x$ is often used instead of $f_x(x, y)$, and the symbol $\partial f/\partial y$ is used instead of $f_y(x, y)$.

Since $f_x(x, y)$ and $f_y(x, y)$ are also functions of two variables, we may compute their partial derivatives. The corresponding notation for such *second order* partial derivatives is indicated below.

$$\frac{\partial}{\partial y}\left(\frac{\partial f}{\partial x}\right) = f_{xy}(x, y) = \frac{\partial^2 f}{\partial y\, \partial x}$$

$$\frac{\partial}{\partial x}\left(\frac{\partial f}{\partial y}\right) = f_{yx}(x, y) = \frac{\partial^2 f}{\partial x\, \partial y}$$

$$\frac{\partial}{\partial x}\left(\frac{\partial f}{\partial x}\right) = f_{xx}(x, y) = \frac{\partial^2 f}{\partial x^2}$$

$$\frac{\partial}{\partial y}\left(\frac{\partial f}{\partial y}\right) = f_{yy}(x, y) = \frac{\partial^2 f}{\partial y^2}$$

Example 2 Compute all four second order partial derivatives of

$$f(x, y) = xy + \sin xy + x^2 + 5x \ln y.$$

Since

$$f_x(x, y) = y + y \cos xy + 2x + 5 \ln y,$$

$$f_y(x, y) = x + x \cos xy + \frac{5x}{y},$$

we have

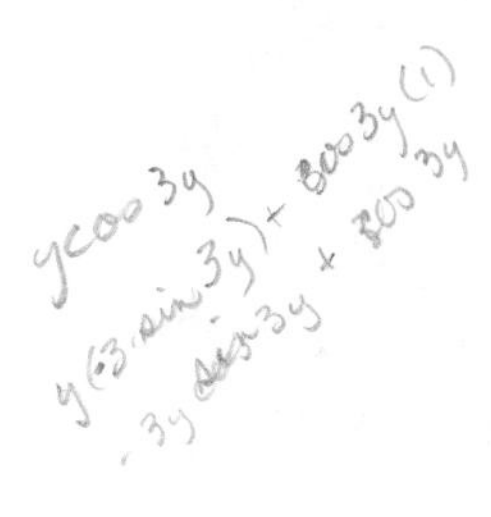

$$f_{xx}(x, y) = \frac{\partial}{\partial x}(f_x(x, y)) = -y^2 \sin xy + 2,$$

$$f_{xy}(x, y) = \frac{\partial}{\partial y}(f_x(x, y)) = 1 + \cos xy - xy \sin xy + \frac{5}{y},$$

$$f_{yy}(x, y) = \frac{\partial}{\partial y}(f_y(x, y)) = -x^2 \sin xy - \frac{5x}{y^2},$$

$$f_{yx}(x, y) = \frac{\partial}{\partial x}(f_y(x, y)) = 1 + \cos xy - xy \sin xy + \frac{5}{y}. \qquad \|$$

Of course one can easily extend the notation to higher order partial derivatives, for example,

$$\frac{\partial^3 f}{\partial y\, \partial y\, \partial x} = f_{xyy}(x, y) = \frac{\partial}{\partial y}\left(\frac{\partial}{\partial y}\left(\frac{\partial f}{\partial x}\right)\right).$$

There is a simple extension of the idea of partial derivatives to functions of more than two variables.

Definition 15.1.3

> The partial derivative of $f: R^n \to R$ with respect to x_i at the point $(a_1, a_2, \ldots, a_n)$ is denoted by $f_{x_i}(a_1, a_2, \ldots, a_n)$ and is defined as
>
> $$f_{x_i}(a_1, a_2, \ldots, a_n) = \lim_{h \to 0} \frac{f(a_1, a_2, \ldots, a_i + h, \ldots, a_n) - f(a_1, a_2, \ldots, a_i, \ldots, a_n)}{h}.$$

The notation $\partial f/\partial x_i$ is also used for f_{x_i} and the calculations are very similar to the previous computations. When calculating f_{x_i} we simply differentiate with respect to x_i, while treating all remaining variables as constants.

Example 3 Compute $f_{x_3}(4, 2, -1, 7)$ if

$$f(x_1, x_2, x_3, x_4) = x_1x_2 + x_2x_3 + x_3x_4.$$

$$f_{x_3}(x_1, x_2, x_3, x_4) = x_2 + x_4, \quad \text{and so}$$

$$f_{x_3}(4, 2, -1, 7) = 2 + 7 = 9.$$

||

The reader has no doubt noticed in Example 2 that $f_{xy}(x, y) = f_{yx}(x, y)$. That is not accidental, as the next theorem indicates. The proof of the theorem appears in many advanced calculus texts.

Theorem 15.1.1

> If $f_{xy}(x, y)$ and $f_{yx}(x, y)$ are continuous in some small disc centered at the point (x_0, y_0), then
>
> $$f_{xy}(x_0, y_0) = f_{yx}(x_0, y_0).$$

Exercises 15.1

In Exercises 1–18, find (a) $\dfrac{\partial f}{\partial x}$, (b) $\dfrac{\partial f}{\partial y}$, (c) $\dfrac{\partial^2 f}{\partial x^2}$, (d) $\dfrac{\partial^2 f}{\partial y^2}$, (e) $\dfrac{\partial^2 f}{\partial x\, \partial y}$.

1. $f(x, y) = 2x + 3y + 5$.
2. $f(x, y) = 6x^2 + 3y^3 + 5$.
3. $f(x, y) = x^2y + 2xy + 1$.
4. $f(x, y) = x^3y^2 + xy + 1$.
5. $f(x, y) = (x^2 + y^2)^{1/2}$.
6. $f(x, y) = (x^2 - y^2)/(x^2 + y^2)$.
7. $f(x, y) = \sin(x + y) + 2x$.
8. $f(x, y) = (x + y)^{10} + 3x$.
9. $f(x, y) = \ln|xy| + \sin x$.
10. $f(x, y) = \sin(xy) + \ln|x|$.
11. $f(x, y) = x \tan(xy)$.
12. $f(x, y) = x \cos y + y \cos x$.
13. $f(x, y) = \arctan(y/x)$.
14. $f(x, y) = e^{x+y} \cos(x - y)$.
15. $f(x, y, z) = xyz + e^{xz} + \arcsin(xy)$.
16. $f(x, y, z) = e^{xy} + xy^2 + \arcsin(yz)^2$.
17. $f(x, y, z) = (x + yz)^4 + \sin(xyz)$.
18. $f(x, y, z) = (x^2 + y^2)^{-1/2} + 3x + zx$.

In Exercises 19–22, compute the slope of the curve formed by the intersection of the graph of the given equation with a vertical plane that is parallel to the y-axis and passes through the given point.

19. $z = xe^y$, $(1, 1)$.
20. $z = x \sin xy$, $(0, 1)$.
21. $z = \arctan(xy)$, $(1, 1)$.
22. $z = e^{xy} \sin x$, $(0, 1)$.
23. Compute f_{x_i} if $f: R^n \to R$ is given by

$$f(x_1, x_2, \ldots, x_n) = x_1 + x_2{}^2 + x_3{}^3 + \cdots + x_n{}^n.$$

24. Compute f_{x_i} if $f: R^n \to R$ is given by $f(\mathbf{X}) = \mathbf{X} \cdot \mathbf{X}$.
25. Compute $f_{x_i x_j}$ if $f: R^n \to R$ is given by $f(\mathbf{X}) = \mathbf{X} \cdot \mathbf{X}$.
26. Compute $f_{x_i x_i}$ if $f: R^n \to R$ is given by $f(\mathbf{X}) = \mathbf{X} \cdot \mathbf{X}$.
27. Find (a) $f_x(x, y)$ and (b) $f_y(x, y)$ if $f(x, y) = x^y$.
28. Find (a) $f_x(x, y)$ and (b) $f_y(x, y)$ if $f(x, y) = x^{y+2x}$.
29. Let $F(x, y) = f(x - ay) + g(x + ay)$ and suppose that the second partial derivatives exist. Show that $F_{yy}(x, y) = a^2 F_{xx}(x, y)$ for all points (x, y).

15.2 Chain Rules

In our previous discussion of functions of a single variable we found that the chain rule afforded a simple method of calculating the derivative of a composite function. The situation concerning functions of several variables is somewhat more complicated, but our aim is the same: differentiation of composite functions.

Before attempting to derive a chain rule we shall need to prove the analog of The Mean Value Theorem for functions of two variables. Let $f(x, y)$ be a function of two variables. If we fix $y = y_0$ and let $F(x) = f(x, y_0)$, then F is a function of the single variable x. So if F has a derivative for all x such that $a \le x \le b$, we can use The Mean Value Theorem (Theorem 3.6.2) to obtain

$$\frac{F(b) - F(a)}{b - a} = F'(c), \tag{1}$$

where $a < c < b$. Since $f_x(x, y_0)$ is obtained by differentiating $f(x, y)$ with respect to x while holding y constant at y_0, we have

$$F'(c) = f_x(c, y_0).$$

Then substitution into (1) produces this theorem:

Theorem 15.2.1

If $f_x(x, y_0)$ exists for all x such that $a \le x \le b$, then for some c such that $a < c < b$,

$$\frac{f(b, y_0) - f(a, y_0)}{b - a} = f_x(c, y_0).$$

Now suppose that $x = g(t)$, $y = h(t)$, and $z = f(x, y)$, as illustrated in Figure 15.2.1. Since z is actually a function of the single variable t, that is, $z = f(g(t), h(t))$,

Figure 15.2.1

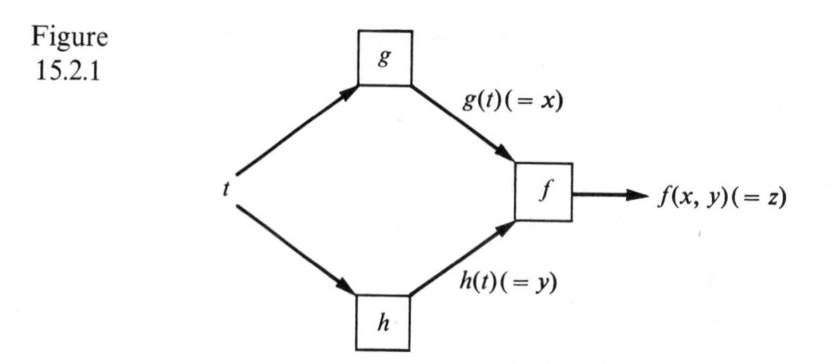

it is reasonable to attempt to find df/dt in terms of dg/dt, dh/dt, $\partial f/\partial x$, and $\partial f/\partial y$. By definition

$$\frac{df}{dt} = \lim_{k\to 0} \frac{f(g(t+k), h(t+k)) - f(g(t), h(t))}{k}$$

$$= \lim_{k\to 0} \left[\frac{f(g(t+k), h(t+k)) - f(g(t), h(t+k)) + f(g(t), h(t+k)) - f(g(t), h(t))}{k}\right].$$

Application of the sum rule for limits, gives

$$\frac{df}{dt} = \lim_{k\to 0} \frac{f(g(t+k), h(t+k)) - f(g(t), h(t+k))}{k} + \lim_{k\to 0} \frac{f(g(t), h(t+k)) - f(g(t), h(t))}{k}, \tag{2}$$

if both limits exist.

For the moment we will concentrate our efforts on the first of those two limits. If $g(t+k) - g(t) \neq 0$, we may write the first limit as

$$\lim_{k\to 0} \left[\frac{f(g(t+k), h(t+k)) - f(g(t), h(t+k))}{g(t+k) - g(t)} \cdot \frac{g(t+k) - g(t)}{k}\right], \tag{3}$$

or, assuming that the separate limits exist,

$$\lim_{k\to 0} \frac{f(g(t+k), h(t+k)) - f(g(t), h(t+k))}{g(t+k) - g(t)} \cdot \lim_{k\to 0} \frac{g(t+k) - g(t)}{k}.$$

Since the second of those two limits is readily seen to be $g'(t)$, we concentrate on

$$\lim_{k\to 0} \frac{f(g(t+k), h(t+k)) - f(g(t), h(t+k))}{g(t+k) - g(t)}.$$

Applying Theorem 15.2.1, with $y_0 = h(t+k)$, $b = g(t+k)$, and $a = g(t)$ we have

$$\frac{f(g(t+k), h(t+k)) - f(g(t), h(t+k))}{g(t+k) - g(t)} = f_x(c, h(t+k)),$$

where c is between $g(t+k)$ and $g(t)$. Consequently

$$\lim_{k\to 0} \frac{f(g(t+k), h(t+k)) - f(g(t), h(t+k))}{g(t+k) - g(t)} = \lim_{k\to 0} f_x(c, h(t+k)).$$

Since we have assumed that g is a differentiable function, g must also be continuous. Consequently, since c is trapped between $g(t + k)$ and $g(t)$, c must approach $g(t)$ as k approaches 0. In addition, the assumption that h is continuous causes $h(t + k)$ to approach $h(t)$ as k approaches 0. Thus, if f_x is continuous at $(g(t), h(t))$, we have

$$\lim_{k\to 0} \frac{f(g(t + k), h(t + k)) - f(g(t), h(t + k))}{g(t + k) - g(t)} = f_x(g(t), h(t)).$$

The substitution of those results into (3) then produces

$$\lim_{k\to 0} \frac{f(g(t + k), h(t + k)) - f(g(t), h(t + k))}{k} = f_x(g(t), h(t))g'(t).$$

The first of the two limits appearing in (2) has been evaluated.

The evaluation of the second limit appearing in (2) is an easier task. We note that if $h(t + k) - h(t)$ is not 0, we have

$$\lim_{k\to 0} \frac{f(g(t), h(t + k)) - f(g(t), h(t))}{k} = \lim_{k\to 0} \frac{f(g(t), h(t + k)) - f(g(t), h(t))}{h(t + k) - h(t)} \cdot \frac{h(t + k) - h(t)}{k}.$$

Then, if both limits exist,

$$\lim_{k\to 0} \frac{f(g(t), h(t + k)) - f(g(t), h(t))}{k}$$

$$= \lim_{k\to 0} \frac{f(g(t), h(t + k)) - f(g(t), h(t))}{h(t + k) - h(t)} \cdot \lim_{k\to 0} \frac{h(t + k) - h(t)}{k}. \tag{4}$$

Let $m = h(t + k) - h(t)$. Then, since h is a continuous function,

$$\lim_{k\to 0} m = \lim_{k\to 0} h(t + k) - \lim_{k\to 0} h(t) = h(t) - h(t) = 0.$$

Substitution into (4) gives

$$\lim_{k\to 0} \frac{f(g(t), h(t + k)) - f(g(t), h(t))}{k}$$

$$= \lim_{m\to 0} \frac{f(g(t), h(t) + m) - f(g(t), h(t))}{m} \cdot \lim_{k\to 0} \frac{h(t + k) - h(t)}{k}.$$

Thus

$$\lim_{k\to 0} \frac{f(g(t), h(t + k)) - f(g(t), h(t))}{k} = f_y(g(t), h(t))h'(t),$$

and the second of the two limits appearing in (2) has been evaluated. Consequently we have

$$\frac{df}{dt} = f_x(g(t), h(t))g'(t) + f_y(g(t), h(t))h'(t).$$

Slightly more sophisticated techniques would allow us to omit the assumptions that $h(t + k) - h(t) \neq 0$ and $g(t + k) - g(t) \neq 0$. Additional work would also permit us to omit the hypothesis that $f_x(x, y)$ is continuous. This *chain rule* is of sufficient importance that we shall make formal note of the result and the hypothesis that leads to it.

Theorem 15.2.2

> If both $x = g(t)$ and $y = h(t)$ are differentiable at t and $f_x(g(t), h(t))$ and $f_y(g(t), h(t))$ both exist, then
> $$\frac{df(x, y)}{dt} = \frac{df}{dt} = f_x(g(t), h(t))g'(t) + f_y(g(t), h(t))h'(t).$$

Since $x = g(t)$ and $y = h(t)$, we can state that result in a more easily remembered form:

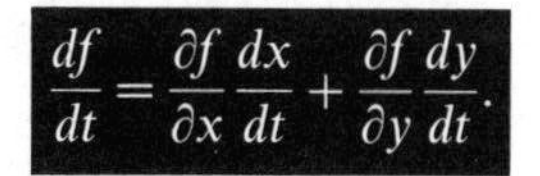

$$\frac{df}{dt} = \frac{\partial f}{\partial x}\frac{dx}{dt} + \frac{\partial f}{\partial y}\frac{dy}{dt}.$$

Example 1 | Find df/dt if $f(x, y) = x \sin y$, where $x = t^2 + 2t + 1$ and $y = \ln t$.

Substitution into

$$\frac{df}{dt} = \frac{\partial f}{\partial x}\frac{dx}{dt} + \frac{\partial f}{\partial y}\frac{dy}{dt}$$

yields

$$\frac{df}{dt} = (\sin y)(2t + 2) + (x \cos y)\left(\frac{1}{t}\right)$$

$$= 2(t + 1) \sin y + \frac{x}{t} \cos y.$$

Of course if we want that result in terms of t alone, we substitute $t^2 + 2t + 1$ and $\ln t$ for x and y, respectively, to get

$$\frac{df}{dt} = 2(t + 1) \sin(\ln t) + \frac{t^2 + 2t + 1}{t} \cos(\ln t).$$ ||

Example 2 | A particle moves in a circular motion in the plane in such a way that its position is given by $x = \sin t$ and $y = \cos t$, for any time t. A force of magnitude $f(x, y) = x^2 + y^2 + 2xy$ is exerted on the particle at the point (x, y). Find an expression for the rate of change of the magnitude of the force exerted on the particle with respect to time when $t = 1$.

Since

$$\frac{df}{dt} = \frac{\partial f}{\partial x}\frac{dx}{dt} + \frac{\partial f}{\partial y}\frac{dy}{dt},$$

we have

$$\frac{df}{dt} = 2(x + y) \cos t - 2(x + y) \sin t.$$

Then, since $x = \sin 1$ and $y = \cos 1$ when $t = 1$, we have, if $t = 1$,

$$\frac{df}{dt} = 2(\sin 1 + \cos 1) \cos 1 - 2(\sin 1 + \cos 1) \sin 1$$

$$= 2(\cos^2 1 - \sin^2 1).$$ ||

Often full knowledge of the functions involved is unnecessary, as the next example indicates.

Example 3 Suppose that a flat metal sheet is heated so that its temperature in degrees at the point (x, y) is given by

$$T(x, y) = 20x^2 + 2xy - 2y - 8x + 30.$$

A temperature sensing device is moved over the plate in such a way that it moves through the origin with a velocity having x-component 2 ft/min and y-component 1 ft/min. Find the rate of change of the temperature recorded by the sensing device as it passes through the origin.

Note that although x and y are functions of time, the specific functions are not given. However, we do know that $dx/dt = 2$ ft/min and $dy/dt = 1$ ft/min when $x = y = 0$.

Then, since
$$\frac{dT}{dt} = \frac{\partial T}{\partial x}\frac{dx}{dt} + \frac{\partial T}{\partial y}\frac{dy}{dt},$$

we have
$$\frac{dT}{dt} = (40x + 2y - 8)(2) + (2x - 2)(1).$$

Consequently, when $x = y = 0$, we have $dT/dt = -18$ deg/min. ||

There is an additional more general form of the chain rule that is often useful. Consider a function f of variables $y_1, y_2, \ldots, y_m$, say $z = f(y_1, y_2, \ldots, y_m)$. Also suppose that each of the y's is in turn a function of the variables $x_1, x_2, \ldots, x_n$, say $y_i = g_i(x_1, \ldots, x_n)$ for $i = 1, 2, \ldots, m$. The diagram of Figure 15.2.2 illustrates the situation. Then, in a manner similar to the proof of Theorem 15.2.2, it can be proved that

$$\frac{\partial f}{\partial x_1} = \frac{\partial f}{\partial y_1}\frac{\partial y_1}{\partial x_1} + \frac{\partial f}{\partial y_2}\frac{\partial y_2}{\partial x_1} + \cdots + \frac{\partial f}{\partial y_m}\frac{\partial y_m}{\partial x_1},$$

$$\frac{\partial f}{\partial x_2} = \frac{\partial f}{\partial y_1}\frac{\partial y_1}{\partial x_2} + \frac{\partial f}{\partial y_2}\frac{\partial y_2}{\partial x_2} + \cdots + \frac{\partial f}{\partial y_m}\frac{\partial y_m}{\partial x_2}.$$

Figure 15.2.2

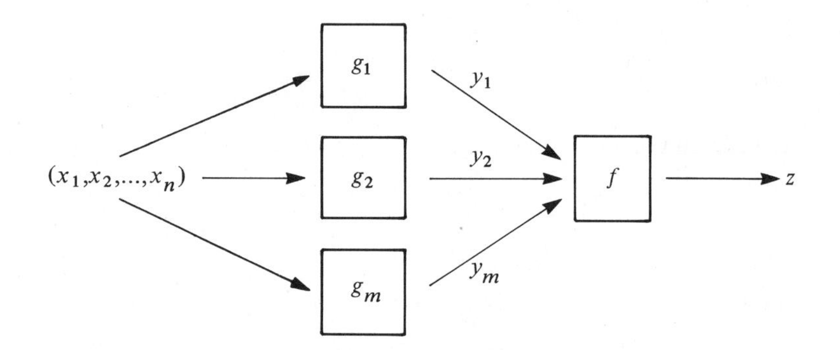

In general,

$$\frac{\partial f}{\partial x_i} = \frac{\partial f}{\partial y_1}\frac{\partial y_1}{\partial x_i} + \frac{\partial f}{\partial y_2}\frac{\partial y_2}{\partial x_i} + \cdots + \frac{\partial f}{\partial y_m}\frac{\partial y_m}{\partial x_i}, \qquad i = 1, 2, \ldots, n. \tag{5}$$

Example 4 Find $\dfrac{\partial f}{\partial x_1}$ if $f(y_1, y_2, y_3) = y_3 + 2y_1 \sin y_2$ and $y_1 = x_1 \sin x_2$, $y_2 = 2 \tan x_1 + x_3$, and $y_3 = {x_1}^2 + 2x_2 - 3{x_3}^{-1}$.

In this case

$$\frac{\partial f}{\partial x_1} = \frac{\partial f}{\partial y_1}\frac{\partial y_1}{\partial x_1} + \frac{\partial f}{\partial y_2}\frac{\partial y_2}{\partial x_1} + \frac{\partial f}{\partial y_3}\frac{\partial y_3}{\partial x_1}. \tag{6}$$

Then

$$\frac{\partial f}{\partial y_1} = 2 \sin y_2, \qquad \frac{\partial f}{\partial y_2} = 2y_1 \cos y_2, \qquad \frac{\partial f}{\partial y_3} = 1,$$

$$\frac{\partial y_1}{\partial x_1} = \sin x_2, \qquad \frac{\partial y_2}{\partial x_1} = 2 \sec^2 x_1, \qquad \frac{\partial y_3}{\partial x_1} = 2x_1.$$

Substitution into (6) gives

$$\frac{\partial f}{\partial x_1} = 2 \sin y_2 \sin x_2 + 4y_1 \cos y_2 \sec^2 x_1 + 2x_1. \qquad \|$$

In many cases, equation (5) can be simplified. Suppose for example that $z = f(x, y)$, where $x = g(u, v)$ and $y = h(u, v)$. Then

$$\frac{\partial f}{\partial u} = \frac{\partial f}{\partial x}\frac{\partial x}{\partial u} + \frac{\partial f}{\partial y}\frac{\partial y}{\partial u},$$

$$\frac{\partial f}{\partial v} = \frac{\partial f}{\partial x}\frac{\partial x}{\partial v} + \frac{\partial f}{\partial y}\frac{\partial y}{\partial v}.$$

Example 5 Find both $\partial f/\partial u$ and $\partial f/\partial v$ if $f(x, y) = x^2 + y$, $x = u \sin v$, and $y = uv$.

Since

$$\frac{\partial f}{\partial u} = \frac{\partial f}{\partial x}\frac{\partial x}{\partial u} + \frac{\partial f}{\partial y}\frac{\partial y}{\partial u},$$

we have, in this particular case,

$$\frac{\partial f}{\partial u} = 2x \sin v + v.$$

Similarly,

$$\frac{\partial f}{\partial v} = 2xu \cos v + u. \qquad \|$$

Exercises 15.2

In Exercises 1–12, use a chain rule to find df/dt. When a value of t is given, find df/dt for that value of t.

1. $f(x, y) = x + 2x^2y, \quad x = 3t + 5, \quad y = t^2, \quad t = 1.$
2. $f(x, y) = \sin x + \sin y, \quad x = t, \quad y = t^2, \quad t = 1.$
3. $f(x, y) = y \ln|x|, \quad x = \cos t, \quad y = t, \quad t = \pi.$
4. $f(x, y) = \tan(xy), \quad x = 3t^2 + 1, \quad y = t^3.$
5. $f(x, y) = x \arctan y, \quad x = \ln t, \quad y = \sec t.$
6. $f(x, y) = \arcsin y, \quad x = \sec^3 t, \quad y = t, \quad t = 1.$
7. $f(x, y) = e^x + e^y, \quad x = t, \quad y = t.$
8. $f(x, y) = xe^{xy}, \quad x = \sin t, \quad y = \cos t.$
9. $f(x, y, z) = x^2 + xy^2 + z, \quad x = \sin t, \quad y = \cos t, \quad z = at.$
10. $f(x, y, z) = \ln(xyz), \quad x = e^t, \quad y = e^{2t}, \quad z = e^{3t}, \quad t = 0.$
11. $f(\mathbf{X}) = |\mathbf{X}|^2, \quad \mathbf{X} = t(1, 1, \dots, 1).$
12. $f(\mathbf{X}) = |\mathbf{X}|^2, \quad \mathbf{X} = (1, t, t^2, \dots, t^n).$
13. Use a chain rule to compute dz/dt, where $z = \arctan(x/y)$, $y = \cos t$, $x = \sin t$. By substituting for x and y, simplify the expression for dz/dt. How do you explain the result?

In Exercises 14–21, use a chain rule to find $\partial f/\partial u$ and $\partial f/\partial v$. Where values of u and v are given, also compute the partial derivatives at the specified point.

14. $f(x, y) = x + y^2$, where $x = u + v$, $y = \ln u$, $u = v = 1$.
15. $f(x, y) = x^3y^5$, where $x = u - v$, $y = u + v$, $u = 0$, $v = 1$.
16. $f(x, y) = x^2 + y^2$, where $x = u \cos v$, $y = u \sin v$, $v = \pi/2$, $u = 2$.
17. $f(x, y) = \sin(xy)$, where $x = uv$ and $y = u$.
18. $f(x, y, z) = z \sin x + z \cos y$, where $x = u^2 + v$, $y = v^3$, $z = uv^2$.
19. $f(x, y, z) = xyz$, where $x = \sin u$, $y = v$, $z = u + v$, $u = \pi/2$, $v = 0$.
20. $f(x, y, z) = xz \ln|y|$, where $x = uv$, $y = e^u$, $z = ue^v$, $u = v = 1$.
21. $f(x, y, z) = x^2 + y^2 + z^2$, where $x = \cos u$, $y = \sin u$, $z = uv$.
22. Find $\dfrac{\partial f}{\partial u}, \dfrac{\partial f}{\partial v}$, and $\dfrac{\partial f}{\partial w}$ if $f(x, y, z) = xy + xz - z^2$, where $x = \sin u$, $y = u^2 + 3u - w$, and $z = u \tan v$.
23. Find $\dfrac{\partial f}{\partial \rho}, \dfrac{\partial f}{\partial \phi}$, and $\dfrac{\partial f}{\partial \theta}$ if $f(x, y, z) = x^2 + y^2 + z^2$ and $x = \rho \sin \phi \cos \theta$, $y = \rho \sin \phi \sin \theta$, and $z = \rho \cos \phi$.
24. A particle is moving in xy-plane in such a way that at time t its position is given by $x = t^2$ and $y = t^3$. A force of magnitude $(x^2 + y^2)^{-1}$ is exerted on the particle at the point (x, y). Find the rate of change of the magnitude of the force on the particle at time $t = 1$.
25. A force of magnitude $(x^2 + y^2)^{-1}$ is exerted on any particle at (x, y). A particle moves through the point $(1, 2)$ in such a way that the x-component of its velocity is 3 units per second and the y-component is -1 unit per second. Find the rate of change of the magnitude of the force as the particle passes through $(1, 2)$.

26. Let $z = f(x, y)$, where $x = r\cos\theta$ and $y = r\sin\theta$. Show that

$$\left(\frac{\partial z}{\partial x}\right)^2 + \left(\frac{\partial z}{\partial y}\right)^2 = \left(\frac{\partial z}{\partial r}\right)^2 + \frac{1}{r^2}\left(\frac{\partial z}{\partial \theta}\right)^2.$$

27. Let $z = f(u, v, w)$, where $u = x - y$, $v = y - z$, and $w = z - x$. Show that

$$\frac{\partial f}{\partial x} + \frac{\partial f}{\partial y} + \frac{\partial f}{\partial z} = 0.$$

28. Let $z = f(x - y, y - x)$. Show that $\dfrac{\partial f}{\partial x} + \dfrac{\partial f}{\partial y} = 0$.

15.3 The Gradient

We begin with the consideration of a simple geometric problem with far-reaching consequences. A line L is said to be perpendicular or *normal* to a surface S at point (a, b, c) if the line passes through (a, b, c) and is perpendicular to the plane tangent to S at (a, b, c). Any vector parallel to L is said to be normal to the surface S at (a, b, c).

Since the line L is perpendicular to the plane tangent to S at (a, b, c), we know that L must be perpendicular to every line in that tangent plane. Consequently, a vector normal to S at (a, b, c) must also be perpendicular to every line in the tangent plane. In particular, a normal vector must be perpendicular to the lines L_1 and L_2 that are tangent to the surface in the planes $x = a$ and $x = b$, respectively. The lines L_1 and L_2 are illustrated in Figure 15.3.1.

Figure 15.3.1

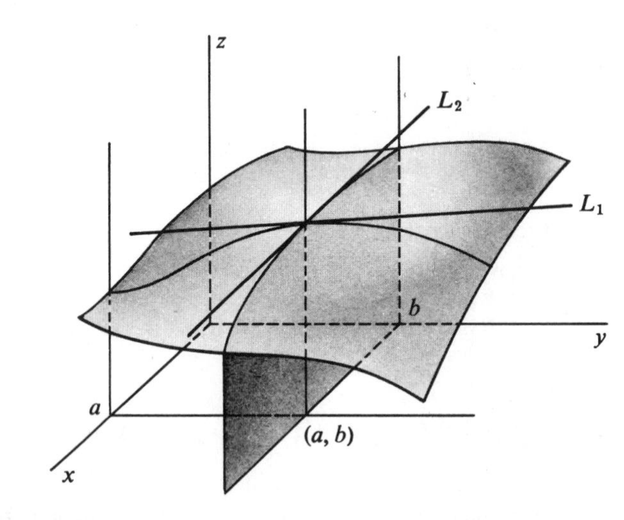

Suppose that the surface under consideration is given by the equation $g(x, y, z) = k$, where k is a constant. It is shown in many advanced calculus texts that if $g_z(a, b, c) \neq 0$, then $g(x, y, z) = k$ determines z as a function of x and y near the point (a, b). We shall assume that $g_z(a, b, c) \neq 0$, and so we have $z = f(x, y)$ for (x, y) sufficiently close to (a, b). Then concentrating on the plane $x = a$, we obtain the situation illustrated in Figure 15.3.2, where an increase of $f_y(a, b)$ corresponds to a unit change in y.

Figure 15.3.2

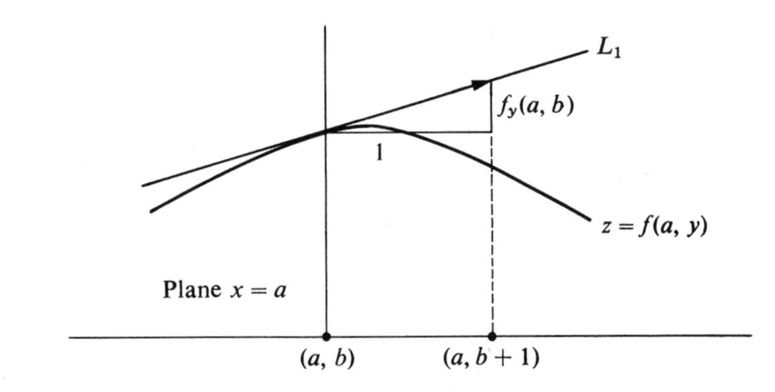

Thus the vector $(0, 1, f_y(a, b))$ is parallel to the line L_1. In a similar fashion we can show that the line L_2 in the plane $y = b$ is parallel to the vector $(1, 0, f_x(a, b))$. It is a simple matter now to find a vector that is perpendicular to the lines L_1 and L_2. We need only find a vector perpendicular to the vectors $(1, 0, f_x(a, b))$ and $(0, 1, f_y(a, b))$. The following vector will serve our purpose:

$$(1, 0, f_x(a, b)) \times (0, 1, f_y(a, b)) = \begin{vmatrix} \mathbf{i} & \mathbf{j} & \mathbf{k} \\ 1 & 0 & f_x(a, b) \\ 0 & 1 & f_y(a, b) \end{vmatrix}$$
$$= (-f_x(a, b), \ -f_y(a, b), 1).$$

Therefore, a vector that is normal to the surface $z = f(x, y)$ at the point on the surface where $x = a$ and $y = b$ is

$$(-f_x(a, b), \ -f_y(a, b), 1). \tag{1}$$

The problem as originally stated, however, gave the surface as $g(x, y, z) = k$, and it was assumed that we could solve for z to obtain $z = f(x, y)$. In general it will be worthwhile to express the normal vector in terms of the function g, which can be accomplished by using a chain rule.

Since $$g(x, y, z) = k \quad \text{and} \quad z = f(x, y),$$

we have

$$g(x, y, f(x, y)) = k. \tag{2}$$

By the chain rule the partial derivative of $g(x, y, f(x, y))$ with respect to x is given by

$$\frac{\partial g(x, y, f(x, y))}{\partial x} = \frac{\partial g(x, y, z)}{\partial x}\frac{\partial x}{\partial x} + \frac{\partial g(x, y, z)}{\partial y}\frac{\partial y}{\partial x} + \frac{\partial g(x, y, z)}{\partial z}\frac{\partial z}{\partial x}.$$

Since x and y are independent of each other, $\dfrac{\partial y}{\partial x} = 0$. Then differentiating both sides of (2) with respect to x we obtain

$$\frac{\partial g}{\partial x} + \frac{\partial g}{\partial z}\frac{\partial f}{\partial x} = 0.$$

Similarly, on differentiation with respect to y, we obtain

$$\frac{\partial g}{\partial y} + \frac{\partial g}{\partial z}\frac{\partial f}{\partial y} = 0.$$

Since we have assumed that $\partial g/\partial z \neq 0$ at (a, b, c), we have

$$\frac{\partial f}{\partial x} = -\frac{\partial g}{\partial x}\bigg/\frac{\partial g}{\partial z} = -\frac{g_x}{g_z}, \quad \text{and}$$

$$\frac{\partial f}{\partial y} = -\frac{\partial g}{\partial y}\bigg/\frac{\partial g}{\partial z} = -\frac{g_y}{g_z}, \quad \text{at } (a, b, c).$$

Substitution of those results into (1) then shows that the vector

$$\left(\frac{g_x(a, b, c)}{g_z(a, b, c)}, \frac{g_y(a, b, c)}{g_z(a, b, c)}, 1\right) \tag{3}$$

is normal to the surface $g(x, y, z) = k$ at the point (a, b, c) where $c = f(a, b)$. Multiplication by the nonzero constant $g_z(a, b, c)$ will produce a vector parallel to the vector given in (3). Hence the vector

$$(g_x(a, b, c), \quad g_y(a, b, c), \quad g_z(a, b, c))$$

is normal to the graph of $g(x, y, z) = k$ at (a, b, c). The negative of that vector is

$$(-g_x(a, b, c), \quad -g_y(a, b, c), \quad -g_z(a, b, c)),$$

which is also normal to the surface but it points in the opposite direction.

The vector-valued function $\nabla g(x, y, z)$ defined by

$$\nabla g(x, y, z) = (g_x(x, y, z), \quad g_y(x, y, z), \quad g_z(x, y, z))$$

is called the *gradient* of $g(x, y, z)$. The symbol ∇ is often read as "del," and the gradient of g is also read as "del g." From the discussion above we have the following result.

Theorem 15.3.1

> The gradient and its negative, $\pm\nabla g(a, b, c)$, are perpendicular to the surface $g(x, y, z) = k$ at the point (a, b, c).

Example 1 Find a unit vector that is normal to the surface $xy + 2xz - 6 = 0$ at the point $(1, 2, 2)$.

The gradient of the function $g(x, y, z) = xy + 2xz - 6$ is

$$\nabla g(x, y, z) = (y + 2z, x, 2x),$$

and so the vector $\nabla g(1, 2, 2) = (6, 1, 2)$ is in the proper direction. To obtain the desired unit vector we need only divide by the length, $\sqrt{41}$, of $(6, 1, 2)$. Consequently $(6, 1, 2)/\sqrt{41}$ is a unit vector that is normal to the given surface at $(1, 2, 2)$. Certainly $-(6, 1, 2)/\sqrt{41}$ is also a unit vector that is normal to the surface at $(1, 2, 2)$. ||

Example 2 Find an equation of the plane that is tangent to the surface $xy + 2xz - 6 = 0$ at the point $(1, 2, 2)$.

Since the vector $\nabla g(1, 2, 2) = (6, 1, 2)$ is normal to the tangent plane, we can use Theorem 14.6.4. Hence, an equation of the tangent plane is

$$6(x - 1) + (y - 2) + 2(z - 2) = 0. \qquad \|$$

For future work it will be useful to find $\sec\theta$, where θ is the acute angle between the xy-plane and the plane tangent to the surface $z = f(x, y)$ at a point (x_0, y_0, z_0). The angle θ is by definition the acute angle between the z-axis and a normal to the tangent plane.

If an equation of the surface is given by $g(x, y, z) = 0$, then $\pm\nabla g(x_0, y_0, z_0)$ are vectors that are normal to the surface at the point (x_0, y_0, z_0). Since $\pm(0, 0, 1)$ are unit vectors in the direction of the z-axis, and since θ is the angle between these two vectors, we have

$$\cos\theta = \pm\frac{\nabla g(x_0, y_0, z_0) \cdot (0, 0, 1)}{|\nabla g(x_0, y_0, z_0)|}. \tag{4}$$

Since

$$\nabla g(x_0, y_0, z_0) = (g_x(x_0, y_0, z_0), g_y(x_0, y_0, z_0), g_z(x_0, y_0, z_0)),$$

we have

$$|\nabla g(x_0, y_0, z_0)| = [g_x^2(x_0, y_0, z_0) + g_y^2(x_0, y_0, z_0) + g_z^2(x_0, y_0, z_0)]^{1/2}$$

and

$$\nabla g(x_0, y_0, z_0) \cdot (0, 0, 1) = g_z(x_0, y_0, z_0).$$

Substitution into (4) gives

$$\cos\theta = \pm\frac{g_z(x_0, y_0, z_0)}{[g_x^2(x_0, y_0, z_0) + g_y^2(x_0, y_0, z_0) + g_z^2(x_0, y_0, z_0)]^{1/2}}.$$

Hence

$$\sec\theta = \frac{1}{\cos\theta} = \pm\frac{[g_x^2(x_0, y_0, z_0) + g_y^2(x_0, y_0, z_0) + g_z^2(x_0, y_0, z_0)]^{1/2}}{g_z(x_0, y_0, z_0)}. \tag{5}$$

In the particular case under discussion, Equation (5) can be simplified somewhat. In this case an equation of the surface is given by

$$z = f(x, y), \quad \text{or}$$

$$z - f(x, y) = 0.$$

So we take

$$g(x, y, z) = z - f(x, y).$$

Thus

$$g_x(x, y, z) = -f_x(x, y)$$

$$g_y(x, y, z) = -f_y(x, y)$$

$$g_z(x, y, z) = 1.$$

Hence Equation (5) becomes

$$\sec\theta = \pm\sqrt{f_x^2(x_0, y_0) + f_y^2(x_0, y_0) + 1}.$$

Then, since θ is an acute angle, $\sec\theta$ is positive, and we must use the plus sign:

$$\sec\theta = \sqrt{f_x^2(x_0, y_0) + f_y^2(x_0, y_0) + 1}.$$

Example 3 | Show that the plane tangent to $f(x, y) = x^2 + xy + y^2 + 3y$ at $(1, -2, -3)$ is parallel to the xy-plane.

The tangent plane is parallel to the xy-plane if and only if $\theta = 0$. That is, if and only if $\sec\theta = 1$.

Since $f_x(x, y) = 2x + y$ and $f_y(x, y) = x + 2y + 3$, we have $f_x(1, -2) = 0$ and $f_y(1, -2) = 0$. Consequently, $\sec\theta = \sqrt{0 + 0 + 1} = 1$, and so the tangent plane is parallel to the xy-plane. ||

The gradient function need not be limited to functions of three variables, but can be extended in an obvious way. Since the extension will be useful in subsequent sections, we make the following definition.

Definition 15.3.1

Let $g: R^n \to R$. Then the *gradient of g* is defined as

$$\nabla g(\mathbf{X}) = \left(\frac{\partial g(\mathbf{X})}{\partial x_1}, \ldots, \frac{\partial g(\mathbf{X})}{\partial x_n}\right).$$

It should be noted that the domain and range of ∇g are subsets of R^n.

Example 4 | Compute ∇g if $g(\mathbf{X}) = \mathbf{X} \cdot \mathbf{X}$, where $\mathbf{X} = (x_1, x_2, \ldots, x_n)$.

$$\begin{aligned} g(\mathbf{X}) &= (x_1, x_2, \ldots, x_n) \cdot (x_1, x_2, \ldots, x_n) \\ &= x_1^2 + x_2^2 + \cdots + x_n^2. \end{aligned}$$

Hence

$$\frac{\partial g}{\partial x_i} = 2x_i, \qquad \text{for all } i,$$

and so

$$\nabla g(\mathbf{X}) = (2x_1, 2x_2, \ldots, 2x_n) = 2\mathbf{X}.$$ ||

The gradient can be used to express the chain rules of the previous section in an easily remembered form. For example, the equation

$$\frac{\partial f}{\partial u} = \frac{\partial f}{\partial x_1}\frac{\partial x_1}{\partial u} + \frac{\partial f}{\partial x_2}\frac{\partial x_2}{\partial u} + \frac{\partial f}{\partial x_3}\frac{\partial x_3}{\partial u}$$

can be written as

$$\frac{\partial f}{\partial u} = \left(\frac{\partial f}{\partial x_1}, \frac{\partial f}{\partial x_2}, \frac{\partial f}{\partial x_3}\right) \cdot \left(\frac{\partial x_1}{\partial u}, \frac{\partial x_2}{\partial u}, \frac{\partial x_3}{\partial u}\right).$$

We let

$$\left(\frac{\partial x_1}{\partial u}, \frac{\partial x_2}{\partial u}, \frac{\partial x_3}{\partial u}\right) = \frac{\partial \mathbf{X}}{\partial u}.$$

Then, since

$$\left(\frac{\partial f}{\partial x_1}, \frac{\partial f}{\partial x_2}, \frac{\partial f}{\partial x_3}\right) = \nabla f,$$

we have
$$\frac{\partial f}{\partial u} = \nabla f \cdot \frac{\partial \mathbf{X}}{\partial u}.$$

Since we have used the notation ∇f for the gradient of a function f, some mention of the symbol ∇ is appropriate. If one views ∇ as the "vector"
$$\nabla = \left(\frac{\partial}{\partial x_1}, \frac{\partial}{\partial x_2}, \ldots, \frac{\partial}{\partial x_n}\right),$$
then quite naturally we have
$$\nabla f = \left(\frac{\partial}{\partial x_1}, \frac{\partial}{\partial x_2}, \ldots, \frac{\partial}{\partial x_n}\right) f = \left(\frac{\partial f}{\partial x_1}, \frac{\partial f}{\partial x_2}, \ldots, \frac{\partial f}{\partial x_n}\right).$$
A word of warning is in order, however. Since ∇ is *not* an ordered n-tuple of real numbers, ∇ is not actually a vector. On the other hand, we shall find that certain formulas are more easily remembered by considering ∇ to behave like a vector.

Exercises 15.3

In Exercises 1–20, compute the gradient of the given function. If a point is specified, compute the gradient at the given point.

1. $f(x, y) = xy + x^2 + y^2$, $(1, -1)$.
2. $f(x, y) = 2xy + x$.
3. $f(x, y) = x + ye^x$.
4. $f(x, y) = x \sin y + e^{xy}$, $(1, 0)$.
5. $f(x, y) = \cos(x - y) + 2x$.
6. $f(x, y) = \ln|xy|$.
7. $f(x, y) = x \tan(xy)$.
8. $f(x, y) = \arctan(y/x)$.
9. $f(x, y, z) = x^2 + y^2 + z^2$.
10. $f(x, y, z) = x \sin(yz)$.
11. $f(x, y, z) = xy + yz$, $(1, 0, -1)$.
12. $f(x, y, z) = xy + \sin yz$.
13. $f(x, y, z) = x \cos xy + ye^z$.
14. $f(x, y, z) = e^{xyz} + xy$, $(1, 2, 0)$.
15. $f(x_1, x_2, x_3, x_4) = x_1x_2x_3x_4$.
16. $f(x_1, x_2, x_3, x_4) = x_3 + x_1x_2 + x_4^2$.
17. $f(x_1, x_2, x_3, x_4) = \ln(x_1x_2x_3x_4)$.
18. $f(x_1, x_2, x_3, x_4) = x_1^2 + x_2^2 + x_3^2 + x_4^2$.
19. $f(\mathbf{X}) = |\mathbf{X}|$.
20. $f(\mathbf{X}) = 1/|\mathbf{X}|$.
21. Find two unit vectors that are normal to the surface $2x^2 + y^2 + z^2 = 4$ at the point $(1, 1, 1)$.
22. Find an equation for the plane that is tangent to the surface $xyz = 12$ at the point $(-2, -1, 6)$.
23. Find an equation for the plane that is tangent to the surface $z = x^2 + y^2$ at the point $(1, 2, 5)$.
24. Find an equation for the plane that is tangent to the surface $z^2 = x^2 + y^2$ at the point $(1, 1, \sqrt{2})$.
25. Let $f(\mathbf{X}) = 1/|\mathbf{X}|$ for $\mathbf{X}$ in R^3. Show that $\nabla f(\mathbf{X}) = -\mathbf{X}/|\mathbf{X}|^3$ and $|\nabla f(\mathbf{X})| = 1/|\mathbf{X}|^2$. (The function f is called the potential function for the gravitational force due to a point mass at the origin.)

15.4 Directional Derivatives

In our study of partial derivatives we noted that the rate of change of a function f at a point (a, b) depends on the direction along which the point (a, b) is approached. When the directions are parallel to the coordinate axes, the concept of partial derivative arises. Many times it is important to consider rates of change of a function in other directions. Such considerations will lead to the concept of directional derivatives.

Suppose we wish to study the rate of change of $f(x, y)$ at the point (a, b) in the direction of the unit vector $\mathbf{U}$ in the xy-plane that makes an angle α, $0 \leq \alpha < 2\pi$ with the positive x-axis, as illustrated in Figure 15.4.1.

Figure 15.4.1

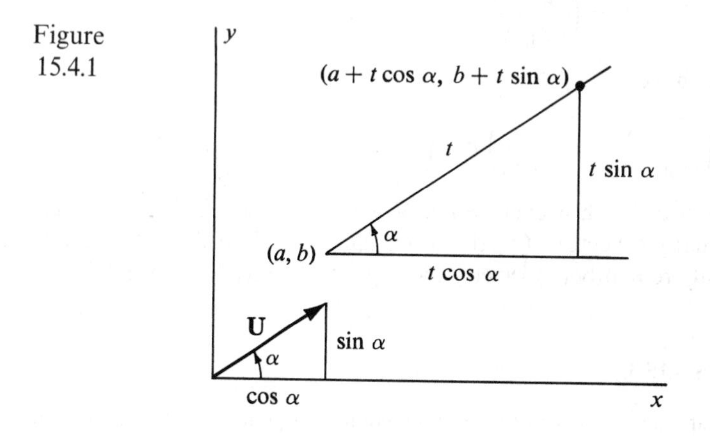

This is equivalent to a study of the rate of change of the function $f(x, y)$ at (a, b), where (x, y) is required to be of the form $(a + t \cos \alpha, b + t \sin \alpha)$ and t is any real number. Thus we are seeking the derivative of $f(a + t \cos \alpha, b + t \sin \alpha)$ with respect to t when $t = 0$, that is, when $(a + t \cos \alpha, b + t \sin \alpha)$ is the point (a, b).

Figure 15.4.2 shows a cross section of the surface $z = f(x, y)$ formed by the plane perpendicular to the xy-plane, through the point (a, b), and parallel to $\mathbf{U}$. It is important to realize that $f(a + t \cos \alpha, b + t \sin \alpha)$ is first differentiated with respect to t, and then t is set equal to 0; not vice versa. This leads to the following definition.

Figure 15.4.2

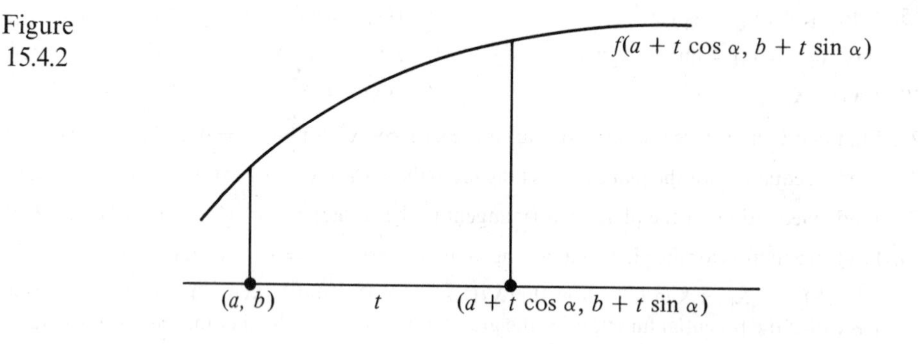

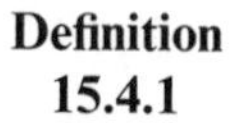

Definition 15.4.1

The directional derivative of $f(x, y)$ at (a, b) in the direction of the unit vector $\mathbf{U}$ that makes an angle α with the positive x-axis is denoted by $D_{(\alpha)}f(a, b)$ and defined as

$$D_{(\alpha)}f(a, b) = \frac{d}{dt} f(a + t \cos \alpha, b + t \sin \alpha), \qquad \text{for } t = 0. \qquad (1)$$

If we wish to emphasize the role of the vector **U**, the notation $D_{\mathbf{U}}\, f(a, b)$ will often be used.

Fortunately the chain rule affords a relatively simple method for calculating directional derivatives. Recall that if $x = g(t)$ and $y = h(t)$ and if both partials of f exist, then

$$\frac{d}{dt} f(g(t), h(t)) = f_x(g(t), h(t))g'(t) + f_y(g(t), h(t))h'(t).$$

In our case

$$x = g(t) = a + t\cos\alpha \quad \text{and} \quad y = h(t) = b + t\sin\alpha.$$

Hence

$$g'(t) = \cos\alpha \quad \text{and} \quad h'(t) = \sin\alpha, \quad \text{and so}$$

$$\frac{d}{dt} f(a + t\cos\alpha, b + t\sin\alpha) = f_x(a + t\cos\alpha, b + t\sin\alpha)\cos\alpha$$
$$+ f_y(a + t\cos\alpha, b + t\sin\alpha)\sin\alpha.$$

Now, setting $t = 0$ and substituting in (1) above, we get

$$D_{\mathbf{U}}\, f(a, b) = D_{(\alpha)} f(a, b) = f_x(a, b)\cos\alpha + f_y(a, b)\sin\alpha. \tag{2}$$

Note that since the unit vector **U** in the direction α can be expressed as

$$\mathbf{U} = (\cos\alpha, \sin\alpha),$$

we can express (2) as

$$D_{\mathbf{U}}\, f(a, b) = D_{(\alpha)} f(a, b) = (f_x(a, b), f_y(a, b)) \cdot \mathbf{U}.$$

Then, using the fact that $\nabla f(a, b) = (f_x(a, b), f_y(a, b))$, we have

$$D_{\mathbf{U}}\, f(a, b) = \nabla f(a, b) \cdot \mathbf{U},$$

which is often written in this compact form:

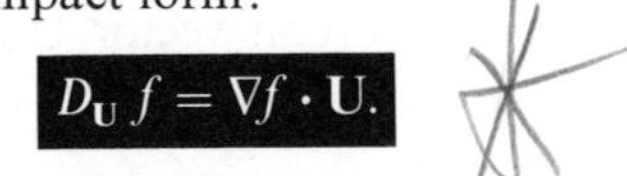

$$D_{\mathbf{U}}\, f = \nabla f \cdot \mathbf{U}.$$

That equation can be extended in a direct way to functions of several variables. If, for example, $f: R^n \to R$ and **U** is a unit vector of R^n, then the directional derivative of f in the direction of **U** is defined by

$$D_{\mathbf{U}}\, f(\mathbf{X}) = \nabla f(\mathbf{X}) \cdot \mathbf{U}.$$

In words, the directional derivative of f in the direction of the unit vector **U** is simply the dot product of the gradient of f and **U**.

Note that if $\mathbf{X}_0$ is a fixed vector, the value of $D_{\mathbf{U}}\, f(\mathbf{X}_0)$ depends only on **U**.

Example 1 | Compute the directional derivative of $f(x, y, z) = xy + xe^y + z^2$ at the point (1, 2, 3) in the direction of the vector $\mathbf{V} = (-1, 3, 2)$.

The unit vector **U** in the direction of **V** is

$$\mathbf{U} = \frac{\mathbf{V}}{|\mathbf{V}|} = \frac{1}{\sqrt{14}}(-1, 3, 2).$$

Thus, since $\nabla f(x, y, z) = (y + e^y, x + xe^y, 2z)$, we have

$$\nabla f(1, 2, 3) = (2 + e^2, 1 + e^2, 6), \quad \text{and so}$$

$$\begin{aligned} D_{\mathbf{U}} f(1, 2, 3) &= (2 + e^2, 1 + e^2, 6) \cdot \frac{1}{\sqrt{14}}(-1, 3, 2) \\ &= \frac{1}{\sqrt{14}}(-2 - e^2 + 3 + 3e^2 + 12) \\ &= \frac{1}{\sqrt{14}}(13 + 2e^2). \end{aligned}$$

||

Example 2 Compute the directional derivative of the function $f(x, y) = xy^2 + yx^2$ at $(1, 2)$ in the direction of a unit vector making an angle α with the positive x-axis.

The unit vector $\mathbf{U}$ making an angle α with the positive x-axis is $\mathbf{U} = (\cos\alpha, \sin\alpha)$. Since

$$\nabla f(x, y) = (y^2 + 2yx, 2xy + x^2),$$

we have

$$\nabla f(1, 2) = (8, 5)$$

and so

$$\begin{aligned} D_{(\alpha)} f(1, 2) &= D_{\mathbf{U}} f(1, 2) \\ &= (8, 5) \cdot \mathbf{U} \\ &= (8, 5) \cdot (\cos\alpha, \sin\alpha) \\ &= 8\cos\alpha + 5\sin\alpha. \end{aligned}$$

||

Example 2 illustrates the fact that the directional derivative of a function at a specified point is a function of the unit vector $\mathbf{U}$ alone. It is of interest in many applications of calculus to find those directions in which a function is increasing or decreasing most rapidly.

The directional derivative of a function $f: R^n \to R$ at $\mathbf{X}_0$ in the direction of the unit vector $\mathbf{U}$ is given by

$$D_{\mathbf{U}} f(\mathbf{X}_0) = \nabla f(\mathbf{X}_0) \cdot \mathbf{U}.$$

Now let θ be the angle between the vectors $\mathbf{U}$ and $\nabla f(\mathbf{X}_0)$. Then

$$\cos\theta = \frac{\nabla f(\mathbf{X}_0) \cdot \mathbf{U}}{|\nabla f(\mathbf{X}_0)|\,|\mathbf{U}|}.$$

But $|\mathbf{U}| = 1$. So $\nabla f(\mathbf{X}_0) \cdot \mathbf{U} = |\nabla f(\mathbf{X}_0)| \cos\theta$. Therefore,

$$D_{\mathbf{U}} f(\mathbf{X}_0) = |\nabla f(\mathbf{X}_0)| \cos\theta.$$

Clearly then the maximum value of $D_{\mathbf{U}} f(\mathbf{X}_0)$ occurs when $\theta = 0$, that is, when $\mathbf{U}$ is in the direction of $\nabla f(\mathbf{X}_0)$. Also, the maximum value is $|\nabla f(\mathbf{X}_0)|$. Consequently the maximum value of $D_{\mathbf{U}} f(\mathbf{X}_0)$ as $\mathbf{U}$ varies is in the direction of $\nabla f(\mathbf{X}_0)$ with magnitude $|\nabla f(\mathbf{X}_0)|$. The gradient of a function is simply a vector whose magnitude is the maximal value of the directional derivative at the point and whose direction is the direction of that maximal directional derivative.

We have presented two interpretations of the gradient. In the first, the gradient is a vector that is normal to the graph of $g(x, y, z) = k$, and in the second, the gradient points in the direction of the maximal directional derivative of $g(x, y, z)$. Actually those two ideas are simply different ways of viewing the same phenomenon. Consider a function g such that $g: R^3 \to R$, say $g(x, y, z) = w$. Since the domain of g is R^3, we cannot sketch a graph of g. However, we can indicate the *level surfaces* of g. That is, we can sketch those surfaces in R^3 where g is a constant. In Figure 15.4.3 we indicate portions of several level surfaces of g. Now suppose that (a, b, c) is on the surface $g(x, y, z) = k_2$, that is, $g(a, b, c) = k_2$. Then $\nabla g(a, b, c)$ is a vector that is normal to the graph of $g(x, y, z) = k_2$ at (a, b, c), and it also points in the direction of most rapid change of the function $g(x, y, z) = w$. Intuitively, to change $g(x, y, z)$ most rapidly we move as rapidly as possible away from a level surface. Consequently we move in a direction that is normal to the level surface.

Figure 15.4.3

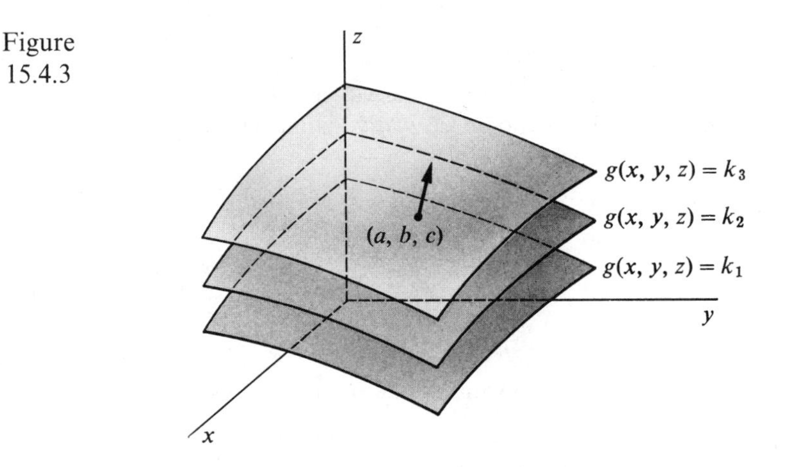

As a specific example, consider the function $g(x, y, z) = x^2 + y^2 + z^2$. As shown in Figure 15.4.4, the level surfaces of the function are spheres centered at the origin. In this case, $\nabla g(0, \sqrt{2}, \sqrt{2}) = (0, 2\sqrt{2}, 2\sqrt{2})$ is a vector that is normal to the graph of the surface

Figure 15.4.4

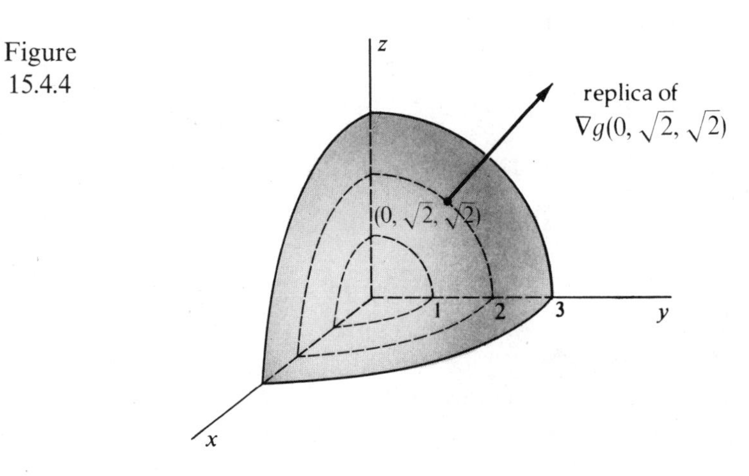

$x^2 + y^2 + z^2 = 1$. Moreover $\nabla g(1, 1, 0)$ points in the direction of the most rapid change of the function $g(x, y, z) = x^2 + y^2 + z^2$.

Care must be exercised to compute the gradient of the correct function, as is illustrated in the next example.

Example 3 Let $f(x, y) = x^2 + y^3$.
(a) Find the direction and magnitude of the maximal directional derivative at (3, 1).
(b) Find a vector that is normal to the graph of $z = x^2 + y^3$ at (3, 1, 10).

(a) The gradient of $f(x, y)$ is a vector in the direction of the maximal rate of change of $f(x, y)$. Moreover the magnitude of the gradient is the maximal rate of change. Since $f(x, y) = x^2 + y^3$,

$$\nabla f(x, y) = (2x, 3y^2).$$

So

$$\nabla f(3, 1) = (6, 3),$$

and the magnitude of the maximal directional derivative at (3, 1) is $\sqrt{6^2 + 3^2} = 3\sqrt{5}$. The direction of the maximal directional derivative is in the direction of the unit vector

$$\mathbf{U} = \frac{(6, 3)}{|(6, 3)|} = \frac{1}{\sqrt{45}}(6, 3) = \frac{1}{\sqrt{5}}(2, 1).$$

(b) Here we seek a vector that is normal to the graph of $z = x^2 + y^3$ at (3, 1, 10). Let

$$g(x, y, z) = x^2 + y^3 - z.$$

Then we know that $\pm\nabla g(3, 1, 10)$ is normal to the surface

$$g(x, y, z) = 0 \quad \text{or} \quad x^2 + y^3 - z = 0$$

at (3, 1, 10). Consequently

$$\pm\nabla g(x, y, z) = \pm(2x, 3y^2, -1), \quad \text{and so}$$

$$\pm\nabla g(3, 1, 10) = \pm(6, 3, -1)$$

are normal to the given surface at (3, 1, 10). ||

Exercises 15.4

In Exercises 1–16, calculate the directional derivative of each function at the given point and in the indicated direction.

1. $f(x, y) = x^2y^4$, (1, 1), $\mathbf{U} = (1/\sqrt{5})(1, 2)$.
2. $f(x, y) = x^2y^4$, (1, 1), $\mathbf{U} = (1/\sqrt{10})(3, -1)$.
3. $f(x, y) = x^2y^3$, (2, 1), $\alpha = \pi/4$.
4. $f(x, y) = x + xy^2$, (1, 1), $\alpha = 2\pi/3$.
5. $f(x, y) = x^2 + 2xy + y^2$, (2, 1), $\mathbf{U} = (1/\sqrt{2})(1, 1)$.
6. $f(x, y) = xe^{xy}$, (1, 2), $\alpha = 3\pi/4$.

7. $f(x, y) = xe^{xy}$, $(1, 2)$, $\alpha = -\pi/4$.
8. $f(x, y) = x^3y^2$, $(-1, 1)$, $\mathbf{U} = (1/2)(\sqrt{3}, 1)$.
9. $f(x, y) = e^x \cos y$, $(0, 0)$, $\alpha = 2\pi/3$.
10. $f(x, y) = \ln(x^2 + y^2)$, $(3, 2)$, $\mathbf{U} = (-1/\sqrt{2})(1, 1)$.
11. $f(x, y) = x^3 + 3xy + 10$, $(0, 1)$, $\alpha = 3\pi/4$.
12. $f(x, y) = x^3 + 3y^2$, $(-1, -2)$, $\alpha = \pi/6$.
13. $f(x, y, z) = xy - yz + x^2y$, $(-1, 1, 2)$, $\mathbf{U} = (1/\sqrt{3})(1, 1, -1)$.
14. $f(x, y, z) = x \cos xy + ye^z$, $(1, \pi/2, 0)$, $\mathbf{U} = (1/\sqrt{3})(-1, -1, 1)$.
15. $f(\mathbf{X}) = \mathbf{X} \cdot \mathbf{X}$, $(1, 1, 1, 1)$, $\mathbf{U} = (1/\sqrt{7})(1, -1, 2, -1)$.
16. $f(\mathbf{X}) = |\mathbf{X}|$, $(0, 0, 0, 1)$, $\mathbf{U} = (0, 0, 1, 0)$.

In Exercises 17–20, find the magnitude and direction of the maximum value of the directional derivative of the indicated functions.

17. $f(x, y) = x^2 + (\sqrt{3}/2)y^2$ at $(1/2, 1)$.
18. $f(x, y) = x^2 + 4y^2 - 2x$ at $(5, 1)$.
19. $f(x, y) = x^2 \sin xy$ at $(0, 1)$.
20. $f(x, y) = e^{xy}$ at $(1, 1)$.
21. If $f(x, y) = xe^{xy}$, (a) find the direction and magnitude of the maximal directional derivative at $(2, 0)$, and (b) find a vector that is normal to the graph of $z = xe^{xy}$ at $(2, 0, 2)$.
22. If $f(x, y) = \sin(xy)$, (a) find the direction and magnitude of the maximal directional derivative at $(2, \pi/12)$, and (b) find a vector that is normal to the graph of $z = \sin(xy)$ at $(2, \pi/12, 1/2)$.
23. Show that $D_{(\mathbf{E}_i)}f(x_1, x_2, \ldots, x_n) = f_{x_i}(x_1, x_2, \ldots, x_n)$ if $\mathbf{E}_i = (0, 0, \ldots, 0, 1, 0, \ldots, 0)$ where the 1 appears in the ith place.
24. Verify that $D_{(\alpha)}f(a, b) = -D_{(\alpha+\pi)}f(a, b)$ and interpret the result geometrically.
25. Let f be symmetric in x and y; that is, let f be a function such that $f(x, y) = f(y, x)$. What are the possible directions of the gradient of f at the point (a, a)?
26. Show that there are at least two unit vectors $\mathbf{U}_1$ and $\mathbf{U}_2$ such that the directional derivatives $D_{(\mathbf{U}_1)}f$ and $D_{(\mathbf{U}_2)}f$ are 0 at (a, b).

15.5 Local Extrema of Functions of Two Variables

Previously we used derivatives of functions of a single variable as an aid in locating local maxima and minima of functions. In this section we shall find that partial derivatives can be used for a similar purpose.

Before we can proceed with finding local extrema of functions of two variables we need a few basic definitions.

Definition 15.5.1

> For each positive number r, the *r-neighborhood* of a point (a, b) is the set of all points (x, y) in the xy-plane that are less than a distance r from (a, b).

Geometrically, an r-neighborhood of (a, b) is simply the disc of radius r centered at (a, b). Note, however, that an r-neighborhood does not include points a distance r from (a, b); that is, the disc does not include its "boundary."

Definition 15.5.2

> A point (a, b) is an *interior point* of a set S if there is some r-neighborhood of (a, b) that is contained in S.

Example 1 Consider the set S consisting of all points (x, y) such that $1 < x \leq 2$ and $0 \leq y < 1$, as illustrated in Figure 15.5.1.

Then, for example, the point (1.5, .75) is an interior point of S since the 1/4-neighborhood of (1.5, .75) is contained in S. In a similar fashion it can be shown that any point (x, y) of S with $x \neq 2$ and $y \neq 0$ is an interior point of S. The points with $x = 1$ or 2, or $y = 0$ or 1, are boundary points of S according to the next definition. ||

Definition 15.5.3

> A point (a, b) is a *boundary point* of a set S if every r-neighborhood of (a, b) contains both points in S and points not in S.

Example 2 Consider the set S consisting of all points (x, y) such that $1 < x \leq 2$ and $0 \leq y < 1$.

That set is pictured in Figure 15.5.1. The point (2, 1/2) is a boundary point of S since every r-neighborhood of (2, 1/2) contains both points in S and outside S. Similarly, the point (1, 1/2) is a boundary point of S even though the point itself is not in S. It is left as an exercise to show that a point of S is either an interior point of S or a boundary point of S. ||

Figure 15.5.1

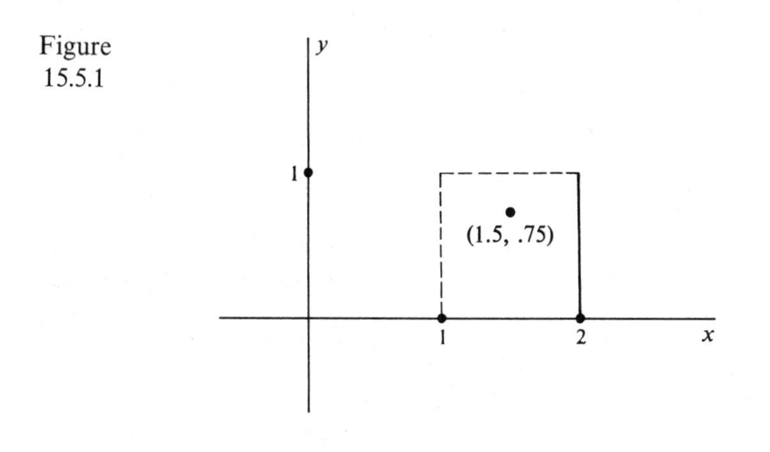

Definition 15.5.4

> A function $f(x, y)$ is said to have a *local maximum* at (a, b) if there is some r-neighborhood N of (a, b) such that $f(a, b) \geq f(x, y)$ for all points (x, y) in N and in the domain of f. If $f(x, y)$ has a local maximum at (a, b), $f(a, b)$ is called a *local maximum value* of $f(x, y)$, and the point $(a, b, f(a, b))$ is called a *local maximum point* of $f(x, y)$.

The definition of a *local minimum* is completely analogous and is left as an exercise.

Definition 15.5.5

A function $f(x, y)$ is said to have an *absolute maximum* (or an *absolute minimum*) at (a, b) if $f(a, b) \geq f(x, y)$ (or $f(a, b) \leq f(x, y)$) for all (x, y) in the domain of f. If $f(x, y)$ has an absolute maximum (absolute minimum) at (a, b), then $f(a, b)$ is called the *absolute maximum (minimum) value* of $f(x, y)$ and $(a, b, f(a, b))$ is called an *absolute maximum (minimum) point.*

Obviously every absolute extreme point is also a local extreme point. We shall concentrate for the time being on developing a method of finding local extreme points.

It is clear that if $f(x, y)$ has a local maximum at (a, b), then the function $f(x, b)$ has a local maximum at a. That fact is illustrated in Figure 15.5.2. Hence, if $df(x, b)/dx$ exists, and if (a, b) is an interior point of the domain of $f(x, y)$, then the derivative $df(x, b)/dx$ must be 0 at $x = a$. However, by definition,

$$\frac{d}{dx} f(x, b) = f_x(x, b).$$

Figure 15.5.2

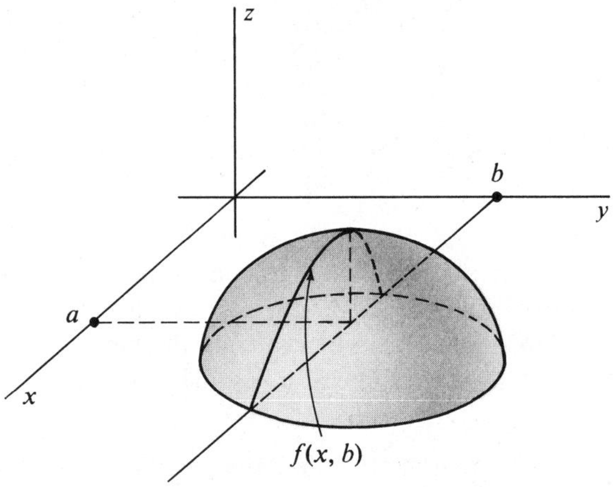

Thus, if $f(x, y)$ has a local maximum at (a, b) and if $f_x(a, b)$ exists, then $f_x(a, b) = 0$. In this case $f(a, y)$ will also have a local maximum at b; hence we also find that $f_y(a, b) = 0$. Since the same sort of statements hold for local minima, we have the following theorem.

Theorem 15.5.1

Let (a, b) be an interior point of the domain of f. If $f(x, y)$ has a local maximum or a local minimum at (a, b), and if $f_x(a, b)$ and $f_y(a, b)$ exist, then

$$f_x(a, b) = f_y(a, b) = 0.$$

Since $\nabla f(a, b) = (f_x(a, b), f_y(a, b))$, and since $(f_x(a, b), f_y(a, b)) = 0$ if and only if $f_x(a, b) = 0$ and $f_y(a, b) = 0$, the theorem can be stated in the following more compact form.

Theorem 15.5.2

> Let (a, b) be an interior point of the domain of f. If $f(x, y)$ has a local maximum or a local minimum at (a, b), and if $\nabla f(a, b)$ exists, then $\nabla f(a, b) = 0$.

As in the case of functions of a single variable, we must be careful to realize what the theorem does *not* say. For example, we cannot deduce that $f(x, y)$ has a local maximum or minimum at (a, b) simply because $\nabla f(a, b) = 0$. Figure 15.5.3 illustrates a "saddle point" at which both partial derivatives are zero and yet $f(x, y)$ has neither a local maximum nor a local minimum at (a, b).

Figure 15.5.3

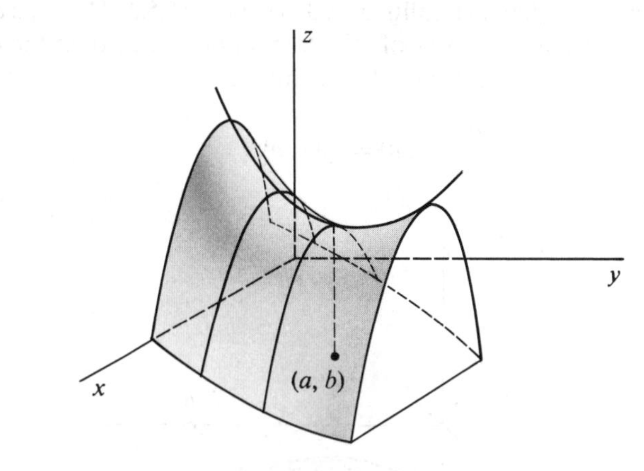

However, since a point in the domain of f is either an interior point or a boundary point, we can make use of Theorem 15.5.1 to assert that a local maximum or minimum of a function $f(x, y)$ can occur only where one of the partial derivatives does not exist, or where both partials are zero, or at a boundary point.

Example 3 Use Theorem 15.5.1 to find the possible local maximum and minimum points of $f(x, y) = x^3y - 3xy + 5$.

In this case,

$$f_x(x, y) = 3x^2y - 3y \quad \text{and} \quad f_y(x, y) = x^3 - 3x.$$

Since those partial derivatives exist for all points (x, y), we need only list the points for which both $f_x(x, y) = 0$ and $f_y(x, y) = 0$. We have

$$f_x(x, y) = 3x^2y - 3y = 3y(x - 1)(x + 1).$$

Thus $f_x(x, y)$ is 0 when $y = 0$ or $x = 1$ or $x = -1$. Similarly,

$$f_y(x, y) = x(x^2 - 3),$$

and so $f_y(x, y)$ is 0 when $x = 0$ or $x = \sqrt{3}$ or $x = -\sqrt{3}$.

Consequently the points where both f_x and f_y are 0 are $(-\sqrt{3}, 0)$, $(0, 0)$, and $(\sqrt{3}, 0)$. The domain of $f(x, y)$ in this case is the entire plane. Hence there are no boundary points, and we have obtained a complete list of *possible* local maximum and minimum points. Of course the list might include points that are neither local maxima nor local minima. For example, at a saddle point both partial derivatives are zero; yet in view of the following definition, a saddle point is neither a local maximum nor a local minimum. ||

Definition 15.5.6

Let (a, b) be a point at which $\nabla f(a, b) = 0$. Consider the curves formed by the intersection of the graph of $z = f(x, y)$ with the vertical planes through the point (a, b). If some of the curves so formed have local maximums while others have local minimums, then (a, b) is a *saddle point* of f.

We call a point (x, y) where $f_x(x, y) = f_y(x, y) = 0$ or where one of the partial derivatives does not exist a *critical point*. Example 3 raises an interesting question concerning critical points. Can we find a method of classifying a critical point as a local maximum or a local minimum?

Suppose that (a, b) is an interior point of the domain of $f(x, y)$ at which $f_x(a, b) = f_y(a, b) = 0$. To find out whether $f(x, y)$ has a local maximum, local minimum, or saddle point at (a, b), we shall examine the behavior of $f(x, y)$ when the point (x, y) is restricted to various lines through the point (a, b). Figure 15.5.4 illustrates the restriction of $f(x, y)$ to several lines through (a, b).

Figure 15.5.4

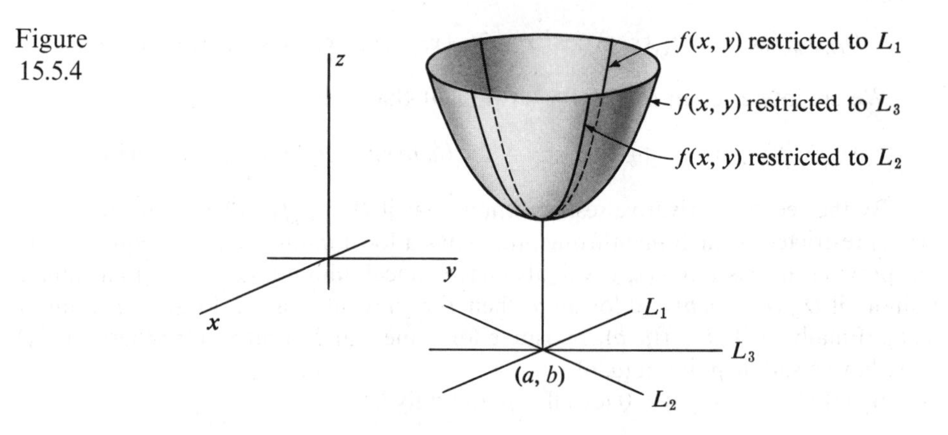

If, as in Figure 15.5.4, $f(x, y)$ restricted to each line L has a local minimum, then the function $f(x, y)$ itself must have a local minimum at (a, b). On the other hand, if $f(x, y)$ restricted to each line L has a local maximum, then $f(x, y)$ has a local maximum at (a, b).

Finally if $f(x, y)$ restricted to some lines has a local maximum, while $f(x, y)$ restricted to other lines has a local minimum, then $f(x, y)$ must have a saddle point at (a, b). This situation is pictured in Figure 15.5.3, where $f(x, y)$ restricted to a line parallel to the x-axis has a local maximum, and $f(x, y)$ restricted to a line parallel to the y-axis has a local minimum.

Our aim is to apply the second derivative test to $f(x, y)$ restricted to a line through (a, b) in order to determine if $f(x, y)$ restricted to this line has a maximum or minimum.

The first derivative of $f(x, y)$ restricted to a line making an angle α with the positive x-axis is simply the directional derivative of $f(x, y)$ in the direction α. Hence the first derivative is

$$D_{(\alpha)}f(x, y) = f_x(x, y)\cos\alpha + f_y(x, y)\sin\alpha.$$

This first derivative is again a function of two variables x and y. To obtain the required second derivative we take the directional derivative of $D_{(\alpha)}f(x, y)$ in the direction α. So we consider

$$\begin{aligned} D_{(\alpha)}D_{(\alpha)}f(x, y) &= \frac{\partial}{\partial x}(D_{(\alpha)}f(x, y))\cos\alpha + \frac{\partial}{\partial y}(D_{(\alpha)}f(x, y))\sin\alpha \\ &= \cos\alpha\frac{\partial}{\partial x}(f_x(x, y)\cos\alpha + f_y(x, y)\sin\alpha) \\ &\quad + \sin\alpha\frac{\partial}{\partial y}(f_x(x, y)\cos\alpha + f_y(x, y)\sin\alpha). \end{aligned}$$

Hence

$$\begin{aligned} D_{(\alpha)}D_{(\alpha)}f(x, y) = f_{xx}(x, y)\cos^2\alpha &+ f_{yx}(x, y)\sin\alpha\cos\alpha \\ &+ f_{xy}(x, y)\sin\alpha\cos\alpha + f_{yy}(x, y)\sin^2\alpha. \end{aligned}$$

Then, if $f(x, y)$ has continuous second order partial derivatives, $f_{xy}(x, y) = f_{yx}(x, y)$, and so

$$D_{(\alpha)}D_{(\alpha)}f(x, y) = f_{xx}(x, y)\cos^2\alpha + 2f_{xy}(x, y)\sin\alpha\cos\alpha + f_{yy}(x, y)\sin^2\alpha.$$

We are interested in $D_{(\alpha)}D_{(\alpha)}f(x, y)$ at (a, b), that is, in

$$D_{(\alpha)}D_{(\alpha)}f(a, b) = f_{xx}(a, b)\cos^2\alpha + 2f_{xy}(a, b)\sin\alpha\cos\alpha + f_{yy}(a, b)\sin^2\alpha. \tag{1}$$

By the second derivative test we know that if $D_{(\alpha)}D_{(\alpha)}f(a, b) > 0$ for all α, then $f(x, y)$ restricted to each line through (a, b) has a local minimum. Consequently, by our previous discussion, $f(x, y)$ will also have a local minimum at (a, b). In a similar fashion, if $D_{(\alpha)}D_{(\alpha)}f(a, b) < 0$ for all α, then $f(x, y)$ would have a local maximum at (a, b). Finally, if $D_{(\alpha)}D_{(\alpha)}f(a, b)$ is positive for some α and negative for others, $f(x, y)$ must have a saddle point at (a, b).

By (1), $D_{(\alpha)}D_{(\alpha)}f(a, b) > 0$ for all α, if and only if,

$$f_{xx}(a, b)\cos^2\alpha + 2f_{xy}(a, b)\sin\alpha\cos\alpha + f_{yy}(a, b)\sin^2\alpha > 0 \tag{2}$$

for all α.

Inequality (2) is to hold for all α, and we first examine those α for which $\sin\alpha = 0$. In this case we must have

$$f_{xx}(a, b)\cos^2\alpha > 0.$$

But $\cos\alpha \neq 0$, since $\sin\alpha = 0$; hence

$$f_{xx}(a, b) > 0. \tag{3}$$

If $\sin\alpha \neq 0$, we may divide by the positive number $\sin^2\alpha$ to convert (2) into the form

$$f_{xx}(a, b)\cot^2\alpha + 2f_{xy}(a, b)\cot\alpha + f_{yy}(a, b) > 0 \qquad \text{for all } \alpha.$$

Consequently we seek conditions which will insure that the quadratic expression

$$f_{xx}(a, b)\,u^2 + 2f_{xy}(a, b)\,u + f_{yy}(a, b) \tag{4}$$

is always positive.

Since $f_{xx}(a, b) > 0$, the graph of the quadratic in (4) opens upward. Figure 15.5.5 indicates two possibilities. In Figure 15.5.5(a) the quadratic is always positive, while in Figure 15.5.5(b) it is not always positive. Consequently, to insure that the quadratic in (4) is always positive, it is sufficient that the quadratic have no real zeros, that is, have a negative discriminant. So it is sufficient that

$$f_{xy}^2(a, b) - f_{xx}(a, b)\,f_{yy}(a, b) < 0. \tag{5}$$

Figure 15.5.5

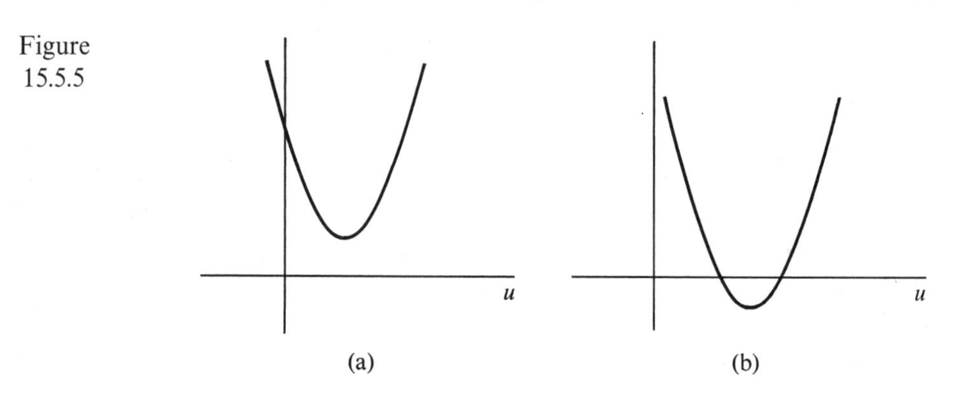

Hence we have shown that if $f_x(a, b) = f_y(a, b) = 0$ and if (3) and (5) hold, then $f(x, y)$ has a local minimum at (a, b).

Reversal of the appropriate inequalities will establish similar results for the case of a local maximum or a saddle point. Those cases are left as exercises. In summary, then, we have this result:

Theorem 15.5.3

Let (a, b) be an interior point of the domain of $f(x, y)$ at which $f_x(a, b) = f_y(a, b) = 0$, and suppose that $f(x, y)$ has continuous second order partial derivatives at (a, b). Then

(i) If $f_{xx}(a, b) > 0$ and $f_{xy}^2(a, b) - f_{xx}(a, b)f_{yy}(a, b) < 0$, then $f(x, y)$ has a local minimum at (a, b);

(ii) If $f_{xx}(a, b) < 0$ and $f_{xy}^2(a, b) - f_{xx}(a, b)f_{yy}(a, b) < 0$, then $f(x, y)$ has a local maximum at (a, b);

(iii) If $f_{xy}^2(a, b) - f_{xx}(a, b)f_{yy}(a, b) > 0$, then $f(x, y)$ has a saddle point at (a, b).

Note that the theorem affords no conclusions if

$$f_{xy}^2(a, b) - f_{xx}(a, b)f_{yy}(a, b) = 0.$$

Example 4 Find and classify all critical points of the function

$$f(x, y) = x^2 - 12y^2 - 4y^3 + 3y^4.$$

Since $$f_x(x, y) = 2x$$

and $$f_y(x, y) = -24y - 12y^2 + 12y^3 = 12y(y-2)(y+1),$$

we find that (0, 0), (0, 2), and (0, −1) are the critical points of $f(x, y)$. Also

$$f_{xy}(x, y) = 0, \quad f_{xx}(x, y) = 2, \quad f_{yy}(x, y) = -24 - 24y + 36y^2.$$

Thus, $f_{xx}(x, y) > 0$ for all points (x, y). Moreover, since

$$f_{xy}{}^2(0, 2) - f_{xx}(0, 2)f_{yy}(0, 2) = -144 < 0 \quad \text{and}$$

$$f_{xy}{}^2(0, -1) - f_{xx}(0, -1)f_{yy}(0, -1) = -72 < 0,$$

$f(x, y)$ has a local minimum at both (0, 2) and (0, −1). On the other hand, since

$$f_{xy}{}^2(0, 0) - f_{xx}(0, 0)f_{yy}(0, 0) = 48 > 0,$$

it follows that $f(x, y)$ has a saddle point at (0, 0). ||

Example 5 A rectangular box is to hold a fixed volume V. Show that the surface area has a local minimum when the box is a cube.

Let x, y, and z be the dimensions of the rectangular box. We must show that a local minimum of the surface area

$$A = 2xy + 2yz + 2xz$$

under the condition $V = xyz$ occurs when $x = y = z$. Since $V = xyz$, then $z = V/xy$. Substitution of that into our area formula produces

$$A = 2xy + 2Vx^{-1} + 2Vy^{-1}.$$

Differentiating, we get

$$\frac{\partial A}{\partial x} = 2y - 2Vx^{-2} \quad \text{and} \quad \frac{\partial A}{\partial y} = 2x - 2Vy^{-2}.$$

Setting $\partial A/\partial x = 0$ and $\partial A/\partial y = 0$, we obtain

$$2y - 2Vx^{-2} = 0,$$

$$2x - 2Vy^{-2} = 0.$$

Simultaneous solution of those equations yields $x = V^{1/3}$ and $y = V^{1/3}$. Consequently the only extreme value of A occurs when $x = V^{1/3}$ and $y = V^{1/3}$. Using the fact that $z = V/xy$ we also obtain $z = V^{1/3}$. Since

$$\frac{\partial^2 A}{\partial y\, \partial x} = 2, \quad \frac{\partial^2 A}{\partial x^2} = 4Vx^{-3}, \quad \text{and} \quad \frac{\partial^2 A}{\partial y^2} = 4Vy^{-3},$$

$$\left(\frac{\partial^2 A}{\partial y\, \partial x}\right)^2 - \frac{\partial^2 A}{\partial x^2} \cdot \frac{\partial^2 A}{\partial y^2} = 4 - 16 < 0 \quad \text{and} \quad \frac{\partial^2 A}{\partial x^2} = 4 > 0,$$

when $x = y = V^{1/3}$. Thus the surface area has a local minimum when $x = y = V^{1/3}$, that is, when the box is a cube. ||

Exercises 15.5

In Exercises 1–26, find and classify all critical points.

1. $f(x, y) = 2x^2 + 2y^2 - x + 2y - 1$.
2. $f(x, y) = x^2 + 2y^2 - 2x + 8y + 5$.
3. $f(x, y) = xy - 2x + 3y$.
4. $f(x, y) = x^2 - xy + y^2$.
5. $f(x, y) = x^2 + 4xy - y^2$.
6. $f(x, y) = x^2 + 4xy + y^2 + 3x + 10$.
7. $f(x, y) = x^3 + y^2 - 3x$.
8. $f(x, y) = x^3 + y^3 - 3xy$.
9. $f(x, y) = x^3 - 12xy^2 + y^3 + 45y$.
10. $f(x, y) = x^3 + 12x^2 + y^2 + 36x - 4y$.
11. $f(x, y) = 3x^2 - 4xy + 5y^2 - 2x + 3y + 7$.
12. $f(x, y) = (x - 5)^2 + (y + 2)^2$.
13. $f(x, y) = (x + y)^2 + x^4$.
14. $f(x, y) = x^4 + y^4 + xy - (x^2 + y^2)$.
15. $f(x, y) = 2y^3 + 3y^2 - 12y - x^2 + 2x$.
16. $f(x, y) = 3x^2 + 2y^2 - 2y + 5$.
17. $f(x, y) = 2y^3 - 3x^2 - 3xy + 9x$.
18. $f(x, y) = (y - x^2)(y - 2x^2)$.
19. $f(x, y) = 3y^3 - yx^2 + y$.
20. $f(x, y) = xy - 1/x - 1/y$.
21. $f(x, y) = (x - y)/(x + y)$.
22. $f(x, y) = xy/(x^2 + y^2)$.
23. $f(x, y) = \cos y + \sin(x + y)$.
24. $f(x, y) = \sin x + \sin y + \cos(x + y)$.
25. $f(x, y) = x + y \sin x$.
26. $f(x, y) = x^3y^2(6 - x - y)$.
27. Suppose that the relation between production z, the number of man hours x, and the amount of capital y is given by $z = f(x, y)$. The ratio z/x can be thought of as the average production per man hour. Show that $\partial z/\partial x = z/x$ when the average production per man hour is at a maximum.
28. Prove part (ii) of Theorem 15.5.3.
29. Prove part (iii) of Theorem 15.5.3.
30. Show that every point of a set S is either an interior point or a boundary point.

15.6 Absolute Maxima and Minima

As with functions of a single variable, we are often faced with the problem of finding the absolute maximum or minimum points of a function of two variables. Although the techniques of the preceding section can be used to determine the local maximum and minimum points, some additional theory is necessary to determine the absolute extrema. We shall begin with a few basic notions.

Recall that one basic property of functions of a single variable is that a continuous function on a *closed* interval actually attains an absolute maximum and minimum on that interval. The concept of a closed interval is of little use when discussing functions of two variables. So, to begin, we must generalize the concept of a closed interval.

Since an interval is closed if and only if it contains its boundary (or end) points, we are led to the following definition.

Definition 15.6.1

> A subset S of the plane is *closed* if it contains all its boundary points.

Example 1 | Consider the disc D illustrated in Figure 15.6.1.

Clearly D consists of all points (x, y) such that $x^2 + y^2 < 1$. The disk D is *not* closed because the point (0, 1) is a boundary point of D but is not in D. Indeed, every point $(x, y,)$ with $x^2 + y^2 = 1$ has that property; so D contains none of its boundary points. Note, however, that if a set is missing only *one* of its boundary points, the set is not closed. ||

Figure 15.6.1

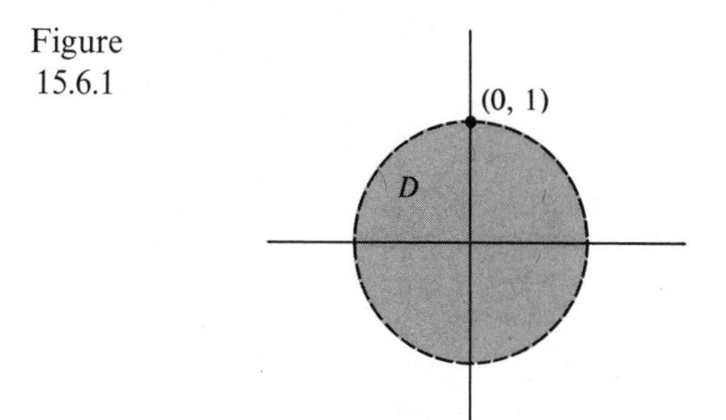

Example 2 | The disc D consisting of all points (x, y) with $x^2 + y^2 \leq 1$ is closed. We note that (a, b) is a boundary point of D if and only if $a^2 + b^2 = 1$. Since the circle $x^2 + y^2 = 1$ is in D, it follows that D contains all its boundary points. Hence D is closed. ||

With one more definition we shall be ready to proceed with our consideration of absolute maxima and minima.

Definition 15.6.2

> A subset S of the plane is *bounded* if it is contained within some disk centered at the origin.

Example 3 | The graph of the line $x + y = 0$ is not bounded since it is not contained within any disc centered at the origin. ||

Example 4 | The set of all points (x, y) such that $|x| \leq 1$ and $|y| \leq 2$, shown in Figure 15.6.2, is bounded since it is contained within a disc of radius 3 centered at the origin. ||

Figure 15.6.2

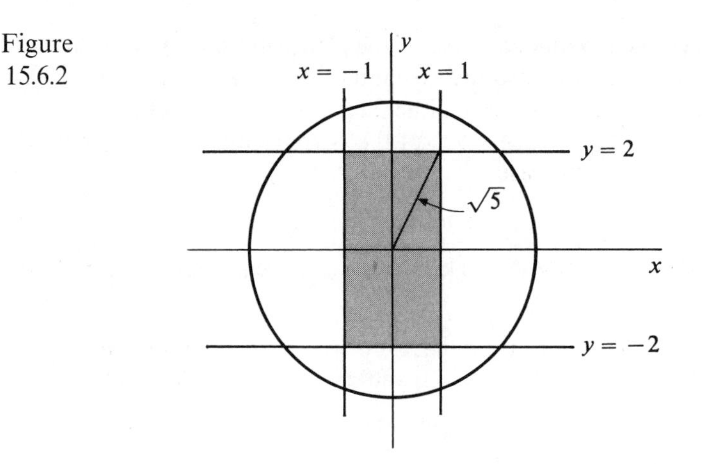

Now that we have the definitions of closed and bounded we can state the following theorem, whose proof appears in some advanced calculus texts.

Theorem 15.6.1

A function that is continuous on a closed, bounded subset S of the plane attains an absolute maximum value and an absolute minimum value on S. That is, there are points (a, b) and (a', b') in S such that $f(a, b) \geq f(x, y)$ and $f(a', b') \leq f(x, y)$, for all points (x, y) in S.

Example 5

The function $f(x, y) = [1 - (x^2 + y^2)]^{-1}$ is continuous on the set S of all (x, y) such that $x^2 + y^2 < 1$. Note, however, that $f(x, y)$ attains no absolute maximum on S since $f(x, y)$ can be made arbitrarily large by taking $x^2 + y^2$ sufficiently close to 1, that is, by taking a point (x, y) in S sufficiently close to the boundary of S. This example indicates that Theorem 15.6.1 is not applicable to sets that are not closed. ||

The application of Theorem 15.6.1 to maximum-minimum problems is illustrated in the following examples. It is important to realize that a local maximum or minimum can occur only where $f_x(x, y) = f_y(x, y) = 0$, or where at least one of those partial derivatives fails to exist, or at a boundary point of the domain of f. As in the case of functions of a single variable it is important to remember that *the boundary of the region in question must be considered separately.*

Example 6

Find the absolute maximum and minimum points of the function

$$f(x, y) = 3y - x + 6$$

on the rectangle determined by $0 \leq x \leq 2$ and $1 \leq y \leq 4$.

Since the rectangle is a closed, bounded subset of the plane, and since f is continuous, Theorem 15.6.1 assures us that an absolute extremum of each kind must occur. We know that the absolute extrema must occur either at a point (a, b) for which $f_x(a, b) = f_y(a, b) = 0$, or at a point where at least one of the partials fails to exist, or on the boundary of the rectangle. In this case

$$f_x(x, y) = -1 \quad \text{and} \quad f_y(x, y) = 3,$$

for all points interior to the rectangle. Therefore the extrema must occur on the boundary of the rectangle.

We now can restrict our attention to the behavior of $f(x, y)$ on the boundary: $x = 0$, $1 \le y \le 4$; $x = 2$, $1 \le y \le 4$; $y = 1$, $0 \le x \le 2$; and $y = 4$, $0 \le x \le 2$. We have

$$f(x, y) = 3y + 6, \qquad \text{when } x = 0 \text{ and } 1 \le y \le 4;$$

$$f(x, y) = 3y + 4, \qquad \text{when } x = 2 \text{ and } 1 \le y \le 4;$$

$$f(x, y) = -x + 9, \qquad \text{when } y = 1 \text{ and } 0 \le x \le 2;$$

$$f(x, y) = -x + 18, \qquad \text{when } y = 4 \text{ and } 0 \le x \le 2.$$

The restriction of f to the appropriate intervals is indicated in Figure 15.6.3. Since these restrictions of f are linear, their absolute extrema must occur at an endpoint of their intervals of definition, that is, at (0, 1), (2, 1), (2, 4), or (0, 4).

Figure 15.6.3

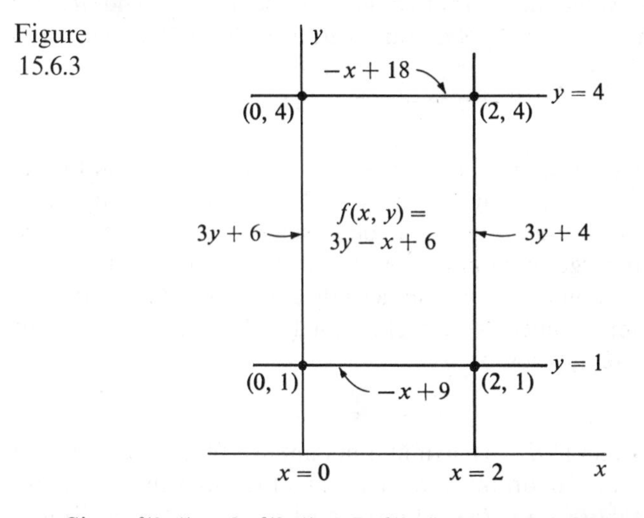

Since $f(0, 1) = 9$, $f(2, 1) = 7$, $f(2, 4) = 16$, and $f(0, 4) = 18$, it follows that (2, 1, 7) is the absolute minimum point and (0, 4, 18) is the absolute maximum point of $f(x, y)$ on the rectangle. ||

Example 7 Find the absolute maximum and the absolute minimum values of the function $f(x, y) = 4y^2 - 2x^2 + 12x - 7$ on the disc $x^2 + y^2 \le 16$.

Since
$$f_x(x, y) = -4x + 12 \quad \text{and} \quad f_y(x, y) = 8y,$$
the only point where both partials are zero is (3, 0).

On the boundary of the disk, $x^2 + y^2 = 16$, or $y^2 = 16 - x^2$. Substitution into $f(x, y) = 4y^2 - 2x^2 + 12x - 7$ then gives

$$f(x, y) = 4(16 - x^2) - 2x^2 + 12x - 7$$

or

$$f(x, y) = -6x^2 + 12x + 57,$$

if $x^2 + y^2 = 16$.

Since $x^2 + y^2 = 16$, we must have $-4 \le x \le 4$. Thus we are to find the extrema of

$$g(x) = -6x^2 + 12x + 57, \qquad \text{for } -4 \le x \le 4.$$

The extrema can exist only where $g'(x) = 0$ or at one of the endpoints $x = \pm 4$. In this case,

$$g'(x) = -12x + 12,$$

so $g'(x) = 0$ only when $x = 1$. The x values that are of interest are therefore $x = 1$, ± 4. On the boundary, $y^2 = 16 - x^2$; so the points of interest are $(1, \sqrt{15})$, $(1, -\sqrt{15})$, $(4, 0)$, and $(-4, 0)$. Those points, along with the point $(3, 0)$ where both partials are 0, are plotted in Figure 15.6.4. The absolute maximum and absolute minimum must occur at one of those points. By direct calculation we find

$$f(-4, 0) = -2 \cdot 16 - 12 \cdot 4 - 7 = -87,$$

$$f(4, 0) = -2 \cdot 16 + 12 \cdot 4 - 7 = 9,$$

$$f(3, 0) = -2 \cdot 9 + 12 \cdot 3 - 7 = 11,$$

$$f(1, \sqrt{15}) = 4 \cdot 15 - 2 + 12 - 7 = 63,$$

$$f(1, -\sqrt{15}) = 4 \cdot 15 - 2 + 12 - 7 = 63.$$

Figure 15.6.4

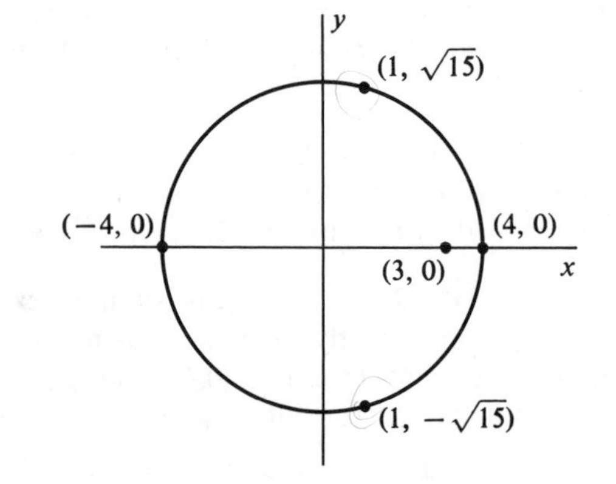

Consequently the absolute minimum value of -87 occurs at $(-4, 0)$; and the absolute maximum value of 63 occurs at both $(1, \sqrt{15})$ and $(1, -\sqrt{15})$. ||

Example 8 Find the absolute maximum and the absolute minimum values of the function

$$f(x, y) = -x^2 + 2y^2 + x - 4y + 2$$

on the triangle determined by $x \ge 0$, $y \ge 0$, and $x + y \le 2$.

The triangle in question is illustrated in Figure 15.6.5. Since $f_x(x, y) = -2x + 1$ and $f_y(x, y) = 4y - 4$, the only point where both partials are zero is the point $(1/2, 1)$. Since the partials exist everywhere, we need only further consider the boundary of the triangle.

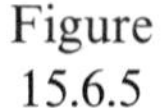

Figure 15.6.5

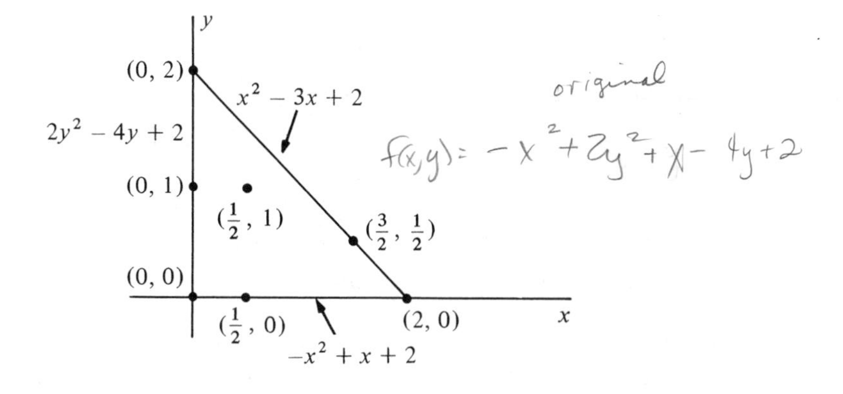

The restrictions of $f(x, y)$ to the sides of the triangle are given by

$$f(x, y) = 2y^2 - 4y + 2, \text{ when } x = 0 \text{ and } 0 \le y \le 2;$$

$$f(x, y) = -x^2 + x + 2, \text{ when } y = 0 \text{ and } 0 \le x \le 2;$$

$$f(x, y) = x^2 - 3x + 2, \text{ when } x + y = 2 \text{ and } 0 \le x \le 2.$$

These restrictions are indicated in Figure 15.6.5. The function $f(x, y) = 2y^2 - 4y + 2$, $0 \le y \le 2$, has a zero derivative at $y = 1$. Consequently, it can attain its absolute extreme values only when $y = 1$ or at the endpoints of the interval $0 \le y \le 2$. Thus the points $(0, 0)$, $(0, 1)$, and $(0, 2)$ are possible points where $f(x, y)$ attains an absolute extreme.

The function $f(x, y) = -x^2 + x + 2$, $0 \le x \le 2$, can attain its absolute extreme values either where its derivative is zero, that is, at $x = 1/2$, or at the endpoints of the interval $0 \le x \le 2$. Thus the points $(0, 0)$, $(1/2, 0)$, and $(2, 0)$ are also possibilities.

Finally, the function $f(x, y) = x^2 - 3x + 2$, $0 \le x \le 2$, can attain its extreme values only at $x = 3/2$, $x = 0$, or $x = 2$. Consequently, since for this restriction we have $x + y = 2$, we must also consider the points $(3/2, 1/2)$, $(0, 2)$, and $(2, 0)$. Thus, $(0, 0)$, $(1/2, 0)$, $(2, 0)$, $(3/2, 1/2)$, $(0, 2)$, $(0, 1)$, and $(1/2, 1)$ are all the points at which the function

$$f(x, y) = -x^2 + 2y^2 + x - 4y + 2, \quad x \ge 0,\ y \ge 0,\ x + y \le 2$$

could attain its absolute extreme values.

Direct calculation yields:

$$f(0, 0) = 2, \quad f\left(\frac{1}{2}, 0\right) = \frac{9}{4}, \quad f(2, 0) = 0, \quad f\left(\frac{3}{2}, \frac{1}{2}\right) = -\frac{1}{4},$$

$$f(0, 2) = 2, \quad f(0, 1) = 0, \quad f\left(\frac{1}{2}, 1\right) = \frac{1}{4}.$$

Clearly then the absolute maximum value of 9/4 is attained at (1/2, 0), and the absolute minimum value of $-1/4$ is attained at (3/2, 1/2). ||

Exercises 15.6

In Exercises 1–20, find the absolute maximum and absolute minimum points of the given function on the given set.

1. $f(x, y) = -2x + y - 10, \quad -2 \le x \le 0, \quad 1 \le y \le 2.$
2. $f(x, y) = x + y + 1, \quad |x| \le 5, \quad |y| \le 1.$
3. $f(x, y) = 3x + 4y - 8, \quad x \le 1, \quad y \ge 0, \quad y \le x.$
4. $f(x, y) = 2y - 7x, \quad y \le 1, \quad x \ge 0, \quad y \ge x.$
5. $f(x, y) = x^2 + 2y, \quad |x| \le 1, \quad |y| \le 1.$
6. $f(x, y) = x^2 + x + y, \quad -1 \le x \le 0, \quad 0 \le y \le 1.$
7. $f(x, y) = x^2 + 3x + 2y, \quad x \ge 0, \quad y \ge 0, \quad x + y \le 4.$
8. $f(x, y) = x^2 - y^2 - 2x, \quad x \ge 0, \quad y \ge 0, \quad x + 2y \le 2.$
9. $f(x, y) = x + 7y^3, \quad |x| \le 2, \quad |y| \le 1.$
10. $f(x, y) = x^2 - x - y + 10, \quad 0 \le x \le 1, \quad 0 \le y \le 1.$
11. $f(x, y) = x^2 + y^2 - 4xy + 6x, \quad 3x \ge y, \quad x \le 2, \quad y \ge 0.$
12. $f(x, y) = x^3 - x^2y, \quad x \ge -1, \quad x + y \le 1, \quad y \ge -1.$
13. $f(x, y) = x^2 + y^2 + 14, \quad x^2 + y^2 \le 1.$
14. $f(x, y) = x^2 + y^2 + 2x + 14, \quad x^2 + y^2 \le 1.$
15. $f(x, y) = x^2 - x - 2y^2 + y + 6, \quad 0 \le x \le 1, \quad 0 \le y \le 2.$
16. $f(x, y) = y^2 + 2x + 7, \quad x^2 + y^2 \le 4.$
17. $f(x, y) = 2 - x^2 - 4y, \quad x^2 + y^2 \le 9.$
18. $f(x, y) = 3x^2 + 2y^2 - 2y + 1, \quad x^2 + y^2 \le 4.$
19. $f(x, y) = 4y^2 + 2x - 3, \quad x^2/4 + y^2 \le 1.$
20. $f(x, y) = -9x^2 - 4y + 7, \quad x^2 + y^2/9 \le 1.$
21. Show that the function $f(x, y) = x^2 - y^2 + 2xy - 3$ has no absolute maximum or absolute minimum.
22. A manufacturer wants to construct a closed rectangular box with a volume of 12 cubic feet. What dimensions should be selected to minimize the cost if the top and bottom cost twice as much per square foot as the sides?
23. The profit a firm makes by producing x units of one item and y units of a second item is given by

$$P(x, y) = 100x + 130y - 10xy - x^2 - 13.$$

In addition, plant capacity dictates that $x \le 10$ and $y \le 5$. How many items of each type should be produced to maximize the profit?

24. A firm manufactures flashlights at a cost of \$1.00 per flashlight and flashlight batteries at a cost of 25 cents per battery. If the sale price for flashlights is x cents each and for batteries y cents each, the expected weekly sales are $42(10)^4/(x^2y)$ flashlights and $58(10)^4/(xy^2)$ batteries. What should the prices be for a maximum profit?

25. A firm produces a single product that is sold in two different markets. If x and y are the amounts sold per year in the two markets, let

$$p(x) = 5(10)^4 - 5x$$

$$q(y) = 4(10)^4 - 3y$$

be the prices in cents per unit at which the product is sold in each market. If w units are produced, the production cost in cents is

$$c(w) = 10^7 + 8(10)^3 w + 5w^2.$$

Determine the x and y that maximize the profit.

26. In Exercise 25 assume that production facilities are limited so that no more than 2000 units can be produced per year. What values of x and y will now maximize the profit?

27. A firm sells its product in two separate markets. Let $C(w)$ be the cost of producing w items. Let $R(x)$ be the revenue from the sale of x items in one market and $T(y)$ be the revenue from the sale of y items in the second market. Show that $dR/dx = dT/dy$ when maximum profit is attained.

28. Economists use utility functions to measure the usefulness to a given consumer of quantities of given products. In particular, the Cobb-Douglas utility function, which has the form $z = Mx^a y^b$, measures the usefulness z of quantities x and y of two given products. Normally the constants a, b, and M satisfy the inequalities $0 < a < 1$, $0 < b < 1$, and $M > 0$.

    As a particular example, let $a = 1/3$, $b = 2/3$, and $M = 3$; and suppose that one unit of the first product costs \$5.00, while one unit of the second costs \$3.00. If a consumer has a total of \$60.00 to spend, find how the money should be spent to maximize the utility.

29. Show that the function

$$f(x, y) = ax + by + c, \quad 0 \le x \le x_1, \quad 0 \le y \le y_1$$

must attain its absolute maximum at one of the vertices of the rectangle $0 \le x \le x_1, 0 \le y \le y_1$. How could you generalize the result?

30. When two firms are in competition, the profit that each realizes is dependent on its production and on its competitor's production. Let p_1 and p_2 be the profits of the two firms, and let x and y be their respective productions. In addition, assume that

$$p_1 = 36x - 2x^2 - 3y^2 + 1500, \quad \text{and}$$

$$p_2 = 48y - 3y^2 - x^2 + 3000.$$

    (a) If each firm acts to maximize its own profit, find x, y, p_1, p_2, and $p_1 + p_2$.
    (b) If the firms work together to maximize $p_1 + p_2$, find x, y, p_1, p_2, and $p_1 + p_2$.

31. Use the methods of this section to find the shortest distance from the origin to the plane with equation $x + y + z = a$.

32. Show that the rectangular box having fixed volume and minimal surface area is a cube.

33. Show that the largest (in volume) rectangular box that can be inscribed in a sphere is a cube.

34. What are the relative dimensions of a rectangular box that has an open top and minimal surface area? The volume of the box is to be one cubic foot.

35. Divide the number N into three parts such that the product of the parts is a maximum.

36. Find the minimum distance between the parabolas $2y = x^2$ and $y = -(x - 9)^2$.

Brief Review of Chapter 15

1. Partial Derivatives

The partial derivatives of a function $f(x, y)$ at (a, b) are defined thus:

$$f_x(a, b) = \lim_{h \to 0} \frac{f(a+h, b) - f(a, b)}{h},$$

$$f_y(a, b) = \lim_{h \to 0} \frac{f(a, b+h) - f(a, b)}{h},$$

In general, if $f: R^n \to R$, the partial of f with respect to x_i at $(a_1, a_2, \ldots, a_n)$ is defined as follows:

$$f_{x_i}(a_1, a_2, \ldots, a_n) = \lim_{h \to 0} \frac{f(a_1, a_2, \ldots, a_i + h, \ldots, a_n) - f(a_1, a_2, \ldots, a_n)}{h}$$

Partial derivatives may be calculated by differentiating with respect to one variable while treating the others as if they were constants.

2. Chain Rules

Chain rules enable us to compute derivatives of composite functions:

(a) If $x = g(t)$, $y = h(t)$, and $f(x, y) = z$, then

$$\frac{df}{dt} = \frac{\partial f}{\partial x}\frac{dx}{dt} + \frac{\partial f}{\partial y}\frac{dy}{dt}.$$

(b) If $x = g(u, v)$, $y = h(u, v)$, and $f(x, y) = z$, then

$$\frac{\partial f}{\partial u} = \frac{\partial f}{\partial x}\frac{\partial x}{\partial u} + \frac{\partial f}{\partial y}\frac{\partial y}{\partial u}, \quad \text{and}$$

$$\frac{\partial f}{\partial v} = \frac{\partial f}{\partial x}\frac{\partial x}{\partial v} + \frac{\partial f}{\partial y}\frac{\partial y}{\partial v}.$$

Those rules have straightforward extensions to functions of more than two variables.

3. The Gradient

If $g: R^n \to R$, the gradient of g is defined as

$$\nabla g(\mathbf{X}) = (g_{x_1}(\mathbf{X}), \ldots, g_{x_n}(\mathbf{X})).$$

Note that both the domain and range of ∇g are subsets of R^n.

Geometrically the gradient of $g: R^3 \to R$ at (a, b, c), that is, $\nabla g(a, b, c)$, is a vector which is normal to the surface $g(x, y, z) = 0$ at (a, b, c).

4. Directional Derivatives

If $f: R^n \to R$ and $\mathbf{U}$ is a unit vector of R^n, then the directional derivative of f in the direction of $\mathbf{U}$ is

$$D_{\mathbf{U}} f(\mathbf{X}) = \nabla f(\mathbf{X}) \cdot \mathbf{U},$$

which represents the rate of change of f in the direction of $\mathbf{U}$ at the point $\mathbf{X}$. If $f: R^2 \to R$, the directional derivative of f at (a, b) in a direction making an angle α with the positive x-axis is given by

$$D_{(\alpha)} f(a, b) = f_x(a, b) \cos \alpha + f_y(a, b) \sin \alpha.$$

The gradient of f at $\mathbf{X}$ is a vector whose magnitude is the maximal value of the directional derivative at $\mathbf{X}$. Moreover, the direction of the gradient is the direction of that maximal directional derivative.

5. Extrema of Functions of Two Variables

Definitions of the following terms are necessary: (a) r-neighborhood, (b) interior point, (c) boundary point, (d) local maximum and local minimum, and (e) absolute maximum and absolute minimum.

A local maximum or minimum of a function $f(x, y)$ can occur only
(a) at a point (a, b) where $\nabla f(a, b) = 0$, or
(b) at a point where at least one partial derivative does not exist, or
(c) at a boundary point of the domain of $f(x, y)$.
Remember that $\nabla f(a, b) = 0$ does not imply that $f(x, y)$ has a local maximum or minimum at (a, b); however, a list including all points of (a), (b), and (c) above will be a complete list of possible points where the function might assume a local maximum or minimum.

If (a, b) is an interior point of the domain of f and $\nabla f(a, b) = 0$, the following second derivative test is an aid in sorting out local maxima, local minima, and saddle points.
(a) If $f_{xx}(a, b) > 0$ and $f_{xy}^2(a, b) - f_{xx}(a, b) f_{yy}(a, b) < 0$, then $f(x, y)$ has a local minimum at (a, b).
(b) If $f_{xx}(a, b) < 0$ and $f_{xy}^2(a, b) - f_{xx}(a, b) f_{yy}(a, b) < 0$, then $f(x, y)$ has a local maximum at (a, b).
(c) If $f_{xy}^2(a, b) - f_{xx}(a, b) f_{yy}(a, b) > 0$, then $f(x, y)$ has a saddle point at (a, b).
A function $f(x, y)$ that is continuous on a closed, bounded subset S of the plane attains an absolute maximum and an absolute minimum on S. To find the absolute extrema of f on S, it is only necessary to make a complete list of possible local extrema (all points where $\nabla f = 0$, or where a partial derivative does not exist, or on the boundary of S). An equation for the boundary of S can often be used to reduce $f(x, y)$ to a function of a single variable. By finding the local extrema of that function of a single variable we obtain the possible extrema of $f(x, y)$ on the boundary of S. As a final step, the function f is evaluated at each of the points obtained above. In this way we can find the absolute maximum points and the absolute minimum points of $f(x, y)$.

Technique Review Exercises, Chapter 15

1. Compute $f_x(x, y)$, $f_y(x, y)$, and $f_{xy}(x, y)$, if $f(x, y) = \sin x + x\tan(xy)$.
2. Compute $\partial f/\partial x_i$ if $f(\mathbf{X}) = 1/|\mathbf{X}|$.
3. Use a chain rule to compute df/dt when $t = 0$, if $f(x, y) = \ln|x^2 - 3xy|$, $x = 2t + 1$, and $y = \cos t$.
4. Use a chain rule to compute $\partial f/\partial u$ when $u = 2$ and $v = \pi$, if $f(x, y) = xy^2 - 3xy$, $x = u\cos v$, and $y = u\sin v$.
5. Find a unit vector that is perpendicular to the graph of $x^2 - 3xy + y^2 = z$ at $(2, -1, 11)$.
6. Compute the directional derivative of $f(x, y, z) = x(yz - z^2)$ in the direction of the vector $(3, 0, 2)$ at the point $(1, 0, -1)$.
7. Find the magnitude and direction of the maximum value of the directional derivative of $f(x, y, z) = x(yz - z^2)$ at $(1, 0, -1)$.
8. Find and classify all critical points of $f(x, y) = x^3 + y^2 - 3x$.
9. Find the absolute maximum and absolute minimum points of $f(x, y) = x^2 - 12x + 3y^2$ for $x^2 + y^2 \leq 25$.
10. An open top box is to have a volume of 24 cubic feet. Find the dimensions that minimize the surface area of the box.

Additional Exercises, Chapter 15

Section 15.1

In Exercises 1–6, find (a) $\dfrac{\partial f}{\partial x}$ and (b) $\dfrac{\partial f}{\partial y}$.

1. $f(x, y) = x^2y + 3xy^2$.
2. $f(x, y) = (x^2 + y^2)\ln(x/y)$.
3. $f(x, y) = \dfrac{x - y}{x + y}$.
4. $f(x, y) = x\ln y$.
5. $f(x, y, z) = xy^2 + e^{-x^2}y - xe^z$.
6. $f(x, y, z) = e^{xy}\cos x - \tan xz$.
7. In Exercise 1 find $\partial^2 f/\partial x^2$ and $\partial^2 f/\partial y^2$.
8. In Exercise 4 find $\partial^2 f/\partial x^2$ and $\partial^2 f/\partial y^2$.
9. In Exercise 1 find $\partial^2 f/\partial x\,\partial y$ and $\partial^2 f/\partial y\,\partial x$.
10. In Exercise 4 find $\partial^2 f/\partial x\,\partial y$ and $\partial^2 f/\partial y\,\partial x$.
11. Compute f_{x_i} if $f: R^n \to R$ is given by $f(\mathbf{X}) = |\mathbf{X}|^4$.
12. Compute f_{x_i} if $f: R^n \to R$ is given by $f(\mathbf{X}) = e^{|\mathbf{X}|}$.

Section 15.2

In Exercises 13–16, use the chain rule to find df/dt. When a value of t is given, find df/dt at that point.

13. $f(x, y) = x^2 + y^2$, $x = \sin t$, $y = \cos t$.
14. $f(x, y) = \cos x - \sin y$, $x = t$, $y = t^2$, $t = 0$.

15. $f(x, y, z) = \sec(xyz)$, $x = t$, $y = e^t$, $z = t^2$, $t = 0$.

16. $f(x, y, z) = x^2 + y^2 + \ln|x|$, $x = \sin t$, $y = \cos t$, $z = e^t$, $t = \pi/2$.

In Exercises 17–20, use a chain rule to find $\partial f/\partial u$ and $\partial f/\partial v$. Where a value of u and v are given, also compute the partial derivatives at the specified point.

17. $f(x, y) = x + y^2 + x^2$, $x = u \cos v$, $y = u \sin v$, $u = 1$, $v = \pi/2$.

18. $f(x, y) = x^2y^4$, $x = \ln u$, $y = e^v$.

19. $f(x, y, z) = z \tan x + ze^y$, $x = u^2$, $y = uv$, $z = uv^3$.

20. $f(x, y, z) = \ln|xyz|$, $x = uv$, $y = u \sin v$, $z = v^2$.

Section 15.3

In Exercises 21–24, compute the gradient of the given function. If a point is specified, compute the gradient at the given point.

21. $f(x, y) = (x - y)^3$, $(2, 1)$.

22. $f(x, y) = xe^{xy}$, $(1, 1)$.

23. $f(x, y, z) = z \ln|xy|$.

24. $f(\mathbf{X}) = |\mathbf{X}|^2$, $(1, 1, \ldots, 1)$.

25. Find two unit vectors normal to the surface $z = 17 - 5x^2 - 4y^2$ at $(-1, 1, 8)$.

26. Find an equation for the plane tangent to the surface $z = 17 - 5x^2 - 4y^2$ at $(-1, 1, 8)$.

Section 15.4

In Exercises 27–32, calculate the directional derivative of each function at the given point and in the indicated direction.

27. $f(x, y) = x^2 + y^2$, $(1, 1)$, $\alpha = \pi/6$.

28. $f(x, y) = x \tan(xy)$, $(1, \pi/4)$, $\mathbf{U} = (2, 3)/\sqrt{13}$.

29. $f(x, y) = x \ln(x^2 + y^2)$, $(1, 0)$, $\mathbf{U} = (3, -1)/\sqrt{10}$.

30. $f(x, y) = \arcsin(xy)$, $(0, 1)$, $\alpha = \pi/4$.

31. $f(\mathbf{X}) = |\mathbf{X}|^2$, $(2, 2, 2, 2)$, $\mathbf{U} = (1, 2, 1, -1)/\sqrt{7}$.

32. $f(\mathbf{X}) = -7|\mathbf{X}|$, $(1, -1, 1, -1)$, $\mathbf{U} = (1, 1, 1, 1)/2$.

Section 15.5

In Exercises 33–42, find and classify all critical points.

33. $f(x, y) = 2x^2 + y^2$.

34. $f(x, y) = x^2 + xy - y^2 + x$.

35. $f(x, y) = x^2 - y^2$.

36. $f(x, y) = x^3 + 6xy + y^3$.

37. $f(x, y) = xe^{-x(1+y)}$.

38. $f(x, y) = x^3 + y^3 + 3x^2y - x - y - 1$.

39. $f(x, y) = x^2y^3 + xy^2 + y - 1$.

40. $f(x, y) = x^4 + y^4 + 16x - y - 5$.

41. $f(x, y) = 6x^2 + 2y^2 - 24x + 36y + 2$.

42. $f(x, y) = x^2 + y^2 + (x - y)^2$.

Section 15.6

In Exercises 43–50, find the absolute maximum value and the absolute minimum value of the given function on the given set.

43. $f(x, y) = 3x - 4y + 14$, $1 \le x \le 4$, $2 \le y \le 6$.

44. $f(x, y) = 2x - y + 80$, $5 \le y \le 15$, $x \le 30$, $x \ge y$.

45. $f(x, y) = x^2 + y^2 + 2y - 10, \quad x^2 + y^2 \leq 1.$

46. $f(x, y) = -2x^2 + x + y^2 - y - 5, \quad 0 \leq x \leq 1, \quad 0 \leq y \leq 2.$

47. $f(x, y) = y^2 - y - x - 5, \quad 0 \leq x \leq 1, \quad 0 \leq y \leq 1.$

48. $f(x, y) = x^2 + 2y - 3, \quad x^2 + y^2 \leq 4.$

49. $f(x, y) = y^2 + 4x - 10, \quad x^2 + y^2 \leq 9.$

50. $f(x, y) = 4x + 9y^2 - 3, \quad x^2 + 9y^2 \leq 9.$

Challenging Problems, Chapter 15

1. A number of sightings are taken on a fixed object. Due to experimental error, its position is recorded as $(x_1, y_1), (x_2, y_2), \ldots, (x_n, y_n)$ during the n observations. The point $(\bar{x}, \bar{y})$ that yields a local minimum of

$$\sum_{i=1}^{n} [D((x_i, y_i), (\bar{x}, \bar{y}))]^2$$

is often taken as the best possible approximation of the position. Find $(\bar{x}, \bar{y})$ in terms of $x_1, y_1, x_2, y_2, \ldots, x_n, y_n$. (This is an example of the method of "least squares.")

2. The n points $(x_1, y_1), (x_2, y_2), \ldots, (x_n, y_n)$ are observations of an experiment. Suppose that theoretically those points are to lie on a straight line, but due to experimental error they fail to have that property. Find the straight line $y = mx + b$ that minimizes

$$f(m, b) = \sum_{i=1}^{n} (mx_i + b - y_i)^2.$$

That straight line *best fits* the data in the "least squares" sense.

3. A function f is called homogenous of degree n if for every positive number k, $f(kx, ky) = k^n f(x, y)$. Show that if f is homogenous of degree n, then

$$x\frac{\partial f}{\partial x} + y\frac{\partial f}{\partial y} = nf(x, y).$$

This result is called Euler's Theorem on homogenous functions. (*Hint*: Take the derivative of both sides of $f(kx, ky) = k^n f(x, y)$.

4. To show that it need not be the case that $f_{xy}(a, b) = f_{yx}(a, b)$, consider the following function

$$f(x, y) = \begin{cases} xy\dfrac{x^2 - y^2}{x^2 + y^2} & \text{if} \quad (x, y) \neq (0, 0) \\ 0 & \text{if} \quad (x, y) = (0, 0). \end{cases}$$

Use the definitions of partial derivatives to show:
(i) $f_x(0, y) = -y$. (ii) $f_{xy}(0, 0) = -1$. (iii) $f_y(x, 0) = x$. (iv) $f_{yx}(0, 0) = 1$.

5. Let f be a function of the two variables x and y. Show that $f_x = f_y$ if and only if f can be written as a differentiable function of the sum of $x + y$. (*Hint*: Let $u = x + y$ and $v = x - y$. Write f as a function of u and v. Then compute f_v.)

6. Let $w = f(x, y, z)$, where $x = r\cos\theta$ and $y = r\sin\theta$. If all the second order partial derivatives of f are continuous, show that

$$\frac{\partial^2 f}{\partial x^2} + \frac{\partial^2 f}{\partial y^2} + \frac{\partial^2 f}{\partial z^2} = \frac{\partial^2 f}{\partial r^2} + \frac{1}{r}\frac{\partial f}{\partial r} + \frac{1}{r^2}\frac{\partial^2 f}{\partial \theta^2} + \frac{\partial^2 f}{\partial z^2}.$$

16

Multiple Integration

16.1 Double Integration

The concept of definite integration of a function of a single variable can be extended to functions of two or more variables. In this section we shall be concerned with the integration of functions of two variables over a bounded region R.

Before beginning this extension of the integration process let's recall the major steps involved in defining the definite integral of a function of a single variable, $\int_a^b f(x)dx$.

1. The interval $a \leq x \leq b$ is subdivided into n subintervals, each of width h. This is accomplished by introducing the points $a = x_0 < x_1 < x_2 < \cdots < x_{n-1} < x_n = b$ in such a way that the difference between every pair of successive points is h, that is,

$$x_i - x_{i-1} = h = \frac{b-a}{n}, \qquad \text{for } i = 1, 2, \ldots, n.$$

2. An arbitrary point x_i^* is then selected from the ith subinterval. Thus $x_{i-1} \leq x_i^* \leq x_i$, for $i = 1, 2, \ldots, n$.
3. The function f is evaluated at each point x_i^*, and each of these values is multiplied by h. The resulting products are added, and finally the limit of that sum is taken as h approaches 0 to obtain $\int_a^b f(x)dx$. That is,

$$\int_a^b f(x)dx = \lim_{h \to 0} \sum_{i=1}^{n} f(x_i^*)h.$$

The motivation for the steps above was the area problem. That is, we arrived at those steps by trying to determine the area bounded by the graph of $y = f(x)$ and the x-axis between a and b. When $f(x) \geq 0$, this can be thought of as the area of the region under the curve $y = f(x)$ and above the 1-dimensional region consisting of the interval $a \leq x \leq b$.

We shall generalize our previous work and motivation to define the *double integral*, $\int_R f(x, y)dA$, of a continuous function $f(x, y)$ over a 2-dimensional, bounded region R in the xy-plane.

The change from a function of one variable to a function of two variables will lead us to change the 2-dimensional area problem to the 3-dimensional volume problem. The volume problem is the problem of determining the volume of the solid under the surface $z = f(x, y)$ and above the 2-dimensional region R in the xy-plane, where $f(x, y) \geq 0$ on R. Figure 16.1.1 shows this sort of solid.

To solve the problem we shall get approximations to the volume by slicing the solid into smaller and smaller columns and then we shall use the approximations to obtain the exact volume.

Figure 16.1.1

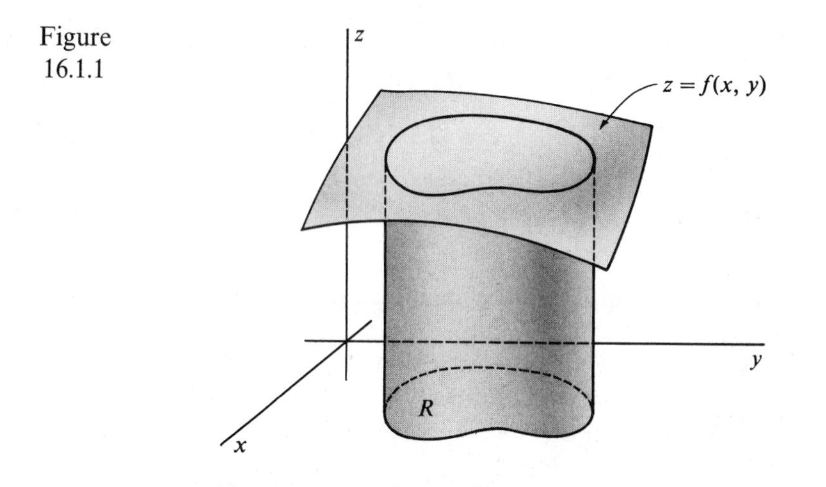

The first step is to divide the region R into subregions. Since R is bounded, it lies between lines of the form $x = a$ and $x = b$ and lines of the form $y = c$ and $y = d$, as shown in Figure 16.1.2.

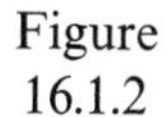
Figure 16.1.2

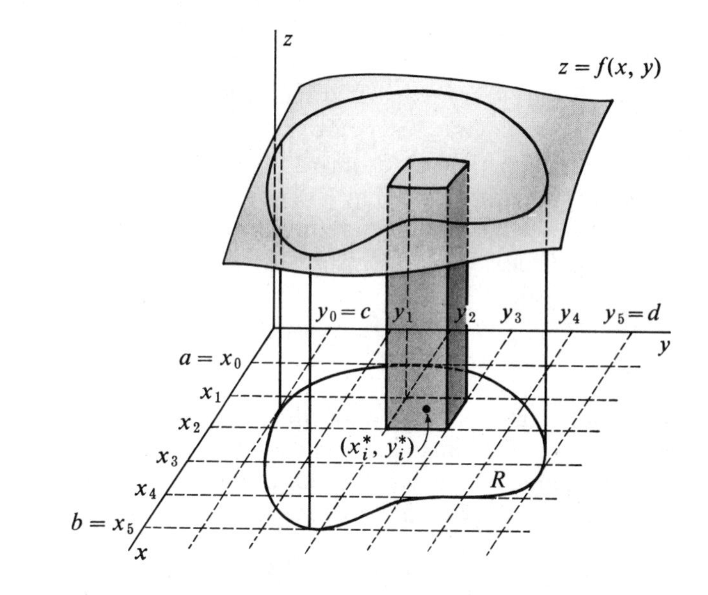

We divide the interval $a \le x \le b$ into n equal subintervals of width h, and the interval $c \le y \le d$ into n equal subintervals of width k. This is accomplished by introducing the points

$$a = x_0 < x_1 < x_2 < \cdots < x_{n-1} < x_n = b$$

on the x-axis and the points

$$c = y_0 < y_1 < y_2 < \cdots < y_{n-1} < y_n = d$$

on the y-axis, so that the difference between successive points on the x-axis is h and on the y-axis is k. That is,

$$x_i - x_{i-1} = h = \frac{b-a}{n} \quad \text{and} \quad y_i - y_{i-1} = k = \frac{c-d}{n}.$$

By drawing lines through those points parallel to the x- and y-axes, R is divided into subregions. This is illustrated in Figure 16.1.2 for a small value of n, namely $n = 5$.

The columns into which the solid is sliced are those directly above the subregions. Altogether there are a certain number of subregions, say m, where $m \geq n$, and hence there are m columns. The subregions are numbered in some order from 1 to m. The ith column is shown in Figure 16.1.2. For each i from 1 to m, we chose an arbitrary point (x_i^*, y_i^*) in the ith subregion. The approximate height of the ith column is $f(x_i^*, y_i^*)$. Thus, if we let A_i be the area of the ith subregion, the approximate volume of the ith column is $f(x_i^*, y_i^*)A_i$. Consequently an approximation to the total volume is

$$f(x_1^*, y_1^*)A_1 + f(x_2^*, y_2^*)A_2 + \cdots + f(x_m^*, y_m^*)A_m.$$

This sum is often written more compactly as follows:

$$\sum_{i=1}^{m} f(x_i^*, y_i^*)A_i.$$

Although this might not be a very good approximation when n is small, it becomes better as n gets larger. In fact, when $f(x, y)$ is continuous on the closed region R and the boundary of R is "reasonably behaved," the limit of this sum exists as n increases without bound. Moreover, this limit is independent of the choices of the points (x_i^*, y_i^*) in their respective subregions, and equals the exact volume. That is, the exact volume is

$$\lim_{n \to \infty} \sum_{i=1}^{m} f(x_i^*, y_i^*)A_i. \qquad (1)$$

Although our motivation is for the case where $f(x, y) \geq 0$ on R, the next definition will be given without that restriction. In the more general situation where $f(x, y)$ may also be negative, the limit in (1) is the volume of the part of the solid above the region R minus the volume of the part of the solid below R.

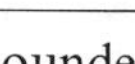

Definition 16.1.1

Let $f(x, y)$ be a function defined over a bounded region R in the xy-plane. Choosing the symbols as above, we call

$$\lim_{n \to \infty} \sum_{i=1}^{m} f(x_i^*, y_i^*)A_i$$

the *double integral* of $f(x, y)$ over the region R, and we denote it by

$$\int_R f(x, y)dA.$$

It should be noted that although the subregions we used to define the double integral were obtained from a rectangular "mesh," this is not necessary. We could have used subregions of any shape. The only restrictions on the subregions are that they completely cover R, that they overlap at most on their boundaries, and that the largest dimension of every subregion approaches zero as the number of subregions increases without bound.

The next theorem indicates the meaning of the words "reasonably behaved" as applied to the boundary of the region R. The proof of the theorem is beyond our present scope and can be found in some advanced calculus books.

Theorem 16.1.1

> Let $f(x, y)$ be continuous on a bounded closed region R, whose boundary consists of a finite number of functions of the form $y = p(x)$ and $x = q(y)$, which have continuous derivatives. Then $\int_R f(x, y)dA$ exists and is independent of the way the points (x_i^*, y_i^*) are chosen in their respective subregions.

Example 1 Use the definition of the double integral to evaluate $\int_R 7\, dA$, where R is the rectangle determined by the intervals $0 \le x \le 2$ and $3 \le y \le 6$.

In this case, $f(x, y) = 7$ for all points (x, y) of R. Thus $f(x_i^*, y_i^*) = 7$ for all i. Also, each A_i is a rectangle of dimensions h and k, where

$$h = \frac{b-a}{n} = \frac{2}{n} \quad \text{and} \quad k = \frac{d-c}{n} = \frac{3}{n},$$

and $m = n^2$. Therefore $A_i = hk = 6/n^2$. Consequently,

$$\int_R 7\, dA = \lim_{n\to\infty} \sum_{i=1}^{n^2} \left(7 \cdot \frac{6}{n^2}\right).$$

Since this is a sum of n^2 terms, all of which are the same, we obtain

$$\int_R 7\, dA = \lim_{n\to\infty} \left(7 \cdot \frac{6}{n^2} \cdot n^2\right) = \lim_{n\to\infty} (7 \cdot 6) = 42.$$

It should be clear that only rather trivial double integrals, such as the one in Example 1, can be conveniently evaluated directly from the definition. What we require and will soon develop is some sort of analog of the Fundamental Theorem of Calculus. That is, we require a theorem that will enable us to compute double integrals in a straightforward manner.

A good way to remember that the volume under the surface $z = f(x, y)$ and above the region R in the xy-plane is equal to $\int_R f(x, y)dA$ is to sketch a picture where a typical column of volume has height $f(x, y)$ and base area dA, as in Figure 16.1.3. Then think of $\int_R$ as summing up the volumes of the columns throughout the region R to get $\int_R f(x, y)dA$. Remember, however, that this is only an intuitive development.

Figure 16.1.3

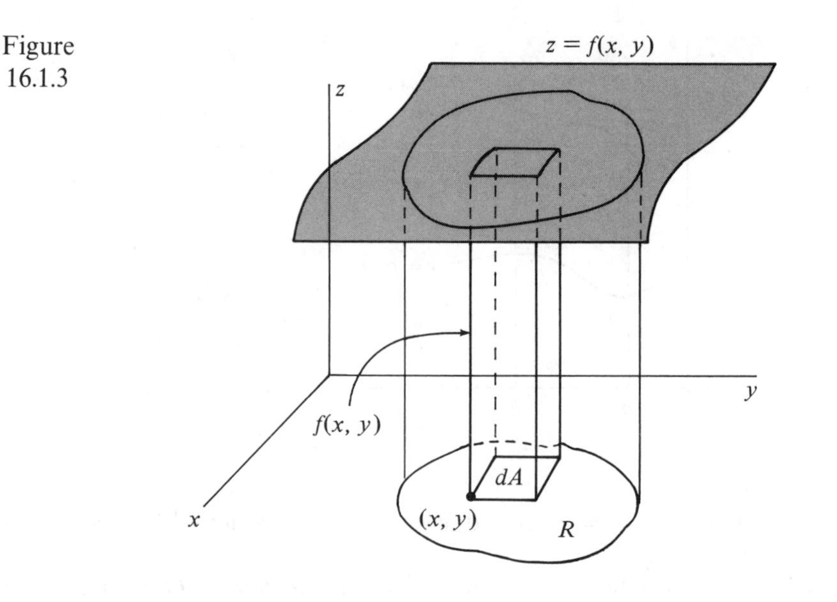

Exercises 16.1

1. Write out in detail the definition of the double integral $\int_R f(x, y)dA$.

For Exercises 2–6, use the definition of the double integral to find the volume of the solid under the graph of $z = f(x, y)$ and above the region R in the xy-plane.

2. $f(x, y) = 3$, R is the rectangle determined by $1 \leq x \leq 5$ and $2 \leq y \leq 10$.
3. $f(x, y) = 6$, R is the rectangle determined by $3 \leq x \leq 10$ and $4 \leq y \leq 9$.
4. $f(x, y) = y/2$, R is the rectangle determined by $2 \leq x \leq 6$ and $1 \leq y \leq 4$.
5. $f(x, y) = x/2$, R is the rectangle determined by $0 \leq x \leq 5$ and $2 \leq y \leq 10$.
6. $f(x, y) = 0$, R is the rectangle determined by $7 \leq x \leq 12$ and $-5 \leq y \leq 0$.

16.2 Iterated Integrals; The Fundamental Theorem for Double Integrals

By considering another approach to the volume problem we shall obtain a technique for evaluating double integrals that is analogous to The Fundamental Theorem of Calculus discussed in Section 4.4. In fact, we shall use The Fundamental Theorem of Section 4.4 to obtain the result.

A second approach to the volume problem is contained in Section 5.1. We found there that if the area $A(x)$ of each cross section made by a plane perpendicular to the x-axis is known for $a \leq x \leq b$ and is a continuous function of x, then the volume between $x = a$ and $x = b$ is

$$\int_a^b A(x)dx.$$

Let R be a region in the xy-plane whose left and right boundaries are the lines $x = a$ and $x = b$. Moreover, assume that the upper and lower boundaries of R

Figure 16.2.1

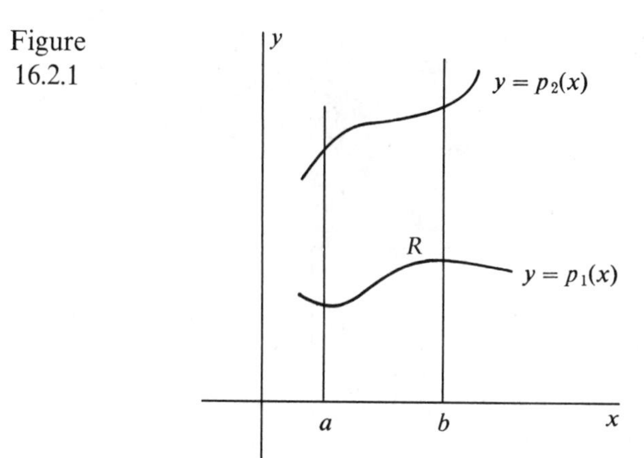

are the continuous curves $y = p_2(x)$ and $y = p_1(x)$. The region R is pictured in Figure 16.2.1.

We shall find the volume of the solid under the surface $z = f(x, y)$ and above the region R.

Intuitive Development. We subdivide the interval $a \leq x \leq b$ into small subintervals each having width dx. This process yields a subdivision of the solid into a number of thin slices, a typical one being shown in Figure 16.2.2.

Figure 16.2.2

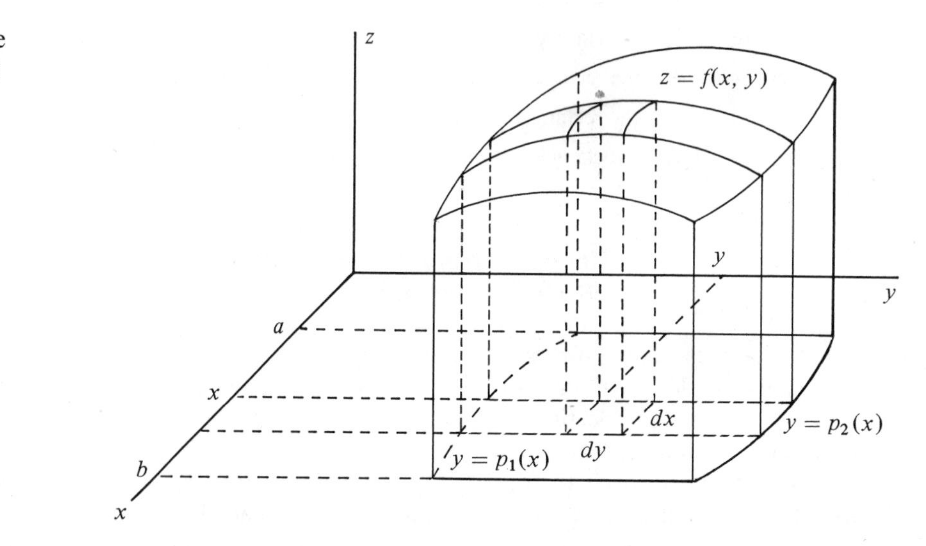

First we get the volume of the indicated slice, noting that for that slice, x is held constant. The slice is divided up into small columns, a typical one having dimensions dx and dy at its rectangular base, as pictured, and height $f(x, y)$. Each column has an approximate volume of $f(x, y)dy\,dx$. Use of the integral sign to sum up the volumes of these columns from $y = p_1(x)$ to $y = p_2(x)$ gives the volume of the slice to be

$$\left(\int_{p_1(x)}^{p_2(x)} f(x, y)dy\right)dx.$$

where x is held constant. Now, to get the total volume, V, we use another integral to sum up the volumes of the slices from $x = a$ to $x = b$. We then get

$$V = \int_a^b \left(\int_{p_1(x)}^{p_2(x)} f(x, y)dy\right)dx.$$

Since x is held constant for $\int_{p_1(x)}^{p_2(x)} f(x, y)dy$, we would evaluate it by looking for a function $F(x, y)$ such that $\dfrac{\partial F}{\partial y} = f(x, y)$. Then

$$\int_{p_1(x)}^{p_2(x)} f(x, y)dy = F(x, y)\Big|_{y=p_1(x)}^{y=p_2(x)} = F(x, p_2(x)) - F(x, p_1(x)).$$

We then complete the integration by calculating the definite integral

$$\int_a^b [F(x, p_2(x)) - F(x, p_1(x))]dx.$$

More Rigorous Development. We assume that on R, f is continuous and $f(x, y) \geq 0$. This situation is illustrated in Figure 16.2.3, where the area of the cross section cut out by a plane at a fixed distance x_0 from the yz-plane is $A(x_0)$.

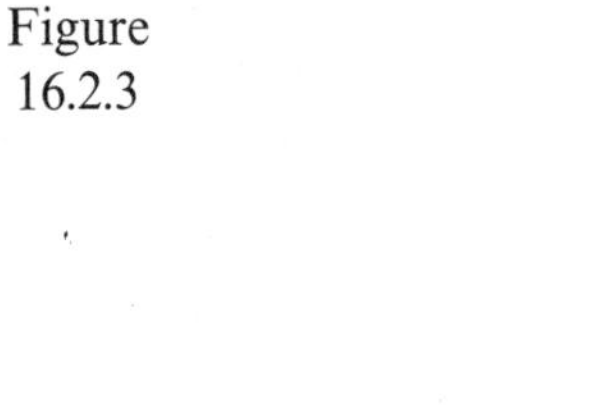

Figure 16.2.3

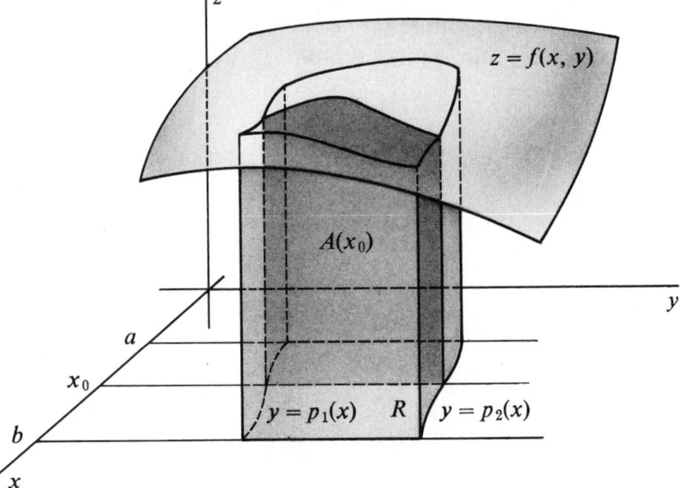

First we determine $A(x_0)$. Note that $A(x_0)$ is the area of the region under the curve $z = f(x_0, y)$ between the vertical lines $y = p_1(x_0)$ and $y = p_2(x_0)$ in the plane $x = x_0$. Thus

$$A(x_0) = \int_{p_1(x_0)}^{p_2(x_0)} f(x_0, y)dy. \tag{1}$$

Using the Fundamental Theorem of Calculus from Section 4.4, we see that if $F(x_0, y)$ is any function of y such that $dF/dy = f(x_0, y)$, then

$$A(x_0) = F(x_0, p_2(x_0)) - F(x_0, p_1(x_0)). \tag{2}$$

Since (2) holds for any x_0 for which $a \leq x_0 \leq b$, we can arrive at a convenient notation by using x in place of x_0. However, we should be careful to realize that up to this point, the x used in place of x_0 in (1) and (2) must be treated as a constant. In terms of x we have

$$A(x) = \int_{p_1(x)}^{p_2(x)} f(x, y)dy = F(x, p_2(x)) - F(x, p_1(x)),$$

for each x such that $a \leq x \leq b$. Since x is treated as a constant, $F(x, y)$ is any function such that

$$\frac{\partial F(x, y)}{\partial y} = f(x, y).$$

It can be proved that since $f(x, y)$, $p_1(x)$, and $p_2(x)$ are continuous, $A(x)$ is continuous. Now that we have $A(x)$, we may use the cross-sectional area method of Section 5.1 to obtain the volume V under consideration. We obtain

$$V = \int_a^b A(x)dx = \int_a^b [F(x, p_2(x)) - F(x, p_1(x))]dx.$$

Thus

$$V = \int_a^b \left(\int_{p_1(x)}^{p_2(x)} f(x, y)dy \right) dx, \tag{3}$$

where the integral inside the parenthesis is evaluated first, treating x as a constant.

Integrals of the type in (3) are of considerable importance and hence are given the name indicated in the next definition.

Definition 16.2.1

An *iterated integral* is an integral of the form

$$\int_a^b \left(\int_{p_1(x)}^{p_2(x)} f(x, y)dy \right) dx,$$

where the integral inside the parenthesis is evaluated first, treating x as a constant.

Similarly we define iterated integrals of the form

$$\int_c^d \left(\int_{q_1(y)}^{q_2(y)} f(x, y)dx \right) dy.$$

For convenience, we usually omit the parenthesis when writing an iterated integral and we write simply

$$\int_a^b \int_{p_1(x)}^{p_2(x)} f(x, y)dy\, dx \quad \text{or} \quad \int_c^d \int_{q_1(y)}^{q_2(y)} f(x, y)dx\, dy.$$

Example 1 | Evaluate $\int_2^3 \int_{2x}^{x^2} 30x^2\, dy\, dx$.

We recall that since x is first treated as a constant, we need to find a function $F(x, y)$ such that $F_y(x, y) = 30x^2$. The function $F(x, y) = 30x^2y$ serves this purpose; so we have

$$\begin{aligned}\int_2^3 \int_{2x}^{x^2} 30x^2\, dy\, dx &= \int_2^3 30x^2y\Big|_{2x}^{x^2} dx\\ &= \int_2^3 [30x^2(x^2) - 30x^2(2x)]dx\\ &= \int_2^3 (30x^4 - 60x^3)dx\\ &= (6x^5 - 15x^4)\Big|_2^3\\ &= 291.\end{aligned}$$

||

Recall that in Section 16.1 we were able to express volumes as double integrals. In this section certain volumes are expressed as iterated integrals. By taking these two approaches we now have the next theorem for the case where $f(x, y) \geq 0$ on R. The theorem provides a technique for evaluating double integrals and is also true without the restriction that $f(x, y) \geq 0$. We shall not give the more general proof, which, however, is very similar to the proof given. The more general proof merely requires careful consideration of the fact that when $f(x_i^*, y_i^*)$ is negative, $f(x_i^*, y_i^*)A_i$ is also negative.

Theorem 16.2.1

The Fundamental Theorem of Calculus for Double Integrals, First Form

Let R be the region in the xy-plane described by $a \leq x \leq b$ and $p_1(x) \leq y \leq p_2(x)$, where $p_1(x)$ and $p_2(x)$ are continuous. If $f(x, y)$ is continuous on R, then

$$\int_R f(x, y)dA = \int_a^b \int_{p_1(x)}^{p_2(x)} f(x, y)dy\, dx.$$

Example 2 | Use a double integral to find the volume of the tetrahedron bounded by the three coordinate planes and the plane $2x + y + 2z = 4$.

A sketch of the graph of the plane $2x + y + 2z = 4$ is indicated in Figure 16.2.4. The surface $2x + y + 2z = 4$ intersects the xy-plane in the line $y = 4 - 2x$, and so the region R is the triangle bounded by the x-axis, the y-axis, and that line. Thus $p_1(x) = 0$, $p_2(x) = 4 - 2x$, $a = 0$, and $b = 2$, as indicated in the figure. Solving $2x + y + 2z = 4$ for z produces $z = f(x, y) = -x - (y/2) + 2$. Therefore the volume V we wish to find is

$$V = \int_R f(x, y)dA = \int_0^2 \int_0^{4-2x} \left(-x - \frac{y}{2} + 2\right)dy\, dx.$$

Figure 16.2.4

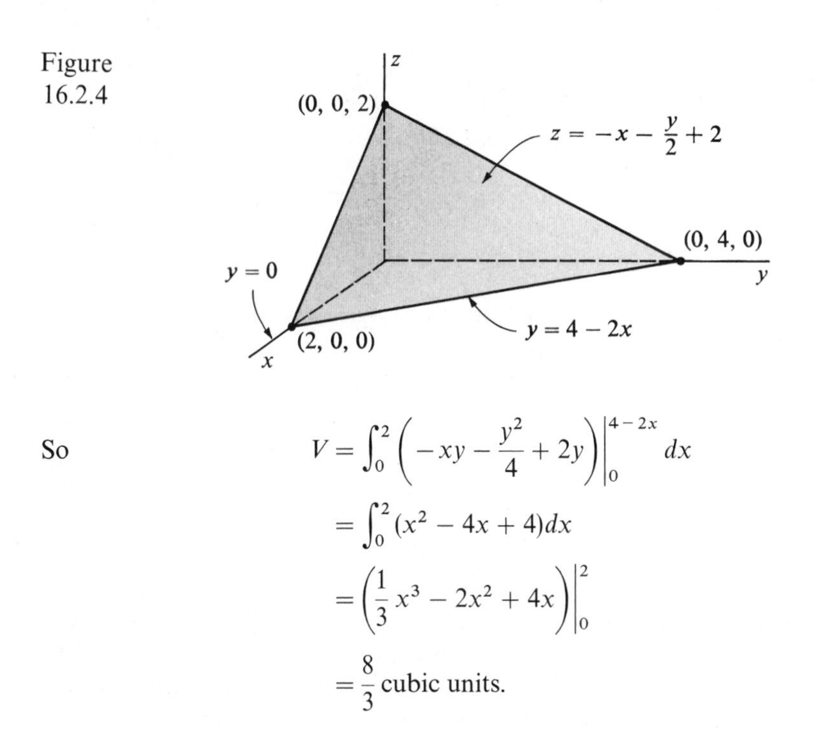

So

$$V = \int_0^2 \left(-xy - \frac{y^2}{4} + 2y\right)\Bigg|_0^{4-2x} dx$$

$$= \int_0^2 (x^2 - 4x + 4)dx$$

$$= \left(\frac{1}{3}x^3 - 2x^2 + 4x\right)\Bigg|_0^2$$

$$= \frac{8}{3} \text{ cubic units.}$$

||

By interchanging the roles of x and y, the proof of Theorem 16.2.1 can be used to establish the second form of The Fundamental Theorem.

Theorem 16.2.2

The Fundamental Theorem of Calculus for Double Integrals, Second Form

Let R be the region in the xy-plane described by $c \le y \le d$ and $q_1(y) \le x \le q_2(y)$, where $q_1(x)$ and $q_2(x)$ are continuous. If $f(x, y)$ is continuous on R, then

$$\int_R f(x, y)dA = \int_c^d \int_{q_1(y)}^{q_2(y)} f(x, y)dx\, dy.$$

(The region R referred to in this theorem is illustrated in Figure 16.2.5.)

Figure 16.2.5

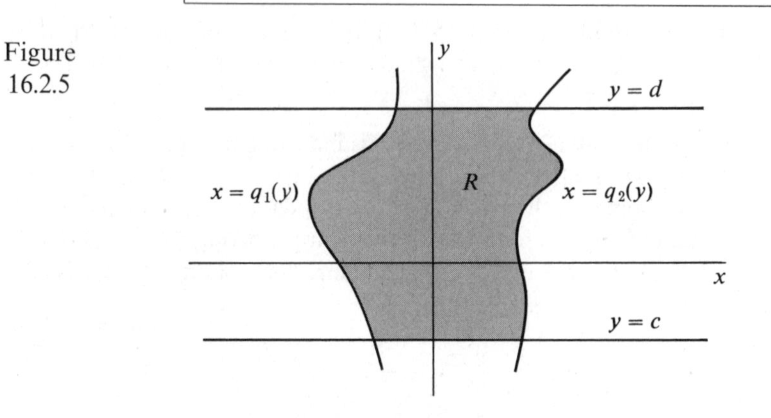

Thus we can express the double integral in two ways as an iterated integral. In one case the integration is first with respect to y and then with respect to x, and in the second case the order of the integration is reversed.

Example 3 The volume of Example 2 can be written also as an iterated integral in which the first integration is with respect to y. From Figure 16.2.4 we see that in this case, $c = 0$ and $d = 4$, while $q_1(y) = 0$ and $q_2(y) = (4 - y)/2$. Thus the volume is also equal to

$$\int_0^4 \int_0^{(4-y)/2} \left(-x - \frac{y}{2} + 2\right) dx\, dy. \qquad ||$$

Sometimes it is more convenient to integrate in one order rather than the other, as is illustrated in the next example.

Example 4 Find the volume of the solid under the cylinder $x^2 + z^2 = 1$ and above the triangle in the xy-plane bounded by $y = x$, $y = 2x$, and $x = 1$.

Figure 16.2.6

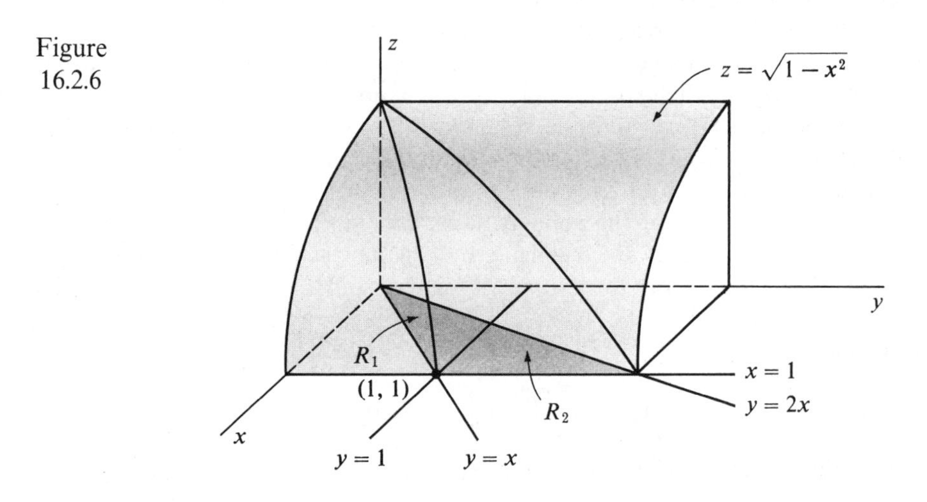

A sketch of the solid appears in Figure 16.2.6, where the region R in the xy-plane is shaded. If we integrate first with respect to y, we obtain

$$\begin{aligned}
\int_0^1 \int_x^{2x} \sqrt{1 - x^2}\, dy\, dx &= \int_0^1 (\sqrt{1 - x^2})y\Big|_x^{2x} dx \\
&= \int_0^1 (1 - x^2)^{1/2} x\, dx \\
&= -\frac{1}{3}(1 - x^2)^{3/2}\Big|_0^1 \\
&= \frac{1}{3} \text{ cubic units.}
\end{aligned}$$

However, if we integrate first with respect to x, we must break R into two separate regions R_1 and R_2 on each side of the line $y = 1$, as shown in the figure. The total volume we are trying to find is then

$$\int_{R_1} \sqrt{1 - x^2}\, dA + \int_{R_2} \sqrt{1 - x^2}\, dA.$$

Using two separate regions is necessary because one boundary of R in the x-direction is made up of two parts. One part has equation $x = y$ and the other part has equation $x = 1$. Writing the integrals as iterated integrals we obtain

$$\int_0^1 \int_{y/2}^{y} \sqrt{1 - x^2}\, dx\, dy + \int_1^2 \int_{y/2}^{1} \sqrt{1 - x^2}\, dx\, dy.$$

Note also that both of those integrals are more difficult to evaluate than in the case where we first integrated with respect to y, since they both require the substitution $x = \sin\theta$. ||

Sometimes the nature of R makes it necessary to use two or more separate subregions regardless of the order of integration. Also, in some cases, even when both orders of integration can be used without subdividing R, one order of integration produces an integral that is considerably easier to evaluate than the other.

Another point to be realized is that we can interchange the roles of y and z or the roles of x and z in all of the results of this section. In particular, we can define the double integral in the expected way for a function $y = f(x, z)$ over a region R in the xz-plane, and we can find the volume of the solid to the left of the surface $y = f(x, z)$ and to the right of the region R in the xz-plane. Similar statements hold for a function $x = f(y, z)$ over a region R in the yz-plane.

To find the volume of the solid between two surfaces $z = f_1(x, y)$ and $z = f_2(x, y)$ above a region R in the xy-plane, we can subtract the volume of the part under the lower surface from the volume of the part under the upper surface. That is, if $f_1(x, y) \le f_2(x, y)$ for all (x, y) in R, then the volume of the solid between those surfaces and above R is

$$\int_R f_2(x, y)dA - \int_R f_1(x, y)dA.$$

However, that difference can also be expressed as one double integral, as the next theorem indicates.

Theorem 16.2.3

> Let $f_1(x, y)$ and $f_2(x, y)$ be continuous on a closed bounded region R, whose boundary consists of a finite number of functions that have the form $y = p(x)$ or $x = q(y)$ and have continuous derivatives. Then
>
> $$\int_R f_2(x, y)dA \pm \int_R f_1(x, y)dA = \int_R [f_2(x, y) \pm f_1(x, y)]dA.$$

Proof. By Theorem 16.1.1, both integrals on the left side of the equation exist. Therefore, using the notation of Section 16.1, the following limits which define those two integrals exist:

$$\lim_{n\to\infty} \sum_{i=1}^{m} f_1(x_i^*, y_i^*)A_i \quad \text{and} \quad \lim_{n\to\infty} \sum_{i=1}^{m} f_2(x_i^*, y_i^*)A_i.$$

Hence, using The Limit Theorem, we have

$$\begin{aligned}
\lim_{n\to\infty} \sum_{i=1}^{m} f_2(x_i^*, y_i^*)A_i \pm \lim_{n\to\infty} \sum_{i=1}^{m} f_1(x_i^*, y_i^*)A_i
&= \lim_{n\to\infty} \left[\sum_{i=1}^{m} f_2(x_i^*, y_i^*)A_i \pm \sum_{i=1}^{m} f_1(x_i^*, y_i^*)A_i \right]. \\
&= \lim_{n\to\infty} \sum_{i=1}^{m} [f_2(x_i^*, y_i^*) \pm f_1(x_i^*, y_i^*)]A_i.
\end{aligned}$$

Consequently, according to the definition of double integrals, we have the desired conclusion:

$$\int_R f_2(x, y)dA \pm \int_R f_1(x, y)dA = \int_R [f_2(x, y) \pm f_1(x, y)]dA.$$

Example 5 Set up an *iterated integral* for the volume of the solid bounded by the upper part of the sphere $x^2 + y^2 + z^2 = 6$ and the paraboloid $x^2 + y^2 = z$.

Figure 16.2.7

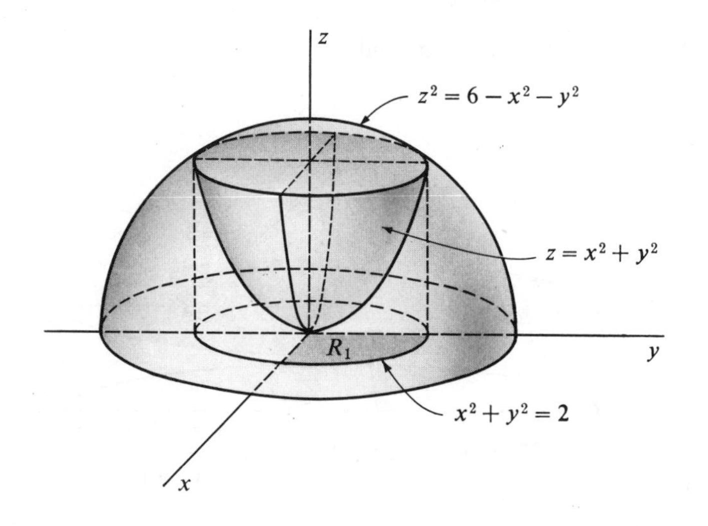

A sketch of the surfaces appears in Figure 16.2.7. First we find the curve of intersection of the surfaces $x^2 + y^2 + z^2 = 6$ and $x^2 + y^2 = z$. Replacing $x^2 + y^2$ by z in the first equation yields $z + z^2 = 6$ or $z^2 + z - 6 = 0$, for which the solutions are $z = -3$ and $z = 2$. Since $x^2 + y^2 = z$, it is clear that z cannot be negative; and hence we keep only $z = 2$ to obtain the intersection

$$x^2 + y^2 = 2, \qquad z = 2.$$

So the intersection is on the cylinder $x^2 + y^2 = 2$ at a height 2 units above the xy-plane. Thus the whole solid lies above the interior of the circle $x^2 + y^2 = 2$ in the xy-plane, and R is the disc $x^2 + y^2 \leq 2$. Consequently the volume is

$$\int_R (\sqrt{6 - x^2 - y^2} - x^2 - y^2)dA,$$

which, by the symmetry of the solid, is four times the volume above the part of R in the first quadrant. That region, R_1, is shaded in the figure. Thus the volume is

$$4 \int_{R_1} (\sqrt{6 - x^2 - y^2} - x^2 - y^2)dA = 4 \int_0^{\sqrt{2}} \int_0^{\sqrt{2-x^2}} (\sqrt{6 - x^2 - y^2} - x^2 - y^2)dy\, dx.$$

This integral will be evaluated in Section 16.5 when we have developed additional techniques. ||

When a solid is partly above and partly below the xy-plane, we must be careful to integrate over the proper region R to obtain the volume. If the upper boundary of the solid has equation $z = f_2(x, y)$ and the lower boundary of the solid has equation $z = f_1(x, y)$, we must integrate $f_2(x, y) - f_1(x, y)$ over R to obtain the volume.

Example 6 | Set up, but do not evaluate, an iterated integral for the volume of the solid bounded above by the cone $x^2 + y^2 = (z - 3)^2$, $z \leq 3$, and below by the plane $z = -2$.

Figure 16.2.8 shows the indicated solid.

Figure 16.2.8

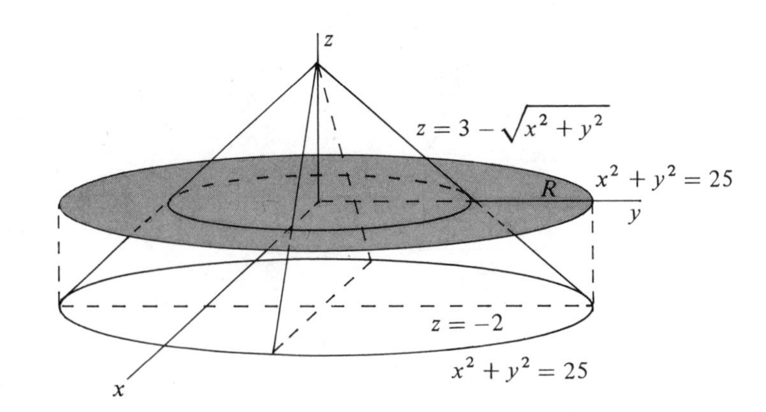

As shown in the figure, the plane $z = -2$ intersects the cone in the circle $x^2 + y^2 = (-2 - 3)^2 = 25$. The region R is the shaded disk shown. Solving the equation of the cone for z we get $z = 3 \pm \sqrt{x^2 + y^2}$. The upper boundary of the solid is the lower portion of the cone, and hence has equation $z = 3 - \sqrt{x^2 + y^2}$. The lower boundary has equation $z = -2$. Thus, we must integrate the following over R:

$$3 - \sqrt{x^2 + y^2} - (-2) = 5 - \sqrt{x^2 + y^2}.$$

Hence, the volume is

$$\int_{-5}^{5} \int_{-\sqrt{25-x^2}}^{\sqrt{25-x^2}} (5 - \sqrt{x^2 + y^2})dy\, dx,$$

or, using the symmetry of the solid, we get

$$4 \int_{0}^{5} \int_{0}^{\sqrt{25-x^2}} (5 - \sqrt{x^2 + y^2})dy\, dx. \qquad \|$$

Let us consider $\int_R f(x, y)dA$ when $f(x, y) = 1$ for all points (x, y) of R. In this case we get the volume of a solid above the region R that has constant height of 1. Since the number of units of this volume is the same as the number of units of the area of R, we can interpret $\int_R 1\, dA$ as the area of R. Thus double integration can also be used to find areas.

Example 7 | Use double integration to find the area of the region bounded by $y = x^2$ and $x = y^2$.

The curves intersect at (0, 0) and (1, 1), as shown in Figure 16.2.9. The desired area is then

$$\int_R 1\, dA = \int_0^1 \int_{x^2}^{\sqrt{x}} dy\, dx$$

$$= \int_0^1 (x^{1/2} - x^2)dx = \frac{1}{3} \text{ sq. units.} \qquad \|$$

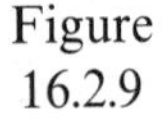
Figure 16.2.9

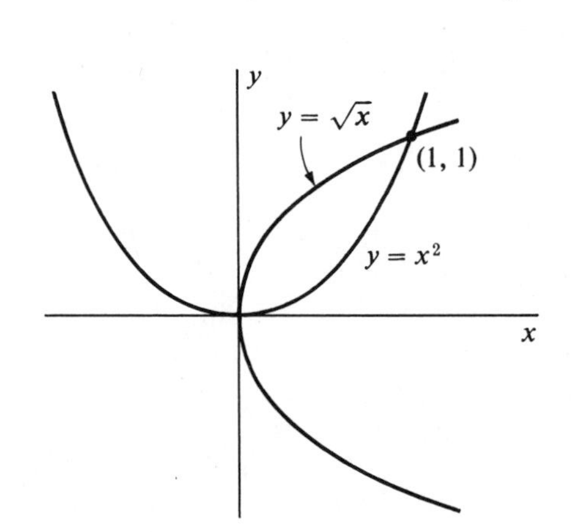

The definition of the double integral, Definition 16.1.1, can be extended to include certain cases where the region R is not bounded. For example, assume that $p_1(x) \leq p_2(x)$ when $x \geq a$, and let R be the unbounded region that is to the right of the vertical line $x = a$ and between the continuous curves $y = p_2(x)$ and $y = p_1(x)$. (See Figure 16.2.10.)

Figure 16.2.10

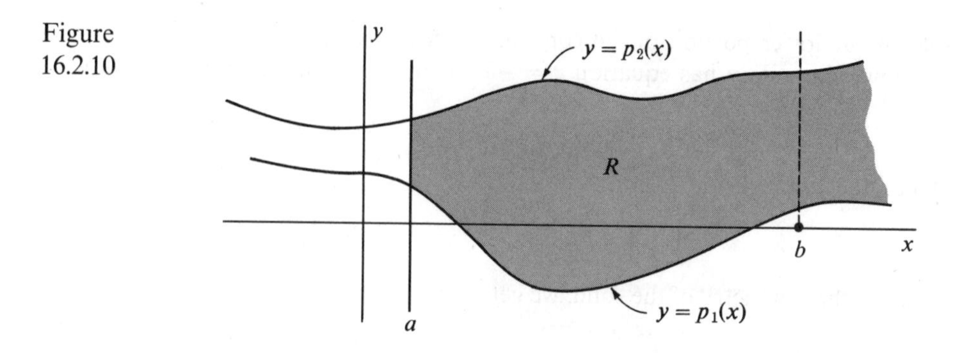

As in the case of improper integrals of functions of a single variable, we note that Definition 16.1.1 applies to the part R_b of the region R of Figure 16.2.10 that lies between the vertical lines $x = a$ and $x = b$, since that part of R is bounded. If we assume also that $f(x, y)$ is continuous, we can use the first form of The Fundamental Theorem to obtain

$$\int_{R_b} f(x, y)dA = \int_a^b \int_{p_1(x)}^{p_2(x)} f(x, y)dy\, dx.$$

We then define

$$\begin{aligned}\int_R f(x, y)dA &= \int_a^\infty \int_{p_1(x)}^{p_2(x)} f(x, y)dy\, dx \\ &= \lim_{b\to\infty} \int_a^b \int_{p_1(x)}^{p_2(x)} f(x, y)dy\, dx\end{aligned}$$

when that limit exists; otherwise we do not define $\int_R f(x, y)dA$.

Example 8 Let $f(x, y) = x + 2y$ and let R be the region above the x-axis, below the curve $y = x^{-3}$, and to the right of the vertical line $x = 1$. Find $\int_R f(x, y)dA$.

Using the results above, we obtain

$$\begin{aligned}\int_R (x + 2y)dA &= \lim_{b\to\infty} \int_1^b \int_0^{x^{-3}} (x + 2y)dy\, dx \\ &= \lim_{b\to\infty} \int_1^b (xy + y^2)\Big|_0^{x^{-3}} dx \\ &= \lim_{b\to\infty} \int_1^b (x^{-2} + x^{-6})dx \\ &= \lim_{b\to\infty} \left(-b^{-1} - \frac{b^{-5}}{5} + 1 + \frac{1}{5}\right) \\ &= \frac{6}{5}.\end{aligned}$$

||

The second form of The Fundamental Theorem can be used to define $\int_R f(x, y)dA$ where R is an unbounded region of the sort illustrated in Figure 16.2.11.

Figure 16.2.11

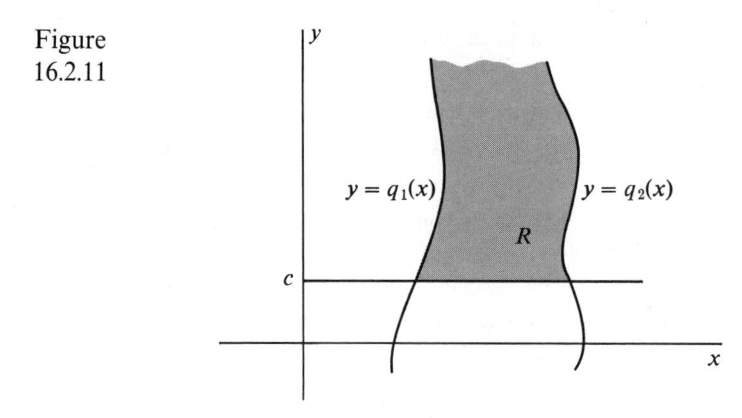

If the limit exists, we have

$$\int_R f(x, y)dA = \int_c^\infty \int_{q_1(y)}^{q_2(y)} f(x, y)dx\,dy$$
$$= \lim_{d\to\infty} \int_c^d \int_{q_1(y)}^{q_2(y)} f(x, y)dx\,dy,$$

for a function $f(x, y)$ continuous on R.

Exercises 16.2

Evaluate the integrals in Exercises 1–16.

1. $\int_{-1}^{3} \int_{x-1}^{x^2} 6xy\,dy\,dx.$

2. $\int_1^2 \int_2^{x^2} 12xy^{-2}\,dy\,dx.$

3. $6\int_0^2 \int_y^3 (x^2 + y^2)dx\,dy.$

4. $\int_0^5 \int_{5-y}^{\sqrt{25-y^2}} x\,dx\,dy.$

5. $\int_0^{2\pi} \int_0^2 r^3\,dr\,d\theta.$

6. $\int_0^{2\pi} \int_0^2 r^3 \cos^2\theta\,dr\,d\theta.$

7. $\int_0^\pi \int_0^{1+\cos\theta} r^2 \sin\theta\,dr\,d\theta.$

8. $\int_0^{\pi/2} \int_1^{\cos\theta} 3r^2\,dr\,d\theta.$

9. $\int_1^2 \int_0^{\pi/2} r^2 \sin\theta\,d\theta\,dr.$

10. $\int_2^3 \int_0^{\pi/4} r\cos\theta\,d\theta\,dr.$

11. $\int_1^2 \int_y^{\sqrt{y}} 18x^2y\,dx\,dy.$

12. $\int_1^3 \int_{y^2}^{y} \frac{x}{x^2 + y^2}\,dx\,dy.$

13. $\int_{-1}^2 \int_0^x y(x^2 + y^2)^{1/2}\,dy\,dx.$

14. $\int_1^2 \int_x^{x^2} \frac{x^2}{y^2}\,dy\,dx.$

15. $\int_1^2 \int_x^{x^2} \frac{1}{x^3y^3}\,dy\,dx.$

16. $\int_2^3 \int_{-y}^{y} x(x^2 + y^2)^2\,dx\,dy.$

17. Find the volume of the tetrahedron bounded by the coordinate planes and the plane $x + y + 2z = 6$.

18. Find the volume of the tetrahedron bounded by the coordinate planes and the plane $x - 2y - z = 10$.

19. Set up, but do not evaluate, an iterated integral for the volume of the solid bounded by the surface $x^2 + 4y^2 + z = 16$ and the xy-plane.

20. Set up, but do not evaluate, an iterated integral for the volume of the solid bounded by the plane $x + y + 2z = 6$, the cylinder $x^2 + y^2 = 9$, and the xy-plane.

21. Find the volume of the solid that is nearest the origin and bounded by the surfaces $y = x$, $y = x^2$, the plane $x + y + 2z = 6$, and the plane $z = 0$.

22. Find the volume of the solid bounded by the cylinders $x^2 + y^2 = 4$ and $y^2 + z^2 = 4$.

23. Set up, but do not evaluate, an iterated integral for the volume of the ellipsoid $x^2 + 4y^2 + 9z^2 = 36$.

24. Use double integration to find the area of the region bounded by $y = x^2$ and $y = x + 6$:
(a) By integrating first with respect to y;
(b) By integrating first with respect to x.

25. Use double integration to find the area of the region bounded by $x = y^2$ and $x + 2y = 8$:
(a) By integrating first with respect to x;
(b) By integrating first with respect to y.

26. Find the volume of the solid above the plane $z = 3$ and under the paraboloid $x^2 + y^2 + z = 4$.

27. Find the volume of the solid above the plane $z = -5$ and under the paraboloid $x^2 + y^2 + z = 4$.

28. Set up, but do not evaluate, one or more iterated integrals for the volume of the solid bounded by the cone $9x^2 + y^2 = (z - 2)^2$ and the planes $z = 0$ and $z = -4$.

29. Set up, but do not evaluate, one or more iterated integrals for the volume of the solid bounded by the planes $z = 0$ and $z = -12$ and the paraboloid $4x^2 + y^2 + z = 4$.

30. Set up, but do not evaluate, an iterated integral for the volume of the solid bounded by the paraboloids $x^2 + 9y^2 + z = 7$ and $x^2 + 9y^2 - z = 11$.

31. Set up, but do not evaluate, an iterated integral for the volume of the solid bounded by the paraboloids $4x^2 + y^2 - z = 7$ and $4x^2 + y^2 + z = 1$.

32. Set up, but do not evaluate, an iterated integral for the volume of the smaller solid above the xy-plane, under $x^2 + 4y^2 + z = 6$, and above the cone $x^2 + 4y^2 = z^2$.

33. Set up, but do not evaluate, an iterated integral for the volume of the upper solid bounded by $x^2 + 2y^2 + z^2 = 6$ and the paraboloid $x^2 + 2y^2 = z$.

34. Set up, but do not evaluate, an iterated integral for the volume of the solid that is to the right of the plane $y = 2$ and that is bounded by the sphere $x^2 + y^2 + z^2 = 29$.

35. Let $f(x, y) = x^{-2}y$, and let R be the region above the x-axis, below the line $y = 2$, and to the right of the line $x = 3$. Find $\int_R f(x, y)dA$.

36. Let $f(x, y) = 6x^{-5}y$, and let R be the region above the x-axis, below the graph of $y = x^2$, and to the right of the line $x = 2$. Find $\int_R f(x, y)dA$.

37. Let $f(x, y) = 24y^{-5}x^2$, and let R be the region that is above the line $y = 1$ and that is bounded by the y-axis and the line $y = 2x$. Find $\int_R f(x, y)dA$.

38. Let $f(x, y) = e^{-y^2}$, and let R be the region bounded by the positive y-axis and the line $y = x$. Find $\int_R f(x, y)dA$.

In Exercises 39–46, reverse the order of integration.

39. $\int_0^2 \int_{y^2}^4 f(x, y)dx\, dy.$

40. $\int_0^3 \int_{y^2}^9 f(x, y)dx\, dy$

41. $\int_0^3 \int_0^{(2/3)\sqrt{9-x^2}} f(x, y)dy\, dx.$

42. $\int_0^5 \int_0^{(3/5)\sqrt{25-x^2}} f(x, y)dy\, dx.$

43. $\int_{-3}^3 \int_0^{(2/3)\sqrt{9-x^2}} f(x, y)dy\, dx.$

44. $\int_{-5}^5 \int_0^{(3/5)\sqrt{25-x^2}} f(x, y)dy\, dx.$

45. $\int_{-2}^3 \int_{y^2}^{y+6} f(x, y)dx\, dy.$

46. $\int_{-5}^4 \int_{y^2}^{20-y} f(x, y)dx\, dy.$

Evaluate each of the integrals.

47. $\int_2^{\infty}\int_0^{x-4}(4x+6y)dy\,dx.$

48. $\int_1^{\infty}\int_0^{x-1}x^{-1}y\,dy\,dx.$

49. $\int_0^{\infty}\int_0^{2y}xe^{-y^3}\,dx\,dy.$

50. $\int_0^{\infty}\int_0^{y^2}42x^2e^{-y^7}\,dx\,dy.$

16.3 Surface Area

Although we have already discussed areas of flat surfaces, that is, areas of planar regions, we have not considered areas of surfaces that do not lie in a plane. We shall now consider the area of the part of a surface that lies above a bounded, closed region R in the xy-plane.

First let's consider a nonvertical plane H and the *acute* angle θ between H and the xy-plane, as shown in Figure 16.3.1. The angle θ is, by definition, the angle between a normal to the plane H and a normal to the xy-plane. That is, θ is the angle between the z-axis and a normal to the plane H.

Figure 16.3.1

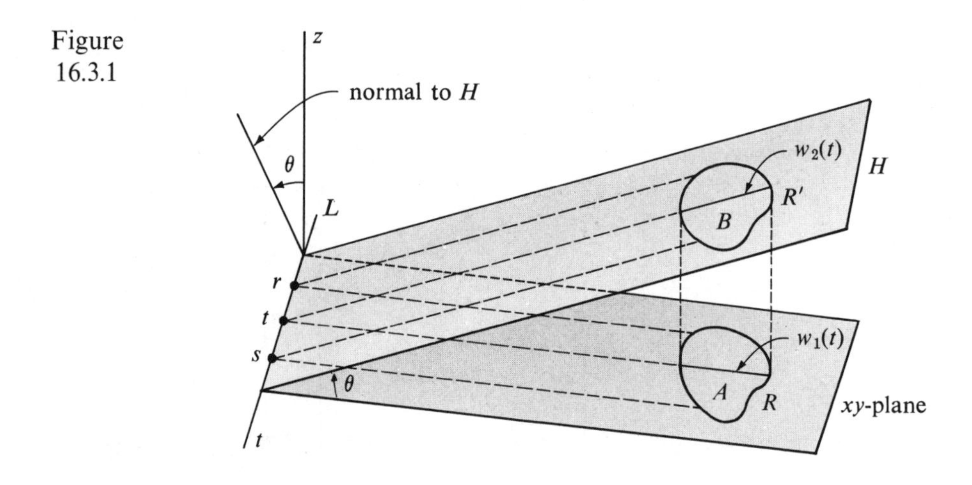

Let A be the area of a region R in the xy-plane, and let B be the area of the region R' vertically above R in the plane H. Let L be the line of intersection of H and the xy-plane. Let t be a coordinate axis on L as indicated, let $w_1(t)$ be the width of the region R, and let $w_2(t)$ be the width of the region R' at t. Then, as in Section 4.6,

$$A=\int_r^s w_1(t)dt \quad \text{and} \quad B=\int_r^s w_2(t)dt.$$

Figure 16.3.2

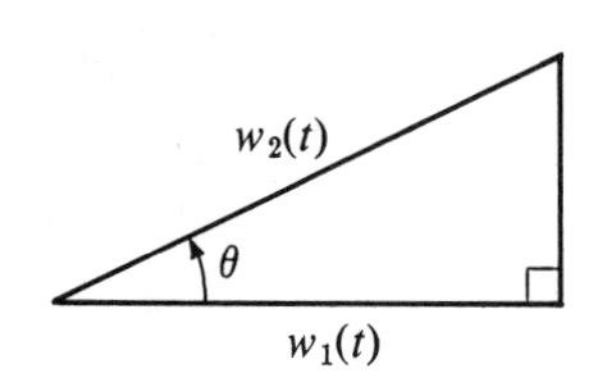

However, because the angle between $w_1(t)$ and $w_2(t)$ is θ, we have $w_2(t) = w_1(t) \sec\theta$. Therefore, since $\sec\theta$ is a constant,

$$B = \int_r^s w_2(t)dt = \int_r^s w_1(t) \sec\theta \, dt$$
$$= \sec\theta \int_r^s w_1(t)dt = A \sec\theta.$$

Thus, $B = A \sec\theta$, and so the larger the acute angle θ, the larger the area B.

Our objective is to determine the surface area, S, of that part of the graph of $z = f(x, y)$ that lies above a closed and bounded region R in the xy-plane. We shall assume that f and its partial derivatives are continuous on R. Although our development is for the case where $f(x, y) \geq 0$ on R, slight modifications will produce the same formula in the general case.

Intuitive Development. Subdivide the region R into m nonoverlapping subregions R_i, $i = 1, \ldots, m$.

Let dA be the area of a typical subregion R_i of R, and let dS be the area of the portion of S that lies above that subregion, where the coordinates are as shown in Figure 16.3.3.

Figure 16.3.3

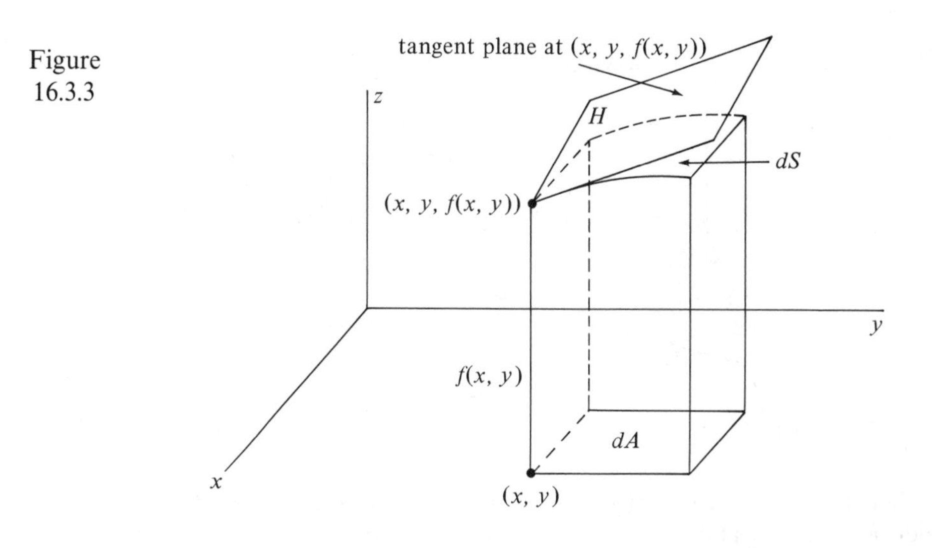

Let H be the tangent plane to the surface $z = f(x, y)$ at the point $(x, y, f(x, y))$. The area of the portion of H that lies above the subregion R_i is an approximation to dS. Thus, $(\sec\theta)dA$ approximates dS, where θ is the acute angle between the plane H and the xy-plane. From our previous work in Section 15.3 we know that

$$\sec\theta = \sqrt{f_x^2(x, y) + f_y^2(x, y) + 1},$$

and so we approximate dS by

$$\sqrt{f_x^2(x, y) + f_y^2(x, y) + 1} \, dA.$$

Use of $\int_R$ to sum up those areas over R gives

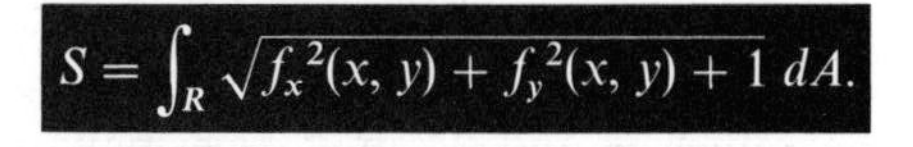

$$S = \int_R \sqrt{f_x^2(x, y) + f_y^2(x, y) + 1}\, dA.$$

More Rigorous Development. We first subdivide the region R into m nonoverlapping subregions R_i, $i = 1, \ldots, m$.

In order to state our assumptions that will lead to the formula S we shall need the following.

1 Let T_i be that portion of the graph of $z = f(x, y)$ that lies above R_i.
2 Let S_i be the area of T_i.
3 Consider the tangent plane to T_i that makes the smallest acute angle with the xy-plane. Let p_i be the area of the portion of this plane that lies above R_i.
4 Consider the tangent plane to T_i that makes the largest acute angle with the xy-plane. Let P_i be the area of the portion of this plane that lies above R_i.

Now we may state our first assumption concerning surface area:

$$p_i \leq S_i \leq P_i.$$

We also assume that surface area is additive. That is, if the region R is divided into nonoverlapping subregions, the total surface area above R is the sum of the surface areas over the separate subregions.

Those two assumptions lead to a double integral for the total surface area of the part of the graph of $z = f(x, y)$ that lies above R. We first subdivide R into m subregions by using equally spaced points on the x- and y-axes, as we did in our consideration of the volume problem. Then, using the assumption of additivity of surface area, we have

$$S = \sum_{i=1}^{m} S_i. \tag{1}$$

To proceed we shall first consider $\sec\theta$, where θ is the acute angle between the xy-plane and a tangent plane to the surface $z = f(x, y)$ at a point (x, y, z). The angle θ is by definition the acute angle between the z-axis and a normal to the tangent plane. As in Section 15.3, we note that

$$\sec\theta = \sqrt{f_x^2(x, y) + f_y^2(x, y) + 1}.$$

Since $f_x(x, y)$ and $f_y(x, y)$ are continuous on R,

$$\sqrt{1 + f_x^2(x, y) + f_y^2(x, y)}$$

is also continuous on the closed region R_i, and hence has an absolute maximum value M and an absolute minimum value m over R_i. But for $0 \leq \theta \leq \pi/2$, $\sec\theta$ is increasing, and so M occurs for the largest value of θ and m occurs for the smallest value of θ. Thus by our first assumption we have

$$mA_i \leq S_i \leq MA_i, \quad \text{or} \quad m \leq \frac{S_i}{A_i} \leq M.$$

However, since the continuous function

$$\sqrt{1 + f_x^{\,2}(x, y) + f_y^{\,2}(x, y)}$$

assumes both the values m and M in R_i, we may apply The Intermediate Value Theorem for Continuous Functions of Two Variables (Theorem 14.8.3) to conclude that there is a point (x_i^*, y_i^*) in R_i such that

$$\frac{S_i}{A_i} = \sqrt{1 + f_x^{\,2}(x_i^*, y_i^*) + f_y^{\,2}(x_i^*, y_i^*)}.$$

So

$$S_i = \sqrt{1 + f_x^{\,2}(x_i^*, y_i^*) + f_y^{\,2}(x_i^*, y_i^*)}A_i.$$

Hence, using (1), we obtain

$$S = \sum_{i=1}^{m} \sqrt{1 + f_x^{\,2}(x_i^*, y_i^*) + f_y^{\,2}(x_i^*, y_i^*)}A_i.$$

Then, since S is independent of n,

$$S = \lim_{n\to\infty} S = \lim_{n\to\infty} \sum_{i=1}^{m} \sqrt{1 + f_x^{\,2}(x_i^*, y_i^*) + f_y^{\,2}(x_i^*, y_i^*)}A_i.$$

But, by definition, the last expression is the double integral

$$\int_R \sqrt{1 + f_x^{\,2}(x, y) + f_y^{\,2}(x, y)}dA.$$

We therefore have

$$S = \int_R \sqrt{1 + f_x^{\,2}(x, y) + f_y^{\,2}(x, y)}dA. \tag{2}$$

Example 1 | Find the area of the triangle cut from the plane $3x - y - z = -6$ by the coordinate planes.

Here $z = 3x - y + 6$ and $\partial z/\partial x = 3$, $\partial z/\partial y = -1$. The region R over which the surface area lies is the triangle formed by the line $y = 3x + 6$ and the x- and y-axes as indicated in Figure 16.3.4. The surface area is thus equal to

$$\int_{-2}^{0}\int_{0}^{3x+6} \sqrt{1 + 3^2 + (-1)^2}\, dy\, dx = \int_{-2}^{0} \sqrt{11}(3x + 6)dx = 6\sqrt{11}. \qquad \|$$

Figure 16.3.4

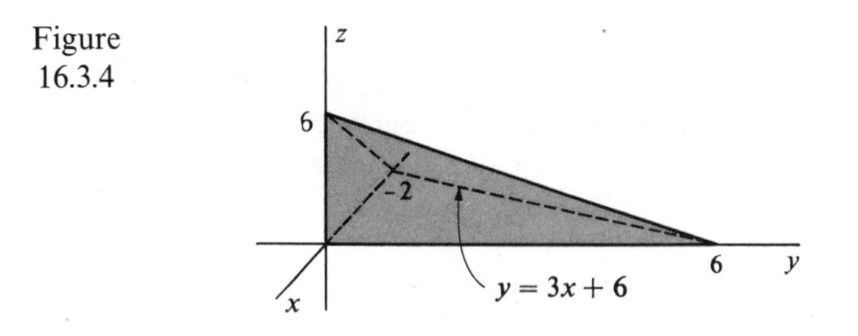

Vector notation can also be used to express Equation (2). If we use

$$g(x, y, z) = z - f(x, y),$$

then

$$\nabla g(x, y, z) = (-f_x(x, y), -f_y(x, y), 1)$$

and

$$|\nabla g(x, y, z)| = \sqrt{f_x^2(x, y) + f_y^2(x, y) + 1}.$$

Substitution into Equation (2) then yields

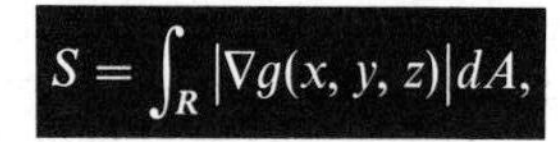

$$S = \int_R |\nabla g(x, y, z)| dA,$$

where $g(x, y, z) = z - f(x, y)$.

Or we can use the vector equation

$$\mathbf{X} = (x, y, f(x, y))$$

for the surface to express the integral for surface area. To accomplish this, note that

$$\frac{\partial \mathbf{X}}{\partial x} \quad (1, 0, f_x(x, y)), \quad \text{and}$$

$$\frac{\partial \mathbf{X}}{\partial y} = (0, 1, f_y(x, y)).$$

Therefore

$$\frac{\partial \mathbf{X}}{\partial x} \times \frac{\partial \mathbf{X}}{\partial y} = \begin{vmatrix} \mathbf{i} & \mathbf{j} & \mathbf{k} \\ 1 & 0 & f_x(x, y) \\ 0 & 1 & f_y(x, y) \end{vmatrix} = (-f_x(x, y), -f_y(x, y), 1).$$

So

$$\left| \frac{\partial \mathbf{X}}{\partial x} \times \frac{\partial \mathbf{X}}{\partial y} \right| = \sqrt{f_x^2(x, y) + f_y^2(x, y) + 1}.$$

Consequently, on substitution into

$$S = \int_R \sqrt{1 + f_x^2(x, y) + f_y^2(x, y)} dA,$$

we have

$$S = \int_R \left| \frac{\partial \mathbf{X}}{\partial x} \times \frac{\partial \mathbf{X}}{\partial y} \right| dA,$$

where $\mathbf{X} = (x, y, f(x, y))$ is a vector equation for the surface. ||

Exercises 16.3

1. Find the area of the triangle cut from the plane $3x - y + 2z = 7$ by the coordinate planes.
2. Find the area of the triangle cut from the plane $2x + y - z = 4$ by the coordinate planes.
3. Set up, but do not evaluate, an iterated integral for the area of the part of the surface $21 - z = x^2 + y^2$ lying above the plane $z = 5$.
4. Set up, but do not evaluate, an iterated integral for the area of the part of the surface $2z = 3x^{2/3} + 3y^{2/3}$ that lies above the triangle in the xy-plane bounded by $y = 0$, $y = x$, and $x = 1$.

5. Set up, but do not evaluate, an iterated integral for the surface area of the part of the sphere $x^2 + y^2 + z^2 = 25$ above the plane $z = 3$.
6. Set up, but do not evaluate, an iterated integral for the area of the part of surface $9 - z = x^2 + y^2$ above the xy-plane.
7. Set up, but do not evaluate, iterated integrals for the surface area of the part of the cylinder $x^2 + z^2 = 9$ that is above the region in the xy-plane bounded by $y = 2x$, $y = 3x$, and $x = 2$: (a) by integrating first with respect to y, (b) by integrating first with respect to x.
8. Set up, but do not evaluate, iterated integrals for the surface area of the part of the sphere $x^2 + y^2 + z^2 = 400$ that is above the region in the xy-plane bounded by $y = 2x$ and $y + 8 = x^2$: (a) where the integration is first with respect to y, (b) where the integration is first with respect to x.
9. Find the area of the part of the surface $z^2 = 9x^2 + 9y^2$ above the xy-plane and below the plane $z = 6$.
10. Find the area of the part of the surface $x^2 + z^2 = y^2$ above the xy-plane and between the planes $y = 0$ and $y = 7$.

16.4 Triple Integration

To extend a concept from functions of two variables to functions of three variables is considerably less difficult than the generalization from functions of one variable to functions of two variables. This is especially true for integrals. However, since it is impossible to graph a function of three variables, we can't rely on our usual illustrations of integrals.

Let T be a bounded closed 3-dimensional region and let $f(x, y, z)$ be a continuous function on T. Since T is bounded, it lies inside a rectangular box determined by $a \leq x \leq b$, $c \leq y \leq d$, and $r \leq z \leq s$, for fixed numbers a, b, c, d, r, and s. As usual we subdivide T into subregions by taking on the coordinate axes equally spaced points that subdivide each of the intervals. The points are taken as $a = x_0 < x_1 < x_2 < \cdots < x_n = b$, $c = y_0 < y_1 < y_2 < \cdots < y_n = d$, and $r = z_0 < z_1 < z_2 < \cdots < z_n = s$. Through each of those points we pass planes that are perpendicular to the axis on which the point lies. Those planes divide the box enclosing T into n^3 smaller rectangular, box-shaped subregions, and divide T into a finite number $m(\leq n^3)$ of subregions. In each of these subregions of T, we take an arbitrary point. Numbering the subregions in some order, we take an arbitrary point (x_i^*, y_i^*, z_i^*) in the ith subregion, for each i from 1 to m.

We then multiply each $f(x_i^*, y_i^*, z_i^*)$ by V_i, the volume of the ith subregion, and we add to get

$$f(x_1^*, y_1^*, z_1^*)V_1 + f(x_2^*, y_2^*, z_2^*)V_2 + \cdots + f(x_m^*, y_m^*, z_m^*)V_m,$$

or more compactly,

$$\sum_{i=1}^{m} f(x_i^*, y_i^*, z_i^*)V_i.$$

The triple integral $\int_T f(x, y\ z)dV$ of f over T is then defined as

$$\int_T f(x, y, z)dV = \lim_{n\to\infty} \sum_{i=1}^{m} f(x_i^*, y_i^*, z_i^*)V_i,$$

The next theorem indicates that triple integrals exist for continuous functions over regions with reasonably behaved boundaries. The proof of the theorem can be found in some advanced calculus tests.

Theorem 16.4.1

> Let $f(x, y, z)$ be continuous on a closed and bounded 3-dimensional region T whose boundary consists of a finite number of continuous functions of the form $z = p(x, y)$, $y = q(x, z)$, and $x = r(y, z)$. Then $\int_T f(x, y, z)dV$ exists and is independent of the way the points (x_i^*, y_i^*, z_i^*) are chosen in their respective subregions.

As in the case of single and double integrals, there is a Fundamental Theorem that enables us to evaluate many triple integrals. As before, a kind of iterated integral is the tool. Although we shall not prove the result, we state it as a theorem.

Theorem 16.4.2

> Let the solid T be bounded above by a surface $z = f_2(x, y)$ and bounded below by a surface $z = f_1(x, y)$, as illustrated in Figure 16.4.1. Also let the solid T lie over a region R in the xy-plane described by $a \le x \le b$ and $p_1(x) \le y \le p_2(x)$. Then
>
> $$\int_T f(x, y, z)dV = \int_a^b \int_{p_1(x)}^{p_2(x)} \int_{f_1(x, y)}^{f_2(x, y)} f(x, y, z)dz\,dy\,dx.$$

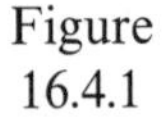
Figure 16.4.1

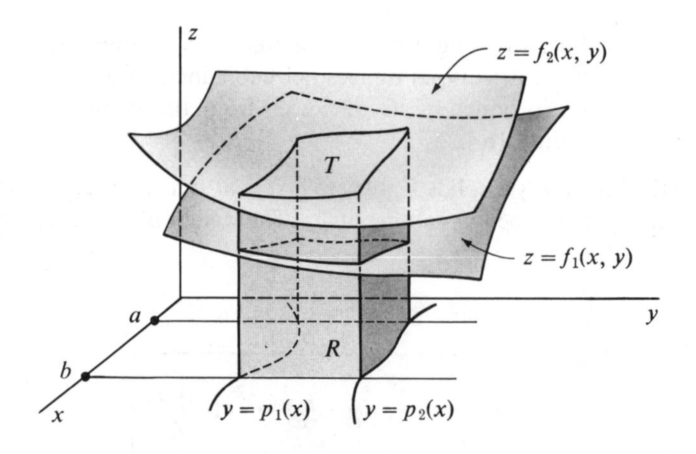

The notation on the right side of that equation means that first we find

$$\int_{f_1(x, y)}^{f_2(x, y)} f(x, y, z)dz$$

by integrating with respect to z, between $f_1(x, y)$ and $f_2(x, y)$, considering both x and y as constants. Then the result is integrated with respect to y, between $p_1(x)$ and $p_2(x)$, treating x as a constant. Finally, that is integrated with respect to x, between a and b.

By interchanging the roles played by x, y, and z and the requirements on the boundary of T, we can express Theorem 16.4.2 in five other forms corresponding to the five other orders of integration. For example, with suitable restrictions, the triple integral can be expressed in the form

$$\int_T f(x, y, z)dV = \int_r^s \int_{g_1(z)}^{g_2(z)} \int_{k_1(y, z)}^{k_2(y, z)} f(x, y, z)dx\,dy\,dz.$$

Example 1 Find $60 \int_T xy\,dV$, where T is the solid bounded by $z = 2$ and $z = x$ above the triangle determined by the x-axis, the y-axis, and the graph of $x + y = 1$ in the xy-plane.

$$\begin{aligned}
60 \int_T xy\,dV &= 60 \int_0^1 \int_0^{-x+1} \int_x^2 xy\,dz\,dy\,dx \\
&= 60 \int_0^1 \int_0^{-x+1} xyz\Big|_x^2\,dy\,dx \\
&= 60 \int_0^1 \int_0^{-x+1} (2xy - x^2y)dy\,dx \\
&= 60 \int_0^1 \left(xy^2 - \frac{x^2y^2}{2}\right)\Bigg|_0^{-x+1} dx \\
&= 60 \int_0^1 \left[x(x^2 - 2x + 1) - \frac{x^2(x^2 - 2x + 1)}{2}\right]dx \\
&= 4.
\end{aligned}$$

||

Example 2 Find the weight of a rectangular box of dimensions 3″ × 5″ × 2″, if the density, in ounces per cubic inch, at any point in the box is the product of its distances from the coordinate planes, as in Figure 16.4.2.

We divide the box into subsolids T_i, $i = 1, 2, \ldots, m$, and we take an arbitrary point (x_i^*, y_i^*, z_i^*) in T_i. Let V_i be the volume of T_i. The weight of T_i is then approximately

Figure 16.4.2

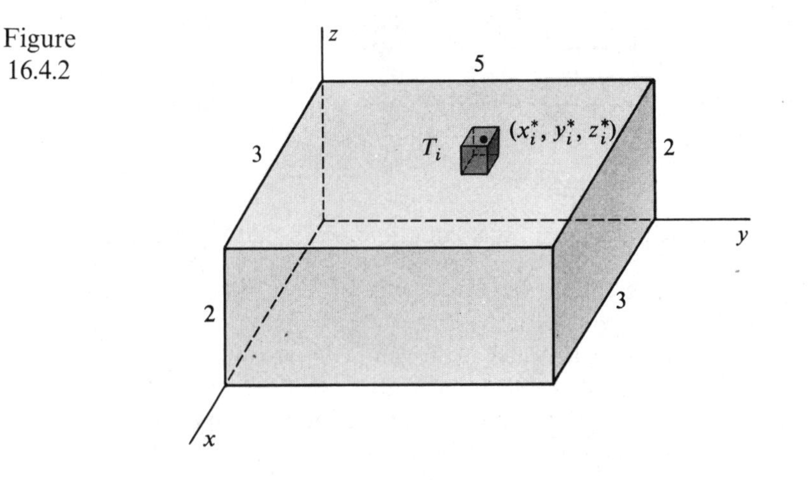

equal to $x_i^* y_i^* z_i^* V_i$. Thus the weight of the whole box is approximately

$$\sum_{i=1}^{m} x_i^* y_i^* z_i^* V_i.$$

It can be shown that the total weight is then the limit of this sum as $m \to \infty$. Hence, the total weight is

$$W = \int_T xyz\, dV.$$

Therefore,

$$\begin{aligned} W &= \int_0^3 \int_0^5 \int_0^2 xyz\, dz\, dy\, dx \\ &= \int_0^3 \int_0^5 2xy\, dy\, dx \\ &= \int_0^3 25x\, dx \\ &= \frac{225}{2} \text{ oz.} \end{aligned}$$

||

In Section 16.2, we saw that $\int_R 1\, dA$ could be used to obtain the area of R. Similarly, triple integrals can be used to obtain volumes. In fact, when the density of T is 1, the number of units of the weight of T is equal to the number of units of the volume of T. Thus $\int_T 1\, dV$ equals the volume of T.

Example 3 | Set up, but do not evaluate, a triple integral for the volume V of the solid bounded by the upper part of the sphere $x^2 + y^2 + z^2 = 6$ and the paraboloid $x^2 + y^2 = z$.

Referring to Figure 16.4.3, we obtain

$$V = 4 \int_0^{\sqrt{2}} \int_0^{\sqrt{2-x^2}} \int_{x^2+y^2}^{\sqrt{6-x^2-y^2}} dz\, dy\, dx.$$

Figure 16.4.3

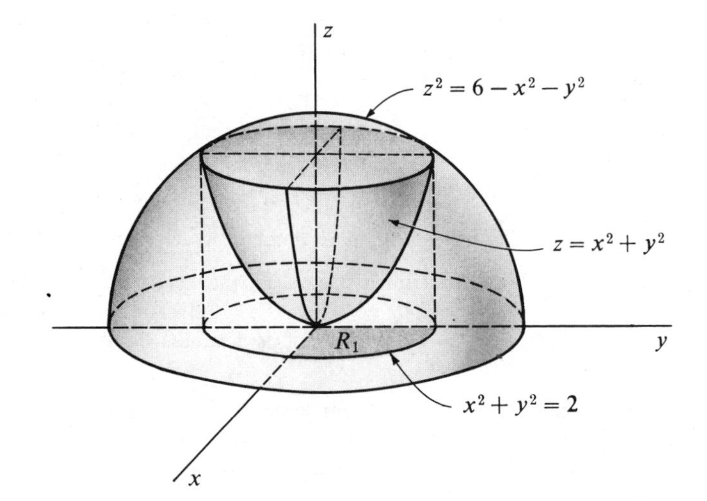

Here, for convenience, we used the symmetry of the figure to express the integration over the part of the volume that occurs where x, y, and z are all positive. ||

Exercises 16.4

In Exercises 1–6, evaluate $\int_T f(x, y, z)dV$.

$z = 12xy^2$

1. $f(x, y, z) = 12xy^2z$ and T is the solid bounded by the planes with equations $x = -1$, $x = 2$, $y = 1$, $y = 3$, $z = 2$, $z = 5$.
2. $f(x, y, z) = x + y - 2z$ and T is the solid bounded by the planes $x = 0$, $x = 2$, $y = -1$, $y = 2$, $z = 2$, $z = 3$.
3. $f(x, y, z) = 6x + 8yz$ and T is the cylinder bounded by the graphs of $x^2 + y^2 = 4$, $z = 3$, and $z = 5$.
4. $f(x, y, z) = x$ and T is the solid bounded by the graphs of $z = 4 - y$, $z = 0$, and $y = x^2$.
5. $f(x, y, z) = 6x + 8yz$ and T is the solid bounded by the graphs of $y = (4 - x^2)^{1/2}$, $y = 0$, $z = 3$, and $z = 5$.
6. $f(x, y, z) = y$ and T is the tetrahedron bounded by the plane $x + 2y + z = 4$ and the coordinate planes.

In Exercises 7–12, set up, but do not evaluate, iterated integrals equal to $\int_T f(x, y, z)dV$ for each given function f and each solid T.

7. $f(x, y, z) = xz$ and T is the solid bounded by the upper part of the ellipsoid $9x^2 + y^2 + z^2 = 90$ and the paraboloid $9x^2 + y^2 = z$.
8. $f(x, y, z) = 2xy^2z$ and T is the solid bounded by the plane $x + y + 2z = 6$, the cylinder $x^2 + 9y^2 = 9$, and the xy-plane.
9. $f(x, y, z) = x^2yz$ and T is the solid above the plane $z = 5$ and under the paraboloid $x^2 + 4y^2 + z = 21$.
10. $f(x, y, z) = 7yz$ and T is the solid bounded by the paraboloids $x^2 + y^2 + z = 13$ and $x^2 + y^2 - z = 5$.
11. $f(x, y, z) = xyz$ and T is the solid bounded by the paraboloid $x^2 + y^2 + 2z = 35$ and the upper part of the cone $x^2 + y^2 = z^2$.
12. $f(x, y, z) = 7yz$ and T is the solid bounded by the cone $9x^2 + 4y^2 = 4z^2$ and the plane $z = 3$.

Evaluate each integral in Exercises 13–18.

13. $\int_1^2 \int_1^x \int_0^y 8x \, dz \, dy \, dx$.
14. $\int_1^2 \int_0^x \int_1^y 72y \, dz \, dy \, dx$.
15. $\int_0^2 \int_1^y \int_1^z 10xy \, dx \, dz \, dy$.
16. $\int_0^1 \int_1^{z^2} \int_1^x 2xz \, dy \, dx \, dz$.
17. $\int_{\pi/6}^{\pi/2} \int_0^{\sin\theta} \int_0^{r\sin\theta} 3r\cos\theta \, dz \, dr \, d\theta$.
18. $\int_0^{\pi/3} \int_0^{\pi/4} \int_0^{\sin\theta} 4\sin\phi\cos\theta \, dr \, d\theta \, d\phi$.
19. Set up, but do not evaluate, an iterated integral for the weight of the tetrahedron bounded by the coordinate planes and the plane $x + 2y + 3z = 12$, if the density, in ounces per cubic unit, at any point is three times the product of its distances from the xz-plane and the yz-plane.
20. A metal shaft shaped as the graph of the cylinder $x^2 + z^2 = 3, 0 \le y \le 6$, has a cost (in dollars per cubic unit) at each point of its interior, equal to twice the square of the distance from the point to the axis of the shaft. Set up an iterated integral to find the total cost of the shaft.

21. Express as an iterated integral, but do not evaluate:

$$\int_T (x - y)dV,$$

where T is the smaller solid bounded by the paraboloid $z = x^2 + y^2$ and the sphere $x^2 + y^2 + z^2 = 20$.

22. Express as an iterated integral, but do not evaluate:

$$\int_T (x^2 + y^2 - z)dV,$$

where T is the solid bounded by $x^2 + z^2 = 9$ and $x^2 + y^2 = 9$.

In Exercises 23–26, set up, but do not evaluate, an iterated triple integral for the volume of the solid bounded by the surfaces given.

23. The coordinate planes and the plane $x + y + 2z = 6$.
24. The surfaces $y = x$ and $y = x^2$, and the planes $x + y + 2z = 6$ and $z = 0$. (Note that there are two such solids. Get the integral only for the one where $z \geq 0$).
25. The xy-plane and the surface $x^2 + y^2 + z = 4$.
26. The cylinders $x^2 + y^2 = 4$ and $y^2 + z^2 = 4$.

16.5 Transformation of Multiple Integrals

The substitution rule for integrals of a single variable gave us a very powerful technique for simplifying certain definite integrals. Likewise, a change of coördinate system resulted in a simplification of many problems. Actually these two techniques, substitution and change of coordinates, are very closely related. In this section we shall consider the corresponding 2-dimensional problem of changing from our standard xy-coordinate system to an arbitrary new uv-coordinate system, where the two coordinate systems are related by the equations $x = f(u, v)$ and $y = g(u, v)$.

For example, if the new coordinate system is the polar coordinate system, we may take u to be r and v to be θ. In this case,

$$x = r\cos\theta \quad \text{and} \quad y = r\sin\theta;$$

so

$$f(r, \theta) = r\cos\theta \quad \text{and} \quad g(r, \theta) = r\sin\theta.$$

Suppose we are given a double integral $\int_R F(x, y)dA$ in the xy-system. Our problem then is to use the relations $x = f(u, v)$ and $y = g(u, v)$ to express this integral in the uv-system.

That is, we must get a double integral $\int_{R'} H(u, v)dA'$ such that

$$\int_R F(x, y)dA = \int_{R'} H(u, v)dA'$$

where the regions R and R' are related by the functions $x = f(u, v)$ and $y = g(u, v)$. We shall assume, in fact, that each point of R corresponds to a single point of R', and vice versa; that is, we assume that there is a *one-to-one* (often written as 1–1) *correspondence* between the points of R and those of R'. That correspondence is pictured in Figure 16.5.1. Since a more rigorous development is exceptionally long and involved, we shall give only the intuitive development.

Figure 16.5.1

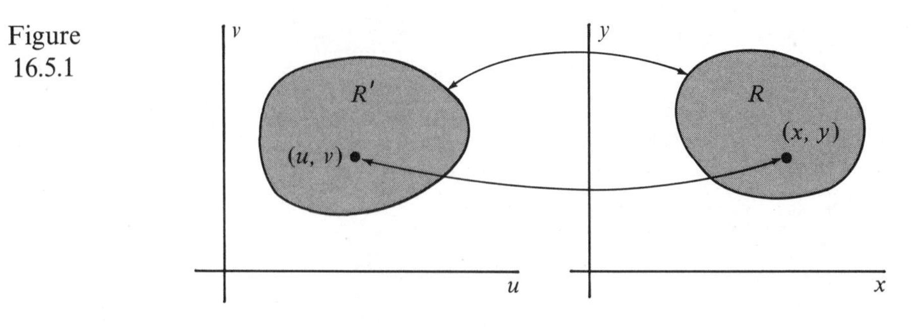

Intuitive Development. To determine the new integral we first subdivide the region R' into small rectangular subregions. The relationship $x = f(u, v)$, $y = g(u, v)$ then gives us a subdivision of R into corresponding subregions, as indicated in Figure 16.5.2, where typical corresponding subregions S' and S have the respective areas dA' and dA.

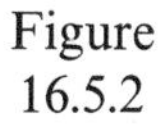
Figure 16.5.2

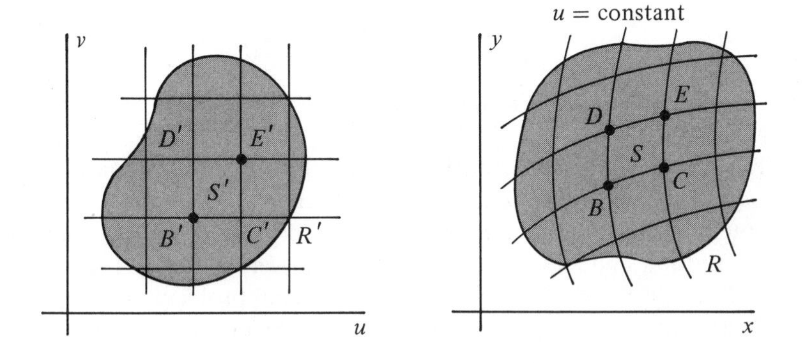

We first concentrate on the relationship between dA' and dA. Consider the rectangle S' having dimensions du and dv with corners that have the coordinates indicated in Figure 16.5.3(a). Then the relationship $x = f(u, v)$, $y = g(u, v)$ determines the coordinates of the points B, C, and D, as shown in Figure 16.5.3(b).

Figure 16.5.3

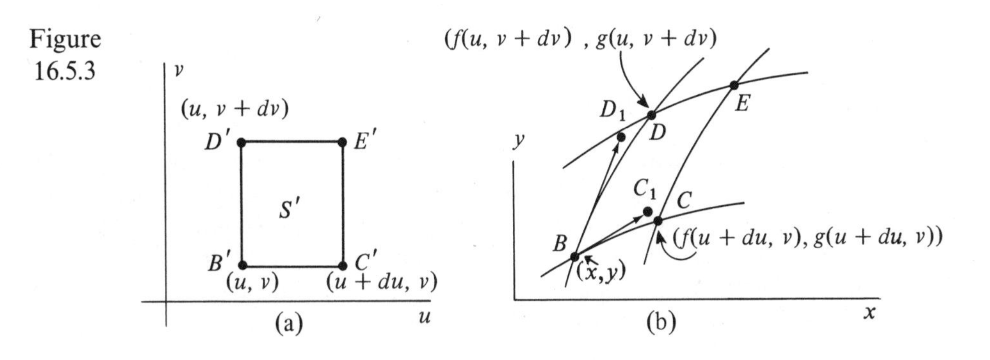

When we move from the point B' to C' in the uv-plane, u changes to $u + du$ while v remains fixed. This corresponds to moving from B to C in the xy-plane. We approximate C by approximating the change in the coordinates (x, y). When u changes to $u + du$ and v remains fixed, $\frac{\partial f}{\partial u}\,du$ approximates the change in x and $\frac{\partial g}{\partial u}\,du$ approximates the change in y. Thus, the point C_1 with coordinates $\left(x + \frac{\partial f}{\partial u}\,du,\ y + \frac{\partial g}{\partial u}\,du\right)$ approximates the point C. Similarly, the point D_1 with coordinates $\left(x + \frac{\partial f}{\partial v}\,dv,\ y + \frac{\partial g}{\partial v}\,dv\right)$ approximates the point D.

Thus, the area of the parallelogram with three of its vertices at B, D_1, and C_1 approximates the area dA of S. We use the vectors whose replicas are shown in Figure 16.5.4 to obtain that approximation to dA. Note that the z-component of each vector is 0 since we are in the xy-plane.

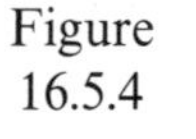

Figure 16.5.4

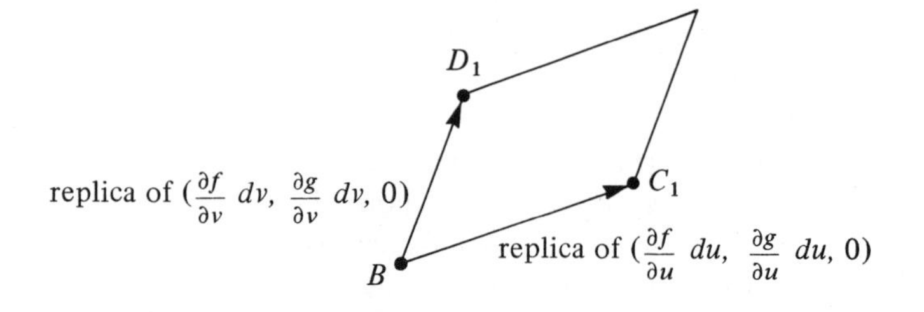

Hence, dA is approximately equal to the absolute value of cross product of those vectors. Now the cross product is equal to

$$\begin{vmatrix} \mathbf{i} & \mathbf{j} & \mathbf{k} \\ \dfrac{\partial f}{\partial u}\,du & \dfrac{\partial g}{\partial u}\,du & 0 \\ \dfrac{\partial f}{\partial v}\,dv & \dfrac{\partial g}{\partial v}\,dv & 0 \end{vmatrix} = \mathbf{k}\left(\frac{\partial f}{\partial u}\frac{\partial g}{\partial v} - \frac{\partial f}{\partial v}\frac{\partial g}{\partial u}\right) du\,dv.$$

Thus, since $dA' = du\,dv$,

$$dA \approx \left|\frac{\partial f}{\partial u}\frac{\partial g}{\partial v} - \frac{\partial f}{\partial v}\frac{\partial g}{\partial u}\right| dA',$$

and so $$F(x, y)dA \approx F(f(u, v), g(u, v))\left|\frac{\partial f}{\partial u}\frac{\partial g}{\partial v} - \frac{\partial f}{\partial v}\frac{\partial g}{\partial u}\right| dA'.$$

We sum up these values over their corresponding regions R and R' with a double integral to get

$$\int_R F(x, y)dA = \int_{R'} F(f(u, v), g(u, v))\left|\frac{\partial f}{\partial u}\frac{\partial g}{\partial v} - \frac{\partial f}{\partial v}\frac{\partial g}{\partial u}\right| dA'.$$

The quantity

$$\frac{\partial f}{\partial u}\frac{\partial g}{\partial v}-\frac{\partial f}{\partial v}\frac{\partial g}{\partial u},$$

which can also be expressed as the determinant

$$\begin{vmatrix} \dfrac{\partial f}{\partial u} & \dfrac{\partial g}{\partial u} \\ \dfrac{\partial f}{\partial v} & \dfrac{\partial g}{\partial v} \end{vmatrix}$$

is called the *Jacobian* of the transformation $x = f(u, v)$, $y = g(u, v)$ and is often written as

$$\frac{\partial(f, g)}{\partial(u, v)}$$

Intuitively *the Jacobian is the factor we must multiply the element of area dA' by in order to obtain the element of area dA.*

Thus we obtain the following theorem.

Theorem 16.5.1

Let the transformation $x = f(u, v)$ and $y = g(u, v)$ map the region R' in uv-space to the region R in xy-space in 1–1 fashion. Assume also that f and g have continuous first partial derivatives and that the Jacobian is not zero. Then

$$\int_R F(x, y)\,dA = \int_{R'} F(f(u, v), g(u, v))\left|\frac{\partial(f, g)}{\partial(u, v)}\right| dA'.$$

A good way to remember that formula is to think of going from $\int_R F(x, y)\,dA$ to $\int_{R'} F(f(u, v), g(u, v))\left|\frac{\partial(f, g)}{\partial(u, v)}\right| dA'$ by replacing R by R', x by $f(u, v)$, y by $g(u, v)$, and the element of area dA by the element of area $\left|\frac{\partial(f, g)}{\partial(u, v)}\right| dA'$.

Example 1 Transform the integral

$$\int_0^1 \int_x^1 xy\,dy\,dx$$

by using the relations $x = u + v$ and $y = u - v$.

Here, $f(u, v) = u + v$ and $g(u, v) = u - v$; so the Jacobian of the transformation is

$$\frac{\partial f}{\partial u}\frac{\partial g}{\partial v}-\frac{\partial f}{\partial v}\frac{\partial g}{\partial u} = (1)(-1) - (1)(1) = -2.$$

Also, we see from the given integral that R is the region shown in Figure 16.5.5. The boundaries of R are $x = 0$, $y = 1$, and $y = x$. Thus, in terms of u and v, the boundaries are $u + v = 0$, $u - v = 1$, and $u - v = u + v$, or

$$v = -u, \quad v = u - 1, \quad \text{and} \quad v = 0.$$

Figure 16.5.5

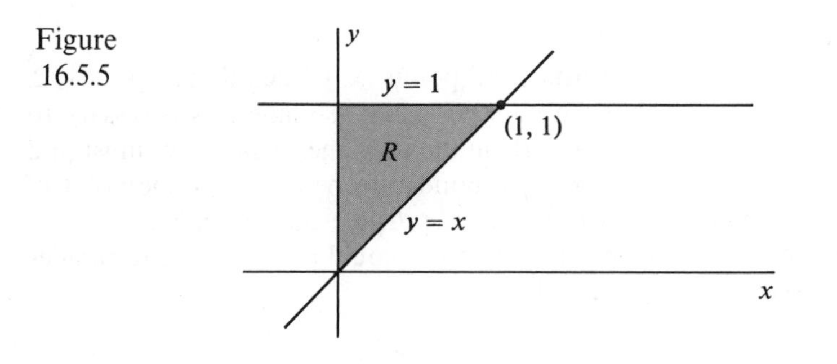

The region R' is indicated in Figure 16.5.6. Using Theorem 16.5.1 we obtain

$$\int_R F(x, y)dA = \int_{R'} F(u + v, u - v)|-2|dA'', \quad \text{or}$$

$$\int_0^1 \int_x^1 xy\, dy\, dx = 2 \int_{-1/2}^{0} \int_{-v}^{v+1} (u^2 - v^2)du\, dv. \qquad \|$$

Figure 16.5.6

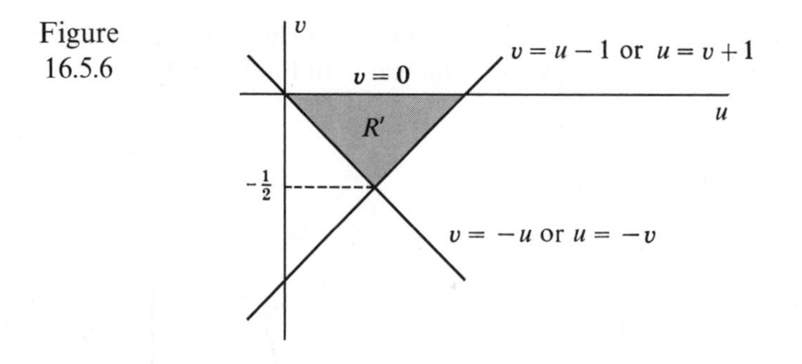

Example 2 Evaluate

$$\int_R (\sqrt{6 - x^2 - y^2} - x^2 - y^2)dA, \tag{1}$$

where R is the set of all points (x, y) such that $x^2 + y^2 \leq 2$.

In Example 5 of Section 16.2 we found that the integral may be expressed as

$$\int_{-\sqrt{2}}^{\sqrt{2}} \int_{-\sqrt{2-x^2}}^{\sqrt{2-x^2}} (\sqrt{6 - x^2 - y^2} - x^2 - y^2)dy\, dx.$$

However, instead of evaluating this integral directly, we shall find it simpler to transform integral (1) by means of the relations

$$x = u \cos v \quad \text{and} \quad y = u \sin v.$$

In this case

$$f(u, v) = u \cos v \quad \text{and} \quad g(u, v) = u \sin v.$$

So the Jacobian of the transformation is

$$\begin{aligned}\frac{\partial f}{\partial u}\frac{\partial g}{\partial v} - \frac{\partial f}{\partial v}\frac{\partial g}{\partial u} &= (\cos v)(u \cos v) - (-u \sin v)(\sin v)\\ &= u \cos^2 v + u \sin^2 v\\ &= u.\end{aligned}$$

The region R in the xy-plane consists of all points (x, y) such that $x^2 + y^2 \leq 2$. To transform an integral over R into an integral in the uv-plane it is necessary to find a corresponding region of integration, R', in the uv-plane. That is, we must find some region R' so that each point in R corresponds to one and only one point of R', and vice versa. Sometimes there will be several possible choices for R'.

In our particular case, for example, the region R' could be any of the rectangles determined by the following sets of inequalities:

$$\begin{aligned} 0 \leq u \leq \sqrt{2}, &\qquad 0 \leq v < 2\pi;\\ -\sqrt{2} \leq u \leq 0, &\qquad 0 \leq v < 2\pi;\\ 0 \leq u \leq \sqrt{2}, &\qquad 2\pi < v \leq 4\pi;\\ -\sqrt{2} \leq u \leq 0, &\qquad \pi \leq v < 3\pi;\\ -\sqrt{2} \leq u \leq \sqrt{2} &\qquad 0 \leq v \leq \pi.\end{aligned}$$

In fact, there are infinitely many choices for R'. A convenient choice is the rectangle R' determined by $0 \leq u \leq \sqrt{2}$ and $0 \leq v < 2\pi$, as indicated in Figure 16.5.7.

Figure 16.5.7

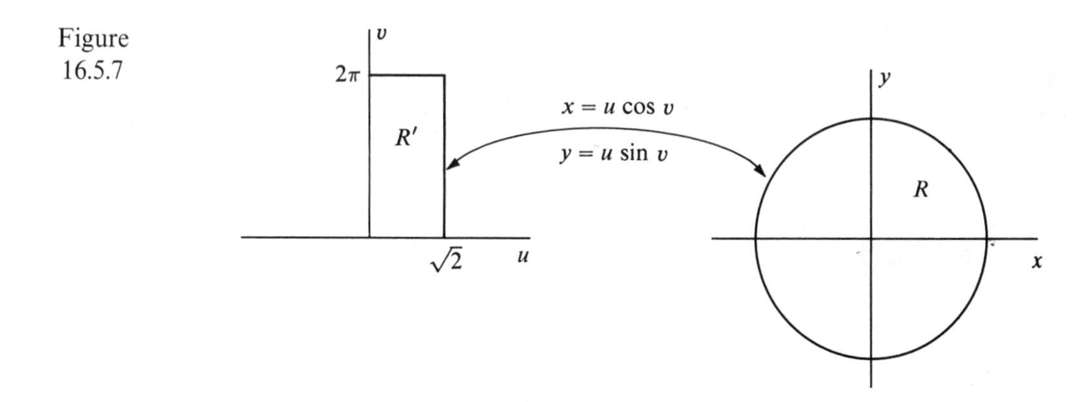

Consequently, since $x^2 + y^2 = u^2$ and since the Jacobian of the transformation is u, we have

$$\int_R (\sqrt{6 - x^2 - y^2} - x^2 - y^2)\,dA = \int_{R'} (\sqrt{6 - u^2} - u^2)|u|\,dA'.$$

Expressing the integral over R' as an iterated integral, we obtain

$$\int_R (\sqrt{6 - x^2 - y^2} - x^2 - y^2)\,dA = \int_0^{\sqrt{2}} \int_0^{2\pi} (\sqrt{6 - u^2} - u^2)|u|\,dv\,du.$$

However, $|u| = u$ over R'; so we get

$$\int_0^{\sqrt{2}} \int_0^{2\pi} (\sqrt{6-u^2} - u^2)u \, dv \, du = 2\pi \int_0^{\sqrt{2}} (\sqrt{6-u^2} - u^2)u \, du$$

$$= 2\pi \left[-\frac{1}{3}(6-u^2)^{3/2} - \frac{u^4}{4} \right]\Bigg|_0^{\sqrt{2}}$$

$$= 2\pi \left[-\frac{8}{3} - 1 + 2\sqrt{6} \right]$$

$$= 2\pi \left(2\sqrt{6} - \frac{11}{3} \right).$$

To summarize, we have

$$\int_R (\sqrt{6-x^2-y^2} - x^2 - y^2) dA = 2\pi \left(2\sqrt{6} - \frac{11}{3} \right). \qquad \|$$

Exercises 16.5

For Exercises 1–12, calculate the Jacobian of the indicated transformation.

1. $x = 3u - 4v, \quad y = 2u + 5v.$
2. $x = u + 2v - 1, \quad y = 6u - 3v + 2.$
3. $x = e^{2u}, \quad y = \sin 3v.$
4. $x = 2 \cos 5v, \quad y = \tan 2u.$
5. $x = u^2v^3, \quad y = 3u^3v^4.$
6. $x = ue^{3v}, \quad y = v^2e^{2u}.$
7. $x = \cot uv, \quad y = \cos uv.$
8. $x = u^v, \quad y = \ln uv.$
9. $x = u, \quad y = v.$
10. $x = v \sin u, \quad y = v \cos u.$
11. $x = e^u, \quad y = e^v.$
12. $x = u \ln v, \quad y = v \ln u.$

For Exercises 13–20, use the given transformation to transform the integral to an iterated integral in u and v.

13. $\int_R (x + y)dA$, where R is the region bounded by $y = 2x$, $x = 1$, and $y = 0$. Use $x = u - v$, $y = 2v$.
14. $\int_R (3x + 2y)dA$, where R is the region bounded by $y = -x$, $y = 2$, and $x = 0$. Use $x = 3u + v$, $y = u - v$.
15. $\int_R \left(\frac{x^2}{16} + \frac{y^2}{9}\right)dA$, where R is the set of all points (x, y) for which $\frac{x^2}{16} + \frac{y^2}{9} \le 1$. Use $x = 4u$, $y = 3v$.
16. $\int_R (x^2 + y)dA$, where R is the set of all points (x, y) for which $x^2 + y^2 \le 4$. Use $x = 2 + 3u$, $y = 1 - 2v$.
17. $\int_{-2}^{0} \int_{-x-2}^{0} (xy + 3)dy \, dx$, using $x = u - v$, $y = u + v$.
18. $\int_{-3}^{0} \int_{x+3}^{3} (x + 2y)dy \, dx$, using $x = 2u - v$, $y = u - 2v$.
19. $\int_0^4 \int_{\sqrt{y}}^{2} y \, dx \, dy$, using $x = 2u$, $y = -v$.
20. $\int_0^1 \int_{y^2}^{\sqrt{y}} x \, dx \, dy$, using $x = 3v$, $y = 2u$.

16.6 Double Integrals in Polar Coordinates

As an application of Theorem 16.5.1 we shall consider the problem of transforming an integral in rectangular coordinates to polar coordinates. In this case we have

$$x = r\cos\theta \quad \text{and} \quad y = r\sin\theta,$$

that is

$$f(r, \theta) = r\cos\theta \quad \text{and} \quad g(r, \theta) = r\sin\theta.$$

Consequently the Jacobian of the transformation is

$$\begin{aligned}\frac{\partial(f, g)}{\partial(r, \theta)} &= \frac{\partial f}{\partial r}\frac{\partial g}{\partial \theta} - \frac{\partial f}{\partial \theta}\frac{\partial g}{\partial r} \\ &= (\cos\theta)(r\cos\theta) - (-r\sin\theta)\sin\theta \\ &= r\cos^2\theta + r\sin^2\theta \\ &= r.\end{aligned}$$

Therefore, by Theorem 16.5.1 we have

$$\int_R F(x, y)\,dA = \int_{R'} F(r\cos\theta, r\sin\theta)|r|\,dA', \tag{1}$$

where a polar region R' is mapped in 1–1 fashion by $x = r\cos\theta$ and $y = r\sin\theta$ onto the region R in rectangular coordinates.

Example 1 Evaluate the following integral by using a transformation to polar coordinates:

$$\int_R \sqrt{x^2 + y^2}\,dA,$$

where R is the set of all points interior to the unit circle centered at the origin.

The points of the region R in the xy-plane correspond in 1–1 fashion to the points of the rectangle determined by $0 \le r \le 1$ and $0 \le \theta < 2\pi$ in the $r\theta$-plane. Thus use of equation (1) along with $x^2 + y^2 = r^2$ yields

$$\int_R \sqrt{x^2 + y^2}\,dA = \int_0^{2\pi}\int_0^1 |r|\,|r|\,dr\,d\theta.$$

Of course, since $0 \le r \le 1$, we have $|r| = r$; and so

$$\int_R \sqrt{x^2 + y^2}\,dA = \int_0^{2\pi}\int_0^1 r^2\,dr\,d\theta = \frac{2\pi}{3}. \qquad \|$$

The reader is warned to distinguish clearly between the $r\theta$-plane and the usual polar coordinates that are superimposed on the xy-plane as a matter of convenience. For example the *rectangle* determined by $1 \le r \le 3$ and $\pi/4 \le \theta \le \pi/2$ in the $r\theta$-plane corresponds to a sector of an annulus in the xy-plane, as pictured in Figure 16.6.1. Although Figure 16.6.1(b) illustrates a subset of the xy-plane, points of this plane are often *labeled* with their corresponding (r, θ) coordinates, as in Figure 16.6.2.

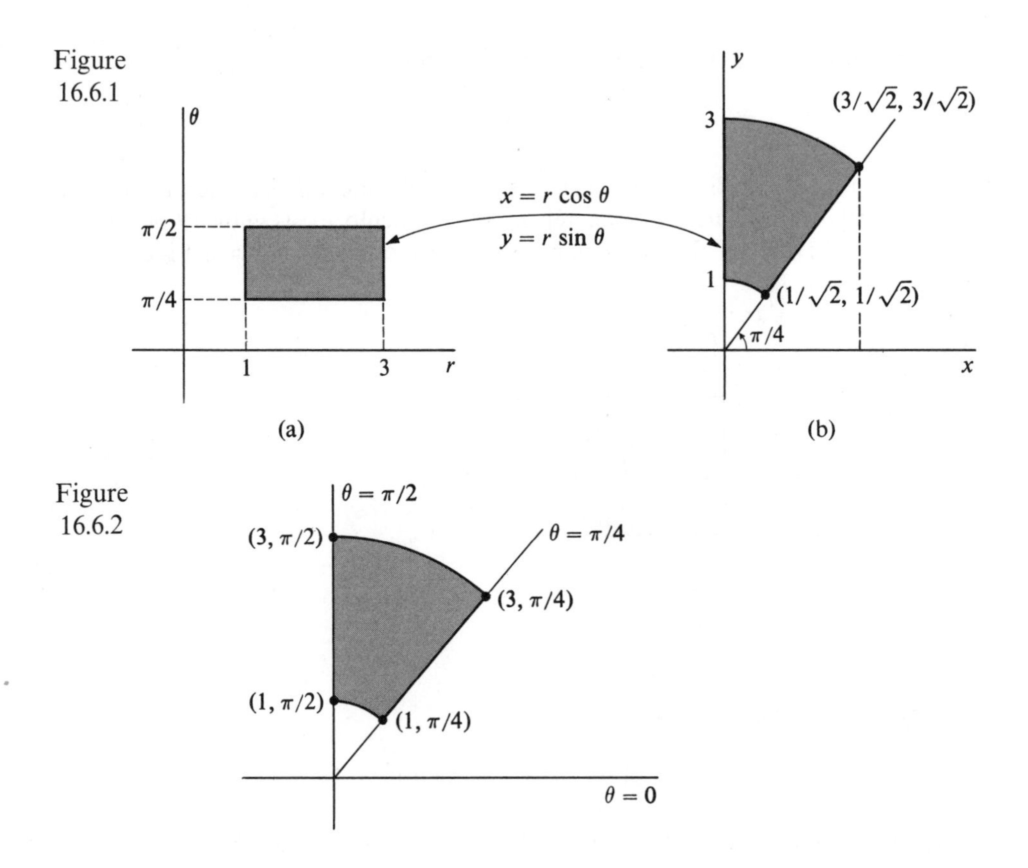

Figure 16.6.1

Figure 16.6.2

When points in the xy-plane are so labeled, we refer to this as the polar coordinate system. Consequently the polar coordinate system may be obtained by renaming the points in the xy-plane with their corresponding names from the $r\theta$-plane. Of course this correspondence of names is determined by the relations

$$x = r\cos\theta \quad \text{and} \quad y = r\sin\theta.$$

With these facts in mind we shall again derive the formula for area using an iterated integral in polar coordinates. Suppose that we wish to use polar coordinates to find the area A of the region illustrated in Figure 16.6.3, where $f(\theta) \geq 0$ and $\alpha \leq \theta \leq \beta$.

Figure 16.6.3

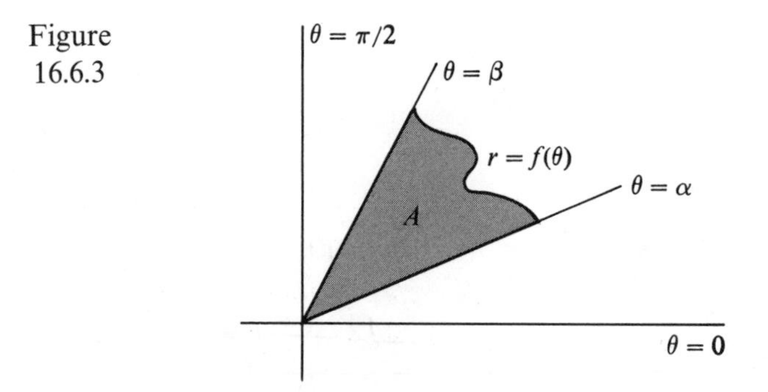

In terms of a double integral the area is given by

$$A = \int_R dA.$$

However, it could be difficult to express that double integral as an iterated integral in rectangular coordinates. Fortunately no such difficulty exists in the $r\theta$-plane. The region R' in the $r\theta$-plane that corresponds in 1–1 fashion to R is shown in Figure 16.6.4.

Figure 16.6.4

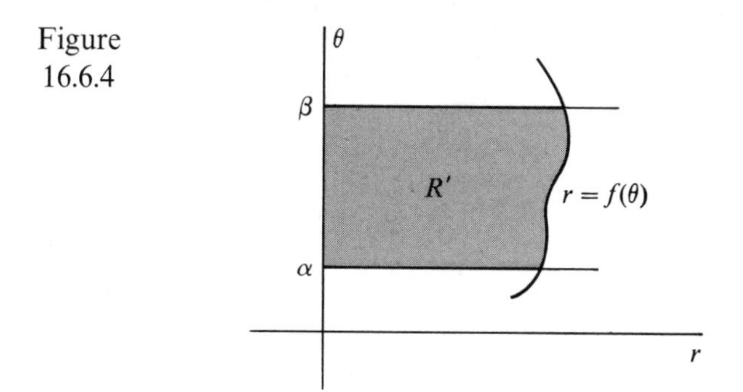

Use of equation (1) gives

$$A = \int_R dA = \int_{R'} |r|\, dA'.$$

On replacing the last double integral with an iterated integral we get

$$A = \int_\alpha^\beta \int_0^{f(\theta)} |r|\, dr\, d\theta.$$

A good way to remember that formula for $r > 0$ is to think of the typical element of area dA as being the area of the region bounded by circles of radii r and $r + dr$ and angles of magnitude θ and $\theta + d\theta$, as in Figure 16.6.5. Thus dA is approximately the area of a rectangle of dimensions $r\, d\theta$ and dr; that is, think of dA as being equal to $r\, dr\, d\theta$. Then think of $\int_\alpha^\beta \int_0^{f(\theta)}$ as summing up those elements of area to get

$$A = \int_\alpha^\beta \int_0^{f(\theta)} r\, dr\, d\theta, \qquad \text{when } r > 0.$$

Figure 16.6.5

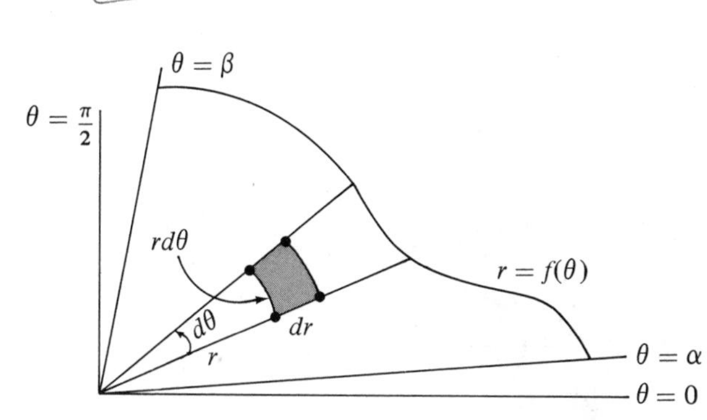

Similarly, we shall also obtain a formula for volume as an iterated integral in polar coordinates. We consider the volume V under a surface $z = f(x, y)$ and above a region R in the xy-plane of the type pictured in Figure 16.6.6, where the polar coordinate system is superimposed on the xy-plane. Since

$$V = \int_R F(x, y)dA,$$

we obtain

$$V = \int_{R'} F(r\cos\theta, r\sin\theta)|r|dA',$$

where R' is the corresponding region in the $r\theta$-plane. As an iterated integral this becomes

$$V = \int_{\theta_1}^{\theta_2} \int_{f_1(\theta)}^{f_2(\theta)} F(r\cos\theta, r\sin\theta)|r|dr\, d\theta.$$

Figure 16.6.6

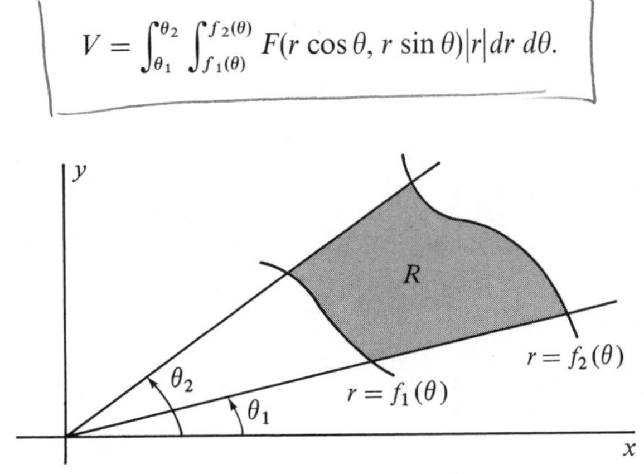

Example 2 | Use polar coordinates to calculate the volume of the solid under the part of the paraboloid $z = x^2 + y^2$ and above the region R indicated in Figure 16.6.7.

Figure 16.6.7

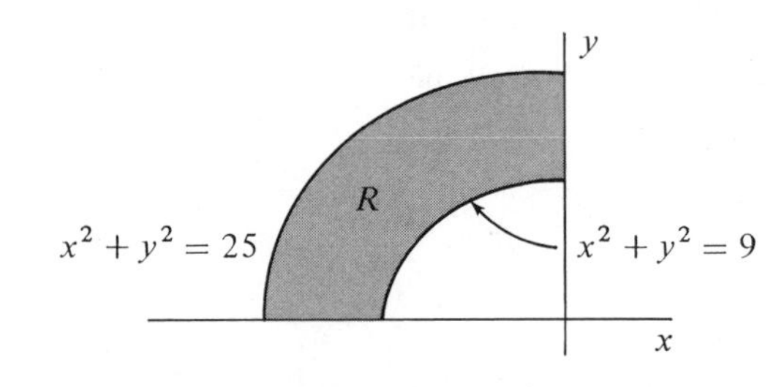

Since $F(x, y) = x^2 + y^2$, we have $F(r\cos\theta, r\sin\theta) = r^2$. The boundaries of R expressed in polar coordinates are $r = 3$, $r = 5$, $\theta = \pi/2$, and $\theta = \pi$. Thus, the volume V is

$$V = \int_{\pi/2}^{\pi} \int_3^5 r^2\, r\, dr\, d\theta$$

$$= \int_{\pi/2}^{\pi} \frac{r^4}{4}\Big|_3^5 d\theta = \int_{\pi/2}^{\pi} 136\, d\theta$$

$$= 68\pi \text{ cubic units.}$$ ||

Exercises 16.6

In Exercises 1–12, transform the given integral to an iterated integral in polar coordinates and evaluate the resulting integral.

1. $\int_R e^{(x^2+y^2)}\,dA$, where R is bounded by the unit circle centered at the origin.
2. $\int_R (x^2+y^2)^{3/2}\,dA$, where R is the set of points (x, y) for which $x^2+y^2 \le 2$ and $y \ge 0$.
3. $\int_R x(x^2+y^2)^{1/2}\,dA$, where R is the set of points (x, y) for which $1 \le x^2+y^2 \le 4$ and $0 \le y \le x$.
4. $\int_R (x^2+y^2)dA$, where R is bounded by the circle $(x-1)^2+y^2=1$.
5. $\int_0^6 \int_0^{\sqrt{36-x^2}} (x^2+y^2)dy\,dx$.
6. $\int_0^5 \int_0^x \sqrt{x^2+y^2}\,dy\,dx$.
7. $\int_0^6 \int_0^{\sqrt{6x-x^2}} (x^2+y^2)dy\,dx$.
8. $\int_0^1 \int_0^{\sqrt{1-x^2}} e^{x^2+y^2}\,dy\,dx$.
9. $\int_0^2 \int_{-\sqrt{4-y^2}}^{\sqrt{4-y^2}} e^{4-x^2-y^2}\,dx\,dy$.
10. $\int_0^1 \int_0^y x\,dx\,dy$.
11. $\int_0^3 \int_0^{\sqrt{9-x^2}} \sqrt{9-x^2-y^2}\,dy\,dx$.
12. $\int_{-1}^1 \int_{-\sqrt{1-x^2}}^{\sqrt{1-x^2}} \sqrt{4x^2+4y^2+1}\,dy\,dx$.
13. Find the surface area of the paraboloid $z = x^2+y^2-4$ that lies below the xy-plane.
14. Find the surface area of the portion of the hemisphere $x^2+y^2+z^2=8$, $z \ge 0$ that is cut out by the cone $x^2+y^2=z^2$.

In Exercises 15–18, find the area of the given region by using a double integral.

15. The region inside $r = \sin 2\theta$.
16. The region inside $r = 2 + \sin\theta$.
17. The region inside $r = 1 + \sin\theta$ and outside $r = 1$.
18. The region inside $r = 2 + \cos\theta$ and outside $r = 1 + \cos\theta$.

In Exercises 19–22, find the volume of the given solid by transforming a double integral to polar coordinates.

19. The solid bounded by $x^2+y^2=4$, $z=-1$, and $z=2$.
20. The solid bounded by $x^2+y^2=16$, $z=0$, and $z=(25-x^2-y^2)^{1/2}$.
21. The solid bounded by $x^2+y^2=9$, $z=0$, and $z=x^2+y^2$.
22. The smaller solid bounded by $x^2+y^2+z^2=8$ and $z=(x^2+y^2)^{1/2}$.

16.7 Transformation of Triple Integrals

The problem of transforming a triple integral

$$\int_T F(x, y, z)dV$$

by means of equations $x = f(u, v, w)$, $y = g(u, v, w)$, and $z = h(u, v, w)$ is closely related to the corresponding problem for double integrals. In fact, the proof of Theorem 16.5.1 may be extended to this new case. Since the extension is involved and tedious, we shall only give an intuitive development of the basic results.

The problem is to obtain a triple integral $\int_{T'} H(u, v, w)dV'$ such that

$$\int_T F(x, y, z)dV = \int_{T'} H(u, v, w)dV',$$

where the regions T and T' are related by the functions $x = f(u, v, w)$, $y = g(u, v, w)$, and $z = h(u, v, w)$ in a 1–1 correspondence.

To obtain the new integral we must find the relationship between dV and dV'.

Intuitive Development. We first subdivide the region T' into a number of small rectangular boxes. The 1–1 correspondence given by $x = f(u, v, w)$, $y = g(u, v, w)$, and $z = h(u, v, w)$ then yields a corresponding subdivision of the region T.

Consider a typical subregion T_i' of T' having dimensions du, dv, and dw, and having the particular corner points with coordinates as indicated in Figure 16.7.1(a). Note that if dV' denotes the volume of T_i', we have $dV' = du\, dv\, dw$.

Figure 16.7.1

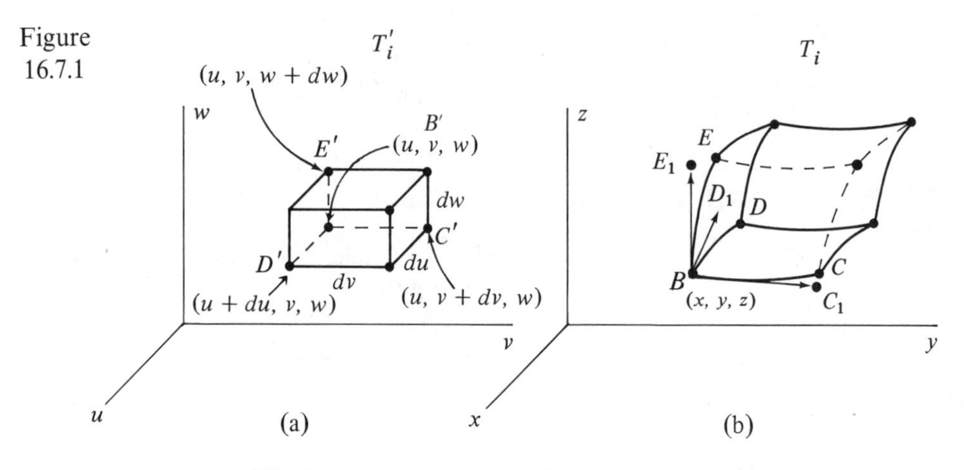

The 1–1 correspondence determines the subregion of V that corresponds to T_i'. Call that subregion T_i. In particular, the 1–1 correspondence determines the points B, C, D, and E of T_i that correspond to the corner points B', C', D', and E' of T_i', as shown in Figure 16.7.1.

When we move from the point B' to C' in uvw-space, v changes to $v + dv$, while u and w remain fixed. This corresponds to moving from B to C in xyz-space. We approximate C by approximating the change in the coordinates (x, y, z). When v changes to $v + dv$, and u and w remain fixed, $\dfrac{\partial f}{\partial v}dv$ approximates the change in x, $\dfrac{\partial g}{\partial v}dv$ approximates the change in y, and $\dfrac{\partial h}{dv}dv$ approximates the change in z. Thus, the point C_1 with coordinates $\left(x + \dfrac{\partial f}{\partial v}dv,\ y + \dfrac{\partial g}{\partial v}dv,\ z + \dfrac{\partial h}{\partial v}dv\right)$ approximates the point C. Similarly, the points D_1 and

E_1 with respective coordinates $\left(x + \frac{\partial f}{\partial u} du, y + \frac{\partial g}{\partial u} du, z + \frac{\partial h}{\partial u} du\right)$ and $\left(x + \frac{\partial f}{\partial w} dw, y + \frac{\partial g}{\partial w} dw, z + \frac{\partial h}{\partial w} dw\right)$ respectively approximate the points D and E.

Thus, the volume of the parallelepiped with four of its vertices at B, C_1, D_1, and E_1 approximates the volume dV. We use the vectors whose replicas are shown in Figure 16.7.2 to obtain that approximation to dV. Hence, from the results of Section 14.5, dV is approximated by the absolute value of

$$\begin{vmatrix} \frac{\partial f}{\partial u} du & \frac{\partial g}{\partial u} du & \frac{\partial h}{\partial u} du \\ \frac{\partial f}{\partial v} dv & \frac{\partial g}{\partial v} dv & \frac{\partial h}{\partial v} dv \\ \frac{\partial f}{\partial w} dw & \frac{\partial g}{\partial w} dw & \frac{\partial h}{\partial w} dw \end{vmatrix}.$$

Figure 16.7.2

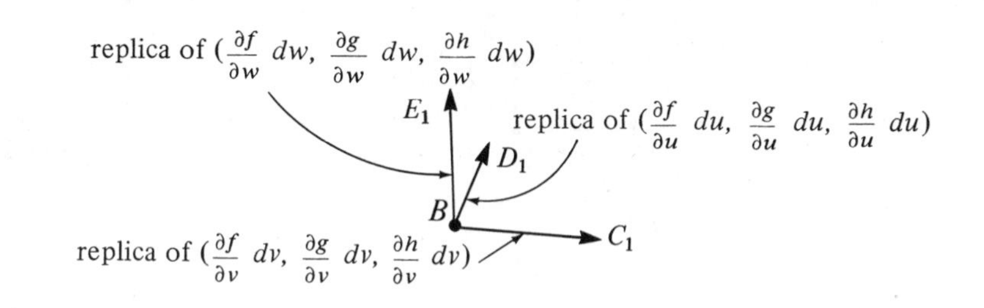

Factoring out the $du\, dv\, dw$, we get

$$dV \approx \left|\frac{\partial(f, g, h)}{\partial(u, v, w)}\right| du\, dv\, dw,$$

where

$$\frac{\partial(f, g, h)}{\partial(u, v, w)} = \begin{vmatrix} \frac{\partial f}{\partial u} & \frac{\partial g}{\partial u} & \frac{\partial h}{\partial u} \\ \frac{\partial f}{\partial v} & \frac{\partial g}{\partial v} & \frac{\partial h}{\partial v} \\ \frac{\partial f}{\partial w} & \frac{\partial g}{\partial w} & \frac{\partial h}{\partial w} \end{vmatrix}.$$

Thus, since $dV' = du\, dv\, dw$,

$$dV \approx \left|\frac{\partial(f, g, h)}{\partial(u, v, w)}\right| dV';$$

and so

$$F(x, y, z)dV \approx F(f(u, v, w), g(u, v, w), h(u, v, w))\left|\frac{\partial(f, g, h)}{\partial(u, v, w)}\right| dV'.$$

We sum up these values over the regions T and T' to get

$$\int_T F(x, y, z)dV = \int_{T'} F(f(u,v, w), g(u, v, w), h(u, v, w))\left|\frac{\partial(f, g, h)}{\partial(u, v, w)}\right| dV'.$$

The expression $\frac{\partial(f, g, h)}{\partial(u, v, w)}$ is called the *Jacobian* of the transformation $x = f(u, v, w)$, $y = g(u, v, w)$, $z = h(u, v, w)$. Note that it is the generalization to 3-dimensions of the Jacobian of Section 16.5. As in the two-dimensional case, *the Jacobian is the factor we must multiply the element of volume dV' by in order to obtain the element of volume dV.*

We are now prepared to state the major theorem of this section.

Theorem 16.7.1

Let the transformation $x = f(u, v, w)$, $y = g(u, v, w)$, and $z = h(u, v, w)$ map the region T' in uvw-space to the region T in xyz-space in 1–1 fashion. Assume also that f, g and h have continuous first partial derivatives and that the Jacobian is not zero. Then

$$\int_T F(x, y, z)dV = \int_{T'} F(f(u, v, w), g(u, v, w), h(u, v, w))\left|\frac{\partial(f, g, h)}{\partial(u, v, w)}\right| dV'.$$

As before, to help remember that formula think of going from the left side of the equation to the right side by replacing the element of volume dV by $\left|\frac{\partial(f, g, h)}{\partial(u, v, w)}\right|$ times the element of volume dV', and making the other indicated replacements.

Example 1

Let T be the tetrahedron bounded by the planes $x = 0$, $y = 0$, $z = y$, and $x + z = 2$. Use the transformation

$$x = f(u, v, w) = 2w,$$

$$y = g(u, v, w) = u,$$

$$z = h(u, v, w) = u + v$$

to change $\int_T xyz\, dV$ to an iterated triple integral in u, v, and w.

The boundary of T' comes from the boundary $x = 0$, $y = 0$, $z = y$, and $x + z = 2$ of T by using $x = 2w$, $y = u$, and $z = u + v$. From $x = 0$ we get $w = 0$; from $y = 0$

we get $u = 0$; from $z = y$ we get $u + v = u$ or $v = 0$; and from $x + z = 2$ we get $2w + u + v = 2$. Thus T' is bounded by

$$u = 0, \quad v = 0, \quad w = 0, \quad \text{and} \quad u + v + 2w = 2.$$

Thus T' is the tetrahedron sketched in Figure 16.7.3. The Jacobian of the transformation is

$$\frac{\partial(f, g, h)}{\partial(u, v, w)} = \begin{vmatrix} 0 & 0 & 2 \\ 1 & 0 & 0 \\ 1 & 1 & 0 \end{vmatrix} = 2, \quad \text{and so}$$

$$\int_T xyz\, dV = \int_{T'} 2wu(u + v)(2) dV' = 4 \int_0^2 \int_0^{2-u} \int_0^{(2-u-v)/2} uw(u + v) dw\, dv\, du. \quad \|$$

Figure 16.7.3

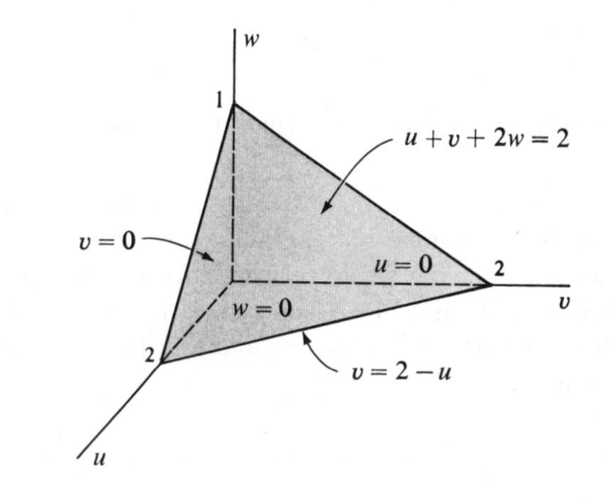

Example 2 Use the transformation $x = r \cos\theta$, $y = r \sin\theta$, and $z = aw$ to evaluate the integral

$$\int_0^a \int_0^1 \int_0^{\sqrt{1-y^2}} \left(x^2 + y^2 + \frac{z^2}{a^2}\right) dx\, dy\, dz,$$

where $a > 0$.

Since $f(r, \theta, w) = r \cos\theta$, $g(r, \theta, w) = r \sin\theta$, and $h(r, \theta, w) = aw$, the Jacobian of the transformation is

$$\frac{\partial(f, g, h)}{\partial(r, \theta, w)} = \begin{vmatrix} \cos\theta & -r \sin\theta & 0 \\ \sin\theta & r \cos\theta & 0 \\ 0 & 0 & a \end{vmatrix} = a(r \cos^2\theta + r \sin^2\theta) = ar.$$

Next we must determine a region T', in $r\theta w$-space that corresponds to the region of integration in the xyz-space. This task is relatively easy since the relation of the xy-plane to the $r\theta$-plane is simply the relation between rectangular and polar coordinates. Consequently the region T, pictured in Figure 16.7.4 is mapped 1–1 onto the rectangular box $0 \le r \le 1$, $0 \le \theta \le \pi/2$, $0 \le w \le 1$ in $r\theta w$-space.

Figure 16.7.4

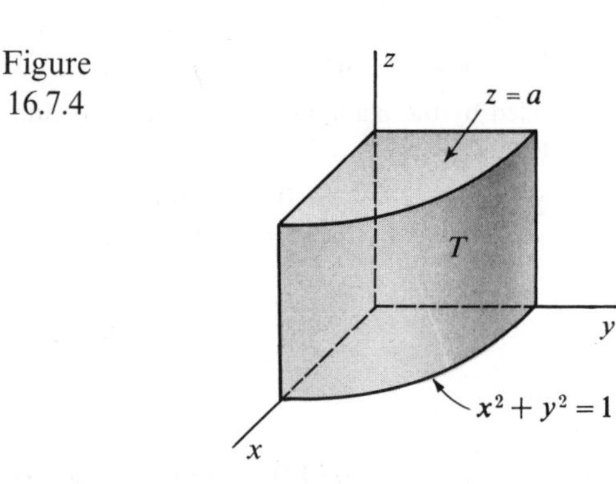

Making the appropriate substitutions, we have

$$\int_0^a \int_0^1 \int_0^{\sqrt{1-y^2}} \left(x^2 + y^2 + \frac{z^2}{a^2}\right) dx\, dy\, dz$$

$$= \int_0^1 \int_0^{\pi/2} \int_0^1 \left(r^2 \cos^2\theta + r^2 \sin^2\theta + \frac{a^2w^2}{a^2}\right) |ar| dr\, d\theta\, dw.$$

Since both a and r are positive, we may express that integral as

$$\int_0^1 \int_0^{\pi/2} \int_0^1 (r^2 + w^2) ar\, dr\, d\theta\, dw = a \int_0^1 \int_0^{\pi/2} \left(\frac{1}{4} + \frac{w^2}{2}\right) d\theta\, dw$$

$$= \frac{a\pi}{2} \int_0^1 \left(\frac{1}{4} + \frac{w^2}{2}\right) dw$$

$$= \frac{a\pi}{2}\left(\frac{1}{4} + \frac{1}{6}\right) = \frac{5a\pi}{24}. \qquad \|$$

Exercises 16.7

In Exercises 1–6, calculate the Jacobian of the given transformation.

1. $x = au, \quad y = bv, \quad z = cw.$
2. $x = u + v, \quad y = v + w, \quad z = u + w.$
3. $x = u \cos v, \quad y = u \sin v, \quad z = w.$
4. $x = e^u, \quad y = u + v + w, \quad z = e^v.$
5. $x = v \tan u, \quad y = vw, \quad z = e^u.$
6. $x = u - w, \quad y = u + 2v, \quad z = 3u - v.$

In Exercises 7–12, use the given transformation to express the integral as an iterated triple integral in u, v, and w. Do not evaluate your result.

7. $\int_T \left(\frac{x^2}{a^2} + \frac{y^2}{b^2} + \frac{z^2}{c^2}\right) dV$, where T is the region consisting of all points (x, y, z) for which $\frac{x^2}{a^2} + \frac{y^2}{b^2} + \frac{z^2}{c^2} \leq 1$. Use $x = au$, $y = bv$, and $z = cw$.

8. $\int_T 4xz(x + y)dV$, where T is the tetrahedron bounded by the planes $x = 0$, $y = 0$, $z = 0$, and $x + y + 2z = 2$. Use $x = v$, $y = w - v$, and $z = u/2$.

9. $\int_T (3x + 6y)dV$, where T is the tetrahedron bounded by the planes $x = 4$, $y = x$, $z = 0$, and $z = y$. Use $x = 4 - u/3$, $y = 4 - u/3 - w/3$, $z = v/3$.

10. $\int_0^2 \int_{x-3}^{x-5} \int_{x-y+1}^{x-y-2} 12x(x-y)dz\,dy\,dx$. Use $x = w/2$, $y = w/2 - v$, and $z = v - u$.

11. $\int_T (xy + z)dV$, where T is the tetrahedron bounded by the planes $x = 0$, $y = 0$, $z = 0$, and $x + y + z = 10$. Use $x = 2w$, $y = 2v - 2u$, $z = 10 - 2v$.

12. $\int_T xy\,dV$, where T is the tetrahedron bounded by the planes $x + y = 0$, $x - y = 0$, $y + z = 0$, and $7x - 2y + 3z = 12$. Use $x = (1/2)(u + v)$, $y = (1/2)(u - v)$, and $z = w - (1/2)(u + v)$.

16.8 Cylindrical Coordinates

Let (x, y, z) be a point in space. The cylindrical coordinates of that point are determined by using polar coordinates in the xy-plane and the usual z-coordinate in the third dimension. This process is illustrated in Figure 16.8.1.

Figure 16.8.1

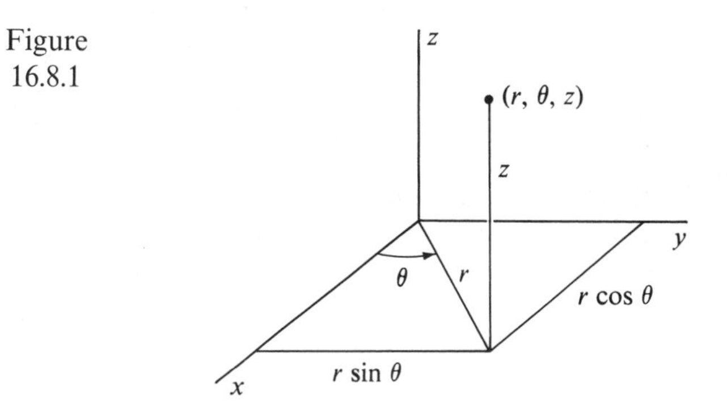

From the diagram above it is clear that the relationship between rectangular and cylindrical coordinates is given by

$$x = r\cos\theta, \quad y = r\sin\theta, \quad \text{and} \quad z = z.$$

Consequently the Jacobian of the transformation from rectangular to cylindrical coordinates is

$$\frac{\partial(x, y, z)}{\partial(r, \theta, z)} = \begin{vmatrix} \cos\theta & -r\sin\theta & 0 \\ \sin\theta & r\cos\theta & 0 \\ 0 & 0 & 1 \end{vmatrix} = r.$$

Thus, by Theorem 16.7.1, an integral may be transformed from rectangular to cylindrical coordinates by use of the equation

$$\int_T F(x, y, z)dV = \int_{T'} F(r\cos\theta, r\sin\theta, z)|r|dV',$$

where T' is a region in $r\theta z$-space that is mapped 1–1 onto the region T.

Intuitively, in cylindrical coordinates the element of volume dV is the volume of a region between two coaxial cylinders, as shown in Figure 16.8.2. Approximating dV as the volume of a rectangular box we obtain

$$dV = r\,dr\,d\theta\,dz = r\,dV'.$$

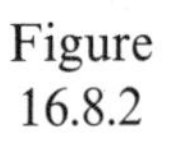

Figure 16.8.2

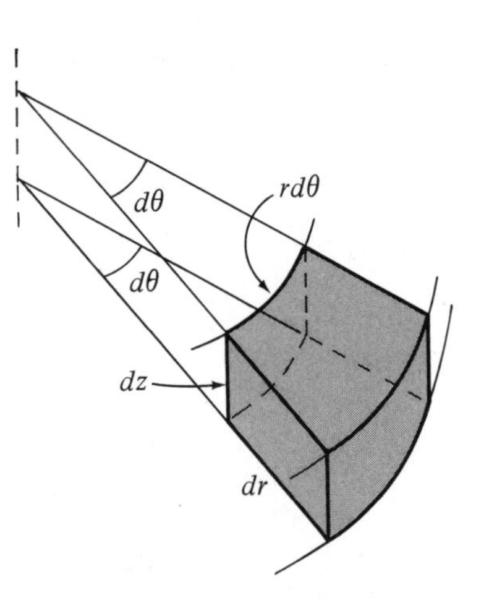

Example 1 Let T be a right circular cylinder whose height is 3 and whose base is the disc with radius 2 and center at the origin. Transform $\int_T (x^2 + y^2)z\, dV$ to cylindrical coordinates and evaluate the result.

The points of the region T correspond 1–1 to the points of the rectangular box determined by $0 \leq r \leq 2$, $0 \leq \theta < 2\pi$, and $0 \leq z \leq 3$ in $r\theta z$-space. So,

$$\begin{aligned}\int_T (x^2 + y^2)z\, dV &= \int_0^3 \int_0^{2\pi} \int_0^2 (r^2 \cos^2\theta + r^2 \sin^2\theta)z|r|dr\, d\theta\, dz \\ &= \int_0^3 \int_0^{2\pi} \int_0^2 r^3\, z\, dr\, d\theta\, dz = 4 \int_0^3 \int_0^{2\pi} z\, d\theta\, dz \\ &= 8\pi \int_0^3 z\, dz = 36\pi.\end{aligned}$$

||

As in our discussion of polar coordinates we must take care to distinguish between xyz-space and $r\theta z$-space. A possible confusion arises from the usual practice of labeling points in xyz-space with their $r\theta z$-names. For example, the point (1, 1, 1) in xyz-space may be labeled as $(\sqrt{2}, \pi/4, 1)$, and the plane $y = x$ labeled as $\theta = \pi/4$. Figure 16.8.3 illustrates this technique.

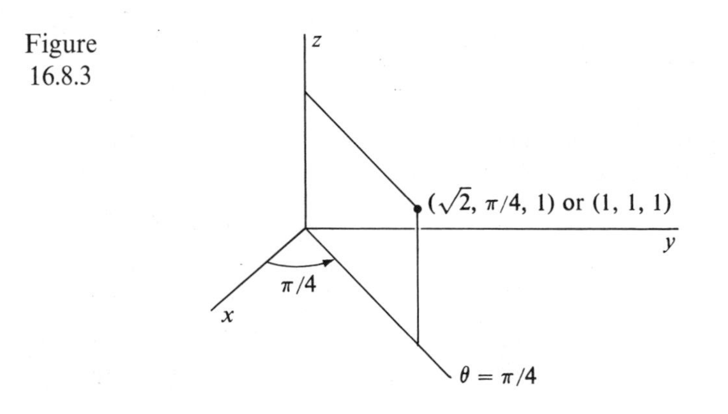

Figure 16.8.3

To demonstrate the power of these ideas we shall find the volume of the region T in xyz-space illustrated in Figure 16.8.4(a). That is, we shall find the volume of the solid that is above the region bounded by $r = g(\theta)$, $r = f(\theta)$, $\theta = \alpha$, $\theta = \beta$, and that is below the graph of $z = h(r, \theta)$. Though the description of T in terms of x, y, and z might be quite complicated, the corresponding region T' in $r\theta z$-space is easy to describe. (See Figure 16.8.4(b).)

Figure 16.8.4

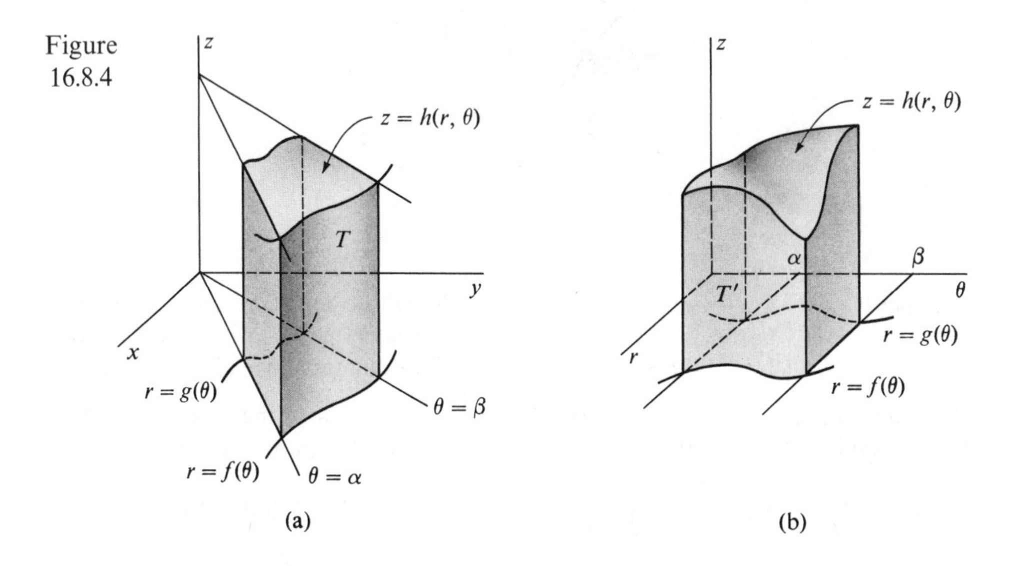

Of course the volume V of T is given by the triple integral $V = \int_T dV$. A transformation of that integral to cylindrical coordinates gives

$$V = \int_{T'} |r| dV',$$

which, by reference to Figure 16.8.4, can be expressed as the iterated integral

$$V = \int_{\alpha}^{\beta} \int_{g(\theta)}^{f(\theta)} \int_{0}^{h(r,\,\theta)} |r| dz\, dr\, d\theta.$$

Completion of the first part of the integration yields

$$V = \int_{\alpha}^{\beta} \int_{g(\theta)}^{f(\theta)} |r| h(r, \theta) dr\, d\theta.$$

Example 2 Transform the integral $\int_0^1 \int_0^1 \int_0^{\sqrt{1-y^2}} z\, dx\, dy\, dz$ to cylindrical coordinates and evaluate the result.

First note that

$$\int_0^1 \int_0^1 \int_0^{\sqrt{1-y^2}} z\, dx\, dy\, dz = \int_T z\, dV,$$

where T is that portion of the cylinder pictured in Figure 16.8.5. The points of the region T correspond in 1–1 fashion to the points of the region determined by $0 \le$

Figure 16.8.5

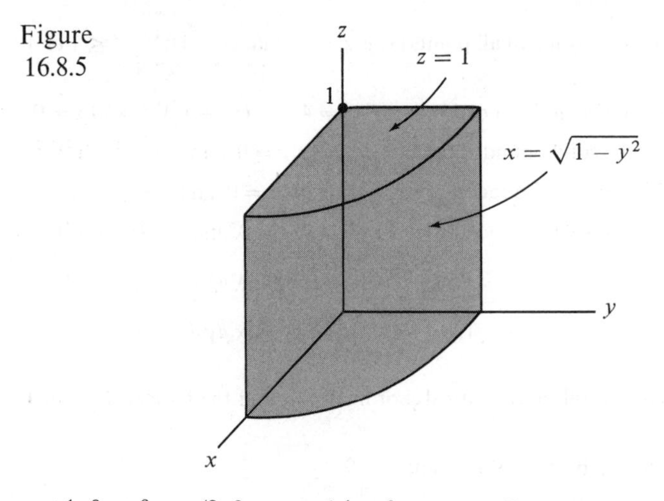

$r \leq 1, 0 \leq \theta \leq \pi/2, 0 \leq z \leq 1$ in $r\theta z$-space. So,

$$\int_0^1 \int_0^1 \int_0^{\sqrt{1-y^2}} z \, dx \, dy \, dz = \int_0^1 \int_0^{\pi/2} \int_0^1 z|r| dr \, d\theta \, dz$$

$$= \frac{1}{2} \int_0^1 \int_0^{\pi/2} z \, d\theta \, dz$$

$$= \frac{\pi}{4} \int_0^1 z \, dz = \frac{\pi}{8}. \qquad \|$$

Example 3 | Find the volume of the solid that lies above the region that is bounded by the spiral $r = \theta$, $0 \leq \theta \leq \pi/2$, and the lines $\theta = 0$, $\theta = \pi/2$, and that lies below the graph of $z = e^{\theta^3}$.

From the discussion above we have

$$V = \int_0^{\pi/2} \int_0^{\theta} |r| e^{\theta^3} \, dr \, d\theta.$$

Then, since $r > 0$, we have $|r| = r$ and so

$$V = \int_0^{\pi/2} \frac{1}{2} \theta^2 e^{\theta^3} \, d\theta = \frac{1}{6} e^{\theta^3} \Big|_0^{\pi/2}$$

$$= \frac{1}{6} [e^{(\pi/2)^3} - 1] \text{ cubic units.} \qquad \|$$

Exercises 16.8

In Exercises 1–12, evaluate the given integral by using a transformation to cylindrical coordinates.

1. $\int_T z(x^2 + y^2)^{1/2} \, dV$, where T is the right circular cylinder consisting of all points (x, y, z) satisfying $x^2 + y^2 \leq 3$ and $-1 \leq z \leq 2$.
2. $\int_T (z^2 + x^2 + y^2) dV$, where T is the region of Exercise 1.
3. $\int_T \frac{z}{x^2 + y^2 + 1} \, dV$, where T is the region of Exercise 1.

4. $\int_T z\,dV$, where T is the region consisting of all points (x, y, z) such that $(x-1)^2 + y^2 \le 1$ and $0 \le z \le 5$.

5. $\int_T (x^2 + y^2)^{1/2}\,dV$, where T is the solid bounded by $x^2 + y^2 = 4$, $z = (x^2 + y^2)^{1/2}$, and $z = 0$.

6. $\int_T (x^2 + y^2)^{1/2}\,dV$, where T is the solid bounded by $x^2 + y^2 = 4y$, $z = 0$, and $z = (x^2 + y^2)^{1/2}$.

7. $\int_T (x^2 + y^2)^{1/2}\,dV$, where T is the solid bounded by $x^2 + y^2 = 6x$, $z = 0$, and $z = x^2 + y^2$.

8. $\int_T (x^2 + y^2)^{3/2}\,dV$, where T is the solid bounded by $x^2 + y^2 = 9$, $z = 0$, and $z = (x^2 + y^2)^{1/2}$.

9. $\int_{-1}^{1}\int_{0}^{2}\int_{0}^{\sqrt{4-x^2}} dy\,dx\,dz.$

10. $\int_{0}^{1}\int_{0}^{2}\int_{0}^{\sqrt{4-x^2}} xyz\,dy\,dx\,dz.$

11. $\int_{-2}^{0}\int_{0}^{1/\sqrt{2}}\int_{y}^{\sqrt{1-y^2}} xy\,dx\,dy\,dz.$

12. $\int_{0}^{2}\int_{0}^{1}\int_{y\sqrt{3}}^{\sqrt{4-y^2}} x^2y\,dx\,dy\,dz.$

In Exercises 13–20, use iterated integrals in cylindrical coordinates to find the indicated volume or surface area.

13. The volume of the solid bounded by $z = x^2 + y^2$ and $z = 9$.

14. The volume of a hemisphere of radius 5.

15. The volume of the smaller solid bounded by $x^2 + y^2 = 2z$ and $x^2 + y^2 + z^2 = 8$.

16. The lateral surface area of a circular cone of base-radius a and height h.

17. The volume of the solid bounded by $z = x^2 + y^2$ and $z^2 = x^2 + y^2$.

18. The volume of the solid bounded by $x^2 + y^2 = 6y$, $z = 0$, and $z^2 = x^2 + y^2$.

19. The volume of the solid bounded by $x^2 + y^2 = 4y$, $z = 0$, and $z = x^2 + y^2$.

20. The volume of the solid bounded by $x^2 + y^2 = 8x$, $z = 0$, and $z = x^2 + y^2$.

16.9 Spherical Coordinates

The reader has undoubtedly realized that cylindrical coordinates are especially useful for problems that involve symmetry about an axis. However, spherical coordinates are the useful tool in problems that involve symmetry about a point.

Figure 16.9.1

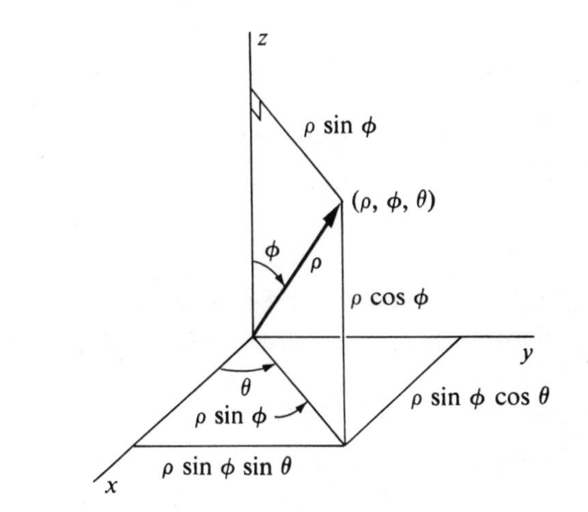

The *spherical coordinates* (ρ, ϕ, θ) of a point are determined as indicated in Figure 16.9.1. In this case ρ is the directed distance of the point from the origin; that is, if ρ is measured in the direction indicated by ϕ and θ, then ρ is positive; and if in the opposite direction, ρ is negative. From this illustration it is clear that the relation between the spherical and rectangular coordinates is given by

$$x = \rho \sin \phi \cos \theta, \quad y = \rho \sin \phi \sin \theta, \quad z = \rho \cos \phi.$$

It is also often useful to note that

$$\rho^2 = x^2 + y^2 + z^2.$$

The power of spherical coordinates lies in the fact that certain common subsets of xyz-space have simple descriptions in $\rho\phi\theta$-space. For example, an equation of a sphere with center at the origin and radius a can be written simply as $\rho = a$; and a cone with vertex at the origin can be expressed in the form $\phi = \alpha$.

Example 1 The region V consisting of all points (x, y, z) such that

$$x^2 + y^2 + z^2 \leq 4, \quad x \geq 0, \quad y \geq 0, \quad z \geq 0$$

is the first octant of a sphere, as pictured in Figure 16.9.2. This region is mapped 1—1 onto the rectangular box determined by $0 \leq \theta \leq \pi/2$, $0 \leq \phi \leq \pi/2$, and $0 \leq \rho \leq 2$. ||

Figure 16.9.2

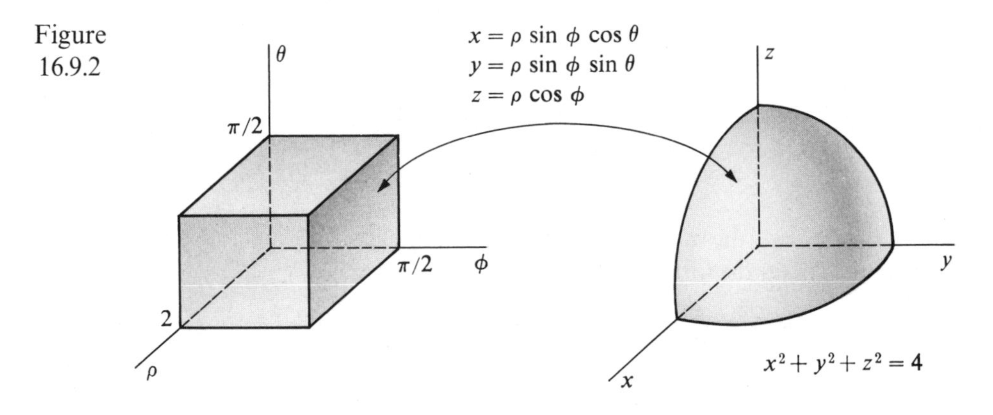

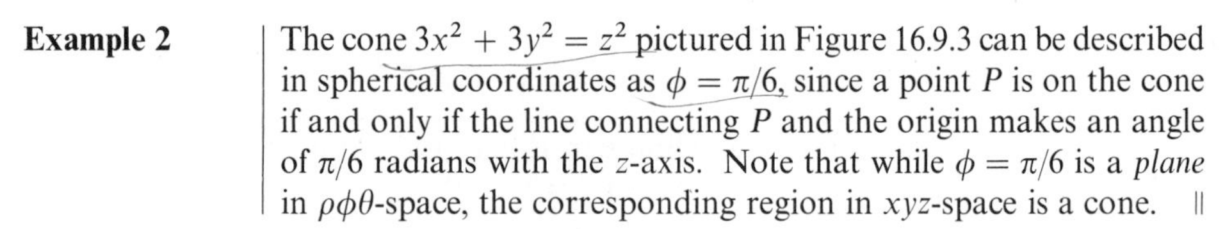

Example 2 The cone $3x^2 + 3y^2 = z^2$ pictured in Figure 16.9.3 can be described in spherical coordinates as $\phi = \pi/6$, since a point P is on the cone if and only if the line connecting P and the origin makes an angle of $\pi/6$ radians with the z-axis. Note that while $\phi = \pi/6$ is a *plane* in $\rho\phi\theta$-space, the corresponding region in xyz-space is a cone. ||

To transform integrals from rectangular to spherical coordinates, we must compute the Jacobian of the transformation. Since

$$x = \rho \sin \phi \cos \theta, \quad y = \rho \sin \phi \sin \theta, \quad \text{and} \quad z = \rho \cos \phi,$$

Figure 16.9.3

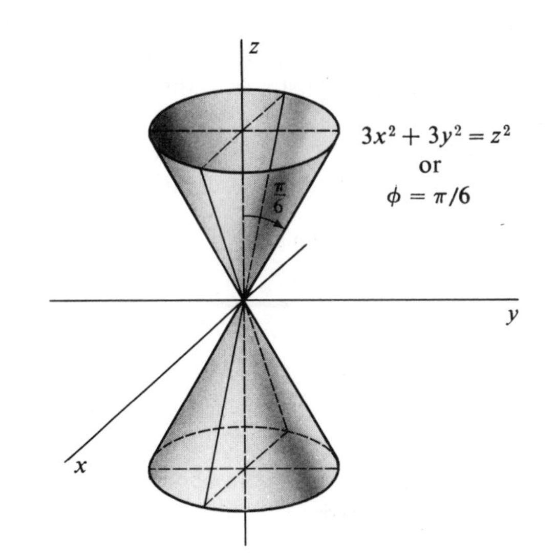

we have

$$\frac{\partial(x, y, z)}{\partial(\rho, \phi, \theta)} = \begin{vmatrix} \sin\phi\cos\theta & \rho\cos\phi\cos\theta & -\rho\sin\phi\sin\theta \\ \sin\phi\sin\theta & \rho\cos\phi\sin\theta & \rho\sin\phi\cos\theta \\ \cos\phi & -\rho\sin\phi & 0 \end{vmatrix}$$

$$= \rho^2\cos^2\phi\cos^2\theta\sin\phi + \rho^2\sin^3\phi\sin^2\theta + \rho^2\cos^2\phi\sin^2\theta\sin\phi + \rho^2\cos^2\theta\sin^3\phi$$

$$= (\rho^2\sin\phi)[\cos^2\phi\cos^2\theta + \sin^2\phi\sin^2\theta + \cos^2\phi\sin^2\theta + \cos^2\theta\sin^2\phi]$$

$$= (\rho^2\sin\phi)[(\cos^2\phi)(\cos^2\theta + \sin^2\theta) + (\sin^2\phi)(\cos^2\theta + \sin^2\theta)]$$

$$= \rho^2\sin\phi.$$

Consequently the transformation of a triple integral from rectangular to spherical coordinates takes the form

$$\int_T F(x, y, z)dV = \int_{T'} F(\rho\sin\phi\cos\theta, \rho\sin\phi\sin\theta, \rho\cos\phi)|\rho^2\sin\phi|dV'.$$

Example 3 Evaluate the integral

$$\int_T \frac{1}{x^2 + y^2 + z^2 + 1}\,dV,$$

where T is the region consisting of all points (x, y, z) for which $x^2 + y^2 + z^2 \leq 4$, $x \geq 0$, $y \geq 0$, and $z \geq 0$.

In Example 1 it was determined that this region T corresponds in 1—1 fashion to the rectangular box determined by $0 \leq \theta \leq \pi/2$, $0 \leq \phi \leq \pi/2$, and $0 \leq \rho \leq 2$. Thus

$$\int_T \frac{1}{x^2 + y^2 + z^2 + 1}\,dV = \int_0^{\pi/2}\int_0^{\pi/2}\int_0^2 \frac{1}{\rho^2 + 1}|\rho^2\sin\phi|d\rho\,d\phi\,d\theta.$$

Since $0 \le \phi \le \pi/2$, we have $\sin\phi \ge 0$; and so the absolute value sign may be omitted. Hence

$$\int_T \frac{1}{x^2+y^2+z^2+1}\,dV = \int_0^{\pi/2}\int_0^{\pi/2}\int_0^2 \frac{\rho^2}{\rho^2+1}\sin\phi\,d\rho\,d\phi\,d\theta$$

$$= \int_0^{\pi/2}\int_0^{\pi/2}\int_0^2 \left(1-\frac{1}{\rho^2+1}\right)\sin\phi\,d\rho\,d\phi\,d\theta$$

$$= \int_0^{\pi/2}\int_0^{\pi/2}(2-\arctan 2)\sin\phi\,d\phi\,d\theta$$

$$= (2-\arctan 2)\int_0^{\pi/2} d\theta$$

$$= \frac{\pi}{2}(2-\arctan 2). \qquad \|$$

Some volumes are more easily calculated when expressed in the spherical coordinate system. The volume V of any region T in xyz-space is given by $V = \int_T dV$. After transforming to the spherical coordinate system we obtain

$$V = \int_{T'} \rho^2|\sin\phi|\,dV',$$

where T' is a region in $\rho\phi\theta$-space that corresponds in 1–1 fashion to T.

Intuitively, in spherical coordinates the element of volume dV is the volume of a region between two concentric spheres as shown in Figure 16.9.4. Approximating dV as the volume of a rectangular box we obtain

$$dV = \rho\sin\phi\,d\theta\,\rho\,d\phi\,d\rho = \rho^2\sin\phi\,dV'.$$

Figure 16.9.4

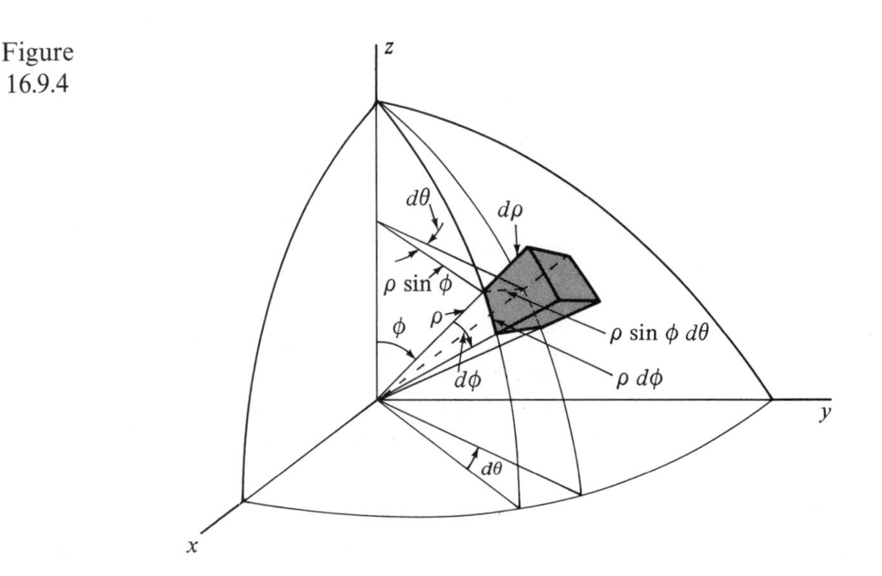

Example 4 | Find the volume V of the region T interior to the sphere $\rho = 2$, exterior to the cone $\phi = \pi/6$, and above the xy-plane.

Since this region is symmetric with respect to each of the three coordinate planes, we illustrate in Figure 16.9.5 only that portion of the region for which $x \geq 0$, $y \geq 0$, and $z \geq 0$. The total volume is four times the illustrated volume. Thus, on substitution into

$$V = 4 \int_{T'} \rho^2 |\sin \phi| dV'$$

where T' is the region determined by $0 \leq \rho \leq 2$, $\pi/6 \leq \phi \leq \pi/2$, and $0 \leq \theta \leq \pi/2$, we obtain

$$\begin{aligned} V &= 4 \int_0^2 \int_{\pi/6}^{\pi/2} \int_0^{\pi/2} \rho^2 |\sin \phi| d\theta \, d\phi \, d\rho \\ &= 2\pi \int_0^2 \int_{\pi/6}^{\pi/2} \rho^2 |\sin \phi| d\phi \, d\rho. \end{aligned}$$

Since $\pi/6 \leq \phi \leq \pi/2$, we have $\sin \phi > 0$; and so the absolute value signs may be omitted. Therefore,

$$\begin{aligned} V &= 2\pi \int_0^2 \int_{\pi/6}^{\pi/2} \rho^2 \sin \phi \, d\phi \, d\rho \\ &= \sqrt{3}\, \pi \int_0^2 \rho^2 \, d\rho \\ &= \frac{8\sqrt{3}\, \pi}{3} \text{ cubic units.} \end{aligned}$$

||

Figure 16.9.5

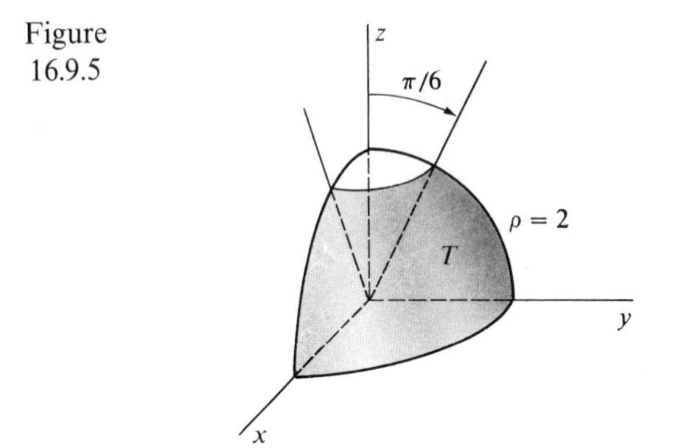

Example 5 The density at each point of a spherical shell of inner radius a and outer radius b is inversely proportional to the square of the distance from the center. Derive an equation for the weight of this spherical shell.

We shall take the origin of our coordinate system to be at the center of the concentric spheres forming the boundaries of the shell T. The equations of the two spheres are given by

$$x^2 + y^2 + z^2 = a^2 \quad \text{and} \quad x^2 + y^2 + z^2 = b^2, \quad a < b.$$

Then the density at each point (x, y, z) of the spherical shell is $k/(x^2 + y^2 + z^2)$, where k is the constant of proportionality.

Then the total weight W is

$$\int_T \frac{k}{x^2 + y^2 + z^2}\, dV.$$

Since the region T and the density are symmetric about the origin, a transformation to spherical coordinates will be profitable. The region T corresponds in 1–1 fashion to the region $a \le \rho \le b$, $0 \le \phi \le \pi$, and $0 \le \theta < 2\pi$ in the $\rho\phi\theta$-coordinate system. Then, since $\rho^2 = x^2 + y^2 + z^2$,

$$\begin{aligned} W &= \int_0^{2\pi} \int_0^{\pi} \int_a^b |\rho^2 \sin\phi| \frac{k}{\rho^2}\, d\rho\, d\phi\, d\theta \\ &= \int_0^{2\pi} \int_0^{\pi} k(b-a)|\sin\phi| d\phi\, d\theta. \end{aligned}$$

Since $\sin\phi \ge 0$ for $0 \le \phi \le \pi$, and since k, a, and b are constants,

$$\begin{aligned} W &= k(b-a) \int_0^{2\pi} \int_0^{\pi} \sin\phi\; d\phi\; d\theta \\ &= k(b-a) \int_0^{2\pi} 2\; d\theta \\ &= 4k\pi(b-a) \text{ units of mass.} \end{aligned}$$

||

Exercises 16.9

In Exercises 1–6, use spherical coordinates to describe the given set in xyz-space.

1. $x^2 + y^2 + z^2 \ge a^2$, $y \ge 0$, where $a > 0$.
2. $b^2 \le x^2 + y^2 + z^2 \le a^2$, $z \ge 0$, where a and b are positive.
3. $x = y$.
4. $x^2 + y^2 \ge z^2$, $z \ge 0$.
5. $x^2 + y^2 \le z^2$, $x^2 + y^2 + z^2 \le 1$, $z \ge 0$.
6. $y = -x$.

In Exercises 7–13, evaluate the given integral by transforming it to spherical coordinates.

7. $\int_T z\, dV$, where T is the unit ball centered at the origin.
8. $\int_T dV$, where T is the region of Exercise 5.
9. $\displaystyle\int_T \frac{dV}{1 + x^2 + y^2 + z^2}$, where T is the region of Exercise 5.
10. $\int_T z^2(x^2 + y^2 + z^2)^{-1/2}\, dV$, where T is the region between spheres of radii 4 and 5 centered at the origin.
11. $\displaystyle\int_0^1 \int_0^{\sqrt{1-x^2}} \int_{\sqrt{x^2+y^2}}^{\sqrt{2-x^2-y^2}} dz\, dy\, dx.$
12. $\displaystyle\int_0^{1/\sqrt{2}} \int_z^{\sqrt{1-z^2}} \int_{-\sqrt{1-z^2-y^2}}^{\sqrt{1-z^2-y^2}} z^2\, dx\, dy\, dz.$
13. $\displaystyle\int_0^3 \int_{-\sqrt{9-x^2}}^0 \int_{\sqrt{x^2+y^2}}^{\sqrt{18-x^2-y^2}} dz\, dy\, dx.$

In Exercises 14–17, use triple integration in spherical coordinates to find the indicated volume.

14. The volume of a sphere of radius r.
15. The volume of the solid bounded above by the sphere $\rho = 8$ and below by the cone $\phi = \pi/3$.
16. The volume of the solid bounded by the plane $\phi = \pi/2$, the sphere $\rho = 10$, and the cone $\phi = \pi/6$.
17. The volume of the region inside the sphere of radius 11 and outside the sphere of radius 10.
18. The density at each point of a ball of radius 7 is equal to twice the distance from the center. Find the weight of the ball.
19. The density at each point of a ball of radius 2 is equal to three times the square of the distance from the center. Find the weight of the ball.
20. Find the volume of the solid bounded by $z = (36 - x^2 - y^2)^{1/2}$ and $z = (3x^2 + 3y^2)^{1/2}$.

Brief Review of Chapter 16

1. The Double Integral

Definition (incomplete).

$$\int_R f(x, y)dA = \lim_{n \to \infty} \sum_{i=1}^{m} f(x_i^*, y_i^*)A_i.$$

(A complete definition must include the meaning of m, n, R, (x_i^*, y_i^*), and A_i.)

Interpretation of Definition. Geometrically, if $f(x, y) \geq 0$ for all points of the bounded region R in the xy-plane, $\int_R f(x, y)dA$ is the volume of the solid under the surface $z = f(x, y)$ and above the region R. When $f(x, y) = 1$,

$$\int_R f(x, y)dA = \int_R 1\, dA = \text{the area of } R.$$

2. The Iterated Integral

Definition. An *iterated integral* is an integral of the form

$$\int_a^b \left(\int_{p_1(x)}^{p_2(x)} f(x, y)dy \right) dx,$$

where the integral inside the parenthesis is evaluated first by treating x as a constant. An iterated integral of the form

$$\int_c^d \left(\int_{q_1(y)}^{q_2(y)} f(x, y)dx \right) dy$$

is defined similarly. For convenience the parenthesis are usually omitted when writing iterated integrals.

Use. Iterated integrals are used to evaluate double integrals using the two forms of The Fundamental Theorem of Calculus for Double Integrals. Let R be the region in the xy-plane described by $a \leq x \leq b$ and $p_1(x) \leq y \leq p_2(x)$, where $p_1(x)$ and $p_2(x)$ are continuous. If $f(x, y)$ is continuous on R, then according to The First Form of The Fundamental Theorem of Calculus for Double Integrals we have

$$\int_R f(x, y)dA = \int_a^b \int_{p_1(x)}^{p_2(x)} f(x, y)dy\,dx.$$

Similarly, according to the second form of The Fundamental Theorem for Double integrals, with the appropriate changes in the restrictions of the first form of the theorem,

$$\int_R f(x, y)dA = \int_c^d \int_{q_1(y)}^{q_2(y)} f(x, y)dx\,dy.$$

3. Surface Area

The area S of the part of the surface $z = f(x, y)$ above a bounded region R in the xy-plane is given by

$$S = \int_R \sqrt{1 + f_x^2(x, y) + f_y^2(x, y)}dA.$$

Letting $g(x, y, z) = z - f(x, y)$, we can express that integral in vector notation as

$$S = \int_R |\nabla g(x, y, z)|dA.$$

or, using the vector equation $\mathbf{X} = (x, y, f(x, y))$, as

$$S = \int_R \left| \frac{\partial \mathbf{X}}{\partial x} \times \frac{\partial \mathbf{X}}{\partial y} \right| dA.$$

4. The Triple Integral

Definition (incomplete).

$$\int_T f(x, y, z)dV = \lim_{n \to \infty} \sum_{i=1}^{m} f(x_i^*, y_i^*, z_i^*)V_i.$$

(A complete definition must include the meaning of m, n, T, (x_i^*, y_i^*, z_i^*), and V_i.)

Evaluation. When the region T is bounded above by a surface $z = f_2(x, y)$ and below by a surface $z = f_1(x, y)$, and when T is over a region in the xy-plane described by $a \leq x \leq b$ and $p_1(x) \leq y \leq p_2(x)$, then

$$\int_T f(x, y, z)dV = \int_a^b \int_{p_1(x)}^{p_2(x)} \int_{f_1(x, y)}^{f_2(x, y)} f(x, y, z)dz\,dy\,dx. \quad (1)$$

Here the inside integration is done first with respect to z, treating both x and y as constants. Then that result is integrated with respect to y, treating x as a constant, and finally the result is integrated with respect to x between a and b. With similar restrictions the triple integral can be evaluated using different orders of integration for the iterated integral on the right side of equation (1).

Uses. Triple integrals can be used to find the weight of an object of varying density. Also, when $f(x, y, z) = 1$,

$$\int_T f(x, y, z)dV = \int_T dV = \text{the volume of } T.$$

5. Transformation of Multiple Integrals

Double Integrals. Let the transformation $x = f(u, v)$ and $y = g(u, v)$ map the region R' in uv-space 1–1 onto the region R in the xy-space. Then

$$\int_R F(x,y)dA = \int_{R'} F(f(u, v), g(u, v))\left|\frac{\partial(f, g)}{\partial(u, v)}\right| dA'. \tag{2}$$

When the transformation is made to polar coordinates $x = r\cos\theta$ and $y = r\sin\theta$, equation (2) becomes

$$\int_R F(x, y)dA = \int_{R'} F(r\cos\theta, r\sin\theta)|r|dA'.$$

Triple Integrals. When the transformation $x = f(u, v, w)$, $y = g(u, v, w)$, $z = h(u, v, w)$ maps T' in uvw-space to the region T in xyz-space in a 1–1 fashion, we have

$$\int_T F(x, y, z)dV = \int_{T'} F(f(u, v, w), g(u, v, w), h(u, v, w))\left|\frac{\partial(f, g, h)}{\partial(u, v, w)}\right| dV'. \tag{3}$$

When the transformation is made to cylindrical coordinates $x = r\cos\theta$, $y = r\sin\theta$, and $z = z$, then equation (3) becomes

$$\int_T F(x, y, z)dV = \int_{T'} F(r\cos\theta, r\sin\theta, z)|r|dV'.$$

When the transformation is to spherical coordinates $x = \rho\sin\phi\cos\theta$, $y = \rho\sin\phi\sin\theta$, and $z = \rho\cos\phi$, then equation (3) becomes

$$\int_T F(x, y, z)dV = \int_{T'} F(\rho\sin\phi\cos\theta, \rho\sin\phi\sin\theta, \rho\cos\phi)|\rho^2\sin\phi|dV'.$$

Technique Review Exercises, Chapter 16

1. Use the definition of the double integral to find the volume of the solid that is under the graph of $z = 20$ and above the rectangle determined by $4 \le x \le 11$ and $3 \le y \le 7$.
2. Find the volume of the solid that is under the surface $z = 8 - x + 2y$ and above the region R in the xy-plane bounded by $y = x$ and $y = x^2 - 6$, using integration first with respect to y.
3. Do Exercise 2 using integration first with respect to x.
4. Set up, but do not evaluate, an iterated integral for the volume of the solid bounded by $z = x^2 + y^2$ and $14 - z = x^2 + y^2$.
5. Find the area of the triangle cut from the plane $5x - 7y + 2z = 9$ by the three coordinate planes.

6. Use double integration to find the area of the region bounded by $x = y^2$ and $2y = x - 15$.
7. Set up, but do not evaluate, an iterated integral for the weight of the solid bounded by the upper part of the sphere $x^2 + y^2 + z^2 = 54$ and the cone $2x^2 + 2y^2 = z^2$. The density in pounds per cubic unit at any point in the solid is the product of its distance from the z-axis and its distance from the xy-plane.
8. Transform the integral
$$\int_0^{1/2} \int_{(x-2)/2}^{-3x/2} (12x - 8y)dy\, dx.$$
by using the relations $x = v/2$, and $y = v/4 - u/2$.
9. Evaluate $\int_0^1 \int_0^{\sqrt{1-x^2}} e^{x^2+y^2}\, dy\, dx$.
10. Let T be the tetrahedron bounded by the planes $z = 0, z = y, x + 3y + z = 2$, and $x + y + z = 1$. Use the transformation $x = w - u - v$, $y = u/2 + v$, and $z = u/2$, to change $4\int_T (xz + yz + z^2)dV$ to an iterated triple integral in u, v, and w.
11. Evaluate by first transforming to cylindrical coordinates, where T is bounded by the cylinder $x^2 + y^2 = 9$ and the planes $z = -2$ and $z = 3$.
$$\int_T z^2(x^2 + y^2)^{1/2}\, dV.$$
12. Transform to an iterated integral in spherical coordinates:
$$\int_{-2}^{0} \int_0^{\sqrt{4-x^2}} \int_{\sqrt{3x^2+3y^2}}^{\sqrt{16-x^2-y^2}} z\, dz\, dy\, dx.$$

Additional Exercises, Chapter 16

Section 16.1

1. Use the definition of the double integral to find the volume of the solid under the graph of $z = 9$ and above the rectangle determined by $-4 \leq x \leq 3$ and $8 \leq y \leq 11$.
2. Use the definition of the double integral to find the volume of the solid under the graph of $z = 3x$ and above the rectangle determined by $1 \leq x \leq 3$ and $4 \leq y \leq 8$.

Section 16.2

In Exercises 3–6, evaluate the integrals.

3. $\int_{-1}^{1} \int_{x^2}^{x} x^2 y\, dy\, dx$.
4. $\int_3^5 \int_{x-1}^{2x-1} xe^{xy}\, dy\, dx$.
5. $\int_2^4 \int_0^y x(x^2 + y^2)^2\, dx\, dy$.
6. $\int_0^{\pi/2} \int_0^{\sin y} \sin y\, dx\, dy$.
7. Find the volume of the tetrahedron bounded by the coordinate planes and the plane $2x + 5y - 2z = 10$.
8. Let R be the region bounded by the graphs of $x = y^2$ and $2y = x - 3$. Find $\int_R xy\, dA$.
9. Set up, but do not evaluate, an iterated integral for the volume of the solid bounded by $x^2 + y^2 = 16$, $z = x^2 + y^2$, and $z = 0$.
10. Set up, but do not evaluate, an iterated integral for the volume of the region bounded by $x^2 + y^2 = 4y$, $z^2 = x^2 + y^2$, and $z = 0$.
11. Evaluate $\int_1^\infty \int_0^{y-3} (4x + y)dx\, dy$.

Section 16.3

12. Find the area of the triangle cut from the plane $3x - 2y - 4z = 12$ by the coordinate planes.
13. Set up, but do not evaluate, an iterated integral for the area of that part of the surface $16 - z = x^2 + y^2$ above the plane $z = 9$.
14. Set up, but do not evaluate, iterated integrals for the surface area of the part of the cone $z^2 = 3x^2 + 3y^2$ that is above the region in the xy-plane bounded by $x = y^2$ and $y = x - 12$: (a) where the integration is first with respect to x, (b) where the integration is first with respect to y.

Section 16.4

15. Evaluate $\int_0^1 \int_z^{z^2} \int_{xz}^{x+z} 12xyz \, dy \, dx \, dz$.
16. Evaluate $\int_2^5 \int_1^4 \int_y^{3y} e^{xy} \, dz \, dx \, dy$.
17. Evaluate $\int_T 10yz \, dV$, where T is the solid bounded by $x^2 + y^2 = 9$, $z = -2$, and $z = 7$.
18. Let T be the solid bounded by $z = (16 - 4x^2 - y^2)^{1/2}$ and $z + 4x^2 + y^2 = 14$, and let the density at any point P of T be twice the distance between P and the z-axis. Set up, but do not evaluate, an iterated integral for the weight of T.

Section 16.5

19. Calculate the Jacobian of the transformation $x = u^2 - v^2$, $y = u^2v^3$.
20. Calculate the Jacobian of the transformation $x = \sin uv$, $y = \cos uv$.
21. Transform $\int_0^3 \int_0^{3-x} xy \, dy \, dx$ to an iterated integral in u and v by using $x = u + v$, $y = u - v$.
22. Transform $\int_0^4 \int_y^4 x^2y \, dx \, dy$ to an iterated integral in u and v by using $x = u + v$, $y = u - v$.

Section 16.6

In Exercises 23–26, transform the given integral to an iterated integral in polar coordinates and evaluate the resulting integral.

23. $\int_R (x^2 + y^2)^{5/2} \, dA$, where R is the region bounded by $x = (9 - y^2)^{1/2}$ and $x = 0$.
24. $\int_R y(x^2 + y^2)^{1/2} \, dA$ where R is the region bounded by $y = (4 - x^2)^{1/2}$, $y = x$, and $x = 0$.
25. $\displaystyle\int_0^3 \int_0^{\sqrt{9-y^2}} (x^2 + y^2)^2 \, dx \, dy$.
26. $\displaystyle\int_1^4 \int_0^y (x^2 + y^2)^{1/2} \, dx \, dy$.
27. Use a double integral to find the area of the region inside $r = 3 + \cos\theta$.

In Exercises 28–30, find the volume of the given solid by transforming to a double integral in polar coordinates.

28. The solid bounded by $x^2 + y^2 = 25$, $z = 1$, and $z = 4$.
29. The solid bounded by $x^2 + y^2 = 36$, $z = 0$, and $z = x^2 + y^2$.
30. The solid bounded by $x^2 + y^2 = 6y$, $z = 0$, and $z = x^2 + y^2$.

Section 16.7

31. Calculate the Jacobian of the transformation $x = u - 2v + w$, $y = 2u + 3v + 2w$, $z = u + 5v - 3w$.
32. Calculate the Jacobian of the transformation $x = e^{u+v}$, $y = e^{uv}$, $z = uvw$.

In Exercises 33–34, use the given transformation to express the integral as an iterated integral in u, v, and w. Do not evaluate your result.

33. $\int_T 8xy(y+z)dV$, where T is the tetrahedron bounded by the planes $x=0$, $y=0$, $z=0$, and $2x+y+z=2$. Use $x=u/2$, $y=v$, $z=w-v$.

34. $\int_0^2 \int_0^{2-y} \int_x^{2-y} xz\,dz\,dx\,dy$. Use $x=u$, $y=2w$, and $z=u+v$.

Section 16.8

In Exercises 35–40, evaluate the given integral by using a transformation to cylindrical coordinates.

35. $\int_{-2}^{3} \int_0^5 \int_0^{\sqrt{25-x^2}} dy\,dx\,dz$.

36. $\int_0^2 \int_0^3 \int_0^{\sqrt{9-y^2}} xyz\,dx\,dy\,dz$.

37. $\int_T z^2(x^2+y^2)^{1/2}\,dV$, where T is the solid bounded by $x^2+y^2=4$, $z=-2$, and $z=3$.

38. $\int_T (x^2+y^2)^{1/2}\,dV$, where T is the solid bounded by $x^2+y^2=25$, $z=(x^2+y^2)^{1/2}$, and $z=0$.

39. $\int_T (x^2+y^2)^{1/2}\,dV$, where T is the solid bounded by $x^2+y^2=8y$, $z=0$, and $z=(x^2+y^2)^{1/2}$.

40. $\int_T (x^2+y^2)^{3/2}\,dV$, where T is the solid bounded by $x^2+y^2=16$, $z=0$, and $z=x^2+y^2$.

41. Find the volume of the solid bounded by $z=x^2+y^2$ and $z^2=4x^2+4y^2$.

42. Find the volume of the solid bounded by $x^2+y^2=4x$, $z=0$, and $z=x^2+y^2$.

Section 16.9

In Exercises 43–46, use spherical coordinates to describe the given set in xyz-space.

43. $x^2+y^2+z^2 \geq 9$, $x \leq 0$.

44. $z^2 \geq x^2+y^2$.

45. $x^2+y^2+z^2 \leq 4$, $z^2 \leq x^2+y^2$.

46. $y \geq 0$, $z^2 \geq x^2+y^2$, $x^2+y^2+z^2 \leq 25$.

In Exercises 47–48, evaluate the given integral by transforming it to spherical coordinates.

47. $\int_T z^2(x^2+y^2+z^2)^{-1/2}\,dV$, where T is the unit ball centered at the origin.

48. $\int_{-2}^{2} \int_0^{\sqrt{4-x^2}} \int_{\sqrt{x^2+y^2}}^{\sqrt{8-x^2-y^2}} (x^2+y^2+z^2)dz\,dy\,dx$.

49. Use triple integration in spherical coordinates to find the volume of the solid bounded by $\phi=\pi/3$, $\phi=\pi/2$, and the sphere $\rho=6$.

50. The density at each point of a right circular cone of altitude $10(3)^{1/2}$ and base-radius 10 is equal to twice the distance from the vertex. Find the weight of the cone.

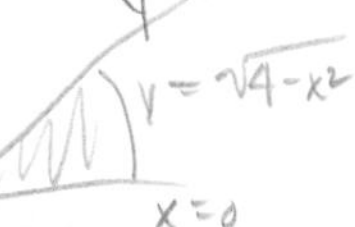

Challenging Problems, Chapter 16

1. It can be proved that $\int_0^\infty e^{(-x^2)}\,dx$ exists. Proceeding under the assumption that it does exist and that the other integrals used exist, show that $\int_0^\infty e^{(-x^2)}\,dx = \sqrt{\pi}/2$ as follows:
   (a) Show that $(\int_0^\infty e^{(-x^2)}\,dx)(\int_0^\infty e^{(-y^2)}\,dy) = \int_0^\infty \int_0^\infty e^{(-x^2)}e^{(-y^2)}\,dx\,dy$.

(b) Express $\int_0^\infty \int_0^\infty e^{-(x^2+y^2)}\, dx\, dy$ in the form $\int_0^\infty \int_0^{\pi/2} f(r, \theta) d\theta\, dr$, and evaluate it over its unbounded region.

(c) Combine the results of (a) and (b).

2. Let R denote the rectangle $a \leq x \leq b$, $a \leq y \leq b$. Show that if f and g are integrable on $a \leq x \leq b$, then

$$\frac{1}{2}\int_R [f(x) - f(y)][g(x) - g(y)]dA$$

$$= (b - a)\int_a^b f(x)g(x)dx - \left(\int_a^b f(x)dx\right)\left(\int_a^b g(x)dx\right).$$

17

Vector Calculus

17.1 Work and Line Integrals

In Section 14.7 we saw that a force field can be fully described as a vector-valued function of position, that is, as a function F such that $F:R^3 \to R^3$. In this section we shall consider the problem of computing the work done by moving a particle along a given path in a given force field. The solution to this problem will lead to the concept of a line integral.

For simplicity we shall consider initially a 2-dimensional force field $F:R^2 \to R^2$ whose y-component is 0 and whose x-component is given by $f_1(\mathbf{X})$, that is, $F(\mathbf{X}) = (f_1(\mathbf{X}), 0)$. We shall find the work done by such a force in moving a particle along a curve C from the point A to the point B. This situation is pictured in Figure 17.1.1.

Figure 17.1.1

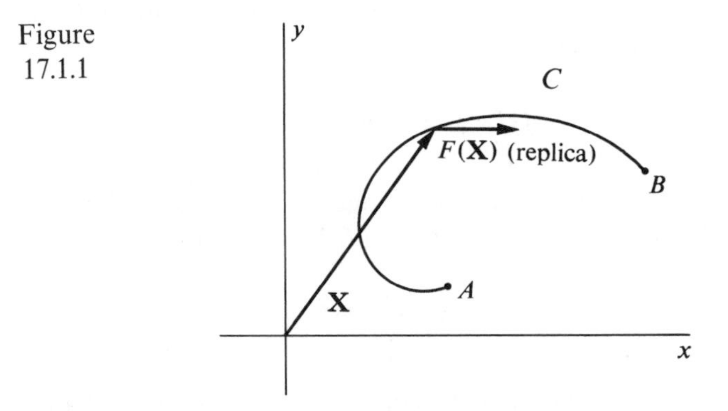

We first subdivide the curve C from A to B into n parts by selecting successive points

$$A = (x_0, y_0), (x_1, y_1), (x_2, x_2), \ldots, (x_i, y_i), \ldots, (x_n, y_n) = B$$

on C. Then, if W_i is the work done in moving the particle from (x_{i-1}, y_{i-1}) to (x_i, y_i), the total work W is given by

$$W = \sum_{i=1}^{n} W_i.$$

Now we concentrate on finding W_i. Since the x-component of the force $f(\mathbf{X})$, is continuous, it has a maximum value M and a minimum value m on that segment of the curve between (x_{i-1}, y_{i-1}) and (x_i, y_i). Moreover, W_i lies between the maximum force M times the distance $x_i - x_{i-1}$ through which it acts, and the minimum force m times that distance. Hence, if $x_i \geq x_{i-1}$, then

$$m(x_i - x_{i-1}) \leq W_i \leq M(x_i - x_{i-1}).$$

If $x_{i-1} \geq x_i$, the above inequalities are reversed; but in either case,

$$m \leq \frac{W_i}{x_i - x_{i-1}} \leq M.$$

But since M and m are the maximal and minimal values of the continuous function $f_1(\mathbf{X})$ on the segment, The Intermediate Value Theorem for continuous functions assures us of a point $X^* = (x_i^*, y_i^*)$ on C between (x_{i-1}, y_{i-1}) and (x_i, y_i) such that

$$\frac{W_i}{x_i - x_{i-1}} = f_1(x_i^*, y_i^*).$$

Hence $$W_i = f_1(x_i^*, y_i^*)(x_i - x_{i-1}).$$

Therefore, the total work W done by the force over the entire curve C from A to B is

$$W = \sum_{i=1}^{n} W_i = \sum_{i=1}^{n} f_1(x_i^*, y_i^*)(x_i - x_{i-1}).$$

However, W is independent of n, and so

$$W = \lim_{n\to\infty} W = \lim_{n\to\infty} \sum_{i=1}^{n} f_1(x_i^{**}, y_i^{**})(x_i - x_{i-1}). \tag{1}$$

In the same way it can be shown that if the x-component of the force is 0 and the y-component is given by $f_2(\mathbf{X})$, then

$$W = \lim_{n\to\infty} \sum_{i=1}^{n} f_2(x_i^*, y_i^*)(y_i - y_{i-1}). \tag{2}$$

In case $F(\mathbf{X}) = (f_1(\mathbf{X}), f_2(\mathbf{X}))$ where neither component is always zero the work done will be the sum of the work done by the forces $(f_1(\mathbf{X}), 0)$ and $(0, f_2(\mathbf{X}))$. Thus

$$W = \lim_{n\to\infty} \sum_{i=1}^{n} f_1(x_i^*, y_i^*)(x_i - x_{i-1}) + \lim_{n\to\infty} \sum_{i=1}^{n} f_2(x_i^{**}, y_i^{**})(y_i - y_{i-1}).$$

Since the limits in (1) and (2) occur quite often in various applications of mathematics, we make the following definition.

Definition 17.1.1

Let $f(x, y)$ be a function that is continuous over a bounded region R in the xy-plane, and let C be a continuous curve entirely inside R with specified initial and terminal points. Using the notation introduced above, the *line integral of f with respect to x along C*, denoted by $\int_C f(x, y)dx$, is defined as

$$\int_C f(x, y)dx = \lim_{n\to\infty} \sum_{i=1}^{n} f(x_i^*, y_i^*)(x_i - x_{i-1}),$$

where the limit is taken so that all differences $x_i - x_{i-1}$ approach 0 as n increases without bound.

By replacing $x_i - x_{i-1}$ with $y_i - y_{i-1}$ in the preceding definition, we get the definition of the line integral $\int_C f(x, y)dy$:

$$\int_C f(x, y)dy = \lim_{n\to\infty} \sum_{i=1}^{n} f(x_i^*, y_i^*)(y_i - y_{i-1}),$$

where, as before, all of the differences $y_i - y_{i-1}$ must approach 0 as n increases without bound.

Consequently, using line integrals, the total work W done in moving a particle along the curve C in the force field

$$F(\mathbf{X}) = (f_1(\mathbf{X}), f_2(\mathbf{X})) = (f_1(x, y), f_2(x, y))$$

is

$$W = \int_C f_1(x, y)dx + \int_C f_2(x, y)dy.$$

Such a sum of line integrals is often expressed this way:

$$W = \int_C f_1(x, y)dx + f_2(x, y)dy.$$

In the next section we shall derive a few of the major properties of line integrals and consider the problem of computing these integrals.

Exercises 17.1

1. Use the definition of the line integral to evaluate $\int_C f(x, y)dx$ where $f(x, y) = 5$ for all (x, y) and C is the graph of $x + y = 2$ with initial point $(0, 2)$ and terminal point $(1, 1)$.
2. Use the definition of the line integral to evaluate $\int_C f(x, y)dy$ where $f(x, y) = x + y$ and C is the graph of $x + y = 5$ with initial point $(6, -1)$ and terminal point $(1, 4)$.
3. Use the definition of the line integral to evaluate $\int_C f(x, y)dy$ where $f(x, y) = x^2 - 2xy + y^2$ and C is the graph of $x - y = -1$ with initial point $(0, 1)$ and terminal point $(1, 2)$.
4. Use the definition of the line integral to evaluate $\int_C f(x, y)dx$ where $f(x, y) = e^{x^2 - y + 1}$ and C is the graph of $y = x^2$ with initial point $(0, 0)$ and terminal point $(1, 1)$.
5. Show that if C has the equation $x = k$, where k is a constant, then

$$\int_C f(x, y)dx = 0$$

for all initial and terminal points on C.
6. Prove that $\int_C kf(x, y)dy = k \int_C f(x, y)dy$.
7. Prove that $\int_C [f(x, y) + g(x, y)]dx = \int_C f(x, y)dx + \int_C g(x, y)dx$.

17.2 Computation of Line Integrals

As in previous discussions of integrals two questions should be considered:

(a) What conditions must we impose on the real-valued function f and on the curve C so that $\int_C f(x, y)dx$ and $\int_C f(x, y)dy$ exist?
(b) Is there some relatively simple method for computing line integrals?

Before answering those questions we shall review parametric equations in light of what we have learned about vector-valued functions. Suppose that a curve C is given by the parametric equations $x = r(t)$, $y = s(t)$, for t in an interval I. Then a point (x, y) is on C if and only if $x = r(t)$ and $y = s(t)$ for some t in I. With vector notation it is possible to write the pair of parametric equations as a single equation.

Figure 17.2.1

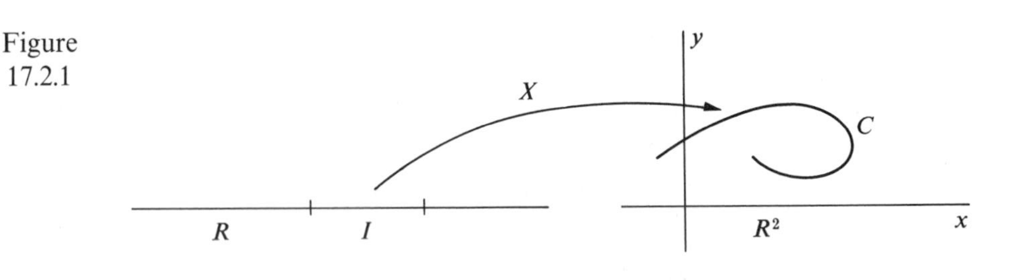

Consider the function $\mathbf{X}: I \to R^2$ where $\mathbf{X}(t) = (r(t), s(t))$. Since (x, y) is on C if and only if $x = r(t)$ and $y = s(t)$ for some t in I, (x, y) is on C if and only if $(x, y) = \mathbf{X}(t)$ for some t in I. This relationship is illustrated in Figure 17.2.1.

We can now answer the questions posed at the beginning of this section. The following theorem, whose proof can be found in most advanced calculus texts, provides sufficient conditions for the existence of line integrals.

Theorem 17.2.1

Let the curve C have the parametric equation $X(t) = (r(t), s(t))$ with initial point $X(t_I) = (r(t_I), s(t_I))$ and terminal point $X(t_T) = (r(t_T), s(t_T))$. Let $r'(t)$, $s'(t)$, and $f(r(t), s(t))$ be continuous for t between t_I and t_T. Then the line integrals

$$\int_C f(x, y)dx \quad \text{and} \quad \int_C f(x, y)dy$$

exist and are independent of the choices of (x_i^*, y_i^*) and of the method of subdivision.

In the following discussion we shall assume that all curves discussed have a parametric equation $X(t) = (r(t), s(t))$ such that $r(t)$ and $s(t)$ are both differentiable.

Computation of line integrals is not a very difficult task. Normally the following theorem can be used to convert a line integral into a standard integral of a single variable.

Theorem 17.2.2

Let the curve C have the parametric equation $X(t) = (r(t), s(t))$ with initial point $(r(t_I), s(t_I))$ and terminal point $(r(t_T), s(t_T))$. Moreover, let $r'(t)$, $s'(t)$, and $f(r(t), s(t))$ be continuous for t between t_I and t_T. Then

$$\int_C f(x, y)dx = \int_{t_I}^{t_T} f(r(t), s(t))r'(t)dt$$

and

$$\int_C g(x, y)dy = \int_{t_I}^{t_T} g(r(t), s(t))s'(t)dt.$$

Proof. Since we know so little about line integrals at this point, it will be necessary to return to Definition 17.1.1 in order to prove this theorem. Definition 17.1.1 requires that the curve C be subdivided into n subcurves. By Theorem 17.2.1, the line integrals exist and are independent of the method of subdivision. If we subdivide the interval $t_I \le t \le t_T$ into n subintervals of equal length by the numbers

$$t_I = t_0 < t_1 < \cdots < t_n = t_T,$$

then the points

$$X(t_0) = (r(t_0), s(t_0)), X(t_1) = (r(t_1), s(t_1)), \ldots, X(t_n) = (r(t_n), s(t_n))$$

yield a subdivision of C.

Next we must select an arbitrary point (x_i^*, y_i^*) on C between $X(t_{i-1})$ and $X(t_i)$. This is easily accomplished by selecting points t_i^*, $i = 1, 2, \ldots, n$, such that $t_{i-1} \leq t_i^* \leq t_i$. Then we can let

$$(x_i^*, y_i^*) = (r(t_i^*), s(t_i^*)).$$

By definition

$$\int_C f(x, y)dx = \lim_{n \to \infty} \sum_{i=1}^{n} f(x_i^*, y_i^*)(x_i - x_{i-1}).$$

So by our choices of x_i^*, y_i^*, x_i, and x_{i-1}, we have

$$\int_C f(x, y)dx = \lim_{n \to \infty} \sum_{i=1}^{n} f(r(t_i^*), s(t_i^*))(r(t_i) - r(t_{i-1})).$$

Thus

$$\int_C f(x, y)dx = \lim_{n \to \infty} \sum_{i=1}^{n} f(r(t_i^*), s(t_i^*))\frac{r(t_i) - r(t_{i-1})}{t_i - t_{i-1}}(t_i - t_{i-1}).$$

Since $r(t)$ is a differentiable function we know by The Mean Value Theorem that there is some t_i^{**} such that $t_{i-1} < t_i^{**} < t_i$ and

$$r'(t_i^{**}) = \frac{r(t_i) - r(t_{i-1})}{t_i - t_{i-1}}.$$

Hence we have

$$\int_C f(x, y)dx = \lim_{n \to \infty} \sum_{i=1}^{n} f(r(t_i^*), s(t_i^*))r'(t_i^{**})(t_i - t_{i-1}).$$

Then, on letting $h = t_i - t_{i-1}$, we obtain

$$\int_C f(x, y)dx = \lim_{n \to \infty} \sum_{i=1}^{n} f(r(t_i^*), s(t_i^*))r'(t_i^{**})h.$$

But by Bliss' Theorem (Theorem 5.6.1) the right side of that equation is a definite integral, and so we get

$$\int_C f(x, y)dx = \int_{t_I}^{t_T} f(r(t), s(t))r'(t)dt.$$

Similarly $$\int_C g(x, y)dy = \int_{t_I}^{t_T} g(r(t), s(t))s'(t)dt.$$

Example 1 | Find $\int_C xy^2\, dy$ where C is the portion of the curve $X(t) = (2t, t^{1/2})$ from $t = 1$ to $t = 4$.

Here $r(t) = 2t$ and $s(t) = t^{1/2}$, and so application of Theorem 17.2.2 gives

$$\int_C xy^2\,dy = \int_1^4 (2t)t\left(\frac{1}{2}t^{-1/2}\right)dt = \int_1^4 t^{3/2}\,dt = \frac{62}{5}. \qquad \|$$

If a line integral exists, its value is not dependent on the particular choice of a parametric equation for C. The integral of the next example is the same as that of Example 1; only the parametric representation for C has been changed.

Example 2 | Find $\int_C xy^2\,dy$, where C is the portion of the curve $X(t) = (2t^2, t)$ from $t = 1$ to $t = 2$.

In this case we take $r(t) = 2t^2$ and $s(t) = t$. Then by Theorem 17.2.2 we have

$$\int_C xy^2\,dy = \int_1^2 2t^2(t^2)(1)dt = 2\int_1^2 t^4\,dt = \frac{62}{5}. \qquad \|$$

Often we must evaluate a line integral over a curve C that is represented by an equation of the form $y = p(x)$ or $x = q(y)$. In this case we first change the equation for the curve to parametric form and then we apply Theorem 17.2.2 as in the preceding examples.

Example 3 | Find $\int_C (x + y^2)dx$, if C is the portion of the parabola $y = x^2$ from the initial point $(1, 1)$ to the terminal point $(2, 4)$.

If we introduce the parameter t by setting $x = t$, then $y = t^2$; and so C is given by $X(t) = (t, t^2)$ with initial point at $t = 1$ and terminal point at $t = 2$. Thus, since $r(t) = t$ and $s(t) = t^2$,

$$\int_C (x + y^2)dx = \int_1^2 (t + t^4)(1)dt = \left(\frac{t^2}{2} + \frac{t^5}{5}\right)\Bigg|_1^2 = \frac{77}{10}. \qquad \|$$

Example 4 | Find $\int_C \sin x\,dx + x^2y^2\,dy$, where C is the portion of the curve $x = y^3$ from $y = -1$ to $y = 1$.

In this case we let $y = t$; then $x = t^3$, and so $X(t) = (t^3, t)$, $-1 \leq t \leq 1$, is an equation for C. Thus

$$\begin{aligned}\int_C \sin x\,dx + x^2y^2\,dy &= \int_{-1}^1 (\sin t^3)\cdot 3t^2\,dt + \int_{-1}^1 t^6\cdot t^2\,dt \\ &= (-\cos t^3)\Big|_{-1}^1 + \frac{1}{9}t^9\Big|_{-1}^1 \\ &= -\cos 1 + \cos(-1) + \frac{2}{9}.\end{aligned}$$

Then, since $\cos 1 = \cos(-1)$, we have

$$\int_C \sin x\,dx + x^2y^2\,dy = \frac{2}{9}. \qquad \|$$

By Theorem 17.2.2 we know that if a curve C has equation $X(t) = (r(t), s(t))$ with initial point $(r(t_I), s(t_I))$ and terminal point $(r(t_T), s(t_T))$, then

$$\int_C f(x, y)dx + g(x, y)dy = \int_{t_I}^{t_T} [f(r(t), s(t))r'(t) + g(r(t), s(t))s'(t)]dt.$$

If we let $X(t) = (r(t), s(t))$ represent C, we have

$$X'(t) = (r'(t), s'(t)).$$

Thus, if we let F be the vector-valued function $F(X) = (f(X), g(X))$, then

$$F(X(t)) = (f(X(t)), g(X(t))) = (f(r(t), s(t)), g(r(t), s(t)));$$

and so we can write

$$F(X(t)) \cdot X'(t) = f(r(t), s(t))r'(t) + g(r(t), s(t))s'(t).$$

Consequently we can state the result of Theorem 17.2.2 in this more compact form:

$$\int_C f(x, y)dx + g(x, y)dy = \int_{t_I}^{t_T} F(X(t)) \cdot X'(t)dt.$$

Line integrals have many properties in common with integrals of a single variable. Parts (i) and (ii) of the following theorem appeared as exercises in the previous section, and parts (iii) and (iv) will appear as exercises in the present section. The proofs of these properties follow directly from the definition of line integrals.

Theorem 17.2.3

When the indicated integrals exist, they have the following properties:

(i) $\int_C kf(x, y)dx = k \int_C f(x, y)dx$, if k is any constant.

(ii) $\int_C [f(x, y) \pm g(x, y)]dx = \int_C f(x, y)dx \pm \int_C g(x, y)dx$.

(iii) If C' is the same as C except that the initial point of C' is the terminal point of C and the terminal point of C' is the initial point of C, then

$$\int_{C'} f(x, y)dx = -\int_C f(x, y)dx.$$

(iv) Let C be a curve with initial point P_1 and terminal point P_2, and let P be any point on C. Let C_1 be the portion of C from P_1 to P, and let C_2 be the portion of C from P to P_2, as in Figure 17.2.2. Then

$$\int_C f(x, y)dx = \int_{C_1} f(x, y)dx + \int_{C_2} f(x, y)dx.$$

Figure 17.2.2

Theorem 17.2.3 also holds when dx is replaced by dy.

Example 5 In Example 4, if the curve C' is the same as C except that it goes from $y = 1$ to $y = -1$, we may apply part (iii) of Theorem 17.2.3 to obtain

$$\int_{C'} \sin x \, dx + x^2 y^2 \, dy = -\frac{2}{9}. \quad \|$$

Example 6 Find $\int_C xy^2 \, dx$, where C is the curve that goes from (0, 0) to (2, 4) along $y = x^2$ and then from (2, 4) and (0, 6) along $y = 6 - x$.

Since the curve C is naturally split into two segments, C_1 along the parabola $y = x^2$, and C_2 along the line $y = 6 - x$, we may apply part (iv) of the preceding theorem to obtain

$$\int_C xy^2 \, dx = \int_{C_1} xy^2 \, dx + \int_{C_2} xy^2 \, dx.$$

Then, using the equation $X_1(t) = (t, t^2)$ for C_1, and $X_2(t) = (t, 6 - t)$ for C_2, we have

$$\begin{aligned}
\int_C xy^2 \, dx &= \int_0^2 t \cdot (t^2)^2 \, dt + \int_2^0 t(6 - t)^2 \, dt \\
&= \int_0^2 t^5 \, dt + \int_2^0 (t^3 - 12t^2 + 36t) dt \\
&= \frac{32}{3} - 44 = -\frac{100}{3}. \quad \|
\end{aligned}$$

Exercises 17.2

In Exercises 1–16, compute the given line integrals.

1. $\int_C (x + y^2) dx + xy \, dy$, where C has the parametric equation $X(t) = (t^2, 5t)$ with initial point (1, 5) and terminal point (9, 15).
2. $\int_C (x + y^2) dx + xy \, dy$, where C has the parametric equation $X(t) = (1 - t, t^2)$ with initial point (1, 0) and terminal point (0, 1).
3. $\int_C \cos y \, dx + e^x \, dy$, where C has the parametric equation $X(t) = (3t, 2t)$ with initial point (0, 0) and terminal point (3, 2).
4. $\int_C y \, dx - x \, dy$, where C has the parametric equation $X(t) = (\sin t, \cos t)$ with initial point (0, 1) and terminal point (1, 0).
5. $\int_C \frac{dx}{x^2 + y^2} + \frac{dy}{x^2 + y^2}$, where C has the parametric equation $X(t) = (\cos t, \sin t)$ with initial point $X(0)$ and terminal point $X(\pi)$.
6. $\int_C dx + (x + y) dy$, where C has the parametric equation $X(t) = (t, t^2)$ with initial point (1, 1) and terminal point (0, 0).
7. $\int_C 2xy \, dx + x^2 \, dy$, where C is the line segment from (0, 0) to (2, 4).
8. $\int_C x^2 y^{1/3} \, dx + x^{1/3} \, y \, dy$, where C is the line segment from (1, 2) to (−3, 1).
9. $\int_C (x^2 + y) dx$, along $y = 2x$ from (2, 4) to (3, 6).
10. $\int_C y \, dx + x \, dy$, along $y = x^2$ from (1, 1) to (2, 4).
11. $\int_C y \, dx + x \, dy$, along $x = \sin y$ from (0, 0) to (0, π).

12. $\int_C (xy + x^2)dy$, along $x = y^3$ from (0, 0) to (8, 2).
13. $\int_C (x + y)dx + dy$, where C is the curve $y = x^{1/2}$ from (0, 0) to (1, 1) and then $y = x^2$ from (1, 1) to (0, 0).
14. $\int_C x^2\,dy + (x + y)dx$, where C is the curve $y = 2x$ from (1, 2) to (2, 4) and then $y = x^2$ from (2, 4) to (3, 9).
15. $\int_C (x^2 + y)dx$, where C is taken in counterclockwise direction around the rectangle whose vertices are (0, 0), (3, 0), (3, 2), and (0, 2).
16. $\int_C (x^2 + y)dx$, where C is taken in counterclockwise direction around the triangle whose vertices are (0, 0), (3, 0), and (3, 1).
17. To illustrate that line integrals are independent of the parametrization, compute $\int_C (x + y)dx + (x - y)dy$ by using the following two parametrizations of C.
    (a) $X(t) = (t, 2t^{1/2})$, with initial point $X(1)$ and terminal point $X(4)$.
    (b) $X(t) = (t^2, 2t)$, with initial point $X(1)$ and terminal point $X(2)$.
18. Find the work done in moving a particle along the curve $X(t) = (t^2, t)$ from $t = 1$ to $t = 3$ under the influence of a force field $F(x, y) = (x + y, y)$.
19. A particle of mass m is moved from the point (2, 3) to the point $(-3, 2)$ in counterclockwise direction along the circle $x^2 + y^2 = 13$. Find the work done in moving the particle against a force that attracts the particle toward the origin with a magnitude of km/d^2, where k is a constant and d is the distance between the particle and the origin.
20. A particle of mass m is always attracted toward the x-axis by a force whose magnitude is km/d, where d is the y-coordinate of the particle. Find the work done against this attracting force when the particle is moved from (1, 1) to (2, 4) along the parabola $y = x^2$.
21. Prove parts (iii) and (iv) of Theorem 17.2.3.

17.3 Green's Theorem

In this section we shall obtain a remarkable theorem that relates line integrals to double integrals and will greatly simplify the calculation of the line integrals along "closed curves." To develop the theorem we shall need the following definitions.

Definition 17.3.1

> A *simple closed curve* is a closed curve that does not cross or touch itself.

In Figure 17.3.1, C_1 and C_2 are simple closed curves, but C_3 and C_4 are not.

Figure 17.3.1

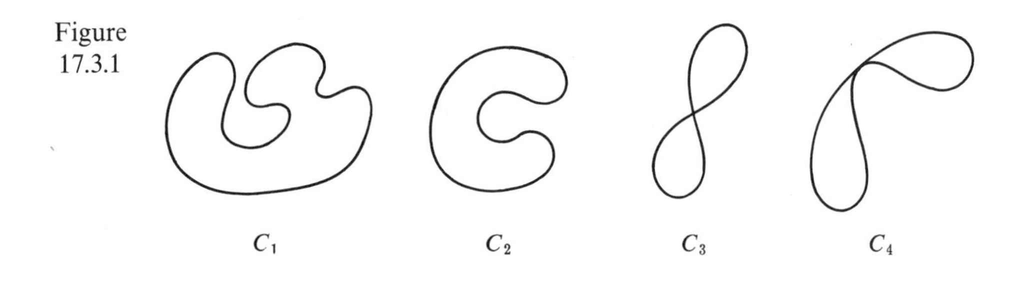

Definition 17.3.2

> Let C be a closed curve in the xy-plane and suppose that the z-axis points upward. The *positive direction* of C is the direction that we should move around C in order to keep the interior of the curve on our left.

In the sequel, all integrals around closed curves will be assumed to be taken in the positive direction unless otherwise stated.

We are now in a position to state and prove Green's Theorem.

Theorem 17.3.1

> *Green's Theorem*
>
> Let $P(x, y)$, $Q(x, y)$, $\partial Q/\partial x$, and $\partial P/\partial y$ be continuous in a bounded region S in the xy-plane; and let C be a simple closed curve which, together with its interior, lies entirely inside S. Let R be the closed region bounded by C. Then
>
> $$\int_C P(x, y)dx + Q(x, y)dy = \int_R \left(\frac{\partial Q}{\partial x} - \frac{\partial P}{\partial y}\right) dA.$$

Proof. First we consider the case illustrated in Figure 17.3.2, where no horizontal or vertical line cuts C in more than two points.

Since R is bounded and no vertical line cuts C in more than two points, R has a left extreme at $x = a$ and a right extreme at $x = b$. Moreover there is an equation $y = p(x)$ for the lower portion of C and an equation $y = q(x)$ for the upper portion. We then have

$$\int_R \frac{\partial P}{\partial y} dA = \int_a^b \int_{p(x)}^{q(x)} \frac{\partial P}{\partial y} dy\, dx = \int_a^b [P(x, q(x)) - P(x, p(x))]dx.$$

On the other hand,

$$\begin{aligned}\int_C P(x, y)dx &= \int_{C_1} P(x, y)dx + \int_{C_2} P(x, y)dx \\ &= \int_a^b P(t, p(t))dt + \int_b^a P(t, q(t))dt \\ &= \int_a^b P(t, p(t))dt - \int_a^b P(t, q(t))dt.\end{aligned}$$

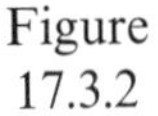
Figure 17.3.2

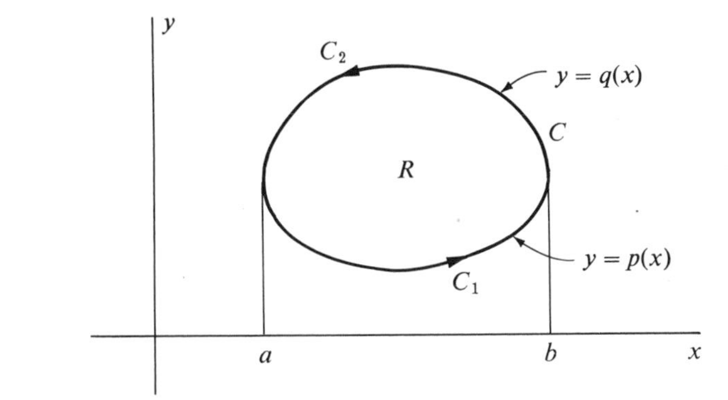

Consequently,

$$\int_C P(x, y)dx = -\int_R \frac{\partial P}{\partial y}\, dA. \tag{1}$$

Similarly, since R is bounded and no horizontal line cuts C in more than two points, R has a lower extreme at $y = c$, an upper extreme at $y = d$, a left portion with an equation $x = g(y)$, and a right portion $x = h(y)$, as shown in Figure 17.3.3.

Figure 17.3.3

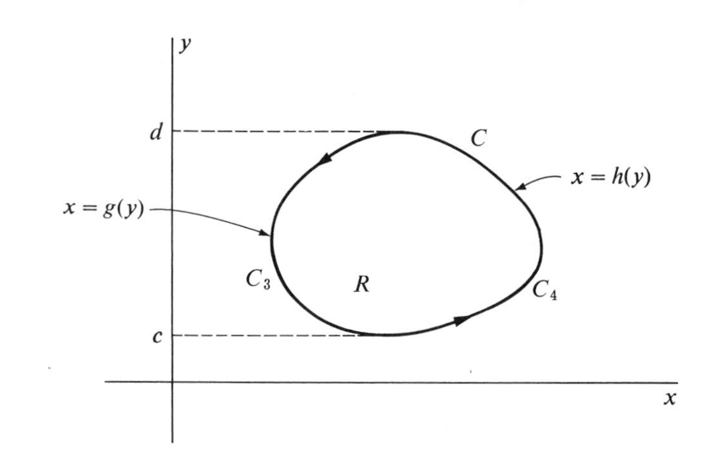

So
$$\int_R \frac{\partial Q}{\partial x}\, dA = \int_c^d \int_{g(y)}^{h(y)} \frac{\partial Q}{\partial x}\, dx\, dy$$
$$= \int_c^d [Q(h(y), y) - Q(g(y), y)]dy.$$

On the other hand,

$$\int_C Q(x, y)dy = \int_{C_4} Q(x, y)dy + \int_{C_3} Q(x, y)dy$$
$$= \int_c^d Q(h(t), t)dt + \int_d^c Q(g(t), t)dt$$
$$= \int_c^d Q(h(t), t)dt - \int_c^d Q(g(t), t)dt.$$

Consequently,

$$\int_C Q(x, y)dy = \int_R \frac{\partial Q}{\partial x}\, dA. \tag{2}$$

Therefore, on combining (1) and (2) we have

$$\int_C P(x, y)dx + Q(x, y)dy = \int_R \left(\frac{\partial Q}{\partial x} - \frac{\partial P}{\partial y}\right) dA.$$

Thus we have established Green's Theorem for simple closed curves C that are intersected by horizontal or vertical lines in at most two points.

Green's Theorem can be easily extended to certain other curves C. For example, even though horizontal or vertical lines intersect C in more than two points, it may be possible to subdivide the region R bounded by C into smaller subregions bounded by curves that are intersected by horizontal or vertical lines in at most two points. For example, the region R of Figure 17.3.4 is divided into three smaller subregions R_1, R_2, and R_3 that are bounded by curves C_1, C_2, and C_3. For each subregion R_i we have

$$\int_{C_i} P\,dx + Q\,dy = \int_{R_i} \left(\frac{\partial Q}{\partial x} - \frac{\partial P}{\partial y}\right) dA, \qquad i = 1, 2, 3,$$

Figure 17.3.4

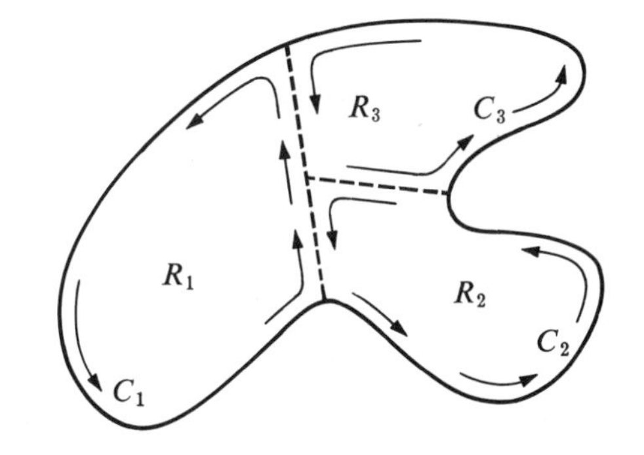

because horizontal and vertical lines cut each C_i in at most two points. However,

$$\int_{C_1} P\,dx + Q\,dy + \int_{C_2} P\,dx + Q\,dy + \int_{C_3} P\,dx + Q\,dy = \int_C P\,dx + Q\,dy,$$

since the integrals on the left go over the dotted portions of the curve once in each direction. Also,

$$\int_{R_1} \left(\frac{\partial Q}{\partial x} - \frac{\partial P}{\partial y}\right) dA + \int_{R_2} \left(\frac{\partial Q}{\partial x} - \frac{\partial P}{\partial y}\right) dA + \int_{R_3} \left(\frac{\partial Q}{\partial x} - \frac{\partial P}{\partial y}\right) dA = \int_R \left(\frac{\partial Q}{\partial x} - \frac{\partial P}{\partial y}\right) dA.$$

and therefore

$$\int_C P\,dx + Q\,dy = \int_R \left(\frac{\partial Q}{\partial x} - \frac{\partial P}{\partial y}\right) dA.$$

The proof of Green's Theorem can be modified to include cases not covered by the given proof. In particular, it is left as an exercise to show that Green's Theorem can be applied to a rectangle whose sides are parallel to the coordinate axes. A proof of Green's Theorem in its more general form can be found in some advanced calculus texts.

Example 1 Use Green's Theorem to compute

$$\int_C x^2 y\,dx + (x^2 + y)dy,$$

where C is the rectangle having vertices $(\pm 1, 0)$ and $(\pm 1, 1)$ and is traversed in the counterclockwise direction.

The situation is pictured in Figure 17.3.5. Since $P(x, y) = x^2y$ and $Q(x, y) = x^2 + y$, we have $\partial Q/\partial x = 2x$ and $\partial P/\partial y = x^2$. Thus

$$\int_C x^2y\,dx + (x^2 + y)dy = \int_R (2x - x^2)dA.$$

Hence

$$\int_C x^2y\,dx + (x^2 + y)dy = \int_0^1 \int_{-1}^1 (2x - x^2)dx\,dy$$

$$= \int_0^1 \left(x^2 - \frac{1}{3}x^3\right)\Bigg|_{-1}^1 dy$$

$$= -\frac{2}{3}.$$

||

Figure 17.3.5

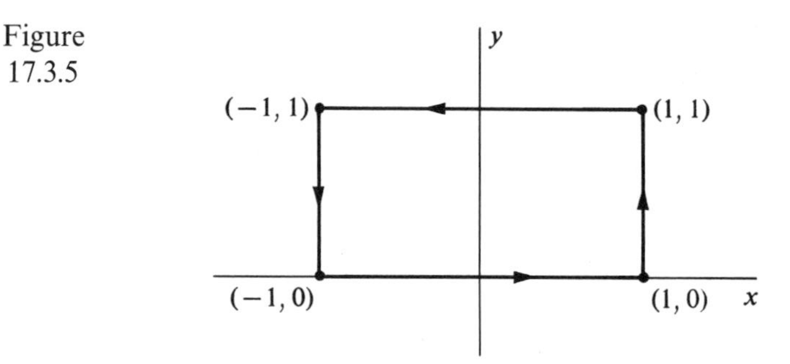

It should be noticed that Green's Theorem can be applied to a region R that is inside a simple closed curve C_1 and outside a simple closed curve C_2. To apply the theorem, however, we must connect C_1 and C_2 by lines (or curves) L_1 and L_2, as shown in Figure 17.3.6.

Figure 17.3.6

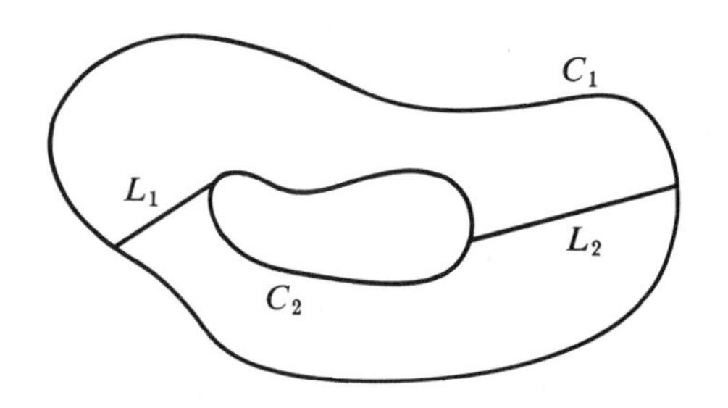

The simple closed curve consisting of L_1, L_2, and the lower parts of C_1 and C_2 will be denoted by C_L. The simple closed curve consisting of L_1, L_2, and the upper parts of C_1 and C_2 will be denoted by C_U. We then integrate over C_U and C_L in the positive direction, as indicated in Figure 17.3.7. Note that by integrating around C_U and C_L in the positive direction we have

(a) integrated over both L_1 and L_2 in each direction,

(b) integrated around C_1 in the positive direction and around C_2 in the negative direction.

Figure 17.3.7

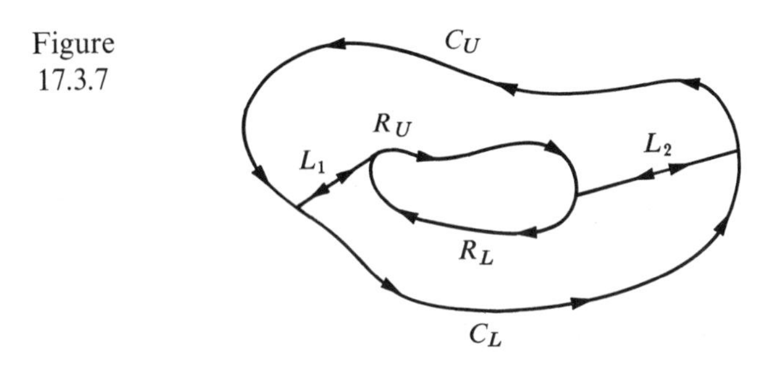

By (a) and (b) above, the integrals over L_1 and L_2 cancel thus,

$$\int_{C_1} P\,dx + Q\,dy - \int_{C_2} P\,dx + Q\,dy = \int_{C_L} P\,dx + Q\,dy + \int_{C_U} P\,dx + Q\,dy.$$

Now we can apply Green's Theorem to the integrals around C_L and C_U. We let the regions bounded by C_L and C_U be R_L and R_U, respectively. Then

$$\int_{C_1} P\,dx + Q\,dy - \int_{C_2} P\,dx + Q\,dy = \int_{R_L} \left(\frac{\partial Q}{\partial x} - \frac{\partial P}{\partial y}\right) dA + \int_{R_U} \left(\frac{\partial Q}{\partial x} - \frac{\partial P}{\partial y}\right) dA.$$

Note that R_L and R_U together make up the region R. Then, since R_L and R_U overlap only along their boundaries,

$$\int_{C_1} P\,dx + Q\,dy - \int_{C_2} P\,dx + Q\,dy = \int_{R} \left(\frac{\partial Q}{\partial x} - \frac{\partial P}{\partial y}\right) dA.$$

Of course this application of Green's Theorem requires that C_1, C_2, and R lie entirely within a region S, where $P(x, y)$, $Q(x, y)$, $\partial Q/\partial x$, and $\partial P/\partial y$ are continuous.

Example 2 Use Green's Theorem to compute

$$\int_{C_1} xy^2\,dx + (x + y^2)dy - \int_{C_2} xy^2\,dx + (x + y^2)dy,$$

where C_1 is the circle with radius 2 and center at the origin, and C_2 is the unit circle with center at the origin.

The curves C_1 and C_2 and the region R are pictured in Figure 17.3.8.

Figure 17.3.8

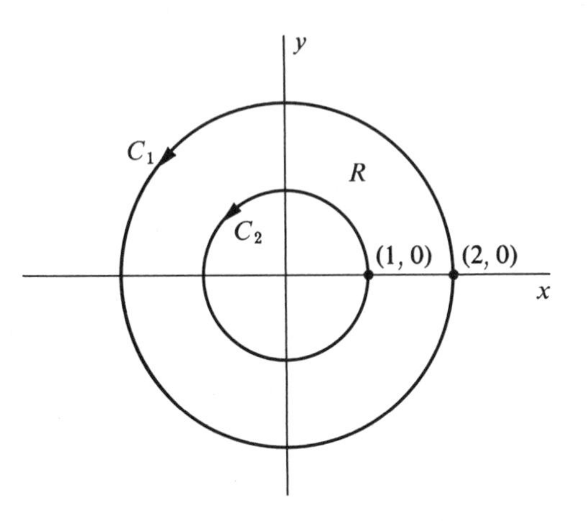

Then, since $P(x, y) = xy^2$ and $Q(x, y) = x + y^2$, we have $\partial Q/\partial x = 1$ and $\partial P/\partial y = 2xy$. Hence

$$\int_{C_1} xy^2\,dx + (x + y^2)dy - \int_{C_2} xy^2\,dx + (x + y^2)dy = \int_R (1 - 2xy)dA.$$

A change to polar coordinates will ease the calculation of the last integral.

$$\begin{aligned}\int_R (1 - 2xy)dA &= \int_0^{2\pi} \int_1^2 (1 - 2r^2 \cos\theta \sin\theta)r\,dr\,d\theta \\ &= \int_0^{2\pi} \left(\frac{1}{2}r^2 - \frac{2}{4}r^4 \cos\theta \sin\theta\right)\Big|_1^2 d\theta \\ &= \int_0^{2\pi} \left(\frac{3}{2} - \frac{15}{2}\cos\theta \sin\theta\right)d\theta \\ &= \frac{3}{2}\theta - \frac{15}{4}\sin^2\theta\Big|_0^{2\pi} \\ &= 3\pi.\end{aligned}$$

Thus $$\int_{C_1} xy^2\,dx + (x + y^2)dy - \int_{C_2} xy^2\,dx + (x + y^2)dy = 3\pi.$$ ||

Exercises 17.3

Use Green's Theorem to compute the integrals in Exercises 1–16.

1. $\int_C xy\,dx + (x + y)dy$, where C is the rectangle with vertices at $(-1, 0)$, $(0, 0)$, $(0, 1)$, and $(-1, 1)$.
2. $\int_C x^2y\,dx + x^2y^2\,dy$, where C is the rectangle with vertices at $(0, 0)$, $(3, 0)$, $(3, 2)$, and $(0, 2)$.
3. $\int_C 2xy\,dx$, where C is the rectangle with vertices $(3, 1)$, $(3, 3)$, $(5, 3)$, and $(5, 1)$.
4. $\int_C x^2y\,dy$, where C is the rectangle with vertices $(3, 1)$, $(3, 3)$, $(5, 3)$, and $(5, 1)$.
5. $\int_C e^x \sin y\,dx + e^x \cos y\,dy$, where C is the triangle whose vertices are at $(1, 1)$, $(1, 0)$, and $(2, 0)$.
6. $\int_C xe^y\,dx + e^x\,dy$, where C is the triangle whose vertices are at $(0, 0)$, $(2, 2)$, and $(2, 0)$.
7. $\int_C x^3\,dy - y^3\,dx$, where C is the closed curve formed by $y = x^3$ and $y = x$, between the points $(0, 0)$ and $(1, 1)$.
8. $\int_C x\,dx + xy^2\,dy$, where C is the closed curve formed by $y = x$ and $y = x^2$, between the points $(0, 0)$ and $(1, 1)$.
9. $\int_C y\,dx + x\,dy$ where C is the circle with parametric equation $X(t) = (\sin t, \cos t)$.
10. $\int_C (x^2 - y^2)dx$, where C is the circle $x^2 + y^2 = 1$.
11. $\int_C (x^3 + y)dy$, where C is the circle $x^2 + y^2 = 1$.
12. $\int_C (x^2 + y^2)dx + (y^2 - x^2)dy$ where C is the circle $x^2 + y^2 = 1$.
13. $\int_C y^3\,dx + (y - x^3)dy$, where C is the circle $x^2 + y^2 = 1$.
14. $\int_C - y^3\,dx - 3xy^2\,dy$, where C is the circle $x^2 + y^2 = 4$.
15. $\int_C x^3y^4\,dx + x^4y^3\,dy$, where C is the circle $x^2 + y^2 = 9$.
16. $\int_C e^x \sin y\,dx + e^x \cos y\,dy$, where C is the circle $x^2 + y^2 = 1$.

17. Can Green's Theorem be used to evaluate the integral

$$\int_C \frac{y}{x^2+y^2}\,dx - \frac{x}{x^2+y^2}\,dy$$

(a) where C is the circle $x^2 + y^2 = 1$?
(b) where C is the triangle with vertices (1, 0), (2, 0), and (2, 3)?

18. Show that if C is a simple closed curve enclosing a region with area A then

$$A = \int_C x\,dy = -\int_C y\,dx.$$

19. Use the result of Exercise 18 to compute the area of the region enclosed by the circle $x^2 + y^2 = r^2$.

20. Use the result of Exercise 18 to compute the area of the region enclosed by the ellipse $a^2x^2 + b^2y^2 = 1$.

21. Let C_1, C_2, and C_3 be the simple closed curves indicated below, and let R be the region between these curves. Suppose Q_x and P_y are continuous in R. Show that

$$\int_{C_1} P\,dx + Q\,dy - \int_{C_2} P\,dx + Q\,dy - \int_{C_3} P\,dx + Q\,dy = \int_R \left(\frac{\partial Q}{\partial x} - \frac{\partial P}{\partial y}\right) dA.$$

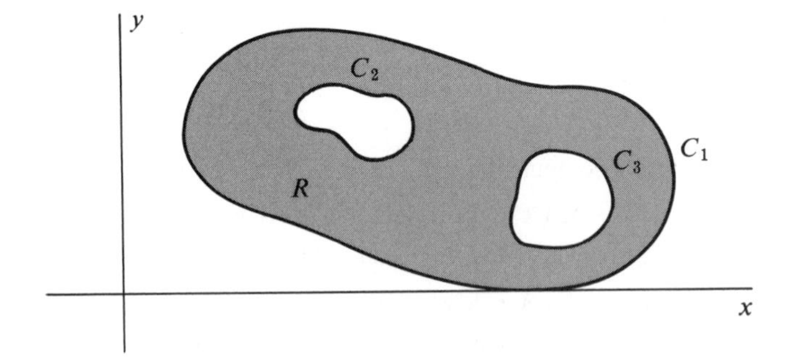

22. A force field $F(x, y) = (f_1(x, y), f_2(x, y))$ is said to be *conservative* if

$$\int_C f_1(x, y)dx + f_2(x, y)dy = 0$$

for every simple closed curve C.

A function $\phi: R^2 \to R$ is called a *potential function* for the force field $F(x, y)$ if

$$F(x, y) = \left(\frac{\partial \phi}{\partial x}, \frac{\partial \phi}{\partial y}\right),$$

where ϕ has continuous second partial derivatives. Prove that if a force field has a potential function, then the field is conservative.

23. Prove that the force field $F(x, y) = (y \cos xy, x \cos xy)$ is conservative. (See Exercise 22.)

24. Is the force field given by $F(x, y) = (y - x, x + y)$ conservative? (See Exercise 22.)

25. Prove Green's Theorem for the case where R is a rectangle having its sides parallel to the coordinate axes.

17.4 Independence of Path

In some cases a line integral $\int_C P\,dx + Q\,dy$ along a path C from a point A to a point B depends on the end points A and B but not on the particular path joining A and B. That phenomenon is sufficiently important to warrant the following definition.

Definition 17.4.1

The line integral $\int_C P\,dx + Q\,dy$ along a path C from a point A to a point B is *independent of path in a region* R if

$$\int_{C_1} P\,dx + Q\,dy = \int_{C_2} P\,dx + Q\,dy$$

for every pair of paths C_1 and C_2 in R from A to B. Figure 17.4.1 pictures this situation.

Figure 17.4.1

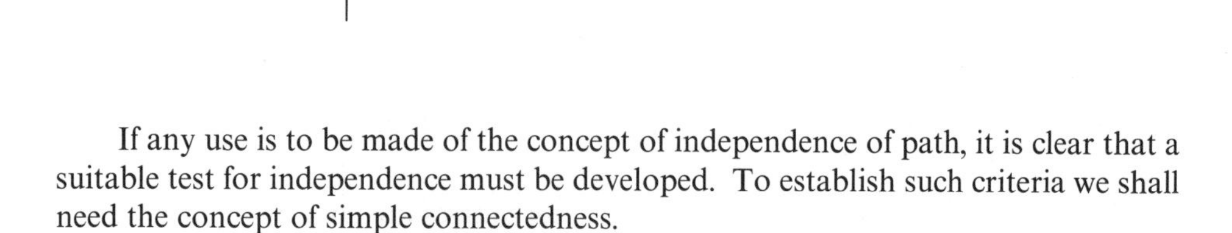

If any use is to be made of the concept of independence of path, it is clear that a suitable test for independence must be developed. To establish such criteria we shall need the concept of simple connectedness.

Definition 17.4.2

A region R is said to be *simply connected* if every simple closed curve in R is the boundary of a region S that is fully contained in R.

For example, the region of Figure 17.4.2(a) is simply connected, but the region illustrated in Figure 17.4.2(b) is not. To see that the second of these two regions is not simply connected just note that the curve C does not bound a region fully contained in R.

The next theorem is a step toward the development of a criterion for independence of path.

Figure 17.4.2

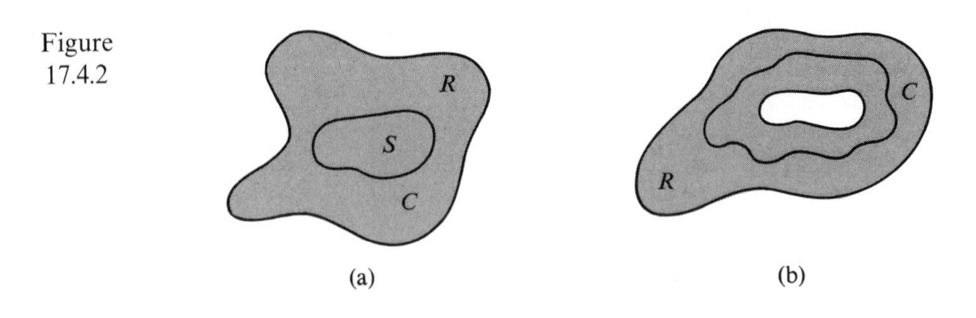

Theorem 17.4.1

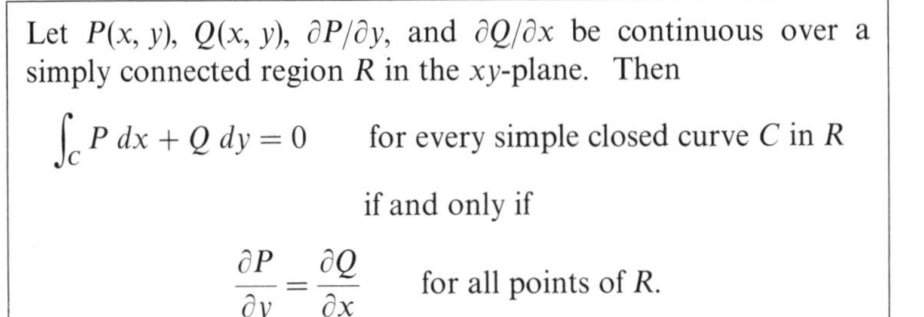

Let $P(x, y)$, $Q(x, y)$, $\partial P/\partial y$, and $\partial Q/\partial x$ be continuous over a simply connected region R in the xy-plane. Then

$$\int_C P\,dx + Q\,dy = 0 \qquad \text{for every simple closed curve } C \text{ in } R$$

if and only if

$$\frac{\partial P}{\partial y} = \frac{\partial Q}{\partial x} \qquad \text{for all points of } R.$$

Proof. First suppose that $\partial P/\partial y = \partial Q/\partial x$ for all points of R, and let C be any simple closed curve in the region R. Since R is simply connected, C bounds a region S fully contained in R. So we may apply Green's Theorem to obtain

$$\int_C P\,dx + Q\,dy = \int_S \left(\frac{\partial Q}{\partial x} - \frac{\partial P}{\partial y}\right) dA = \int_S 0\,dA = 0.$$

Now suppose that $\partial P/\partial y \neq \partial Q/dx$ for some point (x_0, y_0) of R; then

$$\frac{\partial Q}{\partial x} - \frac{\partial P}{\partial y} \neq 0 \quad \text{at} \quad (x_0, y_0).$$

However, since those partial derivatives are continuous in R, there must be some neighborhood N of (x_0, y_0) where $\partial Q/\partial x - \partial P/\partial y \neq 0$; and thus in N the difference of the partial derivatives must be everywhere positive or everywhere negative. Let C be a simple closed curve lying entirely inside N, and let R_1 be the region bounded by C. Then by Green's Theorem

$$\int_C P\,dx + Q\,dy = \int_{R_1} \left(\frac{\partial Q}{\partial x} - \frac{\partial P}{\partial y}\right) dA \neq 0,$$

since $\partial Q/\partial x - \partial P/\partial y$ is everywhere positive or everywhere negative in R_1.

Thus we have shown that if $\partial P/\partial y - \partial Q/\partial x$ is not zero for at least one point is R, then $\int_C P\,dx + Q\,dy \neq 0$ for at least one simple closed curve C in R. Therefore, if $\int_C P\,dx + Q\,dy = 0$ for every simple closed curve C in R, then

$$\frac{\partial P}{\partial y} - \frac{\partial Q}{\partial x} = 0, \quad \text{or} \quad \frac{\partial Q}{\partial x} = \frac{\partial P}{\partial y},$$

for every point of R, and so the theorem has been proved.

One final result is needed.

Theorem 17.4.2

$$\int_C P\,dx + Q\,dy \text{ is independent of path in } R$$

$$\text{if and only if}$$

$$\int_{C_1} P\,dx + Q\,dy = 0 \qquad \text{for every simple closed curve } C_1 \text{ in } R.$$

The proof in one direction is quite easy and is left as an exercise. The other half of the proof, however, would take us quite far afield. The interested reader should refer to a suitable advanced calculus text.

We can now combine the results of Theorem 17.4.1 and Theorem 17.4.2 to obtain the following simple test for independence of path in R.

Theorem 17.4.3

Let $P(x, y)$, $Q(x, y)$, $\partial P/\partial y$, and $\partial Q/\partial x$ be continuous over a simply connected region R in the xy-plane. Then

$$\int_C P\,dx + Q\,dy \text{ is independent of path in } R$$

$$\text{if and only if}$$

$$\frac{\partial P}{\partial y} = \frac{\partial Q}{\partial x} \qquad \text{for all points of } R.$$

In some applications of this theorem the region R may be taken to be the entire plane, as illustrated in the following example.

Example 1 Find where the following integral is independent of path:

$$\int_C (\sin x + \cos y)dx - x \sin y\,dy.$$

In this case, $P(x, y) = \sin x + \cos y$ and $Q(x, y) = -x \sin y$.

So
$$\frac{\partial P}{\partial y} = -\sin y = \frac{\partial Q}{\partial x}$$

for all points (x, y) in the plane. Since the entire plane is simply connected, we can use Theorem 17.4.3 to establish that the given integral is independent of path in the entire plane. ||

Independence of path is commonly used to simplify the calculation of line integrals, as shown in the next example.

Example 2 Compute

$$\int_C (\sin x + \cos y)dx - x \sin y\, dy,$$

where C is the curve having the equation $X(t) = (\cos t, \sin t)$ from $t = 0$ to $t = \pi/2$.

From Example 1 we know that this integral is independent of path; hence we can replace the given path with any path starting at $X(0) = (1, 0)$ and ending at $X(\pi/2) = (0, 1)$. An appropriate path to use is the polygonal path C' consisting of C_1 and C_2, as indicated in Figure 17.4.3. Then

$$\begin{aligned}\int_C (\sin x + \cos y)dx - x \sin y\, dy &= \int_{C_1} (\sin x + \cos y)dx - x \sin y\, dy \\ &\quad + \int_{C_2} (\sin x + \cos y)dx - x \sin y\, dy \\ &= -\int_0^1 \sin t\, dt + \int_1^0 (\sin t + \cos 1)dt \\ &= \cos 1 - \cos 0 - \cos 0 + \cos 1 - \cos 1 \\ &= \cos 1 - 2.\end{aligned}$$

||

Figure 17.4.3

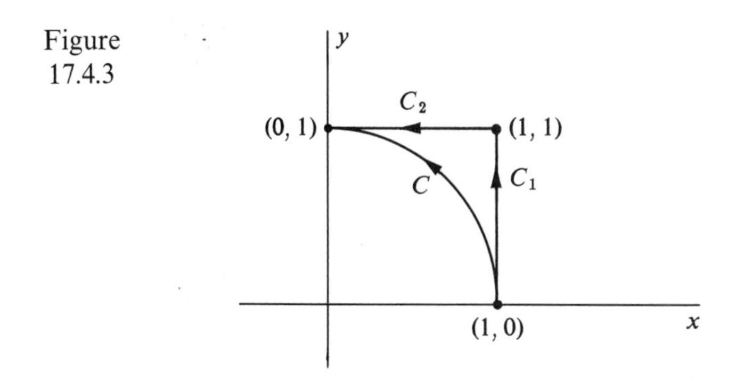

Of course it is possible that an integral may be independent of path only in some subregion of the plane.

Example 3 Find where the following integral is independent of path:

$$\int_C \frac{y}{x^2 + y^2}dx - \frac{x}{x^2 + y^2}dy.$$

In this case $P(x, y) = \dfrac{y}{x^2 + y^2}$ and $Q(x, y) = -\dfrac{x}{x^2 + y^2}$.

So

$$\frac{\partial P}{\partial y} = \frac{x^2 - y^2}{(x^2 + y^2)^2} = \frac{\partial Q}{\partial x}.$$

Consequently P_y and Q_x are equal and continuous, except at the origin where neither exists. An application of Theorem 17.4.3 then indicates that the given integral is independent of path in any *simply connected* subregion of the plane *not* containing the origin. Figure 17.4.4 indicates two possible choices for R. In Figure 17.4.4(b) the region R includes everything but a ray emanating from the origin. ||

Figure 17.4.4

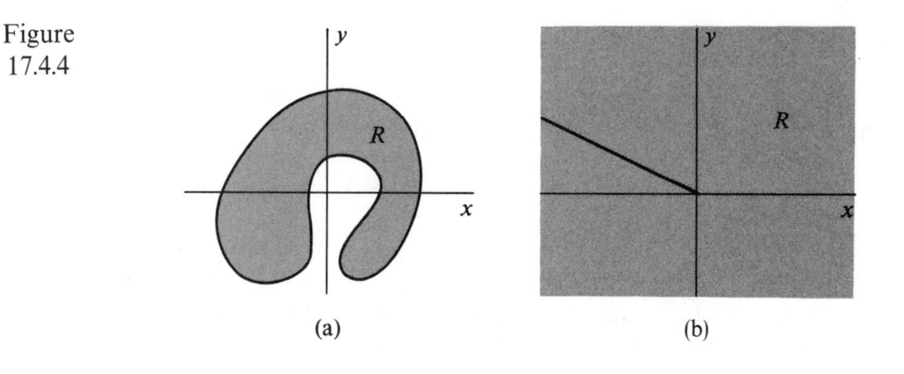

Exercises 17.4

In Exercises 1–12, show that the given integral is independent of path in the entire plane and then evaluate it.

1. $\int_{(0,1)}^{(1,3)} 3x^2y^2\,dx + 2x^3y\,dy.$

2. $\int_{(1,2)}^{(3,6)} 2xy\,dx + (x^2 + y^2)dy.$

3. $\int_{(1,1)}^{(2,2)} 2xe^y\,dx + x^2e^y\,dy.$

4. $\int_{(0,0)}^{(2,3)} ye^{xy}\,dx + xe^{xy}\,dy.$

5. $\int_{(0,1)}^{(-1,0)} e^xy^2\,dx + 2ye^x\,dy.$

6. $\int_{(\pi/2,1)}^{(\pi,-1)} y\sin x\,dx - \cos x\,dy.$

7. $\int_{(0,0)}^{(\pi,\pi)} \cos x\sin y\,dx + \sin x\cos y\,dy.$

8. $\int_{(0,0)}^{(\pi/2,\pi/2)} \sin y\,dx + x\cos y\,dy.$

9. $\int_{(0,1)}^{(1,0)} (2xe^y - e^xy^2)dx + (x^2e^y - 2ye^x)dy.$

10. $\int_{(0,\pi/2)}^{(\pi/2,\pi)} (-y\sin x + \sin y)dx + (\cos x + x\cos y)dy.$

11. $\int_{(0,\pi/2)}^{(\pi/4,\pi)} (e^x\sin y - e^y\sin x)dx + (e^y\cos x + e^x\cos y)dy.$

12. $\int_{(0,0)}^{(1,\pi/2)} 2x\sin y\,dx + x^2\cos y\,dy.$

13. Compute $\int_C e^x\sin y\,dx + e^x\cos y\,dy$, where C is the curve with equation $X(t) = (\sin t, \cos t)$ from $t = 0$ to $t = \pi$.

14. Compute $\int_C (x^2y^3 + xy^2)dx + (x^3y^2 + x^2y)dy$, where C is the curve $y = \ln x$ from $(1, 0)$ to $(e, 1)$.

15. Find where the following integral is independent of path:

$$\int_C \frac{y+1}{(x-1)^2 + (y+1)^2}\,dx + \frac{1-x}{(x-1)^2 + (y+1)^2}\,dy.$$

16. Find where the following integral is independent of path:

$$\int_C \frac{e^x(x+y-1)}{(x+y)^2}\,dx - \frac{e^x}{(x+y)^2}\,dy.$$

17. Let f and g be continuous functions. Prove that any line integral of the following form is independent of path:

$$\int_C f(x)dx + g(y)dy.$$

18. Let C be a simple closed curve. Show that

$$\int_C -\frac{y}{x^2 + y^2}\,dx + \frac{x}{x^2 + y^2}\,dy$$

is either 2π or 0, depending on whether or not the curve C encloses the origin.

19. Prove that if $\int_C P\,dx + Q\,dy$ is independent of path in R, then

$$\int_{C_1} P\,dx + Q\,dy = 0,$$

where C_1 is any simple closed curve in R.

20. Prove Theorem 17.4.3.

21. Prove that work is independent of path in a conservative field. See Exercise 22 of Section 17.3 for the necessary definitions.

17.5 Line Integrals in Three Dimensions

We shall now define line integrals in three dimensions. These definitions will parallel the 2-dimensional case.

Definition 17.5.1

Let $f(x, y, z)$ be continuous over a region V in xyz-space. Let C be a curve entirely inside V with an initial point A and a terminal point B. Divide the path from A to B into n parts by selecting on C the points

$$A = (x_0, y_0, z_0), \quad (x_1, y_1, z_1), \ldots, (x_i, y_i, z_i), \ldots, (x_n, y_n, z_n) = B.$$

Then for each i, take any point (x_i^*, y_i^*, z_i^*) on C between $(x_{i-1}, y_{i-1}, z_{i-1})$ and (x_i, y_i, z_i). The *line integral of f with respect to x along C* is defined as follows:

$$\int_C f(x, y, z)dx = \lim_{n\to\infty} \sum_{i=1}^{n} f(x_i^*, y_i^*, z_i^*)(x_i - x_{i-1}),$$

where the limit is taken so that all differences $x_i - x_{i-1}$ approach 0 as n increases without bound.

By replacing $x_i - x_{i-1}$ by $y_i - y_{i-1}$ and then by $z_i - z_{i-1}$ in Definition 17.5.1, we also obtain the definitions of these line integrals:

$$\int_C f(x, y, z)dy \quad \text{and} \quad \int_C f(x, y, z)dz.$$

Line integrals in three dimensions usually occur in the form

$$\int_C P(x, y, z)dx + \int_C Q(x, y, z)dy + \int_C R(x, y, z)dz,$$

which, for convenience, is often written as

$$\int_C P\,dx + Q\,dy + R\,dz.$$

Sufficient conditions for the existence of 3-dimensional line integrals are given in the next theorem, whose proof can be found in many advanced calculus texts.

Theorem 17.5.1

Let the curve C have parametric equation $X(t) = (q(t), r(t), s(t))$ with initial point $X(t_I) = (q(t_I), r(t_I), s(t_I))$ and terminal point $X(t_T) = (q(t_T), r(t_T), s(t_T))$.

Then, if $q'(t)$, $r'(t)$, $s'(t)$, and $f(q(t), r(t), s(t))$ are all continuous for t between t_I and t_T, the line integrals

$$\int_C f(x, y, z)dx, \quad \int_C f(x, y, z)dy, \quad \text{and} \quad \int_C f(x, y, z)dz$$

exist and are independent of the choices of (x_i^*, y_i^*, z_i^*) and of the method of subdivision.

In the following we shall assume that all curves discussed have a parametric equation $X(t) = (q(t), r(t), s(t))$ such that $q(t)$, $r(t)$, and $s(t)$ are differentiable.

As in the case of line integrals in two dimensions, if the curve C is given in parametric form, the integral can be expressed as an ordinary integral.

Theorem 17.5.2

If the curve C has parametric equation $X(t) = (q(t), r(t), s(t))$ with initial point $(q(t_I), r(t_I), s(t_I))$ and terminal point $(q(t_T), r(t_T), s(t_T))$, and if $q'(t)$, $r'(t)$, and $s'(t)$ are continuous between t_I and t_T, then line integrals can be expressed as ordinary definite integrals of functions of the single variable t as follows:

$$\int_C P(x, y, z)dx = \int_{t_I}^{t_T} P(q(t), r(t), s(t))q'(t)dt,$$

$$\int_C Q(x, y, z)dy = \int_{t_I}^{t_T} Q(q(t), r(t), s(t))r'(t)dt,$$

$$\int_C R(x, y, z)dz = \int_{t_I}^{t_T} R(q(t), r(t), s(t))s'(t)dt.$$

Those three results can be collected into this single equation:

$$\int_C P(x, y, z)dx + Q(x, y, z)dy + R(x, y, z)dz$$

$$= \int_{t_I}^{t_T} [P(q(t), r(t), s(t))q'(t) + Q(q(t), r(t), s(t))r'(t) + R(q(t), r(t), s(t))s'(t)]dt.$$

Example 1 Evaluate the line integral

$$\int_C xyz\,dx + y\,dy + xz\,dz,$$

where C is given by the parametric equations $x = t$, $y = t^2$, $z = t^3$ for $0 \leq t \leq 2$, with initial point $(0, 0, 0)$ and terminal point $(2, 4, 8)$.

An application of Theorem 17.5.2 yields

$$\int_C xyz\,dx + y\,dy + xz\,dz = \int_0^2 t^6\,dt + \int_0^2 2t^3\,dt + \int_0^2 3t^6\,dt = 81\frac{1}{7}. \qquad ||$$

Often the parametric equations for the curve C are not given explicitly, but must be calculated before computing a given integral. To apply Theorem 17.5.2 it is necessary that the curve C be expressed in parametric form. Consequently, if the curve is given in some other way, a parametric equation must first be developed. As in the case of the 2-dimensional line integral, the value of a 3-dimensional line integral is independent of the particular parametric equations used for C.

From our work with the equations of lines in Section 14.6, the line segment connecting the points (x_0, y_0, z_0) and (x_1, y_1, z_1) has parametric equations

$$\begin{aligned} x &= (x_1 - x_0)t + x_0, \\ y &= (y_1 - y_0)t + y_0, \\ z &= (z_1 - z_0)t + z_0, \end{aligned}$$

where $0 \leq t \leq 1$.

Example 2 Evaluate the line integral

$$\int_C (y + z)dx + (x + z)dy + (x + y)dz,$$

where C is the line segment with initial point $(1, 0, 3)$ and terminal point $(-1, 2, 0)$.

By the discussion above, a parametrization of the line segment C is given by

$$\begin{aligned} x &= -2t + 1, \\ y &= 2t, \\ z &= -3t + 3, \end{aligned}$$

where $0 \leq t \leq 1$. Applying Theorem 17.5.2 we have

$$\begin{aligned} \int_C (y + z)dx + (x + z)dy + (x + y)dz &= \int_0^1 [(-t + 3)(-2) + (-5t + 4)(2) + (-3)]dt \\ &= \int_0^1 (-8t - 1)dt = -5. \qquad || \end{aligned}$$

As in the 2-dimensional case, use of vector notation will simplify the statement of Theorem 17.5.2.

If we let $X(t) = (q(t), r(t), s(t))$ be an equation for C, we have

$$X'(t) = (q'(t), r'(t), s'(t)).$$

If we let $F: R^3 \to R^3$ be the vector-valued function given by

$$F(\mathbf{Y}) = (P(\mathbf{Y}), Q(\mathbf{Y}), R(\mathbf{Y})),$$

then $$F(X(t)) = (P(q(t), r(t), s(t)), Q(q(t), r(t), s(t)), R(q(t), r(t), s(t))).$$

Thus $$F(X(t)) \cdot X'(t) = P(q(t), r(t), s(t))q'(t) + Q(q(t), r(t), s(t))r'(t) + R(q(t), r(t), s(t))s'(t).$$

So the result of Theorem 17.5.2 can be stated as follows:

$$\int_C P(X)dx + Q(X)dy + R(X)dz = \int_{t_I}^{t_T} [F(X(t)) \cdot X'(t)]dt. \tag{1}$$

Example 3 Evaluate the line integral

$$\int_C (x + z)dx + x\,dy + (x + y)dz,$$

where C is the circle with equation $X(t) = (\cos t, \sin t, 1)$, initial point $X(0)$, and terminal point $X(2\pi)$.

Since $X(t) = (\cos t, \sin t, 1)$, we have $X'(t) = (-\sin t, \cos t, 0)$; and if $\mathbf{Y} = (x, y, z)$, then $F(\mathbf{Y}) = (x + z, x, x + y)$. Hence

$$F(X(t)) = (\cos t + 1, \cos t, \cos t + \sin t);$$

and so $$F(X(t)) \cdot X'(t) = -(\cos t + 1) \sin t + \cos^2 t.$$

Then use of equation (1) gives

$$\int_C (x + z)dx + x\,dy + (x + y)dz = \int_0^{2\pi} [-(\cos t + 1) \sin t + \cos^2 t]dt$$

$$= \frac{\cos^2 t}{2} + \cos t + \frac{1}{2}t + \frac{1}{4}\sin 2t \bigg|_0^{2\pi} = \pi. \quad \|$$

For considering the independence of path of integrals in three dimensions we need the following analog of Definition 17.4.2.

Definition 17.5.2

A region T in R^3 is called *simply connected* if every simple closed curve C in T bounds a 2-dimensional surface that is contained in T.

For example, R^3 itself, and R^3 with any one point deleted, are simply connected. On the other hand, both R^3 with the x-axis deleted and a torus (solid donut-shaped object) are not simply connected.

The following theorems, which are analogous to results obtained in the 2-dimensional case, hold for line integrals in 3-dimensional space.

Theorem 17.5.3

Let $P(x, y, z)$, $Q(x, y, z)$, $R(x, y, z)$, P_y, P_z, Q_x, Q_z, R_x, and R_y be continuous in a 3-dimensional, simply connected region T. Then

$$\int_C P\,dx + Q\,dy + R\,dz = 0 \qquad \text{for every simple closed curve } C \text{ in } T$$

if and only if

$$P_y = Q_x, \quad Q_z = R_y, \quad \text{and} \quad R_x = P_z \quad \text{for all points of } T.$$

The proof of that theorem will be postponed until we have proved Stokes' Theorem in Section 17.9. However, assuming the result, it is possible to establish the following criteria for a line integral to be independent of path.

Theorem 17.5.4

Let P, Q, R, P_y, P_z, Q_x, Q_z, R_x, and R_y be continuous in a 3-dimensional, simply connected region T. Then

$$\int_C P\,dx + Q\,dy + R\,dz \text{ is independent of path in } T$$

if and only if

$$P_y = Q_x, \quad Q_z = R_y, \quad \text{and} \quad R_x = P_z \quad \text{for all points of } T.$$

Though one of the implications contained in that theorem is easy to prove, the proof of the other implication would take us too far afield. For the entire proof the reader may consult a suitable advanced calculus text.

Example 4 Prove that the integral

$$\int_C (y^3x^2 + y^2x + xz^2)dx + (y^2x^3 + yx^2)dy + x^2z\,dz$$

is independent of path in the entire plane. Compute the value of the integral where C is a path with initial point $(2, 1, 2)$ and terminal point $(-1, -2, 3)$.

We have

$$P(x, y, z) = y^3x^2 + y^2x + xz^2,$$
$$Q(x, y, z) = y^2x^3 + yx^2,$$
$$R(x, y, z) = x^2z.$$

So

$$P_y = 3y^2x^2 + 2yx = Q_x,$$
$$Q_z = 0 = R_y,$$
$$R_x = 2xz = P_z.$$

Then, since the partial derivatives are continuous everywhere in R^3 and R^3 is simply connected, the integral is independent of path in R^3. Hence, to compute the integral we may select any path connecting the two points $(2, 1, 2)$ and $(-1, -2, 3)$. The simplest method of attack is to use the polygonal path pictured in Figure 17.5.1.

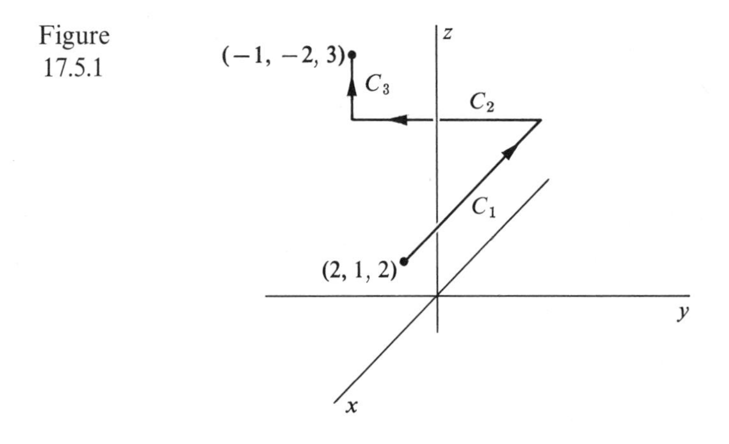

Figure 17.5.1

Parametrically the line segments C_1, C_2, and C_3 are represented as follows:

$$C_1: \begin{cases} x = t \\ y = 1 \\ z = 2 \end{cases} \quad t_I = 2, \quad t_T = -1;$$

$$C_2: \begin{cases} x = -1 \\ y = t \\ z = 2 \end{cases} \quad t_I = 1, \quad t_T = -2;$$

$$C_3: \begin{cases} x = -1 \\ y = -2 \\ z = t \end{cases} \quad t_I = 2, \quad t_T = 3.$$

Breaking the original integral into integrals over C_1, C_2, and C_3, we get

$$\int_C (y^3x^2 + y^2x + xz^2)dx + (y^2x^3 + yx^2)dy + x^2z\,dz$$

$$= \int_2^{-1} (t^2 + t + 4t)dt + \int_1^{-2} (-t^2 + t)dt + \int_2^3 t\,dt$$

$$= -\frac{7}{2}.$$

||

Exercises 17.5

In Exercises 1–10, compute the given line integrals.

1. $\int_C y\,dx + (x + z)dy + z\,dz$, where C has the parametric equation $X(t) = (t, t^2, t^3)$ with initial point (0, 0, 0) and terminal point (1, 1, 1).
2. $\int_C xy\,dx + (y + z)dy + (z + x)dz$, where C has the parametric equation $X(t) = (t^2, t^3, t)$ with initial point (4, 8, 2) and terminal point (1, 1, 1).
3. $\int_C xy\,dx + yz\,dy + z^2\,dz$, where C has the parametric equation $X(t) = (\cos t, \sin t, \sin^2 t)$ with initial point $X(\pi/2)$ and terminal point $X(0)$.
4. $\int_C xe^{yz}\,dx + ye^x\,dy + ye^{yz}\,dz$, where C has the parametric equation $X(t) = (t^2, t, t^3)$ with initial point $X(1)$ and terminal point $X(-1)$.
5. $\int_C \sin x\,dx + x\,dy + e^z\,dz$, where C has the parametric equation $X(t) = (t^3, t^2, t)$ with initial point (0, 0, 0) and terminal point (1, 1, 1).

6. $\int_C e^x\,dx + xz\,dy + xyz\,dz$, where C has the parametric equation $X(t) = (t^4, t^3, t^2)$ with initial point $(1, 1, 1)$ and terminal point $(0, 0, 0)$.
7. $\int_C y^2\,dx + (x + y)dy + (x + z)dz$, where C is the line segment with initial point $(3, 4, 1)$ and terminal point $(-2, 6, 2)$.
8. $\int_C \cos z\,dx + \sin x\,dy + \cos y\,dz$, where C is the line segment with initial point $(2, -3, -1)$ and terminal point $(-1, 2, -2)$.
9. $\int_C x\,dx + y\,dy + z\,dz$, where C is the circle in the plane $z = -2$ with radius 1 and center $(0, 0, -2)$.
10. $\int_C xy\,dx + z\,dy + x^2yz\,dz$, where C is the path from $(3, 2, 1)$ to $(1, 6, 7)$ along straight line segments from $(3, 2, 1)$ to $(3, 6, 1)$ to $(1, 6, 1)$ to $(1, 6, 7)$.
11. Show that $\int_C yz\,dx + xz\,dy + xy\,dz$ is independent of path in R^3, and then evaluate the integral for a curve C with initial point $(-1, 3, 2)$ and terminal point $(2, 1, -1)$.
12. Show that $\int_C (x + z)dx + (y + z)dy + (x + y)dz$ is independent of path in R^3, and then evaluate the integral for a curve C with initial point $(1, 4, -2)$ and terminal point $(-2, 3, -1)$.
13. Show that $\int_C 2xy\,dx + (x^2 + 2yz)dy + (y^2 + 1)dz$ is independent of path in R^3, and then evaluate the integral for a curve C with initial point $(0, 1, -1)$ and terminal point $(1, 2, 0)$.
14. Show that $\int_C 2x^3\,yz\,dz + (z^2x^3 + 2y)dy + 3z^2yx^2\,dx$ is independent of path in R^3, and then evaluate the integral for a curve C with initial point $(0, 0, 0)$ and terminal point $(1, 1, 1)$.
15. Compute $\int_C (y^2 + z^2 + yz)dx + (2yx + xz)dy + (2xz + xy)dz$, where C is the curve with parametric equation $X(t) = (\sin \pi t, \cos \pi t, t)$, initial point $X(1/2)$, and terminal point $X(1)$.
16. Compute $\int_C (x^2y^3 + xy^2 + xz^2)dx + (x^3y^2 + x^2y)dy + zx^2\,dz$, where C is the curve with parametric equation $X(t) = (\ln t, t, t^2)$, initial point $X(1)$, and terminal point $X(2)$.
17. Evaluate the line integral $\int_C y\,dx + dy + x\,dz$, where C is the semicircle with initial point $(3, 0, -2)$, terminal point $(3, 0, 2)$, and equations $x = 3$, $y = (4 - z^2)^{1/2}$.
    (a) Use the parametrization
    $x = 3, \quad z = t, \quad y = (4 - t^2)^{1/2}, \quad -2 \le t \le 2.$
    (b) Use the parametrization
    $x = 3, \quad z = -2 \cos t, \quad y = 2 \sin t, \quad 0 \le t \le \pi.$
    Note that even though different parametrizations are used, the same result is obtained.
18. Evaluate $\int_C 2x\,dx + z\,dy + y\,dz$, where C is a semicircle with initial point $(1, 2, 3)$ and terminal point $(3, 5, 7)$.
19. The components of a force field are $f_1(x, y, z) = ye^{xy}$, $f_2(x, y, z) = xe^{xy}$, and $f_3(x, y, z) = 2z$. Find the work done as a particle is displaced along a semicircular path with initial point $(1, 2, 3)$ and terminal point $(7, -2, -2)$.
20. Let f, g, and h be continuous functions. Prove that any integral of the form

$$\int_C f(x)dx + g(y)dy + h(z)dz$$

is independent of path in R^3.

17.6 The Surface Integral

The concept of a surface integral closely parallels that of a double integral. The only essential change is that instead of integrating a function of two variables over a region in the xy-plane, we shall integrate a function of three variables over some surface.

For example, if we wish to integrate the function $f(x, y, z)$ over a surface S, which is the graph of a function $z = h(x, y)$, we simply subdivide the surface S into n subsurfaces T_i for $i = 1, \ldots, n$. On each of the subsurfaces we select an arbitrary point (x_i^*, y_i^*, z_i^*) at which to evaluate the function f. If S_i is the area of the ith subsurface T_i, we form the sum

$$\sum_{i=1}^{n} f(x_i^*, y_i^*, z_i^*)S_i.$$

If as n increases without bound the largest dimension of each T_i approaches zero, then the limit

$$\lim_{n\to\infty} \sum_{i=1}^{n} f(x_i^*, y_i^*, z_i^*)S_i$$

is called the *integral of* $f(x, y, z)$ *over the surface* S. Figure 17.6.1 indicates the geometric significance of the major steps discussed above.

Figure 17.6.1

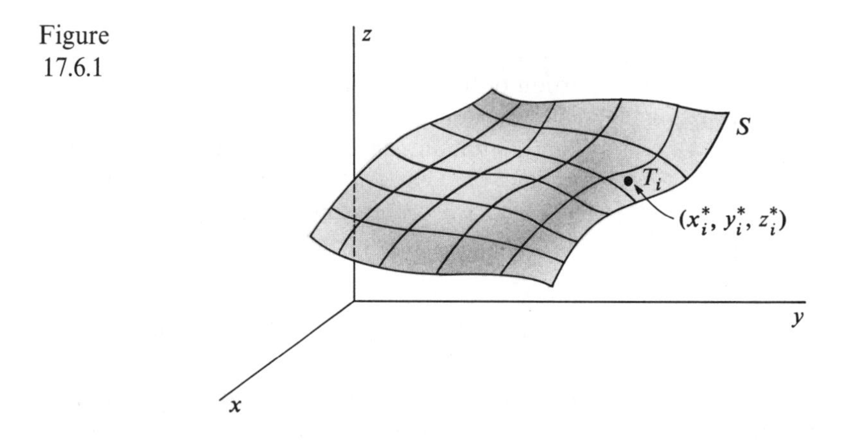

Using the notation $\int_S f(x, y, z)dS$ to denote the surface integral, we can formalize the preceding discussion into a definition.

Definition 17.6.1

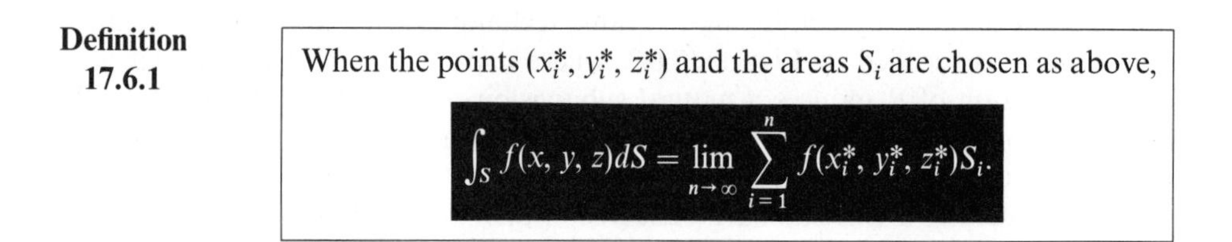

When the points (x_i^*, y_i^*, z_i^*) and the areas S_i are chosen as above,

$$\int_S f(x, y, z)dS = \lim_{n\to\infty} \sum_{i=1}^{n} f(x_i^*, y_i^*, z_i^*)S_i.$$

The same major questions that arose earlier concerning other types of integrals now face us for the surface integral:

1. Under what conditions will the surface integral $\int_S f(x, y, z)dS$ exist and be independent of the choice of the points (x_i^*, y_i^*, z_i^*)?
2. What methods are available for computing surface integrals?

To consider the first question recall that the surface S is the graph of the function $h(x, y)$ for (x, y) in a region R of the xy-plane, as indicated in Figure 17.6.2. An answer to the first question is then given by the following theorem, which is proved in many advanced calculus texts.

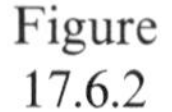
Figure 17.6.2

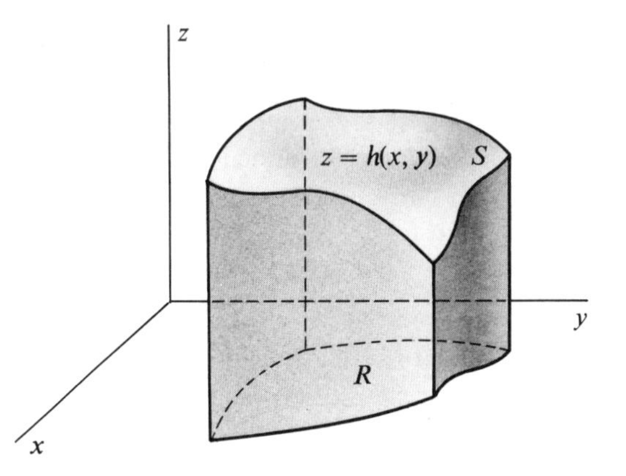

Theorem 17.6.1

> Let S be that portion of the graph of $z = h(x, y)$ that lies above the region R in the xy-plane. If the functions $h(x, y)$, $h_x(x, y)$, and $h_y(x, y)$ are all continuous on the bounded region R and $f(x, y, z)$ is continuous on S, and if the double integral $\int_R dA$ exists, then $\int_S f(x, y, z)dS$ exists and is independent of the choice of the points (x_i^*, y_i^*, z_i^*) and of the method of subdivision.

The reader should refer to Theorem 16.1.1 for conditions that insure the existence of $\int_R dA$.

Actually the hypothesis of Theorem 17.6.1 can be weakened considerably to allow the functions involved to be discontinuous at some point. We shall, from time to time, use this stronger version of the theorem.

We now answer the second question stated earlier; that is, we shall find a method of calculating surface integrals. We begin by subdividing the surface S in a particular manner. Since the region R in the xy-plane is bounded, we may subdivide R into subregions R_i by employing the same techniques used for double integrals. This process is recalled in Figure 17.6.3, where a particular R_i has been labeled. This subdivision of R induces a natural subdivision on the surface S. We take T_i to be that portion of the graph of $z = h(x, y)$ (that is, the surface S) which lies above R_i, as illustrated in Figure 17.6.4.

Consequently, if this standard subdivision of R by subdivision of each of the intervals $a \le x \le b$ and $c \le y \le d$ into n subintervals results in m subregions R_i of R, we may express the surface integral as

$$\int_S f(x, y, z)dS = \lim_{n\to\infty} \sum_{i=1}^{m} f(x_i^*, y_i^*, z_i^*)S_i, \tag{1}$$

where S_i is the area of T_i.

Figure 17.6.3

Figure 17.6.4

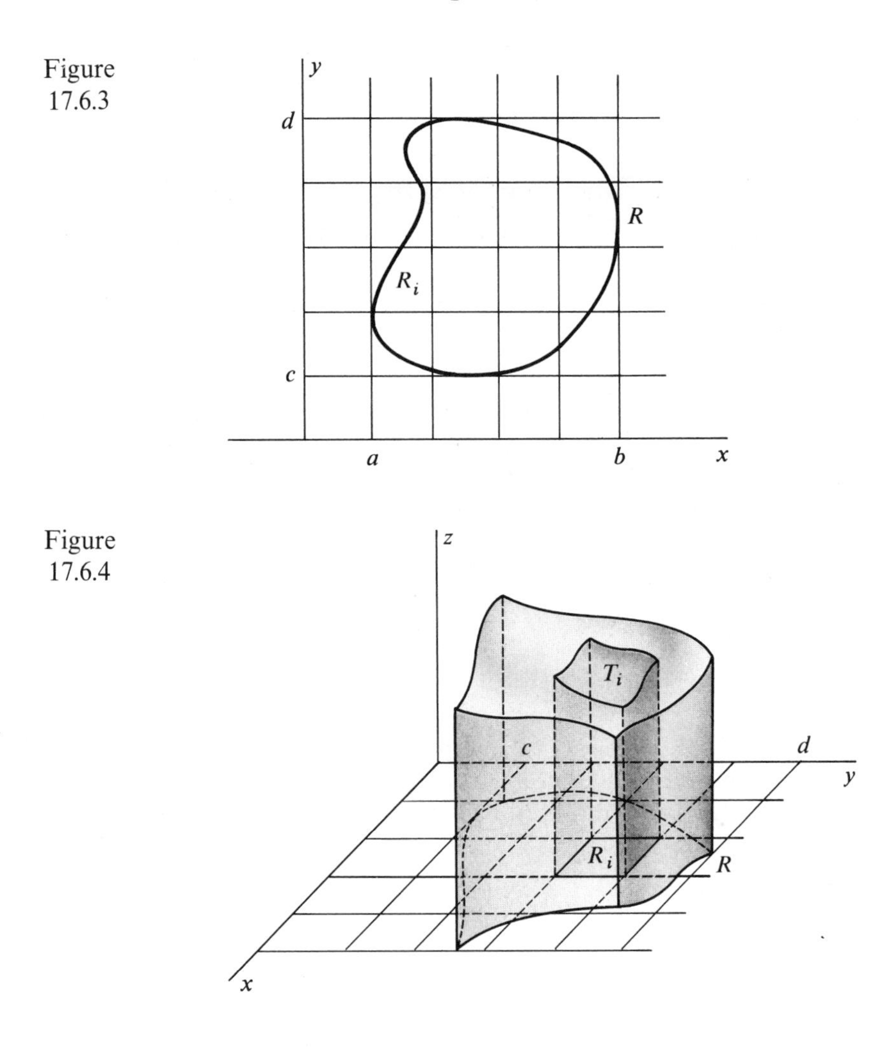

From our discussion of surface area in Section 16.3 we know that there is some point (x_i^*, y_i^*) of R_i such that

$$S_i = \sqrt{1 + h_x^2(x_i^*, y_i^*) + h_y^2(x_i^*, y_i^*)}A_i, \tag{2}$$

where A_i is the area of the subregion R_i.

Since T_i is the graph of $h(x, y)$ above R_i, and since (x_i^*, y_i^*) is in R_i, the point $(x_i^*, y_i^*, h(x_i^*, y_i^*))$ is in T_i. We then evaluate the function f at this particular point (by Theorem 17.6.1 any point of T_i is suitable if h_x and h_y are continuous), and we substitute (2) into (1) to obtain

$$\int_S f(x, y, z)dS = \lim_{n \to \infty} \sum_{i=1}^{m} f(x_i^*, y_i^*, h(x_i^*, y_i^*))\sqrt{1 + h_x^2(x_i^*, y_i^*) + h_y^2(x_i^*, y_i^*)}A_i.$$

The right side of that equation is nothing more than the definition of the double integral of the function $f(x, y, h(x, y))\sqrt{1 + h_x^2(x, y) + h_y^2(x, y)}$ over the region R. Because of its major significance we formalize this result in the following theorem.

Theorem 17.6.2

If the surface S is the graph of $z = h(x, y)$ for (x, y) in the region R of the xy-plane, and if $h_x(x, y)$ and $h_y(x, y)$ are continuous on R, then

$$\int_S f(x, y, z)dS = \int_R f(x, y, h(x, y))\sqrt{1 + h_x^2(x, y) + h_y^2(x, y)}dA.$$

Example 1 Evaluate the surface integral $\int_S xyz\, dS$, where S is the graph of $z = \sqrt{4 - x^2 - y^2}$ for $x \geq 0$ and $y \geq 0$.

In this case $f(x, y, z) = xyz$, $h(x, y) = \sqrt{4 - x^2 - y^2}$, and R is the quarter disk described by $x^2 + y^2 \leq 4$, $x \geq 0$, $y \geq 0$. Then, employing Theorem 17.6.2, we have

$$\begin{aligned}\int_S xyz\, dS &= \int_R xy\sqrt{4 - x^2 - y^2}\sqrt{1 + \frac{x^2}{4 - x^2 - y^2} + \frac{y^2}{4 - x^2 - y^2}}dA \\ &= \int_R 2xy\, dA = 2\int_0^2 \int_0^{\sqrt{4-x^2}} xy\, dy\, dx \\ &= \int_0^2 x(4 - x^2)dx = 4.\end{aligned}$$

||

As a consequence of Theorem 17.6.2 and Section 16.3, we see that the area of the surface S is given by the surface integral $\int_S dS$.

Vector notation can be used to write Theorem 17.6.2 in a different form. If $z = h(x, y)$ is an equation of the surface S, then

$$X(x, y) = (x, y, h(x, y))$$

is a vector equation for S.

Then

$$\frac{\partial X}{\partial x} = (1, 0, h_x(x, y))$$

and

$$\frac{\partial X}{\partial y} = (0, 1, h_y(x, y)).$$

So

$$\frac{\partial X}{\partial x} \times \frac{\partial X}{\partial y} = \begin{vmatrix} \mathbf{i} & \mathbf{j} & \mathbf{k} \\ 1 & 0 & h_x(x, y) \\ 0 & 1 & h_y(x, y) \end{vmatrix} = (-h_x(x, y), -h_y(x, y), 1)$$

and hence

$$\left|\frac{\partial X}{\partial x} \times \frac{\partial X}{\partial y}\right| = \sqrt{h_x^2(x, y) + h_y^2(x, y) + 1}. \tag{3}$$

Moreover

$$f(x, y, h(x, y)) = f(X(x, y))$$

since

$$X(x, y) = (x, y, h(x, y)). \tag{4}$$

Thus, in view of (3) and (4), we can state Theorem 17.6.2 as follows:

Theorem 17.6.3

If the surface S has the vector equation $X(x, y) = (x, y, h(x, y))$ for (x, y) in the region R of the xy-plane, then

$$\int_S f(x, y, z)dS = \int_R f(X)\left|\frac{\partial X}{\partial x} \times \frac{\partial X}{\partial y}\right| dA.$$

Example 2

Evaluate the surface integral $\int_S xy^2z^2\, dS$, where S is the portion of the cone $z = \sqrt{x^2 + y^2}$ above the region R given by $x^2 + y^2 \leq 1$, $x \geq 0$, and $y \geq 0$.

In this case a vector equation for the surface S is

$$X(x, y, z) = (x, y, \sqrt{x^2 + y^2}), \quad x^2 + y^2 \leq 1, \quad x \geq 0, \quad y \geq 0.$$

So

$$\frac{\partial X}{\partial x} = \left(1, 0, \frac{x}{\sqrt{x^2 + y^2}}\right)$$

and

$$\frac{\partial X}{\partial y} = \left(0, 1, \frac{y}{\sqrt{x^2 + y^2}}\right).$$

Hence

$$\frac{\partial X}{\partial x} \times \frac{\partial X}{\partial y} = \left(\frac{-x}{\sqrt{x^2 + y^2}}, \frac{-y}{\sqrt{x^2 + y^2}}, 1\right)$$

and so

$$\left|\frac{\partial X}{\partial x} \times \frac{\partial X}{\partial y}\right| = \sqrt{\frac{x^2}{x^2 + y^2} + \frac{y^2}{x^2 + y^2} + 1} = \sqrt{2}.$$

Then, since $f(x, y, z) = xy^2z^2$,

$$f(X) = f(x, y, \sqrt{x^2 + y^2}) = xy^2(x^2 + y^2).$$

Hence by Theorem 17.6.3,

$$\int_S xy^2z^2\, dS = \int_R xy^2(x^2 + y^2)\sqrt{2}\, dA, \tag{5}$$

where R is the region given by $x^2 + y^2 \leq 1$, $x \geq 0$, and $y \geq 0$.
On transforming (5) to polar coordinates we get

$$\begin{aligned}\int_S xy^2z^2\, dS &= \int_0^{\pi/2}\int_0^1 (r\cos\theta)(r^2\sin^2\theta)r^2\sqrt{2}r\, dr\, d\theta \\ &= \sqrt{2}\int_0^{\pi/2}\int_0^1 r^6\cos\theta\sin^2\theta\, dr\, d\theta \\ &= \frac{\sqrt{2}}{7}\int_0^{\pi/2}\cos\theta\sin^2\theta\, d\theta \\ &= \frac{\sqrt{2}}{7}\frac{\sin^3\theta}{3}\bigg|_0^{\pi/2} \\ &= \frac{\sqrt{2}}{21}.\end{aligned}$$

||

The following two theorems are direct results of Theorem 17.6.2 and the corresponding results for double integrals. The proofs are left as exercises.

Theorem 17.6.4

If all integrals involved exist, then

$$\int_S [f_1(x, y, z) + f_2(x, y, z)]dS = \int_S f_1(x, y, z)dS + \int_S f_2(x, y, z)dS.$$

Theorem 17.6.5

If a surface S is the graph of $z = h(x, y)$ and can be expressed as the union of two surfaces S_1 and S_2 that either don't overlap or only overlap along their boundaries, then

$$\int_S f(x, y, z)dS = \int_{S_1} f(x, y, z)dS + \int_{S_2} f(x, y, z)dS.$$

Often, applications of surface integrals require that the integral be evaluated over a surface S that cannot be expressed as the graph of a function $z = h(x, y)$ because vertical lines intersect the surface in more than one point. For example, we may wish to integrate over the entire sphere $x^2 + y^2 + z^2 = r^2$. To extend the definition to include this case, we take our clue from Theorem 17.6.5.

Definition 17.6.2

Let the surface S be the union of subsurfaces S_i, $i = 1, 2, \ldots, n$, that intersect at most along their boundaries. Then, if each S_i is graph of a function $z = h_i(x, y)$, we define $\int_S f(x, y, z)dS$ as follows:

$$\int_S f(x, y, z)dS = \int_{S_1} f(x, y, z)dS + \int_{S_2} f(x, y, z)dS + \cdots + \int_{S_n} f(x, y, z)dS.$$

Example 3

The density of the hollow sphere $x^2 + y^2 + z^2 = 1$ at any point (x, y, z) is $(x^2 + y^2)$ grams per unit area. Find the weight of the sphere.

The total weight of the sphere is given by

$$W = \int_S (x^2 + y^2)dS, \qquad \text{where } S \text{ is the sphere } x^2 + y^2 + z^2 = 1.$$

Note in this case that S is not the graph of a single function. However, the sphere can be described as the union of the upper hemisphere S_1 and the lower hemisphere S_2, both of which are graphs of functions. S_1 is the graph of $z = (1 - x^2 - y^2)^{1/2}$, and S_2 is the graph of $z = -(1 - x^2 - y^2)^{1/2}$. Using Definition 17.6.2, we have

$$W = \int_{S_1} (x^2 + y^2)dS + \int_{S_2} (x^2 + y^2)dS.$$

Use of Theorem 17.6.2 or Theorem 17.6.3 gives

$$\int_{S_1} (x^2 + y^2) dS = \int_{S_2} (x^2 + y^2) dS$$
$$= \int_R (x^2 + y^2) \sqrt{1 + \frac{x^2}{1 - x^2 - y^2} + \frac{y^2}{1 - x^2 - y^2}}\, dA,$$

where R is the disk in the xy-plane consisting of all points (x, y) for which $x^2 + y^2 \leq 1$. Thus

$$W = 2 \int_R (x^2 + y^2) \sqrt{1 + \frac{x^2}{1 - x^2 - y^2} + \frac{y^2}{1 - x^2 - y^2}}\, dA$$
$$= 2 \int_R \frac{x^2 + y^2}{\sqrt{1 - x^2 - y^2}}\, dA.$$

A transformation to polar coordinates will enable us to evaluate this integral:

$$W = 2 \int_R \frac{x^2 + y^2}{\sqrt{1 - x^2 - y^2}}\, dA = 2 \int_0^{2\pi} \int_0^1 \frac{r^2}{\sqrt{1 - r^2}}\, r\, dr\, d\theta.$$

Now the integration with respect to r is most easily accomplished by use of the substitution $u = (1 - r^2)^{1/2}$. Then $u^2 = 1 - r^2$, $r^2 = 1 - u^2$, and $r\, dr = -u\, du$. Consequently, since $u = 1$ when $r = 0$, and $u = 0$ when $r = 1$, we get

$$W = 2 \int_0^{2\pi} \int_1^0 -(1 - u^2) u^{-1} u\, du\, d\theta$$
$$= 2 \int_0^{2\pi} -\left(u - \frac{1}{3} u^3\right)\Big|_1^0\, d\theta$$
$$= 2 \int_0^{2\pi} \frac{2}{3}\, d\theta = \frac{8\pi}{3}\ \text{g}.$$

||

Exercises 17.6

In Exercises 1–16, evaluate the given surface integral.

1. $\int_S (x + y + z) dS$, where S is the portion of the plane $z = 2x - 4y$ above the region R bounded by $y = 0$, $x = 1$, and $x = 3y$ in the xy-plane.
2. $\int_S x\, dS$, where S is the portion of the plane $x + y + 2z = 4$ above the region R bounded by $x = 1$, $y = 1$, $x = 0$, and $y = 0$.
3. $\int_S (-4x^2 - 16y^2 + z^2) dS$, where S is the surface described in Exercise 1.
4. $\int_S (3x^2 - 4z^2) dS$, where S is the surface described in Exercise 2.
5. $\int_S \cos z\, dS$, where S is the portion of the plane $2z + 6y + 4x = 2$ above the region R given by $0 \leq x \leq 1$ and $-1 \leq y \leq 2$ in the xy-plane.
6. $\int_S e^z\, dS$, where S is the portion of the plane $z = 2 + 2x + 3y$ above the region R given by $1 \leq x \leq 2$ and $2 \leq y \leq 3$ in the xy-plane.
7. $\int_S y\, dS$, where S is the portion of the plane $x + y + z = 1$ inside the cylinder $x^2 + y^2 = 1$.
8. $\int_S \sin z\, dS$, where S is the portion of the plane $z = 4 - x - y$ below the region bounded by the coordinate axes and the lines $x + y = 4$ and $x + y = 5$.

9. $\int_S (x^2 - \sin y + z)dS$, where S is the surface of Exercise 8.

10. $\int_S z\,dS$: (a) where S is the upper half of the sphere $x^2 + y^2 + z^2 = a^2$,
    (b) where S is the whole sphere.

11. $\int_S z\,dS$, where S is the tetrahedron formed by the coordinate planes and the plane $x + y + z = 1$.

12. $\int_S (x^2 + y^2)dS$, where S is the surface of the closed cylinder $x^2 + y^2 = 1$, $0 \le z \le 1$. (*Note*: Include the top and bottom of the cylinder.)

13. $\int_S z\,dS$, where S is the tetrahedron formed by the coordinate planes and the plane $2x + 4y + z = 8$.

14. $\int_S (x + y + e^z)dS$, where S is the surface of Exercise 13.

15. $\int_S e^{x^2+y^2}\,dS$, where S is the portion of the cone $z = (x^2 + y^2)^{1/2}$ above the region R given by $x^2 + y^2 \le 1$ for $x \ge 0$ and $y \le 0$.

16. $\int_S \ln(x^2 + y^2 + z^2)dS$, where S is the portion of the sphere $x^2 + y^2 + z^2 = 5$ above the disk $x^2 + y^2 \le 1$.

17. Prove Theorem 17.6.4.

18. Prove Theorem 17.6.5.

17.7 Steady-State Fluid Flow

In this section we shall use surface integrals to calculate the volume of fluid flowing through a surface per unit time, which is called the *flux* of the fluid through the surface.

We shall restrict our attention to steady-state fluid flow, that is, a flow in which the fluid velocity is dependent only on position and does not change with time. Let $\mathbf{X}$ be the position vector of a point (x, y, z) in the fluid. Then the velocity of the fluid can be described by giving the velocity at the head of each $\mathbf{X}$. This is most easily accomplished by specifying the velocity function

$$G(\mathbf{X}) = (P(\mathbf{X}), Q(\mathbf{X}), R(\mathbf{X}))$$

where $P(\mathbf{X})$, $Q(\mathbf{X})$, and $R(\mathbf{X})$ are the x-, y-, and z-components of the velocity at the head of $\mathbf{X}$. Note then that G is a vector-valued function of the vector $\mathbf{X}$. Thus $G: R^3 \to R^3$.

As a first example let's consider a particularly simple case. Suppose that $G(\mathbf{X}) = (5, 0, 0)$, for all $\mathbf{X}$. Thus the flow has a velocity of five units per unit time in a direction parallel to the x-axis. To calculate the flux of fluid through the rectangle $x = 0$, $0 \le z \le 1$, and $0 \le y \le 2$, it is only necessary to multiply the area of the rectangle by the velocity component perpendicular to the rectangle. Hence, if the time units are in seconds and the length units are in feet, then the flux through the given rectangle is

$$(5 \text{ ft/sec})(2 \text{ ft}^2) = 10 \text{ ft}^3\text{/sec}.$$

We now consider a slightly more general case. Let A be the area of a planar region with unit normal $\mathbf{N}$, and suppose that the velocity of the fluid is $G(\mathbf{X}) = (v_1, v_2, v_3)$, where v_1, v_2, and v_3 are constants. We denote the component vector of the velocity in the direction of $\mathbf{N}$ by $\mathbf{V_N}$. This situation is pictured in Figure 17.7.1.

To obtain the flux through the planar region in the direction of $\mathbf{N}$ we multiply the area A by the magnitude of the vector component in the direction of $\mathbf{N}$.

Figure 17.7.1

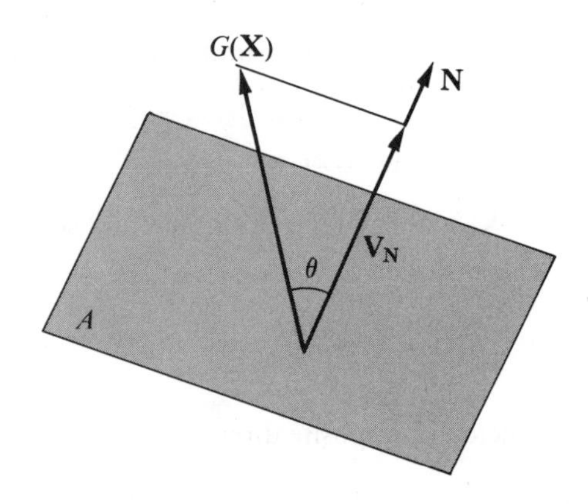

Consequently the flux F through the region is given by

$$F = |\mathbf{V_N}|A.$$

However, if $\mathbf{N}$ is a unit normal, then $|\mathbf{N}| = 1$ and

$$|\mathbf{V_N}| = |G(\mathbf{X})| \cos\theta = |G(\mathbf{X})| \frac{G(\mathbf{X}) \cdot \mathbf{N}}{|G(\mathbf{X})|\,|\mathbf{N}|} = G(\mathbf{X}) \cdot \mathbf{N},$$

and so

$$F = [G(\mathbf{X}) \cdot \mathbf{N}]A$$

To extend the preceding discussion we shall attempt to calculate the flux through a nonplanar bounded surface S with equation $z = f(x, y)$. In addition, we shall allow the velocity $G(\mathbf{X}) = (P(\mathbf{X}), Q(\mathbf{X}), R(\mathbf{X}))$ to vary with position.

In the usual fashion we subdivide the surface S into m smaller subsurfaces S_i. The flux through S will be the sum of the flows per unit time through the subsurfaces S_i.

Let A_i be the area of S_i. Procedures that are now familiar will show that the flux through S_i in the direction of the unit normal $N(\mathbf{X})$ is given by

$$[G(\mathbf{X}_i^*) \cdot N(\mathbf{X}_i^*)]A_i, \tag{1}$$

where $\mathbf{X}_i^* = (x_i^*, y_i^*, z_i^*)$ is the position vector of some point of S_i. It must be emphasized that (1) yields the flux through S_i in the direction of $N(\mathbf{X})$. Choice of the oppositely directed normal would yield a result with the same absolute value but the opposite sign.

To obtain the flux, we shall sum the partial flows per unit time. Thus the total flux is given by

$$F = \sum_{i=1}^{m} [G(x_i^*, y_i^*, z_i^*) \cdot N(x_i^*, y_i^*, z_i^*)]A_i,$$

On taking the limit as m increases without bound, the right-hand side becomes a surface integral over the surface S. Thus

$$F = \int_S [G(x, y, z) \cdot N(x, y, z)]\,dS, \quad \text{or}$$

$$F = \int_S [G(\mathbf{X}) \cdot N(\mathbf{X})]\,dS. \tag{2}$$

To apply (2) we must find an expression for the unit normal $N(\mathbf{X})$ to the surface S being considered. If the surface S has the equation

$$g(\mathbf{X}) = g(x, y, z) = k,$$

then we know that each of $\pm \nabla g(\mathbf{X})$ is normal to the surface, and so the following are unit normals to the surface at $\mathbf{X}$:

$$N(\mathbf{X}) = \pm \frac{\nabla g(\mathbf{X})}{|\nabla g(\mathbf{X})|}.$$

Of course those normals are directed in opposite directions.

Example 1 Find the flux in the outward and upward direction through the northern hemisphere of $x^2 + y^2 + z^2 = 1$, if the fluid velocity is given by $G(x, y, z) = (-y, x, z)$.

Here
$$g(x, y, z) = x^2 + y^2 + z^2;$$

so
$$N(x, y, z) = \pm \frac{(2x, 2y, 2z)}{\sqrt{4x^2 + 4y^2 + 4z^2}} = \pm \frac{(x, y, z)}{\sqrt{x^2 + y^2 + z^2}}.$$

Since $z > 0$ on the northern hemisphere, the plus sign will give the upwardly directed normal. Consequently

$$G(x, y, z) \cdot N(x, y, z) = (-y, x, z) \cdot \frac{(x, y, z)}{\sqrt{x^2 + y^2 + z^2}}$$

$$= \frac{z^2}{\sqrt{x^2 + y^2 + z^2}}.$$

Hence
$$F = \int_S \frac{z^2}{\sqrt{x^2 + y^2 + z^2}} \, dS,$$

where S is the surface $z = \sqrt{1 - x^2 - y^2}$.

By Theorem 17.6.2 or Theorem 17.6.3 we may express that surface integral as a double integral:

$$F = \int_S \frac{z^2}{\sqrt{x^2 + y^2 + z^2}} \, dS$$

$$= \int_R (1 - x^2 - y^2) \sqrt{1 + \frac{x^2}{1 - x^2 - y^2} + \frac{y^2}{1 - x^2 - y^2}} \, dA,$$

where R is the disk given by $x^2 + y^2 \leq 1$. Hence

$$F = \int_R (1 - x^2 - y^2) \frac{1}{\sqrt{1 - x^2 - y^2}} \, dA$$

$$= \int_R \sqrt{1 - x^2 - y^2} \, dA.$$

A transformation to polar coordinates will simplify the computation of that integral:

$$\begin{aligned} F &= \int_0^{2\pi} \int_0^1 \sqrt{1 - r^2}\, r\, dr\, d\theta \\ &= -\frac{1}{2} \int_0^{2\pi} \frac{2}{3}(1 - r^2)^{3/2} \Big|_0^1 \, d\theta \\ &= \frac{1}{2} \cdot \frac{2}{3} \int_0^{2\pi} d\theta \\ &= \frac{2\pi}{3} \text{ cubic units per unit time.} \end{aligned}$$

||

Example 2 | Find the flux in the upward direction through the northern hemisphere of $x^2 + y^2 + z^2 = 1$, if $G(x, y, z) = (-y, x, 0)$.

In this case we again have

$$N(x, y, z) = \frac{(x, y, z)}{\sqrt{x^2 + y^2 + z^2}}.$$

Thus

$$G(x, y, z) \cdot N(x, y, z) = (-y, x, 0) \cdot \frac{(x, y, z)}{\sqrt{x^2 + y^2 + z^2}} = 0,$$

and so

$$F = \int_S [G(\mathbf{X}) \cdot N(\mathbf{X})]\, dS = 0.$$

That is, as much fluid flows through the surface of the hemisphere in one direction as in the other, or the *total* rate of flow through S is 0. ||

Example 3 | Calculate the flux in the outward direction through the surface of the tetrahedron bounded by the coordinate planes and the plane $z + 2x + 2y = 8$, if the fluid velocity is given by

$$G(x, y, z) = (x + y, -z + 2x, y).$$

The surface under consideration is pictured in Figure 17.7.2. Since the surface S is composed of the four subsurfaces S_1, S_2, S_3, and S_4, we have

$$\begin{aligned} F = \int_S [G(\mathbf{X}) \cdot N(\mathbf{X})]\, dS &= \int_{S_1} [G(\mathbf{X}) \cdot N(\mathbf{X})]\, dS + \int_{S_2} [G(\mathbf{X}) \cdot N(\mathbf{X})]\, dS \\ &\quad + \int_{S_3} [G(\mathbf{X}) \cdot N(\mathbf{X})]\, dS + \int_{S_4} [G(\mathbf{X}) \cdot N(\mathbf{X})]\, dS. \end{aligned} \tag{3}$$

On S_1, we have $N(\mathbf{X}) = (-1, 0, 0)$; on S_2, $N(\mathbf{X}) = (0, 0, -1)$; and on S_3, $N(\mathbf{X}) = (0, -1, 0)$.

Since the surface S_4 is the plane with equation $2x + 2y + z = 8$, we may take $g(x, y, z) = 2x + 2y + z$. Then

$$N(\mathbf{X}) = \pm \frac{\nabla g(x, y, z)}{|\nabla g(x, y, z)|} = \pm \frac{(2, 2, 1)}{\sqrt{9}} = \pm \left(\frac{2}{3}, \frac{2}{3}, \frac{1}{3} \right).$$

Figure 17.7.2

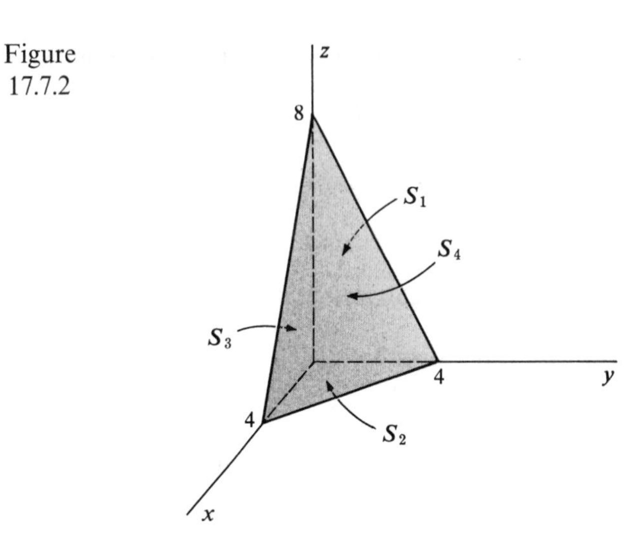

Since the normal is to be outwardly directed, we select the plus sign. Hence, on S_4 we have

$$N(\mathbf{X}) = \left(\frac{2}{3}, \frac{2}{3}, \frac{1}{3}\right).$$

Substitution into equation (3) then gives

$$\begin{aligned} F = &\int_{S_1} [(x + y, -z + 2x, y) \cdot (-1, 0, 0)]dS \\ &+ \int_{S_2} [(x + y, -z + 2x, y) \cdot (0, 0, -1)]dS \\ &+ \int_{S_3} [(x + y, -z + 2x, y) \cdot (0, -1, 0)]dS \\ &+ \int_{S_4} \left[(x + y, -z + 2x, y) \cdot \left(\frac{2}{3}, \frac{2}{3}, \frac{1}{3}\right)\right]dS. \end{aligned}$$

Hence

$$\begin{aligned} F = &-\int_{S_1} (x + y)dS - \int_{S_2} y\, dS - \int_{S_3} (-z + 2x)dS \\ &+ \frac{1}{3}\int_{S_4} (6x + 3y - 2z)dS. \end{aligned}$$

In terms of iterated integrals we have (using Theorem 17.6.2)

$$\begin{aligned} F = &-\int_0^4 \int_0^{8-2y} y\, dz\, dy - \int_0^4 \int_0^{4-y} y\, dx\, dy \\ &-\int_0^4 \int_0^{8-2x} (-z + 2x)dz\, dx \\ &+\frac{1}{3}\int_0^4 \int_0^{4-x} (10x + 7y - 16)3\, dy\, dx. \end{aligned}$$

Thus, $F = 64/3$ cubic units per unit time. ||

The Divergence Theorem of the next section will greatly simplify many of these calculations.

Exercises 17.7

In Exercises 1–8, compute the flux (volume of the flow per unit time) through the given surface S in the upward direction.

1. $G(x, y, z) = (x, y, z)$; S is given by $x + y + z = 3$, $x \geq 0$, $y \geq 0$, $z \geq 0$.
2. $G(x, y, z) = (0, x, y)$; S is given by $2x + 3y + z = 1$, $x \geq 0$, $y \geq 0$, $z \geq 0$.
3. $G(x, y, z) = (2x, -x^2, z - 2x + 2y)$; S is given by $2x + 2y + z = 6$, $x \geq 0$, $y \geq 0$, $z \geq 0$.
4. $G(x, y, z) = (\sin x, y + z, \cos y)$; S is given by $x + y + 3z = 6$, $x \geq 0$, $y \geq 0$, $z \geq 0$.
5. $G(x, y, z) = (y, -x, xy)$; S is given by $z = (x^2 + y^2)^{1/2}$, $x^2 + y^2 \leq 1$.
6. $G(x, y, z) = (-1, 1, z)$; S is given by $z = (x^2 + y^2)^{1/2}$, $x^2 + y^2 \leq 1$.
7. $G(x, y, z) = (x, y, z)$; S is given by $x^2 + y^2 + z^2 = 1$, $z \geq 0$.
8. $G(x, y, z) = (x, y, z)$; S is given by $x^2 + y^2 + z^2 = 1$, $z \leq 0$.

In Exercises 9–15, compute the flux through the given surface in the outward direction.

9. $G(x, y, z) = (x, y, z)$; S is the surface of the tetrahedron formed by the planes $x + y + z = 3$, $x = 0$, $y = 0$, and $z = 0$. (See Exercise 1.)
10. $G(x, y, z) = (0, x, y)$; S is the surface of the tetrahedron formed by the planes $2x + 3y + z = 1$, $x = 0$, $y = 0$, and $z = 0$. (See Exercise 2.)
11. $G(x, y, z) = (2x, -x^2, z - 2x + 2y)$; S is the surface of the tetrahedron formed by the planes $2x + 2y + z = 6$, $x = 0$, $y = 0$, and $z = 0$. (See Exercise 3.)
12. $G(x, y, z) = (\sin x, y + z, \cos y)$; S is the surface of the tetrahedron formed by the planes $x + y + 3z = 6$, $x = 0$, $y = 0$, and $z = 0$. (See Exercise 4.)
13. $G(x, y, z) = (y, -x, xy)$; S is the surface of the region bounded by $z = 1$ and $z = (x^2 + y^2)^{1/2}$. (See Exercise 5.)
14. $G(x, y, z) = (-1, 1, z)$; S is the surface of the region bounded by $z = 1$ and $z = (x^2 + y^2)^{1/2}$. (See Exercise 6.)
15. $G(x, y, z) = (x, y, z)$; S is the sphere $x^2 + y^2 + z^2 = 1$. (See Exercise 7.)
16. Show that if the velocity of flow is constant, then the flux through the sphere $x^2 + y^2 + z^2 = 1$ is 0.
17. Extend the result of Exercise 16 to other surfaces by using an intuitive argument.

17.8 The Divergence Theorem

In our study of line integrals we developed Green's Theorem, which relates a line integral over a closed path to an integral over the simply connected region bounded by the path. In an analogous fashion The Divergence Theorem will relate a surface integral over a closed surface to an integral over the solid bounded by the surface.

Suppose that the solid T in question is bounded above by $z = f_2(x, y)$ and bounded below by $z = f_1(x, y)$, and suppose that the projection of T into the xy-plane is the region W pictured in Figure 17.8.1. The surface S bounding the solid T

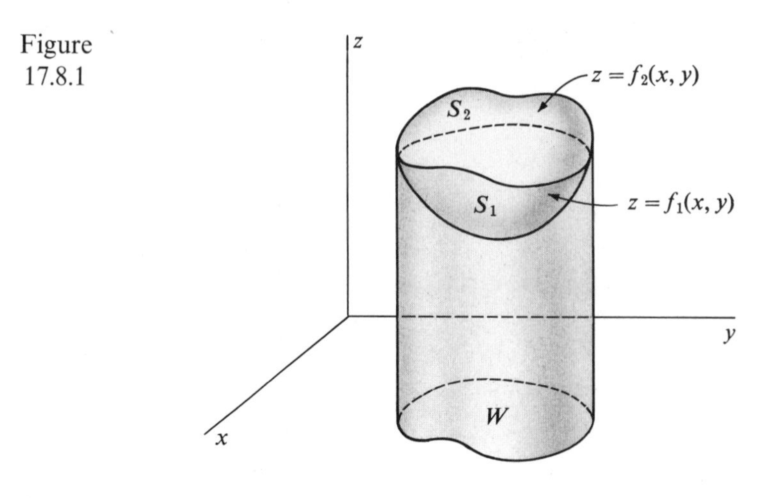

Figure 17.8.1

consists of the graphs of $z = f_1(x, y)$ and $z = f_2(x, y)$, which will be denoted by S_1 and S_2, respectively. Both of those graphs project onto the region W.

As an aid to arriving at our result, we shall suppose that there is a steady-state fluid flow through S. From Section 17.7 we know that the flux through S in an outward direction is given by

$$\int_S [G(\mathbf{X}) \cdot N(\mathbf{X})] dS,$$

where $N(\mathbf{X})$ is the outwardly directed unit normal to S at the point with position vector $\mathbf{X} = (x, y, z)$. Thus, if $G(\mathbf{X}) = (P(\mathbf{X}), Q(\mathbf{X}), R(\mathbf{X}))$ and $N(\mathbf{X}) = (n_1(\mathbf{X}), n_2(\mathbf{X}), n_3(\mathbf{X}))$, we have

$$G(\mathbf{X}) \cdot N(\mathbf{X}) = P(\mathbf{X})n_1(\mathbf{X}) + Q(\mathbf{X})n_2(\mathbf{X}) + R(\mathbf{X})n_3(\mathbf{X}), \quad \text{and so}$$

$$\int_S [G(\mathbf{X}) \cdot N(\mathbf{X})] dS = \int_S [P(\mathbf{X})n_1(\mathbf{X}) + Q(\mathbf{X})n_2(\mathbf{X}) + R(\mathbf{X})n_3(\mathbf{X})] dS. \tag{1}$$

Let's examine the latter integral in (1) more fully. First, since S is the union of the surfaces S_1 and S_2, we know that

$$\int_S R(\mathbf{X})n_3(\mathbf{X}) dS = \int_{S_1} R(\mathbf{X})n_3(\mathbf{X}) dS + \int_{S_2} R(\mathbf{X})n_3(\mathbf{X}) dS. \tag{2}$$

For the moment we concentrate on the computation of $n_3(\mathbf{X})$. In order to do this, note that on S_1 we have $z = f_1(x, y)$; and so

$$N(\mathbf{X}) = N(x, y, z) = \pm \frac{\nabla g(x, y, z)}{|\nabla g(x, y, z)|},$$

where $g(x, y, z) = z - f_1(x, y)$. Thus

$$N(x, y, z) = \pm \frac{\left(-\dfrac{\partial f_1}{\partial x}, -\dfrac{\partial f_1}{\partial y}, 1\right)}{\sqrt{\left(\dfrac{\partial f_1}{\partial x}\right)^2 + \left(\dfrac{\partial f_1}{\partial y}\right)^2 + 1}}.$$

Thus, on S_1 the third component of $N(x, y, z)$, that is n_3, is given by

$$n_3(x, y, z) = \pm \frac{1}{\sqrt{\left(\frac{\partial f_1}{\partial x}\right)^2 + \left(\frac{\partial f_1}{\partial y}\right)^2 + 1}}. \tag{3}$$

In a similar fashion, on S_2 we have $z = f_2(x, y)$ and so

$$n_3(x, y, z) = \pm \frac{1}{\sqrt{\left(\frac{\partial f_2}{\partial x}\right)^2 + \left(\frac{\partial f_2}{\partial y}\right)^2 + 1}}. \tag{4}$$

To obtain n_3 it still remains to choose the correct sign in equations (3) and (4). Figure 17.8.2 illustrates the situation. From this illustration we see that the third component of $N(x, y, z)$, the *outwardly* directed normal, is positive on S_2 and negative on S_1. Thus, refering to equations (3) and (4), we have

$$n_3(x, y, z) = -\frac{1}{\sqrt{\left(\frac{\partial f_1}{\partial x}\right)^2 + \left(\frac{\partial f_1}{\partial y}\right)^2 + 1}} \quad \text{on } S_1, \tag{5}$$

and

$$n_3(x, y, z) = \frac{1}{\sqrt{\left(\frac{\partial f_2}{\partial x}\right)^2 + \left(\frac{\partial f_2}{\partial y}\right)^2 + 1}} \quad \text{on } S_2. \tag{6}$$

Figure 17.8.2

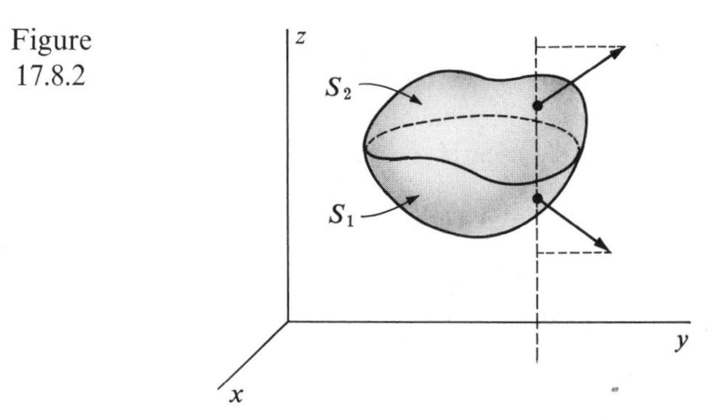

Thus, from (2) we have

$$\int_S R(\mathbf{X})n_3(\mathbf{X})dS$$

$$= -\int_{S_1} \frac{R(\mathbf{X})}{\sqrt{\left(\frac{\partial f_1}{\partial x}\right)^2 + \left(\frac{\partial f_1}{\partial y}\right)^2 + 1}} dS + \int_{S_2} \frac{R(\mathbf{X})}{\sqrt{\left(\frac{\partial f_2}{\partial x}\right)^2 + \left(\frac{\partial f_2}{\partial y}\right)^2 + 1}} dS. \tag{7}$$

Now Theorem 17.6.2 can be used to convert those surface integrals to double integrals:

$$\int_{S_1} \frac{R(\mathbf{X})}{\sqrt{\left(\frac{\partial f_1}{\partial x}\right)^2 + \left(\frac{\partial f_1}{\partial y}\right)^2 + 1}} dS = \int_W \frac{R(x, y, f_1(x, y))}{\sqrt{\left(\frac{\partial f_1}{\partial x}\right)^2 + \left(\frac{\partial f_1}{\partial y}\right)^2 + 1}} \sqrt{\left(\frac{\partial f_1}{\partial x}\right)^2 + \left(\frac{\partial f_1}{\partial y}\right)^2 + 1}\, dA$$

$$= \int_W R(x, y, f_1(x, y)) dA.$$

Similarly

$$\int_{S_2} \frac{R(\mathbf{X})}{\sqrt{\left(\frac{\partial f_2}{\partial x}\right)^2 + \left(\frac{\partial f_2}{\partial y}\right)^2 + 1}} dS = \int_W R(x, y, f_2(x, y)) dA.$$

Substitution of those results into equation (7) yields

$$\int_S R(\mathbf{X}) n_3(\mathbf{X}) dS = -\int_W R(x, y, f_1(x, y)) dA + \int_W R(x, y, f_2(x, y)) dA$$

$$= \int_W [R(x, y, f_2(x, y)) - R(x, y, f_1(x, y))] dA.$$

However, since

$$\int_{f_1(x, y)}^{f_2(x, y)} \frac{\partial R}{\partial z} dz = R(x, y, f_2(x, y)) - R(x, y, f_1(x, y)),$$

we have

$$\int_S R(\mathbf{X}) n_3(\mathbf{X}) dS = \int_W \left(\int_{f_1(x, y)}^{f_2(x, y)} \frac{\partial R}{\partial z} dz \right) dA.$$

The solid T bounded by S is bounded below by the graph of $z = f_1(x, y)$ and above by the graph of $z = f_2(x, y)$. Since the projection of T into the xy-plane is W, the integral above on the right is just a triple integral over T:

$$\int_S R(\mathbf{X}) n_3(\mathbf{X}) dS = \int_T \frac{\partial R}{\partial z} dV. \tag{8}$$

By assuming that lines parallel to the x- and y-axes intersect S in at most two points we can show in an analogous fashion that

$$\int_S P(\mathbf{X}) n_1(\mathbf{X}) dS = \int_T \frac{\partial P}{\partial x} dV, \quad \text{and} \tag{9}$$

$$\int_S Q(\mathbf{X}) n_2(\mathbf{X}) dS = \int_T \frac{\partial Q}{\partial y} dV. \tag{10}$$

On adding equations (8), (9), and (10) we obtain

$$\int_S [P(\mathbf{X}) n_1(\mathbf{X}) + Q(\mathbf{X}) n_2(\mathbf{X}) + R(\mathbf{X}) n_3(\mathbf{X})] dS = \int_T \left[\frac{\partial P}{\partial x} + \frac{\partial Q}{\partial y} + \frac{\partial R}{\partial z} \right] dV. \tag{11}$$

The use of vector notation and the following definition can be used to simplify the expression of that result.

Definition 17.8.1

Let $G(\mathbf{X}) = (P(\mathbf{X}), Q(\mathbf{X}), R(\mathbf{X}))$; then the *divergence* of G, written as div G, is defined by

$$\operatorname{div} G = \frac{\partial P}{\partial x} + \frac{\partial Q}{\partial y} + \frac{\partial R}{\partial z}.$$

Note that the domain of div G is a subset of R^3 and the range of div G is a subset of R.

Since $N(\mathbf{X}) = (n_1(\mathbf{X}), n_2(\mathbf{X}), n_3(\mathbf{X}))$, equation (11) can be written as

$$\int_S [G(\mathbf{X}) \cdot N(\mathbf{X})] dS = \int_T \operatorname{div} G \, dV.$$

That result is so important that we formalize it in the following theorem.

Theorem 17.8.1

The Divergence Theorem

Let S be a closed surface bounding a region T. If S is intersected in at most two points by lines parallel to the coordinate axes, then

$$\int_S [G(\mathbf{X}) \cdot N(\mathbf{X})] dS = \int_T \operatorname{div} G \, dV,$$

where $N(\mathbf{X})$ is the outwardly directed unit normal to S.

Actually the conclusion of The Divergence Theorem holds under less restrictive conditions than the ones indicated. Instead of assuming that S is intersected in at most two points by lines parallel to the coordinate axes, we only have to assume that the normal $N(\mathbf{X})$ is continuous on the surface S.

To show the great power of The Divergence Theorem we again consider Example 3 of Section 17.7.

Example 1

Calculate the flux in the outward direction through the surface of the tetrahedron bounded by the coordinate planes and the plane $z + 2x + 2y = 8$, if the velocity of the flow is given by

$$G(x, y, z) = (x + y, -z + 2x, y).$$

The flux is given by

$$F = \int_S [G(\mathbf{X}) \cdot N(\mathbf{X})] dS.$$

So by The Divergence Theorem

$$F = \int_T \left[\frac{\partial}{\partial x}(x + y) + \frac{\partial}{\partial y}(-z + 2x) + \frac{\partial y}{\partial z} \right] dV$$

$$= \int_T dV = \int_0^4 \int_0^{4-x} (8 - 2x - 2y) dy\, dx$$

$$= \frac{64}{3} \text{ cubic units per unit time.}$$

The computation of that result should be compared with the computation in Example 3 of Section 17.7 to see the great difference! ||

Another notation is often used for div G. If we assume that

$$\nabla = \left(\frac{\partial}{\partial x}, \frac{\partial}{\partial y}, \frac{\partial}{\partial z} \right)$$

behaves like a vector, then it is reasonable to define

$$\nabla \cdot G(\mathbf{X}) = \nabla \cdot (P(\mathbf{X}), Q(\mathbf{X}), R(\mathbf{X})) = \left(\frac{\partial}{\partial x}, \frac{\partial}{\partial y}, \frac{\partial}{\partial z} \right) \cdot (P(\mathbf{X}), Q(\mathbf{X}), R(\mathbf{X}))$$

as

$$\nabla \cdot G(\mathbf{X}) = \frac{\partial P}{\partial x} + \frac{\partial Q}{\partial y} + \frac{\partial R}{\partial z}.$$

Thus one often sees $\nabla \cdot G$ written for div G.

The Divergence Theorem also yields an interesting physical interpretation of the divergence of G that explains the use of the term "divergence." Since by The Divergence Theorem the total flow F through S is given by

$$F = \int_T \operatorname{div} G \, dV,$$

the total outward flow is the integral of the divergence of G over the entire region T. Consequently div $G(\mathbf{X})$ may be thought of as the local outflow or divergence of fluid from a small surface surrounding the point with position vector $\mathbf{X}$. If div $G(\mathbf{X})$ is identically 0 in R, then $G(\mathbf{X})$ is called an *incompressible flow*.

Exercises 17.8

In Exercises 1–4, use The Divergence Theorem to compute the surface integral $\int_S [G(\mathbf{X}) \cdot N(\mathbf{X})] dS$.

1. $G(\mathbf{X}) = (x + y, y + z, z + x)$; S is the sphere $x^2 + y^2 + z^2 = 4$.
2. $G(\mathbf{X}) = (z^2 + e^y, x^2 \sin z, 5z + \cos y)$; S is the boundary of the cylinder $x^2 + y^2 \leq 1, 0 \leq z \leq 1$.
3. $G(\mathbf{X}) = k\mathbf{X}$, where k is a real number and S is the sphere $x^2 + y^2 + z^2 = a^2$.
4. $G(\mathbf{X}) = \mathbf{X} \times (1, 1, 1)$; S is the ellipsoid $a^2x^2 + b^2y^2 + c^2z^2 = 1$.
5. Compute the flux in the outward direction through the surface of the tetrahedron formed by the planes $x + y + z = 3$, $x = 0$, $y = 0$, and $z = 0$, if the fluid velocity is given by $G(x, y, z) = (x, y, z)$. Compare the result with Exercise 9 of Section 17.7.

6. Compute the flux in the outward direction through the surface of the tetrahedron formed by the planes $2x + 3y + z = 1$, $x = 0$, $y = 0$, and $z = 0$, if the fluid velocity is given by $G(x, y, z) = (0, x, y)$. Compare the result with Exercise 10 of Section 17.7.
7. Compute the flux in the outward direction through the surface of the tetrahedron formed by the planes $2x + 2y + z = 6$, $x = 0$, $y = 0$, and $z = 0$, if the fluid velocity is given by $G(x, y, z) = (2x, -x^2, z - 2x + 2y)$. Compare the result with Exercise 11 of Section 17.7.
8. Compute the flux in the outward direction through the surface of the tetrahedron formed by the planes $x + y + 3z = 6$, $x = 0$, $y = 0$, and $z = 0$, if the fluid velocity is given by $G(x, y, z) = (\sin x, y + z, \cos y)$. Compare the result with Exercise 12 of Section 17.7.
9. Compute the flux in the outward direction through the surface of the region bounded by $z = 1$ and $z = (x^2 + y^2)^{1/2}$, if the fluid velocity is given by $G(x, y, z) = (y, -x, xy)$. Compare the result with Exercise 13 of Section 17.7.
10. Compute the flux in the outward direction through the surface of the region bounded by $z = 1$ and $z = (x^2 + y^2)^{1/2}$, if the fluid velocity is given by $G(x, y, z) = (-1, 1, z)$. Compare the result with Exercise 14 of Section 17.7.
11. Compute the flux in the outward direction through the sphere $x^2 + y^2 + z^2 = 1$, if the fluid velocity is given by $G(x, y, z) = (x, y, z)$.
12. By calculating each integral separately, verify without use of The Divergence Theorem that

$$\int_S [G(\mathbf{X}) \cdot N(\mathbf{X})]\,dS = \int_T \operatorname{div} G\,dV,$$

where $G(\mathbf{X}) = \mathbf{X}$, and T is the cube $0 \le x \le 1$, $0 \le y \le 1$, and $0 \le z \le 1$.

13. Show that if the velocity components of a fluid are constant, then the total flow through a closed surface is zero.
14. Prove that

$$\nabla \cdot (f\,\nabla g - g\,\nabla f) = f\,\nabla^2 g - g\,\nabla^2 f,$$

where

$$\nabla^2 f(x, y, z) = \frac{\partial^2 f}{\partial x^2} + \frac{\partial^2 f}{\partial y^2} + \frac{\partial^2 f}{\partial z^2}.$$

15. Explain intuitively why an incompressible fluid must have velocity G such that $\operatorname{div} G = 0$.

17.9 Stokes' Theorem

Recall that Green's Theorem relates a line integral about a simple closed curve C in the xy-plane to a double integral taken over the region bounded by C. In the last section we saw one extension of Green's Theorem to three dimensions, The Divergence Theorem. One additional extension of Green's Theorem to three dimensions is possible. This result, called Stokes' Theorem, relates a line integral taken over a simple closed curve C in three dimensions to the surface integral taken over a surface S bounded by C.

Although more general developments of Stokes' Theorem can be found in many advanced calculus tests, we shall restrict ourselves to the case where vertical lines intersect the surface S in no more than one point, so that S is the graph of an equation $z = f(x, y)$ for (x, y) in some region W of the xy-plane. In this case the curve C is simply the graph of $z = f(x, y)$ where (x, y) is restricted to lie on the curve D that bounds W in the xy-plane. The situation is illustrated in Figure 17.9.1.

Figure 17.9.1

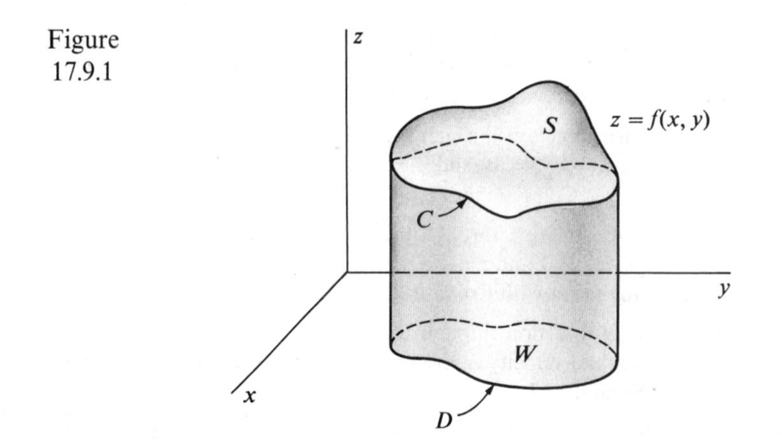

Before stating Stokes' Theorem we shall discuss some notation that will considerably simplify the statement. If we again assume that

$$\nabla = \left(\frac{\partial}{\partial x}, \frac{\partial}{\partial y}, \frac{\partial}{\partial z}\right)$$

behaves like a vector, and we let $F(\mathbf{X}) = (P(\mathbf{X}), Q(\mathbf{X}), R(\mathbf{X}))$, then the following is a vector-valued function of $\mathbf{X}$.

$$(\nabla \times F)(\mathbf{X}) = \begin{vmatrix} \mathbf{i} & \mathbf{j} & \mathbf{k} \\ \frac{\partial}{\partial x} & \frac{\partial}{\partial y} & \frac{\partial}{\partial z} \\ P(\mathbf{X}) & Q(\mathbf{X}) & R(\mathbf{X}) \end{vmatrix}$$

$$= \left(\frac{\partial R}{\partial y} - \frac{\partial Q}{\partial z}, \frac{\partial P}{\partial z} - \frac{\partial R}{\partial x}, \frac{\partial Q}{\partial x} - \frac{\partial P}{\partial y}\right).$$

The vector $\nabla \times F$ is often called the *curl* of F and is often written curl F. Note that the domain of $\nabla \times F$ is a subset of R^3 and the range of $\nabla \times F$ is a subset of R^3.

With the new notation we can state Stokes' Theorem as follows:

Theorem 17.9.1

Stokes' Theorem

If the surface S is bounded by the curve C and has equation $z = f(x, y)$, and if f, P, Q, and R all have continuous first partial derivatives in some region of space containing S, then

$$\int_C P\,dx + Q\,dy + R\,dz = \int_S [(\nabla \times F)(\mathbf{X}) \cdot N(\mathbf{X})]\,dS,$$

where $F(\mathbf{X}) = (P(\mathbf{X}), Q(\mathbf{X}), R(\mathbf{X}))$, and $N(\mathbf{X})$ is the upwardly directed unit normal (the normal with a positive z component) to the surface S at the point with position vector $\mathbf{X}$.

By definition

$$(\nabla \times F)(\mathbf{X}) = \left(\frac{\partial R}{\partial y} - \frac{\partial Q}{\partial z}, \frac{\partial P}{\partial z} - \frac{\partial R}{\partial x}, \frac{\partial Q}{\partial x} - \frac{\partial P}{\partial y}\right).$$

If we let

$$N(\mathbf{X}) = (n_1(\mathbf{X}), n_2(\mathbf{X}), n_3(\mathbf{X})),$$

then

$$(\nabla \times F)(\mathbf{X}) \cdot N(\mathbf{X}) = \left(\frac{\partial R}{\partial y} - \frac{\partial Q}{\partial z}\right) n_1(\mathbf{X}) + \left(\frac{\partial P}{\partial z} - \frac{\partial R}{\partial x}\right) n_2(\mathbf{X}) + \left(\frac{\partial Q}{\partial x} - \frac{\partial P}{\partial y}\right) n_3(\mathbf{X}).$$

Consequently the conclusion of Stokes' Theorem can be stated as

$$\int_C P\,dx + Q\,dy + R\,dz$$

$$= \int_S \left[\left(\frac{\partial R}{\partial y} - \frac{\partial Q}{\partial z}\right) n_1(\mathbf{X}) + \left(\frac{\partial P}{\partial z} - \frac{\partial R}{\partial x}\right) n_2(\mathbf{X}) + \left(\frac{\partial Q}{\partial x} - \frac{\partial P}{\partial y}\right) n_3(\mathbf{X})\right] dS.$$

In the proof of Stokes' Theorem it will be easier to refer to that equation.

Proof. Initially we confine our attention to the line integral $\int_C P(x, y, z)dx$. To evaluate this integral we let

$$x = q(t), \quad y = r(t), \quad t_1 \le t \le t_2$$

be a parametric representation of the boundary D of W. (See Figure 17.9.1.) Then, since C is simply the graph of $z = f(x, y)$, where (x, y) is restricted to D, a parametric representation for C is given by

$$x = q(t), \quad y = r(t), \quad z = f(q(t), r(t)), \quad t_1 \le t \le t_2.$$

Then using Theorem 17.5.2 we have

$$\int_C P(x, y, z)dx = \int_{t_1}^{t_2} P(q(t), r(t), f(q(t), r(t)))q'(t)dt. \tag{1}$$

Now consider the line integral $\int_D P(x, y, f(x, y))dx$. Since $x = g(t)$, $y = r(t)$, $t_1 \le t \le t_2$, is a parametric representation of D, we may apply Theorem 17.2.2 to obtain

$$\int_D P(x, y, f(x, y))dx = \int_{t_1}^{t_2} P(q(t), r(t), f(q(t), r(t)))q'(t)dt. \tag{2}$$

Then a comparison of (1) and (2) indicates that

$$\int_C P(x, y, z)dx = \int_D P(x, y, f(x, y))dx. \tag{3}$$

Since the curve D lies in the xy-plane, we may apply Green's Theorem (Theorem 17.3.1) to get

$$\int_D P(x, y, f(x, y))dx = -\int_W \frac{\partial P(x, y, f(x, y))}{\partial y}\,dA.$$

By the chain rule we have

$$\frac{\partial P(x, y, f(x, y))}{\partial y} = \frac{\partial P(x, y, z)}{\partial y} + \frac{\partial P(x, y, z)}{\partial z}\frac{\partial f}{\partial y}.$$

Hence $\displaystyle\int_D P(x, y, f(x, y))dx = -\int_W \left(\frac{\partial P}{\partial y} + \frac{\partial P}{\partial z}\frac{\partial f}{\partial y}\right) dA,$

and in view of (3),

$$\int_C P(x, y, z)dx = -\int_W \left(\frac{\partial P}{\partial y} + \frac{\partial P}{\partial z}\frac{\partial f}{\partial y}\right) dA. \tag{4}$$

Then, by an application of Theorem 17.6.2, we have

$$-\int_W \left(\frac{\partial P}{\partial y} + \frac{\partial P}{\partial y}\frac{\partial f}{\partial y}\right) dA$$
$$= -\int_S \left(\frac{\partial P}{\partial y} + \frac{\partial P}{\partial z}\frac{\partial f}{\partial y}\right) \frac{1}{\sqrt{1 + f_x^2(x, y) + f_y^2(x, y)}}\, dS.$$

Thus,

$$-\int_W \left(\frac{\partial P}{\partial y} + \frac{\partial P}{\partial z}\frac{\partial f}{\partial y}\right) dA$$
$$= -\int_S \left[\left(\frac{\partial P}{\partial y}\right) \frac{1}{\sqrt{1 + f_x^2(x, y) + f_y^2(x, y)}} + \left(\frac{\partial P}{\partial z}\right) \frac{f_y(x, y)}{\sqrt{1 + f_x^2(x, y) + f_y^2(x, y)}}\right] dS. \tag{5}$$

The *upwardly* directed unit normal to S is given by

$$N(\mathbf{X}) = (n_1(\mathbf{X}), n_2(\mathbf{X}), n_3(\mathbf{X}))$$
$$= \frac{1}{\sqrt{1 + f_x^2(x, y) + f_y^2(x, y)}}(-f_x(x, y), -f_y(x, y), 1),$$

since the upwardly directed unit normal must have a positive third component.

Consequently equation (5) may be written as

$$-\int_W \left(\frac{\partial P}{\partial y} + \frac{\partial P}{\partial z}\frac{\partial f}{\partial y}\right) dA = -\int_S \left(\frac{\partial P}{\partial y} n_3 - \frac{\partial P}{\partial z} n_2\right) dS.$$

Hence from (4) we have

$$\int_C P(x, y, z)dx = \int_S \left(\frac{\partial P}{\partial z} n_2 - \frac{\partial P}{\partial y} n_3\right) dS.$$

This proves Stokes' Theorem as far as it concerns the function $P(x, y, z)$. In a similar way we can show that

$$\int_C Q(x, y, z)dy = \int_S \left(\frac{\partial Q}{\partial x} n_3 - \frac{\partial Q}{\partial z} n_1\right) dS \tag{7}$$

and

$$\int_C R(x, y, z)dz = \int_S \left(\frac{\partial R}{\partial y} n_1 - \frac{\partial R}{\partial x} n_2\right) dS. \tag{8}$$

Addition of (6), (7), and (8) then produces the desired result.

Example 1 Use Stokes' Theorem to evaluate the line integral

$$\int_C (x + z)dx + x\,dy + (x + y)dz,$$

where C is the circle given by the parametric equations $x = \sin t$, $y = \cos t$, $z = 1$, for $0 \le t \le 2\pi$.

The circle C in question bounds the surface S represented by $z = 1$ and $x^2 + y^2 \le 1$. In this case, $N(\mathbf{X}) = (0, 0, 1)$. Consequently, since $P = x + z$, $Q = x$, and $R = x + y$, we have

$$F(x, y, z) = (x + z, x, x + y),$$

and so

$$\nabla \times F = \begin{vmatrix} \mathbf{i} & \mathbf{j} & \mathbf{k} \\ \dfrac{\partial}{\partial x} & \dfrac{\partial}{\partial y} & \dfrac{\partial}{\partial z} \\ x + z & x & x + y \end{vmatrix} = (1, 0, 1).$$

Then by Stokes' Theorem

$$\int_C (x + z)dx + x\,dy + (x + y)dz = \int_S [(1, 0, 1) \cdot (0, 0, 1)]dS$$
$$= \int_S 1\,dS = \pi. \qquad \|$$

The reader should compare this example with Example 3 of Section 17.5, where the given integral is calculated directly.

Example 2 Use Stokes' Theorem to evaluate the line integral

$$\int_C (x + z)dx + (2x - y)dy + (y + z)dz,$$

where C is the triangle cut from the plane $3x + y + 2z = 6$ by the three coordinate planes.

The surface S bounded by C will be taken as the portion of the plane $3x + y + 2z = 6$ bounded by C. The unit normals to that plane are given by

$$\pm\frac{\nabla g(x, y, z)}{|\nabla g(x, y, z)|}, \qquad \text{where } g(x, y, z) = 3x + y + 2z.$$

Thus

$$N(x, y, z) = \pm\frac{1}{\sqrt{14}}(3, 1, 2).$$

Since the normal is to be directed upward, the third component should be positive. So we take

$$N(x, y, z) = \frac{1}{\sqrt{14}}(3, 1, 2).$$

Since

$$F(x, y, z) = (x + z, 2x - y, y + z),$$

we have
$$\nabla \times F = \begin{vmatrix} \mathbf{i} & \mathbf{j} & \mathbf{k} \\ \dfrac{\partial}{\partial x} & \dfrac{\partial}{\partial y} & \dfrac{\partial}{\partial z} \\ x+z & 2x-y & y+z \end{vmatrix} = (1, 1, 2).$$

So by Stokes' Theorem,

$$\int_C (x+z)dx + (2x-y)dy + (y+z)dz = \int_S \left[(1, 1, 2) \cdot \frac{1}{\sqrt{14}}(3, 1, 2) \right] dS$$
$$= \int_S \frac{8}{\sqrt{14}}\, dS = \frac{8}{\sqrt{14}}\, 6\sqrt{\frac{7}{2}} = 24. \qquad \|$$

We are now in a position to give a proof of Theorem 17.5.3.

Theorem 17.5.3

> Let $P(x, y, z)$, $Q(x, y, z)$, $R(x, y, z)$, P_y, P_z, Q_x, Q_z, R_x and R_y be continuous in a 3-dimensional simply-connected region T. Then
>
> $\int_C P\,dx + Q\,dy + R\,dz = 0$ for every simple closed curve C in T
>
> if and only if
>
> $P_y = Q_x$, $Q_z = R_y$, and $R_x = P_z$, for all points of T.

Proof. Let $F(\mathbf{X}) = (P(\mathbf{X}), Q(\mathbf{X}), R(\mathbf{X}))$; Then

$$\nabla \times F = \begin{vmatrix} \mathbf{i} & \mathbf{j} & \mathbf{k} \\ \dfrac{\partial}{\partial x} & \dfrac{\partial}{\partial y} & \dfrac{\partial}{\partial z} \\ P & Q & R \end{vmatrix} = \left(\frac{\partial R}{\partial y} - \frac{\partial Q}{\partial z}, \frac{\partial P}{\partial z} - \frac{\partial R}{\partial x}, \frac{\partial Q}{\partial x} - \frac{\partial P}{\partial y} \right).$$

Thus, $R_y = Q_z$, $P_z = R_x$, and $Q_x = P_y$, if and only if $\nabla \times F = \mathbf{0}$.

Now recall that a region T is simply connected if and only if each simple closed curve C in T is the boundary of a surface S fully contained in T. On applying Stokes' Theorem we have

$$\int_C P\,dx + Q\,dy + R\,dz = \int_S [(\nabla \times F) \cdot N]\,dS.$$

Consequently, if $\nabla \times F = \mathbf{0}$ in T, then $\nabla \times F = \mathbf{0}$ on S and

$$\int_C P\,dx + Q\,dy + R\,dz = 0.$$

The implication in one direction is now established.

Conversely, suppose that one of the three equations $P_y = Q_x$, $Q_z = R_y$, or $R_x = P_z$ does not hold at some point of T. Suppose, for example, that at the point (x_0, y_0, z_0) of T we have $Q_x > P_y$. Then, since Q_x and P_y are continuous in the region T, the expression

Figure 17.9.2

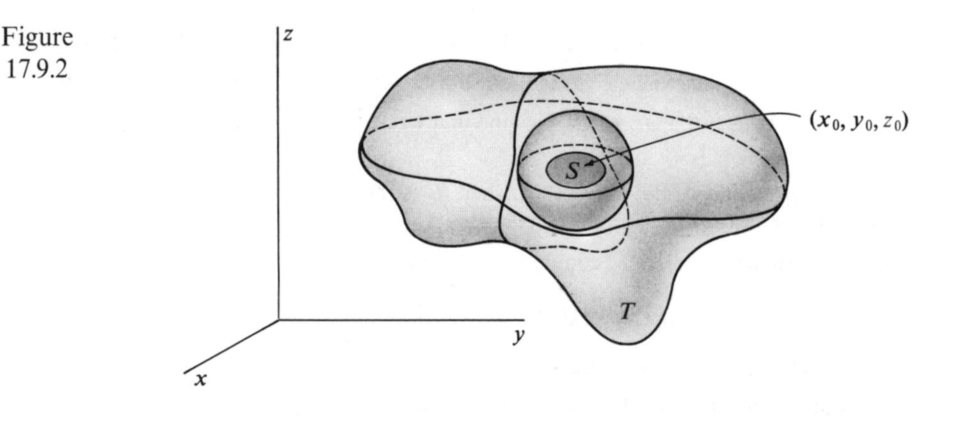

$Q_x(x, y, z) - P_y(x, y, z)$ is positive in some ball centered at (x_0, y_0, z_0). As illustrated in Figure 17.9.2, we can select a horizontal disc S fully contained in the ball where $Q_x - P_y > 0$. Consider the line integral around C, the circle that bounds the disc S. By Stokes' Theorem

$$\int_C P\,dx + Q\,dy + R\,dz$$
$$= \int_S \left[\left(\frac{\partial R}{\partial y} - \frac{\partial Q}{\partial z}\right)n_1 + \left(\frac{\partial P}{\partial z} - \frac{\partial R}{\partial x}\right)n_2 + \left(\frac{\partial Q}{\partial x} - \frac{\partial P}{\partial y}\right)n_3\right] dS.$$

Since S is horizontal, $N(\mathbf{X}) = (0, 0, 1)$. Thus

$$\int_C P\,dx + Q\,dy + R\,dz = \int_S \left(\frac{\partial Q}{\partial x} - \frac{\partial P}{\partial y}\right) dS.$$

Then, since $\dfrac{\partial Q}{\partial x} - \dfrac{\partial P}{\partial y}$ is always positive on S, we have

$$\int_C P\,dx + Q\,dy + R\,dz = \int_S \left(\frac{\partial Q}{\partial x} - \frac{\partial P}{\partial y}\right) dS > 0.$$

and so
$$\int_C P\,dx + Q\,dy + R\,dz \neq 0.$$

A similar argument will establish the result when $Q_x < P_y$ or in the cases where $Q_z \neq R_y$ or $R_x \neq P_z$. This completes the proof of Theorem 17.5.3.

Exercises 17.9

In Exercises 1–7, use Stokes' Theorem to compute the given line integral.

1. $\int_C (x + z)dx + (y + z)dy + \sin z\,dz$, where C is the circle formed by the intersection of $x^2 + y^2 = 4$ and $z = 2$.
2. $\int_C (x + z)dx + (y + z)dy + \sin z\,dz$, where C is the triangle formed by the intersection of the plane $x + 2y + 3z = 6$ with the three coordinate planes.
3. $\int_C (2y + 3x)dx + (\sin y + z)dy + (\cos z + x)dz$, where C is the circle formed by the intersection of the cylinder $x^2 + y^2 = 1$, with the hemisphere $z = (10 - x^2 - y^2)^{1/2}$.

4. $\int_C (y + x^2)dx + z\,dy + x\,dz$, where C is the triangle with vertices (0, 0, 1), (0, 1, 0), and (1, 0, 0).
5. $\int_C y\,dx + z\,dy + x\,dz$, where C is the triangle with vertices (0, 0, 1), (0, 1, 1), and (1, 0, 0).
6. $\int_C 2y\,dx - 2x\,dy + z^2x\,dz$, where C is the unit circle that has center at (0, 0, −1) and lies in the plane $z = -1$.
7. $\int_C (-z - y)dx + (x + z)dy + (x - y)dz$, where C is the curve formed by the intersection of the surface $z = 5 - x^2 - y^2$ with the plane $z = 1$.
8. Illustrate Stokes' Theorem by computing both sides of the equation

$$\int_C P\,dx + Q\,dy + R\,dz = \int_S [(\nabla \times F) \cdot N(\mathbf{X})]dS,$$

where $F(\mathbf{X}) = (z, x, y)$ and S is the surface $x^2 + y^2 + z = 4$, $z \geq 0$.

9. Illustrate Stokes' Theorem by computing both sides of the equation

$$\int_C P\,dx + Q\,dy + R\,dz = \int_S [(\nabla \times F) \cdot N(\mathbf{X})]dS,$$

where $F(\mathbf{X}) = (z, y, -x)$ and S is the surface $x^2 + y^2 + z^2 = 4$, $z \geq 0$.

10. Suppose that a force field is given by a potential function $\phi(x, y, z)$; that is, the force $F(x, y, z) = \nabla\phi(x, y, z)$. If ϕ has continuous second partial derivatives, show that the force field is conservative, that is, the work done by moving a particle along a closed path is 0.
11. Prove Theorem 17.5.3 in the case where $Q_z > R_y$ at some point of V.
12. If f has continuous second partial derivatives, verify that $\nabla \times \nabla f = \mathbf{0}$. That is, curl(grad f) = **0**.
13. **If** F has continuous second partial derivatives, verify that $\nabla \cdot (\ \ \times F) = \mathbf{0}$. That is, div(curl F) = 0.
14. If F has continuous second partial derivatives, verify that $\nabla \times (\nabla \times F) = \nabla(\nabla \cdot F) - \nabla^2 F$, where, if $F(\mathbf{X}) = (P(\mathbf{X}), Q(\mathbf{X}), R(\mathbf{X}))$, then $\nabla^2 F(\mathbf{X}) = (\nabla^2 P(\mathbf{X}), \nabla^2 Q(\mathbf{X}), \nabla^2 R(x))$.
15. Use Stokes' Theorem to prove that $\int_C P(x)dx + Q(y)dy + R(z)dz = 0$, where C is the boundary of some surface given by $z = f(x, y)$.

Brief Review of Chapter 17

1. Definition of Line Integrals in Two Dimensions

Line integrals are defined by the equations

$$\int_C f(x, y)dx = \lim_{n \to \infty} \sum_{i=1}^{n} f(x_i^*, y_i^*)(x_i - x_{i-1})$$

and

$$\int_C f(x, y)dy = \lim_{n \to \infty} \sum_{i=1}^{n} f(x_i^*, y_i^*)(y_i - y_{i-1}),$$

where the curve C has been divided into n parts by the points (x_i, y_i), and (x_i^*, y_i^*) is any point on C between (x_{i-1}, y_{i-1}) and (x_i, y_i). Moreover, the subdivision of C must be done in such a way that all the $x_i - x_{i-1}$ and $y_i - y_{i-1}$ approach 0 as n increases without bound.

2. Computation of Line Integrals

The major tool for computation of line integrals is the following theorem.

Theorem. Let the curve C have the parametric equation $X(t) = (r(t), s(t))$ with initial point $(r(t_I), s(t_I))$ and terminal point $(r(t_T), s(t_T))$. If $r'(t)$, $s'(t)$ and $f(r(t), s(t))$ are continuous for t between t_I and t_T, then

$$\int_C P(x, y)dx + Q(x, y)dy = \int_{t_I}^{t_T} [P(r(t), s(t))r'(t) + Q(r(t), s(t))s'(t)]dt.$$

To apply this theorem the curve C is often broken into curves $C_1, C_2, \ldots, C_k$. The line integrals over the smaller curves $C_1, \ldots, C_k$ are then summed to compute the line integral over C.

Independence of Path. Independence of path is another aid in the computation of some line integrals. If $P_y(x, y) = Q_x(x, y)$ and both partials are continuous on a simply connected region R, then the line integral $\int_C P(x, y)dx + Q(x, y)dy$ is independent of path in the region R. Thus, if the curve C lies in R, it may be replaced with any curve C' in R that connects the given initial and terminal points. Then by choosing a simple curve such as a straight line or polygonal path for C', the calculation of the line integral may be simplified considerably.

If C is a simple closed curve in a region S, where P, Q, P_y, and Q_x are all continuous, we may use *Green's Theorem* to compute the line integral

$$\int_C P(x, y)dx + Q(x, y)dy.$$

According to Green's Theorem

$$\int_C P(x, y)dx + Q(x, y)dy = \int_R [Q_x(x, y) - P_y(x, y)]dA,$$

where R is the region bounded by C.

Green's Theorem may also be applied to a region R between two simple closed curves C_1 and C_2. If C_1 is the outer curve, we have

$$\int_{C_1} P\,dx + Q\,dy - \int_{C_2} P\,dx + Q\,dy = \int_R [Q_x(x, y) - P_y(x, y)]dA.$$

3. Line Integrals in Three Dimensions

The definition of

$$\int_C P(x, y, z)dx + Q(x, y, z)dy + R(x, y, z)dz$$

is very similar to the definition of line integrals in the 2-dimensional case. As in the 2-dimensional case the main tool for the computation of line integrals is the following theorem.

Theorem. Let the curve C have the parametric equation $X(t) = (q(t), r(t), s(t))$ with initial point $(q(t_I), r(t_I), s(t_I))$ and terminal point $(q(t_T), r(t_T), s(t_T))$. Then, if all the functions involved in the integrals are continuous, we have

$$\int_C P\,dx + Q\,dy + R\,dz = \int_{t_I}^{t_T} [P(q(t), r(t), s(t))q'(t) + Q(q(t), r(t), s(t))r'(t) + R(q(t), r(t), s(t))s'(t)]\,dt.$$

The integral $\int_C P\,dx + Q\,dy + R\,dz$ is independent of path in a simply connected region V if and only if $P_y = Q_x$, $Q_z = R_y$, and $R_x = P_z$, for all points of V. In this case the curve C may be replaced by a simpler curve C' so long as both C and C' are both in V and connect the same initial and terminal points.

4. Surface Integrals

The surface integral of $f(x, y, z)$ over the surface S is defined as

$$\int_S f(x, y, z)dS = \lim_{n\to\infty} \sum_{i=1}^{n} f(x_i^*, y_i^*, z_i^*)S_i.$$

For a full explanation of the notation see the discussion in Section 17.6. To compute a surface integral we usually use the following theorem, which converts the surface integral into a double integral.

Theorem. Let the surface S be the graph of $z = h(x, y)$ for (x, y) in the region R of the xy-plane. Then, if $h_x(x, y)$ and $h_y(x, y)$ are continuous on R and $f(x, y, z)$ is continuous on S, we have

$$\int_S f(x, y, z)dS = \int_R f(x, y, h(x, y))\sqrt{1 + h_x^2(x, y) + h_y^2(x, y)}\,dA.$$

To evaluate a surface integral over a surface S that is not a graph of a function we must subdivide S into subsurfaces $S_1, \ldots, S_n$ that are graphs of functions. The surface integral over S is then the sum of the surface integrals over the subsurfaces.

If the velocity of a fluid at (x, y, z) is given by the vector $V(x, y, z)$, then the volume of flow per unit time, or flux, through S in the direction of the unit normal $N(x, y, z)$ is given by the surface integral

$$F = \int_S [V(x, y, z) \cdot N(x, y, z)]dS.$$

5. The Divergence Theorem

Let $F(\mathbf{X}) = (P(\mathbf{X}), Q(\mathbf{X}), R(\mathbf{X}))$. Then the divergence of F is defined by

$$\text{div } F = \frac{\partial P}{\partial x} + \frac{\partial Q}{\partial y} + \frac{\partial R}{\partial z}.$$

The divergence of F is also written as $\nabla \cdot F$. Note that the domain of $\nabla \cdot F$ is a subset of R^3 and the range of $\nabla \cdot F$ is a subset of R.

The Divergence Theorem simplifies the calculation of certain surface integrals. According to the theorem.

$$\int_S [F(\mathbf{X}) \cdot N(\mathbf{X})]dS = \int_T \operatorname{div} F\, dV,$$

where $N(\mathbf{X})$ is the outwardly directed unit normal to the closed surface S, and T is the region bounded by S.

6. Stokes' Theorem

Let $F(\mathbf{X}) = (P(\mathbf{X}), Q(\mathbf{X}), R(\mathbf{X}))$; then the curl of F is defined by

$$\operatorname{curl} F(\mathbf{X}) = \left(\frac{\partial R}{\partial y} - \frac{\partial Q}{\partial z}, \frac{\partial P}{\partial z} - \frac{\partial R}{\partial x}, \frac{\partial Q}{\partial x} - \frac{\partial P}{\partial y}\right).$$

The curl of F is also denoted by $\nabla \times F$, since

$$\operatorname{curl} F = \begin{vmatrix} \mathbf{i} & \mathbf{j} & \mathbf{k} \\ \dfrac{\partial}{\partial x} & \dfrac{\partial}{\partial y} & \dfrac{\partial}{\partial z} \\ P & Q & R \end{vmatrix}.$$

Stokes' Theorem relates a 3-dimensional line integral over a closed curve C to a surface integral over a surface S bounded by C. According to the theorem

$$\int_C P\,dx + Q\,dy + R\,dz = \int_S [(\nabla \times F)(\mathbf{X}) \cdot N(\mathbf{X})]dS,$$

where $F(\mathbf{X}) = (P(\mathbf{X}), Q(\mathbf{X}), R(\mathbf{X}))$ and $N(\mathbf{X})$ is the upwardly directed unit normal to S at the point with position vector $\mathbf{X}$.

Technique Review Exercises, Chapter 17

1. Compute the line integral $\int_C (x + y)dx + x^2\,dy$, where C is the curve with parametric equation $X(t) = (t^2, t^3)$ with initial point $X(0)$ and terminal point $X(2)$.
2. Compute the line integral $\int_C e^{\sqrt{x}}y\,dx + e^y x\,dy$, where C is the graph of $x = y^2$ from $(1, 1)$ to $(4, 2)$.
3. Use Greens' Theorem to compute the line integral $\int_C x^2 y\,dx + xy^2\,dy$, where C is the square determined by the lines $x = \pm 1$ and $y = \pm 1$.
4. Show that the line integral $\int_C yz\,dx + xz\,dy + xy\,dz$ is independent of path in R^3.
5. Compute the line integral $\int_C (x + z)dx + (x + z)dy + (y + z)dz$, where C is the path given parametrically by $X(t) = (\sin^2 t, \cos^2 t, \tan^3 t)$ with initial point $X(0)$ and terminal point $X(\pi/4)$.
6. Evaluate the surface integral $\int_S z\,dS$, where S is the cylinder $z^2 + y^2 = 1, 0 \leq x \leq 1$.
7. Use The Divergence Theorem to compute the flux in an outward direction through the surface of the cube determined by $-1 \leq x \leq 1$, $-1 \leq y \leq 1$, and $-1 \leq z \leq 1$. The fluid velocity is given by $V(\mathbf{X}) = a\mathbf{X}$.

8. Compute the flux in the downward direction through the section of the plane given by $x + 2y + z = 4$, $x > 0$, $y > 0$, $z > 0$. The fluid velocity is given by $V(\mathbf{X}) = a\mathbf{X}$.
9. Use Stokes' Theorem to evaluate the line integral $\int_C -2y\,dx + 2x\,dy + z^2x\,dz$, where C is the circle formed by the intersection of the plane $z = 4$ and the paraboloid $x^2 + y^2 = z$.
10. Let $F:R^3 \to R^3$ and $f:R^3 \to R$. Define the following functions and indicate their range and domain:
    (a) grad $f = \nabla f$
    (b) del $F = \nabla \cdot F$
    (c) curl $F = \nabla \times F$.

Additional Exercises, Chapter 17

Section 17.1

1. Use the definition of the line integral to evaluate $\int_C (x^2 + 2xy + y^2)dx$, where C is the graph of $x + y = 5$ with initial point (1, 4) and terminal point (0, 5).
2. Use the definition of the line integral to evaluate $\int_C e^{x^2+y^2}\,dy$ where C is the quarter of the unit circle centered at the origin with initial point (1, 0) and terminal point (0, 1).

Section 17.2

In Exercises 3–9, compute the given line integral.

3. $\int_C (xy + y^2)dx + y^2\,dy$, C has the parametric equation $X(t) = (t, t^2)$ with initial point (1, 1) and terminal point (2, 4).
4. $\int_C \sin y\,dx + e^x\,dy$, C has the parametric equation $X(t) = (3t, -2t)$ with initial point $X(1)$ and terminal point $X(-1)$.
5. $\int_C y^2\,dx + 2xy\,dy$, where C is the line segment from (3, 1) to $(-2, 2)$.
6. $\int_C x^{1/2}y^3\,dx + xy^{1/3}\,dy$, where C is the line segment from $(2, -1)$ to (0, 0).
7. $\int_C y\,dx + y^2\,dy$, from (0, 0) to $(\pi/4, 1)$ along $y = \tan x$.
8. $\int_C y\,dx + y^2\,dy$, from (0, 0) to $(\pi/4, 1)$ along $x = \pi y^2/4$.
9. $\int_C y^2\,dx + (x + y)dy$, where C is the curve along $y = -x$ from (0, 0) to $(-1, 1)$ and then along $y = x^2$ from $(-1, 1)$ to (0, 0).

Section 17.3

Use Green's Theorem to compute the integrals given in Exercises 10–15.

10. $\int_C y \sin x\,dx + x \cos y\,dy$, where C is the rectangle with vertices at $(-1, -2)$, $(1, -2)$, (1, 1), and $(-1, 1)$.
11. $\int_C (x - y)^2\,dx + (x + y)^2\,dy$, where C is the closed curve formed by $y = x^3$ and $y = x$ between the points (0, 0) and (1, 1).
12. $\int_C e^x y\,dx + e^x\,dy$, where C is the closed curve formed by $y = x^2$ and $y = x$, where $0 \le x \le 1$.
13. $\int_C (x^3 - 3y^2)dy$, where C is the circle $x^2 + y^2 = 4$.
14. $\int_C (x^2 - y^2)\,dx + e^x\,dy$, where C is the circle $x^2 + y^2 = 1$.
15. $\int_C \sin y\,dx + \sin x\,dy$, where C is the triangle formed by the lines $y = -x + 1$, $x = 0$, and $y = 0$.

Section 17.4

In Exercises 16–20, first show that the given integral is independent of path and then evaluate it.

16. $\int_{(1,1)}^{(2,-1)} 2y^3x\,dx + 3x^2y^2\,dy.$

17. $\int_{(1,-1)}^{(-1,1)} \cos y\,dx - x\sin y\,dy.$

18. $\int_{(0,0)}^{(1,-1)} (6y^2x - 3yx^2)dx + (6yx^2 - x^3)dy.$

19. $\int_{(1,2)}^{(-1,3)} (y^2 - 4yx^3)dx + (2xy - x^4 - 1)dy.$

20. $\int_{(0,\pi)}^{(1,\pi/2)} e^{-x}\cos y\,dx + e^{-x}\sin y\,dy.$

21. Where is the integral $\int_C \ln y\,dx + (x/y)dy$ independent of path?

Section 17.5

In Exercises 22–26, compute the given line integrals.

22. $\int_C (x^2 - z)dx + (y^2 - x)dy + (z^2 - y)dz$, where C has the parametric equation $X(t) = (t, t^2, t^3)$ with initial point $X(0)$ and terminal point $X(1)$.

23. $\int_C (x^2 - z)dx + (y^2 - x)dy + (z^2 - y)dz$ where C is the line segment from $(0, 0, 0)$ to $(1, 1, 1)$.

24. $\int_C x\sin y\,dx + \cos y\,dy + (x + y)dz$, where C is the line segment from $(0, \pi/2, 1)$ to $(1, -\pi/2, 0)$.

25. $\int_C z\sin xz\,dx + xz\cos xyz\,dy + x\sin xz\,dz$, where C has the parametric equation $X(t) = (t, t^4, t^2)$ with initial point $(0, 0, 0)$ and terminal point $(1, 1, 1)$.

26. $\int_C e^x\,dx + \sin y\,dy + y\,dz$, where C has the parametric equation $X(t) = (t, t^3, t^2)$ with initial point $(0, 0, 0)$ and terminal point $(1, 1, 1)$.

27. Show that $\int_C 2x\,dx + z^2\,dy + 2yz\,dz$ is independent of path.

28. Show that $\int_C (6y - 2xz)dx + (2yz^3 + 6x)dy + (3y^2z^2 - x^2)dz$ is independent of path.

Section 17.6

Evaluate the given surface integral.

29. $\int_S (x^2y - z)dS$, where S is the portion of the plane $z = x + y$ above the region R bounded by $x = 0$, $y = 0$, and $x + y = 1$ in the xy-plane.

30. $\int_S xz\,dS$, where S is the portion of the plane $x + y + z = 2$ above the region R given by $0 \le x \le 1$ and $0 \le y \le x$.

31. $\int_S (x^2 + y^2)dS$, where S is the surface of Exercise 29.

32. $\int_S \sin z\,dS$, where S is the surface of Exercise 30.

33. $\int_S dS$, where S is that part of the surface of the paraboloid $z = 2 - (x^2 + y^2)$ which is above the xy-plane.

34. $\int_S (x^2 + y^2)dS$, where S is that portion of the cone $z = (x^2 + y^2)^{1/2}$ between $z = 0$ and $z = 1$.

Section 17.7

Compute flux across the given surface in the upward direction.

35. The fluid velocity at the point (x, y, z) is $G(x, y, z) = (zx^2, -z^2y, x - 2z)$, and the surface is that portion of the plane $z = 1$ where $0 \le x \le 1$ and $0 \le y \le 1$.

36. The fluid velocity at the point (x, y, z) is $G(x, y, z) = (\cos x, x + y, \sin z)$, and the surface is given by $x + 2y + z = 3$, $x \ge 0$, $y \ge 0$, $z \ge 0$.

37. The fluid velocity at the point (x, y, z) is $G(x, y, z) = (0, 0, z)$, and the surface is given by $z^2 = x^2 + y^2$, $x^2 + y^2 \le 2$, $z \ge 0$.

38. The fluid velocity at the point (x, y, z) is $G(x, y, z) = (x^3, y^3, z^3)$, and the surface is given by $x^2 + y^2 + z^2 = 1$, $z \geq 0$.

Section 17.8

In Exercises 39–42, use The Divergence Theorem in order to compute the surface integral $\int_S [G(X) \cdot N(X)] dS$.

39. $G(X) = (x + z, x + y, y + z)$, S is the sphere $x^2 + y^2 + z^2 = 4$.

40. $G(X) = (x^2 z, -yz^2, x - 2z)$, S is the surface of the cube given by $0 \leq x \leq 1$, $0 \leq y \leq 1$, $0 \leq z \leq 1$.

41. $G(X) = (xy^2 + 2zy, yz^2 - x^3, x^2 z)$, S is the sphere $x^2 + y^2 + z^2 = 1$.

42. $G(X) = (xy^2 + 2zy, yz^2 - x^3, x^3 z)$, S is the closed surface bounded by $z = -(1 - x^2 - y^2)^{1/2}$ and $z = 0$.

43. The velocity of a fluid at the point (x, y, z) is $G(x, y, z) = (zx^2, -z^2 y, x - 2z)$. Compute the flux in the outward direction through the tetrahedron bounded by $x + y + z = 1$ and the coordinate planes.

44. The velocity of a fluid at the point (x, y, z) is $G(x, y, z) = (2xy + z, y^2, -x - 3y)$. Compute the flux in the outward direction through the tetrahedron bounded by $x + 2y + 3z = 6$ and the coordinate planes.

Section 17.9

Use Stokes' Theorem to compute the given line integral.

45. $\int_C 2y\, dx + 3x\, dy - z^2\, dz$, where C is the circle formed by the intersection of the cylinder $x^2 + y^2 = 9$ and the xy-plane.

46. $\int_C xy\, dx + yz\, dy + xz\, dz$, where C is the triangle with vertices $(1, 0, 0)$, $(0, 1, 0)$, and $(0, 0, 2)$.

47. $\int_C (x^2 - y + z)dx + (\sin y + x)dy + (x - e^z)dz$, where C is the triangle with vertices $(1, 1, 1)$, $(0, 1, 0)$, and $(0, 0, 0)$.

48. $\int_C z\, dx + z\, dy + x\, dz$, where C is the path consisting of the line segments joining the points $(0, 0, 0)$, $(1, 1, 0)$, $(1, 2, 2)$, $(1, 0, 2)$, and $(0, 0, 0)$, in the order indicated.

Challenging Problems, Chapter 17

1. Assume that there is a function $\phi(x, y)$ with continuous second partial derivatives and such that

$$\frac{\partial \phi}{\partial x} = P(x, y) \quad \text{and} \quad \frac{\partial \phi}{\partial y} = Q(x, y).$$

Show:

(i) $\int_{(x_1, y_1)}^{(x_2, y_2)} P\, dx + Q\, dy$ is independent of path.

(ii) $\int_{(x_1, y_1)}^{(x_2, y_2)} P\,dx + Q\,dy = \phi(x_2, y_2) - \phi(x_1, y_1)$.

Statement (ii) is an interesting extension of The Fundamental Theorem of Calculus.

2. Show that the volume V enclosed by a surface S is given by

$$V = \frac{1}{3}\int_S [\mathbf{X} \cdot N(\mathbf{X})]dS.$$

Appendix

Algebraic Formulas

Areas and Volumes

Trigonometric Identities

Differentiation Formulas

Integration Formulas

Algebraic Formulas

1. Quadratic Formula

The solutions to the quadratic equation

$$ax^2 + bx + c = 0, \quad a \neq 0$$

are given by

$$x = \frac{-b \pm \sqrt{b^2 - 4ac}}{2a}.$$

2. Binomial Theorem

If n is a positive integer then

$$(a + b)^n = a^n + na^{n-1}b + \frac{n(n-1)}{2!}a^{n-2}b^2 + \frac{n(n-1)(n-2)}{3!}a^{n-3}b^3 + \cdots + nab^{n-1} + b^n.$$

3. Laws of Exponents

$$a^m a^n = a^{m+n}.$$

$$\frac{a^m}{a^n} = a^{m-n}, \quad a \neq 0.$$

$$(a^m)^n = a^{mn}.$$

$$(ab)^n = a^n b^n.$$

$$a^{p/q} = (\sqrt[q]{a})^p = \sqrt[q]{a^p}.$$

4. Logarithms and Exponential Functions

$$\log_a(xy) = \log_a x + \log_a y.$$

$$\log_a \frac{x}{y} = \log_a x - \log_a y.$$

$$x \log_a y = \log_a(y^x).$$

$$\log_a 1 = 0, \quad \log_a a = 1.$$

$$a^{\log_a x} = x, \quad \log_a(a^x) = x.$$

Areas and Volumes

1. Triangle

Area $= (1/2)bh.$

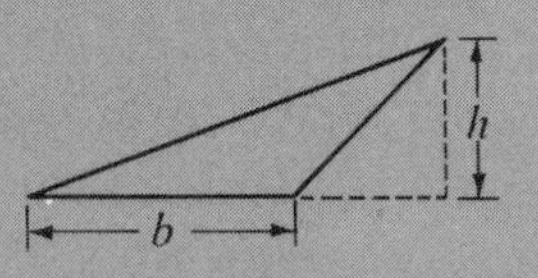

2. Parallelogram

Area $= bh.$

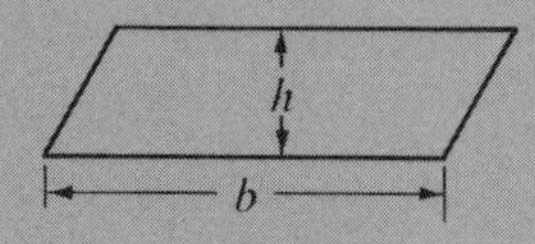

3. Trapezoid

$$\text{Area} = \frac{a+b}{2}h.$$

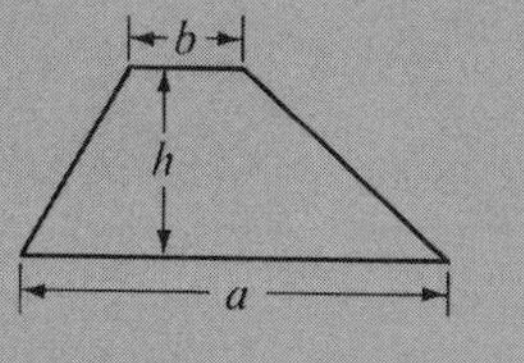

4. Circle

Area $= \pi r^2.$
Circumference $= 2\pi r.$

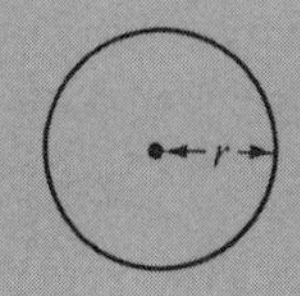

5. Sector of a Circle

Area $= (1/2)r^2\theta$, θ in radians.

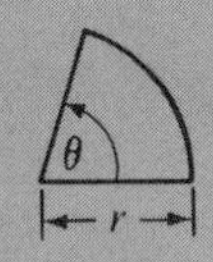

6. Right Circular Cylinder

Volume $= \pi r^2 h.$
Lateral surface area $= 2\pi rh.$

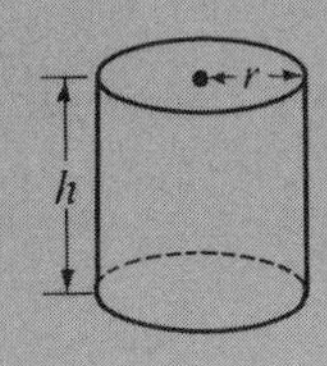

7. Right Circular Cone

Volume $= (1/3)\pi r^2 h$.
Lateral surface area $= \pi rs$.

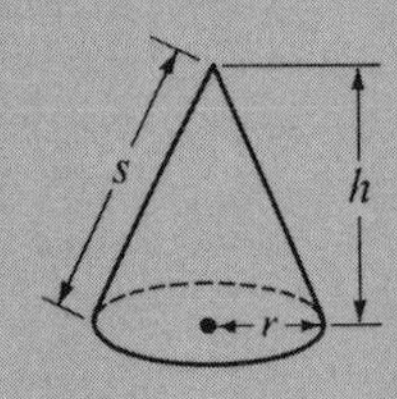

8. Sphere

Volume $= (4/3)\pi r^3$.
Surface area $= 4\pi r^2$.

9. Frustrum of a Right Circular Cone

Volume $= (1/3)\pi h(r_1{}^2 + r_1 r_2 + r_2{}^2)$.
Lateral surface area $= \pi s(r_1 + r_2)$.

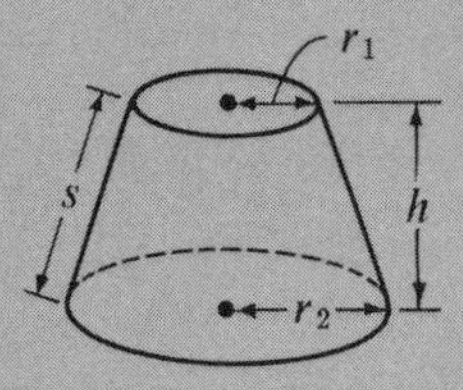

Values of the Trigonometric Functions for Selected Angles.

	0	$\pi/6$	$\pi/4$	$\pi/3$	$\pi/2$	π	$3\pi/2$
$\sin\theta$	0	$1/2$	$\sqrt{2}/2$	$\sqrt{3}/2$	1	0	-1
$\cos\theta$	1	$\sqrt{3}/2$	$\sqrt{2}/2$	$1/2$	0	-1	0
$\tan\theta$	0	$\sqrt{3}/3$	1	$\sqrt{3}$	—	0	—
$\cot\theta$	—	$\sqrt{3}$	1	$\sqrt{3}/3$	0	—	0
$\sec\theta$	1	$2\sqrt{3}/3$	$\sqrt{2}$	2	—	-1	—
$\csc\theta$	—	2	$\sqrt{2}$	$2\sqrt{3}/3$	1	—	-1

Trigonometric Identities

1. $\sin x = \dfrac{1}{\csc x}$

2. $\cos x = \dfrac{1}{\sec x}$

3. $\tan x = \dfrac{1}{\cot x} = \dfrac{\sin x}{\cos x}$

4. $\cot x = \dfrac{1}{\tan x} = \dfrac{\cos x}{\sin x}$

5. $\sin^2 x + \cos^2 x = 1$

6. $1 + \tan^2 x = \sec^2 x$

7. $1 + \cot^2 x = \csc^2 x$

8. $\sin x = \cos\left(\dfrac{\pi}{2} - x\right) = \sin(\pi - x)$

9. $\cos x = \sin\left(\frac{\pi}{2} - x\right) = -\cos(\pi - x)$

10. $\tan x = \cot\left(\frac{\pi}{2} - x\right) = -\tan(\pi - x)$

11. $\cot x = \tan\left(\frac{\pi}{2} - x\right) = -\cot(\pi - x)$

12. $\sin(x \pm y) = \sin x \cos y \pm \cos x \sin y$

13. $\cos(x \pm y) = \cos x \cos y \pm \sin x \sin y$

14. $\tan(x \pm y) = \dfrac{\tan x \pm \tan y}{1 \pm \tan x \tan y}$

15. $\cot(x \pm y) = \dfrac{\cot x \cot y \pm 1}{\cot y \pm \cot x}$

16. $\sin 2x = 2 \sin x \cos x$

17. $\cos 2x = \cos^2 x - \sin^2 x = 2\cos^2 x - 1 = 1 - 2\sin^2 x$

18. $\tan 2x = \dfrac{2 \tan x}{1 - \tan^2 x}$

19. $\cot 2x = \dfrac{\cot^2 x - 1}{2 \cot x}$

20. $\sin^2 x = \dfrac{1 - \cos 2x}{2}$

21. $\cos^2 x = \dfrac{1 + \cos 2x}{2}$

22. $\sin(x + y) - \sin(x - y) = 2 \cos x \sin y.$

Law of Sines.

$$\frac{a}{\sin A} = \frac{b}{\sin B} = \frac{c}{\sin C}$$

Law of Cosines.

$$a^2 = b^2 + c^2 - 2bc \cos A$$

b
C
a
A
B
c

Differentiation Formulas

1. $\dfrac{dc}{dx} = 0$, if c is a constant.

2. $\dfrac{d}{dx}(u \pm v) = \dfrac{du}{dx} \pm \dfrac{dv}{dx}.$

3. $\dfrac{d}{dx}(uv) = u\dfrac{dv}{dx} + v\dfrac{du}{dx}.$

4. $\dfrac{d}{dx}\left(\dfrac{u}{v}\right) = \dfrac{v(du/dx) - u(dv/dx)}{v^2}.$

5. $\dfrac{d(cu)}{dx} = c\dfrac{du}{dx}$, if c is a constant.

6. $\dfrac{du^n}{dx} = nu^{n-1}\dfrac{du}{dx}$, for any real number n.

7. $\dfrac{df(u)}{dx} = \dfrac{df}{du}\dfrac{du}{dx}.$

8. $\dfrac{d}{dx}(\log_a u) = \dfrac{\log_a e}{u}\dfrac{du}{dx}.$

9. $\dfrac{d}{dx}(\ln u) = \dfrac{1}{u}\dfrac{du}{dx}.$

10. $\dfrac{d}{dx}(\log_a |u|) = \dfrac{\log_a e}{u}\dfrac{du}{dx}.$

11. $\frac{d}{dx}(\ln|u|) = \frac{1}{u}\frac{du}{dx}$.

12. $\frac{d}{dx}a^u = a^u \ln a \frac{du}{dx}$.

13. $\frac{d}{dx}e^u = e^u \frac{du}{dx}$.

14. $\frac{d}{dx}\sin u = \cos u \frac{du}{dx}$.

15. $\frac{d}{dx}\cos u = -\sin u \frac{du}{dx}$.

16. $\frac{d}{dx}\tan u = \sec^2 u \frac{du}{dx}$.

17. $\frac{d}{dx}\cot u = -\csc^2 u \frac{du}{dx}$.

18. $\frac{d}{dx}\sec u = \sec u \tan u \frac{du}{dx}$.

19. $\frac{d}{dx}\csc u = -\csc u \cot u \frac{du}{dx}$.

20. $\frac{d}{dx}\arcsin u = \frac{1}{\sqrt{1-u^2}}\frac{du}{dx}$.

21. $\frac{d}{dx}\arccos u = -\frac{1}{\sqrt{1-u^2}}\frac{du}{dx}$.

22. $\frac{d}{dx}\arctan u = \frac{1}{u^2+1}\frac{du}{dx}$.

23. $\frac{d}{dx}\operatorname{arccot} u = -\frac{1}{u^2+1}\frac{du}{dx}$.

24. $\frac{d}{dx}\operatorname{arcsec} u = \frac{1}{|u|\sqrt{u^2-1}}\frac{du}{dx}$.

25. $\frac{d}{dx}\operatorname{arccsc} u = -\frac{1}{|u|\sqrt{u^2-1}}\frac{du}{dx}$.

Integration Formulas

1. $\int f(u)du = F(u) + C$, if $F'(u) = f(u)$.
2. $\int[f(x) \pm g(x)]dx = \int f(x)dx \pm \int g(x)dx$.
3. $\int cf(x)dx = c\int f(x)dx$ if c is a constant.
4. $\int u\,dv = uv - \int v\,du$.
5. $\int u^n\,du = \frac{u^{n+1}}{n+1} + C$.
6. $\int \frac{du}{u} = \ln|u| + C$.
7. $\int a^u\,du = \frac{a^u}{\ln a} + C$.
8. $\int e^u\,du = e^u + C$.
9. $\int \sin u\,du = -\cos u + C$.
10. $\int \cos u\,du = \sin u + C$.
11. $\int \tan u\,du = \ln|\sec u| + C$.
12. $\int \cot u\,du = \ln|\sin u| + C$.
13. $\int \sec u\,du = \ln|\sec u + \tan u| + C$.
14. $\int \csc u\,du = -\ln|\csc u + \cot u| + C$.
15. $\int \sec^2 u\,du = \tan u + C$.
16. $\int \csc^2 u\,du = -\cot u + C$.
17. $\int \sec u \tan u\,du = \sec u + C$.
18. $\int \csc u \cot u\,du = -\csc u + C$.
19. $\int \frac{du}{\sqrt{a^2-u^2}} = \arcsin\frac{u}{a} + C$.
20. $\int \frac{du}{u^2+a^2} = \frac{1}{a}\arctan\frac{u}{a} + C$.
21. $\int \frac{du}{|u|\sqrt{u^2-a^2}} = \frac{1}{a}\operatorname{arcsec}\frac{u}{a} + C$.

22. $$\int \frac{du}{u^2 - a^2} = \frac{1}{2a} \ln \left| \frac{u - a}{u + a} \right| + C.$$

23. $$\int \frac{dx}{(x^2 + a^2)^{k+1}} = \frac{x}{2ka^2(x^2 + a^2)^k} + \frac{2k - 1}{2ka^2} \int \frac{dx}{(x^2 + a^2)^k}.$$

Answers to Odd-Numbered Exercises

Chapter 1

Section 1.2 (page 5)

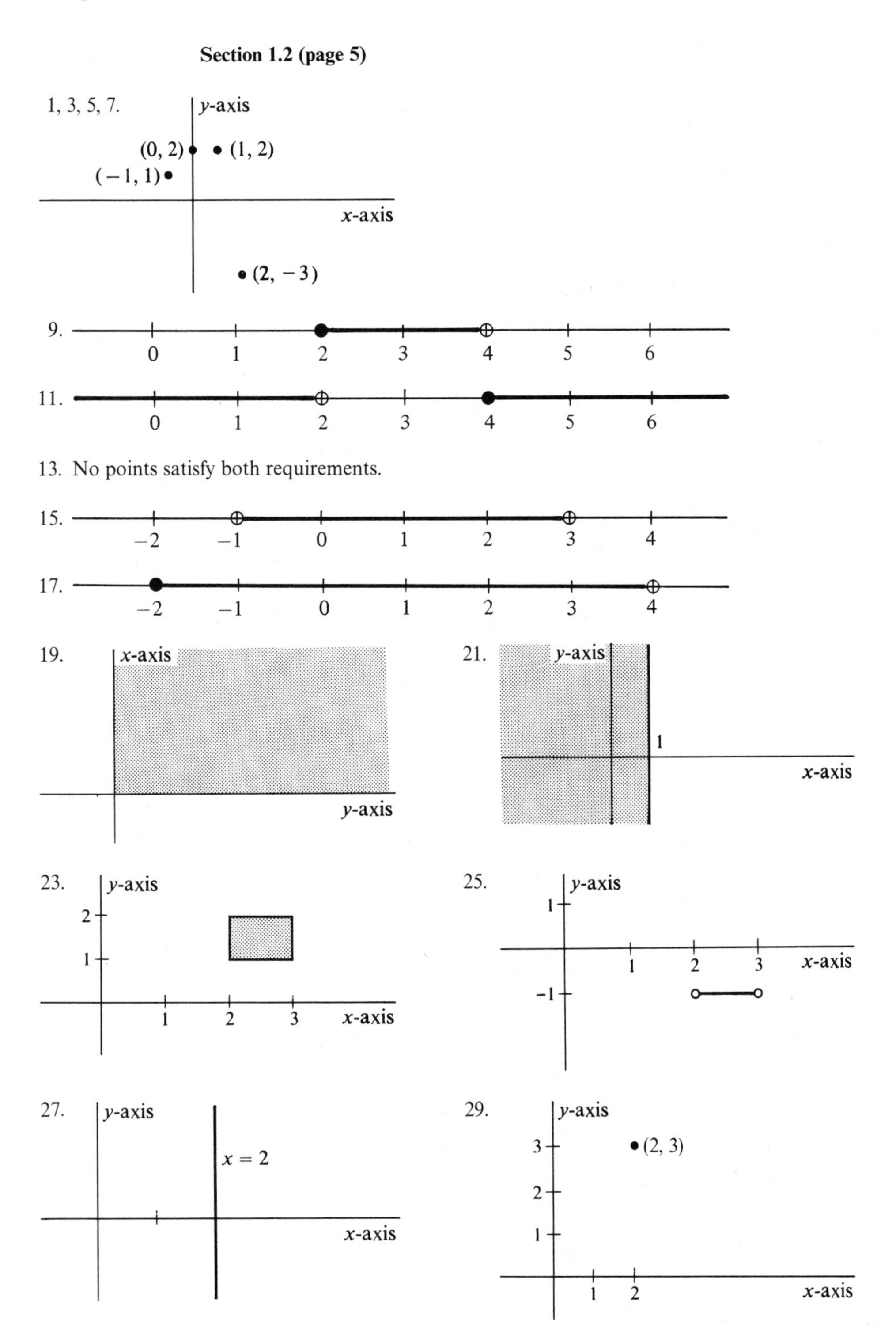

31.

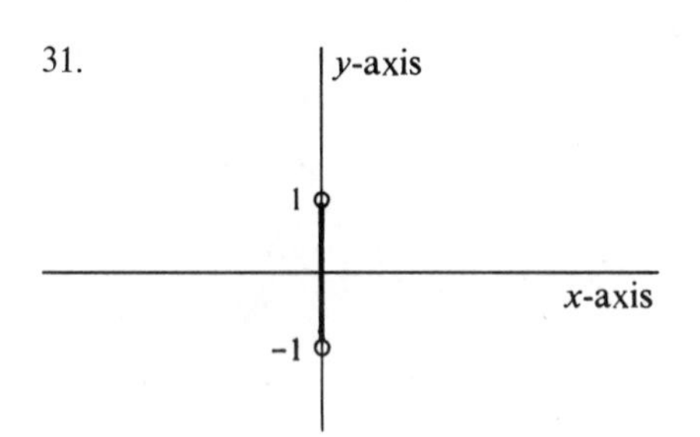

Section 1.3 (page 8)

1. 8
3. 1
5. $a^2 + 4a + 4$
7. $x^4 + 5$
9. 21
11. $x^2 + 8x + 16$
13. $y^6 - 2y^3 + 1$
15. $x^4 + x + 5$
17. $x^6 - 2x^5 + x^4$
19. 64
21. h
23. $2x + h - 2$
25. (a) the set of all real numbers except -2; (b) the set of all real numbers except 0.
27. (a) the set of all x such that $x \geq -2$; (b) the set of all nonnegative real numbers.
29. (a) the set of all x such that $x \neq -2$; (b) the set of nonzero real numbers.
31. (a) the set of all real numbers; (b) the set of all positive real numbers and -3.
33. $f(x) = \begin{cases} 0 & \text{if } 0 \leq x < 1 \\ 1 & \text{if } 1 \leq x < 2 \\ 2 & \text{if } 2 \leq x < 3 \\ 3 & \text{if } 3 \leq x < 4 \\ 4 & \text{if } x = 4 \end{cases}$
35. $A(x) = x(600 - 2x)$
37. $V(h) = (4/3)\pi h^3$
39. $a = 0$, b any value, $c = 0$.

Section 1.4 (page 13)

1. 4
3. 10
5. 11
7. $|x - 3|$
9. $|x + 6|$
11. $|a - b|$
13. The distance between x and 3 is less than 2.
15. The distance between x and -4 is less than or equal to 3.
17. The distance between x and 2 is greater than or equal to 5.
19. The distance between x and -6 is greater than 8.
21. $|x - 4| \leq 3$
23. $|x + 6| > 8$
25. $\sqrt{3}/2$
27. $1/\sqrt{3}$
29. $-\sqrt{3}/2$
31. -2
33. 4
35. -2
37. -2
39. -2
41. $f(x) = \cos x$ or $f(x) = x^2$
43. $f(x) = kx$
45. $f(x) = kx$
47. $f(x) = \log_a x$
53. all x such that $x \leq 0$
55. all x such that $x \geq 0$
57. all x such that $x \leq 0$
59. 0

Section 1.5 (page 18)

1.

3. No breaks.

5. No breaks.

7. No breaks.

9. No breaks.

11. No breaks.

13. Break at $x = 0$.

15. No breaks.

17. No breaks.

19. Break at $x = 0$.

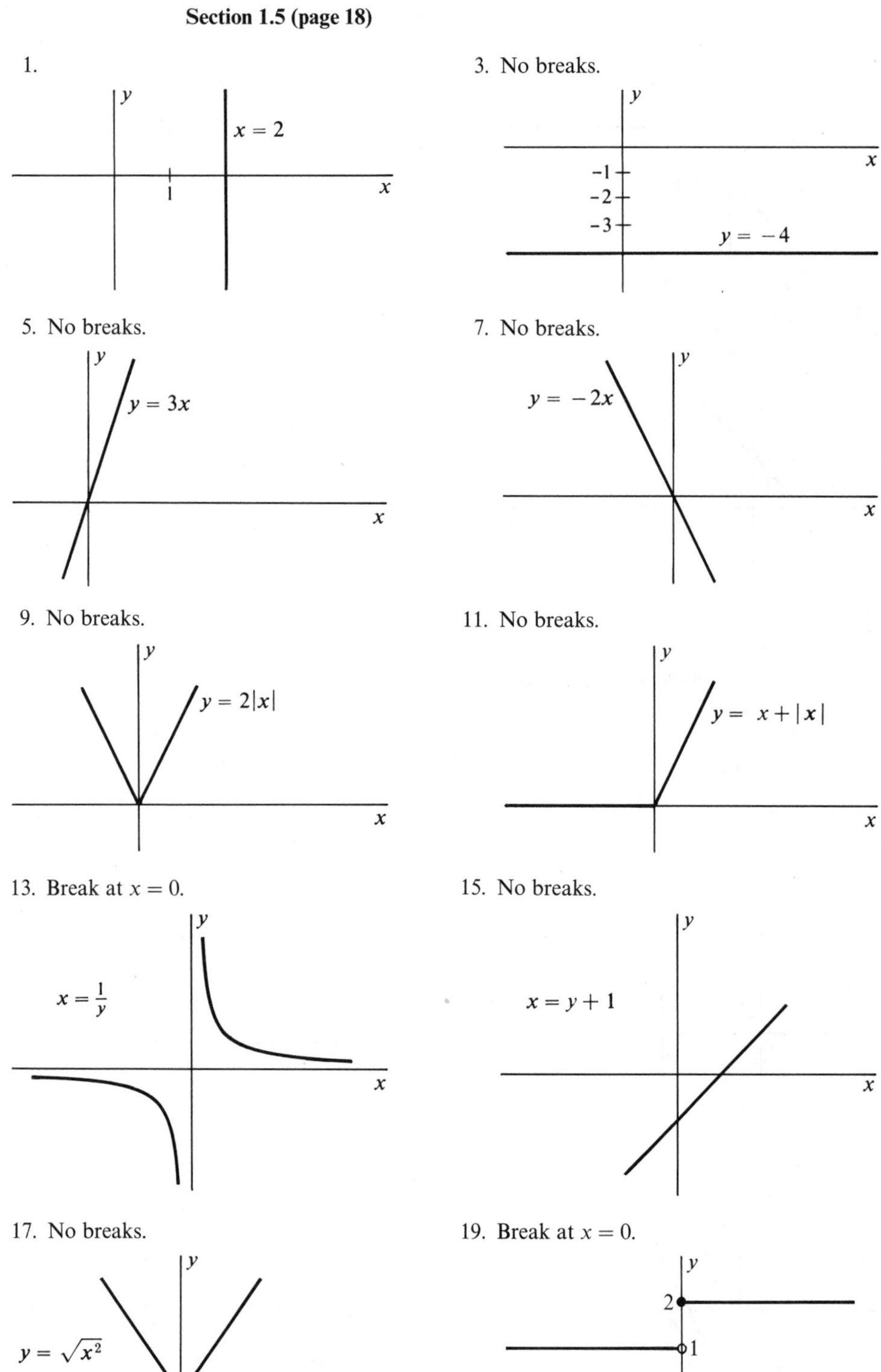

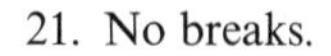
21. No breaks.

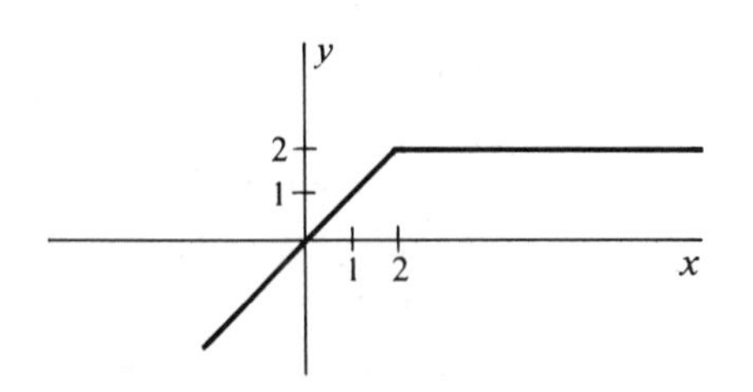

23. Break at $x = 2$.

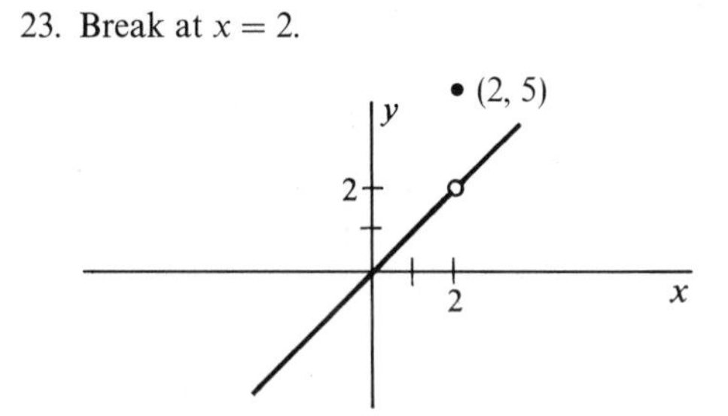

25. Break at $x = 0$.

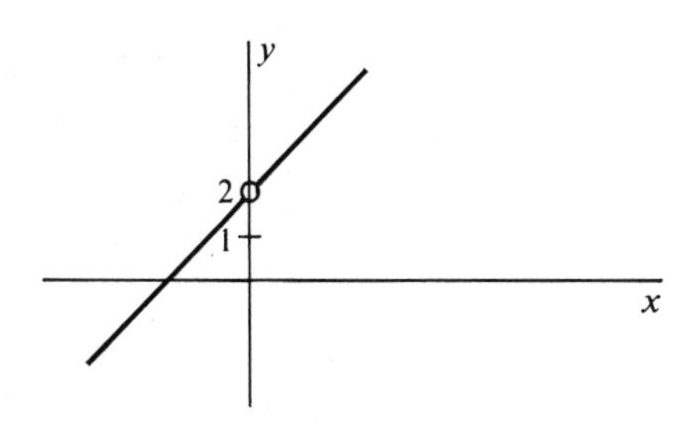

27. No breaks.

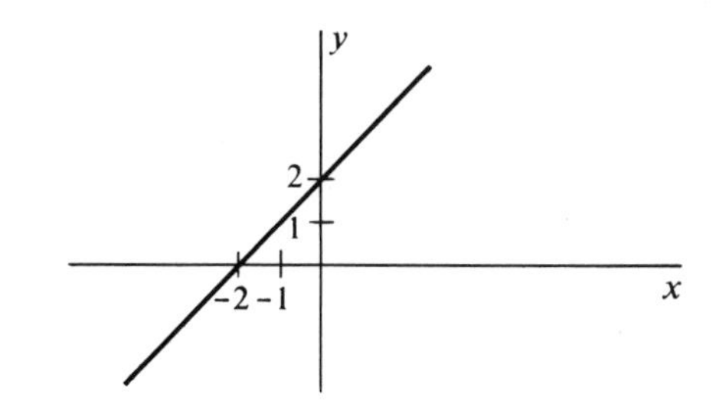

Section 1.6 (page 26)

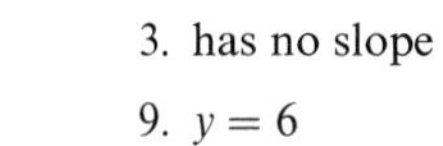

1. $-3/2$
3. has no slope
5. $2x - 5y = 32$
7. $4x - 5y + 19 = 0$
9. $y = 6$
11. $x = -1$
13. $3x + 5y = 1$
15. $y = -2x$
17. $y = -(2/3)x$
19. $y = -5$
21. $x = 6$
29. -1

31.

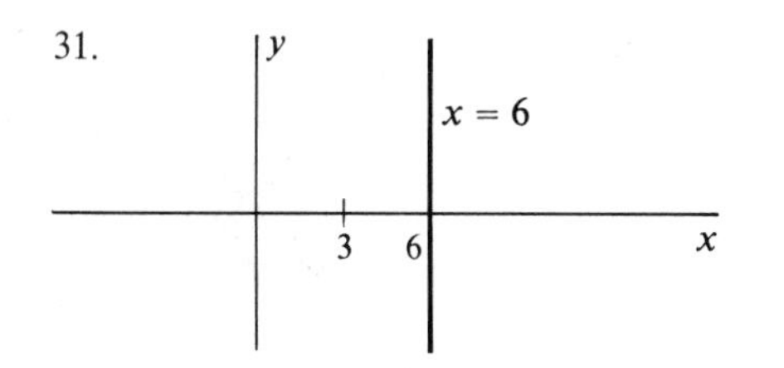

33.

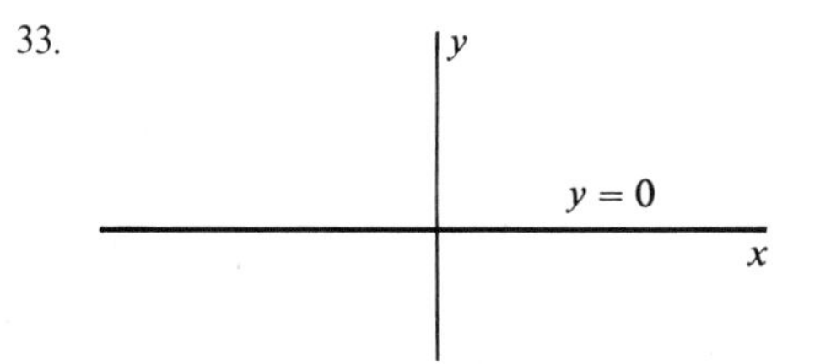

35.

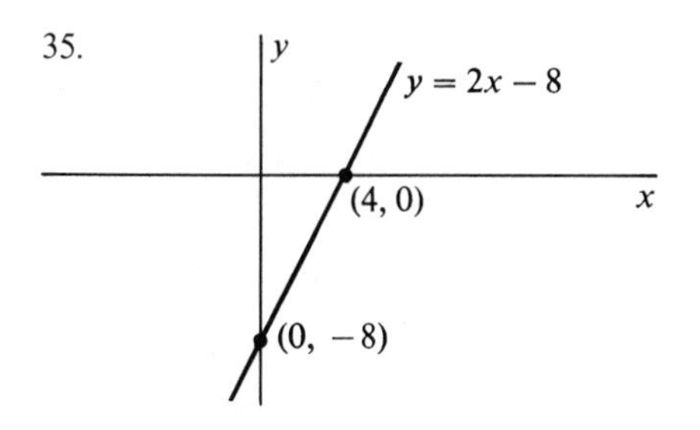

37.

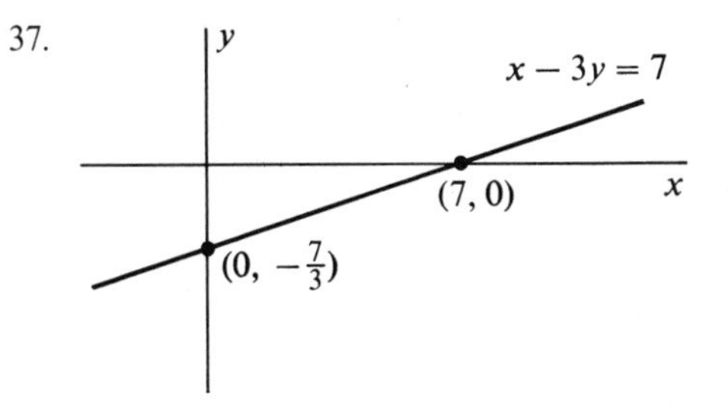

39.

41.

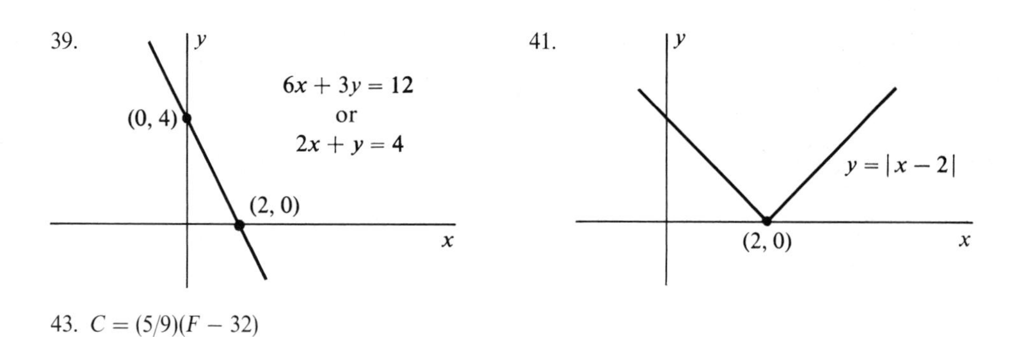

43. $C = (5/9)(F - 32)$

Section 1.7 (page 35)

1. 5
3. 8
5. 13
7. 5, −1
9. $\pm 1/\sqrt{2}$
11. $x^2 + (y - 5)^2 = 36$
13. $(x - 2)^2 + (y + 4)^2 = 50$
15. $x^2 + y^2 = 25$ and $(x - 3)^2 + (y - 1)^2 = 25$
17. $(x + 5)^2 + (y - 5)^2 = 25$ and $(x + 1)^2 + (y - 1)^2 = 1$
19. $\dfrac{x^2}{9} + \dfrac{y^2}{16} = 1$

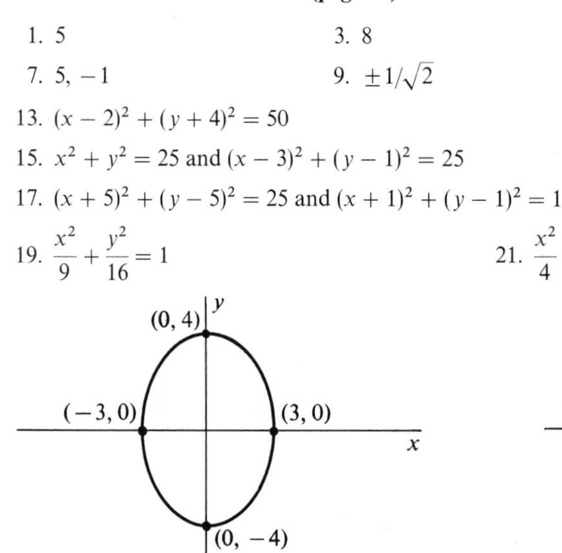

21. $\dfrac{x^2}{4} + \dfrac{y^2}{9} = 1$

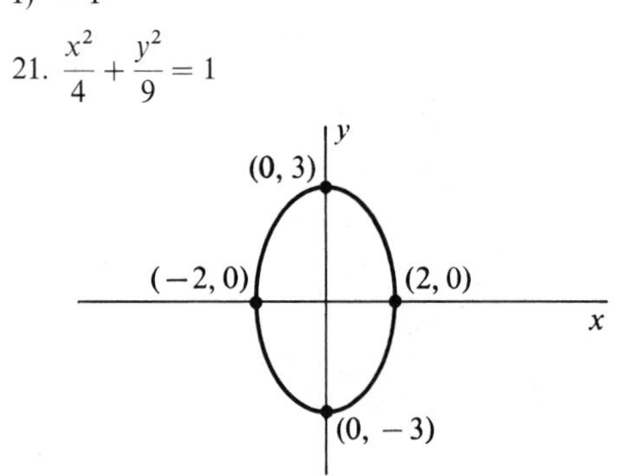

23. $\dfrac{x^2}{9} - \dfrac{y^2}{16} = 1$

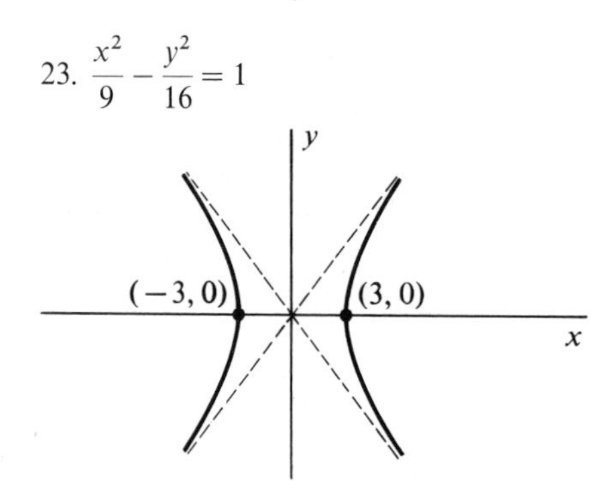

25. $\dfrac{y^2}{4} - \dfrac{x^2}{9} = 1$

27. $12y = x^2$

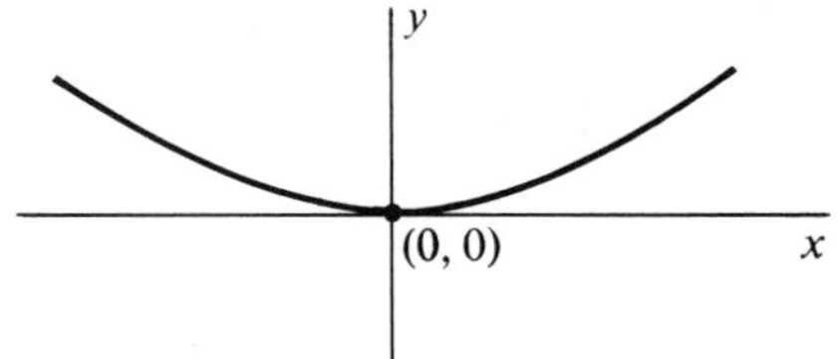

29. $x = 4y^2$

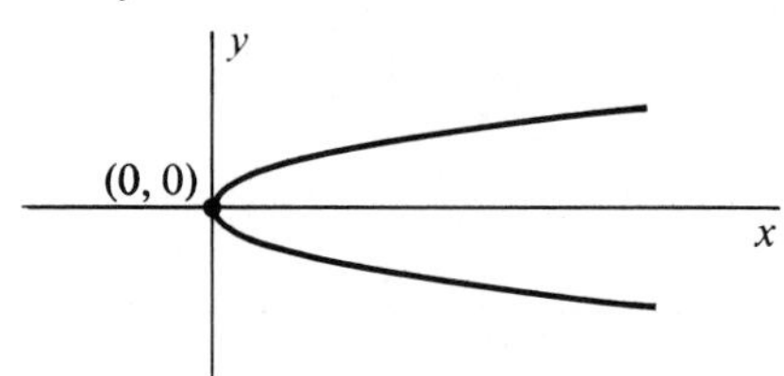

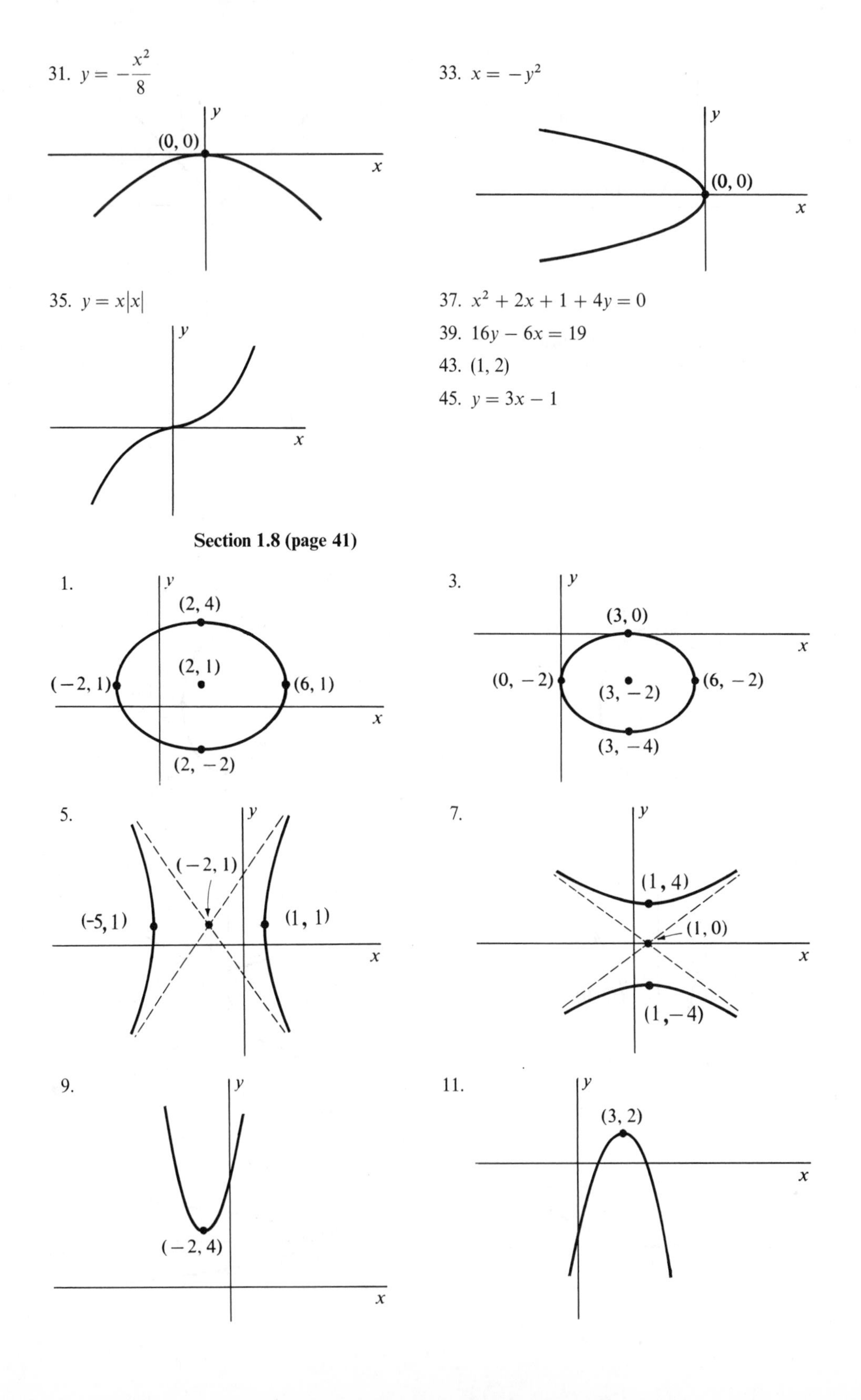
31. $y = -\frac{x^2}{8}$
y
(0, 0)
x
33. $x = -y^2$
y
(0, 0)
x
35. $y = x|x|$
y
x
37. $x^2 + 2x + 1 + 4y = 0$
39. $16y - 6x = 19$
43. (1, 2)
45. $y = 3x - 1$
Section 1.8 (page 41)
1.
y
(2, 4)
(2, 1)
(−2, 1)
(6, 1)
x
(2, −2)
3.
y
(3, 0)
x
(0, −2)
(3, −2)
(6, −2)
(3, −4)
5.
y
(−2, 1)
(-5, 1)
(1, 1)
x
7.
y
(1, 4)
(1, 0)
x
(1, −4)
9.
y
(−2, 4)
x
11.
y
(3, 2)
x

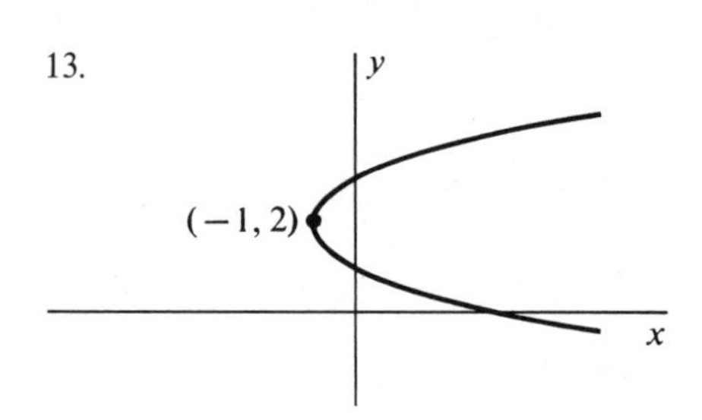

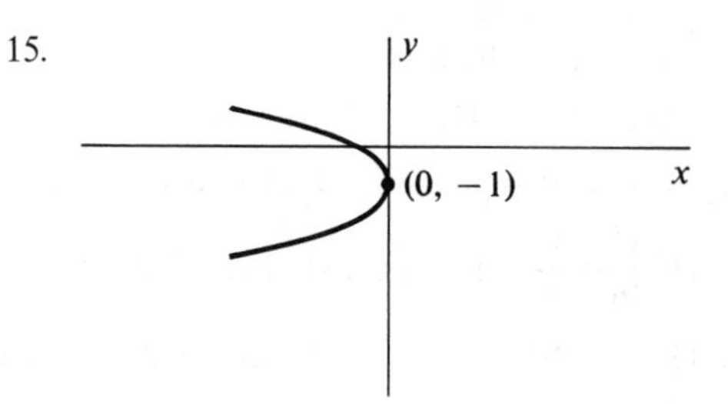

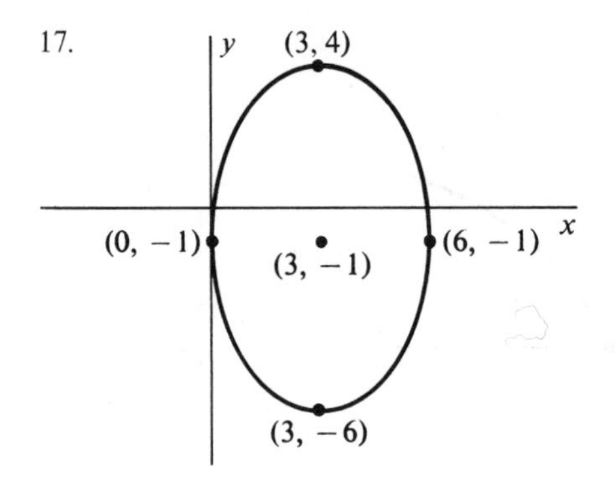

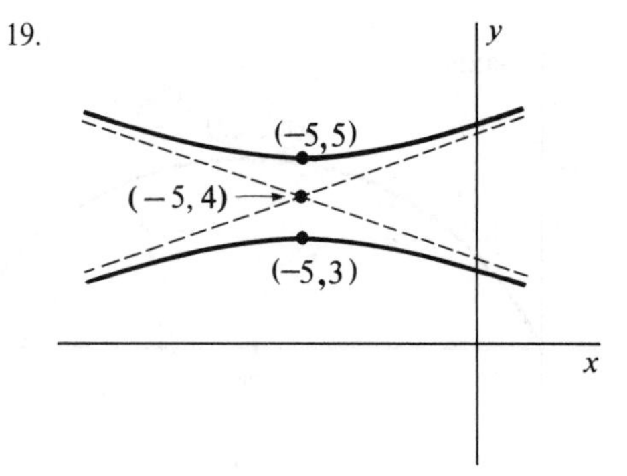

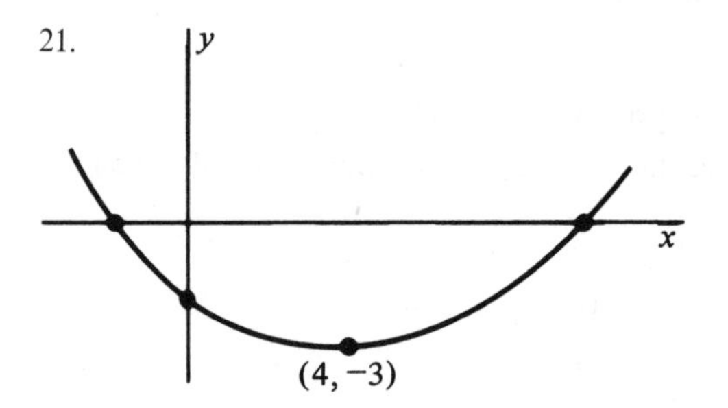

Section 1.9 (page 44)

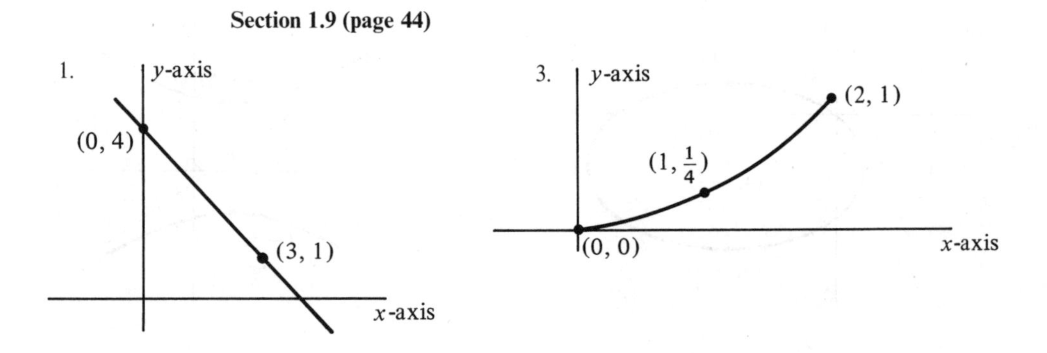

5. $x - 3y = 10$, line
7. $x = 5y^2 - 47y + 108$, parabola
9. $x^2 + y^2 = 1$, $x \geq 0$, right half of circle of radius one centered at the origin.
11. $\frac{x^2}{16} + \frac{y^2}{9} = 1$, $x \geq 0$, right half of an ellipse centered at the origin.
13. $(x-2)^2 + (y-1)^2 = 1$, circle with radius one centered at (2, 1).
15. $x = y^6$, $y \geq 0$

21.

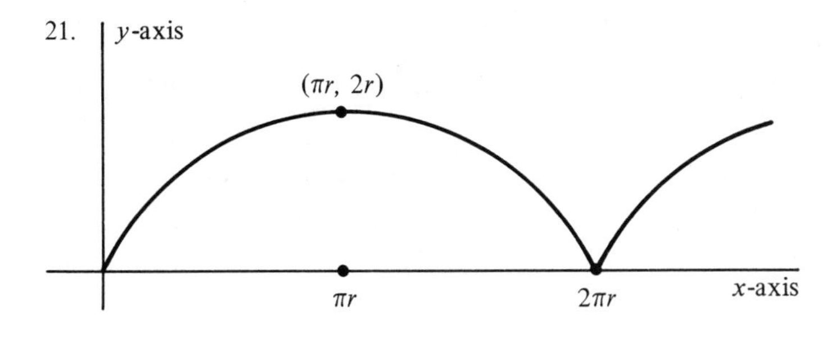

Technique Review Exercises, Chapter 1 (page 49)

1. (a) $x^6 + 3x^3$; (b) 0; (c) $-x^2 - 2x$
2. domain: set of all $x \leq 5$, range: set of all nonnegative real numbers.
3. (a) the distance between x and 5 on the real line; (b) the distance between x and -6 on the real line.

4.

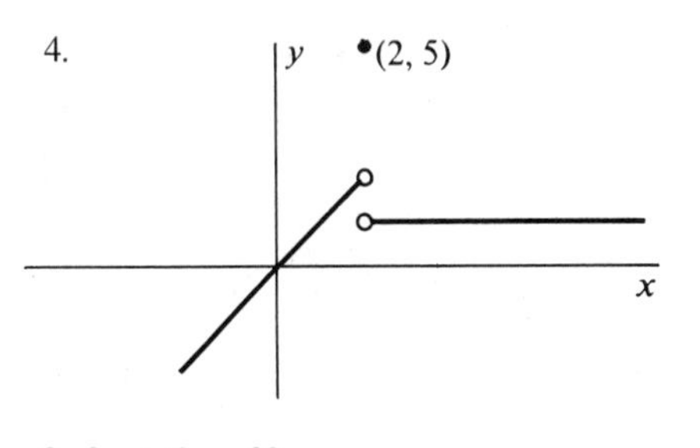

5. $3x - 5y + 21 = 0$
6. $y = -2$
7. $\sqrt{97}$

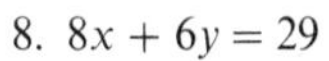

8. $8x + 6y = 29$

9. $(x-5)^2 + (y+1)^2 = 13$

10.

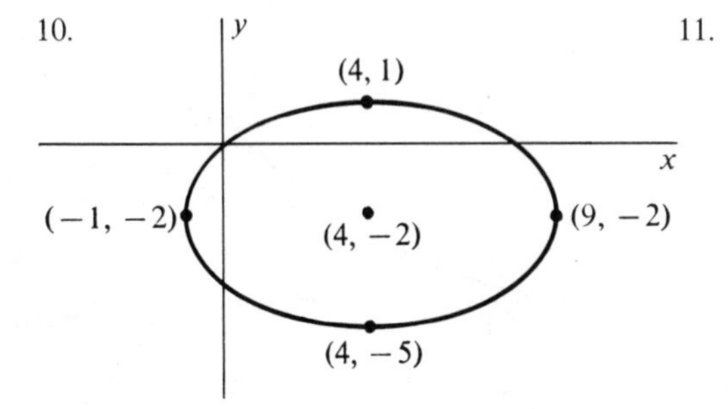

11.

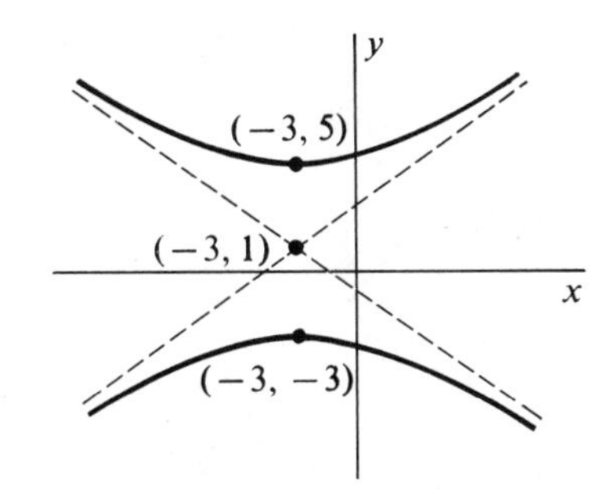

12.

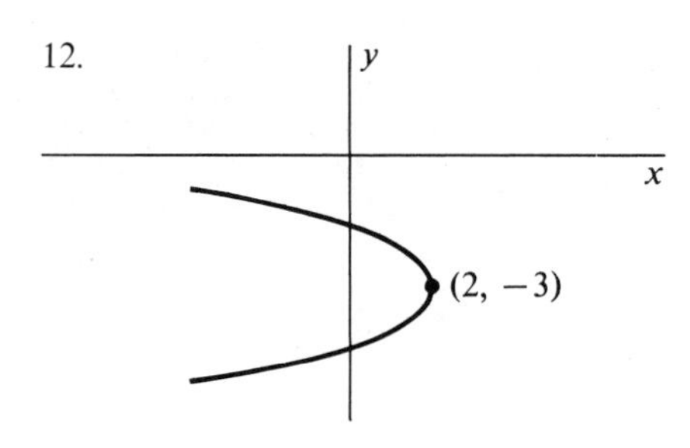

13.

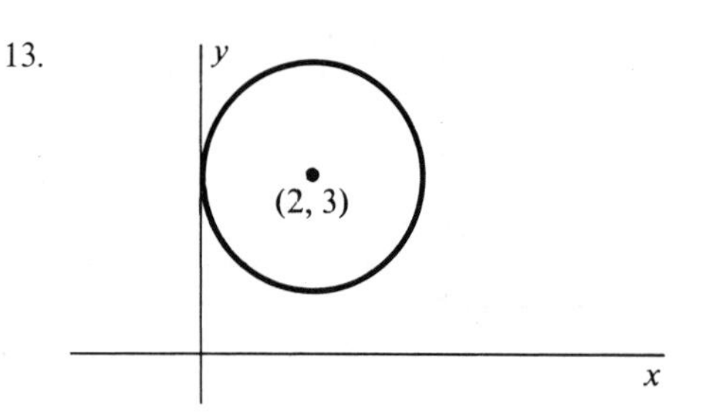

14.

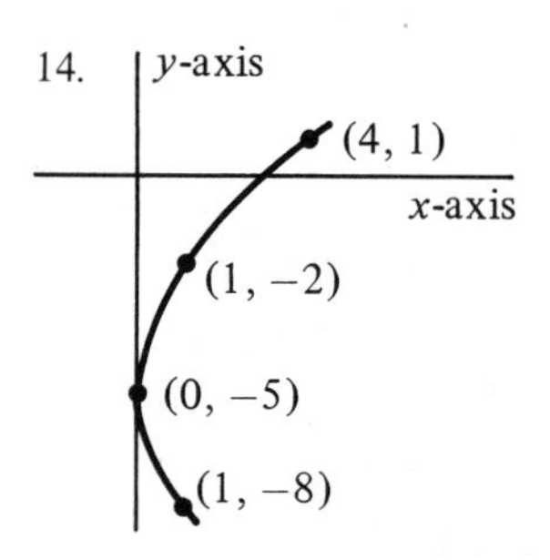

15. $\dfrac{x^2}{4} + \dfrac{y^2}{25} = 1$, ellipse

Additional Exercises, Chapter 1 (page 49)

1. 35
3. $x^4 - 6x^3 + 21x^2 - 36x + 35$
5. $2hx + h^2 - 3h$
7. (a) domain: the set of all real numbers except 2 and -1; (b) range: the set of all real numbers except 0.
9. (a) domain: the set of all real numbers; (b) range: the set of all real numbers ≥ 1.
11. $|x + 7|$
13. The distance between x and -3 is greater than 1.
15. $|x + 10| < 12$
17. $-2/\sqrt{3}$
19. Break at $x = 0$.

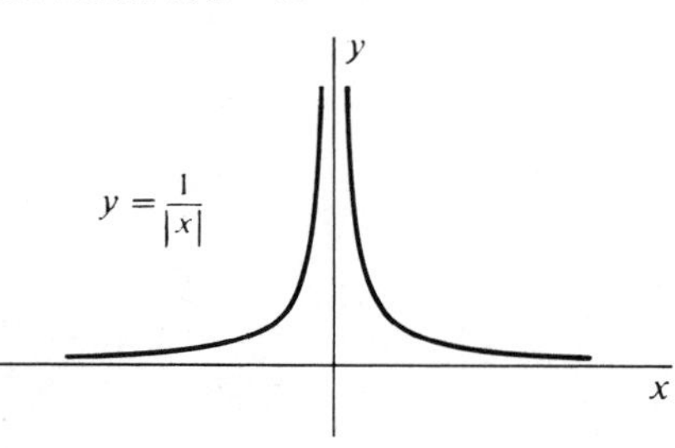

21. Break at $x = 5$.

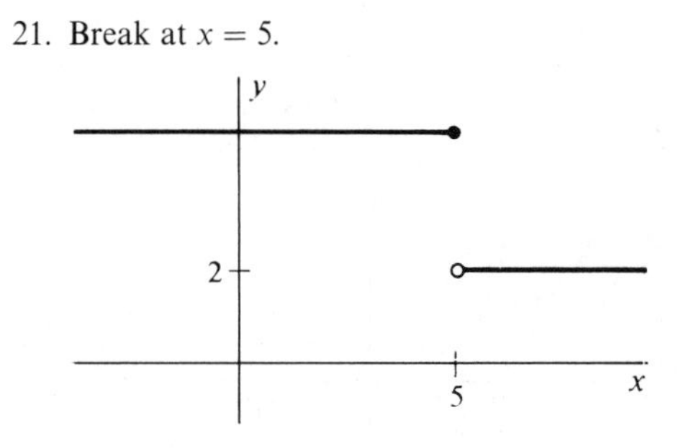

23. Break at $x = -2$.

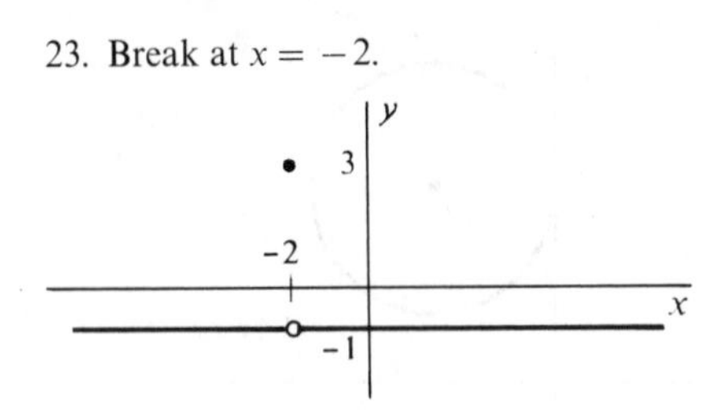

25. Break at $x = 0$.

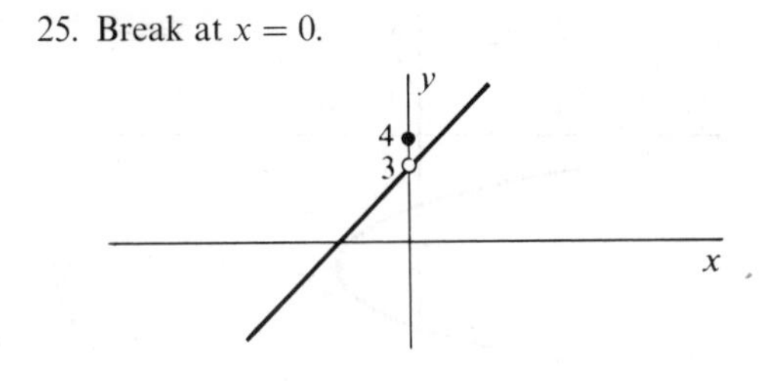

27. $5x + 4y = 17$

29. $7x + 3y = -1$

31. $y = 3$

35. $\sqrt{29}$

37. $(x + 2)^2 + (y - 7)^2 = 109$

39.

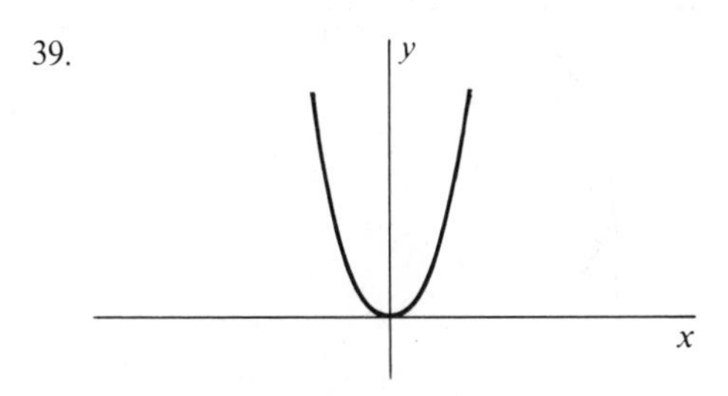

41. $y^2 - 2y - 8x + 33 = 0$

43.

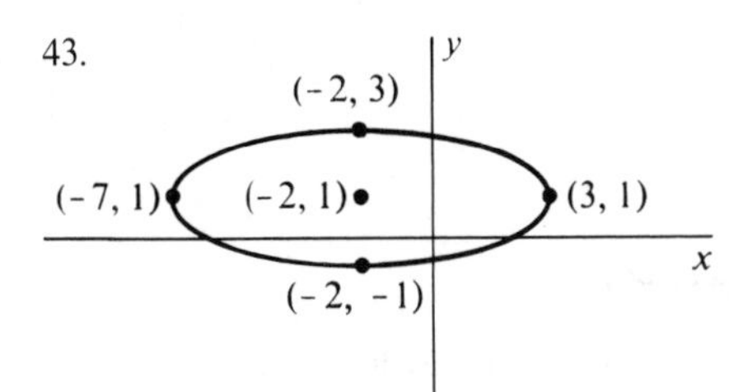

45.

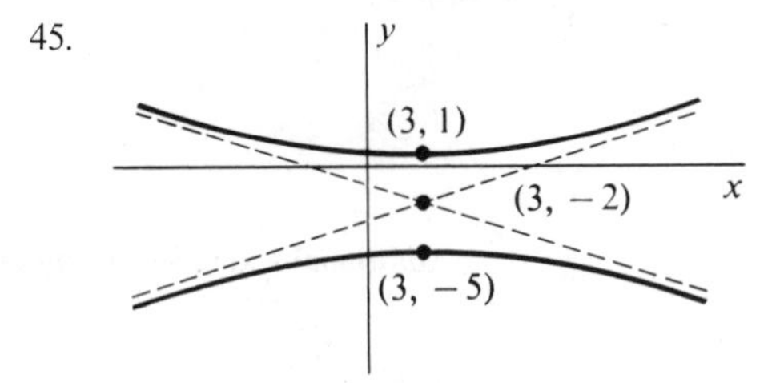

47.

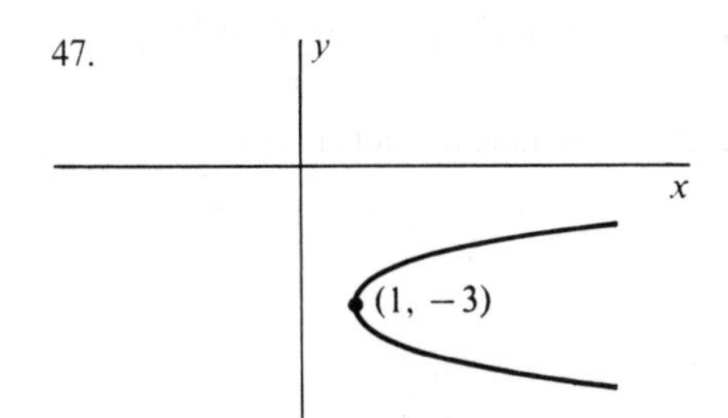

49.

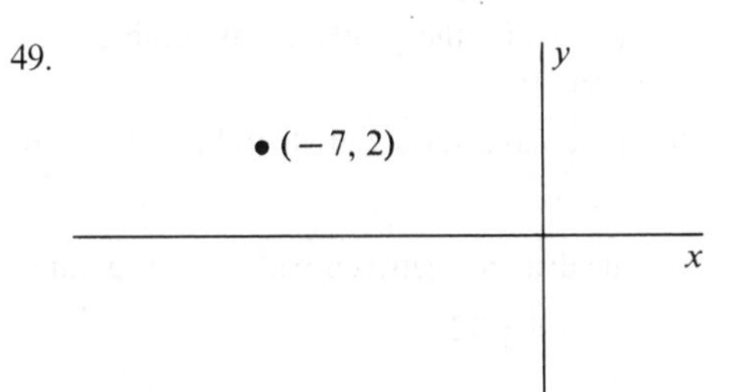

51.

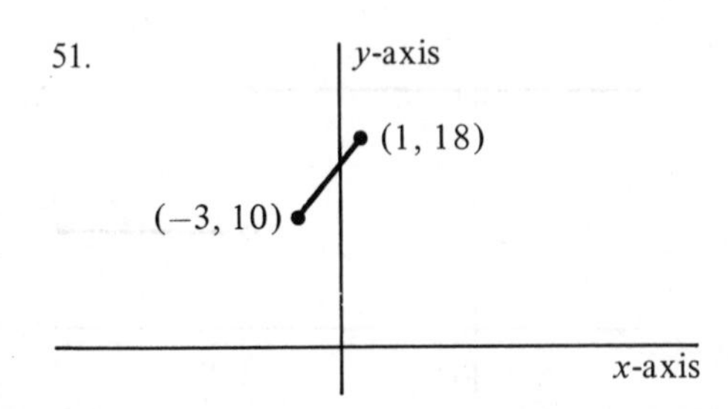

53.

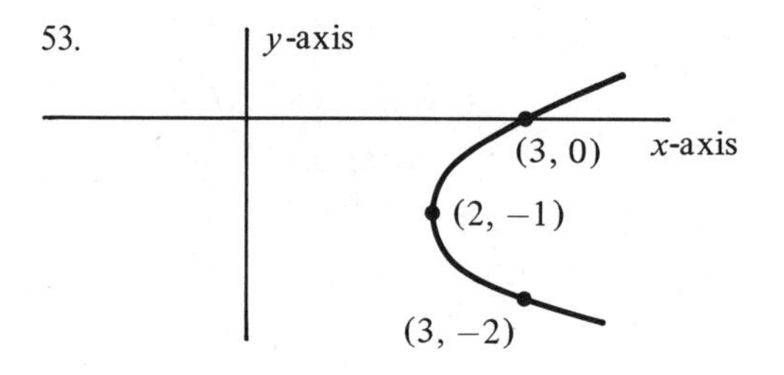

55. $x = 7y^2 + 14y + 10$, parabola

57. $x^2 + \frac{y^2}{4} = 1$, ellipse

59. $(x - 1)^2 + \frac{y^2}{4} = 1$, ellipse

Chapter 2

Section 2.1 (page 58)

1. $y = 6x - 9$
3. $y = 0$
5. $x + y = 4$
7. $y = 5x + 2$
9. $y = x - 1$
11. $y = 2x - 1$
13. $y + 4x = 2$
15. 7
17. 12
19. 3
21. $-1/9$
23. -1
25. 1/2
27. 1
29. 0
31. 0
33. 0
35. $2a + 2$
37. $x + 6y = 25$

Section 2.2 (page 66)

1. 5
3. 4
5. does not exist
7. 1/6
9. $-3/8$
11. 1/5
13. -5
15. 1
17. -2
19. does not exist
21. does not exist
23. 0
25. does not exist
27. 6
29. does not appear to exist
31. 0
33. 2.71828
35. (a) within 1/16 of 3; (b) within $\varepsilon/2$ of 3.
37. (a) within 1/40 of -2; (b) within $\varepsilon/4$ of -2.
39. 1/4

Section 2.3 (page 70)

1. 10
3. -1
5. -5
7. -14
9. $y = x - 9$
11. $y + 4x = 3$
13. $y = 4x - 2$
15. $11x - 2y = 3$

Section 2.4 (page 75)

1. all x
3. all x except 3 and -2
5. all x except 0 and 3
7. all x except -2 and -5
9. all x except -5
11. all x except 0 and 3
13. all x that are not integers and also at $x = 0$
15. $-1/3$
17. 5
19. impossible

Section 2.5 (page 82)

1. $x \geq 3$
3. $-5 \leq x \leq 0$ or $x \geq 2$
5. $x \leq 2$ or $x \geq 3$
7. $-1 \leq x < 0$ or $0 < x \leq 3$ or $x > 5$
9. $-6 < x \leq -2$ or $1 < x \leq 5$
11. $x < -11/2$
13. $-2 < x < -1$
15. $x < 2$ or $x > 3$
17. $x \leq -2$ or $0 \leq x \leq 1$
19. $x < -1$ or $-1 < x \leq 0$ or $3 \leq x < 5$
21. $-6 < x < -3$ or $-1 < x < 1$ or $1 < x < 7$
23. (a) $-1 < x < 3$; (b) $x < -7$ or $-7 < x < -1$ or $x > 3$
25. (a) $x < -6$ or $-2 < x < 0$ or $1 < x < 7$; (b) $-6 < x < -2$ or $0 < x < 1$ or $x > 7$
27. (a) $x < -2$ or $0 < x < 2$; (b) $-2 < x < 0$ or $x > 2$
29. (a) $-3 < x < 0$ or $x > 2$; (b) $x < -3$ or $0 < x < 2$

Section 2.6 (page 90)

1. 1
3. $2x$
5. $-2x$
7. $4x - 4$
9. $3x^2$
11. $-1/x^2$
13. $1/(2\sqrt{x+1})$
15. 0
17. $1 - 2x$
19. $1 + 1/x^2$
21. (a) 4 in/sec, 2 in/sec, -8 in/sec; (b) 5 inches to left of starting point
23. 1/4
25. 1/2

Technique Review Exercises, Chapter 2 (page 91)

1. (a) -3; (b) 1/3
2. (a) 10; (b) 6
3. all x except -2
5.

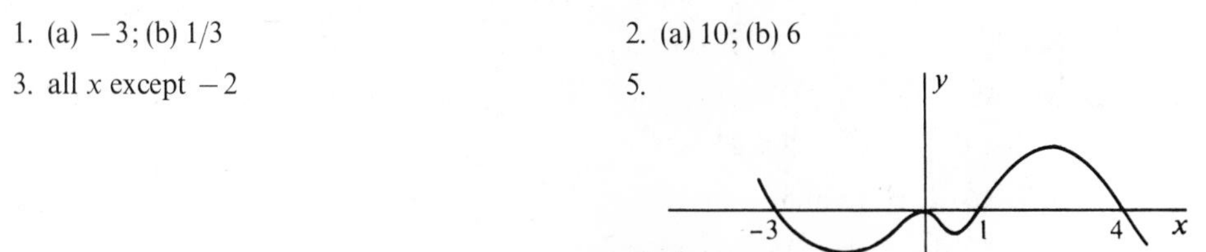

6. $x = -5$ or $-2 < x < 0$ or $1 \le x < 3$
7. $-3 - 2x$
8. $y = 3x + 8$
9. (a) 90 ft/sec; (b) when $t = 50$ sec; (c) 2,500 ft.

Additional Exercises, Chapter 2 (page 92)

1. $4x + y = 7$
3. $7x - y = 9$
5. $y = 3$
7. $y = x + 1$
9. -2
11. 1/8
13. $-1/5$
15. does not exist
17. -2
19. 57
21. -7
23. $2x + y = 4$
25. all x
27. all x except -4 and 2
29. $-1/6$
31. impossible
33. $-4 \le x \le 2$
35. $-5 \le x < -1$ or $1 \le x < 3$
37. $x < -5$ or $x > 2$
39. $-2 \le x < -1$ or $x \ge 5$
41. (a) $-4 < x < 0$ or $x > 2$; (b) $x < -4$ or $0 < x < 2$
43. 8
45. $7 - 4x$
47. 0
49. $-1/3x^2$

Chapter 3

Section 3.1 (page 101)

1. $15x^4 + 6$
3. $-7 + 48x^7$
5. $9x^8 + 12x^3 - 2$
7. $24x^2 + 6x - 7$
9. $2x - 2$
11. $5x^4 + 24x^2 - 4x$
13. $8x^7 + 30x^4 + 18x$
15. $4x^3 - 9x^2$

Section 3.2 (page 106)

1. $30x^4 - 35x^{-6}$
3. $8x^3 + 8x^{-2} + 30x^4$
5. $7x^6 - 5x^4 + 9x^2 - 3$
7. $5x^4 - 15x^2 + 6x$
9. $56x^3 - 12x + 12x^{-2} + 28$
11. $6x^5 + 8x^3 + 2x$
13. $2x - 1$
15. $\dfrac{-8}{(x-3)^2}$
17. $\dfrac{-8x^3}{(2x^4+1)^2}$
19. $\dfrac{3x^2 - 4x - 9}{(x^2+3)^2}$
21. $\dfrac{x^2 - 2x - 5}{(x-1)^2}$
23. $3x^2 + 8x - 11$

Section 3.3 (page 111)

1. $12x^3 - 22x$
3. $[20(x + x^2)^4 - 7 - 2(x + x^2)^{-3}](1 + 2x)$
5. $15(x^{-3} + 1)^{-6}x^{-4}$
7. $6x(x^2 + 4)^5 \dfrac{(x^2+4)^9 - 3(x^2+4)^3 - 6}{((x^2+4)^6 - 1)^2}$

9. $8(2x^3 + 3x + 1)^7(6x^2 + 3)$

11. $-3(4x^{-1} - 2x^{-2})^{-4}(-4x^{-2} + 4x^{-3})$

13. $4(x^3 - 2x + 1)^3(3x^2 - 2) + 12x(x^2 + 5)^5$

15. $\dfrac{-20(x+2)^3}{(x-3)^5}$

17. $\dfrac{6(4x+5)^5}{x^3(x^{-1} + 3x^{-2})^7}(8x^2 + 41x + 30)$

19. $-2(x^2 + 3x)^5(x-1)^{-3} + 5(x^2 + 3x)^4(2x+3)(x-1)^{-2}$

21. $3[(3x^2 + 5x)(x-5)]^2(9x^2 - 20x - 25)$

23. $6x^2(x^2 + 2)^4(x^3 - 1) + 8x(x^2 + 2)^3(x^3 - 1)^2$

25. $[(x^3 + 3)^{10} - 5x^{-2}]^5[120x(x^2 + 3)^9 + 60x^{-3}]$

27. $\dfrac{4}{3t^2 - 7}$

29. $-\dfrac{3t^2 - 2}{2(t-2)^3(t^3 - 2t)^3}$

31. $\dfrac{5(u^3 + 4u)^4(3u^2 + 4)}{4(u - 7u^{-2})^3(1 + 14u^{-3})}$

33. -6

35. $1155/4$

37. $y = 160x + 16$

39. $x + 4y = 8$

41. $0, 2, -2$

43. 15 units/sec

Section 3.4 (page 119)

1. $6x - 10$

3. $6(x^2 + 1)(5x^2 + 1)$

5. $-6(x+1)^{-3}$

7. (a) $x > 2$; (b) $x < 2$; (c) all x; (d) no x

9. (a) $x < -3$; (b) $x > -3$; (c) no x; (d) all x

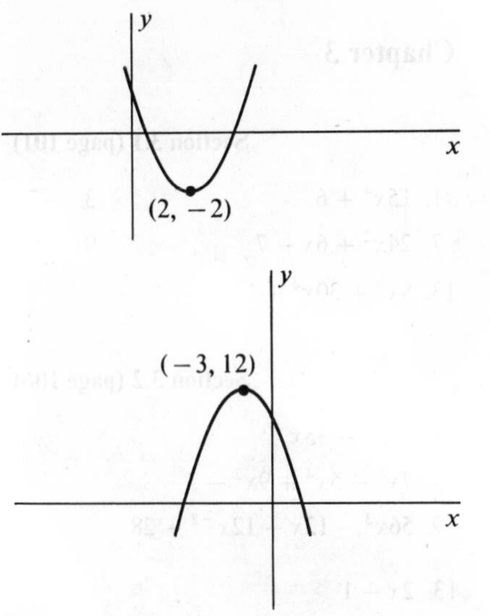

11. (a) $-2 < x < 2$; (b) $x < -2$ or $x > 2$; (c) $x < 0$; (d) $x > 0$

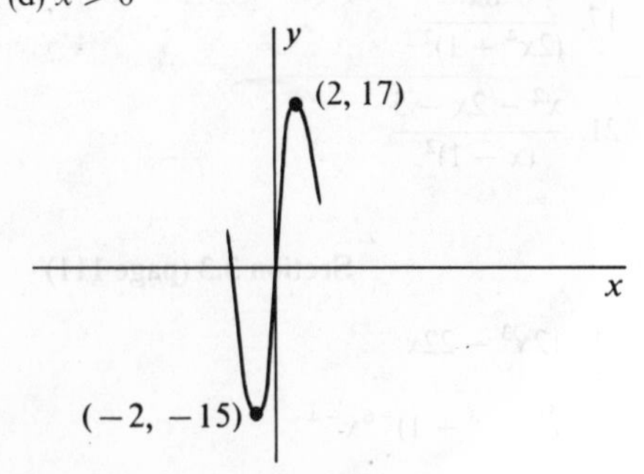

13. (a) $x > 0$; (b) $x < 0$; (c) all $x \neq 0$; (d) no x

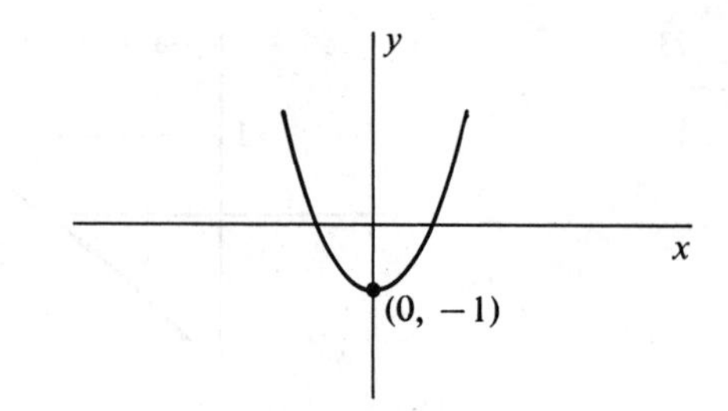

15. (a) $x > 1$; (b) $x < 0$ or $0 < x < 1$; (c) $x < 0$ or $x > 2/3$; (d) $0 < x < 2/3$

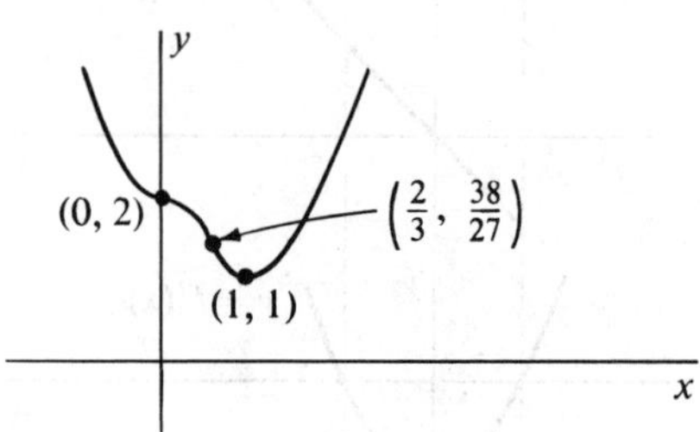

17. (a) $x > 0$; (b) $x < 0$; (c) all x; (d) no x

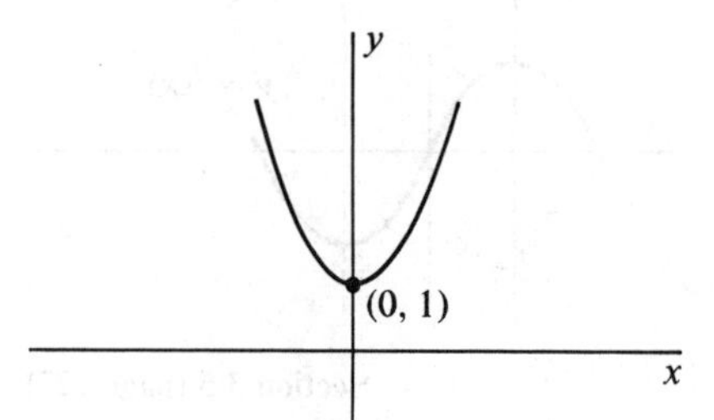

19. (a) $x > 3$; (b) $x < 3$; (c) all $x \neq 3$; (d) no x

21. (a) all x, $x \neq -1$; (b) no x; (c) $x < -1$; (d) $x > -1$

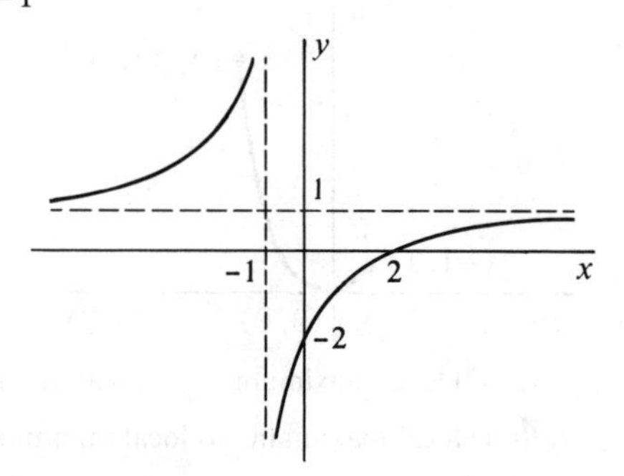

23.

25.

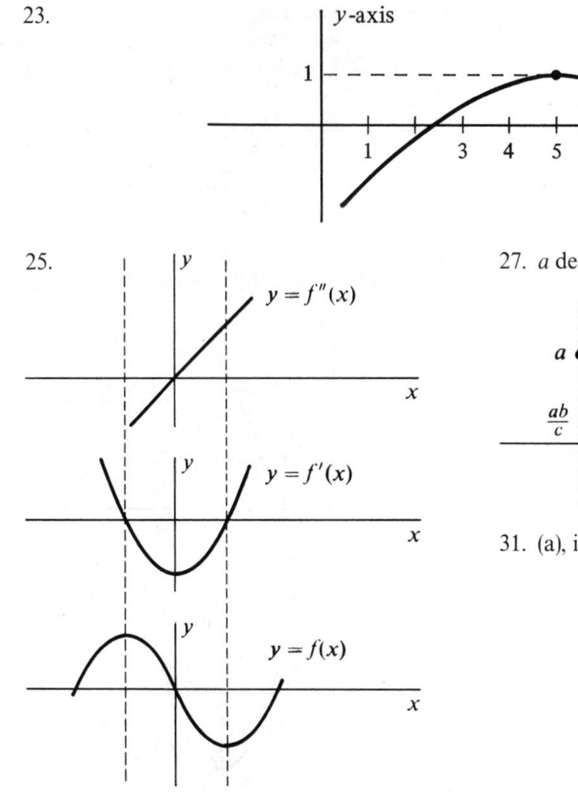

27. a denotes the limit of attainment.

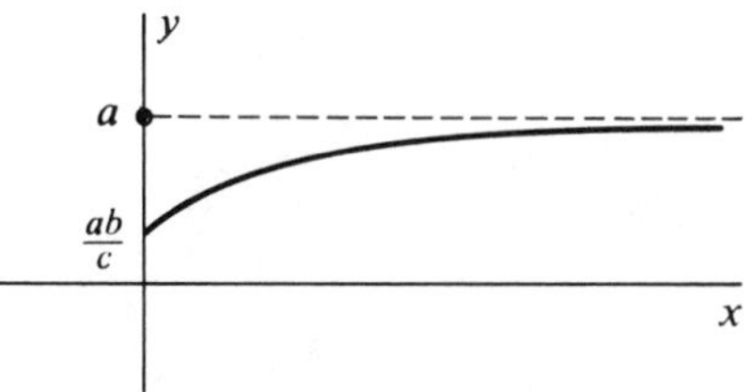

31. (a), iv; (b), vi; (c), ii; (d), v; (e), i; (f), iii.

Section 3.5 (page 127)

1. (2, −11) local minimum, no local maximum
3. (3, −1) local minimum, (5, 7) local maximum
5. (2, −2) local minimum, (0, 2) and (3, −1) local maxima
7. (2, −3) local minimum, (0, 1) local maximum
9. (−3, −62), (0, 1), and (3, −62) local minima, (−1, 2) and (1, 2) local maxima
11. (2, −1/4) local minimum, (1, 0) and (3, −2/9) local maxima
13. no local maximum or minimum
15. (0, 0) local minimum, no local maximum
17.

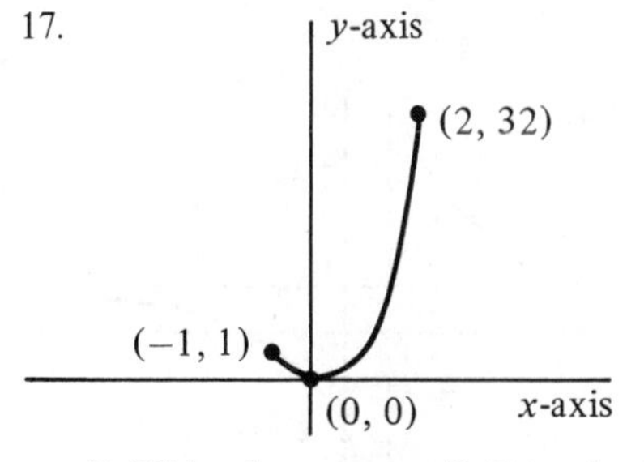

(2, 32) local maximum, (0, 0) local minimum, (−1, 1) local maximum

19. (0, 1) local maximum, no local minimum

Section 3.6 (page 132)

1. $(5, -6)$ absolute maximum, $(3, -10)$ absolute minimum
3. $(0, -1)$ absolute maximum, $(2, -9)$ absolute minimum
5. $(-3, 11)$ absolute maximum, $(0, 2)$ absolute minimum
7. $(4, 208)$ absolute maximum, $(-4, -16)$ absolute minimum
9. $(-1, 14)$ and $(1, 14)$ absolute maxima, $(0, 5)$ absolute minimum
11. no absolute maximum, $(3, 1/9)$ absolute minimum
13. no absolute maximum, $(2, -1/4)$ absolute minimum
15. all points $(k, 7)$ where $0 \le k \le 2$ are absolute maximum and absolute minimum points
17. no absolute maximum or minimum
19. 5
21. $\pm\sqrt{7}$

Section 3.7 (page 136)

1. 225
3. 200 m $\times$ 300 m
5. 18
7. radius $26''/\pi$, length $26''$
9. (a) $x = 1$ or $x = 5$; (b) $x = 18$
11. $20' \times 15'$
13. $\sqrt{a/k}$
15. 25,000
17. $2r/\sqrt{3} \times 2r\sqrt{2}/\sqrt{3}$
19. radius $1/(4\pi)$ km., straightaway 1/4 km.
21. 9
23. 2×3
25. radius $\sqrt[3]{7.2/\pi} \approx 1.32''$, height $\approx 3.95''$
27. $x = 200$ m, $y = 200$ m.

Section 3.8 (page 144)

1. $\dfrac{-3x^2}{5y^4}$
3. $\dfrac{-2xy - 3y^2}{2x^2 + 9xy}$
5. $\dfrac{4x^{3/2}y^{7/2} - y}{x - 6x^{5/2}y^{5/2}}$
7. $\dfrac{2xy}{y^4 + 3x^2}$
9. $\dfrac{5x^{3/5}y^{2/5} - 2y^{2/5}}{3x^{3/5}}$
11. $\dfrac{28x^3(x^4 - y^5)^6 - 15x^2(x^3 + y)^4}{35y^4(x^4 - y^5)^6 + 5(x^3 + y)^4}$
13. $-\dfrac{2xy^4(x^2y^4 - 1)(x^2 + y^2)^4 + 3x}{4x^2y^3(x^2y^4 - 1)(x^2 + y^2)^4 + 3y}$
15. $\dfrac{3(xy + 2)(3x^2y^{-2} + y^2) + 2y(x^3y^{-2} + xy^2)}{6x^3y^{-3}(xy + 2) + 6xy(xy + 2) - 2x(x^3y^{-2} + xy^2)}$
19. $-\dfrac{7}{25y^3}$
21. $\dfrac{3y}{4x^2}$
23. $\dfrac{2y}{x^2}$
25. $16y - 23x + 55 = 0$
27. $y = 0$
29. $(-1, 0)$ and $(1, 0)$ absolute maximum, no absolute minimum
33. $h = 4\sqrt[3]{15}$ ft, $r = \sqrt{2}\sqrt[3]{15}$ ft
35. $x = \dfrac{1}{\sqrt{2}}L$, $y = \dfrac{1}{\sqrt{2}}L$
37. 200 ft along road, then straight to house.

Section 3.9 (page 148)

1. $dy = (14x - 3)dx$
3. $dy = 4x(x^2 - 3)dx$
5. $dy = \dfrac{x^4 + 9x^2 - 6}{(x^2 + 2)^2}dx$
7. $dy = \dfrac{8(x - 2)^2(x + 13)}{(x + 3)^3}dx$
9. $dy = [(7/2)x^{5/2} - x^{-2/3}]dx$
11. $dy = (4/5)(x^2 - 7x^3)^{-1/5}(2x - 21x^2)dx$
13. $-.92$
15. 2.005
17. 4
19. $-.00185$
21. $.048\pi$

Section 3.10 (page 153)

1. $(5x^4 - 3)\cos(x^5 - 3x)$
3. $3x^2 \cos 7x - 7x^3 \sin 7x$
5. $2 \cos 2x \cos 3x - 3 \sin 2x \sin 3x$
7. $15 \sin^2 5x \cos 5x \cos^4 6x - 24 \sin^3 5x \cos^3 6x \sin 6x$
9. $\dfrac{9 \cos 6x \cos 9x + 6 \sin 9x \sin 6x}{\cos^2 6x}$
11. $\sec^2 x$
13. $\sec^2 u \dfrac{du}{dx}$
15. $\sec u \tan u \dfrac{du}{dx}$
17. $-3x^2 \sec^2(5 - x^3)$
19. $(3x^2 - 1)\csc(x - x^3)\cot(x - x^3)$
21. $4 \sec 4x \tan 4x$
23. $10 \sec^2 5x \tan 5x \tan^3 2x + 6 \sec^2 5x \tan^2 2x \sec^2 2x$
25. $\dfrac{-21 \cot^2 7x \csc^2 7x - 10 \cot^3 7x \tan 5x}{\sec^2 5x}$
27. $\dfrac{7 - y \cos xy + \sin x}{x \cos xy + \sec^2 y}$
29. $\dfrac{\sin 2y + 3y \sin 3x - 9}{\cos 3x - 2x \cos 2y}$
31. $\dfrac{-6xy \tan^2(x^2y^2)\sec^2(x^2y^2) + 5y^2 \sec(xy^3)\tan(xy^3)}{6x^2 \tan^2(x^2y^2)\sec^2(x^2y^2) - 15xy \sec(xy^3)\tan(xy^3)}$
33. $12y + 7\pi = (-7\sqrt{3}\pi - 6)\left(x - \dfrac{7\pi}{6}\right)$
35. -1
37. $\cos 2x = \cos^2 x - \sin^2 x$
41. $\dfrac{2}{3\sqrt{3}}$
43. $\dfrac{\pi}{2}$
45. $2\pi\left(1 - \sqrt{\dfrac{2}{3}}\right)$

Section 3.11 (page 158)

1. $3t^2 - 1, 6t$
3. $-t^{-2} + (1/2)t^{-1/2}, 2t^{-3} - (1/4)t^{-3/2}$
5. $3t^2 + 6t, 6t + 6$
7. $160 - 32t$; (a) 160 ft/sec; (b) 5 sec; (c) 500 ft
9. $-1/6$ cm/min
11. 48π in^2/min
13. $2\sqrt{3}/5$ cm^2/hr
15. $-4/3$ ft/min
17. $-24/5$ m/sec
19. $61\pi/16$ ft^3/yr

21. $\dfrac{45}{\sqrt{409}}$ ft/sec ≈ 2.23 ft/sec

23. $(20/3)\pi$ ft^3/yr

25. $r = \dfrac{1}{2\pi}$

Technique Review Exercises, Chapter 3 (page 162)

1. $20x^3 - 6x - 28x^{-8} + 2x^{-3}$

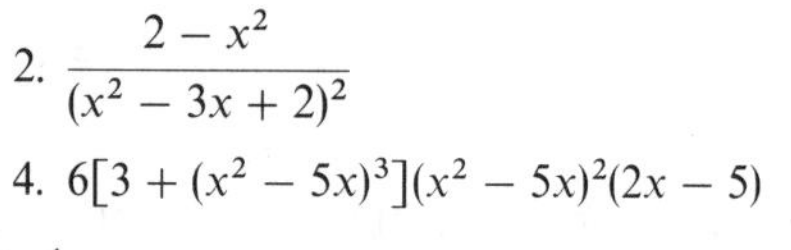

2. $\dfrac{2 - x^2}{(x^2 - 3x + 2)^2}$

3. $x^8(x + 3)^3(13x + 17)$

4. $6[3 + (x^2 - 5x)^3](x^2 - 5x)^2(2x - 5)$

5. increase: $-1 < x < 1$, decrease: $x < -1$ or $x > 1$
   concave up: $x < 0$, concave down: $x > 0$
   critical points: $(-1, 1)$ local minimum,
   $(1, 5)$ local maximum
   inflection point: $(0, 3)$

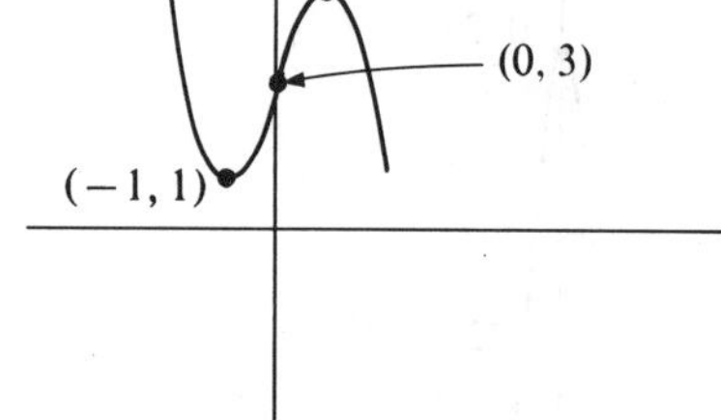

6. local maxima: $(0, 2)$ and $(3, 2)$, local minima: $(-2, -18)$ and $(2, -2)$

7. absolute maximum value: 2185, absolute minimum value: -2

8. 30 yds $\times$ 60 yds

9. $-5/24$ in/min

10. $\dfrac{3x^2 - 4x^3y^2}{2x^4y - 4y^3}$

11. $\pi/20$ m^3

12. (i) $5 \cos x \cos 5x - \sin x \sin 5x$
    (ii) $(3x^2 - 3) \sec^2(x^3 - 3x)$
    (iii) $2 \sin(3x) \cos(3x) \cos^3(7x) - 21 \sin^2(3x) \cos^2(7x) \sin(7x)$

Additional Exercises, Chapter 3 (page 163)

1. $32x^3 - 21x^2$

3. $2x - 5$

5. $6x^5 - 16x^3 + 8x$

7. $-44x^{-5} + 24x^3 + 14x^{-3}$

9. $5x^4 + 20x^3 - 9x^2 - 30x$

11. $\dfrac{8}{(x + 1)^2}$

13. $\dfrac{x^4 + 3x^2 + 4x}{(x^2 + 1)^2}$

15. $(x^2 + 1)(7x^6 + 4x^3 + 3x^2) + 2x(x^3 + 1)(x^4 + 1)$

17. $[18(x^{-1} + x^{-2})^2 + 6(x^{-1} + x^{-2})](-x^{-2} - 2x^{-3})$

19. $4(x^2 - 3x)^3(2x - 3)(2x + 7)^5 + 10(x^2 - 3x)^4(2x + 7)^4$

21. $\dfrac{dy}{dx} = \dfrac{3t^2 - 2}{7t^6 + 3t^{-2}}$

23. $\dfrac{dy}{dx} = \dfrac{3u^2 + 10u}{7u^6 - 9u^{-4}}$

25. $\dfrac{2(2x - 1)(x + 3) - 10(x + 3)^2}{(2x - 1)^6}$

l x; (d) no x

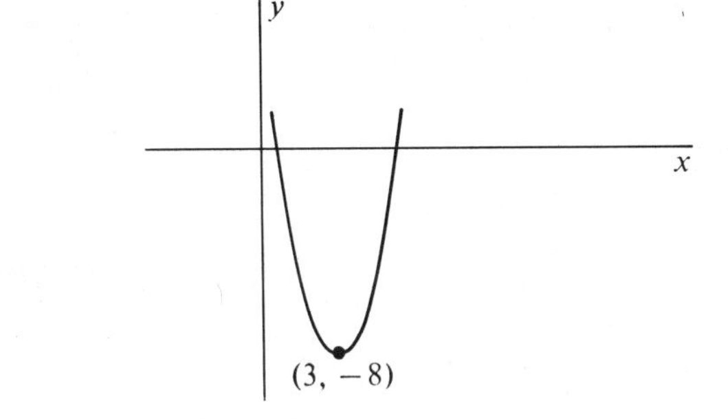

29. (a) $x < 0$ or $x > 4$; (b) $0 < x < 4$; (c) $x > 2$; (d) $x < 2$

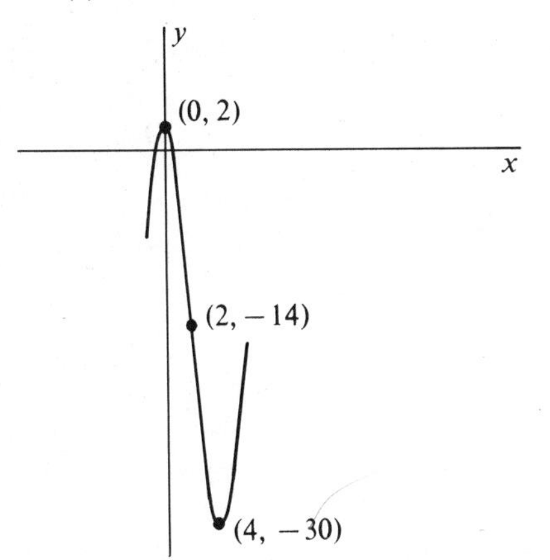

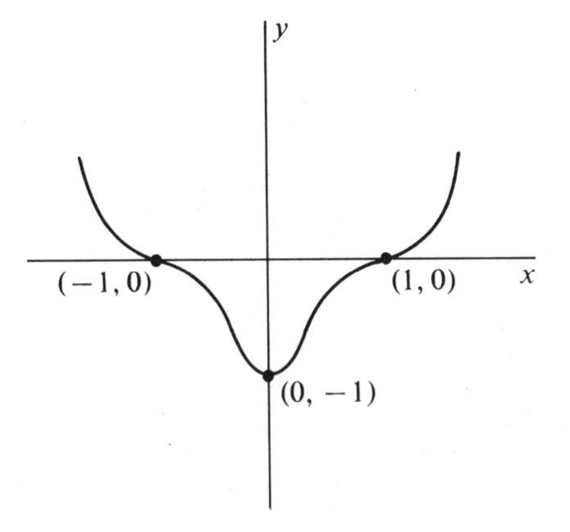

31. (a) $0 < x < 1$ or $x > 1$;
    (b) $x < -1$ or $-1 < x < 0$;
    (c) $x < -1$ or $-1/\sqrt{5} < x < 1/\sqrt{5}$ or $x > 1$;
    (d) $-1 < x < -1/\sqrt{5}$ or $1/\sqrt{5} < x < 1$

33. $(-1, 8)$ local maximum, no local minimum

35. $(5, 38)$ local maximum, $(0, -7)$ local minimum

37. $(-5, -2)$ and $(-1, -10)$ local maximum, $(-2, -11)$ local minimum

39. none

41. $(3, 24)$ absolute maximum, $(0, 3)$ absolute minimum

43. $(1, 8)$ absolute maximum, $(-2, -1)$ absolute minimum

45. none

47. none

49. $100' \times 100'$

51. $r = h/\sqrt{2}, h = r\sqrt{2}$

53. $\dfrac{2y - 3x^2y^3}{4x^3y^2 - 4x}$

55. $\dfrac{6y^{3/5}}{x^{2/5}y^{3/5} + 6x^{2/5}}$

59. $dy = \dfrac{7dx}{(x+4)^2}$

61. $dy = [3(x+2)^2(x^7 - 5x^3)^4 + 4(x+2)^3(x^7 - 5x^3)^3(7x^6 - 15x^2)]dx$

63. 2.0075

65. $2x \sin(3x - 1) + 3x^2 \cos(3x - 1)$

67. $-\sin x \tan x^2 + 2x \cos x \sec^2 x^2$

69. $2 \sin x \cos x \tan(x^2 - 7x) + (2x - 7) \sin^2 x \sec^2(x^2 - 7x)$

71. 96π cm/min

73. -80 ft^2/sec

Chapter 4

Section 4.1 (page 176)

1. $60h^2$
3. $84h^2$
5. $2h^2n(n+1)$
7. $2h^2n(n-1)$
9. $30h^3$
11. 45
13. 0
15. 16
17. 9
19. does not exist

Section 4.2 (page 181)

1. 16 sq units
3. -25 sq units
5. 16 sq units
7. 110 sq units
9. 4 sq units
11. 64 units^2

Section 4.3 (page 188)

1. $3x^2/2 + C$
3. $4x^3/3 + C$
5. $3x^4 + C$
7. $2x + C$
9. C
11. $x^6/6 + C$
13. $x^2/2 + 3x + C$
15. $x^2/2 - x^3/3 + C$
17. 15 sq units
19. 12 sq units
21. 45 sq units
23. 15/4 sq units
25. 1872 sq units
27. -5 sq units
29. 17/2 sq units
31. -15 sq units
33. $s = -16t^2 + 8t + 24$, 3/2 sec, -40 ft/sec
35. $s = -16t^2 - 8t + 24$, 1 sec, -40 ft/sec
37. $\sqrt{6}$ sec, $-32\sqrt{6}$ ft/sec
39. 300 ft

Section 4.4 (page 191)

1. 21/2
3. 196
5. 20
7. 320
9. 15/2
11. 318
13. 4
15. 6
17. -113
19. 0
21. 0
23. 8

Section 4.5 (page 195)

1. $-21/2$
3. 0
5. 0
7. -20
9. 1062
11. 0
13. 29
15. 19/3
17. 19/2
19. 5
21. $8\frac{1}{6}$
23. 11/3
25. 49
31. (a) 0; (b) 0

Section 4.6 (page 200)

1. $21/2$ or $10\frac{1}{2}$ sq units
3. 16 sq units
5. 21 sq units
7. 8 sq units
9. 10 sq units
11. 22 sq units
13. 97 sq units
15. 78 sq units
17. 57 sq units
19. 30 units^2
21. $11/6$ units^2
23. 36 units^2
25. $9/2$ sq units
27. 8 sq units
29. $1/6$ sq units
31. $125/6$ sq units
33. $4/3$ sq units

Section 4.7 (page 207)

1. $5x + C$
3. $t^4/4 + C$
5. $3/8$
7. $2x^{1/2} + C$
9. $5x - x^3/3 + 4x^6/3 + C$
11. $-(1/6)(3 - t^2)^3 + C$
13. $9x - 2x^3 + x^5/5 + C$
15. $-(1/4)(1 - v)^4 + C$
17. $(1/3)(x^2 + 4)^{3/2} + C$
19. $(1/24)(x^4 + 8x^2)^6 + C$
21. $\sqrt{14} - \sqrt{2}$
23. $5/64$
25. $(1/27)(3u + 5)^9 + C$
27. $-(1/3)(9 - x^2)^{3/2} + C$
29. $(1/2)(2 - 5^{1/3})$
31. $h(x)$

Section 4.8 (page 210)

1. 6
3. $19/3$
5. 10
7. 29
9. 28
11. 4
13. $\sqrt{31/3}$
15. $\sqrt[3]{25/4}$
17. $2/\sqrt{3}$
19. 80°C
21. (a) 50; (b) 70; (c) 65; (d) 11,700
23. (a) $5g$ ft/sec; (b) $28g/5$ ft/sec

Section 4.9 (page 212)

1. $-1/2 \cos 2x + C$
3. $\tan t + C$
5. $1/2 \sec 2x + C$
7. $1/2 \tan(x^2) + C$
9. $1/3 \sec v^3 + C$
11. $\dfrac{\sin^4 x}{4} + C$
13. $-1/6 \cos^6 x + C$
15. $-\dfrac{\csc^8 x}{8} + C$
17. $\dfrac{\sec^3 t}{3} + C$
19. $-\dfrac{7 \cos^2 x}{2} + \cos^3 x + C$
21. 1 unit^2
23. 0
25. $1/2 \sin 4\pi^2$ units^2

Technique Review Exercises, Chapter 4 (page 214)

1. 56 sq units
2. -30 sq units
4. $-145/2$
5. (a) $41\frac{1}{4}$ sq units; (b) $43\frac{1}{4}$ sq units

6. $4\frac{1}{2}$ sq units
7. $(1/33)(x^3 - 7)^{11} + C$
8. $(1/16)(2x^4 + 1)^{-2} + C$
9. $(1/5)x^5 + (3/2)x^4 + 3x^3 + C$
10. 21/96800
11. 8/3
12. 26

Additional Exercises, Chapter 4 (page 215)

1. 102
3. -4
5. $9x^2 + C$
7. $2x^5 + C$
9. 147 sq units
11. 36 sq units
13. 480 sq units
15. -4
17. 95
19. 0
21. 8
23. 0
25. -4
27. $49\frac{1}{2}$
29. 244 sq units
31. $8\frac{1}{2}$ sq units
33. 400 sq units
35. 1/4 sq units
37. 12
39. $\frac{x^4}{4} - \frac{x^8}{2} + 2x^3 + C$
41. $(1/3)(x^2 - 1)^{3/2} + C$
43. $(1/27)(x^6 - 3x)^9 + C$
45. 7/9
47. -5
49. $\sqrt[3]{10}$
51. $(1/7)\sin 7x + C$
53. $-(1/8)\cot 8x + C$
55. $-(1/3)\cos(t^3 + 7) + C$
57. $\frac{\tan^6 u}{6} + C$

Chapter 5

Section 5.1 (page 230)

1. $(1/3)(5^{3/2} - 3^{3/2})$ units^3
3. $1\frac{1}{15}$ units^3
5. 2/3 units^3
7. $\pi r^2 h/3$
9. 625π units^3
11. (i) $\frac{768\pi}{7}$ units^3 (ii) $\frac{96\pi}{5}$ units^3 (iii) $\frac{384\pi}{5}$ units^3 (iv) $\frac{1104\pi}{7}$ units^3
13. (i) $\frac{72\pi}{5}$ units^3 (ii) $\frac{16\pi}{3}$ units^3 (iii) $\frac{484\pi}{15}$ units^3 (iv) $\frac{52\pi}{3}$ units^3

Section 5.2 (page 236)

1. 17,250 ft-lbs
3. 88,000 ft-lbs
5. $6666\frac{2}{3}$ ft-lbs
7. 9 kg-cm
9. 40 in-lbs
11. $\frac{kM}{232{,}320{,}000}$ ft-lbs
13. $8000\pi + 8000/3$ ft-lbs
15. 14,400 ft-lbs
17. 357,930 ft-lbs

Section 5.3 (page 241)

1. 42 units
3. $(2/243)[325^{3/2} - 82^{3/2}]$ units
5. $(2/3)(8 - 2\sqrt{2})$ units
6. 66 units

9. $144\frac{5}{6}$ units
11. $7\frac{11}{16}$ units
13. $(8/27)(10^{3/2} - 1)$ units
15. $7\frac{1}{3}$ units
17. $\sqrt{2}(e^{\pi} - 1)$ units
19. 99/2 units
21. 6 units
23. $\frac{1}{3}\int_0^3 \sqrt{\frac{81 - 5x^2}{9 - x^2}}\,dx$

Section 5.5 (page 246)

1. 192
3. 100
5. 1152
7. $469\frac{1}{3}$
9. $106\frac{2}{3}$
11. 9
13. $47\frac{1}{3}$

Section 5.7 (page 253)

1. 36π units^3
3. 36π units^3
5. 72π units^3
7. 90π units^3
9. 54π units^3
11. 108π units^3
13. $8\pi/3$ units^3
15. $6\pi/7$ units^3
17. $7\pi/6$ units^3
19. $63\pi/2$ units^3
21. $207\pi/5$ units^3
23. $\pi \int_0^{1/4} 3\sqrt{1 - 4y}\,dy$

Section 5.8 (page 258)

1. $24\pi\sqrt{5}$ units^2
3. $15\pi\sqrt{5}$ units^2
5. $(\pi/27)[(145)^{3/2} - 10^{3/2}]$ units^2
7. $35\pi/16$ units^2
9. $(\pi/6)(27 - 5^{3/2})$ units^2
11. $13\pi/3$ units^2
13. $(\pi/6)(17^{3/2} - 1)$ units^2
15. $13\sqrt{5}\pi$ units^2
17. $15\sqrt{5}\pi$ units^2

Section 5.9 (page 263)

1. (4/3, 1)
3. (9/20, 9/20)
5. $(3\sqrt{5}/8, 3)$
7. (2/5, 1/2)
9. (3/5, 12/35)
11. $(b/3, a/3)$
13. (0, 5/9)
15. (121/63, 85/63)
17. $\bar{x} = \frac{3}{4}\int_{-1}^{1} x(1 - x^2)dx = 0; \bar{y} = \frac{3}{2}\int_0^1 y\sqrt{1 - y}\,dy$
19. $567\pi/32$ units^3
21. $2\pi/5$ units^3
23. $2\pi^2ar^2$ units^3

Section 5.10 (page 266)

1. $20{,}833\frac{1}{3}$ lbs
3. $5208\frac{1}{3}$ lbs
5. 4000 lbs
7. 5859 lbs
9. $166\frac{2}{3}$ lbs
11. 1029 lbs

Section 5.11 (page 269)

1. 312
3. 2459
5. 22,747
7. 47,147
9. 3797
11. 126

Technique Review Exercises, Chapter 5 (page 272)

1. 16/15 units^3
2. (a) 24π units^3; (b) 24π units^3
3. 380,000 ft-lbs
4. 976/27 units
5. $(2/3)(5^{3/2} - 2^{3/2})$ units
6. 9000
7. $(\pi/3)(2^{3/2} - 1)$ units^2
8. (3/4, 3/10)
9. $266\frac{2}{3}$ lbs
10. 47,224

Additional Exercises, Chapter 5 (page 272)

1. $34\frac{2}{15}$
3. $152\pi\sqrt{3}/5$ units^3
5. $875\pi/3$ units^3
7. 250π units^3
9. 875π units^3
11. 60,600 ft-lbs
13. 16 in-lbs
15. 95, 510 ft-lbs
17. 416/3 units
19. 43/12 units
21. 66 units
23. $3\sqrt{2}$ units
25. $2\pi|a|$ units
27. $468\frac{3}{4}$
29. $20\sqrt{3}$
31. 8π units^3
33. 52π units^3
35. (a) 32π units^3; (b) 60π units^3
37. $153\pi/5$ units^3
39. $45\pi/2$ units^3
41. $(42\pi\sqrt{14} - 120\pi\sqrt{2})$ units^2
43. $13\pi/9$ units^2
45. $(\pi/27)(10^{3/2} - 1)$ units^2
47. (1, 4)
49. (1/2, 5/12)
51. 36π units^3
53. 90,000 lbs
55. 9000 lbs
57. (a) 12,526; (b) 24,829
59. (a) 13,090; (b) 16,830

Chapter 6

Section 6.1 (page 281)

1. 2
3. 1/8
5. 27
7. 4/9

9. 11.

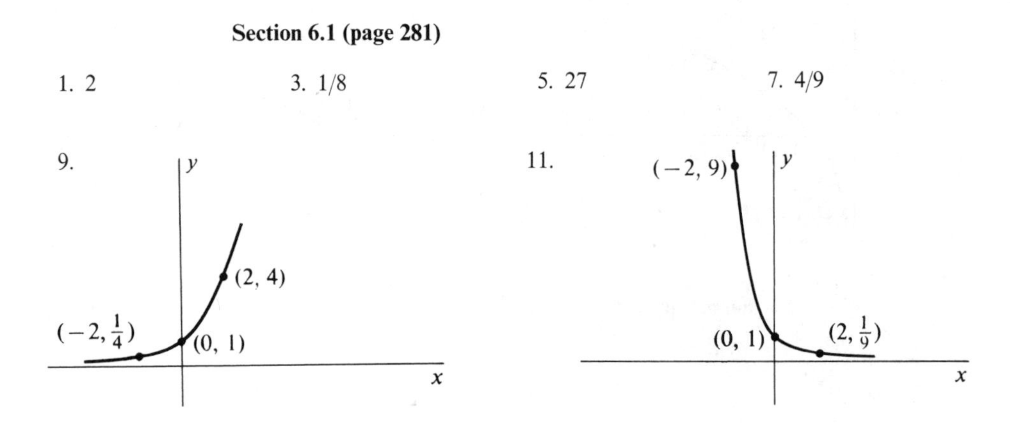

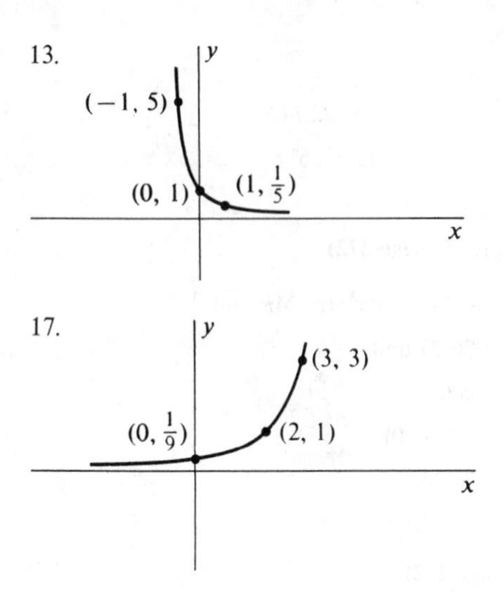

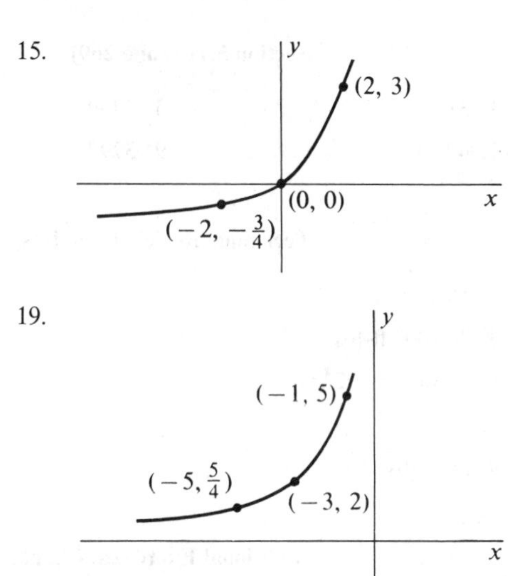

Section 6.2 (page 286)

1. $(y-5)/2,\ 1 \le y \le 19$
3. $-\sqrt{y+4},\ y \ge -4$
5. $\sqrt{y+4},\ y \ge -4$
7. $y^2+5,\ 0 < y < 5$
9. $g(y) = \begin{cases} y/2 & \text{if } 0 \le y < 2 \\ 5-y & \text{if } 2 \le y \le 4 \end{cases}$
11. $|x|,\ x \ge 0;\ |x|,\ x \le 0$
13. $(x-3)^2,\ x \ge 3;\ (x-3)^2,\ x \le 3$
15. $|5-x|,\ x \ge 5;\ |5-x|,\ x \le 5$
17. $6-x^4,\ x \ge 0;\ 6-x^4,\ x \le 0$
19. $x^2-3,\ x \ge 0;\ x^2-3,\ x \le 0$
21. 1/5
23. 1/11
25. 16/61
27. 1/2

Section 6.3 (page 288)

1. -2
3. -3
5. 3
7. 4
9. 5
11. 17

13.

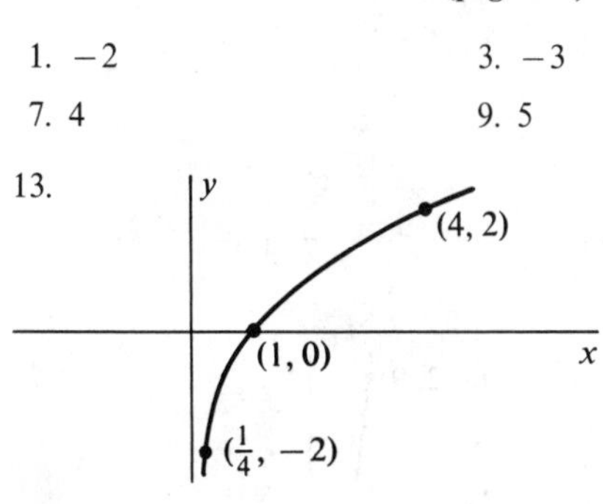

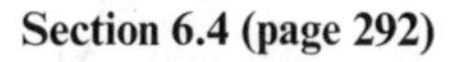

15.

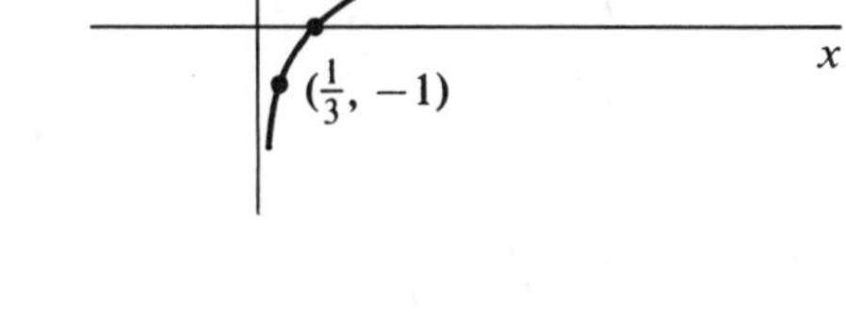

Section 6.4 (page 292)

1. $\dfrac{6x^5+6x}{x^6+3x^2+5}$
3. $\dfrac{x^2}{x-1}\log e + 2x\log(x-1)$

5. $\dfrac{4x}{4x+5}\log_5 e + \log_5(4x+5)$

7. $\dfrac{2x-3}{x^2-3x+2}$

9. $\dfrac{3(3x^2-3)}{x^3-3x+1}$

11. $\dfrac{3x^2}{(x^3+2)\ln|x^3+2|}$

13. $\ln 3$

15. $(1/2)\ln|x^2+3|+C$

17. $(1/2)\ln|x^2-6x+7|+C$

19. $-(1/4)\ln|1-x^4|+C$

21. $\ln|\ln x|+C$

23. 4

25. $(1/2)\ln^2 x + C$

27. local minimum at $(1/e, -1/e)$; no local maximum; no inflection points

29. $(1/4)\ln(73/8)$

31. $1/\sqrt{e}$

Section 6.5 (page 297)

1. $2\sqrt{2}x(x^2+9)^{\sqrt{2}-1}$

3. $2x(3x^4-1)^{1/\sqrt{3}} + \dfrac{12x^5}{\sqrt{3}}(3x^4-1)^{1/\sqrt{3}-1}$

5. $7^x \ln 7$

7. $(10t-3)e^{5t^2-3t+2}$

9. $3x^2e^{x^3+8}$

11. $6x^2(x^3+2)e^{(x^3+2)^2}$

13. $x^x(1+\ln x)$

15. $(x+1)^{2x-1}\left[\dfrac{2x-1}{x+1}+2\ln(x+1)\right]$

17. $(x+5)^4(x-1)^3(x+3)^2\left[\dfrac{4}{x+5}+\dfrac{3}{x-1}+\dfrac{2}{x+3}\right]$

19. $\dfrac{(x^2+5)^3(1-x)^5}{(x^3+1)^2(x+4)}\left[\dfrac{6x}{x^2+5}-\dfrac{5}{1-x}-\dfrac{6x^2}{x^3+1}-\dfrac{1}{x+4}\right]$

21. $\dfrac{(2x+7)^{3/2}(x^2+1)^{3/4}}{(x+2)^2}\left[\dfrac{3}{2x+7}+\dfrac{3x}{2(x^2+1)}-\dfrac{3}{x+2}\right]$

23. $10{,}000\left(\dfrac{6}{5}\right)^{7/2} \approx 18{,}929$

25. $10\left(\dfrac{1}{2}\right)^{.017647} \approx 9.88$ g

27. $\dfrac{\ln(3/73)}{\ln(63/73)} \approx 21.7$ minutes

29. $5600\,\dfrac{\ln 5}{\ln 2} \approx 13{,}000$ years

31. $-\dfrac{1}{2e}$

33. 50,000; 6250; 3125; $\dfrac{dN}{dt} = -\dfrac{(\ln 2)(50{,}000)}{3}$ when $t=0$

$\dfrac{dN}{dt} = -\dfrac{(\ln 2)(25{,}000)}{3}$ when $t=3$

Section 6.6 (page 300)

1. $(1/3)e^{3x}+C$

3. $-e^{-t}+C$

5. 0

7. $(1/3)e^{x^3+3x^2}+C$

9. $-(1/2)e^{x^{-2}}+C$

11. $124/(3\ln 5)$

13. $2(1-e^{-1})$

15. $x-(2/3)e^{-3x}-(1/6)e^{-6x}+C$

17. $-(1/2)e^{-x^2} + C$

19. $\ln|5 - e^{-x}| + C$

21. $2/3$ units^2

23. $\sqrt{\frac{2}{e}}$ units^2

Section 6.7 (page 304)

1. $\frac{x^8 \ln|x|}{8} - \frac{x^8}{64} + C$

3. $x^3e^x - 3x^2e^x + 6xe^x - 6e^x + C$

5. $-\frac{x}{2}e^{-2x} - \frac{1}{4}e^{-2x} + C$

7. $\frac{x^2}{2}\ln^2|x| - \frac{x^2}{2}\ln|x| + \frac{x^2}{4} + C$

9. $-x^{-1}\ln|x| - x^{-1} + C$

11. $\frac{x^2}{3}e^{3x} - \frac{2}{9}xe^{3x} + \frac{2}{27}e^{3x} + C$

13. $x \ln x - x + C$

15. $x^5e^x - 5x^4e^x + 20x^3e^x - 60x^2e^x + 120xe^x - 120e^x + C$

17. $\frac{x^4 - x^3}{2}e^{2x} - \frac{4x^3 - 3x^2}{4}e^{2x} + \frac{6x^2 - 3x}{4}e^{2x} - \frac{12x - 3}{8}e^{2x} + \frac{3}{4}e^{2x} + C$

19. $-\frac{x^7 + 3x^4}{2}e^{-2x} + \frac{7x^6 + 12x^3}{4}e^{-2x} - \frac{21x^5 + 18x^2}{4}e^{-2x} + \frac{105x^4 + 36x}{8}e^{-2x} -$
$\frac{105x^3 + 9}{4}e^{-2x} + \frac{315x^2}{8}e^{-2x} - \frac{315x}{8}e^{-2x} + \frac{315}{16}e^{-2x} + C$

21. $\pi(e^2 - 3e^{-2} + 2)$ units^2

Section 6.8 (page 308)

11. $(2x + 5)\cosh(x^2 + 5x)$

13. $9\cosh^2(3x + 2)\sinh(3x + 2)$

15. $-(8x + 7)\operatorname{csch}^2(4x^2 + 7x)$

17. $-5(2x - 1)^{-1/2}\operatorname{sech}^5\sqrt{2x - 1}\tanh\sqrt{2x - 1}$

19. $3e^{2x}\sinh(3x) + 2e^{2x}\cosh(3x)$

21. $-\cosh x \operatorname{csch}^2 x + \sinh x \coth x$

23. $(1/2)\cosh(2x + 4) + C$

25. $\sinh(\ln x) + C$

27. $(1/2)\tanh^2 x + C$

29. $(1/2)(\cosh 1 - 1)$

31. $(1/4)\sinh^4 x + C$

33. $x^2\sinh x - 2x\cosh x + 2\sinh x + C$

35. $\sinh(e^x) + C$

37. $\frac{1}{3}\left[\frac{1}{\sinh^3 1} - \frac{1}{\sinh^3 2}\right]$

39. $\ln|\sinh x| + C$

41. $\frac{b}{b^2 - a^2}\sinh ax \sinh bx - \frac{a}{b^2 - a^2}\cosh ax \cosh bx + C$

Technique Review Exercises, Chapter 6 (page 312)

1. $-\sqrt{y - 3}, 4 \le y \le 52$

2. (a) $(6x - 2)e^{(3x^2 - 2x + 1)}$; (b) $(-3x)6^{1-3x}\ln 6$

3. $x^{3x-1}\left[\frac{3x - 1}{x} + 3\ln x\right]$

4. $\frac{15x^2}{x^3 - 2}$

5. $-(1/2)e^{2/x} + C$

6. $\frac{x^4}{2}\ln x - \frac{x^4}{8} + C$

7. $\frac{x^2}{3}e^{3x} - \frac{2x}{9}e^{3x} + \frac{2}{27}e^{3x} + C$

8. $(1/6)\ln|3x^2 + 5| + C$

9. $6x\sinh^2(x^2 + 5)\cosh(x^2 + 5)$

10. $(1/28)\sinh^4(7x) + C$

Additional Exercises, Chapter 6 (page 312)

1. 16

3. 4

5.

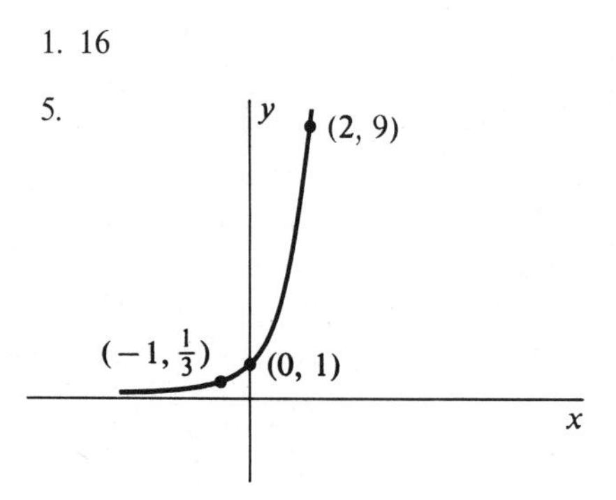

7.

$(-2, 9)$ y (0, 1) $(1, \frac{1}{3})$ x

9. $\frac{y + 1}{6}$, $-7 \le y \le 17$

11. $\sqrt{y - 5}$, $y \ge 5$

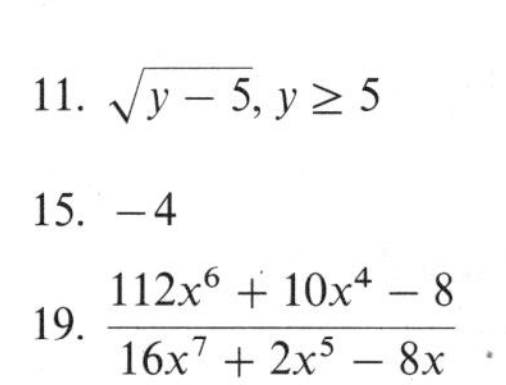

13. -4

15. -4

17.

y $(\frac{1}{5}, 1)$ (1, 0) x $(5, -1)$

19. $\frac{112x^6 + 10x^4 - 8}{16x^7 + 2x^5 - 8x}$

21. $\frac{\log_3 e}{x - 7}$

23. $\frac{4(-1 - 3x^2)}{2 - x - x^3}$

25. $-\ln 4$

27. $\frac{-1}{\ln x} + C$

29. $16^x \ln 16$

31. $9x^{9x}(1 + \ln x)$

33. $\frac{(x + 17)^4}{(x - 6)^5(x - 8)^9}\left[\frac{4}{x + 17} - \frac{5}{x - 6} - \frac{9}{x - 8}\right]$

35. $(1/15)e^{15x} + C$

37. $(1/4)e^{x^4 + 5} + C$

39. $-(1/3)e^{1/x^3} + C$

41. $\frac{1 - e^{-1}}{3}$

43. $\frac{xe^{4x}}{4} - \frac{e^{4x}}{16} + C$

45. $\frac{x^{11}}{11}\ln|x| - \frac{x^{11}}{121} + C$

47. $\frac{x^5}{5}\ln|x^3| - \frac{3}{25}x^5 + C$

49. $\frac{x^3}{3}e^{3x} - \frac{x^2}{3}e^{3x} + \frac{2}{9}xe^{3x} - \frac{2}{27}e^{3x} + C$

51. $4\sinh(4x - 3)$

53. $5(2x-1)\sinh^4(x^2-x)\cosh(x^2-x)$
55. $-8x\tanh^3(2-x^2)\,\text{sech}^2(2-x^2)$
57. $-(1/2)\cosh(5-x^2)+C$
59. $(1/2)\,\text{sech}(3-2x)+C$

Chapter 7

Section 7.1 (page 320)

1. 1
3. 9/2
5. 4
7. -7
9. 3/5
11. 0
13. does not exist
15. $7\cos 7x$
17. $12\sin^3 3x\cos 3x$
19. $(2x+1)\cos(x^2+x-1)$
21. $\dfrac{-\cos x}{\sin^2 x}$
23. $e^{\sin x}\cos x$
25. $x^{\sin x}\left(\dfrac{\sin x}{x}+\cos x\ln x\right)$
27. $\theta=\pi/4$
29. $\dfrac{2}{\sqrt{1+\pi^2}}\approx .61$ ft
31. $\pi/2$

Section 7.2 (page 323)

1. $2x\sin(3-x^2)$
3. $6\cos^2(5-2x)\sin(5-2x)$
5. $12x\sec^2(3x^2-1)\tan(3x^2-1)$
7. $(2x-1)^{-1/2}\sec^2\sqrt{2x-1}$
9. $-2(e^x+1)\csc^2(e^x+x)\cot(e^x+x)$
11. $3e^{2x}\sec 3x\tan 3x+2e^{2x}\sec 3x$
13. $-6x\cot^2(x^2-1)\csc^2(x^2-1)$
15. $15\sec^4 3x\tan^4 3x+6\sec^2 3x\tan^6 3x$
17. $-3\csc 3x\cot 3x\,e^{\csc 3x}\,dx$
19. $5\tan 5x\,dx$
21. $-5\sin 5x\,dx$
23. $(2\sec^3 2x+2\sec 2x\tan^2 2x)dx$
29. 30 mph
31. 19.16 ft
33. the value of θ, $0\le\theta<\pi/2$ for which $\tan\theta=\mu$
35. $h=\sqrt{2}$, $r=\sqrt{2}/2$
37. $\cos 2x=\cos^2 x-\sin^2 x$
39. $x=k\pi$, $k=0,\pm1,\pm2,\ldots.$

Section 7.3 (page 328)

1. $-(1/5)\cos 5x+C$
3. $(1/2)\ln|\sec x^2|+C$
5. $-\cos(\ln x)+C$
7. $-(1/3)\cos^3 x+C$
9. $\tan x+C$
11. $-x\cos x+\sin x+C$
13. $-(1/5)\cot(5x-11)+C$
15. $(1/5)\ln|\sec 5x+\tan 5x|+C$
17. $(1/5)\ln 2$
19. $(1/12)\tan^6 2x+C$
21. $(\sqrt{2}-1)/3$
23. $(1/5)\sec^5 x+C$
25. $-e^{\cos x}+C$
27. $-\dfrac{x}{2}\cos 2x+\dfrac{1}{4}\sin 2x+C$
29. $-x^2\cos x+2x\sin x+2\cos x+C$
31. $-(1/3)\cos^3(\ln x)+C$

33. $4/3$ units^2
35. $2\pi^2$ units^3
37. $1/4$ ft-lbs

Section 7.4 (page 333)

1. $\pi/3$
3. $\pi/6$
5. $-\pi/3$
7. $\pi/3$
9. $3\pi/4$
11. $\pi/6$
13. 5
15. 0
17. $5\pi/6$
19. $2\pi/3$
21. $\sqrt{3}/2$
23. $2\sqrt{6}$
25. $-2\sqrt{6}$
27. $\sqrt{101}$
29. $\pi/8$
31. $(1/5)\arcsin(y/3),\ -3 \le y \le 3$
33. $(1/2)\sin(y/3),\ -3\pi/2 \le y \le 3\pi/2$
35. $(1/3)[2 - \operatorname{arcsec}(y/5)],\ y \le -5$ or $y \ge 5$
37. $(1/2)\arcsin 2y,\ -1/2 \le y \le 1/2$

Section 7.5 (page 336)

1. $\dfrac{5}{\sqrt{1-25x^2}}$
3. $\dfrac{-2}{x\sqrt{x^{-4}-1}}$
5. $\dfrac{e^x}{e^{2x}-2e^x+2}$
7. $\dfrac{-1}{|x|\sqrt{x^2-1}\operatorname{arccsc} x}$
9. $\dfrac{3e^{\arctan 3x}}{9x^2+1}$
11. $\dfrac{3x^2}{\sqrt{1-x^6}}$
13. $\dfrac{1}{\sqrt{e^{2x}-1}}$
15. 0; $\arcsin x + \arccos x$ is a constant.
17. $-1/2$; $\operatorname{arccot}(\sec x + \tan x) = -x/2 +$ a constant
19. $x\arctan x - (1/2)\ln(x^2+1) + C$
21. $x\arccos x - (1-x^2)^{1/2} + C$
23. $10\sqrt{3}$ ft

Section 7.6 (page 338)

1. $\pi/(3\sqrt{3})$
3. $(1/6)\arctan(3x/2) + C$
5. $(1/2)\operatorname{arcsec}(e^x/2) + C$
7. $(1/4)\operatorname{arcsec}(3x/4) + C$
9. $\arcsin(x/5) + C$
11. $(1/6)\operatorname{arcsec}(x^2/3) + C$
13. $(1/3)\arcsin((3\ln x)/2)$
15. $-(16-x^2)^{1/2} + C$
17. $(1/4)\arctan(x^2/2) + C$
19. $2\arctan\sqrt{x} + C$
21. $\arctan(\ln x) + C$
23. $\pi/2$ units^2
25. $\pi^2/4$ units^3

Technique Review Exercises, Chapter 7 (page 340)

1. $3\sec(3x-1)\tan(3x-1)$
2. $4x\cot(1-x^2)\csc^2(1-x^2)$
3. $-6(x^3-2x)(3x^2-2)\cos^2(x^3-2x)^2\sin(x^3-2x)^2$
4. $-(1/2)\cos(x^2+3) + C$
5. $(1/2)\ln|\sec(2x-3)| + C$
6. $(1/2)\ln|\csc(1-2x)+\cot(1-2x)| + C$
7. $(1/2)\tan(x^2-2) + C$
8. $(1/5)\sec 5x + C$
9. (*a*) $\pi/3$; (b) $-\pi/4$
10. $\dfrac{2x}{\sqrt{10x^2-x^4-24}}$
11. $\dfrac{6\arctan(3x)}{1+9x^2}$

12. $\dfrac{2x}{(x^2+1)\sqrt{x^4+2x^2}}$

13. $\arcsin(x/4)+C$

14. $(1/3)\arctan(x^3)+C$

15. $(1/2)\operatorname{arcsec}(x^2)+C$

Additional Exercises, Chapter 7 (page 341)

1. $1/2$
3. does not exist
5. 1
7. $12\cos 12x$
9. $28\sin^6 4x\cos 4x$
11. $3(\ln^2(8x))/x$
13. $-3x^2\sin(x^3-2)$
15. $4x\tan(x^2-1)\sec^2(x^2-1)$
17. $-3e^{2x}\csc 3x\cot 3x+2e^{2x}\csc 3x$
19. $10x\csc^5(3-x^2)\cot(3-x^2)$
21. $-(1/15)\cos 15x+C$
23. $(1/3)\ln(\cos(2-x^3))+C$
25. $(1/3)\sec x^3+C$
27. $-(1/3)\ln|\csc x^3+\cot x^3|+C$
29. $-\dfrac{x}{3}\cos 3x+\dfrac{1}{9}\sin 3x+C$
31. $-\pi/6$
33. 0
35. $(1/4)\arccos(y/2),\ -2\le y\le 2$
37. $\dfrac{10}{\sqrt{1-100x^2}}$
39. $\dfrac{2}{\sqrt{e^{4x}-1}}$
41. $\dfrac{2x}{\sqrt{1-x^4}\arcsin(x^2)}$
43. $\arcsin(x/7)+C$
45. $(1/3)\operatorname{arcsec}(2x)+C$
47. $\operatorname{arcsec} e^x+C$
49. $(1/7)\arctan(e^{2x}/7)+C$

Chapter 8

Section 8.1 (page 352)

1. $2[(1/5)(x+3)^{5/2}-(2/3)(x+3)^{3/2}]+C$
3. $3[(1/11)(x+5)^{11/3}-(5/4)(x+5)^{8/3}+(22/5)(x+5)^{5/3}]+C$
5. $(1/4)[(1/3)(4x-1)^{3/2}+7(4x-1)^{1/2}]+C$
7. $2(\sqrt{3}-1/\sqrt{3}-\sqrt{2}+1/\sqrt{2})$
9. $(1/5)(2e^{2x}\sin x-e^{2x}\cos x)+C$
11. $(2e^{\pi}+1)/5$
13. $(1/4)e^{2x}(\sin 2x-\cos 2x)+C$
15. $(3/10)[e^{3x}\cos x+(1/3)e^{3x}\sin x]+C$
17. 0
19. $(5/21)\cos 2x\sin 5x-(2/21)\sin 2x\cos 5x+C$
21. $\dfrac{x}{2}(\sin(\ln x)-\cos(\ln x))+C$
23. $-x^2\cos x+2x\sin x+2\cos x+C$
25. $\pi-2$
27. $\dfrac{be^{ax}}{a^2+b^2}\left(\dfrac{a}{b}\sin bx-\cos bx\right)+C$
29. $(1/2)(\sec x\tan x+\ln|\sec x+\tan x|)+C$
31. $-\dfrac{1}{4x}\sqrt{4-x^2}+C$
33. $\ln|x+2|+4(x+1)^{-1}-2(x+2)^{-2}+C$
35. $2/5$ square units
37. $2\pi^2$ cubic units

Section 8.2 (page 356)

1. $(1/5)\sin^5 x - (1/7)\sin^7 x + C$
3. $-(1/3)\cos^3 x + (1/5)\cos^5 x + C$
5. $1/3$
7. $(1/2)[(1/3)\sec^3 2x - \sec 2x] + C$
9. $(1/4)\sec^4 x - \sec^2 x + \ln|\sec x| + C$
11. $\tan x + (2/3)\tan^3 x + (1/5)\tan^5 x + C$
13. $(2/3)\tan^{3/2} x + (2/7)\tan^{7/2} x + C$
15. 0
17. $(1/5)[-\sin^{-1} 5x - 3\sin 5x + \sin^3 5x - (1/5)\sin^5 5x] + C$
19. $-(1/5)[(1/3)\cot^3 5x + (1/5)\cot^5 5x] + C$
21. $(1/4)\cot^{-4} x + C$
23. $\pi/4$
25. $(x/16) - (1/64)\sin 4x + (1/48)\sin^3 2x + C$
27. $5\pi/128$
29. $\pi^2/4$ cubic units

Section 8.3 (page 360)

1. $\ln\left|\frac{1}{2}\sqrt{4+x^2} + \frac{x}{2}\right| + C$
3. $\frac{9}{2}\left[\arcsin\frac{x}{3} + \frac{1}{9}x\sqrt{9-x^2}\right] + C$
5. $\frac{1}{2\sqrt{2}} - \frac{1}{3}$
7. $-(9/4)\sqrt{9-2x^2} + (1/12)(9-2x^2)^{3/2} + C$
9. $2\sqrt{5}/45$
11. $-\frac{x}{16\sqrt{x^2-16}} + C$
13. $\frac{7}{8}\arcsin\frac{4x}{\sqrt{7}} + \frac{x}{2}\sqrt{7-16x^2} + C$
15. $\ln\left|\frac{\sqrt{41}}{4} + \frac{5}{4}\right| - \ln|\sqrt{2}+1|$
17. $-(16/3)(-x^2-14x-45)^{3/2} + (8/5)(-x^2-14x-45)^{5/2} - (1/7)(-x^2-14x-45)^{7/2} + C$
19. $(1/5)(x^2-6x+13)^{5/2} - (4/3)(x^2-6x+13)^{3/2} + C$
21. $\ln|x + \sqrt{x^2-16}| + C$
23. $(1/2)\ln|x^2 + \sqrt{x^4-25}| + C$
25. 2π square units
29. $\frac{1}{2}\ln\left|\tan\left(\frac{x}{2}\right)\right| - \frac{1}{4}\tan^2\left(\frac{x}{2}\right) + C$
31. $\tan\frac{x}{2} + C$

Section 8.4 (page 364)

1. $\frac{1}{2}\arctan\frac{x}{2} + C$
3. $\frac{1}{6}\ln\left|\frac{x-3}{x+3}\right| + C$
5. $-\frac{1}{4\sqrt{3}}\ln\left|\frac{x-2\sqrt{3}}{x+2\sqrt{3}}\right| + C$
7. $\frac{1}{\sqrt{19}}\arctan\frac{x}{\sqrt{19}} + C$
9. $\frac{3}{2}\ln|x^2+16| - \frac{7}{4}\arctan\frac{x}{4} + C$
11. $-2\sqrt{4-(x-3)^2} - 5\arcsin\left(\frac{x+3}{2}\right) + C$
13. $\frac{1}{2}\arctan\frac{x+2}{2} + C$
15. $-\frac{1}{8}\ln\left|\frac{x-1}{x+7}\right| + C$
17. $\frac{10}{\sqrt{27}}\arctan\frac{2x-3}{\sqrt{27}} + C$
19. $3\ln|x+2+\sqrt{(x+2)^2-4}| + C$

21. $5 \arcsin\left(\frac{x-3}{5}\right) + C$

23. $\ln\sqrt{10} - \ln\sqrt{13} + (2/3)[\arctan(2/3) - \arctan(1/3)]$

25. $-\frac{1}{2}\ln|20 - x^2 + 4x| + \frac{1}{4\sqrt{6}}\ln\left|\frac{x-2-2\sqrt{6}}{x-2+2\sqrt{6}}\right| + C$

27. $\ln\frac{7}{9} + \frac{4}{3\sqrt{3}}(\arctan(1/\sqrt{3}) + \arctan(1/3\sqrt{3}))$

29. $-(1/2)\ln|7 - x^2 - 6x| + C$

31. $(3/4)\ln|3 + 4x^2 - x| - \frac{17\sqrt{47}}{94}\arctan\frac{\sqrt{47}}{47}(8x-1) + C$

33. $-(5 - x^2 - 4x)^{1/2} + \arcsin\frac{x+2}{3} + C$

35. $5(x^2 - 4x - 5)^{1/2} + (16/5)\ln|x - 2 + \sqrt{x^2 - 4x - 5}| + C$

37. $\pi\left(\ln\frac{13}{8}\right) + \frac{\pi^2}{2} - 2\pi\arctan\frac{3}{2}$ cubic units

Section 8.5 (page 367)

1. $\frac{1}{\sqrt{45}}\ln\left|\frac{-2\tan(x/2) + 7 - \sqrt{45}}{-2\tan(x/2) + 7 + \sqrt{45}}\right| + C$

3. $\frac{2}{\sqrt{33}}(\arctan(3/\sqrt{33}) + \arctan(4/\sqrt{33}))$

5. $\frac{x}{6(x^2+3)} + \frac{1}{6\sqrt{3}}\arctan(x/\sqrt{3}) + C$

7. $\frac{x}{28(x^2+7)^2} + \frac{3}{28}\left[\frac{x}{14(x^2+7)} + \frac{1}{14\sqrt{7}}\arctan(x/\sqrt{7})\right] + C$

13. $(1/3)\cos^2 x \sin x + (2/3)\sin x + C$

15. $(1/4)\cos^3 x \sin x + (3/8)\cos x \sin x + (1/2)x + C$

Section 8.6 (page 374)

1. $-\ln 12$

3. $\ln\left|\frac{(x+3)^3(x-4)}{x}\right| + C$

5. $x - 4\ln|x+1| + C$

7. $\ln\left|\frac{x(x+2)^3}{x-2}\right| + C$

9. $2\ln 4/3 - 5/2$

11. $2\ln|x+1| + \frac{1}{x+1} + C$

13. $\frac{1}{2}\ln|x^2+1| + \frac{1}{2(x^2+1)} + \frac{x}{2(x^2+1)} + \frac{1}{2}\arctan x + C$

15. $\frac{1}{2}x^2 + x + \ln\frac{(x-5)^2}{|x+3|} + C$

17. $\frac{1}{3}x^3 + 2x - \ln|(x-2)(x+2)^6| + C$

19. $\ln(|x^2 + 2x + 5|\,|x - 1|^3) - \dfrac{3}{2}\arctan\dfrac{x+1}{2} + C$

21. $1 + \ln(18/49)$

23. $\arctan\dfrac{x}{2} - \dfrac{1}{2(x^2+4)} + C$

25. $\ln|x - 1| + \dfrac{2}{x^2+2} + \dfrac{x}{4(x^2+2)} + \dfrac{1}{4\sqrt{2}}\arctan\dfrac{x}{\sqrt{2}} + C$

27. $\dfrac{1}{25}\ln|3x| - \dfrac{1}{25}\ln|3x+5| + \dfrac{1}{5(3x+5)} + C$

31. $\dfrac{1}{k(a-m)}\left(\ln\left|\dfrac{x-m}{a-x}\right|\right) + C$, the population decreases with increasing time.

Section 8.7 (page 378)

1. $(1/2)\ln|x^2 + 10x| + C$

3. $x + \ln\left|\dfrac{x-2}{x+2}\right| + C$

5. $\dfrac{1}{6}\ln\left|\dfrac{x-6}{x}\right| + C$

7. $\pi/8$

9. $-x^2\cos x + 2x\sin x + 2\cos x + C$

11. $(1/3)x^3 - x^2 + 4x - 7\ln|x+2| + C$

13. $\ln(4/3)$

15. $2\ln|x+3| - \dfrac{1}{2}\ln|x^2+2x+3| + \dfrac{5}{\sqrt{2}}\arctan\dfrac{x+1}{\sqrt{2}} + C$

17. 0

19. $\dfrac{x}{18(x^2+9)} + \dfrac{1}{54}\arctan\dfrac{x}{3} + C$

21. $e^{\sin x} + C$

23. $(1/4)\ln(4 + e^{4x}) + C$

25. $(1/7)x^7\ln 2x - (1/49)x^7 + C$

27. $12/5$

29. $(1/3)x^3 - x^2 + x + \ln\dfrac{|x+7|^3}{|x-5|^2} + C$

31. $\ln|\tan x| + C$

33. $(1/2)x - (1/20)\sin 10x + C$

35. $\dfrac{x}{2\sqrt{2-x^2}} + C$

37. $\pi/6$

39. $x\sec x - \ln|\sec x + \tan x| + C$

41. $(1/2)\arctan x^2 + C$

43. $(1/7)\sin 7x - (1/21)\sin^3 7x + C$

45. $(1/9)(3x^2+1)^{3/2} + C$

47. $-2\sqrt{x}\cos\sqrt{x} + 2\sin\sqrt{x} + C$

49. $3e^{\sqrt[3]{t}}(x^{2/3} - 2x^{1/3} + 2) + C$

51. $(x+3)\ln\sqrt{x+3} - \dfrac{x+3}{2} + C$

53. $m_0\pi/3$ units/sec

Technique Review Exercises, Chapter 8 (page 381)

1. $(3/2)x^2 + x + (42/5)\ln|x-3| + (3/5)\ln|x+2| + C$

2. $\dfrac{1}{2}\left(\sqrt{2x+1} + \dfrac{1}{\sqrt{2x+1}}\right) + C$

3. $(1/3)\sin^3 x - (2/5)\sin^5 x + (1/7)\sin^7 x + C$

4. $(1/8)(t - (1/4)\pi \sin 4\pi t) + C$

5. $(1/2)\arctan 1/2.$

6. $\frac{9}{2}\left(\frac{\pi}{6} - \frac{\sqrt{3}}{4}\right)$

7. $\frac{\sec^4 2x}{4}\left(\frac{\sec^2 2x}{3} - \frac{1}{2}\right) + C$

8. $\arcsin\frac{x-2}{3} + C$

9. $-(1/8)(\sin 3x \sin x + 3\cos x \cos 3x) + C$

10. $(1/2)\ln|x^2 - 4x - 5| + (1/3)\ln\left|\frac{x-5}{x+1}\right| + C$

11. $2e^{(x+1)^{1/2}} + C$

Additional Exercises, Chapter 8 (page 381)

1. $(2/15)(32\sqrt{2} - 13)$

3. $(1/10)(e^x \cos 3x + 3e^x \sin 3x) + C$

5. $-1/2$

7. $-(1/2)x\cos 2x + (1/4)\sin 2x + C$

9. $(1/12)\tan^6(x^2) + C$

11. $2/63$

13. $-(4/17)\csc^{17/4} x + (8/9)\csc^{9/4} x - 4\csc^{1/4} x + C$

15. $(2/7)\tan^{7/2} x + (2/11)\tan^{11/2} x + C$

17. $\ln|\sqrt{9 + x^2} + x| + C$

19. $-(1/3)(7 - x^2)^{3/2} + C$

21. $\ln\left(\frac{\sqrt{21} + 3}{6}\right)$

23. $[26 - (x-4)^2]^{3/2}[26/3 - (1/5)(26 - (x-4)^2)] + C$

25. $\frac{1}{4}\ln\left|\frac{x-5}{x-1}\right| + C$

27. $\frac{1}{2}\ln|x^2 - 6x + 5| + \frac{3}{3}\ln\left|\frac{x-5}{x-1}\right| + C$

29. $\frac{3}{2}\ln|4 - 2x + x^2| + \frac{7}{\sqrt{3}}\arctan\frac{x-1}{\sqrt{3}} + C$

31. $\frac{1}{4}\ln|3 + 4x^2 - 8x| + \frac{1}{2}\ln\left|\frac{2x-3}{2x-1}\right| + C$

33. $\arcsin\left(\frac{x-3}{7}\right) + C$

35. $\frac{2}{\sqrt{3}}\arctan\left(\frac{2\tan(x/2) + 1}{\sqrt{3}}\right) + C$

37. $\frac{1}{160} + \frac{1}{16}\left(\frac{\pi}{4} - \arctan\frac{1}{2}\right)$

39. $\frac{x}{16(x^2+4)^2} + \frac{3}{128}\left(\frac{x}{x^2+4} + \frac{1}{2}\arctan\frac{x}{2}\right) + C$

41. $\frac{1}{3}\ln\left|\frac{x-2}{x+1}\right| + C$

43. $(1/2)x^2 + 2x + (27/4)\ln|x-3| + (1/4)\ln|x+1| + C$

45. $\ln\left(\frac{160}{7}\right) + \frac{1}{\sqrt{6}}\left(\arctan\frac{2}{\sqrt{6}} - \arctan\frac{1}{\sqrt{6}}\right)$

47. $(1/4)\ln|x| + (3/4)\ln|x-2| - (3/2)(x-2)^{-1} + C$

49. $\ln\sqrt{2} - 1/4$

51. $\frac{1}{2\sqrt{5}}\arctan\left(\frac{x}{\sqrt{5}}\right) - \frac{x}{2(x^2+5)} + C$

53. $(1/2)\ln|x^2 + 4x + 5| - 2\arctan(x+2) + C$

55. $x - (1/9)\ln|x| + (14/9)\ln|x-3| - (13/9)\ln|x+3| + C$

Chapter 9

Section 9.2 (page 391)

1. 2.236
3. $-1.\overline{30}$
5. 1.656
7. 1.227
9. -960.204
13. $E_1 \le 1, E_2 \le 1/4, E_3 \le 1/64$
15. $E_1 \le 1, E_2 \le 1/3, E_3 \le 1/27$
17. $E_1 \le 1, E_2 \le 6, E_3 \le 216$
19. $c_{i+1} = \dfrac{c_i{}^2 + a}{2c_i}$
21. 1.4691

Section 9.3 (page 394)

1. 4.8125
3. 3.1875

Section 9.4 (page 398)

1. $1.68\overline{3}$
3. 1.1038
5. .2469
7. 2.079
9. $-\dfrac{1}{2}\left(\dfrac{12\sqrt{2}}{35} + \dfrac{1}{3}\right)$
11. $-.4855$
13. $\dfrac{1}{2}\left(\dfrac{\sqrt{2}}{2} + \dfrac{\sqrt{13}}{2} + \sqrt{5} + \dfrac{\sqrt{29}}{2} + \dfrac{\sqrt{10}}{2}\right)$
15. $E_4 \le 2/3$
17. $E_4 \le .1302$
19. $E_{10} \le 9/200$
21. (a) 33; (b) 104
23. 15
25. $\dfrac{\pi}{4}(1 + \sqrt{2} + \sqrt{6})$
27. 19.5 m^3
29. 24.35 m

Section 9.5 (page 404)

1. 1.622
3. 1.1051
5. .2263
7. 2.0188
9. $-\dfrac{1}{3}\left(\dfrac{1}{3} + \dfrac{24\sqrt{2}}{35}\right)$
11. .4205
13. $(1/6)(\sqrt{2} + 2\sqrt{13} + 2\sqrt{5} + 2\sqrt{29} + \sqrt{10})$
15. $E_4 \le 24/45$
17. $E_4 \le .01356$
19. $E_{10} \le .00162$
21. (a) 12; (b) 20
23. 8
25. $\dfrac{\pi^3}{12}(1 + \sqrt{2}), \dfrac{\pi^7}{92{,}160}$
27. 660 kg-m
29. 230 ft^2

Section 9.6 (page 409)

1. $-2 - 6x$
3. $5 - 8(x - 1)$
5. $7 - 2x$
7. $5 - 2(x - 1)$
9. $\ln 2 + (1/2)(x - 2)$
11. $1 + x$

13. $1 + 2(x - \pi/4)$

15. no linear approximation exists

17. $x - 1$

19. $1 + x$

21. 2

23.

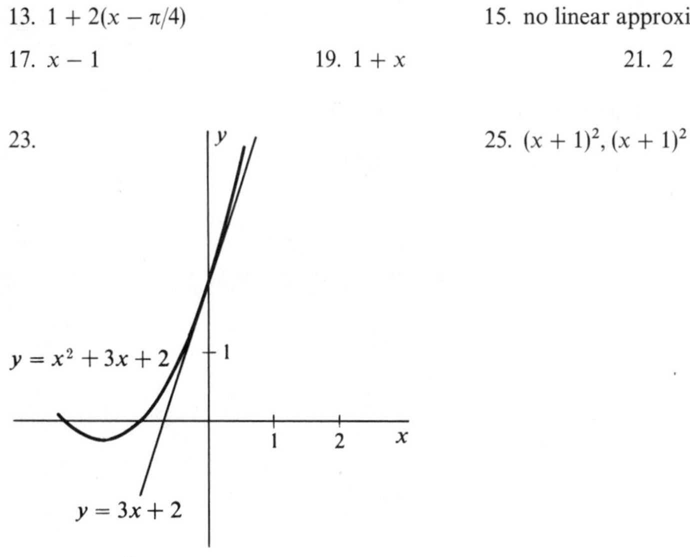

25. $(x + 1)^2, (x + 1)^2$

27. $ex^2/2$

29. .98

31. .00019

Section 9.7 (page 418)

1. $7 - x + x^4, 0$

3. $-3 + x^3 - 3x^4, 2|x|^5$

5. $x - \frac{1}{3!}x^3, \frac{1}{5!}|x|^5$

7. $1 + x + \frac{1}{2!}x^2 + \frac{1}{3!}x^3 + \frac{1}{4!}x^4, \frac{e}{5!}|x|^5$

9. $2^{1/3} + \frac{1}{3 \cdot 2^{2/3}}x - \frac{1}{3^2 \cdot 2^{2/3} \cdot 2!}x^2 + \frac{5}{3^3 \cdot 2^{5/3} \cdot 3!}x^3 - \frac{5 \cdot 8}{3^4 \cdot 2^{8/3} \cdot 4!}x^4, \frac{22}{3^6}|x|^5$

11. $-\frac{1}{3} - \frac{1}{3^2}x - \frac{1}{3^3}x^2 - \frac{1}{3^4}x^3 - \frac{1}{3^5}x^4, \frac{|x|^5}{64}$

13. $1 + x^2 + (1/2)x^4, (13/5)e|x|^5$

15. $-2 - 13(x - 1) - 9(x - 1)^2 + (x - 1)^3 + 2(x - 1)^4, 0$

17. $-\frac{\pi}{2}(x - 1) + \frac{\pi^3}{48}(x - 1)^3, \frac{\pi^5}{2^5 5!}|x - 1|^5$

19. $1 + \frac{1}{2}(x - 1) - \frac{1}{8}(x - 1)^2 + \frac{1}{16}(x - 1)^3 - \frac{5}{128}(x - 1)^4, \frac{7(2)^{1/2}}{16}|x - 1|^5$

21. $(x - 1) + (1/2)(x - 1)^2 - (1/6)(x - 1)^3 + (1/12)(x - 1)^4, (4/5)|x - 1|^5$

27. $4.3(10)^{-13}$

Technique Review Exercises, Chapter 9 (page 421)

1. $c_{i+1} = \frac{c_i^2 + 3}{2c_i}$

2. 1.732, .125

3. 1.6875

4. 5.625

5. $5.41\overline{6}$

6. (a) .75; (b) .028

7. $1 - x + \frac{x^2}{2} - \frac{x^3}{3!} + \frac{x^4}{4!}, \frac{e}{5!}|x|^5$

Additional Exercises, Chapter 9 (page 421)

1. 2.45
3. 2.321
5. 1.153
11. 2.1875
13. 1.5625
15. .6970
17. $.25(.5 + e^{.0625} + e^{.25} + e^{.5625} + .5e)$
19. .01042
21. 37
23. .6933
25. $\frac{.25}{3}(1 + 4e^{.0625} + 2e^{.25} + 4e^{.5625} + e)$
27. .000521
29. 6
31. $-(x - \pi/2)$
33. no linear approximation
35. $(1/2)(x - \pi/2)^2$
37. $-(x - \pi/2) + \frac{1}{3!}(x - \pi/2)^3$
39. $1 - x^2 + (1/3)x^4$
41. $29 + 44(x - 2) + 27(x - 2)^2 + 8(x - 2)^3 + (x - 2)^4$
43. $(1/120)|x - \pi/2|^5$
45. $2x^5/15$

Chapter 10

Section 10.1 (page 431)

1. ∞
3. 0
5. 0
7. 0
9. 0
11. does not exist
13. 5
15. -3
17. 10
19. -2
21. 0
23. $-\infty$
25. 2/5
27. 1
29. $\lim_{n\to\infty} f_n(r) = 1, \lim_{r\to 1} f_n(r) = 1$
31.

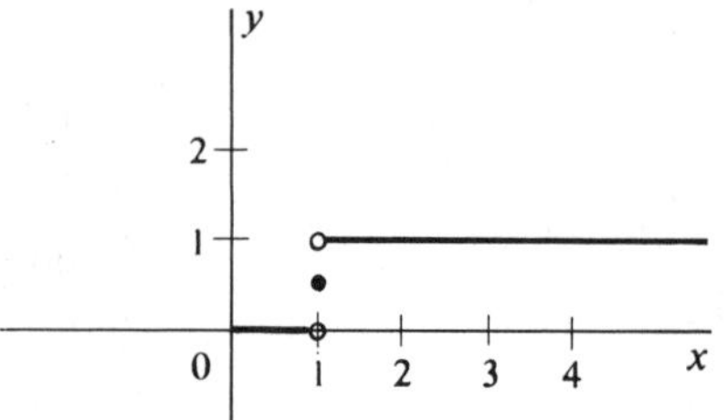

Section 10.2 (page 436)

1. 1/2
3. 3
5. $-1/4$
7. 5
9. $-3/2$
11. 0
13. ∞
15. 0
17. 0
19. 0
21. $-\infty$
23. 3
25. 0
27. 1
29. 0
31. 3/2
33. 2/25

Section 10.3 (page 439)

1. 0
3. 0
5. 1/2
7. a
9. 0
11. 0
13. 1
15. 1
17. 1
19. 0
21. $1/e$
23. e^6
25. $\ln 3 - \ln 2 = \ln(3/2)$
27. e^2

Section 10.4 (page 448)

1. 2
3. diverges
5. diverges
7. $-\dfrac{5 \cdot 3^{2/5}}{2}$
9. $\pi/4$
11. $-3\pi/2$
13. diverges
15. diverges
17. diverges
19. -1
21. $(1/6)\ln 3$
23. $(1/2)\ln(\sqrt{2} + 1)$
25. diverges
27. diverges
29. diverges
31. diverges
33. $\pi/2$
35. $\pi/2$
37. 77/4
41. $16,666.67
43. $1/t$
45. $\dfrac{1}{1 + t^2}$
47. $\dfrac{1}{t - a}$

Technique Review Exercises, Chapter 10 (page 450)

1. (a) $-7/2$; (b) 0; (c) 1/2
2. (a) 9/2; (b) 0
3. (a) ∞; (b) 1

Additional Exercises, Chapter 10 (page 450)

1. $-7/10$
3. does not exist
5. 3/2
7. 2
9. 0
11. does not exist
13. 0
15. 1/3
17. $2/\pi$
19. 0
21. 1
23. 0
25. 0
27. does not exist
29. ∞
31. ∞
33. 0
35. diverges
37. $\pi/4$
39. diverges

Chapter 11

Section 11.1 (page 459)

1. 2/3
3. 0
5. ∞
7. 1
9. 0
11. diverges
13. 0
15. 0
17. 0

19. diverges
21. 5
25. (a) $n/(n+1)$; (b) 1, 2; (c) 1; (d) 1/2, 0; (e) 1/2; (f) 1
27. (a) 5; (b) 5, 6; (c) 5; (d) 4, 5; (e) 5; (f) 5
29. (a) $(3/2)+(-1)^n(1/2)$; (b) 2, 5; (c) 2; (d) 0, 1; (e) 1; (f) does not exist
31. (a) $a_n = \begin{cases} \dfrac{1}{2^{(n-1)/2}} & n \text{ odd} \\ -\dfrac{1}{2^{(n-2)/2}} & n \text{ even} \end{cases}$; (b) 1, 2; (c) 1; (d) −1, −2; (e) −1; (f) 0
33. (a) n; (b) no upper bound; (c) no least upper bound; (d) 0, 1; (e) 1; (f) does not exist
35. (a) $a_n = \begin{cases} 1 & n \text{ odd} \\ \dfrac{n}{n+2} & n \text{ even} \end{cases}$; (b) 1, 2; (c) 1; (d) 0, 1/2; (e) 1/2; (f) 1
37. $\{N, (1.2)N, (1.2)^2N, (1.2)^3N, \ldots, (1.2)^nN, \ldots\}$, diverges

Section 11.2 (page 466)

1. diverges
3. 15/2
5. diverges
7. 3/4
9. diverges
11. diverges
13. diverges
15. $\dfrac{r^2}{1-r^2}$, $r^2 < 1$, and diverges if $r^2 \geq 1$
17. $1/(1-2x)$
19. 1
21. 0
23. $s_n = 1 - \dfrac{1}{n+1}$, 1
25. $6 + 12(.7) + 12(.7)^2 + 12(.7)^3 + \cdots + 12(.7)^n + \cdots = 34$ ft
27. $\dfrac{100(M-m)}{100-R}$
29. 2/3

Section 11.3 (page 472)

1. diverges
3. converges
5. converges
7. diverges
9. diverges
11. diverges
13. diverges
15. converges
17. diverges
19. diverges
21. converges
23. diverges
25. converges
27. converges
29. diverges
31. diverges
33. converges
35. converges

Section 11.4 (page 476)

1. diverges
3. diverges
5. diverges
7. diverges
9. converges
11. converges
13. converges
15. diverges
17. diverges

19. converges
21. diverges
23. diverges
25. diverges
27. converges
29. converges
31. converges
33. yes
35. no
37. yes

Section 11.5 (page 482)

1. converges
3. diverges
5. converges
7. diverges
9. converges
11. converges
13. converges
15. converges
17. .625
19. 917/1000
21. diverges
23. conditionally convergent
25. absolutely convergent
27. diverges
29. conditionally convergent
31. yes
33. yes
35. yes

Section 11.6 (page 487)

1. converges
3. converges
5. diverges
7. converges
9. converges
11. diverges
13. diverges
15. diverges
17. converges
19. converges
21. diverges
23. converges
25. diverges
27. absolutely convergent
29. converges
31. converges
33. converges
35. converges
37. converges
39. diverges

Section 11.7 (page 493)

1. all x, radius of convergence ∞; $1 + x + \frac{x^2}{2!} + \frac{x^3}{3!} + \frac{x^4}{4!} + \cdots$, radius of convergence ∞.

3. all x, radius of convergence ∞; $\frac{3x^2}{2!} + \frac{6x^5}{3!} + \frac{9x^8}{4!} + \frac{12x^{11}}{5!} + \cdots + \frac{3(n-1)x^{3n-4}}{n!} + \cdots$, radius of convergence ∞.

5. $|x| < 7$, radius of convergence 7; $\frac{3}{7} + \frac{5 \cdot 2}{7^2}x + \frac{7 \cdot 3}{7^3}x^2 + \frac{9 \cdot 4}{7^4}x^3 + \cdots = \sum_{n=1}^{\infty} \frac{(2n+1)n}{7^n}x^{n-1}$, radius of convergence 7.

7. $-3 \le x \le 3$, radius of convergence 3; $\sum_{n=1}^{\infty} (-1)^n \frac{nx^{n-1}}{3^n n^{3/2}}$, radius of convergence 3.

9. $-6 \le x < -4$, radius of convergence 1; $\sum_{n=0}^{\infty} (x+5)^n$, radius of convergence 1.

11. $-2 < x < 4$, radius of convergence 3; $\sum_{n=1}^{\infty} (-1)^n \frac{n^2}{3^n}(x-1)^{n-1}$, radius of convergence 3.

13. $x = 0$, radius of convergence 0; no derivative series.

15. $-9 \leq x \leq 1$, radius of convergence 5; $\sum_{n=1}^{\infty} \frac{(x+4)^{n-1}}{5^{n+1}n}$, radius of convergence 5.

17. $x = 5$, radius of convergence 0; no derivative series.

19. $-4 < x \leq -2$, radius of convergence 1; $\sum_{k=1}^{\infty} (-1)^k \sqrt{k}(x+3)^{k-1}$, radius of convergence 1.

21. $x > 1$ or $x < -1$, no radius of convergence.

23. $1 \leq x < 2$, radius of convergence 1/2; $\sum_{n=1}^{\infty} 2(2x-3)^{n-1}$, radius of convergence 1/2.

25. all $x < 1/e$, radius of convergence ∞, $\sum_{n=1}^{\infty} \frac{n^4}{n!} x^{n-1}$, radius of convergence ∞.

27. $1/(1-x)$ for $|x| < 1$

29. $(1-x)^{-2}$ for $|x| < 1$

31. $-\ln|1-x|$ for $|x| < 1$

33. $2(1-x)^{-3}$ for $|x| < 1$

35. e^x

Section 11.8 (page 501)

1. (a) $1 + x^3$; (b) $2 + 3(x-1) + 3(x-1)^2 + (x-1)^3$; (c) $3(x+1) - 3(x+1)^2 + (x+1)^3$. All series converge to $f(x)$ for all x. Uniqueness requires a fixed value of a.

3. $1 + 2(x - \pi/4) + 2(x - \pi/4)^2 + \cdots$.

7. $1 - \frac{2^2}{2!}x^2 + \frac{2^4}{4!}x^4 - \frac{2^6}{6!}x^6 + \cdots$, for all x.

9. $x^2 - \frac{1}{3!}x^4 + \frac{1}{5!}x^6 - \frac{1}{7!}x^8 + \cdots$, for all x.

11. $\sum_{n=0}^{\infty} (-x)^n$, $|x| < 1$

13. $x + x^2 + \left(\frac{1}{2!} - \frac{1}{3!}\right)x^3 + \left(\frac{1}{5!} - \left(\frac{1}{2!}\cdot\frac{1}{3!}\right) + \frac{1}{4!}\right)x^5 + \left(\frac{2}{5!} - \frac{1}{(3!)^2}\right)x^6 + \cdots$, for all x.

15. $1 - x^2 + x^4 - x^6 + \cdots + (-1)^n x^{2n} + \cdots$, for all $|x| < 1$.

21. .333

23. 0.9318

25. $\arcsin x = x + \frac{1}{2\cdot 3}x^3 + \frac{1\cdot 3}{2\cdot 4\cdot 5}x^5 + \frac{1\cdot 3\cdot 5}{2\cdot 4\cdot 6\cdot 7}x^7 + \cdots$, for $|x| < 1$

$(1-x^2)^{-3/2} = 1 + \frac{1\cdot 3}{2}x^2 + \frac{1\cdot 3\cdot 5}{2\cdot 4}x^4 + \frac{1\cdot 3\cdot 5\cdot 7}{2\cdot 4\cdot 6}x^6 + \cdots$, for $|x| < 1$

Technique Review Exercises, Chapter 11 (page 505)

1. (a) diverges; (b) 0
2. diverges
3. diverges
4. 1/3
5. converges
6. converges
7. $-5 \leq x < -1$
8. conditionally convergent
9. $1 - x^3 + \frac{x^6}{2!} - \frac{x^9}{3!} + \cdots + (-1)^n \frac{x^{3n}}{n!} + \cdots$ for all x.
10. $1 + 2x - \frac{1}{2!}x^2 - \frac{2}{3!}x^3 + \frac{1}{4!}x^4 + \frac{2}{5!}x^5 - \frac{1}{6!}x^6 + \frac{2}{7!}x^7 + \cdots$ for all x.

Additional Exercises, Chapter 11 (page 506)

1. 3
3. does not exist
5. 5/13
9. 1/35
11. diverges
13. diverges
15. $1/(e^2 - 1)$
17. converges
19. converges
21. diverges
23. converges
25. converges
27. converges
29. conditionally convergent
31. absolutely convergent
33. absolutely convergent
35. converges
37. absolutely convergent
39. diverges
41. $-4 < x < -2$
43. all x
45. $-1 \le x \le 1$
47. $\sum_{n=0}^{\infty} x^n$
49. $\ln 2 + (1/2)x + x^2 + 0$

Chapter 12

Section 12.1 (page 516)

1, 3, 5, 7, 9, 11.

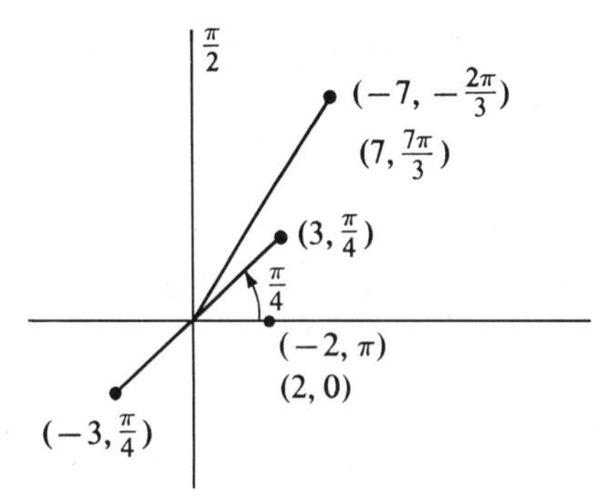

13, 15, 17, 19.

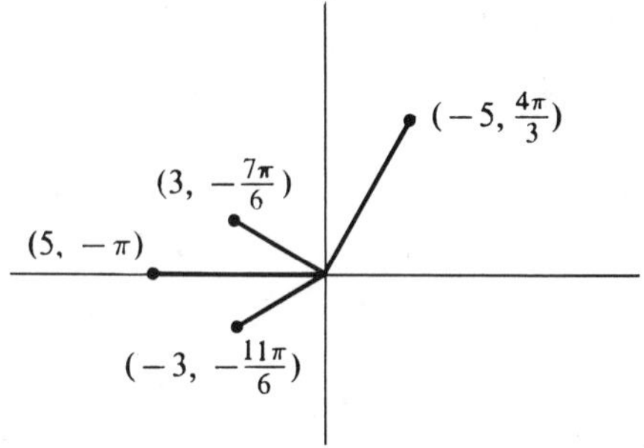

Section 12.2 (page 522)

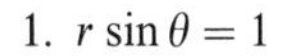
1. $r \sin\theta = 1$

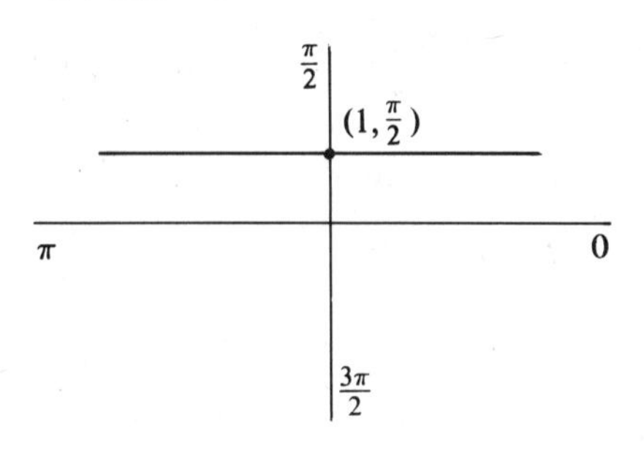

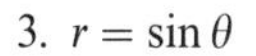
3. $r = \sin\theta$

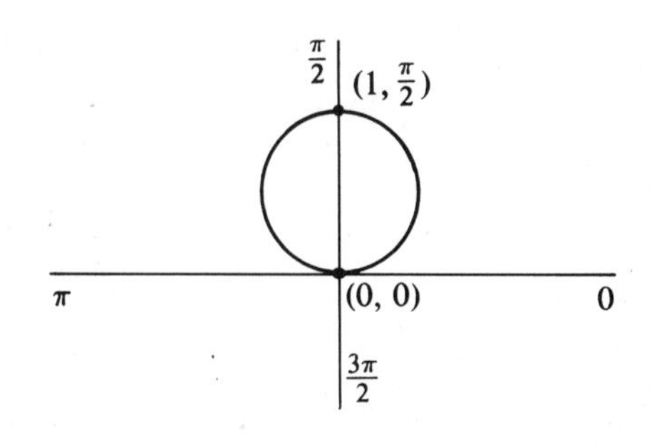

5. $r = 1 - \cos\theta$

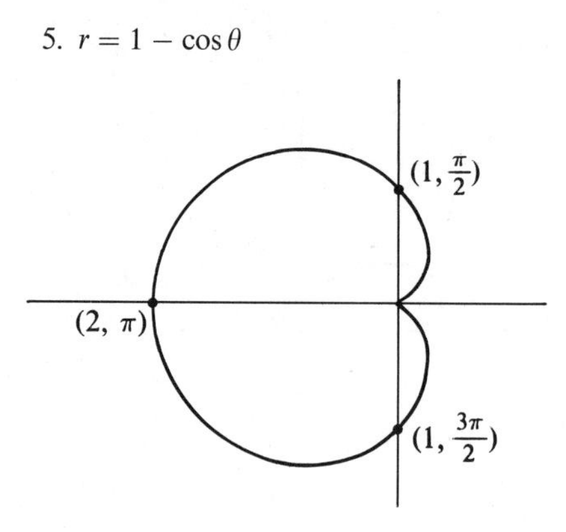

7. $r = 4 - \sin\theta$

$(3, \frac{\pi}{2})$
$(4, \pi)$
$(4, 0)$
$(5, \frac{3\pi}{2})$

9. $r = \cos 3\theta$

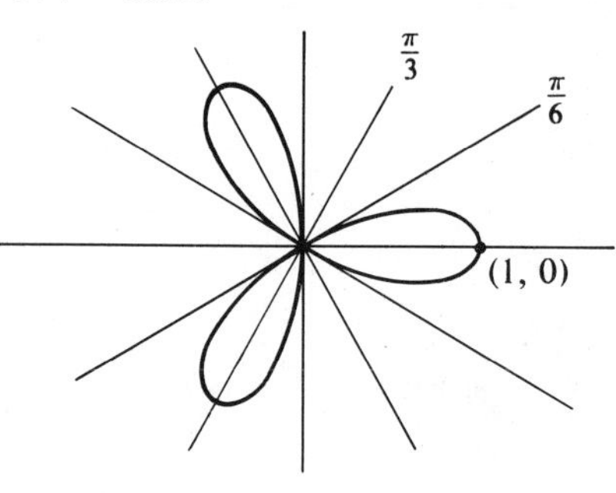

11. $r = 2 + 3\cos\theta$

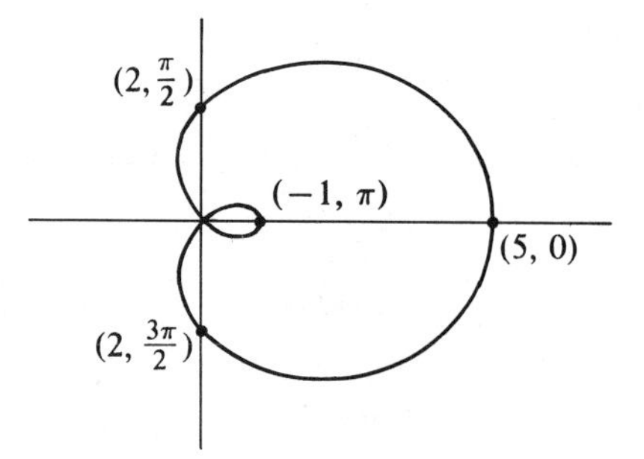

13. $r = \sin 3\theta$

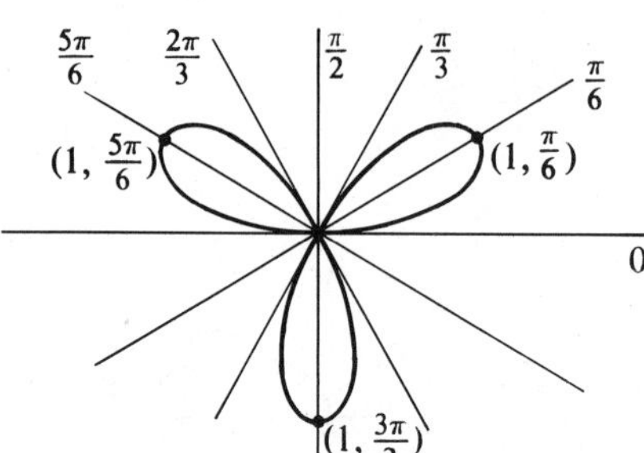

15. $r = \cos(\theta/2)$

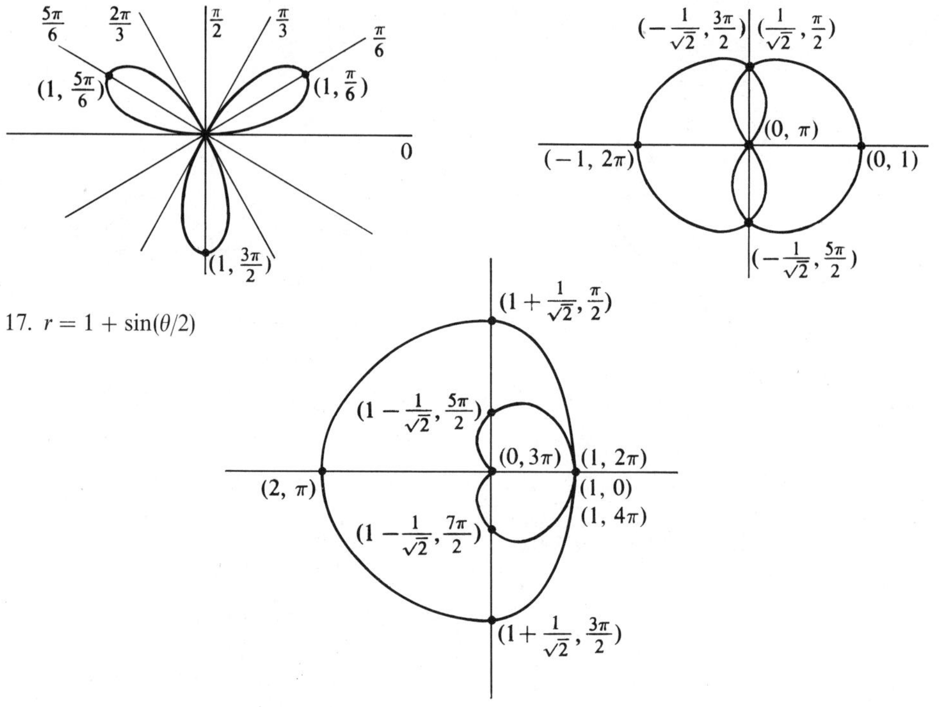

17. $r = 1 + \sin(\theta/2)$

19. $r = e^{\theta/\pi}, \theta \geq 0$

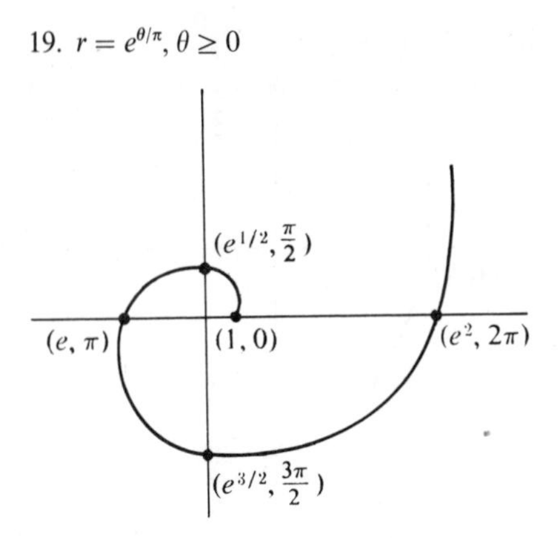

21. $r^2 = a^2 \cos 2\theta, a > 0$

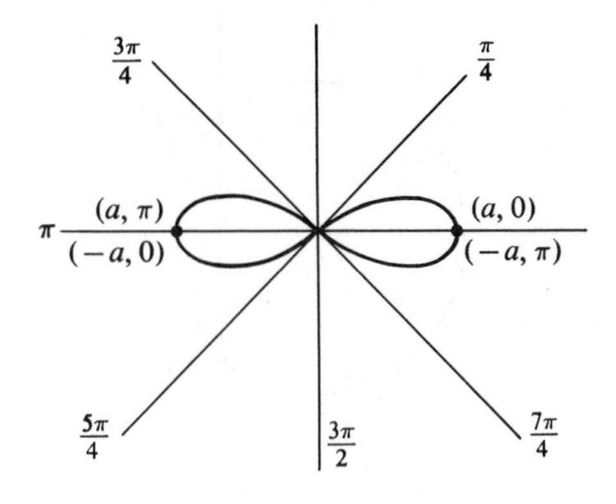

25. $r = a \sec\theta - b$

Section 12.3 (page 525)

1. $r^2 = 8$
3. $r^2 = 4r \cos\theta$
5. $r \sin\theta = 2r \cos\theta + r^2$
7. $r \cos\theta = -1$
9. $r^2 = 2r \cos\theta - 3$
11. $r^4 = 2r^2 \sin\theta \cos\theta = r^2 \sin 2\theta$
13. $r^6 = r^2 \cos^2\theta$
15. $r^2 + 7r \cos\theta = 3r \sin\theta$
17. $x^2 + y^2 = 64$
19. $x + y = 1$
21. $y = -5$
23. $x^2 + y^2 = 2x, x^2 + y^2 \neq 0$
25. $(x^2 + y^2)^{3/2} = y$
27. $(x^2 + y^2)^2 = x^2 - y^2$
29. $y^2 = 9 - 6x$

31.

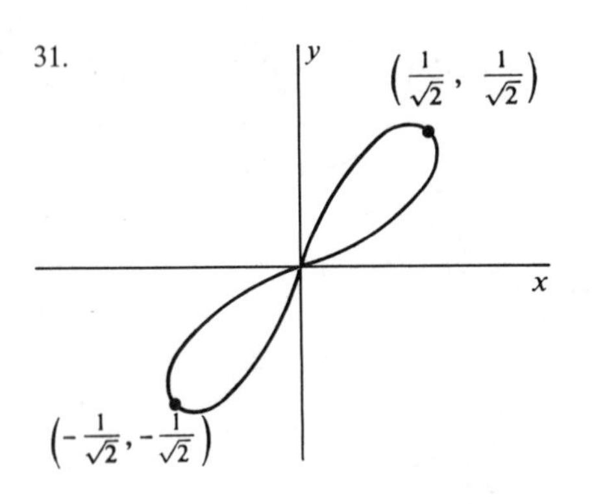

33.

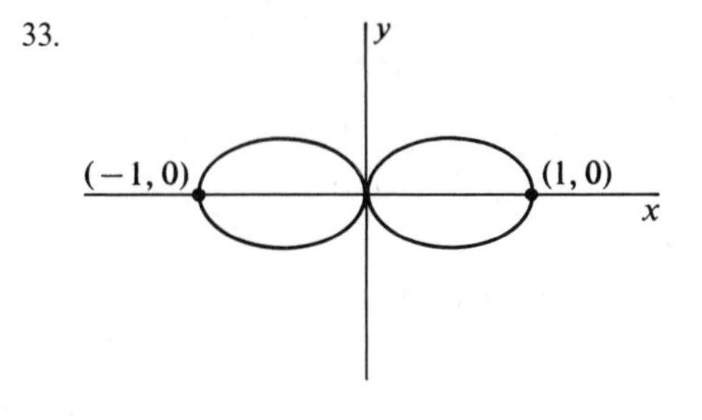

Section 12.4 (page 530)

1. $\frac{1}{8}\left(\frac{\pi}{2} + 1\right)$ square units
3. $(1/12)(e^{3\pi} - 1)$ square units
5. $9\pi/2$ square units
7. $3\pi/2$ square units
9. $\pi/3$ square units
11. $2\pi + 3\sqrt{3}/2$ square units

13. $2 + \pi/4$ square units
15. $8\pi/3 + \sqrt{3}$ square units
17. $\pi - 3\sqrt{3}/2$ square units
19. $1 - \pi/4$ square units
21. $\dfrac{a\pi^2}{48}(6 + \pi)$

Section 12.5 (page 533)

1. 10π units
3. 4 units
5. $(1/3)[(4 + \pi^2)^{3/2} - 8]$ units
7. 9π units
9. $242\sqrt{26}/5$ units
11. 12π units
13. 3 units
15. 8 units
17. $3\pi/2$ units

Technique Review Exercises, Chapter 12 (page 535)

2. $\sqrt{x^2 + y^2} = 3 + 5y/\sqrt{x^2 + y^2}$
4. $\dfrac{\sqrt{5}}{2}(e^{4\pi} - 1)$ units
5. 8 units
6. 9 square units

Additional Exercises, Chapter 12 (page 535)

1, 3, 5.

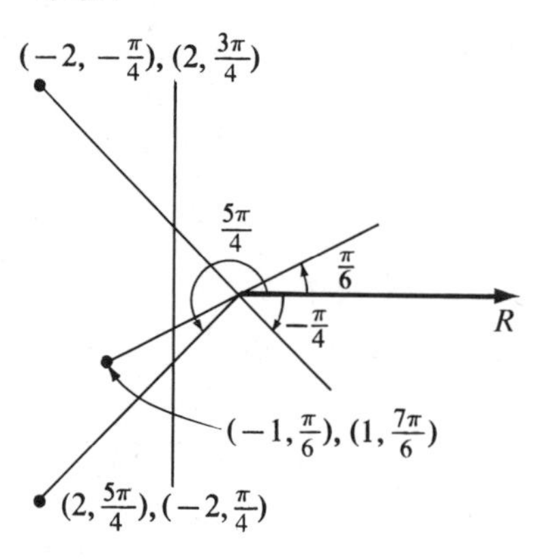

7.

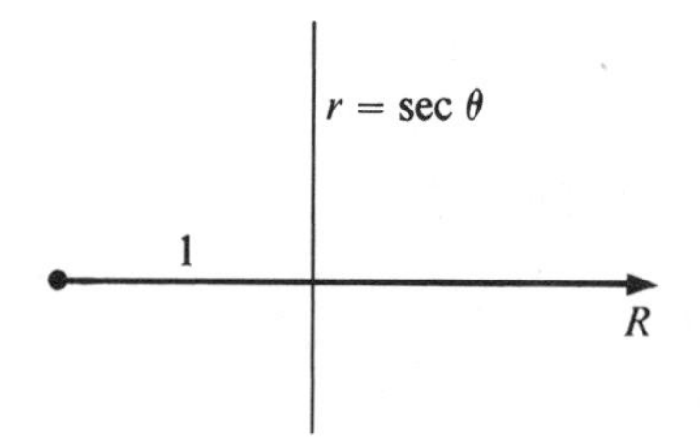

9.

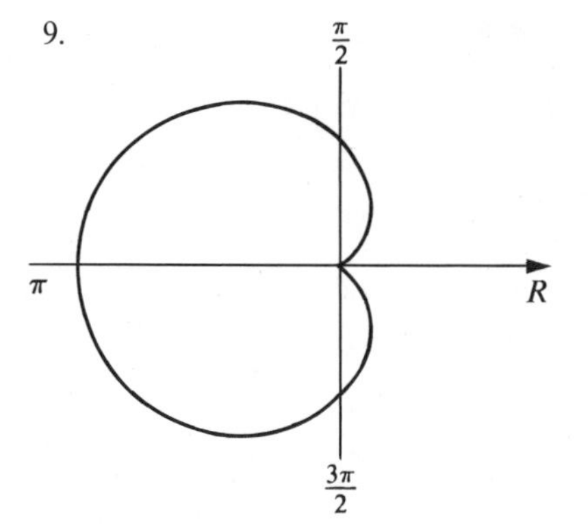

11.

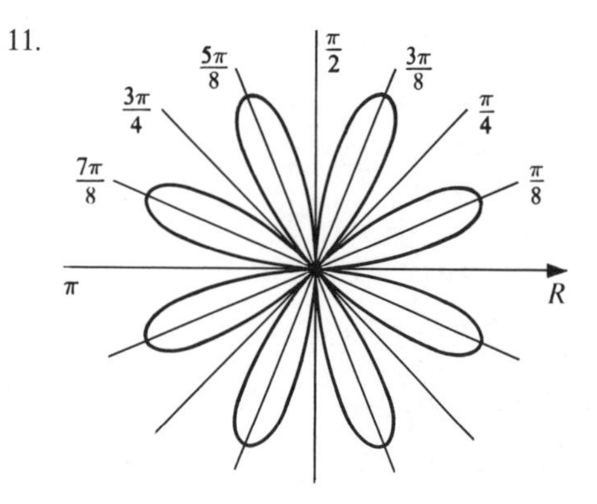

13.

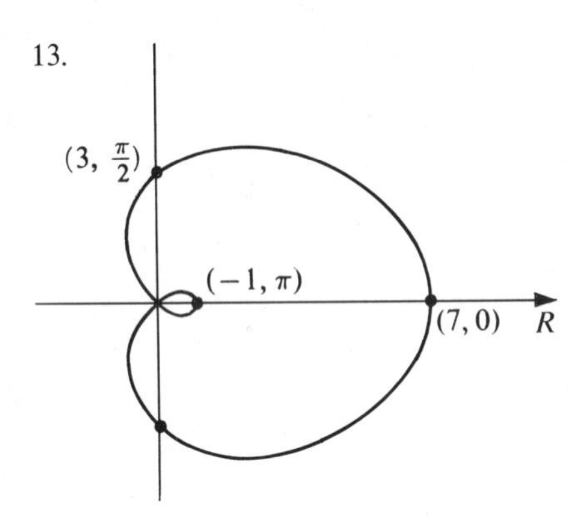

15.

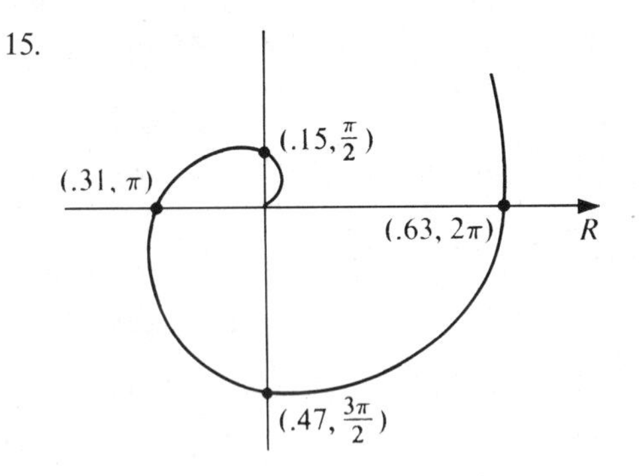

17. $r^2 = 2r \cos\theta$

19. $r^2 - 4r\cos\theta + 3 = 0$

21. $y = \dfrac{1}{\sqrt{3}}x$

23. $x - y = 1$

25. $(1/8)(e^{3\pi} - e^{2\pi})$ square units

27. $9\pi/2$ square units

29. $7\pi/2$ square units

31. $\pi - 3\sqrt{3}/2$ square units

33. $\dfrac{8}{3a}[(1 + a^2\pi^2)^{3/2} - 1]$ units

35. 8 units

37. $\dfrac{1}{2}\left(\pi + \dfrac{3\sqrt{3}}{4}\right)$ units

39. 2 units

Chapter 13

Section 13.1 (page 541)

11. $y = 2e^{-kx}$

13. $y = e^{2x} - (5/4)e^{x} + (1/4)e^{-3x}$

15. $y = e^{-2x} - e^{-4x}$

17. $\lambda = \pm\sqrt{C}$

Section 13.2 (page 545)

1. $y = -1/(x^2 + C)$

3. not separable

5. $y = \dfrac{C}{x}e^{-x^2/2}$

7. $\arctan x = \ln(\sin y) + C$

9. $|y + 3| = \dfrac{7}{|x - 1|}$

11. $y = 2x^5 + 6$

13. $y = \tan(e^x + C)$

15. $N = Ce^{kt}$

17. $x = C \cdot 2^{-t/90}$

19. $x = 20 + 580(14/29)^{t/30}$

21. $x = \dfrac{NCe^{Nkt}}{1 + Ce^{Nkt}}$

23. $f(t) = t^{(\ln 3)/(\ln 2)}$

25. $x = Ce^{kt}$

Section 13.3 (page 550)

1. $y = (1/8)x^3 + (2/7)x^2 + Cx^{-5}$, $y = (1/8)x^3 + (2/7)x^2 - (23/56)x^{-2}$
3. $s = e^{\sin t}(t + C)$
5. $y = x^2[-(2/3)x^{-3} + C]$
7. not linear
9. $y = x^2 - 2x + 2 + Ce^{-x}$, $y = x^2 - 2x + 2 + 3e^{-x}$
11. $y \cos x = x \sin x + \cos x + C$, $y \cos x = x \sin x + \cos x - \pi/4\sqrt{2}$
13. not linear
15. $x = y^2(y^2 + C)$
17. $y = [Ce^x - x - 2]^2$
19. $x = 50(1 - e^{-t/10})$ lbs
21. $y = (2/3)Ax^{-1} + Cx^2$
23. $C - 2 \ln r$

Section 13.4 (page 555)

1. $x = (1/4)t^2 + C_1 \ln|t| + C_2$
3. $(2/3)y^{3/2} = x + 2/3$
5. $y = (1/2)x^2 + C_1x^3 + C_2$
7. $y = -(2/3) \ln|x| + (2/9)x^3 - (2/9)$
9. $y = \sin x$
11. $s = (1/2)t^2 - (1/2)$
13. $y = -2x + \dfrac{1}{C_1} \ln \left|\dfrac{1 + C_1x}{1 - C_1x}\right| + \dfrac{2}{C_1} \arctan C_1x + C_2$
15. $s = (\ln|t|)^2 + (C_1 - t) \ln|t| + 2t + C_2$
17. $y = 0$
19. $x = 5 \sin(t/5 + \pi/2)$

Section 13.5 (page 560)

1. $a_0[1 - x^3 + (1/2)x^5 - \cdots] + [x - (3/4)x^4 + (9/28)x^7 + \cdots]$
3. $a_0[1 + x^2 + (1/2)x^4 + \cdots]$
5. $a_0 + a_1x + (3/2)a_0 x^2 + (3/2)a_1x^3 + \cdots$
7. $a_0 + a_1x + \dfrac{a_0}{2}x^2 + \dfrac{a_1}{3}x^3 + \cdots$
9. $a_0 + a_1x + \dfrac{a_0}{3 \cdot 2}x^3 + \cdots$
11. $(x - 1) + 2(x - 1)^2 + (7/3)(x - 1)^3 + (13/6)(x - 1)^4 + \cdots$
13. $a_0 + a_1(x - 1) - \dfrac{a_0}{2}(x - 1)^2 - \dfrac{a_1}{3}(x - 1)^3 + \dfrac{a_0}{8}(x - 1)^4 + \cdots$

Section 13.6 (page 563)

1. $y = x + 1, 0 \le x \le 1/3$
   $y = 1.25x + .91\overline{6}, 1/3 \le x \le 2/3$
   $y = 1.381x + .829, 2/3 \le x \le 1$
3. $y = 1, 0 \le x \le 1/3$
   $y = x/9 + 26/27, 1/3 \le x \le 2/3$
   $y = 3x/7 + 142/189, 2/3 \le x \le 1$
5. $y = x, 0 \le x \le 1$
   $y = \sqrt{.99}x + (.1 - \sqrt{.0099}), .1 \le x \le .2$
   $y = \sqrt{.9999 - \sqrt{.000396}}x + (1 + \sqrt{.0099} - \sqrt{.03920 - \sqrt{.0000006}}), .2 \le x \le .3$

Technique Review Exercises, Chapter 13 (page 564)

1. $y = x^{-2}(x - 1)$
2. $x = \dfrac{1}{\cos t - 2t - C}$
3. $y = (1/9)x^3 - (1/3)\ln|x| - (1/9)$
4. $x = (C_5 t + C_6)^{2/3}$
5. $a_0 + a_1 x + 2a_0 x^2 + a_1 x^3 + (4/3)a_0 x^4 + \cdots$
6. $y = x + 1$ for $0 \le x \le .1$
   $y = 1.2x + .98$ for $.1 \le x \le .2$
   $y = 1.42x + .936$ for $.2 \le x \le .3$

Additional Exercises, Chapter 13 (page 564)

7. $y = x$
9. $y = 1$
11. $y = Ce^{(x^2/2) - x}$
13. $y = Ce^{(x^3/3) + x}$
15. $x = \dfrac{e^{t^2} - 1}{1 + e^{t^2}}$
17. not separable
19. $y = x^{-3}[(1/7)x^7 + (3/5)x^5 + C]$
21. $s = \sin t + C \cos t$
23. $s = (1/3) + Ce^{3t - 2/2}$
25. $y = (-x - 2 + Ce^{x/2})^2$
27. $y = 1$
29. $\sqrt{C_1}y + \sqrt{C_1 y^2 - 1} = \pm x + C_2$
31. $y = (1/25)x^5 + C_1 \ln|x| + C_2$
33. $s = -\dfrac{1}{C_1}\left(t + \dfrac{1}{C_1}\ln|tC_1 - 1|\right) + C_2$
35. $y = C\left(1 + \dfrac{x^2}{2} + \dfrac{x^4}{8} + \cdots\right) - 1$
37. $y = a_0 + \dfrac{a_0}{2}x + \dfrac{(2 + a_0)}{8}x^2 + \dfrac{2 - 7a_0}{48}x^3 + \cdots$
39. $y = a_0 + a_1 x + \dfrac{1}{2}x^2 - \dfrac{a_1}{6}x^3 + \cdots$
41. $y = a_0 + a_1 x + \dfrac{1 - a_0}{6}x^3 + \dfrac{1 - a_0 - a_1}{12}x^4 + \cdots$
43. $y = a_0 + (1 + a_0)x + (1 + a_0)x^2 + (2/3)(1 + a_0)x^3 + \cdots$
45. $y = 1$ for $0 \le x \le \pi/6$
    $y = (1/2)x + 1 - \pi/12$ for $\pi/6 \le x \le \pi/3$
    $y = \sin\left(\dfrac{4\pi}{\pi + 12}\right)(x - \pi/3) + \pi/12 + 1$ for $\pi/3 \le x \le \pi/2$

Chapter 14

Section 14.1 (page 571)

1. (1, 0, 0), (1, 2, 0), (0, 2, 0), (0, 0, 3), (1, 0, 3), (0, 2, 3)
3. (3, 6, 1), (1, 6, 1), (1, 7, 1), (1, 7, 4), (3, 7, 4), (3, 6, 4)
5. (−3, 3, −2), (−3, −1, −2), (−1, −1, −2), (−1, −1, 5), (−1, 3, 5), (−3, 3, 5)

7. the xy-plane.
9. a plane parallel to the xz-plane and 10 units to the right of it.
11. a plane parallel to the yz-plane and 2 units behind it.
13. a plane parallel to the xy-plane and 1 unit below it.
15. a plane through the x-axis, perpendicular to the yz-plane, and cutting the yz-plane in the line $y = z$.
17. the z-axis.
19. the y-axis.
21. the point $(1, -2, 0)$.
23. two planes parallel to the xz-plane, each being 2 units away from the xz-plane.
25. two planes, each parallel to the yz-plane, one being 5 units in front of and one being 1 unit behind the yz-plane.
27. two planes, each parallel to the xy-plane, one being 7 units above and the other being 1 unit below the xy-plane.
31. 172 units^3

Section 14.2 (page 573)

1. $\sqrt{17}$
3. $\sqrt{57}$
5. $(x-1)^2 + (y+2)^2 + (z-3)^2 = 49$
7. $(x+4)^2 + (y-2)^2 + (z-3)^2 = 17$
9. $(x-1)^2 + (y-2)^2 + (z-4)^2 = 11$
11. $(x-4)^2 + (y+1)^2 + (z-3)^2 < 4$
13. $(x+4)^2 + (y-2)^2 + (z-3)^2 > 17$
15. $(x-1)^2 + (y+2)^2 + (z-3)^2 < 49$ and $(x+4)^2 + (y-2)^2 + (z-3)^2 > 17$
17. $x^2 + z^2 = 16 - 8y$
19. $x + 5y - 2z = -12$

Section 14.3 (page 580)

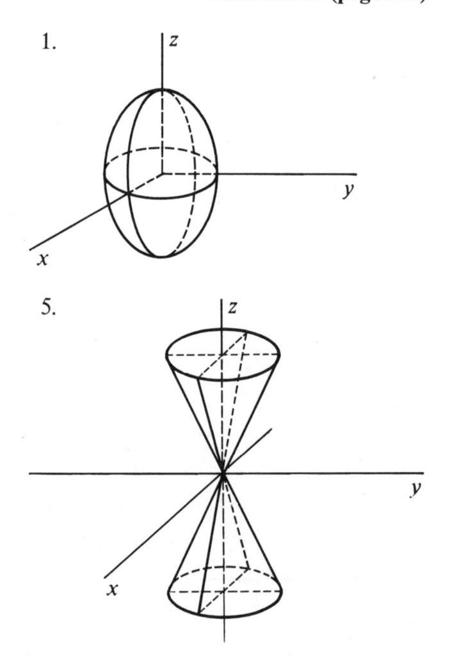

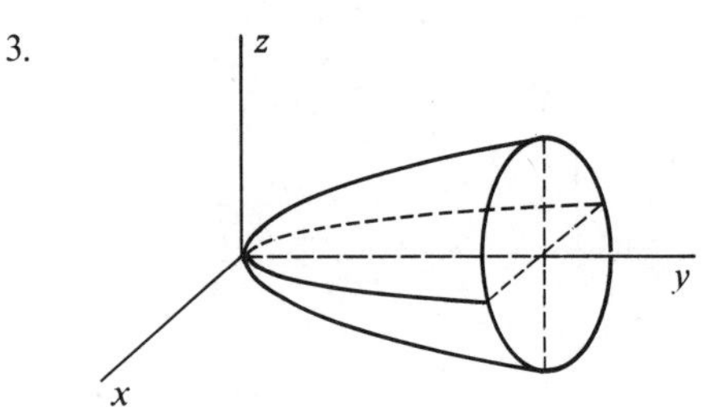

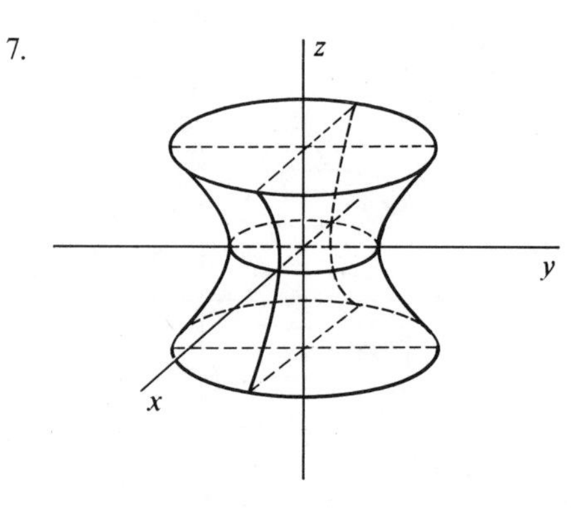

9.

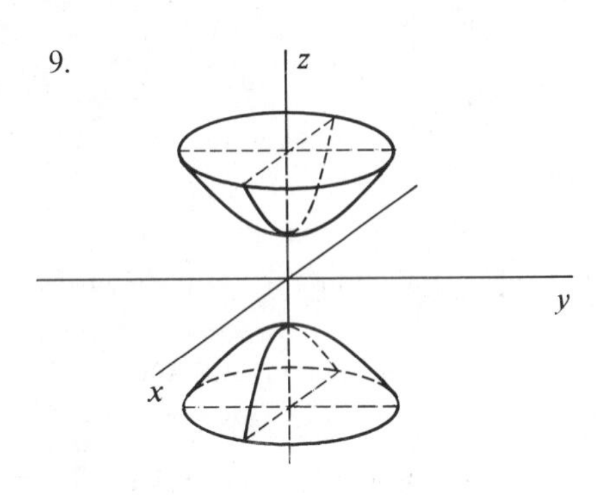

11.

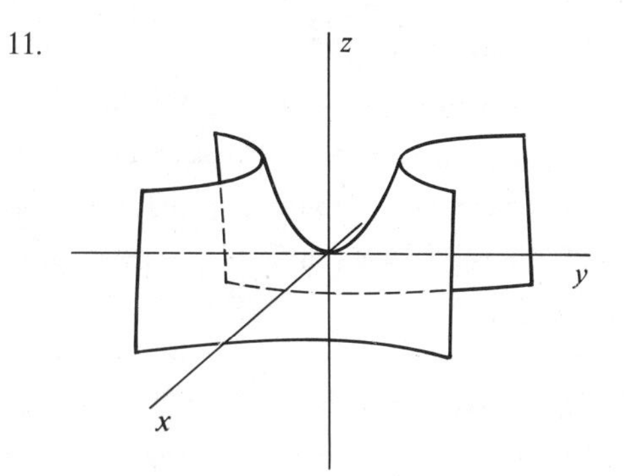

13.

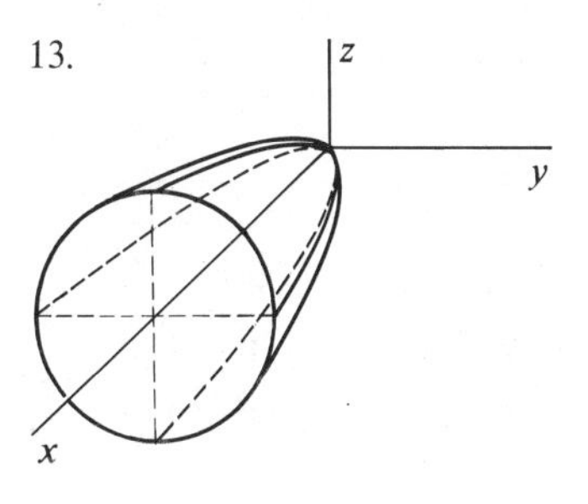

15.

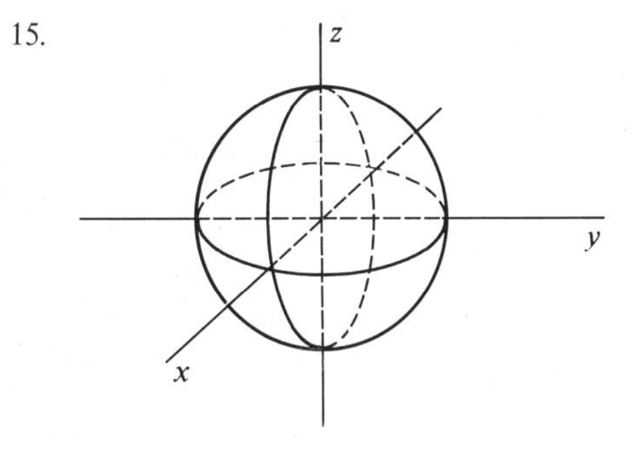

17.

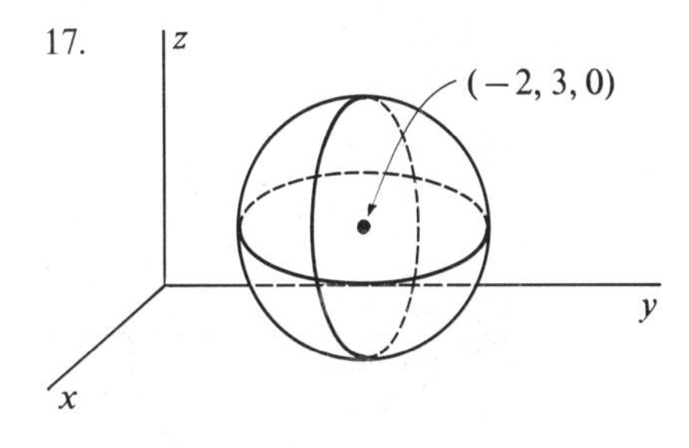

19.

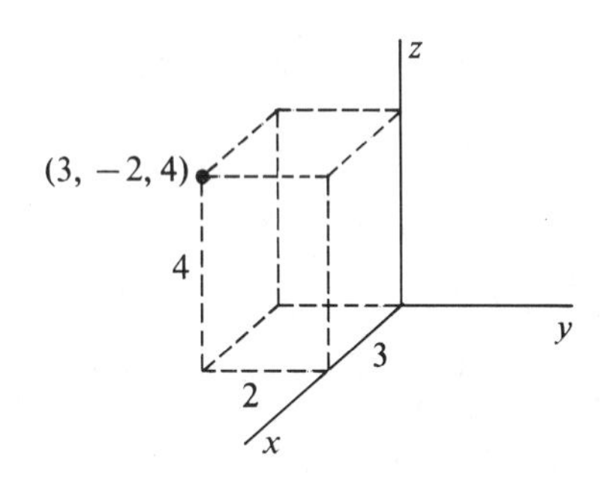

21.

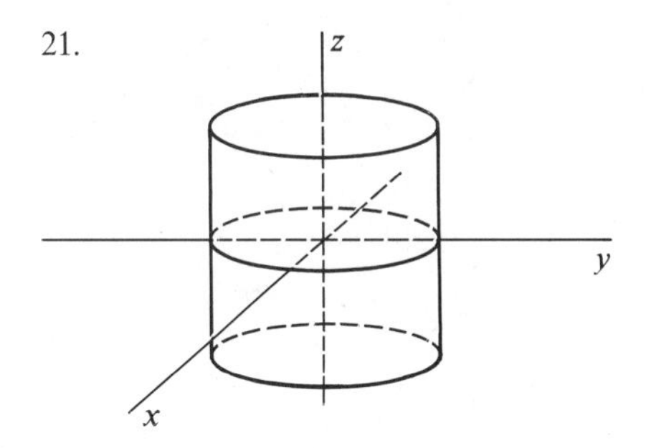

23.

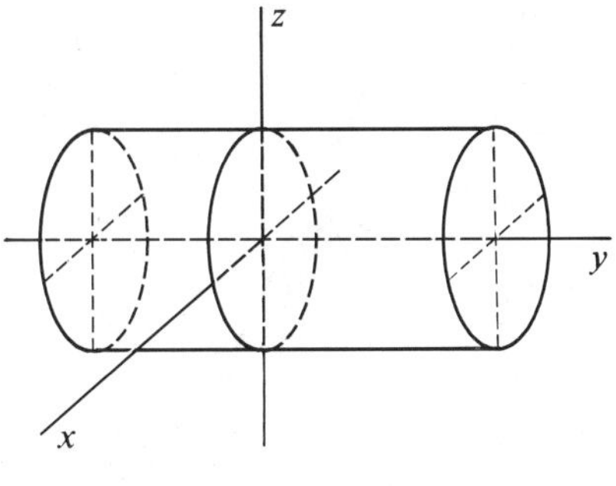

25.

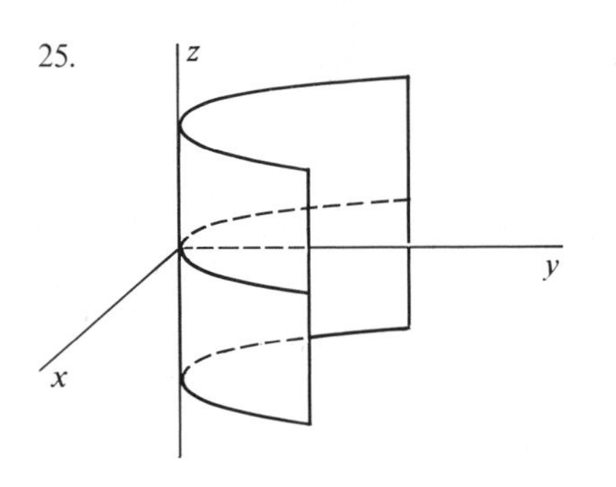

27.

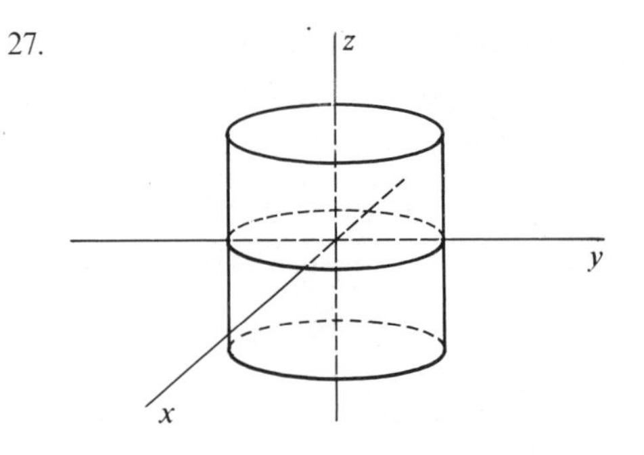

29.

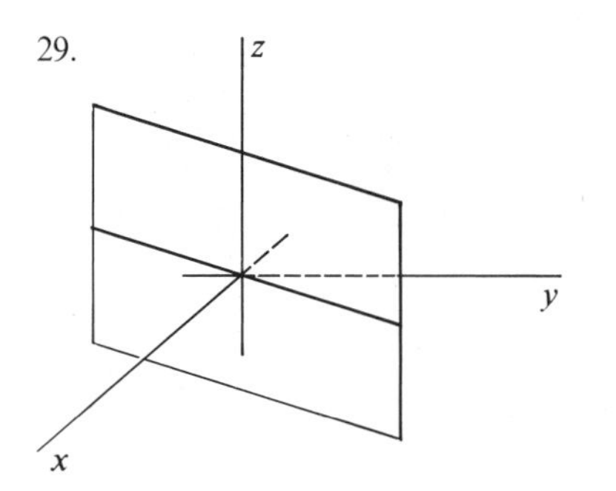

31.

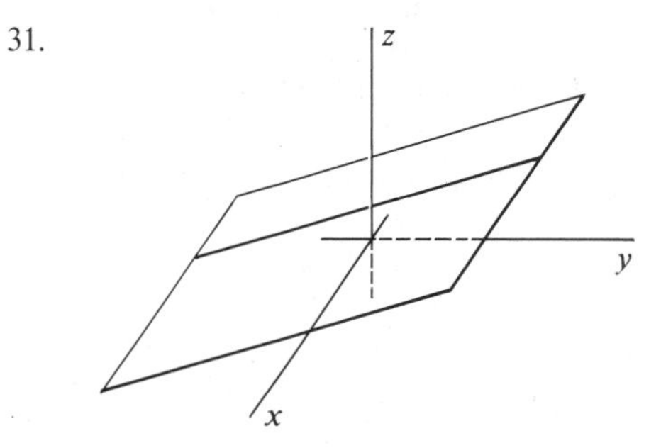

33.

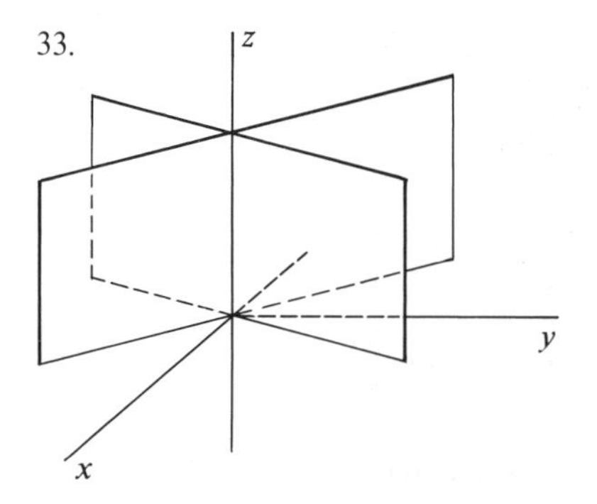

Section 14.4 (page 588)

1. $(7, 11, -2)$
3. $(-4, 8, 20, 12)$
5. $(6, 2, -1)$
7. $(30, 18)$
9. $(-9, -21)$
11. $(-20, 10, 5)$
13. $(-3, -4, 1)$
15. $(3/2, -3/2)$
17. $x = 6, y = 5$
19. $x = 3/13, y = 31/13$
21. $x = 3/2, y = 8$
23. $x = 3, y = 7/3$
25. $2\sqrt{13}$
27. 3
29. $\sqrt{30}$
33. $(1/3, -2/3, 2/3)$
35. $(3/7, 2/7, -6/7)$

Section 14.5 (page 596)

1. (a) 0; (b) $(0, -1, 0)$; (c) $(0, 1, 0)$; (d) $\sqrt{2}$; (e) $\sqrt{2}$; (f) $\pi/2$; (g) $(0, 0, 0)$
3. (a) 0; (b) $(0, 18, 9)$; (c) $(0, -18, -9)$; (d) $3\sqrt{6}$; (e) $3\sqrt{6}$; (f) $\pi/2$; (g) $(0, 0, 0)$

5. (a) 4; (b) $(2\sqrt{6}, -2\sqrt{6}, 0)$; (c) $(-2\sqrt{6}, 2\sqrt{6}, 0)$ (d) $2\sqrt{2}$; (e) $2\sqrt{2}$; (f) $\pi/3$; (g) $(1/2)(-1, -1, \sqrt{6})$

9. $(-11, 2, 5)$

11. $(-7, 7, -21)$

27. $(1/2)\sqrt{222}$

31. 172 units^3

Section 14.6 (page 604)

1. $x = 6 + 2t, y = 1, z = -2 + 5t$

3. $x = 2, y = -2 + 5t, z = 3$

5. (a) $x = 8 - 3t, y = -2 + 6t, z = 4 - 5t$; (b) $\dfrac{8-x}{3} = \dfrac{y+2}{6} = \dfrac{4-z}{5}$

7. (a) $x = 3 - 5t, y = 5t, z = -1 + 8t$; (b) $\dfrac{3-x}{5} = \dfrac{y}{5} = \dfrac{z+1}{8}$

9. $x = 6 + 2t, y = 1, z = -2 + 5t, 0 \le t \le 1; x = 2, y = -2 + 5t, z = 3, 0 \le t \le 1; x = 8 - 3t, y = -2 + 6t, z = 4 - 5t, 0 \le t \le 1; x = 3 - 5t, y = 5t, z = -1 + 8t, 0 \le t \le 1$.

11. $x = 2 + 4t, y = -1 - t, z = 4 + 3t$

13. $x = 8 + 5t, y = -2 + 2t, z = 3 - t$

15. $\dfrac{x-1}{1} = \dfrac{y+2}{2} = \dfrac{z-5}{4}$

17. $x - y = 2$

19. $(5, 5, 3)$

21. $(-3, -10, 3)$

25. $5x + 6y + z = 7$

27. $x + 6z + 3 = 0$

29. $x + 6y + 2z = 36$

31. $3x - 5y + 7z + 43 = 0$

33. $6x - 3y + 2z = 6$

35. $-2x + y = 5$

37. $\pi/3$

Section 14.7 (page 610)

1. $(2t, 3t^2, 5, 0)$

3. $(2e^{2t}, \cos t, 1/t)$

5. $(6 \sin 3t \cos 3t, -6 \sin 3t \cos 3t, -3t^{-4})$

7. range: all nonnegative real numbers; domain: all points (x, y) with $|x| \ge |y|$

9. range: $0 \le f(x, y) \le 3$; domain: all points (x, y) with $x^2 + y^2 \le 9$

11. range: all nonnegative real numbers; domain: all points (x, y) with $|x| \ge |y|$, but $x \ne y$

13. range: all nonnegative real numbers; domain: all points (x, y) with $x \ge 0$ and $y \ge 0$

15. the velocity vector is always directed tangentially to the path of the point

17. (a) $\mathbf{V} = (-2 \sin t, 2 \cos t, 5)$, $\mathbf{A} = (-2 \cos t, -2 \sin t, 0)$, $|\mathbf{V}| = \sqrt{29}$, $|\mathbf{A}| = 2$; (b) $(-2, 0, 5\pi)$, $(0, -2, 5)$, $(2, 0, 0)$; (c) the path looks like a vine winding around the cylinder $x^2 + y^2 = 4$, getting higher as it winds.

19. $(3t^3 - 20, 4t^3 - 33, 3t^2 + 5t - 20)$

21. (a) $\mathbf{V} = (-\omega \sin \omega t, -\omega \sin \omega t, \omega\sqrt{2} \cos \omega t)$, $\mathbf{A} = (-\omega^2 \cos \omega t, -\omega^2 \cos \omega t, -\omega^2\sqrt{2} \sin \omega t)$; (b) $(0, 0, -\sqrt{2})$, $(3\pi, 3\pi, 0)$, $(0, 0, 9\pi^2\sqrt{2})$; (c) a circle of radius $\sqrt{2}$ centered at the origin and in the plane $x = y$

Section 14.8 (page 616)

1. (a) $(8, -3, 27, 10)$; (b) 90

3. (a) 3/8; (b) does not exist

5. (a) 9; (b) does not exist

7. the point $(0, 0)$

9. all points not on the line $y = 3x$ and also at $(0, 0)$ and $(2, 6)$

Technique Review Exercises, Chapter 14 (page 620)

1. a plane parallel to the xz-plane and 10 units to the left of the xz-plane
2. $\sqrt{38}$
3. $(x-2)^2+(y+3)^2+(z-1)^2=29$
4.

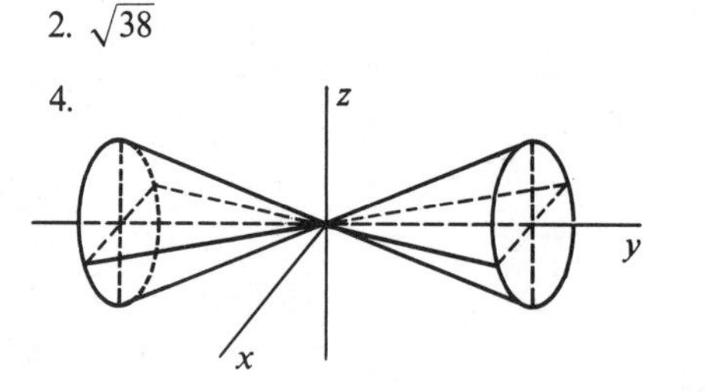

5. 2
6. Their dot product is zero so they are perpendicular.

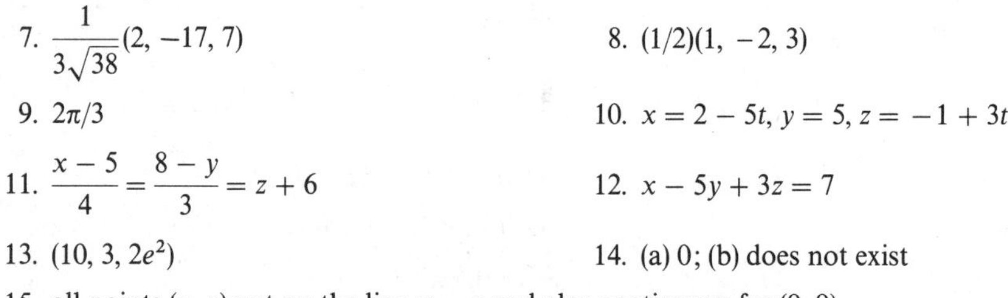

7. $\dfrac{1}{3\sqrt{38}}(2, -17, 7)$
8. $(1/2)(1, -2, 3)$
9. $2\pi/3$
10. $x=2-5t$, $y=5$, $z=-1+3t$
11. $\dfrac{x-5}{4}=\dfrac{8-y}{3}=z+6$
12. $x-5y+3z=7$
13. $(10, 3, 2e^2)$
14. (a) 0; (b) does not exist
15. all points (x, y) not on the line $x=y$ and also continuous for $(0, 0)$.

Additional Exercises, Chapter 14 (page 621)

1. a plane parallel to the yz-plane and 7 units in front of it
3. a plane perpendicular to the xy-plane and intersecting the xy-plane in the line $y=2x$
5. $\sqrt{134}$
7. $(x-5)^2+(y-6)^2+(z+2)^2=81$
9. $(x-4)^2+(y+1)^2+(z-3)^2=30$
11. $(x+1)^2+(y-7)^2+(z-9)^2>110$

13.

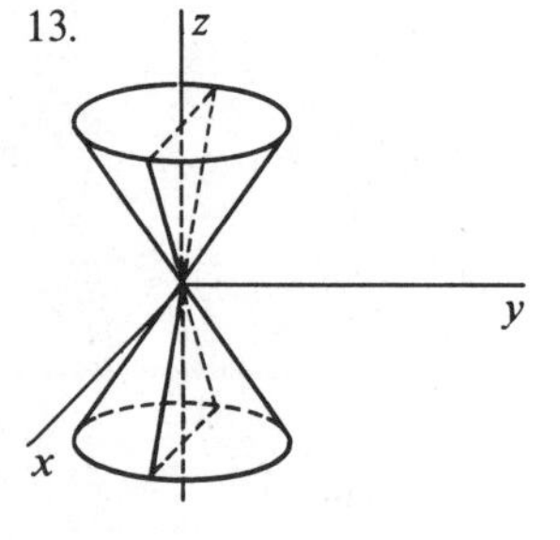

15.

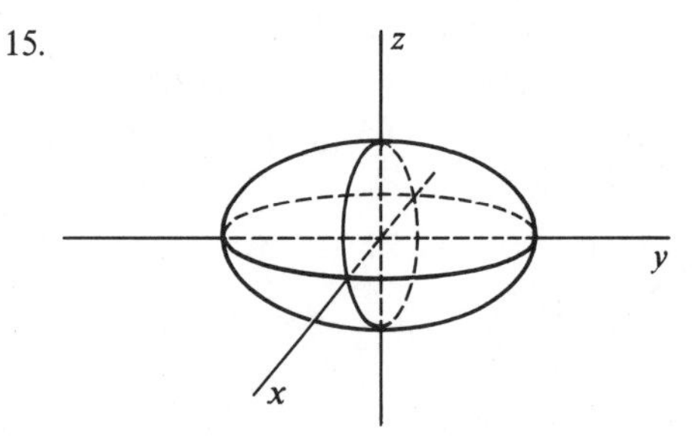

17.

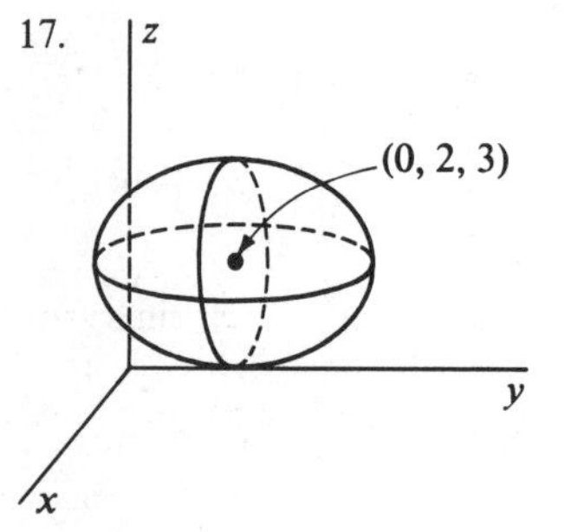

19.

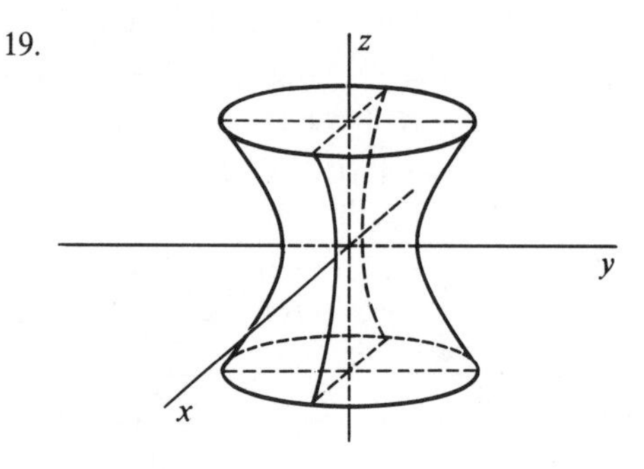

21.

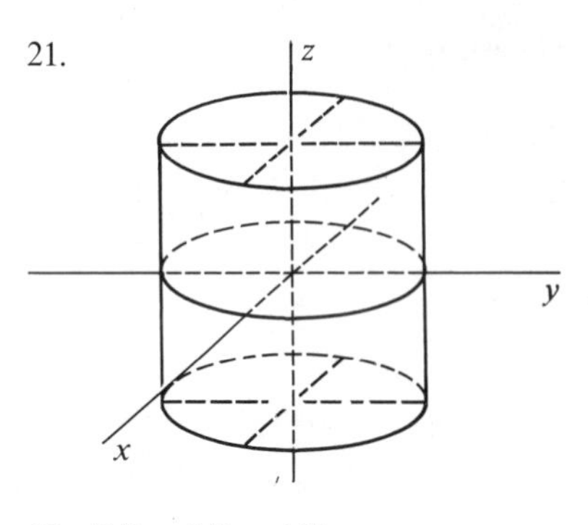

23. $(4/3, -7/3, -4/3)$

25. $2\sqrt{29}$

29. 0

31. $\dfrac{4}{21}(2, -1, 4)$

33. $\dfrac{1}{\sqrt{35}}(-1, 5, 3)$

35. $x = 5 - 4t,\ y = -7 + 9t,\ z = 2 - 5t$

37. $x = 8 + 4t,\ y = 5 - 7t,\ z = 1 + 3t$

39. $2x - 3y - 5z = 0$

41. $5x + 2y - z = 17$

43. $(3\cos 3t, 7e^{7t}, 4t^3)$

45. $(e^t + 4 - e^2, 2t^2 - 9, t^3 - 5)$

47. (a) 5; (b) 0

Chapter 15

Section 15.1 (page 628)

1. (a) 2; (b) 3; (c) 0; (d) 0; (e) 0
3. (a) $2xy + 2y$; (b) $x^2 + 2x$; (c) $2y$; (d) 0; (e) $2x + 2$
5. (a) $x(x^2 + y^2)^{-1/2}$; (b) $y(x^2 + y^2)^{-1/2}$; (c) $y^2(x^2 + y^2)^{-3/2}$; (d) $x^2(x^2 + y^2)^{-3/2}$; (e) $-xy(x^2 + y^2)^{-3/2}$
7. (a) $\cos(x + y) + 2$; (b) $\cos(x + y)$; (c) $-\sin(x + y)$; (d) $-\sin(x + y)$; (e) $-\sin(x + y)$
9. (a) $\dfrac{1}{x} + \cos x$; (b) y^{-1}; (c) $-x^{-2} - \sin x$; (d) $-y^{-2}$; (e) 0
11. (a) $\tan(xy) + xy\sec^2(xy)$; (b) $x^2\sec^2(xy)$; (c) $2y\sec^2(xy) + 2xy^2\sec^2(xy)\tan(xy)$; (d) $2x^3\sec^2(xy)\tan(xy)$; (e) $2x\sec^2(xy) + 2x^2y\sec^2(xy)\tan(xy)$
13. (a) $-y(x^2 + y^2)^{-1}$; (b) $x(x^2 + y^2)^{-1}$; (c) $2xy(x^2 + y^2)^{-2}$; (d) $-2xy(x^2 + y^2)^{-2}$; (e) $(-x^2 + y^2)(x^2 + y^2)^{-2}$
15. (a) $yz + ze^{xz} + y(1 - x^2y^2)^{-1/2}$; (b) $xz + x(1 - x^2y^2)^{-1/2}$; (c) $z^2e^{xz} + xy^3(1 - x^2y^2)^{-3/2}$; (d) $x^3y(1 - x^2y^2)^{-3/2}$; (e) $z + (1 - x^2y^2)^{-3/2}$
17. (a) $4(x + yz)^3 + yz\cos(xyz)$; (b) $4z(x + yz)^3 + xz\cos(xyz)$; (c) $12(x + yz)^2 - (yz)^2\sin(xyz)$; (d) $12z^2(x + yz)^2 - (xz)^2\sin(xyz)$; (e) $12z(x + yz)^2 + z\cos(xyz) - xyz^2\sin(xyz)$

19. e

21. $1/2$

23. ix_i^{i-1}

25. 0

27. (a) yx^{y-1}; (b) $x^y \ln x$

Section 15.2 (page 635)

1. 355
3. 0
5. $(\arctan y)\frac{1}{t} + \frac{x}{1+y^2}\sec t \tan t$
7. $e^x + e^y$
9. $(2x + y^2)\cos t - 2xy \sin t + a$
11. $2x_1 + 2x_2 + \cdots + 2x_n$
13. 1
15. $3x^2y^5 + 5x^3y^4,\ -3x^2y^5 + 5x^3y^4,\ -2,\ -8$
17. $yv \cos(xy) + x\cos(xy),\ yu\cos(xy)$
19. $yz \cos u + xy,\ xz + xy,\ 0,\ \pi/2$
21. $-2x \sin u + 2y \cos u + 2zv,\ 2zu$
23. $\frac{\partial f}{\partial \rho} = 2\rho,\ \frac{\partial f}{\partial \phi} = 0,\ \frac{\partial f}{\partial \theta} = 0$
25. $-2/25$

Section 15.3 (page 641)

1. $(1, -1)$
3. $(1 + ye^x, e^x)$
5. $(-\sin(x - y) + 2, \sin(x - y))$
7. $(\tan(xy) + xy \sec^2(xy), x^2 \sec^2(xy))$
9. $(2x, 2y, 2z)$
11. $(0, 0, 0)$
13. $(\cos xy - xy \sin xy,\ -x^2 \sin xy + e^z,\ ye^z)$
15. $(x_2x_3x_4, x_1x_3x_4, x_1x_2x_4, x_1x_2x_3)$
17. $(1/x_1, 1/x_2, 1/x_3, 1/x_4)$
19. $\frac{1}{|\mathbf{X}|}\mathbf{X}$
21. $\pm(2, 1, 1)/\sqrt{6}$
23. $2x + 4y - z = 5$

Section 15.4 (page 646)

1. $2\sqrt{5}$
3. $8\sqrt{2}$
5. $6\sqrt{2}$
7. $\sqrt{2}e^2$
9. $-1/2$
11. $-3/\sqrt{2}$
13. $-4\sqrt{3}$
15. $2/\sqrt{7}$
17. magnitude 2, direction is direction of unit vector $(1/2)(1, \sqrt{3})$
19. magnitude 0, any direction
21. (a) magnitude $\sqrt{17}$, direction is direction of the unit vector $\frac{1}{\sqrt{17}}(1, 4)$; (b) $(1, 4, -1)$
25. direction of the unit vector $\frac{1}{\sqrt{2}}(1, 1)$

Section 15.5 (page 655)

1. $(1/4, -1/2)$, local minimum
3. $(-3, 2)$, saddle point
5. $(0, 0)$, saddle point
7. $(1, 0)$ local minimum; $(-1, 0)$ saddle point
9. $(2, 1)$ and $(-2, -1)$ are saddle points.
11. $(2/11, -5/22)$, local minimum
13. $(0, 0)$, local minimum
15. $(1, -2)$ local maximum; $(1, 1)$ saddle point
17. $(2, -1)$ local maximum; $(9/8, 3/4)$ saddle point
19. $(1, 0)$ and $(-1, 0)$ are saddle points.

21. no critical points

23. $(l\pi + \pi/2, n\pi)$ is a local maximum if l and n are even;
$(l\pi + \pi/2, n\pi)$ is a local minimum if l is even and n is odd;
$(l\pi + \pi/2, n\pi)$ is a saddle point if l is odd.

25. $(k\pi, (-1)^{k+1})$; $k = 0, \pm 1, \pm 2, \ldots$ are all saddle points.

Section 15.6 (page 661)

1. $(-2, 2, 4)$ absolute maximum point
$(0, 1, -9)$ absolute minimum point

3. $(1, 1, -1)$ absolute maximum point
$(0, 0, -8)$ absolute minimum point

5. $(1, 1, 3)$ and $(-1, 1, 3)$ absolute maximum points
$(0, -1, -2)$ absolute minimum point

7. $(4, 0, 28)$ absolute maximum point; $(0, 0, 0)$ absolute minimum point

9. $(2, 1, 9)$ absolute maximum point
$(-2, -1, -9)$ absolute minimum point

11. $(2, 0, 16)$ absolute maximum; $(0, 0, 0)$ and $(2, 4, 0)$ absolute minimum

13. $(x, y, 15)$ where $x^2 + y^2 = 1$, absolute maximum points
$(0, 0, 14)$ absolute minimum point

15. $(0, 1/4, 49/8)$ and $(1, 1/4, 49/8)$ absolute maximum points
$(1/2, 2, -1/4)$ absolute minimum point

17. $(0, -3, 14)$ absolute maximum point
$(\sqrt{5}, 2, -11)$ and $(-\sqrt{5}, 2, -11)$ absolute minimum points

19. $(1, \sqrt{3}/2, 2)$ and $(1, -\sqrt{3}/2, 2)$ absolute maximum points
$(-2, 0, -7)$ absolute minimum point

23. $x = 10$, $y = 5$

25. $x = 1600$, $y = 1000$

31. $a/\sqrt{3}$

35. $N/3, N/3, N/3$

Technique Review Exercises, Chapter 15 (page 665)

1. $f_x(x, y) = \cos x + \tan(xy) + xy\sec^2(xy)$
$f_y(x, y) = x^2 \sec^2(xy)$
$f_{xy}(x, y) = 2x\sec^2(xy) + x^2 y \sec^2(xy)\tan(xy)$

2. $-\dfrac{x_i}{|\mathbf{X}|^3}$

3. 1

4. 0

5. $\pm\dfrac{1}{\sqrt{114}}(7, -8, -1)$

6. $1/\sqrt{13}$

7. magnitude $\sqrt{6}$, direction is the direction of the unit vector $\dfrac{1}{\sqrt{6}}(-1, -1, 2)$

8. $(1, 0, -2)$ local minimum point

9. $(-3, -4, 93)$, $(-3, 4, 93)$ absolute maximum points; $(5, 0, -35)$ absolute minimum point

10. $2\sqrt[3]{6} \times 2\sqrt[3]{6} \times \sqrt[3]{6}$

Additional Exercises, Chapter 15 (page 665)

1. (a) $2xy + 3y^2$; (b) $x^2 + 6xy$
3. (a) $2y/(x+y)^2$; (b) $-2x/(x+y)^2$
5. (a) $y^2 - 2xye^{-x^2} - e^z$; (b) $2xy + e^{-x^2}$
7. $\dfrac{\partial^2 f}{\partial x^2} = 2y, \dfrac{\partial^2 f}{\partial y^2} = 6x$
9. $\dfrac{\partial^2 f}{\partial x\,\partial y} = 2x + 6y, \dfrac{\partial^2 f}{\partial y\,\partial x} = 2x + 6y$
11. $4x_i|\mathbf{X}|^2$
13. 0
15. 0
17. $\dfrac{\partial f}{\partial u} = (1+2x)\cos v + 2y\sin v$

    $\dfrac{\partial f}{\partial v} = -(1+2x)u\sin v + 2yu\cos v, 2, -1$
19. $\dfrac{\partial f}{\partial u} = 2uz\sec^2 x + vze^y + v^3(\tan x + e^y)$

    $\dfrac{\partial f}{\partial v} = uze^y + 3uv^2(\tan x + e^y)$
21. $(3, -3)$
23. $(z/x, z/y, \ln|xy|)$
25. $\pm(-10, 8, 1)/\sqrt{165}$
27. $\sqrt{3} + 1$
29. $6/\sqrt{10}$
31. $12/\sqrt{7}$
33. (0, 0, 0) is a local minimum point
35. (0, 0, 0) is a saddle point
37. no critical points
39. no critical points
41. $(2, -9, -184)$ local minimum point
43. 18 absolute maximum value, -7 absolute minimum value
45. -7 absolute maximum value, -11 absolute minimum value
47. -5 absolute maximum value, $-6\frac{1}{4}$ absolute minimum value
49. 3 absolute maximum value, -22 absolute minimum value

Chapter 16

Section 16.1 (page 675)

3. 210 units^3
5. 50 units^3

Section 16.2 (page 687)

1. 348
3. 124
5. 8π
7. 4/3
9. 7/3
11. $(96\sqrt{2} - 12)/7 - 186/5$
13. $17(5\sqrt{2} - 1)/12$
15. 9/256
17. 18 units^3
19. $4\int_0^4 \int_0^{\sqrt{(16-x)^2/2}} (16 - x^2 - 4y^2)\,dy\,dx$
21. 17/40 units^3
23. $\dfrac{8}{3}\int_0^6 \int_0^{(1/2)\sqrt{36-x^2}} \sqrt{36 - x^2 - 4y^2}\,dy\,dx$
25. (a) 36 units^2; (b) 36 units^2
27. $81\pi/2$ units^3

29. $\int_0^2 \int_0^{2\sqrt{4-x^2}} (16 - 4x^2 - y^2)dy\,dx - 4\int_0^1 \int_0^{2\sqrt{1-x^2}} (4 - 4x^2 - y^2)dy\,dx$

31. $8\int_0^1 \int_0^{2\sqrt{1-x^2}} (4 - 4x^2 - y^2)dy\,dx$

33. $4\int_0^{\sqrt{2}} \int_0^{(1/2)\sqrt{4-2x^2}} (\sqrt{6 - x^2 - 2y^2} - x^2 - 2y^2)dy\,dx$

35. 2/3

37. 1

39. $\int_0^4 \int_0^{\sqrt{x}} f(x, y)dy\,dx$

41. $\int_0^2 \int_0^{(3/2)\sqrt{4-y^2}} f(x, y)dx\,dy$

43. $\int_0^2 \int_{-(3/2)\sqrt{4-y^2}}^{(3/2)\sqrt{4-y^2}} f(x, y)dx\,dy$

45. $\int_0^4 \int_{-\sqrt{x}}^{\sqrt{x}} f(x, y)dy\,dx + \int_4^9 \int_{x-6}^{\sqrt{x}} f(x, y)dy\,dx$

47. 451/896

49. 2/3

Section 16.3 (page 693)

1. $49\sqrt{14}/12$ units^2

3. $4\int_0^4 \int_0^{\sqrt{16-x^2}} \sqrt{1 + 4x^2 + 4y^2}\,dy\,dx$

5. $4\int_0^4 \int_0^{\sqrt{16-x^2}} 5(25 - x^2 - y^2)^{-1/2}\,dy\,dx$

7. (a) $\int_0^2 \int_{2x}^{3x} 3(9 - x^2)^{-1/2}\,dy\,dx$; (b) $\int_0^4 \int_{y/3}^{y/2} 3(9 - x^2)^{-1/2}\,dx\,dy + \int_4^6 \int_{y/3}^{2} 3(9 - x^2)^{-1/2}\,dx\,dy$

9. $4\sqrt{10}\pi$ units^2

Section 16.4 (page 698)

1. 1638

3. 0

5. $341\frac{1}{3}$

7. $\int_{-1}^1 \int_{-3\sqrt{1-x^2}}^{3\sqrt{1-x^2}} \int_{9x^2+4y^2}^{\sqrt{90-9x^2-y^2}} zx\,dz\,dy\,dx$

9. $\int_{-4}^4 \int_{-(1/2)\sqrt{1-x^2}}^{(1/2)\sqrt{16-x^2}} \int_5^{21-x^2-4y^2} x^2yz\,dz\,dy\,dx$

11. $\int_{-5}^5 \int_{-\sqrt{25-x^2}}^{\sqrt{25-x^2}} \int_{\sqrt{x^2+y^2}}^{(1/2)\sqrt{35-x^2-y^2}} xyz\,dz\,dy\,dx$

13. 9

15. 4

17. 31/160

19. $\int_0^{12} \int_0^{(12-x)/2} \int_0^{(12-x-2y)/3} 3xy\,dz\,dy\,dx$

21. $\int_{-2}^2 \int_{-\sqrt{4-x^2}}^{\sqrt{4-x^2}} \int_{x^2+y^2}^{\sqrt{20-x^2-y^2}} (x - y)dz\,dy\,dx$

23. $\int_0^6 \int_0^{6-x} \int_0^{(6-x-y)/2} dz\,dy\,dx$

25. $4\int_0^2 \int_0^{\sqrt{4-x^2}} \int_0^{4-x^2-y^2} dz\,dy\,dx$

Section 16.5 (page 705)

1. 23

3. $6e^{2u}\cos 3v$

5. $-3u^4v^6$

7. 0

9. 1

11. e^{u+v}

13. $2\int_0^1 \int_{2v}^{v+1} (u + v)du\,dv$

15. $12\int_{-1}^1 \int_{-\sqrt{1-u^2}}^{\sqrt{1-u^2}} (u^2 + v^2)dv\,du$

17. $2\int_{-1}^0 \int_u^{-u} (u^2 - v^2 + 3)dv\,du$

19. $-\int_{-4}^0 \int_{(1/2)\sqrt{-v}}^1 v\,du\,dv$

Section 16.6 (page 710)

1. $\pi(1 - e^{-1})$

3. $15/4\sqrt{2}$

5. 162π

7. $243\pi/4$

9. $(e^4 - 1)\pi/2$

11. $9\pi/2$

13. $(\pi/6)(17^{3/2} - 1)$ units^2
15. $\pi/2$ units^2
17. $2 + \pi/4$ units^2
19. 12π units^3
21. $81\pi/2$ units^3

Section 16.7 (page 715)

1. abc
3. u
5. $ve^u \tan u$
7. $\int_{-1}^{1} \int_{-\sqrt{1-u^2}}^{\sqrt{1-u^2}} \int_{-\sqrt{1-u^2-v^2}}^{\sqrt{1-u^2-v^2}} (u^2 + v^2 + w^2)|abc| dw\, dv\, du$
9. $\frac{1}{27} \int_0^{12} \int_0^{12-u} \int_0^{12-u-v} (36 - 3u - 2w) dw\, dv\, du$
11. $8 \int_0^5 \int_0^v \int_0^u (4vw - 4uw - 2v + 10) dw\, du\, dv$

Section 16.8 (page 719)

1. $3^{3/2}\pi$
3. $3\pi \ln 2$
5. 8π
7. 41,472/25
9. 2π
11. 1/8
13. $81\pi/2$ units^3
15. $\frac{2\pi}{3}(8^{3/2} - 14)$ units^3
17. $\pi/6$ units^3
19. 24π units^3

Section 16.9 (page 725)

1. the set of all points (ρ, ϕ, θ) where $\rho \geq a$, $0 \leq \phi \leq \pi$, and $0 \leq \theta \leq \pi$
3. the set of all points (ρ, ϕ, θ) where $\theta = \pi/4$
5. the set of all points (ρ, ϕ, θ) where $0 \leq \rho \leq 1$ and $0 \leq \phi \leq \pi/4$
7. 0
9. $\pi\left(1 - \frac{\pi}{4}\right)(2 - \sqrt{2})$
11. $\frac{\pi}{3}(\sqrt{2} - 1)$
13. $9\pi(\sqrt{2} - 1)$
15. $\frac{512\pi}{3}$ units^3
17. $\frac{1324}{3}\pi$ units^3
19. $\frac{384\pi}{5}$ units

Technique Review Exercises, Chapter 16 (page 728)

1. 560 units^3
2. $72\frac{11}{12}$ units^3
3. $72\frac{11}{12}$ units^3
4. $8 \int_0^{\sqrt{7}} \int_0^{\sqrt{7-x^2}} (7 - x^2 - y^2) dy\, dx$
5. $81\sqrt{78}/140$ units^2
6. $85\frac{1}{3}$ units^2
7. $4 \int_0^{3\sqrt{2}} \int_0^{\sqrt{18-x^2}} \int_{\sqrt{2x^2+2y^2}}^{\sqrt{54-x^2-y^2}} z\sqrt{x^2 + y^2}\, dz\, dy\, dx$
8. $\int_0^2 \int_0^{u/2} (v + u) dv\, du$
9. $(e - 1)\pi/4$
10. $\int_0^{1/2} \int_0^{1-2v} \int_1^{2-u-2v} uw\, dw\, du\, dv$
11. 210π
12. $\int_{\pi/2}^{\pi} \int_0^{\pi/6} \int_0^4 \rho^3 \cos\phi \sin\phi\, d\rho\, d\phi\, d\theta$

Additional Exercises, Chapter 16 (page 729)

1. 189 units^3
3. $-207/35$
5. $2709\frac{1}{3}$
7. 25/3 units^3
9. $4\int_0^4 \int_0^{\sqrt{16-x^2}} (x^2 + y^2) dy\, dx$
11. does not exist
13. $4\int_0^{\sqrt{7}} \int_0^{\sqrt{7-x^2}} \sqrt{1 + 4x^2 + 4y^2}\, dy\, dx$
15. $-277/720$
17. 0
19. $6u^3v^2 + 4uv^4$
21. $2\int_0^{3/2} \int_{-u}^{u} (u^2 - v^2) dv\, du$
23. $3^7\pi/7$
25. $243\pi/4$
27. $19\pi/2$ units^2
29. 648π units^3
31. -28
33. $2\int_0^2 \int_0^{2-u} \int_v^{2-u} uvw\, dw\, dv\, du$
35. $125\pi/4$
37. $560\pi/9$
39. 384π
41. $8\pi/3$ units^3
43. the set of all points (ρ, ϕ, θ) where $\rho \geq 3$ and $-\pi/2 \leq \theta \leq \pi/2$
45. the set of all points (ρ, ϕ, θ) where $\rho \leq 2$ and $\pi/4 \leq \phi \leq 3\pi/4$
47. $\pi/3$
49. 72π units^3

Chapter 17

Section 17.1 (page 737)

1. 5
3. 1

Section 17.2 (page 742)

1. 1540
3. $(3/2)\sin 2 + (2/3)(e^3 - 1)$
5. -2
7. 16
9. 34/3
11. 0
13. 1/3
15. -6
17. 31/2
19. 0

Section 17.3 (page 749)

1. 3/2
3. -32
5. 0
7. 2/5
9. 0
11. $3\pi/4$
13. $-3\pi/2$
15. 0
17. (a) no; (b) yes
19. πr^2

Section 17.4 (page 755)

1. 9
3. $4e^2 - e$
5. -1
7. 0
9. 2

11. $\frac{1}{\sqrt{2}}e^{\pi} - \frac{1}{\sqrt{2}}e^{\pi/2} - e^{\pi/4} - 1$

13. $-2\sin 1$

15. any simply connected region not containing $(1, -1)$

Section 17.5 (page 761)

1. 19/10
3. $-1/4$
5. $2/5 + e - \cos 1$
7. $-341/3$
9. 0
11. 4
13. 4
15. $-1/4$
17. 12
19. $e^{-14} - e^2 - 5$

Section 17.6 (page 769)

1. $5\sqrt{21}/18$
3. $-2\sqrt{21}/9$
5. $\frac{\sqrt{14}}{6}(-\cos 7 + \cos 2 + \cos 5 - \cos 4)$
7. 0
9. $\sqrt{3}((341/12) + \sin 5 - \sin 4)$
11. $(1/6)(2 + \sqrt{3})$
13. $32(2 + \sqrt{21})$
15. $\frac{\sqrt{2}\pi}{4}(e - 1)$

Section 17.7 (page 775)

1. 13.5 cubic units per unit time.
3. 13.5 cubic units per unit time.
5. 0 cubic units per unit time.
7. 2π cubic units per unit time.
9. 13.5 cubic units per unit time.
11. 27 cubic units per unit time.
13. 0 cubic units per unit time.
15. 4π cubic units per unit time.

Section 17.8 (page 780)

1. 32π
3. $4ka^3\pi$
5. 13.5 cubic units per unit time.
7. 27 cubic units per unit time.
9. 0 cubic units per unit time.
11. 4π cubic units per unit time.

Section 17.9 (page 787)

1. 0
3. -2π
5. -1
7. 8π
9. 0

Technique Review Exercises, Chapter 17 (page 791)

1. 2648/35
2. $6e^2 - 3e$
3. 0
5. $7/2 - 3\pi/4$
6. 0
7. $24a$ cubic units per unit time.
8. $-16a$ cubic units per unit time.
9. 16π

10. (a) grad $f = \nabla f = \left(\frac{\partial f}{\partial x}, \frac{\partial f}{\partial y}, \frac{\partial f}{\partial z}\right)$, domain a subset of R^3, range a subset of R^3.

(b) Let $F(\mathbf{X}) = (P(\mathbf{X}), Q(\mathbf{X}), R(\mathbf{X}))$. Then

$$\text{del } F = \nabla \cdot F = \frac{\partial P}{\partial x} + \frac{\partial Q}{\partial y} + \frac{\partial R}{\partial z},$$ domain a subset of R^3, range a subset of R.

(c) Let $F(\mathbf{X}) = (P(\mathbf{X}), Q(\mathbf{X}), R(\mathbf{X}))$;

$$\text{curl } F = \nabla \times F = \begin{vmatrix} \mathbf{i} & \mathbf{j} & \mathbf{k} \\ \frac{\partial}{\partial x} & \frac{\partial}{\partial y} & \frac{\partial}{\partial z} \\ P(\mathbf{X}) & Q(\mathbf{X}) & R(\mathbf{X}) \end{vmatrix},$$

domain a subset of R^3, range a subset of R^3.

Additional Exercises, Chapter 17 (page 792)

1. -25
3. $30\frac{19}{20}$
5. -11
7. $(1/3) + (1/2)\ln 2$
9. $1/30$
11. $8/15$
13. 12π
15. 0
17. $-2\cos 1$
19. -15
21. In any simply connected region that does not intersect the line $y = 0$.
23. $-1/2$
25. $1 - \cos 1 + (4/7)\sin 1$
29. $-19\sqrt{3}/60$
31. $\sqrt{3}/6$
33. $13\pi/3$
35. $-3/2$ cubic units per unit time
37. $4\pi\sqrt{2}/3$ cubic units per unit time
39. 32π
41. $4\pi/5$
43. $-1/3$ cubic units per unit time
45. 9π
47. 1

Index

Differentiation Formulas

1. $\frac{dc}{dx} = 0$, if c is a constant.

2. $\frac{d}{dx}(u \pm v) = \frac{du}{dx} \pm \frac{dv}{dx}$.

3. $\frac{d}{dx}(uv) = u\frac{dv}{dx} + v\frac{du}{dx}$.

4. $\frac{d}{dx}\left(\frac{u}{v}\right) = \frac{v(du/dx) - u(dv/dx)}{v^2}$.

5. $\frac{d(cu)}{dx} = c\frac{du}{dx}$, if c is a constant.

6. $\frac{du^n}{dx} = nu^{n-1}\frac{du}{dx}$, for any real number n.

7. $\frac{df(u)}{dx} = \frac{df}{du}\frac{du}{dx}$.

8. $\frac{d}{dx}(\log_a u) = \frac{\log_a e}{u}\frac{du}{dx}$.

9. $\frac{d}{dx}(\ln u) = \frac{1}{u}\frac{du}{dx}$.

10. $\frac{d}{dx}(\log_a |u|) = \frac{\log_a e}{u}\frac{du}{dx}$.

11. $\frac{d}{dx}(\ln |u|) = \frac{1}{u}\frac{du}{dx}$.

12. $\frac{d}{dx}a^u = a^u \ln a \frac{du}{dx}$.

13. $\frac{d}{dx}e^u = e^u\frac{du}{dx}$.

14. $\frac{d}{dx}\sin u = \cos u\frac{du}{dx}$.

15. $\frac{d}{dx}\cos u = -\sin u\frac{du}{dx}$.

16. $\frac{d}{dx}\tan u = \sec^2 u\frac{du}{dx}$.

17. $\frac{d}{dx}\cot u = -\csc^2 u\frac{du}{dx}$.

18. $\frac{d}{dx}\sec u = \sec u \tan u\frac{du}{dx}$.

19. $\frac{d}{dx}\csc u = -\csc u \cot u\frac{du}{dx}$.

20. $\frac{d}{dx}\arcsin u = \frac{1}{\sqrt{1-u^2}}\frac{du}{dx}$.

21. $\frac{d}{dx}\arccos u = -\frac{1}{\sqrt{1-u^2}}\frac{du}{dx}$.

22. $\frac{d}{dx}\arctan u = \frac{1}{u^2+1}\frac{du}{dx}$.

23. $\frac{d}{dx}\operatorname{arccot} u = -\frac{1}{u^2+1}\frac{du}{dx}$.

24. $\frac{d}{dx}\operatorname{arcsec} u = \frac{1}{|u|\sqrt{u^2-1}}\frac{du}{dx}$.

25. $\frac{d}{dx}\operatorname{arccsc} u = -\frac{1}{|u|\sqrt{u^2-1}}\frac{du}{dx}$.